Springer Series in Materials Science

Volume 337

The Springer Series in Materials Science covers the complete spectrum of materials research and technology, including fundamental principles, physical properties, materials theory and design. Recognizing the increasing importance of materials science in future device technologies, the book titles in this series reflect the state-of-the-art in understanding and controlling the structure and properties of all important classes of materials.

Christos G. Aneziris · Horst Biermann
Editors

Multifunctional Ceramic Filter Systems for Metal Melt Filtration

Towards Zero-Defect Materials

 Springer

Editors
Christos G. Aneziris
Institute of Ceramics, Refractories
and Composite Materials
TU Bergakademie Freiberg
Freiberg, Sachsen, Germany

Horst Biermann
Institute for Materials Engineering
TU Bergakademie Freiberg
Freiberg, Sachsen, Germany

ISSN 0933-033X ISSN 2196-2812 (electronic)
Springer Series in Materials Science
ISBN 978-3-031-40929-5 ISBN 978-3-031-40930-1 (eBook)
https://doi.org/10.1007/978-3-031-40930-1

This work was supported by German Research Foundation (Deutsche Forschungsgemeinschaft, DFG) (169148856).

This Springer imprint is published by the registered company Springer Nature Switzerland AG
The registered company address is: Gewerbestrasse 11, 6330 Cham, Switzerland

Paper in this product is recyclable.

Preface

Safety for road and railway vehicles as well as aircrafts requires highly stressable steel, iron, aluminum and magnesium-based components. During the production process, contaminations within the metal melt can occur resulting in defects in the form of inclusions. It is difficult or even sometimes impossible to reduce or remove those inclusions. The Collaborative Research Center 920 (CRC 920) 'Multifunctional filters for metal melt filtration — a contribution towards zero defect materials' focussed on research in a new generation of metal qualities — also during recycling — via melt filtration with superior mechanical properties for use in lightweight structures and high-demand construction materials. The Collaborative Research Centre 920 funded by the German Research Foundation (DFG) ran from 2011 to 2023 at the Technische Universität Bergakademie Freiberg, Germany (project number: 169148856). The chapters contained in this book provide an overview of the most important results of the projects of the CRC in the completed funding periods and at the same time present current, in some cases, still unpublished results. This book is the final publication of the CRC 920.

The aim of the CRC 920 is an enormous reduction of non-metallic inclusions in the metal matrix by the use of intelligent filter materials as well as filter systems with a functionalized filter surface. Especially, in the last 4 years, a new generation of combined refining filter systems was the focus. The metal melt comes first in contact with the reactive carbon-bonded filters which generate, e.g. in case of steel CO gas bubbles in the melt as well as activate gas bubbles on the surface of the inclusions. As a result, a kind of flotation of the inclusions towards the slag on the surface of the melt takes place. Further, the high reactivity as well as the gas bubbles contribute to the agglomeration of the fine inclusions to big clusters. These clusters flow due to buoyancy forces to the surface of the melt and are captured by slag or are filtrated on the surface of active filters, which do not form gas bubbles but provide the same chemistry as the inclusions on their functionalized surfaces for a sufficient adhesion and as a result for a sufficient filtration of the inclusions. Modelling activities mainly focus on the contributions of the gas bubbles and on the reactive layers formed in situ on the surface of the reactive filters. Moreover, they generate codes with respect to the thermomechanical and functional properties of the filters for a 3D printing of filter structures which are then end shaped and produced for instance with the aid of a robot-assisted flame spraying technique.

The book is thematically divided into four sections, (i) filter materials and filter material processing, (ii) modelling of filter structures and filter systems, (iii) filtration efficiency, filter material properties and properties of metals after cleaning and (iv) industrial transfer projects. These four sections focus in a comprehensive way on selected results covering the entire chain of filtration, from the research of processing of the filter materials, the understanding of filtration mechanisms and kinetics, the contribution of the gas phase and especially the formation of in situ collecting layers as well as gas collecting suboxides up to mechanical properties of the final metallic products by means of improving ductility, strength and fatigue lives. A total of 19 scientific projects with more than 70 funded researchers were working together by bridging different disciplines in the last 12 years.

Especially, in Chap. 1, carbon filter materials and micro- and macro-structuring are investigated with the aid of cold- and hot-coating techniques, 3D processing approaches as well the possibility to use environmentally friendly binder systems. Chapter 5 supports the structural and thermodynamic properties with the aid of Raman and DFT investigations and in Chap. 2 the confocal laser scanning high-temperature microscopy is applied to investigate the interaction of filter materials with endogenous and exogenous inclusions of steel melts. Oxide filters for aluminum melts are investigated in Chap. 3 and material thermodynamics of Chap. 4 explain the interactions between filter materials and melts. Chapter 7 contributes to the understanding of interactions between filters and melts without a melt flow in an SPS/FAST sintering facility and Chap. 6 supports the studies with XRD, TEM and modelling of the carbon structures. Chapter 8 deals with possible approaches for the Fe removement in Al–Si-alloys and Chap. 9 is dealing with the application of lithium-containing filter materials for reducing the H_2-impact in Al-alloys. Chapter 10 investigates filter materials based also on MgAlON for cleaning Mg-alloys. The macro- and microstructures before and after metal casting have been evaluated with the aid of X-ray tomography. These data support also the modelling and especially the particle transport inside open cell foam filters in Chap. 13 accompanied by heat transport and diffusion processes during melt filtration in Chap. 14. Water-based modelling and models presented in Chap. 12 contribute to a better understanding of filtration and especially to the micro-processes of agglomeration by variation of filters and particle properties as well as process parameters. Agglomeration of non-metallic inclusions is studied in Chap. 15 by hot investigations of wetting behaviour with the aid of a high-temperature atomic force microscope. Chapter 16 is modelling and evaluating the chemo-thermomechanical behaviour of filter materials and structures as well as supports with model-based design the 3D printing of filter structures via the virtual 3D prototyping presented in Chap. 18. Thereby the melt filtration in continuous casting is the focus of Chap. 17 which supports the registration of filtration efficiency of active and reactive filters in a unique steel casting simulator in Chap. 19. Chapter 20 explores the decopperization of steel melts and in Chap. 21 the interactions and wetting behaviour between molten iron and carbon-bonded filters are illustrated with the aid of a hot stage microscope. The high-temperature strength, the form stability and the determination of the temperature-dependent fracture of carbon-bonded filters and filter materials are presented in the Chaps. 22 and 23. Chapter 24 deals with the influence of internal defects on the fatigue lives of steel and aluminum alloys. Chapter 25 presents the analysis of detrimental inclusions in steel and aluminum, complemented also by numerical and statistical modelling of positions of inclusions based on thermography data from ultrasonic fatigue experiments in Chap. 26. In Chap. 27 the mechanical characterization of the impact of inclusions continues based on the temperature- and strain rate-dependent strength, deformation and toughness behaviour of metallic products. The influence of casting characteristics is explored in Chap. 28 for the aluminum casting and in Chap. 29 for cast iron with nodular graphite. Chapter 30 is investigating exchangeable filter systems for the continuous casting in a real industrial tundish supported by numerical simulation in Chap. 31. Chapter 33 is exploring new approaches of applying filtration systems during ingot casting and especially by functionalizing feeders, spider bricks and starter casting tubes. Chapter 32 deals with the precipitation of iron-containing phases in aluminium alloys with the aid of filtration approaches and Chap. 33 explores the impact of filter composite structures with fibres in each structure to remove inclusions with a size of less than 50 μm in aluminum melts. Chapter 34 shows the detrimental impact of inclusion on steel products even at higher operation temperatures.

As speaker and deputy speaker, we would like to thank all current and former academic members of the Collaborative Research Centre 920 as well as our partners from the industrial consulting board for their constant support. The successful work would not have been possible without the dedicated cooperation of all scientists who worked on or supported the projects. We would also like to thank other contributors, all technical and administrative staff as well as the countless students for their outstanding cooperation.

Due to the excellent scientific work, the CRC 920 has been able to produce many scientific qualification theses, from habilitation and doctorate degrees to students' theses. In this 12 years-scientific journey, numerous scientific careers have been established and many graduates have been shaped scientifically. This young talent work was also promoted within the framework of a graduate school, which taught many soft skills in addition to professional qualifications.

We would also like to thank the public relations team, especially Prof. Dr. A. Geigenmüller, which made the scientific results available to a broad public and thus made an important contribution to the reputation of the Technische Universität Bergakademie Freiberg. This public relations work has also interested many students in the special research areas of the Collaborative Research Centre and thus drawn their attention to the university's courses of study.

Our special thanks go to the German Research Foundation for the trust it has placed in us. Our special thanks go to Dr. A. Winkler, Dr. C. Petry and Dr. M. Beilein from the Collaborative Research Centres Department, Dr. X. Molodova, Dr. K. Sick and Dr. C. Schuster from the Materials Science and Engineering Department and Mrs. P. Hammel, Mrs. C. Niebus and Mrs. J. Winkel from the administration centre.

Finally, we are also deeply indebted to all the experts who have followed our work with interest and supported us:

Prof. Dr. A. Benz, Darmstadt, Prof. Dr. J. Bill, Stuttgart, Prof. Dr.-Ing. G. Brenner, Clausthal, Prof. Dr.-Ing. H. J. Christ, Siegen, Prof. Dr. O. Guillon, Jülich, Prof. Dr.-Ing. S. Heinrich, Hamburg-Harburg, Prof. Dr.-Ing. S. Kabelac, Hannover, Prof. Dr.-Ing. E. Kerscher, Kaiserslautern, Prof. Dr. P. Kratzer, Duisburg-Essen, Prof. Dr.-Ing. W. Krenkel, Bayreuth, Prof. Dr. S. Kuhnt, Dortmund, Prof. Dr.-Ing. H. Pfeifer, Aachen, Prof. Dr.-Ing. P. D. Portella, BAM Berlin, Prof. Dr. F. Riehle, Braunschweig, Prof. Dr.-Ing. A. Roosen, Erlangen, Prof. Dr.-Ing. E. Schmidt, Wuppertal, Prof. Dr.-Ing. B. Scholtes, Kassel, Prof. Dr.-Ing. L. Singheiser, Jülich, Prof. Dr.-Ing. P. Schumacher, Leoben, Prof. Dr.-Ing. B. Skrotzki, BAM Berlin, Prof. Dr.-Ing. P. Steinmann, Erlangen, Prof. Dr. U. K. Woggon, Berlin.

We would also like to highlight the support of the members of the Senate Committee of the German Research Foundation Prof. Dr. P. Greil, Erlangen, Prof. Dr. J. Rödel, Darmstadt and Prof. Dr. M. Scheffler, Magdeburg.

Finally, we would like to thank deeply Mrs. A. Penzkofer and especially Dr. U. Fischer, whose work in the office of the Collaborative Research Centre ensured the smooth running of all financial aspects and the organization of all events.

After 12 years of CRC 920, we can summarize: The never-ending scientific journey to Ithaka is still going on.

Freiberg, Germany   Prof. Dr.-Ing. Christos G. Aneziris
                                                                                                       Speaker

                                                                                Prof. Dr.-Ing. Horst Biermann
                                                                                                Deputy Speaker

Contents

Contributors

Martin Abendroth Institute of Mechanics and Fluid Dynamics, Technische Universität Bergakademie Freiberg, Freiberg, Germany

Christos G. Aneziris Institute of Ceramics, Refractories and Composite Materials, Technische Universität Bergakademie Freiberg, Freiberg, Germany

Amjad Asad Institute of Mechanics and Fluid Dynamics, Technische Universität Bergakademie Freiberg, Freiberg, Germany

Andreas Baaske Thyssenkrupp Steel Europe AG, Duisburg, Germany

Katrin Bauer Institute of Mechanics and Fluid Dynamics, Technische Universität Bergakademie Freiberg, Freiberg, Germany

Benedict Baumann Foundry Institute, Technische Universität Bergakademie Freiberg, Freiberg, Germany

Hanka Becker Institute of Materials Science, Technische Universität Bergakademie Freiberg, Freiberg, Germany

Horst Biermann Institute of Materials Engineering, Technische Universität Bergakademie Freiberg, Freiberg, Germany

Benjamin Bock-Seefeld Institute of Ceramics, Refractories and Composite Materials, Technische Universität Bergakademie Freiberg, Freiberg, Germany

Nora Brachhold Institute of Ceramics, Refractories and Composite Materials, Technische Universität Bergakademie Freiberg, Freiberg, Germany

Simon Brehm Institute of Theoretical Physics, TU Bergakademie Freiberg, Freiberg, Germany

Erica Brendler Institute of Analytical Chemistry, Technische Universität Bergakademie Freiberg, Freiberg, Germany

Michael Budnitzki Institute for Advanced Simulation (IAS-9: Materials Data Science and Informatics), Forschungszentrum Jülich GmbH, Jülich, Germany

Sarah Daus Institute of Mechanical Process Engineering and Mineral Processing, Technische Universität Bergakademie Freiberg, Freiberg, Germany

Cornelius Demuth Institute of Thermal Engineering, Technische Universität Bergakademie Freiberg, Freiberg, Germany

Lisa Ditscherlein Institute of Mechanical Process Engineering and Mineral Processing, Technische Universität Bergakademie Freiberg, Freiberg, Germany

Claudia Dommaschk Foundry Institute, Technische Universität Bergakademie Freiberg, Freiberg, Germany

Milan Dopita Department of Condensed Matter Physics, Charles University, Prague, Czech Republic

Steffen Dudczig Institute of Ceramic, Refractories and Composite Materials, Technische Universität Bergakademie Freiberg, Freiberg, Germany

Olga Fabrichnaya Institute of Material Science, Technische Universität Bergakademie Freiberg, Freiberg, Germany

Beate Fankhänel Institute of Nonferrous Metallurgy and Purest Materials, Technische Universität Bergakademie Freiberg, Freiberg, Germany

Tobias Michael Fieback Institute of Thermal Engineering, Technische Universität Bergakademie Freiberg, Freiberg, Germany

Undine Fischer Institute of Ceramics, Refractories and Composite Materials, Technische Universität Bergakademie Freiberg, Freiberg, Germany

Patrick Gehre Institute of Ceramics, Refractories and Composite Materials, Technische Universität Bergakademie Freiberg, Freiberg, Germany

Lisa-Marie Heisig Institute of Thermal Engineering, Technische Universität Bergakademie Freiberg, Freiberg, Germany

Sebastian Henschel Institute of Materials Engineering, Technische Universität Bergakademie Freiberg, Freiberg, Germany

Cameliu Himcinschi Institute of Theoretical Physics, Technische Universität Bergakademie Freiberg, Freiberg, Germany

Daniel Hoppach Institute of Mechanical Process Engineering and Mineral Processing, Technische Universität Bergakademie Freiberg, Freiberg, Germany

Jana Hubálková Institute of Ceramics, Refractories and Composite Materials, Technische Universität Bergakademie Freiberg, Freiberg, Germany

Mariia Ilatovskaia Institute of Materials Science, Technische Universität Bergakademie Freiberg, Freiberg, Germany

Bernhard Jung Institute for Informatics, Technische Universität Bergakademie Freiberg, Freiberg, Germany

Sven Karrasch Thyssenkrupp Steel Europe AG, Duisburg, Germany

Andreas Keßler Foundry Institute, Technische Universität Bergakademie Freiberg, Freiberg, Germany

Bjoern Kiefer Institute of Mechanics and Fluid Dynamics, Technische Universität Bergakademie Freiberg, Freiberg, Germany

Thomas Kirste Federal-Mogul Nürnberg GmbH, a Tenneco Group Company, Nürnberg, Germany

Kevin Koch Institute of Materials Engineering, Technische Universität Bergakademie Freiberg, Freiberg, Germany

Jens Kortus Institute of Theoretical Physics, Technische Universität Bergakademie Freiberg, Freiberg, Germany

Michael Koster Institute for Advanced Simulation (IAS-9: Materials Data Science and Informatics), Forschungszentrum Jülich GmbH, Jülich, Germany

Jakob Kraus Institute of Theoretical Physics, Technische Universität Bergakademie Freiberg, Freiberg, Germany

Lutz Krüger Institute of Materials Engineering, Technische Universität Bergakademie Freiberg, Freiberg, Germany

Meinhard Kuna Institute of Mechanics and Fluid Dynamics, Technische Universität Bergakademie Freiberg, Freiberg, Germany

Henry Lehmann Institute for Informatics, Technische Universität Bergakademie Freiberg, Freiberg, Germany

Andreas Leineweber Institute of Materials Science, Technische Universität Bergakademie Freiberg, Freiberg, Germany

Alexander Malik Institute of Mechanics and Fluid Dynamics, Technische Universität Bergakademie Freiberg, Freiberg, Germany

Katrin Markuske Institute of Thermal Engineering, Technische Universität Bergakademie Freiberg, Freiberg, Germany

Miguel A. A. Mendes Instituto de Engenharia Mecânica, Instituto Superior Técnico, Universidade de Lisboa, Lisboa, Portugal

Roman Morgenstern Federal-Mogul Nürnberg GmbH, a Tenneco Group Company, Nürnberg, Germany

Mykhaylo Motylenko Institute of Materials Science, Technische Universität Bergakademie Freiberg, Freiberg, Germany

Sebastian Neumann Institute of Mechanics and Fluid Dynamics, Technische Universität Bergakademie Freiberg, Freiberg, Germany

Jan Nicklas Institute of Mechanical Process Engineering and Mineral Processing, Technische Universität Bergakademie Freiberg, Freiberg, Germany

Urs A. Peuker Institute of Mechanical Process Engineering and Mineral Processing, Technische Universität Bergakademie Freiberg, Freiberg, Germany

David Rafaja Institute of Materials Science, Technische Universität Bergakademie Freiberg, Freiberg, Germany

Subhashis Ray Institute of Thermal Engineering, Technische Universität Bergakademie Freiberg, Freiberg, Germany

Stephan Roth Institute of Mechanics and Fluid Dynamics, Technische Universität Bergakademie Freiberg, Freiberg, Germany

Shyamal Roy Institute for Advanced Simulation: Materials Data Science and Informatics, Forschungszentrum Juelich GmbH, Juelich, Germany

Anton Salomon Institute of Materials Science, Technische Universität Bergakademie Freiberg, Freiberg, Germany

Stefan Sandfeld Institute for Advanced Simulation: Materials Data Science and Informatics, Forschungszentrum Juelich GmbH, Juelich, Germany

Christiane Scharf Institute of Nonferrous Metallurgy and Purest Materials, Technische Universität Bergakademie Freiberg, Freiberg, Germany

Ekaterina Schmid Freiberger Compound Materials GmbH, Freiberg, Germany

Gert Schmidt Institute of Ceramics, Refractories and Composite Materials, Technische Universität Bergakademie Freiberg, Freiberg, Germany

Alexander Schmiedel Institute of Materials Engineering, Technische Universität Bergakademie Freiberg, Freiberg, Germany

Johannes Paul Schoß Foundry Institute, Technische Universität Bergakademie Freiberg, Freiberg, Germany

Leandro Schöttler Deutsche Edelstahlwerke Specialty Steel GmbH & Co. KG, Siegen, Germany

Alina Schramm Institute of Nonferrous Metallurgy and Purest Materials, Technische Universität Bergakademie Freiberg, Freiberg, Germany

Christina Schröder Institute of Iron and Steel Technology, Technische Universität Bergakademie Freiberg, Freiberg, Germany

Matthias Schwarz Deutsche Edelstahlwerke Specialty Steel GmbH & Co. KG, Siegen, Germany

Rüdiger Schwarze Institute of Mechanics and Fluid Dynamics, Technische Universität Bergakademie Freiberg, Freiberg, Germany

Mikhail Seleznev Institute of Materials Engineering, Technische Universität Bergakademie Freiberg, Freiberg, Germany

Andreas Seupel Institute of Mechanics and Fluid Dynamics, Technische Universität Bergakademie Freiberg, Freiberg, Germany

Andy Spitzenberger Institute of Mechanics and Fluid Dynamics, Technische Universität Bergakademie Freiberg, Freiberg, Germany

Michael Stelter Institute of Nonferrous Metallurgy and Purest Materials, Technische Universität Bergakademie Freiberg, Freiberg, Germany

Enrico Storti Institute of Ceramic, Refractories and Composite Materials, Technische Universität Bergakademie Freiberg, Freiberg, Germany

Michal Szucki Foundry Institute, Technische Universität Bergakademie Freiberg, Freiberg, Germany

Shahin Takht Firouzeh Institute of Mechanics and Fluid Dynamics, Technische Universität Bergakademie Freiberg, Freiberg, Germany

Martin Thümmler Institute of Materials Science, Technische Universität Bergakademie Freiberg, Freiberg, Germany

Dimosthenis Trimis Division of Combustion Technology, Engler-Bunte-Institute, Karlsruhe Institute of Technology, Karlsruhe, Germany

Claudia Voigt Institute of Ceramics, Refractories and Composite Materials, Technische Universität Bergakademie Freiberg, Freiberg, Germany

Olena Volkova Institute of Iron and Steel Technology, Technische Universität Bergakademie Freiberg, Freiberg, Germany

Ruben Wagner Institute of Materials Engineering, Technische Universität Bergakademie Freiberg, Freiberg, Germany

Xingwen Wei Institute of Iron and Steel Technology, Technische Universität Bergakademie Freiberg, Freiberg, Germany

Anja Weidner Institute of Materials Engineering, Technische Universität Bergakademie Freiberg, Freiberg, Germany

Eric Werzner Institute of Thermal Engineering, Technische Universität Bergakademie Freiberg, Freiberg, Germany

Tony Wetzig Institute of Ceramics, Refractories and Composite Materials, Technische Universität Bergakademie Freiberg, Freiberg, Germany

Gotthard Wolf Foundry Institute, Technische Universität Bergakademie Freiberg, Freiberg, Germany

Xian Wu Institute of Materials Engineering, Technische Universität Bergakademie Freiberg, Freiberg, Germany

Rhena Wulf Institute of Thermal Engineering, Technische Universität Bergakademie Freiberg, Freiberg, Germany

Chapter 1
Carbon-Bonded Filter Materials and Filter Structures with Active and Reactive Functional Pores for Steel Melt Filtration

Benjamin Bock-Seefeld, Patrick Gehre, and Christos G. Aneziris

1.1 Introduction

Solid, nonmetallic inclusions considerably impair the mechanical properties of cast steel products. While exogenous inclusions can be removed by slag or casting systems, the removal of endogenous inclusions within micrometer scale constitutes a major challenge. In order increase the quality of the steel product, ceramic foam filters (CFF) such as zirconia filters are applied in the foundry. Besides calming the flow conditions from turbulent to more laminar, the CFF effect an inclusion deposition on the filter surface. However, zirconia filters exhibit a low creep resistance at elevated temperatures, resulting in the deformation of the filter structures and hence to a decrease of the flow rates. For this reason, the present study focused on the development of carbon-bonded alumina filters (Al_2O_3-C), since the carbon matrix provides high refractoriness, improved creep resistance and a low wettability by molten steel. Instead of the pitches and phenolic resins predominantly used in the refractory industry, alternative binder systems were investigated to reduce the amount of carcinogenic substances released during pyrolysis.

The interactions taking place between the Al_2O_3-C filter and the steel melt can be described as follows. Due to the presence of carbon, the "reactive" filter material initializes a carbothermal reaction with liquid iron as a catalyst, whereby Al_2O_3-C reacts with gases dissolved in molten steel. As a result, CO as well as gaseous alumina suboxides emerge and a polycrystalline alumina layer with entrapped inclusions is formed on the filter surface. The gases effect an inclusion flotation and an agglomeration of fine inclusions to bigger clusters by activating nano-bubbles on the

B. Bock-Seefeld (✉) · P. Gehre · C. G. Aneziris
Institute of Ceramics, Refractories and Composite Materials, Technische Universität
Bergakademie Freiberg, Agricolastr. 17, 09599 Freiberg, Germany
e-mail: Benjamin.bock@ikfvw.tu-freiberg.de

© The Author(s) 2024
C. G. Aneziris and H. Biermann (eds.), *Multifunctional Ceramic Filter Systems
for Metal Melt Filtration*, Springer Series in Materials Science 337,
https://doi.org/10.1007/978-3-031-40930-1_1

1

inclusion surface, which move up in the steel melt towards the slag due to buoyancy forces. These reactions proceed until a dense polycrystalline alumina layer is formed and thus no further interaction between Al_2O_3-C and molten steel takes place. At this point, the "reactive" filtration behavior turns into an "active" behavior, whereby predominantly inclusions and inclusion clusters with a similar chemical composition as the in-situ formed layer are attracted, deposited and sintered on the filter surface [1].

Considering the strong impact of the filter surface chemistry on the interaction between the filter and molten steel, the application of functional coatings on the filter surface constitutes a great opportunity to further increase filtration efficiency. Besides an increase in the filter reactivity, the filter surface chemistry can be adapted to the prevailing inclusion composition and thus encourage the "active" filtration. Therefore, functional coatings were explored and their impact on the Al_2O_3-C filter properties was characterized.

Another important issue in this work was the development of novel filter manufacturing methods. Al_2O_3-C filters are commonly manufactured by the replica technique, using reticulated polyurethane foams with a random, anisotropic structure as filter template. In order to enable the fabrication of reproducible filter structures with well-defined geometries, several approaches from the field of additive manufacturing were devised within this work.

1.2 Experimental Details

The following section provides an overview of the experimental setups and conditions to characterize the filter materials and filter structures.

1.2.1 Rheology

The rheological behavior of ceramics slips was measured by the rheometer devices RheoStress RS 150 (Haake, Germany) and MARS 60 (Thermo Fisher Scientific, Germany), using a coaxial measuring system of the types Z40 DIN and CC40 DIN respectively. Thereby, the dynamic viscosity was determined by the linear increase of the shear rate from 0.1 up to 1000 s^{-1}, which was held constant for 90 s. Afterwards, the shear rate was decreased with the same rate to 0.1 s^{-1}.

1.2.2 Residual Carbon Content

In order to evaluate the qualitative and quantitative residual carbon content (RCC) after the pyrolysis, 20 mg of ground carbon-bonded alumina based on Carbores®P were analyzed utilizing a carbon analyzer A RC 412 (Leco, USA). The heating regime was performed from 150 to 900 °C with a heating rate of 50 K min^{-1} under oxidizing atmosphere.

1.2.3 Cold Crushing Strength and Open Porosity

The determination of the cold crushing strength (CSS) was conducted using the universal testing machine TT28100 (TIRA GmbH, Germany) with a 5 kN pressure cell. A displacement speed of 3 mm min^{-1} was selected until a force drop of 80% was reached, which terminated the experiment. Furthermore, the ball-on-three-balls method (B3B) was used to determine the flexural strength of discs with a dimension of approx. 0.5 mm in thickness and 8 mm in diameter. The open porosity was examined by a mercury porosimeter of the type PASCAL series (Porotec, Germany).

1.2.4 Analysis of the Macro- and Microstructure

For the evaluation of the filter structure, a microfocus X-ray computer tomograph CT-ALPHA (ProCon X-ray, Germany), operating at 150 kV and 60 µA, was applied to examine the filter macrostructure. The software VG Studio Max 2.2 (Volume Graphics, Germany) was utilized for the visualization. A digital light microscope VHX-2000 D (Keyence, Belgium) equipped with a VH-Z20R objective was utilized to examine the filter surface. A scanning electron microscope (SEM) of the type ESEM FEG XL30 (FEI, Netherlands) configured with EDX (EDAX-Ametek, USA) was used to analyze the microstructure of filter fragments sputtered with carbon.

1.2.5 Hot Stage Microscope

In order the analyze the wetting behavior of filter materials by molten steel, the macroscopic Young contact angle was characterized in a hot stage microscope, as described by Aneziris et al. [2]. For this purpose, filter substrates and a cylindrical 42CrMo4 sample (10 mm in height and diameter) were placed in the tube and heated up to 1540 °C under argon atmosphere using a heating regime of 10 K min^{-1} and a dwell time of 30 min at maximum temperature. Afterwards, the samples were cooled with 10 K min^{-1}. The contact angles were determined by the software ImageJ (Version: 1.6.0.24).

1.2.6   Thermal Shock Behavior

Impingement tests were conducted to examine the filters' thermal shock behavior. For each filter fixed in a furan-bonded sand mold, 9 kg steel was molten at 1650 °C and poured on the filter from a height of approx. 0.5 m. The experiment was considered successful if the filter neither broke nor exhibited critical defects.

1.3   Sample Preparation and Experimental Results

In this work, three main studies were examined. The first study is the development of Al_2O_3-C filters based on the replica technique, whereby binder systems with a reduced amount of carcinogenic components were focused. In the second study, the functionalization of the Al_2O_3-C filter surface was investigated. Besides the evaluation of different coating compositions, three different application techniques were examined. The third study concerns the fabrication of alternative filter structures by additive manufacturing. The following section provides an overview of the materials, sample preparation methods as well as experimental results.

1.3.1   Carbon-Bonded Alumina Filters

Carbon-Bonded Alumina Filters Based on Carbores®P

In a first approach, the production of Al_2O_3-C filters composed of alumina, fine natural graphite, carbon black powder as well as modified coal tar pitch Carbores®P was explored. In comparison to other pitches, Carbores®P is characterized by a lower B(a)P value (<500 ppm), a softening point of 200 °C and a RCC of approx. 85 wt%. The carbon bonds are obtained by the graphitization of an intermediate, liquid pitch-mesophase, which is formed during the pyrolysis below 1000 °C [3]. To determine the interactions between the individual raw materials, the impact of three compositions (AC1, AC2 and AC3) presented in Table 1.1 on the filter producibility were analyzed.

The filter manufacturing was performed in multi-stage processes based on the replica technique. For this purpose, the raw materials, deionized water as well as the additives polypropylene glycol (wetting agent), ligninsulfonate (temporary binder), polycarboxylate ether (dispersing agent) and polyalkylene glycolether (antifoam agent) were admixed in a high shear Hobart-type mixer (ToniTechnik, Germany) to obtain the impregnation, centrifugation and spraying slips. These slips were applied on reticulated polyurethane (PU) foams with a 10 pores per inch (ppi) macrostructure, according to the process routes shown in Fig. 1.1. Afterwards, the Al_2O_3-C filters were placed in a steel retort embedded in pet coke (MÜCO Mücher & Enstipp

Table 1.1 Slip compositions for carbon-bonded alumina filters, in wt% [4, 5]

Material	Raw material		Recipe				
	Type	Supplier	AC1	AC2	AC3	AC4	AC5
Solids							
Al_2O_3	Martoxid MR 70	Martinswerk, Germany	68	67	66	66	66
Graphite	AF 96/97	Graphit Kropfmühl, Germany	12	10	8	8	8
Carbon black	Luvomaxx N-991	Lehmann & Voss & Co., Germany	10	8	6	6	6
Coal tar pitch	Carbores@P	Rütgers, Germany	10	15	20	–	–
Lactose	Mivolis Milchzucker	Mivolis, Germany	–	–	–	16.7	16.7
Tannin	Quebracho-Extrakt	Otto Dille, Germany	–	–	–	3.3	3.3
Additives							
Ligninsulfonat[a]	T11B	Otto-Dille, Germany	1.5	1.5	1.5	1.5	1.5
Polypropylene glycol[a]	PPG P400	Sigma–Aldrich, Germany	0.8	0.8	0.8	0.8	0.8
Polycarboxylate ether[a]	Castament VP 95 L	BASF, Germany	0.3	0.3	0.3	0.3	0.3
Polyalkylene glycolether[a]	Contraspum K 1012	Zschimmer & Schwarz, Germany	0.1	0.1	0.1	0.1	0.1
TiO_2[a]	TR	Crenox GmbH, Germany	–	–	–	0.5	0.5
SiO_2[a]	RW-Filler	RW silicium, Germany	–	–	–	4.0	4.0
Al[a]		Carl Roth, Germany	–	–	–	0.1	0.1
Hexamethylenetetramine[b]		Alfa Aesar, Germany	–	–	–	10	10
P-doped n-silicon or SiC[b]		Silchem, Germany	–	–	–	–	5
Total solid content							
Impregnation slip			81.7	82.3	83.0	78.0	78.0
Centrifugation slip			–	–	65.0	–	–
Spraying slip			74.1	75.6	76.1	70.0	70.0

[a] Related to amount of solids
[b] Related to lactose/tannin content

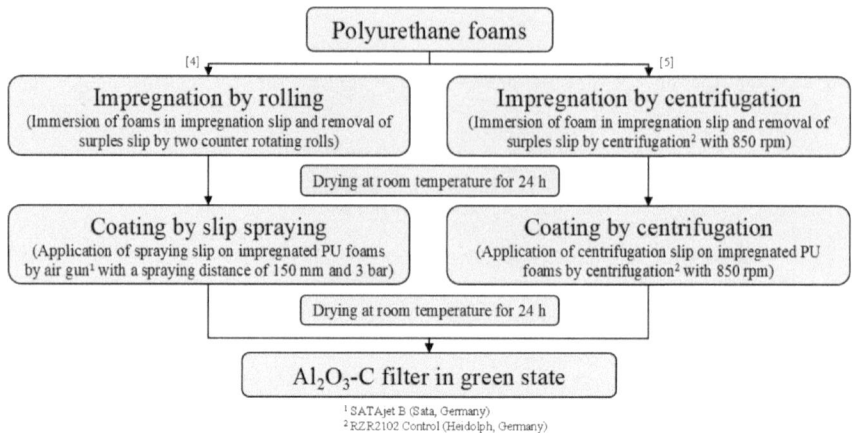

Fig. 1.1 Manufacturing routes for carbon-bonded alumina filters

GmbH, Germany) and fired at 800 °C for 3 h. The heating regime was carried out with 1 K min^{-1} and 30 min dwell times each 100 K.

The comparison of the impact of the slip compositions on the filter producibility revealed a considerable effect of the amount of Carbores®P on the slip properties. It was observed that an increase of the Carbores®P content related to the total solid content led to a considerable decrease in the water amount, which is required for the preparation of coating slips with the same viscosity. This was attributed to the fact that Carbores®P possesses a smaller specific surface area compared to alumina, graphite and carbon black and thus a higher packing density was achieved. However, pure Carbores®P exhibited a shear-thickening behavior, wherefore a minimum amount of graphite and carbon black must be added to obtain stable and shear-thinning slips for the coating procedure [4].

Figure 1.2 presents the filter structure as well as the cross-section of a hollow filter strut after the pyrolysis. Due to the decomposition of the PU foam during the heat treatment, a triconcave cavity with sharp edges was formed, which passed through the entire filter structure. The wall thickness of the filter struts ranged between 300 and 400 μm when the "rolling and spraying" preparation was used. Although the appearance of the AC1, AC2 and AC3 filter structures were similar to each other, the filter composition exerted a significant impact on the filter properties.

As presented in Table 1.2, the RCC increased with decreasing alumina content due to the higher amount of carbon sources. Furthermore, the increase of the Carbores®P content related to the total solid content increased the volumetric shrinkage, whereas the open porosity decreased. Both phenomena were attributed to the high temperature behavior of Carbores®P. On the one hand, the higher amount of Carbores®P caused a higher release of organics bond within the Carbores®P structure and the emergence of a higher amount of liquid pitch-mesophase, which both contributed to a higher volumetric shrinkage and a reduced occurrence of defects. On the other hand, the higher amount of the liquid pitch-mesophase effected a higher infiltration rate of

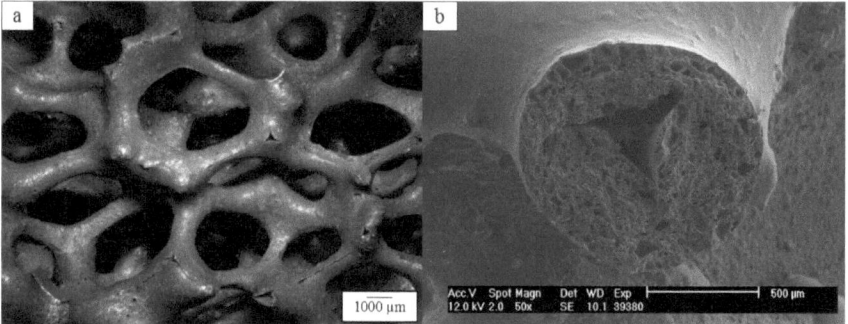

Fig. 1.2 **a** AC3 filter structure and **b** cross-section of a filter strut [4] after pyrolysis

Table 1.2 Filter properties of AC1, AC2 and AC3 after pyrolysis [4, 6]

Recipe	AC1	AC2	AC3
Residual carbon content [wt%]	29.3	29.6	30.1
Volume shrinkage [%]	2.2	3.8	4.7
Open porosity [%]	34.2	32.7	30.2
Cold crushing strength [MPa]	0.07	0.09	0.16[a]/0.21[b]

[a] "Rolling + spraying" method
[b] "Centrifugation + centrifugation" method

pores and thus decreased the porosity. Also a considerable increase of the CCS with increasing Carbores®P content was determined. Besides the decrease of defects and porosity, a higher cross-linking degree of the particles was obtained, which improved the structure bonding and the mechanical filter properties [4]. These effects could be further amplified by increasing the Carbores®P content up to 30 wt%. However, a higher Carbores®P content would impair the coating slip properties and raise the B(a)P value, which is not expedient for the filter production. Considering the improved mechanical properties as well as the higher RCC, which is essential for the carbothermal reaction, AC3 turned out to be the most suitable Al_2O_3-C filter composition for the steel melt filtration.

Apart from the filter composition, a strong dependence of the filter structure and the mechanical filter properties on the coating procedure was found. While the "centrifugation + centrifugation" method led to a homogenous Al_2O_3-C structure over the entire filter volume, the "rolling + spraying" procedure exhibited a decreasing wall thickness with increasing distance to the filter surface. This was traced back to the fact that the inner filter struts were obscured by the outer filter struts during spraying and hence impeded the access of the spraying slip to the interior filter volume. In comparison, the immersion of the impregnated foam into the centrifugation slip ensured an even distribution of the slip on the filter surface. Although the median wall thickness was lower in case of "centrifugation + centrifugation" due to the higher water amount used, the more homogeneous application of the coating resulted in a higher CCS (cp. Table 1.2). Especially for larger filter sizes, this effect was intensified and caused a CCS increase of about 2.5 times [6].

Carbon-Bonded Alumina Filters Based on Lactose and Tannin

Since Carbores®P still exceeds the maximum B(a)P value regulated by European Parliament (<50 ppm), an environment-friendly binder system based on lactose and condensed tannin was examined for the Al_2O_3-C filter production. Both substances are used in the pharmaceutical and food industry demanding a high level of environmental compatibility. Due to the reactive nature of condensed tannins, interlinked carbon structures similar to phenolic resins are formed during the pyrolysis, which should ensure the filter stability [7]. In order to evaluate the usability of the environment-friendly binder system, the AC3 filter composition was adjusted, whereby Carbores®P was completely replaced by lactose/tannin (L/T) with a mixing ratio of 5:1. Besides the raw materials of AC3, Hexamethylenetetramine (hardening agent), aluminum powder (antioxidant), micro silica (antifoam agent), rutile as well as P-doped n-silicon or silicon carbide (stabilizer of carbon matrix) were added. The slip compositions (AC4 and AC5) are listed in Table 1.1.

The filter fabrication was conducted by the "rolling + spraying" method, as described in Sect. "Carbon-Bonded Alumina Filters Based on Carbores®P". Afterwards, the filters were hardened in a drying chamber up to 180 °C to initialize the polymerization of tannin monomers and pyrolyzed at 1000 °C under a reducing atmosphere. Both heat treatments corresponded to those of Himcinschi et al. [5].

Figure 1.3 shows the filter structure of AC4. On top of the filter surface, bright spots were detected, which most likely originate from the interactions of SiO_2 and carbon resulting in metallic silicon formation. The filters exhibited a rough surface with a large number of cracks, which were formed during the heat treatments due to the additional admixture of fine solid additives. Consequently, a lower CCS was obtained compared to the AC3 filters, as presented in Table 1.3.

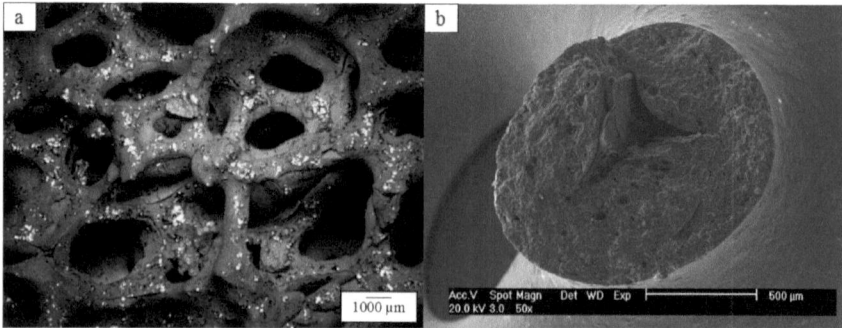

Fig. 1.3 **a** AC4 filter structure and **b** cross-section of a filter strut after pyrolysis [5]

Table 1.3 Filter properties of AC4 and AC5 with n-Si or SiC after pyrolysis [5]

Recipe	AC4	AC5 with n-Si	AC5 with SiC
Open porosity [%]	50.8	54.3	53.8
Cold crushing strength [MPa]	0.07	0.12	0.12

Another fact for the decrease of the CCS consisted in the increase of the open porosity. In comparison to Carbores®P, which converts into a liquid pitch-mesophase during the pyrolysis, L/T creates a rigid carbon matrix similar to phenolic resins. Hence, the pores remained unaltered within the Al_2O_3-C structure and affected the mechanical filter properties. Furthermore, a significant decrease in the RCC content (19.3 wt%) was observed. While Carbores®P has a RCC of 85 wt%, the L/T mixture achieved only 41 wt%. Based on the lower RCC, the carbon matrix evinced a weaker carbon bonding and hence a less stable carbon matrix. However, a positive impact on the filter properties provoked the addition of n-Si and SiC. Besides an improvement of the CSS, the RCC increased up to 20.6 wt%, whereby SiC exerted a slightly higher effect. This observation was attributed to the ability of both substances to emit electrons into the carbon matrix at elevated temperatures. In analogy to the mechanisms described by Yamaguchi et al. [8], the graphitization of the carbon matrix was enhanced resulting in a stronger carbon bonding. In order to evaluate the usability of the AC4 and AC5 filters for steel melt filtration, impingement tests were conducted. The AC5 filters with n-Si or SiC survived the thermal shock, whereas the AC4 filters were destroyed.

Consequently, it can be concluded that the AC5 filter composition with SiC is most suitable for steel melt filtration. Although Al_2O_3-C filters with Carbores®P obtained better filter properties, the new binder system based on lactose and tannin offers a great potential for manufacturing environment-friendly Al_2O_3-C filters.

1.3.2 Functionalization of Carbon-Bonded Alumina Filters Surface

As the deposition of non-metallic inclusions on the filter surface is determined by the attraction force and the adhesion between the filter material and inclusions [9], the functionalization of the Al_2O_3-C filter surface by carbon-free and carbon-containing coatings was analyzed. Therefore, three different coating techniques were investigated.

Surface Functionalization by Slip Spraying

Similar to the established Al_2O_3-C filter production, spray coating was the first technology applied to deposit functional coatings. For this purpose, coating slips were admixed and homogenized in a barrel with alumina grinding balls (d = 1.5 mm). Afterwards, the Al_2O_3-C filters were sprayed with an airgun of the type SATAjet B (Sata, Germany) with a nozzle diameter of 1.5 mm and a spraying distance of approx. 15 cm. The coated filters were placed in an alumina retort embedded in pet coke and sintered at 1400 °C for 3 h. An overview of the coating slip composition is presented in Table 1.4.

In order to enhance the filtration efficiency, pure oxidic coatings composed of alumina ("A", CL 370, Almatis Germany, d_{50} = 2.5 μm), spinel ("S-1", AR 78, Almatis, Germany, d_{50} = 2.0 μm) and mullite ("M", SYMULOX M 72 K0C, Nabaltec, Germany, d_{50} = 5.0 μm) were applied on thermally pretreated and untreated AC3 filters, respectively. After the pyrolysis at 1400 °C, macrocracks emerged within the filter structure of the untreated filter samples. This was attributed to the partial absorption of Carbores®P of the Al_2O_3-C substrate by the oxidic coating due to capillary forces and hence led to a deterioration of the binder matrix. In contrast, Fig. 1.4 shows the thermally pretreated filter samples, which evinced a stable oxidic coating with a thickness of 50 to 80 μm after the pyrolysis. The intermediate layer without solid content (marked with a dashed arrow in Fig. 1.4a) indicated that the coating was not chemically bonded to the Al_2O_3-C substrate, but adhered due to mechanical bonding within the coating. This observation can be explained by the sintering behavior of both media. The Al_2O_3-C substrate underwent a thermal expansion until 850 °C due to the dehydrogenation process resulting in a slight detachment of the coating and the formation at the gap. Afterwards, the oxidic coatings shrank on the filter surface due to further heating and cooling. As a result, the particles interlocked mechanically with each other and compression stresses were generated within the coatings, which ensured the coating stability. Since a higher shrinkage leads to an increase in the mechanical bond as well as to a decrease of the porosity, the alumina-coated Al_2O_3-C filters exhibited the highest CCS of 0.63 MPa, followed by spinel- (0.54 MPa) and mullite-coated (0.47 MPa) Al_2O_3-C filters. Impingement tests revealed that the coated filters withstood the thermal shock of molten steel and

Table 1.4 Slip composition for functional coatings, in wt% [9–16]

Material	Recipe								
	A	M	S-1	S-2	S-3	A-C	S-C	C-C	Nanos
Alumina	100	–	–	71.7	78	93.75	–	–	–
Magnesia	–	–	–	28.3	22	–	–	–	–
Mullite	–	100	–	–	–	–	–	–	–
Spinel	–	–	100	–	–	–	–	–	–
Spinel (Fe–Mn–Mg–Al-O)	–	–	–	–	–	–	93.75	–	–
Calcium aluminate	–	–	–	–	–	–	–	65.1	–
Nano materials	–	–	–	–	–	–	–	–	0.003
Carbores®P	–	–	–	–	–	6.25	6.25	34.9	24.9

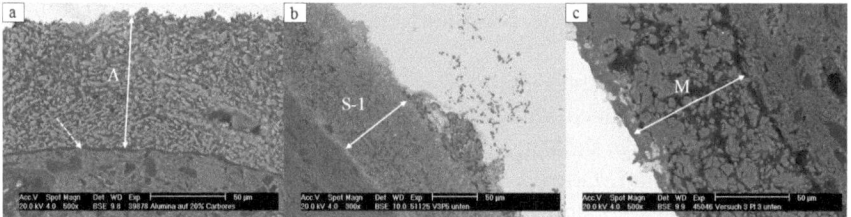

Fig. 1.4 Structure of Al_2O_3-C filters coated with **a** alumina, **b** spinel and **c** mullite

led to a deposition of inclusions on the filter surface, which underlines their suitability for steel melt filtration [9, 17].

However, an excessive shrinkage is negative. On the one hand, too high compressive stresses initiated the formation of cracks and defects, which caused a complete detachment of the coating. On the other hand, the carbothermal reaction and the gas exchange were affected, as the reduction of the porosity contributed to a permeability decrease of the coatings. For this reason, carbon was added to the oxidic coatings due to its flexible structure as well as its influence on the coating properties and analyzed regarding the interactions with the steel melt.

In the first instance, the wettability behavior of molten steel on sintered alumina ("A-C", Martoxid MR 70, Martinswerk, Germany, $d_{50} = 1.8$ μm) coatings with 0, 4 and 30 wt% carbon (after the pyrolysis) was determined between 1500 and 1520 °C within 30 min. Since the steel melt exhibits a poorer wetting on carbon compared to alumina, an increase in the contact angle with increasing carbon content was observed. Additionally, the larger particle size of the carbon materials as well as the higher open porosity formed by the removal of volatiles effected an increase of the coating surface roughness, which resulted in a further increase of the contact angle up to 144°. For all samples, a decrease in the contact angle occurred with increasing wetting time. After 15 min, an almost constant value was detected, which was attributed to the formation of an in-situ formed layer on the coating surface due to the carbothermal reaction. Hence, it was concluded that the addition of carbon enhances the steel melt purification, as a higher contact angle encourages the deposition of solid, non-metallic inclusions due to the more favorable interfacial energy situation [10]. Furthermore, an improvement of the coating structure was achieved. Besides a slight decrease in the shrinkage, the admixture of carbon provoked a higher open porosity due to the release of volatiles during the sintering. For this purpose, carbonaceous coatings (6.25 wt% carbon) based on alumina ("A-C", Martoxid MR 70, Martinswerk, Germany, $d_{50} = 1.8$ μm) and calcium aluminates ("C–C", Kerneos, France) were applied on Al_2O_3-C filters, pyrolized at 1400 °C under reducing atmosphere and brought into contact with molten steel. The results evinced distinct interactions between the coating materials and the steel melt, whereby a significant amount of inclusions was removed due to the "reactive" and "active" filtration mechanisms. Especially calcium aluminates turned out to be most effective due to the reactive nature of CaO under steelmaking conditions and the associated gas bubble formation [11, 12].

In another approach, the feasibility of an in-situ spinel synthesis during the filtration process was investigated. Since the spinel formation is accompanied by a volume expansion, the excessive shrinkage could be counteracted. Therefore, filter samples composed of alumina (Martoxid MR 70, Martinswerk, Germany, $d_{50} = 1.8$ µm), magnesia (Refratechnik, Germany, $d = 0–63$ µm) and with or without carbon sources, respectively, were manufactured according to "S-2" and "S-3" (cp. Table 1.4) and thermally pretreated at 800 °C. Afterwards, the filters were immersed in molten steel at 1640 °C. Preliminary examinations of the sintering behavior revealed that the spinel formation started at a temperature of 925 °C and caused a volume expansion of 0.7% (S-2) and 2.4% (S-3). The spinel formation proceeded with increasing temperature and was almost completed at 1600 °C for S-3, whereby an overall shrinkage of 0.4% was obtained after cooling. In comparison, the presence of carbon inhibited the spinel formation and led to a slightly higher shrinkage of approx. 0.55% in the case of S-2. A contrary development was found after the immersion into the steel melt. Here, the alumina/magnesia mixture of S-2 was completely transferred into the spinel phase, with only a small amount of alumina and magnesia remaining unaltered. It was concluded that the carbothermal reaction lowers the reaction temperature and promotes the in-situ spinel formation under casting conditions [13]. Although the in-situ spinel formation was successfully conducted, the filter samples exhibited lower mechanical properties before the immersion attempt due to the absence of a sintering step. For this purpose, the Al_2O_3-C filter surface was coated with presynthesized spinel compounds $MgAl_2O_4$, $FeAl_2O_4$, $MnAl_2O_4$, $Fe_{0.5}Mg_{0.5}Al_2O_4$ and $Fe_{0.5}Mn_{0.5}Al_2O_4$ ("S-C", AGH University of Science and Technology, Poland) in combination with carbon and sintered at 1400 °C under reducing atmosphere. The qualitative analysis of the phase composition before and after sintering indicated that a partial decomposition of the $FeAl_2O_4$, $MnAl_2O_4$, $Fe_{0.5}Mg_{0.5}Al_2O_4$, and $Fe_{0.5}Mn_{0.5}Al_2O_4$ took place during the pyrolysis most likely due to the low oxygen pressure, which encouraged the decomposition of the spinel compounds at elevated temperatures. $MgAl_2O_4$ was not affected by the decomposition due to higher thermal resistance. The carbonaceous spinel coatings withstood the thermal shock by the immersion into the steel melt and entailed a considerable removal of non-metallic inclusions, whereby metallic manganese formed during the decomposition was assumed to exert a catalytic effect similar to iron [14].

While the oxidic coatings presented were applied to enhance the inclusion deposition rate on the filter surface, the possibility of improving the filter reactivity was focused on within the next step. Therefore, nano-scaled materials were adopted, since their high specific surface could promote the "reactive" filtration behavior. The used raw materials were multi-walled carbon nanotube ("MWCNT", TNM8, Chengdu Organic Chemicals, China) and graphene oxide ("GO", VSCHT Praha, Czech Republic), which were prepared in advance according to a modified Tour's method [18]. For the preparation of the nano materials dispersions, xanthan stock solutions were dissolved in deionized water under high-shear mixing for 15 min to ensure the total hydration of the biopolymer [15, 19]. Afterwards, an appropriate mass of MWCNT was added and the resulting mixtures were ultrasonicated with an ultrasonic probe (Sonopuls HD 2200, 20 kHz, 200 W) for 3 min at 50% amplitude to force unbundling of the nanotubes and promote the interaction with the surfactant molecules. Finally, the suspensions were thoroughly stirred for 30 min to achieve proper homogeneity [20]. The preparation of the GO-based dispersion was performed similarly [16, 21]. For the binding of carbon nanotubes and graphene oxide to the ceramic foams during the heat treatment, a separate Carbores®P dispersion was prepared in deionized water (50 wt%) using horizontal ball milling for 1 h. Furthermore, 0.3 wt% ligninsulfonate (T11B, Otto-Dille, Germany) and 0.1 wt% polyalkylene glycolether (Contraspum K 1012, Zschimmer & Schwarz, Germany) were used to promote the wettability of the pitch powder with the water and to prevent foaming during the preparation. In order to improve the stability of the system and achieve a proper rheological behavior, Xanthan gum (0.3 wt%) was added. After the different dispersions were mixed in various weight ratios and stirred for 5 min, the resulting coating slips ("Nanos", cp. Table 1.2) were applied on Al_2O_3-C filters. Subsequently, the filters were pyrolyzed at 800 °C under a reducing atmosphere for 3 h. The heating was realized with 0.8 K min^{-1}, whereby a heating rate of 0.3 K min^{-1} between 100 and 250 °C was conducted in the case of a GO-coated filter to prevent sudden exfoliation and allow a slow transformation of GO into thermally reduced graphene.

The examination of the mechanical filter properties revealed that an increasing proportion of nano materials related to the Carbores®P content entailed a progressively worse CCS of the filter samples. On the one hand, this observation was attributed to the oxidation of the coating, which cannot be completely avoided during the pyrolysis. Thereby, the oxidation was promoted by the higher surface area, which increased with increasing nano material proportion. On the other hand, the decreasing amount of Carbores®P impaired the crack and porosity cure, achieved by the infiltration of the liquid pitch mesophase and the subsequent graphitization. Nonetheless, the mechanical properties were sufficient to withstand the thermal shock of molten steel at 1650 °C.

Figure 1.5 presents the microstructure of the Al_2O_3-C filters after the pyrolysis functionalized with MWCNT, GO and MWCNT-GO. While the presence of cracks was observed for the GO filters, the GO-MWCNT filters exhibited an apparently smooth surface covered by reduced graphene oxide. In the case of MWCNT, clusters of Carbores®P residuals and MWCNT agglomerates with dimensions of a few microns were found resulting in an irregular filter surface. Since no such large agglomerates were detected in the dispersions before spraying, they likely developed during the pyrolysis [16]. Although a rougher surface should contribute to an increase in the wetting angle between molten steel and filter material, the GO-MWCNT coating exhibited the highest contact angle (>150°), followed by GO and MWCNT [21]. It was presumed that the high specific surface of the nano materials provoked a more intense carbothermal reaction, whereby the roughness increased due to the reaction of the nano materials and oxygen dissolved molten steel. Hence, the application of carbon-based nano materials constitutes an excellent opportunity to improve the filter reactivity and purification degree of molten steel.

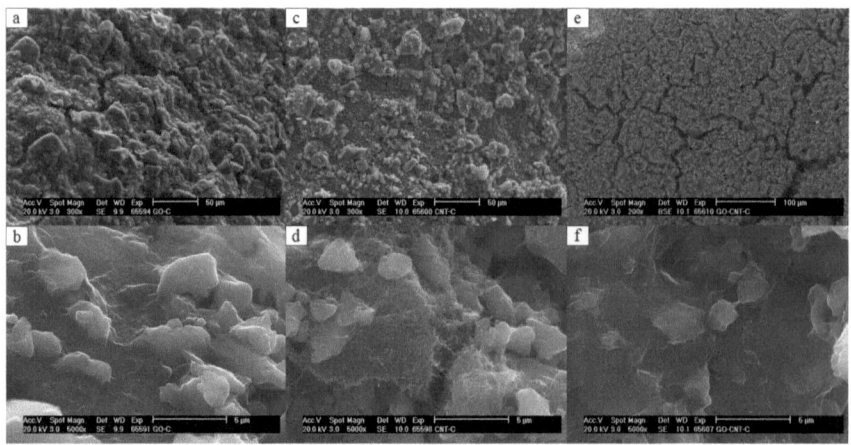

Fig. 1.5 Microstructure of Al_2O_3-C coated with **a, b** graphene oxide, **c, d** multi-walled carbon nanotubes and **e, f** a combination of graphene oxide and multi-walled carbon nanotubes [16]

Surface Functionalization by Flame Spraying

In order to avoid a second pyrolysis step and the associated impairments of the filter and coating structure mentioned in the previous section, the application of functional coatings by flame spraying was investigated. Compared to slip spraying, the flame spraying technique is based on the application of melted ceramic particles on the filter surface, where the particles rapidly solidify, sinter and form a dense coating. As a result, the functional coating reaches immediately its final state after the application and further thermal treatments can be omitted.

For this purpose, sintered rods composed of 100% Al_2O_3 ("A-FS") and Al_2O_3-ZrO_2-TiO_2 (AZT, "90-5-5", "85-5-10" and "80-10-10") were prepared to be utilized as feedstock. The raw materials were mixed with a thermoplastic binder, extruded with a heatable extruder of the type DSE 25 (Brabender, Germany), and sintered at 1500 °C. Furthermore, an Al_2O_3 flexicord feedstock was used, which was composed of an alumina powder material in a paste coextruded with a flexible organic skin.

The A-FS and AZT rods as well as the alumina flexicord were fed to the Master Jet® Gun (Saint-Gobain Coating Solutions, France) to coat the Al_2O_3-C filters or produce free-standing structures. Inside the flame spraying gun, oxygen and acetylene as fuel gas combusted to melt the feedstock. The combustion temperature (max. approx. 3160 °C) depends on the flow rate of the oxygen and acetylene gases and can be controlled by adjusting the gas pressure and the ball height at the flowmeter [22]. During the flame spraying process, compressed air was used to fluidize, transport and at the same time homogenize the melted drops. Thereby, several traverses were necessary until the overlaying splats formed an effective coating that nearly completely covered the substrate.

Before characterizing the flame-spray coated Al_2O_3-C filters, promising freestanding flame-sprayed structures composed of A-FS and AZT were investigated.

Flame-sprayed alumina was characterized by a typical surface with splats of varying morphology such as almost circular discs, pancakes, and splashes of so-called flower structures as well as microcracks. At polished cross-sections, the typical lamellar structure composed of flat Al_2O_3 grains with the highest extension perpendicular to the spray direction, pores and additional phases inside the grains became visible. In addition, grain boundaries were observed in the case of the 90-5-5 and 85-5-10 samples. Based on the 85-5-10 samples, two different microcrack patterns were identified: (1) A primary pattern with major propagating cracks with widths between 150 and 300 nm, which was formed by the rapid cooling/solidification (cooling rate of approx. 10^6 K s^{-1}) was partially interconnected with (2) a secondary pattern with microcracks in the range of 120 nm and less, which resulted from slow crack growth during further cooling of the droplets after their solidification. Thereby, the feedstock composition played an important role. Due to its ability to create multiple phases during the flame spraying process, the 85-5-10 showed a higher level of interconnection between the microcracks compared to A-FS. The A-FS samples exhibited a comparatively low open porosity of 7.1%. The porosity increased with the addition of titania and zirconia (9.7 and 9.3% in sample 90-5-5 and 85-5-10, respectively), most likely due to the incomplete melting of the zirconia particles. Controlled crack propagation experiments of bar-shaped samples yielded a distinct R-curve behavior for A-FS and 80–10-10 samples with initial values of K_{I0} of approx. 0.5 MPa m$^{1/2}$, while plateau values reached about 1.8–2.5 MPa m$^{1/2}$ [23]. The rising portion between those characteristics spanned a range of approximately 100–400 µm of the crack extension. It is assumed that the raising crack resistance was related to the microstructure as a result of energy dissipating processes like crack deflection and crack branching on the one hand and crack wake interactions such as microcrack toughening and crack bridging on the other hand [24]. In order to evaluate the thermomechanical properties, the B3B method was conducted before and after thermal shock trails. It was found that A-FS discs evinced a strength of 83.46 MPa with a Weibull modulus m of 5.48 before and 42.27 MPa with a Weibull modulus m of 1.88 after thermal shock at 1000 °C to room temperature [25]. Since the B3B method is also suitable for flexural tests at elevated temperatures, an increasing ductile behavior was identified with increasing temperature. The determination of the residual bending strengths of band-shaped samples after thermal shock at rising temperature gradients revealed that all A-FS and AZT compositions presented a similar thermal shock behavior with an average strength plateau kept up to a specific temperature gradient, whereby a significant strength loss was registered after a critical temperature gradient was applied. Due to a microcrack toughening effect, the A-FS and AZT compositions demonstrated a slightly increasing strength up to quenching temperature difference in the range of 800 (A-FS) to 900 K (85-5-10). Hence, flame sprayed A-FS and AZT provide outstanding thermal shock resistance.

For this purpose, the application and impact of flame-sprayed A-FS coatings on commercial Al_2O_3-C filters were examined. After four traversing steps, an average coating thickness of 90 µm was obtained. As a very high cooling rate occurred during the solidification of the molten alumina droplets, the formation of a high portion of amorphous phases and metastable transition γ- and η-Al_2O_3 as well as a low portion of corundum was detected [22].

Figure 1.6 presents the filter structure after the coating procedure. It was observed that an irregular and patchy coating predominantly present at the top of the struts and filter parts facing the flame-spray gun arose, since the particles did not flow across the surface of the filter struts due to the rapid cooling. Thereby, the typical structure of overlaid molten droplets became visible. At some positions of the flame-spray coated surface, the subjacent Al_2O_3-C microstructure as well as microcracks crossing the droplets could be seen. The CT analysis revealed a thick, crack-free and dense coating in particular at the outside area of the filter. However, a continuous coating could only be detected in the outer regions of the filter, covering approx. one-third of the filter towards the center. The core of the filter was not well coated, since the outer filter struts obscured the inner part of the filter volume. Regarding the mechanical filter properties, a CCS reduction of approx. 30% was measured. This was traced back to an emerging gap of up to 25 μm between the filter and the coating. Due to the heat transfer from the molten droplets to the filter substrate, its temperature increased to max. 450 °C, expanded during flame spraying and shrank after finishing the spray process. As the molten particles immediately cooled to ambient/ substrate temperature after impacting onto the filter and hence did not shrink, a gap between filter and coating was formed. Nonetheless, the flame-spray coatings survived the immersion attempts in molten steel and resulted in an inclusions deposition due to an improved "active" filtration behavior.

In a further approach, the feasibility of producing filters completely composed of flame-sprayed alumina structures was investigated to encounter the disadvantages of sintered oxide and carbon-bonded filters. Therefore, Al_2O_3-C filters were flame-spray coated with A-FS and subsequently sintered at 800, 1100 and 1400 °C, respectively, to burn the carbon-bonded filter substrate. As long as there is no notable sintering effect, the remaining alumina flame-spray coating covering the decomposed carbon-free residual alumina filter was too thin and weak to act as a reliable load-bearing structure. This led to a negligible CCS of 0.01 and 0.02 MPa of the coated filters after sintering at 800 and 1100 °C, respectively, whereas sintering at 1400 °C led to a higher CCS of 0.16 MPa. Casting tests with molten steel will reveal whether

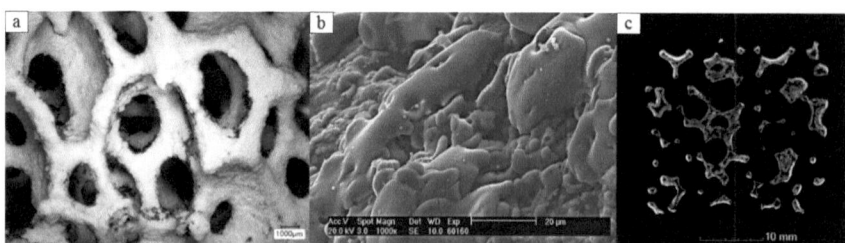

Fig. 1.6 Structure of A-FS coated Al_2O_3-C filters by **a** digital light microscopy, **b** SEM and **c** CT scan [22]

this strength is high enough for the application of such new filters for molten metal filtration.

Surface Functionalization by Electrospinning

Refractory fibers represent a kind of 1D materials with many advantages such as light-weight, good thermal stability, low thermal conductivity, small specific heat and a high surface-to-volume ratio. A large number of advanced techniques have been developed to fabricate 1D nanostructures with well-controlled morphology and chemical composition. Unfortunately, most of these methods are multistep, require high energy or generate secondary products. Electrospinning provides a simple and versatile process capable of generating continuous fibers with diameters ranging from a few tens of nanometers to several micrometers. Based on this, functional fiber coatings produced by electrospinning constitute a promising opportunity to increase the filter reactivity and improve the steel melt purification.

In order to evaluate the feasibility of the coating fabrication by electrospinning, fiber mats composed of magnesium borate ("MB", $Mg_2B_2O_5$) and calcium zirconate ("CZ") were analyzed. For this purpose, the raw materials boron ethoxide ($C_6H_{15}BO_3$) and magnesium ethoxide ($C_4H_{10}MgO_2$) as well as zirconium basic carbonate ($Zr(OH)_2CO_3 \cdot ZrO_2$) and calcium nitrate tetrahydrate ($Ca(NO_3)_2 \cdot 4H_2O$) were used as precursors. Furthermore, polyvinyl-pyrrolidone powder, 2-Methoxyethanol, N,N-dimethylformamide, glacial acetic acid and methanol were utilized as solvents to generate proper fibrous structures. The raw materials were purchased from Sigma–Aldrich (USA), while the solvents were derived from Carl Roth GmbH (Germany). The production of the fibers was performed in a NE100 unit (Inovenso, Turkey). The prepared electrospinning solutions were loaded into a 10 ml syringe connected to a 0.8 mm diameter conductive brass needle and fed at 0.5 ml h^{-1} with a syringe pump (NE-300, New Era Pump Systems, USA). Electrospinning was carried out at room conditions and at a voltage of 15 kV using a DC power supply. An aluminum plate covered with aluminum foil was used as the collector and the distance between nozzle and collector was 10–12 cm. After the process, the as-spun fiber mat was removed from the aluminum foil. The magnesium borate fibers were then pyrolized at 800 °C under a reducing atmosphere for 3 h or at 900 °C in air for 1 h, whereas the calcium zirconate fibers were heat-treated at 800 or 1000 °C in air for 1 h. At least, Al_2O_3-C filters were used instead of the aluminum plate to apply the fibrous coating and sintered at 800 °C under reducing atmosphere for 3 h [26, 35].

The SEM analysis of the MB fibers revealed that a matt of continuous fibers without any droplets or beads was created, whereby the fibers exhibited a main diameter of 500–700 μm. After the pyrolysis at 900 °C in air, the fibers consisted of small particles, which joined together to form a thin filament and are known as "necklace" structure. Thereby, a decrease of the fiber diameter to 300–500 μm took place due to the thermal release of aqueous and organic precursors. The fibers were primarily composed of $Mg_2B_2O_5$ with $Mg_3B_2O_6$ as a secondary phase. In comparison, the pyrolysis under a reducing atmosphere caused a similar shrinkage,

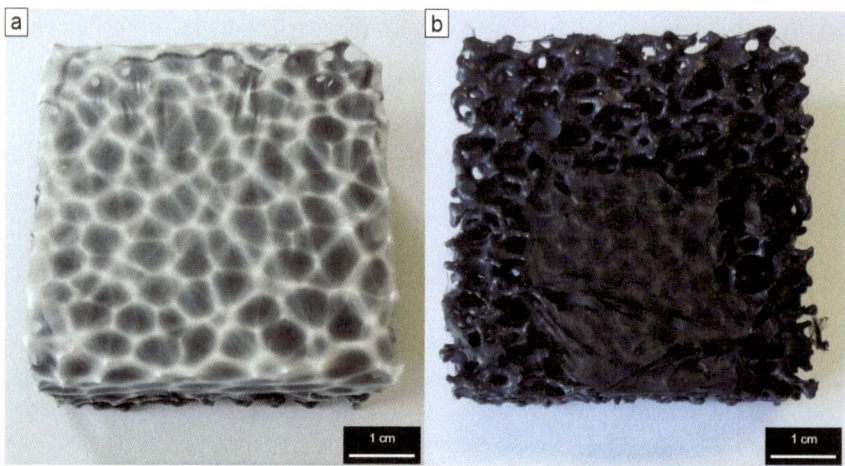

Fig. 1.7 Al_2O_3-C filters coated by electrospinning **a** before and **b** after the pyrolysis

but the occurrence of the "necklace" structure was avoided. However, the specific surface was 10-times higher and the main phase consisted of $Mg_3B_2O_6$ due to the evaporation of boron. The CZ fibers evinced a diameter between 400 and 800 μm in the green state, which was halved after the pyrolysis. Thereby, the formation of a "necklace" structure was not observed. Regarding the phase composition, a higher conversion of the precursors to $CaZrO_3$ was achieved with increasing pyrolysis temperature, reaching a purity above 98% at 1000 °C [26, 35].

As presented in Fig. 1.7, the fiber mats were successfully applied to the filter surface. However, the fibers were deposited only on the first conductive surface in reach, but not in the filter pore volume. Due to the shrinkage and lack of bonding between fibers and filter substrate, a detachment and loss of the fiber mat was inevitable after the pyrolysis. Hence, the application of functional coatings by electrospinning is possible, but not expedient for the steel melt filtration.

1.3.3 Generative Manufacturing of Filter Structures

Apart from the chemical composition of the filter structures, the filter design exerts a significant impact on the filter properties and the steel melt purification efficiency. While the previous methods resulted in random, anisotropic filter structures with sharp-edged cavities within the filter struts, the generative filter manufacturing allows the fabrication of well-defined filter geometries, which can be adjusted to each specific application. For this purpose, novel approaches in filter fabrication were developed, which are described in the following sections.

Alginate-Based Robo Gel Casting

In a first approach, the feasibility of producing filter structures with full-strut cross-sections by in-situ gel casting was investigated. Therefore, sustainable alginates were admixed to ceramic slips and cast in an aqueous solutions enriched with divalent ions such as Ba^{2+} or Ca^{2+}. Since alginates are composed of two of the monomers D-mannuronic and L-guluronic which exhibit a ribbon and buckle geometry, a polymerization of the monomers is initialized by the integration of divalent ions in the structure. As a result of the cross-linkage, a stiff gel is formed ensuring the dimensional stability of the ceramic component.

In order to fabricate Al_2O_3-C filter structures, carbonaceous alumina slips based on the AC3 slip composition (cp. Sect. "Carbon-Bonded Alumina Filters Based on Carbores®P") with a total solid content of 50.0, 52.6 and 55.6 wt%, respectively, were admixed with 0.35, 0.50 or 0.65 wt% sodium alginate (C.E. Roeper GmbH, Germany). Furthermore, the additives polyvinyl alcohol (temporary binder, C.T.S, Italy), polyacrylic acid (dispersing agent, Dolapix PC21, Zschimmer & Schwarz, Germany) and polyalkylene glycol ether (antifoam agent, Contraspum K 1012, Zschimmer & Schwarz, Germany) were used. The resulting slips were continuously pumped through a tube with a diameter of 2 mm and a length of 1 m into an aqueous cross-linking solution enriched with 1 wt% $Ba(NO_3)_2$ or $CaCl_2$ using a peristaltic pump of the type Pumpdrive 5206 (Heidolph Instruments, Germany). The tube was fixed to a robot arm (Robot armMover 4, Commonplace Robotics, Germany), which was controlled by the software RobotExpress (Commonplace Robotics, Germany). Based on this setup, droplet beads and filter samples with a periodic grid structure were manufactured and dried at room temperature. Subsequently, the filter samples and the beads were pyrolized at 800 °C for 3 h under a reducing atmosphere. Besides the rheological behavior of the slips, the dimensional stability and mechanical properties of the samples were analyzed.

All slips exhibited a shear-thinning behavior without major thixotropic effects. The dynamic viscosity increased with increasing alginate and total solid content. However, the viscosity increase caused by the increase of alginate content can be counteracted by increasing the water content. Thereby, the addition of 10 wt% water related to the solid content countered an alginate increase of 0.15 wt% to obtain slips with the same rheological behavior. Furthermore, a strong dependency between the filter dimensions and the slip composition was observed. While low alginate and total solid contents led to an increased shrinkage and sagging of the filter struts, the shrinkage in the filter height plateaued at 30% for high contents. Hence, exceedance of the optimal amount would only result in an increase of the viscosity, but not in dimensional stability. Additionally, a radial strut shrinkage up to 13.6% occurred, which has to be taken into account for the modelling of the filter macrostructure. Since the determination of the mechanical properties for all pyrolized bead samples did not reveal any noticeable differences, an alginate content of 0.65 wt% and a total solid content of 55.6 wt% turned out to be the most suitable composition for the alginate-based robo gel casting. In the case of the cross-linking solutions, the utilization of Ba^{2+} ions entailed a stiffer gel structure and increased filter properties

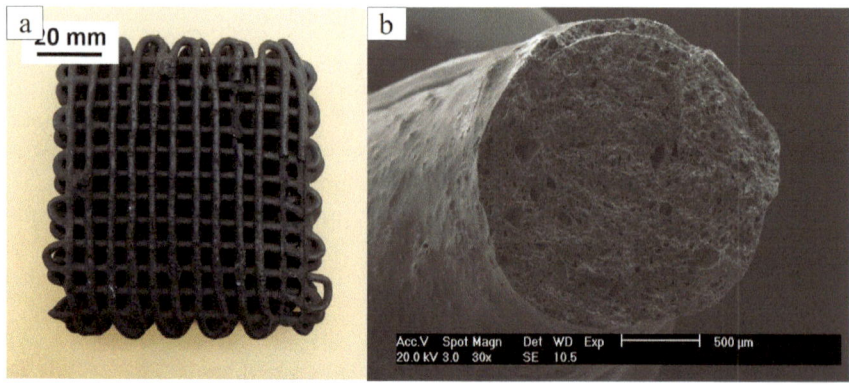

Fig. 1.8 **a** Al_2O_3-C filter structure manufactured by alginate-based robo gel casting with periodic grid pattern and **b** full-strut cross-sections [27]

compared to Ca^{2+} ions. As shown in Fig. 1.8, the filter exhibited a periodic grid structure with full-strut cross-sections. However, as the single slip immediately gels after the contact with the cross-linking solution, the single strands lie on top of each other but are not connected at the knot points. Therefore, the filters were coated by an AC3 spraying slip before the pyrolysis, which avoided a shift of the strands. As a result, Al_2O_3-C filter structures with an improved resistance against external load and a high interaction performance with molten steel were fabricated. Since the procedure is not restricted to dimensional limits, this novel approach enables the feasibility of easy upscaling and thus offers a promising alternative for the fabrication of ceramic filters [28].

Filter Structures Based on Water-Soluble Filter Templates

In a further approach, the adjustment of the filter template geometry was explored. In order to avoid the disadvantages of commercial PU foams mentioned before, filter template geometries with round filter struts were designed, modelled and evaluated regarding their impact on the steel melt filtration efficiency [29, 30]. The filter designs with the highest filtration potential were selected and fabricated using additive manufacturing methods. In particular, fused filament fabrication ("FFF") and selective laser sintering ("SLS") were focused on due to the variety of available raw materials. The examination of the resulting filter templates revealed that the filter templates produced by SLS exhibited a rough surface, whereas the FFF generated templates with a smooth extruded surface. Since a rougher surface encouraged a better adhesion of coating slips to the template, the SLS technique based on the device SnowWhite (Sharebot S.r.l., Italy) with a printing area of $100 \times 100 \times 100$ mm^3 and a 14 W CO_2 laser was considered for further experiments [31]. As most polymers undergo a high volume expansion at evaluated temperatures, which cause macrocracks in the ceramic structure and could lead to a filter breakdown [32], the fabrication of filter structures based on water-soluble raw materials was investigated. Besides an improvement of the filter structure, water-soluble raw materials offer the possibility for recycling in the following processes. Thereby, two variants were conducted to produce either "reactive" or "active" filter structures.

"Reactive" Al_2O_3-C filter structures with 5 ppi macrostructure were manufactured using polyvinyl alcohol ("PVA") filter templates as described by Bock-Seefeld et al. [33]. Since conventional coating slips would lead to premature dissolution of the template and hence to a deformation of the filter structure, the filter templates were dip-coated into an alginate-based slip, which corresponded to the final slip composition mentioned in Sect. "Alginate-Based Robo Gel Casting". Subsequently, the coated filters were placed in an aqueous cross-linking solution for 48 h, which was enriched with 1 wt% $Ba(NO_3)_2$. Simultaneously to the formation of the alginate network which avoided a collapse of the filter structure, the filter template was completely dissolved and removed from the filter structure. However, the dissolution was accompanied by gas bubble formation, which caused breaches and cracks in the Al_2O_3-C filter structure. Nonetheless, the filter structures exhibited sufficient dimensional stability in the green state, which is crucial for further processing. To cure the defects and to increase the coating thickness, the filters were spray-coated twice with alginate-free slips (AC3 spraying slip, cp. Sect. "Carbon-Bonded Alumina Filters Based on Carbores®P") and pyrolyzed at 800 °C for 3 h under a reducing atmosphere.

As shown in Fig. 1.9, most of the macroscopic defects were successfully sealed after the pyrolysis. However, the filter evinced a porous structure, especially in the inner part of the filter strut, which was traced back to the removal of the higher water and alginate content of the alginate-based slip. Due to the multiple application of the coatings, the initial functional pore volume of the filter template diminished. Furthermore, a shrinkage in the filter height of approx. 18% occurred so that the final filter structure almost corresponded to an Al_2O_3-C filter with a 10 ppi macrostructure. After the thermal treatment, the filter structure achieved a CCS of approx. 0.13 MPa, which was slightly lower compared to conventional Al_2O_3-C filters. Although the generation of round cavities within the filter structure should inhibit the growth of critical cracks and hence improve the mechanical filter properties, it was assumed that the comparably larger cavity size superposed the beneficial effect. Furthermore, a decrease in the coating thickness with increasing distance from the filter surface due to the spraying process was observed, which also affected the CCS. Nonetheless, the novel method of Al_2O_3-C filter production using water-soluble PVA templates coated by alginate-based slips provides a promising alternative for "reactive" filter systems [33].

For the production of "active" filter structures, filter templates composed of water-soluble polysaccharide were additively manufactured. The filter templates were coated with Al_2O_3 employing flame spraying, whereby the spraying parameters corresponded to those described in Sect. "Surface Functionalization by Flame Spraying". Although the polysaccharide exhibits a higher thermal resistance compared to polymeric raw materials, the spraying procedure was performed with multiple spray shots pausing for several seconds after each shot to avoid thermal

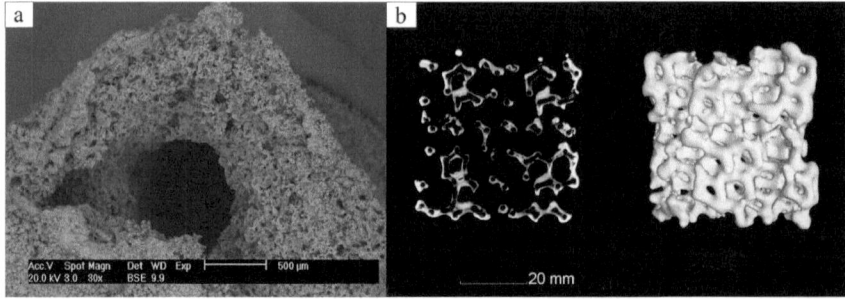

Fig. 1.9 Structure of an Al_2O_3-C filter manufactured by water-soluble PVA filter templates and alginate-based slips by **a** SEM and **b** CT scan [33]

damage. This procedure was repeated from all directions until a dense alumina coating was formed on the filter template struts. Afterwards, the filters were placed in a stirred water bath for 24 h to remove the polysaccharide template. Since the flame-sprayed alumina coating reached its final state after the application, no thermal treatment at high temperatures had to be carried out.

After the removal from the water bath and subsequent drying at room temperature, a dense free-standing filter structure was achieved, which can be used immediately. The coating procedure turned out to be a challenging process, as the spraying have to be carried out by very quick shots to avoid melting the template. Hence, their number and the required coating time increased significantly. The median wall thickness at the filter surface averaged 300 μm. However, the wall thickness decreased with increasing distance from the filter surface, which was already observed in previous studies for flame-sprayed structures (cp. Sect. "Surface Functionalization by Flame Spraying"). Nonetheless, the resulting filter structures evinced good mechanical stability after the complete removal of the polysaccharide template. Due to the excellent thermal shock resistance of flame-sprayed Al_2O_3, the filter withstood the thermal shock by immersion into molten steel at 1650 °C [34]. For this reason, "active" filters based on water-soluble filter templates and flame spraying could play an important role in the purification of steel melts.

1.4 Conclusion

In order to reduce the amount of solid, non-metallic inclusions and hence improve the quality of cast steel products, the present work dealt with the exploration of ceramic foam filters for the steel melt filtration.

In a first approach, a new generation of "reactive" filters based on carbon-bonded alumina was developed, which initialize a carbothermal reaction with molten steel and lead to inclusion flotation due to the formation of gas bubbles. In comparison to other carbon-bonded refractories, using pitches and resins with an increased amount of carcinogenic components, modified coal tar pitch (Carbores®P) and a mixture of lactose and tannin were successfully applied as carbonaceous binder systems. Both binder systems encouraged the environmental sustainability of the carbon-bonded alumina filters, whereby the Carbores®P-containing filters stood out due to their excellent thermomechanical properties and high residual carbon content, which is essential for an improved carbothermal reaction with molten steel. Since the intensity of the interactions between the filter and molten steel is determined by the composition of the filter surface, the functionalization feasibility of the carbon-bonded alumina filter surface by slip spraying, flame spraying and electrospinning was investigated. In terms of slip spraying, oxidic coatings exhibited sufficient thermomechanical properties, so that they can promote an "active" filtration. However, slip-sprayed coatings require a further sintering step, which resulted in an increased shrinkage behavior and hence defect occurrence. Therefore, carbonaceous coatings were developed to counteract these disadvantages. Due to the addition of carbon, a considerable increase in the thermomechanical coating properties was achieved. Furthermore, an increase in the filter reactivity was obtained, whereby carbonaceous coatings composed of calcium aluminate and nano materials showed the greatest potential for steel melt purification. A promising alternative to slip spraying provided the flame spraying technique. Although the application of a homogenous flame-sprayed coating turned out to be a challenging process and is restricted to oxidic coatings due to the high process temperatures, the resulting coating structure exhibited an outstanding thermal shock resistance and improved thermomechanical properties, which is crucial for their application. Since no further thermal treatments are needed, the filters can directly be applied in the process. In contrast, electrospinning did not prove expedient for the filter surface functionalization, as the coatings detached during the production.

Since not only the chemical composition but also the filter design exerts a tremendous impact on the steel melt purification performance, novel filter production approaches based on additive manufacturing were explored. Based on the alginate-based robo gel casting and water-soluble filter templates, "reactive" and "active" filter structures were successfully fabricated. As the filter geometry can be easily adjusted for all steelmaking conditions, these processes open new perspectives for steel melt purification.

Based on the present work, a large variety of filter materials and filter manufacturing techniques for steel melt filtration have successfully been developed. While the individual filter types already lead to a reduction in inclusions, the intelligent design of combined filter systems could result in a new era of high-quality steel products. A detailed characterization of the interactions between the filter systems and molten steel as well their contribution to the removal of solid, non-metallic inclusions will be provided in the following chapters.

Acknowledgements The authors thank the German Research Foundation (DFG) for the financial support of this study within the framework of the Collaborative Research Center 920, subproject A01 (Project-ID 169148856). Furthermore, the authors acknowledge Dr.-Ing. Marcus Emmel, Dr.-Ing Anne Schmidt and Dr.-Ing. Enrico Storti, who made a decisive contribution to this work.

References

1. A. Schmidt, A. Salomon, S. Dudczig, H. Berek, D. Rafaja, C.G. Aneziris, Adv. Eng. Mater. **19**, 1700170 (2017). https://doi.org/10.1002/adem.201700170
2. C.G. Aneziris, F. Homola, D. Borzov, Adv. Eng. Mater. **6**, 562 (2004). https://doi.org/10.1002/adem.200400417
3. B. Rand, B. McEnaney, Br. Ceram. Trans. **84**, 157 (1985)
4. M. Emmel, C.G. Aneziris, Ceram. Int. **38**, 5165 (2012). https://doi.org/10.1016/j.ceramint.2012.03.022
5. C. Himcinschi, C. Biermann, E. Storti, B. Dietrich, G. Wolf, J. Kortus, C.G. Aneziris, J. Eur. Ceram. Soc. **38**, 5580 (2018). https://doi.org/10.1016/j.jeurceramsoc.2018.08.029
6. B. Luchini, J. Hubalkova, T. Wertzig, J. Grabenhorst, J. Fruhstorfer, V.C. Pandolfelli, C.G. Aneziris, Ceram. Int. **44**, 13832 (2018). https://doi.org/10.1016/j.ceramint.2018.04.228
7. C. Biermann, Dissertation, Technische Universität Bergakademie Freiberg, 2016
8. A. Yamaguchi, S. Zhang, J. Yu, J. Am. Ceram. Soc. **79**, 2509 (1996). https://doi.org/10.1111/j.1151-2916.1996.tb09009.x
9. M. Emmel, C.G. Aneziris, J. Mater. Res. **28**, 2234 (2013). https://doi.org/10.1557/jmr.2013.56
10. W. Yan, A. Schmidt, S. Dudczig, T. Wetzig, Y. Wie, Y. Li, S. Schafföner, C.G. Aneziris, J. Eur. Ceram. Soc. **38**, 2164 (2018). https://doi.org/10.1016/j.jeurceramsoc.2017.12.001
11. A. Schmidt, S. Dudczig, A. Salomon, T. Zienert, E.D. Rafaja, C.G. Aneziris, Time dependent interaction between carbon-bonded alumina filters and molten steel, in *UNITECR 2017–15th Biennial Worldwide Congress on Refractories*, Proceeding No. O066 (Santiago, Chile, 2017), pp. 250–253
12. E. Storti, M. Farhani, C.G. Aneziris, C. Wöhrmeyer, C. Parr, Steel Res. Int. **88**, 1700247 (2017). https://doi.org/10.1002/srin.201700247
13. M. Emmel, C.G. Aneziris, F. Sponza, S. Dudczig, P. Colombo, Ceram. Int. **40**, 13507 (2014). https://doi.org/10.1016/j.ceramint.2014.05.033
14. B. Bock, A. Schmidt, E. Sniezek, S. Dudczig, G. Schmidt, J. Szczerba, C.G. Aneziris, Ceram. Int. **45**, 8761 (2019). https://doi.org/10.1016/j.ceramint.2018.11.131
15. E. Storti, S. Dudczig, G. Schmidt, P. Colombo, C.G. Aneziris, J. Eur. Ceram. Soc. **36**, 857 (2016). https://doi.org/10.1016/j.jeurceramsoc.2015.10.036
16. O. Jankovský, E. Storti, K. Moritz, B. Luchini, A. Jiříčková, C.G. Aneziris, J. Eur. Ceram. Soc. **38**, 4732 (2018). https://doi.org/10.1016/j.jeurceramsoc.2018.04.068
17. M. Emmel, C.G. Aneziris, G. Schmidt, D. Krewerth, H. Biermann, Adv. Eng. Mater. **15**, 1188 (2013). https://doi.org/10.1002/adem.201300118
18. D.C. Marcano, D.V. Kosynkin, J.M. Berlin, A. Sinitskii, Z. Sun, A. Slesarev, L.B. Alemany, W. Lu, J.M. Tour, ACS Nano **4**, 4806 (2010). https://doi.org/10.1021/nn1006368
19. E. Storti, M. Emmel, S. Dudczig, P. Colombo, C.G. Aneziris, J. Eur. Ceram. Soc. **35**, 1569 (2015). https://doi.org/10.1016/j.jeurceramsoc.2014.11.026
20. E. Storti, S. Dudczig, A. Schmidt, G. Schmidt, C.G. Aneziris, Steel Res. Int. **88**, 1700142 (2017). https://doi.org/10.1002/srin.201700142
21. O. Jankovský, E. Storti, G. Schmidt, S. Dudczig, Z. Sofer, C.G. Aneziris, Appl. Mater. Today **13**, 24 (2018). https://doi.org/10.1016/j.apmt.2018.08.002
22. P. Gehre, A. Schmidt, S. Dudzig, J. Hubalkova, C.G. Aneziris, N. Child, I. Delaney, G. Rancoule, D. DeBastiani, J. Am. Ceram. Soc. **101**, 3222 (2018). https://doi.org/10.1111/jace.15431
23. M. Neumann, P. Gehre, J. Kuebler, N. Dadivanyan, H. Jelitto, G.A. Schneider, C.G. Aneziris, Ceram. Int. **45**, 8761 (2019). https://doi.org/10.1016/j.ceramint.2019.01.200
24. M. Neumann, P. Gehre, R.I. Nwokoye, H. Jelitto, G.A. Schneider, C.G. Aneziris, Ceram. Int. **47**, 18656 (2021). https://doi.org/10.1016/j.ceramint.2021.03.197
25. M. Neumann, P. Gehre, J. Hubalkova, H. Zielke, M. Abendroth, C.G. Aneziris, J. Therm. Spray Technol. **29**, 2026 (2020). https://doi.org/10.1007/s11666-020-01114-6
26. E. Storti, M. Roso, M. Modesti, C.G. Aneziris, P. Colombo, J. Eur. Ceram. Soc. **36**, 2593 (2016). https://doi.org/10.1016/j.jeurceramsoc.2016.02.049

27. T. Wetzig, Dissertation, Technische Universität Bergakademie Freiberg, 2022
28. T. Wetzig, A. Schmidt, S. Dudczig, G. Schmidt, N. Brachhold, C.G. Aneziris, Adv. Eng. Mater. **22**, 1900657 (2020). https://doi.org/10.1002/adem.201900657
29. E. Werzner, M. Abendroth, C. Demuth, C. Settgast, D. Trimis, H. Krause, S. Ray, Adv. Eng. Mater. **19**, 1700240 (2017). https://doi.org/10.1002/adem.201700240
30. M. Abendroth, E. Werzner, C. Settgast, S. Ray, Adv. Eng. Mater. **19**, 1700080 (2017). https://doi.org/10.1002/adem.201700080
31. A. Herdering, J. Hubalkova, M. Abendroth, P. Gehre, C.G. Aneziris, Interceram **68**, 30 (2019). https://doi.org/10.1007/s42411-019-0013-z
32. A. Herdering, M. Abendroth, P. Gehre, J. Hubalkova, C.G. Aneziris, Ceram. Int. **45**, 153 (2019). https://doi.org/10.1016/j.ceramint.2018.09.146
33. B. Bock-Seefeld, T. Wetzig, J. Hubalkova, G. Schmidt, M. Abendroth, C.G. Aneziris, Adv. Eng. Mater. **24**, 2100655 (2022). https://doi.org/10.1002/adem.202100655
34. A. Herdering, Dissertation, Technische Universität Bergakademie Freiberg, 2022
35. Synthesis and characterization of calcium zirconate nanofibers produced by electrospinning

Chapter 2
In Situ Observation of Collision Between Exogenous and Endogenous Inclusions on Steel Melts for Active Steel Filtration

Christina Schröder, Xingwen Wei, Undine Fischer, Gert Schmidt, Olena Volkova, and Christos G. Aneziris

2.1 Introduction

Non-metallic oxide inclusions are usually undesirable in steel because, mainly by forming clusters, they negatively affect the mechanical properties, especially the strength of the steel, induce cracking under stress and promote their growth. They can be formed both by reoxidation and by the decrease in oxygen solubility during cooling. Carbon-bonded refractory material have a high thermal shock and creep resistance and are used in both the tundish and the casting mold to reduce inclusions from molten steel.

Janiszewski and Kudlinski [1] provided the theoretical basis for the application of active filters. Oxide coatings with the same chemical composition as the inclusions

C. Schröder (✉) · X. Wei · O. Volkova
Institute of Iron and Steel Technology, Technische Universität Bergakademie Freiberg, Leipziger Str. 34, 09599 Freiberg, Germany
e-mail: christina.schroeder@iest.tu-freiberg.de

X. Wei
e-mail: xingwen.wei@iest.tu-freiberg.de

O. Volkova
e-mail: volkova@iest.tu-freiberg.de

U. Fischer · G. Schmidt · C. G. Aneziris
Institute of Ceramics, Refractories and Composite Materials, Technische Universität Bergakademie Freiberg, Agricolastraße 17, 09599 Freiberg, Germany
e-mail: undine.fischer@ikfvw.tu-freiberg.de

G. Schmidt
e-mail: gert.schmidt@ikfvw.tu-freiberg.de

C. G. Aneziris
e-mail: aneziris@ikfvw.tu-freiberg.de

© The Author(s) 2024
C. G. Aneziris and H. Biermann (eds.), *Multifunctional Ceramic Filter Systems for Metal Melt Filtration*, Springer Series in Materials Science 337,
https://doi.org/10.1007/978-3-031-40930-1_2

to be filtered in the melt serve as active filter material and open up the possibility of eliminating semiliquid or liquid inclusions as well. The capillary forces between the particles play a decisive role in the attraction of the inclusions. The principle of attraction between colloidal particles has been extensively studied by Kralchevsky and Nagayama [2, 3] and Danov et al. [4]. In addition to the flotation force caused by their own weight, the particles are affected by the Archimedes force resulting from gravity and capillary forces due to wettability. Thereby, the flotation and immersion forces can be attractive or repulsive. The meniscus slope angles Ψ_1 and Ψ_2 of the particles in the melt are the decisive quantities for a movement of the particles towards or away from each other (Fig. 2.1). The capillary force is attractive when the product $(\sin \Psi 1 \cdot \sin \Psi 2)$ is greater than zero and repulsive when the product $(\sin \Psi 1 \cdot \sin \Psi 2)$ is less than zero. Mainly three factors influence the interaction between the particles: the particle density, the contact angle and the surface tension [5]. The higher the density the stronger gets the capillary attraction. But in reality the particle form clusters, so that an apparent density of particles have to be considered. The capillary force also increases with increasing contact angle of the oxide particles. The influence is big with increasing angles from 100 to ca. 130° but smaller at increasing angle from 130 to 170°. With increasing surface tension of the steel melt, the capillary forces of the particle decrease slightly, also with the presence of sulfur as surfactant element.

To study the clogging tendency against endogenous and exogenous non-metallic inclusions, carbon-bonded nozzles with and without alumina active coatings were tested in a steel casting simulator by Aneziris et al. [6]. 'Denser' as well 'coral-like' clogging areas were found in both cases, whereas the coated nozzles show a stronger clogging. This opens the possible use of the material for filters. The impact of the wetting properties and possible reactions between filter material and melt on the filtration efficiency of solid inclusions are not clear yet. High-temperature investigations at the HT-CLSM provide the opportunity to observe the interaction of endogenous particles with exogenous particles in situ in the melt. Yin et al. [7] were the first to observe in situ the collision and agglomeration behaviour of non-metallic particles in the steel melt of low carbon Al-killed steels. For the case, that one particle,

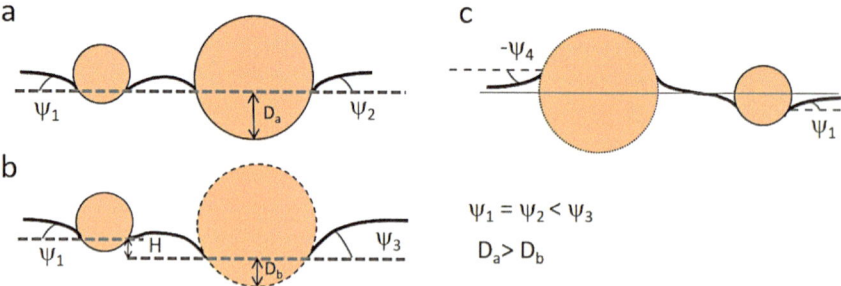

Fig. 2.1 Meniscus slope angles (MSA) of endogenous (smaller) and exogenous (bigger) inclusions. **a** $\Psi_1 = \Psi_2$, **b** higher MSA Ψ_3, lower immersion depth ($D_b < D_a$) and increased high H of the liquid level line of the endogenous particle, **c** repulsion at MSA with reversed sign

guest particle, move to the other, host particle, they calculated the attractive force F as the product of the mass m_1 of the guest particle and its acceleration a_i at each time i.

$$F_i = m_1 \cdot a_i \tag{2.1}$$

In the case that both guest and host particle are moved simultaneously, a revision factor was introduced with

$$m_2/(m_1 + m_2) \tag{2.2}$$

where m_2 is the mass of the larger host particle. The mass of the particles is estimated by its size and the material density.

Yin et al. [7] found that the strength of the capillary attraction forces increases for inclusions and inclusion pairs as follows: It is zero for a liquid/liquid pair < liquid/semiliquid pair < liquid/solid pair < semiliquid/solid pair < solid/solid pair. The solid/solid pair has the strongest attraction force if only the effect of particle morphology is considered. For alumina particle smaller than 3 μm, a long-range strong attraction force over 10^{-16} N was determined in a distance over 10 μm between the particles or their agglomerates. The acting distance is affected by the inclusion size and shape. Mu et al. [8] determined a wide-ranging acting distance of over 150 μm for Al_2O_3 clusters with equivalent radius greater 40 μm and attractive capillary force greater $2 \cdot 10^{-13}$ N. They show, that model calculations using equivalent radius R_k with

$$R_k = \sqrt{A_k/\pi} \tag{2.3}$$

where A_k is the specific area of the particle k, leads to a greater confirmation between calculated attractive force and experimental determined attractive force for inclusion clusters than using the effective radius $R_{k,eff}$ with

$$R_{k,eff} = \frac{P_k}{2\pi} = R_k/\sqrt{CF_k} \tag{2.4}$$

where P_k is the perimeter of the particle k, and CF_k is the particle circularity.

For MgO inclusion pairs and 93%Al_2O_3·7%MgO inclusions pairs Kimura et al. [9] found similar attractive forces at a maximum acting distance of 21–22 μm. They are with $5 \cdot 10^{-18}$–$5 \cdot 10^{-16}$ N approximately a power of ten lower as for Al_2O_3 inclusion pairs. The attractive forces of solid Al_2O_3·SiO_2 pairs is also weaker and their acting length shorter than between alumina particles [7]. In contrast to the alumina particles, their shapes were smoother and their clusters were denser.

In a continuing paper by Yin et al. [10] the collision and agglomeration of particles in the systems CaO-Al_2O_3 and CaO-Al_2O_3-SiO_2 were investigated. The attractive forces between semi-fluid particles of the CaO-Al_2O_3 system are 10^{-14}-10^{-15} N with a range of about 42 μm. If one particle of the pair has an irregular shape (solid particle), the attractive force increases to $8 \cdot 10^{-13}$–$7 \cdot 10^{-14}$ N and the range rises slightly. Larger attractive forces than alumina particles exhibit were determined for solid CaO-80 wt.% Al_2O_3- and for solid $(CaO \cdot Al_2O_3)$-95 wt.% SiO_2 pairs. Both particle pairs have a denser structure as the alumina clusters and so a greater difference of liquid steel surface height between inside and outside of the pair. No attraction was found between totally liquid inclusion particles. SiO_2 lowers the liquidus temperature in CaO-Al_2O_3-SiO_2 inclusions. The attractive forces between solid CaO-Al_2O_3-SiO_2 particles decrease to $<10^{-15}$ N. No capillary attractive forces were observed between liquid CaO-Al_2O_3-SiO_2 inclusions. A summary on the in situ agglomeration observations of non-metallic inclusions at the interface of liquid steel/argon gas and liquid steel/slag is given by Mu et al. [11].

Uemura et al. [12] investigated the filtering mechanism of non-metallic inclusions in steel by ceramic loop filters. According to their research, the filter efficiency depends on the initial oxygen content in addition to the fineness of the filters. Equal filter efficiencies were determined for the $CaO \cdot 6Al_2O_3$, short CA_6, and Al_2O_3 filter materials. Storti et al. [13] studied carbon-bonded filters functionalized with coatings based on calcium aluminates CA_2 and CA_6 in contact with molten steel at 1650 °C. The coating aims to react with the molten steel and generate a liquid layer, that act as an active collector for endogenous inclusions. In addition, the boiling Ca and forming CO move the melt and promote the clustering of fine alumina so that it can rise to the surface of the melt pool. They describe the formation of a thin secondary layer on the with CA_6-coated carbon-bonded alumina filter material after the immersion test. On the other hand, they show in thermodynamic considerations that the Ca activity is low with 10^{-5} to 10^{-4}. However, the analyses of the solid steel showed excellent filtration performances of both compositions.

The aim of the presented investigations within subproject A01 of the CRC 920 was to observe and analyze the interactions of endogenous particles with exogenous non-metallic alumina, magnesia, $MgAl_2O_4$ spinel und CA_6 particles as a possible filter material. For this purpose, various steels were tested for their suitability for high-temperature investigations in the HT-CLSM with respect to a homogeneous melt pool over the experimental period.

2.2 Experimental Details

2.2.1 Preparation of the Steel Samples

For the investigations at the HT-CLSM, steel platelets with a thickness of approx. 1 mm were cut from hot-rolled round bars with a 10 mm diameter. Previous removal of the rolling surface of the round bars by turning reduces the introduction of oxides during the melting test. The sample platelets were metallographically ground and polished so that they have a thickness of approx. 600 μm in the test. Immediately before being placed in the furnace, the steel platelets were polished with 1 μm grit diamond emulsion to minimize the oxide layer on the specimen surface.

For the experiments in the heating microscope, plane-parallel steel cylinders with a diameter of 6 mm and a height of 10 mm were manufactured. In order to reduce the oxide layer on the specimens, the steel specimens were etched in dilute hydrochloric acid directly before the experiment in the heating microscope.

2.2.2 Exogenous Non-metallic Test Material

To investigate the interaction of exogenous non-metallic materials with the melt and its endogenous non-metallic precipitates, individual 50–80 μm particles of the material to be investigated were preferably positioned in the centre of the steel lamina before the start of the melting test. The characteristic properties of the non-metallic test material are presented in Table 2.1.

MgO and CA_6 tablets were prepared for wetting behaviour investigation in the heating microscope. For MgO tablets electro fused magnesia with 97.8 wt.% MgO and 1 wt.% CaO was mixed with 2 wt.% Mg-ligninsulfonate Duriment BV P as temporary pressing additive and 2 wt.% deionized water. The mass was pressed in cylindric samples with 20 mm diameter and 2.5 mm height and a pressure of 25 MPa. After drying at 120 °C for 4 h the tablets were sintered at 1600 °C/1 h under atmosphere. The CA_6-tablets have a diameter of 45 mm and are 7 mm high. They were pressed with the binder dextrin at 120 MPa, dried at 120 °C and sintered at 1600 °C for 5 h. The surface roughness after polishing was 1.57 μm with an open porosity of 17%.

For calculations of the phase fractions during heating and melting of the steel alloys the software ThermoCalc with the thermodynamic database of steels/Fe-alloys Version 8.1 were used. The software FactSage 7.2 with the databases FTmisc 6.3, FToxid 6.3, and FactPS 6.3 were utilised to calculate possible interactions between the refractory material and the oxygen from the passive layer of steel or others oxygen sources.

2.2.3 Experimental Equipment and Experimental Conditions

The HT-CLSM 1LM21H/SVF17SP (Lasertec, Japan) makes it possible to digitally record interactions of molten steel with non-metallic particles in situ in the high temperature range. The principle sketch of the furnace operation is described in an earlier article by Aneziris et al. [14]. The temperature regime of the experiments can be pre-programmed with the aid of the device's own control software. Figure 2.2 shows an example of the temperature control during the tests. 5 working steps have been programmed: In the first stage, heating is performed at 200 K/min to 600 °C to preserve the durability of the halogen lamp. This is followed by an increased heating rate of 400 K/min to avoid precipitate formation during heating up to 1300 °C with a hold time of 30 s to eliminate overheating due to the high heating rate. In steps 3

Table 2.1 Specifics of the non-metallic exogenous materials

Ceramic	Graining	Specific properties	Fabricator
Al_2O_3	Tabular alumina; T60/64	Particle size 50–70 μm, Specific surface 1.13 m^2/g	Almatis, Inc., Leetsdale, PA
MgO	Fused magnesia	Particle size 63–80 μm; include 0.35 wt.% CaO	Qinghua refractories, Dashigao, China
$MgAl_2O_4$	Magnesium aluminate spinel; AR 78	Particle size 60–80 μm >74.0 wt.% Al_2O_3, ca. 22.5 wt.% MgO, ca. 0.24 wt.% CaO, <0.15 wt.% SiO_2, <0.32 wt.% Na_2O, <0.25 wt.% Fe_2O_3; bulk spec. density 3.3 g cm^{-3}, apparent porosity <2.6 vol.%, water absorption <0.8 vol.%	Almatis GmbH, Germany
$CaO·6Al_2O_3$; (CA_6)	Bonite	Particle size 0–0.020 mm, d_{50} = 6.0 μm >99 wt.% CA_6	Kerneos, France
$CaO·6Al_2O_3$; (CA_6)	Bonite	Particle size 500–1000 μm ca. 91 wt.% Al_2O_3, ca. 7.6 wt.% CaO, ca. 0.09 wt.% Fe_2O_3, ca. 0.9 wt.% SiO_2; bulk spec. density 3.0 g cm^{-3}, apparent porosity 9.8 vol.%, water absorption 3.4 vol.%	Almatis GmbH, Germany

and 4 the heating rates are reduced first to 120 K/min up to a temperature of 1420 °C with a holding time of 30 s and to 30 K/min up to a temperature of 1515 °C and a holding time of 6 min.

Step 4 allows observation of partial melting and the interaction between the melt and the non-metallic particle. The heating temperature and the holding time can be manually increased or extended during the test. In program step 5, the sample cools down very quickly by switching off the heating source due to the low molten sample volume. A HeNe laser scans the sample surface for in situ observation at a frame rate of up to 30 frames per second. Particles down to a size of 0.5 μm can be detected in the image. To analyse the particle trajectories of the HT-CLSM videos, image

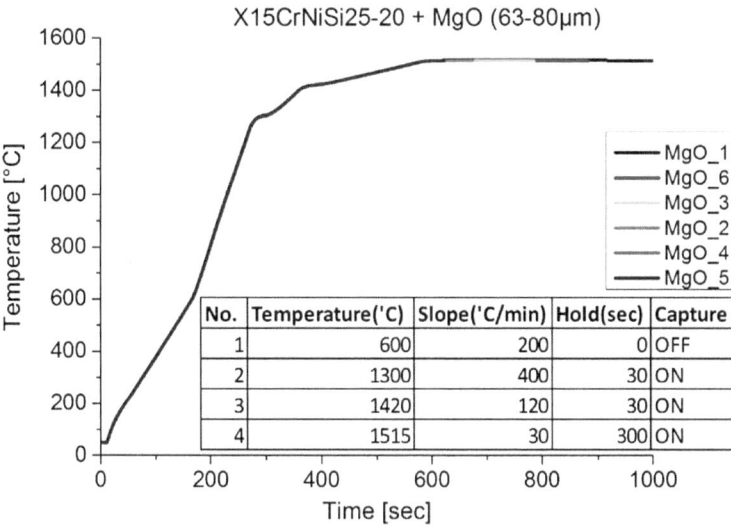

Fig. 2.2 Temperature control during the melting tests in the HT-CLSM using the example of exogenous MgO particles on the steel X15CrNiSi25-20 including the pre-programmed work steps

sequences were selected for automatic motion analysis using WINanalyse software (Mikromak Services, Germany).

The assembly of a heating microscope is shown schematically by Wei et al. [15]. The heating rates in this work were 20 K/min up to 400 °C and 40 K/min above 400 °C up to the test temperature of 1450 °C. Due to its surface tension, the liquid steel forms a drop (sessile drop method). To determine the surface tension of the steel droplet and its contact angle with the ceramic surface, a calculation software is used based on the findings of Rotenberg et al. [16] and described in the work of Chebykin et al. [17].

After the HT-CLSM and heating microscope investigations the micro-structural phase evaluations of the steel discs were carried out at room temperature (RT) at a scanning electron microscopy (SEM) Ultra 55 (Zeiss, Germany) using integrated energy dispersive X-ray spectrometer (EDX) from Edax/Ametek and electron back scattering diffractometer (EBSD). EDAX OIM software was used for data processing. The untreated sample surfaces, as well as vertically cut, grinded and polished samples were investigated.

2.3 Results

2.3.1 Melting Behaviour of Different Steels in HT-CLSM

A prerequisite for observing the interaction of non-metallic particles in the melt is the generation of a melt pool which is maintained as homogeneously as possible over a certain experimental period. Table 2.2 summarizes the chemical composition of three investigated steels. The steels have graded carbon contents.

42CrMo4 is a low-alloyed steel for quenching and tempering used as a base alloy in several CRC 920 projects. According to ThermoCalc calculations Fig. 2.3a, ferritic-austenitic solidification starts at a liquidus temperature T_L of 1487 °C. The primarily formed δ-ferrite is peritectically transformed into austenite in a narrow temperature interval of approx. 5 K. Its solidus temperature T_S is 1406 °C. After solidification, the steel has a completely austenitic structure. Below 750 °C, pearlite formation from the austenitic high-temperature phase begins. Melted (1) and partially resolidified (2) areas are seen on the surface of 42CrMo4 at 1556 °C during HT-CLSM investigation in Fig. 2.3b.

Despite the temperature increase, a part of the melt solidifies again. A strong movement and/or doubling of the grain boundaries is visible, which may indicate the transformation of austenite to δ-ferrite. This observation leads to the following conclusion: Due to the low carbon solubility in δ-ferrite, carbon preferentially diffuses into the melt, reacts with the residual oxygen on the steel surface and is lost. The carbon depletion leads to an increase in the melting temperature. An indirect evidence

Table 2.2 Chemical composition of the studied steels [wt.%], rest Fe

Steel	C	Si	Mn	P	S	Cr	Mo	Ni	Al	Cu	Ti
42CrMo4	0.438	0.26	0.83	0.016	0.031	1.09	0.18	0.21	0.013	0.27	–
X20Cr13	0.216	0.40	0.56	0.283	0.004	12.98	0.15	0.78	0.004	0.06	0.002
X15CrNiSi25-20	0.118	2.38	1.03	0.030	0.027	25.10	0.07	18.80	0.020	0.07	0.020

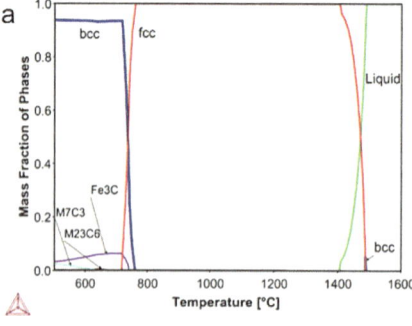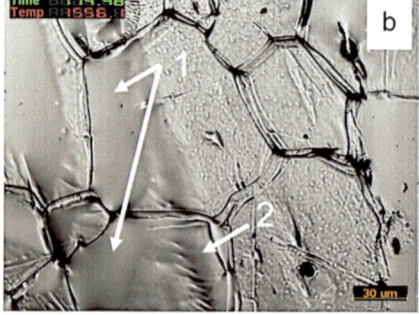

Fig. 2.3 42CrMo4, **a** Phase fraction diagram and **b** HT-CLSM image at 1556 °C

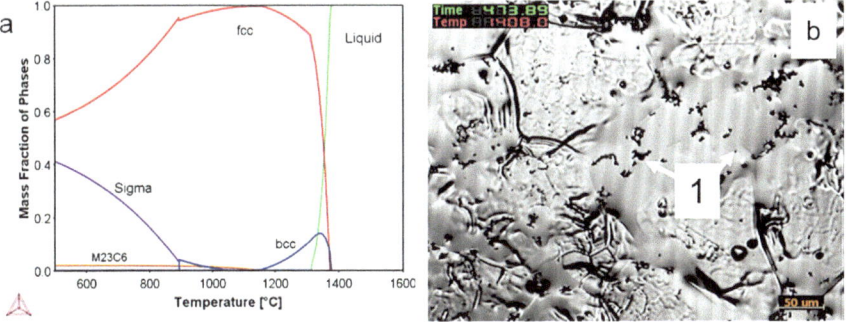

Fig. 2.4 X15CrNiSi25-20, **a** Phase fraction diagram and **b** HT-CLSM image at 1408 °C with partially melted areas and mobile endogenous particles (1)

of this assumption are black deposits on the furnace wall of the HT-CLSM. A uniform melt pool could not be formed without melting the peripheral areas of the steel sample and led to droplet formation due to the surface tension of the steel.

X15CrNiSi25-20 steel is a high-alloy stainless steel that undergoes primary ferritic-austenitic solidification (Fig. 2.4). Compared to 42CrMo4 and X20Cr13, its carbon content is lowered, reducing the tendency for carbon loss during melting. The higher proportions of the alloying elements chromium, nickel and especially silicon lead to the reduction of the liquidus and solidus temperatures to about 1378 and 1300 °C, respectively. 40 ppm oxygen was analysed in the steel.

Due to the reaction of solved aluminium in the steel with oxygen, which is present by reduction of the chromium oxide layer, endogenous particles of approx. 1 μm in size are formed on the surface of the melt. These move freely in the melt, interact with other endogenous particles and form conglomerates. Due to the stability of the melt over an investigation period of >300 s and the possibility of observing the interaction of endogenous particles with endogenous and exogenous particles, the further melting experiments were carried out with steel platelets of the alloy X15CrNiSi25-20.

2.3.2 Structural Changes in the Steel X15CrNiSi25-20 During Heating

In the initial state of the steel, (Mn,Cr,Al,Fe)-sulphides with about 36 at.% Mn, 8 at.% Cr, 3 at.% Fe and 2–7 at% Al and (Cr,Fe,Ni)-carbides are detectable. No sigma phase was observed.

During heating, the reflectivity of the steel plates decreases briefly at about 850 °C and about 1240 °C. The change in reflectance behaviour may be due to new grain formation during a phase transition. It can also be caused by the formation of non-metallic precipitates on the steel surface due to changes in the solubility of the alloying elements during a phase transition. According to ThermoCalc calculations (Fig. 2.4), phase transitions occur in both temperature ranges. Tables 2.3 and 2.4 show the calculated equilibrium phases with their volume fractions and the chemical composition of the phases. The bcc phases are rich in chromium, nickel and silicon and take up lower carbon, manganese and phosphorus contents and slightly higher sulphur and titanium contents compared to the fcc phase. $(Cr,Ni)_3$-phosphide, which appears at 894 °C under equilibrium calculation, is dissolved at 1130 °C in the matrix. Carbon and manganese have a lower solubility in the ferritic high temperature phase and can be enriched at the phase interfaces and the steel surface during the phase transition from fcc to bcc. The Ti and C rich fcc#2 phase takes up sulphur at 1130 °C and forms $Ti_4C_2S_2$ under equilibrium conditions. All phase transitions can influence the reflectance ratios of the steel surface. In order to investigate the phase formations in the steel during the melting and solidification process, steel samples without exogenous inclusions were examined in the HT-CLSM.

Figure 2.5 shows a steel disc after the melting test. The sample has a metallic sheen without increased formation of visible oxide layers. The not yet melted peripheral

Table 2.3 Calculated phases at formation of bcc at 894.5 °C and the phase compositions; rest Fe

Phases at 894.5 °C	Phase fraction	C	Si	Mn	P	S	Cr	Mo	Ni	Al	Cu	Ti
	Vol.%	wt.%	wt.%	wt.%	wt.%	wt.%	wt.%	wt.%	wt.%	wt.%	wt.%	wt.%
Fcc#1	95.0	1E-2	2.4	2.0	1E-2	5E-9	22.5	0.15	18.0	9E-3	0.17	4E-3
Sigma	3.1	–	3.0	0.4	–	–	44.0	0.14	1.9	5E-7	–	7E-5
M23C6	1.7	5.5	–	0.6	–	–	77.0	5.3	0.9	–	–	–
M3P	0.1	–	–	–	15.9	–	40.4	–	24.2	–	1E-5	–
MnS	2E-2	–	–	63.1	–	36.9	3E-7	–	–	–	3E-8	–
Fcc#2	4E-3	18.4	5E-8	1E-3	5E-11	5E-11	3.1	0.2	7E-4	5E-11	2E-7	78.3
bcc	0.0	2E-3	2.6	1.03	9E-4	6E-9	38.9	0.2	6.3	7E-3	2E-2	7E-3

Table 2.4 Calculated phases at transition temperature fcc to bcc at 1127 °C and its compositions; rest Fe

Phases at 1127 °C	Phase fraction	C	Si	Mn	P	S	Cr	Mo	Ni	Al	Cu	Ti
	Vol.%	wt.%	wt.%	wt.%	wt.%	wt.%	wt.%	wt.%	wt.%	wt.%	wt.%	wt.%
Fcc#1	99.6	8E-2	2.4	1.9	3E-2	4E-7	23.9	0.23	17.2	8E-3	0.16	4E-3
M23C6	0.3	5.5	–	0.9	–	–	72.8	1.9	1.9	–	–	–
MnS	2E-2	–	–	63.0	–	36.9	8E-6	–	–	–	6E-7	–
Ti4C2S2	4E-3	8.6	–	–	–	22.9	–	–	–	–	–	68.5
Bcc	0.0	2E-3	2.7	1.2	2E-3	6E-7	33.8	0.3	9.2	8E-3	4E-2	8E-3

area of the sample and the dendritically solidified structures from the periphery to the interior can be clearly seen. In the centre, the residual melt is fine-grained crystalline. In the border area between molten and non-molten material (Fig. 2.5c, d), a ((Mn,Cr,Fe)-S)-rich phase (1) has solidified at the grain boundaries. Fine, non-metallic aluminium oxides have also been deposited preferentially at the grain boundaries (2).

In order to investigate the cause of the radiation intensity loss at approx. 1300 °C, a steel sample was heated to 1300 °C and cooled in 10 cycles at 400 K/min. The steel did not melt during this process. The surface of the steel plate lost its metallic sheen. Drop-like precipitations were formed. In addition, a cluster of excretions was observed (Fig. 2.6). In contrast to the melting test, predominantly (Si,Cr)-rich and occasionally (Cr,Si,Mn)-rich oxidic phases were detected under the given conditions. EDX phase analyses of the spots marked in Fig. 2.6 (1–2.4) are summarised in Table 2.5.

On the one hand, the high-purity argon inert gas (99.999 wt.%) with a residual oxygen content of approx. 2 ppm serves as an oxygen source for oxide formation. Secondly, steel with chromium contents ≥ 11 wt.% forms a chromium oxide passive layer on the surface, which can serve as a further source of oxygen. According to the literature, steels with a chromium oxide coating are scale resistant up to approx. 850 °C. Aluminium oxide coated steels have a higher scaling resistance up to approx. 1000 °C [18]. The passive layers soften at higher temperatures and contract into drops due to their surface tension and reduced adhesion to the polished steel matrix. According to the Richardson-Ellingham diagram, the oxides of the metals Mn, Si, Ti, and Al have lower Gibbs free energies than chromium oxide [19]. In this context, the Gibbs free energy for the oxidation of the metals decreases in the listed order. As the above metals are dissolved in the steel, they can reduce chromium oxide.

Above about 1400 °C, chromium oxide can also be reduced by carbon to form carbon monoxide. The steel matrix absorbs the reduced chromium. In this context, it is interesting to note that a broadening of the grain boundaries was also observed in the case of X15CrNiSi25-20 immediately before the matrix melted. Since, according to the phase quantity diagram Fig. 2.4, ferrite is formed in the high-temperature range and, in addition, chromium stabilizes the bcc ferritic phase, it can be assumed that the manifestation of the grain boundaries is caused by the formation of δ-ferrite. The austenite begins to melt due to its higher carbon content, while the δ-ferrite remains in the solid state at the former grain boundaries.

2.3.3 Interaction of Exogenous Alumina Particles with Molten Steel and Its Endogenous Particles

In contrast to the investigations of Yin et al. [7], who observed particle agglomeration on the completely melted sample surface, in the present experiments the occurrence of local surface flow of the melt could be almost prevented by partial melting within

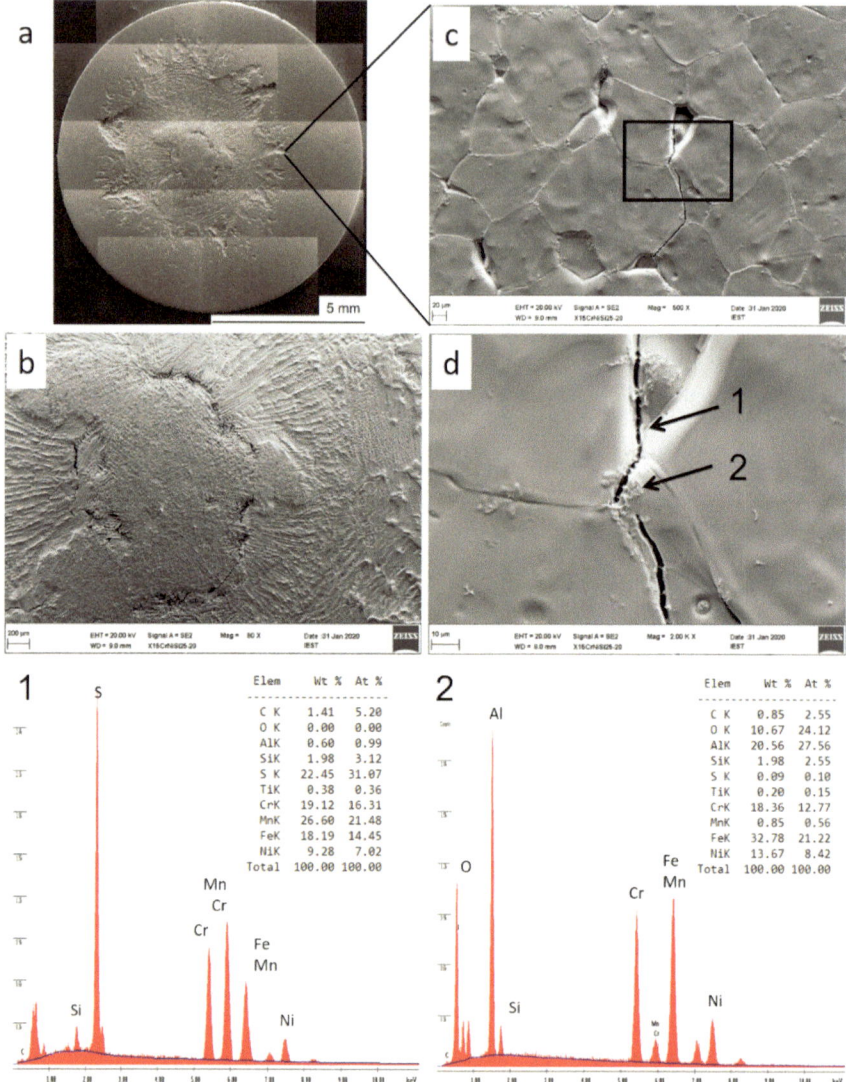

Fig. 2.5 X15CrNiSi25-20 disc after melting test in the HT-CLSM **a** Overall view, **b** dendritic and fine-grained solidification structure, **c** partially molten zone, **d** sulfidic (**1**) and oxidic (**2**) precipitations at the grain boundaries with EDX spectra and analysis

still existing grain boundaries. The rapid collision and agglomeration tendency and growth of alumina clusters due to strong attractive forces over a long range on the surface of the molten steel are confirmed. Figure 2.7 shows an exogenous alumina particle of about 50–70 μm in size on a partially melted steel disc with endogenous particles or particle clusters.

Fig. 2.6 Steel surface after heat treatment at 1300 °C with formation of (Cr,Si,Mn)-rich (1) and (Si,Cr)-rich oxides (2.x)

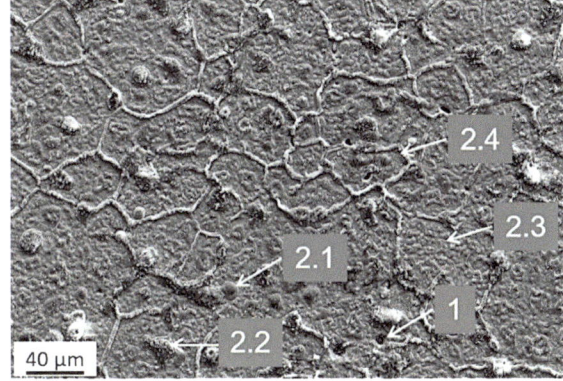

Table 2.5 EDX analyses of the marked oxides on the surface of X15CrNiSi25-20 steel disc after heating treatment at 1300 °C (Fig. 2.6), fraction in at.%

Spot	Si	Cr	Mn	Al	Ni	Ti	O
1	14.07	28.07	12.01	0.84	0.54	0.47	42.39
2.1	27.25	12.66	4.61	–	1.51	–	49.14
2.2	23.05	19.77	8.01	0.53	0.88	0.43	44.33
2.3	18.01	15.21	1.14	0.30	8.47	0.19	31.34
2.4	30.04	7.04	2.53	0.56	1.29	0.18	54.03

Fig. 2.7 Exogenous alumina inclusion surrounded by molten steel in which solid, non-metallic, endogenous inclusions move in the direction of the exogenous inclusion at approx. 1469 °C (exemplary: red curves of particle 14 and 15)

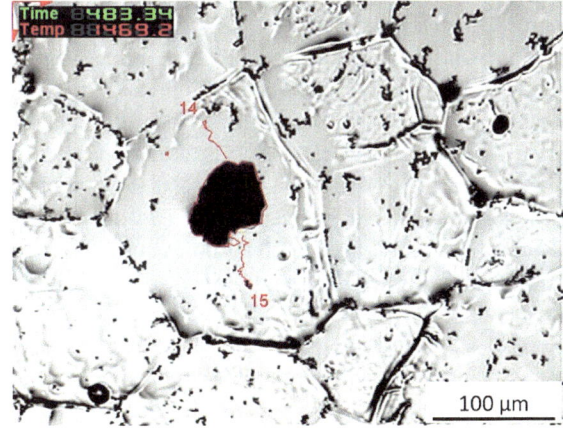

Endogenous alumina particles move over a range of >50 µm to the exogenous host particle. The attractive length is limited by the grain boundaries. The red lines indicate exemplary paths of two endogenous particles. Attractive forces of 10^{-15}–10^{-16} N were determined, which are slightly larger than those obtained by Yin et al. [7] and correspond to the forces of alumina clusters by Mu et al. [8] for particle pairs

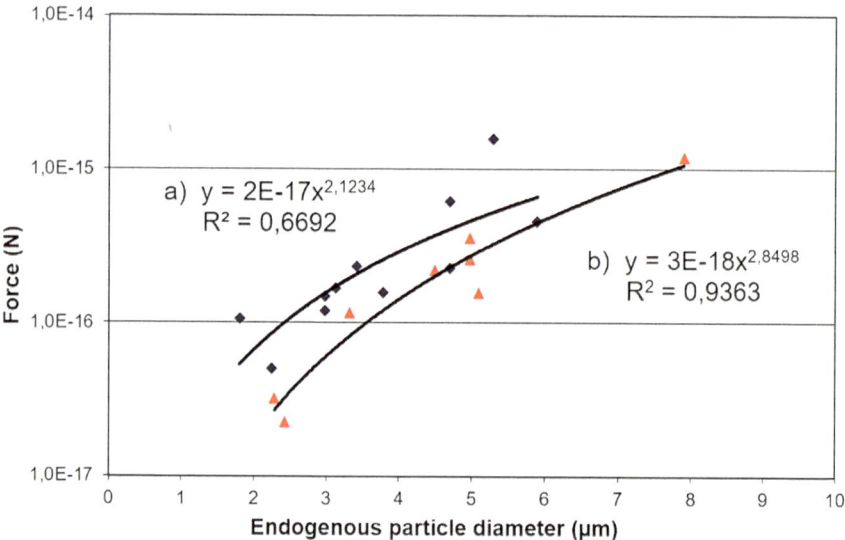

a) $y = 2E\text{-}17x^{2,1234}$
$R^2 = 0,6692$

b) $y = 3E\text{-}18x^{2.8498}$
$R^2 = 0,9363$

Fig. 2.8 Maximum real force as a function of endogenous inclusion: **a** endogenous inclusion toward an exogenous Al_2O_3 particle (blue rhombus) and **b** endogenous inclusions toward an endogenous inclusion (orange triangle) [14]

with nearly the same particle sizes ratio. The influence of the viscosity of the steel melt on the mobility of the non-metallic particles has not yet been investigated. It is to be expected that a lower viscosity with increasing temperature also leads to an increase in particle mobility.

The own investigations show that the attractive forces between the 1–6 μm sized endogenous alumina particles and the exogenous alumina particle are somewhat larger than between the endogenous particles themselves and that the attractive forces increase with increasing particle or cluster size of guest particle (Fig. 2.8) [14]. With increasing examination time at °C the grain boundaries dissolve. At 1515 °C also liquid, round inclusions are seen at the steel surface. The liquid inclusions attach to both solid and liquid particles and likewise form their own loose clusters. They contribute to a densification of the solid inclusion clusters. Titanium oxide was frequently detected. During cooling of the sample, two distinguished phases based on $\alpha\text{-}Al_2O_3$ and Ti_2O_3 were formed in the neighborhood of the alumina particles. No formation of Al_2TiO_5 could be detected.

2.3.4 Interaction of Exogenous Magnesia Particles with Molten Steel and Its Endogenous Particles

A different behavior was observed for magnesia host particles [20]. Figure 2.9a shows an exogenous MgO particle on a steel disc close to the steel melting temperature.

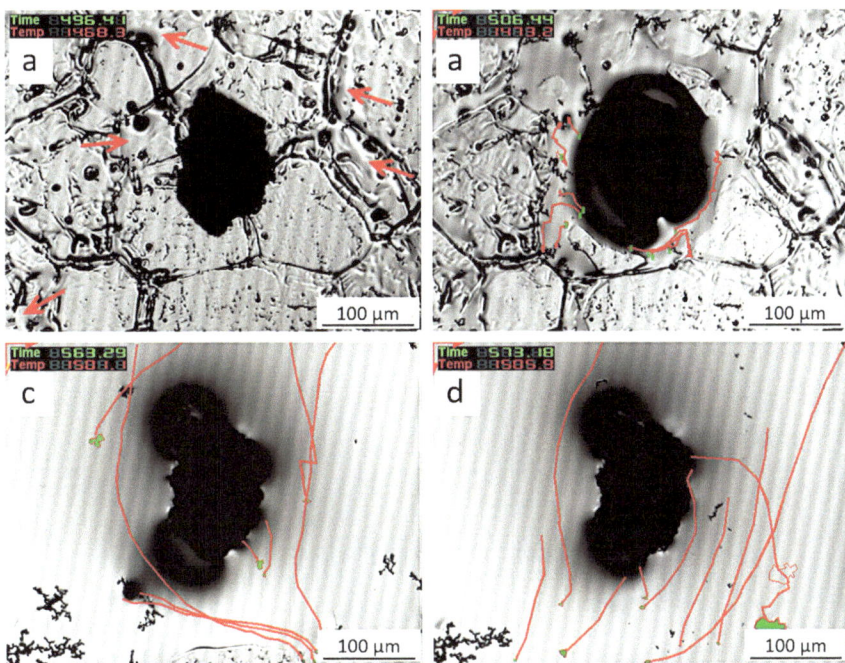

Fig. 2.9 Exogenous MgO particles in contact with molten steel and their endogenous particles **a** first molten areas, initial size of MgO host particle, **b** meniscus formation after contact of MgO with molten steel, **c** partial dissolution of the layer at 1500 °C, **d** red traces document the movement of endogenous particles (green) [20]

First melt has formed in the zones marked with arrows. As soon as the exogenous particle comes into contact with the molten steel, a meniscus with smooth borders forms around the magnesia (Fig. 2.9b). Inside the inner meniscus and the magnesia particle a lighter area was formed and the edges of the MgO particle are visible. In this area moving particles were observed, showing that a liquid phase was built between exogenous particle and steel melt, which increase the perimeter of the magnesia up to 57%. At this stage, no endogenous alumina remains attached to the magnesia. They flow past the exogenous particle at the liquid–liquid phase boundary between the magnesia layer and the melt and move away. Red traces mark the path of individual endogenous particles in the image. With increasing temperature and test duration, the layer becomes unstable and partially dissolves (Fig. 2.9c, d). Same endogenous inclusions that reach the surface of the free MgO adhere only weakly. Others continue to move past the particle.

The maximum real forces determined for endogenous inclusions are summarized in the diagram Fig. 2.10 for the different stages of MgO wetting. For all particle pairs, the attraction force increases with increasing particle size. At about 1420 °C, before contact of the MgO particles with the melt, the endogenous particles exhibit the same attractive forces among themselves as in the experiments with exogenous

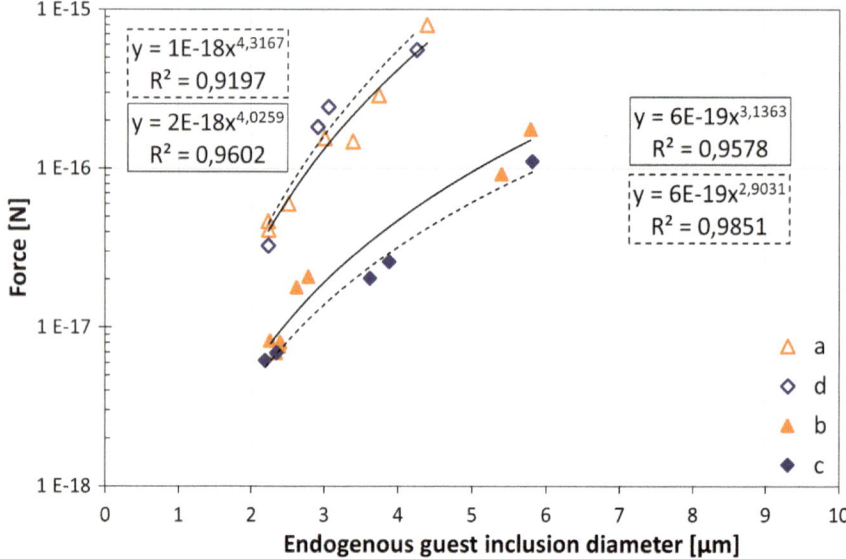

Fig. 2.10 Maximum real force of endogenous particles as function of their inclusion diameter, **a** endogenous toward endogenous inclusion or boundary before the meniscus formation around the MgO at 1417 °C, **b** endogenous toward endogenous inclusion or boundary after the meniscus formation around the MgO at 1482 °C, **c** endogenous particles toward exogenous MgO with liquid layer at 1482 °C, **d** endogenous particles toward exogenous MgO after dissolution of liquid layer at 1505 °C [20]

Al_2O_3 (Fig. 2.10a). After formation of the liquid layer around the exogenous MgO, due to the interaction of the particle with the melt, the attractive forces of the endogenous particles among themselves decrease by about one decimal power (Fig. 2.10b). The attractive forces of the endogenous inclusions towards the exogenous MgO are slightly lower than the forces between the endogenous inclusions (Fig. 2.10c). After partial dissolution of the liquid layer, an increase in the attractive forces was recorded for the interaction of endogenous inclusions to the exogenous particle to values slightly above those of the endogenous particles among themselves at the beginning of melt formation (Fig. 2.10d). SEM/EDX images show that the endogenous particles are enriched in magnesium [20]. Titanium oxide was detected in contact with the Al_2O_3-MgO cluster and at an MgO host particle. The cross-sectioned MgO particle has a layer of two phases on its surface, a (Si,Cr,Mg) rich phase and a magnesia phase enriched with (Ti,Cr and/or Al). The rounded contours of these phases reveal that they were in a liquid or semi-liquid state. It can be assumed that the Ti and Cr enriched phase precipitated from the supersaturated Si rich phase during cooling.

Wetting experiments in the heating microscope also confirm the reaction of the MgO with the molten steel to form a reaction layer [20]. Park et al. [21] describe the dissolution behaviour of MgO inclusions in CaO-Al_2O_3-SiO_2 slags at 1550 °C with the formation of ring-like Al_2O_3·MgO spinel structures when Al_2O_3 rich slags are used and the formation of circularly arranged Ca_2SiO_4 inclusions in (Si,Ca)-rich slags. In both cases, the MgO was surrounded by a liquid phase. In agreement with the observations of Yin et al. [10], this liquid layer prevents the agglomeration of endogenous particles with the exogenous MgO.

2.3.5 Interaction of Exogenous MgO·Al₂O₃ Spinel Particles with Molten Steel and Its Endogenous Particles

As reported in the previous section, a cluster of Al_2O_3 and MgO inclusions is formed during the reaction of exogenous MgO with the molten steel and its endogenous particles. The interaction of exogenous $MgO·Al_2O_3$ spinel particles with the melt was the focus of further investigations.

Figure 2.11a shows an exogenous spinel particle (1) on the steel surface with first molten steel (2) during the melting experiment in the HT-CLSM at 1439 °C. The grain boundaries have thickened in the area where austenite has transformed to δ-Ferrit (3). A strong meniscus forms around the particle once the exogenous spinel particle is contacted by the melt (Fig. 2.11b). Endogenous particles move to the exogenous spinel, but slide along the meniscus and move away again (1, 2). The exogenous particle achieves a higher wettability with increasing temperature and test time (Fig. 2.11c). At the same time, the sharp-edged particle contours of spinel become visible again. Endogenous particles or clusters (3, 4) move to these areas of the exogenous particle, adhere and form clusters. Based on the roundness of the endogenous particles, it can be assumed that a large number of the inclusions are in semiliquid or liquid state (Fig. 2.11d).

It is remarkable that, unlike in experiments with MgO particles, the exogenous spinel particle sinks deeply into the steel matrix, as shown by a cross-section on the SEM in the article by Aneziris et al. [22]. The spinel interacts with the molten steel. Manganese accumulated in the spinel lattice. Alumina was detected at all interfaces towards the steel matrix, especially also under the spinel inclusion. Magnesium aluminate spinel is known as refractory oxide for its high stability [23]. It can absorb non-stoichiometrically many impurities from liquid slag/steel systems. The manufacturer Almatis GmbH gives upper limits for CaO, SiO_2, Na_2O und Fe_2O_3 impurities for the used Spinel AR78 (Table 2.1). However, EBSD examinations of the spinel prior to use did not detect any significant impurities. Another indication of a strong interaction of the steel with the spinel is that the molten steel overflowed some spinel particles and covered them completely. Individual spinel particles were still detected approx. 10 μm below the steel surface. The formation of a liquid layer around the spinel particle was not observed.

Approximately equal real maximum forces of $5·10^{-15}$ N were calculated for the endogenous particles at about 1487 °C when moving along the meniscus, almost independently of their size. These forces are not primarily based on attractive forces, but appear to be influenced by convection and the Marangoni effect [22]. The Marangoni effect results from a gradient in surface tension caused by diffusion processes of

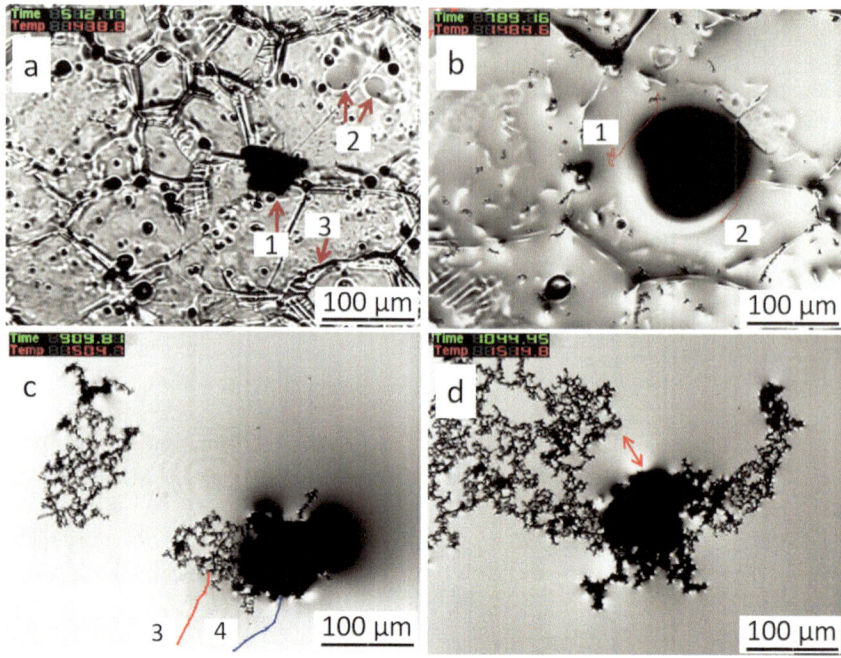

Fig. 2.11 Exogenous MgO·Al$_2$O$_3$ in contact with steel **a** initial size of spinel particle (1), first melt (2); δ-ferrite (3) at 1439 °C, **b** meniscus formation at 1485 °C (endogenous particles flow past the exogenous particle), **c** partial dissolution of meniscus and attraction of particles and **d** cluster detaches from the spinel particle

different materials or temperature differences. It is a possible reason for the observed cyclic acceleration of the endogenous particles on their way to the host particle. The simplified calculation of the attractive force according to Eq. (2.1) is not valid in this case. With the dissolution of meniscus on the exogenous spinel particle at about 1504 °C, maximum real forces of $1 \cdot 10^{-17}$ N to $4 \cdot 10^{-15}$ N were determined for the endogenous particles, Fig. 2.12. As already observed for Al$_2$O$_3$ and MgO host particles, the maximum real force of endogenous particles increases with increasing endogenous particle size. While endogenous particle pairs with a size less than about 3 μm attract each other more strongly, the forces between endogenous Al$_2$O$_3$ and exogenous spinel particles are larger from a particle size of about 3 μm.

2.3.6 Interaction of Exogenous CaO·6Al$_2$O$_3$-Particles with Molten Steel and Its Endogenous Particles

Two different particle classifications were available for the investigations of exogenous CA$_6$ particles with the steel melt X15CrNiSi25-20. The powders with an

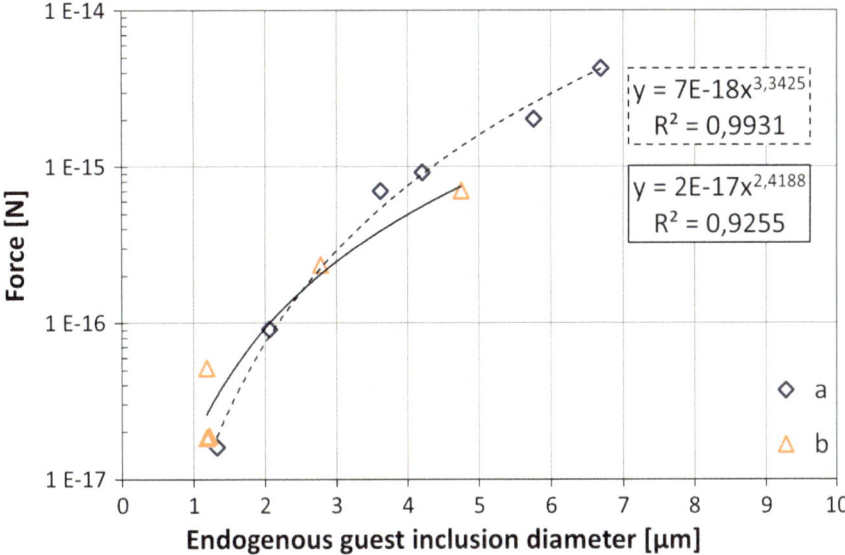

Fig. 2.12 Maximum real forces of the endogenous particles and clusters at approx. 1500 °C during movement **a** to the spinel host particle after removal of the meniscus, **b** to a cluster of endogenous particles

average particle size d_{50} of approx. 6 μm form conglomerates due to adhesion forces. Individual particles could not be separated.

Different phases of the melting of a steel plate with a CA_6 conglomerate can be seen in Fig. 2.13. First liquid areas formed on the steel surface already at 1345 °C, at temperatures approx. 90–120 K lower than in the previous investigations. When the melt comes into contact with the conglomerates, an irregular pulsating movement of the melt is observed, which is attributed to bubble formation. It was also observed that the melt resolidifies in the immediate surroundings of the conglomerate and new grain boundaries form around the conglomerate, Fig. 2.13b. This newly formed δ-ferrite has a higher melting temperature and remains on the conglomerate, even if the steel is already in the molten state in the surrounding area. The observations indicate that oxygen from the conglomerate reacts with the carbon supersaturated steel melt. Upon contact of the melt with the CA_6 conglomerate, μm-sized particles move towards it, but it is not possible to identify whether they are endogenous Al_2O_3 particles or CA_6 particles of the same species.

Further information on the interaction of CA_6 particles with the steel melt is provided by investigations on approx. 500 μm large particles. Figure 2.14 shows part of an exogenous CA_6 particle. At approx. 1485 °C, the first molten areas can be seen (Fig. 2.14a). Even though the temperature is increased further, the melt on the particle solidifies (Fig. 2.14b), similar to the observations with powder conglomerates. A holding time of approx. 30 s at 1515 °C led to renewed melting of the steel

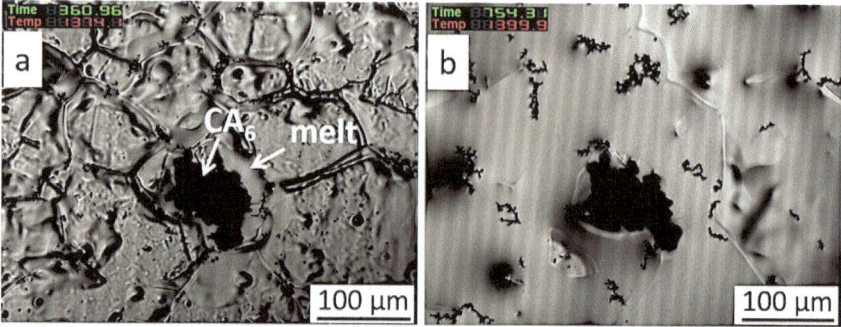

Fig. 2.13 Conglomerate of CA_6 powder on a X15CrNiSi25-20 steel disc **a** 1374 °C: melt on the conglomerate, **b** 1400 °C: last grain boundaries on the conglomerate surrounded by melt

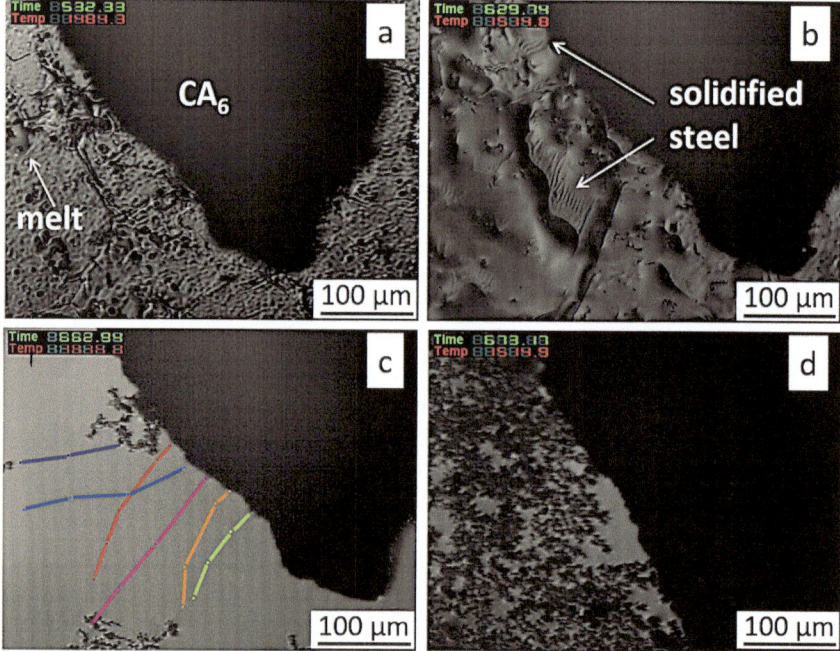

Fig. 2.14 Part of a 500 μm CA_6 particle on a X15CrNiSi25-20 steel disc **a** first molten areas at 1484 °C, **b** solidified steel melt at 1515 °C, **c** traces of particle movement towards the exogenous particle and **d** clusters of endogenous inclusions on the exogenous CA_6 particle

associated with an intensive accumulation of inclusions and clusters on the exogenous particle. Figure 2.14c shows individual paths from endogenous particles to CA_6 and an attached cluster. The maximum observed distance of the attractive forces is approx. 230 μm. After about 40 s, clusters of endogenous particles formed around the CA_6 (Fig. 2.14d). It can be seen that the endogenous particles are more densely bonded to each other than to the exogenous particle.

The maximum calculated attractive forces for endogenous particles and clusters are shown in Fig. 2.15. The attractive forces are several orders of magnitude larger than those of the previously studied exogenous materials. A reason for this is the much larger particle size of CA_6.

A section through a CA_6 particle after the melting test shows that the exogenous particle is not immersed (Fig. 2.16). It lies on a relatively flat solidified steel surface.

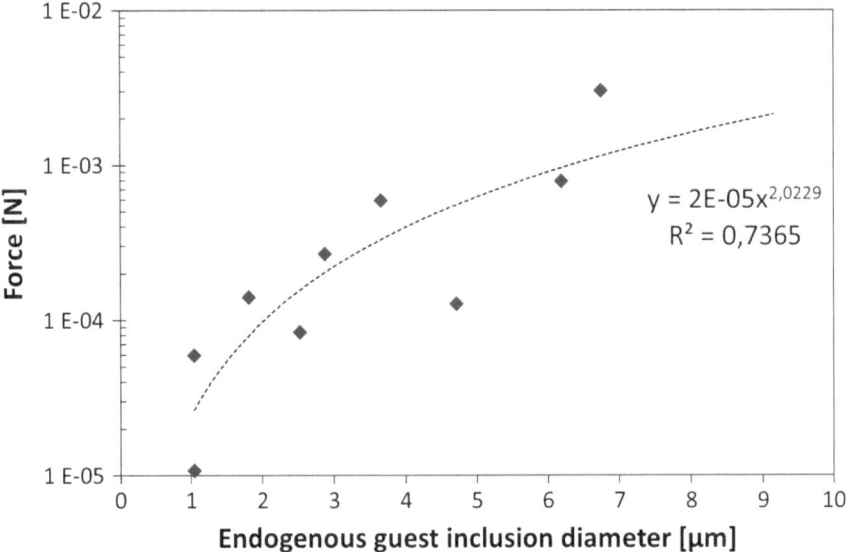

Fig. 2.15 Maximum real forces of the endogenous particles and clusters at approx. 1515 °C during movement to the CA_6 host particle

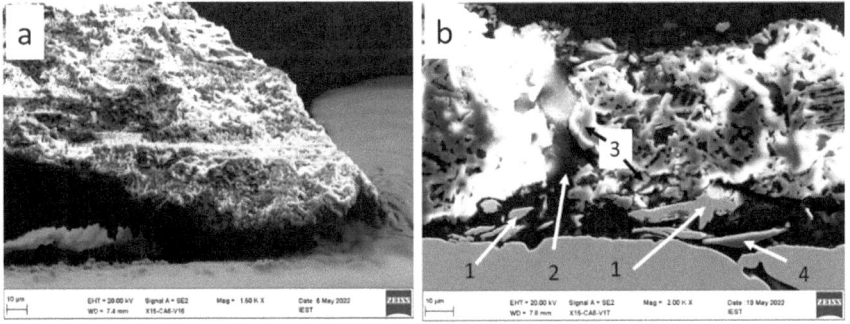

Fig. 2.16 SEM-micrographs of CA_6 particle after melting test **a** particle with cavity and **b** Cross-section with metallic particles (1), a crack (2) through the CA_6 (3) and Al_2O_3 particles (4)

These observations support the thesis that the buoyancy forces acting on the CA_6 are large due to its low density and approximately 10% porosity. Despite the visible dendritic solidification of the steel under the particle, indicating that large portions of the steel were in the molten state, the particle only partially contacts the steel surface. Voids were formed between the particle and the steel and non-metallic particles have deposited on both the steel plate and the exogenous particle. The CA_6 particle exhibits intense cracks (Fig. 2.16b). Both Al_2O_3 and steel particles were detected on the contact surface between exogenous particle and steel. Detailed images of a CA_6 particle after the melting test show a porous body with a brighter oxidic phase enriched in Mn, Si and Cr from the steel [Fig. 2.17a, b (4, 5, 7)].

The pores of CA_6 particle are partially filled with an (Al,Si,Ca)-rich slag (3, 8). The Ca contents of the slag are greater than that in CA_6. The peripheral regions of CA_6 are depleted in Ca (Fig. 2.17c) and enriched in Mn (Fig. 2.17d). This is illustrated by the higher values of the Al/Ca-at.% ratio compared to the ratio in CA_6, which has the value of 12. Also pure alumina was analysed (6). The composition of the phases and their Al/Ca ratios are summarised in Table 2.6.

Melting tests of cylindrical steel samples using the sessile drop method on sintered CA_6 tablets in the heating microscope confirm the formation of an oxide phase (2) on the liquid steel drop (Fig. 2.18). The oxide layers influence the wetting angle and led to divergent results between 115 and 140° at a test temperature of 1450 °C for three tests. Monaghan et al. [24] found wetting angles of 130° at 1450 °C and <110 °C at 1550 °C for iron with a dissolved carbon content of 5 wt.% on CA_6 plates. Wei

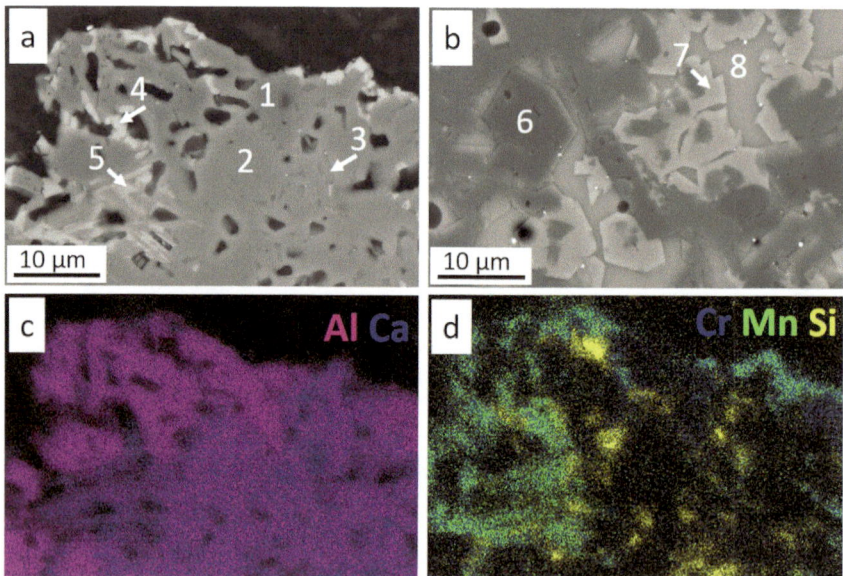

Fig. 2.17 Detail of porous CA_6 particle **a** and **b** SEM images, **c** and **d** element mapping of image **a** with element distribution of **c** Al and Ca and **d** Cr, Mn and Si

Table 2.6 EDX-analysis of areas in Fig. 2.17a, b (fractions in at.%) and Al/Ca ratio

Spot	Al	Ca	Si	Mn	Cr	Al/Ca
1	35.0	2.8	0.3	0.06	0.4	12.6
2	22.6	2.3	0.4	0.06	–	9.7
3	25.7	4.8	7.4	1.7	0.7	5.3
4	22.9	0.8	3.5	5.2	2.8	26.9
5	27.8	1.5	1.2	4.1	1.8	18.6
6	40.0	–	–	–	–	–
7	21.6	0.6	2.7	7.3	0.8	36.0
8	10.6	2.8	13.2	6.1	0.2	3.8

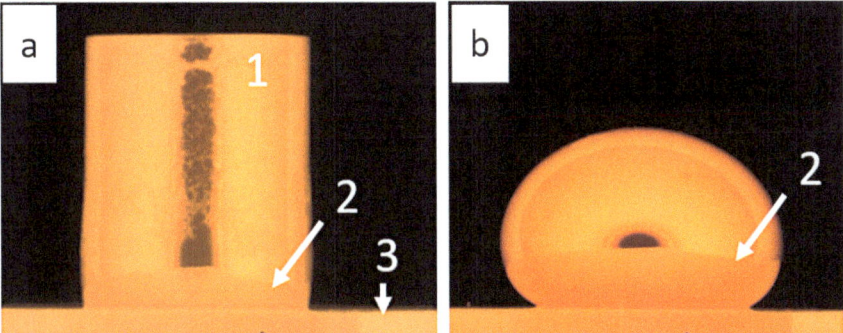

Fig. 2.18 Heating microscope images **a** before melting the steel cylinder (1) on the CA_6-Tablette (3) at 1384 °C and **b** at 1450 °C with oxid layers (2)

et al. [25] determined a contact angle of 134° for electrolytic iron on a CA_6 tablet at 1600 °C. Analog to Yin´s conclusions, the high wetting angles between 100° and 130 lead to high attractive forces also [5].

After the test, individual ceramic particles adhere to the ceramic-steel interface, whereby it cannot be clearly assigned whether the particles were newly formed from the steel matrix or detached from the CA_6 tablet. The liquid (Al,Si,Ca)-rich phase was also observed in the pores of the CA_6 tablets to a depth of approx. 300 µm (Fig. 2.19). While the slag at a depth of approx. 160 µm has a Ca content of 5 at.%, the Ca content at a depth of approx. 90 µm is <1 at.%. Additionally, in contact with the liquid phase, the CA_6 is depleted in Ca and enriched in Cr. The CA_6 particles that were in contact with the slag have smoother boundaries than the unaffected CA_6 (Fig. 2.19b, c). Small dendrites grow from their boundaries to the slag (Fig. 2.19d).

It is possible, that the liquid oxide from the slag solidified at the former CA6 grain boundaries. After immersion tests, Storti [13] already found whiskers on the CA6-coated surface of carbon-bonded alumina filter material. Jiang et al. [21] report a steel/slag reaction in which solid oxidic particles transferred into lower melting point inclusions in presence of only 0.0002 wt.% dissolved Ca. Raviraj et al. [26]

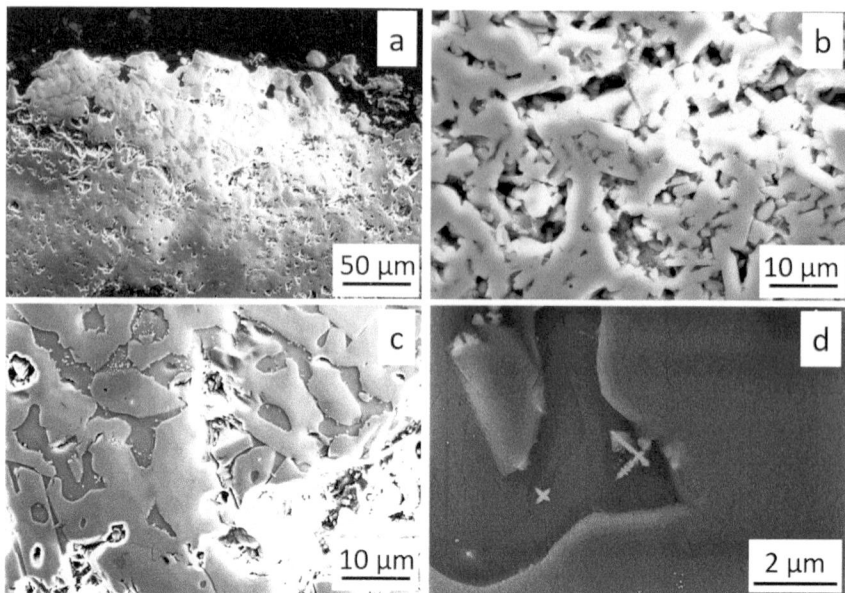

Fig. 2.19 CA_6 tablet after contact with molten steel **a** ceramic particles on the surface, **b** uninfluenced CA_6 **c** infiltrated CA_6 and **d** dendrites growth

observed the reduction of a Si-Ca-Al rich solid slag with the formation of cavities with embedded inclusions using low and high Al steels. In this case, SiO is reduced to Al_2O_3, unfortunately without taking into account the high CaO content originally present in the slag.

In the binary Al_2O_3-CaO diagram (Fig. 2.20), CA and CA_6 are in equilibrium with CA_2 at about 1300 °C. As the CaO content increases, the melting point of the phases drops from 1766 to 1370 °C. The liquid silicon oxide, which is formed by reducing the Cr_2O_3 passive layer on the steel surface and is also present in minute quantities in the exogenous CA_6 particle, can contribute to a further reduction in the melting temperature. FactSage calculations of the steel with the addition of 10^{-2} g each of Cr_2O_3, Al_2S_3, FeS and MnS detected as sulphides on the steel surface, 40 ppm oxygen and 1 g CA_6 related to 100 g steel lead to the formation of 0.36 g liquid slag with 69 wt.% Al_2O_3, 23 wt.% CaO, 4 wt.% SiO_2, 2 wt.% Ti_2O_3 and approx. 1 wt.% (TiO_2 and CrO). CA_6 no longer exists as an equilibrium phase. Calculations without Cr_2O_3 and sulphides with 40 and 400 ppm oxygen at 1515 °C lead to a small reduction of CA_6 under the formation of slag from Al_2O_3, SiO_2, Ti_2O_3, TiO_2, CaO, MnO, CrO and FeO.

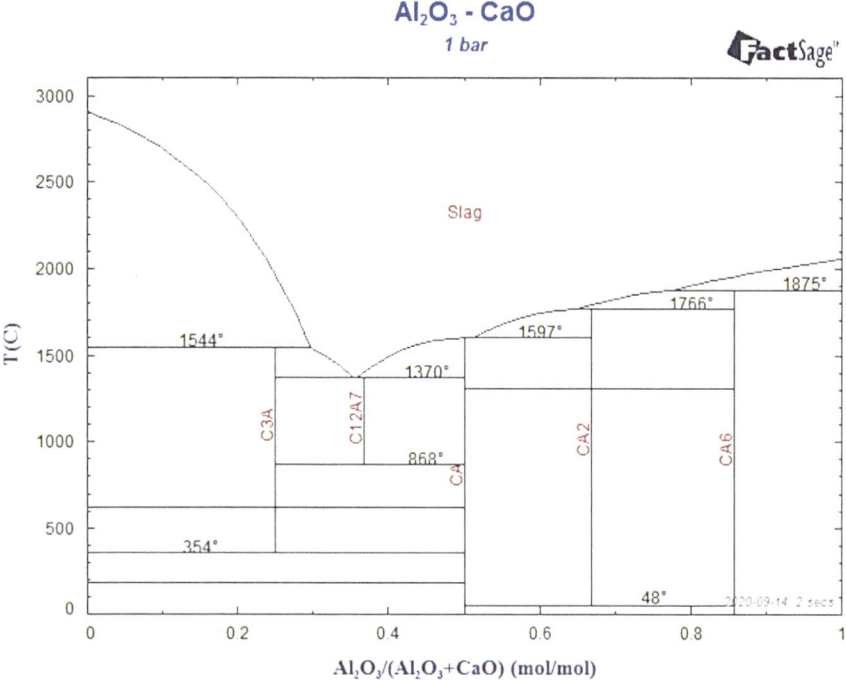

Fig. 2.20 Binary Al₂O₃-CaO diagram

2.4   Conclusion

The trend lines of the maximum real forces of endogenous Al₂O₃ particles interacting with exogenous Al₂O₃, MgO und MgAl₂O₄ particles are summarized in Fig. 2.21. Accordingly, endogenous Al₂O₃ particles smaller than about 2.5 μm in diameter exhibit the largest forces to the conspecific material. Clusters with a particle diameter greater than 4 μm based on its area are more strongly accelerated to magnesia. The endogenous alumina particles with size of 2.5 μm and larger were most strongly attracted to the exogenous spinel particle. However, the endogenous Al₂O₃ particles formed into clusters detached from the spinel in the progress of the experiment (Fig. 2.11d).

In contrast to the investigations presented above, the initial material of CA₆ had a very small particle size of approx. 6 μm or a much larger particle size of approx. 500 μm. A direct comparison of the attractive forces to the other refractory materials was therefore not possible. However, it could be shown that, analogous to the observations of Storti et al. [13], a liquid phase forms on the surface of the CA₆ particle in contact with the melt. An (Al,Si,Ca)-rich liquid phase was also detected in the pores of the coarsely ground CA₆ refractory material. Since this phase has more Ca in deeper areas than in the peripheral areas, it can be assumed that the dissolved Ca evaporates. The cracks in the CA₆ particle and the cavities under the particle after

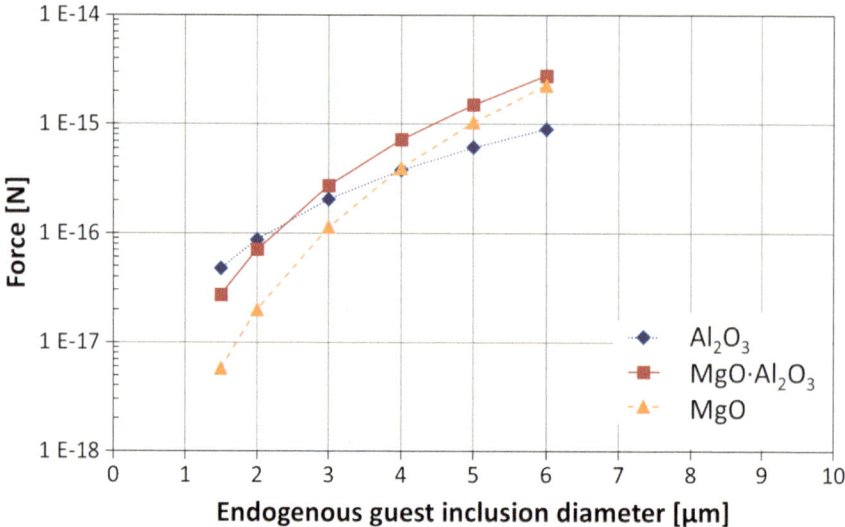

Fig. 2.21 Trend lines of the calculated maximum attractive forces between exogenous approx. 50 μm refractory particles and endogenous alumina particles as a function of their particle size

cooling of the sample could be due to vapour phase formation. On the other hand, the formation of an (Al,Mn,Cr)-rich oxide is observed at the particle surface and inside the particle in direct contact with the liquid phase. The deposition of alumina, manganese and chromium oxide from the molten steel can help to increase the purity of the steel.

2.5 Summary

Refractory materials are used in metallurgy to protect ladles and crucibles due to their high temperature resistance. They are also used as filter materials to remove/reduce non-metallic inclusions and contribute to increasing the degree of purity of steels. In the present work, the interactions of Al_2O_3, MgO, $MgAl_2O_4$ and CA_6 refractories with molten steel X15CrNiSi25-20 were investigated. The steel X15CrNiSi25-20 is an aluminium-killed stainless steel for quenching and tempering. Its carbon content is low, which minimises the tendency for carbon loss during the melting process. Since the ferritic high-temperature phase forms only to a small extent, a relatively stable melt pool is possible over a sufficiently long time interval to observe the interaction of endogenous and exogenous particles during initial contact with the melt. Endogenous aluminium oxides form on steel surface during melting. In this process, the dissolving chromium oxide passivation layer serves as oxygen source.

Exogenous alumina refractory does not react with the steel melt. The particle remains unchanged in its chemical composition. Already at the first contact of the molten steel with the exogenous alumina, the endogenous alumina particles move towards the larger exogenous particle and adhere to it. In contrast, chemical reactions were observed in the interaction of exogenous $MgAl_2O_4$ spinel, MgO and CA_6 particles with the molten steel. They lead to the formation of a strong meniscus in the case of the spinel and to the formation of a liquid phase on the MgO and CA_6 particles, respectively. Contact of the endogenous inclusions with the exogenous refractory at the surface of the molten steel was observed only after degradation of the meniscus or dissolution of the liquid interface in the HT-CLSM. CA_6 has a special feature. In the porous CA_6 particle, a liquid (Al,Si,Ca)-rich phase forms in the voids. In addition, the formation of (Al,Mn,Cr)-rich oxides was observed. The CA_6 particle is depleted of Ca. The emission of Ca can contribute to greater bath movement in the molten steel and lead to new slag being able to penetrate the porous particle.

The maximum forces determined between the endogenous alumina particles and the tested refractories depend on the size of the endogenous inclusions. As the endogenous particle size increases, the maximum real forces acting on the particles increase. Endogenous particles smaller than about 2.3 μm were most strongly attracted to conspecific exogenous particles. The maximum real forces were largest for larger endogenous alumina particles interacting with exogenous spinel. At the same time, it was observed that the adhesion of endogenous particles to the spinel was low. A direct comparison to the attractive forces using CA_6 as host particle was not possible due to the different particle classification.

Acknowledgements The authors would like to thank Ms. Grahl for metallographic sample preparation and Mr. Fischer for technical assistance with equipment maintenance. The studies were carried out with financial support from the German Research Foundation (DFG) within the framework of the Collaborative Research Center CRC 920 "Multi-Functional Filters for Metal Melt Filtration–A Contribution toward Zero Defect Materials" (Project-ID 169148856) subproject A01.

References

1. K. Janiszewski, Z. Kudliński, Removal of liquid non metallic inclusion from molten steel using method filtration. Metal 1–9 (2006). Retrieved from http://metal2014.tanger.cz/files/proceedings/metal_06/indexe.htm
2. P.A. Kralchevsky, V.N. Paunov, N.D. Denkov, I.B. Ivanov, K. Nagayama, J. Colloid Interface Sci. **155**(2), 420–437 (1993). https://doi.org/10.1006/jcis.1993.1056
3. P.A. Kralchevsky, K. Nagayama, Langmuir **10**(1), 23–36 (1994). https://doi.org/10.1021/la0013a004
4. K.D. Danov, B. Pouligny, P.A. Kralchevsky, Langmuir **17**(21), 6599–6609 (2001). https://doi.org/10.1021/la0107300
5. H. Yin, H. Shibata, T. Emi, J. KIM, Capillary attraction of solid particles at gas/steel melt interface. HTC **97**, 380–388 (1997)
6. C.G. Aneziris, S. Dudczig, J. Hubálková, M. Emmel, G. Schmidt, Ceram. Int. **39**(3), 2835–2843 (2013). https://doi.org/10.1016/j.ceramint.2012.09.055
7. H. Yin, H. Shibata, T. Emi, M. Suzuki, In-situ" observation of collision, agglomeration and cluster formation of alumina inclusion particles on steel melts. ISIJ Int. **37**(10), 936–945 (1997)

8. W. Mu, N. Dogan, K.S. Coley, Metall. Mater. Trans. B **48**(5), 2379–2388 (2017). https://doi.org/10.1007/s11663-017-1027-4
9. S. Kimura, K. Nakajima, S. Mizoguchi, Metall. Mater. Trans. B **32**(1), 79–85 (2001). https://doi.org/10.1007/s11663-001-0010-1
10. H. Yin, H. Shibata, T. Emi, M. Suzuki, ISIJ Int. **37**(10), 946–955 (1997). https://doi.org/10.2355/isijinternational.37.946
11. W. Mu, N. Dogan, K.S. Coley, JOM **70**(7), 1199–1209 (2018). https://doi.org/10.1007/s11837-018-2893-1
12. K. Uemura, M. Takahashi, S. Koyama, M. Nitta, ISIJ Int. **32**(1), 150–156 (1992). https://doi.org/10.2355/isijinternational.32.150
13. E. Storti, M. Farhani, C. G. Aneziris, C. Wöhrmeyer, C. Parr, *Steel Res. Int.* **88**(11), 1700247 (2017). https://doi.org/10.1002/srin.201700247
14. C.G. Aneziris, C. Schroeder, M. Emmel, G. Schmidt, H.P. Heller, H. Berek, Metall. Mater. Trans. B **44**(4), 954–968 (2013). https://doi.org/10.1007/s11663-013-9828-6
15. X. Wei et al., J. Eur. Ceram. Soc. **42**(11), 1676–4685 (2022). https://doi.org/10.1016/j.jeurceramsoc.2022.04.058
16. Y. Rotenberg, L. Boruvka, A.W. Neumann, J. Colloid Interface Sci. **93**(1), 169–183 (1983). https://doi.org/10.1016/0021-9797(83)90396-X
17. D. Chebykin, T. Dubberstein, H.-P. Heller, O. Fabrichnaya, O. Volkova, Ceram. Int. **48**(3), 3771–3778 (2022). https://doi.org/10.1016/j.ceramint.2021.10.160
18. W. Weißbach, M. Dahms, C. Jaroschek, in *Werkstoffe und ihre Anwendungen: Metalle, Kunststoffe und mehr*, ed. by W. Weißbach, M. Dahms, C. Jaroschek (Springer Fachmedien, Wiesbaden, 2018), pp. 463–487. https://doi.org/10.1007/978-3-658-19892-3_12
19. M. Hasegawa, in *Treatise on Process Metallurgy*, ed. by S. Seetharaman, (Elsevier, Boston, 2014), pp. 507–516. https://doi.org/10.1016/B978-0-08-096986-2.00032-1
20. C. Schröder, U. Fischer, A. Schmidt, G. Schmidt, O. Volkova, C. G. Aneziris, Adv. Eng. Mater. **19**(9), 1700146. https://doi.org/10.1002/adem.201700146
21. J.-H. Park, I.-H. Jung, H.-G. Lee, ISIJ Int. **46**(11), 1626–1634 (2006). https://doi.org/10.2355/isijinternational.46.1626
22. C.G. Aneziris et al., Adv. Eng. Mater. **15**(12), 1168–1176 (2013). https://doi.org/10.1002/adem.201300155
23. I. Ganesh et al., Ceram. Int. **28**(3), 245–253 (2002). https://doi.org/10.1016/S0272-8842(01)00086-4
24. B. Monaghan, M. Chapman, S. Nightingale, The wetting of liquid iron carbon on aluminate minerals formed during coke dissolution in iron. *Faculty of Engineering—Papers (Archive)* (2009). https://ro.uow.edu.au/engpapers/498
25. X. Wei et al., J. Eur. Ceram. Soc. **42**(5), 2535–2544 (2022). https://doi.org/10.1016/j.jeurceramsoc.2022.01.011
26. A. Raviraj et al., Metall. Mater. Trans. B **52**(2), 1154–1163 (2021). https://doi.org/10.1007/s11663-021-02091-z

Chapter 3
Ceramic Filter Materials and Filter Structures with Active and Reactive Functional Pores for the Aluminum Melt Filtration

Claudia Voigt, Tony Wetzig, Jana Hubálková, Patrick Gehre, Nora Brachhold, and Christos G. Aneziris

3.1 Aluminum Filtration–State of the Art

3.1.1 Non-metallic Inclusions in Aluminum

Due to its low density and the high strength per mass, aluminum is a valued construction material in areas where weight plays a significant role. In the course of development, ever-increasing demands are being placed on the mechanical performance of the material, which depends to a large extent on the degree of purity of the aluminum. When considering the degree of purity of aluminum, a distinction between non-metallic inclusions, gases (for example hydrogen) and trace elements (for example

C. Voigt (✉) · T. Wetzig · J. Hubálková · P. Gehre · N. Brachhold · C. G. Aneziris
Institute of Ceramics, Refractories and Composite Materials, Technische Universität
Bergakademie Freiberg, Agricolastraße 17, 09599 Freiberg, Germany
e-mail: claudia.voigt@ikfvw.tu-freiberg.de

T. Wetzig
e-mail: tony.wetzig@ikfvw.tu-freiberg.de

J. Hubálková
e-mail: jana.hubalkova@ikfvw.tu-freiberg.de

P. Gehre
e-mail: patrick.gehre@ikfvw.tu-freiberg.de

N. Brachhold
e-mail: nora.brachhold@ikfvw.tu-freiberg.de

C. G. Aneziris
e-mail: christos.aneziris@ikfvw.tu-freiberg.de

© The Author(s) 2024
C. G. Aneziris and H. Biermann (eds.), *Multifunctional Ceramic Filter Systems
for Metal Melt Filtration*, Springer Series in Materials Science 337,
https://doi.org/10.1007/978-3-031-40930-1_3

57

alkalis) is made. The focus of this chapter is on the filtration of the non-metallic inclusions.

The non-metallic inclusions affect the castability, the mechanical properties and the workability of the casting. The castability is lowered by a deterioration of the flowability of the aluminum melt [1, 2]. The reduction of the mechanical properties (tensile strength, fracture strain) of aluminum is due to the different physical properties (density and coefficient of expansion) of aluminum and the non-metallic inclusions and their presence as two separate phases without bonding to each other [3, 4]. Such differences lead to critical stress fields during cooling processes or when external stresses are applied. In addition, non-metallic inclusions worsen machinability and workability, as they have a higher hardness than aluminum. The non-metallic inclusions lead to a reduction in the surface quality of bright products. Furthermore, it has been observed that a higher content of non-metallic inclusions in the aluminum melt causes a higher porosity in the cast product [5]. Due to the various negative effects of non-metallic inclusions, efforts are being made to minimize their content.

Non-metallic inclusions can for example be classified in terms of the inclusion shape. The non-metallic inclusions can be present as individual particles, agglomerates or films. Figure 3.1 shows examples of aluminum oxides, spinels and magnesium oxides occurring as agglomerates or films in cast aluminum.

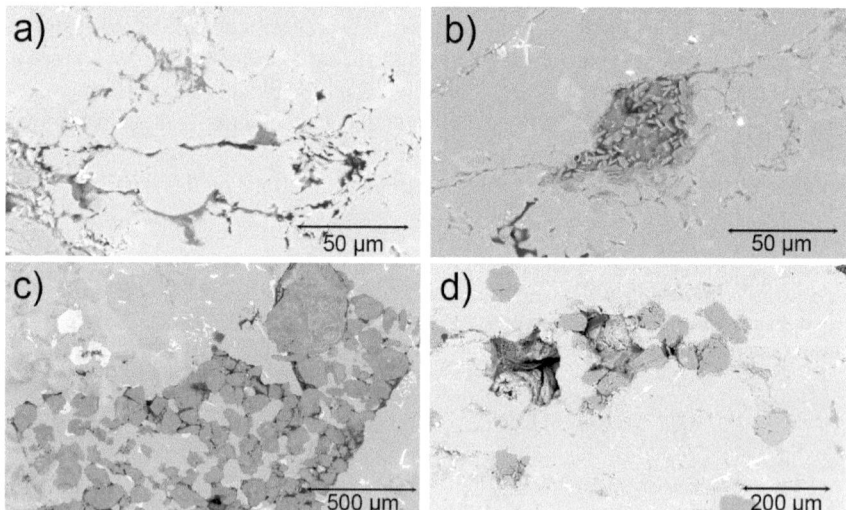

Fig. 3.1 SEM micrographs of typical inclusions: **a** Al_2O_3 films, **b** Al_2O_3 agglomerate, **c** $MgAl_2O_4$ agglomerate, and **d** SiO_2 agglomerate

3.1.2 Filtration of Aluminum Melts

In order to minimize the inclusion content, two approaches must be followed:

- Avoidance of inclusion formation,
- Removal of existing inclusions.

Avoidance of inclusions is achieved, by smart design of the casting system. The correct design of the casting system avoids turbulence, which would lead to the creation of new aluminum surfaces in contact with the surrounding atmosphere and thus to the formation of oxide films [6].

There are various methods for the removal of non-metallic inclusions such as purge gas treatment, sedimentation and filtration whereby only the topic of filtration is dealt with in the following. Since the 1960s, the foundry industry has used filters to reduce the inclusion content of cast metal melts. The filters for metal casting improve the quality of the melt by filtering out the non-metallic inclusions and also by equalizing and calming the melt stream. Calming the melt has the advantage of reducing or even preventing the rupture of the oxide skin on the surface of the melt stream, resulting in less few new oxide films [7]. According to Dam et al. the filtration efficiency depends on the following parameters [8]:

- Melt parameters (temperature, viscosity and composition),
- Process parameters (filtration rate, melt pretreatment and structure of the casting system),
- Inclusion parameters (chemism, structure, size and number),
- Filter parameters (geometry, ratio of solid to open porosity, chemism, wetting behavior, size and distribution of functional macropores).

3.1.3 Method for the Determination of the Non-metallic Inclusion's Concentration in Aluminum

The determination of the concentration of non-metallic inclusions is not only an important quality test for aluminum but it is also essential for the evaluation of filters and their filtration efficiency. However, such determinations are complicated due to the inhomogeneous distribution and low concentration of non-metallic inclusions. The important methods for determining inclusion concentration can be divided in two groups [9]:

- Metallographic examinations,
- Indirect determination of inclusions.

Metallographic examinations of cast aluminum do not lead to reliable statements about the purity of the aluminum due to the low concentration and the inhomogeneous distribution of the inclusions. For this reason, methods have been developed in which the inclusions are enriched in the sample. The most commonly used methods are the PoDFA (Porous Disk Filtration Apparatus) and the LAIS (Liquid Aluminum Inclusion Sampler) methods. In the PoDFA and LAIS methods, liquid aluminum is forced or drawn through a narrow-mesh filter at a defined temperature, concentrating the inclusions on a screen. To force the molten aluminum through the filter, the PoDFA method uses positive pressure while the LAIS method uses negative pressure. After the melt has cooled, the filter together with the filter cake, is ground and can be examined metallographically. Metallographic examination of the filter cake allows to gain information about the chemistry, structure, and number of inclusions [10].

In the field of indirect determination of inclusions (often referred to as physical methods), the inclusions are not evaluated directly, but conclusions about the number of inclusions present are drawn using other measurable quantities such as the flow rate per time or the change in conductivity. The best-known indirect analysis methods (Prefil, LiMCA and ultrasound) are briefly presented below. The Prefil method (Pressure filtration) belongs to the online methods, as it enables a melt quality level to be determined in the shortest possible time. For this purpose, liquid aluminum is filled into the provided crucible of the Prefil® Footprinter analyzer and forced through a close-meshed filter at a defined temperature and overpressure [11]. The non-metallic inclusions remain on the filter as a filter cake and increase the flow resistance during filtration. The aluminum that has flowed through the filter is collected in a second crucible, which is placed on a balance. The calculated time-dependent change in flow rate provides an indication of the purity of the aluminum melt in terms of non-metallic inclusions. An aluminum melt with many non-metallic inclusions results in a higher filter cake, which increases the flow resistance and decreases the flow rate over time. The Prefil method allows a rough evaluation of the impurity level of the melt, but not a fine differentiation of the purity level.

The LiMCA method (Liquid Metal Cleanliness Analyzer) also belongs to the online methods and is based on the principle of a Coulter Counter, which uses the change in the average electrical conductivity between two electrodes to determine the number and size of inclusions with a conductivity different from that of the molten metal. For this purpose, an electrically insulated sample tube made of borosilicate glass with an integrated electrode and an opening (diameter selected depending on the analysis task) is immersed in the molten aluminum. The electrical conductivity is continuously measured between the electrode integrated in the sample tube and an electrode in the melt. When a non-metallic inclusion is transported through the opening in the sample tube, the absolute resistance changes and thus a voltage drop can be detected. The number of non-metallic inclusions can be determined by the number of voltage drops, while the size of the inclusions can be determined by the magnitude and variation of the voltage drop [10].

The ultrasonic method can be used to determine inclusions in both liquid aluminum and solidified aluminum. In both cases, the principle of the pulse-echo method is applied [12]. Achard et al. [13] presented a working ultrasonic system for liquid aluminum. In the solidified state, ultrasonic testing for aluminum is a recognized technique for detecting inclusions and voids. Here, the detection limit of the inclusion size is approximately 0.2–0.5 wavelengths of the sound in the material under examination. Taking the values for aluminum into account, the minimum detection limit in aluminum (20 MHz and 3130 m/s) is about 30 μm. Determining the exact location, size and differentiation between pore and inclusion is a very complex and difficult task [14].

3.1.4 Filters for Aluminum Melt Filtration

For the reduction of non-metallic inclusions, various types of filters are used, which differ in terms of their structure and chemical composition, for example, fabric filters, perforated filters, ceramic foam filters, tubular filters and packed bed filters are used in the industry. Tubular filters and packed bed filters are only applied in continuous casting, as they have to be preheated. Ceramic foam filters are used in continuous casting and mold casting, while fabric filters and sieve cores are used only in mold casting [7].

Ceramic foams for aluminum melt filtration often consist of aluminum oxide or silicon carbide. These ceramic materials have been selected on the basis of economic, process engineering and thermomechanical aspects, but not with regard to the filtration properties. In the technical literature, only sparse information can be found on the influence of the chemism of the filter surface on the filtration properties of aluminum. When searching the literature, the term "active filtration" stands for:

- Approach that improves filtration via the reaction of filter surface with inclusion or contamination [15].
- Approach that improves filtration by allowing inclusions to remain on the filter surface due to adhesion at high temperatures [16].
- Approach with filter surfaces made of refractory materials, which has a similar chemism to the inclusions to be filtered out and have different wetting behaviors [17].

3.1.5 Wetting Behavior of Ceramic Materials with Liquid Aluminum

The determination of the wetting behavior of liquid aluminum on ceramic materials is a great challenge and a frequently studied phenomenon. Bao et al. [18] showed the wide variation (between 45 and 167°) of the measured wetting angles in the Al/ Al_2O_3 system. The possible reasons for the high scatter of the results are the various influencing parameters such as the roughness [19, pp. 58–61], the volumetric mass [19, p. 73], the temperature [20], infiltration of aluminum into the substrate [19, p. 107], dwell time and phase composition [21], alloy composition [22] and formation of oxide skin. In investigations of the wetting behavior of molten aluminum, the formation of the oxide skin has a significant influence on the obtained results and will therefore be explained in some detail below.

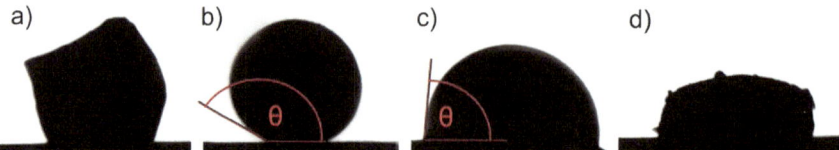

Fig. 3.2 Photographs of sessile drop tests: **a** aluminum at room temperature, **b** molten aluminum–poor wetting, **c** molten aluminum–good wetting, and **d** due to a pronounced oxide skin, only internally molten aluminum

Even at the lowest oxygen partial pressures (10^{-49} bar), an oxide skin forms on the aluminum droplet due to the high affinity of aluminum to oxygen. Not only oxygen from the air leads to oxidation of the aluminum, but also the presence of gaseous water. According to Eberhardt [23], the oxidation of aluminum single crystals is four times faster under the influence of water vapor than purely under oxygen from the air.

The oxide skin prevents the formation of the true aluminum/ceramic interface. Depending on the thickness of the oxide skin, two effects can occur. If the oxide skin is very thick (for example, at a high oxygen partial pressure), it forms a kind of shell over the already molten aluminum, and this prevents droplet formation (see Fig. 3.2). If the oxide skin is only a few nanometers thick, it stays deformable. Thus, a droplet can form but the oxide skin leads to a wetting angle which is almost independent of the temperature and masks possible effects of different substrate materials or of alloying elements [22].

Laurent et al. [24] and Eustathopoulos et al. [19, pp. 234–238] have shown that under the effect of a sufficiently good vacuum, droplet formation of the aluminum occurs at temperatures greater than 950 °C. According to the literature, it is assumed that liquid aluminum (Al) reacts with alumina (Al_2O_3) and forms gaseous aluminum(I) oxide (Al_2O).

The prerequisite for droplet formation is a thin oxide layer, in which cracks form during heating and the oxide skin ruptures as a result. The resulting suboxide Al_2O would be gaseous at the experimental temperatures and can leave the aluminum droplet, causing the oxide skin to degrade. Droplet formation of the aluminum under vacuum only takes place at sufficiently high reaction rates from an experimental temperature of >950 °C [19, pp. 234–238]. It follows that a determination of the contact angle using the conventional sessile drop technique (also called contact heating mode) is difficult to implement at typical casting temperatures. For the conventional sessile drop technique, a piece of the metal is placed on a substrate in a furnace and subsequently contact heated to the required temperature. During heating and holding time the drop shape is recorded for the determination of the contact angle.

Compared to contact times during the filtration process, relatively long holding times are required to determine the contact angles by the conventional sessile drop technique. It must be taken into account that evaporation of the aluminum also

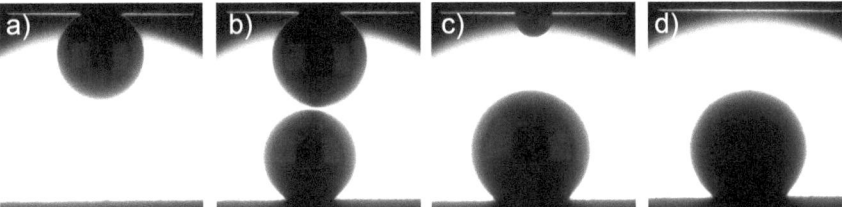

Fig. 3.3 Sequence of photographs of the sessile drop experiment with capillary purification of an AlSi7Mg droplet, **a** the first aluminum drop at the capillary, **b** the first droplet deposited at the substrate and the second droplet formed at the capillary, **c** a large drop formed after merging of two droplets) and **d** the droplet at the substrate after 30 min holding time at the test temperature [25]

changes the contact angle due to the reduction of the droplet volume [19, p. 110–111]. In order to circumvent the required high temperatures and the long holding times, a modified method was developed, in which an aluminum sample is melted separately from the substrate. The aluminum is then pressed through a nozzle with the aid of a piston, and falls by gravity onto the substrate, see Fig. 3.3. The oxide skin remains in the nozzle. This procedure is called sessile drop measurement with capillary purification. The disadvantage of this method is the large influence of the drop mass and the height of fall on the wetting angle [21, 22].

3.2   Insights in the Aluminum Filtration of the CRC 920

The subproject A02 of the Collaborative Research Center 920 deals with the filtration of aluminum melts and examines the influence of different filter parameters on the filtration efficiency. In the first step, the influence of the filter surface chemistry is analyzed, followed by the influence of the wetting behavior of substrates and aluminum melt. Hereinafter, the impact of the nano-functionalized filters and the so far unknown influencing factor of the filter roughness were examined. Finally, the filtration behavior and wetting properties of carbon-bonded alumina filter surfaces were investigated.

3.2.1   Influence of the Filter Surface Chemistry on the Filtration

The influence of the filter surface chemistry was determined with long-term filtration tests at Constellium (Voreppe, France) with an aluminum quantity of 730 kg AlSi7Mg for every tested filter. The filtration pilot plant consists of a melting furnace, a casting channel made of refractory material with an integrated receptacle for the filter and a crucible to collect the liquid aluminum after filtration [26]. The system allows to use two LiMCA probes (ABB Inc., Quebec, Canada) to determine the concentration of non-metallic inclusions in molten aluminum. The filters in form of truncated pyramids with the dimensions of 120 × 120 mm, 90 × 90 mm and height of 50 mm were utilized. Four different filter surface chemistries were tested: Al_2O_3 (alumina), $MgAl_2O_4$ (spinel), $3Al_2O_3 \cdot 2SiO_2$ (mullite), and TiO_2 (rutil) [27].

In order to be able to determine the influence of the chemical composition of the filter surface on the filtration efficiency, the inclusion concentration of the initial melt of the different tests must be comparable as the initial degree of impurity has an influence on the filtration efficiency [28].

The comparison of the inclusion concentration was made using the N20 values of the LiMCA probe, which provide the total inclusion concentrations as a function of time for all inclusions with a particle size between 20 and 300 μm. The LiMCA measurements at the filter entrance showed a typical behavior with a decrease in the N20 values at the beginning of the measurement and an increase at the end of the measurement, see Fig. 3.4. The drop at the beginning of the measurement was due to the reduction of the inclusion concentration caused by the settling of the inclusions. When the last part of the aluminum melt was poured, the settled inclusions were also poured off resulting in the increase of the N20 values. Furthermore, comparable numbers of inclusions for all four tested filters have been determined, see Fig. 3.4 [27].

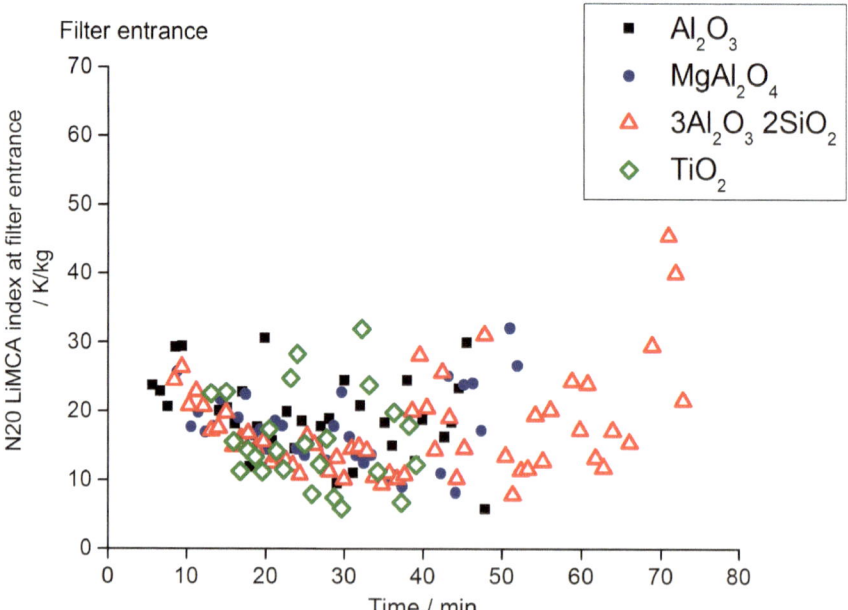

Fig. 3.4 LiMCA index–total inclusion content at the filter entrance as a function of filtration time [27]

By measuring the N20 values at the filter entrance and the filter exit, it is possible to calculate the total filtration efficiency whereby the N20—N300 values in K/kg were used. The $3Al_2O_3$-$2SiO_2$ filter showed the best total filtration efficiency of 89.1% followed by the Al_2O_3 filter with 83.0% and the $MgAl_2O_4$ filter with 83.3%. The TiO_2 filter showed the lowest total filtration efficiency (79.9%).

Moreover, the filtration efficiency was evaluated as a function of inclusion size. Larger inclusions possess a greater damaging effect in solidified aluminum. Therefore, it is of great importance to consider separately the large inclusions, which are outnumbered in terms of quantity, otherwise they would be lost in the large number of finer inclusions [15]. For the evaluation of the filtration efficiency as a function of inclusion size are the differences in settling behavior, inertia and agglomeration behavior of differently sized inclusions [15, 29].

When comparing the filtration efficiencies as a function of inclusion size, differences between the tested filters were observed, see Fig. 3.5. The Al_2O_3 and $MgAl_2O_4$ filters showed very stable filtration behavior over the entire inclusion size range with filtration efficiencies higher than 75%. From an inclusion size larger than 40 μm, the filtration efficiency increased to 90% and larger. 100% filtration efficiency was achieved from an inclusion size larger than 70 μm (Al_2O_3) and 90 μm ($MgAl_2O_4$), respectively [27].

The $3Al_2O_3$·$2SiO_2$ filter showed a quite different filtration behavior in dependence of inclusion size. The lowest filtration efficiency of all tests was measured for inclusions larger than 110 μm for the filter coated with mullite. In the range of inclusion sizes smaller than 60 μm, however, it had the best filtration efficiencies compared to the other three filters. The filtration efficiencies ranged from 93 to 99%

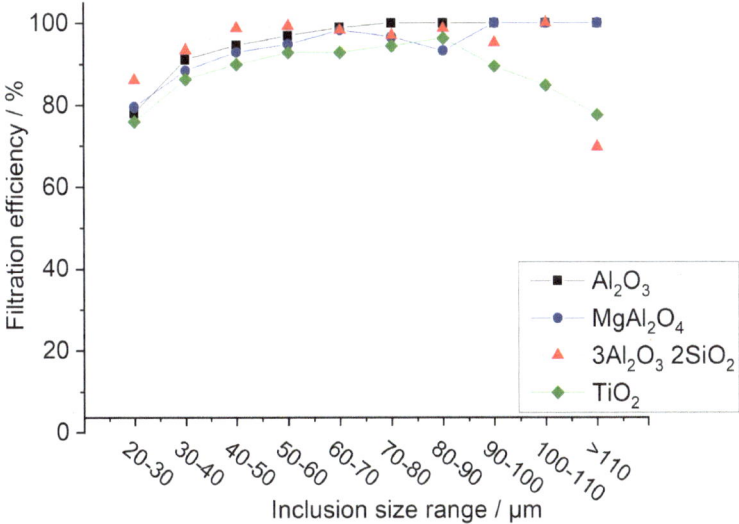

Fig. 3.5 Filtration efficiency depending on filter material and inclusion size [27]

for the inclusion size of 30–60 μm. For inclusions larger than 110 μm, the efficiencies decrease to about 60%. The reason for this decrease is not known, yet. The TiO_2 filter showed the lowest filtration efficiency for almost all inclusion size classes [27].

The LiMCA results provided information regarding the number and inclusion size but did not yield information on the shape and chemistry of the inclusions. For this reason, an examination of the cast filters was performed using scanning electron microscopy (SEM) and energy-dispersive X-ray spectroscopy (EDX). The detected inclusions consisted of:

- Aluminum and oxygen, probably as a corundum phase,
- Aluminum, silicon and oxygen, probably as a mullite phase,
- Aluminum, magnesium and oxygen, probably as a spinel phase,
- Silicon and carbon, probably as a silicon carbide phase.

Figure 3.6 shows SEM micrographes of various typical inclusions, which appeared as particles and as films [27].

In order to be able to determine whether the chemistry of the filter surface had an influence on the chemical composition of the filtered inclusions, the detected inclusions were broken down with regard to their composition, see Fig. 3.6. No significant differences were found with regard to the chemistry of the inclusions as a function of the filter surface. All four inclusion types were detected in comparable concentrations in all tested filters. Further information about the sample preparation,

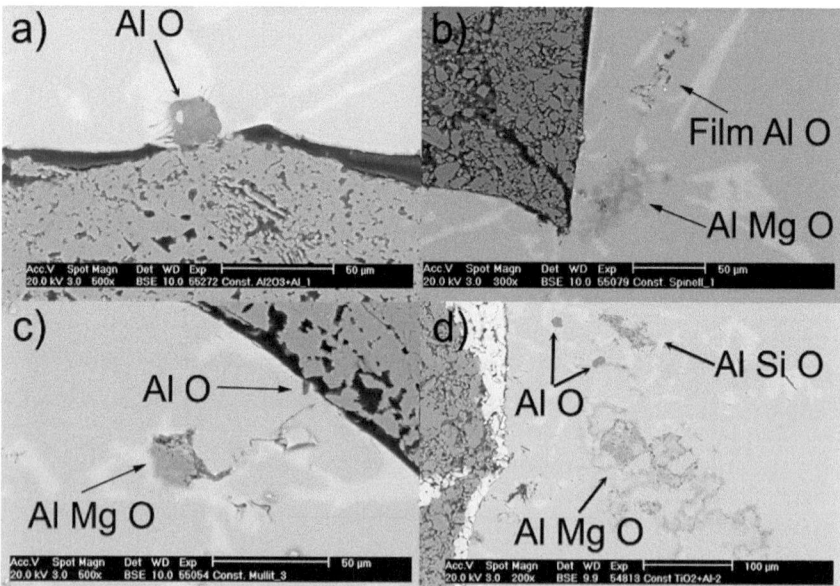

Fig. 3.6 SEM micrograph of the alumina skeletons with different coatings **a** Al_2O_3, **b** $MgAl_2O_4$, **c** $3Al_2O_3$ $2SiO_2$, and **d** TiO_2 after the casting trials [27]

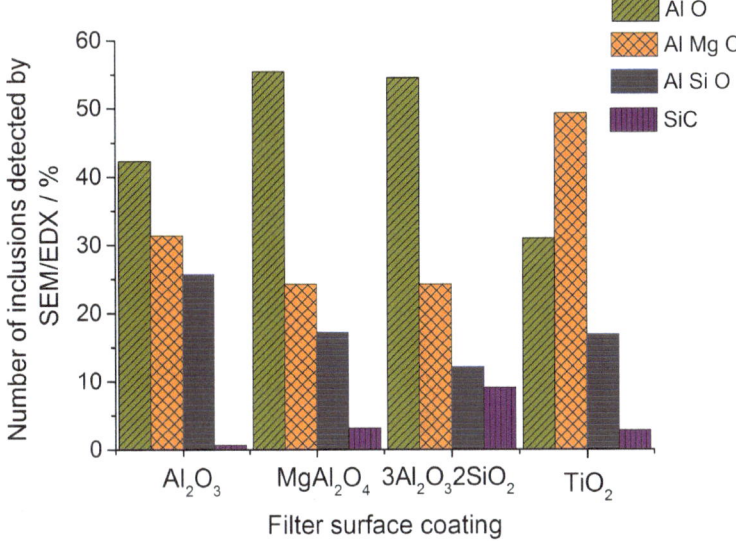

Fig. 3.7 Chemistry of inclusions compared to the filter material [27]

parameter of experiments and additional analysis can be found elsewhere [27, 30, 31] (Fig. 3.7).

3.2.2 Influence of the Wetting Behavior on the Filtration

For the sessile drop measurements at 730 °C (the temperature of the filtration trials, see Sect. 3.2.1), a sessile drop apparatus with a capillary purification at Foundry Research Institute (Krakow, Poland) were used. The test substrate and aluminum were simultaneously, but separately heated under vacuum with 8 K/min. At the temperature of 730 °C, one or two droplets were extruded out of the graphite syringe and dropped on the preheated ceramic substrate. The extrusion took place at a vacuum smaller than $1 \cdot 10^{-5}$ mbar. After dropping, the measurement took 30 min which is comparable to the filtration trials lasting for 40 to 76 min. Four different substrates coated with Al_2O_3 (alumina), $MgAl_2O_4$ (spinel), $3Al_2O_3 \cdot 2SiO_2$ (mullite), and TiO_2 (rutil) were tested, whereby the substrates were coated with the same slurry which was used for the preparation of the filters to ensure a comparable surface quality of the sessile drop substrates and the filter [25].

The contact angles between the AlSi7Mg droplets and the tested substrates as a function of time are presented in Fig. 3.8 and show clear differences. The contact angles vary during the dropping process due to the fall and impingement of the drops on the substrate. One minute after the dropping, the wetting angle is very stable and the differences between the wetting angles directly after dropping and after a holding

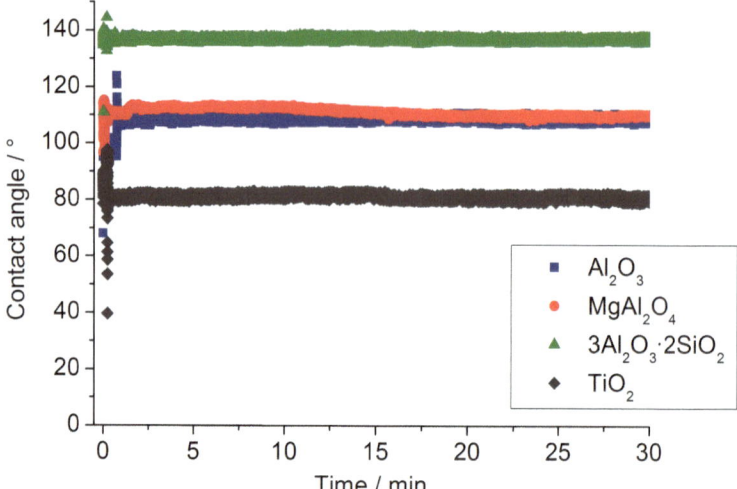

Fig. 3.8 Contact angle measured by sessile drop measurement combined with capillary purification technique [25]

time of 30 min are smaller than 6° for each substrate. The highest contact angle after the holding time of 30 min was measured for $3Al_2O_3 \cdot 2SiO_2$ (136°), followed by $MgAl_2O_4$ (113°) and Al_2O_3 (108°). The lowest wetting angle of 81° was observed for the TiO_2 substrate [25].

In the next step, the evaluated total filtration efficiencies of the filtration trial performed at Constellium [27], see Sect. 3.2.1, were regressed on the measured wetting angles. The coefficient of determination of $R^2 = 0.92$ shows a high goodness of fit between the contact angles and the total filtration efficiency (Fig. 3.9). According to the correlation the following relationship is valid: the larger the contact angle the higher is the filtration efficiency [25].

This finding contrasts with assumptions of Bao et al. [32] stating that a better wetting (lower contact angle) causes an increased convergence of the aluminum melt to the filter wall which increases the probability of a true contact between inclusion and the filter wall.

As the contact angles not only depend on the surface chemistry but also on the surface roughness of the substrates, the roughness of the sessile drop substrates was evaluated. The arithmetic surface roughness S_a shows clear differences whereby the $3Al_2O_3 \cdot 2SiO_2$ substrate ($S_a = 15.8 \, \mu m \pm 1 \, \mu m$) possessed the highest roughness followed by Al_2O_3 ($S_a = 2.7 \, \mu m \pm 0.2 \, \mu m$) and $MgAl_2O_3$. ($S_a = 2.3 \, \mu m \pm 0.1 \, \mu m$). The TiO_2 substrates with a $S_a = 1.0 \, \mu m \pm 0.1 \, \mu m$ have a very smooth surface [25].

The differences in the roughness between the tested materials can be traced back to the different particle size distributions used for the preparation of the coating slurries. The different particle sizes are observable with the help of scanning electron microscopy (SEM) investigations of the filter surfaces, see Fig. 3.10. The

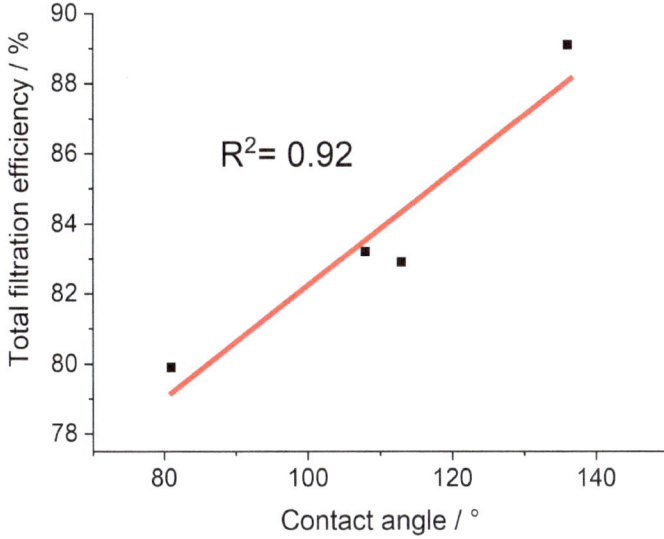

Fig. 3.9 Total filtration efficiency in dependence on the measured contact angle between AlSi7Mg and substrates made of the filter material

$3Al_2O_3 \cdot 2SiO_2$ filter possesses large particles at the surface having a large impact on the roughness value. In contrast, the TiO_2 filter surface appears very smooth with scale-like morphology, see Fig. 3.10 [25]. It can be concluded that the roughness measurements show significant differences whereby the higher the roughness the higher the contact angle and the higher is the total filtration efficiency. Further information about the sample preparation, parameters of experiments and additional analysis can be found elsewhere [25].

3.2.3 Influence of the Nano-functionalized Filters on the Filtration

The contact angle between the aluminum melt and ceramic substrate is influenced by chemistry and phase composition of the substrate material and by the surface roughness of the substrate. In that context, the individual contribution of each parameter on the filtration behavior is of great interest. This question was investigated with the help of nano-functionalized coatings which show differences in the wetting behavior in dependence on the particle size (influencing the surface energy) with at the same time comparable roughness. Thus, the impact of nano-functionalized coatings on the wetting behavior and filtration efficiency in comparison to the Al_2O_3 reference material (see Sect. 3.2.1) was studied.

Al_2O_3 nano-powder with $d_{50} = 80$ nm of IoLiTec Ionic Liquids Technologies GmbH was used as a basis to produce a water-based slurry for nano-functionalization

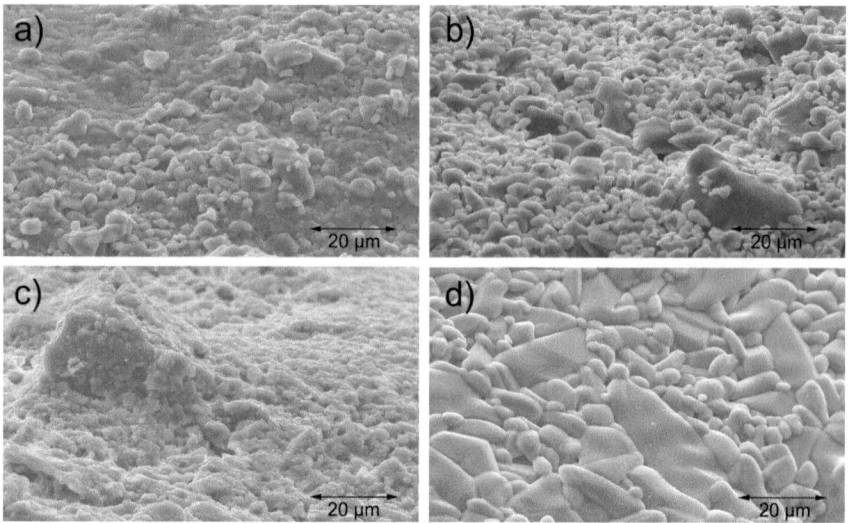

Fig. 3.10 SEM micrographs of filter surfaces before filtration trials: **a** Al_2O_3, **b** $MgAl_2O_4$, **c** $3Al_2O_3 \cdot 2SiO_2$ and **d** TiO_2 [25]

with a solid content of 9 wt%. The resulting slurry was mixed with the help of alumina grinding balls on a drum roller for 24 h. After mixing, the slurry was sonicated by means of an ultrasonic homogenizer (Sonopuls HD 2200, 20 kHz, 200 W) for 3 min with a 50% amplitude at 100 Hz.

Pre-sintered Al_2O_3 substrates and Al_2O_3 filters (30 ppi) were dip-coated using the nano-functionalization slurry. Subsequently, the samples were liberated from excessive coating slurry by means of centrifugation and left for drying. Finally, the dried filters and substrates were sintered at 800 °C and 1600 °C [33].

First, the contact angle between the aluminum melt and substrates with and without nano-functionalization was measured at the Institute of Nonferrous Metallurgy and Purest Materials (TU Bergakademie Freiberg, Germany), following the conventional sessile drop technique. The substrate carrying 60 ± 2 mg AlSi7Mg aluminum alloy (Trimet Aluminium AG, Germany) was placed in the furnace and the chamber was evacuated for 90 min to a pressure of $p \leq 1.5 \cdot 10^{-5}$ mbar. Subsequently, the furnace was heated to 950 °C at 350 K/min with a final dwell time of 180 min. The average pressure after this procedure was $7.4 \pm 0.4 \cdot 10^{-6}$ mbar [33].

In order to measure the surface roughness of the substrates, an area of 1500×1400 μm^2 in the center of the sample was analyzed by means of a VK-X laser scanning microscope (Keyence, Japan) at a magnification of 20x. The waviness was removed (with a cut-off wavelength λ_c of 2.5 mm) prior to the determination of the surface roughness S_a. The two nano-coated substrates showed comparable surface roughness values S_a of 2.2 μm (nano 1600 °C) and 2.5 μm (nano 800 °C), which was slightly lower than that of the Al_2O_3 reference substrate amounting to 2.9 μm [33].

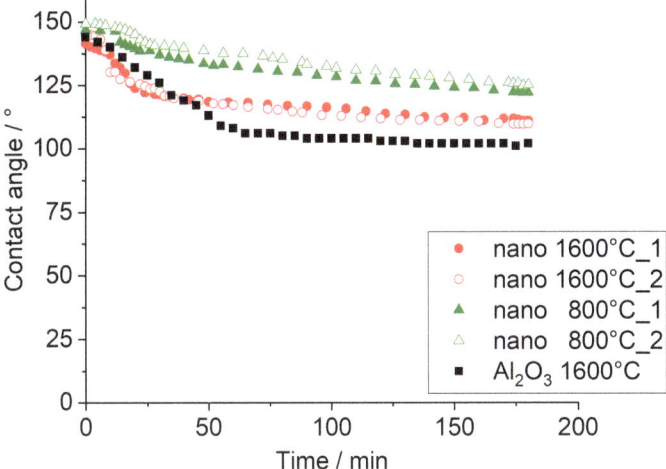

Fig. 3.11 Contact angles measured by sessile drop measurement as a function of time [33]

Figure 3.11 shows the characteristic curve progression of the contact angle measurements under vacuum. Directly after reaching the measuring temperature of 950 °C, high contact angles of >140° were observed followed by a steady decrease of the values indicating the decomposition of the oxide skin on the aluminum. After approximately 60 min, the slope decreased. No stable limit value was reached for nano-coated substrates within the testing period. Consequently, the contact angles were recorded after 180 min. Repeated measurements showed a high reproducibility [33].

For nano-coated substrates, a sintering temperature of 1600 °C resulted in a reduced contact angle of $\theta = 111° \pm 1°$ compared to substrates sintered at 800 °C ($\theta = 124° \pm 2°$). All nano-coated samples showed higher contact angles than the Al_2O_3 reference substrates ($\theta = 101° \pm 2°$). According to the corresponding surface roughness investigation, roughness could be excluded as an influencing factor for this behavior. The increased contact angles observed for nano-coated samples with lower sintering temperatures can be explained by the higher specific surface area of the coating. SEM analyses (see Fig. 3.12) supported this assumption by revealing differences in average particle size on filter surfaces sintered at 800 and 1600 °C [33].

Finally, filtration trials were performed in a pilot filtration plant at Hydro Aluminium Rolled Products GmbH (Germany). The determination of the inclusion levels were carried out before and after the addition of 1.25 kg of AlTi3B1 (AMG, UK) grain refiner (at $t \approx 45$ min). The pilot filtration line used for the long-term trials includes a gas-fired melting furnace with three chambers (1.5 t capacity), a launder system, a filter box and a lifting pump. In the main furnace chamber, 1.3 t of wrought Al99.5 aluminum alloy ingots were melted followed by manual removal of the dross. With a small cascade upstream from the filter box, the aluminum melt was

Fig. 3.12 SEM micrographes of the tested filter surfaces with **a** Al$_2$O$_3$ reference **b** nano-powder sintered at 800 °C; **c** nano-powder sintered at 1600 °C [33]

pumped into the elevated launder system during the tests. After passing the filter, the melt flowed back to the melting furnace in a cascade resulting in a continuous pumping loop. Within this system, two LiMCA II units were utilized (ABB Ltd., Canada) [33].

The filtration effect in the measuring period before the addition of the grain refiner was clearly recognizable as significant decrease of the N20 LiMCA indexes (number of inclusions with a size between 20 and 300 μm in thousand inclusions per kilogram aluminum) measured at the filter exit compared to the N20 values of the filter entrance, see Fig. 3.13. The filtration behavior of the reference filter and the nano-filter sintered at 800 °C were comparable in that regard. The average N20 indexes for the reference sample was at the filter entrance approx. 6.9 k/kg and 0.20 k/kg at the filter outlet. For the nano-filter sintered at 800 °C, the N20 indexes amounted to 9.1 k/kg at the filter entrance and 0.23 k/kg at the filter exit filter. The filtration efficiencies derived from that data are 97.0% (Al$_2$O$_3$ reference) and 97.5% (nano-filter), indicating that the nano-coating had no significant impact on the filtration effect [33].

According to the literature [34], addition of the AlTi3B1 grain refiner results in a decreased filtration efficiency, i.e. the N20 level at the filter entrance decreases and increases at the filter exit. Similar observations were made in this study, see Fig. 3.13 [33].

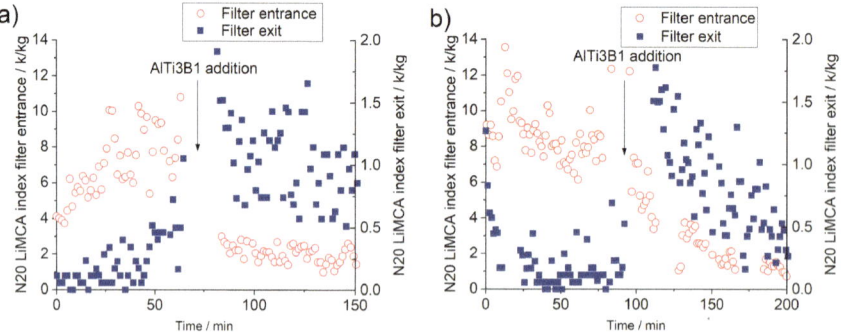

Fig. 3.13 N20 LiMCA indexes from the long-term filtration trials for **a** the Al$_2$O$_3$ reference filter and **b** nano-coated filter [33]

According to Laé et al. [35], grain refinement is assumed to reduce the filtration rate by preventing the formation of bridges in the filter consisting of non-metallic inclusions. No hints for such a phenomenon could be observed. The true origin of this behavior and possible explanations for the decreasing N20 value at the filter entrance remain disputed, making further analysis necessary.

Based on the gathered data, the mean inclusion size was calculated and plotted as a function of time, see Fig. 3.14. The application of the Al_2O_3 reference filter and the nano-coated filter at 800 °C resulted in a comparable average inclusion size. In both cases, the mean inclusion size was almost identical at the filter entrance and exit. The scatter of the data at the filter exit was elevated, mainly due to the lower number of inclusions. For both filter types, the addition of the AlTi3B1 grain refiner resulted in a reduction of the mean inclusion size from approx. 31 μm to approx. 25 μm [33]. Until now, there is no clear explanation for this observation.

In conclusion, nano-functionalization did not enhance the filtration efficiency of the applied filters. Considering the comparable starting conditions, i.e. the alumina as filter surface material, surface roughness, functional pore size and experimental conditions, the trials indicated that measured contact angles between metal melt and filter material did not impact the filtration behavior in a significant manner. Consequently, the observed positive correlation between measured apparent contact angles and filtration efficiency described by Voigt et al. [25] did not originate from changes in the intrinsic contact angle but from the different roughness of the materials [33]. Further information about the sample preparation, parameters of experiments and additional analysis can be found elsewhere [33].

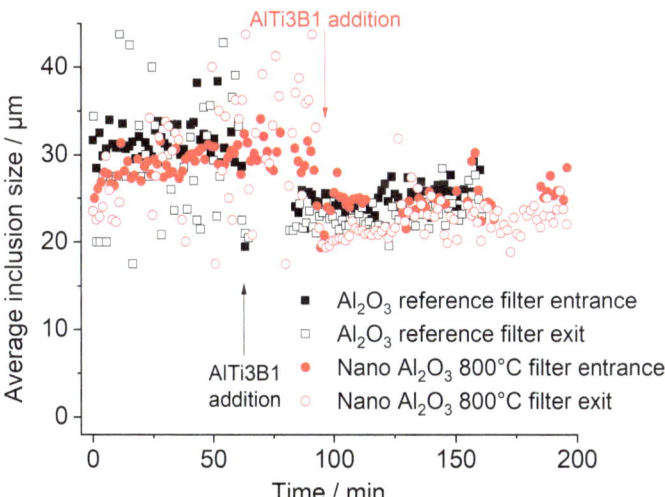

Fig. 3.14 Mean inclusion size during the long-term filtration trials as a function of time and the addition of grain refiner [33]

3.2.4 Influence of Filters with Enhanced Roughness on the Filtration

To evaluate the influence of filters of the roughness, a filter with enhanced roughness was applied in the pilot filtration plant at Hydro Aluminum Rolled Products GmbH (Germany). The underlying test procedure is described in Sect. 3.2.3. The roughness of the filter was increased by adding raw materials with larger particle size distributions. The coating slurry was prepared by mixing 50% alumina raw material P6 (Almatis, Germany) with a median particle size of $d_{50} = 30$ µm and 50% CT 3000 SG (Almatis, Germany) with deionized water. Figure 3.15 shows the filter surface of the sample with enhanced roughness in comparison to the Al_2O_3 reference filter [36].

The filtration behavior of the filter with enhanced roughness was very different from that of the Al_2O_3 reference filter, see Fig. 3.16. The mean N20 LiMCA index after the filter was 0.04 k/kg and thus slightly lower than for the Al_2O_3 reference filter (0.2 k/kg). Furthermore, a strong decline of the N20 LiMCA index from around 12 k/kg to around 3.5 k/kg was observed at the filter entrance with enhanced roughness. The N20 LiMCA index plateaued in the last 14 min of the measurement sequence, see Fig. 3.16 [36].

The strong decline of the N20 LiMCA index at the filter entrance of the rough filter might indicate a higher filtration efficiency of the rough filter resulting in an effective reduction of the inclusion concentration. In that case, the removal of inclusions could be higher than the number of new inclusions formed by the two aluminum melt cascades. Bergin et al. [37] did not observe an increase in filtration efficiency during tests with rough filters in a similar filtration pilot plant. In subsequent SEM investigations (see Fig. 3.17), small light inclusions of the grain refiner were found in close proximity to the Al- and O-based inclusions. Grain refiner particles docking at the inclusions could have resulted in the formation of small clusters with increased density and enhanced settling behavior.

Further information about the sample preparation, parameter of experiments and additional analysis can be found elsewhere [36].

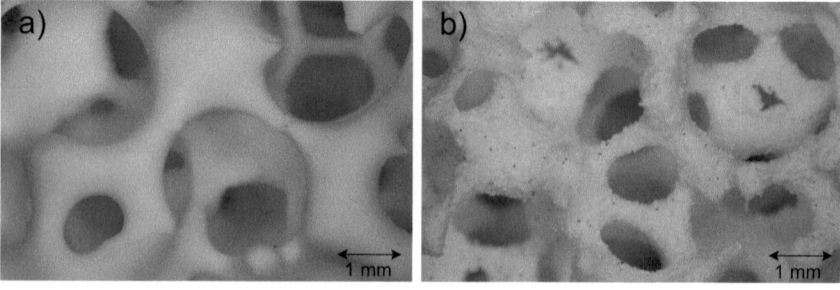

Fig. 3.15 Digital micrographs of filter structures **a** Al_2O_3 reference filter **b** filter with enhanced roughness

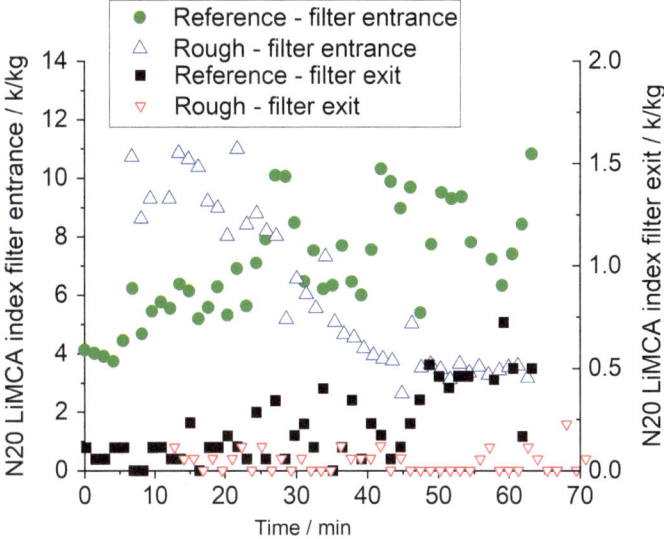

Fig. 3.16 N20 LiMCA indexes from the filtration trials with different roughness

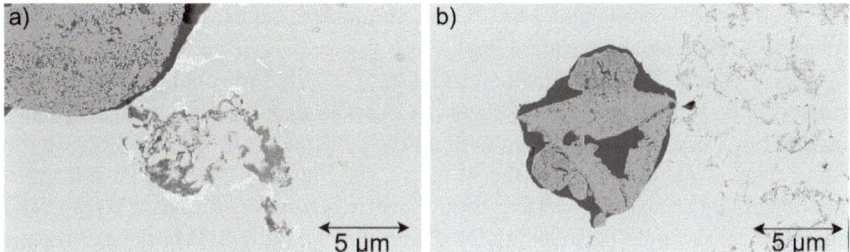

Fig. 3.17 SEM micrographs of the tested filter surfaces with **a** Al_2O_3 reference filter **b** filter with enhanced roughness

3.2.5 Influence of Carbon Containing Filter Surfaces on the Filtration

Alumina (Al_2O_3) or Silicon carbide (SiC) are the most common filter materials used for ceramic foam filters applied in the filtration of aluminum and aluminum alloys [7]. In comparison, zirconia (ZrO_2) or carbon bonded alumina (Al_2O_3-C) filters are typical for the filtration of steel melts. Al_2O_3-C is a versatile refractory material due to its low thermal expansion, negligible shrinkage after thermal treatment and enhanced slagging resistance originating from low wetting between carbon bonded alumina and metallic melts. A major disadvantage of Al_2O_3-C is its low resistance against oxidation at high temperatures.

In order to determine the influence of the carbon-bonded alumina coatings in comparison to the Al_2O_3 reference material used as reference material, its wetting behavior and filtration performance were investigated. For that purpose, carbon-bonded alumina substrates and ceramic foam filters were manufactured.

In order to ensure comparability between the filtration behavior of the Al_2O_3-C filter and the reference alumina filter, the functional pore size distribution of the filter should be similar. The main challenge in this regard was the difference in shrinkage behavior due to the very different nature of the thermal treatment. Contrary to the Al_2O_3-C, alumina foams exhibits a shrinkage of larger than 5% during thermal treatment. In order to avoid differences in the functional pore size distribution, pure Al_2O_3 skeleton filters were used as a basis for all experiments and coated with Al_2O_3 or Al_2O_3-C slurry instead.

The Al_2O_3-C base material consisted of 66 wt% Al_2O_3 Martoxid MR-70 (Martinswerk, Germany), 20 wt% coal pitch Carbores P (Rütgers, Germany), 6.3 wt% carbon black N991 (Lehmann & Voss & Co., Germany), 7.7 wt% graphite AF 96–97 (Graphit Kropfmühl, Germany). The slurry composition was based on Emmel et al. [38]. Dipping and subsequent centrifugation was applied to coat the presintered alumina skeleton filters. The coating slurry was also used to cast substrates for the wetting tests in a plastic mold with plaster base. After drying at room temperature, the carbon-bonded substrates and filters with carbon-bonded coating were thermally treated at 800 °C (in a steel retort) and 1400 °C (in an alumina retort) filled with pet coke providing a reducing atmosphere [39].

At the Foundry Research Institute (Krakow, Poland), the sessile drop measurements were performed at 730 °C by means of a sessile drop apparatus with a capillary purification. Placing the aluminum melt on the substrate as droplet resulting in the rolling away of the droplet from the substrate, see Fig. 3.18. The repulsive interaction indicated a strong non-wetting behavior between Al_2O_3-C and AlSi7Mg. The apparent contact angle between Al_2O_3-C and the rolling AlSi7Mg droplet (determined from three last images of Fig. 3.18) amounted to $(157 \pm 1)°$. In comparison, the contact angle between Al_2O_3 and AlSi7Mg yielded only 108° [25]. Due to the rolling off of the droplet from the substrate, an investigation of interactions between the Al_2O_3-C and aluminum was not possible.

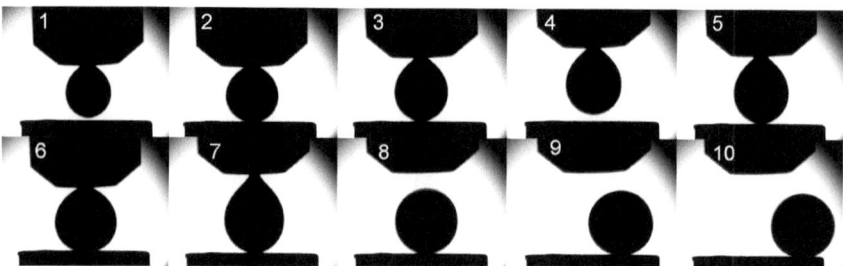

Fig. 3.18 Sequence of photographs of the sessile drop experiment on Al_2O_3-C substrate with capillary purification of AlSi7Mg alloy [39]

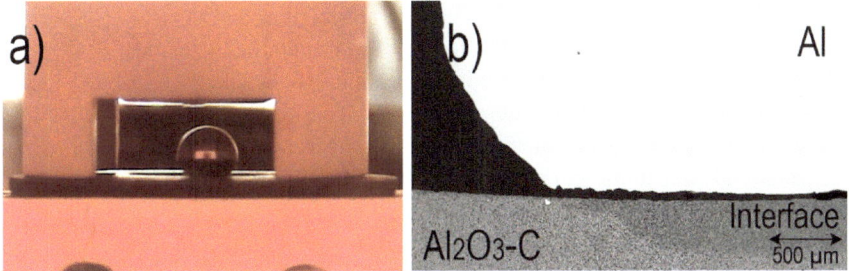

Fig. 3.19 Sessile drop experiment **a** photograph of the sessile drop setup after dropping, **b** SEM micrograph (BSE mode) of Al_2O_3-C/aluminum couple

A simple sessile drop setup with a boron nitride dropping unit (Henze Boron Nitride Products AG, Germany) was used for further tests [40]. The buildup of the dropping unit comprises a hopper for melting the aluminum and a steel plunger pulling the molten aluminum through a bottleneck in order to retain the oxide skin. The dropping unit loaded with a cylindrical piece of Al99.7 (Rheinfelden, Germany) was placed on the substrate inside a hot stage microscope, (see Fig. 3.19a). The system was heated to a temperature of 730 °C at 10 K/min followed by a holding time of 10 min. In order to minimize the oxygen level, the chamber of the hot stage microscope was constantly flushed with argon during the thermal treatment [41].

After the cooling, the droplet mass was calculated as the difference between the determined mass of the substrate/Al couple and the original mass of the ceramic substrate. The droplet mass of 300–600 mg was significantly higher than in the previous tests (<100 mg). The higher droplet mass significantly increases the impact of gravity on the contact angle. However, the new trials enabled the formation of Al_2O_3-C/Al couples for subsequent SEM and EDX analyses to investigate the interactions between aluminum melt and Al_2O_3-C. No adhesion of the Al droplet to the substrate was observed by SEM investigation of the interface. However, there was a pronounced reaction zone (dark area in the center right) in the Al_2O_3-C substrate, (see Fig. 3.19b). In EDX analyses, the two different zones exhibited only slight differences in the chemical composition, i.e. the reaction zone showed a higher silicon peak and a lower carbon peak in comparison with the unreacted zone. The origin of the detected silicon is unknown since pure aluminum (99.7%) was used for the measurements [41].

Two short-term filtration trial series were conducted at the metal foundry Georg Herrmann Metallgiesserei (Muldenhütten, Germany). The applied aluminum alloy AlSi7Mg (EN AC-42100) from Rheinfelden Alloys (Rheinfelden, Germany) was made from 50% ingots and 50% scrap, i.e. aluminum recycled from solidified feeders and runners, in order to introduce non-metallic inclusions. After melting and skimming, the AlSi7Mg alloy was cast into a combined sand/steel mold. The casting system consisted of a joint sprue, four horizontal runners with filter chambers (made by green sand) and four vertical steel molds [41]. The aluminum from the steel molds was disjoined from the feeders and analyzed. A cold Porous Disk Filtration Apparatus analysis (PoDFA) was performed by HOESCH Metallurgical Service (Niederzier, Germany) in order to investigate the content of non-metallic inclusions.

The first short-term filtration trial lasted about 16 s and the four steel molds showed an equal filling of the aluminum melt. The PoDFA index determined by metallographic examination of the filter cake is given as area (of inclusions per kilogram analyzed aluminum and is a measure correlating with the total number of non-metallic inclusions). Consequently, a lower PoDFA index indicates better melt quality. In the framework of the PoDFA analyses, Al_2O_3 films, carbides, magnesium oxide, spinel, refractory material, iron and manganese oxides as well as grain refiners were found in the castings. The determined PoDFA indexes were 0.106 mm^2/kg for the Al_2O_3-C filter sample and 0.246 mm^2/kg for the Al_2O_3 reference filter sample in trial 1, indicating that Al_2O_3-C filters had an improved filtration efficiency [39].

In short-term filtration trial 2, the PoDFA indexes were once again lowest for the Al_2O_3-C filters, i.e. 0.482 mm^2/kg for the Al_2O_3-C filter fired at 800 °C and 0.618 mm^2/kg for the Al_2O_3-C filter fired at 1400 °C. The rough Al_2O_3 reference filter resulted in the highest PoDFA index of trial series 2 with 1.06 mm^2/kg. SEM analyses revealed large inclusions on the surface of applied filters containing mainly Al, Mg, Si and O. The Al_2O_3-C coating showed no structural changes after the filtration trial which were visible by SEM [39].

For further investigation, long-term filtration trialslong-term filtration trials were performed in a pilot filtration line at Hydro Aluminium AS primary aluminum plant (Sunndal, Norway). The system comprises a melting furnace, a launder system, a filter box, and a lifting pump and is equipped with two LiMCA II units (ABB Ldt., Canada). Approximately 8 mT of wrought aluminum alloy 6082 (main alloying elements of Si ~ 0.95%, Fe ~ 0.2%, Mn ~ 0.6% and Mg ~ 0.65%) were melted and circulated in the loop by the lifting pump in each trial. Every 10 min, 4 kg of compacted aluminum sawdust chips were added behind the lifting pump in order to introduce inclusions into the melt. In total, two long-term filtration trials were performed. One trial with a Al_2O_3 reference filter and one trial with carbon bonded alumina (Al_2O_3-C) filter. In both cases, the aluminum melt temperature at the priming procedure was 740 °C [41].

For both filters, the N20 values of the LiMCA measurements showed comparable time-dependent behavior at the filter entrance and exit, see Fig. 3.20. The N20 values of the filter entrance decreased steadily at the beginning of the trial and stabilized in the second half of the test. An initial decline of the N20 values at the filter exit was observed. Furthermore, the scatter of the N20 values at the filter exit was elevated in comparison to data at the filter entrance the filter. Using the N20 data from stable filtration areas, the filtration efficiency was calculated with 91.2 ± 6.7% for the Al_2O_3 filter (from 23.1 to 45 min) and with 88.8 ± 12.3% for the Al_2O_3-C filter (from 15 to 25.9 min), indicating comparable filtration behavior [41].

According to intensive SEM investigations, nearly no Al_2O_3 films, magnesium oxide, spinel, carbides or refractory material were found, indicating that only few inclusions were captured within the filters. It can be concluded that introducing a comparable level of typical inclusions by adding compacted aluminum sawdust chips was not very effective. Closer examination of the Al_2O_3-C coating revealed

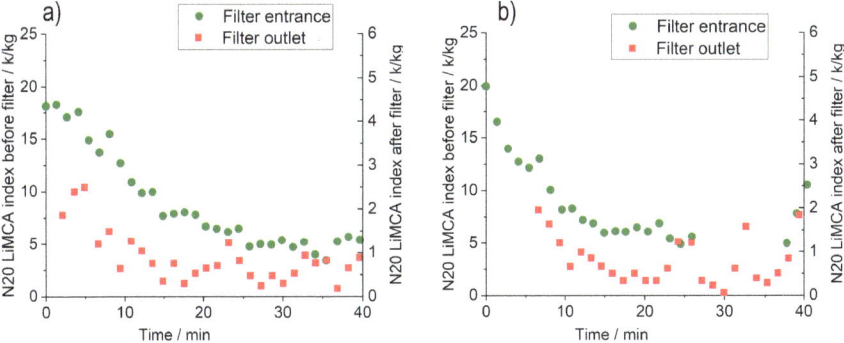

Fig. 3.20 Results of the long-term filtration trial **a** Al_2O_3 reference filter and **b** Al_2O_3-C filter

microstructural changes compared to as coked filters (Fig. 3.21). The Fig. 3.21b Position 1 shows a very homogenous, dense, and compact coating without many pores or larger carbon structures. In contrast, the Fig. 3.21b Position 2 shows large pores within the Al_2O_3-C coating. The large pores of as coked filter are filled with carbon raw materials such as graphite. Figure 3.21b shows both structures in proximity. Apparently, only coating areas with low thickness experienced the homogenous and dense structures. According to EDX analyses, denser coating areas exhibited lower carbon peaks implicating that the carbon partly disappeared, leaving pores or dense structures consisting mainly of alumina behind. The temperature during the filtration trials did not exceed 750 °C, i.e., it was significantly lower than the sintering temperature required for densification of alumina. Further in-depth investigations are necessary for clarification [41].

In conclusion, the Al_2O_3-C/Al couples for SEM and EDX analyses were successfully generated with the aid of the sessile drop trials with a simple dropping unit. Despite missing adhesion between Al_2O_3-C substrate and the aluminum droplet, a reaction zone with a thickness of approximately 500 μm was observed in the substrate material. In short-term filtration trials, Al_2O_3-C filter showed better PoDFA indexes

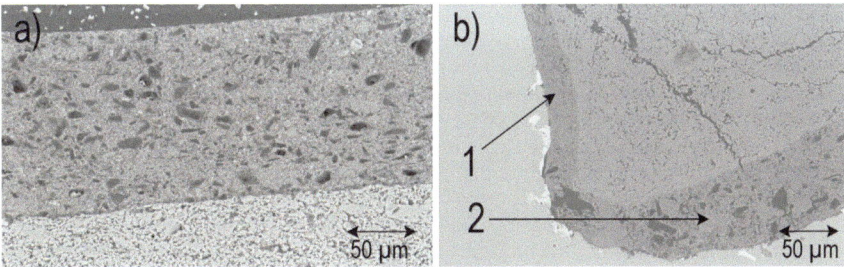

Fig. 3.21 SEM micrographs (BSE mode) of Al_2O_3-C filter **a** before and **b** after long-term filtration trial

compared to the Al_2O_3 reference filter. In contrast, LiMCA results from the long-term filtration trialslong-term filtration trials did reveal no differences in the filtration effect. Few typical non-metallic inclusions were detected in the solidified aluminum via extensive SEM investigations for both Al_2O_3 and Al_2O_3-C filters. Furthermore, the SEM analyses revealed partial microstructural transformation of the Al_2O_3-C coating from heterogenous, porous layers into a dense, compact layers. To conclude, Al_2O_3-C in the presented form is not suitable for long-term filtration tasks [41]. Further investigations are necessary to address the filtration behavior and stability of Al_2O_3-C filters during filtration. Further information about the sample preparation, parameter of experiments and additional analysis you can find elsewhere [39, 41].

Acknowledgements The authors would like to thank the German Research Foundation (DFG) for supporting these investigations as part of the Collaborative Research Centre 920 "Multi-Functional Filters for Metal Melt Filtration—A Contribution towards Zero Defect Materials" (Project-ID 169148856) subproject A02.

References

1. Y.D. Kwon, Z.H. Lee, Mater. Sci. Eng. A **360**, 372–376 (2003). https://doi.org/10.1016/S0921-5093(03)00504-5
2. M.D. Sabatino, L. Arnberg, S. Rørvik, A. Prestmo, Mater. Sci. Eng. A **413–414**, 272–276 (2005). https://doi.org/10.1016/j.msea.2005.08.175
3. L. Liu, F.H. Samuel, J. Mater. Sci. **33**, 2269–2281 (1998). https://doi.org/10.1023/A:1004331219406
4. F.H. Samuel, H. Liu, A.M. Samuel, Metall. Mater. Trans. A **24A**, 1631–1645 (1993). https://doi.org/10.1007/BF02646602
5. C. Tian, J. Law, J. van der Touw, M. Murray, J.-Y. Yao, D. Graham, D. St. John, J. Mater. Process. Technol. **122**, 82–93 (2002). https://doi.org/10.1016/S0924-0136(01)01229-8
6. X. Dai, X. Yang, J. Campbell, J. Wood, Mater. Sci. Eng. A **354**, 315–325 (2003). https://doi.org/10.1016/S0921-5093(03)00021-2
7. R.A. Olson III, L.C.B. Martins, Adv. Eng. Mater. **7**(4), 187–192 (2005). https://doi.org/10.1002/adem.200500021
8. L.N.W. Damoah, L. Zhang, Mater. Trans. B, **41B**, 886–907 (2010). https://doi.org/10.1007/S11663-010-9367-3
9. D. Doutre, B. Gariépy, J.P. Martin, G. Dubé, Light Metals 296–304 (1985). https://doi.org/10.1007/978-3-319-48228-6_36
10. T.A. Utigard, I. Sommerville, Cleanliness of aluminum and steel: a comparison of assessment methods. Light Metals 951–956 (2005)
11. A.A. Simard, F. Dallaire, J. Proulx, P. Rochette, Cleanliness measurement benchmarks of aluminum alloys obtained directly at-line using the prefil-footprinter instrument. Light Metals 739–744 (2000)
12. T.L. Mansfield, Ultrasonic technology for measuring molten aluminum quality. Light Metals 305–311 (1982)
13. J.-L. Achard, N. Ramel, G. Beretta, P.-Y. Menet, J. Prigent, P. Le Brun: Batscan™, Con-stellium in-melt ultrasonic inclusion detector: industrial performance. Light Metals 936–943 (2020)
14. K. Schiebold, *Zerstörungsfreie Werkstoffprüfung—Ultraschallprüfung*. Springer–Verlag, Berlin, Heidelberg, 2015)

15. H. Görner, M. Syvertsen, E.J. Øvrelid, T.A. Engh, AlF3 as an aluminium filter medium. Light Metals 939–944 (2005)
16. M. Zhou, D. Shu, Metall. Mater. Trans. A **34A**, 1183–1191 (2003). https://doi.org/10.1007/s11 661-003-0138-5
17. K. Oosumi, Y. Nagakura, R. Masuda, M. Yanagawa, Kobelco Technol. Rev. **24**, 32–36 (2001). https://doi.org/10.1002/9781118788073.ch84
18. S. Bao, K. Tang, A. Kvithyld, M. Tangstad, T.A. Engh, Metall. Mater. Trans. B **42B**, 1358–1366 (2011). https://doi.org/10.1007/s11663-011-9544-z
19. N. Eustathopoulos, M.G. Nicholas, B. Drevet, Wettability at high temperatures. (Pergamon, 1999)
20. S.K. Rhee, J. Am. Ceram. Soc. **53**(7), 386–389 (1970). https://doi.org/10.1111/j.1151-2916. 1970.tb12138.x
21. P. Shen, H. Fujii, T. Matsumoto, K. Nogi, J. Am. Ceram. Soc. **87**(7), 1265–1273 (2004). https:// doi.org/10.1111/j.1151-2916.2004.tb06376.x
22. J.-G. Li, Ceram. Int. **20**, 391–412 (1994). https://doi.org/10.1016/0272-8842(94)90027-2
23. W. Eberhardt, E. Kunz, Surf. Sci. **75**(4), 709–720 (1978). https://doi.org/10.1016/0039-602 8(78)90188-7
24. V. Laurent, D. Chatain, C. Chatillon, N. Eustathopolos, Acta Mater. **36**(7), 1797–1803 (1988). https://doi.org/10.1016/0001-6160(88)90248-9
25. C. Voigt, L. Ditscherlein, E. Werzner, T. Zienert, R. Nowak, U.A. Peuker, N. Sobczak, C.G. Aneziris, Mater. and Des. **150**, 75–85 (2018). https://doi.org/10.1016/j.matdes.2018.04.026
26. H. Duval, C. Riviere, A. Laé, P. Le Brun, J.-B. Guillot, Simulations of aluminum filtration including lubrication effect in three-dimensional foam microstructure. Light Metals 645–650 (2007)
27. C. Voigt, E. Jäckel, F. Taina, T. Zienert, A. Salomon, G. Wolf, C.G. Aneziris, P. le Brun, Metall. Mater. Trans. B **48B**, 497–505 (2017). https://doi.org/10.1007/s11663-016-0869-5
28. G.F.A. Acosta, E.A.H. Castillejos, Metall. Mater. Trans. B **31B**, 503–514 (2000). https://doi. org/10.1007/s11663-000-0156-2
29. E. Werzner, M.A. Mendes, S. Ray, D. Trimis, Adv. Eng. Mater. **15**(12), 1307–1314 (2013). https://doi.org/10.1002/adem.201300465
30. C. Voigt, T. Zienert, P. Schubert, C.G. Aneziris, J. Hubálková, J. Am. Ceram. Soc. **97**, 2046–2053 (2014). https://doi.org/10.1111/jace.12977
31. C. Voigt, E. Jäckel, J. Hubálková, C.G. Aneziris, Adv. Eng. Mater. **15**(12), 1197–1205 (2013). https://doi.org/10.1002/adem.201300111
32. S. Bao, M. Syvertsen, A. Kvithyld, T. Engh, Trans. Nonferrous Met. Soc. China **24**, 3922–3928 (2014). https://doi.org/10.1016/S1003-6326(14)63552-4
33. C. Voigt, B. Fankhänel, B. Dietrich, E. Storti, M. Badowski, M. Gorshunova, G. Wolf, M. Stelter, C.G. Aneziris, Metall. Mater. Trans. B **51B**, 2371–2380 (2020). https://doi.org/10. 1007/s11663-020-01900-1
34. N. Towsey, W. Schneider, H.-P. Krug, A. Hardman, N.J. Keegan, Light Metals 291–295 (2001). https://doi.org/10.1007/978-3-319-48228-6_35
35. E. Laé, H. Duval, C. Rivière, P. Le Brun, J.-B. Guillot, Light Metals 753–758 (2006). https:// doi.org/10.1007/978-3-319-48228-6_34
36. C. Voigt, B. Dietrich, M. Badowski, M. Gorshunova, G. Wolf, C.G. Aneziris, Light Metals 1063–1069 (2019). https://doi.org/10.1007/978-3-030-05864-7_130
37. A. Bergin, C. Voigt, R. Fritzsch, S. Akhtar, L. Arnberg, C.G. Aneziris, R.E. Aune, Light Metals 640–648 (2022). https://doi.org/10.1007/978-3-030-92529-1_84
38. M. Emmel, C.G. Aneziris, Ceram. Int. **38**, 5165–5173 (2012). https://doi.org/10.1016/j.cer amint.2012.03.022

39. C. Voigt, J. Hubálková, T. Zienert, B. Fankhänel, M. Stelter, A. Charitos, C.G. Aneziris, Materials **13**(18), 20203962 (2020). https://doi.org/10.3390/ma13183962
40. P. Malczyk, T. Zienert, F. Kerber, C. Weigelt, S.O. Sauke, H. Semrau, C.G. Aneziris, Materials **13**(21), 4737 (2020). https://doi.org/10.3390/ma13214737
41. C. Voigt, J. Hubálková, A. Bergin, R. Fritzsch, S. Akhtar, R. Aune, C. G. Aneziris, Light Metals 626–632 (2022). https://doi.org/10.1007/978-3-030-92529-1_82

Chapter 4
Thermodynamic Assessment as a Tool for Modeling Interactions at the Interface Between Ceramic Filter and Melt

Mariia Ilatovskaia and Olga Fabrichnaya

4.1 Introduction

Phase diagrams are widely used in materials research and engineering to understand the interrelationship of composition, microstructure, and process conditions. However, most industrial materials are complex, consisting of more than two or three components. In this case, the direct application of phase diagrams is less representative. On the other hand, only limited data (on phase diagrams) are available for many multicomponent systems. Computational methods such as CALPHAD (CALculation of PHAse Diagrams) are therefore employed as a powerful tool to model thermodynamic properties for each phase and simulate multicomponent multi-phase relations in complex systems [1, 2].

In the context of the Collaborative Research Center 920 (CRC 920), which deals with creating a new generation of metal having improved qualities (e.g. superior mechanical properties) through melt filtration, computational thermodynamics can provide valuable information on the phase behavior in the filter material over a wide temperature range. The formation of stable and metastable phases during the interaction of filter material, coatings, as well as inclusions in molten steel or aluminum alloy can be predicted. Moreover, relevant energy effects and basic information about the physical and chemical parameters of industrial processes can be provided.

As long as two filtration processes have been proposed, for steel and aluminum melt, two main independent complex filter systems can be distinguished and considered separately. For the steel melt filtration, Al_2O_3 filters with different coatings including carbon-bonded ceramics such as Al_2O_3 or spinel $MgAl_2O_4$ have been suggested [3, 4]. Interaction of coatings with 42CrMo4 steel and its purification

M. Ilatovskaia · O. Fabrichnaya (✉)
Institute of Materials Science, Technische Universität Bergakademie Freiberg, Gustav-Zeuner-Str. 5, 09599 Freiberg, Germany
e-mail: fabrich@ww.tu-freiberg.de

© The Author(s) 2024
C. G. Aneziris and H. Biermann (eds.), *Multifunctional Ceramic Filter Systems for Metal Melt Filtration*, Springer Series in Materials Science 337,
https://doi.org/10.1007/978-3-031-40930-1_4

were indicated [5, 6]. Therefore, thermodynamic modeling of the complex spinel $(Fe,Mg,Mn)Al_2O_4$ system is necessary. A modification of Al_2O_3 filter surface using TiO_2 and ZrO_2 additives have also been suggested for the steel filtration [7, 8]. Besides this, different metastable Al_2O_3 phases have been suggested as coatings for the filtration process and therefore the description of the metastable Al_2O_3 phases should be also implemented. Thus, the first main focus of this work is the development of thermodynamic database for the ceramic filter materials consisting of α-Al_2O_3, metastable γ-, δ-, k-, and θ-Al_2O_3, oxides originating from steel oxidation, i.e. FeO, Fe_2O_3, MnO, Mn_2O_3, and oxides used for filter surface modification MgO, TiO_2, and ZrO_2.

For the aluminum melt filtration, carbon-free Al_2O_3 have been suggested as filter materials. The influence of different filter surface chemistries (Al_2O_3, spinel $MgAl_2O_4$, mullite $3Al_2O_3 \cdot 2SiO_2$, SiO_2, and TiO_2) has been discussed [9, 10]. Moreover, the thermodynamic description of the aluminum alloy itself is of interest to control and modify the morphology of Fe-containing intermetallic phases. Therefore, the focuses in this case are on development of thermodynamic databases for the ceramic filter materials (Al_2O_3, MgO, SiO_2, TiO_2) and aluminum alloy (Al, Fe, Mg, Si).

It is clear that thermodynamic descriptions of some systems are important for both processes, whether it is the filtration of steel or aluminum melt. These are, for example, the binary Al_2O_3–MgO and Al_2O_3–TiO_2 systems and the ternary Al_2O_3–MgO–TiO_2 system. Since the CALPHAD method provides a structured and block approach, the derived thermodynamic description of any binary or ternary system can be easily implemented into a complex database. The only limitation is the consistence of the merged databases.

In the beginning of the project, there were very few related thermodynamic descriptions available. They were mainly based on critical assessment of binary and ternary systems available in literature. Phase relations and melting behavior in the FeO–Fe_2O_3–MgO–SiO_2 [11] and MgO–Al_2O_3–SiO_2 [12] systems were described thermodynamically at 1 bar. However, different models were used in Refs. [11] and [12] to describe the liquid phase. Other related systems were available and should be considered within the project, they are MgO–Al_2O_3 [13, 14], Fe–Mn–O [15], Al_2O_3–SiO_2 [16] etc. The liquid phases in these cases were described by the ionic two-sublattice model, so the descriptions could be easily implemented into the derived datasets in case of consistency. The opposite situation is applied to the FeO–Fe_2O_3–MgO–SiO_2 [17], MnO–Al_2O_3 [18], MgO–Al_2O_3–SiO_2 [19] systems where modified quasi-chemical model was applied for the liquid phase and due to inconsistency it cannot be implemented into database utilizing two-sublattice partially ionic liquid model. However, descriptions of the solid phases based on the compound energy formalism in the form of sublattice model could be used. Moreover, commercial databases (e.g. FactSage, Thermo-Calc) were also available, however not all required ternary oxide systems were described at the beginning of the project. Commercial databases have been expanded over the years, but further development of these databases for the aims of the present work is not possible because the thermodynamic parameters are not available to users. Since there was a great gap in describing many related systems (even binaries) in commercial databases and in literature, precise material design was not possible. Moreover, besides to the existing datasets, consideration of new experimental data on phase relations and thermodynamics is essential. Therefore, the CRC 920 required the development of self-consistent multicomponent multi-phase oxide database that would provide necessary information for simulation of the metal melt filtration process.

4.2 The CALPHAD Approach

The total Gibbs free energy is minimized to determine a state of chemical equilibrium. The CALPHAD (CALculation of PHAse Diagrams) method [1, 20, 21] has been originated as an algorithm utilizing models developed for the Gibbs energy, G, of phases in low-order systems as a function of temperature, pressure and composition, starting from pure compounds and binaries. These models describe the Gibbs energy of each phase in a system, including stable and metastable temperature and composition ranges. By fitting a set of critically evaluated and selected experimental data (for example phase diagram information, thermodynamic values) and theoretical predictions, it is possible to calculate optimal model parameters reproducing experimental data within uncertainty limits, followed by storage in a database. Thus, the CALPHAD method is a hierarchy method that requires to develop thermodynamic descriptions from unary to higher-order systems. Then the derived thermodynamic databases can be used to solve, among other things, practical issues of many technological processes. The algorithm of the CALPHAD approach is shown schematically in Fig. 4.1.

Although the CALPHAD method is convenient and efficient for predicting and estimating thermodynamic properties and phase equilibrium data, it is unable to predict the existence of a phase unless it was experimentally confirmed and included in the thermodynamic assessment. The latter highlights the need for accurate experimental data, because optimized parameters in a thermodynamic assessment are dependent on experimental data.

4.2.1 Experimental Data

The quality of a thermodynamic description is determined by how well nature of phases corresponds to the applied models, e.g. how the crystal structure, short range order, species present in the phase, site occupancies etc. are accounted and how well the description can reproduce experimental data. Consequently, high quality experimental data are essential to develop advanced CALPHAD descriptions. Their availability and critical assessment make the thermodynamic description of the system more efficient and invaluable in describing real processes and in the design of new

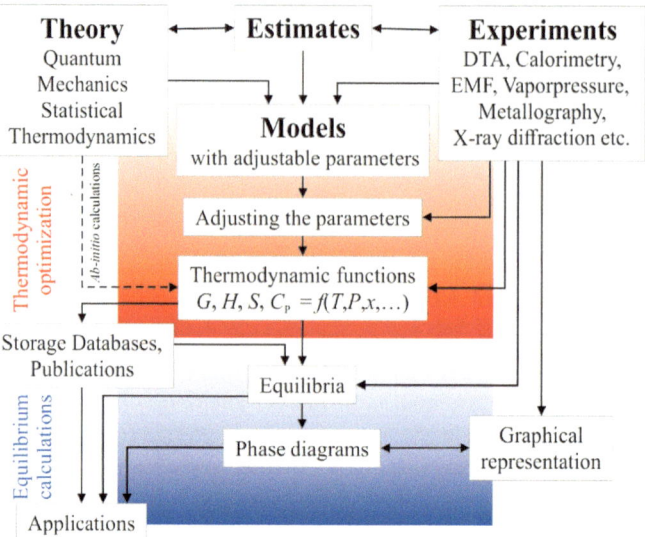

Fig. 4.1 Schematic view of the CALPHAD approach

materials. In general, the experimental data used to optimize thermodynamic parameters can be distinguished as phase diagram data and thermodynamic data [1]. A summary of the methods applied to ensure the reliable experimental data and their evaluation criteria can be found elsewhere [2], while in-detailed information on many methods used for the determination of phase diagram data is provided by Zhao [22]. Only the techniques directly applied within the current project are discussed in detail below. The corresponding methods are differential thermal analysis, X-ray diffraction, and scanning electron microscopy (phase diagram data) and scanning differential calorimetry (heat capacity and enthalpy of transformation data).

Sample Preparation

Samples of the Al–Fe system were prepared by levitation melting followed by casting in a cold copper mold [23] and arc melting of pure metals. The arc melting furnace (AM furnace, Edmund Bühler GmbH, Hechingen, Germany) was three times evacuated to $1 \cdot 10^{-5}$ atm and backfilled with argon prior melting to prevent oxidation of the samples. The as-cast samples were encapsulated in fused silica glass tubes under Ar. The argon atmosphere was adjusted to normal pressure at target temperature. The samples were then annealed under specified temperature–time conditions and either cooled in a furnace to room temperature to get a homogeneous microstructure applicable for the heat capacity measurement or quenched (RHTV 120/300/18, Nabertherm, Germany) into cold water to get the phase diagram data.

Samples of the oxide systems were prepared by three different methods: solid-state reaction, co-precipitation routine, and co-hydrolysis. Al_2O_3–MgO and TiO_2–SiO_2 samples were prepared by solid-state reaction from high-purity oxide powders. To reduce the time to reach equilibrium at a target temperature, co-precipitation routine and co-hydrolysis (for SiO_2-containing samples) [24] followed by thermal decomposition of aqueous solutions were mainly used. When the samples were prepared by co-precipitation, the initial chemicals dissolved in distilled water were mixed at predetermined ratios and then added dropwise to an aqueous solution of ammonia NH_4OH with the pH value maintained above 9.0 during the process. To grow the precipitated particles and divide the working solutions into two parts, filtrate and filtride, the suspensions were heated up and held at 333 K for 1–2 h followed by filtration. To check composition of the obtained samples, filtrates (to check completeness of the co-precipitation) and precipitates dissolved in diluted solution of H_2SO_4 were controlled by inductively coupled plasma optical emission spectrometry (ICP-OES). In case of a significant deviation of the sample composition from the nominal detected by ICP-OES (mostly in case of TiO_2-containing samples), the filtration stage was replaced by evaporation. The precipitates after filtration or substances after evaporation were dried at 353 K for 1–3 days. By grinding, the powder samples were then subjected to one- or two-stage calcination at 673–1073 K for 2–5 h in air. The powders obtained were ball-milled, pressed into tablets at 300 MPa (tablet size is 5 or 8 mm in diameter and about 2–3 mm in height), and annealed under specified temperature–time conditions in air in Pt-crucibles in a muffle furnace (NABERTHERM, Germany) in order to achieve a pseudo-equilibrium state followed by furnace-cooling. The annealing duration varied depending on the sintering temperature and microstructure development suitable for SEM/EDX quantification. The MnO-containing sample tablets were placed in sealed (under Ar) quartz tubes together with a solid piece of graphite and then annealed. Graphite is required to maintain a reducing atmosphere inside the tube so that the low oxygen potential keeps manganese in its $+ 2$ state.

The SiO_2-containing samples were prepared by co-hydrolysis as follows. The metal–organic precursors of every desired metal were first separately dissolved in isopropyl alcohol and then mixed and stirred for 2 h with preliminary prepared solutions of (H_2O + $(CH_3)_2CHOH$ + NH_4OH) at a fixed concentration of the components. After, the ready solutions were slowly mixed together in the presence of NH_4OH to maintain the pH above 9.0 during the process and to ensure complete precipitation of the corresponding hydroxides. The precipitate was visible as a cloudy white solution. The precipitate was sequentially decanted, evaporated, dried, and calcined to obtain the desired mixed oxide powders. The further procedure is similar to that described above for samples prepared by co-precipitation.

Phase Diagram Data

Structural Investigation

Powder-X-ray diffraction (P-XRD) was carried out to determine relative phase fractions in the powder samples obtained after prolonged annealing at target temperatures. The lattice parameters and crystal structures of the phases were also verified. The powdered samples were investigated at room temperature using an URD63 X-ray diffractometer (Seifert, FPM, Freiberg, Germany) working in Bragg–Brentano reflection geometry and equipped with a graphite monochromator at CuKα radiation ($\lambda = 1.5418$ Å). Thin layers of powder samples were sedimented with a drop of ethanol on monocrystalline silicon substrate with (510) orientation which gives a "zero-background" at 2θ of $15°$–$110°$.

Qualitative and quantitative analyses of the P-XRD patterns were performed by Rietveld analysis using MAUD software [25, 26]. ICSD (Inorganic Crystal Structure Database, 2017, Karlsruhe, Germany) [27] was used for interpretation of the powder diffraction patterns. The method makes it possible to determine the site occupancy parameters by analyzing the polycrystalline samples. Each solid phase has its own characteristic diffraction pattern as a function of intensity depending on the diffraction angle 2θ.

Quantitative analysis of the P-XRD patterns of the SiO_2-rich samples subjected to melting can be complicated by a broad background peak in the range of 2θ between 18 and $25°$ due to the non-crystalline glass formation.

Microstructural Investigation

Microstructural investigations by scanning electron microscopy (SEM) including secondary (SE) and backscattered electrons (BSE) imaging, energy dispersive X-ray spectrometry (EDX), and electron backscatter diffraction (EBSD) were carried out on polished cross-sections of the solidified/quenched samples after prolonged annealing at the target temperatures. The sample microstructures were analyzed using LEO 1530 Gemini (Zeiss, Germany) equipped with an EDX detector (Bruker AXS Mikroanalysis GmbH, Germany) or/and JEOL JSM 7800F (Tokyo, Japan) equipped with an EDX/WDX detector and with an EDAX Hikari Super EBSD system (Octane Elite, EDAX Inc., Berwyn, PA, USA). Both microscopes were equipped with a field emission cathode, used at the acceleration voltage of 20 kV and working distance of 8.0–11.3 mm.

EDX was used to check the chemical compositions of the samples, to determine the compositions of solid phases which also means phase identification, and to estimate the composition of the liquid resulting from partial melting at a given temperature or after melting during an invariant reaction. An experimental uncertainty of EDX measurement is around 2–4 at.%. For SEM/BSE imaging, specimens (cross-sections) must be electrically conductive, at least at the surface, and electrically grounded to prevent the accumulation of electrostatic charge at the surface when scanned by the electron beam. Therefore, specimens (especially less- or nonconductive oxide samples) were well ground and polished followed by sputter coating with an ultrathin graphite layer. In some cases, metal coatings (a thin film of copper or silver paint on the specimen surface) were additionally used.

Thermal Analysis

Temperatures of the solid-state transformations and melting behavior were investigated by differential thermal analysis (DTA). DTA is a technique in which the temperature difference between a substance and a thermally inert reference material is measured as a function of the temperature (or time), while the substance and reference material are subjected to a controlled temperature program. During identical thermal cycles (i.e. same cooling or heating program), any temperature difference between sample and reference is registered so that any changes in the sample, either exothermic or endothermic, relative to the inert reference in terms of thermal voltage (in μV) can be detected. Note that an empty crucible is often used as a reference. DTA can be accompanied by thermogravimetric analysis (TG) which makes it possible to record the mass change of a sample as a function of the temperature (or time) during thermal cycles.

DTA measurements were performed on (i) TG–DTA SETSYS Evolution-1750 (SETARAM Instrumentation, France) in air or inert atmosphere (Ar, He) using B-type tri-couple DTA rod (PtRh 6%/30% thermocouple) and open Pt crucibles (Pt/Rh 10%) and (ii) TG–DTA SETSYS Evolution-2400 (SETARAM Instrumentation, France) using W5-type DTA rod (WRe 5%/26% thermocouple) and open W crucibles and employing a permanent inert He flow. The heating and cooling curves were recorded at the rates of 10 and 30 K·min^{-1}, respectively. In case of the metallic sample investigation, ceramic crucibles (mostly Al_2O_3) were used for SETSYS Evolution-1750.

Temperature calibration of both devices was regularly done using well-known melting points of pure reference substances. Temperature calibration of SETSYS Evolution-1750 (in case of ceramic crucibles application) was performed using Al, Ag, Au, Cu and Ni. Temperature calibration of SETSYS Evolution-2400 was carried out using melting points of Al, Al_2O_3, and temperature of solid-state transformation in $LaYO_3$ previously measured using SETSYS Evolution-1750.

The temperatures of the phase transformations were recorded on DTA curve in heating mode and were determined as an onset point, i.e. the intersection of the baseline with the linear extrapolation (tangent) from the largest slope of the DTA curve.

Thermodynamic Data

Heat Capacity Measurement

Differential scanning calorimetry (DSC) is widely applied for the heat capacity (C_P) measurements. The heat capacities of individual phases and compounds were measured (i) by an instrument DSC 8000 (Perkin Elmer, Inc., USA; inert Ar or He flow; heating rate of 10 K/min; Pt/Rh crucible, with an alumina inlay in case of metallic samples) in the temperature range from 235 to 675 K (the whole temperature range was divided into short intervals of 100–150 K) and (ii) by a DSC device Pegasus 404C (NETZSCH, Germany; inert Ar flow; heating rate of 10 K/min; Pt/Rh crucible, with an alumina inlay in case of metallic samples) in the temperature range from 623 to 1400 K. The classical three-step method [28] using a continuous heating mode of 10 K·min^{-1} was applied to measure specific heat capacity as follows:

1. Baseline measurement taking empty both sample and reference crucibles to prevent any negative effects during DSC experiments,
2. Standard material measurement to account for the calibration factor caused by the difference in heat transfer, and
3. Unknown sample measurement.

For ceramic samples, both measurement systems were calibrated using a certified sapphire (Alfa Aesar, Karlsruhe, Germany) as a standard; the mass (84.1 mg) and the radius (5 mm) of the sample pellets were the same as for the standard material. For metallic samples, the calibration substances were selected according to the temperature range of interest: copper for the range of 100–320 K, molybdenum for the range of 300–673 K, and platinum for the range 573–1473 K. Each calibration measurement with a preceding measurement with empty crucibles and the sample pellets measurement were carried out three times with an average uncertainty of 3%. Note that the C_p measurements at temperature above 1200 K (by described DSC instrument) could be less reliable because of the increase in heat radiation that decreases the registered signal. This limitation was considered for fitting of experimental data using the Maier–Kelley equation [29] in the form as

$$C_P = a + bT + cT^{-2} \tag{4.1}$$

where a, b, and c are the parameters and T is temperature in K.

Apart from obtaining experimental C_P data, an algorithm predicting the trend of heat capacity with temperature based on zero-Kelvin properties was derived by Zienert and Fabrichnaya [30]. The algorithm was also implemented as a set of Python classes scripts called *cp-tools*, which is described elsewhere [31]. Given the available thermophysical data at low temperature ($T > 0$ K), the developed algorithm can also be used for prediction the melting point of substances.

4.2.2 Models and Gibbs Free Energies

Since the Gibbs energy is described as a function of temperature, pressure, and composition, the Gibbs energy of a stoichiometric compound (or a stable end-member of a phase) is defined as follows

$$G(N, P, T) = \Delta_f H^\circ_{298.15} + \int_{298.15}^{T} C_p dT - T \left(S^\circ_{298.15} + \int_{298.15}^{T} \frac{C_p}{T} dT \right)$$

$$+ \int_{1}^{P} V dP \tag{4.2}$$

where $\Delta_f H^\circ_{298.15}$ is the enthalpy of formation of the compound at a standard state of 298.15 K and 1 bar, $S^\circ_{298.15}$ is the standard entropy of the compound at 298.15 K and 1 bar, T is temperature in K, and C_p is the heat capacity at constant pressure.

For isobaric conditions, the pressure contribution $\int_1^P V dP$ can be omitted from Eq. (4.2) and it is not considered in thermodynamic assessments within the current project.

According to the NIST database [32], specific heat capacity at constant pressure is often expressed as a power series expansion in T of the type

$$C_P = a + bT + cT^{-2} + dT^2 + eT^{-3} + \dots \tag{4.3}$$

where a, b, c, d, and e are the parameters.

Solid solutions and stoichiometric phases with homogeneity ranges are assumed to be substitutional and the compound energy formalism (CEF) [33] is suggested in this case to model thermodynamic properties and to describe the composition and temperature dependences of the Gibbs energy. This means that a mathematical expression such as the CEF is more general than the actual physical model and can be applied to various constituents with different behavior in a phase. It has been shown that the CEF is well suited to model solid solutions with two or more distinct sub-lattices. Furthermore, it allows for cations and anions of different valances to mix in different sub-lattices, corresponding to the structure of a solid solution [1].

The simplest substitutional model presents the case when species are mixing in one possible site as $(A, B)_1$. This model is applied for the disordered solid solution phases bcc_A2 and fcc_A1 and for the liquid phase of the Al–Fe system as $(Al, Fe)_1$. The Gibbs energy is then

$$G_m = \sum_{i}^{n} x_i {}^\circ G_i + RT \sum_{i}^{n} x_i \ln(x_i) + {}^E G_m + {}^{mag} G_m \tag{4.4}$$

where x_i is the mole fraction of component i, $°G_i$ is the Gibbs energy of end-member, R is the ideal gas constant, ^EG_m is the excess Gibbs energy, and $^{mag}G_m$ is the magnetic contribution to the Gibbs energy due to the magnetic ordering. The first term in Eq. (4.4) corresponds to the mechanical mixture, the second term is the contribution from the configurational entropy of mixing for the solution.

The binary excess energy can be expressed by

$$^EG_m = \sum_{i=1}^{n-1}\sum_{j>i}^{n} x_i x_j L_{ij} \tag{4.5}$$

where i and j are the components and L_{ij} is the regular-solution parameter representing the interaction energy between i and j. The regular-solution parameter can be expressed by Redlich–Kister equation [34]

$$L_{ij} = \sum_{v=0}^{k} (x_i - x_j)^v \cdot {}^vL_{ij} \tag{4.6}$$

where $^vL_{ij}$ can be temperature-dependent.

For element or compound having magnetic ordering, an additional term $^{mag}G_m$ accounting magnetic contribution to the Gibbs energy is included in Eq. (4.4) according to Ingen [35] and Hillert and Jarl [36]

$$^{mag}G_m = RT\ln(\beta + 1)f(\tau), \tag{4.7}$$

where β is the average magnetic moment, and τ is the ratio T/T_{cr} where T_{cr} is the critical temperature (the Curie temperature for ferromagnetic materials or the Neel temperature for antiferromagnetic materials). For the binary Al–Fe system, for example, the concentration dependencies of T_{cr} and β are expressed as

$$T_{cr} = x_{Al}T_{Al}^0 + x_{Fe}T_{Fe}^0 + x_{Al}x_{Fe}T_{cr,i}^{Al,Fe}, \tag{4.8}$$

$$\beta = x_{Al}\beta_{Al}^0 + x_{Fe}\beta_{Fe}^0 + x_{Al}x_{Fe}\beta_i^{Al,Fe}, \tag{4.9}$$

where $T_{cr,i}^{Al,Fe}$ and $\beta_i^{Al,Fe}$ are the parameters to be optimized.

For phases where species can mix on two sublattices, a two-sublattices model in the form of compound energy formalism should be used to describe the Gibbs energy of the phase. This is also applicable to describe the liquid phase using a partially ionic two-sublattice model [37, 38]. The model can be written as $(i_1, i_2, \ldots)_{a_1}(j_1, j_2, \ldots)_{a_2}$, where i and j are the constituents on the first and the second sublattices, respectively. The Gibbs energy is then expressed using the site fraction of the constituents y_i instead of the mole fraction as

$$G_m = \sum_i \sum_j y_i^{a_1} y_j^{a_2} \cdot {}^\circ G_{i:j} + RT \left(a_1 \cdot \sum_i y_i \ln(y_i) + a_2 \cdot \sum_j y_j \ln(y_j) \right)$$
$$+ {}^E G_m \tag{4.10}$$

The excess Gibbs energy is derived in a similar way as for the substitutional model

$$ {}^E G_m = y_{i_1} y_{i_2} y_{j_1} \cdot L_{i_1,i_2:j_1} \tag{4.11}$$

with

$$L_{i_1,i_2:j_1} = \sum_{v=0}^{n} (y_{i_1} - y_{i_2})^v \cdot {}^v L_{i_1,i_2:j_1} \tag{4.12}$$

The Gibbs energy for models with three and more sublattices can be described in a similar way giving more complicated the excess Gibbs energy terms.

Note that if the Neumann–Kopp rule applies (the heat capacity data of a compound is unavailable and it is an average of the heat capacity of the constituting elements), the Gibbs energy, $G_{A_a B_b}$, of a stoichiometric phase $A_a B_b$ is modeled as follows

$$G_{A_a B_b} = aGHSER_A + bGHSER_B + v_1 + v_2 T \tag{4.13}$$

where $GHSER_i$ is the Gibbs energy of the component i given relative to the enthalpy of i in its stable structure, H_i^{SER}; v_1 and v_2 are parameters to be optimized.

4.2.3 Optimization

The optimization methodology of the CALPHAD method can be simplified to the following steps:

1. Collection and critical evaluation of all available experimental data (crystallographic data for phases, phase equilibria data, thermodynamic data),
2. Selection of thermodynamic model for phase description considering its crystallographic information on Wyckoff positions and constituent distribution,
3. Determination of the parameters to be optimized. Accounting for the temperature dependence of the parameters for the end-members and the introduction of mixing parameters for the Gibbs energy description of the solution phases,
4. Optimization of the thermodynamic parameters considering all available experimental and theoretical data and setting their weights of influence,
5. Saving the derived Gibbs energy expressions with the optimized parameters into thermodynamic databases,
6. Calculation of phase diagrams and various phase equilibria using the thermodynamic databases.

4.3 Databases Development and Their Application

The main objective of the research from the very beginning was a development of thermodynamic databases which may be used as a tool for modeling interactions at the interface between the ceramic filter and the melt. As long as two filtration processes were proposed, for steel and aluminum melt, two independent complex filter systems had to be investigated followed by their thermodynamic description. The following filter systems to be developed can be highlighted:

- Database development for Al-melt processing

  - Database development for oxide systems $MgO-Al_2O_3$, $TiO_2-Al_2O_3$, $SiO_2-Al_2O_3$ [12], $MgO-Al_2O_3-SiO_2$ [12], $Al_2O_3-MgO-TiO_2$ to model interactions between coatings and inclusions in Al-melt,
  - Oxide systems TiO_2-SiO_2, $Al_2O_3-TiO_2-SiO_2$, $MgO-TiO_2-SiO_2$ to develop databases serving as bridges linking important sub-systems,
  - Complex Al–Ti–Mg–Si–O database with an emphasis on oxide system $Al_2O_3-MgO-TiO_2-SiO_2$ to predict phase relations between complex functional filter system, coatings, and inclusions,
  - Complex metallic Al–Fe–Mg–Si database to predict phase relations between complex functional filter system, coatings, inclusions and molten Al-based alloy.

- Database development for steel processing

  - Database development for oxide systems $MgO-Al_2O_3$, $FeO_x-Al_2O_3$, $MnO_x-Al_2O_3$ to model interactions between spinel $MgAl_2O_4$ and Al_2O_3 coating system and nonmetal inclusions. Thermodynamic database for steel is available. Therefore, development of thermodynamic description of oxide system makes it possible to simulate interaction between coatings and steel.
  - Introduction of carbon and oxycarbides into the systems for accounting influence of carbon on phase relations in carbon containing coatings.
  - Database development for the $TiO_2-Al_2O_3$, TiO_2-ZrO_2 [39], $Al_2O_3-TiO_2-ZrO_2$ systems to model interactions with inclusions and between coating and filter in case of $Al_2O_3 + TiO_2/ZrO_2$ filter system.

Below the systems and the corresponding thermodynamic databases are discussed in detail. Note that some of the binaries that are key sub-systems for both filter systems are discussed once in the order they are mentioned. In case of the aluminum melt filtration process, thermodynamic description of the Al-based metallic systems is also discussed. Possible applications of the derived thermodynamic databases are also discussed in line with the experimental observations within the whole project.

4.3.1 Database Development for Al-Melt Processing

In context of the project, spinel $MgAl_2O_4$ was suggested as one of the possible coatings for Al_2O_3–C active filter for steel and for Al_2O_3 filter for Al purification processes [83]. Therefore, thermodynamic properties of spinel and phase relations in the MgO–Al_2O_3 system had to be considered. The thermodynamic parameters for the MgO–Al_2O_3 system from the description of Hallstedt [13] were reassessed by Zienert and Fabrichnaya [41] based on new evaluation of literature data and own experimental investigation using XRD, SEM/EDX, and DTA. Available experimental data on inversion degree for stoichiometric spinel $MgAl_2O_4$ were critically analyzed and considered for the parameters optimization. Melting of spinel was investigated using high temperature DTA. The calculated phase diagram of the MgO–Al_2O_3 system and the temperature dependence of degree of inversion for spinel are shown in Fig. 4.2a and b, respectively, along with the literature data.

Later, the influence of different filter surface chemistries (Al_2O_3, spinel $MgAl_2O_4$, mullite $3Al_2O_3 \cdot 2SiO_2$, SiO_2, and TiO_2) on the properties of foam filters were discussed [9] and the corresponding filtration efficiency was evaluated [40]. All investigated filter coatings exhibited quite high filtration efficiency. It was shown that the TiO_2 coatings reacted with Al alloy melt [40]. The wettability of AlSi7Mg alloy on oxide ceramics (Al_2O_3, $MgAl_2O_4$, mullite, and TiO_2) was measured by Fankhänel et al. [10] and contact angles were found higher than 90° for all investigated ceramics. In case of TiO_2, it was found that the contact angle was quickly decreasing with the time and then leveled out at 101°. The investigation of reaction zone indicated reduction of TiO_2, presence of Ti and Si, and formation of Al_2O_3 layer. Moreover, rutile coatings deposited on corundum were supposed to filtrate actively and reactively spinel $MgAl_2O_4$ and Al_2O_3 inclusions present in Al alloy

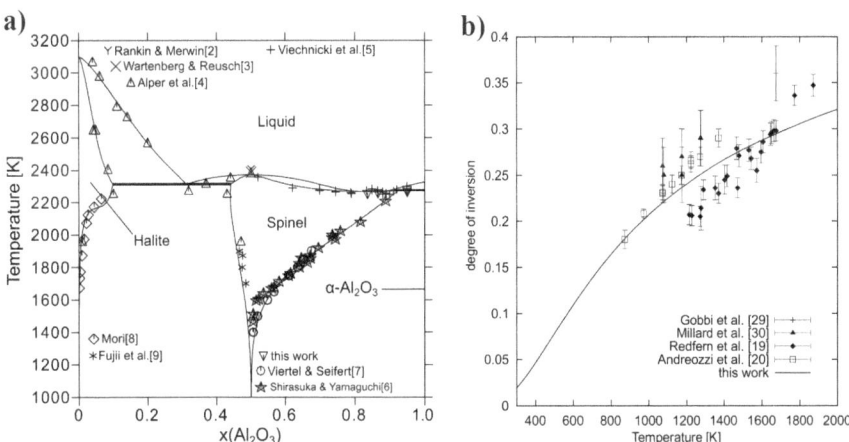

Fig. 4.2 **a** Calculated phase diagram of the MgO–Al_2O_3 system and **b** temperature dependence of degree of inversion for spinel $MgAl_2O_4$ [41]

melt. Therefore, the thermodynamic database had to be developed for the Al–Ti–O system for modelling interaction between Al melt and TiO_2 coatings on Al_2O_3 filters.

A thermodynamic assessment of the Al–Ti–O system was done by Ilatovskaia et al. [42, 43] using available binary databases and new experimental data regarding the Al_2O_3–TiO_2 system. It should be noted that the thermodynamic description of the Ti–O system by Hampl and Schmid-Fetzer [44] was accepted for solid phases, but two-sublattice partially ionic model was used for the liquid phase and therefore the parameters for liquid were assessed, as well as the parameters for Magneli phase $Ti_{20}O_{39}$ were slightly modified [42]. The thermodynamic descriptions for the Ti–Al [45] and Al–O [47] were accepted without modifications. Available experimental data on cation disordering (degree of inversion) for pseudobrookite Al_2TiO_5 were critically analyzed and accounted in the parameters optimization [50]. The phase relations in the Al_2O_3–TiO_2 system were verified experimentally. The stability range of Al_2TiO_5 was experimentally studied by Ilatovskaia et al. [42]. The Al_2TiO_5 formation was determined at 1553 K as low phase stability limit. Peritectic character of Al_2TiO_5 melting was established in the Al_2O_3–TiO_2 system at 2123 K by the high-temperature DTA experiments followed by SEM/EDX investigation of the obtained microstructures. The calculated phase diagram of the Al_2O_3–TiO_2 system and the temperature dependence of degree of inversion for stoichiometric pseudobrookite Al_2TiO_5 are shown in Fig. 4.3a and b, respectively, along with the literature data.

The derived database for the Al–Ti–O system [42] was used to predict phase relations between functional rutile coatings and molten Al alloy. By thermodynamic simulations, it was shown that TiO_2 should be reduced by molten aluminum forming Al_2O_3, while the reduced titanium should react with Al to form Al_3Ti. The results were presented and compared with annealing experiments of Salomon et al. [48] confirming formation of Al_3Ti and Al_2O_3. Presence of small amount of Ti_2O_3 was

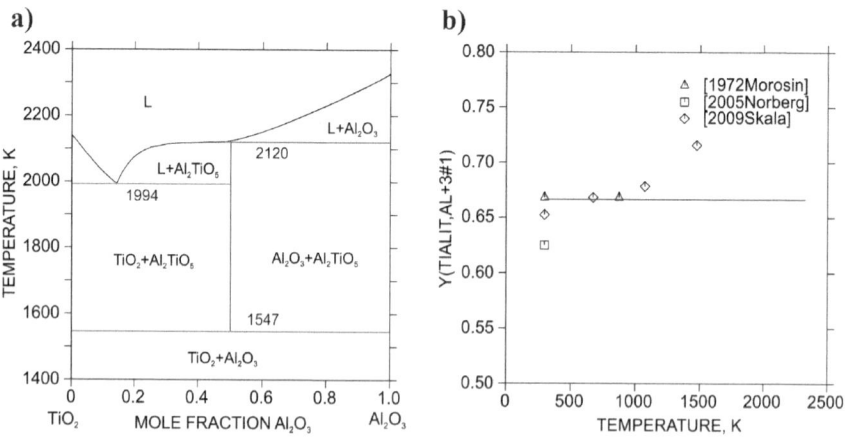

Fig. 4.3 a Calculated phase diagram of the Al_2O_3–TiO_2 system and **b** temperature dependence of degree of inversion for pseudobrookite Al_2TiO_5 [50]

indicated experimentally and it was shown to decrease with the time of annealing. Therefore, Ti_2O_3 was intermediate product forming during TiO_2 reduction.

The experiments performed using Al alloy containing Mg and Si indicated formation of $MgTiO_3$ phase and absence of Al_3Ti phase [48]. However, $Ti_5(Si,Al)_3$, $Ti(Al,Si)_3$ and ternary phases could form. Therefore, thermodynamic database of the complex Al–Ti–Mg–Si–O system is necessary to elucidate influence of Mg content in Al alloys to interface reactions between TiO_2 coating and Al-based alloy melt. By step-by-step approach, the thermodynamic descriptions for the oxide parts of the Al_2O_3–MgO–TiO_2, Al_2O_3–TiO_2–SiO_2, and MgO–TiO_2–SiO_2 systems had to be derived for further joint usage. The thermodynamic description of the forth key oxide ternary MgO–Al_2O_3–SiO_2 sub-system was carried out by Fabrichnaya et al. [12], however the liquid phase model should be changed and the parameters need to be optimized.

Moreover, the thermal shock tests of the Al_2O_3-rich $MgAl_2O_4$-spinel ceramics were done [49]. It was found that the addition or in situ formation of Al_2TiO_5 could enhance the performance of the Al_2O_3-rich spinel refractories by improving their thermal shock resistance.

Preliminarily, as a key binary sub-system, thermodynamic assessment of the MgO–TiO_2 system was carried out based on own phase diagram data and experimental thermodynamic data [43]. Available data on the cation disordering of the intermediate compounds Mg_2TiO_4 and $MgTi_2O_5$ as well as the measured heat capacities data of the intermediate phases (Mg_2TiO_4, $MgTiO_3$, and $MgTi_2O_5$) were also took into account during optimization of the thermodynamic parameters.

Phase relations in the Al_2O_3–MgO–TiO_2 system were investigated in detail using DTA followed by SEM/EDX and XRD investigations [43]. The solid-state reaction, Al_2O_3 + TiO_2 + Spinel s.s. = Pseudobrookite s.s., was indicated at 1433 K by XRD study of the stepwise annealed samples. Two series of continuous solid solutions in the Al_2O_3–MgO–TiO_2 system were found: $MgAl_2O_4$–Mg_2TiO_4 (spinel s.s.) and Al_2TiO_5–$MgTi_2O_5$ (pseudobrookite s.s.) that agrees with the literature data [46]. A miscibility gap for the spinel join with maximum at 1648 ± 20 K [46] was confirmed. Two invariant reactions on the liquidus projection, eutectic L = $MgTiO_3$ + Psbk + Sp and transitional-type L + Al_2O_3 = Sp + Psbk, were found at 1875 K and 2006 K, respectively, using DTA followed by SEM/EDX microstructure investigations. The experimental data were then used to develop the thermodynamic description of the Al_2O_3–MgO–TiO_2 system [50]. The calculated liquidus projection and sub-solidus isothermal section at 1550 K of the Al_2O_3–MgO–TiO_2 system are shown in Fig. 4.4a and b, respectively.

The thermodynamic description of the pseudo binary TiO_2–SiO_2 system was derived based on own experimental study using XRD, SEM/EDX, and DTA as well as available literature data [51]. The reactions occurring in the system including liquid immiscibility were investigated by DTA experiments in air or under He atmosphere. The calculated phase diagram of the TiO_2–SiO_2 system under normal and reduced oxygen conditions is shown in Fig. 4.5a.

Later on, considering the newly obtained results for the TiO_2–SiO_2 system, phase equilibria in the Al_2O_3–TiO_2–SiO_2 system were investigated experimentally in air

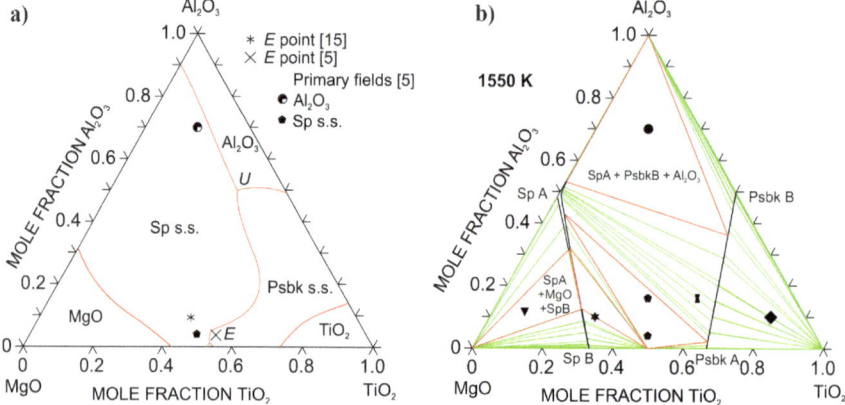

Fig. 4.4 Calculated liquidus projection **a** and sub-solidus isothermal section at 1550 K **b** of the Al$_2$O$_3$–MgO–TiO$_2$ system [50]

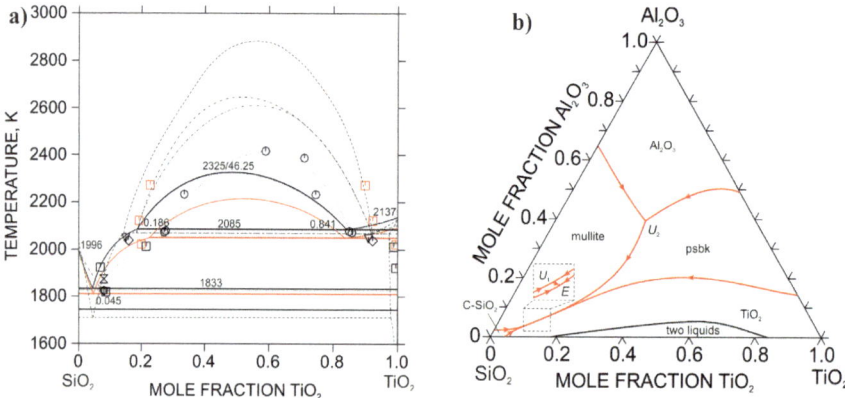

Fig. 4.5 **a** Calculated phase diagram of the TiO$_2$–SiO$_2$ system at p(O$_2$) = 0.21 bar (black) and p(O$_2$) = 10$^{-2.5}$ bar (red) and **b** calculated liquidus projection of the Al$_2$O$_3$–TiO$_2$–SiO$_2$ system [51]

by Ilatovskaia et al. [52]. Solid-state phase relations were characterized using XRD and SEM/EDX and the solid-state invariant reaction SiO$_2$ + Al$_2$TiO$_5$ ↔ Al$_6$Si$_2$O$_{13}$ + TiO$_2$ at 1743 K was observed. The invariant reactions occurring on the liquidus projection of the system were determined by DTA followed by SEM/EDX examination of the obtained microstructures. Consequently, the experimentally observed data were considered for thermodynamic assessment of the system [51]. The calculated liquidus projection is shown in Fig. 4.5b.

Phase relations in the MgO–TiO$_2$–SiO$_2$ system were investigated in air in the range of 1500–1900 K using XRD, SEM/EDX, and DTA [85]. Similar to the ternary oxide systems described above, solid-state reaction occurring at 1625 K was observed, MgSiO$_3$ + TiO$_2$ = SiO$_2$ + MgTi$_2$O$_5$. Moreover, among seven invariant reactions

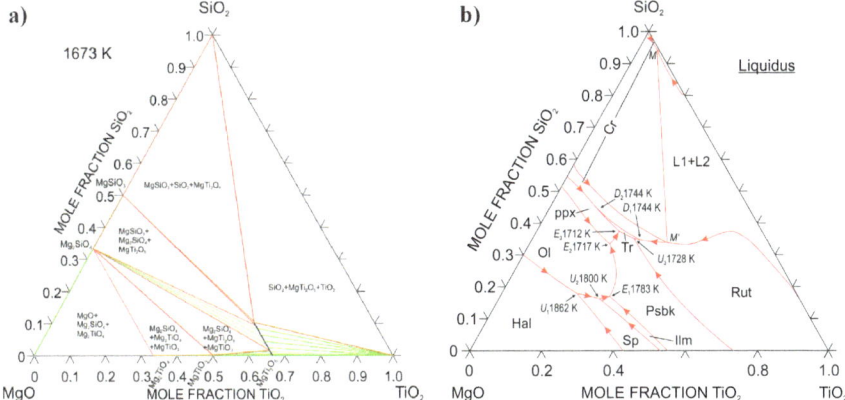

Fig. 4.6 Calculated **a** isothermal section at 1673 K and **b** liquidus projection of the MgO–TiO$_2$–SiO$_2$ system

occurring on the liquidus projection, three of eutectic type were verified experimentally: L = Mg$_2$SiO$_4$ + MgTiO$_3$ + MgTi$_2$O$_5$ (E$_1$) at 1822 K, L = MgSiO$_3$ + Mg$_2$SiO$_4$ + MgTi$_2$O$_5$ (E$_2$) at 1704 K, and L = MgSiO$_3$ + SiO$_2$ + MgTi$_2$O$_5$ (E$_3$) at 1690 K. During the thermodynamic optimization of the system, the Gibbs energy of the MgTi$_2$O$_5$ phase was deliberately adjusted by considering SiO$_2$ solubility to make it more stable than MgSiO$_3$ in order to keep the solid-state reaction. Therefore, the extension of MgTi$_2$O$_5$ into the ternary system is more pronounced compared to the experimental observation of an insignificant solubility of SiO$_2$ in MgTi$_2$O$_5$. Note that similar approach was applied for the Gibbs energy of Al$_2$TiO$_5$ against Al$_6$Si$_2$O$_{13}$ in the Al$_2$O$_3$–TiO$_2$–SiO$_2$ system due to the solid-state reaction. The calculated isothermal section at 1673 K and liquidus projection of the MgO–TiO$_2$–SiO$_2$ system are shown in Fig. 4.6.

To model the interactions between ceramic filter material and Al melt in a proper way, the thermodynamic assessment of the Al-based alloy itself had to be also undertaken. The Fe content in Al-based alloys is low, but the intermetallic compounds forming in the Al–Fe and Al–Fe–Si systems substantially influence mechanical properties of alloys. Despite a number of thermodynamic assessments for the Al–Fe system [53], thermodynamic functions of intermetallic compounds were simplistically modelled using the Neumann–Kopp rule as sum of heat capacities for elements. Therefore, the thermodynamic reassessment of the Al–Fe system was undertaken using newly obtained experimental data. Heat capacity of Fe$_2$Al$_5$ (η) phase was experimentally measured between 235 and 1073 K using differential scanning calorimetry (DSC) [54]. It was found that at temperatures between 523 and 553 K phase transformation occurred in the Fe$_2$Al$_5$ phase. It was assumed that the low temperature modification of this phase has an ordered structure. The heat capacity of Fe$_2$Al$_5$ was measured at high temperatures up to 1390 K and at low temperatures between 120 and 823 K [55]. The other intermetallic phases Fe$_2$Al (ζ) and Fe$_4$Al$_{13}$ (θ) were

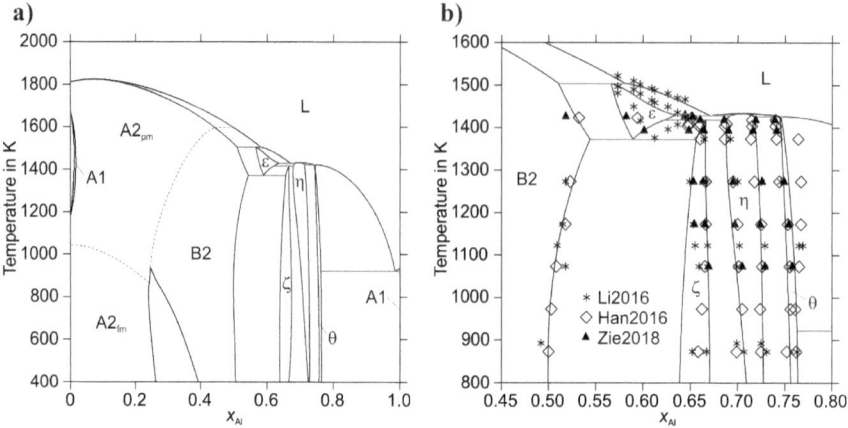

Fig. 4.7 Calculated phase diagram of the Al–Fe system **a** along with experimental data **b**

studied from 210 K to melting points as well as B2 ordered phase [55]. The measurements for B2 phase were compared with high temperature adiabatic calorimetry data and good agreement confirmed reliability of C_p measurements at high temperatures. Based on the measured data for heat capacity of intermetallic compounds and their homogeneity ranges together with experimental information from literature thermodynamic parameters of the Al–Fe system were assessed [56]. Later on, the thermodynamic description of the Fe_2Al_5 (η) phase was reconsidered accounting newly reported data on the crystal structure of the phase [57]. Moreover, in the assessment [56], the model inconsistent with the crystal structure was used to describe the Fe_5Al_8 (ϵ) phase structure which led to the stabilization of Fe_5Al_8 at temperatures above 2300 K. Therefore, the thermodynamic parameters of the η, ϵ, and θ phases in the Al–Fe system were reassessed [86]. The calculated phase diagram of the Al–Fe system is presented in Fig. 4.7.

A thermodynamic dataset for the description of the Al–Fe–Si system for the temperature range below 1073 K was derived by Zienert and Fabrichnaya [58]. The system includes at least 11 known ternary compounds which descriptions were combined from several resources, i.e. $Al_{13}Fe_4$ [59], τ_2, τ_4, τ_5, and τ_6 [60], τ_1, τ_3, τ_7, τ_8, τ_{10}, and the liquid phase [61]. Thermodynamic parameters of these phases were optimized to reproduce available experimental data. The $Al_{64}Fe_{26}Si_{10}$-τ_{11} phase was modelled for the first time in Ref. [58]. With the derived dataset, two reactions were predicted, (i) $\tau_{11} + \tau_2 + \tau_3 = \tau_{10}$ at 795.2 °C and (ii) $\tau_{11} = \tau_1 + \tau_{10} + Al_{13}Fe_4$ at 683.7 °C. However, the real nature of these transitions between τ_{10} and τ_{11} are still unclear and further experimental investigations are necessary. The calculated isothermal section of the Al–Fe–Si system at 727 °C is shown in Fig. 4.8a. The thermodynamic descriptions of pure element were accepted from SGTE Unary Database [62]. Note that the heat capacities of intermediate phases were described using the Neumann–Kopp rule and therefore heat capacities based on more realistic

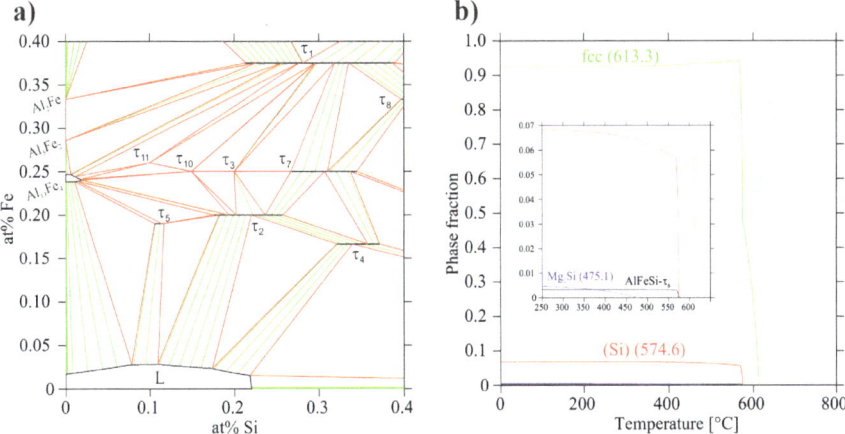

Fig. 4.8 Calculated **a** isothermal sections of the Al-rich side of the Al–Fe–Si system at 727 °C, and **b** phase fraction diagram of Al-alloy A356 [58]

assumption or based on experimental data are desirable. Regarding this, a single-phase sample of AlFeSi-τ_4 was produced and the heat capacity was measured in the single-phase region between 640 and 760 °C. It was shown that use of the Neumann–Kopp rule results in unphysical kink in the temperature dependence of heat capacity of compounds containing a low melting element, i.e. Al. Besides, the heat capacity of the investigated ternary AlFeSi-τ_4 phase was calculated using DFT method [63]. The calculated results are in a very good agreement with the experimental findings and the reasons for deviation from the Neumann–Kopp rule were discussed.

The description of the Al–Mg–Si system is accepted from Ref. [64]. Two other ternary systems Al–Mg–Fe and Si–Mg–Fe were not assessed due to the lack of data and simply combined from the binary descriptions [58]. It should be noted that no ternary phases were found in these systems. Ternary descriptions were then combined into the quaternary description of Al–Mg–Fe–Si.

In the frame of thermodynamic database development for metallic Al-based alloy the solidification behavior of the A356 alloy was investigated experimentally (DTA, XRD, SEM/EDX, SEM/EBSD) and theoretically [58]. It was found experimentally that the AlFeMgSi-π phase solidifies in A356 alloy, however the thermodynamic description of the phase was not included into the dataset for the Al–Fe–Mg–Si system due to the lack of experimental data on the phase. The derived dataset was then used to model phase relations in Al-alloy A356. The calculated phase fraction diagram of the A356 alloy composition is presented in Fig. 4.8b.

As an application issue, the combined datasets of the MgO–Al_2O_3 and of the Al–Fe–Mg–Si systems were used to predict phase relations at the interface between AlSi7Mg0.6 alloy (A356) and alumina filters. The formation of magnesia-alumina-spinel was indicated by thermodynamic calculations. The results were shown and compared with SPS melting experiments provided by Salomon et al. [65]. Moreover, the interaction of molten AlSi7Mg0.6 alloy with mullite and amorphous silica were investigated using SPS [66]. It was found that amorphous SiO_2 as well as SiO_2 originating from mullite decomposition were reduced by Al and Mg. The combined datasets of the Al_2O_3–Fe_2O_3–FeO and Al–Fe–Si–Mg systems were used to predict interactions during contact of molten AlSi7Mg0.6 alloy with mullite and amorphous silica. The formation of Al_2O_3, $MgAl_2O_4$ spinel and metallic Si were predicted by thermodynamic modelling. The results of calculation were in good agreement with experimental data of SPS melting [66]. In detail description of the Al_2O_3–Fe_2O_3–FeO system is given in the next section.

This advanced thermodynamic description could be introduced into already published description of the Al–Mg–Si–Fe–Mn system [67]. The thermodynamic description of this system (including also Cr) could be used for modeling support of experimental study of Dietrich et al. [68] who investigated influence of Mn and Cr additives to microstructures formed during solidification of Al-alloy melt.

4.3.2 Database Development for Steel Processing

Since the carbon-bonded alumina materials have been proposed as a promising next generation ceramic filter material for steel melt filtration [3], the possibility of a carbothermic reaction between alumina and carbon was discussed by means of thermodynamic calculations considering the Fe–Al_2O_3–Al corner of the Al–Fe–O system [69]. It was also observed experimentally that the equilibrium between alumina and liquid iron is unstable due to the constant formation of CO, and alumina dissolves in liquid iron, increasing the concentration of aluminum in the melt.

A further detailed thermodynamic description of the Al_2O_3–Fe_2O_3–FeO system was developed by Dreval et al. [70] taking into account all available literature data. Available experimental data on phase equilibria at different temperatures and oxygen partial pressures, calorimetric and vapor pressure data, the degree of inversion of the spinel phase as well as an extension of homogeneity range in spinel phase by the dissolution of Al_2O_3 and Fe_2O_3 were critically reviewed. Thermodynamic modelling of spinel phase based on compound energy formalism allowed taking into account inversion degree in the $FeAl_2O_4$–Fe_3O_4 solid solution and deviation from stoichiometry in the direction of Al_2O_3 and Fe_2O_3. The calculated isothermal section at 1523 K and phase relations as dependence of $\log p_{O_2}$ on metal ration Fe/(Fe + Al) at 1553 K in the Al_2O_3–Fe_2O_3–FeO system are shown in Fig. 4.9.

An excellent example of the implementation of these databases is thermodynamic simulation of the interaction between carbon-free and carbon-bonded Al_2O_3 and Armco iron [71]. With the combined use of the databases for Fe–Al_2O_3–Al [69], Al_2O_3–Fe_2O_3–FeO [70], and Al_2O_3–Al_4C_3–AlN [72], an increase in Al-containing species in the gas phase with a decrease in oxygen partial pressure and the formation of solid Al_4C_3 and Al_4O_4C at a low oxygen partial pressure were predicted. Those were also discussed considering experimental data, and in this case, further whiskers formation was observed [71]. In another work, the combined database for the Al–Mg–Fe–O system [23, 70, 73] was used to simulate the interaction of Al_2O_3–$MgAl_2O_4$

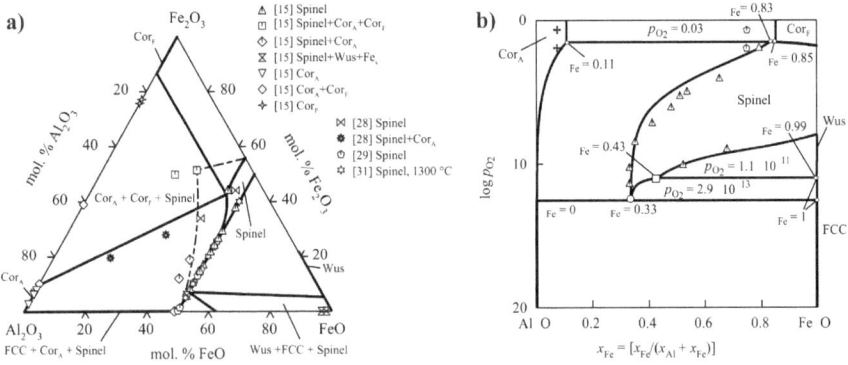

Fig. 4.9 Calculated isothermal sections of the Al_2O_3–Fe_2O_3–FeO system at **a** 1523 K and **b** 1553 K as dependence of $\log p_{O_2}$ on metal ratio [70]

substrate and Armco iron. The results of calculations showed that liquid Fe is enriched mainly in aluminum and Fe is negligible dissolved in $MgAl_2O_4$ at a low oxygen partial pressure which was subsequently confirmed experimentally [74].

Subsequently, carbon-bonded alumina filters with active or reactive coatings for steel melt filtration were discussed. In the case of reactive filters based on carbon bonded MgO, the assumption of gaseous Mg particles was supported by experimental findings of produced secondary MgO fibers on the filter surface [4]. Moreover, the combination of alumina and magnesia in carbon-bonded filters results in the in situ formation of spinel $MgAl_2O_4$ which acts as an active coating and is suitable to remove inclusions with the same crystal structure. Also, the formation of spinel leads to volumetric expansion that would counteract the shrinkage phenomenon [75, 76]. The thermodynamic description of the Mg–Al–O–C system would be particularly useful in this context. As an important part, the thermodynamic description of the oxide MgO–Al_2O_3 system was discussed above.

To optimize the reactions taking place during filtration and their effects on the steel melt purification in more detail, the spinel $(Fe,Mn,Mg)Al_2O_4$ coating system applied on Al_2O_3–C filters was further investigated [77]. It was shown that the subsequent interaction of coating with molten steel leads to the development of multicrystal structures on the filter surface, which stem from interfacial reactions between coating materials, molten steel, and inclusions. Therefore, the thermodynamic database had to be developed by combining data on spinel phases of $MgAl_2O_4$ [41], $FeAl_2O_4$ [70], and $MnAl_2O_4$. The thermodynamic description of the Al_2O_3–MnO system was developed by Ilatovskaia and Fabrichnaya [78] to collect the remaining information requested. Phase relations were investigated experimentally in the range of 1373–2053 K using XRD, SEM/EDX, and DTA and the heat capacity measurements of stoichiometric $MnAl_2O_4$ were carried out at 250–873 K using DSC. The results obtained were considered along with the available literature data on the degree of inversion of stoichiometric spinel $MnAl_2O_4$, standard entropy, and its Gibbs energy of formation from oxides. The calculated phase diagram of the Al_2O_3–MnO system

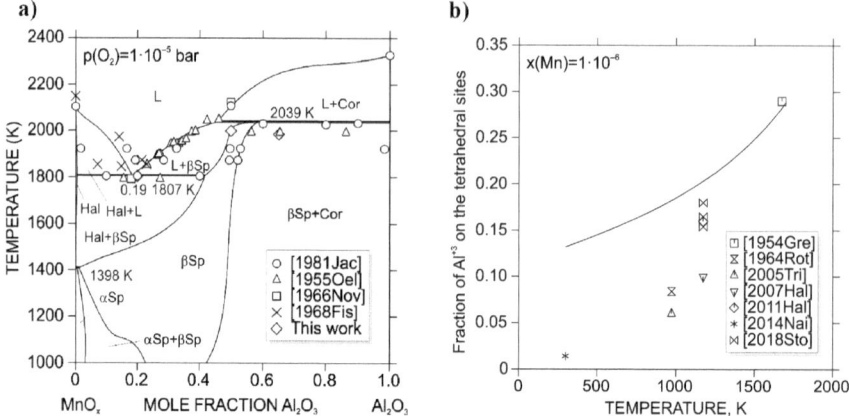

Fig. 4.10 a Calculated phase diagram of the Al_2O_3–MnO system and **b** temperature dependence of degree of inversion for stoichiometric spinel $MnAl_2O_4$ [78]

and the temperature dependence of degree of inversion for stoichiometric spinel $MnAl_2O_4$ are shown in Fig. 4.10a and b, respectively, along with the literature data.

Gehre et al. [7] found that TiO_2 and ZrO_2 additions to the alumina-based refractories improve the thermal shock resistance of alumina and increase its corrosion resistance against molten coal slags. The formation of a protective spinel $MgAl_2O_4$ layer, which stops the slag penetration and the corrosion process, was shown to occur due to the reaction of alumina matrix with MgO of the slag in the presence of titania and zirconia.

Moreover, a modification of Al_2O_3 surface using TiO_2 and ZrO_2 additives was suggested for the steel filtration [79]. Enhanced thermal stability of Al_2TiO_5 provided by $ZrTiO_4$ phase was indicated [80]. Therefore, the thermodynamic database of the Al_2O_3–TiO_2 system had to be extended by including ZrO_2. As a separate work, a thermodynamic description of the TiO_2–ZrO_2 system using the CALPHAD approach was developed based on own experimental data and from the literature [39]. Experimental and theoretical work on the system Al_2O_3–ZrO_2 was available after Fabrichnaya et al. [81] and Lakiza et al. [82]. The derived thermodynamic descriptions of the pseudo binary TiO_2–ZrO_2 and Al_2O_3–ZrO_2 systems were further accepted for thermodynamic assessment of the high-order system.

The CALPHAD assessment of the Al_2O_3–TiO_2–ZrO_2 system was done by Ilatovskaia et al. [84]. The formation of t-ZrO_2–$ZrTiO_4$–Al_2TiO_5 and $ZrTiO_4$–Al_2TiO_5–TiO_2 assemblages was indicated by thermodynamic calculations. Isothermal sections were calculated and the results of calculations for the system were checked experimentally by phase equilibration in the temperature range of 1530–1893 K. Using DTA and equilibration experiments two solid-state reactions at 1593 K and 1648 K were found. Three invariant eutectics were found on the liquidus surface in the range of 1909–1978 K using DTA followed by SEM/EDX microstructure investigation. Obtained experimental results were used for assessment

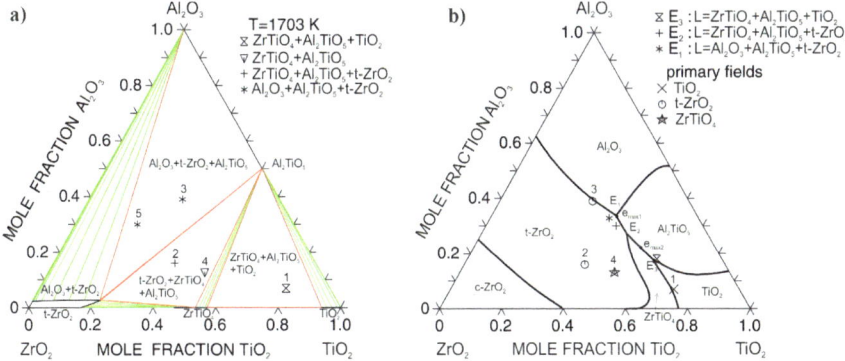

Fig. 4.11 Calculated **a** isothermal sections of the Al_2O_3–TiO_2–ZrO_2 system at 1703 K and **b** liquidus projection [84]

of thermodynamic parameters. The calculated isothermal section at 1703 K as well as the liquidus projection of the Al_2O_3–TiO_2–ZrO_2 system are shown in Fig. 4.11.

The corrosion of carbon containing refractories based on titania-zirconia doped alumina for steel casting was investigated [79]. It was shown that 9 vol.% carbon content is enough to form a functioning matrix material for which the wetting was reduced and corrosion reactions inhibited from the beginning. The calculations using expanded thermodynamic database of the Al_2O_3–TiO_2–ZrO_2–C system could provide new information for a final evaluation of refractories for steel ingot casting considering the oxygen content of the melt.

4.4 Conclusion

Since its launch the project has achieved remarkable results on thermodynamic modeling of the systems related to the metal melt filtration process. Advanced methods of computational thermodynamics were applied for development of a complex databases used to predict and understand mechanisms of chemical reactions occurring in the filter material, coatings, inclusions, molten metal and in-between. Moreover, control and optimization as well as pondering new solutions of technological issues could be suggested.

Considering two independent metal melt filter processes, for steel and aluminum melt, two complex datasets were developed. For Al-melt filtration process, the complex Al–Ti–Mg–Si–O database with an emphasis on oxide part of Al_2O_3–MgO–TiO_2–SiO_2 in order to predict phase relations between complex functional filter system, coatings, and inclusions was developed. Furthermore, the thermodynamic database for metallic Al–Fe–Mg–Si system used in combination with oxide database make it possible to simulate phase relations between filter system and molten Al-based alloy. For the steel filtration process, the complex dataset related to the Al_2O_3–MgO coating system applied on Al_2O_3–C filters was developed. Iron oxides were included into the database to account their presence as inclusions and interactions with coatings. An independent database for the Al_2O_3–TiO_2–ZrO_2 system was also developed for the modeling of interaction in the Al_2O_3 coatings modified by Ti and Zr oxides. Regarding development of thermodynamic datasets for oxides, five binary and four ternary systems were investigated experimentally and assessed thermodynamically using the CALPHAD approach for applications in cooperation with other sub-projects.

Acknowledgements The authors gratefully acknowledge the Deutsche Forschungsgemeinschaft (DFG, German Research Foundation) for financial support of the investigations, which were part of the Collaborative Research Center Multi-Functional Filters for Metal Melt Filtration–A Contribution towards Zero Defect Materials (Project-ID 169148856–SFB 920, subproject A03). Furthermore, the authors would like to thank Dr. Peter Franke, Dr. Liya Dreval, and Dr. Tilo Zienert for scientific support and co-working in the sub-project A03 in the frame of CRC 920.

References

1. H.L. Lukas, S.G. Fries, B. Sundman, *Computational Thermodynamics, The Calphad Method*, (Cambridge University press, Cambridge, UK, 2007). https://doi.org/10.1017/CBO9780511804137
2. U.R. Kattner, High temperature-high pressure. **49**, 31 (2020). https://doi.org/10.32908/hthp.v49.853
3. M. Emmel, C.G. Aneziris, Ceram. Int. **38**, 5165 (2012). https://doi.org/10.1016/j.ceramint.2012.03.022
4. C.G. Aneziris, S. Dudczig, M. Emmel, H. Berek, G. Schmidt, J. Hubalkova, Adv. Eng. Mater. **15**, 46 (2013). https://doi.org/10.1002/adem.201200199
5. S. Dudczig, C.G. Aneziris, M. Dopita, D. Rafaja, Adv. Eng. Mater. **15**(12), 1177 (2013). https://doi.org/10.1002/adem.201300121
6. T. Zienert, O. Fabrichnaya, J. Eur. Ceram. Soc. **35**, 1317 (2015). https://doi.org/10.1016/j.jeurceramsoc.2014.10.033
7. P. Gehre, C.G. Aneziris, M. Klinger, M. Schreiner, M. Neuroth, Fuel **150**, 252 (2015). https://doi.org/10.1016/j.fuel.2015.02.024
8. C.G. Aneziris, S. Dudczig, N. Gerlach, H. Berek, D. Veres, Adv. Eng. Mater. **12**(6), 478 (2010). https://doi.org/10.1002/adem.201000037
9. C. Voigt, T. Zienert, P. Schubert, C.G. Aneziris, J. Hubalkova, J. Am. Ceram. Soc. **97**(7), 2046 (2014). https://doi.org/10.1111/jace.12977
10. B. Fankhänel, M. Stelter, C. Voigt, C.G. Aneziris, Adv. Eng. Mater. **19**, 1700084 (2017). https://doi.org/10.1002/adem.201700084
11. O.B. Fabrichnaya, Calphad **24**(2), 113 (2000). https://doi.org/10.1016/S0364-5916(00)00018-3
12. O. Fabrichnaya, A. Costa e Silvas, F. Aldinger, Z. Metallkd. **95**, 793 (2004). https://doi.org/10.1515/ijmr-2004-0148
13. B. Hallstedt, J. Am. Ceram. Soc. **75**(6), 1497 (1992). https://doi.org/10.1111/j.1151-2916.1992.tb04216.x

14. H. Mao, M. Selleby, B. Sundman, Calphad **28**(3), 307 (2004). https://doi.org/10.1016/j.cal phad.2004.09.001
15. L. Kjellqvist, M. Selleby, J. Phase Equilib. Diffus. **31**(2), 113 (2010). https://doi.org/10.1007/s11669-009-9643-6
16. H. Mao, M. Selleby, B. Sundman, J. Am. Ceram. Soc. **88**(9), 2544 (2005). https://doi.org/10.1111/j.1551-2916.2005.00440.x
17. S.A. Decterov, I.-H. Jung, A.D. Pelton, J. Am. Ceram. Soc. **85**(12), 2903 (2002). https://doi.org/10.1111/j.1151-2916.2002.tb00554.x
18. I.-H. Jung, Y.-B. Kang, S.A. Decterov, A.D. Pelton, Met. Mater. Trans. B **35**(2), 259 (2004). https://doi.org/10.1007/s11663-004-0027-3
19. I.-H. Jung, S.A. Decterov, A.D. Pelton, J. Phase Equilib. Diffus. **25**(4), 329 (2004). https://doi.org/10.1007/s11669-004-0151-4
20. N. Saunders, A.P. Miodownik, *CALPHAD (Calculation of Phase Diagrams) A Comprehensive Guide*, vol. 1 (Elsevier Science Ltd., Oxford, UK, 1998)
21. Z.-K. Liu, Y. Wang, *Computational Thermodynamics of Materials* (Cambridge University Press, Cambridge, UK, 2016)
22. J.C. Zhao, *Methods for Phase Diagram Determination.* (Elsevier, Oxford, 2017). https://doi.org/10.1017/CBO9781139018265
23. X. Li, A. Scherf, M. Heilmaier, F. Stein, J. Phase Equilib. Diff. **37**, 162 (2016). https://doi.org/10.1007/s11669-015-0446-7
24. E. Fidancevska, V. Vassilev, J. Chem. Technol. and Metall. **45**(4), 421 (2010)
25. L. Lutterotti, S. Matthies, H.R. Wenk, CPD Newsletter **21**, 14 (1999)
26. L. Lutterotti, M. Bortolotti, G. Ischia, I. Lonardelli, H.R. Wenk, Z. Kristallogr. Supl. **26**, 125 (2007). https://doi.org/10.1017/S0885715613001346
27. M. Hellenbrandt, Crystallogr. Rev. **10**, 17 (2004). https://doi.org/10.1080/08893110410001664882
28. G.D. Gatta, M.J. Richardson, S.M. Sarge, S. Stølen, Pure Appl. Chem. **78**(7), 1455 (2006). https://doi.org/10.1351/pac200678071455
29. C.G. Maier, K.K. Kelley, J. Am. Chem. Soc. **54**(8), 3242 (1932). https://doi.org/10.1021/ja01347a029
30. T. Zienert, O. Fabrichnaya, CALPHAD **65**, 177 (2019). https://doi.org/10.1016/j.calphad.2019.01.017
31. T. Zienert, Software X **9**, 244 (2019). https://doi.org/10.1016/j.softx.2019.02.002
32. NIST—National Institute of Standards and Technology. NIST chemistry web book. (2017). https://webbook.nist.gov/
33. B. Sundman, J. Agren, J. Phys. Chem. Solids **42**, 297 (1981). https://doi.org/10.1016/0022-3697(81)90144-X
34. O. Redlich, T. Kister, Ind. Eng. Chem. **40**(2), 345 (1948). https://doi.org/10.1021/ie50458a036
35. G. Inden, Z. Metallkde **68**, 529 (1977). https://doi.org/10.1515/ijmr-1975-661003
36. M. Hillert, M. Jarl, Calphad **2**(3), 227 (1978). https://doi.org/10.1016/0364-5916(78)90011-1
37. M. Hillert, B. Jansson, B. Sundman, J. Ågren, Metal. Trans. **A16**, 261 (1985). https://doi.org/10.1007/BF02816052
38. B. Sundman, CALPHAD **15**, 109 (1991). https://doi.org/10.1016/0364-5916(91)90010-H
39. I. Saenko, M. Ilatovskaia, G. Savinykh, O. Fabrichnaya, J. Am. Ceram. Soc. **101**(1), 386 (2017). https://doi.org/10.1111/jace.15176
40. C. Voigt, E. Jäckel, F. Taino, T. Zienert, A. Salomon, G. Wolf, C.G. Aneziris, P. LeBrun, Metal. Mat. Trans. B **48B**, 497 (2017). https://doi.org/10.1007/s11663-016-0869-5
41. T. Zienert, O. Fabrichnaya, CALPHAD **40**, 1 (2013). https://doi.org/10.1016/j.calphad.2012.10.001
42. M. Ilatovskaia, G. Savinykh, O. Fabrichnaya, J. Phase Equilib. Diffus. **38**, 175 (2017). https://doi.org/10.1007/s11669-016-0509-4
43. M. Ilatovskaia, I. Saenko, G. Savinykh, O. Fabrichnaya, J. Am. Ceram. Soc. **101**, 5198 (2018). https://doi.org/10.1111/jace.15748

44. M. Hampl, R. Schmid-Fetzer, Int. J. Mater. Res. **106**(5), 439 (2015). https://doi.org/10.3139/146.111210
45. V.T. Witusiewicz, A.A. Bondar, U. Hecht, S. Rex, T.Ya. Velikanova, J. Alloys Compd **465**, 64 (2009). https://doi.org/10.1016/j.jallcom.2007.10.061
46. P. Boden, F.P. Glasser, Trans. J. Br. Ceram. Soc. **72**(5), 215 (1973)
47. B. Hallstedt, J. Phase Equil. **14**(6), 662 (1993). https://doi.org/10.1007/BF02667878
48. A. Salomon, C. Voigt, O. Fabrichnaya, C.G. Aneziris, D. Rafaja, Adv. Eng. Mater. **19**(9), 1700106 (2017). https://doi.org/10.1002/adem.201700106
49. K. Moritz, C.G. Aneziris, Ceram. Int. **42**, 14155 (2016). https://doi.org/10.1016/j.ceramint.2016.06.037
50. M. Ilatovskaia, O. Fabrichnaya, J. Alloys Compd. **790**, 1137 (2019). https://doi.org/10.1016/j.jallcom.2019.03.046
51. M. Ilatovskaia, O. Fabrichnaya, J. Phase Equilib. Diffus. **43**, 15 (2022). https://doi.org/10.1007/s11669-021-00935-4
52. M. Ilatovskaia, F. Bärthel, O. Fabrichnaya, Ceram. Int. **46**, 29402 (2020). https://doi.org/10.1016/j.ceramint.2020.05.103
53. B. Sundman, I. Onuma, N. Dupin, U.R. Kattner, S.G. Fries, Acta Mater. **57**(10), 2896 (2009). https://doi.org/10.1016/j.actamat.2009.02.046
54. T. Zienert, L. Amirkhanyan, J. Seidel, R. Wirnata, T. Weissbach, T. Gruber, O. Fabrichnaya, J. Kortus, Intermetallics **77**, 14 (2016). https://doi.org/10.1016/j.intermet.2016.07.002
55. T. Zienert, A. Leineweber, O. Fabrichnaya, J. Alloys Compd. **725**, 848 (2017). https://doi.org/10.1016/j.jallcom.2017.07.199
56. T. Zienert, O. Fabrichnaya, J. Alloys Compd. **743**, 795 (2018). https://doi.org/10.1016/j.jallcom.2018.01.316
57. H. Becker, A. Leineweber, Intermetallics **93**, 251 (2018). https://doi.org/10.1016/j.intermet.2017.09.021
58. T. Zienert, O. Fabrichnaya, Adv. Eng. Mater. **15**(12), 1244 (2013). https://doi.org/10.1002/adem.201300113
59. R. Kolby, *COST 507–System Al-Fe-Si.* (European Communities, 1998), pp. 319–321
60. L. Eleno, J. Vezely, B. Sundman, M. Cieslar, J. Lacaze, Mater. Sci. Forum **649**, 523 (2010). https://doi.org/10.4028/www.scientific.net/MSF.649.523
61. Y. Du, J.C. Schuster, Z.-K. Liu, R. Hu, P. Nash, W. Sun, W. Zhang, J. Wang, L. Zhang, C. Tang, Z. Zhu, S. Liu, Y. Ouyang, W. Zhang, N. Krendelsberger, Intermetallics **16**, 554 (2008). https://doi.org/10.1016/j.intermet.2008.01.003
62. SGTE Unary Database 4.4. http://www.crct.polymtl.ca/sgte/index.php
63. L. Amirkhanyan, T. Weissbach, T. Gruber, T. Zienert, O. Fabrichnaya, J. Kortus, J. Alloys Compd. **598**, 137 (2014). https://doi.org/10.1016/j.jallcom.2014.01.234
64. J. Lacaze, R. Valdes, Monatsh. Chem. **136**(11), 1899 (2005). https://doi.org/10.1007/s00706-005-0385-9
65. A. Salomon, T. Zienert, C. Voigt, E. Jäckel, O. Fabrichnaya, D. Rafaja, C.G. Aneziris, Adv. Eng. Mater. **15**(12), 1206 (2013). https://doi.org/10.1002/adem.201300114
66. A. Salomon, T. Zienert, C. Voigt, M. Dopita, O. Fabrichnaya, C.G. Aneziris, D. Rafaja, Corros. Sci. **114**, 79 (2017). https://doi.org/10.1016/j.corsci.2016.10.023
67. Y. Du, Y.A. Chang, S. Liu, B. Huang, F.Y. Xie, Y. Yang, S.-L. Chen, Z. Metallkde **96**(12), 1351 (2005). https://doi.org/10.3139/ijmr-2005-0235
68. B.G. Dietrich, H. Becker, M. Smolka, A. Kessler, A. Leineweber, G. Wolf, Edv. Eng. Mater. **19**(9), 1700286 (2017). https://doi.org/10.1002/adem.201700161
69. T. Zienert, S. Dudczig, O. Fabrichnaya, C.G. Aneziris, Ceram. Int. **41**, 2089 (2015). https://doi.org/10.1016/j.ceramint.2014.10.004
70. L. Dreval, T. Zienert, O. Fabrichnaya, J. Alloys Compd. **657**, 192 (2016). https://doi.org/10.1016/j.jallcom.2015.10.017
71. X. Wei, A. Yehorov, E. Storti, S. Dudczig, O. Fabrichnaya, C.G. Aneziris, O. Volkova, Adv. Eng. Mater. **24**, 2100718 (2022). https://doi.org/10.1002/adem.202100718

72. D. Pavlyuchkov, O. Fabrichnaya, M. Herrmann, H.J. Seifert, J. Phase Equilib. Diffus. 33, 357 (2012). https://doi.org/10.1007/s11669-012-0073-5
73. I. Saenko, O. Fabrichnaya, J. Phase Equilib. Diffus. 42, 254 (2021). https://doi.org/10.1007/s11669-021-00876-y
74. X. Wei, E. Storti, S. Dudczig, A. Yehorov, O. Fabrichnaya, C.G. Aneziris, O. Volkova, J. Eur. Ceram. Soc. 42, 4676 (2022). https://doi.org/10.1016/j.jeurceramsoc.2022.04.058
75. S. Dudczig, C.G. Aneziris, M. Emmel, G. Schmidt, J. Hubalkova, H. Berek, Ceram. Int. 40, 16727 (2014). https://doi.org/10.1016/j.ceramint.2014.08.038
76. E. Storti, S. Dudczig, M. Emmel, P. Colombo, C.G. Aneziris, Steel Research Int. 87, 1030 (2015). https://doi.org/10.1002/srin.201500446
77. B. Bock, A. Schmidt, E. Sniezek, S. Dudczig, G. Schmidt, J. Szczerba, C.G. Aneziris, Ceram. Int. 45, 4499 (2019). https://doi.org/10.1016/j.ceramint.2018.11.131
78. M. Ilatovskaia, O. Fabrichnaya, J. Alloys Compd. 884, 161153 (2021). https://doi.org/10.1016/j.jallcom.2021.161153
79. J. Fruhstofer, S. Dudczig, P. Gehre, G. Schmidt, N. Brachhold, L. Schöttler, C.G. Aneziris, Steel Research Int. 87(8), 1014 (2016). https://doi.org/10.1002/srin.201600023
80. N. Sarkar, J.G. Park, S. Mazumber, C.G. Aneziris, I.J. Kim, J. Eur. Ceram. Soc. 35, 3969 (2015). https://doi.org/10.1016/j.jeurceramsoc.2015.07.004
81. O. Fabrichnaya, S. Lakiza, C. Wang, M. Zinkevich, C. Levi, F. Aldinger, J. Phase Equilib. Diffus. 27, 343 (2006). https://doi.org/10.1007/s11669-006-0006-2
82. S. Lakiza, O. Fabrichnaya, M. Zinkevich, F. Aldinger, J. Alloys Compd. 420, 237 (2006). https://doi.org/10.1016/j.jallcom.2005.09.079
83. C. Aneziris, S. Dudczig, M. Emmel, Patent: Germany, a08–0523, keramische filter für die metallschmelzefiltration auf der grundlage gängiger metallschmelze-filtergeometrien und verfahren zu ihrer herstellung. (2011)
84. M. Ilatovskaia, G. Savinykh, O. Fabrichnaya, J. Eur. Ceram. Soc. 37, 3461 (2017). https://doi.org/10.1016/j.jeurceramsoc.2017.03.064
85. M. Ilatovskaia, A. Treichel, O. Fabrichnaya, J. Am. Ceram. Soc. 106(12), 7704 (2023). https://doi.org/10.1111/jace.19365
86. M. Ilatovskaia, H. Becker, O. Fabrichnaya, A. Leineweber, J. Alloys Compd. 936, 168361 (2023). https://doi.org/10.1016/j.jallcom.2022.168361

Chapter 5
Structural and Thermodynamic Properties of Filter Materials: A Raman and DFT Investigation

Jakob Kraus, Simon Brehm, Cameliu Himcinschi, and Jens Kortus

5.1 Introduction

Over the course of the three funding periods of the Collaborative Research Center (CRC) 920, subproject A04 has tackled a diverse set of topics centered around analyzing materials which are important in the context of this CRC, i.e., metal melt filtration in order to produce zero defect materials. This is unsurprising, given that the subproject was active in a time span of more than a decade.

The first two funding periods, which were shaped by subproject member Dr. Lilit Amirkhanyan, focused on intermetallics. Intermetallic phases are interesting systems for designing construction materials due to their low weight, their pronounced deformation resistance at elevated temperatures, their extraordinary hardness, and their general resilience against wear, oxidation, and corrosion [1, 2]. Through filtration processes such as those investigated in the CRC 920, the ratio of the components within the intermetallic phase can be regulated and impurities can be removed [1]. Within subproject A04, density functional theory (DFT [3, 4]) was chosen to non-empirically predict thermodynamic properties of intermetallic systems related to melts, such as heat capacities over a wide temperature range [1, 2, 5].

Moreover, research was also conducted on interface reactions and interface energies during this first funding period. Hercynite (Al_2FeO_4) has a spinel structure and is among the oxidic compounds that form at the interface between alumina-based filters

J. Kraus (✉) · C. Himcinschi · J. Kortus
Institute of Theoretical Physics, Technische Universität Bergakademie Freiberg, Leipziger Str. 23, D-09599 Freiberg, Germany

J. Kortus
e-mail: jens.kortus@physik.tu-freiberg.de

S. Brehm
Institute of Theoretical Physics, TU Bergakademie Freiberg, Leipziger Str. 23, D-09599 Freiberg, Germany

© The Author(s) 2024
C. G. Aneziris and H. Biermann (eds.), *Multifunctional Ceramic Filter Systems for Metal Melt Filtration*, Springer Series in Materials Science 337, https://doi.org/10.1007/978-3-031-40930-1_5

and iron-containing melts such as steel melts [6]. The deposited hercynite particles are believed to aid filtration efficiency by acting as anchor points for impurities in the melt [6]. Here, DFT was applied to study possible solid state reaction pathways leading to the formation of hercynite at the interface. Rutile (tetragonal TiO_2), on the other hand, is used as a coating for alumina filters which are employed in the filtration of aluminum melts and melts of aluminum alloys, particularly to deal with oxidic inclusions [7]. For these melts, rutile-coated filters can achieve both satisfactory melt flow rates and a decent filtration efficiency [7], favorable properties which are based on chemical reactions of rutile with the components of the melts. In this case, the aim of DFT was to evaluate the energetics of rutile-containing interfaces that are established because of the aforementioned reactions [7].

In 2017, the first major cooperation between DFT and Raman spectroscopy within the CRC 920 took place [8]. Raman spectroscopy is a non-destructive method to measure the vibrations of solids and molecules via the inelastic scattering of light on matter. Rudolph et al. investigated the thermally induced phase transition from boehmite (γ-AlO(OH)) to corundum (α-Al_2O_3) via the metastable transition phases γ-Al_2O_3, δ-Al_2O_3, and θ-Al_2O_3. This investigation was relevant to the goal of metal melt filtration because γ-Al_2O_3 is considered a promising coating material for corundum-based filters due to its high surface area and its function as a catalyst [8].

A year later, in 2018, there was a clear shift in focus within subproject A04. While carbon-bonded filters had been studied with Raman spectroscopy as part of CRC 920 before in the work of Röder et al. [9], the publication by Himcinschi and coworkers marked the starting point for the investigation of those carbon-bonded filters that use environmentally friendly materials as binders, i.e., lactose and tannin [10]. Compared to the usual pitch binders and phenolic resins, lactose/tannin binders do not lead to the emission of toxic phenol or carcinogenic polycyclic aromatic hydrocarbons like benzo[a]pyrene during either the production or the operation of the respective carbon-bonded filters [10, 11]. These emissions are not only a risk to humans and the environment in general, but provide a source of conflict with current pollution and hazard regulations in the European Union, which also has economic repercussions. However, the thermomechanical stability of the filters based on lactose/tannin binders was found to be not always on the level of state-of-the-art conventional binders, which is why further investigation was necessary [10, 11].

With Dr. Himcinschi joining the subproject as a project leader and Simon Brehm as well as Jakob Kraus replacing Dr. Amirkhanyan as project members, Raman spectroscopy was now an integral tool for addressing the remaining workload. Ex-situ Raman measurements were employed to study both the binder materials lactose, tannin, and Carbores®P themselves as well as carbon-bonded filters using various mixtures of these substances as binders [12]. For all of these systems, the prominent D and G Raman peaks were used to confirm graphitization during heat treatment and also to estimate the size of the resulting graphitic carbon clusters, with DFT playing a supporting role for the identification of minor peaks [12]. Afterwards, lactose and common tannins or tannin building blocks were also investigated in-situ in order to clarify their pyrolysis products and pyrolysis temperature [13]. Again, DFT-calculated Raman spectra were used for substance identification, in this case for the aforementioned products. On the purely theoretical side of things, the combination of self-interaction correction models with solvation was studied in order to possibly gain insights into chemical reactions taking place before the curing step [14]. Furthermore, a reaction mechanism including transition states was proposed for the pyrolysis of gallic acid, an important component of gallotannins [15].

In addition to this work focused on carbon-bonded filters, subproject A04 contributed to a publication investigating the synthesis of magnesium aluminum oxynitride (MgAlON), which has been suggested as a coating material for ceramic filters applied to magnesium, aluminum and steel melts due its oxidation resistance and wetting behavior [16]. However, this publication is not discussed in detail within this chapter.

In the following, some of the computational details for DFT and the experimental details for Raman spectroscopy as applied in this subproject are briefly mentioned, followed by our major findings and a conclusion.

5.2 Methods

5.2.1 Density Functional Theory

For the density functional theory calculations, a variety of open-source codes were employed to calculate the electronic and vibrational properties of investigated systems. In order to treat the solid state species, two codes were utilized, namely the Wien2K code [17], an implementation of the linear augmented plane wave (LAPW [18, 19]) method, and the plane-wave Quantum ESPRESSO code [20–22], which was combined with projector augmented wave (PAW [23]) ab-initio pseudopoten-tials. In both cases, the calculations were performed with the help of generalized gradient approximation exchange–correlation functionals, primarily the functional by Perdew, Burke, and Ernzerhof (PBE [24]). For molecular systems, the PySCF [25, 26], PyFLOSIC [27], ERKALE [28–31], and ORCA [32, 33] codes were used in combination with a multitude of functionals and Gaussian-type orbital basis sets.

5.2.2 Raman Spectroscopy

For Raman measurements, two spectrometers of the company Horiba Jobin Yvon were used, one of which was a HR 800 spectrometer with a frequency-doubled Nd:YAG laser (532 nm), a He–Ne laser (633 nm), and a diode laser (785 nm) as excitation sources. The second spectrometer was a HR 800 UV spectrometer with a He-Cd laser (325 nm and 442 nm). Both spectrometers were equipped with a Peltier-cooled CCD detector.

All measurements were performed in backscattering geometry, i.e., the incident and scattered laser beams were focused and collected by the same objective. For this, × 40 and × 50 objectives were used.

The temperature dependent in-situ measurements were carried out utilizing a Linkam TS 1200 heating chamber. The heating chamber was filled with argon to ensure an oxygen-free atmosphere.

5.3 Results and Discussion

5.3.1 Thermodynamic Properties of Intermetallics

This contribution focused on binary Al–Fe and ternary Al–Fe-Si intermetallic phases. Binary phases under investigation included η-AlFe [5], ε-AlFe [2], and η'-AlFe [34], whereas τ-AlFeSi was a ternary phase of interest [1].

DFT in combination with the quasi-harmonic approximation [35] was applied to binary and ternary intermetallic phases to study their thermodynamic properties, with a focus on calculating accurate heat capacities. Here, Quantum ESPRESSO was used in combination with PAW pseudopotentials and PBE.

Binary Phases: η-AlFe, ε-AlFe, and H'-AlFe

The isobaric heat capacities of η-AlFe (Al_5Fe_2) and ε-AlFe (Al_8Fe_5) were evaluated in [5] and [2], respectively. Moreover, the crystal structure of η'-AlFe (Al_8Fe_3) was studied in [34]. Up to temperatures of 460 K, the DFT-calculated heat capacity of η-AlFe was favorably compared to experimental data taken from Chi et al. [36] and measured via differential scanning calorimetry (DSC) [5]. Here, it is important to stress that DFT-calculated heat capacities are not dependent on any free parameters to be chosen by the user or to be fitted to experiment. In this sense, they are properties evaluated from first principles. As an example for the match, a comparison between the data by Chi et al. and DFT is presented in Fig. 5.1 for temperatures up to 270 K. For higher temperatures, the DFT heat capacity was found to be lower than experiment. This was explained by the DFT structure corresponding to the ordered η-AlFe phase, whereas the disordered phase dominates for higher temperatures and was thus measured in experiment [5]. Nevertheless, the DFT results were qualitatively superior to those yielded by the Neumann–Kopp rule, which predicted a nonphysical local maximum of the heat capacity just below 1000 K.

For ε-AlFe, [2] showed agreement between the DFT heat capacity and the heat capacity as predicted by the code of T. Zienert, in contrast to the Neumann–Kopp rule.

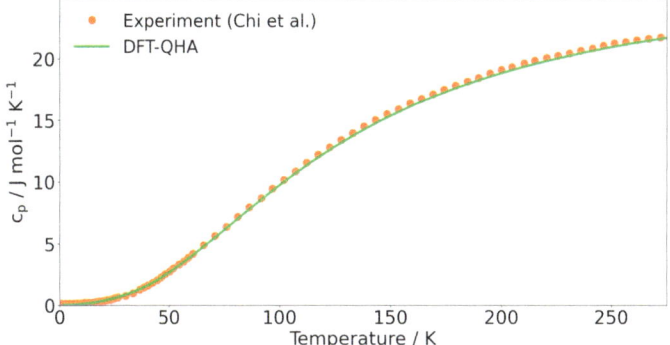

Fig. 5.1 DFT-calculated isobaric heat capacity of η-AlFe (Al_5Fe_2) [5], compared to experimental data by Chi et al. [36]

In this case, only a very small temperature range was accessible to DSC measurements, and these measurements indicated a phase transition above 1250 K. In [34], the η'-AlFe equilibrium crystal structure as predicted by DFT came very close to the data gathered from X-ray diffraction and subsequent Rietveld refinement.

Ternary Phase: τ-AlFeSi

The isobaric heat capacity of τ-AlFeSi (Al_3FeSi_2) was investigated in [1], and the result is shown in Fig. 5.2. According to Fig. 5.2, DFT is accurate with respect to DSC experimental values previously presented in [1] in the measured range of 900–1050 K, as opposed to the Neumann–Kopp rule. Instead, the Neumann–Kopp rule predicts a local maximum of the heat capacity, as was the case for the binary intermetallic phases, and it only somewhat approximates the DFT values in the range of 300–600 K.

5.3.2   Interface Reactions and Interface Energies

DFT was applied to study the formation of hercynite at the α-Al_2O_3‖Fe interface, using LAPW as implemented in the Wien2K code and the PBE exchange–correlation functional. Furthermore, the interface energies for several TiO_2‖α-Al_2O_3 and TiO_2‖$MgTiO_3$ interfaces were evaluated using PAW pseudopotentials in combination with the Quantum ESPRESSO code and, again, the PBE functional.

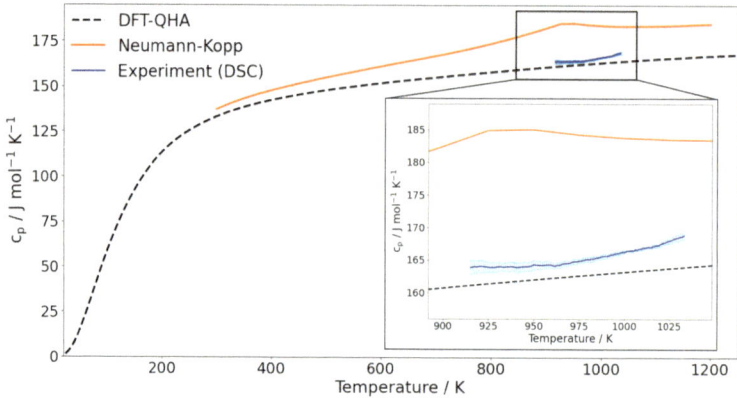

Fig. 5.2 DFT-calculated isobaric heat capacity of τ-AlFeSi (Al_3FeSi_2), compared to experimental data from DSC measurements [1] and the Neumann–Kopp rule. The inset increases the visibility of the mean values and error bars of the DSC measurements

Interface Reactions: The Formation of Hercynite

In [6], several model compounds that play a role in the interaction of an alumina-based filter with a steel melt and the resulting formation of hercynite (Al_2FeO_4) were examined: α-Al_2O_3, FeO, Al_2FeO_4, $AlFe_2O_4$, $AlFeO_3$, AlFe, and Fe (bcc). From the total energies of these systems, reaction energies were calculated, which suggested that the hercynite formation is not based on a direct reaction between α-Al_2O_3 and Fe, but between α-Al_2O_3 and FeO, instead, as shown in Table 5.1, which is inspired by [6].

Interface Energies: Rutile Coatings and Aluminum-Containing Melts

If rutile-coated alumina filters come into contact with aluminum melts, a $TiO_2\|\alpha$-Al_2O_3 interface is established at the solid/liquid phase boundary [7]. Substituting the

Table 5.1 Changes in total energy ΔE for several chemical reactions suspected of occurring at the interface of alumina filter and steel melt [6]. ΔE is given in eV

Chemical reaction	ΔE
$Al_2O_3 + 2\,Fe \rightarrow AlFeO_3 + AlFe$	3.6
$Al_2O_3 + 5/4\,Fe \rightarrow 3/4\,Al_2FeO_4 + 1/2\,AlFe$	2.1
$Al_2O_3 + 11/4\,Fe \rightarrow 3/4\,AlFe_2O_4 + 5/4\,AlFe$	4.3
$Al_2O_3 + 3\,FeO \rightarrow 2\,AlFeO_3 + Fe$	-1.8
$Al_2O_3 + FeO \rightarrow Al_2FeO_4$	-0.2
$Al_2O_3 + FeO + 2\,Fe \rightarrow AlFe_2O_4 + AlFe$	2.8
$Al_2O_3 + 5\,FeO \rightarrow 2\,AlFe_2O_4 + Fe$	-3.2
$Al_2O_3 + 5\,Fe \rightarrow 3\,FeO + 2\,AlFe$	8.9

aluminum melt for the aluminum alloy $AlSi_7Mg_{0.6}$, however, leads to the generation of a $TiO_2\|MgTiO_3$ interface [7]. For certain orientations, both the $TiO_2\|\alpha\text{-}Al_2O_3$ and the $TiO_2\|MgTiO_3$ interfaces were calculated to be energetically favorable, i.e., negative, as presented in [7]. Specifically, the most negative interface energies were found for those interfaces which were also observed in the experiments described in [7]. The negative interface energies were assumed to contribute to the satisfactory adhesion of $\alpha\text{-}Al_2O_3$ and $MgTiO_3$ to rutile coatings [7].

5.3.3 Raman Spectroscopic Characterization of Carbon-Bonded Filters

In the publications reviewed in this part of the chapter, carbon-bonded alumina filters as well as their binders were investigated with Raman spectroscopy.

First, the results of measurements on the pure binders Carbores®P, lactose, and tannin are presented [12]. Afterwards, the results of measurements on the filters are shown [9, 10]. In addition, Raman spectroscopic investigations were carried out on other carbon-bonded materials, and changes in the microstructure or the formation of graphitic structures during mechanical tests at 1500 °C could be demonstrated [37, 38]. However, these results are not discussed in detail here.

Binder Materials

Carbores®P is a product made from coal tar pitch developed and manufactured by the company RÜTGERS. It was developed as an alternative to the classic binders for refractory materials such as phenolic resins and simple pitches, which contain a high amount of phenol or carcinogenic polycyclic aromatic hydrocarbons. With the development of Carbores®P, the amount of harmful substances could be reduced drastically, but it is still above the lawful limit in the European Union at the time of writing [10].

Lactose (sum formula: $C_{12}H_{22}O_{11}$), informally called milk sugar, is one of the most well-known carbohydrates [39]. Lactose consists of the two monosaccharides galactose and glucose, which are connected by a glycosidic bond. In combination with tannins, lactose showed great promise as a binder ingredient for carbon-bonded magnesia refractories [40].

Tannins is the collective term for a group of polyphenolic biomolecules which can be found in the wood, bark, and fruit of certain plants. Tannins are often distinguished into hydrolyzable and condensed tannins [41]. Our focus in this subproject is on hydrolyzable tannins, which themselves can be divided into gallotannins, like tannic acid, and ellagitannins. While gallotannins contain gallic acid as their central building block, ellagitannins have an ellagic acid molecule as a common feature [41]. *Gallic acid* (GA, sum formula: $C_7H_6O_5$) can be described as a benzene ring with three neighboring hydroxyl groups opposite a carboxyl group [42]. GA pyrolysis leads to the formation of pyrogallol and carbon dioxide and occurs at 175–200 °C [43]. *Tannic acid* (TA, nominal sum formula: $C_{76}H_{52}O_{46}$) is a polyphenol, composed of a glucose ring that is linked to five *m*-digallic acid units via ester bonds. Digallic acids and especially GA are major products of TA pyrolysis [44]. TA decomposes starting at temperatures around 190 °C [45]. Notably, the sum formula for TA is mostly nominal, as TA vendors usually offer substances that are actually mixtures of various polygalloyl-glucose compounds [44]. *Ellagic acid* (EA, sum formula: $C_{14}H_6O_8$) is another polyphenol, consisting of two GA molecules esterified with each other and connected by an additional C–C bond. Among the pyrolysis products of EA is 2,2',3,3',4,4'-biphenylhexol, a compound that is equivalent to two pyrogallol molecules linked by the aforementioned C–C bond [46–48]. EA was reported to melt and decomposes in the range of 350–360 °C [49, 50].

Carbon-Bonded Filters

The investigated carbon-bonded alumina filters were produced from dry slip with a total solid content of 78 wt.%. The composition of the slip is presented in Table 5.2. The main part of the raw materials is alumina (Martoxid® MR70) at 66 wt.%, while the binders, Carbores®P, lactose, and tannin, make up 20 wt.% in total.

For different samples, different ratios of the binders were applied. This ratio ranged from Carbores®P only to lactose/tannin only, with the lactose-tannin ratio always being 5 to 1. Besides functioning as binders, these substances also served as carbon sources. Carbon black and graphite were further carbon sources. The additives TiO_2 and Al work as antioxidants [51], SiO_2 and Contraspum® K 1012 work as antifoam agents, Castament® VP 95 L works as a dispersing agent, and ammonium ligninsulfonate functions as a temporary binder and wetting agent. In order to obtain a higher carbon yield, *n*-Si was added to some of the samples [52].

Table 5.2 Dry slip composition for carbon-bonded alumina filters	Raw materials	Mass in %
	Martoxid® MR70	66.0
	Binder	20.0
	Carbon black N 991	6.3
	Graphite AF 96/97	7.7
	Additives	Relative to raw materials (wt.%)
	TiO_2	0.5
	Al	0.1
	SiO_2	4.0
	Castament® VP 95 L	0.3
	Contraspum® K 1012	0.1
	Ammonium ligninsulfonate	1.5
	n-Si	Relative to lactose/tannin (wt.%)
		5.0

Investigation of the Binder Materials

The investigated samples of Carbores®P, lactose, and tannin were fired under reducing atmosphere generated by using a coke bed [12]. The normalized Raman spectra of Carbores®P for annealing temperatures from room temperature (RT) to 1400 °C are shown in Fig. 5.3a. The so-called D peaks at ca. 1350 cm^{-1} and the G peaks at 1600 cm^{-1} are clearly visible on the left side of the figure. The G peak originates from vibrations of sp^2-hybridized carbon atoms, as found in graphite, while the D peak is exclusively seen in disordered carbon systems.

In the spectra from RT to 600 °C, the D peak exhibits shoulders at 1160, 1240, and 1440 cm^{-1} originating from trans-polyacetylene, a chain-like hydrocarbon with alternating single and double bonds [12]. In Fig. 5.3a, these shoulders are marked with arrows. The disappearance of the shoulders at higher temperatures indicates the pyrolysis of trans-polyacetylene.

Moreover, the spectra indicate that the OH groups split off at ca. 400 °C, since the associated band with a maximum at 3350 cm^{-1} disappears above this temperature (see the right part of Fig. 5.3a). From the background slope of the spectra, the hydrogen fraction of the system could be estimated [53]. According to this estimation, the hydrogen fraction decreased from more than 40% at RT to less than 20% at 800 °C. Figure 5.3b shows the D peak becoming narrower and more intense with increasing annealing temperature, indicating an increase in order for the carbon system. Thus, the investigated Carbores®P samples changed chemically and structurally with increasing annealing temperature.

Further information about these changes is provided by the intensity ratios of the D and G peaks I_D/I_G (see the upper part of Fig. 5.4) as well as the position of the

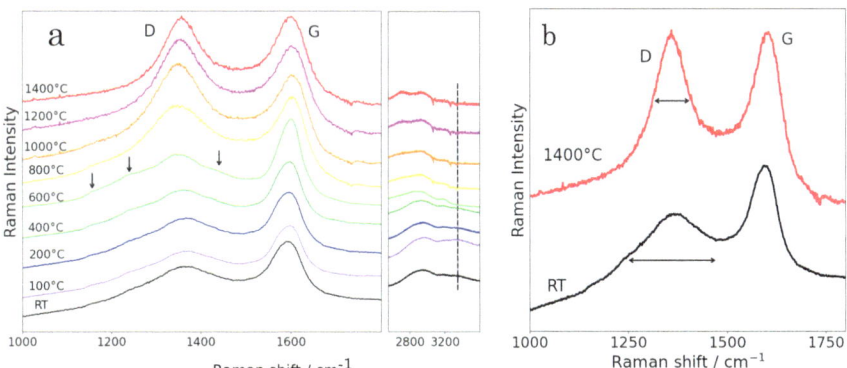

Fig. 5.3 **a** Raman spectra of Carbores®P [12]. The Carbores®P samples were heated up to 1400 °C under reducing atmosphere, with RT representing room temperature. The right part of the subfigure shows the spectral region where -OH vibrations occur, **b** Direct comparison of the spectra of Carbores®P annealed to RT and 1400 °C. The D peaks shows a clear narrowing and an increase with temperature compared to the G peak. The arrows indicating the FWHM are only guides for the eyes and not the actual fit parameters

G peak (given in [12]) as a function of annealing temperature. To determine these parameters, the D and G peaks were fitted with a Lorentz and a Breit–Wigner–Fano function, respectively. An example for the fit process is shown in the inset of Fig. 5.4 for a Carbores®P sample fired at 600 °C. The spectra of lactose and tannin were treated in a similar manner.

For temperatures up to 600 °C, there is hardly any variation in the intensity ratios and peak positions, indicating that the binder systems undergo minimal chemical and structural changes. In the range of 600–1000 °C, the intensity ratio I_D/I_G increases, and the position of the G peak is shifted to higher wavenumbers, indicating a transformation from hydrogen-rich amorphous carbon to nanocrystalline graphite [12].

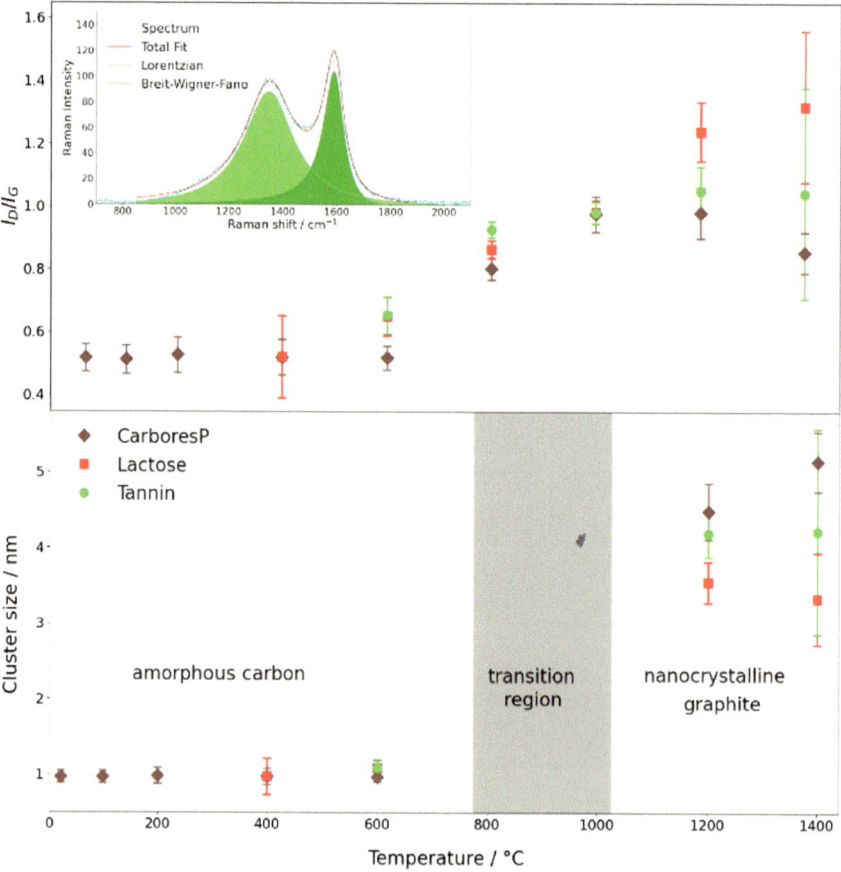

Fig. 5.4 The intensity ratios of the D and G peaks as a function of annealing temperature. An example fit of the D and G peaks is shown in the top left picture for a Carbores®P sample fired at 600 °C. In the lower part, the calculated cluster sizes for Carbores®P, tannin, and lactose are shown. Between 800 °C and 1000 °C, a conversion of amorphous, hydrogen-rich carbon to nanocrystalline graphite occurs [12]

This is associated with a decrease in the content of sp^{3-}hybridized carbon, hydrogen, and chain-like carbon compounds such as trans-polyacetylene.

During this process, the size of graphitic sp^2 carbon ring clusters increased [54]. With the Tuinstra model, these cluster sizes could be estimated from the I_D/I_G ratios [55, 56]. The lower part of Fig. 5.4 shows the results; a significant increase in cluster size with annealing temperature can be seen [12]. These values are in excellent agreement with the XRD results determined by subproject A05 (compare Chap. 6, Fig. 6.4a).

Investigation of Filters with Carbores®P Binder

In the work of Röder et al., carbon-bonded alumina filter with Carbores®P binder fired at temperatures in the range of 800–1600 °C were studied with Raman spectroscopy [9]. The samples exhibited a granularly structured surface. It was possible to distinguish between flakes, which were identified as graphite, and a surrounding matrix part in the sample. The presence of graphite in the sample can be explained by graphite having been added as a raw material (compare Table 5.2). The Raman spectra measured at the matrix part were assigned to nanocrystalline carbon. Additionally, it was shown that the sp^2 carbon clusters in the matrix part grew from around 2.3 nm at 800 °C to roughly 4.4 nm at 1600 °C [9].

Investigation of Filters with Lactose/Tannin Binder and the Influence of Added *n*-Si

Filters with environmentally friendly binders based on lactose/tannin were investigated using Raman spectroscopy for the first time in [10]. Figure 5.5 (left side) shows Raman spectroscopic measurements at RT for Al_2O_3/lactose/tannin samples that were tempered at 200, 300, 400, and 500 °C. Apparently, the photoluminescence decreases with increasing annealing temperature and the aforementioned I_D/I_G ratio, which can also be described as the ratio of the sp^2/sp^3 character of the carbon compounds, increases. As explained earlier, this increase in the I_D/I_G ratio is likely associated with a growth of the sp^2 carbon clusters. A possible mechanism could be the release of some of the OH groups from tannin due to thermal energy, so that the smaller molecular fragments cross-link and form larger aromatic sp^2 carbon clusters. This behavior would agree with the temperature development of the photoluminescence, which indicates a loss of hydrogen in the system [53]. In support of this, Raman spectra presented in [10] show a clear decrease in intensity for the sample that was tempered at 500 °C in the range of -OH vibrations at ca. 3400 cm^{-1}. All the findings detailed here are in excellent agreement with previously reported research performed on the pure binders [12].

In the right part of Fig. 5.5, the Raman spectra of the samples after addition of *n*-doped Si are shown. As can be seen, the photoluminescence intensity is lower for the samples mixed with *n*-Si, except for the sample annealed at 500 °C. Moreover, an

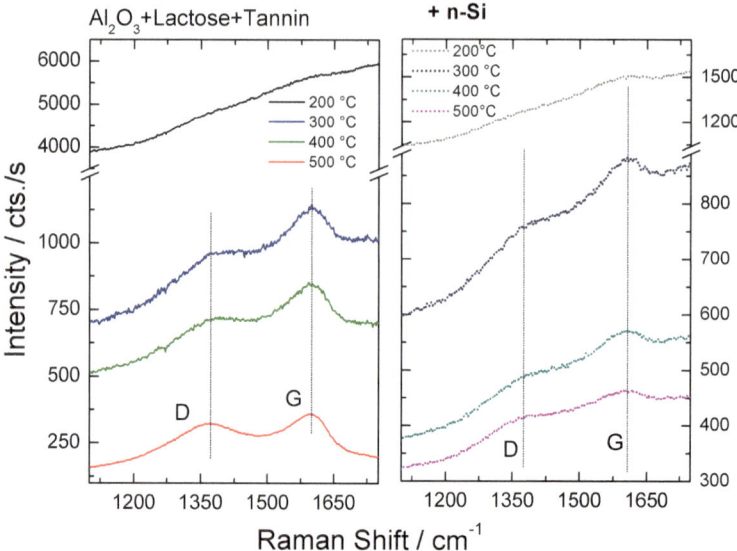

Fig. 5.5 Raman spectra of Al₂O₃/lactose/tannin samples as a function of annealing temperature [10]. Raman spectra of the same samples with added *n*-Si (right)

increase of the I_D/I_G ratio with annealing temperature is observed in these spectra. Such an increase was found by Ferrari et al. to be 'proportional to the number and clustering of rings', indicating a higher degree of order in the amorphous carbon system [56].

For a better visualization of the influence of *n*-Si addition, the Raman spectra, normalized for the same intensity of the D and G peaks, of the coked samples (800 °C) with and without *n*-Si are shown in Fig. 5.6. In addition to the overall lower intensity, the addition of *n*-Si clearly modified the carbon bonding. One can see (marked by the red arrows) that *n*-Si caused a reduction of the intensity of the bands at ca. 1150 cm⁻¹ and a slight increase of the band at ca. 1460 cm⁻¹. These modes were assigned, respectively, to C–C vibrations and C = C vibrations in sp² carbon chains in trans-polyacetylene [57, 58].

In order to achieve a better understanding of the temperature dependence of the chemical processes that take place in the filters with environmentally friendly binders, real-time in-situ Raman measurements were performed as well. The results of these in-situ Raman measurements are discussed in the following.

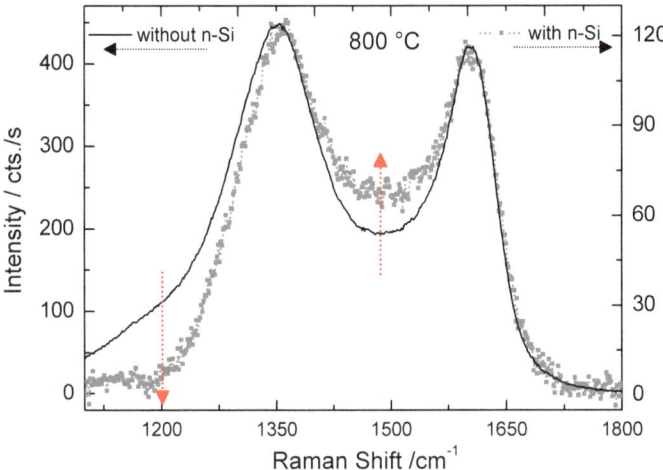

Fig. 5.6 Raman spectrum of a Al_2O_3/lactose/tannin sample coked at 800 °C (solid line) and the Raman spectrum for the sample with *n*-Si (dotted line + symbols) [10]. The added *n*-Si leads to a lower intensity of the band at ca. 1150 cm^{-1} and a higher one for the 1460 cm^{-1} band, here indicated by the red arrows

5.3.4   Raman Spectroscopic In-Situ Study of the Pyrolysis of Tannins

Temperature dependent in-situ Raman measurements of selected representatives of tannins, i.e., GA, TA, and EA were performed to understand their pyrolysis [13]. In Fig. 5.7, spectra of GA and TA as well as their chemical structures are displayed. The investigated samples were placed in a crucible in a heating chamber under argon atmosphere to ensure the absence of oxygen, which avoids a possible oxidation of the sample and thereby enables pyrolysis. The used heating chamber is shown in the inset of Fig. 5.8.

Besides, the Raman spectra of GA from RT up to 700 °C are shown in Fig. 5.8. Except for a broadening of the peaks and a slight shift to lower wavenumbers due to temperature effects, no changes can be observed in the spectra up to 200 °C.

In the temperature range of 250–350 °C, a white substance condensed at the window of the heating chamber. Since the laser light could not penetrate the condensate and reach the actual sample, measurements at these temperatures were not possible. Instead, the focus of the laser beam was set to the condensate at the heating chamber window. The spectrum recorded for the condensate could be assigned to pyrogallol, whose spectrum and structure are shown in Fig. 5.7.

As previously mentioned in this report, pyrogallol is well known as a pyrolysis product of GA [42, 43]. It is formed by the elimination of CO_2 from the GA molecule. Pyrogallol could also be identified as a pyrolysis product of TA. For

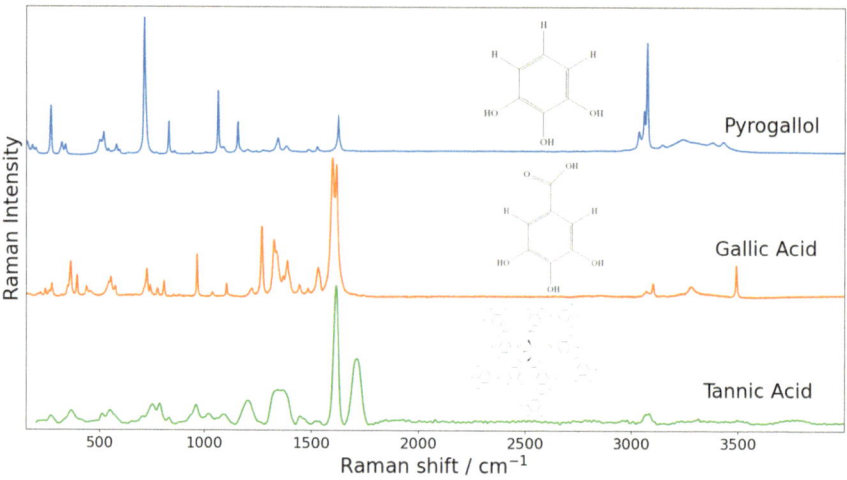

Fig. 5.7 Raman spectra of pyrogallol, GA, and TA measured at RT as well as their Lewis structural formulas [13]

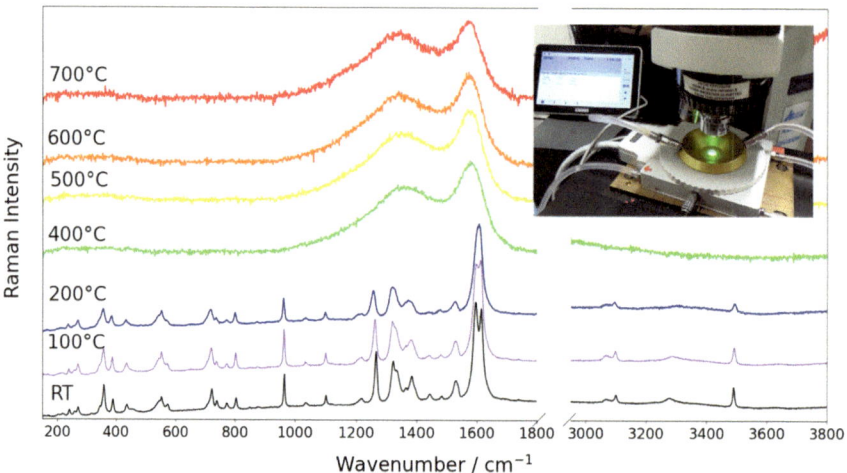

Fig. 5.8 In-situ Raman spectra of GA at different temperatures [13]. The used heating chamber is shown as an inset in the upper right corner

EA, 2,2',3,3',4,4'-biphenylhexol has been reported as a pyrolysis product in literature [46–48], which is also the assumption made in this in-situ study. The pyrolysis product identification was supported by DFT calculations. Moreover, pyrolysis temperature ranges based on Raman measurements could be estimated for GA (225–350 °C), TA (200–325 °C), and EA (150–250 °C).

When the tannins were heated further, they exhibited the Raman spectra of amorphous carbon systems. For GA, these spectra are also shown in Fig. 5.8 for temperatures of 400 °C and above. The spectra contain the characteristic D and G peaks at around 1350 and 1600 cm^{-1}, respectively.

5.3.5   Fermi-Löwdin Orbital Self-Interaction Correction and Solvation

The Fermi-Löwdin orbital self-interaction correction (FLO-SIC [59–62]), which aims to correct prominent failings of DFT, was implemented in the form of PyFLOSIC [27], an extension to the PySCF code. The development of PyFLOSIC made FLO-SIC calculations with a wide variety of basis sets, functionals, and numerical integration grids possible, thus simplifying comparisons to other quantum chemical codes. One such comparison before the development of PyFLOSIC was presented in [63]. The PyFLOSIC code is open-source and freely available on GitHub (https://github.com/pyflosic/pyflosic).

This code, in addition to PySCF and ERKALE, was applied to the AQUA20 test set in order to calculate ionization potentials (IPs) and room-temperature standard enthalpies of formation ($\Delta_f H°(298.15$ K)) for the gas phase and an aqueous solution as simulated with the ddCOSMO [64–70] continuum solvation model [14]. The AQUA20 test set includes carboxylic acids and carboxylate anions, systems that share functional groups with the gallotannins making up part of the environmentally friendly binders for carbon-bonded filters. Moreover, the aqueous solution case was investigated in addition to the gas phase because the binders take the form of a slurry before the curing step during filter production, and the tannins themselves are naturally formed in aqueous polymerization reactions [14].

In [14], the FLO-SIC results were compared to quantum chemical methods, uncorrected DFT using several different functionals, and real-valued self-interaction correction as implemented in the ERKALE code, the latter of which is called RSIC. The FLO-SIC and RSIC values were found to be in good agreement with each other across the board, and gas phase trends previously reported in literature were reproduced. For the aqueous solution, the mean errors (MEs) and mean absolute errors (MAEs) when compared to experiment are presented in Fig. 5.9 for the IPs [14].

According to Fig. 5.9 and the data presented in [14], the DFT approaches COSMO-SCAN and COSMO-B3LYP achieve the closest overall agreement to experiment for the AQUA20 test set, along with COSMO-CCSD(T). For the aqueous standard enthalpy of formation, it was shown that most of the error is caused by the charged species in the test set [14]. Moreover, COSMO-FLO-SIC and COSMO-RSIC improve on the results of the underlying functional, i.e., LDA, for $\Delta_{f,aq} H°(298.15$ K), however, the deviations from experiment actually increase when applying self-interaction correction to aqueous IPs (see Fig. 5.9).

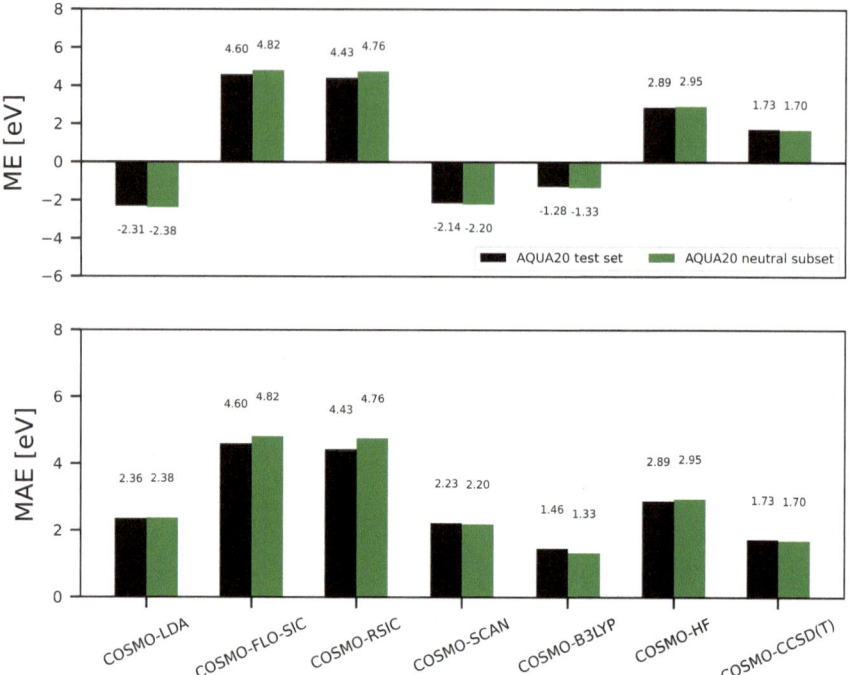

Fig. 5.9 The MEs and MAEs for IP_{aq} in eV, calculated using 7 different quantum mechanical methods combined with COSMO solvation, and applied to the AQUA20 test set [14]

5.3.6 Theoretical Investigation into Gallic Acid Pyrolysis

Thermodynamic and kinetic data on the first two steps of GA pyrolysis, i.e., a decarboxylation followed by a dehydrogenation, were obtained with the help of DFT and quantum chemistry [15], using the ORCA code. The investigated reactions can be written as $GA \rightarrow PG + CO_2$ and $PG \rightarrow OQ + H_2$, with PG representing pyrogallol and OQ representing 3-hydroxy-o-benzoquinone.

For the kinetics, transition states were identified with the help of the climbing image nudged elastic band method (CI-NEB) with subsequent transition state optimization as implemented in ORCA [71–74], employing the PBE functional. Both reactions were found to exhibit two transition states. One of these transition states is related to the rotation of OH groups, and the other one is related to the breaking and forming of bonds. The results of the CI-NEB calculation for the first GA pyrolysis reaction [15] are shown in Fig. 5.10, with Fig. 5.10a displaying the converged minimum energy path and Fig. 5.10b displaying selected images, including the optimized highest-energy transition state.

After applying several DFT functionals and wavefunction methods in order to calculate standard enthalpies of reaction and standard Gibbs energies of reaction, the combination of the first and second pyrolysis reactions was judged to be

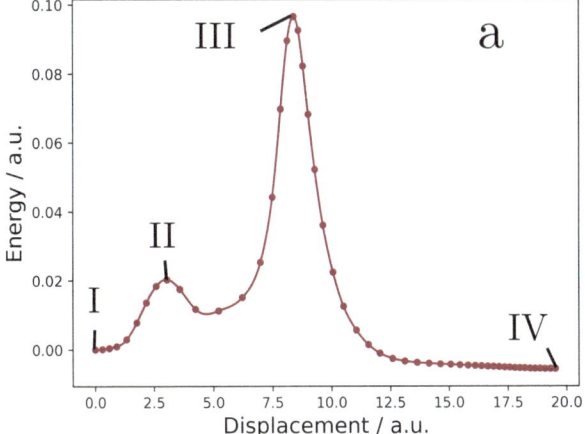

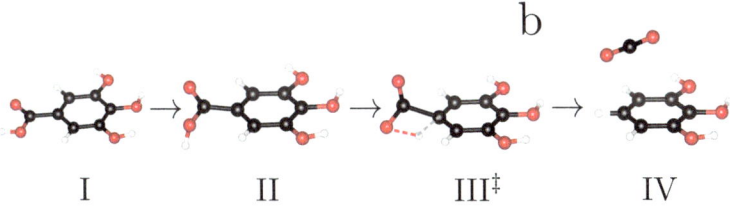

Fig. 5.10 **a** Converged CI-NEB minimum energy path of the first reaction GA → PG + CO$_2$. The x axis shows the cumulative displacement between subsequent images, which is identified as the reaction coordinate. The dots represent the PBE energies of the images relative to the first image, and the line is an interpolating spline. Code for selected images: I–GA, II–rotational transition state, III–converged CI, IV–PG/CO$_2$. This subfigure was created with the help of a Python script by Ásgeirsson [75], **b** Selected images, with III‡ representing the converged transition state for which III acted as a starting point. Color code: white–H atoms, black–C atoms, red–O atoms. The solid lines represent fully formed bonds, the dotted lines represent bonds in the process of being broken or formed. This figure was previously published in [15] and is reused with permission from the publisher

endothermal, and it is predicted to change from endergonic to exergonic between 500 and 750 °C. The second reaction, the dehydrogenation of PG, was identified as the rate-determining step of GA pyrolysis, with reaction rate constants below 1 s^{-1} for temperatures below 1250 K [15].

5.3.7 Thermally Induced Formation of Boehmite

Raman spectroscopy was used to understand the structural changes that occur during the thermally induced transformation of boehmite (γ-AlO(OH)) via the metastable transition phases γ-Al$_2$O$_3$, δ-Al$_2$O$_3$, and θ-Al$_2$O$_3$ to corundum (α-Al$_2$O$_3$)

[8]. Figures 5.11 and 5.12 show characteristic Raman spectra of samples at RT that were annealed in air for 20 h at different temperatures. Figure 5.11 shows the Al-O vibrations range, while Fig. 5.12 corresponds to the –OH vibrations measured at higher Raman shifts.

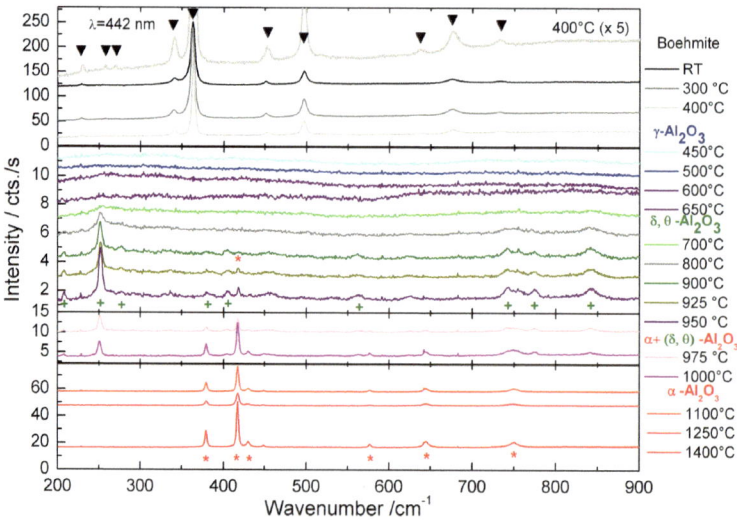

Fig. 5.11 Raman spectra of boehmite (γ-AlO(OH)) in the Al-O vibration range as a function of annealing temperature. They show the transition from boehmite to corundum (α-Al$_2$O$_3$) with the formation of metastable transition phases γ-Al$_2$O$_3$, δ-Al$_2$O$_3$, and θ-Al$_2$O$_3$

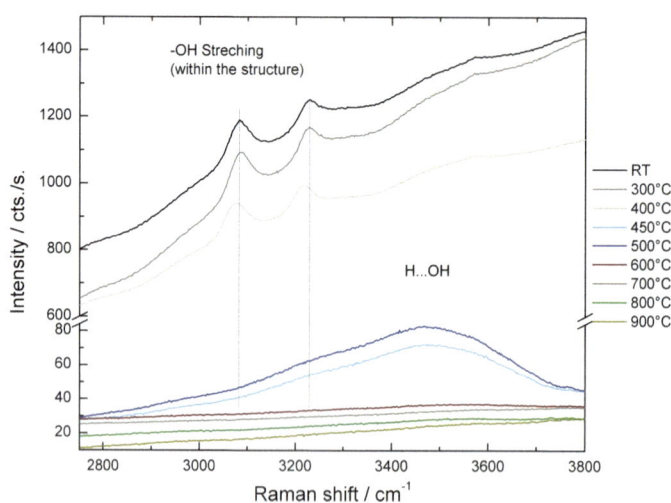

Fig. 5.12 Raman spectra of spectra of boehmite (γ-AlO(OH)) in the -OH vibration range as a function of annealing temperature

Temperature Range RT – 400 °C

Up to a temperature of 400 °C, the Raman modes marked by triangles in Fig. 5.11 were detected for boehmite at 229 cm^{-1}, 257 cm^{-1} (weak), 270 cm^{-1} (weak), 341 cm^{-1}, 363 cm^{-1}, 451 cm^{-1}, 500 cm^{-1}, 638 cm^{-1} (weak), 676 cm^{-1}, and 732 cm^{-1}, indicating that the boehmite structure is stable up to this temperature. The positions of the peaks agree with Doss et al. [76]. In the higher wavenumbers region (Fig. 5.12), the modes at 3083 cm^{-1} and 3227 cm^{-1} can be assigned to the -OH stretching vibration within the boehmite structure [77].

Temperature Range 450–650 °C

The Raman spectra measured for γ-Al2O3 show no peaks in the low wavenumber range (Fig. 5.11). The absence of Raman activity was explained in literature [78] by the distribution of the cation vacancies preferentially to the tetrahedral sites and of the Al^{3+} ions preferentially to the octahedrally coordinated sites in the regular spinel structure. In the higher wavenumbers region (Fig. 5.12), the samples annealed at 450 and 500 °C exhibit a large broad band at ca 3500 cm^{-1}, which corresponds to hydrogen bonded -OH stretching vibrations, and this band is assigned to adsorbed water. A possible reason could be water resorption, as the samples were examined ex-situ. A weak band at ca. 3760 cm^{-1} can be attributed to a terminal -OH stretching vibration in Al–OH (isolated surface hydroxyl groups) [79]. Above 600 °C, the peaks corresponding to water vibrations (H-OH or isolated -OH) are drastically reduced in intensity for all spectra (Fig. 5.12).

Temperature Range 700–1400 °C

At 700 °C, modes of a new phase (peaks marked by + in Fig. 5.11) become visible in the Raman spectra. The positions of the peaks are in agreement with the resonance Raman data for δ-Al$_2$O$_3$ [80] and θ-Al$_2$O$_3$ [78, 81] as measured with UV excitation. In literature, there is no clear distinction between the Raman bands from δ-Al$_2$O$_3$ and θ-Al$_2$O$_3$ phases [82]. At 1000 °C, the α-Al$_2$O$_3$ starts to form. Above this temperature, the seven expected Raman modes (marked by * in Fig. 5.11) of corundum [83] are detected at: 378 cm^{-1} (E$_g$), 418 cm^{-1} (A$_{1g}$), 431 cm^{-1} (E$_g$), 449 cm^{-1} (E$_g$), 577 cm^{-1} (E$_g$), 644 cm^{-1} (A$_{1g}$), 751 cm^{-1} (E$_g$).

In addition to the results recovered by Raman spectroscopy, DFT calculations using PAW pseudopotentials, Quantum ESPRESSO and the PBE functional were carried out as well [8]. The calculated enthalpies suggested that a negative pressure would lead to the γ-Al$_2$O$_3$ phase being more stable than both θ-Al$_2$O$_3$ and α-Al$_2$O$_3$, effectively decelerating the phase transition from boehmite to corundum.

5.4   Conclusion

This summary gave an insight how density functional theory calculations and Raman spectroscopic measurements were used to investigate filter materials and thereby can help to improve them. It could be shown that DFT and Raman spectroscopy are valuable additions to traditional methods in materials science.

Over more than a decade, DFT was used to calculate and study a wide range of physical properties for different systems. For example, the heat capacity of the systems η-AlFe, ε-AlFe, and τ-AlFeSi were calculated. The results showed a better agreement with experimental data compared to heat capacities calculated by the Neumann–Kopp rule.

Furthermore, the reaction energies of the formation of hercynite (Al_2FeO_4) at the interface during the filtration of a steel melt with an Al_2O_3 based filter were calculated by DFT. The results showed that hercynite is probably not a product of the reaction of α-Al_2O_3 with Fe, but rather with FeO. Another property determined by DFT were the interface energies for rutile (TiO_2) coatings in aluminum melts and $AlSi_7Mg_{0.6}$. For the interfaces $TiO_2\| \alpha$-Al_2O_3 and $TiO_2\|MgTiO_3$, the most negative energies were calculated for the orientations observed in experiment.

A major research field of the subproject A04 was the investigation of carbon-bonded alumina filters. In several publications, these filters and the binders Carbores®P, lactose, and tannin were studied with Raman spectroscopy. Raman spectra for different binder compositions and different coking temperatures were recorded and analyzed. Especially the D and G peaks typical for Raman spectra from samples containing sp^2-hybridized carbon were interpreted.

For the pure binders, a decrease of the hydrogen content with increasing coking temperature and an increase of the sp^2 carbon clusters at temperatures above 600 °C were observed. Similar results were found for filters with Carbores®P binder, filters with environmentally friendly lactose/ tannin binder, and filters with a mixture of both. For the filters with added n-Si, the Raman spectra showed a modification of the carbon bonding.

For a better understanding of tannin pyrolysis, temperature dependent in-situ Raman measurement under argon atmosphere were performed. The tannin representatives gallic acid, ellagic acid, and tannic acid were studied this way. Raman spectra of the pyrolysis intermediates and end products of the examined substances were recorded, and pyrolysis temperature ranges were determined. Furthermore, the pyrolysis, i.e., the decarboxylation and dehydrogenation, of gallic acid was investigated with DFT, identifying transition states for both reactions.

Besides investigating carbon-bonded filters, the formation of corundum (α-Al_2O_3) from boehmite (γ-AlO(OH)) via γ-Al_2O_3, δ-Al_2O_3, and θ-Al_2O_3 was monitored in a temperature range from room temperature up to 1400 °C with Raman spectroscopy. Each phase showed a clearly distinguishable spectrum, so an identification of these phases as well as their corresponding temperature ranges was possible.

Acknowledgements Funded by the Deutsche Forschungsgemeinschaft (DFG, German Research Foundation)–Project ID 169148856 – SFB 920, subprojects A01, A03, A04, A05, A07, C06, S03.

The authors thank the ZIH in Dresden and the URZ in Freiberg for computational time and support. The computations at the URZ were performed on the compute cluster of the Faculty of Mathematics and Computer Science of the TU Bergakademie Freiberg, funded by the DFG–Project ID 397252409. Further, we would like to thank numerous previous members of the Institute of Theoretical Physics who have been involved in this research topic at some stage or have been available for discussions and questions: Dr. Lilit Amirkhanyan, Dr. Torsten Weißbach, Dr. René Wirnata, Dr. Torsten Hahn, Dr. Christian Röder, Dr. Steve Schmerler, Dr. Simon Liebing, Dr. Sebastian Schwalbe, M.Sc. Lenz Fiedler, and M.Sc. Nebahat Bulut. We would also like to thank Birgit Ostermay for technical support of the Raman measurements.

References

1. L. Amirkhanyan, T. Weißbach, T. Gruber, T. Zienert, O. Fabrichnaya, J. Kortus, J. Alloys Compd. **598**, 137 (2014). https://doi.org/10.1016/j.jallcom.2014.01.234
2. L. Amirkhanyan, Ph.D. thesis, TU Bergakademie Freiberg, 2017. https://d-nb.info/122610 083X
3. P. Hohenberg, W. Kohn, Phys. Rev. **136**(3B), B864 (1964). https://doi.org/10.1103/PhysRev. 136.B864
4. W. Kohn, L.J. Sham, Phys. Rev. **140**(4A), A1133 (1965). https://doi.org/10.1103/PhysRev.140. A1133
5. T. Zienert, L. Amirkhanyan, J. Seidel, R. Wirnata, T. Weißbach, T. Gruber, O. Fabrichnaya, J. Kortus, Intermetallics **77**, 14 (2016). https://doi.org/10.1016/j.intermet.2016.07.002
6. L. Amirkhanyan, T. Weißbach, J. Kortus, C.G. Aneziris, Ceram. Int. **40**(1), 257 (2014). https:// doi.org/10.1016/j.ceramint.2013.05.132
7. A. Salomon, L. Amirkhanyan, C. Ullrich, M. Motylenko, O. Fabrichnaya, J. Kortus, D. Rafaja, J. Eur. Ceram. Soc. **38**(16), 5590 (2018). https://doi.org/10.1016/j.jeurceramsoc.2018.07.052
8. M. Rudolph, A. Salomon, A. Schmidt, M. Motylenko, T. Zienert, H. Stöcker, C. Himcinschi, L. Amirkhanyan, J. Kortus, C.G. Aneziris, D. Rafaja, Adv. Eng. Mater. **19**(9), 1700141 (2017). https://doi.org/10.1002/adem.201700141
9. C. Röder, T. Weißbach, C. Himcinschi, J. Kortus, S. Dudczig, C.G. Aneziris, J. Raman Spectrosc. **45**(1), 128 (2014). https://doi.org/10.1002/jrs.4426
10. C. Himcinschi, C. Biermann, E. Storti, B. Dietrich, G. Wolf, J. Kortus, C.G. Aneziris, J. Eur. Ceram. Soc. **3**(16), 5580 (2018). https://doi.org/10.1016/j.jeurceramsoc.2018.08.029
11. X. Wu, A. Weidner, C.G. Aneziris, H. Biermann, Ceram. Int. **48**(1), 148 (2022). https://doi. org/10.1016/j.ceramint.2021.09.090
12. S. Brehm, C. Himcinschi, J. Kraus, B. Bock-Seefeld, C. Aneziris, J. Kortus, Adv. Eng. Mater. **24**(2), 2100544 (2022). https://doi.org/10.1002/adem.202100544
13. S. Brehm, J. Kraus, C. Himcinschi, J. Kortus, J. Raman Spectrosc. **53**(8), 1361 (2022). https:// doi.org/10.1002/jrs.6376
14. J. Kraus, S. Schwalbe, K. Trepte, J. Kortus, Adv. Eng. Mater. **24**(2), 2100572 (2022). https:// doi.org/10.1002/adem.202100572
15. J. Kraus, J. Kortus, J. Comput. Chem. **43**(15), 1023 (2022). https://doi.org/10.1002/jcc.26865
16. A. Schramm, M. Thümmler, O. Fabrichnaya, S. Brehm, J. Kraus, J. Kortus, D. Rafaja, C. Scharf, C.G. Aneziris, Crystals **12**(5), 654 (2022). https://doi.org/10.3390/cryst12050654
17. P. Blaha, K. Schwarz, F. Tran, R. Laskowski, G.K.H. Madsen, L.D. Marks, J. Chem. Phys. **152**(7), 074101 (2020). https://doi.org/10.1063/1.5143061
18. O.K. Andersen, Phys. Rev. B **12**(8), 3060 (1975). https://doi.org/10.1103/PhysRevB.12.3060
19. D.D. Koelling, G.O. Arbman, J. Phys. F: Met. Phys. **5**(11), 2041 (1975). https://doi.org/10. 1088/0305-4608/5/11/016

20. P. Giannozzi, S. Baroni, N. Bonini, M. Calandra, R. Car, C. Cavazzoni, D. Ceresoli, G.L. Chiarotti, M. Cococcioni, I. Dabo et al., J. Phys. Condens. Matter **21**(39), 395502 (2009). https://doi.org/10.1088/0953-8984/21/39/395502

21. P. Giannozzi, O. Adreussi, T. Brumme, O. Bunau, M.B. Nardelli, M. Calandra, R. Car, C. Cavazzoni, D. Ceresoli, M. Cococcioni et al., J. Phys. Condens. Matter **29**(46), 465901 (2017). https://doi.org/10.1088/1361-648X/aa8f79

22. P. Giannozzi, O. Baseggio, P. Bonfà, D. Brunato, R. Car, L. Carnimeo, C. Cavazzoni, S. De Gironcoli, P. Delugas, F. Ferrari Ruffino et al., J. Chem. Phys. **152**(15), 154105 (2020). https://doi.org/10.1063/5.0005082

23. P.E. Blöchl, Phys. Rev. B **50**(24), 17953 (1994). https://doi.org/10.1103/PhysRevB.50.17953

24. J.P. Perdew, K. Burke, M. Ernzerhof, Phys. Rev. Lett. **77**(18), 3865 (1996). https://doi.org/10.1103/PhysRevLett.77.3865

25. Q. Sun, T.C. Berkelbach, N.S. Blunt, G.H. Booth, S. Guo, Z. Li, J. Liu, J.D. McClain, E.R. Sayfutyarova, S. Sharma, et al., Wiley Interdiscip. Rev. Comput. Mol. Sci. **8**(1), e1340 (2018). https://doi.org/10.1002/wcms.1340

26. Q. Sun, X. Zhang, S. Banerjee, P. Bao, M. Barbry, N.S. Blunt, N.A. Bogdanov, G.H. Booth, J. Chen, Z.-H. Cui et al., J. Chem. Phys. **153**(2), 024109 (2020). https://doi.org/10.1063/5.0006074

27. S. Schwalbe, L. Fiedler, J. Kraus, J. Kortus, K. Trepte, S. Lehtola, J. Chem. Phys. **153**(8), 084104 (2020). https://doi.org/10.1063/5.0012519

28. J. Lehtola, M. Hakala, A. Sakko, K. Hämäläinen, J. Comput. Chem. **33**(18), 1572 (2012). https://doi.org/10.1002/jcc.22987

29. S. Lehtola, H. Jónsson, J. Chem. Theory Comput. **9**(12), 5365 (2013). https://doi.org/10.1021/ct400793q

30. S. Lehtola, H. Jónsson, J. Chem. Theory Comput. **10**(12), 5324 (2014). https://doi.org/10.1021/ct500637x

31. S. Lehtola, ERKALE. (2013), http://github.com/susilehtola/erkale. Accessed 15 Jan 2022

32. F. Neese, Wiley Interdiscip. Rev. Comput. Mol. Sci. **2**(1), 73 (2012). https://doi.org/10.1002/wcms.81

33. F. Neese, Wiley Interdiscip. Rev. Comput. Mol. Sci. **8**(1), e1327 (2018). https://doi.org/10.1002/wcms.1327

34. H. Becker, L. Amirkhanyan, J. Kortus, A. Leineweber, J. Alloys Compd. **721**, 691 (2017). https://doi.org/10.1016/j.jallcom.2017.05.336

35. M.T. Dove, *Introduction to lattice dynamics*, 1st edn. (Cambridge University Press, Cambridge, 1993)

36. J. Chi, X. Zheng, S.Y. Rodriguez, Y. Li, W. Gou, V. Goruganti, K.D.D. Rathnayaka, J.H. Ross Jr., Phys. Rev. B **82**(17), 174419 (2010). https://doi.org/10.1103/PhysRevB.82.174419

37. J. Solarek, C. Himcinschi, Y. Klemm, C.G. Aneziris, H. Biermann, Carbon **122**, 141 (2017). https://doi.org/10.1016/j.carbon.2017.06.041

38. H. Zielke, T. Wetzig, C. Himcinschi, M. Abendroth, M. Kuna, C.G. Aneziris, Carbon **159**, 324 (2020). https://doi.org/10.1016/j.carbon.2019.12.042

39. G.M. Westhoff, B.F.M. Kuster, M.C. Heslinga, H. Pluim, M. Verhage, Ullmann's Encycl. Ind. Chem. **1** (2014). https://doi.org/10.1002/14356007.a15_107.pub2

40. C. Biermann, Ph.D. thesis, TU Bergakademie Freiberg, 2016. https://d-nb.info/1106500865

41. K. Khanbabaee, T. Van Ree, Nat. Prod. Rep. **18**(6), 641 (2001). https://doi.org/10.1039/B101061L

42. E. Ritzer, R. Sundermann, Ullmann's Encycl. Ind. Chem. **18**, 493 (2000). https://doi.org/10.1002/14356007.a13_519

43. H. Fiege, H.-W. Voges, T. Hamamoto, S. Umemura, T. Iwata, H. Miki, Y. Fujita, H.-J. Buysch, D. Garbe, W. Paulus, Ullmann's Encycl. Ind. Chem. **26**, 521 (2000). https://doi.org/10.1002/14356007.a19_313

44. M. Mattonai, E. Ribechini, J. Anal. Appl. Pyrolysis **135**, 242 (2018). https://doi.org/10.1016/j.jaap.2018.08.029

45. S. Nam, M.W. Easson, B.D. Condon, M.B. Hillyer, L. Sun, Z. Xia, R. Nagarajan, RSC Adv. **9**(19), 10914 (2019). https://doi.org/10.1039/C9RA00763F
46. R.W. Hemingway, W.E. Hillis, Tappi J. **54**(6), 933 (1971)
47. G.C. Galletti, P. Bocchini, Rapid Commun. Mass Sp. **9**(3), 250 (1995). https://doi.org/10.1002/rcm.1290090315
48. J.C. Del Río, A. Gutiérrez, F.J. González-Vila, F. Martín, J. Anal. Appl. Pyrolysis **49**(1), 165 (1999). https://doi.org/10.1016/S0165-2370(98)00099-0
49. K.B. Kobakhidze, M.D. Alaniya, Chem. Nat. Compd. **40**(1), 89 (2004). https://doi.org/10.1023/B:CONC.0000025477.18086.8b
50. I.S. Movsumov, D.Y. Yusifova, T.A. Suleimanov, V. Mahiou-Leddet, G. Herbette, B. Baghdikian, E.E. Garayev, E. Ollivier, E.A. Garayev, Chem. Nat. Compd. **52**(2), 324 (2016). https://doi.org/10.1007/s10600-016-1631-6
51. C.G. Aneziris, J. Hubálková, R. Barabas, J. Eur. Ceram. Soc. **27**(1), 73 (2007). https://doi.org/10.1016/j.jeurceramsoc.2006.03.001
52. V. Stein, C.G. Aneziris, J. Ceram. Sci. Technol. **5**(2), 115 (2014). https://doi.org/10.4416/JCST2013-00036
53. B. Marchon, J. Gui, K. Grannen, G.C. Rauch, J.W. Ager, S.R.P. Silva, J. Robertson, IEEE Trans. Magn. **33**(5), 3148 (1997). https://doi.org/10.1109/20.617873
54. M. Dopita, M. Emmel, A. Salomon, M. Rudolph, Z. Matěj, C.G. Aneziris, D. Rafaja, Carbon **81**, 272 (2015). https://doi.org/10.1016/j.carbon.2014.09.058
55. F. Tuinstra, J.L. Koenig, J. Chem. Phys. **53**(3), 1126 (1970). https://doi.org/10.1063/1.1674108
56. A.C. Ferrari, J. Robertson, Phys. Rev. B **61**(20), 14095 (2000). https://doi.org/10.1103/PhysRevB.61.14095
57. E. Mullazzi, G.P. Brivio, E. Faulques, S. Lefrant, Solid State Commun. **46**(12), 851 (1983). https://doi.org/10.1016/0038-1098(83)90296-X
58. A.C. Ferrari, J. Robertson, Phys. Rev. B **63**(12), 121405 (2001). https://doi.org/10.1103/PhysRevB.63.121405
59. M.R. Pederson, A. Ruzsinszky, J.P. Perdew, J. Chem. Phys. **140**(12), 121103 (2014). https://doi.org/10.1063/1.4869581
60. M.R. Pederson, J. Chem. Phys. **142**(6), 064112 (2015). https://doi.org/10.1063/1.4907592
61. M.R. Pederson, T. Baruah, Adv. At. Mol. Opt. Phys. **64**, 153 (2015). https://doi.org/10.1016/bs.aamop.2015.06.005
62. Z.-H. Yang, M.R. Pederson, J.P. Perdew, Phys. Rev. A **95**(5), 052505 (2017). https://doi.org/10.1103/PhysRevA.95.052505
63. S. Schwalbe, T. Hahn, S. Liebing, K. Trepte, J. Kortus, J. Comput. Chem. **39**(29), 2463 (2018). https://doi.org/10.1002/jcc.25586
64. A. Klamt, G. Schüürmann, J. Chem. Soc. Perkin Trans. **2**(5), 799 (1993). https://doi.org/10.1039/p29930000799
65. E. Cancès, Y. Maday, B. Stamm, J. Chem. Phys. **139**(5), 054111 (2013). https://doi.org/10.1063/1.4816767
66. F. Lipparini, B. Stamm, E. Cancès, Y. Maday, B. Mennucci, J. Chem. Theory Comput. **9**(8), 3637 (2013). https://doi.org/10.1021/ct400280b
67. F. Lipparini, G. Scalmani, L. Lagardère, B. Stamm, E. Cancès, Y. Maday, J.-P. Piquemal, M.J. Frisch, B. Mennucci, J. Chem. Phys. **141**(18), 184108 (2014). https://doi.org/10.1063/1.4901304
68. F. Lipparini, L. Lagardère, G. Scalmani, B. Stamm, E. Cancès, Y. Maday, J.-P. Piquemal, M.J. Frisch, B. Mennucci, J. Phys. Chem. Lett. **5**(6), 953 (2014). https://doi.org/10.1021/jz5002506
69. B. Stamm, E. Cancès, F. Lipparini, Y. Maday, J. Chem. Phys. **144**(5), 054101 (2016). https://doi.org/10.1063/1.4940136
70. B. Stamm, L. Lagardère, G. Scalmani, P. Gatto, E. Cancès, J.-P. Piquemal, Y. Maday, B. Mennucci, F. Lipparini, Int. J. Quantum Chem. **119**(1), e25669 (2019). https://doi.org/10.1002/qua.25669
71. G. Mills, H. Jónsson, Phys. Rev. Lett. **72**(7), 1124 (1994). https://doi.org/10.1103/physrevlett.72.1124

72. G. Mills, H. Jónsson, G.K. Schenter, Surf. Sci. **324**(2), 305 (1995). https://doi.org/10.1016/0039-6028(94)00731-4
73. G. Henkelman, H. Jónsson, J. Chem. Phys. **113**(22), 9978 (2000). https://doi.org/10.1063/1.1323224
74. V. Ásgeirsson, B.O. Birgisson, R. Bjornsson, U. Becker, F. Neese, C. Riplinger, H. Jónsson, J. Chem. Theory Comput. **17**(8), 4929 (2021). https://doi.org/10.1021/acs.jctc.1c00462
75. V. Ásgeirsson, Neb_visualize (2020), https://github.com/via9a/neb_visualize. Accessed 15 Jan 2022
76. C.J. Doss, R. Zallen, Phys. Rev. B **48**(21), 15626 (1993). https://doi.org/10.1103/PhysRevB.48.15626
77. H.D. Ruan, R.L. Frost, J.T. Kloprogge, J. Raman Spectrosc. **32**(9), 745 (2001). https://doi.org/10.1002/jrs.736
78. G. Mariotto, E. Cazzanelli, G. Carturan, R. Di Maggio, P. Scardi, J. Solid State Chem. **86**(2), 263 (1990). https://doi.org/10.1016/0022-4596(90)90142-K
79. H.-S. Kim, P.C. Stair, J. Phys. Chem. A **113**(16), 4346 (2009). https://doi.org/10.1021/jp811019c
80. Z. Wu, H.-S. Kim, P.C. Stair, S. Rugmini, S.D. Jackson, J. Phys. Chem. B **109**(7), 2793 (2005). https://doi.org/10.1021/jp046011m
81. H. Kim, K.M. Kosuda, R.P. Van Duyne, P.C. Stair, Chem. Soc. Rev. **39**(12), 4820 (2010). https://doi.org/10.1039/C0CS00044B
82. G. Deo, F.D. Hardcastle, M. Richards, I.E. Wachs, A.M. Hirt, in *Novel Materials in Heterogeneous Catalysis*, ed. by R.T.K. Baker, L.L. Murrell (Washington DC, 1990)
83. S.P.S. Porto, R.S. Krishnan, J. Chem. Phys. **47**(3), 1009 (1967). https://doi.org/10.1063/1.1711980

Chapter 6
Temperature-Induced Changes in the Microstructure of the Metal Melt Filters and Non-metallic Inclusions

Martin Thümmler, Milan Dopita, Mykhaylo Motylenko, Anton Salomon, Erica Brendler, and David Rafaja

6.1 Introduction

The most prominent inclusions, which are present in the cast components made of steels or aluminum alloys, are aluminum oxides. These inclusions deteriorate dramatically the mechanical properties of the cast components and reduce significantly their lifetime [1–3]. Commonly, the inclusions are classified as exogenous or as endogenous according to their origin. Exogenous inclusions are present already in the starting material. Exogenous alumina inclusions appear typically in form of thermodynamically stable corundum (α-Al_2O_3). Endogenous inclusions form during

M. Thümmler (✉) · M. Motylenko · A. Salomon · D. Rafaja
Institute of Materials Science, Technische Universität Bergakademie Freiberg, Gustav-Zeuner Straße 5, 09599 Freiberg, Germany
e-mail: martin.thuemmler@iww.tu-freiberg.de

M. Motylenko
e-mail: mykhaylo.motylenko@ww.tu-freiberg.de

A. Salomon
e-mail: anton.salomon@ww.tu-freiberg.de

D. Rafaja
e-mail: rafaja@iww.tu-freiberg.de

M. Dopita
Department of Condensed Matter Physics, Charles University, Ke Karlovu 5, 12116 Prague, Czech Republic
e-mail: milan.dopita@matfyz.cuni.cz

E. Brendler
Institute of Analytical Chemistry, Technische Universität Bergakademie Freiberg, Leipziger Straße 29, 09599 Freiberg, Germany
e-mail: erica.brendler@chemie.tu-freiberg.de

© The Author(s) 2024
C. G. Aneziris and H. Biermann (eds.), *Multifunctional Ceramic Filter Systems for Metal Melt Filtration*, Springer Series in Materials Science 337,
https://doi.org/10.1007/978-3-031-40930-1_6

the casting process, for instance as a consequence of a local oversaturation of the melt by oxygen [4]. Endogenous alumina inclusions consist usually of metastable aluminum oxides, mainly of γ-Al$_2$O$_3$, δ-Al$_2$O$_3$ and θ-Al$_2$O$_3$, in particular in the early stages of the formation process [5].

Functional filters must be capable of removing these inclusions from the metal melt either by attaching the oxide particles to the filter surface or by dissolving them in a chemical reaction between the filter surface and the melt. The attachment of the inclusions is facilitated by the epitaxial processes, which occur preferentially at the interfaces of the counterparts having similar crystal structures. For the second process, a selective reactivity of the functionalized filter material with specific elements is needed. In both cases, a high efficiency of the filtration process requires a good adhesion of the non-metallic inclusions and/or reaction products to the functionalized filter surface in order to inhibit their spalling and the contamination of the metallic melt. For these reasons, the filter surface is functionalized by compounds, which have similar crystal structure and similar chemical composition as the non-metallic inclusions [6, 7]. Furthermore, the metal melt filters have to withstand extreme conditions, in particular high thermal shock and the contact with highly corrosive environments at high temperatures [8].

One of the well-established materials used for production of the metal melt filters is the carbon-bonded alumina (Al$_2$O$_3$-C). Dudczig et al. [9], Zienert et al. [10] and Salomon et al. [11] have shown that the Al$_2$O$_3$-C filters react with liquid iron and with the oxygen solved in the steel melt in a complex reaction, which helps to decrease the oxygen concentration in the steel and to remove the alumina inclusions from the melt. Another positive characteristics of the carbon-bonded alumina filters is their good thermo-shock resistance, which is further improved by the addition of a high melting coal tar resin or pitch [12] to the binder phase. In the current developments, the coal tar resin or pitch are replaced by environmentally friendlier and less toxic substitutes like lactose or tannin [13].

At high temperatures, the carbon binders undergo structural and microstructural changes that are accompanied by changes in the materials properties, which can negatively affect the filtration process. In this chapter, the thermally induced microstructure changes and phase transformations in the coal tar pitch, tannin and lactose are discussed. The other materials under study, namely γ-Al$_2$O$_3$, δ-Al$_2$O$_3$ and θ-Al$_2$O$_3$ are considered for production of functional coatings, because their structure is either identical or closely related to the crystal structure of the metastable alumina phases within the metallic melt, which are present in endogenous inclusions.

6.2   Carbon Binders for Ceramic Filter Bodies

The most prominent carbon-containing binders used in the 'Collaborative Research Centre 920' for production of the ceramic foam filters are the high melting coal tar pitch 'Carbores P' (Rütgers), the tannin-rich 'Quebracho extract' (Otto Dille) and conventional lactose. All of them consist mainly of carbon and hydrogen. Tannin and lactose contain, in addition to C and H, different amounts of oxygen, and are considered as environmental friendly alternative to the high melting coal tar pitches. At high temperatures, these compounds are expected to produce nanocrystalline graphite with a strongly distorted crystal structure, which serves as thermoshock resistant binder.

6.2.1 Characterization of the Microstructure of Carbon Binders

The X-ray diffraction (XRD) experiments have shown that turbostratic carbon is a substantial part of 'Carbores P'. Turbostratic carbon can be described as a stack of bent graphene layers having different mutual rotations with respect to the c axis, different shifts along the a and b axes, and noticeable fluctuations of the stacking distance as compared to hexagonal (2H) graphite (Fig. 6.1).

In the reciprocal space, a single graphene layer produces infinitely extended rods along the c^* axis (Fig. 6.2a). The size of the rods in the a^* and b^* directions is inversely proportional to the lateral size of the graphene layer. This reciprocity of the size of the objects in the direct and reciprocal space is a generally valid phenomenon, which is described in many textbooks on X-ray diffraction, e.g., in [14]. A three-dimensionally ordered crystal structure of graphite produces reciprocal lattice points, which size is reciprocally proportional to the size of the graphite crystallite (Fig. 6.2a) in the respective direction. The stacking disorder in turbostratic carbon causes a broadening

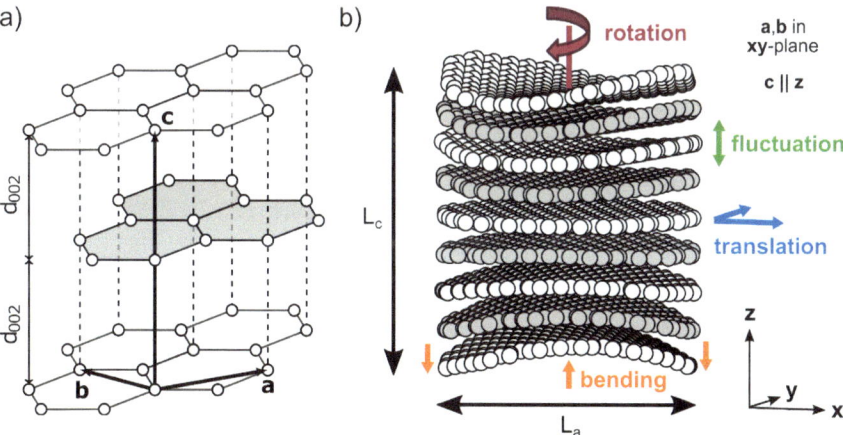

Fig. 6.1 Structure model of 2H graphite (**a**) and turbostratic carbon consisting of a perturbed stack of laterally terminated graphene layers (**b**). The structural perturbations are described by the rotation, translation and bending of individual graphene layers and by the fluctuation of the interlayer distances

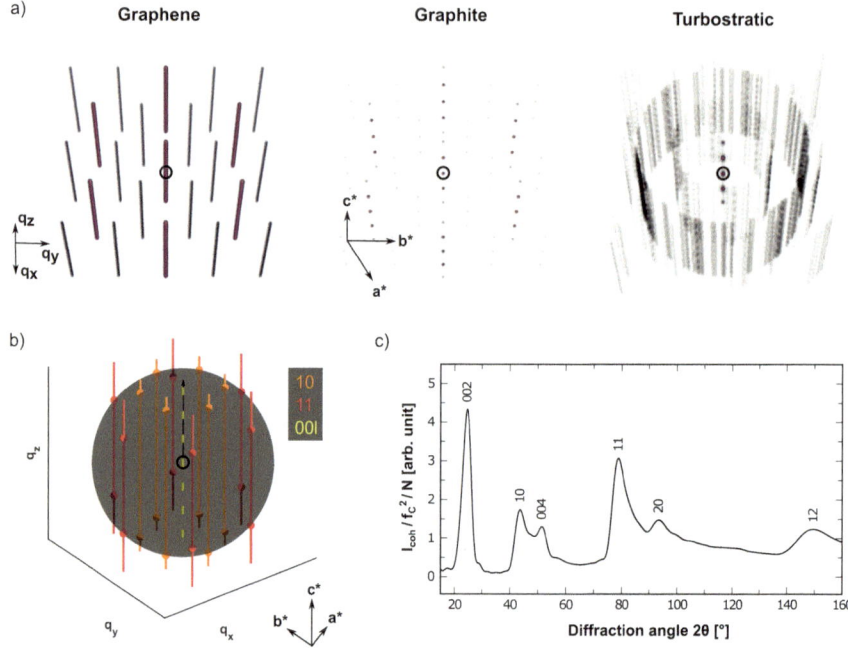

Fig. 6.2 **a** Intensity distributions simulated for graphene, 2H graphite and turbostratic carbon with mutually rotated layers (cf. Fig. 6.1). **b** Intersections of the intensity rods of turbostratic carbon with the surface of the Ewald sphere. For simplicity, only rods corresponding to the bands {10} and {11} are shown. **c** Powder XRD pattern simulated for CuKα radiation. q_x, q_y and q_z are Cartesian components of the diffraction vector q. a^*, b^*, c^* are the reciprocal lattice vectors of graphite. In panels **a** and **b**, the origin of the reciprocal space is highlighted by open circle

of the reciprocal lattice points along the c^* axis and the mutual rotations of individual graphene layers a rotation of the rods around this axis.

When the reciprocal space of a powder sample of turbostratic carbon is scanned in an XRD experiment, the 'scattered intensities' located at the intersections of the broadened reciprocal lattice points with the Ewald sphere (Fig. 6.2b) are integrated according to the powder pattern power theorem, see, e.g., [15]. This integration reveals diffracted intensity for example as a function of the diffraction angle (2θ), cf. Fig. 6.2c. As the reciprocal lattice points 00l of turbostratic carbon form a chain of local intensity maxima, which is located along the c^* axis (Fig. 6.2a), the Ewald sphere intersects these points always only in the c^* direction. Consequently, the corresponding intensity maxima in the powder XRD pattern are broadened symmetrically (Fig. 6.2c). Their width is inversely proportional to the size of the turbostratic graphite clusters in the c direction (L_c in Fig. 6.1b) and scales linearly with the magnitude of the diffraction vectors, if the stacking (interplanar) distances vary. Other reciprocal lattice 'points' than 00l are strongly elongated along the c^* axis by the shift of individual graphene layers along the a and b directions and by their mutual rotation around the c axis (Fig. 6.1b), which leads to an overlap of the reflections hkl with the same indices h and k but different l. These reciprocal lattice points are approached by the Ewald sphere in the $h\boldsymbol{a}^* + k\boldsymbol{b}^*$ direction, which causes an asymmetrical broadening of the diffraction lines with non-zero indices h and k and the formation of strongly asymmetrical $\{hk\}$ bands (Fig. 6.2c).

Within the kinematical X-ray diffraction theory, the total intensity scattered coherently by a cluster of atoms results from the interference of the electromagnetic waves scattered by individual atoms [15]:

$$I_{\text{coh}}(\boldsymbol{q}) \propto |F|^2 = \sum_m f_m \exp(i\boldsymbol{q}\boldsymbol{r}_m) \sum_n f_n \exp(-i\boldsymbol{q}\boldsymbol{r}_n) \tag{6.1}$$

In Eq. (6.1), $\boldsymbol{q}$ is the diffraction vector and F the structure factor (neglecting anomalous absorption and dispersion) of the cluster. The magnitude of the diffraction vector ($|\boldsymbol{q}| \equiv q$) is related to the diffraction angle (2θ) like $q = 4\pi \sin\theta/\lambda$, where λ is the X-ray wavelength. Individual atoms are characterized by their atomic scattering factors (f) and positions ($\boldsymbol{r}$). The integration of the scattered intensity over the Ewald sphere, which describes the X-ray scattering on an ensemble of randomly oriented clusters (on a powder sample), leads to the Debye equation [16]:

$$\begin{aligned} I_{\text{coh}}(q) &= \sum_m \sum_n \frac{f_m f_n}{4\pi r_{mn}^2} \int_0^\pi \int_0^{2\pi} \exp(iqr_{mn}\cos\psi)r_{mn}^2 \sin\psi \, d\phi d\psi \\ &= \sum_m \sum_n f_m f_n \frac{\sin(qr_{mn})}{qr_{mn}} \quad \text{with } r_{mn} = |\boldsymbol{r}_m - \boldsymbol{r}_n| \end{aligned} \tag{6.2}$$

In contrast to the general formulation of the X-ray scattering within the kinematical approximation [Eq. (6.1)], the Debye Eq. (6.2) takes into account all possible orientations of the cluster in the direct space and operates with the interatomic distances (r_{mn}) instead of with the atomic positions. A principal drawback of the Debye equation is a high number of the atomic pairs, $(N^2 - N)/2$, , which scales almost quadratically with the number of the atoms in the cluster (N). Thus, it was assumed that individual graphene layers have the same structure and the same lateral dimensions, and that they are only randomly rotated around the c direction. In such a case, Eq. (6.2) can be replaced by [17, 18]:

$$\frac{I_{\text{coh}}}{f_C^2 N} = i(0, q) + 2\sum_{k=1}^{P-1}\left(1 - \frac{k}{P}\right)i(k, q) \tag{6.3}$$

A consequence of this assumption is that the clusters of turbostratic carbon are approximated by cylindrical objects having the diameter L_a and the height L_c (Fig. 6.1b). In Eq. (6.3), f_C is the atomic scattering factor of carbon, P the number of graphene layers, $i(0, q)$ describes the X-ray scattering by a single layer and $i(k, q)$ is the Warren-Bodenstein integral [19, 20]:

$$i(k,q) = \frac{16N}{\pi L_a^2 Pq} \int_{\frac{1}{2}kc}^{\sqrt{L_a^2 + \frac{1}{4}k^2c^2}} \left(\arccos(u) - u\sqrt{1 - u^2} \right) \sin(qr)\mathrm{d}r \qquad (6.4)$$

With $u = \sqrt{r^2 - 0.25k^2c^2}/L_a$, this approximation contains only intuitive physical parameters. The parameter $c = 2d_{002}$ represents the mean out-of-plane lattice parameter of the turbostratic carbon (Fig. 6.1). Another parameter of the model is the in-plane lattice parameter (a) of graphene, which defines the distances between the carbon atoms within the graphene layers. The calculation of the one-dimensional integral from Eq. (6.4) and the one-dimensional sum from Eq. (6.3) is significantly faster than the 'two-dimensional' summation in Eq. (6.2), because both operations [Eqs. (6.3) and (6.4)] scale linearly with the number of graphene layers. Finally, the scattered intensity from Eq. (6.3) was corrected for static displacements of carbon atoms in c direction (u_c) using the Debye–Waller factor $DW_{00l} = \exp\left(-\langle u_c^2 \rangle q^2\right)$ [18]. This correction covers the bending of the graphene layers and the fluctuations of the interlayer distances.

6.2.2 Thermally Induced Changes in the Microstructure of Carbon Binders

The capability of the whole powder pattern fitting based on the theoretical background from Sect. 6.2.1 was first tested on 'Carbores P' [18, 21] and later applied to the new binder materials that are based on tannin (Quebracho extract) and lactose. In order to examine their temperature behavior, the binders were annealed for 3 h at the temperatures up to 1400 °C in reducing atmosphere ($CO/CO_2 + N_2$) to prevent their combustion.

After annealing at high temperatures, the XRD patterns of all carbon binders show the presence of the bands {10} and {11} (Fig. 6.3), which are the typical diffraction features of turbostratic carbon with a low number or with poorly ordered graphene layers, cf. Fig. 6.2. In the XRD patterns of 'Carbores P', also the diffraction lines 00l are visible, which indicate the formation of a turbostratic carbon structure with a higher number of parallel graphene layers.

The 'Quebracho extract' (denoted as tannin in Fig. 6.3) that was heat-treated above 600 °C contains crystalline phases, which were identified as Na_2CO_3 and CaS stemming from the impurities present in the natural product. The presence of Na, Ca, O, S (and Mg) was confirmed by energy dispersive X-ray spectroscopy (EDX) in the scanning electron microscope (SEM). Additionally, SEM revealed that the grains containing Na_2CO_3 and CaS are attached to the surface of agglomerated carbon clusters. Lactose was crystalline below 400 °C. It contains α-lactose monohydrate ($C_{12}H_{24}O_{12}$) at room temperature and the corresponding anhydride ($C_{12}H_{22}O_{11}$) that forms at temperatures in between 90 and 140 °C. The heat treatment of lactose caused excessive foaming, which was reduced by addition of TiO_2 powder to the samples heated above 400 °C. Consequently, the XRD patterns from Fig. 6.3 contain the

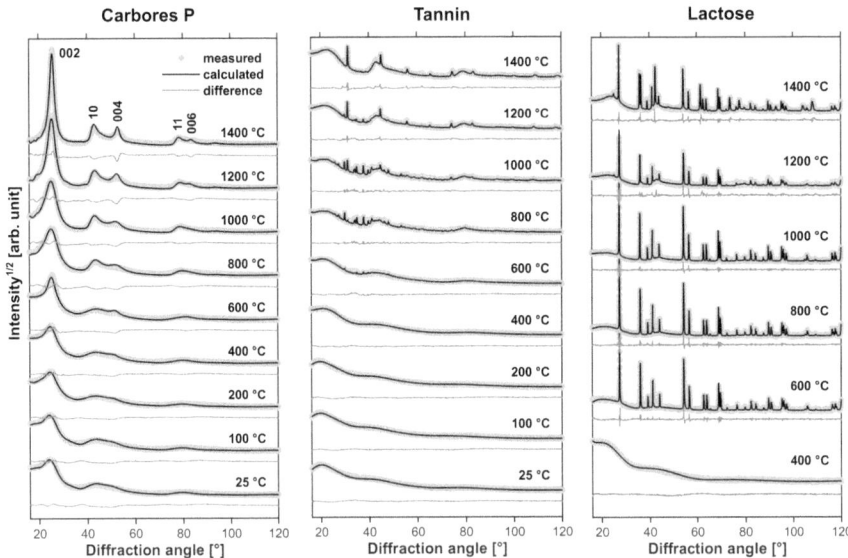

Fig. 6.3 XRD patterns of different carbon binders measured *post mortem* after annealing and their Rietveld fits. The XRD measurements were carried out with the CuKα radiation ($\lambda = 0.15418$ nm)

diffraction lines from two TiO_2 modifications (anatase and rutile) and from TiN, which was produced in the reaction of TiO_2 with carbon and nitrogen: $TiO_2 + 2C + \frac{1}{2}N_2 \rightarrow TiN + 2CO$.

Most of the parameters of the structure model used for the quantitative description of the turbostratic graphite, i.e., L_a, L_c, $\langle u_c^2 \rangle$, a and c (Fig. 6.4a-e), could be determined from the XRD patterns shown in Fig. 6.3. These parameters were complemented by the measurement of the gas pressure in a high-temperature (HT) chamber used for *in situ* annealing experiments (5 Kmin^{-1}) (Fig. 6.4f), by the *ex situ* measurement of the mass loss (Fig. 6.4g) and by the chemical analysis using carrier gas hot extraction (CGHE). These measurements revealed the carbon content (Fig. 6.4h) and the [H]/[C] ratio (Fig. 6.4i). A rapid increase of the gas pressure in the constantly evacuated chamber (cf. Fig. 6.4f) points to an intense formation of gaseous substances like H_2O, CO, CO_2, CH_4 etc. at the respective temperature.

In addition, the layer size L_a was verified by Raman spectroscopy from the intensity ratio of the D (≈ 1355 cm^{-1}) and G (≈ 1575 cm^{-1}) modes [22]. These results (Chap. 5, Fig. 5.4) are in good agreement with the values obtained by XRD (Fig. 6.4a).

Coal Tar Pitch (Carbores P)

In comparison with other binder materials, 'Carbores P' has the highest carbon content of 94(1)wt% (Fig. 6.4h) and the lowest [H]/[C] ratio of 0.53(5) (Fig. 6.4i) already in the initial state. The carbon content and the [H]/[C] ratio stay more or less

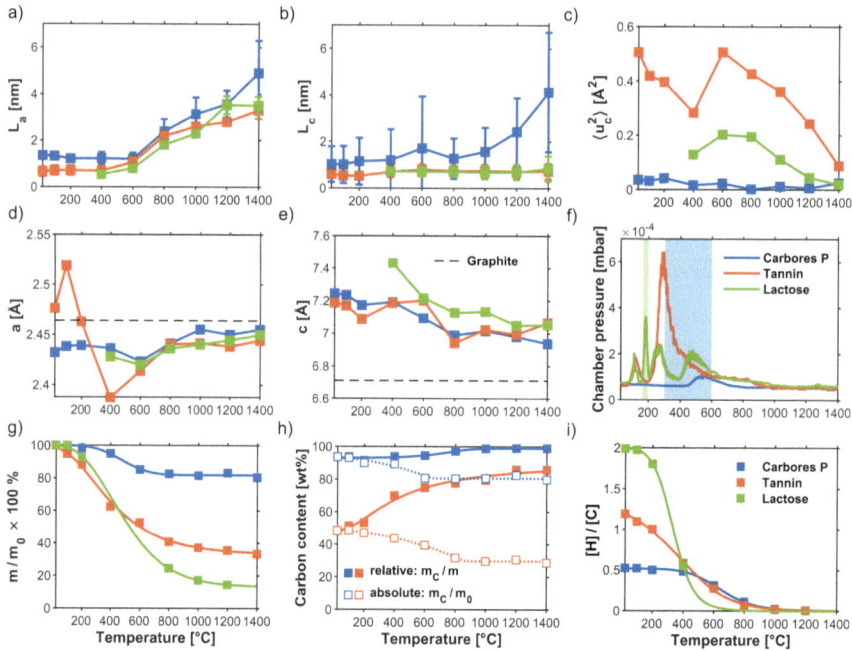

Fig. 6.4 Mean cluster size (**a**, **b**), mean square displacement of the carbon atoms in the c direction (**c**) and the lattice parameters a and c (**d**, **e**) obtained from the Rietveld refinement [18, 21]. The 'error bars' for the cluster size represent the standard deviation of the utilized lognormal distribution. The horizontal dashed lines in panels (**d**) and (**e**) indicate the lattice parameters of the 2H graphite. **f** Chamber pressure during the HTXRD experiments. The marked areas indicate the temperature ranges, in which lactose melts (green bar) and 'Carbores P' softens (blue bar). Mass loss (**g**), carbon content (**h**) and the hydrogen to carbon ratio (**i**) determined by CGHE

constant up to approx. 400 °C. The carbonization process, which is generally associated with an increase of the relative carbon content (Fig. 6.4h) in organic materials upon heating, starts in 'Carbores P' above 600 °C. The carbonization process is typically accompanied by the growth of individual carbon layers (visible as an increase of L_a, Fig. 6.4a), and by the dehydration reactions and by the elimination of point defects [23, 24], which result in an increase of the lattice parameter a [18]. After the graphene sheets are formed, the lattice parameter a approaches the in-plane lattice parameter of graphite [25] (cf. Fig. 6.4d). An onset of the thermally induced graphitization of 'Carbores P' is documented by the formation of the 00l peaks (Fig. 6.3, cf. Fig. 6.2c) and by the increase of L_c above 800 °C (Fig. 6.4b). Since the main components of the coal tar pitch [24] are large polycyclic aromatic hydrocarbons (PAHs), which possess nearly a perfect planar molecular structure, the displacement of carbon atoms in c direction is negligibly small (Fig. 6.4c). However, the lattice parameter c (Fig. 6.4e) is still larger than the corresponding intrinsic lattice parameter of graphite, even after annealing at 1400 °C. Thus, the crystal structure of graphite formed in calcined 'Carbores P' stays highly turbostratic in the whole temperature range under study. It should be noted that the typical graphitization temperatures are between 1600 to 3000 °C [23, 24].

Since 'Carbores P' contains mainly carbon and hydrogen, the gases, which form during the high-temperature treatment and which were responsible for the increase of the pressure (Fig. 6.4f) in the evacuated chamber used for annealing, are H_2 and hydrocarbons [23]. Additional consequences of the formation of these gases are the mass loss (Fig. 6.4g) and the decrease of the [H]/[C] ratio (Fig. 6.4i). However, whereas the mass loss is almost finished at 600 °C, the [H]/[C] ratio decreases up to the annealing temperature of approximately 1000 °C. As it can be assumed that the formation of H_2 causes mainly a decrease of the [H]/[C] ratio, while the formation of hydrocarbons is also responsible for mass loss, these results suggest that the formations of H_2 and hydrocarbons occur at different annealing temperatures. This is confirmed by the decrease of the absolute carbon content, which also indicated a formation of hydrocarbons only up to 600 °C (Fig. 6.4h). The correlation between the mass loss, the carbon content, the change in the [H]/[C] ratio (Fig. 6.4g-i), the cluster growth (Fig. 6.4a,b) and the consolidation of the lattice parameters (Fig. 6.4d,e) confirms that the carbonization process is associated with the removal of hydrogen from the graphene sheets. Thus, the carbonization process results in the increase of the in-plane lattice parameter a and leads finally to the growth of the graphene sheets in the lateral direction. As already mentioned above, the carbonization of 'Carbores P' starts at 600 °C. At this annealing temperature, the main mass loss, which is associated with the production and evaporation of hydrocarbons, is already almost completed.

The formation of volatile phases is finished at approximately 1000 °C. Above this temperature, the relative carbon content in annealed 'Carbores P' reaches nearly 100% (Fig. 6.4h). The in-plane lattice parameter of turbostratic graphite approaches the lattice parameter of 2H graphite. In contrast to the carbonization of 'Carbores P', which is almost finished above 1000 °C, its graphitization is still in progress. The clusters of turbostratic graphite grow. The out-of-plane lattice parameter (c) approaches the lattice parameter c of graphite, but it is still far from its intrinsic value, because the van der Waals bonds between the neighboring graphene layers are not established yet in the turbostratic graphite.

Tannin (Quebracho Extract)

The [H]/[C] ratio measured in the utilized 'Quebracho extract' ([H]/[C] = 1.19) is slightly higher than the value of 1.06 expected for the main component profisetinidin ($C_{17}H_{18}O_5$), cf. Fig. 6.4i. However, as the [H]/[C] ratio decreased already at the annealing temperatures below 200 °C, it can be assumed that the excess of hydrogen in the 'Quebracho extract' stems from adsorbed water. This assumption is substantiated by a significant mass loss during the annealing in this temperature range (Fig. 6.4g). The released water vapor is also responsible for the first increase of the pressure in the evacuated annealing chamber (Fig. 6.4f). The second pressure increase is caused by the thermal decomposition of profisetinidin and by the evaporation of catechol, resorcinol and other organic compounds [26] having their boiling points between 200 and 300 °C. The thermal decomposition of profisetinidin is accompanied by cross-linking reactions involving hydroxyl groups (polycondensation). These cross-link reactions are responsible for large variations of the in-plane lattice parameter a observed in samples annealed below 600 °C (Fig. 6.4d). In the early stages of the annealing process, the chaotic cross-link formation does not lead to the formation of carbon layers, which would be larger than the original size of the organic molecules, cf. Fig. 6.4a. Still these cross-links are a prerequisite of the subsequent layer growth and the main carbonization process above 600 °C that proceeds like in 'Carbores P'. During this heating period, the diffraction bands {10} and {11} become more pronounced.

In contrast to 'Carbores P', the parallel arrangement of graphene sheets and the formation of turbostratic graphite in calcined tannin is inhibited even at the highest annealing temperatures, as it can be seen from the absence of pronounced diffraction lines $00l$ (Fig. 6.3). At the positions of these XRD lines, only very broad maxima of the diffuse scattering were observed. This means that the parameters of the microstructure model, which are typically determined from the positions, shape and intensities of the diffraction lines $00l$, i.e., the vertical size of turbostratic carbon clusters (L_c), the out-of-plane lattice parameter (c) and the displacement of carbon atoms in the c direction ($\langle u_c^2 \rangle$), are no reliable quantities, as their values are expected to correlate strongly. Still, the significantly higher values of $\langle u_c^2 \rangle$ (Fig. 6.4c) obtained for tannin (and lactose), as compared with 'Carbores P', and the consequent decrease of $\langle u_c^2 \rangle$ at higher temperatures are consistent with the average molecular structure of the utilized binder materials. In contrast to 'Carbores P', the hexagonal rings containing carbon atoms are not planar in profisetinidin and in α-lactose [27, 28], but they flatten with increasing layer size L_a during the carbonization process.

The results of the XRD analysis of calcined tannin, which are summarized in Fig. 6.4, can only be interpreted in such a way that no real turbostratic graphite is formed. The vertical size of the turbostratic graphite clusters having certain periodic ordering in the c direction stays below 1 nm, which corresponds to 2–3 nearly parallel graphene-like layers. As lactose showed a strong inhibition of the graphitization process as well (see below), the impurities (Na, Ca, O and S) mentioned above cannot be the primary reason for the retarded graphitization of tannin. It seems more likely, that unfavorable cross-links between strongly tilted graphene-like layers prevent a further parallel arrangement up to 1400 °C or higher.

Lactose

The [H]/[C] ratio of 1.99 (Fig. 6.4i) determined using CGHE in the starting sample of α-lactose monohydrate ($C_{12}H_{24}O_{12}$) is in a good agreement with the expected value, [H]/[C] = 2. The annealing of $C_{12}H_{24}O_{12}$ leads to a rapid reduction of the [H]/[C] ratio and to the extensive mass loss (Fig. 6.4g), which are accompanied by the production of gaseous compounds (Fig. 6.4f). The first transformation step is the release of crystal water and the phase transition of α-lactose monohydrate to the

α-lactose anhydride ($C_{12}H_{22}O_{11}$). This step is completed at 200 °C, where the [H]/[C] has decreased to 1.80 (Fig. 6.4i).

Upon further annealing, the melt is caramelized. The caramelization process is accompanied by condensation reactions between different hydroxyl groups, which lead to the formation of non-planar carbon layers containing hydrogen, oxygen and other point defects, which resemble small, cross-linked and highly perturbed PAHs [29]. This process is most pronounced around 270 °C and is completed around 400 °C. Consequently, the XRD signal typical for hexagonally coordinated carbon was observed only in the XRD patterns of the lactose samples, which were annealed at 400 °C and above. In samples annealed at the temperatures below 400 °C, no noteworthy amount of carbonized lactose was detected by XRD. The graphitization of lactose proceeds similarly to the graphitization of tannin. The release of hydrogen from the perturbed PAHs and the formation of graphene sheets are documented by the increase of the in-plane lattice parameter a and by the increase of the layer size L_a, which start above 600 °C (Fig. 6.4a, d). In analogy with tannin, the formation of a turbostratic structure in carbonized lactose is strongly inhibited.

In comparison with 'Carbores P' and tannin, lactose shows the highest mass loss (~87%), which is accompanied by excessive foaming. These features facilitate formation of a highly porous carbon binder phase, which may negatively affect the mechanical properties of the synthesized filters. In this context, it should be mentioned that the liquid phase sintering in lactose happens at much lower temperatures (Fig. 6.4f) than, e.g., in coal tar pitches. Thus, it cannot contribute significantly to the improvement of the mechanical properties of sintered refractory composites.

Comparison of the Carbon Binders from the Microstructural Point of View

Traditionally, the coal tar pitch is used as a favored source of carbon for production of the carbon binders in thermoshock-resistant refractories. However, as coal tar contains many carcinogenic polycyclic aromatic hydrocarbons, alternatives are sought. Tannin and lactose are considered as possible substitutes. From the microstructural point of view, these alternative binders contain less carbon than coal tar. Additional elements form frequently gaseous phases, which leave the original compound and facilitate the formation of highly porous structures that may negatively affect the overall yield strength of the filter. The graphitization of tannin and lactose is strongly inhibited. As the coefficient of the thermal expansion of graphite is much smaller in the crystallographic direction a than in the crystallographic direction c [30], the lack of the parallel ordering of the graphene sheets and the formation of the graphite structure may improve the thermal shock resistance of the carbon binder.

Therefore, the replacement of certain amount of the coal tar pitch in refractory materials by alternative carbon binders is reasonable, because it reduces the amount of carcinogenic polycyclic aromatic hydrocarbons and improves the thermoshock resistance of the refractories. Still, some problems must be solved. One of them is the absence of a liquid-like phase in the binders with high tannin content. As the presence of a liquid phase usually facilitates the sintering process, the refractories containing binders with a high tannin content must be sintered for a longer time or at higher temperatures. Furthermore, the sources of tannin with a lower amount of impurities and additional phases should be preferred. A principal problem of the carbon binders produced from lactose is the foaming. However, it can be reduced by addition of ceramic powders or fully suppressed by annealing in vacuum.

6.3 Metastable Alumina Phases for Functional Filter Coatings

Metastable alumina phases are promising materials for functionalization of the metal melt filters, because they are supposed to have the same chemical composition and a similar crystal structure like the endogenous inclusions, which form during the casting of steels and aluminum alloys [5]. In general, the alumina phases can be divided in two groups according to their oxygen sublattice [31]. The first group including the thermodynamically stable corundum (α-Al_2O_3) possesses an approximately hexagonal close packed (h.c.p.) sublattice. The second group comprises metastable alumina phases, which crystallize with an approximately cubic close packed (c.c.p.) sublattice.

6.3.1 Structure of γ-Al₂O₃

One of the most popular metastable aluminum oxides is γ-Al_2O_3. Although this phase was extensively investigated in the past, there are still discussions about its crystal structure [32–36], which are motivated mainly by the presence of crystal structure defects [37–39]. In an ideal spinel having the space group (SG)$Fd\bar{3}m$, the anions occupy the Wyckoff positions $32e$, the trivalent cations the octahedral sites $16d$ and the divalent cations the sites $8a$ [40]. In γ-Al_2O_3, the Wyckoff positions $8a$ have also to be occupied by trivalent cations (Fig. 6.5). To preserve the charge neutrality and the cation to anion ratio of 2/3, $2.\bar{6}$ of the 24 cations sites must remain vacant.

A typical feature of the highly defective crystal structure of γ-Al_2O_3 is a highly anisotropic broadening of diffraction spots and lines, which is observed in selected area electron diffraction (SAED) and XRD patterns (Fig. 6.6). Diffraction lines with $h/2$, $k/2$ and $l/2$ being all even or odd, e.g., 004, 222, 440 and 444, which stem primarily from the fully occupied c.c.p. anion sublattice (Wyckoff sites $32e$), are much less broadened than the other diffraction lines, e.g., 111, 113, 220, 224, 331, 333 and 115, to which solely the scattering on the cation sublattice contributes. This anisotropic line broadening was explained by the presence of non-conservative antiphase boundaries (APBs) having the domain boundaries (00l) and the domain

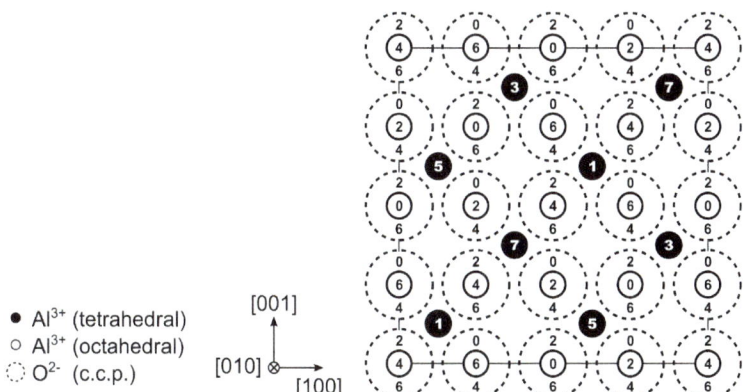

Fig. 6.5 Projection of the idealized cubic spinel type structure ($F d\overline{3} m$) of γ-Al$_2$O$_3$. The numbers represent the y-coordinate of the respective ion in multiples of $a/8$, where a is the lattice parameter. Reprint with a friendly permission of IUCr [41]

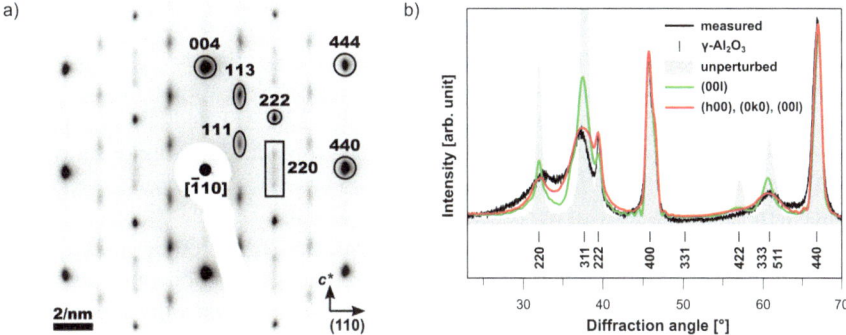

Fig. 6.6 SAED (**a**) and XRD (**b**) patterns of γ-Al$_2$O$_3$ that was produced by annealing highly crystalline boehmite (γ-AlOOH) in air for 20 h at 600 °C. For simplicity, the diffraction spots and lines were indexed utilizing the cubic spinel-type phase ($F d\overline{3} m$). The XRD measurement is complemented by simulated powder XRD patterns of unperturbed γ-Al$_2$O$_3$ and of γ-Al$_2$O$_3$ with non-conservative APBs located on a single family of the lattice planes (green) and on the crystallographically equivalent lattice planes (red). Reprint with friendly permission of IUCr [41]

shift $\frac{1}{4}\langle 10\overline{1}\rangle$ (Fig. 6.7a). These particular defects keep the c.c.p. oxygen sublattice intact but introduce a disorder on cation sublattice [41].

The kind of the APBs and in particular the shift vector were determined using the phase shift factor

$$A_{\boldsymbol{h}}(\boldsymbol{r}) = \exp(2\pi i \, \boldsymbol{h} \cdot \boldsymbol{r}), \tag{6.5}$$

which must be different from unity for heavily broadened reflections [41]. In Eq. (6.5), $\boldsymbol{r}$ is the shift vector and $\boldsymbol{h}$ the vector containing the diffraction indices hkl. The shift

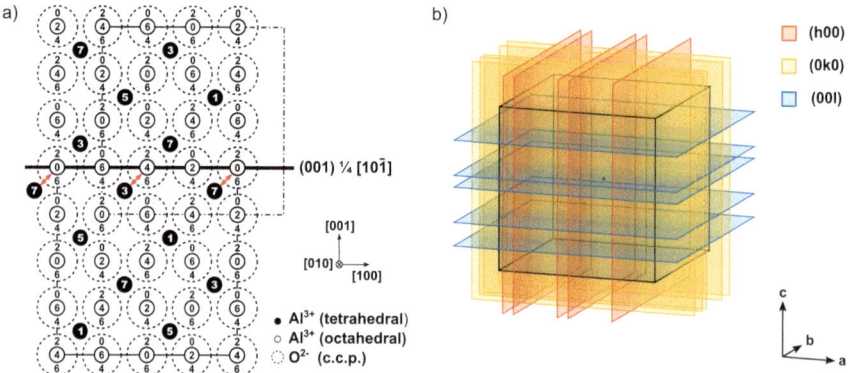

Fig. 6.7 **a** Non-conservative APB (001) $\frac{1}{4}[10\bar{1}]$ (cf. Figs. 6.5 and 6.8) with partially occupied cation positions. Unfavorable nearest neighbors are connected by red arrows. **b** Separation of a cubic nanocrystallite into cuboidal nanodomains by APBs located on crystallographically equivalent lattice planes. Reprint with friendly permission of IUCr [41]

factors calculated for the domain boundary (00*l*) and the non-conservative domain shift along the crystallographically equivalent directions $\frac{1}{4}\langle 10\bar{1}\rangle$ are summarized in Table 6.1. One can see that the reflections 222, 400 and 440 are not broadened by these APBs, while the broadening of the reflections 220, 311 and 333 depends on the shift direction. Furthermore, the reflection 220 is broadened for more equivalent shift directions than the reflection 311 and only in case of non-conservative APBs, which agrees very well with the observation (Fig. 6.6a).

For simulation of the SAED and XRD patterns, the JEMS routine [42] and the Debye equation [Eq. (6.2)] were employed, respectively. The atomic positions were first generated for undisturbed cubic nanocrystallites terminated by the lattice planes {100} that had the size of 11 nm. The lattice parameter of γ-Al$_2$O$_3$ was 7.942(5)

Table 6.1 Phase shift factors [Eq. (6.5)] for APBs on the lattice planes (00*l*). The reflections are broadened, when the phase shift factor is equal to –1. Corresponding reflections stem solely from the cation sublattice and are highlighted in bold. Other reflections originate from the c.c.p. sublattice. Adopted from [41]

hkl	Conservative (00*l*) $\frac{1}{4}\langle 110\rangle$		Non-conservative (00*l*) $\frac{1}{4}\langle 10\bar{1}\rangle$			
	$\frac{1}{4}[110]$	$\frac{1}{4}[\bar{1}10]$	$\frac{1}{4}[10\bar{1}]$	$\frac{1}{4}[\bar{1}0\bar{1}]$	$\frac{1}{4}[01\bar{1}]$	$\frac{1}{4}[0\bar{1}\bar{1}]$
220	1	1	−1	−1	−1	−1
311	1	−1	−1	1	1	−1
222	1	1	1	1	1	1
400	1	1	1	1	1	1
333	−1	1	1	−1	1	−1
440	1	1	1	1	1	1

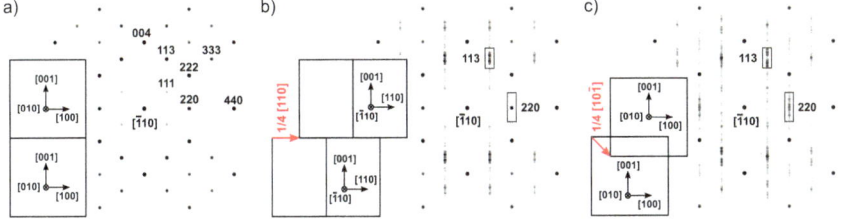

Fig. 6.8 The SAED patterns simulated for the idealized unperturbed spinel type structure of γ-Al$_2$O$_3$ (**a**), conservative APBs $(00l)\frac{1}{4}\langle110\rangle$ (**b**) and non-conservative APBs $(00l)\frac{1}{4}\langle10\bar{1}\rangle$ (**c**). The corresponding mutual shift of two γ-Al$_2$O$_3$ unit cells is shown in the respective inset (cf. Table 6.1). Adopted from [41]

Å. In order to simulate APBs, parts of the nanocrystallites were shifted along the respective shift vector (Fig. 6.8 and Table 6.1) [41]. Near the non-conservative APBs, this shift produces a certain amount of unfavorable nearest neighbors (Fig. 6.7a). Thus, the required vacancies preserving the charge neutrality were located in the vicinity of these APBs – either on the corresponding octahedral or on the tetrahedral sites. The SAED patterns simulated for undisturbed γ-Al$_2$O$_3$ and for γ-Al$_2$O$_3$ with APBs $(00l)\frac{1}{4}\langle110\rangle$ and $(00l)\frac{1}{4}\langle10\bar{1}\rangle$ [38, 43] are shown in Fig. 6.8. In addition to the phase shift factors (Table 6.1), the comparison of the SAED simulations with the measured SAED pattern from Fig. 6.6a confirms that only the non-conservative APBs $(00l)\frac{1}{4}\langle10\bar{1}\rangle$ cause the observed broadening of the diffraction spot 220.

The corresponding XRD patterns for the non-conservative APBs (Fig. 6.7a) on the lattice planes $(00l)$ (green) and on all equivalent planes (red) are depicted in Fig. 6.6b. This comparison reveals that the APBs have to be located on all equivalent planes forming a complex 3D structure of cuboidal nanodomains as depicted in Fig. 6.7b. In order to reduce the long computing time, which is typically required when the Debye equation [Eq. (6.2)] is applied for calculation of the XRD patterns from atomic clusters containing around 1 million atoms, the routine 'cuDebye' was developed [44]. The utilization of 2816 parallel operating CUDA cores (GTX 980 TI) in combination with atomic operations, which make storage and post processing of the calculated distances for a frequency evaluation obsolete, reduce the overall computation time to less than 5 min. This is a consequence of the Debye equation [Eq. (6.2)] as the prevention of reoccurring summands significantly increase the calculation speed. Additional line broadening stems from the nanocrystalline character of γ-Al$_2$O$_3$ under study, which had a crystallite size of about 11 nm. The crystallite size was determined from the width of the diffraction lines related to the oxygen sublattice. The shoulder of the diffraction line 400 (Fig. 6.6b) indicates a slight tetragonal distortion of the cubic elementary cell. The corresponding c/a ratio was 0.985. This tetragonal distortion is usually described by the SG $I4_1/amd$ instead of $Fd\bar{3}m$ [41, 45].

6.3.2 Thermally Induced Phase Transformation of γ-Al₂O₃

At higher temperatures, metastable γ-Al_2O_3 transforms to the thermodynamically stable corundum (α-Al_2O_3) following the phase transformation path [31]:

$$\gamma \rightarrow \delta \rightarrow \theta \rightarrow \alpha. \tag{6.6}$$

Although this transformation path is generally accepted, there are still some uncertainties regarding the transformation process and individual transition temperatures. The main issues are related to the formation of the intermediate phases δ-Al_2O_3 and θ-Al_2O_3 [46–49] and to their defect crystal structure [38, 39, 41, 50], as it is assumed that the phase transitions $\gamma \rightarrow \delta \rightarrow \theta$ lead primarily to a redistribution and possibly to an ordering of crystal structure defects in Al_2O_3. In this context, γ-Al_2O_3 is described as a cubic ($Fd\bar{3}m$) or a tetragonal ($I4_1/amd$) spinel [45, 51] and θ-Al_2O_3 as a monoclinic ($C2/m$) crystal structure [52]. The orientation relations between these structures are depicted in Fig. 6.9. The δ-Al_2O_3 phase possesses an intermediate crystal structure between γ-Al_2O_3 and θ-Al_2O_3, which cannot be described unambiguously from the crystallographic point of view [41, 47, 50, 53]. Initially, δ-Al_2O_3 was described as a supercell consisting of three cubic γ-Al_2O_3 unit cells stacked in the c direction with $a_\delta = a_\gamma$ and $c_\delta = 3a_\gamma$ [54]. Based on the tetragonal description of γ-Al_2O_3, δ-Al_2O_3 exhibits the space group $P\bar{4}m2$ [49, 55] with $a_\delta = a_\gamma/\sqrt{2}$ and $c_\delta = 3a_\gamma$. In both cases, the orientation relation towards the c.c.p. sublattice remains the same as in the parent γ-Al_2O_3 unit cell. Since the initial tetragonal distortion is quite small (cf. Sect. 6.3.1), the cubic model for γ-Al_2O_3 ($Fd\bar{3}m$) is utilized for simplicity in the following text.

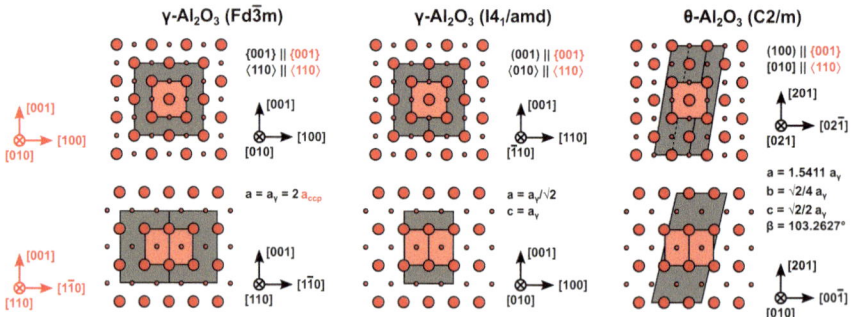

Fig. 6.9 Orientation relationships between cubic and tetragonal γ-Al_2O_3 (SG $Fd\bar{3}m$ and $I4_1/amd$), and θ-Al_2O_3 (SG $C2/m$). The oxygen atoms in idealized unperturbed c.c.p. sublattice are plotted in red. The lattice parameter a_γ refers to the cubic description (cf. Fig. 6.5). The larger circles refer to oxygen anions lying within the projection planes (010) (top) and (110) (bottom)

Identification of δ-Al$_2$O$_3$

A complementary technique used for the phase identification in defect-rich alumina phases is the nuclear magnetic resonance spectroscopy (NMR) (frequently in conjunction with the magic-angle spinning, MAS), as it is sensitive to the local coordination of the ^{27}Al cations [56, 57]. As pointed out in Sect. 6.3.1, the crystal structure of γ-Al$_2$O$_3$ ($Fd\overline{3}m$) requires $8/3 = 2.\overline{6}$ cation vacancies per unit cell, which must be distributed over the tetrahedral ($8a$) and octahedral ($16d$) spinel sites. When all vacancies are located on the eight tetrahedral sites, then $(8/3)/8 = 1/3$ of these sites stay vacant. Thus, $1 - 1/3 = 2/3$ of the tetrahedral sites are occupied by Al^{3+} ions, while all octahedral positions are fully occupied by Al^{3+}. As for γ-Al$_2$O$_3$ the NMR resolves the relative site occupancy and not the occupancy of individual sites, the minimum fraction of the tetrahedrally (AlO$_4$) coordinated Al^{3+} ions in γ/δ-Al$_2$O$_3$ is equal to $(2/3 \times 8)/(2/3 \times 8 + 1 \times 16) = 25\%$. When all vacancies are located on the sixteen octahedral sites, then $1 - 1/6 = 5/6$ of these sites are occupied by Al^{3+}. Thus, the maximum fraction of the tetrahedrally (AlO$_4$) coordinated Al^{3+} ions in γ/δ-Al$_2$O$_3$ is $(1 \times 8)/(1 \times 8 + 5/6 \times 16) = 37.5\%$, s. Table 6.2. In ideal θ-Al$_2$O$_3$ (SG $C2/m$, structure type β-Ga$_2$O$_3$), which has the same number of tetrahedral and octahedral positions (both sites are $4i$), the fraction of tetrahedrally (AlO$_4$) coordinated Al^{3+} ions seen by NMR must be 50% [52]. The remaining 50% of the Al^{3+} ions are located at the octahedral sites. Therefore, during the phase transformation γ-Al$_2$O$_3$ → θ-Al$_2$O$_3$, at least 12.5% ($= 50$–37.5%) of the Al^{3+} ions in the parent γ-Al$_2$O$_3$ phase must move from the octahedral to the tetrahedral interstitial (non-spinel) sites. Due to the structural similarity of δ-Al$_2$O$_3$ with γ-Al$_2$O$_3$, only the intermediate states having the fraction of tetrahedrally coordinated cations below or equal to 37.5% (and showing superstructure reflections in the XRD pattern) are classified as δ-Al$_2$O$_3$ [53].

Table 6.2 Fractions of tetrahedrally (AlO$_4$) and octahedrally (AlO$_6$) coordinated Al^{3+} ions determined by ^{27}Al MAS NMR as function of the annealing temperature (T). The corresponding amounts of vacancies per unit cell residing on the Wyckoff sites $8a$ and $16d$ of γ-Al$_2$O$_3$ that are summarized on the right-hand side of the table satisfy the equations

$[AlO_4] = (8 - N_{8a}^v)/(24 - N_{8a}^v - N_{16d}^v); \; N_{8a}^v + N_{16d}^v = 8/3$

T [°C]	AlO$_4$ [%]	AlO$_6$ [%]	N_{8a}^v	N_{16d}^v
Ideal	37.5	62.5	0.00	2.67
500	33.7	66.3	0.81	1.86
600	34.7	65.3	0.60	2.07
700	35.8	64.2	0.36	2.31
800	37.3	62.7	0.04	2.63
900	37.3	62.7	0.04	2.63

Transformation of γ-Al_2O_3 to δ-Al_2O_3

The relative distributions of the Al^{3+} cations and the respective vacancies over the Wyckoff sites $8a$ and $16d$ of γ-Al_2O_3 that were determined from the ^{27}MAS NMR data are summarized in Table 6.2. As expected, the amount of vacancies on octahedral sites ($16d$) is significantly larger than the amount of vacancies on tetrahedral sites ($8a$). Still, about 30% of the vacancies are located on a tetrahedral site ($8a$) in the initial stage of the γ/δ-Al_2O_3 formation (at 500 °C). With increasing annealing temperature, some octahedral cations (AlO_6) migrate to the tetrahedral sites (AlO_4). When the AlO_4 fraction approaches 37.5% (around 800 °C), almost all vacancies are located on the octahedral sites $16d$ [58–60]. As the measured fraction of the tetrahedrally coordinated Al^{3+} (Table 6.2) does not exceed 37.5%, it can be concluded that no θ-Al_2O_3 has formed up to 900 °C. Still, the diffraction patterns in Fig. 6.10 showing superstructure reflections were labeled as θ-Al_2O_3 because of the unambiguous structure of δ-Al_2O_3 [50, 53].

According to the SAED patterns (Fig. 6.10a, b), the $\gamma \rightarrow \delta$ transformation starts at about 700 °C. At 800 °C, a partially ordered superstructure forms. The superstructure reflections become much more pronounced above 800 °C. The average distance

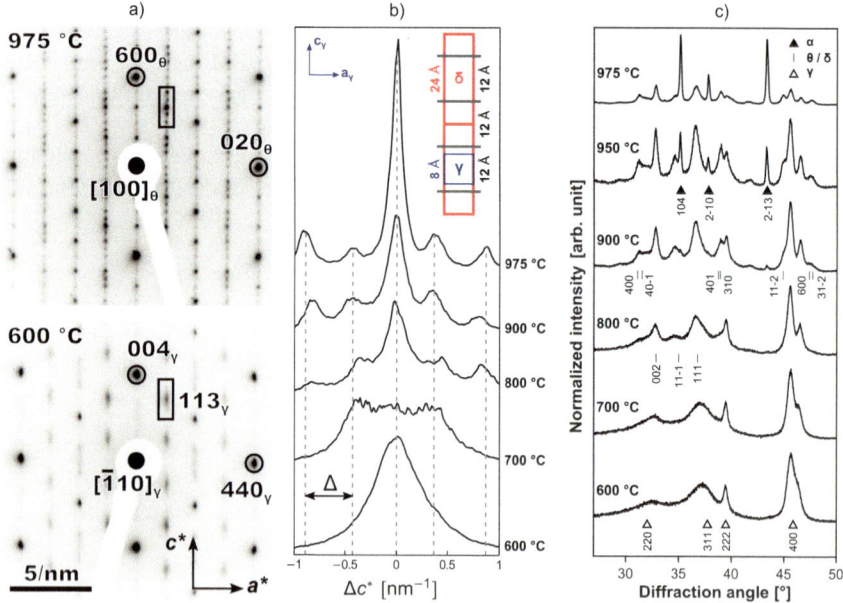

Fig. 6.10 Evolution of the diffraction patterns with increasing annealing temperature. **a** SAED patterns with zone axes parallel to the [$\bar{1}10$] direction in the c.c.p. sublattice (cf. Fig. 6.9). **b** The formation of a superstructure along the c direction is visible from the splitting of the diffraction spot 113_γ. The model of the supercell (red) consisting of three elementary cells γ-Al_2O_3 (blue) is shown in the inset. **c** Measured XRD patterns. Diffraction lines from θ/δ-Al_2O_3 are indexed in the space group $C2/m$ (θ-Al_2O_3). Adopted from [50]

between these satellite reflections ($\Delta = c_\delta^* = 0.42 \, \text{nm}^{-1}$, Fig. 6.10b) indicates the formation of a supercell with the lattice parameter c_δ of approx. 23.8 Å. In the c direction, this elementary cell is approximately tripled in comparison with γ-Al_2O_3 $Fd\bar{3}m$) having the lattice parameter $a_\gamma = 7.942$ Å, which is consistent with the initially published structures of δ-Al_2O_3 [49, 54, 55].

In γ-Al_2O_3, the APBs are distributed more or less randomly. The vacancies are preferentially located near the planar defects, which prevents an unfavorable proximity of Al^{3+} cations. With increasing temperature, the planar defects start to rearrange, and order periodically. This process is facilitated by the migration of Al^{3+} cations, which results in the shift of the vacancies from the tetrahedral sites to the octahedral sites that are close to APBs (cf. Fig. 6.7a). A perfect ordering of the APBs (and thus a perfect ordering of the vacancies) would resemble the formation of ideal (tetragonal) δ-Al_2O_3, which could be distinguished from γ-Al_2O_3 and θ-Al_2O_3 clearly. A simplified example of such ordering is schematically depicted in the inset of Fig. 6.10b. This model shows two elementary cells of δ-Al_2O_3 (red). Each δ-Al_2O_3 supercell has the lattice parameter of $c \approx 24$ Å, consists of three elementary cells of γ-Al_2O_3 (blue) with the lattice parameter $a_\gamma \approx 8$ Å and contains two equidistantly spaced APBs (gray). The distance of the APBs (≈ 12 Å) is close to the distance between APBs, which was determined from the pronounced anisotropy of the XRD line broadening [41]. However, the probability that such an ideal state is observed in an XRD (or SAED) experiment is low, because the same periodic rearrangement of APBs in all diffracting nanocrystallites is rather improbable. Furthermore, the equivalent shift vectors (Table 6.1) and the fact that the APBs can be also found on the equivalent lattice planes (cf. Figs. 6.6b and 6.7b) suggest a more complex arrangement of APBs than it is depicted in the inset (Fig. 6.10b). The variety of the APB arrangements is also a reason for the large number of structural approximations of δ-Al_2O_3 that can be found in literature [48, 49, 53–55].

The ongoing ordering of APBs and vacancies leads to a more pronounced tetragonal distortion of the c.c.p. oxygen sublattice. This can be seen on the further splitting of the diffraction line 004_γ with increasing annealing time and on the appearance of the diffraction lines $11\bar{2}_\theta$ and 600_θ at the temperatures around 900 °C. The latter does not imply the existence of θ-Al_2O_3 as an additional phase to γ/δ-Al_2O_3 (see Sect. "Identification of δ-Al_2O_3" and Fig. 6.10).

The above findings are consistent with the results of the multi-quantum magic-angle spinning (MQMAS) NMR, which show clearly the splitting and sharpening of the AlO_4 peak for γ-Al_2O_3 calcinated at 900 °C (Fig. 6.11) [56, 61]. The presence of at least two sharp AlO_4 peaks can be associated with the presence of differently coordinated Al^{3+} cations on the tetrahedral sites. Unlike in an ideal spinel structure, where the Al^{3+} coordination is always the same, in a structure with partially ordered vacancies, the Al^{3+} cations located in near APBs can have different coordination than the Al^{3+} cations, which are far from these boundaries. This can also apply for Al^{3+} cations in nanocrystallites with a more pronounced tetragonal distortion of the c.c.p. anion sublattice.

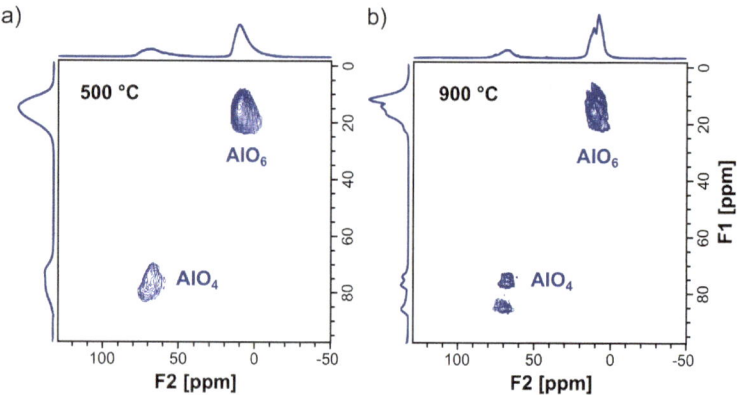

Fig. 6.11 Results of the ^{27}Al MQMAS NMR measurements on calcinated highly crystalline boehmite. Axis F1 represents the chemical shift of ^{27}Al but with higher resolution as in the common MAS NMR spectrum. The F2 axis shows the quadrupole interaction

6.3.3 Stabilization of γ-Al₂O₃

As illustrated above, γ-Al₂O₃ is a metastable phase, which undergoes a phase transition to thermodynamically stable corundum at high temperatures (Fig. 6.10c). The transition temperature is typically between 1000 and 1200 °C [54], but it depends strongly on the kind and on the microstructure of the starting material. For special technical applications, an extension of the temperature stability range of γ-Al₂O₃ is desired. As shown in [50], the use of a reducing environment, e.g., the presence of a carbon binder in the ceramic foam filters, has a positive effect on the improvement of the thermal stability of the metastable alumina phases. It postpones the formation of α-Al₂O₃ of about 50 K. Another approach, which can be applied to stabilize γ-Al₂O₃ to higher temperatures, is based on the inhibition of the migration of Al cations and thus on the hindering of the vacancy ordering. The migration of the Al cations and the ordering of the cation vacancies, which are the fundamental mechanisms accompanying the phase transformations in alumina, can be hindered, when Al₂O₃ is doped with elements having a different valence and coordination. One of the most promising dopants is silicon [62].

In order to quantify the impact of the doping of Al₂O₃ by silicon on the thermal stability of the metastable alumina phases, the nanocrystalline boehmite, which was used as a starting material also in this part of the study, was synthesized using a sol–gel process [63]. In the doped sample, approx. 5% of the Al^{3+} cations were replaced by Si^{4+} cations through the addition of tetraethyl orthosilicate (TEOS) [64]. The differential thermal analysis (DTA) (Fig. 6.12) revealed that the formation of the thermodynamically stable α-Al₂O₃ in the doped sample is shifted to a higher temperature (approx. 180 °C) as compared with the non-doped sample. The first endothermal peak in the DTA curve at approx. 150 °C (Fig. 6.12) corresponds to the loss of adsorbed water, whereas the second endothermal peak at approx. 450 °C

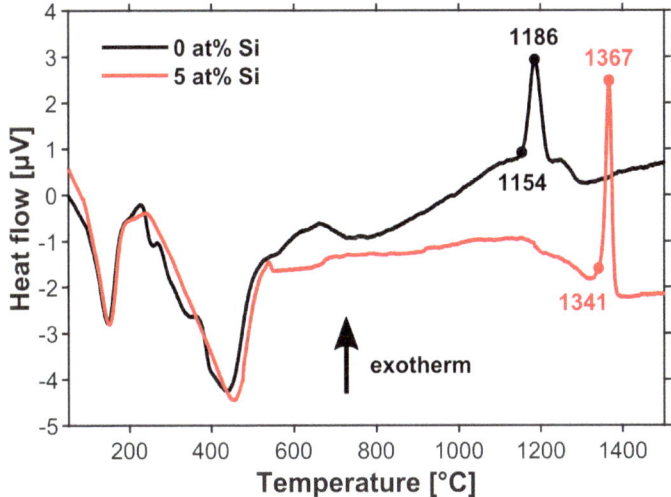

Fig. 6.12 DTA curves of non-doped and Si-doped nanocrystalline boehmite powders. The onset temperatures and the temperatures of the exothermal peaks related to the formation of α-Al₂O₃ are labeled

indicates the structural collapse of boehmite and the subsequent formation of γ-Al$_2$O$_3$ [50, 65].

The XRD measurements confirmed the same phase transformation path (boehmite $\rightarrow$ γ-Al$_2$O$_3$ $\rightarrow$ δ/θ-Al$_2$O$_3$ $\rightarrow$ α-Al$_2$O$_3$) for non-doped and Si-doped samples (Fig. 6.13a). However, these measurements have shown that in the doped sample the phase transformations γ-Al$_2$O$_3$ $\rightarrow$ δ/θ-Al$_2$O$_3$ and δ/θ-Al$_2$O$_3$ $\rightarrow$ α-Al$_2$O$_3$ occur at a higher temperature than in the non-doped one. The results of the XRD analysis were complemented by the results of the ^{27}Al MAS NMR analysis (Fig. 6.13b), which revealed that the fraction of tetrahedrally coordinated Al^{3+} ions (AlO$_4$) increased and the fraction of octahedrally coordinated Al^{3+} ions (AlO$_6$) decreased after annealing at higher temperatures, i.e., when γ-Al$_2$O$_3$ transformed to δ-Al$_2$O$_3$ (see Sect. "Transformation of γ-Al$_2$O$_3$ to δ-Al$_2$O$_3$").

The AlO$_5$ coordination visible in Fig. 6.13b is related to the surface states, which are typically reported for nanocrystalline materials [61]. In the samples synthesized using the sol–gel process, the sizes of the γ-Al$_2$O$_3$ crystallites were 3.1 nm (in the doped sample) and 4.3 nm (in the non-doped sample), as estimated using the Scherrer equation [67] from the integral width of the diffraction line 400$_\gamma$. The net relative fractions of the AlO$_4$ and AlO$_6$ coordinations that were calculated by neglecting the AlO$_5$ fraction are summarized in Table 6.3. The fractions of tetrahedrally coordinated Al^{3+} ions, which are higher than the maximum (ideal) values calculated for γ/δ-Al$_2$O$_3$, indicate the presence of a mixture of δ-Al$_2$O$_3$ and θ-Al$_2$O$_3$. The second part of Table 6.3 displays the concentrations of corresponding vacancies that are located on the Wyckoff sites 8a and 16d of γ-Al$_2$O$_3$. Negative concentrations of 8a

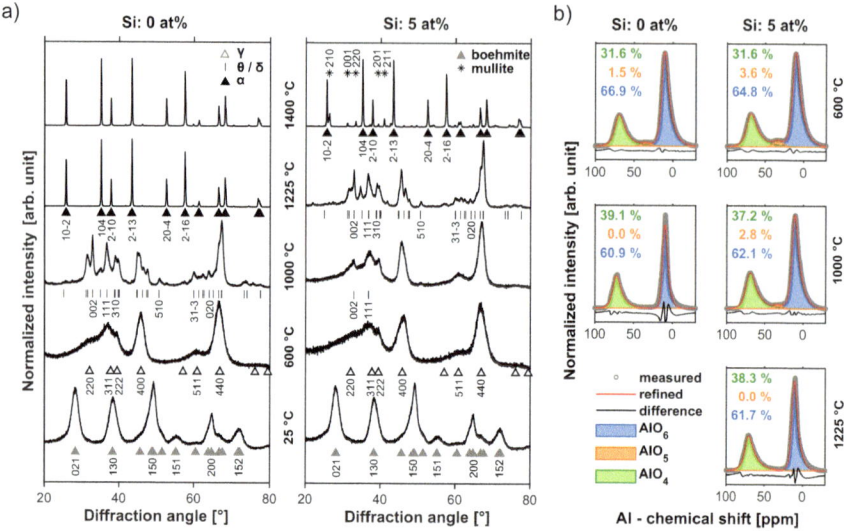

Fig. 6.13 The XRD patterns (**a**) (cf. Fig. 6.10c) and ^{27}Al MAS NMR spectra (**b**) of heat-treated alumina samples obtained by calcinating non-doped and Si-doped nanocrystalline boehmite. The refinement of the NMR spectra was performed using 'dmfit' [57, 66]

vacancies refer to the cations (Al^{3+} and/or Si^{4+}) located on the interstitial (tetrahedral non-spinel) sites.

After annealing at 1000 °C, the non-doped Al$_2$O$_3$ contained θ-Al$_2$O$_3$ domains. This can be seen in the XRD pattern (Fig. 6.13a) and concluded from the fraction of the tetrahedrally coordinated Al^{3+} ions, which exceeds 37.5% (AlO$_4$ in Table 6.3). Consequently, the amount of the 8a vacancies becomes negative. Still, no significant amount of α-Al$_2$O$_3$ has formed at this temperature. First at 1225 °C, Al$_2$O$_3$ in the non-doped sample transformed almost completely to corundum (α-Al$_2$O$_3$). As the Al^{3+} ions in α-Al$_2$O$_3$ are octahedrally coordinated, the NMR recognized the AlO$_6$ coordination only (Table 6.3, 1250 °C).

In the Si-doped γ-Al$_2$O$_3$, the positions of the Si^{4+} ions must be clarified first. The ^{29}Si MAS NMR revealed that Si^{4+} ions are located exclusively on a tetrahedral site (8a, 8b or 48f [41]) and that they are surrounded by the Al^{3+} cations (as second nearest neighbors), which is in a good agreement with the results of Mardkhe et al. [64]. If the Si^{4+} ions occupy the spinel sites 8a ($x_{8a}^{Si} = 1$), then the maximum fraction of AlO$_4$ is 34.9% (non-highlighted values in Table 6.3). If the Si^{4+} ions are located on the non-spinel sites 8b or 48f ($x_{8a}^{Si} = 0$), then the maximum fraction of AlO$_4$ is 40.1% (highlighted values in Table 6.3). Both fractions were calculated for 5 at% Si, i.e., for one Si^{4+} ion ($N_γ^{Si} = 1$) allotted to twenty Al^{3+} ions ($N_γ^{Al} = 20$). If one Si^{4+} ion replaces one Al^{3+} ion on the Wyckoff sites 8a, the charge neutrality is achieved when the unit cell of γ-Al$_2$O$_3$ contains three vacancies ($N_{8a}^v + N_{16d}^v = 9/3$) instead of $8/3 = 2.\overline{6}$ for non-doped γ-Al$_2$O$_3$ (s. Sect. 6.3.1). The main consequence of the Si^{4+} distribution over the spinel and non-spinel sites is that the formation of θ-Al$_2$O$_3$

Table 6.3 Fractions of the tetrahedrally (AlO_4) and octahedrally (AlO_6) coordinated Al^{3+} ions determined using the ^{27}Al MAS NMR (cf. Fig. 6.13b and Table 6.2). The non-highlighted values were calculated assuming that Si^{4+} ions occupy the spinel sites $8a$ ($x^{Si}_{8a} = 1$). The values printed in bold were calculated for Si^{4+} located on the tetrahedral non-spinel sites $8b$ and $48f$ [41] ($x^{Si}_{8a} = 0$). The concentrations of corresponding vacancies in γ-Al_2O_3 that are located on the Wyckoff sites $8a$ and $16d$ (second part of the table) were calculated analogously to the vacancy concentrations given in Table 6.2. In the Si-doped sample, the relationship between the amount of vacancies and the fraction of tetrahedrally coordinated Al^{3+} ions takes the form

$$[AlO_4] = \left(8 - N^v_{8a} - x^{Si}_{8a}N^{Si}_\gamma\right)\Big/\left(24 - N^v_{8a} - N^v_{16d} - x^{Si}_{8a}N^{Si}_\gamma\right);\ N^v_{8a} + N^v_{16d} = 24 - N^{Al}_\gamma - x^{Si}_{8a}N^{Si}_\gamma$$

x_{Si}	0%	5%	0%	5%	0%	5%	0%	5%
$T[°C]$	$AlO_4\ x^{Si}_{8a}=1/0$ [%]		$AlO_6\ x^{Si}_{8a}=1/0$ [%]		N^v_{8a}		N^v_{16d}	
Ideal	37.5	34.9/**40.1**	62.5	65.1/**59.9**	0.00	0.00/**0.00**	2.67	3.02/**4.06**
600	32.1	32.8 **36.1**	67.9	67.2	1.16	0.42/**1.47**	1.51	2.60
1000	39.1		60.9	63.9	−0.34	−0.25/**0.80**	3.01	3.27
1250	0.3	38.3	99.7	61.7		−0.68/**0.37**		3.70

can be concluded without doubt only when the amount of tetrahedrally coordinated Al^{3+} exceeds 40.1%, which was observed in none of the Si-doped samples under study.

The negative amounts of the cation vacancies sitting on the sites $8a$ were only obtained assuming that all Si^{4+} ions are located at the sites $8a$ (non-highlighted values in Table 6.3 for the samples annealed at 1000 °C and above). In this case, the fraction of the tetrahedrally coordinated Al^{3+} ions measured using ^{27}Al MAS NMR exceeds the predicted ideal value of 34.9%. However, the XRD pattern of the Si-doped sample annealed at 1000 °C (Fig. 6.13a) did not show any indication of the θ-Al_2O_3 formation, which would produce pronounced superstructure satellites like in Fig. 6.10c above 900 °C. As the negative amount of the cation vacancies at the $8a$ sites is related to the formation of θ-Al_2O_3 (s. Sect. "Identification of δ-Al_2O_3"), the non-existence of θ-Al_2O_3 in the XRD pattern reveals that the Si^{4+} cations are located also on the interstitial (non-spinel) sites up to 1225 °C. This may hinder the migration of Al^{3+} cations to the non-spinel sites, which is required for the phase transformation of γ/δ-Al_2O_3 to θ-Al_2O_3. In contrast to the Si-doped sample, the ^{27}Al MAS NMR performed on the non-doped sample revealed that the Si-free Al_2O_3 transforms partially to θ-Al_2O_3 already at 1000 °C.

In the Si-doped specimen, also the formation of α-Al_2O_3 is shifted to a higher temperature. At 1225 °C, almost no α-Al_2O_3 was identified by XRD (Fig. 6.13a). As a dominant phase, α-Al_2O_3 was first observed in the doped sample that was annealed at 1400 °C (Fig. 6.13a). This specimen contained mullite ($2Al_2O_3 \cdot SiO_2/$

$3Al_2O_3 \cdot 2SiO_2$, SG *Pbam* [68]) as a minor phase that formed during the reaction of alumina with added silicon. The delay of the α-Al_2O_3 formation has primarily kinetic reasons. As the solubility of silicon in corundum is extremely low [69], Si must leave the parent metastable phase (δ/θ-Al_2O_3), before this phase can transform to α-Al_2O_3. This process requires an additional migration or diffusion of silicon, which shifts the formation of α-Al_2O_3 to a higher temperature.

6.4 Conclusions

This chapter was devoted to the description of temperature-induced changes in the microstructure of carbon binders produced from coal tar pitch, tannin and lactose, and in the microstructure of alumina phases produced by calcination of boehmite. For the crystal structure and microstructure characterizations of the carbon binders, a fast computer routine based on the Debye equation was developed. This routine reveals the essential parameters of the turbostratic graphite, namely the lattice parameters a and c, the size of the graphite blocks in the respective direction and the degree of the atomic disorder. It was shown, how these parameters can be used for description of the kinetics of the graphitization process in different carbon-containing materials. For description of the phase transformations in alumina phases, atomistic models were developed that served as a basis for the explanation of the transformation mechanisms and as a basis for the systematic stabilization of the metastable phases through the obstruction of the transformation mechanisms.

Acknowledgements The authors thank Zdeněk Matěj for the implementation of the Warren-Bodenstein approach into the MStruct routine, Claudia Voigt, Benjamin Bock, Diane Hübgen, Katrin Becker, Astrid Leuteritz and Beate Wahl for sample preparation or measurements as well as Christian Schimpf, Marius Wetzel, Hanka Becker and Andreas Leineweber for fruitful discussions. Thomas Hammer and Wiebke Hadwich supported this work by their student projects. The German Research Foundation (DFG) is acknowledged for funding the Collaborative Research Center 920 (project number 169148856) in the frame of subproject A05.

References

1. A. Schmidt, A. Salomon, S. Dudczig, H. Berek, D. Rafaja, C.G. Aneziris, Adv. Eng. Mater. **19**, 1700170 (2017). https://doi.org/10.1002/adem.201700170
2. M. Seleznev, S. Henschel, E. Storti, C.G. Aneziris, L. Krüger, A. Weidner, H. Biermann, Adv. Eng. Mater. **22**, 1900540 (2020). https://doi.org/10.1002/adem.201900540
3. R. Wagner, A. Schmiedel, S. Dudczig, C.G. Aneziris, O. Volkova, H. Biermann, A. Weidner, Adv. Eng. Mater. **24**, 2100640 (2022). https://doi.org/10.1002/adem.202100640
4. C.G. Aneziris, C. Schroeder, M. Emmel, G. Schmidt, H.P. Heller, H. Berek, Metall. Mater. Trans. B **44**, 954 (2013). https://doi.org/10.1007/s11663-013-9828-6
5. K. Wasai, K. Mukai, A. Miyanaga, ISIJ Int. **42**, 459 (2002). https://doi.org/10.2355/isijinternational.42.459

6. B. Bock, A. Schmidt, E. Sniezek, S. Dudczig, G. Schmidt, J. Szczerba, C.G. Aneziris, Ceram. Int. **45**, 4499 (2019). https://doi.org/10.1016/j.ceramint.2018.11.131
7. M. Emmel, C.G. Aneziris, G. Schmidt, D. Krewerth, H. Biermann, Adv. Eng. Mater. **15**, 1188 (2013). https://doi.org/10.1002/adem.201300118
8. T. Wetzig, A. Baaske, S. Karrasch, N. Brachhold, M. Rudolph, C.G. Aneziris, Ceram. Int. **44**, 23024 (2018). https://doi.org/10.1016/j.ceramint.2018.09.105
9. S. Dudczig, C.G. Aneziris, M. Emmel, G. Schmidt, J. Hubalkova, H. Berek, Ceram. Int. **40**, 16727 (2014). https://doi.org/10.1016/j.ceramint.2014.08.038
10. T. Zienert, S. Dudczig, O. Fabrichnaya, C.G. Aneziris, Ceram. Int. **41**, 2089 (2015). https://doi.org/10.1016/j.ceramint.2014.08.038
11. A. Salomon, M. Motylenko, D. Rafaja, Adv. Eng. Mater. **24**, 2100690 (2022). https://doi.org/10.1002/adem.202100690
12. M. Emmel, C.G. Aneziris, Ceram. Int. **38**, 5165 (2012). https://doi.org/10.1016/j.ceramint.2012.03.022
13. C. Himcinschi, C. Biermann, E. Storti, B. Dietrich, G. Wolf, J. Kortus, C.G. Aneziris, J. Eur. Ceram. Soc. **38**, 5580 (2018). https://doi.org/10.1016/j.jeurceramsoc.2018.08.029
14. L.V. Azároff, *Elements of X-Ray Crystallography*. (McGraw-Hill, New York, 1968), isbn: 9780070026674
15. B.E. Warren, *X-Ray Diffraction*. (Dover Publications, New York, 1990), isbn: 9780486663173
16. P. Debye, Ann. Phys. **351**, 809 (1915). https://doi.org/10.1002/andp.19153510606
17. M. Dopita, M. Rudolph, A. Salomon, M. Emmel, C.G. Aneziris, D. Rafaja, Adv. Eng. Mater. **15**, 1280 (2013). https://doi.org/10.1002/adem.201300157
18. M. Dopita, M. Emmel, A. Salomon, M. Rudolph, Z. Matěj, C.G. Aneziris, D. Rafaja, Carbon **81**, 272 (2015). https://doi.org/10.1016/j.carbon.2014.09.058
19. B.E. Warren, P. Bodenstein, Acta Cryst. **18**, 282 (1965). https://doi.org/10.1107/S0365110X65000609
20. B.E. Warren, P. Bodenstein, Acta Cryst. **20**, 602 (1966). https://doi.org/10.1107/S0365110X66001464
21. Z. Matěj, A. Kadlecová, M. Janeček, L. Matějová, M. Dopita, R. Kužel, Powder Diffr. **29**, S35 (2014). https://doi.org/10.1017/S0885715614000852
22. S. Brehm, C. Himcinschi, J. Kraus, B. Bock-Seefeld, C. Aneziris, J. Kortus, Adv. Eng. Mater. **24**, 2100544 (2022). https://doi.org/10.1002/adem.202100544
23. I. Mochida, S.-H. Yoon, W. Qiao, J. Braz. Chem. Soc. **17**, 1059 (2006). https://doi.org/10.1590/S0103-50532006000600002
24. C.A. Rottmaier, Dissertation, Universität Erlangen-Nürnberg, 2007, urn:urn:nbn:de:bvb:29-opus-11781
25. P. Trucano, R. Chen, Nature **258**, 136 (1975). https://doi.org/10.1038/258136a0
26. Z. Sebestyén, E. Jakab, E. Badea, E. Barta-Rajnai, C. Şendrea, Zs. Czégény, J. Anal. Appl. Pyrolysis 138, 178 (2019). https://doi.org/10.1016/j.jaap.2018.12.022
27. National Library of Medicine (PubChem). cid: 101373967, https://pubchem.ncbi.nlm.nih.gov. Accessed 1 March 2022
28. National Library of Medicine (PubChem), cid: 84571. https://pubchem.ncbi.nlm.nih.gov. Accessed 1 March 2022
29. I.F. Myronyuk, V.I. Mandzyuk, V.M. Sachko, V.M. Gun'ko, Nanoscale Res. Lett. **11**, 508 (2016). https://doi.org/10.1186/s11671-016-1723-z
30. D.K.L. Tsang, B.J. Marsden, S.L. Fok, G. Hall, Carbon **43**, 2902 (2005). https://doi.org/10.1016/j.carbon.2005.06.009
31. P. Euzen, P. Raybaud, X. Krokidis, H. Toulhoat, J.-L. Le Loarer, J.-P. Jolivet, C. Froidefond, in *Handbook of Porous Solids*, ed. by F. Schüth, K. S. W. Sing, J. Weitkamp (Wiley-VCH Verlag GmbH, Weinheim, 2002), p. 1591.https://doi.org/10.1002/9783527618286.ch23b
32. M. Digne, P. Raybaud, P. Sautet, B. Rebours, H. Toulhoat, J. Phys. Chem. B **110**, 20719 (2006). https://doi.org/10.1021/jp061466s
33. A.E. Nelson, M. Sun, J. Adjaye, J. Phys. Chem. B **110**, 20724 (2006). https://doi.org/10.1021/jp0616720

34. G. Paglia, C.E. Buckley, A.L. Rohl, J. Phys. Chem. B **110**, 20721 (2006). https://doi.org/10.1021/jp061648m
35. H.P. Pinto, R.M. Nieminen, S.D. Elliott, Phys. Rev. B **70**, 125402 (2004). https://doi.org/10.1103/PhysRevB.70.125402
36. M. Sun, A.E. Nelson, J. Adjaye, J. Phys. Chem. B **110**, 2310 (2006). https://doi.org/10.1021/jp056465z
37. J.M. Cowley, Acta Cryst. **6**, 53 (1953). https://doi.org/10.1107/S0365110X53000119
38. A. Dauger, D. Fargeot, Radiat. Eff. **74**, 279 (1983). https://doi.org/10.1080/00337578308218421
39. V.P. Pakharukova, D.A. Yatsenko, E. Yu., Gerasimov, A.S. Shalygin, O.N. Martyanov, S.V. Tsybulya, J. Solid State Chem. **246**, 284 (2017). https://doi.org/10.1016/j.jssc.2016.11.032
40. K.E. Sickafus, J.M. Wills, N.W. Grimes, J. Am. Ceram. Soc. **82**, 3279 (1999). https://doi.org/10.1111/j.1151-2916.1999.tb02241.x
41. M. Rudolph, M. Motylenko, D. Rafaja, IUCrJ **6**, 116 (2019). https://doi.org/10.1107/S2052252518015786
42. P. Stadelmann, *JEMS* v3.67 (Saas-Fee, 2012). https://www.jems-swiss.ch. Accessed 2018
43. S.V. Tsybulya, G.N. Kryukova, Phys. Rev. B **77**, 024112 (2008). https://doi.org/10.1103/PhysRevB.77.024112
44. M. Rudolph, *cuDebye* v1.5 (Freiberg, 2018). https://github.com/Martin-Rudolph/cuDebye. Accessed 2019
45. G. Paglia, C.E. Buckley, A.L. Rohl, B.A. Hunter, R.D. Hart, J.V. Hanna, L.T. Byrne, Phys. Rev. B **68**, 144110 (2003). https://doi.org/10.1103/PhysRevB.68.144110
46. L. Kovarik, M. Bowden, D. Shi, N.M. Washton, A. Andersen, J.Z. Hu, J. Lee, J. Szanyi, J.-H. Kwak, C.H.F. Peden, Chem. Mater. **27**, 7042 (2015). https://doi.org/10.1021/acs.chemmater.5b02523
47. L. Kovarik, M. Bowden, A. Genc, J. Szanyi, C.H.F. Peden, J.H. Kwak, J. Phys. Chem. C **118**, 18051 (2014). https://doi.org/10.1021/jp500051j
48. I. Levin, D. Brandon, J. Am. Ceram. Soc. **81**, 1995 (1998). https://doi.org/10.1111/j.1151-2916.1998.tb02581.x
49. G. Paglia, C.E. Buckley, A.L. Rohl, R.D. Hart, K. Winter, A.J. Studer, B.A. Hunter, J.V. Hanna, Chem. Mater. **16**, 220 (2004). https://doi.org/10.1021/cm034917j
50. M. Rudolph, A. Salomon, A. Schmidt, M. Motylenko, T. Zienert, H. Stöcker, C. Himcinschi, L. Amirkhanyan, J. Kortus, C.G. Aneziris, D. Rafaja, Adv. Eng. Mater. **19**, 1700141 (2017). https://doi.org/10.1002/adem.201700141
51. R.-S. Zhou, R.L. Snyder, Acta Cryst. B **47**, 617 (1991). https://doi.org/10.1107/S0108768191002719
52. G. Yamaguchi, I. Yasui, W.-C. Chiu, Bull. Chem. Soc. Jpn **43**, 2487 (1970). https://doi.org/10.1246/bcsj.43.2487
53. L. Kovarik, M. Bowden, J. Szanyi, J. Catal. **393**, 357 (2021). https://doi.org/10.1016/j.jcat.2020.10.009
54. B.C. Lippens, J.H. de Boer, Acta Cryst. **17**, 1312 (1964). https://doi.org/10.1107/S0365110X64003267
55. Y. Repelin, E. Husson, Mater. Res. Bull. **25**, 611 (1990). https://doi.org/10.1016/0025-5408(90)90027-Y
56. L. Kovarik, M. Bowden, A. Andersen, N.R. Jaegers, N. Washton, J. Szanyi, Angew. Chem. **132**, 21903 (2020). https://doi.org/10.1002/ange.202009520
57. J.-B. d'Espinose de Lacaillerie, C. Fretigny, D. Massiot, J. Magn. Reson. **192**, 244 (2008). https://doi.org/10.1016/j.jmr.2008.03.001
58. R. Dupree, M.H. Lewis, M.E. Smith, Phil. Mag. A **53**, L17 (1986). https://doi.org/10.1080/01418618608242816
59. R. Prins, Angew. Chem. Int. Ed. **58**, 15548 (2019). https://doi.org/10.1002/anie.201901497
60. C. Wolverton, K.C. Hass, Phys. Rev. B **63**, 024102 (2000). https://doi.org/10.1103/PhysRevB.63.024102

61. C. Vinod Chandran, C.E.A. Kirschhock, S. Radhakrishnan, F. Taulelle, J.A. Martens, E. Breynaert, Chem. Soc. Rev. **48**, 134 (2019). https://doi.org/10.1039/C8CS00321A
62. B.E. Yoldas, J. Mater. Sci. **11**, 465 (1976). https://doi.org/10.1007/BF00540927
63. G. Mariotto, E. Cazzanelli, G. Carturan, R. Di Maggio, P. Scardi, J. Solid State Chem. **86**, 263 (1990). https://doi.org/10.1016/0022-4596(90)90142-K
64. M.K. Mardkhe, B. Huang, C.H. Bartholomew, T.M. Alam, B.F. Woodfield, J. Porous Mater. **23**, 475 (2016). https://doi.org/10.1007/s10934-015-0101-z
65. R. Tettenhorst, Clays Clay Miner. **28**, 373 (1980). https://doi.org/10.1346/CCMN.1980.028 0507
66. D. Massiot, F. Fayon, M. Capron, I. King, S.L. Calvé, B. Alonso, J.-O. Durand, B. Bujoli, Z. Gan, G. Hoatson, Magn. Reson. Chem. **40**, 70 (2002). https://doi.org/10.1002/mrc.984
67. P. Scherrer, Nachr. Ges. Wiss. Gottingen, Math.-Phys. Kl. **1918**, 98 (1918). https://eudml.org/doc/59018
68. S. Ďurovič, Chemical Papers (Chemické Zvesti) 23, 113 (1969), https://www.chempap.org/?id=7&paper=5979
69. V. Prostakova, D. Shishin, M. Shevchenko, E. Jak, Calphad **67**, 101680 (2019). https://doi.org/10.1016/j.calphad.2019.101680

Chapter 7
Interface Reactions Between the Metal Melt and the Filter Surface Activated by a Spark Plasma Sintering Process

Anton Salomon, Mykhaylo Motylenko, Martin Thümmler, and David Rafaja

7.1 Introduction

The understanding of the interface reactions between metal melts containing inclusions and the filter ceramic surfaces is essential for the development of functional materials for active and reactive filtration of metallic melts. On the laboratory scale, the filtration processes and the deposition of inclusions in ceramic filters are typically analyzed using impingement trials [1–3] or casting simulators [4], before the filters are subjected to prototype casting on industrial scale [5]. However, the results of such experiments are sometimes hardly to interpret. The main reason is that these experiments activate several simultaneous processes, which involve heterogeneous reactions between the metallic melt and the surface of the functional ceramics, production of inclusions in the melt and their deposition on the surface of the functional ceramics, and finally the solid state reactions and diffusion processes between the products of the heterogeneous reactions and the functionalized filter material. Furthermore, these processes are superimposed by the melt flow and/or by the macroscopic convection. Although the traditional laboratory experiments reveal important information about the thermal shock behavior [1, 6], melt flow rate [5, 7] and filtration efficiency [5, 7, 8] of the filters, which is usually complemented by the amount and chemical composition of the deposited non-metallic inclusions, they cannot substitute an in-depth investigation of the fundamental mechanisms of the interface and bulk reactions, which is required for a targeted development of the metal melt filters.

This chapter illustrates how the Spark Plasma Sintering (SPS) can be employed as a generally applicable but more controlled method to produce interface and reaction layers between metal melts and filter ceramics for advanced analysis

A. Salomon · M. Motylenko · M. Thümmler · D. Rafaja (✉)
Institute of Materials Science, Technische Universität Bergakademie Freiberg,
Gustav-Zeuner-Straße 5, 09599 Freiberg, Germany
e-mail: rafaja@iww.tu-freiberg.de

© The Author(s) 2024
C. G. Aneziris and H. Biermann (eds.), *Multifunctional Ceramic Filter Systems for Metal Melt Filtration*, Springer Series in Materials Science 337,
https://doi.org/10.1007/978-3-031-40930-1_7

of filtration processes. The SPS technology, which was originally developed for powder compaction and sintering [9], offers a high variability of process parameters, controlled heating and cooling rates, and widely adjustable dwell temperatures, holding times and atmospheres. In this work, SPS was utilized to melt the 42CrMo4 steel, aluminum and AlSi7Mg alloy that were in a direct contact with selected refractories like alumina, mullite, silicon oxide, titanium oxide, carbon-bonded alumina and carbon-bonded magnesia.

In contrast to other techniques, which bring molten metals in contact with the functionalized metal melt filters, the metal melting in a SPS apparatus minimizes the macroscopic melt flow, which is usually responsible for damage or even for removal of the newly formed reaction products and layers from the filter surface. Furthermore, the SPS allows working with extremely variable sample geometry including planar metal-ceramic interfaces in combination with bulk filter materials up to the powder mixtures of metal and ceramics with a huge contact area between the counterparts, with short reaction diffusion paths and with the finite geometry of the diffusion couples. The samples with planar geometry were utilized to analyze the sequence and morphology of new phases formed at the metal-ceramic interface. The experiments done on powder mixtures assisted in the identification and description of phases with narrow homogeneity ranges, and produced equilibrium-state samples for comparison with the thermodynamic calculations. Thus, the SPS-based melting technique provides complementary results with respect to the classical casting experiments or impingement tests, and helps to elucidate the reaction steps, which are frequently inaccessible, when the established methods are used.

7.2 Adaptation of the Spark Plasma Sintering for Metal Melting

The Spark Plasma Sintering or Field Assisted Sintering Technology (SPS/FAST) is based on resistive Joule heating through the pulsed electric direct current (DC) that passes electrically conducting graphite tools and/or the sample, if the sample is electrically conducting as well [9]. The Joule heating by the graphite die or directly by the sample itself allows for high heating rates. The typical current used for SPS/FAST is in the kA range. For sample having a diameter of about 2 cm, this current corresponds to the current density of approx. 1 kA/cm^2. Although the voltage on the sample is relatively low (max. 10 V), the heating power is still about 30 kW. The sintering atmosphere is usually vacuum or some inert gas like nitrogen or argon [9]. As the SPS/FAST process was developed mainly for compacting and short time sintering of refractories, the standard SPS/FAST tools are not constructed for an extensive melting of one of the components. During the fast melting experiments, however, the metal has to be molten. Therefore, new sample environments must be developed for such experiments, because the molten metal would leak from the sample environment and react quickly with the carbon present in the graphite tools.

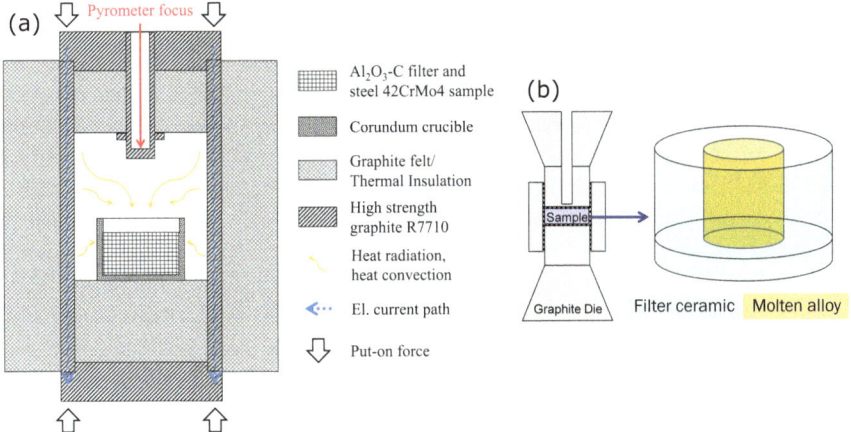

Fig. 7.1 Scheme of the high-speed furnace **a** and tool-in-tool setup **b** used for the sample production ((**b**) adopted from [11])

The first development is a corundum crucible, which is filled with the functionalized filter material and with the metal powder to be molten. The crucible itself is located in a tube made of high strength graphite, which serves as a heater (Fig. 7.1a). The outer diameter of the tube is about 100 mm, the thickness of the walls about 10 mm [10]. In this "high speed furnace", the samples are heated indirectly by the heat irradiated from the heater and by the heat convection. Because of the indirect heating, the heating rate is below 200 K min^{-1}, significantly faster than in a conventional furnace but slower than usual SPS. Still, this tool is useful for special experiments with large samples that should be kept at the dwell temperature for long holding times, for instance during the investigation of the metal infiltration into the real filter structures. A drawback of this tool is the expected inaccuracy of the temperature measurement by a pyrometer, which is focused to a graphite finger that is not in a direct contact with the sample under study.

The second sample environment is based on a "tool-in-tool" setup [11], in which the metal powder to be molten is placed within a crucible that is made from or coated by the functionalized filter ceramics. The ceramic parts were cut from suspension nozzles or produced by slip casting according to the procedures presented in references [1, 2, 7, 8, 12–14]. The size of the ceramic crucible was usually 20 mm in diameter with an internal cavity of 5 to 10 mm. The whole inserts were located at the position, which corresponds to the sample position in a standard SPS experiment (Fig. 7.1b). Due to the small size of the inserts and the direct contact between the heater and the sample, very high heating rates up to 1500 K min^{-1} were achieved [3]. As this setup offered much higher heating rates and an easier handling than the previous one, it was used for the majority of the SPS experiments.

7.3   Methods of Structure and Microstructure Analysis

On the microscopic scale, the solidified products of the chemical reactions between the molten metals and the surface of the ceramic filter were analyzed using scanning electron microscopy with energy dispersive X-ray spectroscopy (SEM/EDX) and electron backscatter diffraction (SEM/EBSD). The elemental analysis using SEM/EDX was complemented by the electron probe microanalysis with wavelength dispersive X-ray spectroscopy (EPMA/WDX). For SEM/EDX/EBSD, a high-resolution SEM LEO-1530 (Carl Zeiss AG, Germany) with field-emission cathode, an EDX detector (Bruker AXS) and a Nordlys II EBSD detector (HKL Technology) was used. The SEM imaging was performed using secondary electrons, back-scattered electrons or in a combined mode. The SEM/EDX/EBSD experiments were carried out at an acceleration voltage of 20 kV. The working distance for EBSD was 15 mm, the tilting angle of the sample 70° and the step size 0.3 μm. For identification of the Kikuchi patterns and for the evaluation of the measured data, the software package Channel 5 (HKL Technology) was used. For the EPMA/WDX measurements, an electron probe microanalyzer JXA-8230 SuperProbe (Jeol GmbH, Germany) with five-crystal spectrometers was used. The EPMA scans were performed with the step size of 0.5 μm.

The phase compositions of the solidified samples were analyzed using a Bragg–Brentano diffractometer URD 63 (Freiberger Praezisionsmechanik) that was equipped with a sealed X-ray tube with copper anode and with a curved graphite monochromator located in front of a scintillation detector. The X-ray diffraction (XRD) patterns patterns were collected between $2\theta = 20°$ and $150°$ with the step size of $\Delta 2\theta = 0.04°$, and with the counting time of 10 s per step. The phase composition of the samples was quantified by using the Rietveld method [15, 16] implemented in the computer program MAUD [17].

On the nanoscale, the samples were characterized using transmission electron microscopy (TEM), selected area electron diffraction (SAED) and energy dispersive X-ray spectroscopy (EDX/TEM). These analyses were done in a JEM 2200 FS transmission electron microscope (JEOL Ltd., Japan) at an acceleration voltage of 200 kV. The TEM samples were prepared by the focused ion beam method (FIB) with a Helios NanoLab 600i (FEI, USA) in form of thin slices.

7.3.1   Reactions Between Molten Steel and Corundum-Based Refractories with Different Carbon Contents

In analogy to the established entry nozzles [18–21], the newly developed carbon-bonded alumina (Al_2O_3-C) filters [8, 13, 14] are expected to react with molten steels and to form new interface layers between the metallic melt and the filter surface. Already the formation of the interface layer should significantly contribute to the removal of unwanted oxygen or other contaminating elements [10]. The interface

Table 7.1 Chemical composition of the utilized steel alloy 42CrMo4 as provided by the powder vender

Element	C	Cr	Mo	Si	Mn	Fe
Mass%	0.4	1.0	0.2	0.2	0.7	Balance

layer itself has to attract and to embed non-metallic inclusions. Generally, the chemical composition and the phase composition of the layer are considered as crucial factors influencing the agglomeration of non-metallic inclusions and their adherence to the filter surface [7, 22]. The carbon additions should improve the high-temperature mechanical properties of the alumina filters [23–26], their thermal shock resistance (via higher thermal conductivity and low thermal expansion) [8, 25] and their resistance against crack initiation and propagation [26]. In this section, the formation of a secondary alumina layer at the interface between the Al_2O_3-C filters with different carbon contents and molten steel 42CrMo4 is described. The chemical composition of the steel is summarized in Table 7.1. The reaction experiments presented here were performed using the tool-in-tool setup with the respective carbon-containing alumina ceramic acting as the reaction vessel.

7.3.2 Time-Dependent Layer Growth and Reaction Scheme on the Microscopic Level

According to the thermodynamic model proposed by Zienert et al. [27], Al_2O_3 is partially decomposed by liquid iron. The decomposition of alumina is facilitated by the presence of carbon in the reaction zone, which also reacts with dissolved oxygen to CO and/or CO_2 [27, 28]. The CO/CO_2 gas leaves partially the reaction zone. The formation and evaporation of CO/CO_2 decreases the carbon concentration in the reaction zone and decelerates the decomposition of Al_2O_3. Nevertheless, if sufficient amount of carbon is present in the filter material, alumina is permanently dissolved in the liquid iron [27–29], which leads to an increase of the concentration of aluminum in the melt.

The microstructural consequences of a short-time reaction between the carbon binder and the dissolved oxygen stemming from the decomposed alumina are illustrated in Fig. 7.2. The contact region between the solidified steel and Al_2O_3(-C) becomes carbon deficient or even carbon-free. The thickness of the carbon-depleted zone does not exceed 40 – 50 μm and is almost independent of the dwell time. Liquid iron, which contributes substantially to the decomposition of alumina, can be observed in form of solidified droplets in the filter constituents containing simultaneously alumina and carbon. In the functional filter coating containing no carbon (Fig. 7.2a), the solidified droplets were observed only in the carbon-bonded alumina filter struts. In the functionalized filter coatings containing carbon (Fig. 7.2b and c), the solidified droplets were observed already in the functional coating. Increased

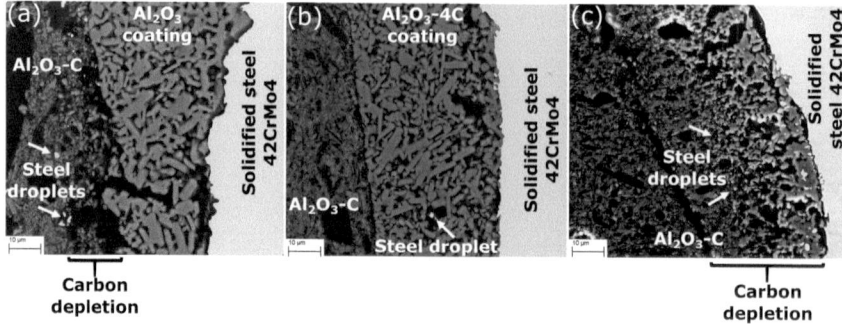

Fig. 7.2 SEM micrographs (BSE contrast) of functionalized Al₂O₃-C filters that were exposed to molten steel 42CrMo4 in the SPS apparatus at 1600 °C for 1 min. The functional coatings consist of corundum without carbon (**a**), Al_2O_3-C with 4 mass% C (**b**), and Al_2O_3-C with 30 mass% C (**c**) (adopted from [29])

concentrations of Al and O in the solidified steel melt were revealed by the SEM/EDS measurements [11, 29]. Figure 7.3a shows the grooves stemming from the escape of the CO/CO_2 bubbles, which were formed during the reaction of free oxygen with the carbon binder. The escape of the CO/CO_2 bubbles is the main reason for the decarburization of the C-bonded Al_2O_3 filter ceramic.

The results of these experiments confirmed that the local concentration of carbon is a crucial factor influencing the rate of the concurrent reactions, namely alumina dissolution and the formation of secondary corundum. Although these reactions occur in contact with liquid iron, the dissolution of Al_2O_3 is facilitated by the presence of carbon, which reacts with free oxygen, while the formation of secondary corundum is assisted by the absence of carbon. For this reason, secondary corundum forms preferentially at the filter surface, i.e., in the carbon-depleted zone, where a sufficiently

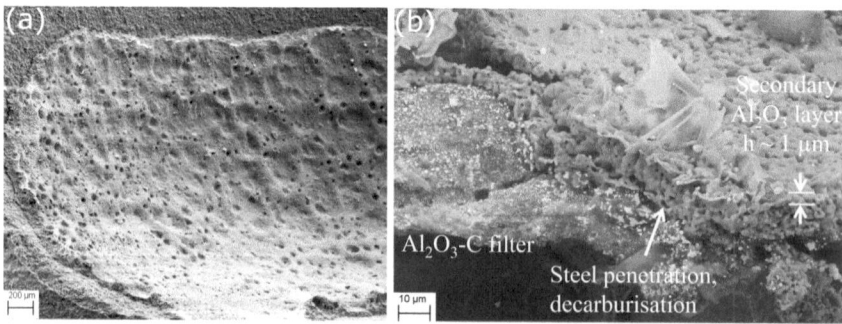

Fig. 7.3 SEM micrographs (SE contrast) of the surface of Al_2O_3-C with 30 mass% of C after SPS at 1600 °C: Layer formed on top of the ceramic body after 1 min with grooves stemming from the escape of CO/CO_2 bubbles (**a**); penetration of the steel melt into the filter, decarburization of Al_2O_3-C and formation of a secondary Al_2O_3 layer with a thickness below 1 μm after 10 min (**b**) ((**a**) adopted from [11])

high amount of dissolved aluminum and oxygen is present (Fig. 7.3b) [27, 29]. The formation of secondary corundum was observed mainly in samples that were kept at high temperature for a longer time (Fig. 7.4). Besides the dwell time, the rate of the CO/CO_2 gas formation is another important factor controlling the secondary corundum formation, because it affects the local concentration of carbon in the filter ceramics.

Also the morphology of the Al_2O_3(-C) filters, their porosity, and the morphology of secondary corundum are additional important factors influencing both reaction processes [27, 29]. At a sufficiently high local concentration of carbon, the secondarily formed corundum can be reduced like the primary corundum in the filter, if it forms only small and separated crystallites that can be soaked by liquid iron containing carbon from the binder [27, 29]. On the other hand, compact and large grains of secondary corundum, which are in contact with carbon-depleted steel melt (Fig. 7.4a-c), stay stable. Hence, the formation of a dense, impenetrable layer of secondary corundum provides an efficient barrier for carbon diffusion from the Al_2O_3-C filter and iron penetration into the filter, and thus inhibits the decomposition of Al_2O_3 [29]. Still, the reaction layer serves as a docking site for non-metallic inclusions contained in the melt (Fig. 7.5).

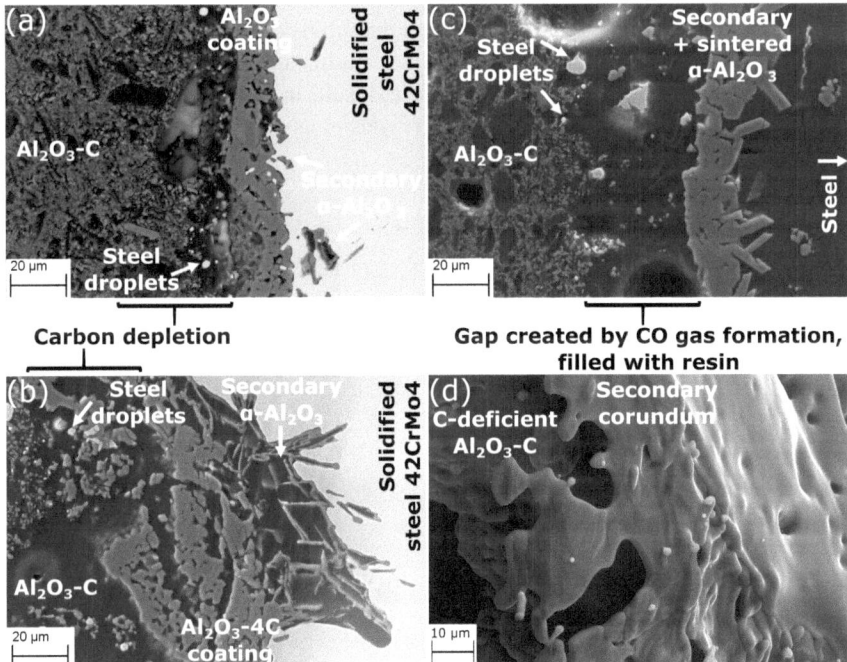

Fig. 7.4 SEM micrographs (SE contrast) of functionalized Al_2O_3-C filters that were exposed to molten steel 42CrMo4 in SPS apparatus at 1600 °C for **a–c** 30 min and **d** for 5 min. The functional coatings consist of **a** corundum without carbon, **b** Al_2O_3-C with 4 mass% C, and **c, d** Al_2O_3-C with 30 mass% C (adopted from [29])

Fig. 7.5 SEM micrograph (BSE/SE) of agglomerated Al_2O_3 inclusions deposited on the Al_2O_3-C filter after 5 min at 1600 °C. The secondary Al_2O_3 layer "bridges" between the agglomerate and the filter

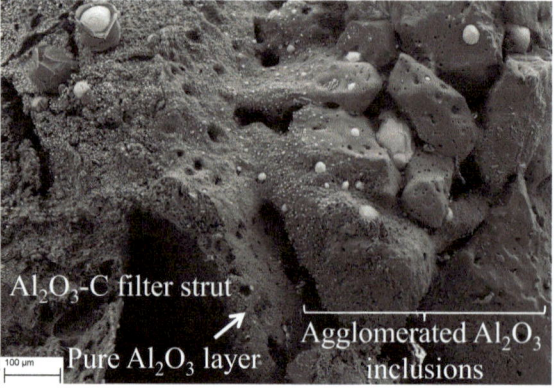

The majority of the inclusions consisted of aluminum oxides. Other endogenous inclusions contained MnS (SG $Fm\bar{3}m$) [11]. According to Sims and Dahle [30], the MnS precipitates can be classified as type II and III. They were found at the high-angle grain boundaries, free surface of the steel droplets and at the interfaces formed between the ceramics and the steel. In addition, MnS in form of a thin film covered partly the Al_2O_3 inclusions (Fig. 7.6). The formation of the MnS film was explained by the heterogeneous nucleation of the sulfide on oxide surfaces and by the supersaturation of Mn and S in the areas of final solidification [31]. Unfortunately, the described MnS formation on inclusion surfaces results in so-called duplex inclusions that are detrimental for the mechanical properties of any (cast) metallic part [32].

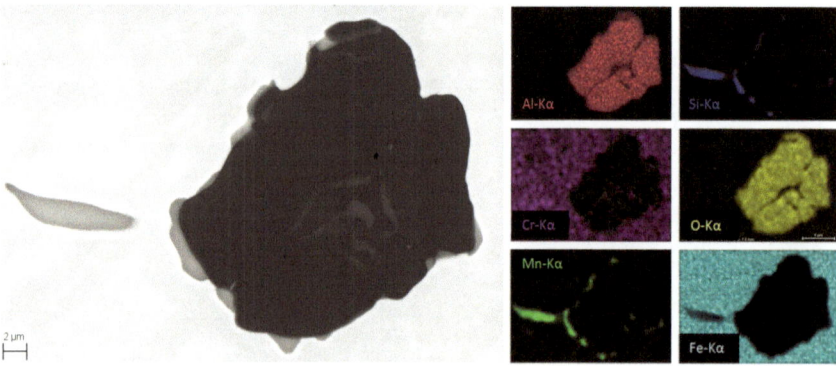

Fig. 7.6 SEM micrograph (BSE contrast) of an Al_2O_3 particle within solidified 42CrMo4 steel (bright gray) partly covered with MnS (dark gray) after 1 min at 1600 °C, and EDX element distribution maps on the right (adopted from [11])

7.3.3   Interface Reactions Preceding the Secondary Corundum Formation

High efficiency of the filtration processes requires a good adhesion of the non-metallic inclusions and/or reaction products to the functionalized filter surface in order to inhibit their spalling and the contamination of the metallic melt. It is assumed that materials having the same chemical composition or similar crystal structures possess better adhesion than fully incompatible compounds. As the secondary corundum grows on the filter surface that contains primary corundum, a local epitaxial growth was considered. However, the SEM/EBSD measurements, which were carried out on the metal melt filter covered by a functional Al_2O_3-C coating with 4 mass% C, revealed that the secondary corundum grows in form of interconnected, almost single-crystalline platelets, but without any pronounced orientation relationship to the corundum substrate (Fig. 7.7). This result motivated further studies, which should elucidate the early stages of the chemical reactions at the interface between the functionalized filter surface and the metallic melt that precede the growth of secondary corundum. These studies were carried out on functional Al_2O_3 coatings containing no carbon, 4 mass% C or 30 mass% C that were treated for a short time (1 – 2 min) at 1600 °C in the SPS apparatus or in a steel-casting simulator. The chemical composition of the interface layers was analyzed using EDX in TEM. The phase composition of the interlayers was concluded from the SAED patterns.

The chemical analysis of the interlayers using EDX confirmed the presence of iron, aluminum and oxygen, which are involved in the carbothermic reaction, as well as the presence of the alloying elements from the steel, mainly silicon [13, 29]. The SAED analysis revealed that the interlayers are almost amorphous (Figs. 7.8 and 7.9). Still, the contrasts observed in the TEM micrographs indicated fluctuations in the chemical and possibly in the phase composition, including the presence of nanocrystalline phases. Detailed analysis of the SAED patterns (Fig. 7.10) disclosed that the nanocrystalline phases contain wuestite (FeO, SG $Fm\overline{3}m$), spinel-like phases with the chemical composition $A^{2+}B_2^{3+}O_4^{2-}$ (A and B being Fe, Al, Si and/or Mg contaminant) and the space group $Fd\overline{3}m$ and garnet-like structures, most probably $Fe_3Al_2(SiO_4)_3$ (SG $Ia\overline{3}d$) [29]. After longer reaction times of a few minutes, the nanocrystalline oxide phases are replaced by corundum. A possible transient phase is metastable γ-Al_2O_3, which possesses a spinel-like crystal structure (SG $Fd\overline{3}m$) containing highly mobile structural vacancies [33]. The structural vacancies facilitate the necessary fast exchange of metallic (cationic) species to remove iron and/or silicon from the spinel-like structure of $A^{2+}B_2^{3+}O_4^{2-}$.

When the seeds of secondary corundum in the interlayer are in a direct contact with the primary corundum from the functionalized coating, the secondary corundum grows epitaxially on the primary corundum (regions 1 and 2 in Fig. 7.11). Platelets of the secondary corundum, which grow from aluminum and oxygen dissolved in the steel melt on the amorphous interlayer, do not develop any pronounced orientation relationship to the filter surface (region 3 in Fig. 7.11), because the amorphous interlayer inhibits the epitaxial growth. Whereas the amorphous layers form on the surface

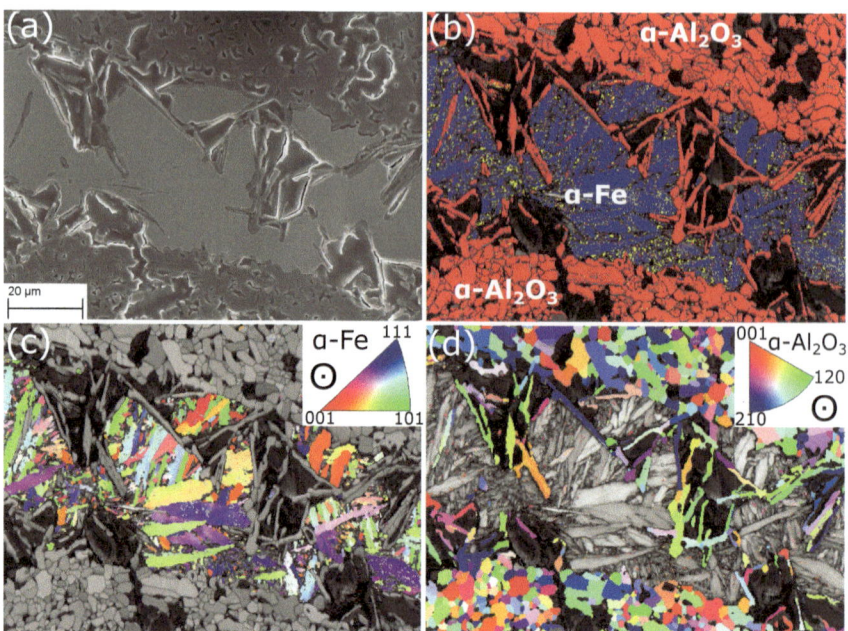

Fig. 7.7 **a** SEM micrograph (SE contrast) of the interface between the solidified steel 42CrMo4 and the functional Al_2O_3-C coating with 4 mass% C after the reaction for 30 min at 1600 °C. **b** Phase map, **c** local orientations of the grains, and **d** local orientations of corundum. Yellow spots in panel (**b**) are MnS inclusions with an fcc crystal structure. The orientation distribution maps in panels (**c**) and (**d**) are related to the direction perpendicular to the image plane (adopted from [29])

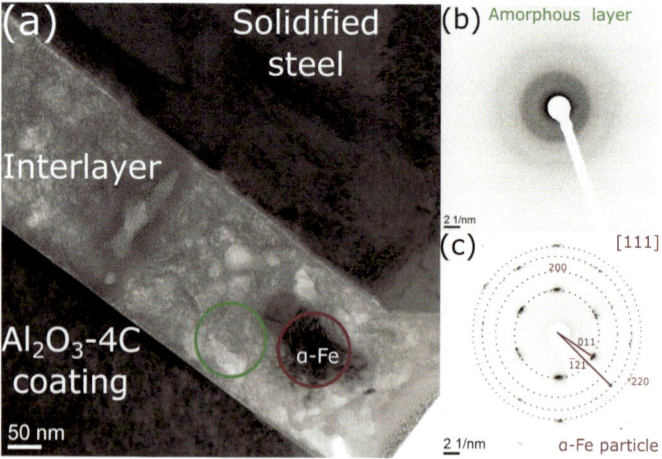

Fig. 7.8 **a** TEM micrograph of an interlayer formed between solidified steel 42CrMo4 and an α-Al_2O_3 coating with 4 mass% C. The reaction experiment was conducted in SPS apparatus at the dwell time of 1 min at 1600 °C. **b** SAED pattern of the interface layer. **c** SAED of an α-Fe particle embedded in the amorphous layer (adopted from [29])

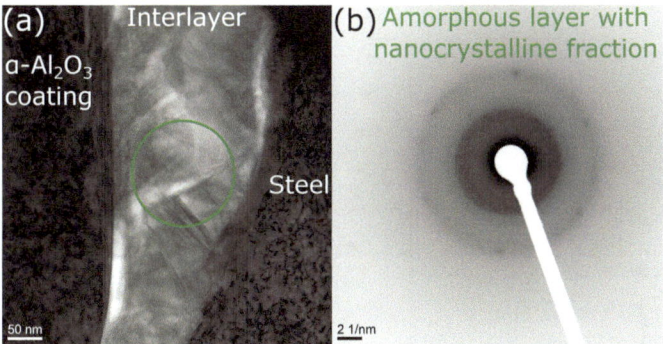

Fig. 7.9 a TEM micrograph of an interface layer formed between solidified steel 42CrMo4 and a carbon-free functional corundum coating. The reaction experiment was conducted in a steel-casting simulator. The immersion time was 2 min at 1600 °C. **b** SAED pattern of the interface layer (adopted from [29])

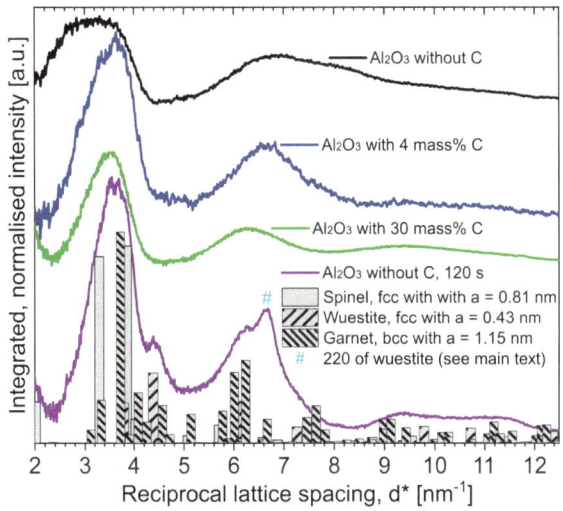

Fig. 7.10 Diffracted intensities obtained by integrating the SAED patterns of the observed Fe–O-(Al–Si)-containing interlayers in the azimuthal direction and plotted versus the reciprocal lattice spacing, calculated according to $d^* = 1/d = \sqrt{h^2 + k^2 + l^2}/a$ for the respective lattice parameter a and the diffraction indices hkl. Theoretical peak positions and the diffracted intensities are shown in a bar chart with differently shaded bars at the bottom of the figure for a spinel structure with a lattice parameter of about 0.81 nm, for fcc wuestite with a lattice parameter of about 0.43 nm and a garnet phase with a lattice parameter of about 1.15 nm. The diffracted intensities were calculated using kinematical diffraction theory assuming random orientation of crystallites (adopted from [29])

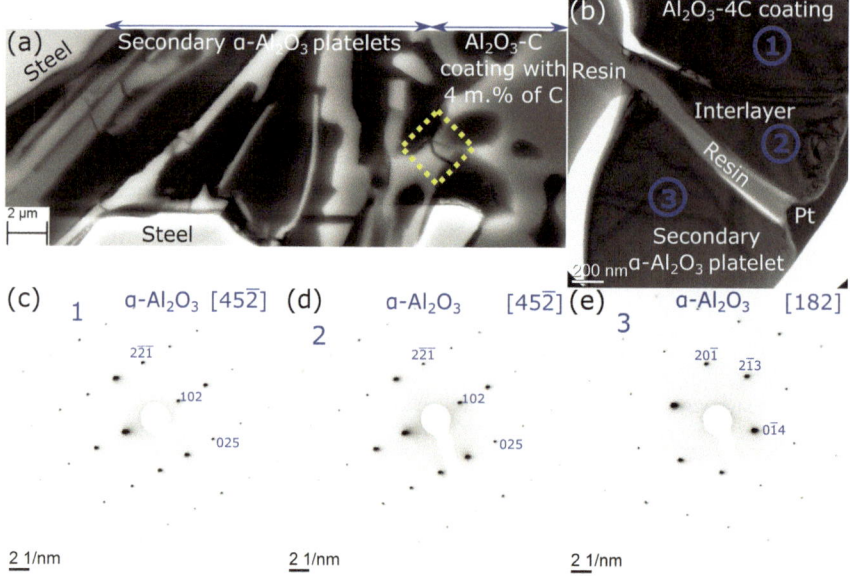

Fig. 7.11 **a** SEM micrograph of a transition region between the Al$_2$O$_3$-C functional coating containing 4 mass% C and the α-Al$_2$O$_3$ platelets showing the position of the FIB sample (highlighted area). The reaction time was 30 min at 1600 °C. **b** TEM micrograph showing an Al$_2$O$_3$ interlayer grown on the surface of the functional coatings and an Al$_2$O$_3$ platelet. The corresponding SAED patterns are displayed in panels (**c**)–(**e**) (adopted from [29])

of all Al$_2$O$_3$-C filters independently of their carbon content, the secondary corundum possesses different morphologies, which depend on the local carbon supply and thus on growth kinetics [29], as discussed in the previous section.

7.4 Reactions Between Molten Steel and Carbon-Bonded Corundum Coated with Carbon-Bonded Magnesia

Carbon-bonded magnesia (MgO-C) combines low thermal expansion and high thermal conductivity of the graphite binder with a low wettability of the composite against slags and metal melts [34]. In analogy to Al$_2$O$_3$-C (cf. previous sections), MgO in MgO-C is decomposed and CO/CO$_2$ gas is formed, when the refractory is brought in contact with liquid steel [14, 27, 34–37]. Dissolved magnesium and oxygen form secondary MgO, which deposits as a dense, thin layer at the steel/ refractory interface [36, 38, 39], analogously to the formation of secondary Al$_2$O$_3$. After a longer reaction time, MgAl$_2$O$_4$ whisker-like fibers (SG $Fd\overline{3}m$) formed at the interface between the MgO-C coating and the Al$_2$O$_3$-C filter substrate (Fig. 7.12). The fibers grow during a vapor–liquid-solid process [4, 37]. The reactants are Mg,

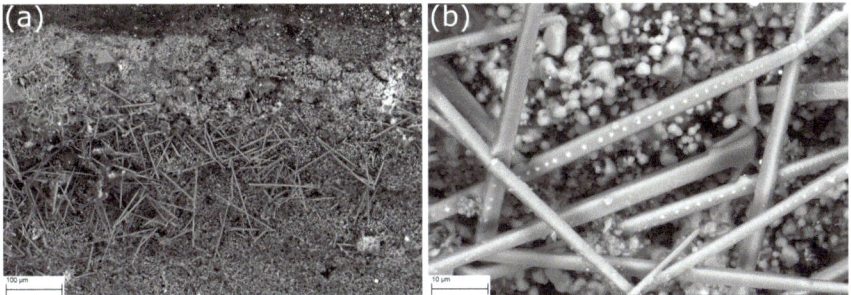

Fig. 7.12 SEM micrograph (BSE) of the contact area between MgO-C coating and Al_2O_3-C substrate ceramic after 60 min at 1600 °C. **a** Fibre formation observed in the contact area; **b** enlarged cut-out of (**a**) showing iron incorporation (bright spheres) within the fibres (adopted from [37])

Al and O that are dissolved in the liquid steel penetrating the porous MgO-C coating [40].

In addition to $MgAl_2O_4$ whiskers, a dense layer of $MgAl_2O_4$ formed at the former MgO-C/steel interface (Fig. 7.13a) [4, 37]. The Al enrichment on the edges of the pre-existing MgO grains of the MgO-C coating produced $MgO/MgAl_2O_4$ core/rim structures within the filter ceramic coating [37]. Vice versa, $MgAl_2O_4$ formed also on the edges or rims of exogenous Al_2O_3 inclusions that were present in the steel (Fig. 7.13b).

The SPS/FAST experiments helped to confirm the assumed "reactive" behavior of the carbon-bonded magnesia coatings deposited on the carbon-bonded alumina filter substrates. The $MgAl_2O_4$ formation on the filter surface and on exogenous inclusions was accompanied by a reduction of the oxygen concentration in the steel. The dense layer of $MgAl_2O_4$ is expected to represent a limiting factor for a further penetration of the molten steel into the interior of the filter material. Thus, a compact $MgAl_2O_4$ layer will slow down the reaction kinetics in analogy with a compact Al_2O_3 coating discussed above. In contrast, the $MgAl_2O_4$ whiskers act against the shrinkage of the

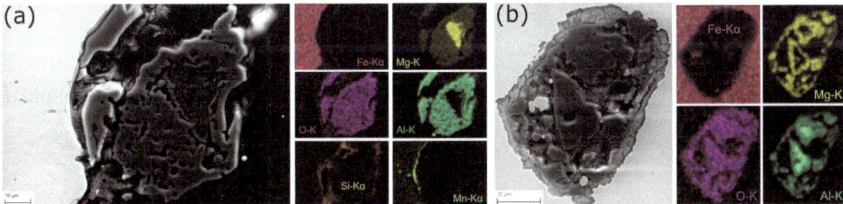

Fig. 7.13 **a** SEM micrograph (BSE) of the steel/MgO-C interface (crosscut) after 60 min at 1600 °C with corresponding EDX elemental mapping. $MgAl_2O_4$ formation on the rim of an MgO grain, Fe–Si–O-rich (fayalite) formation and MnS collection (only Mn shown) is indicated. **b** SEM micrograph (SE contrast) of an Al_2O_3 cluster found within the 42CrMo4 steel after 1 min at 1600 °C and corresponding EDX elemental maps (adopted from [37])

filter during the filtration process [6] and could also be used to increase the strength of the functionalized filter ceramic [4, 37, 40].

7.4.1 Reactions Between Molten Aluminum or AlSi7Mg0.6 Alloy and Selected Carbon-Free Oxide Coatings

Corundum foam filters are used as a standard tool for the aluminum melt filtration [2]. Still, alternative functional materials with specific wettability [41], selective reactivity and interactivity with respect to certain impurities and inclusions are sought. This contribution focusses on mullite ($3Al_2O_3 \cdot 2SiO_2$, SG *Pbam*), amorphous SiO_2 and rutile (TiO_2, SG $P4_2/mnm$), which are considered for production of functional coatings. The main phenomena under study were the reaction of the functional coatings with molten aluminum and molten AlSi7Mg0.6 alloy, formation of the reaction layers on the filter surface and the mechanisms of the adhesion between the reaction layers and the functionalized filter surface. Chemical composition of the AlSi7Mg0.6 alloy is presented in Table 7.2. The majority of the experiments discussed in this subchapter were performed in the SPS apparatus using the tool-in-tool setup. Additional experiments were done via impingement tests [1–3] and sessile drop experiments [41]. The temperature of the melt was always 750 °C.

7.4.2 Reaction Between AlSi7Mg0.6 and Corundum

For short dwell time (1 min), no reaction layers were detected on the surface of the corundum filters. The results of the SPS treatment and the impingement tests were identical, as for both methods the liquid AlSiMg0.6 did not penetrate into the almost dense ceramics. For the SPS-treated samples, slightly better wetting of the corundum filter by the molten aluminum alloy was observed. This phenomenon was attributed to the destruction of the thin oxide layer formed on the surface of original AlSi7Mg0.6 particles in the SPS melting process. The oxide layer is disrupted through the percolation of the electric current and removed by the reducing conditions in the SPS process, which are established by the application of the graphite tooling [3].

Table 7.2 Chemical composition of the aluminum alloy powder used for the melt production as specified by the manufacturer

Alloy	Elemental composition [mass%]							Particle size	Source
	Al	Si	Mg	Fe	Cu	Mn	Zn		
AlSi7Mg0.6	Balance	7.07	0.61	0.09	0.09	0.02	0.28	45 – 100 μm	TLS Technik

Fig. 7.14 SEM micrograph (Inlens/SE detector) of the alumina/aluminum alloy interface after 60 min at 750 °C, the contamination due to the EPMA line scan is visible, with corresponding EPMA line scan results (adopted from [3])

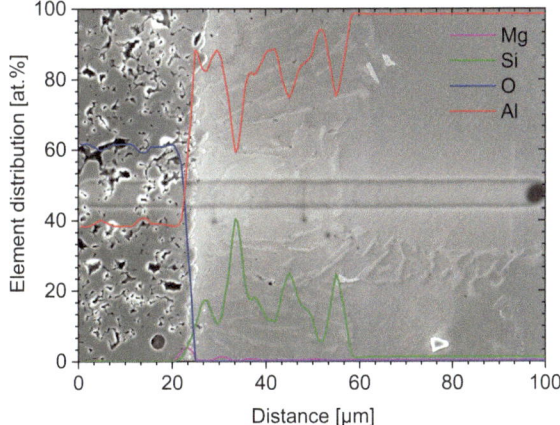

After a longer holding time (30 – 60 min), accumulation of Mg at the surface of the corundum filter was observed (Fig. 7.14). Although no distinct reaction layer was detected, the accumulation of Mg at the Al_2O_3/AlSi7Mg0.6 interface is regarded as a preliminary stage of the $MgAl_2O_4$ formation. Thermodynamic calculations supported this hypothesis and revealed that $MgAl_2O_4$ is the stable phase in the whole temperature range between 20 to 1000 °C [3]. According to the thermodynamic simulation, the formation of the $MgAl_2O_4$ spinel would, in the equilibrium state, fully consume the 0.6 mass% of Mg, which are available in AlSi7Mg0.6 [42, 43]. Consequently, no Mg-containing intermetallic phases, e.g., Mg_2Si (SG $Fm\overline{3}m$), should form. Silicon contained in the melt should precipitate upon cooling. These results of the thermodynamic simulation were proven experimentally. No Mg_2Si was found, while Si formed needles in the solidified melt (Fig. 7.14).

7.4.3 Reaction of AlSi7Mg0.6 with Amorphous SiO₂ and Mullite

It is known from literature [44–47] that aluminum silicates, e.g., mullite ($3Al_2O_3 \cdot 2SiO_2$), decompose in contact with molten aluminum, and form alumina and silica. Alumina usually crystallizes as corundum (α-Al_2O_3), while silica is reduced by liquid aluminum to silicon, which dissolves in the melt. This reaction results in additional formation of Al_2O_3 [44], which occurs in form of metastable alumina phases like η-, θ- and/or γ-Al_2O_3 [48–50]. The presence of Mg in the aluminum alloy is expected to alter the interfacial reactions and to produce additional phases like $MgAl_2O_4$ spinel and/or $MgSiO_3$ pyroxene.

Individual steps of the reaction between AlSi7Mg0.6 melt and SiO_2 or mullite were visualized by model experiments that were performed with compacted powder samples containing 50 mol% of the aluminum alloy and 50 mol% of SiO_2 or mullite.

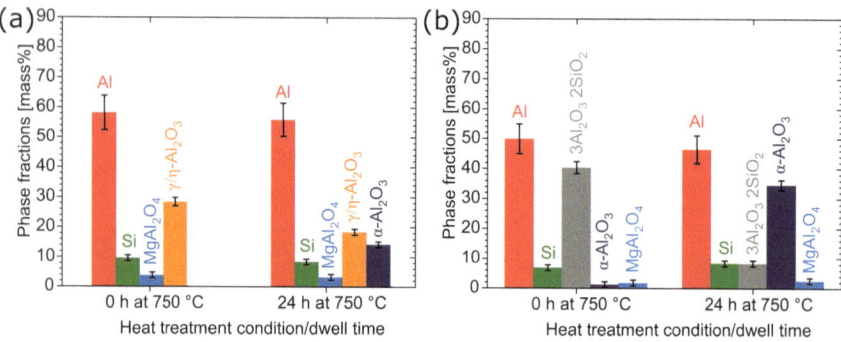

Fig. 7.15 **a** Phase composition of the powder mixture of AlSi7Mg0.6 and amorphous SiO$_2$ after heat treatment at 750 °C without dwell (0 h) and after 24 h at 750 °C (24 h). **b** Phase composition of the powder mixture of AlSi7Mg0.6 and mullite after heat treatment at 750 °C without dwell (0 h) and after 24 h at 750 °C (24 h) (adopted from [44])

A short-time reaction between molten AlSi7Mg0.6 alloy and amorphous SiO$_2$ at 750 °C led to the decomposition of SiO$_2$, and to the formation of Si, γ/η-Al$_2$O$_3$ and MgAl$_2$O$_4$ in the solidified sample (Fig. 7.15a, 0 h). Note that Fig. 7.15 includes only crystalline phases, because the phase fractions were determined using XRD. With longer reaction time (24 h at 750 °C), the amount of Al$_2$O$_3$ increases, and a part of metastable alumina (γ/η-Al$_2$O$_3$) transforms to corundum (α-Al$_2$O$_3$).

Reaction experiments, which were carried out with an alumina plate coated with amorphous SiO$_2$, revealed that Mg is attracted to the filter wall in the initial stages of the reaction process (Fig. 7.16a). However, as no magnesium silicate was found in the powder mixtures sintered for short time (Fig. 7.15a), it can be concluded that the functional SiO$_2$ coating dissolves rather than a magnesium silicate, e.g., MgSiO$_3$, forms. Free silicon forms Si precipitates, which were observed in the solidified AlSi7Mg0.6 melt (Fig. 7.16b). Oxygen reacts with Al and Mg to Al$_2$O$_3$ and to MgAl$_2$O$_4$ (Fig. 7.15a). In the planar sample, the coexistence of these compounds led to the formation of MgAl$_2$O$_4$ precipitates embedded in an Al$_2$O$_3$ layer (left part of Fig. 7.16b). The secondary Al$_2$O$_3$ layer is separated from the solidified melt by an Mg-rich and O-depleted interlayer (central part of Fig. 7.16b).

The reaction between molten AlSi7Mg0.6 and mullite leads to the formation of Si, Al$_2$O$_3$ and MgAl$_2$O$_4$ (Fig. 7.15b) as well. In this case, however, alumina exists in the thermodynamically stable form of corundum already in the initial stages of the reaction process. The formation of corundum instead of the metastable alumina phases is facilitated by the presence of α-Al$_2$O$_3$, which is a product of the mullite decomposition [44–47]. Figure 7.17 depicts the Si needles in the solidified AlSi7Mg0.6 melt and illustrates the preferential diffusion of Mg into the mullite coating along the pore and grain boundaries.

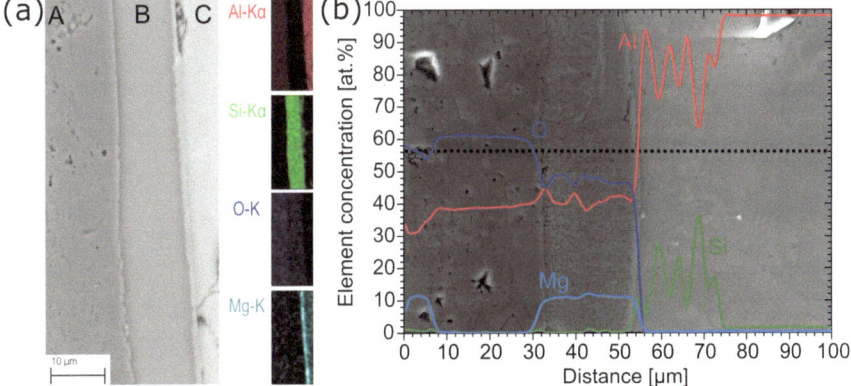

Fig. 7.16 a Left: SEM micrograph (BSE contrast) of the amorphous SiO_2 coating (B) on the Al_2O_3 substrate (A) after 1 min dwell at 750 °C in contact with the solidified AlSi7Mg0.6 alloy (C); Right: Results of EDX element mapping of the area on the left showing Mg enrichment on the interface SiO_2/alloy. **b** SEM micrograph (SE contrast) and overlaid EPMA line scan track (cross-centred black dashed line) with corresponding quantitative elemental analysis showing the α-Al_2O_3 substrate with an $MgAl_2O_4$ precipitate (left side) covered with a newly formed, Mg-enriched, Al- and O-containing and Si-free layer (middle) that replaces the SiO_2 coating after contact with the AlSi7Mg0.6 alloy (right side) at 750 °C for 30 min. The composite structure grows columnar-like (adopted from [44])

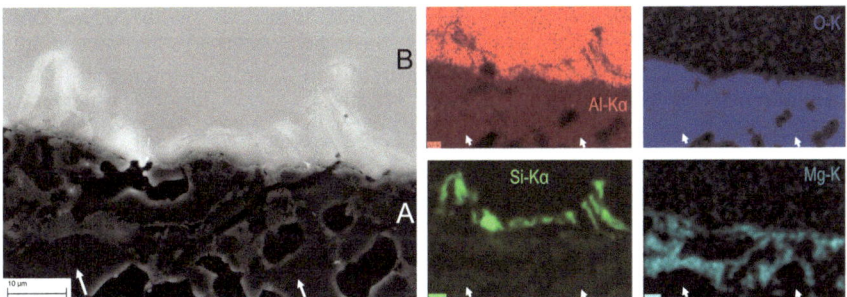

Fig. 7.17 Left: SEM micrograph (BSE contrast) of the mullite coating (**A**) after 30 min dwell at 750 °C in contact with the solidified AlSi7Mg0.6 alloy (**B**), pores and cracks (black) in the coating; Right: Results of EDX element mapping of the area on the left side showing Mg enrichment and Si depletion in the coating and silicon needles in the alloy. The centres of two large grains showing the stoichiometric composition of mullite ($3Al_2O_3 \cdot 2SiO_2$) are marked with white arrows (adopted from [44])

7.4.4 Reaction Between Al and AlSi7Mg0.6 Melts and TiO$_2$

In contact with molten Al or aluminum alloys, TiO_2 is reduced in analogy to SiO_2. Typical reaction products are Ti_3O_5, Ti_2O_3, TiO and Ti [12, 51–57]. Metallic Ti is dissolved in the melt. In our SPS experiments that were performed with the

Fig. 7.18 Phase composition of the powder mixture of pure Al and TiO$_2$ after heat treatment at 750 °C without dwell (0 h) and after 24 h at 750 °C (24 h)

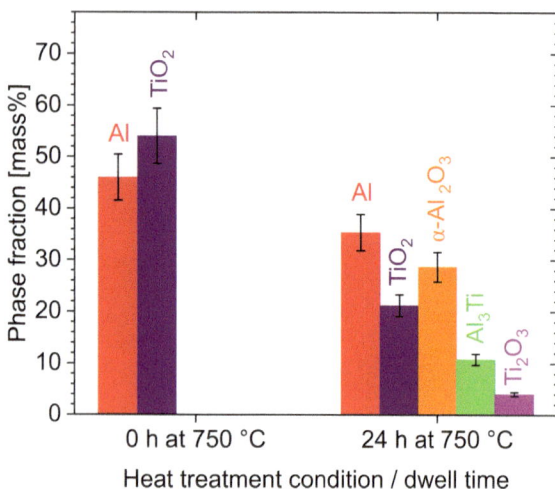

powder mixtures of Al and TiO$_2$ (rutile, SG$P4_2/mnm$), α-Al$_2$O$_3$ and Ti$_2$O$_3$ with the corundum crystal structures (SG$R\bar{3}c$) and Al$_3$Ti (SG $I4/mmm$) were found after 24 h at 750 °C (Fig. 7.18). In literature, also the formation of intermetallic phases like (Al,Si)$_3$Ti (SG$I4/mmm$), Al$_{60}$Si$_{12}$Ti$_{28}$ or Ti$_7$Al$_5$Si$_{12}$ [12, 52, 55, 58, 59] was reported. Ti$_2$O$_3$ present in our samples is a product of the TiO$_2$ reduction. α-Al$_2$O$_3$ is a product of the reaction of Al with released oxygen. The formation of corundum (α-Al$_2$O$_3$) is possibly facilitated by the presence of Ti$_2$O$_3$ having the same crystal structure like α-Al$_2$O$_3$. Al$_3$Ti formed, because the Ti concentration in the melt exceeded locally the solubility limit for Ti in Al (< 1 at% at 750 °C).

The experiments carried out with planar samples disclosed that α-Al$_2$O$_3$ grows in form of a compact layer on the surface of the functional TiO$_2$ coating (Fig. 7.19). The analysis of the Al$_2$O$_3$/TiO$_2$ interface using SEM/EBSD revealed that Al$_2$O$_3$ grows on TiO$_2$ frequently with the orientation relationship $(001)_{TiO_2} \parallel (100)_{Al_2O_3}$ and $[010]_{TiO_2} \parallel [001]_{Al_2O_3}$ (Fig. 7.20). The round brackets stand for parallel lattice planes, the square brackets for parallel crystal axes. The slight misorientations (Fig. 7.20b) compensate the differences in the interatomic distances. The pronounced orientation relationship between α-Al$_2$O$_3$ and TiO$_2$ is a consequence of the similarity of their crystal structures (Fig. 7.20c, orientation relationships were plotted using VESTA 3 [60]) and of the possible presence of Ti$_2$O$_3$ as a thin interlayer. Such heteroepitaxial growth improves the adhesion of Al$_2$O$_3$ to TiO$_2$ significantly. Moreover, a compact corundum layer impedes the direct contact between molten aluminum and the TiO$_2$ coating, which inhibits a further reduction of rutile (and other titanium oxides) and the production of free titanium. The lack of reduced titanium retards or even hinders the formation of Al$_3$Ti.

In contact with molten AlSi7Mg0.6 alloy, TiO$_2$ forms quite quickly MgTiO$_3$ (SG $R\bar{3}$), cf. Fig. 7.21a. Additional oxygen, which is required for this reaction, is supplied by the melt. Silicon present in the AlSi7Mg0.6 melt forms Si precipitates.

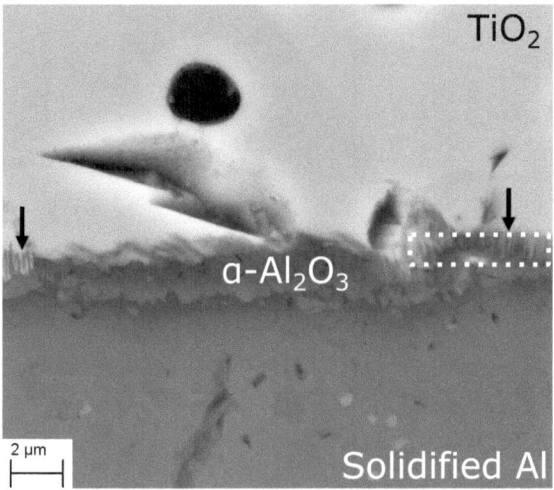

Fig. 7.19 SEM micrograph (BSE contrast) of the rutile coating brought in contact with molten Al at 750 °C for 300 min. The main reaction layer contains corundum (α-Al_2O_3). The white-dotted box shows the position of a FIB lamella, which was investigated by TEM and SAED (cf. Fig. 7.25) in order to explain the nature of the stripes marked by black arrows (adopted from [52])

The highly variable thickness of the $MgTiO_3$ layer (Fig. 7.21b) suggests that the diffusion of Mg and O into TiO_2 is accelerated, when the TiO_2 coating contains cracks, voids or grain boundaries. The analysis of the $MgTiO_3$/TiO_2 interface using SEM/EBSD revealed that the grains of magnesium titanate and rutile possess frequently an orientation relationship $(100)_{MgTiO_3} \parallel (001)_{TiO_2}$ and $[001]_{MgTiO_3} \parallel [010]_{TiO_2}$ (Fig. 7.22). Round brackets denote the lattice planes, while square brackets denote the crystal axes. This orientation relationship was confirmed by a local analysis of the $MgTiO_3$/TiO_2 interface using SAED in TEM (Fig. 7.23). The ab initio simulations using the density functional theory [52] revealed that the above orientation relationship reduces the total energy of the $MgTiO_3$/TiO_2 interface in comparison with the total energy of the individual bulk components.

Diffraction contrasts visible in the TEM micrograph (Fig. 7.23a) explained the slight misorientation between $MgTiO_3$ and TiO_2, which was first concluded from the results of the SEM/EBSD analysis (Fig. 7.22b). The diffraction contrasts stem from geometrically necessary misfit dislocations, which are distributed almost equidistantly along the $MgTiO_3$/TiO_2 interface and which compensate the lattice misfit between $MgTiO_3$ and TiO_2. These dislocations produce small angle grain boundaries that are visible by SEM/EBSD as slight departures from the ideal orientation relationship.

Furthermore, the local analysis of the reaction zone between TiO_2 and the solidified AlSi7Mg0.6 melt using TEM and SAED contributed essentially to the understanding of the reaction kinetics in this system. Within the original TiO_2 coating, bands of Ti_2O_3 (SG $R\overline{3}c$, corundum type) and $MgTiO_3$ (SG $R\overline{3}$) having

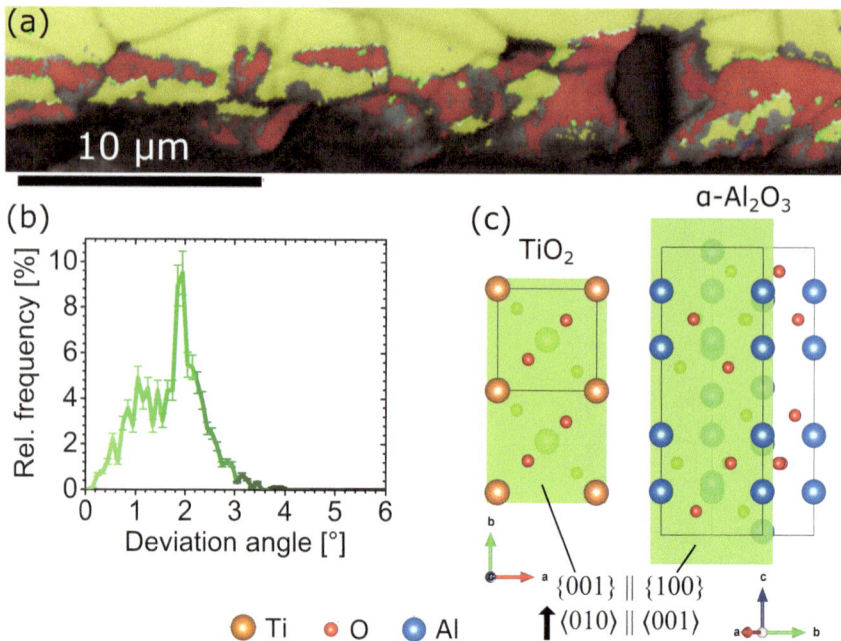

Fig. 7.20 **a** EBSD phase map of the TiO₂/Al interface after 60 min at 750 °C. Rutile is plotted in yellow, corundum in red. Black areas within the colorized region are pores, non-indexed bottom region is solidified aluminum. The green lines mark the interfaces between rutile and corundum crystallites having the orientation relationship $(001)_{TiO_2} \| (100)_{Al_2O_3}$ and $[010]_{TiO_2} \| [001]_{Al_2O_3}$. A histogram of the local deviations from this orientation relationship is shown in (**b**). **c** Model of rutile and corundum in the above orientation relationship plotted using VESTA 3 [60] (adopted from [52])

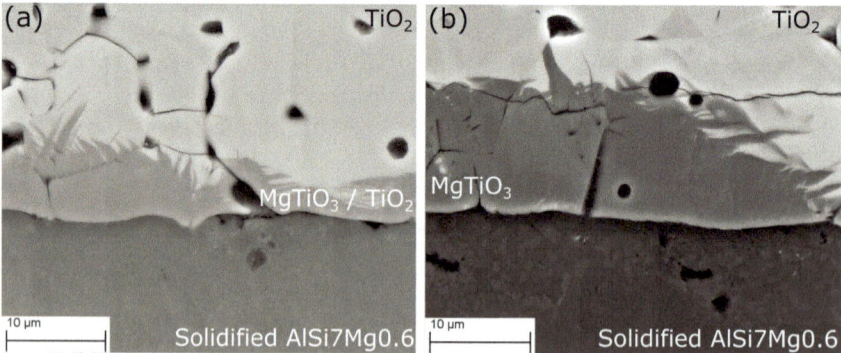

Fig. 7.21 SEM micrographs (BSE contrast) of the rutile coating brought in contact with molten AlSi7Mg0.6 alloy for 60 min (**a**) and 300 min (**b**) at 750 °C. At the interface between the coating and the liquid alloy, a MgTiO₃ layer formed (adopted from [52])

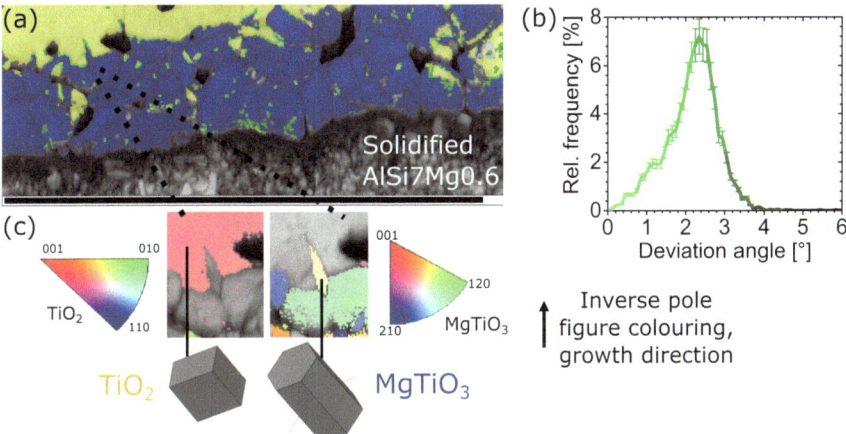

Fig. 7.22 Results of the EBSD analysis performed on the rutile coating after 60 min at 750 °C. **a** Phase map showing the distribution of rutile (yellow) and MgTiO$_3$ (blue). Pores/voids are reproduced in black, the grey region at the bottom is the solidified alloy AlSi7Mg0.6. The green lines mark the interfaces between the TiO$_2$ and MgTiO$_3$ grains having the orientation relationship $(100)_{MgTiO_3} \| (001)_{TiO_2}$ and $[001]_{MgTiO_3} \| [010]_{TiO_2}$. A histogram of the local deviations from this orientation relationship is shown in (**b**). The above orientation relationship is illustrated in figure (**c**) (adopted from [52])

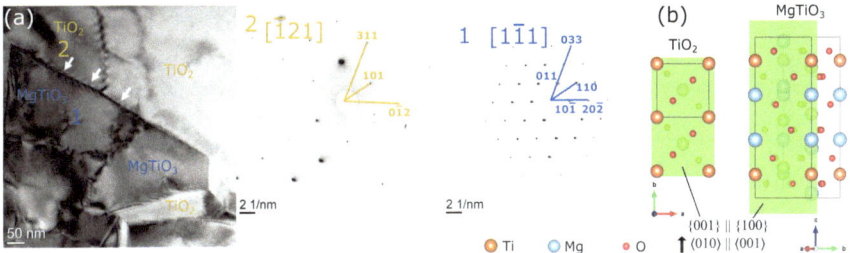

Fig. 7.23 **a** A TEM micrograph of the MgTiO$_3$/TiO$_2$ interface after 60 min at 750 °C. A phase boundary with geometrically necessary misfit dislocations is marked by white arrows. The orientation relationship $(100)_{MgTiO_3} \| (001)_{TiO_2}$ and $[001]_{MgTiO_3} \| [010]_{TiO_2}$ shown in panel **b** was verified by SAED. Crystal structures were created with the VESTA 3 software [60] (adopted from [52])

a pronounced heteroepitaxial orientation relationship $(001)_{Ti_2O_3} \| (001)_{MgTiO_3}$ and $[100]_{Ti_2O_3} \| [100]_{MgTiO_3}$ at their interfaces were detected (Fig. 7.24). The thickness of the individual stripes was about 200 nm. The formation of such structures is facilitated by a local oxygen deficiency. When a sufficient amount of magnesium but a low amount of oxygen diffuse into the TiO$_2$ coating, the oxygen that is required for the formation of MgTiO$_3$ is produced by the reduction of TiO$_2$, which leads to the formation of Ti$_2$O$_3$. Note that Ti$_2$O$_3$ and MgTiO$_3$ are miscible in a broad range of the Ti and Mg concentrations [61] and that MgTiO$_3$ belongs to the group of ilmenites, which can accommodate various metallic species in their crystal structure,

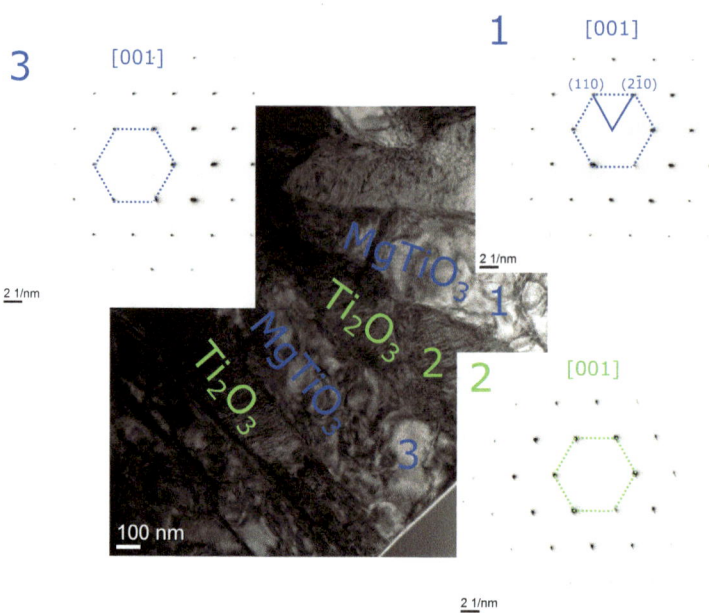

Fig. 7.24 TEM and SAED of the MgTiO$_3$/Ti$_2$O$_3$ interfaces of the TiO$_2$ coating that was in contact with the AlSi7Mg0.6 melt for 300 min at 750 °C. The growth direction is perpendicular to the plane of the image (adopted from [52])

e.g., Fe and Mn in addition to or instead of Mg and Ti [62]. Thus, the MgTiO$_3$/Ti$_2$O$_3$ composite layer can incorporate a variety of alloying or foreign elements.

When TiO$_2$, Mg and O come into contact with alumina, e.g., near the interface between the Al$_2$O$_3$ filter wall and the functional rutile coating, MgAl$_2$O$_4$ spinel forms as an additional phase (Fig. 7.25). Whereas the titanium magnesium oxides exist in the form of ilmenite (MgTiO$_3$, SG$R\bar{3}$) and spinel (Mg(Ti$_{1-x}$Mg$_x$)O$_4$, SG$Fd\bar{3}m$) [63], MgAl$_2$O$_4$ exists only in the spinel form. Furthermore, Mg(Ti$_{1-x}$Mg$_x$)O$_4$ and MgAl$_2$O$_4$ are not miscible. Therefore, a complex microstructure consisting of the TiO$_2$, MgTiO$_3$ and MgAl$_2$O$_4$ phases with segregated Ti and Al atoms forms in the reaction zone containing Ti, Mg, O and Al (Fig. 7.25). Nevertheless, the SAED patterns (Fig. 7.25) indicated that these phases possess a pronounced orientation relationship, which was described as $(100)_{TiO_2} \| (11\bar{1})_{MgAl_2O_4} \| (001)_{MgTiO_3}$ and $[001]_{TiO_2} \| [011]_{MgAl_2O_4} \| [120]_{MgTiO_3}$ and which is substantiated by the crystal structure models depicted in Fig. 7.26. Also in this case, the heteroepitaxy of the neighboring phases is believed to improve the adhesion of the reaction layers formed during the filtration process.

The layer formation in each case provided evidence for the "reactive" interaction between metal melt and the functional filter ceramic coating, i.e. rutile, consuming magnesium and dissolved detrimental oxygen.

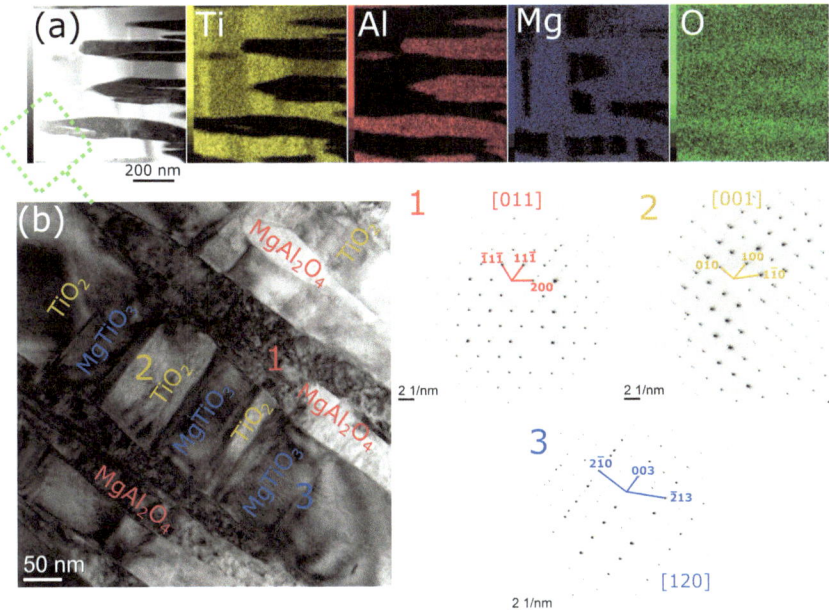

Fig. 7.25 Element maps (**a**) and TEM micrograph (**b**) of the stripes from Fig. 7.19. Individual phases were assigned using a combination of chemical analysis (EDX) and SAED. The viewing direction is perpendicular to the sample surface and corresponds to the growth direction of corundum into rutile (adopted from [52])

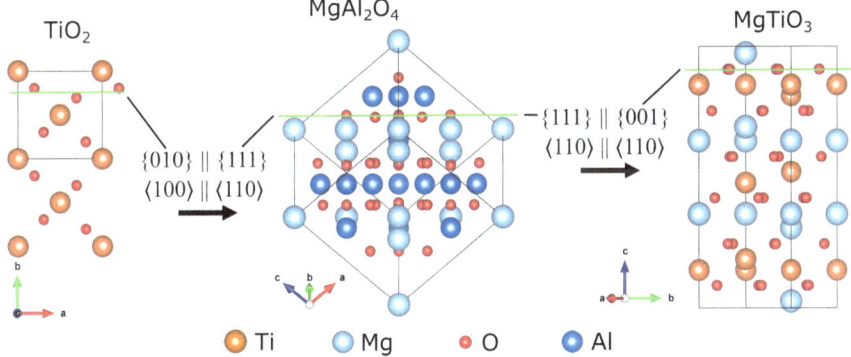

Fig. 7.26 Crystal structure models of TiO_2, $MgAl_2O_4$ and $MgTiO_3$ mutually oriented according to the orientation relationships that were identified using SAED (cf. Fig. 7.25). The parallel planes highlighted in green mark close-packed oxygen sublattice planes in each structure (highly distorted in rutile), arrows mark the corresponding parallel directions. The crystal structures were created with VESTA 3 [60] (adopted from [52])

The binding mechanism between the newly formed magnesium titanate layer and rutile was also investigated systematically to make sure that no spalling or peel-off of $MgTiO_3$ could contaminate the filtered alloy melt. The $MgTiO_3$ layer growth at the interface between the AlSi7Mg0.6 melt and rutile depends strongly on the local orientation of the TiO_2 grains.

7.5   Summary and Conclusions

In this contribution, the reaction processes between functionalized metal melt filters and molten 42CrMo4 steel, pure aluminum melt and molten AlSi7Mg0.6 alloy were studied. The samples were produced in a Spark Plasma Sintering apparatus with modified tooling. This experimental setup allowed very fast heating of the samples, controlled melting of the metals and the investigation of reaction processes at constant reaction temperatures and without convection of the melt.

The initial products of the reaction between molten 42CrMo4 steel and carbon-bonded corundum (Al_2O_3-C) filters were dissolved aluminum and oxygen, CO/CO_2 gas and an amorphous reaction layer, which contained Fe, Al, O as well as the alloying elements and impurities from the steel. After a longer reaction time, nanocrystalline wuestite, garnets, spinel-like phases including metastable alumina, and secondary corundum formed. The production of CO/CO_2, the formation of reaction layers and the growth of secondary corundum contributed significantly to the reduction of the oxygen content in the steel melt. When the secondary corundum grows directly on the primary corundum from the metal melt filter, the growth is epitaxial. The habitus of the secondary corundum (attached platelets or compact layers) depends on the local availability of carbon in the reaction zone, on the carbon concentration in the Al_2O_3-C filter and on the filter morphology.

When the carbon-bonded corundum is coated with carbon-bonded magnesium oxide, $MgAl_2O_4$ spinel forms in a carbothermic reaction between corundum from the Al_2O_3-C substrate and MgO. This reaction is facilitated by the presence of iron from the molten 42CrMo4 steel. Iron that penetrates into the porous MgO-C coating promote the formation of $MgAl_2O_4$ whiskers at the interface between the MgO-C coating and the Al_2O_3-C substrate. These whiskers are supposed to strengthen the filter ceramics, which is an additional benefit to the removal of oxygen from the melt, to the increased flotation of oxygen and inclusions present in the melt by the CO/CO_2 bubbles and to growth of secondary $MgAl_2O_4$ as docking sites for further deposition of endogenous inclusions.

In contact with molten aluminum, functional coatings made from corundum stay intact. Still, they serve as substrate for secondary corundum, which forms from aluminum and oxygen dissolved in the melt. In contrast to Al_2O_3, SiO_2 dissolves completely. Reduced silicon is solved in the melt. Free oxygen reacts with Al to metastable alumina (γ/η-Al_2O_3), which transforms later to corundum (α-Al_2O_3). Mullite ($3Al_2O_3 \cdot 2SiO_2$) decomposes to α-Al_2O_3 and SiO_2. SiO_2 dissolves quickly; α-Al_2O_3 serves as substrate for secondary corundum. Consequently, no metastable alumina phases are formed, when mullite is brought in contact with molten Al. In functional coatings made from rutile, TiO_2 is reduced by molten Al to Ti_2O_3. Free oxygen (including the original oxygen present in the melt) reacts with Al to corundum.

The reactions between molten AlSi7Mg0.6 alloy and the functional coatings made from corundum, mullite and SiO_2 are accompanied by the formation of $MgAl_2O_4$, which is a product of the reaction of Mg and O with Al_2O_3. The other reactions and processes are the same like for the Al melt. Mullite decomposes into corundum and SiO_2. SiO_2 dissolves. Reduced Si is solved in the melt. Oxygen reacts with Al and Mg to Al_2O_3 and $MgAl_2O_4$. On the surface of the corundum and mullite filters, where corundum is present (in mullite as a product of the $3Al_2O_3 \cdot 2SiO_2$ decomposition), Al_2O_3 crystallizes as corundum. If Al_2O_3 cannot grow directly on corundum substrate, e.g., on the surface of a SiO_2 coating, metastable alumina phases are formed first. The TiO_2 coatings react with Mg and O from the melt to $MgTiO_3$. If the local concentration of Mg is higher than the local concentration of O, TiO_2 is reduced to Ti_2O_3. When $MgTiO_3$ comes in contact with Al_2O_3 that stems, e.g., from the skeleton of the filter, $MgAl_2O_4$ and TiO_2 or Ti_2O_3 are formed.

Our study illustrated the importance of the spinel-like structures, which were found in all filter/melt combinations under consideration. In all cases, the spinel-like phases (in particular $MgAl_2O_4$) accommodated oxygen and impurities from the melts. Furthermore, the spinel-like phases are expected to dock the inclusions having the spinel structure, e.g., the metastable alumina phases. For these reasons, the spinel-like phases are considered as possible candidates for production of the functional coatings covering the metal melt filters.

Another finding of this study concerns the role of the epitaxial or heteroepitaxial growth of secondary coatings during the filtration process. Although many of the (primary) functional coatings under study decomposed in contact with the metal melt, the solved elements were captured in oxides, which formed in a secondary "deposition" process. It was shown that almost all oxides involved in this process are able to grow by mutual heteroepitaxy. This was proven for Al_2O_3, Ti_2O_3 (both having the corundum structure), TiO_2 (rutile), $MgAl_2O_4$ (spinel) and $MgTiO_3$ (ilmenite).

Acknowledgements The former CRC 920 project collaborators are highly acknowledged for their contributions: Milan Dopita, Tilo Zienert, Lilit Amirkhanyan, Marcus Emmel, Claudia Voigt, Eva Jäckel and Anne Schmidt. Furthermore, we thank Christiane Ullrich, Steffen Dudczig, Jana Hubalkova, Brigitte Bleiber, Dietrich Heger, Beate Kutzner, Christian Schimpf, Diane Hübgen, Katrin Becker, Astrid Leuteritz, Karin Müller and Galina Savinykh for their support with production, preparation and analysis of the samples. The electron probe microanalyzer was funded by the German Research Foundation in the frame of the Major Research Instrumentation Program under the project number 395240765. The German Research Foundation (DFG) is acknowledged for funding the Collaborative Research Centre 920—Project-ID 169148856, subproject A06.

References

1. C. Voigt, T. Zienert, P. Schubert, C.G. Aneziris, J. Hubalkova, J. Am. Ceram. Soc. **97**, 2046 (2014). https://doi.org/10.1111/jace.12977
2. C. Voigt, E. Jäckel, C.G. Aneziris, J. Hubalkova, Ceram. Int. **39**, 2415 (2013). https://doi.org/10.1016/j.ceramint.2012.09.001
3. A. Salomon, T. Zienert, C. Voigt, E. Jäckel, O. Fabrichnaya, D. Rafaja, C.G. Aneziris, Adv. Eng. Mat. **15**, 1206 (2013). https://doi.org/10.1002/adem.201300114
4. S. Dudczig, C.G. Aneziris, M. Emmel, G. Schmidt, J. Hubalkova, H. Berek, Ceram. Int. **40**, 16727 (2014). https://doi.org/10.1016/j.ceramint.2014.08.038
5. C. Voigt, E. Jäckel, F. Taina, T. Zienert, A. Salomon, G. Wolf, C.G. Aneziris, P. Le Brun, Metall. Mater. Trans. B **48**, 497 (2016). https://doi.org/10.1007/s11663-016-0869-5
6. M. Emmel, C.G. Aneziris, F. Sponza, S. Dudczig, P. Colombo, Ceram. Int. **40**, 13507 (2014). https://doi.org/10.1016/j.ceramint.2014.05.033
7. M. Emmel, C.G. Aneziris, J. Mater. Res. **28**, 2234 (2013). https://doi.org/10.1557/jmr.2013.56
8. M. Emmel, C.G. Aneziris, Ceram. Int. **38**, 5165 (2012). https://doi.org/10.1016/j.ceramint.2012.03.022
9. O. Guillon, J. Gonzalez-Julian, B. Dargatz, T. Kessel, G. Schierning, J. Räthel, M. Herrmann, Adv. Eng. Mat. **16**, 830 (2014). https://doi.org/10.1002/adem.201300409
10. H.U. Kessel, J. Hennicke, Interceram **56**, 164 (2007)
11. A. Salomon, M. Emmel, S. Dudczig, D. Rafaja, C.G. Aneziris, Adv. Eng. Mat. **15**, 1235 (2013). https://doi.org/10.1002/adem.201300119
12. A. Salomon, C. Voigt, O. Fabrichnaya, C.G. Aneziris, D. Rafaja, Adv. Eng. Mat. **19**, 1700106 (2017). https://doi.org/10.1002/adem.201700106
13. A. Schmidt, A. Salomon, S. Dudczig, H. Berek, D. Rafaja, C.G. Aneziris, Adv. Eng. Mat. **19**, 1700170 (2017). https://doi.org/10.1002/adem.201700170
14. C.G. Aneziris, S. Dudczig, M. Emmel, H. Berek, G. Schmidt, J. Hubalkova, Adv. Eng. Mat. **15**, 46 (2013). https://doi.org/10.1002/adem.201200199
15. H.M. Rietveld, Acta Crystallogr. **22**, 151 (1967). https://doi.org/10.1107/S0365110X67000234
16. H.M. Rietveld, J. Appl. Crystallogr. **2**, 65 (1969). https://doi.org/10.1107/S0021889869006558
17. L. Lutteroti, S. Matthies, H.R. Wenk, CPD Newsletter (IUCr) May 1999, 21. http://www.mx.iucr.org/iucr-top/comm/cpd/Newsletters/no21may1999/art17/art17.htm
18. S.N. Singh, Metall. Trans **5**, 2165 (1974). https://doi.org/10.1007/BF02643930
19. A.S. Kondrat'ev, V.N. Popov, L.M. Aksel'rod, M.R. Baranovskii, S.A. Suvorov, N.B. Tebuev, Refractories **31**, 384 (1991). https://doi.org/10.1007/BF01281545
20. R. Dekkers, B. Blanpain, P. Wollants, F. Haers, C. Vercruyssen, B. Gommers, Ironmak. Steelmak. **29**, 437 (2002). https://doi.org/10.1179/030192302225004584
21. J. Poirier, Metall. Res. Technol. **112**, 410 (2015). https://doi.org/10.1051/metal/2015028
22. M. Emmel, C.G. Aneziris, G. Schmidt, D. Krewerth, H. Biermann, Adv. Eng. Mat. **15**, 1188 (2013). https://doi.org/10.1002/adem.201300118
23. X. Wu, Y. Ranglack-Klemm, J. Hubálková, J. Solarek, C.G. Aneziris, A. Weidner, H. Biermann, Ceram. Int. **47**, 3920 (2020). https://doi.org/10.1016/j.ceramint.2020.09.255
24. J. Solarek, C. Himcinschi, Y. Klemm, C.G. Aneziris, H. Biermann, Carbon **122**, 141 (2017). https://doi.org/10.1016/j.carbon.2017.06.041
25. H. Zielke, T. Wetzig, C. Himcinschi, M. Abendroth, M. Kuna, C.G. Aneziris, Carbon **159**, 324 (2020). https://doi.org/10.1016/j.carbon.2019.12.042
26. M. Neumann, T. Wetzig, J. Fruhstorfer, V. Lampert, H. Jelitto, G.A. Schneider, C.G. Aneziris, Ceram. Int. **46**, 11198 (2020). https://doi.org/10.1016/j.ceramint.2020.01.141
27. T. Zienert, S. Dudczig, O. Fabrichnaya, C.G. Aneziris, Ceram. Int. **41**, 2089 (2015). https://doi.org/10.1016/j.ceramint.2014.10.004
28. R. Khanna, S. Kongkarat, S. Seetharaman, V. Sahajwalla, ISIJ Int. **52**, 992 (2012). https://doi.org/10.2355/isijinternational.52.992
29. A. Salomon, M. Motylenko, D. Rafaja, Adv. Eng. Mat. **24**, 2100690 (2021). https://doi.org/10.1002/adem.202100690

30. C.E. Sims, F.B. Dahle, Trans. Am. Foundrymen's Assoc. **46**, 65 (1938)
31. H. Ohta, H. Suito, ISIJ Int. **46**, 480 (2006). https://doi.org/10.2355/isijinternational.46.480
32. S. Henschel, J. Gleinig, T. Lippmann, S. Dudczig, C.G. Aneziris, H. Biermann, L. Krüger, A. Weidner, Adv. Eng. Mat. **19**, 1700199 (2017). https://doi.org/10.1002/adem.201700199
33. M. Rudolph, M. Motylenko, D. Rafaja, IUCrJ **6**, 116 (2019). https://doi.org/10.1107/S20522 52518015786
34. M.-A. Faghihi-Sani, A. Yamaguchi, Ceram. Int. **28**, 835 (2002). https://doi.org/10.1016/S0272-8842(02)00049-4
35. M. Bavand-Vandchali, F. Golestani-Fard, H. Sarpoolaky, H.R. Rezaie, C.G. Aneziris, J. Eur. Cer. Soc. **28**, 563 (2008). https://doi.org/10.1016/j.jeurceramsoc.2007.07.009
36. M. Ahmadi Najafabadi, M. Hirasawa, M. Sano, ISIJ Int. **36**, 1366 (1996). https://doi.org/10.2355/isijinternational.36.1366
37. A. Salomon, M. Dopita, M. Emmel, S. Dudczig, C.G. Aneziris, D. Rafaja, J. Eur. Cer. Soc. **35**, 795 (2015). https://doi.org/10.1016/j.jeurceramsoc.2014.09.033
38. V. Brabie, Steel Res. **68**, 54 (1997). https://doi.org/10.1002/srin.199700542
39. S. Jansson, V. Brabie, P. Jönsson, Ironmak. Steelmak. **33**, 389 (2006). https://doi.org/10.1179/174328106X113977
40. Z. Xie, F. Ye, J. Wuhan Univ. Technol. Mat. Sci. Ed. **24**, 896 (2009). https://doi.org/10.1007/s11595-009-6896-1
41. C. Voigt, L. Ditscherlein, E. Werzner, T. Zienert, R. Nowak, U. Peuker, N. Sobczak, C.G. Aneziris, Mater. Design **150**, 75 (2018). https://doi.org/10.1016/j.matdes.2018.04.026
42. T. Zienert, O. Fabrichnaya, Adv. Eng. Mat. **15**, 1244 (2013). https://doi.org/10.1002/adem.201 300113
43. L. Dreval, T. Zienert, O. Fabrichnaya, J. Alloys Compds. **657**, 192 (2016). https://doi.org/10.1016/j.jallcom.2015.10.017
44. A. Salomon, T. Zienert, C. Voigt, M. Dopita, O. Fabrichnaya, C.G. Aneziris, D. Rafaja, Corros. Sci. **114**, 79 (2017). https://doi.org/10.1016/j.corsci.2016.10.023
45. K.J. Brondyke, J. Am. Ceram. Soc. **36**, 171 (1953). https://doi.org/10.1111/j.1151-2916.1953.tb12860.x
46. S. Afshar, C. Allaire, JOM **48**, 23 (1996). https://doi.org/10.1007/BF03222938
47. A.L. Yurkov, I.A. Pikhutin, Refract. Ind. Ceram. **50**, 212 (2009). https://doi.org/10.1007/s11148-009-9184-x
48. A.E. Standage, M.S. Gani, J. Am. Ceram. Soc. **50**, 101 (1967). https://doi.org/10.1111/j.1151-2916.1967.tb15049.x
49. C. Marumo, J.A. Pask, J. Mater. Sci. **12**, 223 (1977). https://doi.org/10.1007/BF00566262
50. P. Mossino, D. Vallauri, F.A. Deorsola, L. Pederiva, R. Dal Maschio, G. Scavino, I. Amato, Metallurgia Italiana **97**, 25 (2005)
51. M. Ilatovskaia, G. Savinykh, O. Fabrichnaya, J. Phase Equ. Diff. **38**, 175 (2017). https://doi.org/10.1007/s11669-016-0509-4
52. A. Salomon, L. Amirkhanyan, C. Ullrich, M. Motylenko, O. Fabrichnaya, J. Kortus, D. Rafaja, J. Eur. Cer. Soc. **38**, 5590 (2018). https://doi.org/10.1016/j.jeurceramsoc.2018.07.052
53. I. Gheorghe, H.J. Rack, Mater. Sci. Technol. **18**, 1079 (2002). https://doi.org/10.1179/026708 302225005990
54. P. Shen, H. Fujii, K. Nogi, Acta Mater. **54**, 1559 (2006). https://doi.org/10.1016/j.actamat.2005.11.024
55. S. Avraham, P. Beyer, R. Janssen, N. Claussen, W.D. Kaplan, J. Eur. Cer. Soc. **26**, 2719 (2006). https://doi.org/10.1016/j.jeurceramsoc.2005.06.024
56. J. Pan, J.H. Li, H. Fukunaga, X.G. Ning, H.Q. Ye, Z.K. Yao, D.M. Yang, Compos. Sci. Technol. **57**, 319 (1997). https://doi.org/10.1016/S0266-3538(96)00127-3
57. Z.-C. Chen, T. Takeda, K. Ikeda, Compos. Sci. Technol. **68**, 2245 (2008). https://doi.org/10.1016/j.compscitech.2008.04.006
58. I. Tsuchitori, N. Morinaga, H. Fukunaga, J. Jpn. I. Met. **59**, 331 (1995). https://doi.org/10.2320/jinstmet1952.59.3_331

59. S. Avraham, W.D. Kaplan, J. Mater. Sci. **40**, 1093 (2005). https://doi.org/10.1007/s10853-005-6922-4
60. K. Momma, F. Izumi, J. Appl. Crystallogr. **44**, 1272 (2011). https://doi.org/10.1107/S0021889811038970
61. A.B. Sheikh, J.T.S. Irvine, J. Solid State Chem. **103**, 30 (1993). https://doi.org/10.1006/jssc.1993.1075
62. C. Klein, C.S. Hurlbut, Jr., Manual of mineralogy (after James D. Dana), 21st ed., (New York: Wiley, 1993), pp. 380–381, ISBN 047157452X
63. B.A. Wechsler, R.B. von Dreele, Acta Cryst. B **45**, 542 (1989). https://doi.org/10.1107/S010876818900786X

Chapter 8
Dealing with Fe in Secondary Al-Si Alloys Including Metal Melt Filtration

Hanka Becker and Andreas Leineweber

8.1 The Crystal Structures of the Relevant Intermetallic Phases

The intermetallic phases occurring during the various treatments of the secondary Al-Si alloys are introduced in the following. The focus is put on the phases α_h-Al–Si–Fe, α_c-Al–Si–(Fe,Mn,Cr) and β-Al–Si–Fe and δ-Al–Si–Fe which are most relevant in case of Al–Si–Fe alloys with Mn and Cr. The characteristics of these phases are summarized in Table 8.1. Note that in (additional) presence of other transition metal elements other or further Fe-containing intermetallic phases may occur.

8.1.1 α_h-Al–Si–Fe and α_c-Al–Si–(Fe, Mn, Cr)

In pure Al–Si–Fe alloys, the hexagonal α_h-Al–Si–Fe phase (space group $P6_3/mmc$) with a composition given by Al7.1Fe2Si develops [4]. However, in presence of as small amounts as > 0.03 wt% Cr [5] or > 0.3 wt% Mn [5–7] the cubic α_c-Al–Si–(Fe,Mn,Cr) phase forms instead of α_h with approximate compositions between $Al_9Mn_2Si_{1.8}$ [8], $Al_{20}(Fe,Mn)_5Si_2$ [9] and $Al_{15}(Fe,Mn)_3Si_2$ [10]. It has space group $Im\bar{3}$ at low ratios of Mn or Cr to Fe content or $Pm\bar{3}$ at high ratios of Mn or Cr to Fe content [9, 11].

Both the structures of α_h and α_c can be derived from periodic arrangements of Mackay icosahedra (MI), supplemented by further atoms (Fig. 8.1a,b). Thereby, the ideal icosahedral symmetry of such clusters is, of course, broken (Fig. 8.1c). In α_c

H. Becker (✉) · A. Leineweber
Institute of Materials Science, Technische Universität Bergakademie Freiberg,
Gustav-Zeuner-Straße 5, 09599 Freiberg, Germany
e-mail: Hanka.Becker@iww.tu-freiberg.de

© The Author(s) 2024

C. G. Aneziris and H. Biermann (eds.), *Multifunctional Ceramic Filter Systems for Metal Melt Filtration*, Springer Series in Materials Science 337,
https://doi.org/10.1007/978-3-031-40930-1_8

Table 8.1 Summary of the intermetallic phases α_h, α_c, β and δ forming in hypoeutectic Al–Si alloys with Fe and Mn. The Greek letter that is commonly assigned to the phase and a frequently used formula is given next to space group and exemplary lattice parameters. Note that this table is meant to present an overview of the relevant phases and cannot cover variation of lattice parameters within the homogeneity ranges. Within the notes line further formulas for the phases and solubilities are given if available. The mentioned characteristic morphologies and kinetic preferences refer to solidification of hypoeutectic Fe-containing Al-Si casting alloys

Phase (Label)	Space group	Lattice parameters
α_h-$Al_{7.1}Fe_2Si$	$P6_3/mmc$	$a = 12.346$ Å, $c = 26.210$ Å [4]

Used formulas: $Al_{7.1}Fe_2Si$ [4], Al_8Fe_2Si [12, 13]
Homogeneity range: Narrow [4], Relative solubility of Mn: $Fe_{0.97}Mn_{0.03}$ [4]
Morphology: Large particles containing enclosures of Al [14]
Kinetic preference: Slow nucleation, mostly forming as primary phase, can be easily suppressed under high cooling rates [15]

α_c-$Al_9Mn_2Si_{1.8}$	$Pm\bar{3}$	$a = 12.68$ Å [8]

Used formulas: $Al_9Mn_2Si_{1.8}$ [8], $Al_{20}Fe_4MnSi_2$ [9], $Al_{15}(Fe,M)_3Si_2$ [10], $Al_{34.1}(Fe,M)_{9.5}Si_6$ [6]
Crystal structure: Disordering for Mn/Fe ratios < 0.66 to $Im\bar{3}$ [9, 11],
Homogeneity range:
broad range of Si and of (Fe,Mn,Cr): Al: 68.8 [6] to 75 at% [10], Si: 7.4 [10] to 14 at% [8], Fe, Mn, Cr: 15 [10] to 19.2 at% [6]
Relative solubilities of Fe, Cr, Mn[8.1]: $Fe_{0.97}Mn_{0.03}$ to Fe_0Mn_1 [7], $Fe_{0.88}Cr_{0.12}$ [16] to $Fe_{0.51}Cr_{0.49}$ [17]
Morphology: Bulk facetted polyhedral, hopper, coarse and fine dendrites [16, 18]
Kinetic preference: Nucleates easily at all cooling rates during all stages of solidification (primary, co-dendritic, co-eutectic) in sufficient presence of e.g. Mn and Cr [15] and on various surfaces (if present) [19]

β-$Al_{4.5}FeSi$[2]	$A12/a1$ [20, 21] $A2_1/e11$ [22] $I4_1/acd$ [23]	$a = 6.1676$ Å, $b = 6.1661$ Å, $c = 20.8093$ Å, $\beta = 91.0°$ [20]

Used formulas: $Al_{4.5}FeSi$ [10, 20–22, 24], Al_5Fe_2Si [25]
Homogeneity range: Relative solubility of Mn: $Fe_{0.87}Mn_{0.13}$ [26]
Morphology: Plate-shaped [15]
Kinetic preference: Unimpaired nucleation; can be suppressed under high cooling rates in sufficient presence of Mn and Cr in favour of α_c or δ phase [15]

δ-Al_3FeSi_2	$I4/mcm$	$a = 6.061$ Å, $c = 9.525$ Å [27]

Used formulas: Al_3FeSi_2 [10, 26, 27], Al_4FeSi_2 [28]
Morphology: Plate-shaped [15, 26]
Kinetic preference: Forms in co-existence with β phase in the same particles with increasing δ phase fraction with increasing cooling rates [15]

[1] Experiments of the kinetics of intermetallic phase formation on different substrates hints at the fact that under special conditions the α_c phase forms as metastable phase in the ternary Al-Si-Fe system without the presence of further metallic element e.g. transition-metal elements
[2] See text for explanation of the simultaneous presence of three structures

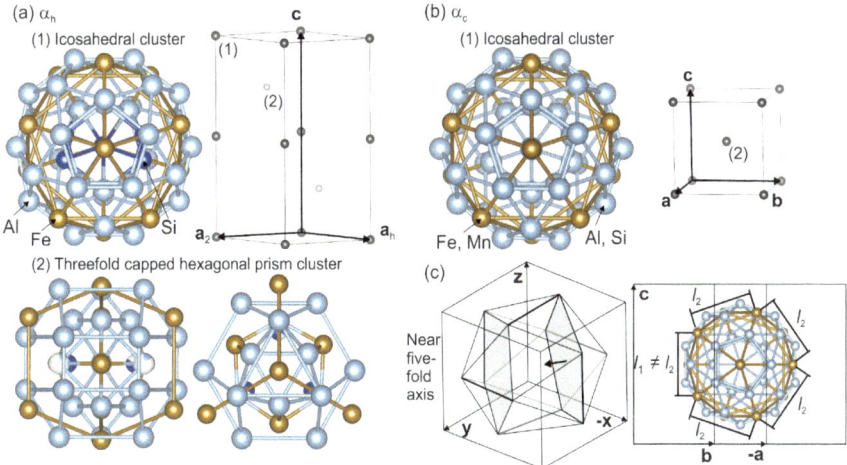

Fig. 8.1 Characteristic clusters and their arrangement in **a** α_h-Al–Si–Fe phase and the **b** α_c-Al–Si–(Fe, Mn, Cr) phase based on the structure descriptions in [4, 8]. Mackay icosahedra and threefold capped hexagonal prism clusters compose the α_h phase. For α_c-Al–Si–(Fe, Mn, Cr) only a packing of Mackay icosaedra is considered. The viewing directions are along the near five-fold axis in case of the distorted icosahedral clusters. The threefold capped hexagonal prism cluster is shown in side view and in top view. The clusters are located on the indicated lattice points (dark and light gray) in the schematic unit cells. Additional sites of Al/Si atoms located in space between neighbor clusters are not shown. The inscription of that distorted cluster type into the unit cell of α_c in (**c**). The inscription of that cluster type into the unit cell of α_c and the distortion are highlighted in (**c**), Copyright Elsevier. The illustration in (**c**) is based on an illustration in [29]. Here and in the following: Crystal structure presentations are created using VESTA [30]

the MI form a bcc arrangement. Additional sites of Al/Si atoms are located between the second-nearest neighbor clusters [9].

In α_h the MI form a primitive hexagonal arrangement, where ordered occupation of 1/2 of the trigonal prismatic interstices by further clusters leads to a doubling of the c axis. There $P6_3/mmc$ symmetry results by the latter clusters occupying the (1/3,2/3,1/4) and (2/3,1/3,3/4) positions forming an AB stacking and the MI end up on the octahedral sites. It is worth to stress that the α_h and α_c phase are clearly distinct phases with different properties and should be clearly distinguished during microstructure analysis.

The α_h phase forms during primary crystallization from a Fe-containing Al–Si alloy melt under slow cooling rates in a shape of large polyhedra that can contain inclusions of Al [14, 15]. In co-dendritic and co-eutectic stages of solidification[1] and at high cooling rates the evolution of the α_h phase is suppressed [15]. The α_c phase forms in Fe-containing Al–Si alloys containing further transition metal elements especially Mn and Cr. Furthermore, experiments on the kinetics of intermetallic phase formation from ternary Fe-containing Al–Si alloys on different substrates reveals that under special conditions the α_c phase forms as metastable phase in the ternary

---

[1] For explanation of stages of solidification of Fe-containing Al–Si cast alloys see [67].

Al–Si–Fe system. This might be attributed to the possible presence of icosahedral clusters already in the melt [31, 32]. Depending on the external conditions, e.g. elemental composition and cooling rate [18], the α_c phase evolves into a broad variety of morphologies (Chinese script, bulk facetted polyhedra etc.) [16, 18]. The morphologies of the α_c phase can be evaluated e. g. on the basis of EBSD data. Challenges arising from pseudosymmetry issues during indexing of EBSD patterns of the α_c phase are scrutinized in [29].

8.1.2   β-Al–Si–Fe and δ-Al–Si–Fe

The structural characteristics of the β-Al–Si–Fe and δ-Al–Si–Fe phase in this section have been detailed in depth in [15, 22, 23]. The atomic structure of the β and δ phase consists of basic layers built from bicapped antiprisms with Al and Si atoms at the corners and with Fe atoms in the center (Fig. 8.2a). Within these layers, the bicapped square antiprisms share common edges, i.e. two (Al, Si) atoms. In the crystal structure of the β phase, pairs of these layers are condensed by sharing (Al, Si) atoms at the caps, thus, forming double layers (Fig. 8.2b,c). Stacking of these double layers by relative layer shifts of $\mathbf{a}/2 + \mathbf{c}_d$ or $\mathbf{b}/2 + \mathbf{c}_d$ allows four positions of the double layers, A, B, C, D, where $\mathbf{c}_d$ is a vector perpendicular to unit cell basis vectors $\mathbf{a}$ and $\mathbf{b}$. $|\mathbf{c}_d|$, thereby, is the distance between the double layers.

The δ phase is structurally closely related to the β phase, but thermodynamically β and δ are clearly distinct phases [23]. The atomic structure of the δ phase is composed of fully condensed basic layers by sharing all cap atoms. The δ phase is described in space group $I4/mcm$ or space group $Pbcn$ when ordering of Al and Si atoms is considered (Fig. 8.2c) [27]. The β phase has an ideal chemical composition of $Al_{4.5}FeSi$ [10, 20, 21, 23, 33, 34]. The δ phase is typically described with an ideal chemical composition Al_3FeSi_2 [27].

Several crystal structure variants for the β phase have been reported. All these structures can be described using near-tetragonal lattice parameters that can be grouped into monoclinic [10, 20, 33] e.g. with space group $A12/a1$ [20, 21], orthorhombic [24, 25] or tetragonal [24, 33] structures which provide lattice parameter c either around 20.8 Å [20, 21, 24, 25] or 41.6 Å [10, 24, 33]. Based on SEM, TEM including SAED and HAADF and DFT calculations, that series of structures was rationalized to a more complete image considering several hierarchical levels of the crystal structure [22].

- **Polytypes**

From the stacking of the A, B, C and D double layers, ordered polytypes or disordered structures arise which were elaborated in [23]. The ordered polytypes that form from a periodic stacking of the layers by $\mathbf{b}/2 + \mathbf{c}_d - \mathbf{b}/2 + \mathbf{c}_d$ or $\mathbf{b}/2 + \mathbf{c}_d - \mathbf{a}/2 + \mathbf{c}_d - \mathbf{b}/2 + \mathbf{c}_d - \mathbf{a}/2 + \mathbf{c}_d$ result in an AB sequence or an ABCD sequence. The former ideally shows orthorhombic $Aeam$ symmetry (with a pseudotetragonal unit cell shape) and the latter ideal tetragonal $I4_1/acd$ symmetry (Fig. 8.2b). However, none of these ideal

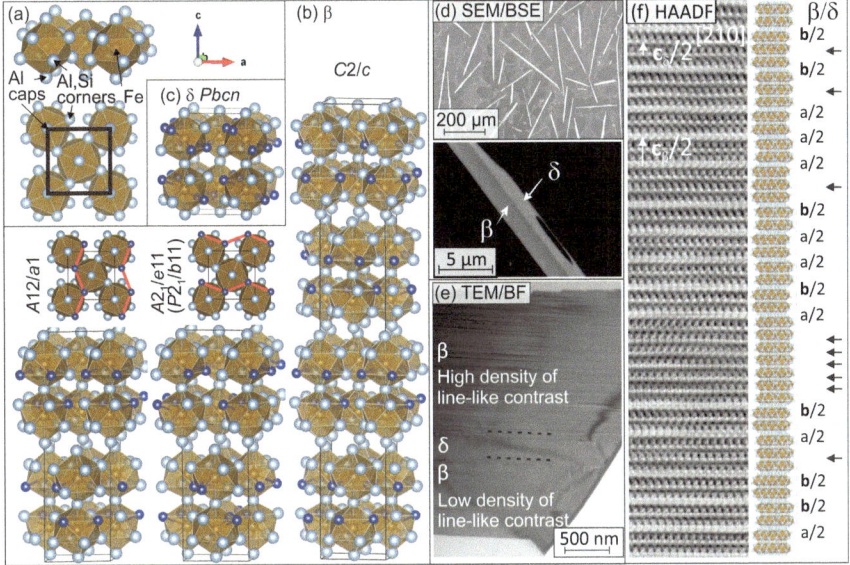

Fig. 8.2 a A basic layer of bicapped square antiprisms composing the crystal structure of the β and δ phases. The coordinate system that also applies in **b, c** is indicated. **b** Ordering of Al versus Si atoms in the unit cells of three ordered structures of β [22]. Two AB polytypes and an ABCD polytype of the stacking of double layers in β phase are shown. The double layers are shifted with respect to each other by $\mathbf{b}/2 + \mathbf{c}_d$ and $\mathbf{a}/2 + \mathbf{c}_d$ respectively. **c** Unit cell of the δ phase in which all cap atoms are condensed illustrated for ordered Al versus Si atoms as reported in [27]. **d** SEM/BSE image of the appearance of the plate-shaped β and δ phase in the Al7.1Si1.5Fe alloy solidified with a cooling rate of 1.4 K/s. **e** Appearance of the line-like contrast in bright field (BF) TEM and in **f** high-angle annular dark field (HAADF) images at different scales of imaging of a lamella from a plate-shaped particle ([15], Copyright Elsevier). Note the irregular sequence of β and δ stacking and non-periodic stacking of double layers in β which are highlighted by the atomic model

structures explains the clearly observed unit cell distortions away from a tetragonal or orthorhombic metric, nor some additional diffraction peaks hinting at further ordering effects. For the purpose of orientationally robust indexing e.g. of EBSD patterns of the β phase with locally varying stacking sequence including also disorderedly stacked regions, an artificial, partially disordered tetragonal crystal structure with *I4/mmm* symmetry has been devised and successfully applied in practice, e.g. upon indexing of EBSD data from δ + β containing microstructures [23, 35].

- **Ordering of the Al Versus Si atoms**

Reconciling experimental evidence from literature and own studies in combination with new theoretical evidence from DFT calculations allowed to address these unsolved microstructure interpretations in [22]. Differently Al versus Si ordered model structures have been subjected to DFT calculations. Analysis of the relaxed lattice parameters imply that unit cell distortions away from an orthorhombic or tetragonal metric are generally enforced by the symmetry induced by the atomic arrangement of Al versus Si atoms. Two monoclinic structures differing solely by the orientation of the ordering in the double layers with respect to the AB stacking provide the lowest-energy. These two structures can be described in space groups.

- $A12/a1$ with lattice parameters $a = 6.1667$ Å, $b = 6.1683$ Å, $c = 20.7791$ Å and $\beta = 91.51°$ (in agreement with structures proposed in previous literature) and
- $A2_1/e11$ (or $P2_1/b11$ with doubled set of atomic sites) with lattice parameters $a = 6.1739$ Å, $b = 6.1602$ Å, $c = 20.7840$ Å and $\alpha = 88.63°$.

The monoclinic angle deviates by about 1.3–1.6° from 90°, which is compatible with results from powder-X-ray diffraction analysis. The ordering in these two lowest-energy variants can be transferred to the ABCD stacking of double layers, also leading to monoclinic symmetry. Thus, previously unexplained irreconcilable sets of additional diffraction peaks comply with the three ordered structures. Three different crystal structures coexisting simultaneously at constant composition of β-$Al_{4.5}FeSi$ are unexpected under thermodynamic equilibrium and can be non-equilibrium effects.

- **Defect structure**

Various disordered stacking and Al versus Si ordering variants of the β can be considered as defect structure. In characteristically plate-shaped particles (Fig. 8.2d, e, f) neighboring extended phase regions of double layer stacked β phase and full condensed basic layer stacking of δ phase exist. Furthermore, next to these regions, short successive additionally condensed basic layer sequences are frequently present in the β phase. Such δ-like stacking sequences are typical planar defects of the β phase. Their presence can be related to compensation of excess Si, as implied by the higher Si content of the δ phase (Al_3FeSi_2) [22].

These structural details, on the one hand, can obscure the microstructure characterization when the full complexity is not considered during the microstructure analysis, e. g. when additional superstructure reflections would remain unidentified or misleadingly suggest the presence of further phases. On the other hand, some characterization methods might not resolve all details, so that it is recommended to use artificially higher symmetries in the microstructure analysis. The method and the structural model should address the same level of resolution.

8.2 Thermodynamic and Kinetic Considerations on the Phase Formation During Solidification

8.2.1 Melt Conditioning Under Influence of Mn, Cr and Mg

The results and conclusions highlighted in this section mainly reported in [14] outline the state of melt conditioning based on new results and literature. The current limitations and chances for dealing with Fe in Al–Si alloys by removal of Fe on the basis of melt conditioning and metal melt filtration are presented. Melt conditioning addresses the intended adjustment of the melt or semi-liquid state for a specific purpose. Here, the formation of primary, Fe-containing, intermetallic phases, the so-called sludge, in liquid aluminium is the precondition for removal of Fe by metal melt filtration. To maximally reduce the residual content of Fe in the Al melt by separating off those primary, Fe-containing, intermetallic particles, the melt has to be specifically conditioned for that purpose according to the parameters described in the following:

- The *modification of the alloy composition* by addition of further elements like Mn and Cr can promote binding Fe in intermetallic phases by an increased tendency to form sludge [36–40] and can reduce the solubility of Fe in the Al melt [41–45]. Furthermore, these elements can change the type of primary phases constituting the sludge, i.e. α_c, α_h, β and/or other phases [14, 15, 44, 45] and, thus, influence the achievable Fe removal.
- The amount of sludge increases with decreasing *conditioning temperature*. Thereby, the remaining amount of Fe and the added elements as Mn and Cr in the melt decreases [41, 46, 47]. In order to remove a high fraction of Fe, the conditioning temperature should be chosen closely above the solidification temperature of the (Al)-solid solution $T_{Al\text{-}s}$.
- The *conditioning time* should be sufficient to achieve a maximum (equilibrium) volume fraction of primary, intermetallic particles. The heat treatment time of 2 h was chosen according to [42, 44–46] to achieve the equilibrium volume fraction of primary particles.
- A coarse compact polyhedral *morphology* with a large particle size is preferred for the purpose of Fe removal by separation of primary, intermetallic particles from the Al melt from technological point of view. Such a coarse morphology is known to form as for the α_c and also the α_h phase under slow cooling or longer heat treatment [16, 18, 39, 44, 45, 48]. Note that various interfaces, especially naturally occurring oxide films are able to promote heterogeneous nucleation of the β, the α_c and the α_h phases [19, 49, 50].

The primary particles form between the liquidus temperature T_L and the onset temperature of the (Al)-solid solution $T_{Al\text{-}s}$. These temperatures as well as the type(s) of the primary phase(s) depend on the alloy composition. Information about these temperatures and phases is required to define the conditioning parameters for the melt conditioning. Three approaches can be evaluated to access these parameters: the sludge factor approach, thermodynamic calculations and thermal analysis.

Sludge factor: Conditions for the formation of primary, intermetallic particles, the sludge, can be estimated based on the sludge factor SF [36–40]

$$SF = w_{Fe} + 2w_{Mn} + 3w_{Cr}. \tag{8.1}$$

w are the mass percentages of the elements in the index. The sludge factor can be used to estimate the effect of the transition-metal elements on the sludge formation predominantly on the basis of the weighted sum of the mass percentages of the metal elements Fe, Mn and Cr. The effect of Mn is twice and that of Cr three times as large as that of Fe. The critical temperature below which sludge forms was proposed to depend solely on the sludge factor:

$$T_{sludge} = (86.7°C \cdots SF/wt\%) + 506°C \text{ [39] based on [37]}, \qquad (8.2)$$

$$T_{sludge} = (44°C \cdots SF/wt\%) + 552°C \text{ [39] based on [36]}. \qquad (8.3)$$

Also another relationship for the critical temperature in dependence on the Fe content has been suggested:

$$T_{sludge} = (34.2°C \cdots w_{Fe}/wt\%)^2 + 645.7°C \text{ [51]}. \qquad (8.4)$$

According to [36] such dependences represent a kind of liquidus temperature and in case of equilibrium conditions the sludge temperature corresponds to the liquidus temperature. Therefore, in the following, the sludge temperature T_{sludge} is referred to as liquidus temperature T_L.

Thermodynamic calculations: A more advanced approach to access T_L and T_{Al-s} are thermodynamic calculations e.g. by using the CALPHAD method [52]. Additionally, these calculations provide information about the type(s) of primary phase(s) that form and provide information about the phase formation sequence during solidification. However, the current thermodynamic databases for thermodynamic calculations evidently require some fine-tuning to predict the effect of modifying elements as Mn and Cr correctly. That current limitation but also possible kinetic effects that cause a deviation from equilibrium conditions, make experimental investigations e.g. by thermal analysis necessary to achieve information on T_L, T_{Al-s} and evolving phases.

Thermal analysis: Thermal analysis data e.g. differential thermal analysis (DTA) curves allow experimentally evaluating reactions occurring during melting of alloys. Occurrence of such reactions is revealed by endothermic signals in DTA data. From data evaluation reaction temperatures e.g. T_L and T_{Al-s} can be extracted. The primary phases are identified based on their coarse morphology as compared to all other phases present in the microstructure. Thus, the formation of the primary phases can be assigned to the occurrence of the endothermic signal observed in the DTA signal corresponding to final melting of the alloy at T_L. T_{Al-s} can be derived from the high intensity signal in the DTA curve corresponding to melting of the main microstructural component Al in the alloy.

The three approaches have been applied to obtain the temperatures T_L and T_{Al-s} for the alloys of the compositions Al7.1Si($1.5-x_M$)Fe(x_M)M with M = Mg, Mn, Cr and x_M = 0, 0.3, 0.6, 0.9, 1.2 and 1.5 at%. The numbers preceding the elements indicate molar fractions. The resulting temperatures and calculated or observed primary phases are shown in Fig. 8.3.

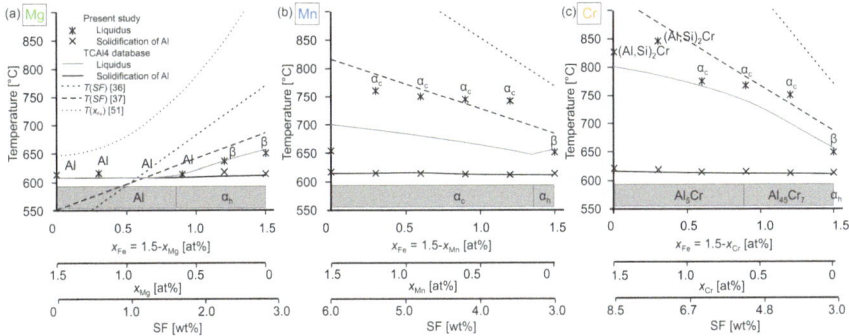

Fig. 8.3 Experimentally determined temperatures (DTA) of melting of the (Al)-solid solution $T_{\text{Al-s}}$ and the liquidus temperature T_L are indicated as data points for the different Al7.1Si(1.5-x_M)Fe(x_M)M alloys with **a** M = Mg, **b** M = Mn and **c** M = Cr. The lines indicate the corresponding temperature predicted by thermodynamic calculations (TCAL4 database [53]) and on the basis of the sludge factor (SF) approach or Fe content (Eqs. 8.2–8.4). The kind of primary phase predicted by thermodynamic calculations is indicated at the bottom (gray), whereas the experimentally encountered primary phase is indicated at the data points. The figure was published in [14], Copyright Springer

- **Solidification of the (Al)-solid solution:**

According to the results from the DTA experiments on the Al7.1Si(1.5-x_M)Fe(x_M)M alloys, solidification of the (Al)-solid solution starts closely below 620 °C. The thermodynamic calculations and DTA experiments show good agreement for $T_{\text{Al-s}}$. Thus, the process window for the melt-conditioning temperature can be reasonably well predicted by the CALPHAD method [52] using the TCAL4 database [53]. Hence, holding the melts at 620 °C, as applied here for melt conditioning, is the reasonable lowest possible for potential development of primary, intermetallic phases without solidification of the (Al)-solid solution during melt conditioning.

- **Solidification of the primary, intermetallic particles**

In contrast to $T_{\text{Al-s}}$, the liquidus temperatures T_L and primary phases obtained from DTA experiments, thermodynamic calculations or the sludge factor approach agree to only some limited extend.

In the alloys with Mg, the deviation is less than 8 K for all Mg-containing alloys including the alloys with $x_{\text{Mg}} \geq 0.6$ at% and with $x_{\text{Mg}} \leq 0.3$ at%. Note that, in the former, the (Al)-solid solution is the primary phase. In the latter, the β phase is the primary phase in the experiment while the α_h phase results from thermodynamic calculations. The discrepancy of the formed primary intermetallic phase can be understood in relation to the retarded nucleation of the α_h phase if the cooling rates are sufficiently high, as obviously the 0.5 K/s applied for cooling in the DTA experiments. Note that, after the actual melt-conditioning treatment of 2 h at 620 °C, the α_h phase is observed as primary phase (Fig. 8.4).

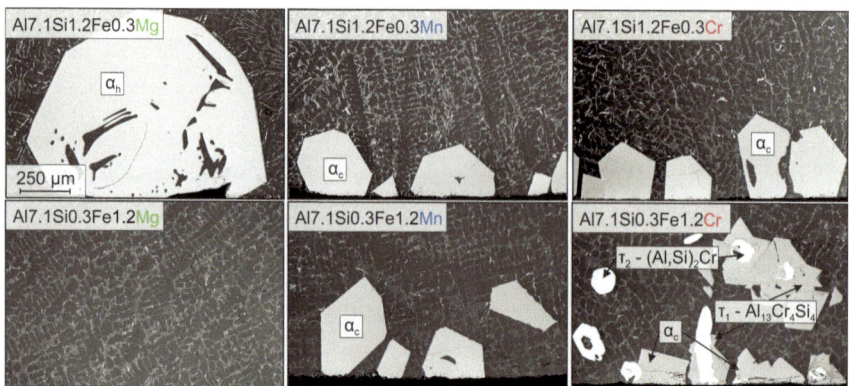

Fig. 8.4 SEM/BSE images of the indicated alloy compositions after melt conditioning at 620 °C for 2 h and quenching. For corresponding qualitative EDS maps see [14]. The figure was recomposed from parts of a figure in [14], Copyright Springer

For Mn- and Cr-containing alloys, the observed type(s) of primary phase(s) after DTA experiments and melt conditioning agree, however, partly differ from CALPHAD based predictions. The appearance of the primary phases in the alloys quenched after melt conditioning is illustrated in Fig. 8.4.

In alloys with Mn, the difference of the thermodynamically predicted and observed liquidus temperatures is remarkable (>50 K) although the primary phase is the α_c phase in both cases.

In alloys with Cr, the difference of the liquidus temperatures is less pronounced compared to Mn-containing alloys. However, the predicted phases for all compositions differ from the observed phases. The α_c phase is observed as the only primary phase for $x_{Cr} = 0.3, 0.6, 0.9$ at%. For $x_{Cr} = 1.2, 1.5$ at%, the $(Al,Si)_2Cr$, the $Al_{13}Cr_4Si_4$ and the α_c phase occur before the onset of solidification of (Al)-solid solution. The α_c phase is not implemented in the description of the Al–Si–Fe–Cr system in the TCAL4 database [53] while the $(Al,Si)_2Cr$ is implemented but not predicted for the present alloy compositions. Furthermore, the observed $(Al,Si)_2Cr$ phase should generally dissolve during solidification and should not be present together with Al in the solid state [54]. Instead, experimentally not observed formation of Al_5Cr and $Al_{45}Cr_7$ is predicted by thermodynamic calculations employing the TCAL4 database [53].

It can be concluded that the databases for prediction of the formation of primary intermetallic phases by CALPHAD generally need some further fine-tuning e.g. by implementation of the α_c phase in the Al–Si–Fe–Cr system which is a true quaternary phase in this system not existing in any of the ternary subsystems.

In view of the sludge factor approach, the agreement of the liquidus temperatures T_L is surprisingly good for the temperature estimation from Eq. 8.3 [36] with experimental values as long as an intermetallic phase is the primary phase (Fig. 8.3). The sludge factor approach was initially introduced for Cu-containing, secondary Al–Si alloys with similar Si content. The agreement to the temperature estimations from Eqs. 8.2 and 8.4 [37, 51] is less good which might be related to a remarkably higher Si content in the alloys in [51] compared to the present alloy compositions. Furthermore, it has been observed that Cu has a minor and Mg has no effect on the temperature of sludge formation [36] leading to the good agreement of the sludge factor approach in the present Cu-free Al–Si alloys. From this observation, it can be further assumed that the sludge factor approach can be applied to commercial Al–Si alloys in presence of alloying elements that do not take part in the sludge formation like Zn, Ni, Pb, Sn and Ti [36, 38] or when the amount of the element is negligible. However, in case of elements like Sc, B and V that promote the α_c-phase formation [1], or if the effect of elements is unknown like Be, Sr, Co, Mo and S the sludge factor approach in form of Eqs. 8.1–8.4 should be used with care.

- **Fe and transition-metal reduction in the residual melt after its conditioning**

The solubility of the elements Fe, Mn, Cr and Mg in the liquid Al at 620 °C determines the remaining minimum amount of these elements in the alloy after separation of the sludge from the liquid metal. These amounts, determined by EDS area measurements, are summarized in the quasi-ternary (Al,Si)–Fe–M projection of the Al-rich corner of the corresponding quaternary phase diagram (Fig. 8.5a) for the alloy series Al7.1Si(1.5-x_M)Fe(x_M)M with M = Mg, Mn, Cr and Al7.1Si1.5Fe$x_M M$ with M = Mn, Cr where x_M = 0, 0.3, 0.6, 0.9, 1.2 and 1.5. at% in both alloy series.

For the evaluation of the success of Fe removal often relative Fe reductions are specified. However, those values depend on the initial Fe content. Consequently, it is recommended to use absolute elemental contents e. g. of the remaining melt in

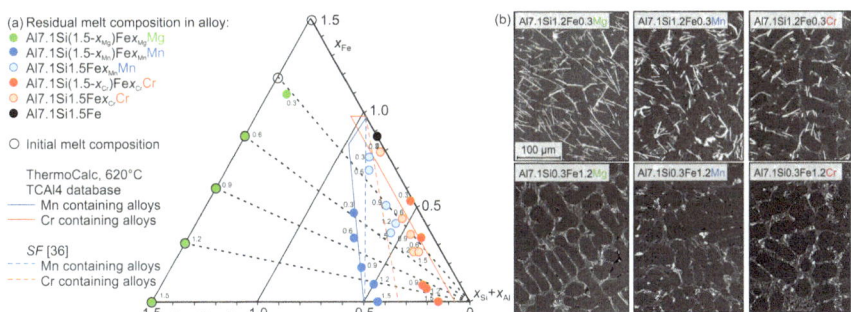

Fig. 8.5 Characteristics of the residual melt quenched after melt conditioning at 620 °C. **a** Projection of the (Al–Si)–Fe–M phase diagram with M = Mg, Mn and Cr for concentrations higher than 98.5 at% Al and Si showing the initial melt compositions and the residual melt compositions. Small numbers next to symbols of the residual melt composition indicate xM of the initial melt composition. Results from calculations based on the TCAL4 database [53] and semi-empirical estimations related to the sludge factor are shown. **b** SEM/BSE images of the microstructure of the solidified remaining melt. The brightest particles indicate Fe-containing phases and dark gray indicates the (Al)-solid solution. Intermediate gray belongs to Si in the Al–Si eutectic. The figure was partially reproduced and modified based on a figure in [14], Copyright Springer

preference or in addition to the relative reductions. That facilitates the comparison of the success of melt conditioning for Fe removal with other studies.

In the Mg-containing alloys with $x_{Mg} \geq 0.6$ at%, no solid phase was present at 620 °C and the chemical composition after melt conditioning equals the initial chemical alloy composition. In the other alloys, the formation of primary, Fe-containing, Mg-free intermetallic particles reduces the Fe content in the remaining material. In the ternary Al7.1Si1.5Fe alloy, 0.88 at% Fe remain dissolved in the liquid metal. The phase responsible for binding Fe in the alloys with $x_{Mg} \leq 0.3$ at% is the α_h phase. Note that Mg as typical additional alloying element of Al–Si cast alloys is not bound by sludge particles and remains in the melt.

In case of alloys with Mn and Cr, the Fe content and the total transition-metal content in the remaining melt decreases with increasing $x_{Mn}/(x_{Mn} + x_{Fe})$ and $x_{Cr}/(x_{Cr} + x_{Fe})$ ratios in the alloys (Fig. 8.5a). That is attributed to the decreased ability of the melt to dissolve Mn and Cr. In presence of Mn and Cr, the α_c phase is responsible for binding Fe and also Mn and Cr in agreement with [14, 41, 44–46]. Furthermore, a smaller amount of Cr than Mn for the same initial Mn and Cr content remains in the melt related to additionally binding Cr in the $(Si,Al)_2Cr$ and the $Al_{13}Cr_4Si_4$ phase which, however, do not bind Fe in relevant amounts.

The optimum Mn and Cr content for Fe reduction sensitively depends on the actual alloy composition [41–43]. Next to the experimentally determined residual melt composition, the chemical composition of the melt at 620 °C from thermodynamic calculations and from the sludge factor approach are illustrated in Fig. 8.5a. In the latter case, the minimum content of Fe in the melt is reached according to [36] when the sludge factor equals 1. In agreement with experimental observation, own and literature, reported in [14, 41–45], the solubility of Fe decreases with increasing $x_{Mn}/(x_{Fe} + x_{Mn})$ ratio and $x_{Cr}/(x_{Fe} + x_{Cr})$ ratio. However, the optimum ratio with respect to the residual transition-metal content is controversial. While [43] achieved minimum $x_{Fe} + x_{Mn}$ contents as low as 0.3 at% for an initial $x_{Mn}:x_{Fe}$ ratio of 3:1, in the present investigation, a continuous decrease of the transition-metal content towards the ternary Al–Si–Mn and Al–Si–Cr alloy composition was observed. In addition to the chemical composition, the residual Fe, Mn and Cr content in the melt decreases with decreasing conditioning temperature [41, 46, 47] and approaches the lowest achievable content after a critical holding time. Then, the maximum amount of Fe and Mn are bound in primary, intermetallic particles [42–44, 46] and their maximum volume fraction is reached. Consequently, a lower residual Fe and Mn content in the present study was achieved compared to similar alloys if these were conditioned at higher temperatures [41, 42, 46] or shorter times [42, 46].

Figure 8.5b depicts the intermetallic phases formed with the Fe, Mn and Cr which have remained in the melt despite optimum melt-conditioning at 620 °C and quenching. These intermetallic phases have formed during the co-dendritic and co-eutectic stage and are located in the Al–Si eutectic regions. Corresponding to the remaining contents of Fe, Mn and Cr, the volume fraction of intermetallic particles is lower in alloys with initially high $x_{Mn}/(x_{Mn} + x_{Fe})$ and $x_{Cr}/(x_{Cr} + x_{Fe})$ ratios. In case of the Mg-containing alloys, the reduced amount of the Fe-containing, intermetallic particles reflects the initially lower Fe content with increasing $x_{Mg}/(x_{Mg} + x_{Fe})$ ratio, although the Fe content is not ($x_{Mg} \geq 0.6$ at%) or not remarkably ($x_{Mg} = 0.3$ at%) changed in presence of Mg. In Mn- and Cr-containing alloys, the reduced amount of the Fe-containing, intermetallic phases is attributed to the remarkably reduced Mn and Cr level, simultaneously to the reduced Fe level remaining in the melt after melt conditioning. Furthermore, although depending to some extent on the cooling rate, the morphology of the Mn- and Cr-containing, intermetallic particles is irregular and roundish compared to the plate-shaped particles in the eutectic region in alloys with low initial $x_{Mn}/(x_{Mn} + x_{Fe})$ and $x_{Cr}/(x_{Cr} + x_{Fe})$ ratios.

8.2.2 Effect of Different Cooling Rates and Mn on the Phase Formation

In this section results and conclusions from [15] about the microstructures that have formed under different cooling rates and $x_{Mn}/(x_{Mn} + x_{Fe})$ ratios in Al7.1Si(1.5-x_{Mn})Fe(x_{Mn})Mn with $x_{Mn} = 0$, 0.3, 0.375, 0.6 and 0.75 at% and a focus on the Fe-containing, intermetallic phases is presented. Mn is the most frequently used element for modification of the microstructure. Depending on the casting technique, relevant cooling rates range from 1–10 K/s in permanent mold casting to 10–100 K/s in high pressure die casting [55]. Modification of intermetallic particles into harmless microstructural components in order to deal with Fe impurities in secondary Al–Si casting alloys requires detailed understanding of the solidification path in presence of modifying elements and non-equilibrium solidification. The present data is further analyzed to provide a systematic understanding on the solidification path for a broad range of Fe- and Mn-containing Al–Si casting alloys. Thus, it is shown how seemingly controversial results of intermetallic phases in the microstructures of related alloys observed in other studies well integrate to give a consistent picture when considering the effect of cooling rate and presence of further elements especially Mn. Solid-state phase transformations that could occur during subsequent heat treatments are beyond the scope of the presented investigations. No signs of such transformation have been observed in [15], except of marginal transformation of the α_h phase into the β phase.

The microstructures of the Al7.1Si(1.5-x_{Mn})Fe(x_{Mn})Mn alloys solidified with cooling rates of 0.05, 1.4, 11.4 and 200 K/s are shown in Fig. 8.6. The type of phases corresponding to the intermetallic particles with different morphology and location within the microstructure and within the sample are indicated. The volume fractions of the intermetallic particles in the solidified microstructures are presented in Fig. 8.7. The intermetallic particles have been quantitatively analyzed of in view of specific morphologies and type of the phases.

- **Mn-free Al7.1Si1.5Fe alloy**

In the Mn-free alloy, at the edge of the slowest cooled samples, where the alloy is in contact with the surrounding oxide film, coarse polyhedral particles of the α_h phase are present. Plate-shaped particles appear associated with the Al–Si eutectic, i.e.

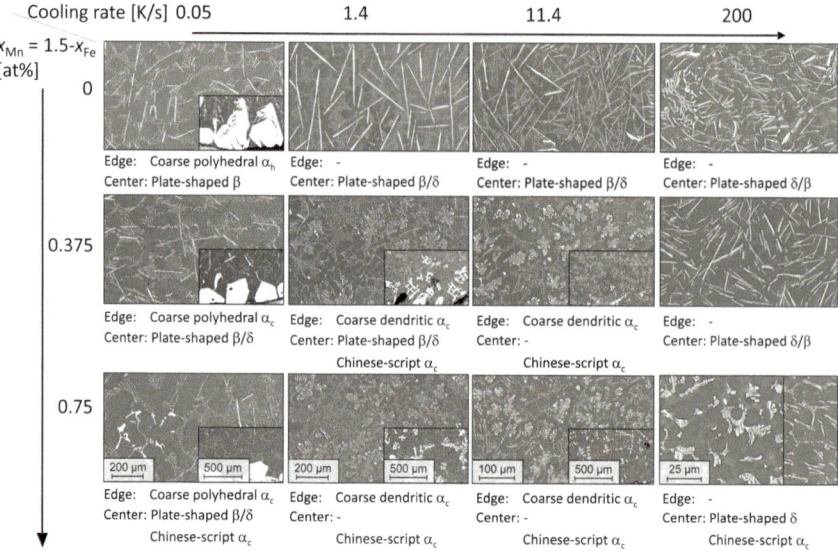

Fig. 8.6 Evolution of the microstructure of the Al7.1Si(1.5-x_{Mn})Fe(x_{Mn})Mn alloys depending on the Fe and Mn content and the cooling rate in SEM/BSE contrast. The images correspond to the bulk of the solidified material while the insets show the primary particles located at the edge of the samples. The intermetallic phases appear white, Si in light gray and Al in gray. Note the change of the magnification depending on the cooling rate. The figure was recomposed from parts of a figure in [15], Copyright Elsevier

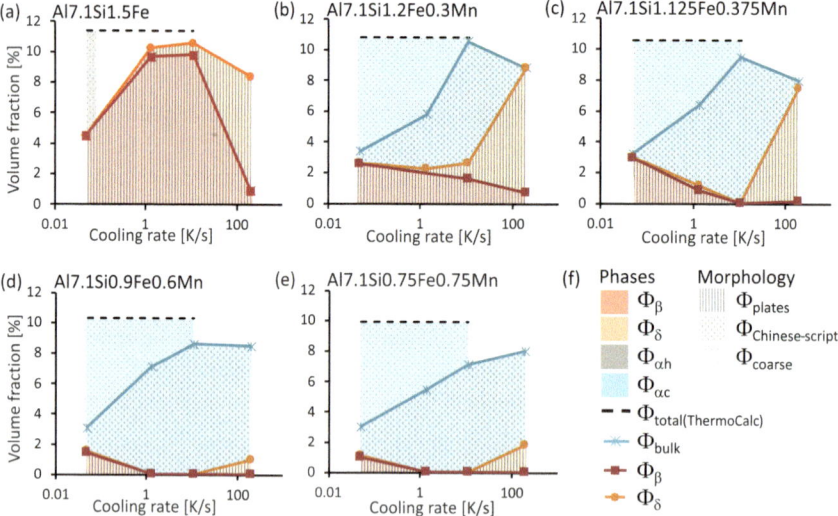

Fig. 8.7 Volume fraction Φ of the occurring intermetallic phases and volume fraction Φ of characteristic morphologies of the intermetallic phases depending on the applied cooling rates for the investigated alloys. Note that the lines and shaded areas are meant as guide to the eye. The figure was published in [15], Copyright Elsevier

interdendritic, region, but partly reach into the Al-dendrites. All other higher cooling rates have led to microstructures without any α_h-phase particles which is attributable to retarded nucleation of the α_h phase. Only plate-shaped particles occur. The plate-shaped particles in samples solidified with intermediate cooling rates appear within the Al-dendrites. In samples solidified with the highest cooling rate, the plate-shaped particles are located within the Al–Si eutectic region. These plate-shaped particles consist mainly of the β phase after solidification with low cooling rates and of the δ phase after solidification with high cooling rates. The observed suppression of the α_h phase and the formation of the δ phase under high cooling rates contradict thermodynamically calculated solidification paths for equilibrium and Scheil solidification conditions which predict the α_h and β phase or solely β phase.

- **Mn-containing Al7.1Si(1.5-x_{Mn})Fe(x_{Mn})Mn alloys**

In Mn-containing alloys, at the edge of the samples cooled up to intermediate cooling rates, coarse polyhedral and coarse dendritic particles of the α_c phase form during solidification. After solidification with low cooling rates plate-shaped particles accompany the coarse particles located in association with the Al–Si eutectic region. Under the highest cooling rates, plate-shaped particles also form, however, with smaller size and within the Al–Si eutectic. At intermediate cooling rates Chinese-script particles have formed being located within the Al-dendrites. The α_c-phase particles form as the only intermetallic microstructural component above a Mn-content dependent cooling rate. The relevant range of cooling rates for the solely α_c-phase formation increases with increasing Mn content. In presence of Mn, the discrepancy to the thermodynamically calculated solidification paths is less pronounced. The equilibrium calculation by trend agrees with the solidification at low cooling rates. Scheil solidification calculations are in accordance with the solidification at low to intermediate cooling rates. However, the formation of the δ phase is not predicted by the thermodynamic calculations similar to the Mn-free Al–Si alloys.

These observed deviations from the calculated equilibrium and Scheil solidification path substantiate the reported importance of kinetic effects during solidification of secondary Al–Si alloys [19, 56–58]. These kinetic effects can include:

- suppression of phases which are expected to occur according to equilibrium and Scheil solidification conditions e.g. due to retarded nucleation,
- occurrence of phases which are not expected to occur according to equilibrium and Scheil solidification conditions and
- formation of metastable phases which are not present in the equilibrium phase diagrams (not observed in the investigation in [15]).

The influence of such kinetic effects during non-equilibrium solidification cannot be fully accounted for by Scheil solidification conditions or alternatively by setting a certain phase or several phases dormant. Therefore, it is suggested to construct so-called *apparent liquidus projections* (Fig. 8.8) which provide a tool to estimate the phase formation during solidification. Their construction is based on the experimental microstructures, which are analyzed according to the present phases and their relative location in the microstructures and in the sample.

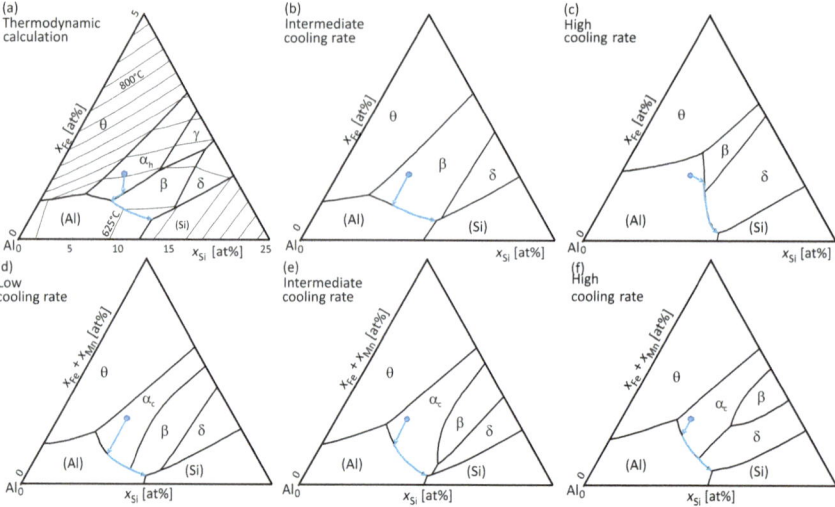

Fig. 8.8 A Liquidus projection of the Al–Fe–Si system based on thermodynamic calculations using the TCAl4 database [53] within the ThermoCalc [52] software. Thick black lines border primary phase areas. Thin black lines in **a** indicate contour lines of the temperature in 25 °C steps. Apparent liquidus projections of the Al–Fe–Si system for solidification conditions with **b** intermediate cooling rates and **c** high cooling rates which comply with the observed solidification microstructures. Apparent liquidus projection of the Al–Si–Fe–Mn system for some fixed Mn level ($x_{Mn}/(x_{Fe} + x_{Mn})$ approx. 0.5) being compatible with the as-solidified microstructures formed under solidification conditions with **d** low cooling rates, **e** intermediate cooling rates and **f** high cooling rates. The blue circles represent the presently investigated alloy composition and blue lines schematically represent the solidification path. Note that the scale of x_{Fe} and x_{Si} is different. Note that (**b-f**) represent apparent liquidus projections which are meant to rationalize changes in the solidification paths with increasing cooling rate. Therefore, no numbers are given at the scale. The figure was published as two separate figures in [15], Copyright Elsevier

As obvious from the apparent liquidus projections, the kinetic effects on the microstructure formation in secondary Al–Si alloys are pronounced. It can be assumed that two aims are followed by addition of further elements and Mn and solidification with a high cooling rate: avoiding sludge formation and suppressing the formation of plate-shaped particles. However, some trade-offs must be considered when the parameters are optimized.

The general alloy composition including the Si content and the total transition metal content, as well as the $x_{Mn}/(x_{Mn} + x_{Fe})$ ratio, determine the potentially forming phases. Aiming at an optimized microstructure with a low intermetallic phase fraction, without plate-shaped particles and without sludge particles, Mn should be present, but the $x_{Mn}/(x_{Mn} + x_{Fe})$ ratio and total transition metal content should be as low as required and the minimum cooling rate should be chosen according to the chemical composition of the alloy.

8.3 Interaction of Different Filter Materials with the Transition Metal Containing Al–Si Alloys

The following sections elucidate the effect of oxide- and carbon-containing filter materials in contact with Al7.1Si1.5Fe alloy on the Fe-removal efficiency in order to produce high-quality, secondary Al–Si alloys. Details exploring the utilization of filter materials to support the Fe removal can be found in [59]. The basic interaction effects investigations were investigated on flat substrates of Al_2O_3, mullite ($3Al_2O_3 \cdot 2SiO_2$), spinel ($MgAl_2O_4$), silicon carbide (SiC) and carbon bonded alumina (Al_2O_3-C).

Sessile-drop experiments have been carried out at 950 °C with the alloy in fully liquid state and at 625 °C with the alloy in the melt conditioned state (see Sect. 8.2.1) to assess the interaction behavior. For the interaction with Al7.1Si and Al7.1Si0.75Fe0.75Mn alloys as well as interaction experiments in small scale crucibles see [59].

Evaluation of the effectiveness of filter materials for Fe-removal requires information about:

(1) the contact of substrate and alloy melt in view of wetting, infiltration, reactive chemical interaction and formation of new layers at the interface,
(2) the effect of the filter material on the alloy melt composition concerning the desired removal of impurity Fe and possibly further transition metal elements, but also changes of the chemical concentration of intended alloying elements and contamination by external elements from the filter material,
(3) the effect on the formation of primary Fe-containing intermetallic particles regarding the type of phase and its amount which might include nucleation impacts and
(4) the position of primary Fe-containing intermetallic particles at the interface to the substrate or within the droplet volume.

The interaction characteristics and the consequences for the alloy are illustrated in Fig. 8.9 and summarized in Table 8.2.

(1) **Contact region of substrate and alloy melt**

The investigated alloy-substrate combinations can be classified into two main groups regarding the wetting behavior in view of the final reaction product in contact region with the alloy melt. The reaction equations with the individual substrates are listed in Table 8.2. One group contains the substrate materials Al_2O_3, mullite and spinel. Characteristic for these substrate materials is that the final reaction product in contact with the alloy melt is Al_2O_3. The final contact angles are in the range of 104 to 119° after melt conditioning at 625 °C in agreement with [60]. The other group contains SiC and Al_2O_3-C. In these systems, the final reaction product Al_4C_3 is in contact with the alloy melt. Consequently, final contact angles are in the range of 71 to 79°.

Hence, the final wettability mainly depends on the wettability of the liquid melt on the new reaction product [60, 61]. Despite the same reaction product, the spreading

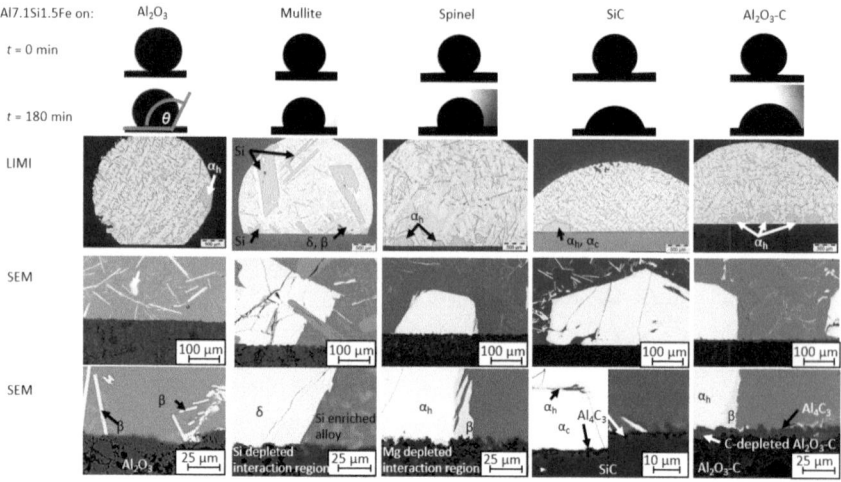

Fig. 8.9 Silhouette images of the Al7.1Si1.5Fe alloy on the bulk substrate (indicated at top) directly after reaching the 950 °C defining time t = 0 min and after holding at 950 °C for 60 min and 180 min. LIMI images from the cross-sections of the droplets. The phase constituting primary coarse particles is indicated. The contact angle θ has been indicated in one example. SEM-BSE images of the interaction region solidified from sessile drop experiments after 180 min at 950 °C. The figure was published in [59], Copyright Wiley

Table 8.2 Summary of the interaction characteristics of the substrate material with the Fe-containing Al-Si alloys. The table was reproduced from [59], Copyright Wiley

	Al$_2$O$_3$	Mullite	Spinel	SiC	Al$_2$O$_3$-C
Reaction	Non-reactive	3 SiO$_2$(s) + 4 [Al](l) → 3 [Si](l) + 2 Al$_2$O$_3$(s)	3 MgAl$_2$O$_4$(s) + 2 [Al](l) → 3 [Mg](l) + 4 Al$_2$O$_3$	3 SiC(s) + 4 [Al](l) → 3 [Si](l) + Al$_x$C$_y$(s)	3 C(s) + 4 [Al](l) → Al$_4$C$_3$(s)
Contact angles (60 min, 625 °C, Al7.1Si1.5Fe)	115°	104°	119°	72°	84°
Phase in contact to the alloy melt	Al$_2$O$_3$	Al$_2$O$_3$ (main component)	Al$_2$O$_3$ (main component)	Al$_4$C$_3$	Al$_4$C$_3$
Alloy composition	As-initial	Si enrichment[8.1]	Mg enrichment[8.1]	Fe, Cr, V, Ni, Al and Ti	0.9 at% Fe
Fe-containing primary intermetallic	α_h	β, δ	α_h	α_c, α_h	α_h
Location of primary intermetallic	Non-specific	At substrate	At substrate	Non-specific	At substrate

[1] While the Si enrichment was clearly measurable, the Mg enrichment was marginal

kinetics are governed by the reactive interaction at the triple line of the droplet to the substrate [61–63]. That temperature–time dependence might, as one aspect, causes the slightly different contact angles.

(2) **Consequences for the alloy melt composition**

The Al−Si melt in contact with Al_2O_3 represents a non-reactive, low-wetting system acting as a reference for the reactive systems with the other substrates. Consequently, no contamination of the melt after the interaction has been recognized. After melt conditioning the composition of the remaining melt, without primary Fe-containing α_h phase, is in agreement with the Fe-content after melt conditioning in Al_2O_3 crucibles (Sect. 8.2.1).

In the reactive systems with SiO_2 in $3Al_2O_3 \cdot 2SiO_2$ mullite and with the $MgAl_2O_4$ spinel, the Al alloy melt reduces the oxides forming Al_2O_3. Thereby, Si enriches in significant large amounts in the melts while Mg enriches in the melt only to a minor extent.

While the interaction of the melt with SiC and Al_2O_3-C leads, in both cases to a dense layer of Al_4C_3 carbide, the consequences for the alloy were different. The SiC substrate material is an example for the effect of impurities in the substrate material. Solution of transition-metal elements from spots that are rich in transition metal elements Fe, Cr, V, Ni, and Ti lead to a detectable modification of the solidified microstructure. A detectable enrichment of Si or C in the melt has not been observed. In the case of the Al_2O_3-C substrate, no contamination of the alloy has been observed. Note that contamination of aluminum alloys with C is undesirable. It could lead to the formation of Al_4C_3 particles in the bulk microstructure which are known to deteriorate the mechanical properties [64, 65]. However, the solubility of C at 960 °C in molar fraction is approximately $6.4 \cdot 10^{-4}$ [66] and thus, might not be relevant for the significant formation of Al_4C_3 during solidification. However, spalling off of Al_4C_3 particles from the reaction layer into the melt must be avoided. Furthermore, it has been suggested by [64] that the presence of Si in the Al-Si melt prevents the formation of the carbide layer. This was not observed in the present study although the resulting carbide layer is thinner on SiC than on Al_2O_3-C after sessile drop experiments.

(3) **Primary Fe-containing intermetallic particles**

The type of primary Fe-containing intermetallic particles is directly related to the alloy composition after the interaction experiment. As expected for the inert system with Al_2O_3, the primary phase is the α_h phase. The α_h phase is also the primary phase in contact to the $MgAl_2O_4$ spinel and Al_2O_3-C. In these systems the introduced impurity content is very low and the elements do not affect the type of primary phase. In case of $3Al_2O_3 \cdot 2SiO_2$ mullite, the remarkable increase of the Si content in the alloy, leads to the primary formation of β and δ phase. Although the introduced impurity content of Fe, Cr, V, Ni, and Ti from the SiC substrate is small, their effect is clearly detectable in the microstructure. The transition elements, best known for the modifying effect of small amounts of Cr [5], change the type of intermetallic phase in dependence on the total amount of the element and the cooling rate from α_h

and β to α_c [15]. Although such a microstructural modification towards the presence of α_c is often desirable [1], it should be applied as intended and controlled action.

(4) Position of primary Fe-containing intermetallic particles

For effective filtration, the specific attachment of primary particles on the substrates is beneficial. The observed location of α_h particles in the alloy on Al_2O_3 filter material and α_h/α_c particles in the alloy on SiC filter material is non-specific. In the alloys on Al_2O_3-C few large primary α_h particles are attached to the interface with the non-reactive Al_2O_3 filter material. In the case of $3Al_2O_3 \cdot 2SiO_2$ mullite and the $MgAl_2O_4$ spinel and especially Al_2O_3-C primary α_h particles are specifically attached to the substrate. Thereby, this attachment of the α_h particles to Al_2O_3–C with the carbide layer is associated with oriented growth of the carbide layer.

In conclusion, under the tested filter materials, Al_2O_3-C is the most promising candidate as a filter material. Although in case of Al_2O_3-C, the Fe content in the remaining melt of approx. 0.9 at% Fe is the same as in the remaining melt with contact to inert Al_2O_3 substrate (as expected from thermodynamic calculations). The reactive interaction of the intermetallic towards the Al_2O_3-C and the resulting attachment to this substrate could possibly be beneficially utilized to increase the Fe-reduction efficiency for secondary Al-Si alloys. Furthermore, the interaction forming a dense thin layer Al_4C_3 does not contaminate the Al–Si alloy.

Acknowledgements This study was financially supported by the German Research Foundation within the Collaborative Research Centre 920 "Multi-Functional Filters for Metal Melt Filtration—A Contribution towards Zero Defect Materials" (Project-ID 169148856) in the subproject A07. The authors are grateful for the financial support for the transmission electron microscope investigations related to activities within the centre for research-based innovation SFI Manufacturing in Norway, partial funding by the Research Council of Norway under contract number 237900 and financial support by the Research Council of Norway to the NORTEM project (197405). One of the authors, H. Becker likes to thank for the financial support by the Department of Materials Science and Engineering at NTNU during her visit in Trondheim (Norway) and by the Department of Civil and Mechanical Engineering at DTU during her visit in Lyngby (Denmark).

References

1. L. Zhang, J. Gao, L.N.W. Damoah, D.G. Robertson, Miner. Process. Extr. Metall. Rev. **33**, 99 (2012). https://doi.org/10.1080/08827508.2010.542211
2. D. Raabe, D. Ponge, P.J. Uggowitzer, M. Roscher, M. Paolantonio, C. Liu, H. Antrekowitsch, E. Kozeschnik, D. Seidmann, B. Gault, F. De Geuser, A. Deschamps, C. Hutchinson, C. Liu, Z. Li, P. Prangnell, J. Robson, P. Shanthraj, S. Vakili, C. Sinclair, L. Bourgeois, S. Pogatscher, Prog. Mater. Sci. **128**, 100947 (2022). https://doi.org/10.1016/j.pmatsci.2022.100947
3. J.A. Taylor, Mater. Sci. Forum **689**, 429 (2011). https://doi.org/10.4028/www.scientific.net/MSF.690.429
4. J. Roger, F. Bosselet, J.C. Viala, J. Solid State Chem. **184**, 1120 (2011). https://doi.org/10.1016/j.jssc.2011.03.025
5. D. Munson, J. Inst. Metals **95**, 217 (1967)

6. P. Orozco-González, M. Castro-Román, J. López-Cuevas, A. Hernández-Rodríguez, R. Muñiz-Valdez, S. Luna-Álvarez, C. Ortiz-Cuellar, Rev. metal. **47**, 453 (2011). https://doi.org/10.3989/revmetalm.1068
7. J.E. Tibballs, R.L. Davis, B.A. Parker, J. Mater. Sci. **24**, 2177 (1989). https://doi.org/10.1007/BF02385438
8. M. Cooper, K. Robinson, Acta Cryst. **20**, 614 (1966). https://doi.org/10.1107/S0365110X6600149X
9. M. Cooper, Acta Crystallogr. **23**, 1106 (1967). https://doi.org/10.1107/S0365110X67004372
10. G. Phragmen, J. Inst. Met. **77**, 489 (1950). https://doi.org/10.1007/s11669-007-9036-7
11. H.J. Kim, T.Y. Park, S.W. Han, H.M. Lee, J. Cryst. Growth **291**, 207 (2006). https://doi.org/10.1016/j.jcrysgro.2006.02.006
12. D. Ferdian, C. Josse, P. Nguyen, N. Gey, N. Ratel-Ramond, P. de Parseval, Y. Thebault, B. Malard, J. Lacaze, L. Salvo, Metall. Mater. Trans. A **46**, 2814 (2015). https://doi.org/10.1007/s11661-015-2917-1
13. W. Khalifa, F.H. Samuel, J.E. Gruzleski, Metall. Mater. Trans. A **34**, 807 (2003). https://doi.org/10.1007/s11661-003-0116-y
14. H. Becker, A. Thum, B. Distl, M.J. Kriegel, A. Leineweber, Metall. Mater. Trans. A **49**, 6375 (2018). https://doi.org/10.1007/s11661-018-4930-7
15. H. Becker, T. Bergh, P.E. Vullum, A. Leineweber, Y. Li, Materialia **5**, 100198 (2019). https://doi.org/10.1016/j.mtla.2018.100198
16. A. Fabrizi, G. Timelli, I.O.P. Conf, Ser. Mater. Sci. Eng. **117**, 012017 (2016). https://doi.org/10.1088/1757-899X/117/1/012017
17. Z. Zhou, Z. Li, Y. Xie, X. Wang, Y. Liu, Z. Long, F. Yin, Int. J. Mater. Res. **106**, 470 (2015). https://doi.org/10.3139/146.111202
18. T. Gao, Y. Wu, C. Li, X. Liu, Mater. Lett. **110**, 191 (2013). https://doi.org/10.1016/j.matlet.2013.08.039
19. W. Khalifa, F.H. Samuel, J.E. Gruzleski, H.W. Doty, S. Valtierra, Metall. Mater. Trans. A **36**, 1017 (2005). https://doi.org/10.1007/s11661-005-0295-9
20. V. Hansen, B. Hauback, M. Sundberg, C. Rømming, J. Gjønnes, Acta Crystallogr. B **54**, 351 (1998). https://doi.org/10.1107/S0108768197017047
21. C. Rømming, V. Hansen, J. Gjønnes, Acta Crystallogr. B **50**, 307 (1994). https://doi.org/10.1107/S0108768193013096
22. H. Becker, N. Bulut, J. Kortus, A. Leineweber, J. Alloys Compd. **911**, 165015 (2022). https://doi.org/10.1016/j.jallcom.2022.165015
23. H. Becker, T. Bergh, P.E. Vullum, A. Leineweber, Y. Li, J. Alloy. Compd. **780**, 917 (2019). https://doi.org/10.1016/j.jallcom.2018.11.396
24. J.G. Zheng, R. Vincent, J.W. Steeds, Philos. Mag. A **80**, 493 (2000). https://doi.org/10.1080/01418610008212063
25. G.J.C. Carpenter, Y. Le Page, Scr. Metall. Mater. **28**, 733 (1993). https://doi.org/10.1080/01418610008212063
26. M.V. Kral, P.N.H. Nakashima, D.R.G. Mitchell, Metall. and Mater. Trans. A. **37**, 1987 (2006). https://doi.org/10.1007/s11661-006-0141-8
27. C. Gueneau, C. Servant, F. d'Yvoire, N. Rodier, Acta Crystallogr. C **51**, 177 (1995). https://doi.org/10.1107/S0108270194009030
28. J.M. Yu, N. Wanderka, A. Rack, R. Daudin, E. Boller, H. Markötter, A. Manzoni, F. Vogel, T. Arlt, I. Manke, J. Banhart, Acta Mater. **129**, 194 (2017). https://doi.org/10.1016/j.actamat.2017.02.048
29. H. Becker, A. Leineweber, Mater Charact **141**, 406 (2018). https://doi.org/10.1016/j.matchar.2018.05.013
30. K. Momma, F. Izumi, J. Appl. Crystallogr. **44**, 1272 (2011). https://doi.org/10.1107/S0021889811038970
31. M. Rappaz, P. Jarry, G. Kurtuldu, J. Zollinger, Metall. Mater. Trans. A **51**, 2651 (2020). https://doi.org/10.1007/s11661-020-05770-9

32. C. Galera-Rueda, M.L. Montero-Sistiaga, K. Vanmeensel, M. Godino-Martínez, J. Llorca, M.T. Pérez-Prado, Addit. Manuf. **44**, 102053 (2021). https://doi.org/10.1016/j.addma.2021.102053
33. P.J. Black, Philosoph. Magaz. J. Sci. **46**, 401 (1955). https://doi.org/10.1080/147864404085 20573
34. C.M. Fang, Z.P. Que, Z. Fan, J. Solid Statre Chem. **299**, 122199 (2021). https://doi.org/10.1016/j.jssc.2021.122199
35. S. Martin, D. Sulik, X.F. Fang, H. Becker, A. Leineweber (2022). https://doi.org/10.2139/ssrn.4039449
36. J. Gobrecht, Schwereseigerungen von Eisen, Mangan und Chrom in Aluminium-Silicium-Gußlegierungen (Teil 1). Giesserei **62**, 263 (1975)
37. J.L. Jorstad, Understanding "Sludge." Die Cast Eng. **30**, 30 (1986)
38. R. Dunn, Aluminum melting problems and their influence on furnace selection. Die Cast Eng. B **9**, 8 (1965)
39. G. Timelli, S. Capuzzi, A. Fabrizi, J. Them. Anal. Calorim. **123**, 249 (2016). https://doi.org/10.1007/s10973-015-4952-y
40. M.M. Makhlouf, D. Apelian, Casting characteristics of aluminum die cast alloys, work performed under contract No. DEFC07–99ID13716 prepared for US department of energy office of industrial technologies prepared by the advanced casting research center Worcester Polytechnic Institute, pp. 1–46 (2002). https://doi.org/10.2172/792701
41. H.I. de Moares, J.R. de Oliveira, D.C.R. Espinosa, J.A.S. Tenório, Mater. Trans. **47**, 1731 (2006). https://doi.org/10.2320/matertrans.47.1731
42. A. Flores-V, M. Sukiennik, A.H. Castillejos-E., F.A. Acosta-G., J.C. Escobedo-B., Intermetallics **6**, 217 (1998). https://doi.org/10.1016/S0966-9795(97)00073-3
43. J.E. Tibballs, J.A. Horst, C.J. Simensen, J. Mater. Sci. **36**, 937 (2001). https://doi.org/10.1023/A:1004815621313
44. B.G. Dietrich, H. Becker, M. Smolka, A. Keßler, A. Leineweber, G. Wolf, Adv. Eng. Mater. **19**, 1700161 (2017). https://doi.org/10.1002/adem.201700161
45. J.P. Schoß, H. Becker, A. Keßler, A. Leineweber, G. Wolf, Adv. Eng. Mater. 2100695 (2021). https://doi.org/10.1002/adem.202100695
46. P. Ashtari, K. Tetley-Gerard, K. Sadayappan, Can. Metall. Quart. **51**, 75 (2012). https://doi.org/10.1179/1879139511Y.0000000026
47. S.W. Kim, U.H. Im, H.C. Cha, S.H. Kim, J.E. Jang, K.Y. Kim, Ch. Foundry **10**, 112 (2013). https://doi.org/10.1179/1879139511Y.0000000026
48. A. Bjurensted, D. Casari, S. Seifeddine, R.H. Methiesen, Acta Mater. **130**, 1 (2017). https://doi.org/10.1016/j.actamat.2017.03.026
49. X. Cao, J. Campbell, Metall. Mater. Trans. A **34**, 1409 (2003). https://doi.org/10.1007/s11661-003-0253-3
50. D.N. Miller, L. Lu, A.K. Dahle, Metall. Mater. Trans. B **37**, 873 (2006). https://doi.org/10.1007/BF02735008
51. S.G. Shabestari, Mater. Sci. Eng. A **383**, 289 (2004). https://doi.org/10.1016/j.msea.2004.06.022
52. H.L. Lukas, S.G. Fries, B. Sundman, Cambridge University Press, pp. 1–324 (2007). https://doi.org/10.1016/j.msea.2004.06.022
53. TCAL4 – TCS Al-based alloy database, Version 4.0 http://www.thermocalc.com/media/19847/dbd_tcal40_extended_info.pdf
54. L.F. Mondolfo, Aluminum alloys: structure and properties, London, Boston, Butterworth, pp. 487–488 (1976). https://doi.org/10.1016/C2013-0-04239-9
55. D. Shimosaka, S. Kumai, F. Casarotto, S. Watanabe, Mater. Trans. **52**, 920 (2011). https://doi.org/10.2320/matertrans.L-MZ201125
56. L.A. Narayanan, F.H. Samuel, J.E. Gruzleski, Metall. Mater. Trans. A **25**, 1761 (1994). https://doi.org/10.1007/BF02668540
57. L. Bäckerud, G. Chai, J. Tamminen, Aluminum-Silicon alloys, Chapter 5. In: Solidification Characteristics of Aluminum Alloys Vol.2:, Foundry Alloys, AFS and Skanaluminium, Oslo, Norway (1990)

58. Y. Langsrud, Key Eng. Mater. **44–45**, 95 (1990). https://doi.org/10.4028/www.scientific.net/KEM.44-45.95

59. H. Becker, B. Fankhänel, C. Voigt, C. Charitos, M. Stelter, C.G. Aneziris, A. Leineweber, Adv. Eng. Mater., 2100595 (2021). https://doi.org/10.1002/adem.202100595

60. B. Fankhänel, M. Stelter, C. Voigt, C.G. Aneziris, Adv. Eng. Mater. **19**, 1700084 (2017). https://doi.org/10.1002/adem.201700084

61. O. Dezellus, N. Eustathopoulos, J. Mater. Sci. **45**, 4256 (2010). https://doi.org/10.1007/s10853-009-4128-x

62. K. Landry, N. Eustahopoulos, Acta Mater. **44**, 3923 (1996). https://doi.org/10.1016/S1359-6454(96)00052-3

63. V. Laurent, C. Rado, N. Eustathopoulos, Wetting kinetics and bonding of Al and Al alloys on u-SiC. Mater. Sci. Eng. A **205**, 1 (1996)

64. S. Bao, A. Kvithyld, T.A. Engh, M. Tangstad, Light Metals TMS **22**, 1930 (2012). https://doi.org/10.1016/S1003-6326(11)61410-6

65. T. Etter, P. Schulz, M. Weber, J. Metz, M. Wimmler, J.F. Löffler, P. J. Uggowitzer, Mater. Sci. Eng., A, **48**, 1 (2007). https://doi.org/10.1016/j.msea.2006.11.088

66. C. Qiu, R. Metselaar, J. Alloy. Compd. **216**, 55 (1994). https://doi.org/10.1016/0925-8388(94)91042-1

67. X. Cao, J. Campbell, Metall. Mater. Trans. A **35**, 1425 (2004). https://doi.org/10.1007/s11661-004-0251-0

Chapter 9
Influence of Reactive Filter Materials on Casting's Quality in Aluminum Casting

Beate Fankhänel, Ekaterina Schmid, and Michael Stelter

9.1 Introduction

The purity of melts plays an essential role in the production of high-quality aluminum castings. Since more than half of the aluminum alloys produced today originate in secondary metallurgy, an increasing number of impurities also find their way into the alloy melts. These impurities can be present in the melts both in dissolved form (e.g. H_2, undesirable alloying elements) and in solid form (e.g. Al_2O_3, MgO, AlN, NaCl, TiB_2). Impurities in solid form can be formed during melting and liquid holding in the melt (endogenous inclusions) or can enter the melt from outside through reactions with the atmosphere or the refractory (exogenous inclusions). Many of these inclusions lead to defects and undesirable macroporosity in the castings. Due to the high solubility of hydrogen in the aluminum melt, its precipitation during solidification in particular leads to increased porosity and a decrease in casting quality. A pronounced hydrogen porosity can thus lead to a reduction in strength, toughness, surface quality and corrosion properties of the components [1].

Even though there are already various ways of reducing the hydrogen content and particulate inclusions of an aluminum melt, solidification in many castings does not always proceed uniformly, so that areas with undesirable macroporosity can still form. In addition, non-metallic inclusions can be generated during melt cleaning by reactions with flushing gases or salt residues. For this reason, it can be advantageous to use filter materials that are specially adapted to the impurities to be removed from

B. Fankhänel (✉) · M. Stelter
Institute of Nonferrous Metallurgy and Purest Materials, Technische Universität Bergakademie Freiberg, Leipziger Str. 34, 09599 Freiberg, Germany
e-mail: beate.fankhaenel@inemet.tu-freiberg.de

E. Schmid
Freiberger Compound Materials GmbH, Am Junger-Löwe-Schacht 5, 09599 Freiberg, Germany
e-mail: ekaterina.schmid@freiberger.com

© The Author(s) 2024
C. G. Aneziris and H. Biermann (eds.), *Multifunctional Ceramic Filter Systems for Metal Melt Filtration*, Springer Series in Materials Science 337,
https://doi.org/10.1007/978-3-031-40930-1_9

the aluminum alloys. The use of ceramic foam filters has been a proven process in aluminum casting for several years [2–5]. With the help of melt filtration, a laminar melt flow is created and solid impurities can be removed. Both improve casting quality. The possibility of coating filters with active materials to achieve improved melt cleaning is becoming increasingly important [6, 7]. Reactive materials could help to force hydrogen precipitation at the beginning of solidification and/or to bind the precipitated hydrogen in order to effectively reduce the hydrogen content of the melt in time.

Although the processing temperatures of aluminum melts are not in extreme ranges, filtration as a cleaning process on an industrial scale can take 30 min or longer, so that a non-negligible interaction between aluminum melt and filter material must be expected. In order to investigate these interactions more in detail, not using the considerable quantities of metal required for a correspondingly long filtration period, laboratory immersion tests and sessile drop experiments were used. In the case of the immersion experiments, compact pieces of the filter materials were immersed in the molten aluminum for a certain time and moved therein. Sessile drop experiments are a method for investigating the wetting behavior of a fluid (molten) media in relation to solid materials as a function of temperature and time. These investigations were completed by casts in the laboratory to obtain an initial impression of the effectiveness of the new filter materials under operating conditions.

9.2 Interaction Between Aluminum, Its Alloys and Reactive Filter Materials

As already mentioned, wetting, roughness [8] and potential chemical reactions between the filter material and the molten metal play a major role on the filtration effect. For this reason, immersion tests were carried out in addition to sessile drop experiments as preparation for the filtration tests. In order to be able to make more realistic statements, substrates were used for the investigations whose surfaces were adapted to those of the filter materials and for whose production the same routine as for the filter production was applied.

9.2.1 Reactivity of Spodumene

While alumina showed a non-reactive behavior towards pure aluminum and selected alloys in all studied cases, a reactive behavior could be detected for other oxides under certain circumstances [9]. However, spodumene ($LiAl(Si_2O_6)$) represents a special type of filter material. Spodumene is a naturally occurring material that is chemically a mixed oxide: $\frac{1}{2}Li_2O \cdot \frac{1}{2}Al_2O_3 \cdot 2SiO_2$. The reason for researching the lithium oxide-containing material as a potentially reactive filter material was the assumption that Li_2O might be able to form a compound with hydrogen dissolved in the aluminum, thus reducing the hydrogen content in the alloy. Lithium salts have been used for several years for melting point reduction in aluminum production and for oxide removal in aluminum welding [10, 11], so that a basic interaction between spodumene as a filter material and the aluminium melt could be expected. Thermodynamic calculations of possible reactions between the liquid aluminum, the Al_2O_3 as the oxide layer and the $LiAl(Si_2O_6)$ as filter material were carried out with FactSage8.1® for different conditions. These calculations expected the formation of $LiAl_5O_8$, $LiAlO_2$ and silicon at the cost of Al_2O_3 in case of a typical casting situation (Eq. 9.1).

$$40Al + 0.1Al_2O_3 + LiAl(Si_2O_6) \rightarrow 37.33Al$$
$$+ 2Si + 0.72LiAl_5O_8 + 0.28\ LiAlO_2$$
$$(\Delta G = -315 - -323\ kJ, 750 - 650\,°C, 1\ bar) \tag{9.1}$$

In the presence of gaseous hydrogen, as released from the molten aluminum during filtration and cooling, the following reaction scenario was calculated (Eq. 9.2):

$$40Al + 0.5H_2 + LiAl(Si_2O_6) \rightarrow 37Al + 2Si$$
$$+ 0.75LiAl_5O_8 + 0.25LiAlH_4 + 0.1Al_2O_3$$
$$(\Delta G = -266 - 281\ kJ, 750 - 650\,°C, 1\ bar) \tag{9.2}$$

Here a hydrogen-containing compound could be formed in addition to the lithium aluminate. However, phases containing lithium and/or hydrogen are undetectable with many analytical methods. Therefore, plasma-based Secondary Neutral Mass Spectroscopy (SNMS) was used as a characterization method for the verification of lithium and hydrogen in the filter materials. This method is based on ion bombardment of the sample and subsequent ionization of the sputtered neutrals [12]. Experiments were performed using the INA-X equipment from SPECS GmbH, Germany [13]. In this configuration an Electron Cyclotron Wave Resonance (ECWR) plasma serves both as source for primary ions and for post ionization. After passing an ion optic, the post-ionized neutrals are separated by a quadrupole mass analyzer and counted by a secondary electron multiplier. The SNMS measurements were done in the so-called High Frequency Mode (HFM) due to the dielectric properties of the investigated samples [14].

SNMS mass spectra (intensity vs. mass number) were plotted for ceramic foam filters containing spodumene (pure $LiAl(Si_2O_6)$ and mixed oxide of 85% Al_2O_3 15% spodumene) as well as for the Al_2O_3 filters as reference. Each of them were measured three times by using a copper mask (5 mm in diameter) positioned on top of the sample. For the measurements, a krypton plasma with 152 W, 4.5 A and a working pressure of $2 \cdot 10^{-3}$ mbar was used. The applied voltage was set to 500 V at a frequency of 91 kHz and a duty cycle of 60%. The distance between the sample surface and the molybdenum aperture during measurements was 1.5 mm. The sputtering time for each mass spectrum was 210 s.

The presence of lithium content within the samples was examined to make sure that the sintering step of the spodumene did not lead to a complete evaporation of the lithium. The SNMS measurements were carried out at the surface of substrates used for the sessile drop measurements. The measured intensity depends not only on the concentration of the atoms, but also on detection factors which are influenced by geometry factors, post-ionization and on transmission factors [15, 16]. For this reason, a quantification of the lithium content would only be possible after referencing the lithium containing components, which is very demanding and time intensive, and has not yet been carried out. However, the qualitative analysis already showed clear lithium peaks indicating the presence of lithium at the surface of sintered lithium containing coatings in comparison to pure alumina substrates without any lithium peaks (Fig. 9.1).

Furthermore, heat treatments of these filter materials under a hydrogen-containing atmosphere were carried out to simulate the contact between the filter material and gaseous hydrogen, which will be released during solidification of the metal melt. For this purpose, heat treatments of spodumene-containing substrates were carried out for 30 to 90 min at temperatures between 700 and 1000 °C under a forming gas atmosphere consisting of 90 vol.% Ar and 10 vol.% H_2 to investigate the ability of the spodumene to react with gaseous hydrogen. Indeed, hydrogen-containing reaction products were detected on the substrate by SNMS measurements. While in the case of untreated substrates almost no hydrogen was measured, the substrates exposed to the hydrogen atmosphere showed clear intensity maxima relatable to hydrogen. It was irrelevant whether the substrate consisted of sintered Al_2O_3 with 15 wt.%

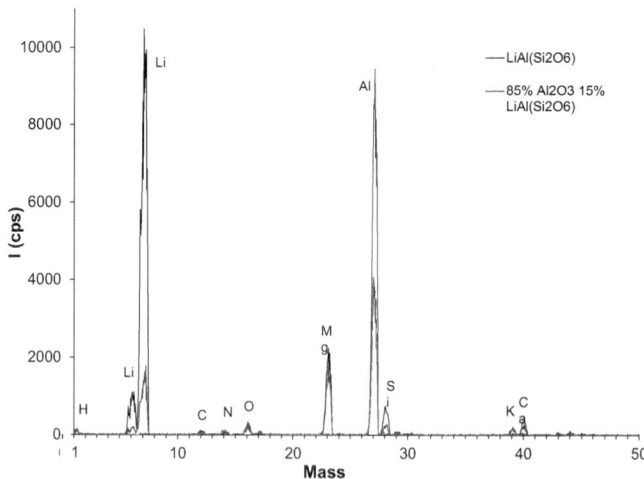

Fig. 9.1 SNMS spectra of $LiAl(Si_2O_6)$-coated Al_2O_3 substrates and mixed oxide substrates in comparison to pure Al_2O_3 substrates

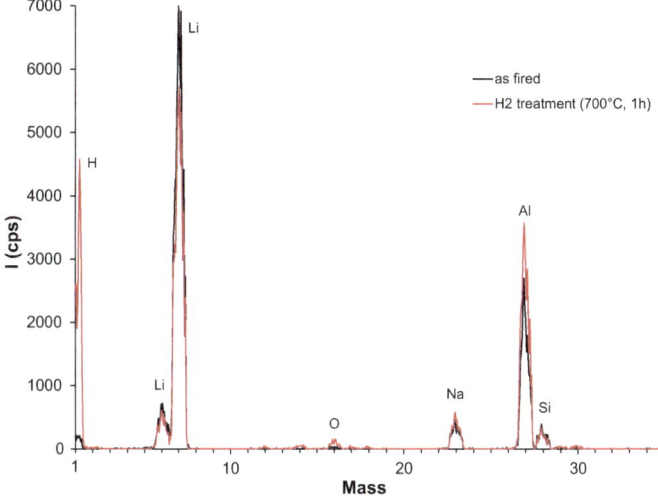

Fig. 9.2 Spodumene coated Al_2O_3 before and after a heat treatment at 700 °C under a hydrogen-containing gas atmosphere (90 vol.% Ar 10 vol.% H_2)

spodumene or of a pure spodumene surface. Similar measurements with pure Al_2O_3 showed hydrogen peaks as well, but to a much lesser extent than with the spodumene-containing samples.

The spectra recorded by SNMS for the element distribution on the sample surface thus indicate that hydrogen is already adsorbed or bound in the filter material from 700 °C onwards (Fig. 9.2).

9.2.2 Sessile Drop Experiments

Procedure

A common method for determining wetting properties is the sessile drop method. The method is based on the evaluation of the shape of liquid droplets on solid surfaces. The shape of the drop is essentially determined from the balance between surface tension and weight. Based on the droplet contour, both the surface tension of the liquid and the contact angle at the triple point of the system can be calculated if the density of the droplet material is known. The basis for the calculations is the Young–Laplace equation (Eq. 9.3), which describes the relationship between the drop shape (r = spherical radius), the surface tension (γ_{lg}) and the pressure difference (Δp) at the phase boundary.

$$\Delta p = \frac{2\gamma_{lg}}{r} \tag{9.3}$$

The influence of gravity on the droplet shape can be neglected for small droplets (m < 100 mg) so that a spherical segment can be assumed as the droplet shape [17]. In this case, it is also possible to calculate the contact angle (θ) using the drop height (h) and the diameter at the drop base (d) (Eq. 9.4):

$$\theta(rad) = 2\tan^{-1}\frac{2h}{d} \qquad (9.4)$$

The existing sessile drop system consists of a 3-zone melting furnace with a maximum working temperature of 1550 °C manufactured by Gero GmbH (Neuhausen, Germany), a vacuum and inert gas system and an optical evaluation unit with CCD camera for computer-aided recording of the melt drop and calculation of the contact angle.

In order to remove the oxide layer on the surface of the aluminum completely and to establish contact between the liquid metal and the substrate surface, a high temperature and a high vacuum pressure are necessary. Only then, the Al_2O_3 at the drop surface is reduced by liquid aluminum to metastable Al_2O. (Eq. 9.5).

$$4Al(l) + Al_2O_3(s) \rightarrow 3Al_2O\ (g) \qquad (9.5)$$

The thus formed Al_2O is gaseous under the above conditions and can be removed via the vacuum pump system. Prior to the sessile drop experiments, the equilibrium partial pressure of Al_2O as a function of temperature was determined for Eq. 9.5 using FactSage® (Fig. 9.3). From the calculation, a temperature of 950 °C and a vacuum pressure of $1.5 \cdot 10^{-5}$ mbar were determined as minimum requirements for the experiments.

Consequently, all sessile drop experiments were carried out according to the following procedure: At first, the metal samples (m < 100 mg) were cut on all of their faces to receive fresh surfaces with oxide layers thinner than 25 Å as described by Bianconi et al. [18] immediately before placing it onto the substrate in the furnace. The furnace was evacuated for 90 min to reach a pressure of $p < 1.5 \cdot 10^{-5}$ mbar. Afterwards the heating process started with a heating rate of 350 K/h up to a maximum temperature of 950 °C, which was held constant for 180 min. At this time, a vacuum pressure of $7.4 \cdot 10^{-6} \pm 0.4 \cdot 10^{-6}$ mbar on average was measured within the furnace. A detailed description of the process and the contact angle calculation is given in [9, 19].

Influence of the Surface Morphology

The wetting behavior of a liquid phase on a solid phase is also influenced by the surface morphology of the solid phase. For this reason, various surface qualities were investigated with regard to their influence on the contact angle between metal and filter material and the resulting wetting statement, which is important for the filtration effect [9]. The classical sessile drop method for determining the contact

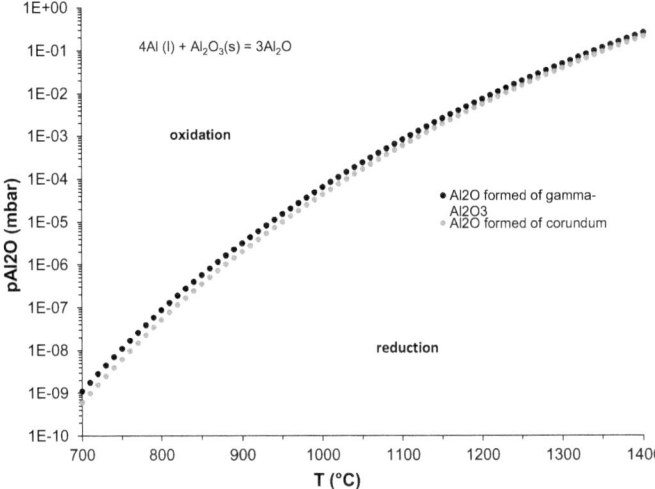

Fig. 9.3 Equilibrium partial pressure of metastable Al_2O as a function of temperature, calculated with FactSage®

angle usually uses dense polished substrates to obtain unobstructed contact between the liquid phase and the solid substrate. The surfaces of the filter struts, on the other hand, are open pored and have a natural roughness. For this reason, the investigations to determine the wetting behavior and the interaction between metal and filter material were extended to include the use of substrates with surfaces comparable to those of the filter struts. Both the filter material in itself as well as the surface roughness of the filter struts determine the effect of the filters during filtration. Both can be reproduced very well by coated substrates. The coating process is a combined dipping and centrifuging process. It has been already used to produce the appropriate foam filters and described in detail by Voigt et al. [6].

As a result, compact substrates are obtained whose surface morphology is comparable to that of the filter struts. Examples are shown in Table 9.1 and Fig. 9.4. It can be seen that the surfaces of the coated substrates are quite comparable with those of the filter struts. In the case of Al_2O_3, there is almost no difference, and also in the case of the other materials the average surface height S_a is similar. The area factor S_{dr} (ISO 25178), which indicates the increase in surface area compared to an ideally smooth surface of the same lateral dimension, differs apart from the alumina from one another. This difference probably is caused by a certain surface roughness of the filter struts in contrast to the pressed tablets. The surface morphology can be enhanced by the coating and cause an increase in the effective surface area.

Sessile drop experiments carried out with these substrates showed that a non-wetting behavior occurs in all cases, with the exception of the spodumene-containing materials, which will be discussed separately in the next section [9, 19]. Filters made of mullite ($3Al_2O_3 \cdot 2SiO_2$), for example, show significantly poorer wetting than other oxidic filter materials only due to their greater surface roughness. Thus, the

Table 9.1 Surface roughness (arithmetic average height S_a, area factor S_{dr} according ISO 25178) of coated substrates compared to filter struts of selected filter materials

Filter material	Coated substrate S_a in μm	S_{dr} in %	Filter struts S_a in μm	S_{dr} in %
Al_2O_3	2 ± 0.2	18 ± 1	4 ± 0.5	17 ± 6
85% Al_2O_3 15% $LiAl(Si_2O_6)$	10 ± 1	65 ± 9	8 ± 2.5	153 ± 13
TiO_2	1 ± 0.1	35 ± 1	6 ± 0.9	99 ± 14
$3Al_2O_3 \cdot 2SiO_2$	13 ± 1	228 ± 6	11 ± 2	379 ± 82

wetting behavior of all investigated oxidic filter materials (Al_2O_3, $3Al_2O_3 \cdot 2SiO_2$, $MgAl_2O_4$, TiO_2) on an AlSi7Mg alloy was comparable. Furthermore, it was shown that all substrates, but the pure Al_2O_3 form reactive systems in presence of the AlSi7Mg alloy under the conditions of the sessile drop experiments. This involved, for example, the reduction of TiO_2 or SiO_2 by the liquid aluminum, which thereby oxidizes itself to Al_2O_3. The silicon thus formed passes into the melt in metallic form, while the TiO_2 reacts to form Ti_2O_3 and with magnesium-containing aluminum alloys to form $MgTiO_3$ [9, 20]. If an aluminum melt comes into contact with a magnesium aluminate ($MgAl_2O_4$), it is depleted of magnesium at the contact zone, which can diffuse into the melt. What remains is a peripheral zone of Al_2O_3. Therefore, it was supposed that the equilibrium value of the contact angle is more dominated by the surface roughness than by one of the investigated materials apart from the alumina. The chemistry of the filter coatings influences only the course of the wetting process in itself.

Since time and temperature are much lower during filtration than during sessile drop experiments, the occurrence of a chemical reaction between the three filter materials ($3Al_2O_3 \cdot 2SiO_2$, $MgAl_2O_4$, and TiO_2) and the melt during casting is, however, rather unlikely. Therefore, the filter materials mentioned here will not exhibit any reactive character during filtration [9].

Nanostructured surfaces represent a special case. Here, the surface roughness is not the determining factor in the contact angle measurement. The wetting behavior is mainly determined by the increased surface energy of the nanostructured surfaces compared to a reference material. Thus, the wetting of Al_2O_3 became worse compared to an AlSi7Mg alloy due to the nanostructuring of the surfaces, while the chemical composition and the surface roughness remained the same. Subsequent filtration tests showed that there is no difference in effect between the filters with nanostructured surfaces and equivalent refractory filters [21].

This means, for non-reactive filter materials only the surface roughness in itself is significant for the filtration process.

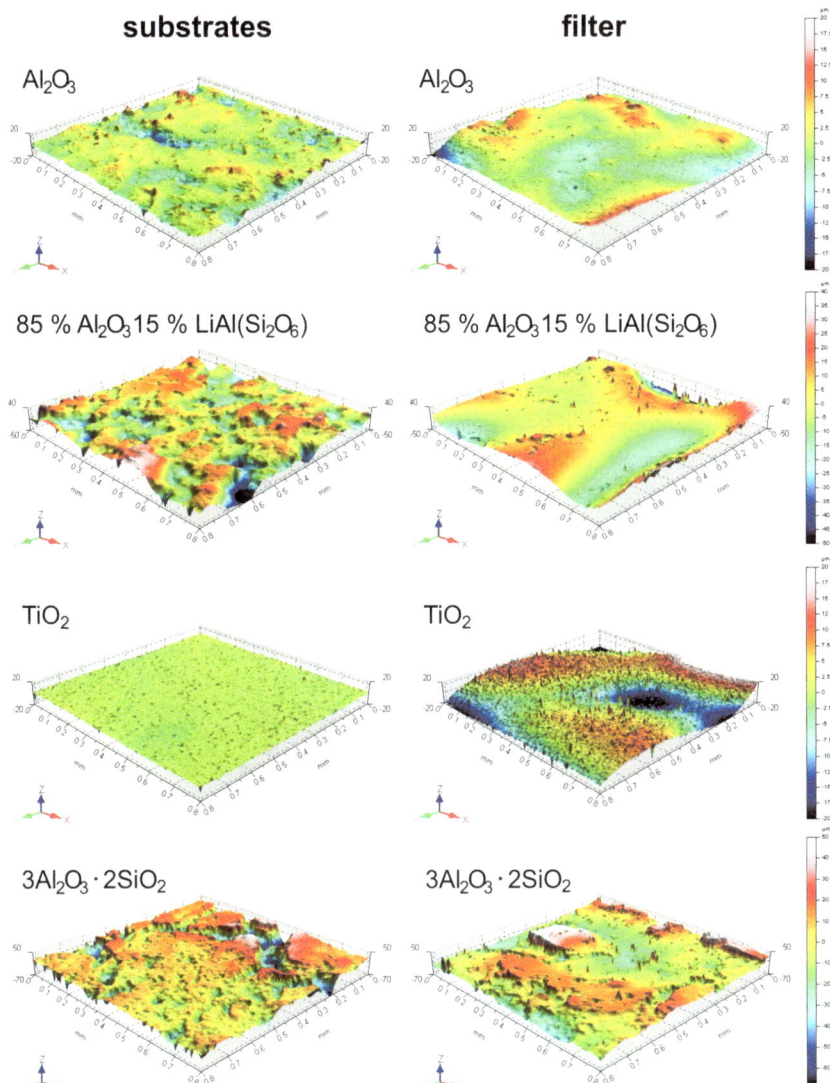

Fig. 9.4 Characteristic surface sections (investigated area: 0.8×0.8 mm) of coated substrates and of the corresponding filter struts; according Table 9.1 (The scale shown is identical for each material.)

Results for Materials Containing Spodumene

Sessile drop experiments confirmed the high reactivity of spodumene as a filter material. In contrast to the other investigated filter materials, the spodumene based substrates showed a completely different wetting behavior. The melt droplet already showed a wetting behavior after a few minutes after deoxidation of the surface and a reaction layer was visible to the naked eye on the spodumene-containing substrates after the experiments (Fig. 9.5). The reaction obviously took place between the LiAl(Si_2O_6) and the metal and/or the oxide on the metal surface. Investigations using Scanning electron microscopy (SEM Ultra55, Zeiss, Germany) showed a reaction zone (Fig. 9.6) consisting of an layered oxide containing aluminum and silicon in varying amounts, where the amount of aluminum is significantly higher (64 ± 2 wt.%) than that of silicon (4 ± 1 wt.%) [19].

Looking at the droplet cross-section of such a sample, comparatively large silicon segregations are noticeable, even in presence of the mixed oxide with only 85 wt.% alumina (Fig. 9.7). The segregations are a result of the reaction of the spodumene with the liquid aluminum, which reduces SiO_2, contained in the spodumene, to metallic silicon (Eqs. 9.1, 9.2).

Since neither lithium nor hydrogen is possible to be detected by EDX-measurements on the SEM, such a reaction layer was analyzed using SNMS. In this

Fig. 9.5 Top view of a typical solidified sessile drop of an AlSi7Mg alloy on a spodumene-coated Al_2O_3-substrate

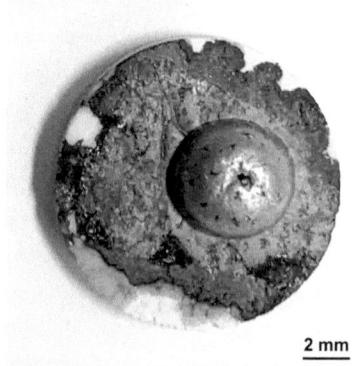

2 mm

Fig. 9.6 Top view of the reaction layer around a AlSi5Mg droplet after the sessile drop experiment on LiAl(Si_2O_6)-coated Al_2O_3 (Reproduced from Fankhänel et al. 2015 [19])

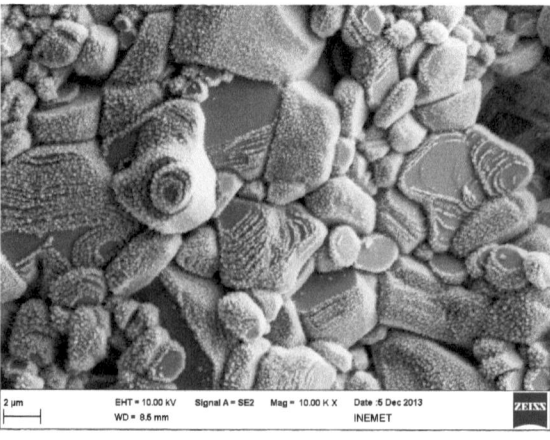

Fig. 9.7 Cross section of an AlSi7Mg droplet after sessile drop experiment on a mixed oxide ceramic (85 wt.% Al_2O_3 15 wt.% $LiAl(Si_2O_6)$)

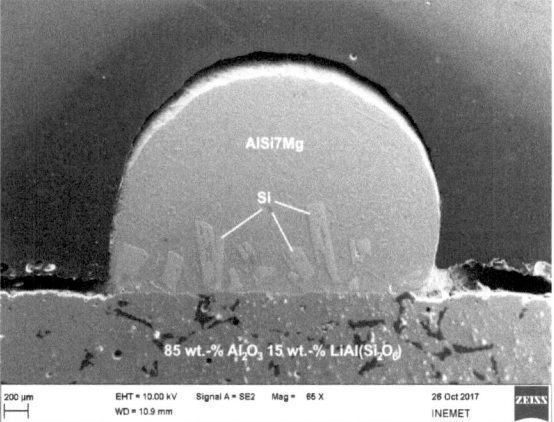

way, it was possible to show that the reaction layer contains lithium and hydrogen in addition to aluminum, silicon and oxygen. (Fig. 9.8). Because the sessile drop experiments were carried out under high vacuum, the source of the hydrogen found in the reaction layer only can be the hydrogen dissolved in the melt. Therefore, a hydrogen-containing reaction product can certainly be assumed.

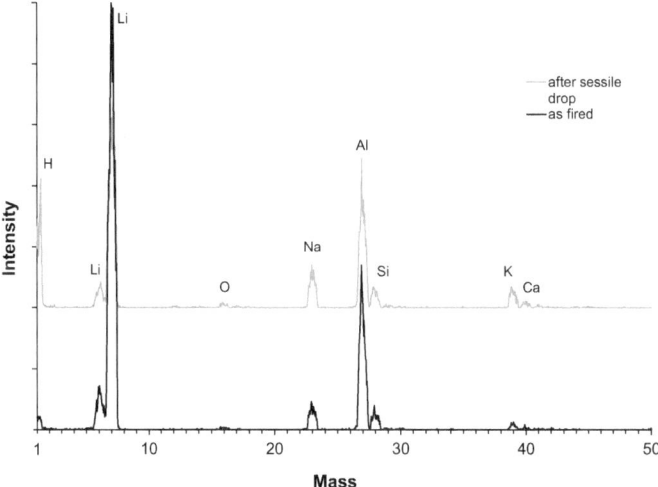

Fig. 9.8 SNMS surface analysis of the a spodumene coated Al_2O_3 substrate before (bottom) and after (top) a sessile drop experiment with pure aluminum at 950 °C under high vacuum

9.2.3 Immersion Tests

In addition to the sessile drop experiments, immersion tests were carried out under environmental conditions. For this purpose, tablets like those used as substrates for the sessile drop experiments were used. The tablets consisted either of 85 wt.% Al_2O_3 and 15 wt.% $LiAl(Si_2O_6)$ or of alumina coated with pure $LiAl(Si_2O_6)$. They were sintered under identical conditions as the filter materials. For the tests, the tablets were immersed in an AlSi7Mg melt and a pure Al melt at 730 °C for 10 or 30 min, respectively.

The reaction marks can already be seen with the naked eye (Fig. 9.9). The surfaces of all tablets showed residues of a reaction product after only a few minutes. Especially the samples coated with spodumene (bottom of Fig. 9.9) showed a strong reactivity, which made it impossible in some cases to separate the metal from the ceramic. But also with the samples consisting of the mixed oxide, 85 wt.% Al_2O_3 and 15 wt.% $LiAl(Si_2O_6)$ (top in Fig. 9.9), reaction traces could be detected in all cases. Already after 10 min exposure time, enrichments of metallic silicon appeared at the interface between an aluminium melt and the filter material. Fragments of the spodumene coatings and oxide particles have been found at the interfaces and at the bottom of the metal caps of the immersion samples. These oxide particles consist of Al, Mg and O according to EDX analysis (Fig. 9.10).

All this indicates a chemical reaction between spodumene and the aluminium melt, even under standard conditions.

9.3 Influencing Hydrogen Porosity in Aluminum Castings

As already mentioned, the production of high-quality aluminum castings requires not only the removal of particulate inclusions but also the reduction of the hydrogen content in the melt. The hydrogen solubility in aluminum is temperature-dependent and is characterized by a solubility leap at the solidification temperature of 660 °C. This means that the solubility for hydrogen in aluminum decreases at the time of solidification by more than one power of ten from 0.7 to 0.04 cm^3/100 g [23]. This leap in solubility is the reason of the pores that are formed in the castings. When the melt cools down, not all of the dissolved hydrogen can be kept in solution. The atomically dissolved hydrogen is released in molecular form and gas pores are formed in the metal. The amount and distribution of gas pores depends on the amount of dissolved gas, which is higher than the solubility at the solidification temperature. The solidification rate, the possibilities for bubble nucleation, and the escape of the gas also play a role. At high cooling rates, more gas can be kept in solution because the melt solidifies quickly and less hydrogen is emitted in a short time. As a result, porosity phenomena are lower because of forming micro-porosity. In a highly contaminated melt, compared to a pure melt, more nuclei are present where gas bubbles can form [24]. This can lead to greater porosity. On the other hand, a high

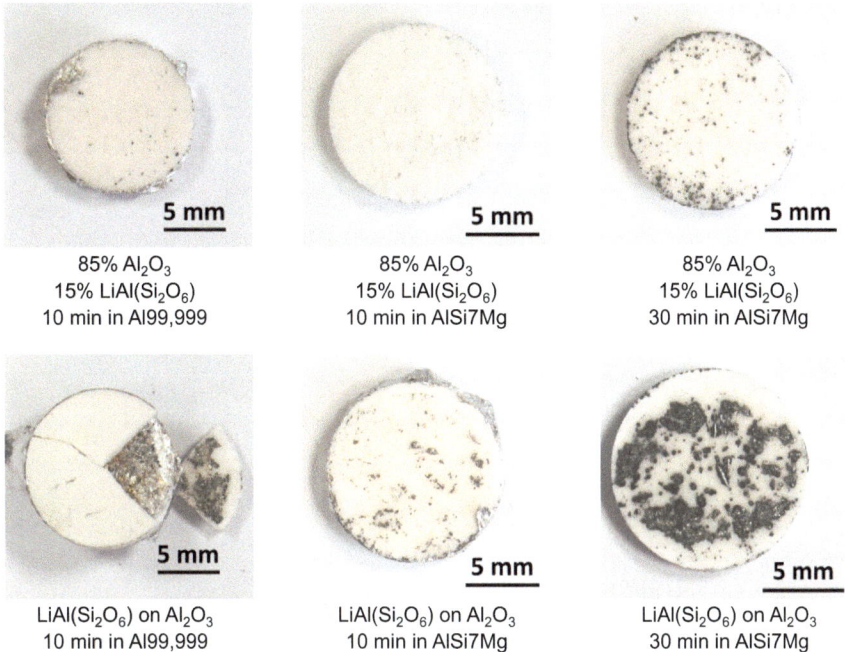

85% Al$_2$O$_3$
15% LiAl(Si$_2$O$_6$)
10 min in Al99,999

85% Al$_2$O$_3$
15% LiAl(Si$_2$O$_6$)
10 min in AlSi7Mg

85% Al$_2$O$_3$
15% LiAl(Si$_2$O$_6$)
30 min in AlSi7Mg

LiAl(Si$_2$O$_6$) on Al$_2$O$_3$
10 min in Al99,999

LiAl(Si$_2$O$_6$) on Al$_2$O$_3$
10 min in AlSi7Mg

LiAl(Si$_2$O$_6$) on Al$_2$O$_3$
30 min in AlSi7Mg

Fig. 9.9 Tablets of 85 wt.% Al$_2$O$_3$ 15 wt.% LiAl(Si$_2$O$_6$) and Al$_2$O$_3$-tablets coated with LiAl(Si$_2$O$_6$) after immersion in an AlSi7Mg or Al melt at 730 °C for 10 min and 30 min respectively (Reproduced from Fankhänel et al. 2019 [22])

number of nuclei often allows faster hydrogen precipitation. A too low number of nuclei can therefore lead to larger pores in the casting. In order to reduce the resulting porosity as effectively as possible, the number of nuclei in the melt must therefore be adapted to the dissolved hydrogen content [25].

Although purging gas treatments and the use of melting salts makes it possible to reduce the proportion of hydrogen and oxidic inclusions, solidification in many castings does not proceed uniformly, so that areas of increased porosity are still formed. As already mentioned, one reason for such an accumulation can be an excess of hydrogen compared to the number of pore nuclei present in the melt, for example. Therefore, it makes sense to find a way not only to remove unwanted inclusions from the melt, which act as pore nuclei, but also to eliminate the gaseous hydrogen released during solidification in time. One solution to this problem can be the use of suitable filter materials that are capable of positively influencing hydrogen precipitation during casting.

In several series of experiments, spodumene (LiAl(Si$_2$O$_6$)) was tested as a reactive filter material. Its influence on the hydrogen porosity of various aluminum alloys was investigated in comparison to non-reactive filter materials. Commercial alloys (AlSi7Mg, EN AC-42000, from Trimet Aluminium AG, Germany and AlCu4Ti, EN AC-21100, from Aluminium Rheinfelden GmbH, Germany) were used for the casts.

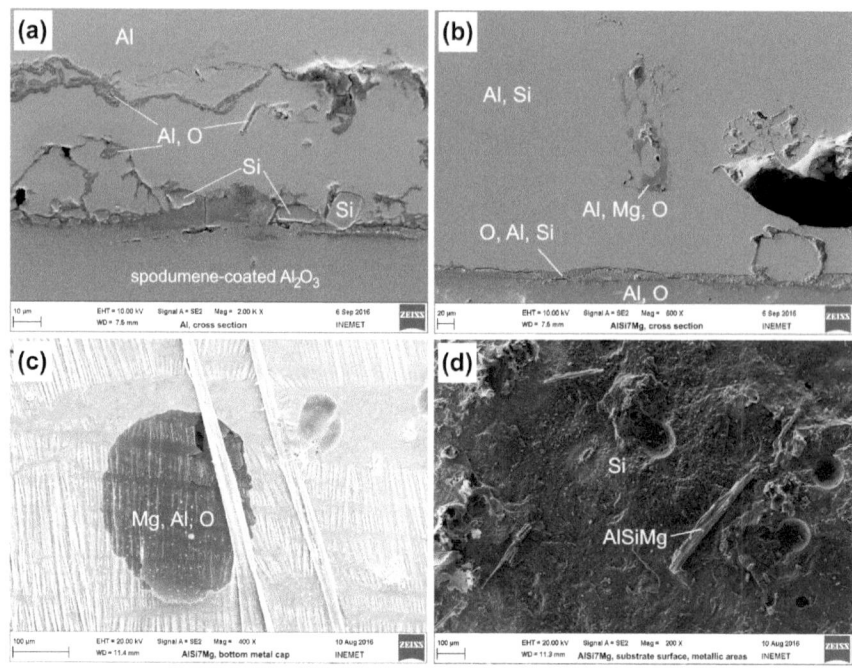

Fig. 9.10 Scanning electron micrographs of selected sample areas of spodumene-coated Al_2O_3-tablets, 30 min immersed in Al99.999 (**a**) and AlSi7Mg (**b-d**) (Reproduced from Fankhänel et al. 2019 [22])

Since the carbon bonded filter materials Al_2O_3-C, which were also investigated, showed no effect on hydrogen porosity [26], they will not be discussed here in more detail.

9.3.1 Casting Method and Pore Analysis

A special mold was made for the casts, which allows the gaseous hydrogen released in the mold during solidification to be almost completely precipitated as pores and thus made visible. The mold reproduces a problematic situation from practice. The wedge-shaped steel mold (Fig. 9.11) was designed in such a way that the upper area complicates feeding due to early solidification. This prevents the gaseous hydrogen from escaping completely after filtration. In the central area of the mold, cooling is much slower, so that an area of increased porosity is formed, which serves as a measure of the precipitated hydrogen. The sprue and the mold were preheated to 350 °C before each casting to prevent premature solidification of the melt. Before casting, the metal was heated to approx. 200 °C above the temperature of solidification to compensate the temperature loss during handling up to the start of casting.

Fig. 9.11 Casting and experimental procedure for the determination of the pore size distribution

The alloys used differ in their solidification behavior. In contrast to the widely used AlSi7Mg (EN AC-42000) alloy, the AlCu4Ti (EN AC-21100) alloy has a solidification range extended by more than 50 °C and tends to form micro-pitting [27]. The first alloy promotes hydrogen precipitation while the latter favors the formation of smaller separated pores at the expense of shrinkage porosity.

The 10 ppi ceramic foam filters used for filtration were fabricated using Schwartzwalder's replica technique as described by Voigt et al. [6, 28]. The basis for all filters were filter skeletons made of pure alumina, which were coated with the desired materials: Al_2O_3, $LiAl(Si_2O_6)$, 85 wt.% Al_2O_3 15 wt.% $LiAl(Si_2O_6)$ (mixed oxide ceramic). Casts without filtration were produced for comparison.

The sprue and the filters were placed on the side of the mold edge so that the liquid metal could flow into the mold in the same way for each casting. All castings were made under the same conditions in terms of handling before and during casting and during cooling to ensure comparability. At least three casts were made and evaluated for each investigated variant.

To evaluate the internal porosity of the castings, polished specimen cross-sections were analyzed by light microscopy (AxioImager.M2m light microscope from Carl Zeiss, Germany). The upper critical area, where the melt solidifies most slowly, was analyzed. The pore radii were determined by automatic image analysis. Each image of a sample cross-section was divided into 33 measuring fields. The pores that appeared dark in the microscope were automatically detected and their area determined. From these values, it was now possible to determine an area-equivalent diameter, which characterizes the size of the pores. Only pores completely contained in the measuring field were taken into account in the measurement. The experimental procedure is shown in Fig. 9.11 and a detailed description of this procedure is given in [22] and [26].

Various characteristic values were used for the evaluation: The total number of pores per examined sample cross-section comprises the pores completely contained in the measuring fields and, together with the total pore area, is a measure of the porosity of a casting. In some specimens, the released hydrogen was not completely precipitated as gas porosity, characterized by the approximately circular pore shape.

Areas of shrinkage porosity were formed in which the hydrogen still remained in the residual melt was precipitated exclusively in the intercrystalline or interdendritic spaces, influenced by the solidification morphology. Such areas are typical for the gravity die casting of complex shaped components, as they are reproduced by the mold used. These areas were not taken into account when determining the number of pores. However, their area was determined separately and included in the calculation of the total pore area to give an impression of the total porosity in the casting.

Selected casting sections were scanned additionally by computer tomography (CT-Alpha from ProCon X-Ray GmbH, Garbsen, Germany) to improve the visualization of the area of the macropores. The upper part of the castings (below the sprue) was analyzed with a voxel size of 40 μm in each case, which means that only pores with diameters larger than the voxel size can be visualized. Due to the beam hardening artifacts, the quantitative evaluation of the pore sizes was not possible.

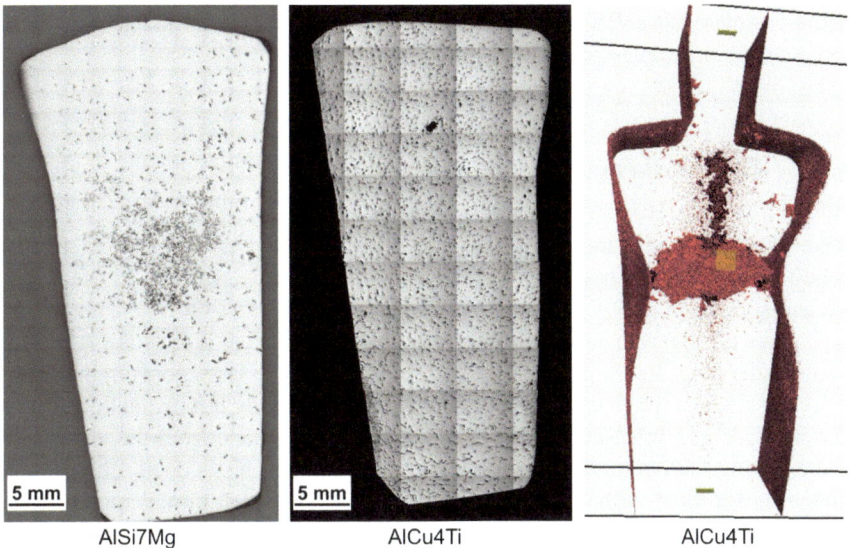

AlSi7Mg   AlCu4Ti   AlCu4Ti

Fig. 9.12 Light microscope images of cross-sections (left and center) and 3D CT image (right) of castings cast without a filter

9.3.2 Pore Structure Without Filtration

Figure 9.12 shows typical cross-sections of castings, which have been cast without filters. The macroporosity can already be seen with the naked eye. In the castings without filters, shrinkage porosity had formed in the central area of the castings. Such a concentration of pores is due to shrinkage of the metal, which also leads to poor dimensional accuracy of the castings. The dendritic shape of the pores in this concentration region indicates rapid solidification of the melt. The rapid solidification initially suppresses hydrogen precipitation from the melt. The hydrogen diffuses into the last solidifying areas and finally precipitates in the interdendritic spaces. There, however, it no longer has the opportunity to form round gas pores. No round pores were visible to the naked eye in the entire cross-sections. The phenomenon can be observed particularly well casting the AlSi7Mg alloy. The alloys used differ in their casting properties and microstructure [27]. As standard casting alloys, hypoeutectic AlSi alloys exhibit good flow and mold filling properties and form a solid solution structure with Si precipitates. AlCuTi alloys, on the other hand, have poor casting properties due to their wide solidification interval and a very dense and fine-grained microstructure. They tend to form micro-pitting. For this reason, small pores at the grain boundaries rather than large gas pores and areas of shrinkage porosity are to be expected here. In the CT analysis, which analyzed a larger sample volume, shrinkage porosity could be detected as well. In general, the castings without filter showed the highest shrinkage and the worst dimensional stability. The upper edges of the castings had collapsed and the internal deficit due to shrinkage was more pronounced for samples produced without a filter.

9.3.3 Al₂O₃-Filters

The pore distribution in the castings produced with the aid of an Al_2O_3-filter appears to be more homogenously distributed over the entire cross-section. In the cross-sections shown in Fig. 9.13, which are typical for the use of Al_2O_3 filters, individual large pores are clearly visible. Especially in the upper casting area of the AlSi7Mg castings, the pores have circular shapes. A high number of large pores in the upper region results in a rough casting surface, which can be easily felt by hand [22]. The gaseous hydrogen released during solidification could escape from the upper side of the casting until complete solidification of the metal. This is due to the general influence of a filter on the melt flow during casting. The melt flow is regulated by using a filter, and fluctuations are prevented. A calming of the melt occurs, as the filter acts as a resistance to flow, and turbulences are avoided. Due to the calmer melt flow through the filter, the hydrogen has time to precipitate from the melt and form large gas bubbles. The area with shrinkage porosity was less pronounced here than in the samples cast without a filter.

The Al_2O_3-filters exhibit a non-wetting character towards aluminum and its alloys. Furthermore, the surface roughness is low compared to the spodumene-containing filter materials. The determined pore area is reduced compared to the samples cast

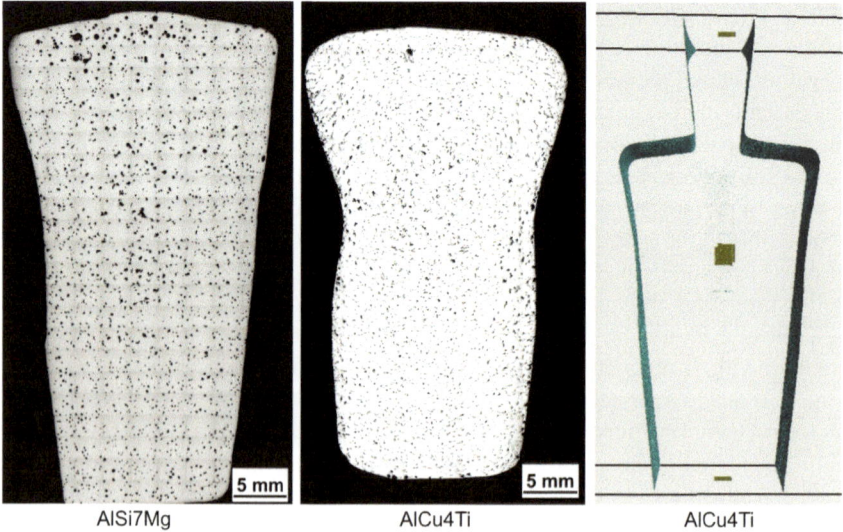

AlSi7Mg   AlCu4Ti   AlCu4Ti

Fig. 9.13 Light microscope images of cross-sections (left and center) and 3D CT image (right) of castings cast with an Al_2O_3-filter

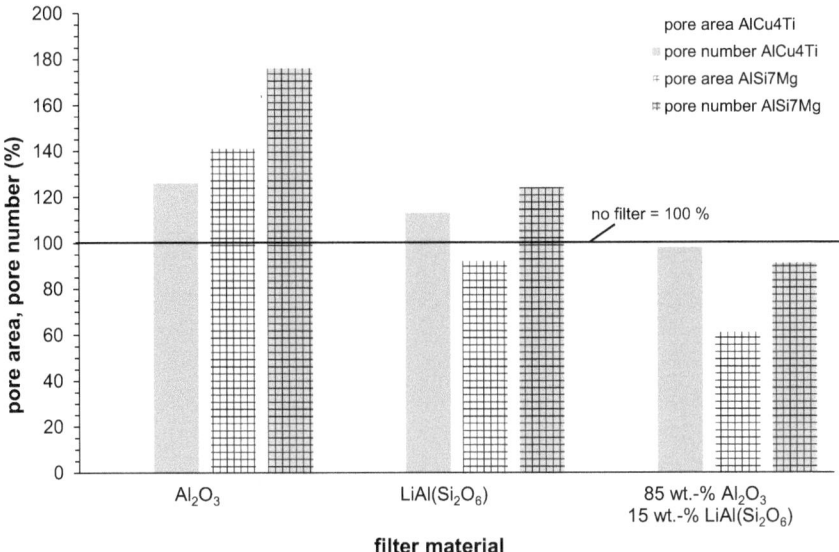

Fig. 9.14 Normalized pore areas and pore numbers evaluated by means of automatic image analysis for three different filter materials and two different aluminum alloys (The values for the cast samples without a filter were set to be 100%.)

without filter while at the same time, the highest number of pores was measured indicating more but smaller pores compared to an unfiltered casting (Fig. 9.14). The main influence of this filter type is the melt calming, which gives the metal more time to fill the mold and thus prolongs the casting process. This leads to enhanced pore formation and stimulates pore growth. In the case of a sufficiently large solidification interval, the gaseous hydrogen thus has the chance to form bubbles and leave the melt before solidification.

9.3.4 LiAl(Si$_2$O$_6$)-Coated Filters

The filter materials containing spodumene showed a different influence on the gas porosity. The filters with a pure spodumene surface lead to a higher pore number compared to the unfiltered castings (Fig. 9.14). But none of the bigger pores can be seen in the cross-sections in case of the AlCu4Ti-alloy when using LiAl(Si$_2$O$_6$)-coated filters (Fig. 9.15). In this case, the released gaseous hydrogen seems almost exclusively precipitated (influenced by the solidification morphology) in the inter-crystalline or inter-dendritic spaces to form many small pores as expected for the AlCu4Ti-alloy, which tends to micro-pitting. In this case, more pores formed a larger pore area. In the case of the AlSi7Mg alloy, however, the porosity appears to be reduced in total. There are still a few larger pores in the last solidified sample region

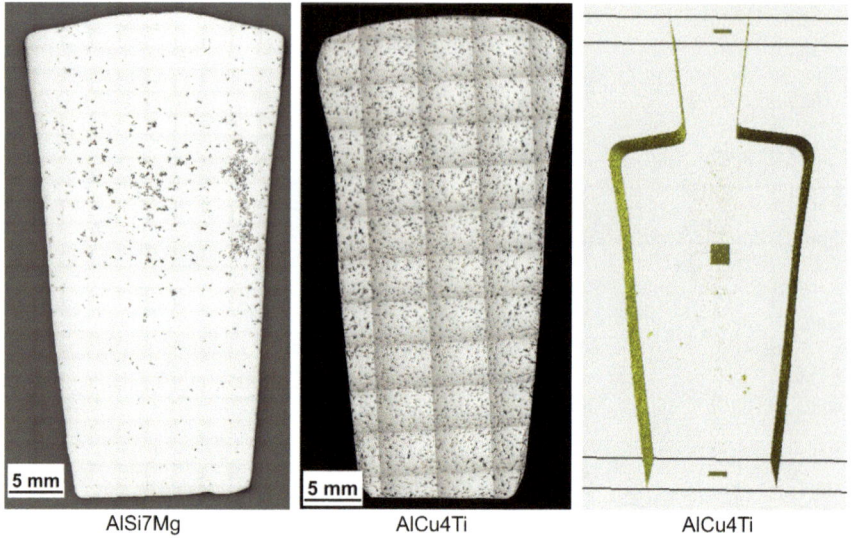

5 mm	5 mm	
AlSi7Mg	AlCu4Ti	AlCu4Ti

Fig. 9.15 Light microscope images of cross-sections (left and center) and 3D CT image (right) of castings cast with a LiAl(Si$_2$O$_6$)-filter

in the center of the castings, but they no longer represent a typical shrinkage porosity compared to the variants already discussed. In fact, for this alloy the pore area is reduced overall, although there are more pores in number, compared to the unfiltered castings.

LiAl(Si$_2$O$_6$) behaves reactively towards aluminum. In addition, high surface roughness increases the effective filter surface area so that more pore nuclei are available, which will influence the gas porosity that is formed after filtration. Therefore, it can be assumed that the LiAl(Si$_2$O$_6$)-coated filter enhances the precipitation of gaseous hydrogen during casting. The rough filter surface generates many nuclei for hydrogen precipitation during filtration due to the melt's interaction with the reactive filter surface. This interaction leads to the formation of many small gas bubbles, which would exhibit a lower tendency to escape from the melt during filtration, compared to larger bubbles. This assumption can explain the large porosity when using the filter with the pure spodumene surface in case of the AlCuTi-alloy, which tends to micro-pitting, and the remaining of lager pores in the center of the AlSi7Mg-specimen, that were not able to leave the melt before solidification.

Interfacial investigations on the filter struts after filtration confirmed that a chemical reaction between the molten metal and the spodumene took place during filtration. Scanning electron micrographs show traces of metallic silicon at the filter/metal interface even in case of the silicon free alloy (Fig. 9.16). This indicates that one of the two reactions described in Eqs. 9.1 and 9.2 took place.

Fig. 9.16 Scanning electron micrograph of the AlCu4Ti/ LiAl(Si_2O_6)-coated filter interface after filtration

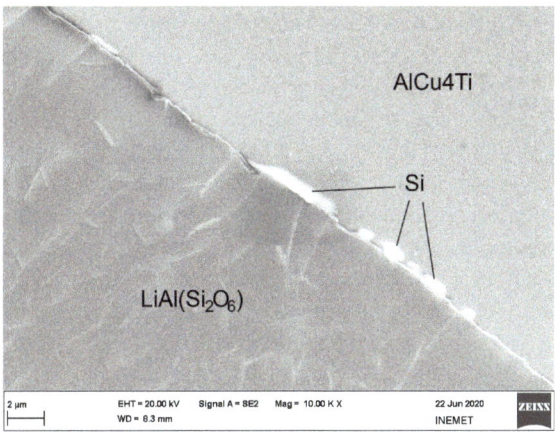

9.3.5 Mixed Oxide Filters

The mixed oxide filters seem to enhance the effect on gas porosity described for the LiAl(Si_2O_6)-coated filters. In Fig. 9.17, which shows characteristic cross-sections of the samples cast with a spodumene-containing mixed oxide filter, larger pores can only be seen in the center of the AlSi7Mg alloy casting.

However, the pores generally appear to be distributed more homogeneously over the entire casting cross-section. The filters with 15 wt.% LiAl(Si_2O_6) cause the

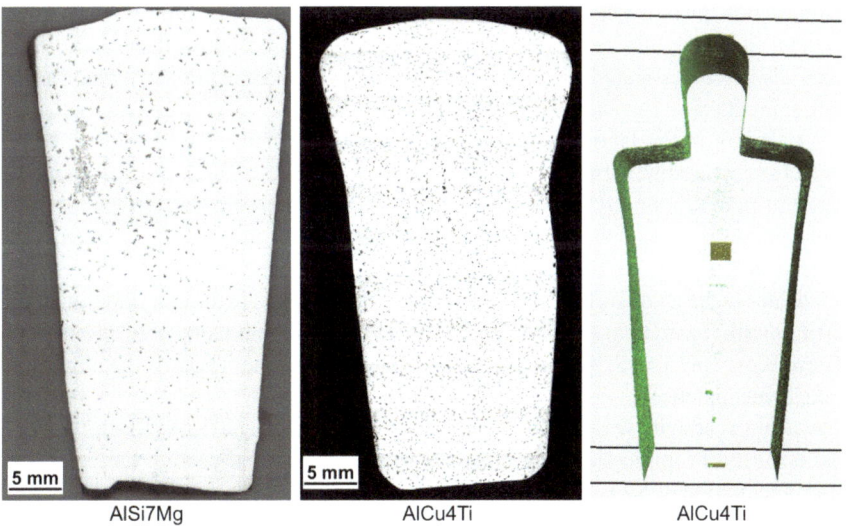

Fig. 9.17 Light microscope images of cross-sections (left and center) and 3D CT image (right) of castings cast with a mixed oxide filter (85 wt.% Al_2O_3 15 wt.% LiAl(Si_2O_6))

formation of fewer pores associated with a significantly reduced pore area in general and therefore a shift of the pore size distribution towards bigger pores (Fig. 9.14).

These filters also feature a rough surface with a large specific surface area. However, the surface consists of non-reactive alumina with reactive parts of spodumene. Hence, in addition to melt calming, this filter can generate supplementary nuclei for hydrogen precipitation, but most likely to a lower extent than in filters with a pure spodumene surface. The additional gaseous hydrogen will create new pores and let some of them grow to larger pore sizes as already discussed. However, if a filter with less spodumene is used, fewer nuclei are present for the early precipitation of gaseous hydrogen, which reduces pore formation and results in an increase in pore size through bubble growth, i.e. by means of coalescence, as explained for the Al_2O_3-filters. However, since more hydrogen is available from the beginning, more hydrogen can escape before solidification. In the end, there is less porosity in general in case of the mixed oxide filters. This could be considered as a combined result of balanced hydrogen nucleation on the one hand and pore growth combined with gas escape during filtration on the other hand.

Unfortunately, it was not possible to measure the hydrogen content in the aluminum melt directly. However, due to identical experimental conditions, it can be assumed that the melts contained comparable hydrogen contents before casting.

9.4 Conclusion

The aim of the work was to investigate the possibility to reduce hydrogen porosity in aluminum castings with the aid of reactive filter materials containing spodumene. To determine this, the reactivity of spodumene compared with other oxidic filter materials was investigated in laboratory tests, and casts were made with two different aluminum alloys.

As a result of the laboratory tests, it was found that under the conditions of the contact angle measurement, only the filters consisting of pure Al_2O_3 did not react with a melt of pure aluminum or an alloy. Chemical reactions were detected in all other filter materials investigated ($MgAl_2O_4$, $3Al_2O_3 \cdot 2SiO_2$, SiO_2, TiO_2, $LiAl(Si_2O_6)$). Despite the partially reactive behavior in the laboratory tests, no negative effects were found later in the analysis of the filters used for filtration and in the examination of the castings with regard to a changed composition of the aluminum melts. In addition, all materials except $LiAl(Si_2O_6)$ exhibited non-wetting behavior towards the aluminum melts investigated.

A strong reactivity was found in contact with the spodumene ($LiAl(Si_2O_6)$), which also occurred in immersion tests under normal pressure and casting temperatures. Here, a reduction of SiO_2 and the formation of metallic silicon occurred. In addition, the formation of new, presumably lithium-containing, aluminosilicates was detected. Under the conditions of the sessile drop experiments, the strong reactivity of the spodumene caused a significant decrease in the contact angle being in direct contact with the molten aluminum. However, the reason for exploring the lithium oxide-containing material spodumene ($LiAl(Si_2O_6)$) as a potential filter material was the assumption that Li_2O might be able to form a compound with the hydrogen dissolved in the aluminum melt and released during the solidification process, and thus reduce the hydrogen content in the alloy. Initial investigations in this regard actually showed an improvement in terms of the pore structure formed within the castings when a filter material containing spodumene was used compared to a pure Al_2O_3 filter. However, it is assumed that the reduction of hydrogen porosity is mainly caused by the early precipitation of hydrogen from the melt stimulated by the chemical reaction and less by the formation of a hydrogen-containing reaction product at the filter surface. In addition, high surface roughness increases the effective filter surface area so that more pore nuclei may be available. This will influence the gas porosity that is formed during filtration. The smallest volume deficits and best dimensional accuracy were observed for castings produced with the spodumene-containing filters. These castings exhibited exact edges and almost no dents.

Acknowledgements The authors gratefully acknowledge the Deutsche Forschungsgemeinschaft (DFG, German Research Foundation) for financial support of the investigations, which were part of the Collaborative Research Center Multi-Functional Filters for Metal Melt Filtration—A Contribution towards Zero Defect Materials (Project-ID 169148856 – CRC 920, subproject C06). Furthermore, the authors would like to thank Dr. Claudia Voigt (subproject A02) for providing the filter materials and for her fundamental research in the field of aluminum filtration, Birgit Witschel (subproject S01) for preparing the sessile drop and filter cross sections and, Dr. Jana Hubálková (subproject S01) for the performance of the CT measurements.

References

1. G.E. Totten, D.S. MacKenzie, *Physical metallurgy and processes*, 1st edn. (Dekker, Hoboken, NJ, 2003)
2. A. Ardekhani, R. Raiszadeh, J. Mater. Eng. Perform. **21**(7), 1352 (2012). https://doi.org/10.1007/s11665-011-9991-3
3. T. Baykara, A.A. Goktas, R. Goren, M. Marsoglu, Prakt. Metallogr. **34**(2), 93 (1997)
4. L.N.W. Damoah, L. Zhang, Metall. Mater. Trans. B **41**(4), 886 (2010). https://doi.org/10.1007/s11663-010-9367-3
5. B. Prillhofer, H. Antrekowitsch, Berg- und Hüttenmännische Monatshefte **152**(2–3), 53 (2007). https://doi.org/10.1007/s00501-007-0274-0
6. C. Voigt, T. Zienert, P. Schubert, C.G. Aneziris, J. Hubálková, J. Am. Ceram. Soc. **97**(7), 2046 (2014). https://doi.org/10.1111/jace.12977
7. M. Zhou, D. Shu, K. Li, W.Y. Zhang, H.J. Ni, B.D. Sun, J. Wang, Metall. Mater. Trans. A **34**(5), 1183 (2003). https://doi.org/10.1007/s11661-003-0138-5
8. C. Voigt, B. Dietrich, M. Badowski, M. Gorshunova, G. Wolf, C.G. Aneziris, in *Light Metals 2019: Proceedings of the Light Metals symposia at the TMS Annual Meeting & Exhibition*, 2019 ed. by C. Chesonis, p. 1063
9. B. Fankhänel, M. Stelter, C. Voigt, C.G. Aneziris, Adv. Eng. Mater. **19**(9), 170084 (2017). https://doi.org/10.1002/adem.201700084
10. R. Bauer, Chem. unserer Zeit **19**(5), 167 (1985). https://doi.org/10.1002/ciuz.19850190505

11. J. Deberitz, G. Boche, Chem. unserer Zeit **37**(4), 258 (2003). https://doi.org/10.1002/ciuz.200 300264
12. H. Oechsner, in *Secondary Neutral Mass Spectrometry (SNMS) and Its Application to Depth Profile and Interface Analysis,* ed. by H. Oechsner. Thin Film and Depth Profile Analysis, vol. 37 (Springer, Berlin, Heidelberg, 1984), p. 63 https://doi.org/10.1007/978-3-642-46499-7_4
13. H. Oechsner, W. Bock, M. Kopnarski, M. Müller, Mikrochim. Acta **133**(1–4), 69 (2000). https://doi.org/10.1007/s006040070074
14. H. Oechsner, Thin Solid Films **341**(1), 105 (1999). https://doi.org/10.1016/S0040-6090(98)015 45-4
15. U.C. Schmidt, M. Fichtner, J. Goschnick, M. Lipp, H.J. Ache, Fresenius J. Anal. Chem. **341**(3–4), 260 (1991). https://doi.org/10.1007/BF00321560
16. H. Jennet, in *Sekundärneutralteilchen-Massenspektrometrie (Plasma-SNMS),* ed. by H. Günzler, A.M. Bahadir, R. Borsdorf, K. Danzer, W. Fresenius, R. Galensa, W. Huber, M. Linscheid, I. Lüderwald, G. Schwedt, G. Tölg, H. Wisser. Analytiker-Taschenbuch, vol 16 (Springer, Berlin Heidelberg New York, 1997), p. 43
17. N. Eustathopoulos, M.G. Nicholas, B. Drevet, *Wettability at high temperatures,* 1st edn. (Pergamon, Amsterdam, New York, 1999)
18. A. Bianconi, R. Bachrach, S. Hagstrom, S. Flodström, Phys. Rev. B **19**(6), 2837 (1979). https://doi.org/10.1103/PhysRevB.19.2837
19. B. Fankhänel, M. Stelter, C. Voigt, C.G. Aneziris, Metall. Mater. Trans. B **46**(3), 1535 (2015). https://doi.org/10.1007/s11663-015-0307-0
20. A. Salomon, C. Voigt, O. Fabrichnaya, C.G. Aneziris, D. Rafaja, Adv. Eng. Mater. **19**(9), 1700106 (2017). https://doi.org/10.1002/adem.201700106
21. C. Voigt, B. Fankhänel, B. Dietrich, E. Storti, M. Badowski, M. Gorshunova, G. Wolf, M. Stelter, C.G. Aneziris, Metall. Mat. Trans. **B51**(5), 2371 (2020). https://doi.org/10.1007/s11 663-020-01900-1
22. B. Fankhänel, S. Grötz, M. Stelter, World of Metallurgy—ERZMETALL **72**(1), 32 (2019)
23. D.L. Zalensas, J.L. Jorstad, *Aluminium casting technology*, 2nd edn. (American Foundrymen's Society, Des Plaines, 1993)
24. G. Drossel, S. Friedrich, W. Huppatz, C. Kammer, W. Lehnert, O. Liesenberg, W. Mader, M. Paul, A. Rudolf, W. Thate, Zeltner S., W. Wenglorz, *Umformen von Aluminium Werkstoffen, Gießen von Aluminium-Teilen, Oberflächenbehandlung von Aluminium, Recycling und Ökologie,* 16th edn. (Aluminium-Verlag Marketing & Kommunikation GmbH, Düsseldorf 2009)
25. B. Fankhänel, M. Stelter, W. Vogel, T. Klug, World of Metallurgy—ERZMETALL **67**(5), 277 (2014)
26. B. Fankhänel, J. Hubálková, C.G. Aneziris, M. Stelter, A. Charitos, Adv. Eng. Mater. **24**(2), 2100579 (2022). https://doi.org/10.1002/adem.202100579
27. C. Kammer, *Grundlagen und Werkstoffe*, 16th edn. (Aluminium-Verlag Marketing & Kommunikation GmbH, Düsseldorf, 2009)
28. C. Voigt, E. Jäckel, C.G. Aneziris, J. Hubálková, Ceram. Int. **39**(3), 2415 (2013). https://doi.org/10.1016/j.ceramint.2012.09.001

Chapter 10
Novel Ceramic Foam Filter Materials for the Filtration of Magnesium Alloy Melts

Alina Schramm, Christos G. Aneziris, and Christiane Scharf

10.1 Introduction

Environmental awareness and protection became a central topic in developing scientific progress, empowering research on lightweight materials suitable for structural components for aerospace and automotive applications to reduce fuel consumption, as well as improving recyclability of the materials. With its low density of $1.74\,g/cm^3$, high specific strength, good castability, machinability, resource abundance, electromagnetic shielding properties and recyclability potential [1, 2], magnesium and its alloys are regarded as promising lightweight materials.

Apart from these positive qualities, there are some disadvantages that hinder their wide-spread application, such as low formability at ambient temperatures, tendency to corrosion, and, most prominently, proneness to high inclusion contents due to high melt reactivity [3]. Measures of melt protection, such as the usage of cover gases, fluxes, and specific closed furnace designs, are therefore required, complicating the foundry process [4]. Non-metallic inclusions, most notably oxide particles, clusters or skins [5], reduce the mechanical properties of magnesium and magnesium alloy castings greatly, acting as a disruption of the metal matrix, prone to crack initiation and propagation. Porosity caused by shrinkage and dissolved gases, such as

A. Schramm (✉) · C. Scharf
Institute of Nonferrous Metallurgy and Purest Materials, Technische Universität Bergakademie Freiberg, Leipziger Straße 34, 09599 Freiberg, Germany
e-mail: Alina.Schramm1@inemet.tu-freiberg.de

C. Scharf
e-mail: Christiane.Scharf@inemet.tu-freiberg.de

C. G. Aneziris
Institute of Ceramics, Refractories and Composite Materials, Technische Universität Bergakademie Freiberg, Agricolastraße 17, 09599 Freiberg, Germany
e-mail: aneziris@ikfvw.tu-freiberg.de

© The Author(s) 2024
C. G. Aneziris and H. Biermann (eds.), *Multifunctional Ceramic Filter Systems for Metal Melt Filtration*, Springer Series in Materials Science 337,
https://doi.org/10.1007/978-3-031-40930-1_10

241

hydrogen, may also be present. Furthermore, unwanted intermetallic particles or flux inclusions lower the corrosion resistance of magnesium-based castings. Therefore, the efficient removal of inclusions by means of purposive metallurgical treatments of metal melts is the key to durable and reliable magnesium products. Apart from flux- and gas-purging treatments, filtration has been proposed as a flux-free alternative for magnesium melt refining.

Earliest attempts at magnesium melt filtration were reported by Unsworth in 1960 [6], using a chilled iron shot bed to remove intermetallic particles from an AZ61 melt. During the 1970ies and 1980ies, further research was performed on the use of stainless steel mesh filters, which were able to retain inclusions down to 0.08 mm in size, improving the ultimate tensile strength of the AZ91 castings to 217 MPa [7, 8].

The usage of ceramic foam filters for magnesium melt filtration first gained attention in the 1990ies through the research of Bakke et al. [9, 10], using AZ91 scrap and therefore bringing the focus of recycling into the picture. The durability of magnesia- and alumina-based ceramic foam filters, their efficiency and deep bed filtration capacity were emphasized in further research [11, 12]. Even without argon sparging of the AZ91 melt, its filtration through MgO- or Al_2O_3-filters improved the ultimate tensile strength and elongation of the castings up to 190.7 MPa/190 MPa and 4.31%/4.05%, respectively, the values for unfiltered AZ91 being 175.3 MPa and 2.74% [13].

Following these promising results for conventional, oxide-based ceramic foams, which had also been successfully applied in aluminum melt filtration [14], the aim of current research was the application of carbon-bonded alumina ceramic foam filters [15] to assess their applicability in magnesium and magnesium alloy melts. With their lower manufacturing temperature, various coatability [16], active and reactive filtration mechanisms, high durability and resistance towards creep deformation, thermal shock and corrosion, these novel filter materials had been successfully applied in the field of refractory and steel melt filtration (cf. [17]).

Carbon-bonded alumina foams were manufactured as described by Emmel and Schwartzwalder et al. [15, 18], equipped with various coatings, such as reaction-sintered MgAlON, and applied in an AZ91 melt [19], noting interface reactions and inclusion contents of the metal. Sessile drop tests were conducted with various ceramic substrates and coatings to determine their wettability by AZ91 [20].

10.2   Manufacturing of Al_2O_3-C Ceramic Foam Filters and Novel MgAlON Coating Material

As carbon-bonded alumina was shown to be a suitable material for refractories and metal melt filters [15], research was focused on advantageous filter manufacturing techniques, which would result in high-quality filter structures with good mechanical properties.

Previous investigations of Luchini et al. [21] were based on the Schwartzwalder replica technique [18] using aqueous slurries containing 66 wt% of Al_2O_3, 20 wt% of the tar pitch binder Carbores P®, 6.3 wt% of carbon black, 7.7 wt% of graphite and 1.5 wt% of ammonium lignin sulfonate, as well as antifoam- and dispersing agents, as described by Emmel et al. [15]. These slurries were applied onto polyurethane foam templates in two coating layers and heat-treated in reducing atmosphere at 800 °C for 3 h. The results of comparative cold crushing tests showed that filters with coatings applied by centrifugation showed homogeneous structures with smoother fracture behavior, as compared to coatings that were applied by means of rolling or spray coating [21].

Additional research was furthermore conducted, putting an emphasis on coating techniques, as well as the slurry preparation routes [22].

Batches of Al_2O_3-C ceramic foam samples with 20 pores per inch (50 × 50 × 20 mm^3) were manufactured according to the Schwartzwalder technique, using combinations of either dipping and rolling, spray coating or centrifugation for the first coating layer and spray coating or centrifugation for the second coating. Two types of Al_2O_3-C slurries were compared for either coating layer. The first type was prepared in a ball mill, using Al_2O_3 grinding balls at 80 rpm for 24 h, the second one was prepared in a high-shear mixer with a mixing duration of 45 min at 100 rpm [22].

After their heat treatment, functional as well as material pores of the ceramic foams were evaluated by means of computer tomography (μCT) analysis, mercury intrusion porosimetry and scanning electron microscopy (SEM). The cold crushing strength of the Al_2O_3-C foams was tested at room temperature at a quasi-static crushing speed of 3 mm/min for 20 samples of each batch.

μCT analysis and mercury intrusion porosimetry showed similar average functional porosities for all manufacturing routes, ranging from 37.7–42.8%. For coatings applied by centrifugation, a porosity gradient of approximately 5% was noted, correlated to the centripetal forces during the coating process. Nonetheless, centrifugation resulted in a more homogeneous foam structure as compared to spray coated foam samples, which showed a tendency towards thinner struts and thicker nodes [21, 22].

With grain sizes $d_{90} < 0.2$ mm, the tar pitch binder Carbores P® is a coarse-grained component, which has a high influence on the rheological and coating properties of the slurry, depending on its milling degree. Filter samples made using only shear-mixed slurry displayed coarse, inhomogeneous structures with large, angular pores up to 140 μm in size, seen during SEM analysis (see Fig. 10.1), resulting in cold crushing strengths around 0.35 MPa (see Fig. 10 in [22]). Additional heat treatments conducted in-between coating steps were shown to be unsuitable, causing shrinkage cracks between the ceramic layers, resulting in frail ceramic foams [22].

Ball-milled slurry showed smaller grain sizes, resulting in coatings that were more homogeneous, finely structured, and had smaller pores after their heat treatment (see Fig. 10.1). Ball-milled second coatings were shown to close flaws in the first layer effectively, proving to be an advantageous manufacturing route, resulting in higher cold crushing strengths around 0.7 MPa. The manufacturing process of the first

Fig. 10.1 SEM micrograph of an Al$_2$O$_3$-C filter strut cross-section after its heat treatment, showing a layer composed of coarse-grained, shear-mixed slurry in the middle *(I)* and a finely-structured coating layer of homogeneous, ball-milled slurry *(II)*

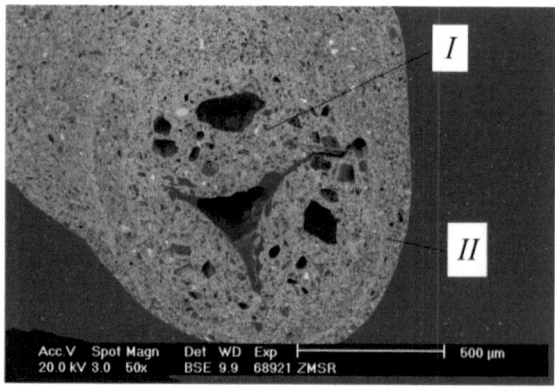

coatings was suggested to have minor influence on the mechanical properties of the foam (see Fig. 10 in [22]).

The use of MgAlON as a coating material for Al$_2$O$_3$-C filters that would be used in magnesium- and aluminum alloy melt filtration is proposed following previous reports on the application of MgAlON as a refractory material [23] and its resistance towards hydration, making the application of aqueous coating slurries possible [24].

Investigations were therefore conducted on pressureless reaction sintering of MgAlON from MgO-, Al$_2$O$_3$- and AlN powders in nitrogen atmosphere at 1500 °C and the wettability of sintered MgAlON by an AlSi7Mg alloy melt at 950 °C at a contact time of 3 h [24]. The range of starting molar fractions of 0.45–0.7 for Al$_2$O$_3$, 0.2–0.45 for MgO and 0.1–0.35 for AlN was selected based on previous syntheses reported in selected literature [25, 26, 26, 28], as well as the phase diagrams proposed by Willems for synthesis temperatures of 1400 °C and 1750 °C [27]. The mixtures were held at 1500 °C for 3 and 6 h, respectively, to investigate the influence of holding times on the synthesis yield. XRD analysis was performed to evaluate phase fractions and conversion degrees to MgAlON in sintered samples [24].

MgAlON formation was shown to increase with increasing fractions of MgO in the starting mixture and amounts of AlN that were below $x_{AlN} = 0.15$. Furthermore, the conversion degree to MgAlON increased with sintering time, reaching an average of 92 wt% after 6 h at 1500 °C, while an average of 76 wt% was reached after 3 h of sintering [24]. The maximum amount of MgAlON (99 wt%) was registered for samples from the MgO-rich corner of the quasi-ternary system MgO-AlN-Al$_2$O$_3$ after 6 h of sintering, with Al$_2$O$_3$:AlN ratios of around 20:6 (cf. Fig. 6a, e in [24]). The compositional changes during sintering and respective thermodynamic calculations [24] were correlated to reactions postulated by Yan et al. for aluminothermic MgAlON synthesis [28]. MgO is hereby consumed first, reacting with Al$_2$O$_3$, forming MgAl$_2$O$_4$. As the homogeneity range of MgAl$_2$O$_4$ expands with increasing temperatures, further dissolution of Al$_2$O$_3$ into the spinel is facilitated. An alumina-rich spinel with cation vacancies is formed, encouraging the dissolution of AlN into its structure.

The AlN-MgAlON dual phase region is estimated to begin around $x_{AlN} = 0.15$ at 1500 °C (cf. Fig. 6 in [24]). The loss of magnesium due to evaporation or parasitic reactions was proven to be marginal during the synthesis, since the lattice parameters did not show a significant decrease as the sintering progressed, showing values around 8.06 Å in the MgO-rich corner of the quasi-ternary system MgO-AlN-Al$_2$O$_3$ (cf. Fig. 7 in [24]).

Carrier gas extraction measurements conducted for sintered samples with the highest (99 wt%) and lowest (76 wt%) amount of MgAlON after 6 h of sintering showed nitrogen contents of 1.48–1.64 wt%, corresponding to relative nitrogen amounts of 1.5 wt% in the MgAlON phase, considering residual amounts of AlN within the respective sample [24].

To evaluate the applicability of the reaction-sintered MgAlON as a coating for filters used in light metal melt filtration, sessile drop tests were conducted at 950 °C using the alloy AlSi7Mg. Samples sintered at 1500 °C for 6 h were selected, containing the lowest (76 wt%) and highest (99 wt%) amount of MgAlON (cf. Fig. 6e [24]). Compared to state-of-the-art filter materials Al$_2$O$_3$ and MgAl$_2$O$_4$, which showed contact angles of 100° and 122°, respectively [29], the MgAlON samples showed remarkably low wettability and reactivity, with contact angles around 145° and 150° [24]. This leads to the conclusion that MgAlON is a promising material for the field of light metal alloy melt filtration.

10.3   Wettability of Ceramic Materials by an AZ91 Magnesium Alloy Melt

The evaluation of wetting behaviors and chemical interactions between the novel filter materials (cf. Sect. 10.2) and magnesium alloy melts are crucial for understanding and evaluating the filtration process and its applicability, as these properties have significant influence on the filtration efficiency. Higher contact angles in sessile drop tests, indicating lower wettability of the ceramic material by the light metal melt, were correlated with a higher filtration efficiency for non-metallic inclusions in filtration tests [30].

Sessile drop tests with magnesium alloys are technologically challenging, given the known high reactivity of the metal. The formation of oxide skins on the metal droplet, as well as impurity segregation, corrosion of ceramic substrates and metal evaporation, lead to significant inaccuracy and complications in the measuring process. Furthermore, the wetting process is practically not calculable, considering the high amount of influencing variables for technical-grade materials and surfaces [31]. Conventional sessile drop testing, in which a piece of metal is molten directly on the ceramic surface and its contact angle measured, does not yield reliable results for magnesium alloys, due to the significant oxide layer on the surface of the metal influencing measurements. Therefore, a metal dropping device and a novel furnace apparatus, optimized for such measurements [32], are used to obtain results for

AZ91 droplets on ceramic substrates consisting of ZrO_2, Al_2O_3, Al_2O_3-C, MgO and MgAlON [20, 24]. The apparatus applies heating elements made of refractory metal, which reduce residual oxygen contents in the protective argon atmosphere and ensure an accuracy of $\pm 2°$ for contact angle measurements between the substrates and metal droplets using a digital high-speed camera at 100 fps and digital image analysis. Capillary purification of the AZ91 droplet is conducted, pressing the metal onto the ceramic substrate through a graphite tube, effectively removing oxide skins and impurities from its surface [32, 33]. The ceramic substrates were unpolished and manufactured using the same materials and heat treatment routes used for ceramic foam filters to ensure comparability with their surfaces. Sessile drop tests were conducted at the common foundry temperature of 680 °C and a contact time of 10 min, the substrates were preheated within the furnace chamber [20].

Before and after the sessile drop tests, the surface roughness of the ceramic substrates was determined by means of a digital optical microscope according to DIN EN ISO 25178. SEM and energy dispersive X-ray (EDX) analysis of contact surfaces and metal droplet cross-sections was conducted to evaluate any traces of interfacial reactions.

Preliminary thermodynamic calculations were performed to evaluate possible interfacial reactions. As anticipated, inertia of MgO towards the magnesium alloy melt was shown, while Al_2O_3, Al_2O_3-C and residual $MgAl_2O_4$ spinel were calculated to react with the metal. For these substrates, the interfacial reaction would result in the formation of MgO and $MgAl_2O_4$ at the interface, along with the dissolution of reduced aluminum in the metal melt. A similar reaction is anticipated for the ZrO_2 substrate, with the formation of MgO, Al- and Zr-containing intermetallics and the enrichment of the alloy with Zr occurring [20]. A reaction of the magnesium alloy melt with MgAlON is not anticipated at 680 °C due to the metastable condition of the ceramic material below 700 °C [37] and its high corrosion resistance towards AZ91 under these experimental conditions [19]. Calculations or simulations of the actual wetting processes of technical materials are not seen as practical, due to many varying influences on them, therefore experimental investigations are considered important trend values to evaluate and plan technological processes [31], such as metal melt filtration.

Thermodynamic calculations were experimentally confirmed by means of SEM and EDX evaluation of the contact surfaces after the sessile drop tests. All AZ91 droplets were found to have a metallic, oxide-free surface, indicating successful protection from oxidation by the novel experimental setup [32]. While no significant differences were found in the surface roughness values of the ceramic substrates, traces of interface reactions were seen for the ZrO_2-, Al_2O_3- and Al_2O_3-C substrates, which were correlated to changes in the registered contact angles that were seen to decrease over time (see Fig. 6 in [20]).

Formed MgO was observed on the ZrO_2 surface that had been in contact with the AZ91 droplet, traces of MgO were also found in proximity to the location of the droplet, suggesting slight evaporation of magnesium during the test, reacting with the ZrO_2 surface after settling. Zirconium was furthermore detected in intermetallic particles on the contact surface and cross-section of the AZ91 droplet, in accordance with the thermodynamic calculation of the reduction of ZrO_2 by the AZ91 melt. This interfacial reaction mirrors the decline of measured contact angles, starting at 114° and reaching 102° after 10 min of contact time (see Fig. 6 in [20]). According to Eustathopoulos and Lee et al. [31, 34], the polycrystalline ZrO_2/AZ91 system can be characterized as wetting and reactive, displaying contact angles entirely below 120°. As the interfacial reaction introduces intermetallics and oxygen into the metal, the surface energy gets decreased, encouraging wetting of the substrate [31]. Therefore, ZrO_2 is not seen as preferable for an application in magnesium alloy melt filtration.

Traces of an interfacial reaction were furthermore found on the surface of the Al_2O_3 substrate after the sessile drop test. Its magnesium content in the absence of metallic particles suggests MgO or $MgAl_2O_4$, as calculated thermodynamically, resulting from the reduction of Al_2O_3 by the AZ91 melt. The cross-section of the AZ91 droplet showed no significant amount of oxides, suggesting that the metal was not contaminated by the contact with the ceramic substrate. The suitability of Al_2O_3 for further research into magnesium alloy filtration is furthermore shown by the contact angles measured, staying above 120° during the test [20], implying non-wettability of the substrate [31, 34]. Staying at 142° for the first 3 min of the test, the contact angles declined towards 126° at 10 min, influenced by the chemical changes and phase formation at the interface, as Al_2O_3 was reduced by the magnesium alloy melt (see Fig. 6 in [20]). MgO in-situ layer formation is furthermore seen as a potential for improved reactive filtration of magnesium alloy melts by Al_2O_3-containing substrates in both carbon-bonded and carbon-free states [19, 20, 35].

The Al_2O_3-C surface showed traces of MgO or $MgAl_2O_4$ formation as well, indicating a similar interface reaction, while no carbide formation was detected, in accordance with the slow kinetics of Al_4C_3 formation at 680 °C. No oxide inclusions were found on the cross-section of the AZ91 droplet, and the grain-refining particles were undisturbed. Despite a contact angle irregularity at 6.5 min of the sessile drop test (see Fig. 6 in [20]), the Al_2O_3-C substrate showed high contact angles around 146–151°, and even after the deviation the contact angles never sank below 120°, indicating non-wettability of the material, as well as its suitability for magnesium alloy melt filtration. The irregularity was suggested to have been caused by an adhering oxide particle, distorting the shape of the AZ91 droplet [20]. A significant contribution towards low wettability of Al_2O_3-C is found in the graphite content of the surface, as its polycrystalline, porous, and micro-faceted structure effectively pins the solid–liquid-vapor triple line [31].

Notably, the highest and most constant contact angles were measured for the MgO and MgAlON substrates at 140° and 150°, respectively (see Fig. 6 in [20]), indicating their non-wettability and inertia towards molten AZ91. No significant chemical changes were found on the surfaces of the substrates after the sessile drop tests, showing no traces of any interfacial reactions. The cross-sections of respective AZ91 droplets were shown to be oxide-free, with the grain-refining intermetallics undisturbed.

While both MgO and MgAlON were shown to be suitable materials for magnesium alloy melt filtration, MgAlON should be pinpointed as a highly promising option, since it was shown to have the advantage of being resistant towards hydration [24]. This ensures the advantage of easier and safer filter manufacturing and coating using aqueous slurries.

10.4 Immersion Testing of Ceramic Foam Filters in AZ91 and Its Interface Reactions

Preliminary short-term immersion tests in an AZ91 melt were carried out, using uncoated and coated carbon-bonded alumina filters (10 pores per inch, $150 \times 25 \times 20$ mm^3) to assess their applicability and interface reactions [35]. The carbon-bonded alumina filter samples were manufactured according to the Schwartzwalder technique [18], using polyurethane templates and aqueous slurries containing Al_2O_3, graphite, carbon black, additives and the tar pitch binder Carbores P®, as described by Emmel et al. in 2012 (cf. Sect. 10.2, [15]). Various coatings were applied onto the carbon-bonded substrates, such as $MgAl_2O_4$, Al_2O_3 [16], carbon-bonded MgO, or carbon nano tubes and alumina nano sheets [36]. The filter samples were immersed in an AZ91 melt for up to 120 s at 680 °C, SF_6 was used as cover gas to protect the melt from oxidation. Every filter sample was retrieved from the melt undamaged (see Fig. 10.2).

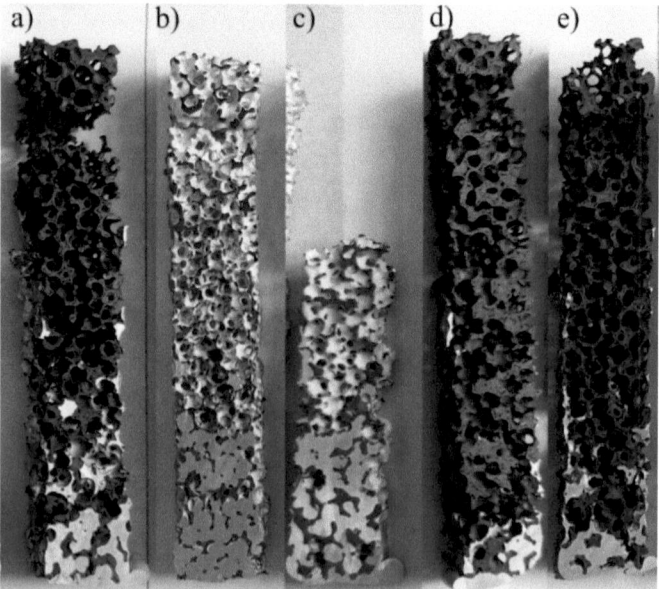

Fig. 10.2 Carbon-bonded alumina foam samples immersed in the AZ91 melt (cf. [35]). Al_2O_3-C, **a** uncoated, **b** Al_2O_3-coated, **c** $MgAl_2O_4$-coated, **d** MgO-C-coated, **e** coated in alumina nano sheets and carbon nano tubes

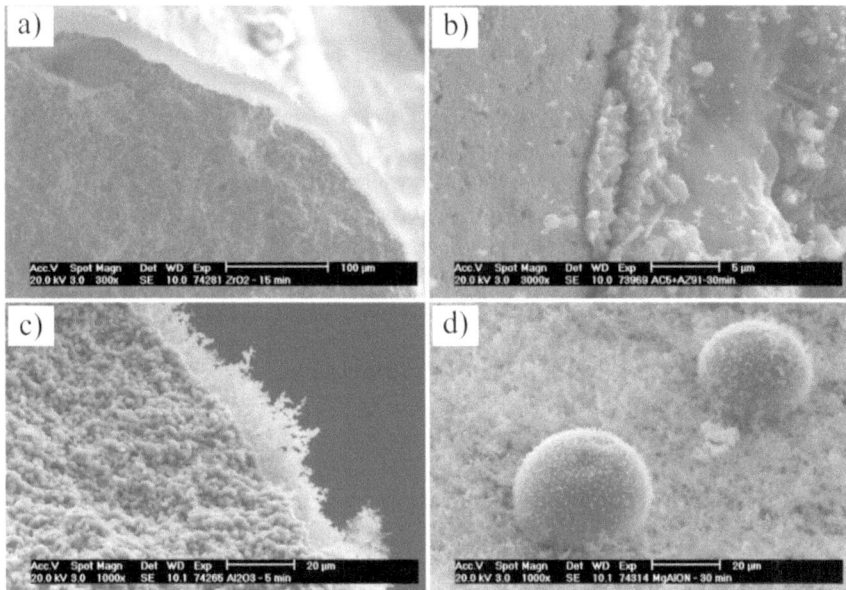

Fig. 10.3 SEM micrographs showing MgO-containing in-situ layers that had formed on ceramic foam surfaces after their contact with molten AZ91. **a** layer on the ZrO_2 filter surface after 15 min of melt contact, **b** layer on the uncoated Al_2O_3-C surface after 30 min of melt contact, **c** layer on the Al_2O_3-coated surface after 5 min of melt contact, **d** layer and AZ91 droplets on the MgAlON-coated surface after 30 min of melt contact

Subsequently, the immersed surfaces of the filter samples were evaluated by means of SEM and EDX analysis. Finely structured, platelet-like in-situ layers (cf. Fig. 10.3) with thicknesses around 2.5 μm were found on the immersed filter surfaces containing Al_2O_3 or $MgAl_2O_4$, while the filter surface underneath kept its structural integrity. EDX analysis suggested its composition of MgO.

These observations were in accordance with thermodynamic calculations, which showed the reduction of Al_2O_3 or $MgAl_2O_4$ contents of the filter surface by the magnesium alloy melt, resulting in the formation of MgO at the interface and the dissolution of metallic aluminum in the melt [35]. Furthermore, the dissolution of carbon or oxygen in the metal melt was calculated to be low, while Al_4C_3 formation is thermodynamically hindered at the given temperature [37].

The formation of CO bubbles, which has been observed during the application of carbon-bonded alumina filters in steel melts and attributed to a cleansing flotation effect [38, 39], was not observed in AZ91, as CO would be immediately reduced by the magnesium alloy melt. Therefore, no flotation effect can be anticipated [35]. Despite this, promising results were achieved in steel melt filtration experiments, using similar reactive ceramic filters, forming in-situ layers that were able to entrap inclusions from the melt [44], therefore great potential for light metal melt filtration is seen in the tested filter materials.

To investigate long-term filter stability and their influence on melt cleanliness, extended immersion tests were carried out with selected coated and uncoated filter samples [19]. Uncoated, as well as Al_2O_3- or MgAlON-coated [24, 40] Al_2O_3-C filters with 10 pores per inch were selected for this purpose. Filter samples made of ZrO_2 were added to the selection as well, following the observation of Wu et al. on ZrO_2 filters being third-most-effective in filtering AZ91, following filters made of MgO and Al_2O_3 [13]. A vertical tube furnace and steel ladle were used to melt the AZ91, the immersion of filters was carried out at 680 °C under protective Ar-0.2 Vol% SF_6 atmosphere. Filter samples were kept at dwell times of 5, 15, 30 and 60 min, a stainless steel strainer was used to remove the oxide layer from the melt surface before each immersion and a batch of fresh AZ91 was used for each type of filter material. AZ91 melt samples and the immersed filters were evaluated regarding melt contamination and interface reactions.

All of the filter samples endured the long immersion times, not showing signs of erosion, dissolution or other damage. SEM analysis of immersed filter surfaces showed in-situ layers that had formed in contact with the AZ91 melt (see Fig. 10.3). EDX data suggested their composition of MgO, small amounts of fluorine and sulfur were detected, originating from a reaction with the protective cover gas mixture, containing SF_6. Occasionally, metallic AZ91 droplets were found adhering to the ceramic filters, their metallic surfaces showed successful melt protection during the immersion process [19].

As expected, the area covered in MgO-containing in-situ layers increased with increasing immersion time, as the interface reactions between the filter material and AZ91 melt progress. For the immersed, uncoated Al_2O_3-C filter, the platelet-like in-situ layer showed thicknesses of approximately 3 μm after an immersion time of 15 min, with the thickness of the layers increasing to 5 μm after 60 min of contact time. The filter areas that were not covered in MgO appeared intact, smooth and undamaged by the melt. For the Al_2O_3-coated filters, denser MgO in-situ layers were observed, reaching thicknesses of 10 and 20 μm after 5 and 60 min of immersion time, respectively.

This can be correlated to the higher amount of Al_2O_3 present in the surface, as compared to Al_2O_3-C, which readily reacts with molten magnesium, forming larger quantities of MgO in the process [19].

Patches of finely structured MgO were also found deposited on the MgAlON-coated filter samples, reaching thicknesses of 15 to 50 μm after 15 and 60 min of contact time, respectively. Considering that MgAlON itself was shown to be non-reactive towards AZ91 (see Sect. 10.3), the MgO can be regarded as inclusion depositions from the melt or the product of reactions with the Al_2O_3-C substrate underneath, which can be exposed through technologically unavoidable firing cracks.

Similar MgO deposits were found on the ZrO_2 substrate. Extensive in-situ layers with thicknesses ranging from 20 to 50 μm for immersion times of 30–60 min, respectively, were found and concluded to result from the reduction of ZrO_2 by the magnesium alloy melt [19]. Furthermore, amounts of magnesium were registered in deeper regions of the filter struts, showing the dissolution of magnesia in ZrO_2, as it is utilized in magnesia-stabilized zirconia [41].

Automated SEM/EDX analysis (ASPEX) was used to evaluate the inclusion contents and compositions of AZ91 samples that had ceramic filters immersed in them for 60 min. 345 mm^2 of each sample cross-section were analyzed, particles larger than 1 μm could be identified [42]. The intermetallic β-phase ($Mg_{17}Al_{12}$) was shown to be unchanged by the contact with filter materials, up to 2 wt% of iron was registered within it, originating from the contact with the steel ladle (cf. [43]), while intermetallics in as-cast AZ91 contained 0.7 wt% of iron. Manganese- and iron-rich intermetallic particles could be described as $Al_8(Mn, Fe)_5$, as reported by Øymo et al. [44]. Zr contamination was negligible in AZ91 that was in contact with ZrO_2. Only traces of sulfur were registered in inclusions, showing that the reaction products of the cover gas component SF_6 remained on the surface of the melt without contaminating it.

666 nitrogen-containing inclusions per 100 mm^2 were registered for the as-cast AZ91 sample, all of the immersed filter materials were successful at reducing the inclusion content down to 81–120 per 100 mm^2, measured for samples that were in contact with ZrO_2 and the Al_2O_3 coating, respectively. Most of the remaining inclusions had sizes below 50 μm [19].

Analysis of oxygen-containing particles showed significantly reduced or unchanged inclusion amounts in AZ91 samples that have been in contact with filter surfaces containing Al_2O_3-C, Al_2O_3 or MgAlON (see Fig. 7 in [19]). For all filter materials, most inclusions were below the size of 20 μm, being around 98, 159 and 301 inclusions per 100 mm^2 respectively, with the as-cast AZ91 sample containing 306 inclusions per 100 mm^2 with sizes up to 50 μm. It should be noted, that while the number of inclusions stayed nearly constant in case of MgAlON, the average size of oxygen-containing inclusions was greatly reduced, being below 10 μm. For the inclusions within the as-cast AZ91 sample, average oxygen contents of 10.3 wt% were registered, AZ91 samples that have been in contact with given ceramic materials showed oxygen amounts of 1.6–2.6 wt% for their inclusions. The overall oxygen content of 0.4 wt% of the as-cast AZ91 sample was reduced to 0.1 wt% through contact with the ceramic filter materials [19].

The only exception from the positive effect of the ceramic materials on the cleanliness of AZ91 was seen for the ZrO_2 filter immersion. The largest inclusion sizes of all immersed AZ91 samples were registered for the material, measuring 39 μm, with a total amount of 521 inclusions per 100 mm^2. This correlates with the observations of Wu et al., 2010, describing the effectiveness of ZrO_2 filters in AZ91 as low, compared to the performance of Al_2O_3 [13]. In contrast to that, coated and uncoated Al_2O_3-C filters were shown to be successfully applicable in further investigations of magnesium alloy melt filtration [19].

10.5   Outlook and Further Research

Preliminary immersion- and wettability tests suggested the possibility of a successful application of coated and uncoated Al_2O_3-C ceramic foams for the filtration of magnesium alloy melts, therefore, filtration tests are the core point of future research. Model, laboratory-scale AZ91 casting experiments were performed in a closed furnace chamber under protective Ar-0.2 Vol% SF_6 cover gas, melting the metal above the uncoated or Al_2O_3-, $MgAl_2O_4$-, or MgAlON-coated Al_2O_3-C foam filters. AZ91 that had flown through the respective filters was collected in a steel ladle to be evaluated using automated SEM analysis. It is notable that all filter samples were retrieved from the filtration experiments without being damaged, after being held at 680 °C for up to 60 min and enduring the highly reactive AZ91 melt. Filtration experiments under industrial conditions are proposed, as well as filtration of various magnesium- and aluminum alloys, and the exploration of further possible filter coatings for active and reactive metal melt filtration.

10.6   Conclusions

Research had been conducted on coated and uncoated carbon-bonded ceramic foams regarding their application as filters for light metal alloy melts, such as AZ91. The following conclusions can be drawn.

With the Schwartzwalder technique applied for the manufacture of Al_2O_3-C filter foams, centrifugation was shown to be an advantageous manufacturing route, resulting in homogenous foam structures with good mechanical properties and cold crushing strengths. Furthermore, the application of ball-milled slurry contributed to homogeneous ceramic microstructures with small, finely distributed pores [22]. The novel ceramic filter coating material MgAlON was successfully manufactured by means of reaction sintering at 1500 °C for 6 h under nitrogen atmosphere. Powder mixtures containing 0.2–0.45 molar fractions of MgO, 0.1–0.35 molar fractions of AlN and 0.45–0.7 molar fractions of Al_2O_3 were used as starting materials. Maximum conversion degrees to MgAlON of 99 wt% were achieved in the MgO-rich corner of the quasi-ternary system MgO-AlN-Al_2O_3 at AlN:Al_2O_3 ratios of around 6:20 and sintering times of 6 h [24].

Preliminary sessile drop measurements between reaction-sintered MgAlON and the aluminum alloy AlSi7Mg showed wetting angles of 145–150°, proving non-wettability and non-reactivity of the ceramic material, and therefore its suitability as a filter coating for light metal melt filtration [24].

Novel sessile drop measurements using molten, capillary-purified AZ91 at 680 °C have proven Al_2O_3, Al_2O_3-C, MgO and MgAlON to be suitable filter materials for magnesium alloy melts. Despite a slight contact angle decrease after 10 min of contact time, induced by their interfacial reaction with AZ91, Al_2O_3-C and Al_2O_3 continued to show contact angles above 130°, implying their non-wetting character.

Both MgO and reaction-sintered MgAlON had constant contact angles around 140° and 150°, respectively, showing their non-reactivity and non-wettability. The only substrate that was reactive and wetted by the AZ91 melt was ZrO_2, showing contact angels around 100° after 10 min of contact time with molten AZ91 [20].

Immersion tests in AZ91 conducted at 680 °C for 10 s to 60 min showed the applicability of Al_2O_3-C filters in their uncoated and Al_2O_3-, $MgAl_2O_4$- and MgAlON-coated modifications. The filters stayed undamaged after the melt contact, interfacial reactions led to the formation of MgO-containing, platelet-like in-situ layers on all alumina-, zirconia-, or spinel-containing surfaces. Originating from the reduction of alumina, spinel, or zirconia by magnesium, these in-situ layers reached thicknesses of approximately 50 μm after contact times of 60 min. Automated SEM analysis of AZ91 samples that were in contact with uncoated, Al_2O_3- or MgAlON-coated Al_2O_3-C ceramic foam filters for 60 min showed a notable decrease in oxidic inclusions larger than 20 μm, suggesting the suitability of these filter materials for improving melt cleanliness. The contact with ZrO_2 lead to an increase in oxidic inclusions in the AZ91 sample, marking the material as unsuitable for its application in magnesium alloy melt filtration [19, 20].

Model gravity filtration experiments using uncoated, as well as Al_2O_3-, $MgAl_2O_4$- and MgAlON-coated Al_2O_3-C filters showed their durability after being flown through by molten AZ91 and held at 680 °C for 60 min.

Acknowledgements   The authors would like to thank the German Research Foundation (DFG) for their support and funding of the Collaborative Research Center 920 "Multifunctional Filters for Metal Melt Filtration—A Contribution Towards Zero Defect Materials" (subproject C06, Project ID: 169148856), making this research on the filtration of magnesium alloys through ceramic foam filters possible.

References

1. E.F. Emley, Principles of Magnesium Technology, First Edition, Pergamon Press Ltd, Oxford, London, Edinburgh, New York, Paris, Frankfurt, Library of Congress Catalog Card No. 64-19590, pp. 567-568 (1966)
2. H. Antrekowitsch, G. Hanko, P. Ebner, Recycling of different types of magnesium scrap, in: Magnesium Technology, Minerals, Metals & Materials Society, ed. by H. L. Kaplan pp. 321–328 (2002)
3. H. Hu, A. Luo, JOM **48**, 47–51 (1996). https://doi.org/10.1007/BF03223103
4. A. Ditze, C. Scharf, Recycling of Magnesium, 1st ed., Papierflieger Verl., Clausthal-Zellerfeld (2008)
5. W.D. Griffiths, N.-W. Lai, Metall. Mater. Trans. A **38**, 190–196 (2007). https://doi.org/10.1007/s11661-006-9048-7
6. W. Unsworth, Control of Intermetallic Particles in Magnesium Alloys by Filtration. Metallurgia (England), Vol. 62, p. 15 (1960)
7. J.N. Reding, Method For Melting and Purifying Magnesium, USA Patent No. 3,843,355 (1974)
8. W. Petrovich, J.S. Waltrip, Flux-Free Refining of Magnesium Die Cast Scrap, Light Metals 1989, pp. 749–755, ed. by R. G. Campbell (Warrendale, PA: TMS, 1989)

9. P. Bakke, A. Nordmark, E. Bathen, T.A. Engh, D. Øymo, Filtration of Magnesium by Ceramic Foam Filters, in: Light Metals, The Minerals, Metals & Materials Society, ed. by E. R. Cutshall, p. 923–935 (1992)

10. P. Bakke, T.A. Engh, E. Bathen, D. Øymo, A. Nordmark, Mater. Manuf. Proc. **9**(1), 111–138 (1994). https://doi.org/10.1080/10426919408934888

11. B.S. You, G.H. Wu, C.D. Yim, Ceramic foam filtration of magnesium alloy melt, in: Magnesium: Proceedings of the 7th International Conference on Magnesium Alloys and Their Applications (Wiley-VCH Verlag GmbH & Co. KGaA, Weinheim) ed. by K.U. Kainer, pp. 221–227 (2007)

12. G.H. Wu, M. Xie, C. Zhai, X. Zeng, Y. Zhu, W. Ding, Purification technology of AZ91 magnesium alloy wastes. Trans. Nonferrous Met. Soc. China **13**(6), 1260–1264 (2003)

13. G.H. Wu, M. Sun, J. Dai, W. Ding, Effects of filter materials on microstructure and mechanical properties of AZ91. China Foundry **7**(4), 400–407 (2010)

14. C. Voigt, B. Dietrich, M. Badowski, M. Gorshunova, G. Wolf, C.A. Aneziris, The Minerals, Metals & Materials Society, ed. by C. Chesonis, Light Metals, pp. 1063–1069 (2019). https://doi.org/10.1007/978-3-030-05864-7_130

15. M. Emmel, C.G. Aneziris, Ceram. Int. **38**(6), 5165–5173 (2012). https://doi.org/10.1016/j.ceramint.2012.03.022

16. M. Emmel, C.G. Aneziris, J. Mater. Res. **28**, 2234–2242 (2013). https://doi.org/10.1557/jmr.2013.56

17. T. Wetzig, B. Luchini, S. Dudczig, J. Hubálková, C.G. Aneziris, Ceram. Int. **44**(15), 18143–18155 (2018). https://doi.org/10.1016/j.ceramint.2018.07.022

18. K. Schwartzwalder, A.V. Sommers, Method of making porous ceramic articles. US Patent 3,090,094 (1961)

19. A. Schramm, V. Recksiek, S. Dudczig, C. Scharf, C.G. Aneziris, Adv. Eng. Mater. **24**(2), 2100519 (2021). https://doi.org/10.1002/adem.202100519

20. A. Schramm, R. Nowak, G. Bruzda, W. Polkowski, O. Fabrichnaya, C.G. Aneziris, J. Eur. Ceram. Soc. **42**(6), 3023–3035 (2022). https://doi.org/10.1016/j.jeurceramsoc.2022.01.040

21. B. Luchini, E. Storti, T. Wetzig, C. Settgast, M. Abendroth, J. Hubálková, V. Pandolfelli, C.G. Aneziris, J. Eur. Ceram. Soc. **39**(8), 2760–2769 (2019). https://doi.org/10.1016/j.jeurceramsoc.2019.02.048

22. A. Schramm, C. Voigt, J. Hubalkova, C. Scharf, C. G. Aneziris, Adv. Eng. Mater. **22**(2) (2019). https://doi.org/10.1002/adem.201900525

23. F. Gao, R. Liu, X.J. Wang, Adv. Mater. Res. **750-752**, 2191–2195 (2013). https://doi.org/10.4028/www.scientific.net/amr.750-752.2191

24. Y. Guo, Z. Ji, J. Phy. Conf. Series **1637**, 012031 (2020). https://doi.org/10.1088/1742-6596/1637/1/012031

25. A. Schramm, M. Thümmler, O. Fabrichnaya, S. Brehm, J. Kraus, J. Kortus, D. Rafaja, C. Scharf, C.G. Aneziris, Crystals **12**, 654 (2022). https://doi.org/10.3390/cryst12050654

26. A. Granon, P. Goeuriot, F. Thevenot, J. Guyader, P. L'Haridon, Y. Laurent, J. Eur. Ceram. Soc. **13**(4), 365–370 (1994). https://doi.org/10.1016/0955-2219(94)90012-4

27. A. Granon, P. Goeuriot, F. Thevenot, J. Eur. Ceram. Soc. **15**(3), 249–254 (1995). https://doi.org/10.1016/0955-2219(95)93946-z

28. Z. Cheng, J.L. Sun, F.S. Li, Z.Y. Chen, L.G. Wan, Adv. Mater. Res. **105**, 769–772 (2010). https://doi.org/10.4028/www.scientific.net/AMR.105-106.769

29. X. Wang, W. Li, S. Seetharaman, Int. J. Mater. Res. **93**, 540–544 (2002). https://doi.org/10.3139/146.020540

30. H.X. Willems, Preparation and properties of translucent gamma-aluminium oxynitride, doctor of philosophy, department of chemical engineering and chemistry, Eindhoven. The Netherlands (1992). https://doi.org/10.6100/IR382898

31. M. Yan, M. Li, Y. Li, L. Li, Y. Sun, S. Tong, J. Sun, Ceram. Int. **43**(17), 14791–14797 (2017). https://doi.org/10.1016/j.ceramint.2017.07.223

32. B. Fankhänel, M. Stelter, C. Voigt, C.G. Aneziris, Adv. Eng. Mater. **19**(9) (2017). https://doi.org/10.1002/adem.201700084

33. C. Voigt, L. Ditscherlein, E. Werzner, T. Zienert, R. Nowak, U. Peuker, N. Sobczak, C.G. Aneziris, Mater. Des. **150**, 75–85 (2018). https://doi.org/10.1016/j.matdes.2018.04.026
34. N. Eustathopoulos, M.G. Nicholas, B. Drevet, Wettability at High Temperatures. Elsevier Science (1999), ISBN: 9780080543789
35. A. Kudyba, N. Sobczak, J. Budzioch, W. Polkowski, D. Giuranno, Mater. Des. **160**, 915–917 (2018). https://doi.org/10.1016/j.matdes.2018.10.028
36. M. Godzierz, A. Olszówka-Myalska, N. Sobczak, R. Nowak, P. Wrześniowski, J. Magnes. Alloy. **9**(1), 156–165 (2021). https://doi.org/10.1016/j.jma.2020.06.006
37. W. Dai, A. Yamaguchi, W. Lin, J. Ommyoji, J. Yu, Z. Zou, J. Ceram. Soc. Jpn. **115**(1337), 42–46 (2007). https://doi.org/10.2109/jcersj.115.42
38. W.E. Lee, S. Zhang, Melt corrosion of refractories, in: Ageing Studies and Lifetime Extension of Materials. ed. by Mallinson L.G., Springer, Boston, MA, (2001). https://doi.org/10.1007/978-1-4615-1215-8_23
39. A. Schramm, B. Bock, A. Schmidt, T. Zienert, A. Ditze, C. Scharf, C.G. Aneziris, Ceram. Int. **44**(14), 17415–17424 (2018). https://doi.org/ https://doi.org/10.1016/j.ceramint.2018.06.207
40. E. Storti, S. Dudczig, A. Schmidt, G. Schmidt, C.G. Aneziris, Steel Res. Int. **88** (2017). https://doi.org/10.1002/srin.201700142
41. T. Etter, P. Schulz, M. Weber, Mat. Sci. Eng. A **448**(1–2), 1–6 (2007). https://doi.org/10.1016/j.msea.2006.11.088
42. A. Schmidt, A. Salomon, S. Dudczig, H. Berek, D. Rafaja, C.G. Aneziris, Adv. Eng. Mater. **19**, 1700170 (2017). https://doi.org/10.1002/adem.201700170
43. T. Zienert, S. Dudczig, O. Fabrichnaya, C.G. Aneziris, Ceram. Int. **41**, 2089–2098 (2015). https://doi.org/10.1016/j.ceramint.2014.10.004
44. S. Dudczig, C.G. Aneziris, M. Emmel, G. Schmidt, J. Hubalkova, H. Berek, Ceram. Int. **40**(10, B), 16727–16742 (2014). https://doi.org/10.1016/j.ceramint.2014.08.038
45. A. Schramm, C.G. Aneziris, B. Fankhänel, M. Thümmler, A. Charitos, Rafaja, D., Beschichtungen und keramische Filter für die Metallschmelzenfiltration, German Patent Application, Patent file number: 10 2020 006 167.2 (2020)
46. D.L. Porter, A.H. Heuer, J. Am. Ceram. Soc. **62**(5–6), 298–305 (1979). https://doi.org/10.1111/j.1151-2916.1979.tb09484.x
47. C. Weigelt, F. Kerber, C. Baumgart, L. Krüger, C.G. Aneziris, Adv. Eng. Mater. **23**, 2001215 (2021). https://doi.org/10.1002/adem.202001215
48. K. Schwerdtfeger, A. Ditze, C.-T. Mutale, Solubility of iron in magnesium-lithium melts. Metall. Mater. Trans. B **33**(6), 929–930 (2002). https://doi.org/10.1007/s11663-002-0077-3
49. D. Øymo, O. Wallevik, D.O. Karlsen, C. Brassard, in: Proc. of the Second Int. Conf. on the Recycling of Metals, American Society for Metals, Amsterdam (1994)

Chapter 11
Qualitative and Quantitative X-ray Tomography of Filter Macrostructures and Functional Components

Jana Hubálková and Christos G. Aneziris

11.1 Introduction

The tomographic methods provide non-destructive 3D insights into the microstructure of materials whereby a variety of radiation, e.g. light [1], Gamma-rays [2] and X-rays [3, 4], as well as particle radiation, e.g. electrons [5], and neutrons [6], can be used as source for tomographic imaging depending on the scale of the samples to be investigated, the desired spatial resolution and the scope of the imaging. Each type of radiation has its own characteristics and interacts on a different way with the matter [7]. When a high resolution on atomic or low nanoscopic scale is required, only destructive methods such as 3D atom probe tomography, TEM and FIB-SEM, respectively, are suitable [7].

The X-rays can basically be generated by industrial/laboratory X-ray tubes or synchrotron light sources. Due to the electrons with a relativistic speed passing through magnetic fields, the synchrotron light sources feature very high photon flux density, high brilliance and monochromatic, almost parallel beam with coherent properties [8]. However, since there is only a restricted number of synchrotron light sources operating imaging facilities, the beamtime is limited and allocated according to the ranking of submitted proposals. The X-ray tomographic imaging can be performed by different contrast modes such as absorption, phase contrast, diffraction, fluorescence, spectroscopy and small angle scattering [7, 8].

In this chapter, specific applications of laboratory, attenuation-based X-ray computed tomography to questions focusing on the research and development of

J. Hubálková (✉) · C. G. Aneziris
Institute of Ceramics, Refractories and Composite Materials, Technische Universität Bergakademie Freiberg, Agricolastr. 17, 09599 Freiberg, Germany
e-mail: jana.hubalkova@ikfvw.tu-freiberg.de

C. G. Aneziris
e-mail: aneziris@ikfvw.tu-freiberg.de

© The Author(s) 2024
C. G. Aneziris and H. Biermann (eds.), *Multifunctional Ceramic Filter Systems for Metal Melt Filtration*, Springer Series in Materials Science 337,
https://doi.org/10.1007/978-3-031-40930-1_11

porous ceramic materials and functional components are discussed. The details on the technology, composition of the investigated samples as well as the applied measuring parameters can be found in the respective reference.

11.2 Qualitative X-ray Computed Tomography

The tomographic investigations presented in this chapter have been performed by means of a conventional attenuation-based X-ray computed tomograph CT ALPHA (ProCon X-Ray, Sarstedt, Germany) using a microfocus X-ray transmission tube FXE-160 (Feinfocus, Garbsen, Germany), a motorized, precision rotation stage M-062.PD (Physik Instrumente, Karlsruhe, Germany) as well as a flat panel detector 79425 K-05 (Hamamatsu, Japan) with a pixel pitch of 50 μm and an photodiode are of 120×120 mm^2, a CMOS flat panel X-ray detector Dexela 1512 (Perkin Elmer Inc. Europe, Walluf, Germany) with a pixel pitch of 75 μm and an active area of 150×120 mm^2 for small samples as well as a large X-ray detector (Perkin Elmer Inc. Europe, Walluf, Germany) with a pixel pitch of 200 μm and a photodiode area of 410×410 mm^2 for large samples. During each X-ray tomographic scan, 1200 radiographs within 360° were acquired with different exposure time as shown in Fig. 11.1.

From the finite number of radiographs, i.e. projections of the analyzed object on the detector, a volume data was reconstructed using the measurement and reconstruction software Volex 6.0 (Fraunhofer EZRT, Fürth, Germany). The reconstructed volume data were visualized and evaluated using the software VG Studio MAX 2.2 (Volume Graphics, Heidelberg, Germany).

Since more than 30 years, the X-ray computed tomography is widely used for 3D imaging of highly porous solids [9], such as polymeric [10, 11], metallic [12, 13], glass [14] and ceramic [15, 16] foams due to their low density and thus favourable X-ray transmission properties.

Therefore, a qualitative X-ray computed tomography shall exemplarily be demonstrated on the open-porous alumina foam structures. The commercially available Al$_2$O$_3$ filters with three different ppi (pores per inch) numbers, i.e. 10 ppi, 30 ppi and 60 ppi were investigated using μCT, see Figs. 11.2, 11.3 and 11.4, respectively.

Fig. 11.1 Radiographs of an alumina foam ceramics with a size of $50 \times 50 \times 25$ mm^3 taken in different angular positions (from left to right: 0, 45, 67.5 and 90°)

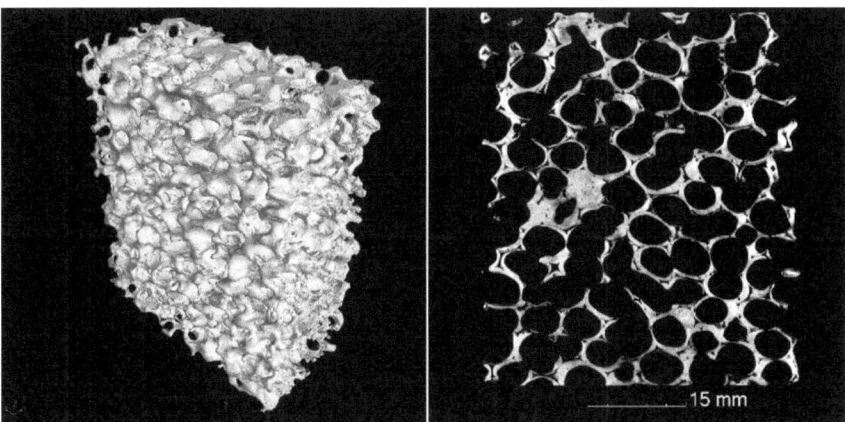

Fig. 11.2 Reconstructed volume image of an Al_2O_3 foam ceramics with a pore density of 10 ppi; 3D isosurface rendering (left) and 2D slice from the middle of the foam (right)

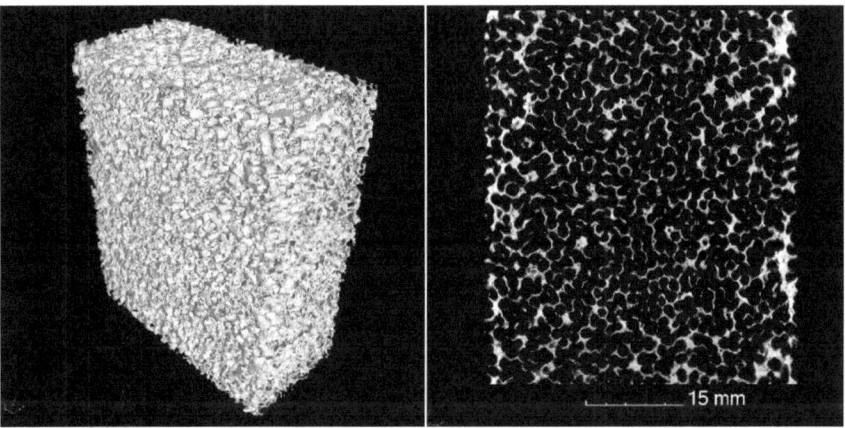

Fig. 11.3 Reconstructed volume image of an Al_2O_3 foam ceramics with a pore density of 30 ppi; 3D isosurface rendering (left) and 2D slice from the middle of the foam (right)

The foam geometry and the pore density have a significant impact on the determination of elastic properties by both static and dynamic methods [17] as well as on the determination of compressive and 3-point bending strength [18].

Furthermore, the reticulated ceramic foam structures can be functionalized with different coating methods. Moritz et al. carried out a feasibility study of electrophoretic deposition of alumina on foam ceramics by varying the voltage and deposition time. Al_2O_3-C foams ($20 \times 15 \times 15$ mm^3) were used as deposition electrode due to their sufficient electrical conductivity. The formation of channel-like pores, the quality of the coating and its adhesion after sintering as well as after thermal shock loading by immersion in molten aluminium were evaluated by means

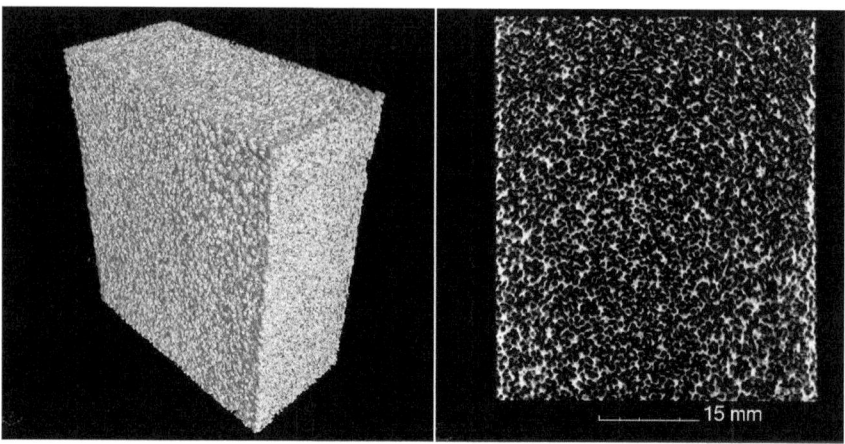

Fig. 11.4 Reconstructed volume image of an Al_2O_3 foam ceramics with a pore density of 60 ppi; 3D isosurface rendering (left) and 2D slice from the middle of the foam (right)

of X-ray computed tomography. It has been proven that the electrophoretic deposition can be successfully applied for coating of Al_2O_3-C foam structures, see Fig. 11.5. The electrophoretic coatings are mainly formed on the outer surface of the substrate foams, but even the outer surface poses the most impacted zone during the metal melt filtration process [19].

Besides electrophoretic deposition also a cold spraying technique or conventional slip casting can be used for coating of foam [20] and dense ceramics by alumina slurry, respectively. Aneziris et al. tested slip casted carbon bonded nozzles with and without active alumina coating regarding their clogging behavior in a steel casting

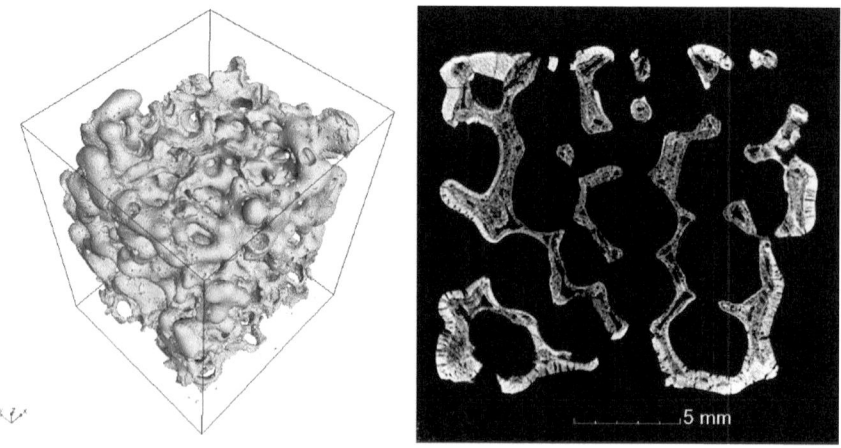

Fig. 11.5 Reconstructed volume images of a electrophoretically coated Al_2O_3-C foam structures; 3D rendering (left) and a 2D slice to visualize the generated pore channels (right)

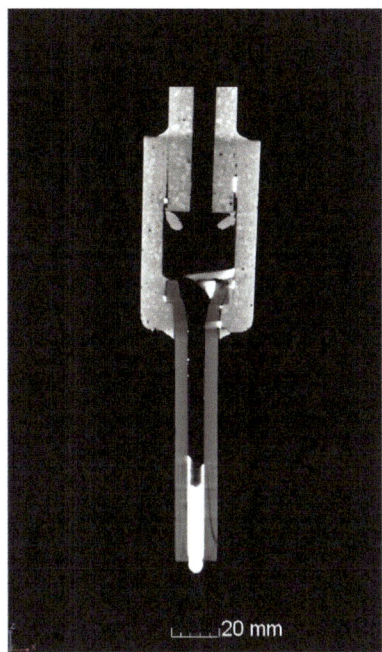

Fig. 11.6 Reconstructed volume image of the nozzle setup after steel casting; 3D rendering with clipping box (left) and 2D slice visualising the adapter in the upper part and the testing nozzle (dark) in the lower part with clogged steel particles on the inner wall and a steel drop (bright) at the exit

simulator consisting of an inductive heated melting unit and an inductive heated tundish unit with two nozzles in the bottom part [21]. The Fig. 11.6 shows the configuration of the nozzle setup comprising the adapter made of coarse-grained spinel-bonded alumina castable with inserted test nozzle made of fine-grained carbon bonded alumina without reactive alumina coating. The CT scans of test nozzle with reactive alumina coating can be found elsewhere [21].

The qualitative X-ray computed tomography is often used for quality control purposes. Moreover, the reconstructed volume images can serve as a reference for target-actual performance comparison to estimate the variance between the computer-aided design and the CT scan of the fabricated object based upon the model. Herdering et al. generated an open cell foam model, printed polyamide (PA) foams thereupon by selective laser sintering (SLS) technique and compared the foam model with the reconstructed isosurface of the printed foam. The SLS additive manufacturing could well reproduce the macrostructure of the computer-aided model, however the strut thickness and the relative density was significantly higher for the printed foams [22].

The Fig. 11.7 shows hybrid manufactured porous alumina ceramics (cubes with an edge length of 25 mm) with a Kelvin structure based on templates 3D printed by SLS method and subsequently flame sprayed with alumina. Instead of usually

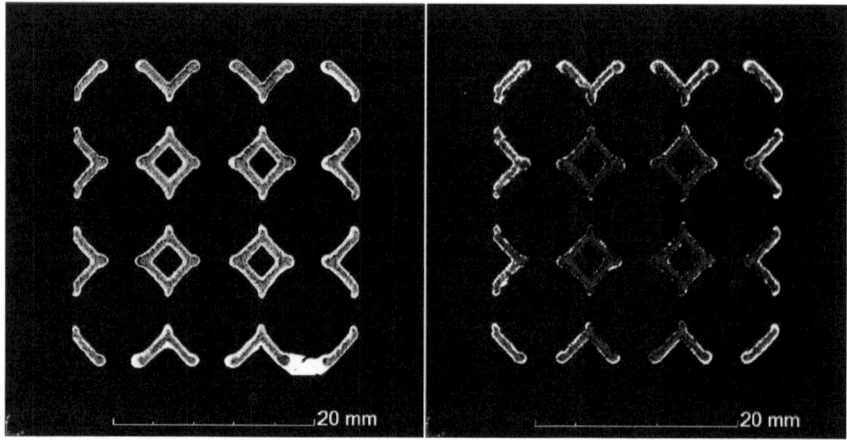

Fig. 11.7 Reconstructed 2D slices of 3D printed environmentally friendly templates with a Kelvin structure based on saccharide (left) and salt (right) after first flame spraying with alumina (bright)

applied polymeric materials (polyamide, polyurethane, polycarbonate) which have to be burned-out, environmentally friendly raw materials were introduced such as saccharide or salt. The new hybrid manufacturing approach overcomes difficulties connected with the thermal treatment of the polymeric materials by solving the template material in water.

The qualitative X-ray computed tomography is very well suited for the investigation of fracture mechanisms as presented in Fig. 11.8. In order to increase the ultimate fracture strain of Al_2O_3-C materials a new approach has been developed and evaluated. An aqueous suspension of graphene oxide (GO) was reinforced with sapphire nanosheets, slip casted to strip-like samples with a size of $25 \times 8 \times 0.2 \text{ mm}^3$ and subjected to 3-point bending loading. An extreme deflection without failure and a delamination as a determining fracture mechanism are clearly visible. The resulting voxel size after reconstruction was 8.6 μm.

Fig. 11.8 Reconstructed 2D slice of strip-like sample based graphene oxide reinforced with sapphire nanosheets after 3-point bending loading

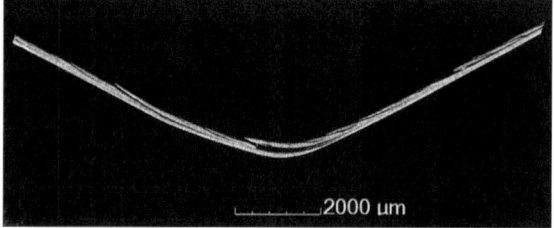

11.3 Quantitative X-ray Computed Tomography

In addition to the visualisation, a quantification of the microstructure in 3D [23, 24] or even in 4D [23] is of major significance. The quantitative analysis of the reconstructed volume data was performed by the software Modular Algorithm for Volume Images MAVI 1.5.3 (Fraunhofer ITWM, Kaiserslautern, Germany).

The quantitative X-ray computed tomography can provide essential 3D structural features and statistical characteristics useful for the development of new technologies and the optimal choice of processing steps, e.g. for the manufacturing of reticulated foam filters [25]. Based on alumina [26] or carbon-bonded alumina [27]. In order to evaluate the homogeneity of the ceramic foam structure, the global and local features were analyzed using the whole filter ($50 \times 50 \times 25$ mm^3) and several ROIs, respectively. Voigt et al. divided the manufactured alumina foam filters virtually in 6 layers along the filters [26] while Luchini et. al constituted five distinct ROIs, one in the center and four in the edges of the foam filter samples [27]. Thus, the homogeneity of the strut thickness distribution, the porosity and the macropore sizes at several filter production steps can be statistically analyzed and used for a direct comparison of conventional rolling/spraying and centrifugation technique.

But not only the relationship between the processing route and the macro and macrostructure but also the correlation between the measurement parameters and the mechanical properties of foam materials, e.g. the influence of the loading plate size, the sample dimensions and the application of a compliant pad on the crushing strength can be evaluated by μCT [28]. The experimental results were validated by mechanical simulations of a crushing strength measurement based on minimum principal stress distribution calculated with FEM.

Moreover, the understanding of the experimental results of the compressive strength measurement can be supported by numerical simulations decoding the crucial determinants. In compliance with the real foam filters, artificial foam geometries were modelled by combination of different relative densities and strut shape factors [29]. A good correlation between the real and virtual strength was obtained and two distinct failure mechanisms depending on the strut shape factor were detected. An advanced experimental–numerical approach taking besides the strut shape and the relative density also pore stretching and polydispersity into account can be used for the prediction of creep deformations of carbon-bonded alumina foams at high temperatures [30]. In order to achieve a high creep resistance and the dimensional stability of a foam filters at high temperature, the stretching direction of the foam should be parallel to the loading direction.

In the run-up to the quantification, the reconstructed volume images must be post-processed in order to extract desired quantitative characteristics, cp. [15]. After the reconstruction of the volume image with a resulting voxel size of 18.9 μm, a region of interest ROI avoiding the edge errors has to be selected, as shown in Fig. 11.9 exemplarily for uncoated carbon bonded alumina foam filters of a cylindrical geometry with a diameter and a height of 20 mm. In the presented case, the volume image was cropped to a size of $300 \times 300 \times 300$ voxels.

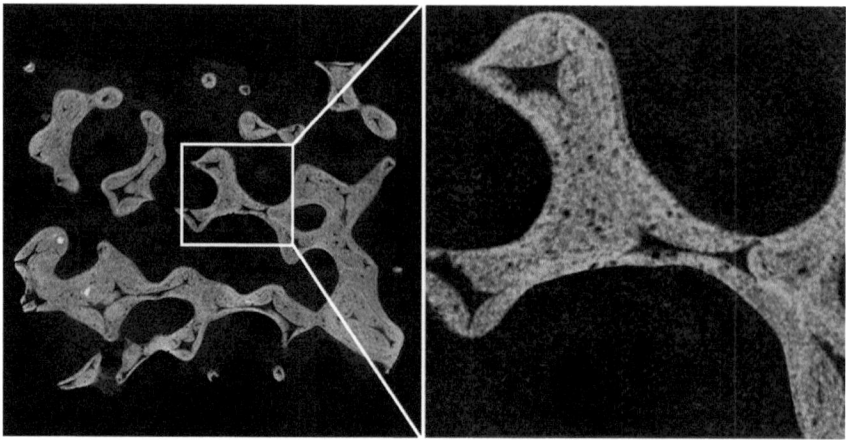

Fig. 11.9 Reconstructed volume image of a carbon bonded alumina foam filter: The original volume image after reconstruction with a size of 1365 × 1365 × 875 without any image processing steps (left) and after cropping with a size of 300 × 300 × 300 voxels (right)

After the binarization step using Otsu´s threshold automatically separating voxels in foreground (white) and background (black) [31], field features were determined, see Fig. 11.10. Subsequently, all pores inside the struts (PU pores, material´s micropores) were eliminated using the morphological closing procedure in order to be able to quantify the strut system as a whole as well as the functional macropores. By complementing the binarization, the background voxels representing the functional macropores can easily be inverted into the foreground voxels and quantified as well.

Afterwards, the morphological procedure called spherical granulometry can be performed on the binarized volume image by successive morphological openings with structuring elements (balls) of increasing size until all foreground voxels disappeared, see Fig. 11.11. As a result, each voxel is assigned to the diameter of the ball belonging to the step where this voxel disappeared for the first time [24].

Thus, by means of the spherical granulometry cumulative distribution functions of both struts and macropores can be established. The spherical granulometry revealed for struts a d_{50}-value of 1.33 mm and a d_{90}-value of 1.82 mm whereas for macropores a d_{50}-value of 2.44 mm and a d_{90}-value of 4.56 mm.

11.4   Ex-Situ and In-Situ Tomographic Investigations

Ex-situ tomographic investigations require neither implementation of the testing equipment inside the μCT device nor restrictions regarding the acquisition time. Therefore, they are widely used for gathering of 3D tomographic data prior and posterior of thermal [32], mechanical [33, 34] or chemical [35] loading events. For ex-situ loading, the investigated sample must be removed from the μCT device and

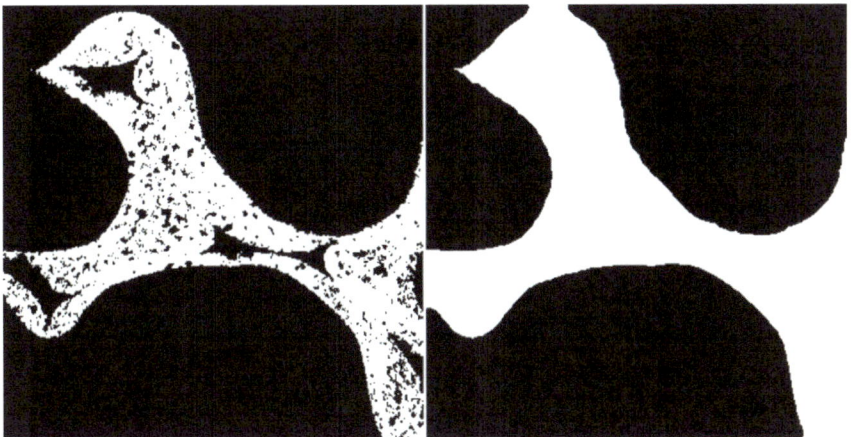

Fig. 11.10 Reconstructed volume image from Fig. 11.9 after binarization with Otsu's threshold (left) and after morphological closing (right)

Fig. 11.11 Reconstructed volume image from Fig. 11.10 after morphological granulometric procedure performed on strut system (left) and on macropores (right)

repositioned on the sample rotation stage upon loading completion [36]. Obviously, the relocation of the sample involves an extrinsic shift of the acquired image data which has to be distinguished from the inherent displacement or alteration induced by intended loading.

The ex-situ tomographic investigations are well suited for localising and identifying the prime cause of failure under extreme loading and environmental conditions not implementable into μCT devices, such as creep experiments at high temperature under inert atmosphere or corrosion experiments in a steel casting simulator.

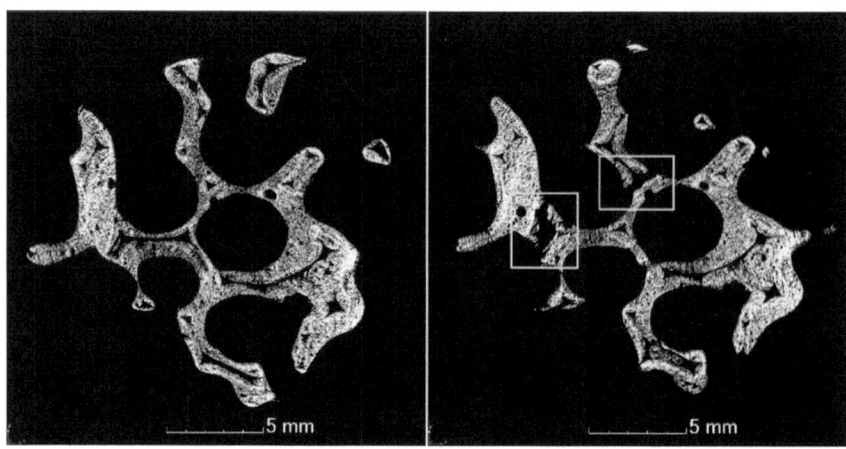

Fig. 11.12 Reconstructed 2D slices from the middle of a cylindrical, carbon-bonded alumina foam sample perpendicular to the loading direction before (left) and after the creep test at 1350 °C in an inert atmosphere (right)

Figure 11.12 compares two different states of a cylindrical, carbon-bonded alumina sample manufactured by the replica technique using a cylindrically shaped polyurethane foam with a diameter of 20 mm and a height of 25 mm [30]. The Al_2O_3-C foam sample was coked at 1400 °C under reducing atmosphere representing the initial state. The initial state was scanned by μCT. The finished, coked foam sample was evacuated and subjected to a creep experiment in an induction heated high temperature testing machine at 1350 °C under argon atmosphere. After cooling, the thermo-mechanically loaded sample was analyzed using the μCT again with the same acquisition parameters. The voxel size after reconstruction was 19.8 μm. The direct comparison of both states at the same position (virtual cross section through the middle of the sample) revealed two cracks originating from the sharp triangular strut pores left by decomposition of the polyurethane foam template.

Figure 11.13 demonstrates another ex-situ investigation of foam filters. Wetzig et al. produced cylindrical carbon-bonded alumina foam filters (10 ppi) with a diameter and a height of 200 mm by replica technique [37]. The polyurethane foam templates were provided with one central and eight circularly arranged macro-channels with a diameter of 40 mm each, allowing flow conditioning during continuous casting of steel. The templates were twice coated with an alumina-carbon slurry, dried and cooked at 800 °C under reducing atmosphere.

The microstructure of the final foam filter was analyzed by μCT using the large PerkinElmer X-ray detector in order to visualize possible radial density and strut thickness gradients and inhomogeneities. The resulting voxel size was 262 μm. Afterwards, the foam filter was tested in a steel casting simulator regarding its thermo-mechanical behavior. The steel melt was cast through the foam filter being placed in a tundish vessel into the copper molds positioned below the tundish vessels. After cooling and opening the steel casting simulator, the removed foam filter with steel

Fig. 11.13 Radiographs of the carbon-bonded alumina foam filter before (left) and after the testing in the steel casting simulator (right)

residues was analyzed by μCT again. As can be seen from Figs. 11.13 and 11.14, the macrostructure of the foam filter remained undamaged but numerous steel residues in the form of spheres and a rim stuck on the surface of the foam filter.

In contrast, for in-situ loading the sample remains in the same position while the whole loading equipment is placed on the rotation stage thus the arising microstructural changes can primarily be attributed to the applied loading. A distinction needs to be drawn between so called interrupted and true in-situ μCT investigations. The interrupted in-situ μCT investigations are carried out discontinuously, i.e. the loading must be interrupted in order to perform a CT scan while maintaining the loading level at a constant value. Obviously, such assumption does not necessary reflect the reality

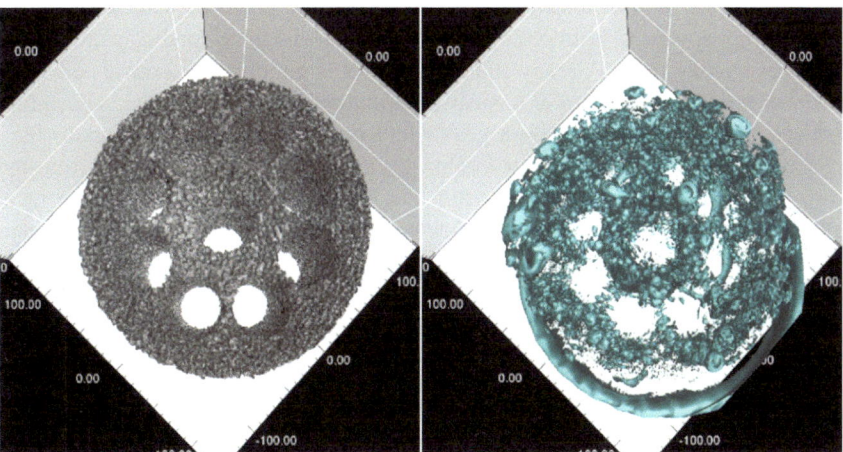

Fig. 11.14 Reconstructed volume rendering of the carbon-bonded alumina foam sample before the testing in the steel casting simulator (left) and the reconstructed isosurface rendering of the steel residues after the testing in the steel casting simulator (right)

since all launched processes are being continued even during the CT scan. In comparison, the true in-situ CT investigations are performed continuously, i.e. the scan time is substantially shorter than the time frame of the unfolding events.

Interrupted in-situ investigations of metallic (316L steel) scaffolds and metalceramic (TRIP-steel/Mg-PSZ) foams and honeycombs under compressive load were successfully performed using lab-based X-ray μCT equipment providing detailed insights into the deformation mechanisms of architectures metallic structures [38] and metal-ceramic composite structures [39].

However, the interrupted in-situ tomographic investigations of brittle foam materials present a particular challenge as a consequence of its sudden failure of a stochastic nature. Therefore, extensive preliminary investigations are necessary.

In the presented study, the interrupted in-situ investigations of white and black glass foams under quasi-static strain-controlled compression loading were performed. The white and black foam glass samples were prepared by foaming of a powdered recycling float glass while adding neutralization foaming agent resp. redox foaming agent at a temperature of 800 °C. The foamed glass samples were cut to cubes with an edge length of 20 mm and underwent in-situ tomographic scans at different compressive strains based on preliminary investigations. After reaching of the intended compressive strain level and subsequent awaiting of force balance, tomographic scans taking approx. 15 min were carried out. The details of the experimental procedure and in-situ μCT parameters can be found elsewhere [40]. For the interrupted in-situ compressive testing, a specially designed in-situ load frame was used [39].

Figure 11.15 shows the reconstructed volume rendering of white foam glass at compressive strains of 0, 1, 3.8 and 5.1%. The tomographic scans underlined that the failure of the white foam glass takes place in a mere brittle mode. At a compressive strain of 1%, no visible changes of the microstructure could be detected, only an elastic compression can be traced by the movement of the lower loading plate against the upper loading plate. Already at a compressive strain of 3.8%, several cell walls in a shear direction failed, i.e. a shear deformation band was formed. The forcedisplacement curve showed a clear peak with a consequent drop of the force.

The compressive behaviour of the black glass foam differs distinctly from the white glass foam. As can be seen from the Fig. 11.16, either a successive local failure nor a formation of a global crack could be identified. A deformation band close to the upper compressive plate perpendicular to the loading direction unfolded. Such quasi-ductile behaviour of the black glass foams can be attributed to the nature of the pores. In contrast to the white foam glass, the pores of the black foam glass are predominantly closed.

The black foam glass exhibited an extraordinary compressive behaviour, i.e. no pronounced force peak could be identified. The force–displacement curve of black foam glass shows a linear increase up to a displacement of approx. 2%. Afterwards, a force plateau arises and the force–displacement curve proceeds constantly up to the last compressive strain of 15%.

By means of DVC (Digital Volume Correlation) software DaVis 10 (LaVision, Göttingen, Germany) the full 3D strain and displacement maps were exemplary

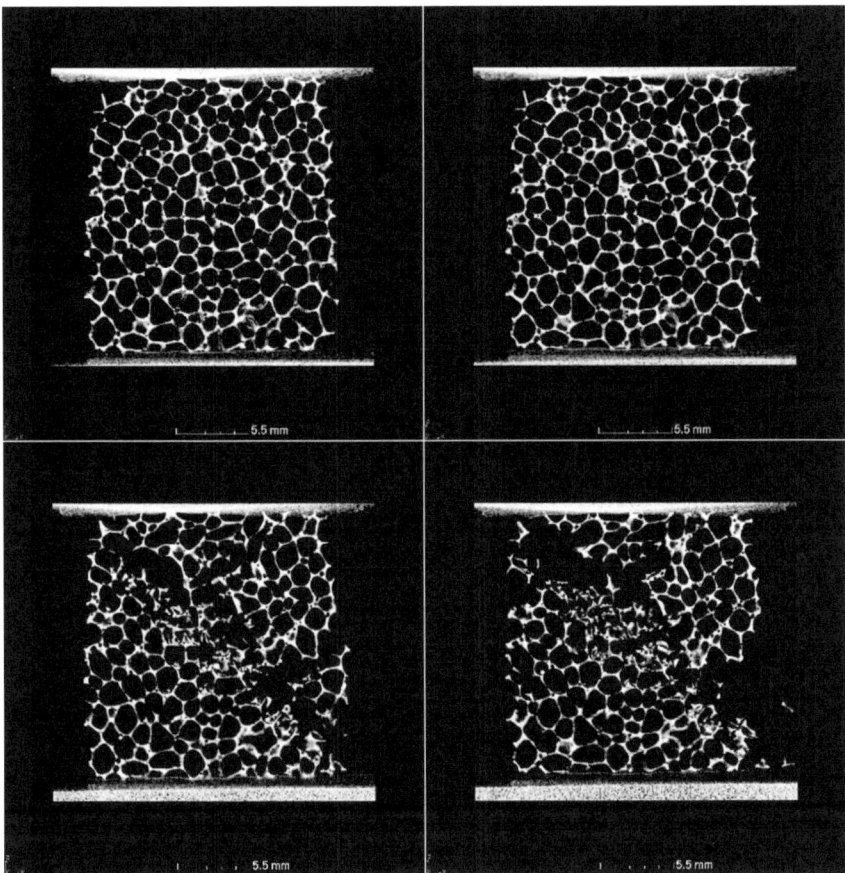

Fig. 11.15 Reconstructed 2D slices of the black foam glass at compressive strains of 0% (top left), 1% (top right), 3.8% (bottom left) and 5.1% (bottom right)

calculated for the black foam glass at a compressive strain of 10% comparing the volume image of the black foam glass in unloaded state (0%) as a reference and in the deformed state. The deformation band was excluded from the DVC calculation due to the extreme alteration (Fig. 11.17).

Nevertheless, the true in-situ X-ray μCT experiments for fast progressing processes, especially for dynamic loading [41, 42], are currently feasible only using synchrotron sources. Recently, dynamic processes during foaming of liquid aluminium such as nucleation, growth and coalescence of the bubbles at high temperatures have been investigated by applying of a continuous X-ray tomoscopy [43]. Otherwise, an automatic tracking algorithm was developed enabling follow up the damage evolution, particularly void nucleation, growth and coalescence during in-situ tensile loading of pure iron material by microtomography [44].

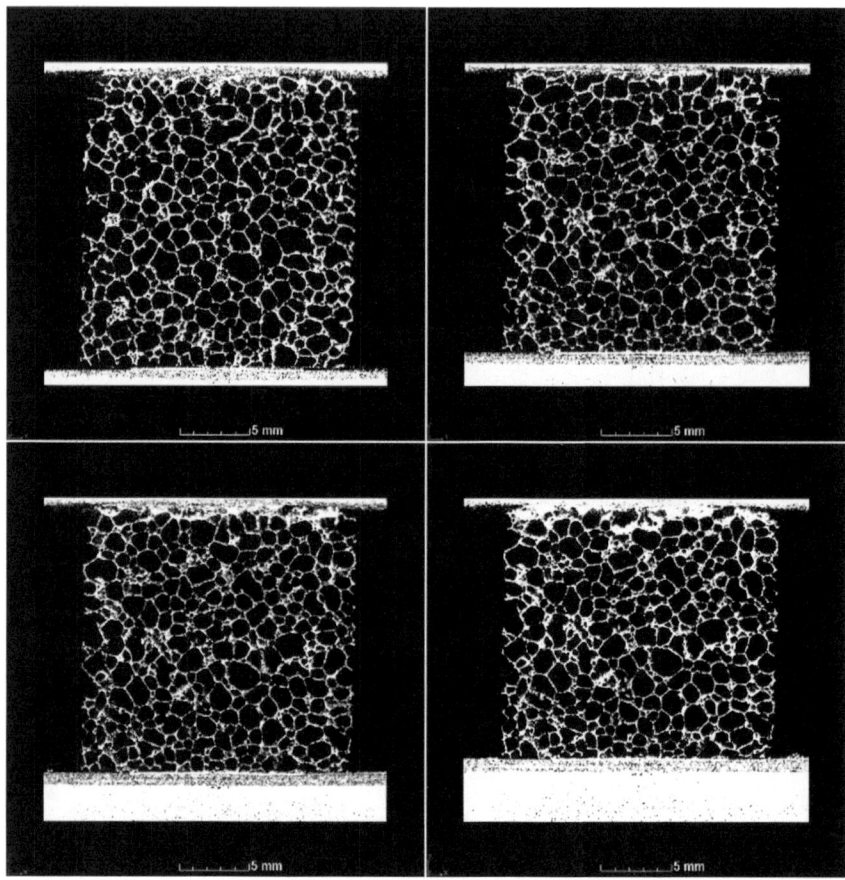

Fig. 11.16 Reconstructed 2D slices of the black foam glass at compressive strains of 0% (top left), 5% (top right), 10% (bottom left) and 15% (bottom right)

11.5 Benefits and Limitations of Laboratory X-ray Computed Tomography

The conventional, attenuation-based X-ray computed tomography has been shown to be a useful tool for assessing of 3D structures as well as detecting of microstructural defects, inhomogeneities, pores and pore size distribution, cracks etc. [45]. The advantages are undoubtedly its non-destructive matter, good availability and possibility to scan real components with large dimensions. However, some limitations still remain. First of all, the spatial resolution is rather limited since X-rays are divergent and the control of the magnification is achieved by adjusting the source to sample distance or by implication the source to detector distance, i.e. the magnification is merely of geometrical nature.

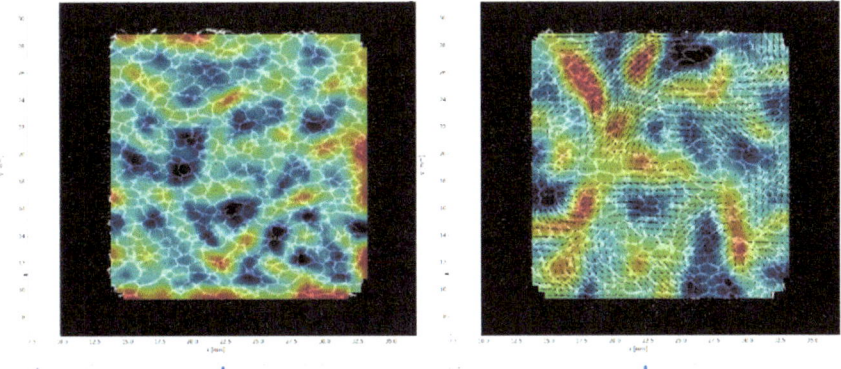

Fig. 11.17 Overlay of the reconstructed 2D slice of the black glass foam at a compressive strain of 10% and the calculated strain (left) and displacement map (right)

The X-ray computed tomography of large components features a tightrope walk between a sufficient transmittance, achievable resolution and incidence of artefacts impairing the reliability and quality of the tomographic images since the X-ray beam leaving the target represents by no means a point source. In fact, a focal spot of a non-negligible size forms on the target resulting in the blurring effects in the acquired radiographs. The spot size is directly proportional to the applied accelerating voltage and current. However, the maximizing spot size along with minimizing the number of projections on the condition that no notable increase of blurriness arises could be beneficial, particularly for high-speed imaging [46].

Furthermore, the conventional, attenuation based X-ray computed tomography is just an imaging technique without any possibility to obtain additional information on chemical or phase composition and distribution. In recent past, novel techniques not only for synchrotron microtomography [47] but also for laboratory X-ray microscopy [48] have been developed.

Nevertheless, the acquired attenuation-based X-ray CT imaging data and extracted 3D/4D quantities can be used as input for a virtual laboratory in order to set realistic boundary conditions, to validate simulation outputs and to establish new, reality-based models of material microstructure as a function of environmental conditions. The steps towards future would comprise the implementation of machine-learning algorithms not only on the acquisition side, e.g. for significant reducing of the acquisition time at required resolution, but also on the image processing side, e.g. for automatizing of binarization and particularly segmentation of structural parts to be evaluated [49].

11.6 Conclusions

The findings of the performed μCT measurements indicate that the attenuation-based, laboratory X-ray microcomputed tomography is a valuable tool for non-destructive analysis of macro and micro structures of different materials and components taking some constraints into account. Obviously, there are some technical (size and resolution of the detector, source-to-detector distance, max. accelerating voltage, max. target power) and physical limitations (polychromatic and cone X-ray beam, focal point on the target) so that the spatial and temporal resolution are constrained while maintaining a reasonable signal-to-noise ratio. Therefore, the size and composition of the objects to be investigated should be chosen with respect to the sufficient penetrability, achievability of expected results and timescale of the processes that unfold during loading.

Acknowledgements The authors would like to acknowledge financial support of the German Research Foundation (DFG) as part (subproject S01) of the Collaborative Research Centre CRC 920 "Multifunctional Filters for Metal Melt Filtration – a Contribution to Zero Defect Materials", project number: 169148856. The investigated samples were kindly provided by the CRC 920 team members.

References

1. B.E. Bouma, J.F. de Boer, D. Huang, I.-K. Jang, T. Yonetsu, C.L. Leggett, R. Leitgeb, D.D. Sampson, M. Sutter, B.J. Vakoc, M. Villiger, M. Wojtkowski, Nat Rev Methods Primers **2**, 79 (2022). https://doi.org/10.1038/s43586-022-00162-2
2. G.A. Johansen, Ind. Tomogr. 197–222 (2015). https://doi.org/10.1016/B978-1-78242-118-4. 00007-1
3. P.J. Withers, C. Bouman, S. Carmignato, V. Knudde, D. Grimaldi, C.K. Hagen, E. Maire, M. Manley, A. Du Plessis, S.R. Stock, Nat. Rev. Methods Primers **1**, 18 (2021). https://doi.org/10. 1038/s43586-021-00015-4
4. S.R. Stock, *Microcomputed Tomography: Methodology and applications*, (CRC Press, Boca Raton, 2019). https://doi.org/10.1201/9780429186745
5. P. Ercius, O. Alaidi, M.J. Rames, G. Ren, Adv. Mat. **27**, 5638–5663 (2016). https://doi.org/10. 1002/adma.201501015
6. N. Kardjilov, I. Manke, A. Hilger, M. Strobel, J. Banhart, Mater. Today **14**, 248–256 (2015). https://doi.org/10.1016/S1369-7021(11)70139-0
7. J. Banhart (ed.), *Advanced Tomographic Methods in Materials Research and Engineering* (Oxford University Press, New York, 2008)
8. A. Bharti, N. Goyal, Fundamental of synchrotron radiations, in *Synchrotron Radiation*, ed. by D. Joseph (Ed.) (IntechOpen, London, 2019). https://doi.org/10.5772/intechopen.82202
9. E. Maire, J. Adrien, C. Petit, Structural characterization of solid foams. C. R. Phys. **15**, 674–682 (2014). https://doi.org/10.1016/j.crhy.2014.09.001
10. G.J. Kemerink, R.J. Lamers, G.R. Thelissen, J.M. van Engelshoven, Med. Phys. **22**, 1445–1450 (1995). https://doi.org/10.1118/1.597568
11. J.G.F. Wismans, J.A.W. van Dommelen, L.E. Govaert, H.E.H. Meijer, Mater. Sci. Forum **638–642**, 2761–2765 (2010). https://doi.org/10.4028/www.scientific.net/MSF.638-642.2761

12. S. Calvo, D.A. Beugre, M. Crine, A. Léonard, P. Marchot, D. Toye, Chem. Eng. Process. **48**, 1030–1039 (2009). https://doi.org/10.1016/j.cep.2009.02.001
13. J.J. Bock, A.M. Jacobi, Mat. Char. **75**, 35–43 (2013). https://doi.org/10.1016/j.matchar.2012.10.001
14. R.C. Atwood, J.R. Jones, P.D. Lee, L.L. Hench, Scr. Mat. **51**, 1029–2033 (2004). https://doi.org/10.1016/j.scriptamat.2004.08.014
15. E. Maire, P. Colombo, J. Adrien, L. Babout, L. Biasetto, J. Eur. Ceram. Soc. **27**, 1973–1978 (2007). https://doi.org/10.1016/j.jeurceramsoc.2006.05.097
16. P. Stochero, E.G. de Moraes, A.C. Moreira, C.P. Fernandes, M.D.M. Innocentini, A.P. Novaes De Oliveira, J. Eur. Ceram. Soc. **40**, 4224–4231 (2020). https://doi.org/10.1016/j.jeurceram soc.2020.05.036
17. J. Grabenhorst, B. Luchini, J. Fruhstorfer, C. Voigt, J. Hubálková, J. Chen, N. Li, Y. Li, C.G. Aneziris, Ceram. Int. **45**, 5987–5995 (2019). https://doi.org/10.1016/j.ceramint.2018.12.069
18. M. Neumann, J. Hubálková, C. Voigt, J. Grabenhorst, C.G. Aneziris, J. Eur. Ceram. Soc. **42**, 2331–2340 (2022). https://doi.org/10.1016/j.jeurceramsoc.2021.12.034
19. K. Moritz. C. Dietze, C. Voigt, J. Hubálková, C.G. Aneziris, Ceram. Int. **45** (2019), 10701–10706. https://doi.org/10.1016/j.ceramint.2019.02.141
20. C. Voigt, T. Zienert, P. Schubert, C.G. Aneziris, J. Hubálková, J. Am. Ceram. Soc. **97**, 2046–2053 (2014). https://doi.org/10.1111/jace.12977
21. C.G. Aneziris, S. Dudczig, J. Hubálková, M. Emmel, G. Schmidt, Ceram. Int. **39**, 2835–2843 (2013). https://doi.org/10.1016/j.ceramint.2012.09.055
22. A. Herdering, M. Abendroth, P. Gehre, J. Hubálková, C.G. Aneziris, Ceram. Int. **45**, 153–159 (2019). https://doi.org/10.1016/j.ceramint.2018.09.146
23. E. Maire, P.J. Withers, Int. Mater. Rev. **59**, 1–43 (2014). https://doi.org/10.1179/1743280413Y.0000000023
24. K. Schladitz, J. Microsc. **243**, 111–117 (2011). https://doi.org/10.1111/j.1365-2818.2011.035 13.x
25. T. Fey, U. Betke, S. Rannabauer, M. Scheffler, Adv. Eng. Mater. **19**, 1700369 (2017). https://doi.org/10.1002/adem.201700369
26. C. Voigt, E. Jäckel, C.G. Aneziris, J. Hubálková, Ceram. Int. **39**, 2415–2422 (2013). https://doi.org/10.1016/j.ceramint.2012.09.001
27. B. Luchini, J. Hubálková, T. Wetzig, J. Grabenhorst, J. Fruhstorfer, V.C. Pandolfelli, C.G. Aneziris, Ceram. Int. **44**, 13832–13840 (2018). https://doi.org/10.1016/j.ceramint.2018.04.228
28. C. Voigt, J. Storm, M. Abendroth, C.G. Aneziris, M. Kuna, J. Hubálková, J. Mater. Res. **28**, 2288–2299 (2013). https://doi.org/10.1557/jmr.2013.96
29. B. Luchini, E. Storti, T. Wetzig, C. Settgast, M. Abendroth, J. Hubálková, V.C. Pandolfelli, C.G. Aneziris, J. Eur. Ceram. Soc. **39**, 2760–2768 (2019). https://doi.org/10.1016/j.jeurceram soc.2019.02.048
30. C. Settgast, Y. Ranglack-Klemm, J. Hubalkova, M. Abendroth, M. Kuna, H. Biermann, J. Eur. Ceram. Soc. **39**, 610–617 (2019). https://doi.org/10.1016/j.jeurceramsoc.2018.09.022
31. N. Otsu, IEEE Trans. on SMC **9**, 62–66 (1979). https://doi.org/10.1109/TSMC.1979.4310076
32. J.-B. le Graverend, J. Adrien, J. Cormier, Mater. Sci. Eng. **695**, 367–378 (2017). https://doi.org/10.1016/j.msea.201703.083
33. R. Lorenzoni, V.N. Lima, T.C.S.P. Figueiredo, M. Hering, S. Paciornik, M. Curbach, V. Mechtcherine, F. de Andrade Silva, Constr. Build. Mater. **323** (2022), 126558. https://doi.org/10.1016/j.conbuildmat.2022.126558
34. T. Leißner, A. Diener, E. Löwer, R. Ditscherlein, K. Krüger, A. Kwade, U.A. Peuker, Adv. Powder Technol. **31**, 78–86 (2020). https://doi.org/10.1016/j.apt.2019.09.038
35. K. van Gaalen, F. Gremse, F. Benn, P.E. McHugh, A. Kopp, T.J. Vaughan, Bioactive Mater. **8**, 545–558 (2022). https://doi.org/10.1016/j.bioactmat.2021.06.024
36. C. Noiriel, F. Renard, C. R. Geosci. **354**, 255–280 (2022). https://doi.org/10.5802/crgeos.137
37. T. Wetzig, B. Luchini, S. Dudczig, J. Hubálková, C.G: Aneziris, Ceram. Int. **44** (2018), 18143–18155. https://doi.org/10.1016/j.ceramint.2018.07.022

38. M. Kachit, A. Kopp, J. Adrien, E. Maire, X. Boulnat, J. Mater. Res. Technol. **20**, 1341–1351 (2022). https://doi.org/10.1016/j.jmrt.2022.08.003

39. H. Berek, M. Oppelt, C.G. Aneziris, X-Ray computer tomography for three-dimensional characterization of deformation and damage processes, in *Austenitic TRIP/TWIP Steels and Steel-Zirconia Composites*. Springer Series in Materials Science, vol. 298, ed. by H., Biermann, C. Aneziris (Springer, Cham, 2020). https://doi.org/10.1007/978-3-030-42603-3_16

40. J. Hubálková, C. Voigt, A. Schmidt, K. Moritz, C.G. Aneziris, Adv. Eng. Mater. **19**, 1700286 (2017). https://doi.org/10.1002/adem.201700286

41. L. Farbaniec, D.J. Chapman, J.R.W. Patten, L.C. Smith, J.D. Hogan, A. Rack, D.E. Eakins, Icarus **159**, 114346 (2021). https://doi.org/10.1016/j.icarus.2021.114346

42. N.K. Bourne, W.U. Mirihanage, M.P. Olbinado, A. Rack, C. Rau, Sci. Rep. **10**, 10366 (2020). https://doi.org/10.1038/s41598-020-67086-3

43. F. García-Moreno, P.H. Kamm, T.R. Neu, F. Bülk, R. Mokso, C.M. Schlepütz, M. Stampanoni, J. Banhart, Net. Commun. **10**, 3762 (2019). https://doi.org/10.1038/s41467-019-11521-1

44. M.A. Azman, C. Le Bourlot, A. King, D. Fabrègue, E. Maire, Mater. Today Commun. **32**, 103892 (2022). https://doi.org/10.1016/j.mtcomm.2022.103892

45. S. Diener, G. Franchin, N. Achilles, T. Kuhnt, F. Rösler, N. Katsikis, P. Colombo, Open Ceram. **5**, 100042 (2021). https://doi.org/10.1016/j.oceram.2020.100042

46. E.A. Zwanenburg, M.A. Williams, J.M. Warnett, Meas. Sci. Technol. **33**, 012003 (2022). https://doi.org/10.1088/1361-6501/ac354a

47. W. Ludwig, S. Schmidt, E.M. Lauridsen, H.F. Poulsen, J. Appl. Crystallogr. **41**, 32–309 (2008). https://doi.org/10.1107/S0021889808001684

48. C. Holzner, L. Lavery, H. Bale, A. Merkle, S. McDonald, P. Withers, Y. Zhang, D. Juul Jensen, M. Kimura, A. Lyckegaard, P. Reischig, E.M. Lauridsen, Microscopy Today, **24**, 34–43 (2016). https://doi.org/10.1017/S1551929516000584

49. P.J. Creveling, J. Fischer, N. LeBaron, M.W. Czabaj, Acta Mat. **201**, 547–560 (2020). https://doi.org/10.1016/j.actamat.2020.10.036

Chapter 12
Metal Melt Filtration in a Water-Based Model System Using a Semi-automated Pilot Plant: Experimental Methods, Influencing Factors, Models

Sarah Daus, Lisa Ditscherlein, Daniel Hoppach, and Urs A. Peuker

12.1 Water Model System

Water models have been applied in a large number of cases to investigate the metal melt flow [1–3]. In addition to lower costs and reduced experimental effort, a water model allows the application of optical measuring methods in the process as well as subsequent analysis methods like computer tomography. Water has a comparable dynamic viscosity to liquid aluminum (1.2 m·Pas at 725 °C), the viscosity of steel melts, however, is significantly higher (5.2 m·Pas at 1600 °C) [4]. The transfer of results is therefore almost exclusively limited to aluminum filtration. Since the impurities predominantly contained in aluminum melts are made of alumina, i.e. corundum and its modifications (mullite and spinel), alumina particles were primarily used as model inclusions. A hydrophobic organic coating, more precisely silane Dynasylan® F8261 from Evonik, Germany, was applied on the surfaces of the hydrophilic model solids (filter and particles) to achieve a similar wettability as in the real system [5]. With this method, contact angles of 104–105° can be measured on very smooth oxidic surfaces. For the experiments, ceramic foam filters (CFF) produced via the Schwartzwalder method were used [6]. These filters came from subprojects A01 and A02 or have been commercially purchased from hofmann ceramic GmbH as they exhibit very low variations in mass and associated strut thickness. To investigate the impact of filter properties, the generated filters differed in pore sizes and surface morphologies. In addition to conventional ceramic foam filters, novel filter structures like the "spaghetti" filters [7] and special filter setups like a packed bed of cubes or filter aids in the form of ceramic fibers were tested [8, 9].

S. Daus (✉) · L. Ditscherlein · D. Hoppach · U. A. Peuker
Institute of Mechanical Process Engineering and Mineral Processing, Technische Universität Bergakademie Freiberg, Agricolastraße 1, 09599 Freiberg, Germany
e-mail: sarah.daus@mvtat.tu-freiberg.de

© The Author(s) 2024
C. G. Aneziris and H. Biermann (eds.), *Multifunctional Ceramic Filter Systems for Metal Melt Filtration*, Springer Series in Materials Science 337,
https://doi.org/10.1007/978-3-031-40930-1_12

12.2 Water-Based Pilot Plant

The filtration experiments were performed at a semi-automated pilot plant that was developed and built at the Institute of Mechanical Process Engineering and Mineral Processing (Fig. 12.1). The plant consists of a stainless steel pressure vessel with a maximum capacity of 80 L. The model melt is prepared by injecting model inclusions into the prefilled vessel. The particles are dispersed in ethanol as a carrier liquid to inhibit agglomeration. The use of an inclined blade stirrer ensures homogeneous distribution of the particles inside the vessel. After starting the filtration test, the model melt passes through an approximately 1 m long vertical pipe section before entering the filter mount where the filter is positioned. Cylindrical filters with a diameter of 40 mm or cuboid filters with a cross-section of up to 50 × 50 mm and a maximum depth of 60 mm can be inserted. To obtain information on the local separation efficiency several thinner filters can be stacked. A filter cloth holder downstream of the mount enables the subsequent gravimetric analysis of the experiments. Flow rate can be set by using a reducer that is placed at the outlet of the pipe. For the flow rate estimation, the filtrate is collected by a weighing vessel. The filtration plant can be operated in three modes: no pressure, constant pressure and constant flow. Flow rate control was achieved by pressurizing the water inside the vessel. After each experiment, both the CFF and the filter cloth were removed and dried for the gravimetric analysis. The mass difference of the filter cloth $m_{f,c}$ can be assumed to be exclusively a result of the particles that passed though the filter without being separated. The filtration efficiency η can then be determined from the overall mass balance with Eq. (12.1):

$$\eta = \frac{m_{p,f}}{m_{p,f} + m_{f,c}} \cdot 100\% \tag{12.1}$$

where $m_{p,f}$ represents the mass of particles that is separated in the filter. Further, the pressure drop across the CFF is another important parameter in metal melt filtration. If the resistance to liquid *metal* flow during the mold fill is too high, there is a risk of melt freezing. The filtration plant was therefore extended by a differential pressure device. A detailed description of the pilot plant can be found in [10].

12.3 Gravimetric Analysis of Filtration

12.3.1 Influence of Filter and Particle Properties

With regards to the design of the model system, one major aim was to evaluate the influence of wetting behavior on the filtration performance. Heuzeroth et al. [11] carried out tests with Al_2O_3 particles with a size fraction of <200 μm and 20 ppi alumina filters at a lab-scale filtration plant. Different combinations of wetting

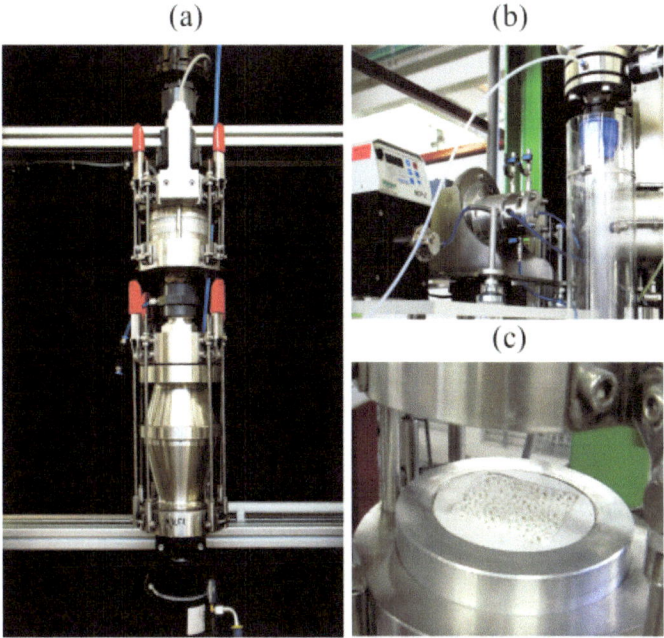

Fig. 12.1 **a** A detailed view of the filter mount and filter cloth mount; **b** micro bubble generator upstream of the filter mount; **c** filter mount with embedded CFF, adapted with permission from [10], 2019 John Wiley and Sons.

states were tested. The samples (p: particles; f: filters) were classified by their surface properties, i.e. either wetted (phil) or not wetted (phob) by the liquid. The results of the gravimetric analysis show a strong correlation between the wetting behavior and the separation performance. The highest mean filtration efficiency was measured for the fully hydrophobic system (p-phob|f-phob) (see Fig. 12.2). One reason for the distinct increase of the filtration performance in the hydrophobic systems is the change in the later described adhesion energy leading to higher attachment probability between solid surfaces after collision. As a result, particles have a higher probability to be deposited permanently after contact with the filter surface.

Moreover, there is a higher tendency to the formation of agglomerates when particle–particle collision occurs. The particle size distribution of the model impurities will consequently shift to larger sizes that can be separated more easily. This effect can be seen in the results for the partly hydrophobic system where only the particles were coated (p-phob|f-phil). The filtration efficiency is still significantly higher than for the cases where the particles are hydrophilic. The increase in filtration efficiency depends on size, shape, and the stability of the formed agglomerates. Because of that, higher standard deviations were observed for the systems where hydrophobic particles were present. Subsequent experiments were conducted in a fully hydrophobic system.

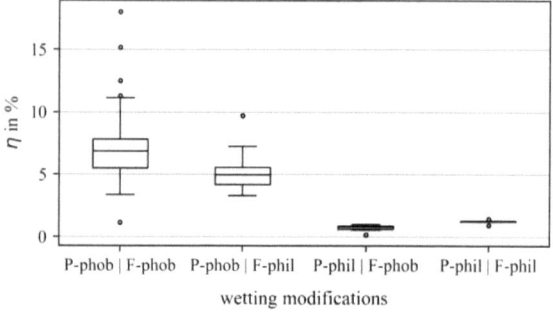

Fig. 12.2 Obtained integral filtration efficiencies as a function of the wetting behavior of the particle and filter medium, adapted with permission from [11], copyright 2015 Elsevier Inc.

Further, the influence of different filter properties was investigated. CFF exhibit large differences in the morphology, depending on the granular components contained in the ceramic slurry. These differences can be characterized by the roughness and influence the particle separation in the filter [12, 13]. The impact of surface roughness on the separation efficiency of the ceramic foam filters was investigated on the macroscale. Four filters with different coating materials were chosen and evaluated regarding their separation efficiency in the model system: rutile, alumina, spinel and mullite. The setup consisted of three stacked 20 ppi filters. Alumina particles with an initial particle size fraction (PSD) of 2–20 μm were used as model inclusions. When the filter material is selected to match the impurities in the real aluminium melt system, chemism influences the separation process. This phenomenology can only be seen to a limited extent in the water model, since only alumina particles are used for these investigations and the solid surfaces have been coated to achieve comparable wetting conditions. Results of the filtration tests show that with higher filter roughness the mean filtration efficiency increases. Thus, the filter coated with TiO_2 and a roughness of 0.3 μm yielded the lowest mean separation efficiency with 9.5% and the filter coated with mullite and a roughness of 1.1 μm the highest with 15.3%. To put the results into perspective, the filters were additionally evaluated regarding their pressure drop at a constant flow rate of about 23 cms^{-1}. All filter types except for the TiO_2 coated one exhibited a similar pressure gradient of about 12 mbar. With increasing roughness of the filter, slightly higher pressure gradients were measured. Only the TiO_2-coated filter had a much lower pressure gradient of 7.9 mbar (Fig. 12.3a) [14].

To isolate the influence of surface roughness, 30 ppi alumina filters with varying arithmetic average roughness between rms = 0.16–0.32 μm (scan area 4^2 μm^2) were tested in an identical setup (Fig. 12.3b). The same influence of roughness was observed than in the previous investigations involving the different coating materials. When the roughness value was doubled from 0.16 to 0.32 μm, the mean separation efficiency increased from 22.9 to 29%.

Regarding the filter structure, the impact of pore size was investigated by varying the ppi number. Filtration experiments were carried out on 10 and 30 ppi industrially manufactured filters (properties summarized in Table 12.1). For every experiment a

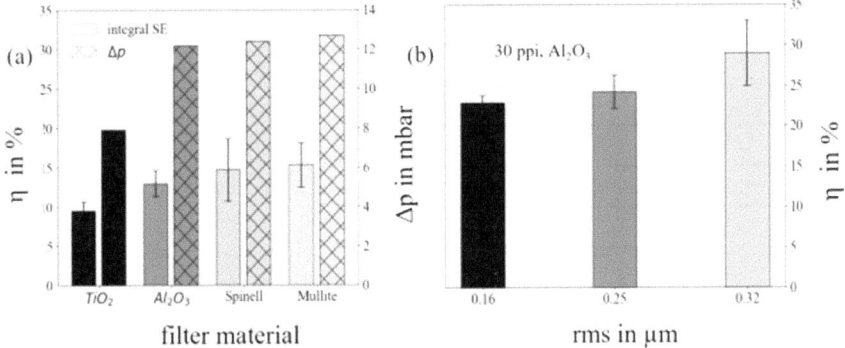

Fig. 12.3 a Integral filtration efficiencies and mean pressure drops at 23 cms^{-1} for 20 ppi filters of different materials; **b** separation efficiency of 30 ppi alumina with increasing roughness [14]

Table 12.1 Properties of the used CFF, adapted with permission from [10], copyright 2019 John Wiley and Sons.		10 ppi	30 ppi
	Single filter dimension in mm	Ø 40 × 13.5	
	Single filter mass in mm	7.5–7.6	
	Medium open porosity	0.77	0.79
	Medium pore diameter in mm	3.98	2.18
	Medium collector thickness in mm	1.48	0.67

stack of four identical cylindrical filters with a diameter of 40 mm and a height of approximately 13.5 mm each were used.

Alumina particles with two different size fractions were chosen as model impurities: A fine fraction (2–20 μm) and a coarser fraction (20–100 μm). Filtration tests were carried out under "constant flow" mode with a predefined flow rate of 3.2 cms^{-1}. Information regarding the state of agglomeration of the added particles was obtained by implementing an online optical measuring system QICPIC (Co. Sympatec GmbH) up and downstream of the filter mount. The setup consists of a liquid cuvette through which the suspension to be measured is pumped via a hose connection. The measuring cuvette contains a viewing window through which a pulsed, expanded laser beam is guided, which passes through a lens system in the measuring device and is detected by a CCD camera. The resulting PSDs upstream of the filter mount were determined at different times during filtration over a measurement period of one minute. The corresponding cumulative volume distributions are shown in Fig. 12.4. Comparisons with the initial PSD show a significant shift in the PSD of the fine fraction (2–20 μm). The size distribution of the coarse fraction (20–100 μm), however, almost remained constant.

Integral filtration efficiencies for both particle fractions and pore densities of the filters are shown in Fig. 12.5a. The reduction of pore size, represented by a higher ppi number, improves the filtration efficiency by a factor of 3 and 4 for the fine

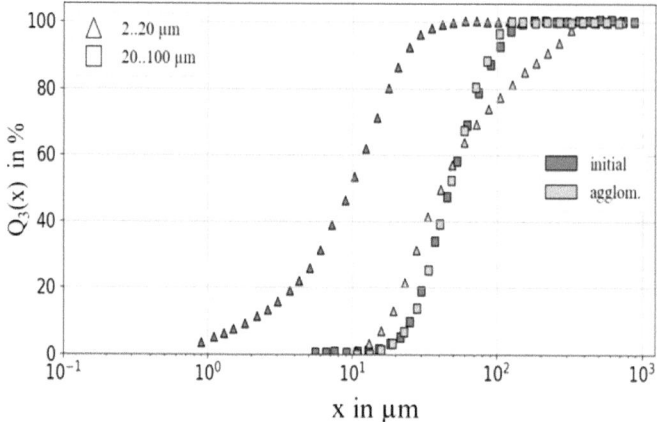

Fig. 12.4 PSD of the applied alumina fractions for the initial state and before entering the CFF (after 7 min), adapted with permission from [10], copyright 2019 John Wiley and Sons.

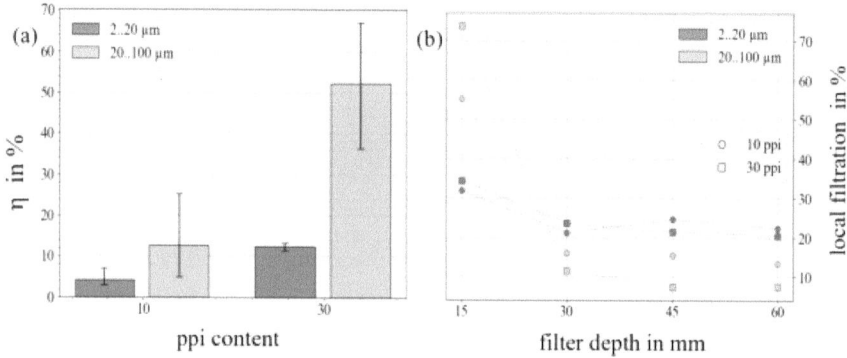

Fig. 12.5 **a** Integral filtration efficiencies for different pore and inclusion sizes with error bars representing the minimum and the maximum values obtained; **b** local filtration efficiencies, [14]

system and the coarse system, respectively. Interestingly, the enhancement of the filtration performance by increasing the primary particle size is in the same order of magnitude than for the reduction of pore diameter. Both effects can be attributed to the higher probability of contact between the particles and the filter surface. When only the integral separation efficiency is considered, the effect of agglomeration seems to be negligible. Even though there are only few differences in the impurity size distribution, the filtration performance is significantly higher for the coarse primary particles. The resulting PSD of 2–20 μm particles, however, still has a considerably higher fine fraction (80% of the particles are <20 μm) as opposed to the resulting 20–100 μm PSD (30% of the particles are <20 μm). Thus, for the 2–20 μm PSD, there is a higher number of small impurities in the model melt which are separated poorly and lead to lower separation efficiencies of the filter [10].

The results for the evaluation of the individual filters show that for all configurations the mass of separated particles was highest in the first filter and then decreased with filter depth (Fig. 12.5b). When the finer particles were used, however, only slightly more particles were separated in the first quarter of the filter depth than in the remaining parts. Additionally, for both pore densities a similar distribution of particles along the filter depth was observed. In case of the coarse particle system, more than half of the particles were deposited in the first filter. The percentage further increases with higher pore densities. Larger particles and agglomerates seem to impinge on the filter struts earlier due to increased interception, sedimentation and inertial effects.

The axial mass separation profiles obtained by the experiments were also compared to values predicted by the logarithmic deep filtration law. The filter coefficient was assumed to be independent of time and calculated using the integral filtration efficiency. While the deposition profile for the fine fraction could be predicted with the coefficient of determination $R^2 > 90\%$, deviations were significantly higher for the larger particles ($R^2 < 75\%$). The differences were mainly caused by the underestimation of particle deposition in the first filter. This indicates that for higher pore densities there is a superposition of depth filtration and surface filtration takes place in the first filter as a result of an inclusion buildup on the filter surface. This effect is more pronounced in the coarse particle system, since the proportion of fine particles <20 μm is significantly lower here.

The filtration of alumina particles (20–100 μm) with a 10 ppi CFF has also been simulated using a pore-scale numerical model by subproject B02 [10]. The model predicts an average volumetric filtration efficiency of 53.4%. This result is significantly higher than the experimentally obtained mean value of 12.7%. One potential reason for the differences is the redispersion of already deposited particles. This behavior is common in deep filtration processes where at first particles or agglomerates primarily get deposited directly on the filter struts. Once a particle layer is formed, the deposition occurs increasingly via the formation of particle aggregates leading to a constriction of pore size. The resulting increase in interstitial velocities leads to detachment of deposited aggregates that can get either redeposited in subsequent pores or get discharged from the bed [15].

12.3.2 Impact of Filter Geometry

The previous results indicate that a decrease in pore size affects the filtration performance significantly. Consequently, the results imply that using a filter with higher ppi number is favorable to separate a wide range of particle sizes. As de-scribed earlier, however, the danger of melt freezing must be considered. The filter in the clean state exhibits a resistance to the melt flow that is increased by deposition of solids in the pores. In the present investigation, the volume of deposits in the filter depth is low compared to the pore volume. Additional pressure loss per unit depth can be described as proportional to the local specific deposit. At the filter surface, however, large particles and agglomerates get trapped once they collide with the filter struts. This leads to a restriction in pore cross-section and can eventually result in local pore blocking. This additional pressure drop due to the blocking filtration is non-linear (Boucher's law) [16]. To keep the pressure gradient low, a more homogeneous distribution of deposits within the filter is desirable. This can be achieved by adjusting the filter geometry. Two different adjustments of the filter geometry were tested: graded filter structures and packed beds. Additionally, the use of ceramic fibers either as a top layer or encapsulated between two filters was investigated and will be discussed in a later chapter in detail. A summary of the filter set ups can be found in Table 12.2.

Graded filter structures have been employed in a wide range of deep filtration applications. The term 'graded' here refers to the pore size, which is typically graded to get smaller in the direction of fluid flow [17]. By combining different pore sizes, it is possible to separate coarse particles first without the risk of pore blockage and simultaneously separate fine particles in the filter depth.

In the case of ceramic foam filters, the influence of graded pore sizes is investigated by stacking thin filters. The objective is to maximize the relative filtration efficiency η_{rel} that is defined as Eq. (12.2):

$$\eta_{rel} = \frac{\eta}{\Delta p_F} \tag{12.2}$$

where Δp_F is the pressure gradient across the filter stack. Like before, Al_2O_3 particles with an initial PSD of 2–20 μm were used as impurities. All filter stacks for the experiments had a rectangular cross section with dimensions of 50×50 mm^2 and a total height of 60 mm. The modified filters were then compared to a reference assembly consisting of three 30 ppi filters of the standard dimensions of $50 \times 50 \times 20$ mm. The assembly is referred to as the "constant" setup and yielded a mean separation efficiency of 22.9% after three experiments. Two different approaches which increase the ppi number over depth were investigated. A "linear" set-up consists of one 20, 30 and 40 ppi filter, each with a depth of 20 mm. The second so called

Table 12.2 Design and results of the tested modified filter geometries, [14]

Set up	ppi-number	Depth in mm	Filtration efficiency in %	Pressure drop in mbar
Graded filters				
Constant	30/30/30	20/20/20	22.9	12
Linear	20/30/40	20/20/20	15.8	10
Exponential	10/20/30/40	20/10/10/20	21.5	8.8
Packed beds (b) on carrier filter (f)				
Pyramid layer	30(f)/–	20/20	10.2	8.1
Cube layer	30(f)/30(b)	20/40	18.2	4.6

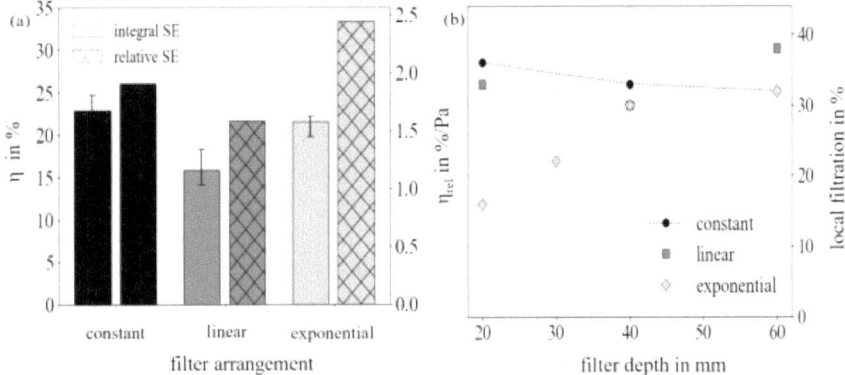

Fig. 12.6 a Integral and relative filtration efficiencies for different filter set-ups with error bars representing the minimum and the maximum values obtained; **b** local filtration efficiencies, [14]

"exponential" filter stack was a 20 mm 10 ppi filter followed up by two 10 mm filters with ppi-numbers 20 and 30 and finished off with a 20 mm 40 ppi filter.

Both set ups lead to a decrease in integral separation efficiency and pressure drop compared to the reference filter. When calculating the relative filtration efficiencies, however, the exponential filter alignment outperforms the reference set up with a constant ppi number (Fig. 12.6). The local distribution shows that with the exponential filter, in contrast to the other two filter alignments, an increase in the separated particles occurs with the filter depth due to the use of higher ppi values.

Two alternative filter concepts were developed and tested: a layer of pyramid-shaped ceramic bodies and a packed bed of ceramic 30 ppi cubes. Both were placed on a 30 ppi ceramic foam filter, the so-called carrier filter. Filtration performance was again compared to a reference filter configuration of two 20 ppi filter with a depth of 20 mm each. The configuration yielded a filtration efficiency of 15.9% at a pressure drop of 8.1 mbar. The height of the pyramid layer was adjusted to meet a reference filter set up. The dimensions of one cube were approximately $16.5 \times 16.5 \times 20$ mm and the number of cubes was selected to match the total volume of the filter material in the reference filter. Consequently, the packed layers of the two filter geometries have a significantly different porosity, which is also reflected in the pressure loss. While the pressure gradient of the filter with the cube bed drops to 4.6 mbar, the pyramid layer leads to an increase in pressure gradient from 8.1 to 10.6 mbar in comparison to the reference filter. Despite the larger pressure drop, the pyramids did not enhance the separation ($\eta = 12.7\%$). One reason for this is the lower surface area of the pyramids compared to the 20 mm filter. A further increase in the surface area by adjusting the number of pyramids is not feasible due to the pressure drop. Using the bed of ceramic cubes, however, led to an increase in filtration efficiency to 18.2% [14].

All the previous filter modification had in common that the basic structure of the ceramic filters remained the same but either the geometry was changed, or filtration aids were added to the filter. One filter geometry that has a completely different structure is the so-called spaghetti filter that was developed by subproject A01 [7]. When producing a spaghetti filter with the same outer dimension as the previously used CFF (50 × 50 × 20 mm), it has a mean surface area of 377 cm^2 which is between that of a 20 ppi (237 cm^2) and a 30 ppi (520 cm^2) filter. Regarding the filtration performance, a single spaghetti filter achieved a mean separation efficiency of 3.6% for the same process conditions as before, whereas the single 20 ppi and 30 ppi filters yielded a separation efficiency of 5.3% and 10.7%, respectively. The decline in filtration performance can be explained by the grid-like structure of the spaghetti filter. Since the cord is laid parallel to each other in layers and each layer is rotated 90° to the previous layer, the tortuosity of the filter is low and only the first and second layers of the struts are directly exposed to the flow [14].

12.3.3 Impact of Microbubbles on the Filtration Process

Besides particulate impurities, gaseous inclusions also occur in aluminum melts. Flotation processes involving micro bubbles (MB) for the purification of melts are reported in literature [18, 19]. Whether this positive impact of microbubbles can also be seen in the filtration process was investigated in the room temperature model system as well. For this purpose, a micro bubble generator was positioned in the pilot plant about 1 m upstream of the filter mount. With the help of a membrane, bubbles in the mid two-digit to mid three-digit micrometer range can be generated. Evaluation of the experiments was performed by means of gravimetric analyses. Even though microbubbles cannot be measured gravimetrically, general tendencies regarding their influence on filtration can be concluded from the results. As a reference experiment, a commercial 30 ppi filter was used to clean the model melt contaminated with Al$_2$O$_3$ particles in the range of 10–20 μm initial PSD at a flow rate of 8.8 cms^{-1}. For this set up, a mean filtration efficiency of 14.8% was achieved. Once microbubbles were added to the system the filtration efficiency dropped to 8.5%. While the separated mass inside the filter remained at about 0.2 g, almost twice as much particulate matter was found in the filter cloth. The additional mass in the filter cloth indicates that the bubbles collect solid impurities that would otherwise not have entered the filter but would have been deposited on the inside of the pipe. Instead of separating the impurities via flotation effects, the heterocoagulates formed are transported into the filter by the flow. However, they are separated more poorly in the filter due to their lower overall density. Measurements with the QICPIC system were also performed to allow an estimation of the separation efficiency based on the volume of the measured microbubbles or heterocoagulates. For this purpose, the model melt was selectively contaminated either solely with particles or additionally with microbubbles. Examples of the obtained projections can be seen in Fig. 12.7. Comparisons between the volumes of the detected objects before and after the filter showed that once microbubbles were added, volume of impurities increased by a factor of 100 in the initial melt. Even though the formation of heterocoagulates led to a decrease of the defect volume

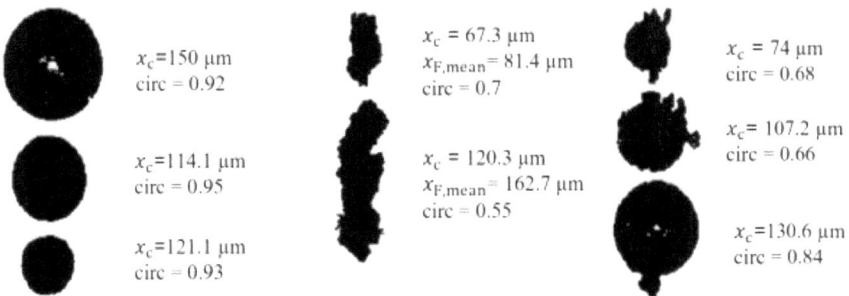

Fig. 12.7 Selected projections of the impurities (from left to right: MB, agglomerates, and heterocoagulates) with their dimensions (equivalent circle diameter: x_c, mean Feret diameter: $x_{F,mean}$) and circularity (circ), with permission from [14]

by 96.6%, the remaining volume of impurities was still about 3.3 times higher than the pure addition of particulate impurities. Further, the bimodal size distributions of the impurities before the filter for the addition of microbubbles indicate that not all particulate impurities attach to a bubble. This is due to the laminar flow condition in the measuring section where the microbubbles are introduced into the system leading to a reduced contact probability. Therefore, it can be concluded that for the real melt filtration process, after treatment of the melt with purge gases, any bubbles in the size range of 250 μm diameter, which also bind particulate contaminants, are separated with a probability of over 90% in a 30 ppi filter. But the remaining MB or heterocoagulates still significantly increase the defect volume in the casting compared to a melt contaminated only with particles [14].

12.4   CT Analysis of a Loaded CFF

The gravimetric analysis enables experiments with different parameters to be compared in terms of separation efficiency. However, it can only give axial information about the particle deposition and is limited to the thickness of one filter. More detailed insight regarding the position of the deposits on the inner filter surface was achieved by performing a CT scan of a 10 ppi filter loaded with 20–100 μm alumina particles. As the model impurities and the filter are both made of the same material, Al_2O_3, and therefore have the same attenuation coefficient, it was challenging to distinguish between the separated particles and the filter matrix. The visualization of the deposition of the alumina particles was achieved by conducting two measurements: an "empty measurement" of the unloaded filter and a measurement of the same loaded filter after the experiment. In a subsequent step, the loaded geometry can be subtracted from the unloaded one, leaving an image stack with the alumina particles deposited on the filter surface. Comparisons between the locally deposited particle masses over the filter depth examined in the CT and by gravimetric evaluation show good agreement. In the gravimetric analysis, an average of 0.266 ± 0.125 g of deposited impurities was determined, which deviates only by 6.5% from the deposited mass of 0.249 g of the filter examined in the CT. If the filter is divided into four sections, as in the gravimetric evaluation routine, the locally deposited masses are within the range of the gravimetric values. Figure 12.8 gives an impression of the particle deposition on the filter surface. 20% of the determined particles are present in large clusters. With increasing filter depth, the size and number of these clusters decreases. This confirms the results of the gravimetric analysis implying that large agglomerates are separated mainly in the first quarter of the filter. A high separation efficiency at a standardized filter depth of approx. 25, 50 and 75% is due to the filter preparation involving the stacking and gluing of the filters. The resulting small gap leads to a flow relaxation in the transition from one filter to the next, as the local porosity increases up to 95% [10].

The impact of the direction of inflow can be investigated assuming an idealized filter strut with a circle diameter d_s by subdividing the surface into angular area fractions and visualizing the mass of deposited particles over an area-normalized angle between 0° and 180°. Here, 0° corresponds to orthogonal deposition of the particles on the upper side of the strut in flow direction, 90° a tangential separation parallel to the direction of inflow and an angle of 180° of orthogonal deposition on the underside of the filter strut. The depiction of depositions via the angle classes shows that the particles are preferentially deposited on the upper side of the strut (Fig. 12.9). The mass of impurities deposited on the filter surface has a plateau between approx. 30° and 66° at which on average 4% of the deposited particles are present per angle class. Subsequently, the local particle masses per angle class drop until reaching a local maximum at 90°. In the further course, a maximum of 1% is locally reached on the bottom side of the strut. Due to the low separation efficiency of $12.7 \pm 7.61\%$ for the 10 ppi filter (Fig. 12.5), very low particle separation probabilities of maximum 0.66% result for an angle class division of 4°. If a distinction is only made between

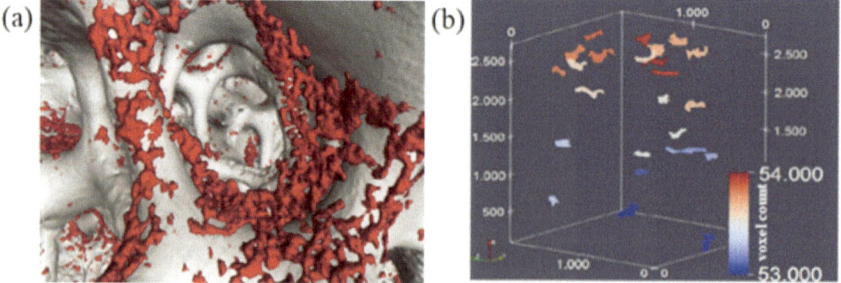

Fig. 12.8 **a** Section of the scanned 10 ppi filter loaded with alumina particles with view from the inflow direction; **b** illustration of the largest deposit clusters over the filter depth with an average individual mass of approx. 0.0017 g. The impurities are represented colorimetrically via a voxel count (voxel edge length of 20 μm), (**a**) with permission from [10], copyright 2019 John Wiley and Sons., (**b**) with permission from [14]

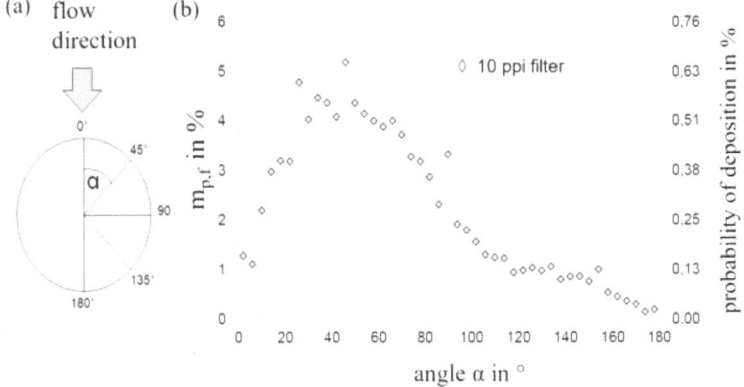

Fig. 12.9 a Representation of the idealized filter strut with chosen angular separation; **b** percentages of the separated masses at the idealized strut. The angle classes are each comprised of 4° with the class center indicated in the graph, with permission from [14]

the top and bottom of the filter strut, the probability of separation is 10.14% on the upper and 2.56% on the lower side [14].

12.5   Viscous Force

In order to investigate the depth filtration process in more detail, the single collector approach can be investigated numerically. Here, only one strut of the complex filter medium is considered and the influence of the characteristic flow on the separation of inclusion particles is estimated. The unsteady flow field is calculated by using the Lattice-Boltzmann method, the strut is assumed to be infinitely long and reduced to a 2D problem, and also, inlet velocity and outlet pressure are kept constant. The particle trajectories are computed by solving the equation of motion in the Langrangian framework; particle–particle interactions and redispersion of separated particles are neglected [20]. Heuzeroth et al. [21] obtain initial data sets for the separation model for different parameter sets (e.g. particle size, main fluid velocities) in which, for example, impact angles and impact velocity vectors are given for each particle. This enables the impact velocity normal to the collector surface u and the impact efficiency to be calculated. However, the hydrodynamic inhibition must also be taken into account, since energy is required to push the fluid phase out of the contact zone. In contrast, the kinetic energy of the particle is acting, which must be correspondingly high in order to allow the particle with radius R to be separated. To additionally consider roughness effects and thus a disturbance of the liquid lamella, the surface roughnesses of inclusion particles and filter surface are included in the calculation of the critical distance (h_{min}, film rupture) in addition to the previously determined minimum distance. The energy balance from viscous energy E_{vis} and kinetic energy of the particle E_{kin} is calculated numerically (Eq. 12.3):

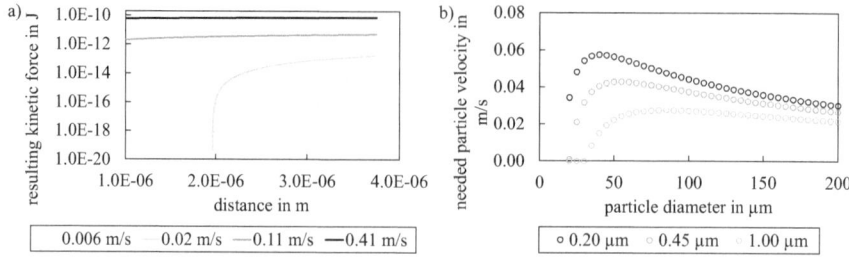

Fig. 12.10 a Resulting kinetic force profiles for different velocities by variation of the distance of a 75 μm particle; **b** needed particle velocity for impact depending on particle size and rms values, adapted with permissions from [21], copyright 2015 Elsevier Inc.

$$E_{vis} = \int_{h_1}^{h_{min}} 6\pi R^2 \upsilon_L \frac{u}{h} dh$$

$$E_{kin} = \int_{h_1}^{h_{min}} \frac{4}{3}\pi R^3 \rho u du \qquad (12.3)$$

υ_L is the kinematic viscosity and ρ the particle density. If the film rupture criterion is reached, the particle is considered to be stably deposited onto the filter strut. If the energy balance, i.e. resulting kinetic energy, falls below 10^{-20} J, it is considered to not have been separated. Figure 12.10 shows the curves of the resulting energy taking into account viscous force as a function of the particle velocity as well as the required particle velocity for different diameters and roughnesses.

In all cases the resulting kinetic energy decreases with decreasing distance, but for the two lower velocities it does not reach the film rupture criterion, i.e. the particle (here 75 μm) is not deposited onto the filter strut. At higher velocities, a film breakage occurs with impact energies of $>10^{-12}$ J on the filter surface. If the roughness of the collector surface increases, the particle size that is not inhibited also rises, since h_{min} is linearly proportional to the roughness. In the case of an influence of viscous force, the required output velocity normal to the collector first increases with larger particle size and then decreases again with further enhanced particle size. The reason for this is the proportionality of the kinetic energy and the viscous force to the particle diameter. If the roughness of the particle and collector increases, the fluid film between the two is more likely to rupture and the required particle velocity decreases.

12.6   Microscopic Scale—Force Measurement and Force Simulation

The atomic force microscope (AFM) is usually the tool of first choice for investigations on the microscale with regard to adhesive forces. With this measuring device, it is possible to scan topographies of the rough surfaces (also used to describe the minimum distance for the calculation of the viscous force) and to carry out force spectroscopic measurements, whereby parameters such as wettability, gas supersaturation or roughness can be varied. Here, too, the water-based model system is used, since the properties of the molten metal (e.g. opacity or temperature) do not permit measurements under real conditions. Spherical, but rough aluminum oxide particles from Denka, Japan with sizes between 15 and 40 µm are used as model inclusion particles. In order to investigate different filter materials, slurries also used in the subprojects A01 and A02 of CRC920 for the manufacturing of ceramic foam filters via Schwartzwalder process were filled into tablet molds, dried and then coked or sintered. The compositions and production parameters can be found in the related publications [22–24]. Schemes of the main modes used as well as exemplary pictures of a colloidal probe cantilever and filter material samples can be found in Fig. 12.11.

To guarantee wetting properties as in the real process and thus to be able to investigate specifically occurring adhesive force effects, alumina particles and filter material samples were hydrophobized with a silane (Dynasylan® F8261), whereby contact angles between 110 and 133° were obtained. These compared to smooth coated oxidic surfaces higher values (104°) are due to additional effects, enhanced by asperities and pores (see Fig. 12.12): Microscopically visible bubbles but also small cap-shaped nanobubbles of various sizes are detected during immersion with the liquid or through gas supersaturation due to temperature change. These bubbles can cause capillary forces when inclusion particles or particle-filter material contacts occur.

It is known from the literature that the three parameters wettability, roughness and gas supersaturation are interconnected [28], but to date it has not been possible to adequately and precisely model the behavior in particle–particle or particle–wall

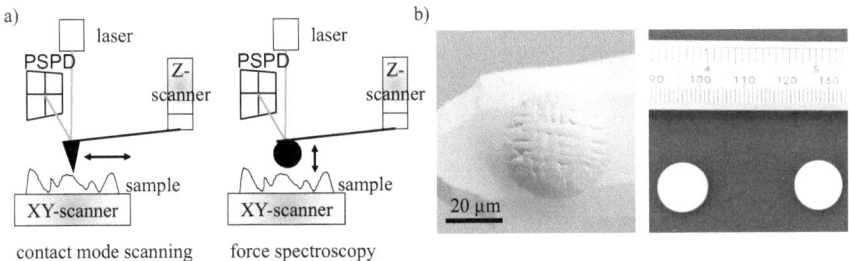

Fig. 12.11 a Schemes of the two used AFM modes (PSPD: position sensitive photo detector); **b** examples of a colloidal probe and filter material samples, [25, 26]

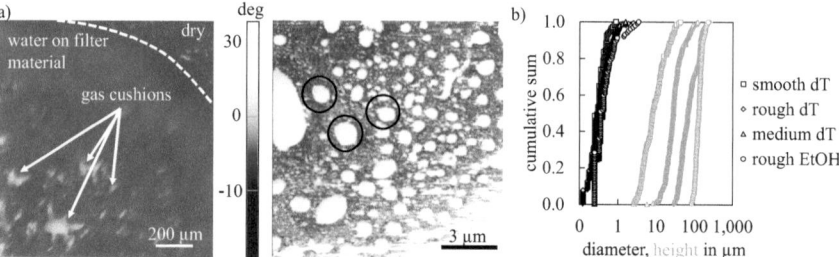

Fig. 12.12 **a** Exemplary images of microscopically visible bubbles; and nanobubbles on an alumina particle, [25], adapted with permissions from [27], copyright 2017 Elsevier Inc.; **b** size distribution for a variety of rough surfaces and gas oversaturation states, [25]

interactions, since common adhesion force models often make simplifications that are only valid for a specific system and that lead, for example, to a significant deviation from the experiment in the case of a change in particle size distribution or material of the filters [23]. It is further problematic that usually only a single adhesion force value is calculated, although literature sources as well as own measurements between particle and particle or particle and filter material clearly show that the values are distributed, for example due to the locally variable roughness [27, 29, 30]. If several adhesion mechanisms are involved, the modelling becomes even more complicated.

It can be assumed that in both model and real system the two main adhesion mechanisms are attractive van der Waals (vdW) interactions and capillary forces [31]. An operating point close to the isoelectric point of the system eliminates a significant contribution of electric double layer forces in the water-based model system, which ultimately leads to better transferability. Van der Waal's interactions are of universal importance, since they act between any combination of molecules and surfaces and can also be expected to take effect in metal melts (here the melt is the intermediate medium) [28]. Using Hamaker's approach, the more precise description by Lifshitz and the simplifications by van Kampen, Ninham and Parsegian by introducing an imaginary response function $\epsilon(i\zeta_n)$, it is possible for ideal geometries to calculate the van der Waals force very precisely with the system specific Hamaker constant [32–34]. The necessary equations, which will also be used later for modelling, are listed below (Eq. 12.4). The zero-frequency component of the Hamaker constant A_H is multiplied by 0.5.

$$A_H = \frac{-3}{2} k_B T \sum_{n=0}^{\infty}{}' \int_{r_n}^{\infty} x \ln\left\{\left[1 - \Delta_{13}\Delta_{23}e^{-x}\right]\left[1 - \overline{\Delta}_{13}\overline{\Delta}_{23}e^{-x}\right]\right\} dx$$

$$\epsilon(i\zeta_n) = 1 + \frac{B}{1 + \zeta\tau} + \sum_{m=1}^{N} \frac{C_m}{1 + \left(\frac{\zeta}{\omega_m}\right)^2 + \frac{g_m\zeta}{\omega_m^2}}$$

$$\zeta_n = \frac{2n\pi k_B T}{\hbar}$$

$$r_n = \frac{2L\zeta_n\sqrt{\epsilon_3}}{c}$$

$$\Delta_{jk} = \frac{\epsilon_j s_k - \epsilon_k s_j}{\epsilon_j s_k + \epsilon_k s_j}$$

$$\overline{\Delta}_{jk} = \frac{s_k - s_j}{s_k + s_j}$$

$$s_k^2 = x^2 + \left(\frac{2L\zeta_n}{c}\right)^2 (\epsilon_k - \epsilon_3) \tag{12.4}$$

Accordingly, only the optical properties of the materials involved are needed (oscillator strength B, damping coefficient τ in the microwave range as well as damping coefficients g_m, characteristic absorption frequencies ω_m and strengths C_m in the IR and UV range, respectively). The dispersion relations in the integral contain the functions Δ_{jk} and $\overline{\Delta}_{jk}$, which define the differences in the dielectric properties of two materials. s_k^2 is an additional damping term describing retardation effects, c represents the speed of light in vacuum, $\hbar$ the reduced Planck constant and L the distance. Since coatings such as the used silane have a significant influence on the Hamaker constant, these must also be taken into account, which is done via the Clausius-Mosotti equation, see Eq. 12.5, and is described in detail by Dagastine et al. [35, 36].

$$\epsilon(0)_{mix} = \sum_i \Phi_i \epsilon(0)_i$$

$$\epsilon(\omega)_{mix} = \frac{1 + 2\sum_i \Phi_i \left(\frac{\epsilon(\omega)-1}{\epsilon(\omega)+2}\right)_i}{1 - \sum_i \Phi_i \left(\frac{\epsilon(\omega)-1}{\epsilon(\omega)+2}\right)_i}$$

$$\Delta_{c_N 1}(c_1, c_2, \ldots, c_{N-1}) = \frac{\Delta_{c_N c_{N-1}} + \Delta_{c_N 1}(c_1, c_2, \ldots, c_{N-2})e^{-t_{N-1}s_{N-1}/L}}{1 + \Delta_{c_N c_{N-1}}\Delta_{c_N 1}(c_1, c_2, \ldots, c_{N-2})e^{-t_{N-1}s_{N-1}/L}} \tag{12.5}$$

Equations for ideal systems that are relevant for the calculation of forces are listed below with particle radius $R = \frac{D}{2}$:

$$F_{vdW}^{sphere-plate} = \frac{-A_H R}{6L^2}$$

$$F_{vdW}^{plate-plate} = \frac{-A_H}{6\pi L^3} \tag{12.6}$$

Capillary forces in the model and real system are caused by small cap-shaped bubbles, so-called nanobubbles. These have mainly been investigated on very smooth, quite homogeneous surfaces; there have only been a few studies on rough surfaces such as ceramic foam filters or inclusion particles. The bubbles can be generated by incomplete immersion with the liquid or melt, by perturbation, by gas supersaturation, e.g. by a temperature gradient, or in steel melt filtration by the formation of CO bubbles during the carbothermal reaction. In the model system, nanobubbles caused by incomplete immersion or gas supersaturation as well as subsequently occurring capillary forces have been investigated so far [22, 37–39]. Even though the quantitative detection of nanobubbles on rough surfaces is more difficult than on smooth surfaces, the bubbles located in roughnesses and pores show a significantly higher stability against detaching from the surface, which ultimately has a positive effect on the separation of the inclusion particles [22]. The calculation of capillary forces caused by small bubbles is difficult due to the strongly varying sizes compared to the model assumptions, a possible higher capillary number due to several small bubbles in the contact area as well as the metastable gas state. Usually, conventional capillary force models are used which differ in their complexity and these are compared with individual measurement curves. The models differ primarily in whether a pressure change is considered and how the capillary is approximated. The pressure change is negligibly small; most models use the toroid approximation [40]. Attard provided the first capillary force model, which was also used in subsequent work, such as the Fritzsche model (Eq. 12.7) [41, 42]:

$$F_{\text{capillary}}^{\text{Attard}} = -\pi r_s p_0 - 2\pi r_s \gamma$$

$$r_s = \frac{3\gamma}{2p_0}\left[1 - \sqrt{1 + \frac{8Rp_0 cos\theta}{9\gamma}}\right] \tag{12.7}$$

Important variables here are the surface tension γ, the contact angle θ, the pressure p_0 and the capillary radius r_s.

If these idealized equations for van der Waals and capillary forces are used to predict the adhesive forces between particulate inclusions and filter material, significant deviations are obtained. In both cases, the adhesive forces calculated in this way are far too large and do not reflect reality. While the wettability is included in the calculations by adjusting the Hamaker constant or the contact angle, the roughness has so far been ignored. The example shown in Fig. 12.13, however, shows a significant influence of this roughness depending on two wetting states.

Under good wetting conditions ("hydrophilic", Fig. 12.13a, i.e. exclusive occurrence of vdW interactions, which can also be assumed to be given by an absence of snap-in events, the median values of the measured normalized adhesive force

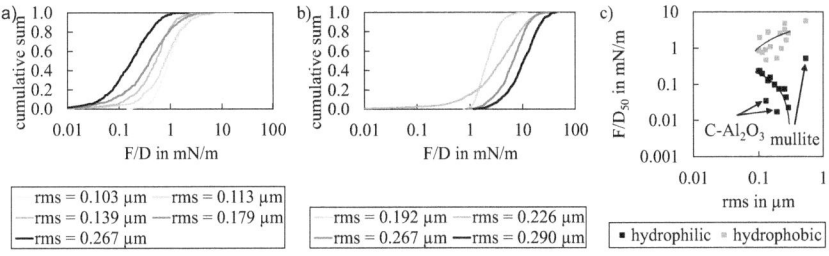

Fig. 12.13 Force distributions for **a** the hydrophilic case; **b** the hydrophobized system; **c** corresponding median normalized forces depending on roughness value rms, [25]

decrease with increasing roughness parameter rms (scan area 2.5^2 μm^2). The roughnesses shown here are significantly larger than those stated in models up to now, but are particularly relevant in the field of engineering. The reason for the decrease in adhesive force with higher rms is a reduction in the contact area and thus an increase in the distance between the model inclusion particle and the filter material sample. If the median value of the normalized adhesive force distribution is plotted against the rms values, a lognormal relationship is obtained. The three outliers AC1, AC95-5 and mullite are due to significant deviations of the Hamaker constants, since, for example, the influence of carbon is not considered in this representation. Instead, a contrary picture emerges when nanobubbles and thus capillary forces occur in addition to vdW (Fig. 12.13b). The cause here can be seen in the superposition by capillary forces, whereby it can be assumed that the stability, size and degree of coverage with nanobubbles increases with increasing roughness. This could be shown on smooth as well as rough surfaces (Si-wafer, filter material, carbon-bonded filter material). In both cases, there are not only single adhesion force distributions, but also overlapping areas. These result from differences in the local roughness (so there would also be a roughness distribution on the surfaces) and the locally different availability and accessibility of the nanobubbles, which are sometimes strongly pinned. A simple fitting of the normalized adhesive force distributions with common mathematical distributions is in most cases significant, i.e. associated with large deviations.

For filter surfaces with normally distributed z-values, Fritzsche's model can be applied as a first approximation, which is similar to Hoffmann's model [43, 44]. Here, a simple description of the roughness via a spherical segment on the plate is assumed. The roughness radius r_A has to be significantly smaller than the particle radius, the distance much smaller than particle radius and roughness radius and the Derjaguin approximation has to be applied. For vdW forces as well as hydrophobic interactions the result is (Fig. 12.14).

$$F_{vdW}^{Fritzsche} = \frac{-A_H}{6}\left[\frac{r_A}{L^2} + \frac{\sqrt{r_A^2 - c_A^2}}{\left(r_A + L - \sqrt{r_A^2 - c_A^2}\right)^2}\right.$$

$$\left. + \frac{\sqrt{R^2 - c_A^2}}{\left(r_A + L + h_A - \sqrt{R^2 - c_A^2}\right)^2}\right]$$

$$F_{hydrophobic}^{Fritzsche} = -2\pi \, w_{phob} e^{\frac{-L-L_0}{\lambda}}\left[\left(\lambda - \sqrt{r_A^2 - c_A^2}\right)e^{\frac{-r_A - \sqrt{r_A^2 - c_A^2}}{\lambda}}\right.$$

$$\left. - (\lambda - r_A) - \left(\lambda - \sqrt{R^2 - c_A^2}\right)e^{\frac{-R - \sqrt{R^2 - c_A^2}}{\lambda}}\right] \tag{12.8}$$

where λ corresponds to the decay length, w_{phob} to the cohesion energy between hydrophobic solid and the liquid, as well as the minimum distance L_0 (0.156 nm). Hydrophobic force is understood here as an additional force due to structuring of water molecules, i.e. not capillary forces. The model presented later shows, however, that there is another, more plausible explanation for this. Hamaker constants from the literature are used. In the Fritzsche model, the relationship between z-values and rms roughness is exploited. Assuming that the radius of the roughness peaks corresponds approximately to the rms roughness, a distribution $f(x)$ of the roughness radii can be calculated with the aid of a normal distribution. The mean asperity angle $\bar{\alpha}$ can be estimated from line profiles; a similar relationship $g(x)$ results, see Eq. (12.9).

$$f(x) = \frac{1}{2}\left[1 + erf\left(\frac{r_A}{rms\sqrt{2}}\right)\right]$$

$$g(x) = \frac{1}{2}\left[1 + erf\left(\frac{\alpha - \bar{\alpha}}{\sigma\sqrt{2}}\right)\right] \tag{12.9}$$

Here, α is the asperity angle and σ the standard deviation. Note that due to the limitations of the Derjaguin approximation, only angles between 0–90° can be applied. Using this method, it is possible, based on AFM topography images and the assumptions made, to then apply these distributions to calculate the modelled adhesive force distribution. Together with experimental results (AFM force spectroscopy), the calculations for hydrophobic and van der Waals are shown in Fig. 12.15.

It can be seen that the model for hydrophobic forces is in the same order of magnitude compared to the experimental data, but there are very strong deviations for the vdW forces. Since the experimentally determined value is used in the calculation of the hydrophobic force, the significance of imprecise model assumptions can be

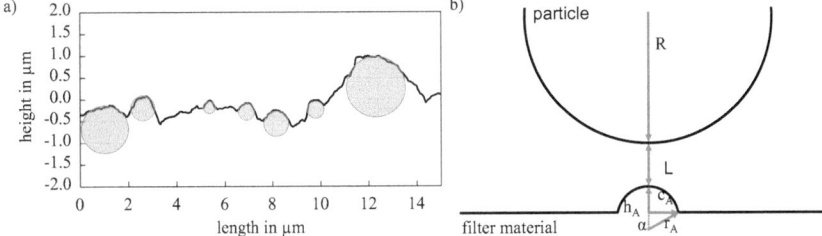

Fig. 12.14 a Line profile of a scanned filter material surface; **b** scheme of the model geometries of particle and surface with cap-shaped roughness, adapted with permissions from [43], copyright 2016 Elsevier Inc.

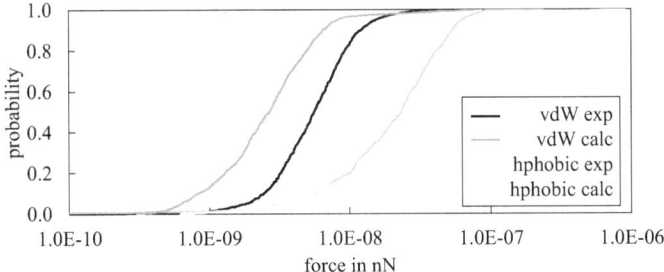

Fig. 12.15 Force distributions measured via AFM of the hydrophilic (vdW exp) and the hydrophobic (hphobic exp) system as well as via Fritzsche's model calculated forces, adapted with permissions from [43], copyright 2016 Elsevier Inc.

reduced. However, for the influencing factors roughness or contact area as well as Hamaker constant, the simple assumptions remain, which are often not valid in this way. An example of this is given in Fig. 12.16a, where also the size dependency of roughness parameters can be seen.

Often there is more than one contact point, which means that the Derjaguin approximation is no longer valid. Also, most of the topography images of the filter material surfaces show significance when tested for normal distribution (likewise lognormal and Weibull distribution), which limits the application of Fritzsche's model. One idea, which originally goes back to Cooper [45], is the virtual contact of two interacting surfaces and the subsequent summation of the pixel pair forces. While Cooper simulated the two surfaces until they matched the experimental values, the model developed in CRC920 uses the AFM topography data of the two surfaces (filter material and model inclusion particles). In a first step, the Dagastine layer model described above is used to determine the correct Hamaker constants, i.e. taking into account silane layers, distance-dependent retardation and—instead of a hydrophobic force—the inclusion of very thin air layers, also known as a water depletion layer. Then the topography data is imported, both surfaces are virtually contacted and the pixel-pair vdW forces are calculated and summed up with the precisely determined,

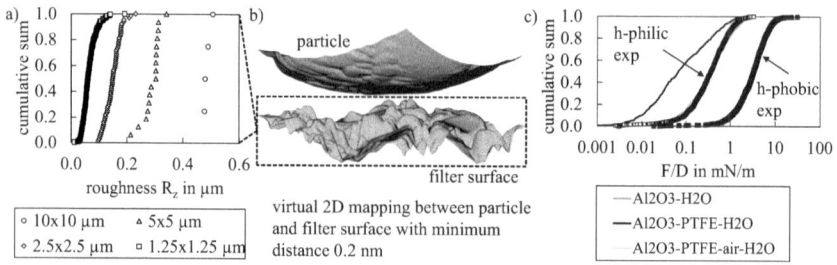

Fig. 12.16 **a** Example of dependence of scan size on a roughness value, here Rz; **b** AFM topography scans of both interacting surfaces; c measured force distributions with corresponding calculated vdW force distributions. Clearly visible is an excellent matching between calculated forces for the hydrophilic system (Al_2O_3-H_2O-Al_2O_3) and the experimental data, but no/only poor matching fir the hydrophobic case (Al_2O_3-PTFE-H_2O-PTFE-Al_2O_3, Al_2O_3-PTFE-air-H_2O-air-PTFE-Al_2O_3), [25, 26], **a** with permission from [10], copyright 2019 John Wiley and Sons.

distance-dependent Hamaker constant. A virtual 2D mapping is used to generate an adhesive force distribution. Figure 12.16c shows that the calculated vdW force data for the hydrophilic surfaces agree very well with the experimental one. In the hydrophobic case, contributions from the silane coating without an air layer are hardly to be expected, since the adhesion force distribution is significantly smaller than the measured values. An incomplete silanization (i.e. the "hydrophilic case") is conceivable, but the positive influence of air layers on the adhesive force is much more significant. Again, it should be mentioned that so far these are exclusively vdW force contributions. The distribution of the experimental results, which lies significantly further to the right at higher values, shows that the capillary forces of the nanobubbles very strongly increase the adhesive forces between particles and filter material (and thus ultimately the separation efficiency). Using the adhesive force data for which a snap-in was detected at the same time, the resulting distribution is then more to the right, so that it can be assumed that there is a further mechanism in addition to vdW and capillary forces due to nanobubbles already present. This can be caused by capillary forces during perturbation or by nanobubbles that are difficult to access, but cannot be detected by measurement so far. With the help of the measurement data and the calculated vdW distributions, the distribution of the capillary forces due to perturbation can also be obtained after estimating the ratios [25].

In order to find the cause for the discrepancy of the capillary force models, AFM topography scans are also used, because nanobubbles could also be detected on the filter material samples. For the sake of completeness, it is also possible to detect the bubbles on the particles. The AFM topography scans can be manipulated in such a way that the bubbles on them can first be removed, the particle and now bubble-free surface are virtually contacted, the bubbles are subsequently added again and capillaries can thus be estimated. This is shown as an example in Figure 12.17.

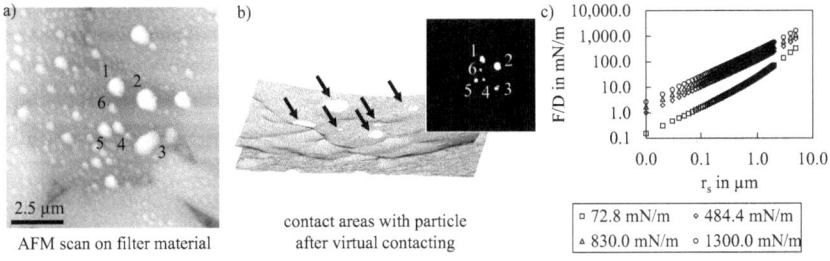

Fig. 12.17 a AFM topography scan of filter material covered with nanobubbles; **b** contacts of particle and bubbles after virtual contacting of particle and filter surface; c dependance of normalized capillary force on capillary radius using Attard's model, [25]

$$V_{\mathrm{NB}} = \pi r_s^2 L - \frac{2}{3}\pi\left[2R^3 - \left(2R^2 + r_s^2\right)\sqrt{R^2 - r_s^2}\right]$$ (12.10)

With the help of Attard's approach, it is not only possible to calculate the contact radius (which could then also be used to estimate the coalescence of bubbles), but also to determine the capillary radius.

The capillary radii are in the range between 4 and 45 nm, which corresponds to $\leq 1/500$ of the cylinder radii of classical capillary force models. In the example with multiple contacts shown, target searches yield $r_{s,i} = (0.056; 0.085; 0.078; 0.031; 0.041; 0.009)\mu m$ and thus $F/D = (0.89 + 1.36 + 1.25 + 0.48 + 0.65 + 0.14)\mathrm{mN/m} = 4.77\,\mathrm{mN/m}$, which fits very well with the experimental results.

It has thus been demonstrated that the two main adhesion mechanisms can be modelled quite precisely. However, a direct transfer to the real process is still connected with some obstacles, since e.g. the data situation for Hamaker constants of ceramic–metal melts is very thin or the results cannot be verified by measurements under real conditions.

Acknowledgements The authors would like to thank Ralf Schünemann, Steffen Scholz, and Thomas Hantusch for their support in the planning and setting up of the filtration plant. Furthermore, we thank Ralf Ditscherlein and Erik Löwer for the CT measurements. The authors gratefully acknowledge the German Research Foundation (DFG) for supporting the Collaborative Research Center CRC 920 (Project ID 169148856–subproject B01).

References

1. F. Acosta G, A. Castillejos E, J. Almanza R, et al., Metall. Mater. Trans. B **26**(1), 159–171 (1995). https://doi.org/10.1007/BF02648988
2. S. Akbarnejad, L.T.I. Jonsson, M.W. Kennedy et al., Metall. Mater. Trans. B **47**(4), 2229–2243 (2016). https://doi.org/10.1007/s11663-016-0703-0

3. A. Asad, K. Bauer, K. Chattopadhyay et al., Metall. Mater. Trans. B **49**(3), 1378–1387 (2018). https://doi.org/10.1007/s11663-018-1200-4
4. R. Brooks, A. Day, R. Andon et al., High temperatures-high pressures **33**(1), 73–82 (2001). https://doi.org/10.1016/10.1068/htwu139
5. Evonik Industries AG, Product information–Dynasylan F 8261. www.dynasylan.com. Accessed 2022
6. K. Schwartzwalder, H. Somers, A. V. Somers, United States Patent No. US3090094A
7. T. Wetzig, A. Schmidt, S. Dudczig et al., Adv. Eng. Mater. **22**(2), 1900657 (2020). https://doi.org/10.1002/adem.201900657
8. D. Hoppach, U. Peuker, Germany Patent No. DE 10 2019 117 513
9. D. Hoppach, U. Peuker, Germany Patent No. DE 10 2018 126 326
10. D. Hoppach, E. Werzner, C. Demuth et al., Adv. Eng. Mater. **22**(2), 1900761 (2020). https://doi.org/10.1002/adem.201900761
11. F. Heuzeroth, J. Fritzsche, U.A. Peuker, Particuology **18**, 50–57 (2015). https://doi.org/10.1016/j.partic.2014.06.001
12. C. Voigt, B. Dietrich, M. Badowski, et al., in *Light Metals* (Springer, 2019), pp. 1063–1069. https://doi.org/10.1007/978-3-030-05864-7_130
13. C. Voigt, E. Jäckel, F. Taina et al., Metall. Mater. Trans. B **48**(1), 497–505 (2017). https://doi.org/10.1007/s11663-016-0869-5
14. D. Hoppach, Dissertation, TU Bergakademie Freiberg (2022)
15. C. Tien, R.M. Turian, H. Pendse, AlChE J. **25**(3), 385–395 (1979). https://doi.org/10.1002/aic.690250302
16. S. Ripperger, W. Gösele, C. Alt, et al., in *Ullmann's Encyclopedia of Industrial Chemistry* (2013) pp. 1–38. https://doi.org/10.1002/14356007.b02_10.pub3
17. K. Sutherland, in *Filtration and Separation*, vol. 45 (2008), pp. 25–28
18. P. L. Brun, F. Taina, C. Voigt, et al., in *Light Metals* (Springer, 2016), pp. 785–789. https://doi.org/10.1007/978-3-319-48251-4_133
19. P. Waite, in *Essential Readings in Light Metals* (Springer, 2016), pp. 51–58
20. E. Werzner, M.A.A. Mendes, S. Ray et al., Adv. Eng. Mater. **15**(12), 1307–1314 (2013). https://doi.org/10.1002/adem.201300465
21. F. Heuzeroth, J. Fritzsche, E. Werzner et al., Powder Technol. **283**, 190–198 (2015). https://doi.org/10.1016/j.powtec.2015.05.018
22. L. Ditscherlein, A. Schmidt, E. Storti et al., Adv. Eng. Mater. **19**(9), 1700088 (2017). https://doi.org/10.1002/adem.201700088
23. C. Voigt, L. Ditscherlein, E. Werzner et al., Mater. Des. **150**, 75–85 (2018). https://doi.org/10.1016/j.matdes.2018.04.026
24. C. Voigt, J. Hubálková, L. Ditscherlein et al., Ceram. Int. **44**(18), 22963–22975 (2018). https://doi.org/10.1016/j.ceramint.2018.09.094
25. L. Ditscherlein, Dissertation, TU Bergakademie Freiberg (2021)
26. L. Ditscherlein, T. Zienert, S. Dudczig et al., Adv. Eng. Mater. **24**(2), 2100634 (2022). https://doi.org/10.1002/adem.202100634
27. P. Knüpfer, L. Ditscherlein, U.A. Peuker, Colloids Surf. A **530**, 117–123 (2017). https://doi.org/10.1016/j.colsurfa.2017.07.056
28. J. N. Israelachvili, *Intermolecular and Surface Forces*. (Academic press, 2011)
29. L. Ditscherlein, U. A. Peuker, presented at the 8. World Congress of Particle Technology (Orlando, 2018) (unpublished)
30. L. K. Ditscherlein, P.; Peuker, U. A., presented at the PARTEC 2019 (Nürnberg, 2019) (unpublished)
31. J. Fritzsche, U.A. Peuker, Procedia Engineering **102**, 45–53 (2015). https://doi.org/10.1016/j.proeng.2015.01.105
32. E. M. Lifshitz, M. Hamermesh, in *Perspectives in Theoretical Physics* (Elsevier, 1992), pp. 329–349. https://doi.org/10.1016/B978-0-08-036364-6.50031-4
33. B. Ninham, V. Parsegian, Biophys. J. **10**(7), 646–663 (1970). https://doi.org/10.1016/S0006-3495(70)86326-3

34. N. Van Kampen, B. Nijboer, K. Schram, Phys. Lett. A **26**(7), 307–308 (1968). https://doi.org/10.1016/0375-9601(68)90665-8
35. R.R. Dagastine, M. Bevan, L.R. White et al., J. Adhes. **80**(5), 365–394 (2004). https://doi.org/10.1080/00218460490465696
36. C. Weber, P. Knüpfer, M. Buchmann et al., Miner. Eng. **167**, 106804 (2021). https://doi.org/10.1016/j.mineng.2021.106804
37. L. Ditscherlein, J. Fritzsche, U. A. Peuker, Colloids Surf. A **497**, 242–250 (2016). https://doi.org/10.1016/j.colsurfa.2016.03.011
38. J. Fritzsche, J. Teichmann, F. Heuzeroth et al., Adv. Eng. Mater. **15**(12), 1299–1306 (2013). https://doi.org/10.1002/adem.201300120
39. F. Heuzeroth, J. Fritzsche, U.A. Peuker, Chem. Ing. Tech. **86**(6), 874–882 (2014). https://doi.org/10.1002/cite.201300183
40. H.-J. Butt, M. Kappl, Adv. Colloid Interface Sci. **146**(1–2), 48–60 (2009). https://doi.org/10.1016/j.cis.2008.10.002
41. P. Attard, Langmuir **16**(10), 4455–4466 (2000). https://doi.org/10.1021/la991258+
42. J. Fritzsche, U.A. Peuker, Colloids Surf. A **509**, 457–466 (2016). https://doi.org/10.1016/j.colsurfa.2016.09.051
43. J. Fritzsche, U.A. Peuker, Powder Technol. **289**, 88–94 (2016). https://doi.org/10.1016/j.powtec.2015.11.057
44. B. Hoffmann, S. Weyrauch, B. Kubier, Chem. Ing. Tech.Ing. Tech. **74**(12), 1722–1726 (2002). https://doi.org/10.1002/cite.200290013
45. K. Cooper, A. Gupta, S. Beaudoin, J. Colloid Interface Sci. **234**(2), 284–292 (2001). https://doi.org/10.1006/jcis.2000.7276

Chapter 13
Simulation of Fluid Flow, Heat Transfer and Particle Transport Inside Open-Cell Foam Filters for Metal Melt Filtration

Eric Werzner, Miguel A. A. Mendes, Cornelius Demuth, Dimosthenis Trimis, and Subhashis Ray

13.1 Introduction

The removal of impurities from liquid metals inside ceramic foam filters (CFF) occurs predominantly through depth filtration, owing to the small size and concentration of inclusions. A deeper understanding of the relevant physical phenomena, involved in this process, is required in order to develop novel filters with improved characteristics. However, due to the high temperature and visual opacity of the melt, as well as due to the complex geometry of CFFs and the stochastic nature of the inclusion capture processes, it is extremely difficult, if not impossible, to obtain detailed information by means of conventional experimental techniques. In contrast to this, numerical simulations can naturally provide information on numerous physical quantities with high temporal and spatial resolution. Furthermore, they are less costly and time-consuming than laboratory trials, enabling comprehensive parametric studies that would be otherwise infeasible to conduct experimentally.

Depth filtration models can be broadly divided into two categories: (1) macroscopic or phenomenological models, which provide correlations or analytical expressions that describe the filtration phenomena in terms of empirical parameters, based

E. Werzner · C. Demuth · S. Ray (✉)
Institute of Thermal Engineering, Technische Universität Bergakademie Freiberg,
Gustav-Zeuner-Str. 7, 09599 Freiberg, Germany
e-mail: ray@iwtt.tu-freiberg.de

M. A. A. Mendes
Instituto de Engenharia Mecânica, Instituto Superior Técnico, Universidade de Lisboa, Av.
Rovisco Pais 1, 1049-001 Lisboa, Portugal

D. Trimis
Division of Combustion Technology, Engler-Bunte-Institute, Karlsruhe Institute of Technology,
Engler-Bunte-Ring 1, 76131 Karlsruhe, Germany

© The Author(s) 2024   301
C. G. Aneziris and H. Biermann (eds.), *Multifunctional Ceramic Filter Systems
for Metal Melt Filtration*, Springer Series in Materials Science 337,
https://doi.org/10.1007/978-3-031-40930-1_13

on either experimental data or theoretical considerations and, (2) pore-scale models, which employ the solution of the governing equations for the melt flow and inclusion dynamics in two- or three-dimensional domains, considering either the true filter geometry or similar computer-generated structures. With the growing capacity and availability of high performance computers, the latter approach is becoming increasingly attractive. The results of pore-scale simulations also contribute to the development of improved macroscopic models, which are often sufficiently accurate for a quick assessment or the comparison of filter structures. An overview of the literature concerning the pore-scale modeling of metal melt filtration inside CFFs was given by Demuth et al. [1] and hence is not repeated here for brevity. This chapter addresses different more advanced issues arising in the detailed modeling of individual physical phenomena that occur during metal melt filtration and presents main results of several recent comprehensive parametric studies, which contributed to an improved understanding of the process and the design of better filter geometries. Further, it shows efficient methods for the determination of relevant effective properties, in particular, the effective thermal conductivity.

This chapter is structured as follows: In the subsequent part of this section, the basic mathematical formulation of the pore-scale model along with references dealing with its implementation are provided. This is followed by a discussion on the flow regimes and the particle capture mechanisms that are encountered during depth filtration inside CFFs. In Sect. 13.2, information on the generation of the geometric models as well as the characterization and scaling of filter geometries are provided. Methods for the determination of effective filter properties, in particular, the effective thermal conductivity, the hydraulic tortuosity, the viscous and inertial permeability coefficients as well as the filtration coefficient are discussed in Sect. 13.3. In the subsequent section, individual modeling issues are addressed. They include the propagation of the melt front during infiltration, the characteristics and modeling of turbulent flow, the agglomeration of inclusions and their accumulation during long-term operation. Section 13.5 presents the results of several numerical investigations dealing with the sensitivity of the filtration process with respect to process conditions and geometry parameters of the filter, followed by a conclusion from this research effort.

13.1.1 Modeling Strategy

In the following, the employed mathematical modeling is briefly outlined for the sake of completeness. For a detailed description of the pore-scale model and its implementation using the lattice-Boltzmann method (LBM), one may refer to Demuth et al. [1, 2] and Lehmann et al. [3].

The simulation of depth filtration inside open-cell foams at the pore-scale is typically carried out using an Euler–Lagrange approach. The flow of liquid metal is governed by the mass and momentum conservation equations for an incompressible, Newtonian fluid, known as:

$$\frac{\partial u_i}{\partial x_i} = 0 \tag{13.1}$$

$$\frac{\partial u_i}{\partial t} + \frac{\partial}{\partial x_j}(u_i u_j) = -\frac{1}{\rho_m}\frac{\partial p}{\partial x_i} + \frac{\partial}{\partial x_j}\left[\nu\left(\frac{\partial u_i}{\partial x_j} + \frac{\partial u_j}{\partial x_i}\right)\right] \tag{13.2}$$

where u, ρ_m, ν and p denote the velocity, the density, the kinematic viscosity and the pressure of the melt, respectively. Additional terms may appear on the RHS of Eq. (13.2), e.g., for the residual stresses when using a turbulence model [2] or when buoyancy-induced convection due to the temperature-dependent density is considered [1]. In the latter case, the following energy conservation equation is also solved:

$$\rho c_p\left[\frac{\partial T}{\partial t} + \frac{\partial}{\partial x_i}(u_i T)\right] = \frac{\partial}{\partial x_i}\left(k\frac{\partial T}{\partial x_i}\right) \tag{13.3}$$

Here, c_p, k and T stand for the specific heat capacity, the thermal conductivity and temperature, respectively. For simplicity, the inclusions are modeled as spherical particles, which are one-way coupled to the flow field. Their trajectories are determined by solving the equation of motion in a Lagrangian reference frame considering the most relevant forces:

$$m_p\frac{dv_i}{dt} = \underbrace{(m_p - m_m)g_i}_{\text{buoyancy}} + \underbrace{m_m\frac{18\nu f s}{d_p^2}(u_i - v_i)}_{\text{drag}}$$

$$+ \underbrace{m_m\frac{Du_i}{Dt}}_{\text{pressure gradient}} + \underbrace{m_m\left(\frac{Du_i}{Dt} - \frac{dv_i}{dt}\right)}_{\text{added mass}} \tag{13.4}$$

where v denotes the velocity of the particles, g stands for the gravitational acceleration and f_S is a Stokes drag correction factor for higher particle Reynolds numbers. Further, $m_p = \rho_p \pi d_p^3/6$ is the mass of a spherical particle with diameter d_p, which displaces a melt volume of mass $m_m = m_p \rho_m/\rho_p$. In the vicinity of the wall, the drag force increases, thereby reducing the probability for wall contact, which is not reflected by the drag term in Eq. (13.4). An approach to correct the obtained collision efficiency during the post-processing of the data was demonstrated by Heuzeroth et al. [4]. Once an inclusion comes in contact with the filter struts, its filtration probability depends on its ability to remain attached. In most numerical studies, a simple adhesion condition is adopted that assumes a particle remains attached once it gets in contact with the strut surface. This seems to be justified in view of the small kinetic energy of the inclusions, the strong capillary forces in metal melts and the possible sintering between the particle and the filter surface, particularly at the high temperatures during steel filtration. The same condition, however, should not be applied to water-model depth filtration experiments inside CFF, for which simulations conducted using the same adhesion condition overpredicted the measured filtration efficiency by more than 53% [5].

The amount of data, generated during detailed pore-scale simulations also poses a challenge for the storage, visualization and analysis of the results. In order to address these issues, methods for the efficient compression and visualization of the large datasets were proposed by Lehmann et al. [6, 7].

13.1.2 Identification of Flow Regimes

In order to reliably characterize the flow regimes for open-cell foams, the Reynolds number according to Ruth and Ma is proposed [8], which is also termed as the Forchheimer number Fo and is defined as:

$$Fo = \frac{u_D k_1}{v k_2} \tag{13.5}$$

where u_D is the superficial average velocity in the direction of the main flow, which is related to the interstitial average velocity $\bar{u}$ and porosity ε according to $u_D = \varepsilon \bar{u}$ and the ratio between viscous and inertial permeabilities, k_1/k_2, is chosen as the length scale. Methods for their determination are discussed in Sect. 13.3.3. Both coefficients appear in the Darcy-Forchheimer law, which describes the flow through porous media at a macroscopic level:

$$\frac{dp}{dx} = \underbrace{-\frac{\mu}{k_1} u_D}_{\text{viscous losses}} \underbrace{- \frac{\rho_m}{k_2} u_D |u_D|}_{\text{inertial losses}} \tag{13.6}$$

It may be mentioned here that k_2 is also known as Forchheimer coefficient. Using Eq. (13.5) and reformulating Eq. (13.6), an expression for the inertial contribution to the total pressure drop, which consists of both viscous and inertial components, is obtained as:

$$\chi \approx \frac{Fo}{1 + Fo} \tag{13.7}$$

Figure 13.1 shows different regimes of flow through porous media as a function of Fo and the corresponding inertial contribution χ. For any flow with $Fo < 0.01$, the inertial contribution is $\chi < 1\%$, meaning that the flow can be considered to be purely viscous. Hence it can be described by the Stokes equation on the pore-scale and Darcy's law on the macro-scale. It is evident that the determination of k_1 using Darcy's law should always be carried out in this regime. For $0.01 < Fo < 1.0$, the flow can be termed inertial but it is still dominated by the viscous effects. Although recirculation zones form in the wake regions of the struts, they still remain stable due to the

Fig. 13.1 Inertial contribution to total pressure drop χ and flow regimes for flow through porous media with respect to Fo. The second abscissa (gray) shows the corresponding superficial velocity u_D for two typical melt filtration systems: the filtration of (a) aluminum alloy AlSi7Mg inside a 30 ppi filter [9] and (b) AISI 4142 steel inside a 10 ppi filter [10]. Coincidentally, the ratio between u_D and Fo is nearly identical for these two systems

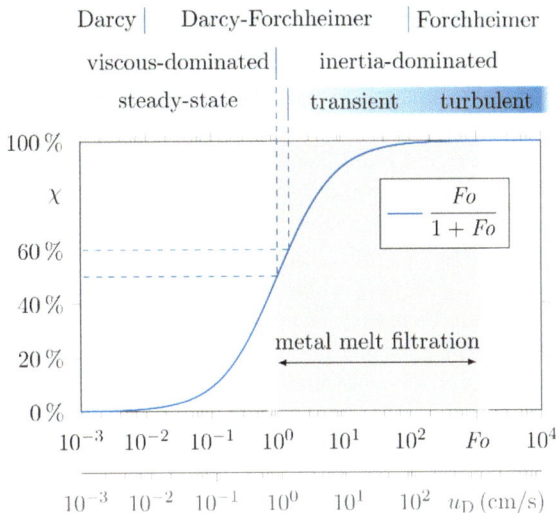

damping effect of viscosity, i.e., the flow can be considered as steady. The transition from steady-state to transient flow was observed to occur for different types of porous media at $\chi \approx 60\%$ [9], which corresponds to $Fo = 1.5$. Despite of its time-dependent nature, the flow still remains laminar in this regime, showing deterministic transients, similar to the vortices of a Kármán vortex street. It is important to note that numerical simulations in this regime require be initialized with a random perturbation of the velocity field in order to enforce the development of the transients, in particularly for regular geometries. As Fo increases further, the secondary flows turn out to be increasingly chaotic and for $Fo > 100$, i.e. $\chi > 99\%$, the flow can be considered fully inertial as well as turbulent.

As indicated in Fig. 13.1, industrial metal melt filtration processes cover a wide range of Fo, since their superficial velocities u_D vary from a few cm/s, e.g., for filtration during continuous casting, up to several m/s in case of gravity casting. While the former can sometimes be simulated as a steady-state problem on relatively coarse grids, the pore-scale simulation of the latter would require extremely high spatial and temporal resolution for capturing all scales of turbulence and the very thin boundary layers. Direct numerical simulations (DNS) of the flow in this regime are limited to small filter sections or periodic unit cells. For larger filter samples, turbulence closures and special wall treatments are required. For the sake of completeness, it may be mentioned that purely viscous flow occurs during centrifugal supergravity filtration using very fine CFFs of 100 pores per inch (ppi) and higher.

13.1.3 Capture Mechanisms

For the characterization of the inclusion behavior, it is important to distinguish between negatively buoyant particles, i.e., inclusions with higher density than the melt $\rho_p/\rho_m > 1$ and the positively buoyant ones, for which $\rho_p/\rho_m < 1$. While the former is typically the case for oxide inclusions in light metals, such as aluminum, the latter typically occurs for heavy metals, e.g., iron or steel melt. The main capture mechanisms of depth filtration and the definitions of the dimensionless numbers characterizing their magnitude are presented in Table 13.1. All dimensionless numbers scale with the particle diameter, either linearly for the case of direct interception or quadratically for inertial impaction and gravitational settling/buoyant rising. While large (negatively buoyant) inclusions are susceptible to all collision mechanisms, the small ones are almost exclusively captured by direct interception. In addition, for large positively buoyant inclusions, the force due to pressure gradient, exerted by the melt, can be significant. It acts towards the centre of curvature of the fluid pathlines [11], pushing inclusions away from the filter wall on the upstream side of the filter strut, thereby significantly reducing the particle capture on the front face. The conventional definition of the Stokes number, however, does not express this difference, discerning the filtration of larger oxide inclusions from light and heavy metals. Therefore, a modified Stokes number is suggested, which takes into account the density difference between the two phases. It is obtained from non-dimensionalization of the equation of motion, considering only the terms representing pressure gradient and Stokes drag, under the assumption of very small particle Reynolds number $Re_p \ll 1$, for which the time derivative of the particle velocity dv/dt approaches the substantial derivative of the fluid velocity Du/Dt, and is presented in Table 13.1. As one can easily recognize, the sign of St_{mod} becomes negative for particles with density ratios $\rho_p/\rho_m < 1$, indicating the marked change in the particle dynamics. The pressure gradient term does not only affect the transport of particles in the wall-normal direction, but also the motion of larger particles inside vortices that occur at higher flow Reynolds numbers. While negatively buoyant particles are flung outward of the vortices due to their inertia, the positively buoyant particles or bubbles are pushed towards their core, resulting in a preferential concentration of inclusions in different regions depending on the density ratio. The resulting local increase in the particle concentration and the relative motion of nearby particles, which drift differently depending on their size, may facilitate the collision of smaller inclusions and the formation of larger agglomerates that are more likely to get captured due to the aforementioned mechanisms of depth filtration. In order to estimate the probability of preferential concentration due to this mechanism, the characteristic time scale of the macroscopic flow, $d_s/\overline{u}$, in the definition of St_{mod} requires to be replaced by a time scale of the turbulent motion. It may be mentioned here that neutrally buoyant particles with $\rho_p/\rho_m = 1$, for which $St_{mod} = 0$, closely follow the streamlines but can still experience a relative motion due to the curved flow field surrounding them.

Table 13.1 Dimensionless numbers characterizing the main mechanisms of particle capture during depth filtration of liquid metals

Collision mechanism	Dimensionless number	
Direct interception	Interception number	$d^* = \frac{d_p}{d_s}$
Inertial impaction/pressure gradient	Modified stokes number	$St_{mod} = \frac{(\rho_p - \rho_m)d_p^2 \bar{u}}{18\mu d_s}$
Gravitational settling/buoyant rising	Gravitational number	$N_G = \frac{|\vec{g}|d_p^2(\rho_p - \rho_m)}{18\mu\bar{u}}$

13.2 Modeling of Geometry

The numerical models of the filter geometries can be either artificially generated or reconstructed from 3D computed tomography (CT) scan images of real CFFs. Conventionally manufactured CFFs often exhibit a macroscopic gradient in their geometric properties, significant anisotropy and random local defects, either hampering the reproducibility of studies or demanding large representative volumes for numerical simulations. A costly repetition of experimental trials and subsequent averaging is necessary in order to obtain reliable results, as required e.g., for sensitivity analyses. Computer-generated random foams are a remedy to this problem, as they can be reproducibly manufactured using 3D printing techniques while capturing the essential geometric features of their real counterparts [4, 9]. They also allow variations of topological and morphological features, enabling comprehensive sensitivity analyses, and can be generated with periodic boundaries with arbitrary size, facilitating handling of the boundary conditions for the numerical simulations. Generation of the artificial random foams, employed in the studies presented later, was elaborately described by Abendroth et al. [12]. Besides random foams, the space-filling unit cells, such as cubic cell [13], Kelvin cell [14], or the Weaire-Phelan structure are also used as models for open cell foams. While in such cases, only a small physical domain needs to be simulated for obtaining the velocity field, the particle deposition can be predicted for a much larger porous medium, by considering an unfolded domain for the particle tracking, as has been shown by Werzner et al. [15]. Figure 13.2 shows different examples of numerical models for open-cell foams.

All structures are stored as volume meshes on a uniform Cartesian grid, consisting of either fluid or solid cubic elements (voxels). In order to improve the accuracy of the curved boundary representation and the computational efficiency of the particle–wall collision checking, the signed Euclidean distance field is employed. It is calculated with sub-voxel accuracy from the voxel mesh after smoothing using a Gaussian kernel.

Ceramic foam filters are self-similar within a certain ppi range, i.e., after appropriate normalization, not only their geometric characteristics, such as the average pore and strut diameter, the specific strut length and surface area, but also the permeability coefficients k_1 and k_2 do not significantly depend on pore count. This is evident

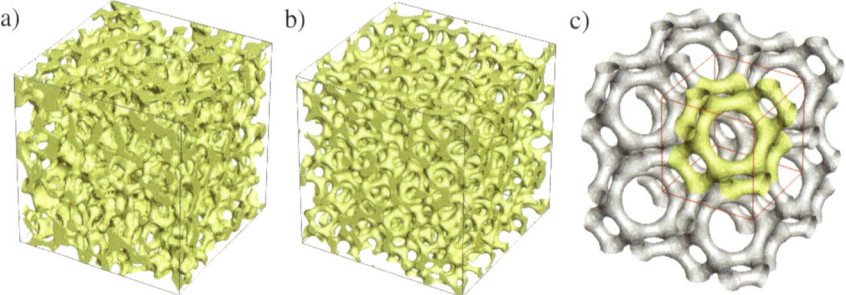

Fig. 13.2 Geometry examples: **a** CT scan reconstruction of a 30 ppi CFF with closed inner porosity and **b** computer-generated random foam with periodic boundaries, both comprising 216 pores, and **c** periodic section of 8 unit cells based on the tetrakaidekahedron (Kelvin cell) containing 16 pores

from Table 13.2, in which normalized geometric and hydraulic properties of 10, 20 and 30 ppi CFFs along with the computer-generated random foams are compared. The choice of an appropriate reference scale depends on the investigated problem. Here, and in previous studies [9, 13, 16], a length, calculated based on the volumetric pore density φ (1/m^3) was chosen, which is defined as follows:

$$L = \frac{1}{\sqrt[3]{\varphi}} \tag{13.8}$$

The inverse of L can be recognized as a more accurate pendant to the typical industrial ppi specification. The advantage of this reference length over other choices is its invariance with respect to morphological parameters, such as porosity or strut shape, and its clear definition. Values of φ and L are presented in Table 13.2.

13.3 Determination of Effective Properties Related to Metal Melt Filtration

Effective properties are required for the macroscopic modeling of the filtration process using the computationally efficient homogenization approach. They also serve to characterize filter geometries with respect to the different transport processes that occur during metal melt filtration.

13.3.1 Effective Thermal Conductivity

The effective thermal conductivity (ETC) is not only important for the characterization of the thermal shock resistance of filter foams but also required for simulations of

Table 13.2 Geometric and hydraulic characteristics of different CFFs and the computer-generated random foam [9, 13]

Filter material			Al₂O₃-C		Al₂O₃			Artificial foam
Nominal pore count		(ppi)	10	20	20	30	30	–
Geometry characteristics								
Pore density	φ	(1/m³)	$1.03 \cdot 10^7$	$3.54 \cdot 10^7$	$1.13 \cdot 10^7$	$3.95 \cdot 10^7$	$4.03 \cdot 10^7$	–
Reference length	L	(mm)	4.60	3.05	4.45	2.94	2.92	–
Porosity		(%)	80%	78%	78%	79%	81%	80%
Spec. surface area	S	(m²/m³)	593	899	655	949	964	–
Spec. projected area	S_P	(m²/m³)	147	248	153	246	242	–
Strut diameter	d_{strut}	(mm)	0.91	0.75	0.86	0.55	0.55	–
Window diameter	d_{win}	(mm)	3.80	2.84	3.55	2.32	2.29	–
Pore diameter	d_{pore}	(mm)	6.05	3.81	6.21	3.85	3.82	–
Normalized geometry parameters								
Surface area	$S \cdot L$		2.73	2.74	2.92	2.79	2.81	2.55
Projected area	$S_P \cdot L$		0.68	0.76	0.68	0.72	0.71	0.65
Strut diameter	d_{strut}/L		0.20	0.25	0.19	0.19	0.19	0.22
Window diameter	d_{win}/L		0.83	0.93	0.80	0.79	0.79	–
Pore diameter	d_{pore}/L		1.32	1.25	1.39	1.31	1.31	1.25

(continued)

Table 13.2 (continued)

Filter material		Al_2O_3-C		Al_2O_3			Artificial foam
Hydraulic properties							
Tortuosity	τ	1.15	1.14	1.13	1.15	1.14	1.11
Viscous permeability	k_1/L^2	$8.23 \cdot 10^{-3}$	$8.08 \cdot 10^{-3}$	$7.98 \cdot 10^{-3}$	$7.62 \cdot 10^{-3}$	$7.93 \cdot 10^{-3}$	$9.15 \cdot 10^{-3}$
Inertial permeability	k_2/L	$4.82 \cdot 10^{-1}$	$4.48 \cdot 10^{-1}$	$6.03 \cdot 10^{-1}$	$4.95 \cdot 10^{-1}$	$4.88 \cdot 10^{-1}$	$7.14 \cdot 10^{-1}$

energy transport using the homogenization approach. Mendes et al. [17]. proposed a simple and effective method for its determination that relies upon the predicted ETC under vacuum condition $k_{\text{eff,s}}$, which is determined for $k_{\text{f}}/k_{\text{s}} = 0$, where the subscripts "f" and "s" stand for the fluid and the solid phases, respectively. The numerical evaluation of $k_{\text{eff,s}}$ requires considerably less computation time compared to solving for the complete porous medium since the calculation in the fluid domain, which constitutes the largest volume fraction equal to the porosity, can be avoided. In order to arrive at the final recommendation, detailed numerical investigations were carried out by considering two artificial and four real open-cell foams. They demonstrated that $k_{\text{eff,s}}$ contains all necessary structural information as far as the ETC is concerned. Therefore, the ETC of the porous medium for any k_{f} can be evaluated from a simplified model, which was proposed as:

$$\tilde{k}_{\text{eff}} = b\tilde{k}_{\text{min}} + (1 - b)\tilde{k}_{\text{max}} \qquad (13.9)$$

where $\tilde{k} = k/k_{\text{s}}$ is the dimensionless thermal conductivity, normalized with respect to k_{s}, while $\tilde{k}_{\text{min}}$ and $\tilde{k}_{\text{max}}$ are the minimum and the maximum bounds of the dimensionless ETC, respectively. The only adjustable parameter b in Eq. (13.9) can be evaluated from $k_{\text{eff,s}}$ as follows:

$$b = \frac{\tilde{k}_{\text{max,s}} - k_{\text{eff,s}}}{\tilde{k}_{\text{max,s}} - \tilde{k}_{\text{min,s}}} \qquad (13.10)$$

where the additional subscript "s" stands for the corresponding values of thermal conductivities, evaluated at vacuum condition, i.e., for $k_{\text{f}}/k_{\text{s}} = 0$. Mendes et al. [17] also demonstrated that the use of lower and upper Hashin-Strikman bounds [18] for $\tilde{k}_{\text{min}}$ and $\tilde{k}_{\text{max}}$, respectively, performs the best for most of the investigated cases. Since Eq. (13.9) is valid for any $\tilde{k}_{\text{f}} = k_{\text{f}}/k_{\text{s}}$, the proposed model can also be used for the indirect evaluation of ETC in presence of hazardous fluids, like liquid metals. An experimental verification of the proposed model was presented by Wulf et al. [19]. They measured the ETC of open-cell FeCrAl alloy metal foams with different porosities and pore dimensions using the transient plane source (TPS) method at room temperature and observed reasonably good agreements with the predictions obtained from the model of Mendes et al. [17].

In principle, $k_{\text{eff,s}}$, required for the determination of b, could also be measured. However, since it is extremely difficult to conduct experiments under true vacuum condition, Mendes et al. [20] proposed to measure the ETC in presence of a commonly available fluid, like air or water, and evaluate the parameter b from an equation similar to Eq. (13.10), indicated by replacing the second suffix "s" by "ref" which stands for the reference fluid instead of the vacuum condition. Their results show that a lower thermal conductivity working fluid, like air, is the most suitable for all investigated cases, although water may also be used if the thermal conductivity of the solid matrix is also sufficiently high.

Subsequently, an extension of this model was also proposed by Mendes et al. [21] consisting of two adjustable parameters those can be determined from the detailed numerical predictions of $k_{\text{eff,s}}$ and $k_{\text{eff,ref}}$, where the latter is evaluated in the presence of a reference fluid with known $\tilde{k}_f$. The model is quite similar to that in Eq. (13.9) except for the fact that $\tilde{k}_{\text{min}}$ is replaced by $\tilde{k}_a$, which is given as:

$$\tilde{k}_a = \left[\frac{a}{\tilde{k}_{\text{min}}} + \frac{1-a}{\tilde{k}_{\text{max}}} \right]^{-1} \qquad (13.11)$$

where a is the second adjustable parameter. However, instead of using the lower and upper Hashin-Strikman bounds [18] the use of equivalent thermal conductivities for the serial and the parallel arrangements of solid and fluid phases were recommended for $\tilde{k}_{\text{min}}$ and $\tilde{k}_{\text{max}}$, respectively. Using these values, the model parameters b and a can be sequentially determined from $k_{\text{eff,s}}$ and $k_{\text{eff,ref}}$, respectively. Quite clearly, the original model of Mendes et al. [17] can be easily retrieved by setting $a = 1$ for which, $\tilde{k}_a = \tilde{k}_{\text{min}}$ is obtained from Eq. (13.11). Performance of the model was investigated by comparing its estimation with the results obtained from the detailed numerical simulations and the predictions obtained from the original model. It was clearly demonstrated that the modified model performs extremely well for all investigated structures, provided the reference fluid is appropriately chosen.

In order to account for the thermal radiation, which is expected inside porous medium in the presence of metal melt at high temperature, Talukdar et al. [22] proposed a 3D numerical model that solves combined conduction–radiation heat transfer employing the finite volume method (FVM), combined with the blocked-off region approach, that is capable of handling the grey–diffusive behavior of the strut surfaces. The model, which is represented by the overall energy balance equation and accounts for the radiative heat flux from the solution of the 3D radiative transfer equation (RTE), uses the voxel based information of the porous medium for representing the geometry that is employed also for solving other conservation equations and hence eliminates the complexity of grid generation. The representative porous media, considered for this investigation, are the cubic cell and the honeycomb structures. The effective thermal conductivities evaluated using the model for different temperatures, emissivities and thermal conductivities of the solid and different absorption coefficients of the fluid were compared with that obtained from FLUENT 6.3 as well as the data available in the literature and good agreements were observed.

After establishing the detailed conduction–radiation model [22], Mendes et al. [23] applied it for evaluating the ETCs of three real foams with different porosities and pore counts. Other than this model [22], Mendes et al. [23] also considered yet another detailed but decoupled model, where pure conduction and pure radiation are separately solved while disregarding the contribution of the other mode of heat transfer, and proposed three simplified one-dimensional (1D) models, based on homogenization approach. The simplified models require information about the ETC due to pure heat conduction $k_{\text{eff,C}}$, calculated from the detailed 3D simulation [17,

20, 21], and the extinction coefficient β, estimated from the 3D CT-scan data using the image processing technique by the projection method [24]. The ETCs predicted by these four models were compared with the data obtained by employing the most accurate detailed model of Talukdar et al. [22], those were considered as reference values in order to determine the accuracy of other simplified models. It was observed that all simplified models perform reasonably well, provided the maximum value of the extension coefficient is chosen, while the alternative detailed and decoupled model consistently underpredicts the ETC for all cases.

An experimental validation of two simplified 1D homogeneous models, proposed by Mendes et al. [23], was presented by Mendes et al. [25], who measured the ETC of two FeCrAl alloy foams with similar porosity but considerably different pore dimensions at temperatures ranging from 130 to 850°C using the panel test technique. The first model (Model-1) of Mendes et al. [25] solves the coupled conduction–radiation heat transfer using the 1D RTE, while their second model (Model-2) employs the extremely simplified Rosseland approximation in order to account for the thermal radiation. As mentioned earlier in the last paragraph, both 1D homogeneous models, however, require $k_{\text{eff,C}}$ and β as inputs. Comparison of the results shows that both simplified models are efficient alternatives to the detailed model for the determination of ETC at high temperatures. Although Model-2 is extremely simple, it may still be used provided the optical thickness of the porous medium is reasonably high. For more accurate predictions, however, Model-1 was recommended by Mendes et al. [25], irrespective of the value of β.

Owing to the manufacturing process, porous foams often exhibit a certain amount of anisotropy and also contain microscopic porosities that cannot be captured by 3D CT-scan images with the available resolution. Although the total porosity ε_T of the foams can be accurately determined by measuring their weights, dimensions and the density of the strut material, the microscopic porosity ε_m, which is extremely sensitive to the manufacturing process and may vary between 5 to 25%, can only be roughly estimated. The macroscopic porosity ε_M, on the other hand, can be evaluated from the measured ε_T and the estimated ε_m as follows:

$$\varepsilon_M = \frac{\varepsilon_T - \varepsilon_m}{1 - \varepsilon_m} \tag{13.12}$$

Increase in the microscopic porosity has two opposing effects as far as the ETCs of porous foams are concerned: (1) it reduces ε_M according to Eq. (13.12) since ε_T remains constant and hence increases the solid fraction or the effective strut volume (that also contains microscopic pores) available for heat conduction and (2) it reduces the thermal conductivity of the modeled struts or the solid fraction, which could be estimated from the upper Hashin-Strikman bound [18]. Considering these issues, Mendes et al. [26] investigated the effects of anisotropy and microscopic porosity on the ETC of 10 and 45 ppi porous foams using both detailed numerical simulations and the simplified model of Mendes et al. [17]. They observed that for the considered foam samples, ETC increases with the increase in ε_m, which, however, may not be the case for other foams owing to the opposing effects of ε_m as mentioned earlier.

Nevertheless, they concluded that the proposed simplified model of Mendes et al. [17] performs extremely well with significantly reduced computational effort and hence may be employed for investigating the sensitivity of ETC on different structural parameters.

The simplifications for modeling the ETC of porous foams, proposed by Mendes et al. [17, 25] in the presence of both conduction and radiation were validated later by Mendes et al. [27]. They measured the ETC of two anisotropic 10 ppi pure alumina large foam samples with almost similar porosity $\varepsilon_T \approx 0.89$ and other geometric parameters using the TPS technique for average foam temperatures T_{av} ranging from 22 to 750°C. The average ETCs of two samples were compared with the predictions obtained using Model-1 of Mendes et al. [25]. The predicted results exhibited the expected effects of anisotropy and estimated microscopic porosity on the directional ETC, as shown in Fig. 13.3. The measured ETC in a particular direction and the predicted ETC in the same direction, obtained after the proposed directionally weighted averaging, compared extremely well for the complete range of investigated temperature. Based on this observation, they concluded that the ETC, measured employing the TPS technique, are to be considered as the directionally averaged value of the ETCs in three mutually perpendicular directions. The investigation also clearly demonstrated the potential of the simplified modeling approach for characterizing even the directional ETC of open-cell foams, based on their structural information that could be obtained solely from the 3D CT-scan data.

Using the modeling tools developed by Mendes et al. [17] and Talukdar et al. [22], the variations in the ETC of a tetrakaidecahedra unit cell as functions of ε, k_s and T_{av} due to combined conduction and radiation were examined by Patel et al. [14]. In addition, they also investigated the effects of surface reflectivity, pore density and porosity on the radiative properties of the structures, for which a pure radiation heat

Fig. 13.3 Directional dependence of predicted ETC and experimental data (lines indicate the model predictions) [27]

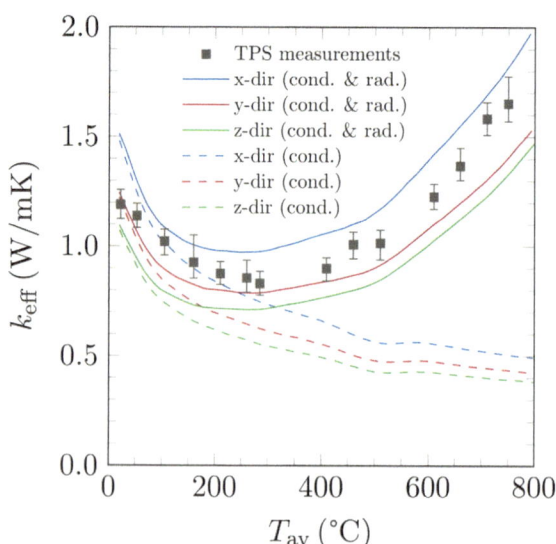

transfer solver was developed and employed. In order to develop useful correlations, three different ETCs were predicted: (1) due to pure heat conduction i.e., without radiation $k_{eff,C}$ [17], (2) due to pure radiation heat transfer i.e., without conduction $k_{eff,R}$ [14] and finally (3) due to combined conduction and radiation k_{eff} [22], which is considered as the reference value for comparison. From the acquired data, three different correlations were proposed with varying degree of complexities. The final correlation, obtained by superposing $k_{eff,R}$ and $k_{eff,C}$ with an adjustable coefficient that takes into account the coupling between radiation and conduction heat transfer, was recommended.

13.3.2 Hydraulic Tortuosity

The probability of contacts between the inclusions and the filter web strongly depends on the tortuosity of the flow pathlines. It can be directly evaluated from the velocity field obtained from pore-scale simulations as follows [28]:

$$\tau = \frac{\langle |u| \rangle}{\langle u_x \rangle} \tag{13.13}$$

Here, the angle brackets $\langle \cdot \rangle$ denote spatial averages, $|u|$ is the velocity magnitude and the direction x stands for the direction of the bulk flow. The hydraulic tortuosity was evaluated for the Stokes flow regime, i.e. at $Fo \ll 1$ and hence it is considered as one of the characteristics of the porous media. Values for different open-cell foams are presented in Table 13.2.

13.3.3 Viscous and Inertial Permeability

The permeability coefficients, appearing in Eq. (13.6), are required for the design of filtration systems, the identification of the flow regime and the simulation of flow through porous media using the homogenization approach. While k_1 can be evaluated in a straightforward manner employing Darcy's law from a single pore-scale simulation in the Stokes regime, the numerical determination of k_2 from a single simulation in the purely inertial regime is computationally challenging for real filter geometries. However, once k_1 is known, an estimate can be obtained by solving the Darcy-Forchheimer law for k_2 using data from a simulation at $Fo \approx 1$, where the inertial contribution to the total pressure drop is already significant but the flow still remains steady. Since the estimation of Fo and hence setting up of such a simulation itself requires the prior knowledge of k_2, its guess value needs to be provided. On the basis of a momentum balance, one can arrive at an extremely simple expression for k_2 of porous media in the limit of high porosity, which reads as [2]:

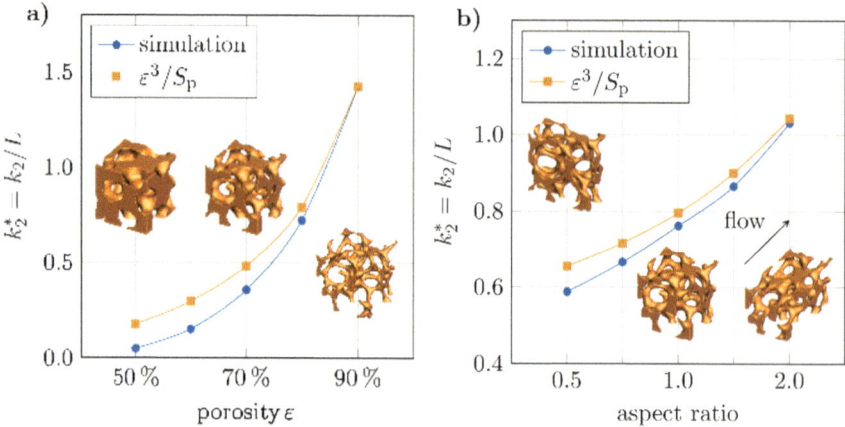

Fig. 13.4 Predictions of dimensionless inertial permeability coefficient k_2^* obtained from detailed pore-scale simulations and the correlation of Eq. (13.14) for artificial open-cell foam structures with **a** different porosity and **b** different aspect ratio at 85% porosity. The simulated foam samples comprises 216 pores

$$k_2 = \frac{\varepsilon}{S_p} \qquad (13.14)$$

where S_p is the cumulative specific surface area, projected in the direction of the main flow, which can be easily estimated from a voxel representation that can be obtained from 3D μCT scan images. Since S_p contains direction-dependent information, it is able to handle the anisotropy, which is present in most CFFs with lower pore count [9]. A comparison of Eq. (13.14) with the predictions obtained from detailed pore-scale simulations for artificial random foams with different porosity and aspect ratio is presented in Fig. 13.4. Accurate correlations for k_1 and k_2, which also account for the effect of tortuosity, were developed by Jorge et al. [13]. Values of k_1 and k_2 for CFFs with pore counts of 10, 20 and 30 ppi are presented in Table 13.2. A model for the prediction of the mechanical behavior of CFFs during filtration employing k_1 and k_2 was proposed by Lange et al. [16].

13.3.4 Filtration Coefficient

Assuming depth filtration to be the dominant mode of particle retention and a homogenous porous medium, the variation in the particle concentration c inside the melt, is governed by:

$$\frac{dc}{dx} = -\lambda c \qquad (13.15)$$

where λ is the filtration coefficient. Solving Eq. (13.15) yields the depth filtration law:

$$\eta(x) = 1 - e^{-\lambda x} \tag{13.16}$$

Here, $\eta(x)$ is defined as the fraction of inclusions with respect to the total amount of suspended inclusions, captured by the filter up to a certain depth x, and is straightforward to obtain from the pore-scale simulation of the filtration process. A simplified deposition model for simulations using the homogenization approach, which also takes into account the variation in λ with the flow velocity was proposed by Asad et al. [10].

In the limit of very high porosity, λ can also be estimated from simple geometric considerations since the filter can then be approximated as a skeleton of thin wires that do not significantly affect the flow field and intercept all inclusions within a distance of $d_P/2$. Assuming that approximately $2/3$ of the available cumulative strut length contributes to the filtration, neglecting the overlapping of the struts in flow direction and introducing the specific strut length l^+, the following simple expression for the filtration coefficient could be obtained:

$$\lambda_{\text{lim}} = \frac{2}{3} d_P l^+ \tag{13.17}$$

13.4 Selected Modeling Issues

In the following, the modeling of different physical phenomena involved in the depth filtration process will be discussed, which have received considerably less attention in the existing literature on metal melt filtration.

13.4.1 Geometry Change During Long-Term Operation

During long-term operation of CFFs, e.g., in the course of continuous casting, or while filtering highly contaminated melts, the accumulation of inclusions, deposited inside the filter is expected to alter its effective geometry. In order to investigate how the change in filter geometry affects the melt flow and the capture of inclusions, a model was developed for the filtration of aluminum during continuous casting inside a 30 ppi CFF [29]. The deposited inclusions are modeled as an impermeable rigid porous sediment, whose porosity was treated as a free parameter with $\varepsilon_{\text{sed}} = 40, 60$ and 80%. For every fluid voxel containing filtered particles, the effective volume of the sediment is calculated. Whenever it exceeds the voxel volume, its state is changed from fluid to solid, resulting in a stepwise relocation of the filter surface. As the real loading process is extremely slow compared to the temporal

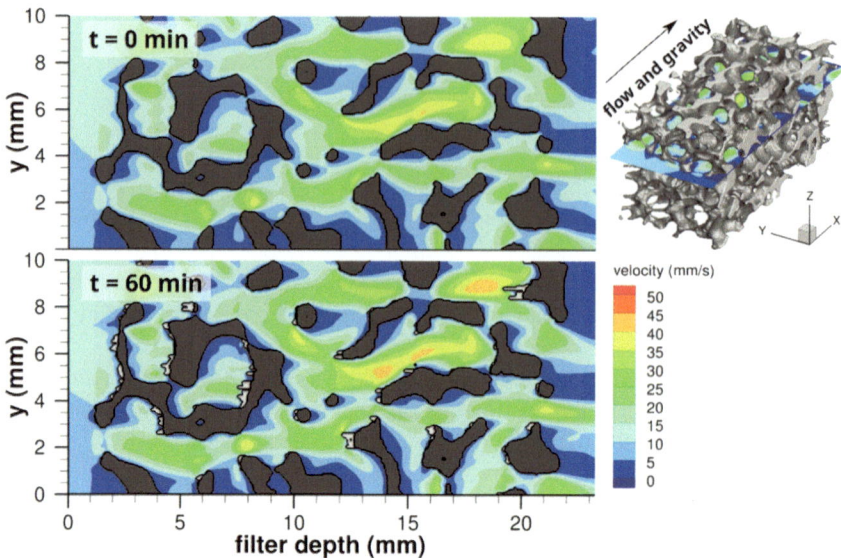

Fig. 13.5 Modification of filter geometry and distribution of velocity magnitude during long-term filtration for a sediment porosity of $\varepsilon_{sed} = 60\%$. A number of 10^7 Al_2O_3 particles with 20 μm diameter were inserted randomly on a plane upstream of the filter. The liquid aluminum was driven with a superficial velocity of 10 mm/s. The filter geometry was reconstructed from a CT scan of a 30 ppi Al_2O_3 foam, obtained with a resolution of 65 μm, which was used also for the LBM simulation. The computational domain was extended by one filter length in the upstream and downstream directions

resolution required for the particle tracking, its simulation would cause prohibitively high computational costs. Therefore, the process was accelerated by imposing a 21-fold increased concentration of impurities. This adjustment is possible as the flow is quasi-steady and since particle-particle interactions are negligible at the low volume fraction of particles inside the melt of 10^{-5} (10 ppm), assumed for the investigated process. In this manner, a physical duration of 60 min could be captured by simulating only 9 average residence times of the melt inside the filter.

Figure 13.5 shows the modification of the filter geometry and the associated change in the velocity distribution assuming a sediment porosity of 60%. As visible from the figure, after 60 min, the filtered particles have formed sediments on the upstream faces of the filter struts, that protrude into the pores and cause a marginal but recognizable increase in local melt velocity. The time-variations of different filter characteristics are presented in Fig. 13.6. The filter loading, which is the fraction of the initially available pore volume occupied by the sediments, remains relatively low within the observed time. Nevertheless, the modification of the filter geometry results in a significant increase in filtration efficiency and also pressure gradient $p\prime$. Depending on sediment porosity, the filtration coefficient λ, which is estimated from the instantaneous filtration efficiency assuming the validity of the depth filtration law, increases between 18% and 76%. While this variation occurs linearly over time

Fig. 13.6 Time-variations of filter loading and different filtration-related characteristics during long-term filter operation assuming different porosities of the deposited sediments ε_{sed}. In the initial stage, the filtration efficiency and related quantities cannot be accurately evaluated as the filter is not yet completely filled with contaminated melt (indicated by dashed lines). Due to the artificial acceleration of the loading process, this phase appears extended in physical time

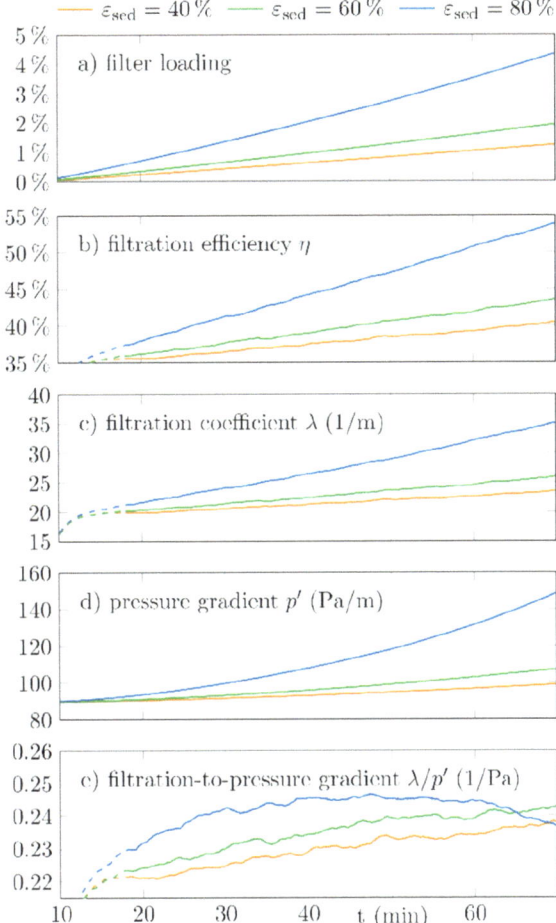

$(\overline{R^2} = 1.00)$, the pressure gradient tends to increase quadratically $(\overline{R^2} = 0.97)$. Hence, in the initial stage, when the change in pressure gradient is small, the filter efficiency in terms of the ratio λ / p' increases with time. For $\varepsilon_{sed} = 80\%$ it attains a maximum after 37 min, when the filter is loaded by 2.5%. This optimal filter loading is also expected for the cases with lower sediment porosity, but could not be observed within the simulated time duration.

13.4.2 Agglomeration

The intentional agglomeration of small inclusions to form larger aggregates, that are more likely to be captured by the filter, may benefit their filtration. Particles can only collide when they experience a relative motion with respect to each other, either due to velocity shear, e.g., in boundary layers and vortices, or due to the difference in their slip velocity that results from a difference in properties, particularly size and density. Furthermore, they must overcome the viscous force [4], which is a result of the squeezed film flow in the gap between two particles. Once collided, the inclusions are expected to adhere to each other, as this state is favorable in terms of surface energy. Furthermore, at higher temperature, sintering takes places that creates a strong bond between the particles. It is obvious that the probability for agglomeration strongly depends also on the particle concentration, which determines the average distance between the particles. Although it is low for most melt filtration systems, i.e., in the order of ppm, the effect of preferential concentration in highly turbulent flows may increase the local concentration to a level, where collisions are more probable to occur. Since an ideal structure for the promotion of agglomeration has to meet different requirements than a filter, composite structures, consisting of several functional layers, have to be considered.

In order to investigate the potential of agglomeration, leading to the improved filtration of smaller inclusions, the filtration of liquid steel was simulated using an agglomeration model. A variety of composite filter structures, consisting of layers of Kelvin cells with 85% porosity, measuring either 5 or 10 mm cell width, as depicted in Fig. 13.7a were tested. The flow of liquid steel (AISI4142) entering with a velocity of $u_D = 10 \text{cm/s}$ was simulated, resulting in $Fo = 10.5$ and 21.0 with corresponding $Re_{\text{strut}} = 160.5$ and 321.0 for the small and large cells, respectively. The simulations were carried out with periodic boundary conditions in the spanwise direction. A number of 10^7 SiO_2 particles were randomly inserted at a rate, which corresponded to an average volumetric concentration of 200 ppm. The collision radius of the agglomerates was calculated according to the closely-packed sphere model, considering an agglomerate porosity of 50%. In order to reduce the computational time for collision checking, the linked-cell method was employed.

As shown in Fig. 13.7b, approximately only 6% of the inclusions form agglomerates. The highest formation of agglomerates is observed for a structure with fine pores, followed by a single layer of large pores (marked in red). Irrespective of the spatial distribution, the amount of agglomeration tends to grow with the fraction of small pores. It was further observed that agglomeration tends to decrease with filter depth, as the number of potential collision partners in the vicinity of an inclusion or agglomerate becomes smaller due to previous agglomeration events and filtration. The vast majority of agglomerates consisted of only two inclusions whereas the largest one comprised 8 inclusions and measured 25 μm, which is still relatively small. Although the agglomerates are more likely to get trapped by a filter strut, particularly due to increased probability of direct interception, their contribution to the overall filtration efficiency (in terms of filtered particle volume) is still lower than their volume fraction among all particles, since they spend less time inside the filter than solitary inclusions. A comparison of the filtration efficiencies, shown in Fig. 13.7a, with those obtained from simulations carried out without the agglomeration model showed an average improvement of only 1%. Hence, agglomeration has a negligible effect on overall filtration, at least for the conditions considered here. The

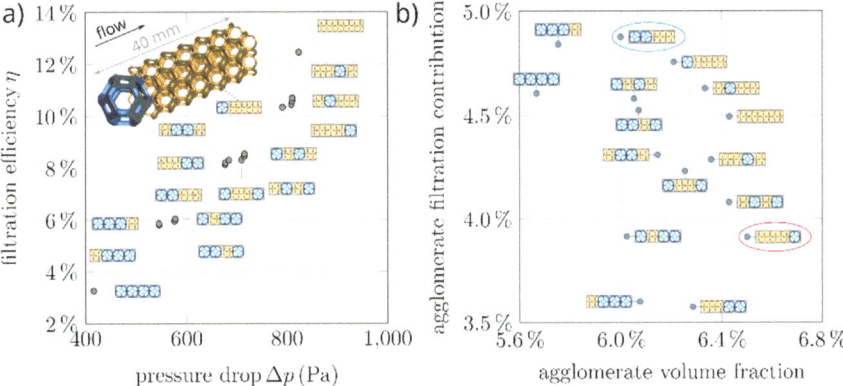

Fig. 13.7 Filtration of SiO_2 inclusions of 10 μm diameter from liquid steel (AISI4142) through composite filters consisting of differently sized Kelvin cells, taking into account the effect of agglomeration: **a** filtration efficiency and pressure drop, **b** contribution of agglomerates to total filtration versus the agglomerate formation, defined as fraction of agglomerates with respect to the total volume of impurities

safe removal of larger agglomerates is crucial, since they are more detrimental to the quality of the cast part than small inclusions. As visible from Fig. 13.7b, composite structures, consisting of a coarse layer followed by a fine one, are performing better in this respect (marked in blue). However, an assessment of the usefulness of composite structures with functionalized zones for intentional agglomeration of smaller inclusions would require further studies. The investigations should also include higher bulk flow velocities in order to study the effect of turbulence as well as particle size distributions and particles with different densities or bubbles.

13.4.3 Infiltration Process

In all numerical studies on liquid metal filtration, known to the authors, the initial stage of the process, during which the melt penetrates into the filter, also termed as priming, is neglected and the structure is initialized as being completely filled with the molten metal. However, this initial phase could be particularly critical as the melt may freeze and hence block the pores or the filter may get damaged due to the thermal shock. Further, Marangoni convection is expected to significantly influence the melt propagation into the porous medium. An additional infiltration resistance follows also from the capillary pressure and the change in momentum, as the flow gets deflected by the struts of the filter. The necessity of reliable priming and the desired high filtration efficiency may pose conflicting requirements on the filter design, e.g., as far as the pore density is concerned. Therefore, the computer-aided development of improved filter geometries should also include the simulation of the initial stage.

For modeling of the infiltration process in the LBM context, a free-surface model was adopted [30]. In this model, the mass and momentum conservation equations are solved only for the fluid phase. The gas–liquid interface is captured using the volume-of-fluid approach and capillary effects are considered by the continuum surface force model. The present implementation handles the wetting behavior using a fictitious extension of the gas-liquid interface into the solid with a prescribed contact angle. While the code was successfully validated for canonical problems, numerical instabilities occurred during simulations of the real infiltration process. The imposition of high Laplace pressure, which follows from the high surface tension coefficient σ of the melt and the small radii of curvature inside the small pores of the CFF, causes a strong distortion of the mesoscopic particle distribution, used by the LBM for representing the fluid state. Furthermore, parasitic currents were observed near the contact line. The identified instabilities could only be avoided by increasing the spatial resolution to a level, at which the strut diameter was resolved with a number of voxels, corresponding to $1.6 \times 10^{-4} Re_{\text{Pore}}/Ca$, where Re_{Pore} is the Reynolds number based on the pore diameter and the capillary number is defined as $Ca = \mu \overline{u}/\sigma$. Since the resulting grid resolution would be prohibitively high for practical simulations of the metal melt filtration, the applicability of the free-surface LBM in this context is limited, unless an adaptive grid refinement of the interface is implemented. Another difficulty arises from the fact that in the LBM, the time step size is indirectly defined through the choice of the sound speed, which serves as a numerical parameter that can be adjusted within certain limits. While the sound speed is often artificially reduced to accelerate the time marching, this practice introduces problems when gravitational acceleration is included, since the large hydrostatic pressure of the liquid metal then introduces a significant density variation along the height, which violates the assumption of weak compressibility, required by the LBM.

Nevertheless, in order to demonstrate the capability of the model and to investigate the effect of the filter coating, the infiltration of aluminum into a CFF was simulated in 2D for different wetting angles at a reduced surface tension, where the metal is fed from the bottom. The results, presented in Fig. 13.8, show that even at reduced σ, the flow behavior is strongly determined by the capillary action. Once the melt touches the solid strut surface, capillary waves are generated at the interface, which oscillate much faster than the average interstitial velocity. At the smaller windows, particularly for the simulation with higher wetting angle, the melt stagnates, causing a redistribution of the melt flow. Once the Laplace pressure is overcome, the melt discharges into the adjacent pore, during which velocities of more than 20 times the inlet velocity as well as large vortices are observed. This discontinuous filling of the pores is also visible from surges in the average pressure, measured on the inlet plane, that are also more pronounced for higher wetting angle.

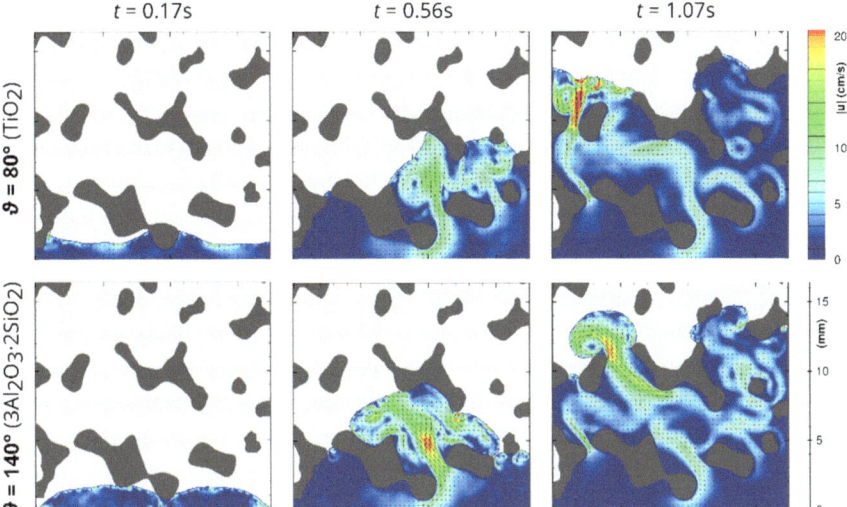

Fig. 13.8 Distribution of the velocity magnitude in the liquid phase during infiltration of Al_2O_3 CFF with a superficial velocity of $u_D = 1cm/s$ at different wetting angles ϑ, measured by Voigt et al. [31] ($Re_{Pore} = 117$, $Ca = 1, 17 \times 10^{-4}$, $Bo = 6, 72$ and direction of gravity $\downarrow$)

13.4.4 Non-isothermal Modeling

Although heat transfer plays a crucial role for the filtration process due to the expected temperature gradients in the melt and the sensitivity of the melt properties, particularly density, viscosity and surface tension, with respect to temperature, its variation is often neglected in numerical simulations and hence the energy conservation equation is not invoked. Demuth et al. [1] developed a model, which couples the LBM for the prediction of the fluid flow with a finite volume method (FVM) for solving the energy equation, considering the temperature-dependent melt density and viscosity. The viability of the model was demonstrated by simulating the filtration of aluminum for conditions as present during continuous casting. They reported the development of buoyancy-induced secondary flows, which led to higher heat losses and a delayed entry of the inclusions. For the considered boundary conditions, however, the effect of temperature-dependent viscosity on the flow field was found to be negligible.

13.4.5 Characterization and Modeling of Turbulence

Selection of a suitable modeling approach for turbulence inside open-cell foams requires a good understanding of its nature. In order to characterize turbulence inside CFFs, a direct numerical simulation (DNS) of flow through an idealized open-cell foam consisting of Kelvin cells was carried out at $Fo = 64$ ($\chi = 98\%$). The simulated unit cell was scaled to dimensions of $5 \times 5 \times 5$mm, as depicted in Fig. 13.9a, and was discretized using a voxel mesh with 4.9 μm resolution, allowing to capture flow details with very fine scales. The corresponding pore density of $\varphi = 1.6 \times 10^7 1/m^3$ and a porosity of 85% are comparable to a 10 ppi CFF. The viscous and inertial permeability were determined before as $k_1 = 1.30 \times 10^{-7} m^2$ and $k_2 = 2.56 \times 10^{-3}$m, respectively, for this structure. Considering the kinematic viscosity of liquid steel at 1600 °C, the corresponding superficial velocity is obtained as $u_D = 61$cm/s. In order to collect spectral information about the turbulence, the velocity time-series were recorded for selected points over 8 average melt residence times. Subsequently, the spectral density of the fluctuation energy in the longitudinal direction $\overline{u'^2}$ was computed and plotted with respect to the wavenumber κ_1, assuming Taylor's frozen flow hypothesis. The spectra, shown in Fig. 13.9b, indicate a significant heterogeneity of turbulence inside the porous structure: in the pore centre, the fluctuation energy is much higher and also exhibits a wider frequency distribution compared to that at the window, where, however, the average flow velocity is 40% higher. This is also indicated by the turbulence intensity, which reached 34% in the pore centre and 15% in the window. As expected for low Reynolds number turbulence, the inertial subrange, for which the spectrum decays with $\kappa_1^{-5/3}$ is narrow and the transition to the dissipative range occurs early. The variations are also in agreement with model spectra for homogenous isotropic turbulence [32], which were calculated on the basis of the turbulent kinetic energy (TKE) and its dissipation rate, evaluated during the simulation. Figure 13.9c shows the distribution of vorticity in z-direction along a xy-plane, marked in Fig. 13.9a. High values are observed particularly on the sides of the struts, where TKE is produced due to the transient flow separation. The magnified view in Fig. 13.9d shows a small vortex pair at the lower end of the energy spectrum, measuring approximately 50 μm, which is comparable to the size of larger inclusions in metal melts. Since these and the smaller scales are impossible to resolve in practical simulations involving real CFFs, their effects have to be taken care of by appropriate turbulence models. For turbulent flows inside open-cell foams with similar or lower Forchheimer numbers, the Large-Eddy Simulation (LES) was adopted [2]. For this purpose, the spatial resolution, dictated by an accurate representation of the intricate strut network, is sufficient to resolve the bulk of the energy-containing scales, allowing an LES to be performed with no significant additional computational costs. In this respect, two-equation turbulence models were shown to overpredict the TKE near the stagnation point [33]. While reconstructing the unresolved velocity fluctuations for the tracking of inclusions, e.g., using the discrete random walk model (DRWM), this would cause an overestimation of the turbulent dispersion. For further details on the implementation of the LES, the reader is referred to Demuth et al. [2], who demonstrated its application for the simulation of filtration of liquid steel in the turbulent regime. They also analyzed the development and character of the turbulent flow inside the filter and estimated the size of the smallest flow structures.

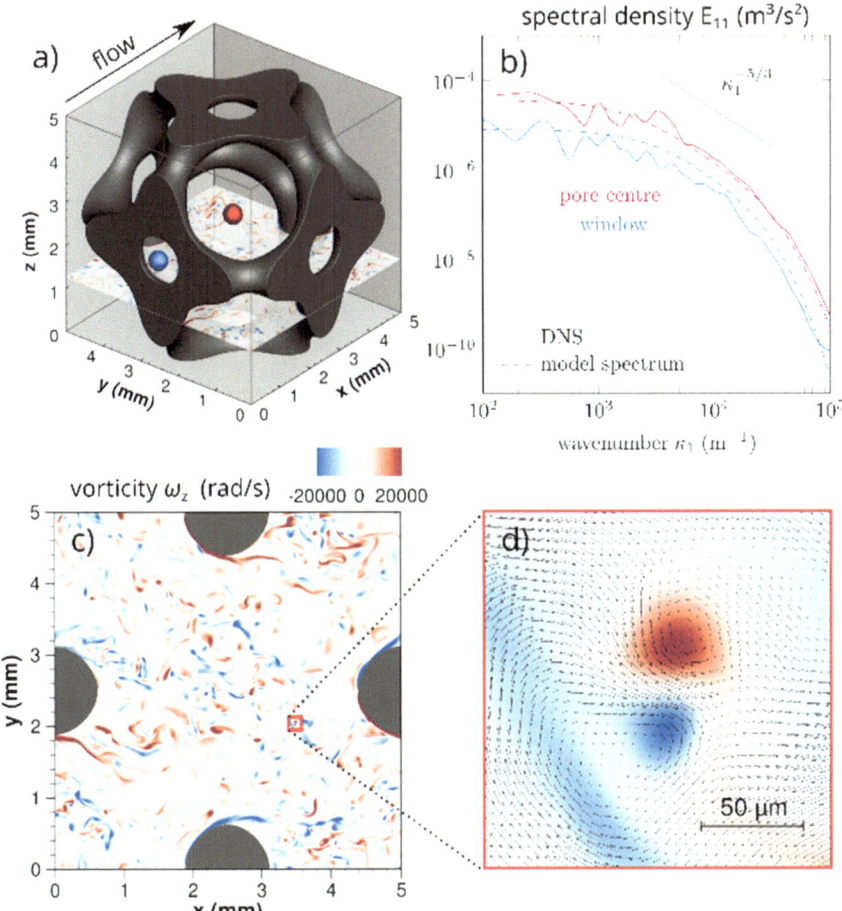

Fig. 13.9 Turbulent flow inside a stack of Kelvin cells mimicking a 10 ppi CFF at $Fo = 64$: **a** simulated periodic unit cell, **b** spectral distribution of longitudinal fluctuation energy for selected locations, **c** distribution of z-vorticity along a plane and **d** magnified section with velocity vectors as seen by a moving observer, showing a small vortex pair

13.5 Sensitivity of the Filtration Process

In the following, the main results of several parametric studies on the sensitivity of the depth filtration process with respect to different process conditions and geometric parameters are presented for the sake of completeness.

As evident from the discussion in Sect. 13.1.3, the filtration efficiency strongly depends on the size of the inclusions. This dependence can be readily observed from Fig. 13.10, which is a result of the simulation of depth filtration of negatively buoyant inclusions inside a stack of 10 ppi filters [5]. While the larger inclusions

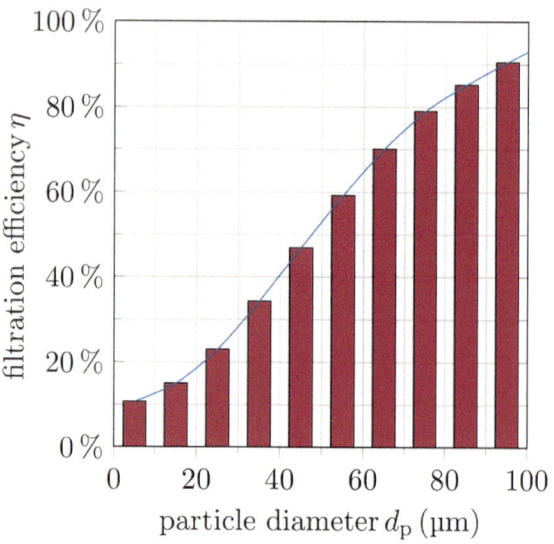

Fig. 13.10 Variation of filtration efficiency for inclusions of different size inside a stack of 10 ppi filters with 54 mm thickness ($\rho_p/\rho_m = 3.96$, $Fo = 2.38$) [5]

are almost certainly filtered, the smaller ones have very low probability of filtration. It has to be mentioned that η should eventually vanish as the size of the inclusions approaches zero, if the Brownian motion is neglected. However, an extremely high spatial resolution or a more accurate boundary treatment would be required in order to capture this behavior. The strong sensitivity of the filtration efficiency with respect to the particle diameter was also observed by Werzner et al. [9, 15], who studied the depth filtration inside CFFs using a 2D idealized filter geometry, consisting of periodic arrays of staggered cylinders, and in 3D computer-generated random open-cell foams.

The bulk velocity of the melt flow affects the individual mechanisms of the filtration process in different ways. For typical melt filtration processes, an increase in the flow rate leads to a higher probability of collisions between the inclusions and the filter struts. This is caused mainly by the fact that, as Forchheimer number increases, the velocity boundary layer becomes thinner, allowing a higher fraction of particles to be intercepted by the struts. This effect was reported by Asad et al. [10], who numerically investigated the depth filtration of non-metallic inclusions with $d_P = 20\mu m$ from liquid steel inside Al_2O_3-C foam filters with pore counts of 10 and 20 ppi, neglecting buoyancy effects. As presented in Fig. 13.11a, the filtration coefficient remains constant for very low velocities and starts to increase above approximately 1 mm/s. Figure 13.11b shows the data in dimensionless form, i.e., after normalization of λ using k_2 as filter length scale and plotted with respect to the Forchheimer number. The data seem to approach two distinct asymptotes in the purely viscous and the inertial flow regimes. The macroscopic pressure gradient $p\prime$ was also normalized with respect to k_2, and its variation was found to agree well with the dimensionless form of Eq. (13.6), as visible from Fig. 13.11d. Demuth et al. [2] performed simulations of a similar process at higher Forchheimer numbers using

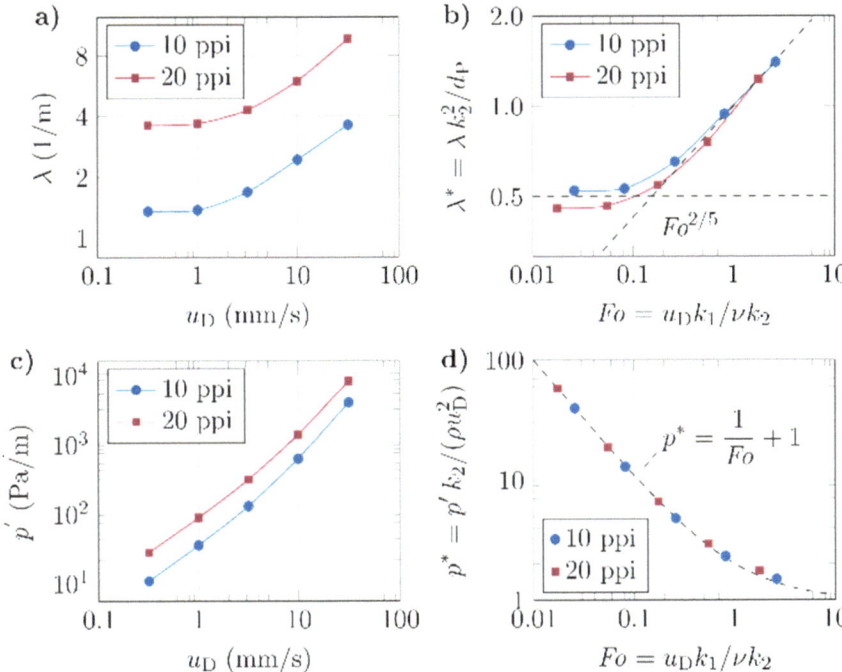

Fig. 13.11 Sensitivity of filtration and pressure gradient with respect to flow velocity during depth filtration of non-metallic inclusions with $d_P = 20\,\mu$m from liquid steel: variation of **a** filtration coefficient and **c** pressure gradient with superficial velocity and variation of **b** dimensionless filtration coefficient and **d** dimensionless pressure gradient with Forchheimer number [10]

an LES turbulence model. They also observed an increase in the filtration efficiency with the bulk flow velocity, which, however, is less pronounced. This difference may be caused by the pressure gradient term in the equation of motion for the particles that was neglected in the study of Asad et al. [10] and which is expected to reduce the inertial impaction for the positively-buoyant particles. The data from both investigations [2, 10], however show that the gain in filtration efficiency that occurs due to the increase in velocity is obtained at the cost of a lower efficiency in terms of filtration-per-pressure gradient. This is expected since the pressure gradient increases quadratically in the inertial regime. For the sake of completeness it may be mentioned here that the contribution of gravitational settling or buoyant rising to the capture of inclusions decreases with flow velocity and hence can be neglected for most filtration processes [15].

The pore count or pore density is one of the most important geometric parameters for characterizing the filtration performance of open-cell foams. Several studies suggest that the filtration coefficient increases quadratically with the linear pore count, i.e. $\lambda \sim \varphi^{2/3}$. This dependence is confirmed not only by the results obtained from two homogenous structures, presented in Fig. 13.7a, but also can be observed from Fig. 13.11a, while taking into account the actual pore density obtained from

3D image analysis [10]. Demuth et al. also observed a substantial increase in λ with the pore count, which, however, remained sub-quadratic. It is also evident from Fig. 13.11c, that the pressure drop increases with pore count. A dimensional analysis of the Darcy-Forchheimer law clearly shows that the viscous flow resistance scales quadratically with the pore count, while the inertial resistance is only linearly dependent. This is also somewhat visible from Fig. 13.11c, which shows that the ratio in pressure gradient p' for the 10 and 20 ppi CFFs becomes smaller as the inertial regime is approached. The linear relationship in the inertial regime is also confirmed by the difference in pressure drop obtained for the two homogenous structures depicted in Fig. 13.7a. Thus, in the inertial regime, an increase in the pore count is beneficial in terms of the ratio λ/p'.

Owing to the manufacturing process, conventional CFFs exhibit a certain anisotropy, which also affects the characteristics related to metal melt filtration. Werzner et al. [9] observed a maximum variation of 21% in the inertial permeability and up to 10% variation in the filtration coefficient for Al_2O_3 CFFs of 20 and 30 ppi pore count, depending on the flow direction. They suggested to exploit the anisotropy of the filters in order to improve the filtration efficiency or the ratio $\lambda/p\prime$.

The effect of several geometric modifications of a random monodisperse open-cell foam, as presented in Sect. 13.2, on the effective properties related to metal melt filtration was thoroughly investigated from comprehensive parametric studies by Werzner et al. [9] and Lehmann et al. [3]. In these studies, the filtration coefficient was assessed according to the conditions similar to a laboratory-scale trial of aluminum filtration inside a 30 ppi CFF during continuous casting [34] while assuming a constant flow rate. The effect of porosity is presented in Fig. 13.12, where Fig. 13.12d shows that the filtration coefficient increases with decreasing porosity. This could be attributed mainly to the increase in interstitial velocity, which leads to thinner boundary layers and higher velocities near the struts, allowing a larger fraction of inclusions to be intercepted. Towards high porosity, λ seems to approach a finite value, which is also in accordance with the discussion presented in Sect. 13.3.4. An evaluation of the filtration coefficient for the high porosity limit according to Eq. (13.17) yields $\lambda_{lim} = 4.2, 8.3$ and $16.6 m^{-1}$ for $d_P = 10, 20$ and $40 \mu m$, respectively. For this evaluation, the dimensionless specific strut length $l^* = l^+ L^2 = 5.39$ of the artificial foam and a reference length $L = 2.94 mm$ for a 30 ppi filter [9] were considered. Although the theoretical model overpredicts the numerical results, its accuracy may still be considered acceptable in view of its simplicity. As may be observed from Fig. 13.12a, b, the viscous and inertial permeabilities tend to infinity in the high porosity regime, which consequently also applies to the ratio $\lambda/p\prime$. Since l^+ scales quadratically with the linear pore count, an increase in the pore count can be regarded as the most effective way to increase λ in the high porosity regime. The increase in λ for lower porosity is also reflected in the variation of hydraulic tortuosity τ, as shown in Fig. 13.12c. The correlation between both these quantities was observed also by Lehmann et al. [3]. The axial strut shape, i.e., whether the struts have a constant diameter along their axis (solid lines) or are tapered towards the middle (dashed lines) do not significantly affect the behavior.

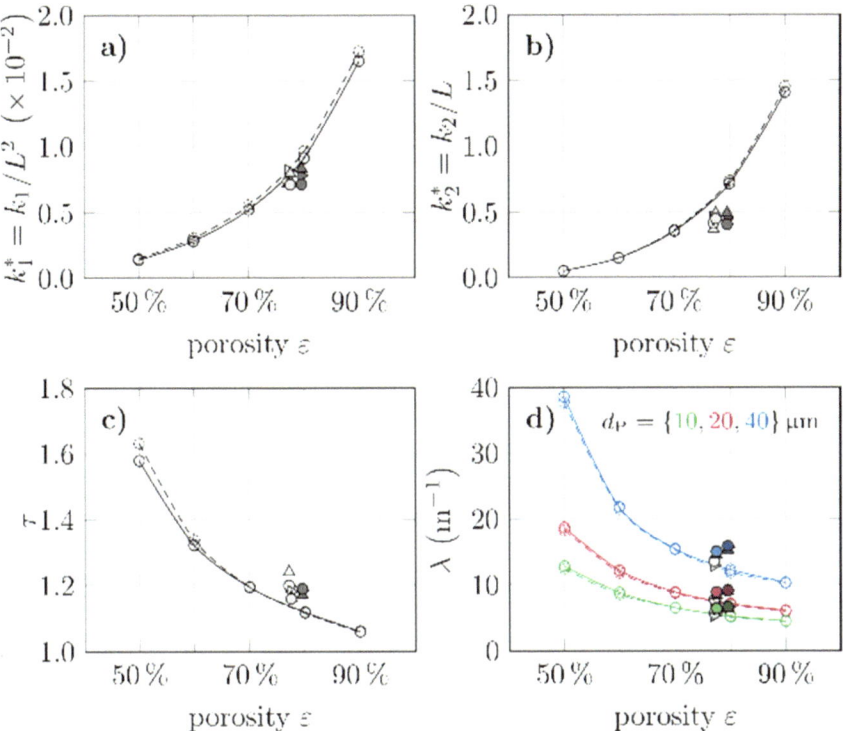

Fig. 13.12 Effect of porosity on effective properties of open-cell foams: **a** dimensionless viscous permeability k_1^*, **b** dimensionless inertial permeability k_2^*, **c** hydraulic tortuosity τ and **d** filtration coefficient λ for Al_2O_3 inclusions of different size employing a scaling, corresponding to a 30 ppi CFF ($L = 2.94$ mm). The solid and dashed lines represent the computer-generated foam with cylindrical and hourglass shaped struts, respectively, whereas the filled symbols are the results for three conventional CFFs, simulated for bulk flow in different directions [9]

With regard to the extended design freedom, enabled by additive manufacturing, Lehmann et al. [3] additionally investigated the effects of closed windows, elliptical strut cross-section and added finger-like struts. Although all modifications eventually lead to an increase in λ, the effect of additional finger-like struts was found to be the most efficient method in terms of filtration per unit pressure drop. This modification could be particularly useful for cases, where an increase in the strut length by increasing the pore count is either difficult or impossible due to the constraints on the manufacturing process.

13.6 Summary and Conclusions

The present investigation contributes to the theoretical characterization, the numerical modeling and the understanding of the depth filtration process, encountered during the removal of non-metallic inclusions from the metal melts using open-cell CFFs. On the basis of theoretical considerations, a modified definition of Stokes number has been proposed, which considers the effect of pressure gradient, allowing to capture the substantial change in the particle dynamics for particle-melt density ratios smaller than unity, which is the case e.g., for steel filtration. Different regimes of fluid flow have been presented in terms of a Reynolds number, which employs the ratio between viscous and inertial permeabilities as the length scale and is believed to indicate the transitions for a broad range of porous media. With regard to effective properties, an extremely simple model for the inertial permeability of CFFs with high porosity has been proposed, which naturally handles the anisotropy and relies only on input parameters that can be easily obtained from CT scan images. Similarly, a simplified model for the prediction of the filtration coefficient in the limit of high porosity has been suggested. Further, recent efforts in the development of simple and detailed methods for the determination of the effective thermal conductivity have been presented. The scale similarity of CFFs between 10 and 30 ppi was shown by normalization of different geometric properties using a length scale derived from the volumetric pore density. As far as the pore-scale simulation of the depth filtration process is concerned, the basic modeling approach was presented and various advanced modeling issues have been demonstrated and discussed. Finally, results of several comprehensive parametric studies on the sensitivity of the filtration process with respect to different process conditions and geometric parameters have been presented and conclusions towards a geometry improvement were drawn.

As evident from the presented results, pore-scale simulations provide good insight into the depth filtration process by giving access to virtually all relevant physical quantities with high temporal and spatial resolution. Nevertheless, for setting up the model itself, a thorough understanding of the involved physical phenomena is required in order to be able to select and parametrize suitable sub-models. The validation of these is often impossible due to the lack of experimental data, which are difficult to obtain for the liquid metal system. Further challenges arise from the computational costs, which usually demand parallel execution on compute clusters and methods for efficient handling of the large amounts of data resulting from such simulations. As the separation step in depth filtration is basically an interface phenomenon, additional care must be taken in order to ensure a sufficient spatial resolution near the filter wall, particularly for the smaller inclusions. High resolution is also required at the melt-gas interface during the infiltration process. While the lattice-Boltzmann method permits the implementation of locally or adaptively refined grids, its simplicity and high parallel efficiency would suffer and a dynamic load balancing may be required. In view of this, other well-established CFD approaches, such as the finite-volume method may be considered equally or more attractive.

The research efforts presented here contribute to the understanding of the filtration process and its sensitivity with respect to geometric characteristics and process conditions, which assists the development of improved filter geometries. It is important to note, however, that besides the removal of unwanted impurities, filters have to also fulfill other tasks that may pose different requirements on geometry. This includes their use for conditioning the turbulent melt flow before it enters the casting mold or the capability for successful priming during the initial stage of filtration under non-ideal conditions, e.g., when a preheating of the filter is not possible. Therefore, these aspects should be addressed in future studies.

Acknowledgements This work was funded by the Deutsche Forschungsgemeinschaft (DFG, German Research Foundation)–Projektnummer 169148856–SFB 920, subproject B02. The authors acknowledge computing time on the compute clusters of the Faculty of Mathematics and Computer Science of Technische Universität Bergakademie Freiberg, operated by the computing center (URZ) and funded by the DFG under grant numbers 397252409 and INST 267 / 159-1 FUGG and would like to thank Oliver Rebentrost and Dieter Simon for the professional support regarding the HPC infrastructure. Further, we would like to thank Jürgen Freitag for his IT support and helpful suggestions and Tommy Flößner for his contributions to the free-surface model.

References

1. C. Demuth, E. Werzner, M.A.A. Mendes, H. Krause, D. Trimis, S. Ray, Adv. Eng. Mater. **19**, 1700238 (2017). https://doi.org/10.1002/adem.201700238
2. C. Demuth, E. Werzner, S. Dudczig, C.G. Aneziris, S. Ray, Adv. Eng. Mater. (2021). https://doi.org/10.1002/adem.202100717
3. H. Lehmann, E. Werzner, A. Malik, M. Abendroth, S. Ray, B. Jung, Adv. Eng. Mater. **24**(2), 2100878 (2021). https://doi.org/10.1002/adem.202100878
4. F. Heuzeroth, J. Fritzsche, E. Werzner, M.A.A. Mendes, S. Ray, D. Trimis, U.A. Peuker, Powder Technol. **283**, 190–198 (2015). https://doi.org/10.1016/j.powtec.2015.05.018
5. D. Hoppach, E. Werzner, C. Demuth, E. Löwer, H. Lehmann, L. Ditscherlein, R. Ditscherlein, U.A. Peuker, S. Ray, Adv. Eng. Mater. **22**, 1900761 (2020). https://doi.org/10.1002/adem.201 900761
6. H. Lehmann, E. Werzner, C. Degenkolb, Simulation Series **48**, 32–39 (2016). https://doi.org/10.22360/SpringSim.2016.HPC.043
7. H. Lehmann, E. Werzner, C. Demuth, S. Ray, B. Jung, Efficient visualization of large-scale metal melt flow simulations using lossy in-situ tabular encoding for query-driven analytics, in *21st IEEE International Conference on Computational Science and Engineering* (Bucharest, Romania, 2018)
8. D. Ruth, H. Ma, Transp. Porous Media **7**, 255–264 (1992). https://doi.org/10.1007/BF0106 3962
9. E. Werzner, M. Abendroth, C. Demuth, C. Settgast, D. Trimis, H. Krause, S. Ray, Adv. Eng. Mater. **19**, 1700240 (2017). https://doi.org/10.1002/adem.201700240
10. A. Asad, E. Werzner, C. Demuth, S. Dudczig, A. Schmidt, S. Ray, C. G. Aneziris, R. Schwarze, Adv. Eng. Mater. **19**, 1700085–n/a, 2017. https://doi.org/10.1002/adem.201700085
11. A. Espinosa-Gayosso, M. Ghisalberti, G.N. Ivey, N.L. Jones, J. Fluid Mech. **783**, 191–210 (2015). https://doi.org/10.1017/jfm.2015.557
12. M. Abendroth, E. Werzner, C. Settgast, S. Ray, Adv. Eng. Mater. **19**, 1700080–n/a (2017). https://doi.org/10.1002/adem.201700080

13. P. Jorge, M.A.A. Mendes, E. Werzner, J.M.C. Pereira, Chem. Eng. Sci. **201**, 397–412 (2019). https://doi.org/10.1016/j.ces.2019.02.010
14. V.M. Patel, M.A.A. Mendes, P. Talukdar, S. Ray, Int. J. Heat Mass Transf. **127**, 843–856 (2018). https://doi.org/10.1016/j.ijheatmasstransfer.2018.07.048
15. E. Werzner, M.A.A. Mendes, S. Ray, D. Trimis, Adv. Eng. Mater. **15**, 1307–1314 (2013). https://doi.org/10.1002/adem.201300465
16. N. Lange, M. Abendroth, E. Werzner, G. Hütter, B. Kiefer, Adv. Eng. Mater. (2021). https://doi.org/10.1002/adem.202100784
17. M.A.A. Mendes, S. Ray, D. Trimis, Int. J. Heat Mass Transf. **66**, 412–422 (2013). https://doi.org/10.1016/j.ijheatmasstransfer.2013.07.032
18. Z. Hashin, S. Shtrikman, J. Appl. Phys. **33**, 3125–3131 (1962). https://doi.org/10.1063/1.1728579
19. R. Wulf, M.A.A. Mendes, V. Skibina, A. Al-Zoubi, D. Trimis, S. Ray, U. Gross, Int. J. Therm. Sci. **86**, 95–103 (2014). https://doi.org/10.1016/j.ijthermalsci.2014.06.030
20. M.A.A. Mendes, S. Ray, D. Trimis, Int. J. Therm. Sci. **79**, 260–265 (2014). https://doi.org/10.1016/j.ijthermalsci.2014.01.009
21. M.A.A. Mendes, S. Ray, D. Trimis, Int. J. Heat Mass Transf. **75**, 224–230 (2014). https://doi.org/10.1016/j.ijheatmasstransfer.2014.02.076
22. P. Talukdar, M.A.A. Mendes, R.K. Parida, D. Trimis, S. Ray, Int. J. Therm. Sci. **72**, 102–114 (2013). https://doi.org/10.1016/j.ijthermalsci.2013.04.027
23. M.A.A. Mendes, P. Talukdar, S. Ray, D. Trimis, Int. J. Heat Mass Transf. **68**, 612–624 (2014). https://doi.org/10.1016/j.ijheatmasstransfer.2013.09.071
24. L.-M. Heisig, K. Markuske, E. Werzner, R. Wulf, T.M. Fieback, Adv. Eng. Mater. (2021). https://doi.org/10.1002/adem.202100723
25. M.A.A. Mendes, V. Skibina, P. Talukdar, R. Wulf, U. Gross, D. Trimis, S. Ray, Int. J. Heat Mass Transf. **78**, 112–120 (2014). https://doi.org/10.1016/j.ijheatmasstransfer.2014.05.058
26. P. Götze, M.A.A. Mendes, A. Asad, H. Jorschick, E. Werzner, R. Wulf, D. Trimis, U. Groß, S. Ray, Spec. Top. Rev. Porous Media **6**, 1–10 (2015). https://doi.org/10.1615/SpecialTopicsRevPorousMedia.v6.i1.10
27. M.A.A. Mendes, P. Götze, P. Talukdar, E. Werzner, C. Demuth, P. Rößger, R. Wulf, U. Groß, D. Trimis, S. Ray, Int. J. Heat Mass Transf. **102**, 396–406 (2016). https://doi.org/10.1016/j.ijheatmasstransfer.2016.06.022
28. A. Duda, Z. Koza, M. Matyka, Phys. Rev. E **84**(3), 036319 (2011). https://doi.org/10.1103/PhysRevE.84.036319
29. E. Werzner, M. Mendes, S. Ray, D. Trimis, Numerical modeling of long-term depth filtration of metal melts inside open-cell ceramic foams, in *Proceedings CellMAT* (Dresden, Germany, 2014)
30. T. Flößner, Numerical modeling of the infiltration of open-cell ceramic foams during metal melt filtration using the lattice-Boltzmann method, Master thesis, Institute of Thermal Engineering, Technische Universität Bergakademie Freiberg, 2017
31. C. Voigt, L. Ditscherlein, E. Werzner, T. Zienert, R. Nowak, U. Peuker, N. Sobczak, C.G. Aneziris, Mater. Des. **150**, 75–85 (2018). https://doi.org/10.1016/j.matdes.2018.04.026
32. S.B. Pope, *Turbulent flows* (Cambridge University Press, Cambridge, UK, 2000)
33. F. Kuwahara, T. Yamane, A. Nakayama, Int. Commun. Heat Mass Transfer **33**, 411–418 (2006). https://doi.org/10.1016/j.icheatmasstransfer.2005.12.011
34. P. L. Brun, F. Taina, C. Voigt, E. Jäckel, C. Aneziris, Assessment of Active Filters for High Quality Aluminium Cast Products, in *Light Metals* (Hoboken, 2016)

Chapter 14
Characterization of Heat Transport and Diffusion Processes During Metal Melt Filtration

Lisa-Marie Heisig, Katrin Markuske, Rhena Wulf, and Tobias Michael Fieback

14.1 Thermophysical Properties of Bulk Materials

The effective thermal conductivity (ETC) of porous media includes heat conduction in the solid and gas phases, convection and radiation. First of all, knowledge of the thermophysical properties of the base materials of the porous media is essential for understanding heat transfer processes, using analytical or numerical models and applying as well as optimizing the porous material. In the following two subchapters, some basic investigations on the thermal properties of two different strut materials of ceramic open-cell foams are presented.

14.1.1 Alumina

In this chapter the experimental procedure to determine the bulk thermal conductivity (BTC) of pure alumina is explained and the main results are presented. The BTC was determined by measuring the thermal diffusivity α, the specific heat capacity c_p and

L.-M. Heisig · K. Markuske (✉) · R. Wulf · T. M. Fieback
Institute of Thermal Engineering, Technische Universität Bergakademie Freiberg,
Gustav-Zeuner-Straße 7, 09599 Freiberg, Germany
e-mail: katrin.markuske@ttd.tu-freiberg.de

L.-M. Heisig
e-mail: lisa-marie.heisig@ttd.tu-freiberg.de

R. Wulf
e-mail: rhena.wulf@ttd.tu-freiberg.de

T. M. Fieback
e-mail: fieback@ttd.tu-freiberg.de

© The Author(s) 2024
C. G. Aneziris and H. Biermann (eds.), *Multifunctional Ceramic Filter Systems for Metal Melt Filtration*, Springer Series in Materials Science 337,
https://doi.org/10.1007/978-3-031-40930-1_14

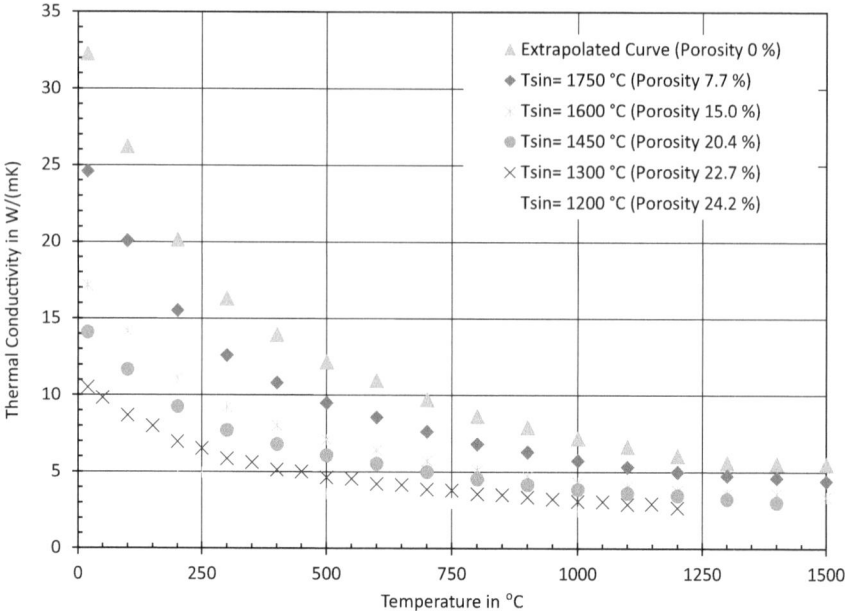

Fig. 14.1 Bulk thermal conductivities (BTC) of alumina samples sintered at different temperatures as well as extrapolated BTC to 0% porosity for temperatures from 20 to 1500 °C, processed data from [1]

the density ρ ($\lambda_B = \rho * \alpha * c_p$) in the range between 20 and 1500 °C. Corresponding to the sintering temperature, the used bulk alumina samples had a certain residual porosity. Since it was aimed to determine the BTC at 0% porosity, measurement results of five different sintering temperatures (1200–1750 °C, porosity ≈ 8–24%) were extrapolated at each measurement temperature. The processed data from [1] are shown in Fig. 14.1.

The BTC of pure alumina strongly decreases with increasing temperature from >30 W/(mK) at 20 °C to approx. 5 W/(mK) at 1500 °C which agrees well with several literature values. The results were applied for the determination of the ETC of alumina open-cell foams, commonly used as high-temperature insulating materials, catalyst carriers or in the foundry for metal melt filtration, using two-phase models.

14.1.2 Carbon-Bonded Alumina

As for the industry, carbon-bonded alumina refractories are very promising materials, their thermophysical properties were intensively investigated in Goetze et al. 2013 [2]. Besides the density and the porosity, the thermal expansion, the thermal diffusivity as well as the specific heat capacity for temperatures up to 800 °C were measured. Samples were produced either by isostatic pressing, uniaxial pressing or slip casting. Furthermore, the content of the coal-tar resin Carbores®P (binding agent) was varied (10, 15, 20%) keeping the general composition (approx. 66% alumina and 34% carbon) nearly constant by adding a corresponding amount of two further carbon modifications (carbon black powder, graphite). The bulk thermal conductivity was calculated as for the alumina samples before. Since the microstructure of the material is affected by the composition as well as by the manufacturing process, an effect of these parameters on several thermophysical properties was detected.

The technical coefficient of thermal expansion α_{techn} generally increases with increasing temperature because of the increasing lattice energy and distance between the atoms. As expected, lower thermal expansion coefficients than for pure alumina were measured. Since the thermal expansion strongly depends on the carbon modification used, differences in the curves of α_{techn} between the different contents of the Carbores®P can be seen and, similarly, the manufacturing procedure also influences the expansion. The specific heat capacity of the carbon-bonded alumina increases with increasing temperature due to enhanced lattice vibrations. The carbon composition or the manufacturing procedure does not significantly affect the measurement results, since heat capacity mainly depends on the chemical composition than on the microstructure. Finally, as expected for solid, crystalline ceramic materials, it was found that the thermal conductivity decreases with increasing temperature. Lower thermal conductivity occurred for higher binding content of Carbores®P traced back to the microstructure and the smaller grain size of the carbon particles. Furthermore, an influence of the manufacturing process as well as an anisotropic behavior for the uniaxial pressed samples were found.

14.2   Effective Thermal Conductivity

In comparison to homogeneous solids, porous media present a particular challenge when determining thermal properties. A key parameter which is often required in designing and optimizing thermal systems including porous media is the effective thermal conductivity (ETC). It results from the homogenization approach assuming equivalent properties for the whole porous medium.

The complex structure can make it difficult, time-consuming, and inaccurate to determine the thermal conductivity of the porous media using analytical approaches or numerical models. Since also experimental validation is always needed, reliable and reproducible measurement of the ETC of such media is mandatory. However, appropriate measurement methods to determine experimentally the ETC are strongly limited. One method that has been proved to be suitable for measuring the thermal properties of porous media is the Transient-Plane-Source (TPS) technique. In the following chapters, first, the measurement method is briefly presented. Thereupon, results of preliminary investigations at room temperature and higher temperatures as well as measurement results for the ETC of several ceramic open-cell foams are shown. Finally, the results of numerical investigations are discussed.

14.2.1 Transient-Plane-Source Technique

The TPS technique proposed by Gustafsson 1991 [3] is a transient measurement method to determine simultaneously the thermal conductivity λ, the thermal diffusivity a and the specific heat capacity c_p. The advantages of this measurement method are mentioned to be the large variety of materials (solid as bulk or thin film, liquid, paste, porous/granular) as well as the wide range of thermal conductivity (0.005–500 W/(mK)) that can be measured, the variable sample size and form and low effort for sample preparation [4, 5].

The sensor consists of an insulated bifilar nickel spiral acting as a heater and a temperature sensor (in form of a resistance thermometer) at the same time. [3–6] The measurements presented below were performed with a Hot Disk TPS 2500 S from the company HotDisk AB. Depending on the temperature range to be investigated sensors with Kapton (RT-250 °C) or Mica (200–750 °C) insulation are used. Furthermore, different sensor diameters (1–59 mm) are available, which are selected according to the structure and thermal conductivity of the sample. The measured, time-dependent resistance change $R(t)$ at the sensor can be described by Eq. (14.1) [3–6].

$$R(t) = R_0[1 + TCR \cdot \Delta T(\tau)] \tag{14.1}$$

A key parameter is the Temperature Coefficient of Resistance (TCR), which is further investigated in the context of high-temperature measurements. To deduce the thermal properties, the experimentally determined temperature increase $\Delta T(\tau)$ is compared with the approximate numerical solution for the temperature increase of a ring source (Eq. (14.2)) [3–6].

$$\Delta T(\tau) = P_0 \left(\pi^{\frac{3}{2}} r \lambda \right)^{-1} D(\tau) \tag{14.2}$$

Further, detailed information on the measurement principle can be found in Gustafsson 1991 [3] or He 2005 [4].

14.2.2 Preliminary Investigations

Room Temperature

The applicability and conditions of application of the TPS method for ceramic open-cell foams at room temperature were intensively investigated by Goetze et al. [7] studying the influence of several key parameters on the ETC. Numerous measurements with pure alumina foams of different pore sizes (10–60 ppi (pores per inch)) using Kapton-insulated sensors with varying sensor sizes were performed. Most importantly, it was found that an appropriate surface preparation, i.e., increasing the contact area between the foam and the sensor by grinding and polishing the samples is inevitable to obtain trustworthy, reproducible data. However, this might a challenging task because of easily breaking struts of the foams with smaller pore sizes. Furthermore, the influence of the measurement time, the heating power and the mechanical load have been found to be as negligible as the anisotropy studied on 10 and 30 ppi foams with vertically and horizontally oriented cells, since differences were only within the measurement uncertainty. Concerning the sensor size, the sensor diameter should be commonly 10 times larger than the mean cell size. With an appropriate sensor (d = 59 mm) for 10 ppi foams, it was not possible to set measurement parameters in order to reach an acceptable characteristic time and penetration depth at the same time due to the relatively high thermal diffusivity of the samples. To achieve reliable measurement results with this sensor by reaching a higher penetrations depth, an elongated sensor connection in combination with bigger samples would be necessary. Measuring 10 and 60 ppi foams the most reliable results were each obtained with a sensor of d = 29 mm. An even smaller sensor (d = 13 mm) leads to higher scattering of measured values.

Additional measurements were performed with commercial alumina foams (10–50 ppi) of adequate sample size using the same three sensor diameters, however, partially with an elongated connection brought to market by the manufacturer in the meantime to ensure sufficient penetration depth. Comparing the standard deviations, it was found that the sensor with d = 30 mm leads to acceptable results while strongly reducing the required sample dimensions from approx. $180 \times 180 \times 60$ mm^3 to $90 \times 90 \times 30$ mm^3 in contrast to the bigger sensor with d = 59 mm. An even smaller sensor would strongly increase measurement uncertainty. For foams ≥ 30 ppi sensors with d = 12 mm or d = 30 mm can be recommended.

High Temperature

Investigating the performance of the TPS method at temperatures up to 750 °C, Goetze et al. [8] found deviations up to 35% between the reference values [9] and the measured effective thermal conductivity of the material Silcal 1100 with an increasing deviation with increasing temperature. Particular high deviations were found around the Nickel Curie temperature, why the range between 350 and 420 °C is excluded from the measurements by the manufacturer.

Numerical simulations indicated that an inhomogeneous temperature field in the furnace does not affect the temperature rise in the sensor and is therefore not responsible for the observed deviations. In contrast, a redefinition of the TCR nearly directly influencing the thermal conductivity has led to a significant reduction of the deviations. A new set of TCR based on measurements with three different materials (Silcal 1100, OM 100 and stainless steel 1.4841) in the range of 20–750 °C was suggested by Goetze et al. [8] by recording the resistance and temperature for different sensor insulations and designs. As a result, deviations from the reference values with the material Silcal 1100 were reduced to ±7%, except in a narrow range around the Nickel Curie temperature.

Further investigations were performed to confirm the TCR proposed by Goetze et al. [8] in comparison to manufacturer values, to enhance the accuracy of high-temperature measurements and to investigate the TCR around the Curie point more

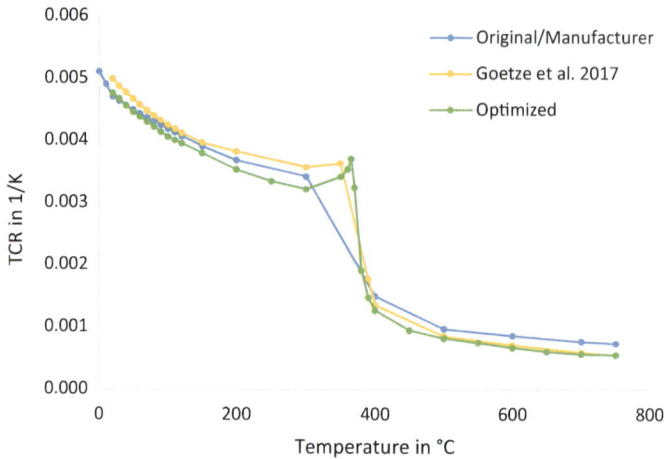

Fig. 14.2 TCR given by the manufacturer, calibrated by Goetze et al. (data taken from [8]) and newly optimized values for temperatures up to 750 °C

in detail. Measurements with four different reference materials (Silcal 1100, Pyroceram, Inconel 600 and Stainless Steel 304) were used to further improve the TCR, achieving deviations from the respective references of max. 6%, mainly below 4% for all four reference materials. By adding additional values, the TCR curve, particularly in the Curie range, was refined. However, a determination of the thermal conductivity at temperatures between 370 °C and 380 °C should be still excluded since deviations in this range strongly increase due to the strong temperature-dependent TCR and a required very accurate temperature measurement in the sample. Figure 14.2 summarizes the original TCR given by the manufacturer in comparison to the values modified by Goetze et al. [8] and the newly optimized TCR.

14.2.3 Measurement Results

With the corrected TCR, measurements with several alumina open-cell foams of different porosities and pore sizes (Fig. 14.3a) and with 10 ppi ceramic open-cell foams made of different materials (Fig. 14.3b) in the temperature range between 25 and 700 °C were conducted. With increasing porosity (65–90%), the effective thermal conductivity of the alumina foams strongly decreases, especially for temperatures <400 °C due to the reduction of solid mass fraction and the conductive heat transfer. While for foams with a porosity of 65 and 80% the thermal conductivity decreases continuously with increasing temperature, at higher porosity (90%).

A minimum in thermal conductivity around 300–400 °C is formed because of the higher void ratio and the resulting increased radiative heat transfer with increasing temperature. The pore size or ppi number has a lower impact on the ETC, both, at

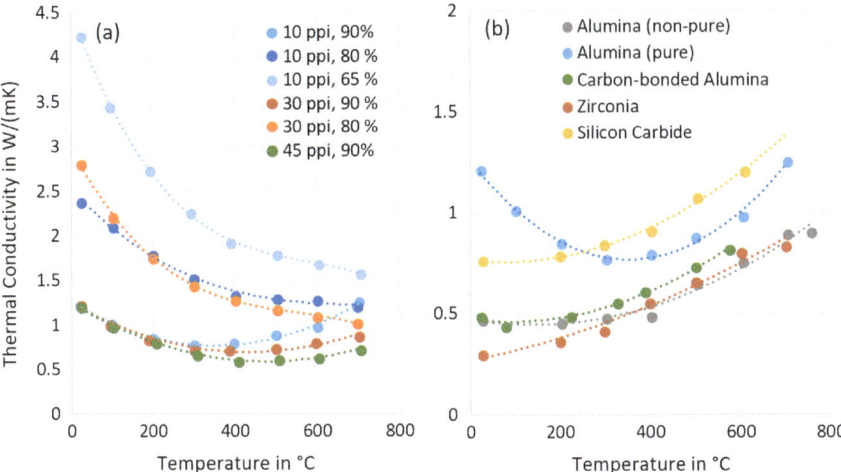

Fig. 14.3 ETC of **a** open-cell ceramic foams made of pure alumina with different porosities and pore ppi and **b** several 10 ppi ceramic open-cell foams made of different materials at temperature up to 700 °C

room temperature and at higher temperatures. With increasing pore size (decreasing ppi) the thermal conductivity at higher temperatures slightly increases because of increasing radiative heat transfer in the bigger pores. Furthermore, the observed minimum (around 300–400 °C with 10 ppi) shifts to higher temperatures with decreasing pore size.

Besides the porosity, the material of the open-cell foam can significantly affect its thermal conductivity. In contrast to the 10 ppi pure alumina foams, the solid thermal conductivity of the base material of all other investigated foams (carbon-bonded alumina, zirconia, silicon carbide, non-pure alumina) material increases with temperature, hence, the curves do not show a minimum of the thermal conductivity.

14.2.4   Models and Prediction

In addition to the experimental determination, a model-based description of the heat transport processes in the foams to numerically calculate the ETC was performed in cooperation with subproject B02 of the CRC 920. The main results are summarized by Mendes et al. [10] by comparing TPS measurements of alumina open-cell foams (10 ppi, porosity 89%) with a numerical 1D homogeneous model considering coupled conductive and radiative heat transfer. Previously the applicability of this model was confirmed by Mendes et al. [11] by comparing different detailed, 3D heterogeneous and simplified, 1D homogeneous models to determine the ETC at high temperatures. It was shown that the homogeneous models can yield errors of less than 10% with strongly reduced computational time and effort. Further validation was demonstrated in Mendes et al. [12] using open-cell metal foams.

To describe the geometry of the foams, 3D CT scans were carried out, which on the one hand enabled an exact structural characterization and on the other hand, after binarization, directly served to create the geometric model for the simulation. Important model parameters, such as the proportion of pure heat conduction through solid and gas phase and the extinction coefficient β were derived from the 3D CT scans. The simplified model is based on the homogenization approach considering a coupled conductive and radiative heat transfer. The 1D steady-state energy conservation equation including radiation as a source term is solved using the finite volume method [10].

The results show the potential of a precise prediction of the ETC only with geometric parameters obtained from the structural characterization of the foams using the 3D CT scans. A major advantage of modeling the heat transfer is that the influence of individual parameters can be analyzed with little effort. It was shown that the samples exhibited a certain anisotropy and, in addition to known influencing variables (such as total porosity and temperature), the microporosity in the struts not captured by the 3D CT scans also plays an important role. Taking the anisotropy of the foams into account the TPS measurements at temperatures up to 750 °C revealed good agreement with the simulations. The measured ETC can be interpreted as an averaged thermal conductivity of the room directions obtained from the simulation [10].

14.3 Radiation

At high-temperature processes like the metal melt filtration, radiative heat transfer can be of high relevance as long as the filter is not yet flowed through, for example during the targeted preheating or the natural heating of the filter by the oncoming melt flow. The temperature achieved through preheating has relevance for the thermal stress of the filter when contacting the melt, as well as to prevent solidification of the melt and thus guarantee a high filtration quality and efficiency.

Quantification of heat transport by radiation requires knowledge of the radiative properties of the material. These are the extinction coefficient, the scattering coefficient and the scattering phase function, which are mostly insufficiently investigated for open-cell ceramic foams so far.

In this chapter a comparison of different, mostly quite simple methods to determine the radiative properties, especially the extinction coefficient, of ceramic filters is presented. A main focus is also set on an experimental investigation of the radiative behavior of several filters by using a Fourier-Transform-Infrared (FTIR) spectrometercombined with an external integrating sphere. Finally, a short outlook of investigations on the radiation heat transfer in the filters is given.

14.3.1   Spectroscopic Measurements

As a basis for the experimental determination of the radiative properties, measurements of hemispherical transmission and reflection were performed with a Bruker Vertex 80v FTIR spectrometer (Fig. 14.4). Since the filters are highly inhomogeneous and normally used sample sizes would not produce representative measurement results, the FTIR spectrometer is equipped with an external integrating sphere (Fig. 14.4) with an inner diameter of 150 mm allowing it to cover a measurement area of about 25 mm. Measurements were performed in the Near-Infrared (NIR, $\lambda \approx 0.9–2.4\ \mu m$) and the Mid-Infrared (MIR, $\lambda \approx 2.0–16\ \mu m$) on different commercial filters used for metal melt filtration and made of alumina (Al_2O_3, 10 and 20 ppi), zirconia (ZrO_2), silicon carbide (SiC) or carbon-bonded alumina (Al_2O_3-C, each 10 ppi). Further information on the measurement parameters and samples can be found in Heisig et al. [13].

14.3.2   Radiative Behavior and Properties

The radiative behavior (transmission, reflection, absorption) as well as the experimentally determined spectral and Rosseland mean extinction coefficients of several ceramic open-cell foams are presented in detail by Heisig et al. [13]. Besides an influence of the ppi value, significant differences in the radiative behavior between light, oxidized ceramics (Al_2O_3, ZrO_2) and dark, carbon-containing materials (SiC, Al_2O_3-C) were demonstrated (see Fig. 14.5).

There are several approaches and methods to determine the radiative properties of porous media which can be either experimentally or numerically based. Furthermore,

Fig. 14.4 FTIR spectrometer Bruker Vertex 80v with external integrating sphere

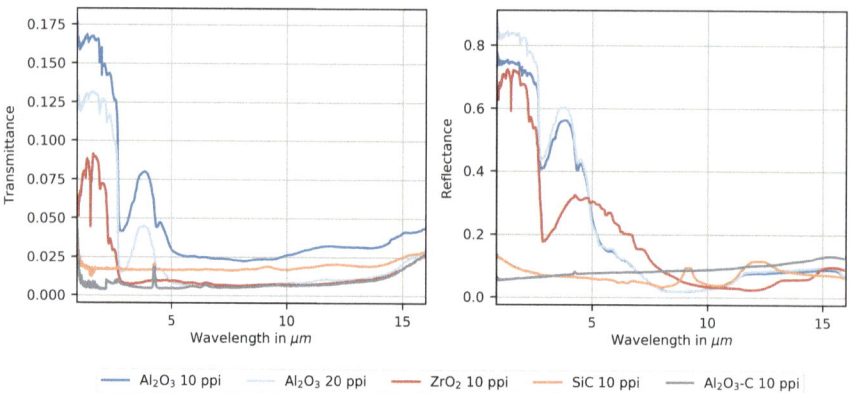

Fig. 14.5 Measured reflectance and transmittance of various ceramic open-cell foams

lots of empirical correlations are proposed in the literature mainly predicting the extinction coefficient.

A simple and often applied way to experimentally determine the extinction coefficient is using ***Bouguer's Law***. Assuming a homogeneous medium only absorbing or scattering radiation isotropically, the spectral extinction coefficient can be derived only from the measured transmittance and the sample thickness. For comparison and practical application, the wavelength-independent Rosseland mean extinction coefficient averaging the weighted spectral extinction according to Planck's radiation distribution over the relevant wavelength spectrum is commonly used [14]. Besides Bouguer's Law, the ***inverse parameter identification technique*** allows to determine radiative properties of the foams by solving the complete radiative transfer equation (RTE), thus considering also non-isotropic scattering. The comparison of simplified methods determining the extinction coefficient of open-cell foams in contrast to the parameter identification method will be a topic of future investigations.

In the practical application, ***empirical correlations*** are often applied, as a fast and simple method to estimate the radiative properties, especially the extinction coefficient avoiding elaborative measurements or numerical simulations. The accuracy and reliability of several predictive correlations were investigated and compared in Heisig et al. [13]. The best agreement with the previously experimentally determined extinction coefficients was achieved with correlations given by Hendricks and Howell [15] as well as Li et al. [16] with a mean deviation of 10 or 16%. Several other considered models lead to significantly higher average deviations up to 43%, since they might be based on limited data and, hence, not universally applicable to all types of ceramic foams. Furthermore, comparing deviations between the different filter materials of all investigated models a significant influence of the specific filter material or certain structural characteristics of the filter type on the extinction coefficient has been detected, not considered in the models so far and further to be investigated.

Finally, another disadvantage of the empirical correlations is that in general a wavelength or temperature dependency of the radiative behavior as it occurs for example with the semi-transparent Al_2O_3 foams is not respected in the models.

Besides empirical correlations an image superposition technique proposed by Loretz et al. [17] referred to as *projection method* in the following offers a simple, fast and more accurate determination of the extinction coefficient. The method is based on Bouguer's Law as well, however, instead of measurements the transmittance is obtained by processing the images of a 3D micro-computed tomography scan of the foam. The superposition method is also applied by Mendes et al. [10, 11]. Deviations from the experimentally determined extinction coefficients are limited to 18% (12% on average). The extinction coefficient is generally overpredicted, since, in contrast to the measurements, forward scattered radiation is not taken as transmission. Quite constant deviations of each filter material indicate that respecting the definite filter structure with the certain geometric characteristics of the different filter materials by using 3D tomography scans leads to strongly reduced deviations. However, a disadvantage is like with the empirical models, that opaque foams are assumed, i.e. the wavelength dependency of Al_2O_3 or ZrO_2 foams at higher temperatures is not considered.

14.3.3 Radiative Heat Transfer

In Mendes et al. [11] the radiative part of the steady-state heat flux in the foams, in addition to the conductive part, is calculated either by solving the complete RTE using the discrete transfer method or by simply applying Rossel and diffusion approximation. This approximation is quite commonly used in the context of open-cell foams. Quite accurate results for the ETC (deviation to detailed model considered as reference: <6%) can be received using the simplified models including radiative heat transfer solved without the approximation. In contrast, using Rosseland approximation leads to an overprediction of the ETC of 15–23%, possibly because the assumption of optically thick media (optical thickness >10) is either not or only just achieved for the investigated open-cell foams. In Mendes et al. [10], only the more accurate, simplified model solving the RTE is used. Nevertheless, the very easily applicable Rosseland approximation can serve as a rough estimation of the radiative transfer and is especially attractive since only the extinction coefficient is needed as a radiative property.

Besides the extinction coefficient two further radiative properties, namely the scattering coefficient or scattering albedo and the scattering phase function are required for solving the RTE, determined in Mendes et al. [10, 11] from suitable approximations: The scattering albedo ω was set equal to the reflectivity of the foam and the phase function was assumed to be the one for large diffuse spheres. The extinction coefficient into different directions was determined by applying the previously mentioned projection method. A strong decrease of the extinction coefficient for foam thicknesses with less than one mean pore diameter was observed, why this range was excluded from the evaluation in Mendes et al. [10]. Over the remaining sample thickness, β_{av}, β_{min} and β_{max} were determined, whereas it was found that results for the ETC fits best using β_{max}. In addition, as mentioned before, a combination of different spectroscopic measurements and the parameter identification method would offer a possibility to experimentally determine all the three radiative properties.

14.4   Convective Heat Transfer

When open-cell foams are flowed through by a fluid of different temperatures, convective heat transfer between the solid and the fluid characterized by the volumetric heat transfer coefficient h_v becomes relevant. In this context, Nusselt (Nu)-Reynolds (Re)-correlations are also often applied as a dimensionless relation between the heat and mass transfer.

Since a big application of ceramic foam filters is the foundry, the main goal is to investigate convective heat transfer between a filter and the metal melt (e.g. aluminum). Up to now, due to the significant simplification mainly experiments with air were conducted as also done by Vijay et al. [18] in the preliminary investigations presented in Sect. 14.4.1. To verify a transferability between the different fluids when using Nu-Re-correlations, h_v during the flow of the liquid metal through the filter should be determined. The progress of the construction of a suitable measurement section for aluminum melt as well as the first results are described in Sect. 14.4.2.

14.4.1   Determination of h_v with Air

Since Vijay et al. [19] found that when performing steady-state experiments thermal dispersion cannot be neglected and the dispersion conductivity k_d is in addition to h_v a second unknown parameter. Therefore, transient experiments using the single-blow method presented in Vijay et al. [18] should serve to determine h_v separately in advance. In a self-built wind tunnel, an air stream provided by a side channel blower was heated by a coil heater and passed through a flow straightener before streaming through an insulated section with the inserted foam. Furthermore, the measurement section is equipped with a thermal mass flow sensor as well as nine thermocouples at the inlet and outlet of the foam. After having a steady-state isothermal flow field in the test section, the inlet temperature was increased to the target temperature and recorded together with the outlet temperature for 20–60 s until thermal equilibrium was nearly reached.

Concerning the numerical evaluation, first, the flow field was determined using a modified Darcy-Forchheimer-Brinkman equation. Subsequently, the 1D energy equations (local thermal non-equilibrium) were solved using a simple implicit scheme. By calculating the instantaneous energy terms of each type of heat transfer in relation to the convection, it was shown that with transient experiments simplifications like neglecting dispersion lead to an error of less than 5%. This proved that the interstitial convection is the dominant mode of heat transfer and neglecting dispersion is permissible. Finally, h_v in dependence of the inlet superficial velocity was presented and compared to the literature for verification of the determined values. For increasing inlet superficial velocities from 1 to 10 m/s h_v increases from ≈ 7 to 22 W/(m^{3}K) for the 10 ppi foam and from ≈ 16 to 51 W/(m^{3}K) for the 30 ppi foam. It can be concluded from the results that smaller pore size (higher ppi number) as well as lower porosity leads to an increased h_v. Partially good agreement with literature values as well as reasons for inconsistencies were given.

14.4.2 Determination of h_v with Aluminum

In general, the experiments with a metal melt are to be carried out in the same way, also using the single-blow method, but with an appropriate adjustment of the experimental setup and the evaluation. A first attempt to create a suitable test rig was presented by Goetze et al. [20]. Using gravity casting the measurement section is a sand mold in which the aluminum is first redirected to flow through the inserted commercial alumina filters from bottom to top to achieve a more stable fluid flow. Between the filter mineral-insulated thermocouples are placed. A cast of the measurement section is shown in Fig. 14.6a.

A rough estimation of the average h_v was given and an average Nu-Re-Correlation was set up using the overall energy balance and assuming the whole process as a set of quasi-steady-state processes. Furthermore, several other, not realistic assumptions like adiabatic walls of the sand mold and hence, a 1D heat transfer, had to be made. For 10 and 30 ppi alumina filters with an achieved fluid velocity of 0.06–0.09 m/s (Re $\approx 500...1200$) average h_v between 780 and 2371 W/(m^{3}K) and average Nu between 0.85 and 2.37 were calculated. It was concluded that Nu increases with increasing flow velocity and with increasing ppi.

Since with the performed measurements, several problems and inaccuracies, like the varying pouring temperature, the melt flow rate that is not constant or adjustable or the low date acquisition rate appeared, general improvements and modifications in the design of the test section were made. Instead of performing gravity casting with a sand mold a steel pipe (see Fig. 14.6b) is placed on a low-pressure furnace, which enables to achieve a more constant casting temperature, a targeted adjustment of the melt velocity via the pressure and a uniform oncoming flow to the filters. Furthermore, the data acquisition rate can theoretically reach 10.000 Hz. Besides the thermocouples, an additional measurement of the melt velocity via an anode–cathode reaction was implemented. Furthermore, to estimate the heat losses of the measurement section as well as the flow profile, before and behind each filter three thermocouples with different distances to the edge were placed. The typical temperature curves at these measurement points are shown in Fig. 14.7 and the position of each thermocouple (TC) can be found in Fig. 14.6b.

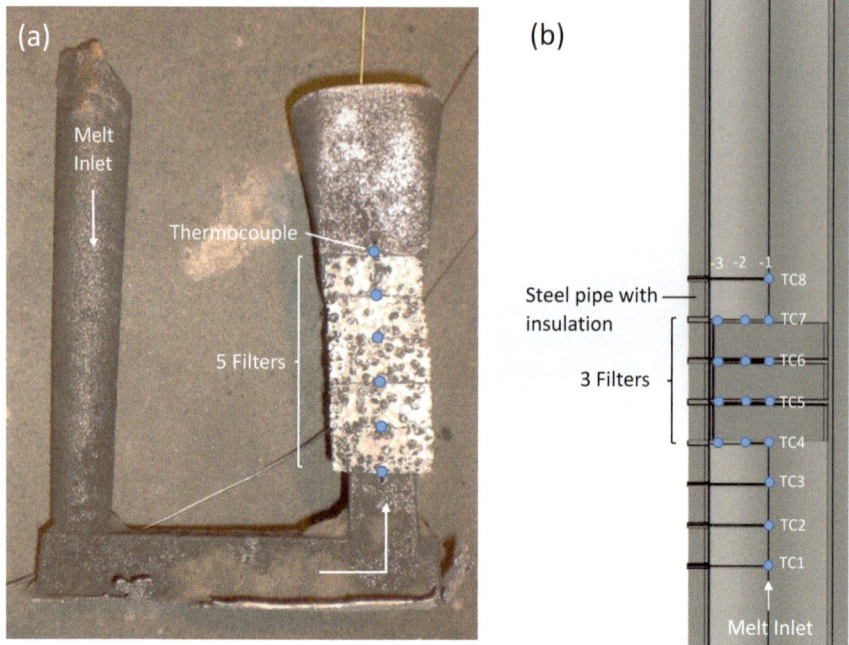

Fig. 14.6 Developed measurement setups for the determination of h_v with aluminum melt, **a** first measurement section for gravity casting, **b** improved measurement section used with a low-pressure furnace

In contrast to the previous evaluation of the measurement data, a more detailed simulation should serve to determine h_v. Since preliminary calculations have shown that the heat losses through the test section to the ambient have a considerable influence on the results, a 2D evaluation of the experiments, i.e. taking into account the radial heat transfer, is aimed. For conductive heat transport, the previously measured, temperature-dependent ETC is used (see Sect. 14.2.3). Using the same 10 and 30 ppi foams as with previous experiments, the first simplified simulations suggest h_v of 10^4 to $4 \cdot 10^5$ W/(m³K) and Nu of 0.5–16, which is significantly higher than the values obtained by Goetze et al. [20]. However, the same dependencies were found: With increasing velocity (Re), Nu increases. Furthermore, Re using the 10 ppi samples (Re ≈ 525–790) are clearly higher because of increased permeability compared to the 30 ppi filters (Re ≈ 180–490).

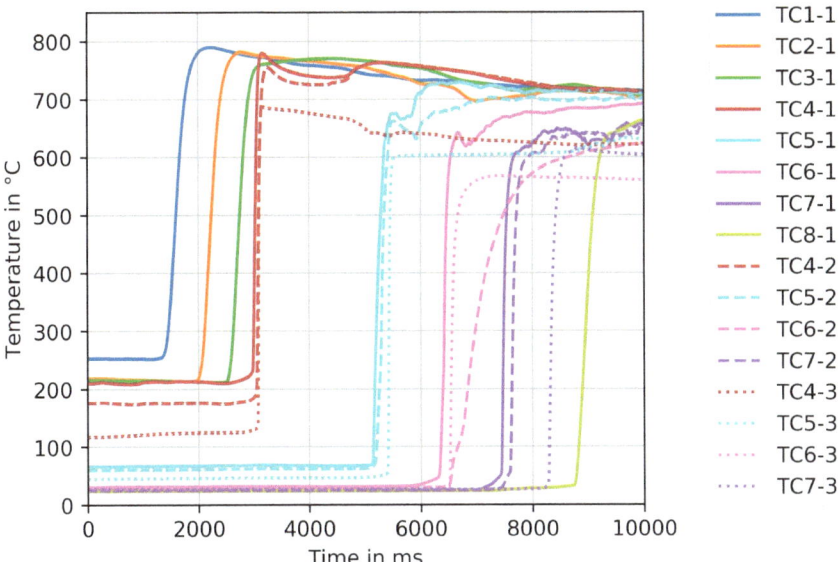

Fig. 14.7 Typical temperature curves received from aluminum experiments using improved measurement setup with a more dimensional temperature measurement

14.5 Solubility and Diffusion Properties of Hydrogen in Metal Melts

Apart from impurities and particle inclusions, especially hydrogen dissolved in molten metals has a negative effect on the mechanical properties, such as tensile strength, elongation and fatigue life of castings [21]. This can be attributed to the difference in hydrogen solubility of liquid and solid aluminum resulting in the formation of gas porosity during solidification. Concrete values of the solubility around the melting temperature vary considerably in the literature, but the individual data indicate a ten- to 20-fold higher solubility of hydrogen in liquid aluminum relative to its solid state [21, 22]. Thus, in order to minimize the hydrogen concentration in molten aluminum alloys, several techniques for melt treatment have been developed, but the understanding of hydrogen solubility and diffusion is insufficient. This issue should now be counteracted by the development of a new measuring apparatus for investigating the solubility and diffusion properties of hydrogen in metal melts. The initial focus was on the system hydrogen-aluminum melt.

After an overview of the initial situation in Sect. 14.5.1, the measurement method and experimental setup are explained in Sect. 14.5.2, followed by a more detailed description of the measurement procedure in Sect. 14.5.3. Finally, the calculation of diffusion coefficients of hydrogen in metal melts based on the kinetics of sorption processes is presented in Sect. 14.5.4.

14.5.1   State of Research

For the experimental determination of hydrogen solubility in aluminum melts, various measurement devices based on indirect or volumetric measuring methods were developed from 1922 until the end of the 1990s. For the volumetric measuring methods, quantities such as the measurement cell volume, dead volume and sample respectively melt volume have to be determined in advance to be able to infer the dissolved hydrogen quantity from the resulting pressure at constant volume or volume change at constant pressure. Most of the reported values of hydrogen solubility in liquid pure aluminum have been determined according to this measurement principle, in particular by means of the so-called "Sieverts' direct absorption method" [23]. Other techniques obtain solubility data from isothermal degassing processes of hydrogen-saturated samples or by measuring the gas content of previously solidified samples. Here, besides the possible loss of gas, the sudden change in hydrogen solubility during the liquidus-solidus transition must also be taken into account.

In general, the indirect determination of solubility data from the previously mentioned measured quantities is associated with numerous sources of error. This causes, together with further uncertainties of the respective method, widely diverging results in the literature [24]. Even at the melting temperature and a hydrogen pressure of 1 atm reported hydrogen solubility in liquid pure aluminum ranges from 0.43 cm³/ 100 g [22] to 0.918 cm³/100 g [21], with discrepancies increasing with increasing temperatures. This in turn leads to significant discrepancies in the results of mathematical models and calculations on melt treatment processes and porosity formation in aluminum products [23].

A very precise measurement principle for sorption investigations, in particular for long equilibrium times, is the gravimetric one, but up to now, this has not been used for the determination of gas solubility in metal melts. A reason for this is certainly that the measuring accuracy of balances in the middle of the twentieth century, when the majority of the experimental investigations were carried out, was comparatively low and thus the small mass of dissolved hydrogen was hardly detectable gravimetrically. Meanwhile, there has been significant progress in the field of weighing technology. Between 1965 and 1995, highly sensitive balances with electromagnetic force compensation were developed for commercial distribution. Using these, the first gravimetric measurements of gas adsorption were already possible for noncorrosive gases at pressures up to a maximum of 15 MPa, but only at temperatures below 175 °C [25]. In order to extend the application range and protect highly sensitive components, a technology was then developed that completely decouples the balance mechanically from the measurement cell. For this purpose, the sample load is transmitted from the measurement cell to an analytical balance placed under ambient conditions by means of a magnetic suspension system. In the sorption analysis, the use of these so-called magnetic suspension balances (MSB) is now well established for precise investigations under demanding conditions and hence intended to be applied to the measurement issue within the collaborative research center. Due to the direct recording of the sorption-related mass changes with the aid of a high-precision microbalance, the main sources of error and uncertainties of the previously used methods are avoided.

14.5.2 Measuring Method

Since at the beginning of the subproject no commercial sorption analyzer with MSB was available for the required temperature range, it was decided to use the thermogravimetric system DynTherm MP-HTIII (instrument type: TGA510), hereafter referred to as TGA, from TA Instruments. The TGA was modified according to the requirements of the measurement task and the instrumental setup was extended by the necessary systems for measurement and control. The adaptations include various aspects such as data acquisition, gas dosing and pressure control, the development of appropriate sample containers and sample preparation, as well as the minimization of environmental influences and explosion protection measures.

In the following, the basic principle of MSB is explained and the technical specifications of the apparatus used are described. For clarification, the schematic structure of the main device is shown in Fig. 14.8.

An electromagnet is mounted on the under-floor weighing attachment of the microbalance. This keeps the permanent magnet located in the upper area of the magnetically neutral coupling housing in a suspended state via a corresponding control device [26]. The permanent magnet is in turn connected to the sample container via a two-part rod. On the upper part of the linkage, between the permanent magnet and the measuring load coupling, there is a sensor core whose vertical position is measured via the inductance of an externally mounted sensor coil. Using a PID controller, the current applied to the electromagnet is adjusted to keep the position of the permanent magnet stable [27]. In this way, the weight force of the sample is transmitted contactlessly to the microbalance in the form of the required magnetic force. In addition, the position of the permanent magnet can be varied with the aid of a superimposed set point controller. This makes it possible to temporarily decouple the measuring load consisting of the sample in the sample container and the lower part of the linkage, which represents the zero point position [27]. In order to record the zero drift of the balance, the permanent magnet is periodically moved from the measuring point position to the zero point position while the measurement is in progress.

Besides the advantages of a magnetic suspension balance resulting from the separation of the measurement cell and the weighing instrument, the temperature control system is of particular importance for the measurement task. It has to thermally decouple the magnetic coupling from the measurement cell and at the same time ensure a precise control of the sample temperature. To achieve this, an electrical high-temperature heating element allows sample temperatures of 50 °C up to 1650 °C to be realized and kept constant inside the Al_2O_3 measurement cell. On the outside, the measurement cell has a thermal insulation and a water-cooled casing, so that

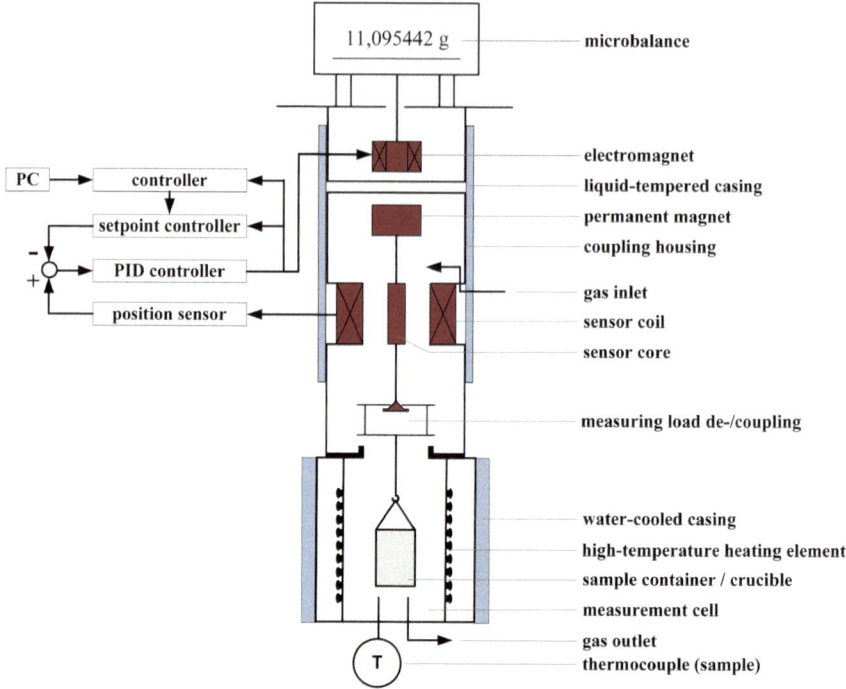

Fig. 14.8 Schematic construction of the applied magnetic suspension balance TGA with associated temperature control system

heat losses to the environment are minimized and a precise regulation of the furnace temperature with defined cooling rates is ensured. The tempering of the magnetic suspension coupling is carried out separately via a liquid-tempered double casing, since in the area of the magnetic force transmission temperatures of max. 258 °C may occur and temperature constancy is of enormous importance.

As the accuracy of the recorded weight contributes significantly to the quality of the measurement results, the TGA was combined with the 66S high-load microbalance of the Sartorius Cubis series, with a resolution of 0.001 mg. In addition, a computer-controlled gas dosing and pressure control unit (GDU) ensures precise pressure regulation and exact reproducibility of the conditions in the measurement cell. As Fig. 14.8 shows, the measurement apparatus has a separate gas inlet and outlet so that a dynamic pressure control can be implemented. By continuously supplying a defined gas flow rate (using thermal mass flow controllers) and regulating the desired pressure in the measurement cell, the conditions remain consistent throughout the entire measurement duration. Whereas with the previously used measurement methods already the loss of the slightest amount of hydrogen due to leakage or diffusion from the individual measurement apparatus led in some cases to significant errors in the calculated hydrogen solubilities, this does not affect the measurement accuracy of the new measurement apparatus.

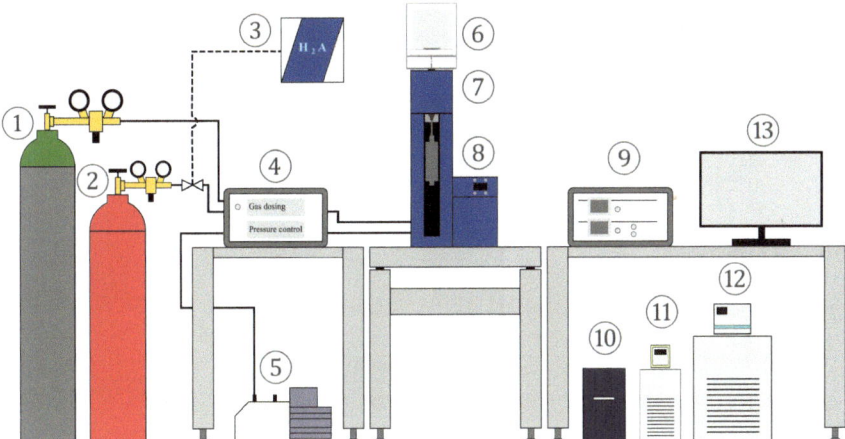

Fig. 14.9 Schematic representation of the measurement setup with the associated components. Therein describes ① Argon, ② Hydrogen, ③ Hydrogen shut-off system, ④ Gas dosing and pressure control unit, ⑤ Vacuum pump (explosion-proof), ⑥ Microbalance, ⑦ Magnetic suspension coupling, ⑧ Measurement cell with heating system, ⑨ Electronic and controller rack, ⑩ Power supply of the heating system, ⑪Temperature control unit of the coupling section, ⑫Recirculating chiller, ⑬Data acquisition

Figure 14.9 illustrates the experimental setup, including gas dosing, magnetic suspension and furnace, as well as several other components to ensure the proper functioning and control of the measurement apparatus.

The presented measurement apparatus enables fully automated long-term measurements in the pressure range of medium vacuum to a maximum of 160 kPa. Besides the time-resolved recording of temperature, pressure and sample weight in a controlled atmosphere, the zero drift of the balance is also recorded for an appropriate correction of the measurement data. This significantly increases the accuracy of the data. The electronic and controller rack ensures that the different positions of the permanent magnet are alternated periodically and that the temperature is regulated according to the previously generated temperature profile during measurements.

14.5.3   Experimental Details

Sample Container

Depending on the target value, different sample containers, also called crucibles, are used to investigate the maximum gas solubility and diffusion coefficients of dissolved gases in metal melts. Since the crucibles are manufactured entirely from high-purity alumina (99.9%) using the additive LCM method, they are suitable for measurement temperatures of up to 1650 °C. This enables a wide measuring range for

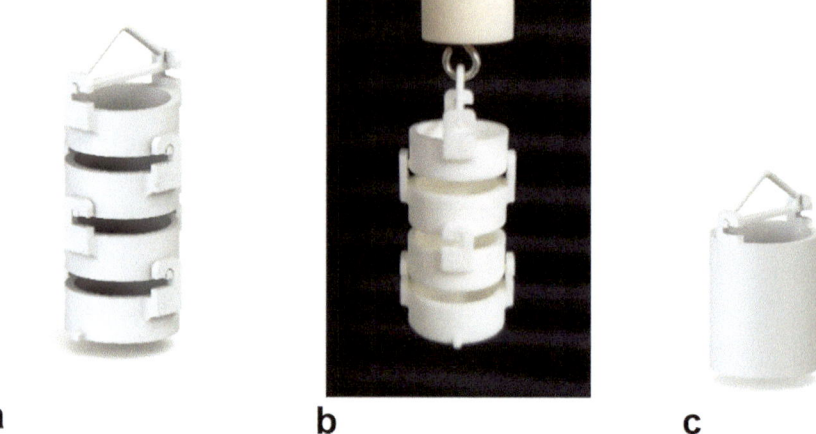

a   **b**   **c**

Fig. 14.10 All-ceramic sample crucibles for measurements of gases in metal melts, **a** model of the multi-layer crucible for solubility measurement, **b** photo of the mounted multi-layer crucible on the measuring apparatus, **c** model of the crucible for the determination of the binary diffusion coefficient

the determination of temperature-dependent sorption properties so that in addition to aluminum melts, molten iron alloys can also be investigated later on.

The crucibles are gas-impermeable, so only the top layer of the melt is accessible to the sorptive gas. In accordance with this, a high ratio of free surface to volume of the metal melt is required for the investigation of the maximum solubility to minimize the measurement time. A sample container system of four flat crucibles stacked above one another at defined distances was developed aiming to increase the sample quantity measured at the same time and thus the measurement sensitivity. Figure 14.10a shows a model of the developed multi-layer sample crucible and Fig. 14.10b illustrates the mounted condition on the measuring apparatus with a photo.

Besides the maximum solubility, the kinetics of the sorption processes are of interest. On the basis of time-resolved measurement data, the diffusion coefficients can be determined by the defined reference plane in the form of the melt's free surface. For this task, a crucible of a larger volume (see Fig. 14.10c) is chosen, so that effects occurring at the surface only influence the kinetics immediately at the beginning. As for liquids, interfacial reactions characteristically proceed faster than the diffusive mass transport of gases in the liquid [28], this method allows the determination of the binary diffusion coefficient.

Measuring Procedure

Aluminum with a purity of 99.999% is used for the investigations so that any influence of alloying elements or impurities is excluded. To determine the maximum solubility, samples with a diameter of 11 mm and a mass of about 1 g are contained by each crucible.

All measurements are performed under defined atmospheres using high purity gases ($\geq$99,999%). Prior to the start of each experiment, the measurement apparatus is evacuated three times and purged with inert gas. Here, argon is chosen as inert and reference gas, since Liu et al. [24] proved helium to be soluble in liquid aluminum. The samples are heated under vacuum or argon atmosphere with a heating rate of 5 K/min from ambient temperature to the desired measurement temperature. Then the atmosphere intended for the investigation as well as the gas flow conditions are set and the sorption measurement itself begins. Afterwards, the recorded data are corrected by baseline subtraction.

Initial Results

Due to the high affinity of molten aluminum to oxygen, the intended measurements on maximum solubility and for determining the diffusion coefficients of hydrogen in aluminum melt had to be deferred. Initial measurements with high-purity aluminum showed an increase in mass when reaching the liquidus temperature, which is attributed to the formation of an oxide layer. A significant impact of the sample temperature was observed, resulting in an increasing mass change ratio with increasing temperature (especially at the beginning of measurements). In contrast, whether the oxide layer formed at ambient conditions was removed or not prior the measurement did not show any significant effect.

These observations revealed the necessity of a lower partial pressure in the vacuum as well as the removal of the oxygen impurities ($\leq$2 ppm mass fraction) from the 5.0 gases used with the aid of special purification cartridges. In this context, the piping system and the sealing of the regularly loosened connections on the furnace were also optimized. Furthermore, a larger inert gas flow rate should ensure that even the slightest oxygen ingress diffusing through existing connections is avoided or removed quickly. If oxide layer formation should occur nevertheless, an O_2-trap based on a material with an even higher affinity to oxygen than aluminum might be installed.

14.5.4 Calculation of Diffusion Coefficients

The characteristically slow diffusion in liquids, and thus also in melts, often limits the overall rate of the occurring processes [28]. Hence, it is assumed that the rate of hydrogen dissolution is also limited by diffusion. This enables the study of diffusive mass transport of gases in metal melts based on the kinetics of sorption processes. A mass diffusion model commonly used in the literature for the diffusion of gases in liquids [29–32] is applied to estimate effective diffusion coefficients for the system hydrogen-aluminum melt on the basis of time-dependent solubility data. Moya et al. [32] even used this model for the estimation of diffusion coefficients out of gravimetrically obtained absorption kinetic curves (of CO_2 in ionic liquids).

In the present work, the mathematical description of the diffusion process is based on the following conditions:

- Reaching the maximum hydrogen solubility in the melt occurs by transient one-dimensional diffusion normal to the free surface.
- The free surface is considered flat; the crucibles are gas impermeable and cylindrical. Thus, the reference surface is constant.
- In the gas phase, no mass transfer coefficient has to be considered, since it consists of pure hydrogen.
- It is a unidirectional diffusion, since the aluminum melt does not transfer into the gas phase.
- No chemical reaction occurs between hydrogen and aluminum.
- The filling level l_0 of the gas-free aluminum sample corresponds to the characteristic length.
- Due to the low solubility [33], the swelling respectively the change of the filling level is negligible for the definition and evaluation of the diffusion coefficients, thus $l = l_0 = l_\infty$. Furthermore, the thermophysical properties do not change by hydrogen dissolution.

Since the diffusive mass transport j of the hydrogen in the melt is a one-dimensional diffusion process without convection in the same direction, the hydrogen concentration c at position z in the liquid aluminum can be described using Fick's law [28, 34].

$$-j_1 = D \cdot \frac{dc_1}{dz} \tag{14.3}$$

Assuming a constant diffusion coefficient D and a constant cross-sectional area over which diffusion occurs, the process can be described using the basic Eq. (14.4) for transient diffusion processes, also referred to as Fick's 2nd law [28, 35].

$$\frac{\partial c}{\partial t} = D \cdot \frac{\partial^2 c}{\partial z^2} \tag{14.4}$$

At the beginning of the measurement, no hydrogen is dissolved in the melt and thus, the entire melt volume has the uniform hydrogen concentration c_0. While this changes over the measurement duration, the hydrogen concentration c_l at the surface of the melt ($z = l$) is constant. These results in the following initial and boundary conditions for the gravimetric investigation method applied [30, 31]:

$$c = c_0 \quad \text{for} \quad 0 < z < l \quad \text{and} \quad t = 0 \quad \text{initial condition}$$

$$c = c_l \quad \text{for} \quad z = l \quad \text{and} \quad t \geq 0 \quad \text{boundary condition}$$

$$\frac{\partial c}{\partial z} = 0 \quad \text{for} \quad z = 0 \quad \text{and} \quad t \geq 0 \quad \text{boundary condition}$$

Under the prevailing conditions, Crank [35] specifies the analytical solution for Eq. (14.4) depending on the filling level l of the crucible as follows (Eq. (14.5)):

$$\frac{M_t}{M_\infty} = 1 - \frac{8}{\pi^2} \cdot \sum_{n=0}^{\infty} \frac{8}{(2n-1)^2} \cdot e^{\frac{-D(2n+1)^2\pi^2 t}{4l^2}}$$ (14.5)

The equation sets the total amount M_t of diffusing gas present in the melt at time t in relation to the corresponding amount M_∞ after infinite time, representing the maximum solubility. According to Crank [35], the solution Eq. (14.5) is accurate to four significant digits for desorption in the range $\frac{M_t}{M_\infty} < \frac{2}{3}$. In addition, the analytical solution (Eq. (14.5)) can be solved graphically for $\frac{Dt}{l^2}$ by appropriate representation based on the relative mass change $\frac{M_t}{M_\infty}$ as a function of sorption time t, and thus the diffusion coefficient can be determined out of the gradient of the plotted sorption data [35].

14.6 Conclusion

In the presented chapter experimental as well as numerical investigations on the heat transfer in open-cell ceramic foams used for metal melt filtration are summarized. Focus was especially set on the determination of important thermal key parameters of the filters.

First, studies concerning the thermophysical properties of alumina and carbon-bonded alumina used as bulk material for open-cell ceramic filters are presented. During preliminary investigations with the TPS method, the influence of different measurement parameters was analyzed and recommendations concerning the sensor and sample size were given. Furthermore, based on measurements with several reference materials, an improvement of the TCR for temperatures up to 750 °C was proposed. Using the optimized TCR, the temperature-dependent ETC of ceramic open-cell filters made of different materials and with different ppi and porosity were presented. In addition to the measurements, also modeling of the ETC was considered.

The radiative behavior of several ceramic open-cell foams was analyzed using an FTIR spectrometer, where the influences of the filter materials and the ppi number were demonstrated. Concerning the determination of the radiative properties, primary of the extinction coefficient, a comparison of the results of the experiments, different empirical correlations and the projection method have been presented. Finally, the calculation of the radiative heat transfer as part of the total heat transfer was examined more in detail.

After successfully characterizing the convective heat transfer in the foams with air, a measuring section was developed for the investigation of convection in the open-cell foams flowed through with molten metal (aluminum) and the determination of h_v during metal melt filtration. Due to the high temperatures and the complicated handling of the metal melt, there were numerous difficulties and necessary adaptations. Finally, the first results for convective heat transfer between alumina filters and aluminum melt could be presented.

A further objective was the development of a new measurement apparatus for the experimental determination of the maximum solubility and diffusion coefficients of hydrogen in metal melts. For this purpose, a thermogravimetric apparatus was combined with a high-load microbalance resolving 0.001 mg precisely by means of the technology of a magnetic suspension coupling. The measurement system was modified according to the requirements of the measurement task and the instrumental setup was extended by the necessary technical measurement and control components and continuously improved. Thus, the new high-temperature sorption apparatus is basically suitable for the determination of hydrogen solubility in metal melts at temperatures of up to 1650 °C in the low pressure range (vacuum to 160 kPa).

The main advantages in comparison to previously used experimental investigation methods are the direct and time-resolved acquisition of the sorption-induced weight changes. In addition to obtaining measurement system-independent equilibrium data, this makes it possible to investigate the kinetics of sorption processes and various other phenomena. Furthermore, the application of magnetic suspension balance technology not only enables measurement at very high sample temperatures, it also increases the accuracy of the data by recording the zero drift during measurement.

Acknowledgements The authors acknowledge the German Research Foundation (DFG) for supporting the Collaborative Research Center CRC 920—Project-ID 169148856, subproject B03. Furthermore, the authors would like to thank the previous contributors to subproject B03 and all those who supported the work.

References

1. P. Goetze, Internal Communication. Collaborative Research Center CRC 920-Project-ID 169148856, subproject B03, 1st founding period
2. P. Goetze, R. Wulf, U. Gross, M. Dopita, D. Rafaja, S. Dudczig, C.G. Aneziris, Y. Klemm, H.B. Biermann, Adv. Eng. Mater. **15**, 12 (2013). https://doi.org/10.1002/adem.201300203
3. S.E. Gustafsson, Rev. Sci. Instrum. **62**, 3 (1991). https://doi.org/10.1063/1.1142087
4. Y. He, Thermochim. Acta **436**, 1–2 (2005). https://doi.org/10.1016/j.tca.2005.06.026
5. Hot Disk AB. Hot Disk Thermal Constants Analyser—Instruction Manual, Revision date: 20 August 2019
6. T. Log, S.E. Gustafsson, Fire Mater. **19**, 1 (1995). https://doi.org/10.1002/fam.810190107
7. P. Goetze, R. Wulf, U. Gross, in *8th World Conference on Experimental Heat Transfer, Fluid Mechanics and Thermodynamics* (Lisbon, Portugal, 2013)
8. P. Goetze, S. Hummel, R. Wulf, T. Fieback, U. Gross, in *Proceedings of the 33rd International Thermal Conductivity Conference (ITCC) and the 21st International Thermal Expansion Symposium (ITES)* ed. by H. Ban (Utah State University, Logan, UT, USA, 2017). https://doi.org/10.12783/tc33-te21/30332
9. H.-P. Ebert, F. Hemberger, Int. J. Therm. Sci. **50**, 10 (2011). https://doi.org/10.1016/j.ijthermalsci.2011.05.007

10. M. A. Mendes, P. Goetze, P. Talukdar, E. Werzner, C. Demuth, P. Rössger, R. Wulf, U. Gross, D. Trimis, S. Ray, Int. J. Heat Mass Transf. **102** (2016). https://doi.org/10.1016/j.ijheatmasstransfer.2016.06.022

11. M. A. Mendes, P. Talukdar, S. Ray, D. Trimis, Int. J. Heat Mass Transf. **68** (2014). https://doi.org/10.1016/j.ijheatmasstransfer.2013.09.071

12. M. A. Mendes, V. Skibina, P. Talukdar, R. Wulf, U. Gross, D. Trimis, S. Ray, Int. J. Heat Mass Transf. **78** (2014). https://doi.org/10.1016/j.ijheatmasstransfer.2014.05.058

13. L.-M. Heisig, K. Markuske, E. Werzner, R. Wulf, T. M. Fieback, Adv. Eng. Mater. 2100723 (2021), https://doi.org/10.1002/adem.202100723

14. R. Siegel, J.R. Howell, *Thermal Radiation Heat Transfer*, 3rd edn. (Hemisphere Publishing Corporation, Washington, DC, 1992)

15. T. J. Hendricks, J. R. Howell, J. Heat Transf. **118** (1996), https://doi.org/10.1115/1.2824071

16. Y. Li, X.-L. Xia, C. Sun, S.-D. Zhang, H.-P. Tan, J. Quant. Spectrosc. Radiat. Transf. **224** (2019), https://doi.org/10.1016/j.jqsrt.2018.11.037

17. M. Loretz, E. Maire, D. Baillis, Adv. Eng. Mater. **10**, 4 (2008). https://doi.org/10.1002/adem.200700334

18. D. Vijay, P. Goetze, R. Wulf, U. Gross, Int. J. Therm. Sci. **98** (2015). https://doi.org/10.1016/j.ijthermalsci.2015.07.013

19. D. Vijay, P. Goetze, R. Wulf, U. Gross, Int. J. Therm. Sci. **98** (2015). https://doi.org/10.1016/j.ijthermalsci.2015.07.017

20. P. Goetze, D. Vijay, E. Jaeckel, R. Wulf, U. Gross, K. Eigenfeld, in *Proceedings of the 15th International Heat Transfer Conference (IHTC-15)* (Kyoto, Japan, 2014). https://doi.org/10.1615/IHTC15.fcv.009167

21. M. Tiryakioğlu, Metals **10**, 3 (2020). https://doi.org/10.3390/met10030368

22. W. Eichenauer, K. Hattenbach, A. Pebler, Int. J. Mater. Res. **52**, 10 (1961). https://doi.org/10.1515/ijmr-1961-521014

23. P.N. Anyalebechi, Mater. Sci. Appl. **13**, 04 (2022). https://doi.org/10.4236/msa.2022.134011

24. H. Liu, M. Bouchard, L. Zhang, J. Mater. Sci. **30**, 17 (1995). https://doi.org/10.1007/BF00361510

25. J.U. Keller, R. Staudt, *Gas Adsorption Equilibria, Experimental Methods and Adsorptive Isotherms* (Springer, New York, 2005)

26. H.W. Lösch, R. Kleinrahm, W. Wagner, Chem. Ing. Tech. **66**, 8 (1994). https://doi.org/10.1002/cite.330660808

27. F. Dreisbach, R. Kleinrahm, H.-W. Lösch, C. Lösch-Will, R. A. H. Seif,W. Wagner, *Patent*, DE 10 2009 009 204 B3 (2010)

28. E.L. Cussler, *Diffusion: Mass Transfer in Fluid Systems*, 3rd edn. (Cambridge Univ. Press, Cambridge, 2009)

29. A. Yokozeki, Int. J. Refrig. **25**, 6 (2002). https://doi.org/10.1016/S0140-7007(01)00066-4

30. M.B. Shiflett, A. Yokozeki, Ind. Eng. Chem. Res. **44**, 12 (2005). https://doi.org/10.1021/ie058003d

31. W. Schabel, I. Mamaliga, M. Kind, Chem. Ing. Tech. **75**, 12 (2003). https://doi.org/10.1002/cite.200390017

32. C. Moya, J. Palomar, M. Gonzalez-Miquel, J. Bedia, F. Rodriguez, Ind. Eng. Chem. Res. **53**, 35 (2014). https://doi.org/10.1021/ie501925d

33. P.N. Anyalebechi, TMS Light Metals (2003)

34. J. M. Zeng, Z. B. Xu, J. He, Adv. Mat. Res. **51** (2008), https://doi.org/10.4028/www.scientific.net/AMR.51.93

35. J. Crank, *The Mathematics of Diffusion*, 2nd edn. (Oxford University Press, Oxford, 1975)

Chapter 15
Microprocesses of Agglomeration, Hetero-coagulation and Particle Deposition of Poorly Wetted Surfaces in the Context of Metal Melt Filtration and Their Scale Up

Jan Nicklas, Lisa Ditscherlein, Shyamal Roy, Stefan Sandfeld, and Urs A. Peuker

15.1 Introduction

This Chapter is dedicated to the microprocesses of agglomeration and hetero-coagulation, which both have the potential to support the removal of non-metallic inclusions from metal melts. The removal of inclusions in the size range between 1 and 10 μm is especially problematic, as their movement is not governed by diffusion and their mass is not big enough for the movement to be fully governed by inertia effects. The attachment of particles to each other after collision leads to the formation of bigger particle identities, so called agglomerates. Similarly, the attachment of particles to bubbles is called hetero-coagulation and leads to the formation

J. Nicklas (✉) · L. Ditscherlein · U. A. Peuker (✉)
Institute of Mechanical Process Engineering and Mineral Processing, Technische Universität Bergakademie Freiberg, Agricolastr. 1, 09599 Freiberg, Germany
e-mail: jan.nicklas@mvtat.tu-freiberg.de

U. A. Peuker
e-mail: urs.peuker@mvtat.tu-freiberg.de

L. Ditscherlein
e-mail: lisa.ditscherlein@mvtat.tu-freiberg.de

S. Roy · S. Sandfeld
Institute for Advanced Simulation: Materials Data Science and Informatics, Forschungszentrum Juelich GmbH, 52425 Juelich, Germany
e-mail: s.roy@fz-juelich.de

S. Sandfeld
e-mail: s.sandfeld@fz-juelich.de

© The Author(s) 2024
C. G. Aneziris and H. Biermann (eds.), *Multifunctional Ceramic Filter Systems for Metal Melt Filtration*, Springer Series in Materials Science 337,
https://doi.org/10.1007/978-3-031-40930-1_15

of bigger identities with reduced density in comparison to the inclusion particles, which supports the cleaning of the melt by flotation of the inclusion particles. For both microprocesses, the likeliness of attachment is influenced by the kinetic energy of the particles and bubbles in the melt, interparticle forces and the surrounding boundary layers. For the hetero-coagulate formation, the deformation of the bubbles also can affect the attachment behavior. Cleaning of the metal melt before the filtration and the subsequent casting process is commonly done by the injection of reactive or inert gases like argon, that are injected by immersion lances or porous plugs. The initial bubble sizes are in the range of 10 mm to 20 mm for porous plugs [1] and in the range of multiple cm for immersion lances and their size increases as the bubbles rise to the surface. The injection of gases does not only serve the removal of inclusions by the flotation and entrainment of inclusion particles but also alters the composition of the melt [1], increases the mixing in the ladle and can decrease the number of gas cavities in the work pieces. Additional to the aforementioned cleaning strategies, the use of active and reactive ceramic foam filters allows an efficient removal of inclusion particles by depth-filtration. The application of reactive foam filters in casting processes leads to the formation of CO-bubbles by a carbothermal reaction between the filter material and the melt. The effect of those bubbles on the filtration efficiency is not fully clear and is neglectable according to numerical simulations of reactive filters in an induction crucible, as the bubbles do not detach from the filter surface [2]. Generally, bubbles inside the melt can partake in the removal of the inclusions by flotation. Additional to the true flotation, which involves particle-bubble-hetero-coagulates, a transfer of inclusions to the slag layer by entrainment has to be expected. The effect of agglomeration of non-metallic-inclusions, mainly alumina, was previously investigated in a number of theoretical and experimental studies in both actual metal melts and a water model system at room temperature. Investigations in the melt are expensive and are limited to relatively rough methods, as the high temperatures and opacity of the melt prevent the precise measurement of forces in the nano- and micro-Newton range, as they occur between micrometer sized inclusions.

15.2 Interparticle Forces in Molten Metal and the Water-Based Model System

The interparticle forces in metal melts are largely unknown and somewhat controversially discussed in the literature. The van der Waals forces cover three types of interactions, the Keesom-, Debye- and London-interactions. The London-interactions are the interactions between two induced dipoles, that are caused by a fluctuation of the charge distributions in the presence of close by molecules and are frequently referred to as dispersion interactions [3]. They are omnipresent in all materials and are the main type of interactions in many systems that have a non-polar character [3]. Many authors have attempted to estimate the van der Waals interactions between inclusion particles in metal melts and two main types of approaches for the estimation of Hamaker constants for metal melt systems are found. The first group of authors makes use of surface energies and utilizes the approaches of Fowkes. The applicability of the Fowkes approach for the determination of the Hamaker-constant for non-metallic inclusion—melt systems remains uncertain because the Fowkes approach only provides reliable results for a selected number of ideal, non-polar liquids [4]. Further it requires the precise measurement of contact angles that later on are used in combination with the Young-Equation, to calculate the surface energies. As the surfaces of inclusions like alumina exhibit high roughness values, it has to be expected that the macroscopic contact angle, obtained from measurements, and the true, microscopic contact angle are deviating. The contact angles of molten metals should be measured in ultra-high vacuum to prevent the formation of an oxide skin. Additionally, chemical reactions between substrate and molten metal are possible. The second group of authors follows along the London theory of dispersion forces to estimate the van der Waals interactions of liquid iron from the available Hamaker-constants for iron at room temperature, which lay in the range 324 to 562 zJ and are well summarized by Gomes de Sousa [5], but ignore any effects of the phase transition. Most notably for this approach are the works of Tanighuchi, who estimated the Hamaker-constant based on a modified Saffman-Turner model, that was applied on agglomeration experiments of alumina, silica and polystyrene latex particles at the IEP [6]. Unfortunately, the London theory of dispersion is formally not valid for interactions in solvents and assumes that the molecules have only 1 ionization potential [3], which is not true for metals. Both approaches have in common, that they require the use of mixing rules with limited applicability for Hamaker-constants and generally a critical discussion of the phase transition, the conducting character of metals, as well as the presence of multiple alloying elements in real metal melts are avoided. A third approach worth mentioning is the use of the Lifshitz and Casimir theories of dispersion in combination with modified oscillator models that have been fitted to optical data of molten and liquid metals. Most commonly Drude-type models are used to describe the dielectric function of metals. The Drude model can be derived from the Lorentz oscillator model for insulators by setting the restoring force equal to zero [7]. The applicability of the Drude model and modified Drude models, as they were suggested by Chen [8] and others to molten metal seems reasonable, as the molten metals lose their electronic band structure when they melt [9]. Daun pointed out that the successful application of the Drude model for molten metallic systems like silver or molten silicon have been reported, but for transition metals like iron and nickel the application still is difficult, due to the d-electron bands, that lead to either unphysical model fits or great deviations from the experimentally determined dielectric functions [9, 10]. The application of such erroneous oscillator models in the Lifshitz- and Casimir-theories would give unpredictable results and only should be attempted with physically sound oscillator models. A possible example of such an application to the liquid metal system is given by Esquivel-Sirvent, who calculated Casimir forces based on dielectric data for gold, mercury and an eutectic indium-gallium alloy, even though extrapolation of the optical data was necessary [11]. The availability of high-quality optical data for molten metal species is very limited. Often the datasets only cover few frequencies and the particularly important part of absorbance spectra is missing. As a consequence of the shortcomings in the theoretical description of the surface forces acting in the melt system, alternative measures of the present surface forces such as contact angle measurements are frequently utilized to characterize the interaction behavior of non-metallic inclusions in melt systems. The Young equation (Eq. 15.1) links the macroscopic contact angle θ to the three energies of the interfaces liquid–gas γ_{lg}, substrate—liquid γ_{sl} and substrate—gas γ_{sg}. The equation underlies the assumption of ideally smooth surfaces and thermodynamic equilibrium between the phases which are rarely present in technical systems.

$$\cos(\theta) = \frac{\gamma_{sg} - \gamma_{sl}}{\gamma_{lg}} \tag{15.1}$$

For molten metal on alumina surfaces, contact angles far above 100° are measured and contact angles as high as 155° are reported [12]. When using water droplets for contact angle measurements, one common convention is the distinction between hydrophilic $\theta < 90°$ and hydrophobic $\theta > 90°$ materials. In analogy for non-aqueous liquids, materials that exhibit contact angles above 90° are considered poorly wetted. This analogy is used in a water-based model system to investigate the behavior of poorly wetted particles during agglomeration, hetero-coagulation and the deposition of particles in ceramic foam filters. Four main factors have to be considered in a model system, the particle wettability, surface tension, viscosity and density. A common choice is the adjustment of the particle wettability by surface functionalization. Within the CRC920 the particle wettability of oxidic model inclusions is altered by silanization with the fluoroalkyl silane Dynasylan F8261 (1H,1H,2H,2H-perfluorooctyltriethoxysilane) from Evonik (Germany), which reacts with the surface OH-groups of oxidic materials, such as alumina, silica and glass [13]. The silanization procedure leads to contact angles of approximately 105° on smooth naturally oxidized Si-wafers and contact angles up to 134° on rough alumina surfaces.

In aqueous systems such as the water model system relevant interparticle forces include capillary forces, van der Waals forces, electric double layer forces and steric effects, as well as the more controversially discussed hydrophobic forces that are observed in poorly wetted particle systems. Those strongly attractive hydrophobic forces have been reported to act in a range of a couple nm up to several hundreds of nm and have been ascribed to organic contaminations, effects of the electric double layer, nanobubbles and other phenomena, of which many have proven to be unphysical. A physically reasonable approach for the description of hydrophobic forces is the separation of the hydrophobic forces into a short range, true hydrophobic part, that is caused by structuring of water molecules close to surfaces with low surface energy and a long-range part that occurs due to capillary bridging between nanobubbles or gas cavities that are pre-existing on the often rough, poorly wetted surfaces [14]. The wettability behavior of alumina surfaces in the metal melt, the expectation of strongly attractive van der Waals forces between the inclusions and the wettability dependent formation of gas cavities [12] suggest a similar surface force behavior to that observed on hydrophobic surfaces in water-based systems, as both the melt system and the water model system show a low affinity between inclusion and liquid phase. The effects of the capillary force are expected to be more pronounced in the melt system due to the higher surface tension. For molten aluminum at 700 °C the surface tension is around $\gamma_{lg} = 865$ mN/m [15], but for water at room temperature only $\gamma_{lg} = 72$ mN/m. The presence of gas adsorption layers in the melt system, seems reasonable when considering the theoretical estimates of strongly attractive van der Waals forces and the extremely high contact angles that the molten metals exhibit on ceramic surfaces. Water is a polar molecule and the adsorbed gas layer in the aqueous systems often is considered as a water depletion layer, which is caused by the reorientation of the dipoles parallel to the poorly wetted, hydrophobic surface as it is energetically more favorable [14].

The Atomic Force Microscope (AFM) is one of the most powerful tools for the characterization of surfaces and surface forces. Measurements with the instrument can provide information on the surface topography, as well as the surface forces of solid–solid and solid–fluid systems. The deflection of a cantilever, a thin beam, is measured via the position of an incident laser beam on a position-sensitive photo diode (PSPD). It can be converted to a force by using Hooke's law [16]. For the determination of the surface topography a soft cantilever with a nanometer thin probe tip is brought into contact with the surface and is moved line wise over it. The measured deflection is used directly to obtain a height-image of the surface from which measures for the characterization of the surface roughness such as the root mean square roughness and the peak to valley roughness can be extracted. The Colloidal-Probe-technique (CP) has frequently been employed in the last decades to investigate particle–particle and particle- bubble interactions by gluing or sintering [17] a commonly spherical three to twenty-five micrometer diameter particle of interest onto a cantilever tip. The CP-cantilevers manufactured by this method are brought into contact with other surfaces to measure the forces acting on the particle.

Alumina surfaces of both, inclusion particles and ceramic foam filters are commonly rough and the roughness has strong implications on the adhesion behavior. Ditscherlein reported root mean square roughness of rms = 0.8 μm for flat alumina samples and rms = 0.08 μm (scan size 3×3 μm^2) for spherical alumina particles [14]. For rough surfaces in the absence of capillary forces a decrease of the adhesion force is observed, that is caused by an altered contact area [18]. The prominent parts of the surfaces function as spacers that prevent the surrounding areas from contact. For the case of similar surfaces, where the van der Waals forces are attractive the roughness itself leads to a reduction of the adhesion force in comparison to smooth surfaces. Laitinen measured the adhesion force between spherical alumina colloidal probe particles and flat alumina substrates, reporting moderately good agreement of the classical Rabinovic model with AFM experiments in air, but did not quantify the influence of capillary condensation and deviations from the ideal spherical geometry on the measurements [19]. The particle adhesion between rough hydrophobized alumina surfaces in water–ethanol mixtures was investigated by Fritzsche [20], who was able to proof the presence of nanobubbles and the therewith connected different capillary interaction mechanisms using CP-FD-spectroscopy. Examples for the four types of FD-curves are shown in Fig. 15.1, of which three have a capillary character

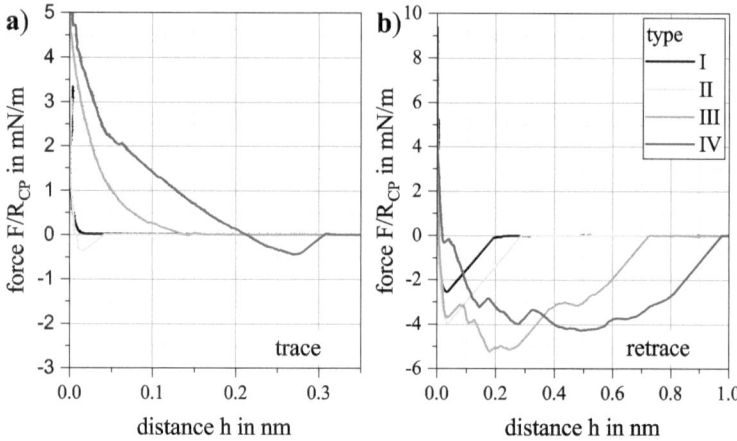

Fig. 15.1 **a** Trace and **b** Retrace FD-curves for the four interaction types observed on poorly wetted alumina [21]: (I) no capillary interaction, (II) snap-in caused by capillary interaction, (III) contact formation with a gas reservoir after retraction and (IV) snap-in in combination with stepwise pull-off as a consequence of the particle roughness

(II-IV) and the remaining type (I) corresponds to van der Waals interactions in the absence of capillary bridging. The FD-curves with capillary character in the most general case show a snap-in (II, IV), at which a sudden increase of the attractive force during the approach phase is measured if sufficiently soft cantilevers are used and which is characteristic for the formation of a capillary bridge. In the subsequent retract phase a snap-off occurs, which marks the loss of contact between the particle and the gas reservoir or surface respectively. Sometimes the contact is not lost at once, but happens in multiple steps with alternating stick and slide phases that are characteristic for the separation of rough particles from gas bubbles and gas reservoirs. The third type of capillary interactions (III) on rough poorly wetted substrates is characterized by a snap-off without the previous occurrence of a snap-in and can be the result of gas nucleation upon perturbation, or of a stable liquid film that does not drain completely during the approach phase, but ruptures upon the retraction of the colloidal probe, which is in line with recent predictions of the Stokes-Reynolds-Young–Laplace model, that predicts the contact formation upon retraction for moderate capillary numbers.

The overall adhesion characteristics of rough, poorly wetted alumina surfaces are resulting from the superimposition of the four types of interactions. The relative frequency f_R for the occurrence of capillary interactions between silanized alumina surfaces follows $f_R = -0.56299 \times \cos(\theta_{liquid}) + 0.37892$, where the wetting angle θ_{liquid} was altered by variation of the ethanol content between 0 and 20 wt.% [20, 21]. The increased forces that are measured in the presence of nanobubbles are caused by capillary bridges that can form upon particle contact. A numerical model for the calculation of capillary forces caused by gaseous capillary bridges was proposed by Fritzsche [22]. It is based on both, the minimization of the surface energies for

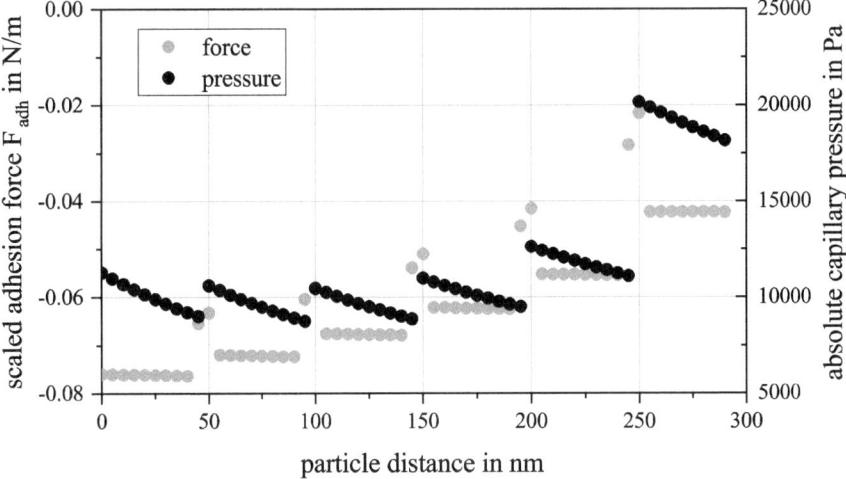

Fig. 15.2 Influence of the interparticle distance on the adhesion force that is caused by the capillary bridge under the assumption that the contact line of the capillary is pinned and jumps into its equilibrium position every 50 nm [22]

the calculation of the three-phase-contact-angle, and on a solution of the discretized Laplace-equation for the calculation of the meniscus outline [22]. Analysis with the model has shown that the total amount of gas molecules in the capillary bridge has only a small influence on the overall adhesion force, which can be explained by the incorporation of the ideal gas law into the set of equations instead of the constant volume constraint that is commonly assumed for liquid capillary bridges [22]. As the pressure increases proportional to the amount of gas molecules, the overall surface area remains almost unaffected which leads to a neglectable influence on the overall adhesion force [22]. Additionally, it was shown that contact line pinning influences the adhesion force when the distance between the particles is altered as shown in Fig. 15.2.

15.3   Modeling of Interparticle and Particle-Bubble Interactions

The short-range contribution of the hydrophobic forces that accounts for the water depletion layers can be modeled as adsorbed gas layers on the particle surface with sub nanometer thickness, using a retarded van der Waals layer model [23] described in Chap. 12. The silane coating on the particle surface is modeled as a PTFE-layer, as no optical data for the silane is available and the perfluorinated silane has chemical similarities to PTFE [24]. The attraction between two PTFE coated particles in theory should be weaker, than for pure alumina, but in experiments strong attraction is observed. The vdW-radius of nitrogen is approximately 0.155 nm and serves as a first estimate for the thickness of a monolayer of adsorbed gas on top of the PTFE layer [24].

The modeled curves for the homo-interaction of (non-)silanized particles and the hetero-interaction of a (non-)silanized alumina particle with an air bubble are shown in Fig. 15.3, for the cases of zero to three layers of adsorbed air. For both cases an increase of the Hamaker-function with an increase of the gas layer thickness is observed. For high values of the thickness of the layers, the Hamaker-function of the silanized particles approach the strongly attractive Hamaker-function of the air–water-air configuration [24]. The attraction between two uncoated alumina particles is greater than for the coated particles in the absence of adsorbed gas. At large distances above approximately 1 nm, the van der Waals force for the unfunctionalized particles is stronger than the forces between the functionalized particles with and without gas layer. For separations below 1 nm the particles with gas layers have much higher attractive Hamaker-function values close to 50 zJ, the Hamaker-function of the pure alumina in contrast plateaus at 37 zJ. In practice this means, that for both, the pure alumina and the silanized alumina agglomeration is likely, but due to the higher attraction at contact between the silanized particles their equilibrium agglomerate size will be bigger because the agglomerates can withstand higher shear rates before the contact between the primary particles in the agglomerate breaks.

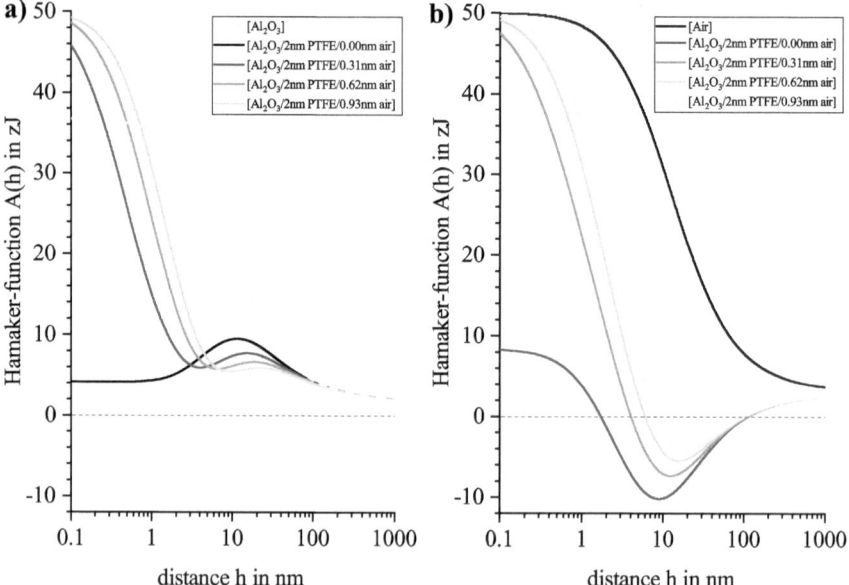

Fig. 15.3 Hamaker-function for the interaction of **a** likewise silanized alumina surfaces in the presence of adsorbed gas layers across water ([Particle]-[H_2O]-[Particle]) and the interaction of silanized alumina surface **b** in the presence of adsorbed gas layers with air across water ([Particle]-[H_2O]-[Air]), calculated with the vdW-layer-model

Similar to the particle–particle interation, the Hamaker-function for a (non-)coated particle interacting with a gas bubble can be calculated by modeling the layers for the first half space that corresponds to the particle and air as the second halfspace to account for the gas bubble. The interaction between the pure alumina and the gas bubble is repulsive at distances above 2 nm, which can easily be seen by the disjoining pressure profiles shown in Fig. 15.4 that correspond to the Hamaker- functions for the particle-bubble interaction shown in Fig. 15.3.

With increasing gas layer thickness the repulsive contribution of the pure alumina—bubble interaction decreases and strong attraction occurs at separations below 7 nm, when the gas layer exceeds a thickness of 0.31 nm. The modeled gas depletion layers on the silanized alumina particles that are used in the water model system all exhibit a strong exponential decay that concides well in range and slope with previously suggested exponential force law for the short-range hydrophobic forces between a hydrophobic particle and a gas bubble [25], that can be converted to a disjoining pressure Π_{dis} by Eq. 15.2.

The decay length parameter in the exponential force law Eq. 15.3 was previously determined in experiments to be approximately 0.3 nm [25]. The pre-exponential factor in Eq. 15.3 depends on the surface tension γ_{lg} and the macroscopic contact angle θ of the particle material with the liquid.

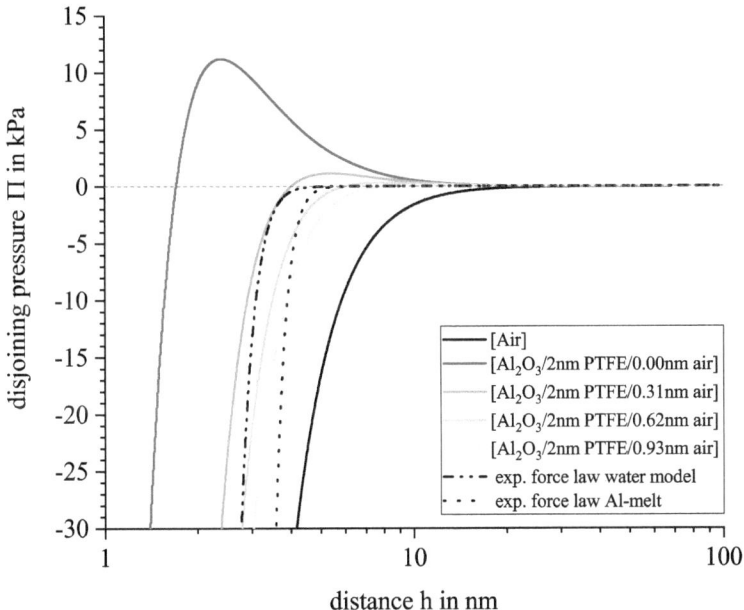

Fig. 15.4 Disjoining pressure for the interaction of silanized alumina surfaces in the presence of adsorbed gas layers with air across water ([Particle]-[H2O]-[Air]), obtained from the Hamaker-functions shown in Fig. 15.3 and approximation with the hydrophobic force law for the water model system and the Al-melt system

$$\Pi_{dis} = -\frac{dE(h)}{dh} \tag{15.2}$$

$$E_{hpb}(h) = -\gamma_{lg}(1 - \cos\theta)e^{\left(-\frac{h}{D_0}\right)} \tag{15.3}$$

As a first approximation for the attractive force between a gas bubble and poorly wetted alumina in molten metal, the surface tension of the melt and its contact angle are combined with the decay length of 0.3 nm for the particle-bubble interaction, as there is no knowledge about the forces in the melt system. From this, a distance dependent disjoining pressure function for the melt-system is approximated, which is used in combination with the Stokes-Reynolds-Young–Laplace-model for particle-bubble interactions, which currently is the most comprehensive model for bubble-particle interactions under conditions met in AFM experiments. Typically, an approach-retrace-cycle is simulated during which the colloidal probe first is moved towards the bubble with constant velocity, until a certain overlap of the colloidal probe with respect to the undisturbed position of the bubble interface is reached. Subsequently the particle is retracted from the bubble again with constant velocity. The validity of the SRYL model has been shown in numerous experimental and theoretical studies for interactions of deformable drop-drop, bubble–bubble, and drop/bubble-particle interactions [26]. It incorporates the key parameters interfacial tension γ_{lg}, fluid viscosity η, contact angle of the bubble θ_b, surface and hydrodynamic forces, as well as the curvatures that determine the collision stability for particle-bubble interactions.

$$\frac{\gamma_{lg}}{r}\frac{\partial}{\partial r}\left(r\frac{\partial h}{\partial r}\right) = \frac{2\gamma_{lg}}{R_n} - \Pi - p \tag{15.4}$$

$$R_n = \left(\frac{1}{R_0} + \frac{1}{R_{CP}}\right)^{-1} \tag{15.5}$$

$$\frac{\partial h}{\partial t} = \frac{1}{12\eta r}\frac{\partial}{\partial r}\left(rh^3\frac{\partial p}{\partial r}\right) \tag{15.6}$$

For the description of the film drainage the augmented Young–Laplace-equation (Eq. 15.4) is used in rewritten form for the film thickness in the AFM configuration. It describes the pressure difference over a curved interface in the presence of surface forces and additional hydrodynamic pressure. The latter is obtained by the Stokes-Reynolds-equation (Eq. 15.6) known from the lubrication theory, which describes creeping flows (Re < 1) in geometries, where one dimension is much smaller than another [26]. The total force between particle and bubble is obtained by integration over the interaction zone via Eq. 15.7 at the apex of the bubble for $r \in [0, r_{max}]$ under the assumption of symmetry boundary condition. The boundary condition at the end of this interaction zone given by Eq. 15.8 was derived from Carnie [27] by matching the inner and outer shape of a drop at the end of the interaction zone. It contains contributions for the forces in the interaction zone, the contact angle behavior $B(\theta_0)$ and optionally the cantilever spring constant k_c.

$$F(t) \cong 2\pi \int_0^{r_{max}} [p(r, t) + \Pi(r, t)]r dr + 2\pi \int_{r_{max}}^{\infty} p(r, t)r dr \qquad (15.7)$$

$$\frac{\partial h(r_{max}, t)}{\partial t} = \frac{dX(t)}{dt} + \left(\frac{1}{k_c} - \frac{1}{2\pi \gamma_{lg}} \left(log \left(\frac{r_{max}}{2R_0} \right) + B(\theta_0) \right) \right) \frac{dF(t)}{dt} \qquad (15.8)$$

$$Ca = \frac{\eta v}{\gamma_{lg}} \qquad (15.9)$$

The scaled total force over the central film thickness calculated with the SRYL-model and the parameters in Table 15.1, are shown in Fig. 15.5 for the room temperature model system in (a) and the Al-melt in (b) for the interaction between a particle with 10 μm diameter and a sessile bubble of size 200 μm diameter. For the water model system, the disjoining pressure is described with the vdW-layer-model for alumina with a 2 nm PTFE-layer and a gas layer of 0.31 nm thickness. Effects of the electrochemical double layer are not considered. The dynamic viscosity and the surface tension are those of water at 25 °C, therefore the capillary number (Eq. 15.9) is between $Ca = 1.85 \times 10^{-7}$ and $Ca = 5.56 \times 10^{-7}$. At approach velocities below 40 μm/s a rupture of the liquid-film between bubble and the functionalized particle is predicted. With increasing approach velocity, a higher repulsive force acts between particle and bubble, caused by the displacement of the liquid in the film. As a consequence, the bubble interface in the interaction zone deforms, which leads to a slower decrease of the central film thickness. This hydrodynamic influence becomes attractive as the particle is retracted from the surface and the liquid has to flow back into the gap, which can favor the contact upon retraction. If the film thickness is reduced below 4 nm the attractive van der Waals forces take over and particle and the bubble get into contact. In the Al-melt system, shown in Fig. 15.5b, the capillary numbers range from $Ca = 1.21*10^{-7}$ to $Ca = 6.06*10^{-7}$, as higher approach velocities between 100 μm/s and 500 μm/s have been chosen to identify the transition region between attachment and no attachment. Approach velocities below 350 μm/s lead to an unavoidable contact formation between particle and bubble, as the high interfacial tension of the melt system favors the attachment.

Table 15.1 Parameters used in the SRYL-simulations for silanized alumina in the water model system (RT-Model) and poorly wetted particles in the aluminum melt (Al-melt)

Parameter	RT-model (T = 25 °C)	Al-melt (T = 700 °C)	Unit
Radius of bubble R_0	100	100	μm
Radius of colloidal probe R_s	5	5	μm
Overlap ΔX	10	10	nm
Interfacial tension γ_{lg} [15]	72	865	mN/m
Dynamic viscosity η [15]	0.89	1.049	mPas
Contact angle θ	120	120	°

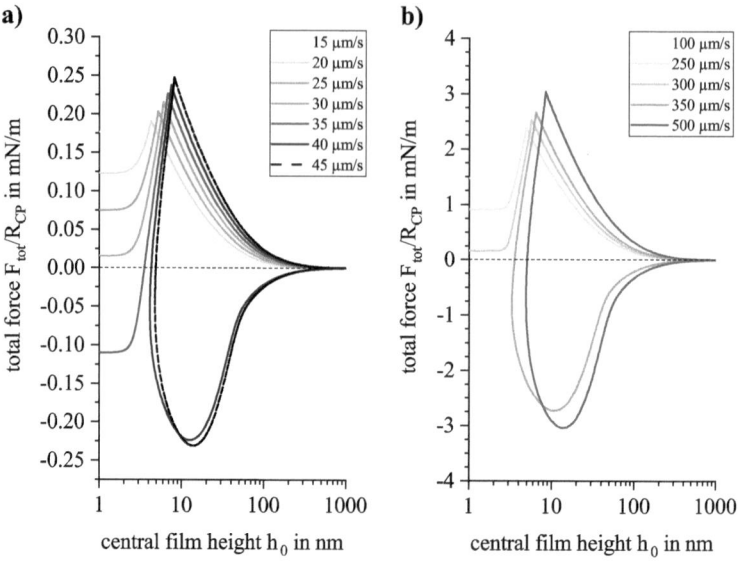

Fig. 15.5 Calculation of the scaled force over the central film thickness h_0 for the interaction of a bubble with $R_0 = 100 \, \mu m$ and a particle with $R_{CP} = 10 \, \mu m$ for **a** $|v| = 15–45 \, \mu m/s$ in the water model system and **b** $|v| = 100–500 \, \mu m/s$ in the Al-melt system

15.4 Particles in the Gas–Liquid-Interface

The behavior of particles in the gas–liquid interface can be studied on a sessile drop/ bubble with the CP-AFM technique. During an approach-retract-cycle as shown schematically in Fig. 15.6, the colloidal probe is driven towards the interface until a defined push-distance of the piezo is reached and is then retracted again from the interface. Knüpfer et al. investigated the case of a particle interacting with a sessile drop in an AFM experiment [13]. As colloidal probe particles both untreated (hydrophilic) and silanized (poorly wetted) silica particles of high sphericity were used to avoid the difficulties involved with the less regular alumina particles. For simplicity first, the interaction with a sessile drop is analyzed, but the overall force-distance-curves observed in experiments with gas bubbles are similar to the force-distance curves measured on a sessile drop. At first no significant force acts on the particle (1) as it approaches the drop until particle and drop are almost in contact. For small enough distances the interface deforms due to the capillary force, until the interface gets into contact with the particle (2). At this point the wetting of the particle begins (3) and the higher the wettability of the particle is, the faster it gets sucked into the interface of the drop.

If the particle wettability is high enough this leads to the so-called snap-in. Reversely for the interaction of the colloidal probe with a gas bubble a de-wetting of the particle occurs, and the velocity of the de-wetting process increases with a decreased particle wettability. For real inclusion particles that attach to gas bubbles

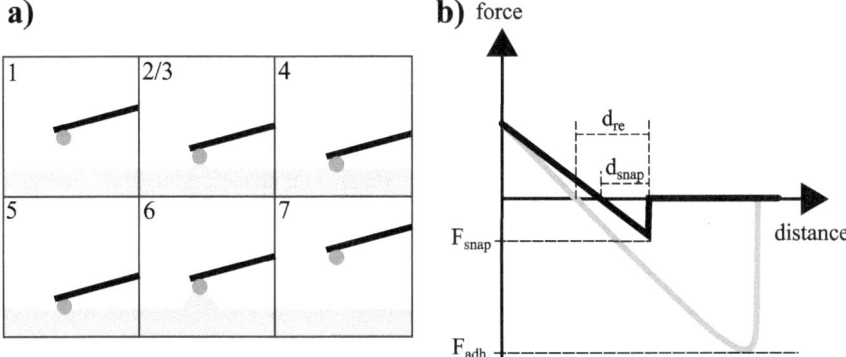

Fig. 15.6 **a** Particle at a gas–liquid interface during an approach-retrace cycle: (1) initial position, no forces act on the CP-particle, (2) snap-in, (3) wetting of the particle, (4) equilibrium position, (5) movement past the equilibrium position, (6) retraction from the interface, (7) particle after the snap-off; **b** Schematic force-distance curve for the interaction of a colloidal probe with a deformable bubble or drop with film rupture

this wettability dependent behavior would translate into an attachment probability that increases with increased three-phase-contact angle. In AFM experiments a strong attractive capillary force is measured at the snap-in. Subsequently the particle is approaching the original position of the interphase line, which leads to a reduction of the attractive force that is exerted on the particle, as the deformation of the interface is reduced. At the point where the force becomes zero, the particle reaches its wetta-bility dependent equilibrium position (4). Any movement of the particle further into the interface results in a repulsive force sensed by the cantilever (5). The distance between the zero crossings of the snap-in and the reaching of the equilibrium posi-tion d_{snap} can be used as a measure for the contact angle upon attachment θ_{at} of the particle in the interface by using Eq. 15.10 where R_{CP} is the radius of the colloidal probe particle and $d = d_{snap}$. For the drop case, θ_{at} is the advancing contact angle and for the bubble case θ_{at} is the receding contact angle. Similarly, the contact angle θ_{de} during the pull-off can be calculated by Eq. 15.10, using the distance between the first-zero crossing of the retrace curve and the position of the snap-in $d = d_{re}$.

$$\theta_{at/de} = \cos^{-1}\left(\frac{R-d}{R}\right) \tag{15.10}$$

The minimum of the force-distance curve marks the detachment or pull-off force, which is defined as the adhesion force F_{adh}. It depends on the size of the colloidal probe particle, the interfacial tension γ_{lg} and the particle wettability, characterized by the contact angle θ_{de} during the pull-off of the particle from the interface. The corresponding capillary force can be approximated by Eq. 15.11, which underlies the assumption of a constant contact angle and a free movement of the contact line and was suggested by Scheludko and Nikolov [13, 28, 29].

$$F_{\mathrm{drop,max}} = -2 \cdot \pi \cdot \gamma_{\mathrm{lg}} \cdot R_{CP} \cdot \cos^2\left(\frac{\theta_{\mathrm{rec}}}{2}\right) \quad \mathrm{Drop}$$

$$F_{\mathrm{bubble,max}} = -2 \cdot \pi \cdot \gamma_{\mathrm{lg}} \cdot R_{CP} \cdot \sin^2\left(\frac{\theta_{\mathrm{adv}}}{2}\right) \quad \mathrm{Bubble} \qquad (15.11)$$

A new model for the calculation of the maximum detachment force, using a blackbox-spring model, was developed by Knüpfer under the assumption of a linear deformation of the interface, which was observed experimentally for the major part of the retraction curve on sessile water droplets with radii in the mm range during the interaction with particles of 30 μm diameter [13]. The model incorporates a free Gibbs energy approach Eq. 15.12 that describes the wetting of the particle but not the deformation of the interface. This leads to the prediction of a linear force trend and the definition of a spring constant with the slope of $2\pi\gamma_{\mathrm{lg}}$. The maximum pull-off force can then be approximated by Eq. 15.13 [13].

$$-F = \frac{dE_{\mathrm{Gibbs}}}{dh} = -2\pi R\gamma_{\mathrm{lg}}\left(1 - \frac{h}{R} + \cos\theta\right) \qquad (15.12)$$

$$F_{max} = 2.33 R\gamma_{lg}(1 + \cos(\theta)) \qquad (15.13)$$

The scaled force of an approach-retract-cycle for a silanized alumina colloidal probe particle with a radius of $R_{CP} = 5$ μm during the interaction with a gas bubble of 120 μm diameter in the water model system is shown in Fig. 15.7 for three approach velocities between 10 and 30 μm/s. The measurements were carried out in an electrolyte solution that contains 1 mM NaCl and 1 mM NaNO$_3$ for the stabilization of the interface and a hydrophobic Si-wafer with a three-phase-contact angle of 105° was used as a substrate for the bubble. The force-distance curves in Fig. 15.7 have been shifted so that the equilibrium position of the particle in the interface during the approach is located at position zero. For all investigated velocities the particle behaves similar in the interface and no systematic dependency of the wetting behavior on the approach- and retrace-velocities is observed. The snap-in parts of the curves for 10 and 20 μm/s coincide directly with each other, but the maximum scaled force at 20 μm/s is 110 mN/m and at 10 μm/s only 60 mN/m, which is a consequence of the stick- and slide phases that the three-phase-contact-line of the gas bubble on the substrate undergoes during the repeated approach-retract-cycles of the AFM-experiment.

At 30 μm/s the observed snap-in distance is a little smaller than for the lower approach velocities, but the depth of the force minimum at the snap in remains unaffected and the difference in the observed snap-in-distances is likely caused by the pinning effects of the gas bubble on the substrate that directly affect the overlap between colloidal probe and initial bubble interface position. The force-distance curves of both, silanized and non-silanized alumina particles have a similar shape and all exhibit a snap-in, that is likely the consequence of gas reservoirs present on the roughness of both, silanized und untreated alumina particles.

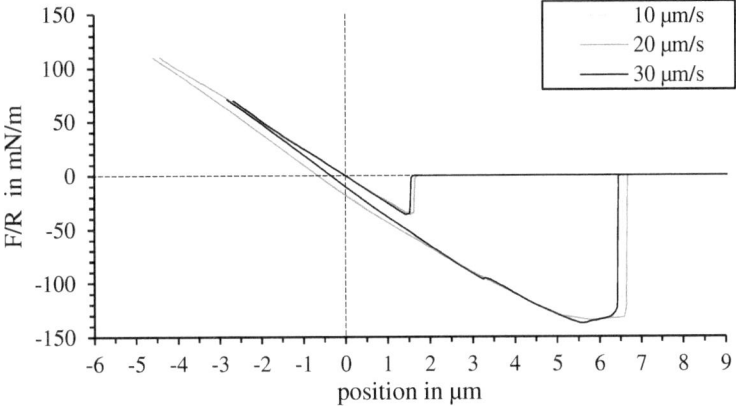

Fig. 15.7 Force distance curves for the interaction between a silanized alumina particle (CP1) and a gas bubble of 120 μm diameter in an electrolyte solution which contains 1 mM NaNO$_3$ and 1 mM NaCl

The scaled pull-off-force and the scaled force at the snap-in for three colloidal probes of each wettability state, silanized-alumina (CP1-CP3) and pure alumina (CP4-CP6), are shown in Fig. 15.8. The diameters of the particles given in Table 15.2 range from 7.6 to 21.4 μm and the measurements were done on three sessile bubbles with diameters between 85 and 120 μm.

The difference in particle wettability is evident from all three, the dynamic contact angles, the scaled force at the snap in and the force of adhesion, as the median values of the silanized colloidal probes CP1 to CP3 exceed the median values for the untreated colloidal probes CP4 to CP6. The static macroscopic contact angle for untreated alumina is $\theta = 90°$ and $\theta = 134°$ for silanized alumina, determined by sessile drop measurements on flat plates. The average of the median receding contact angle of the silanized alumina particles CP1 to CP3 is $\theta_{re} = 50.0°$ and significantly lower than the equilibrium contact angle of $\theta = 134°$. The average advancing contact angle of CP1 to CP3 is $\theta_{adv} = 62.8°$ and also significantly lower than the macroscopic value. Similarly, the average median values $\theta_{re} = 31.3°$ and $\theta_{adv} = 43.8°$ for untreated alumina are lower, which leads to the conclusion that the wetting behavior of particles with slight deviation from an ideal spherical shape and surface roughness is not well described by simple analysis of the force-distance measurements in combination with Eq. 15.10. The median values of the scaled force at the snap-in for silanized alumina lie between 20.5 and 36.8 mN/m and are significantly higher than the median values for the untreated alumina colloidal probes CP4 to CP6 with values of the scaled force between 6.2 mN/m to 13.7 mN/m. The occurrence of the snap-in for pure alumina seems surprising at first glance, but is likely a consequence of the surface roughness and cracks on the particle surface that can host gas reservoirs. Even if initially no gas reservoirs are present on the particle surface, the repeated contacting of the colloidal probe and the gas bubble likely leads to the formation of gas reservoirs that remain in the roughness after the rewetting of the colloidal probe. The scaled adhesion force

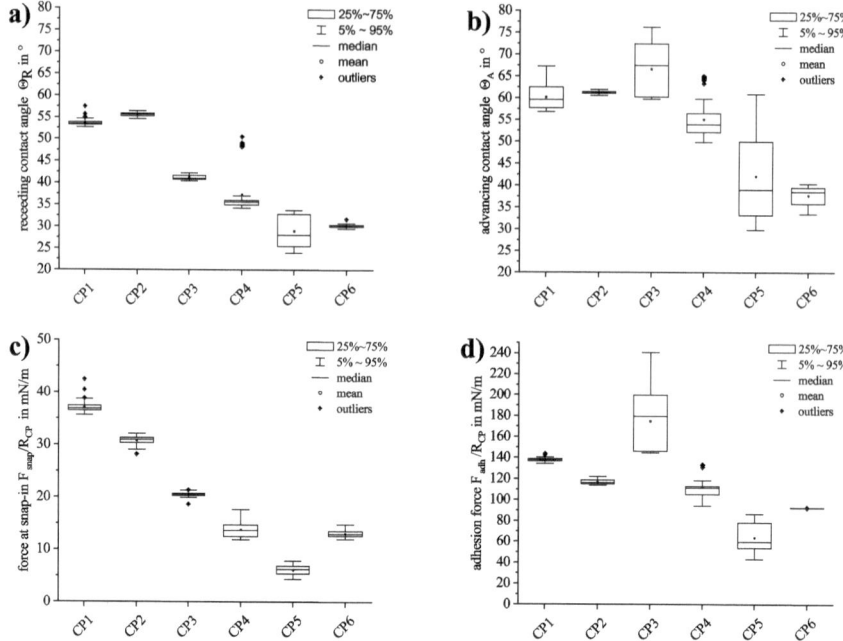

Fig. 15.8 Particle-bubble-interactions of three silanized alumina particles (CP1-CP3) and untreated alumina particles (CP4-CP6) measured on three gas bubbles in a solution of 1 mM NaNO$_3$ with 1 mM NaCl: **a** Receding contact angle; **b** advancing contact angle; **c** scaled force at the snap-in; **d** adhesion force

Table 15.2 Particle diameters, functionalization and spring constants of colloidal probes used in the experimental investigations, as well as the approximate diameters of the sessile bubbles of each experiment

Particle			$\emptyset_{particle}$ in μm	Silanization	k_c in N/m	$\emptyset_{bubble}$ in μm
Al$_2$O$_3$-Sil	CP	1	7.6	Yes	4.0	120
Al$_2$O$_3$-Sil	CP	2	10.9	Yes	6.1	85
Al$_2$O$_3$-Sil	CP	3	15.7	Yes	5.3	93
Al$_2$O$_3$	CP	4	19.2	No	7.0	93
Al$_2$O$_3$	CP	5	21.4	No	6.8	93
Al$_2$O$_3$	CP	6	16.6	No	3.2	93

that has to be overcome to pull the colloidal probe particles out of the bubble interface is less wettability dependent and possibly governed by the particle roughness, as the observed ranges of the pull-off force for both wettability states are overlapping. The median values of the scaled adhesion force for CP1 to CP3 are in the range 116.3 to 179.5 mN/m and for the untreated alumina particles CP4 to CP6 in the range 59.6 to 111.3 mN/m. This result indicates that the likeliness of attachment to a gas bubble is increased with decreased particle wettability and a slight increase of the stability against detachment from the interface for the poorly wetted particle system is observed. Overall the differences in wettability are expected to lead to an increased net-rate of hetero-coagulate formation with decreased particle wettability.

15.5 High Temperature Force Measurements

In contrast to the van der Waals and capillary adhesion mechanisms, the water-based model system does not take into account additional effects such as sintering of the inclusion particles on the filter wall due to the extremely high temperatures, especially in the case of molten steel filtration. A general overview of adhesive mechanisms between two surfaces under elevated temperatures is given by Berbner et al. whereby adhesive forces with material bridges (e.g. chemical reaction, sintering, crystallisation) are usually stronger [30].

Adhesive force measurements under elevated temperatures, especially with the atomic force microscope, are comparatively little known, as the high temperatures can very quickly destroy the expensive sensor technology. Specially shielded devices are used in which the samples can be heated, for example, via a filament or directly by direct current supply. The temperature range investigated in the literature is also significantly lower than in the case of the metal melt filtration process; investigations have so far been carried out in the temperature range between 100 and 200 °C. In an air atmosphere, an increase in adhesive force due to capillary condensation can be observed first, followed by a steady but slight decrease. Even in a nitrogen atmosphere, where no capillary condensation occurs, a slight but steady decrease in adhesive force with increasing temperature can be observed [31–33]. A stronger atomic oscillation energy is assumed to be the cause. If instead the glass transition temperature of the sample surface is reached, in this case polyester, the adhesive force determined via an apparatus similar to the atomic force microscope first begins to rise moderately, then sharply, and drops slightly at even higher temperatures. It is assumed that the highest energy dissipation rate is achieved by viscoelastic behaviour [34]. Using the centrifuge method, it could be shown that gold particles under mild sintering temperatures at 400 °C show an up to 100-fold increase in adhesive strength compared to measurements at 20 °C, due to plastic deformation and surface diffusion at the contact point [35].

In terms of particle interactions, sintering is the diffusion creep under the influence of capillary forces. Sintering only takes place when the sintering potential, i.e. the Gibbs free energy, is negative; the system thus aims to minimise the Gibbs free energy. Material transport takes place at the interface, where different mechanisms can occur depending on the system conditions, such as temperature, material, particle size or heating regime, which are summarised in Table 15.3.

An important point is that the sintering temperature T_{sint} can sometimes be significantly below the melting temperature ($T_{\mathrm{sint}} \approx 0.50 \ldots 0.95 \cdot T_{\mathrm{melt}}$), whereby for individual particles sintering effects also occur in the early range, i.e. at low temperature.

Table 15.3 Parameters of sinter kinetics for different mechanisms

Mechanism	n	m
Viscous flow	2	1
Volume diffusion	6	3
Condensation	3	1
Surface diffusion	7	3
Compressive force sintering	4	2

From the work of Kuczynski and Krupp [36, 37], the force needed to separate two sintered surfaces can be determined with the help of a material pair-specific constant $S(T_{sint})$, the characteristic constants m und n (see Table 15.3), the tensile strength $\sigma_{tensile}$, the particle radius R and, assuming an Arrhenius approach, the force needed to separate two sintered surfaces can be calculated via Eq. 15.14:

$$F = \pi \sigma_{tensile} R^{\frac{2m}{n}} (S_0 t_{sint})^{\frac{2}{n}} e^{-\frac{2E_A}{nk_B T_{sint}}} \tag{15.14}$$

where t_{sint} corresponds to the sintering time, E_A to the activation energy and k_B the Boltzmann constant. This classical equation, which not only results from the tensile strength of the sinter neck, simultaneously considers the neck growth.

With the aid of the UHV SPM 7500 high-temperature scanning force microscope from RHK Technology, it is possible to realise adhesive force measurements up to 800 °C, using a developed method for the production of temperature-stable colloidal probe cantilevers, (see Fig. 15.9b) [17]. Results for the alumina system (Al_2O_3 particles and Al_2O_3-coated Si wafer) are shown in Fig. 15.9c).

On the one hand, the influence of the atmosphere (ambient air and thus a certain humidity increases the adhesive forces compared to vacuum) and the dwell time of the particle on the sample surface (a longer dwell time also corresponds to a higher adhesive force) must be taken into account. A change in the cantilever resonance frequency after the tests at elevated temperatures was not significant. A peak at about 140 °C is noticeable, which can be attributed to burning of residues of the not completely removed dispersant on the particle surface. Up to about 730 °C, the van

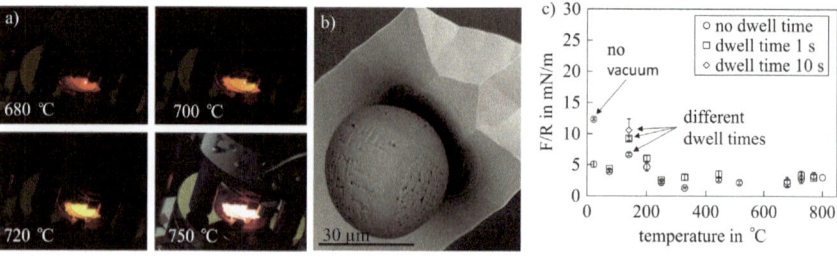

Fig. 15.9 **a** View at the heated samples; **b** temperature stable colloidal probe cantilever made with phenolic resin; **c** mean normalized adhesive force in dependence of sample temperature

der Waals forces decrease slightly, after which they hardly increase. Here, too, as in Lai et al., a stronger fluctuation of the electrons is assumed, which ultimately reduces the attractive interactions. For metal melt filtration, it can therefore be concluded that with pure vdW interactions (i.e. no capillary forces or sintering effects), the adhesion of the inclusion particles to the filter wall is slightly reduced. The opposite is the case with the additional occurrence of sintering effects due to very high temperatures. In order to realise such measurements, a model system had to be used, as temperatures even higher than 800 °C are not feasible. Polystyrene particles were therefore chosen as the model inclusion and temperatures up to a maximum of 240 °C were set [38]. Another characteristic temperature besides the melting point between 240 and 270 °C (depending on the composition) is the glass transition temperature, i.e. the temperature that describes the transition from brittle-elastic to rubber-elastic behaviour; this is about 100 °C for polystyrene. The results are shown in Fig. 15.10.

The SEM image in (a) shows that sintering has already occurred at the selected temperature (here 135 °C) after contact with the hot surface; the previously almost ideally smooth spherical particle shows several rough elevations. At lower temperatures, a flattening could be observed, which fits well with the behaviour when the glass temperature is exceeded (plastic deformation). Compared to theoretically calculated values (vdW forces), the normalised adhesive force values are somewhat lower, which is due to smallest roughnesses. At about 150 °C, a force maximum is achieved at 313 mN/m, which corresponds to more than 20 times the value at 85 °C and can be explained by stronger sintering effects (Fig. 15.10c)). After that, the adhesive force decreases, which is primarily due to more intensive melting and can be observed, for example, through increased ring formation on the particle. Example curves in Fig. 15.10b) also show the changed shape of the force-distance curves, which further support this assumption.

From the experimentally determined averaged adhesive force values and temperature-dependent surface tension values, a simple equation (Eq. 15.15) can finally be approximated to the measured data via a combination of the above-mentioned sinter model and common capillary force models:

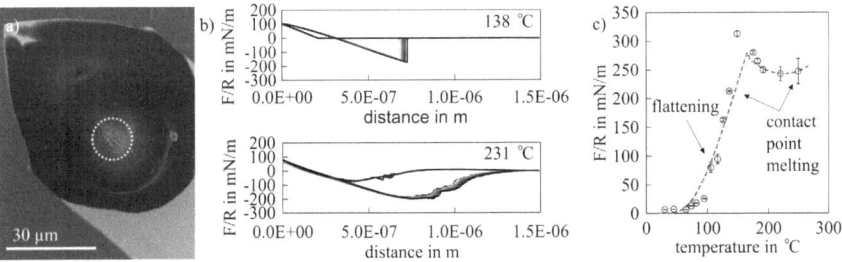

Fig. 15.10 **a** Used polystyrene probe cantilever with characteristic contact zone; **b** exemplary force distance curves showing large shape difference for a temperature just above the glass transition point and just before melting point; **c** mean normalized adhesive forces depending on sample temperature

$$F = 2\pi R\gamma e^{-\frac{C}{T}} \tag{15.15}$$

A radius of 13.075 μm can be chosen for the particles. The surface tension is calculated according to Moreira with $\gamma = f(T in K) = -0.774T + 66.655$ [39]. This results in an activation energy of 1.2 eV for the model particles with $C = \frac{2E_A}{nk_B}$ assuming volume diffusion, which is within a reasonable size range [35].

15.6   Sintering

The high temperatures in the molten metal can lead to sintering of previously agglomerated inclusion particles which results in an increased agglomerate stability against particle redispersion. Sintering of alumina can be studied by atomistic simulations under the condition that an appropriate interatomic potential is chosen, as the quality and predictability of sintering is strongly depended on the performance of the underlying interatomic potentials. The suitability of the four empirical interatomic potentials, Vashishta potential (Vash) [40], Coulomb-Buckingham (CB) potential of matsui type [41], Born–Mayer-Huggins potential (BMH) [42] and the Charge Transfer Ionic + EAM potential (CTIE) [43] was investigated by Roy [44]. Benchmarking of the potentials showed that most material properties predicted by all four potentials are well within the range of the experimental data available in literature, except for the predictions of elastic constants and surface energies by the BMH and CTIE potentials. An accurate prediction of the experimentally reported melting temperature of Al_2O_3 that lies in the range between 2200 and 2350 K is only predicted with the CB potential where the melting temperature is 2340 K. The melting temperatures obtained with the other potentials are significantly higher and range up to 4200 K for the CTIE potential, therefore the homologous temperature of the corresponding potential was used to facilitate an objective comparison of the sintering process with different potentials.

The simulation setup for the MD simulations of the sintering is shown in Fig. 15.11. The sintering progress is evaluated by six parameters, namely the shrinkage ratio, normalized surface area, mean square displacement, neck curvature, the fraction of ions at the neck and the norm of the stretch tensor.

All potentials show an increased amount of sintering with increasing temperature and the sintering is faster for small particles. The Vash potential predicts the material properties after sintering better than the other three potentials, and shows the lowest degree of sintering among the four. The CTIE potential predicts the highest degree of sintering of the potentials considered.

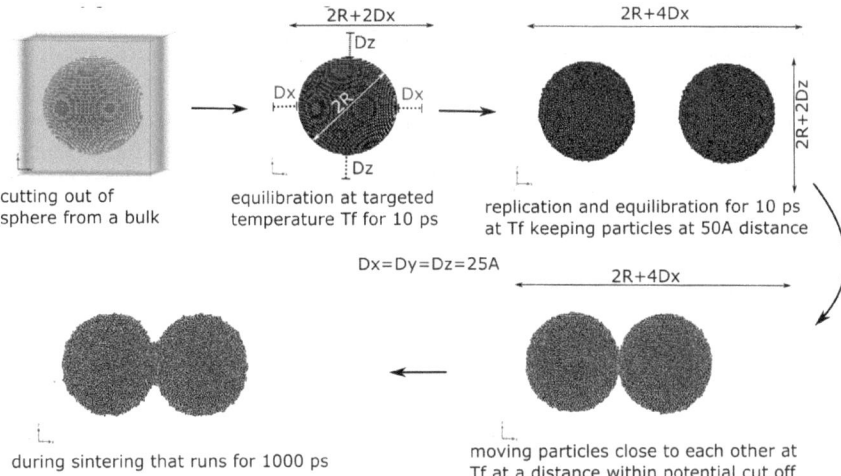

Fig. 15.11 Simulation setup: A particle is cut out from bulk material (shown as semi-transparent), where Dx = Dy = Dz = 25 Å is the smallest distance between particle and the box sides. The particle is heated up to the target temperature Tf and thermalized at Tf for 10 ps each. It is then replicated and equilibrated at Tf. The particles are then brought together at a distance within the potential cut off and at Tf. Subsequently, the system is maintained at the Tf for 1 ns during which sintering can occur under favourable conditions

15.7   Agglomeration and Hetero-coagulation of Poorly Wetted Al₂O₃ Particles

The agglomeration behavior of silanized Al_2O_3 particles under conditions relevant for technical scale processes was investigated in a stirred tank setup by Knüpfer et al., who in addition to the stirred tank experiments provided detailed information on the interplay of flow conditions and surface forces, characterizing the latter by AFM measurements, with special emphasis on the presence of nanobubbles [45]. Nucleation of defined nanobubbles can be triggered by many procedures that include solvent exchange, temperature gradients and perturbation. A solvent exchange procedure is employed in the agglomeration experiments to investigate the influence of nanobubbles on the agglomeration and stability of agglomerates. The presence of nanobubbles can be proved by phase contrast images obtained from AFM as shown in Fig. 15.13b) for nanobubbles on a hydrophobized Al_2O_3 substrate after a solvent exchange with ethanol. The experiments were carried out at the isoelectric point in 0.5 M NaCl solution and with non-spherical Al_2O_3 particles with a maximum particle size of 25 μm. The experimental setup consists of a baffled stirred tank with 4 L volume, a height to diameter ratio of 0.75 and a 6-blade propeller that had half the diameter of the tank. The agglomerate sizes were determined in a bypass by dynamic image analysis, using the QICPIC (Sympatec, Germany). Both, the propeller and the three baffles are required to induce turbulence, which is crucial to enable the collision of inclusions but can also lead to redispersion of agglomerates. An analytical tool for the characterization of the turbulence behavior is the Kolmogorov microscale η, given by Eq. 15.16, which describes the length scale of the smallest eddies and depends on the kinematic viscosity ν and the energy dissipation rate ε. It is assumed that the energy dissipates in the smallest eddies and that the flow within those smallest eddies can be described by a laminar shear flow with a Reynolds Number of $Re = 1$.

$$\eta = \left(\frac{\nu^3}{\varepsilon}\right)^{\frac{1}{4}} \tag{15.16}$$

The local energy dissipation rate determines the influence of turbulence on the agglomeration, but it is experimentally not accessible and resolving the local turbulence in a stirred tank by CFD is still a formidable task, despite today's computational capabilities. Therefore, in most experimental investigations the energy dissipation rate which determines the Kolmogorov micro scale is obtained from either power uptake or torque measurements of the stirrer.

The definition of a global root mean shear rate $\overline{G}$ in Eq. 15.17 as suggested by Camp and Stein [46] and frequently used in other works that are concerned with agglomeration phenomena, can be used to directly relate the equilibrium agglomerate size x to the shear conditions with an allometric fit using Eq. 15.18 [45], as depicted in Fig. 15.13a). The two material specific constants are related to the agglomerate strength and are the agglomerate strength constant C, and the agglomerate strength exponent γ [45].

$$\overline{G} = \left(\frac{\overline{\varepsilon}}{\nu}\right)^{\frac{1}{2}} \tag{15.17}$$

$$x = C\overline{G}^{-\gamma} \tag{15.18}$$

Agglomerates that are smaller than the Kolmogorov length scale are subject to erosion as a consequence of the viscous shear forces in the smallest eddies and therefore redispersion occurs if the adhesion forces are not sufficiently high [45]. This results in the observation of an equilibrium size distribution after 20 min of agglomeration in the room temperature water model system and is shown in Fig. 15.12 for $\overline{G} = 186\,s^{-1}$, which correlates well with the measured adhesion forces of the three investigated wettability states of untreated hydrophilic alumina, silanized hydrophobic alumina and silanized hydrophobic alumina after the solvent exchange, which has the highest adhesion force due to the presence of nanobubbles on the surface.

The agglomeration and dispersion of the agglomerates are fully reversible, as shown by experiments where the agglomeration process is disturbed by an ultrasonic treatment. After the ultrasonic stressing the agglomerates form at the same speed and show the same equilibrium agglomerate sizes, which suggests that the presence of

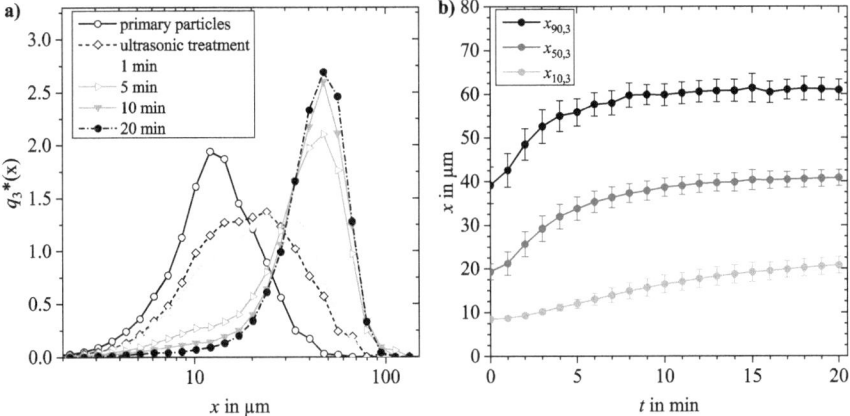

Fig. 15.12 Evolution of the agglomerate size of poorly wetted Al$_2$O$_3$ particles with primary particle size 2 μm to 25 μm for 0.05 wt.% particles in 0.5 M NaCl solution, shown by **a** the log transformed volume weighted density distribution and **b** the volume weighted percentiles

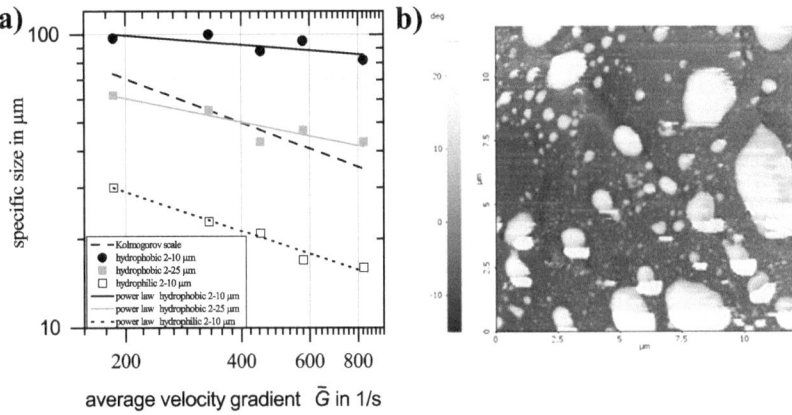

Fig. 15.13 Equilibrium agglomerate size at different shear rates for silanized and untreated Al$_2$O$_3$ (left) and phase contrast image of nanobubbles on a poorly wetted Al$_2$O$_3$ surface after the solvent exchange (right)

the ultrasound does not affect the stability and surface coverage with nanobubbles, that have heights of 50 to 400 nm and diameters in the range 400 to 2000 nm [45].

The effect of aeration on the agglomerate size distribution of spherical silanized Al$_2$O$_3$ is shown in Fig. 15.14. The initial particle size distribution measured after the injection of an ultrasonicated ethanol suspension (t = 0 s) is relatively wide and adjusts to the flow conditions by shifting its modal value to 46 μm at t = 218 s. At this point the number of detected particles is reduced by 40 % relative to the initial number which is a consequence of superimposed effects of agglomeration and deposition of particles in the air–water interface at the top of the tank. After

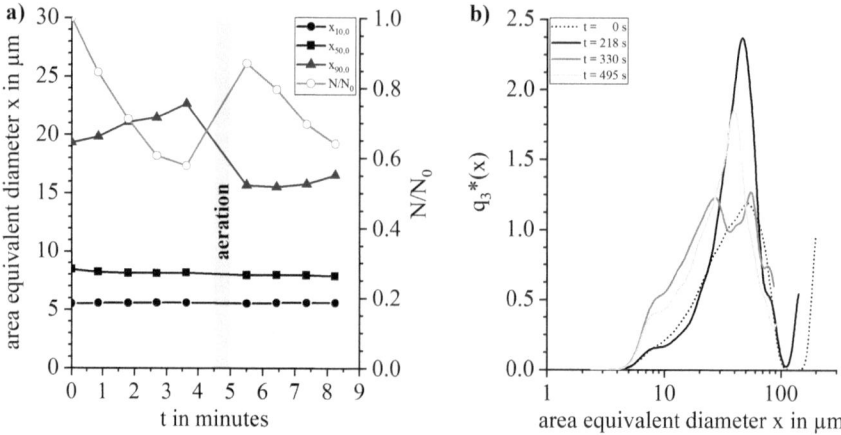

Fig. 15.14 Effect of aeration on the agglomerate size and number of 0.15 wt.% silanized DAW-07 (Denka, Japan) particles with primary particle size of $x_{10,3} = 2.15\,\mu m$, $x_{50,3} = 8.74\,\mu m$ and $x_{90,3} = 24.71\,\mu m$ in 0.75 M NaCl solution. **a** evolution of the percentiles $x_{10,0}$, $x_{50,0}$, $x_{90,0}$ and the dimensionless particle number; **b** log transformed volume weighted density distribution before and after aeration

four and a half minutes the stirred tank is aerated for 30 s with microbubbles from a two-phase-nozzle ($d_{32,bubble} = 145\,\mu m$), which causes a shift to a multimodal density distribution with two main peaks at $27\,\mu m$ and $55\,\mu m$ at t $= 330$ s. An increase of the detected particle number by 51% compared to the particle number before aeration is measured, as well as the increase of the number of particles that is immobilized inside the gas–liquid interface at the top of the stirred tank, quite similar to the slag layer in the real melt. This indicates both, the flotation of particles to the gas–liquid interface, as well as the destruction of existing agglomerates. The majority of bubbles does not form hetero-coagulates and those that do preferably attach to medium sized agglomerates smaller $50\,\mu m$. If the entrainment of the particles in the flow field around the bubbles contributes to the transport of particles to the interface remains unclear, as the redispersion of particles from agglomerates in the bulk is dominant. At t $= 495$ s an increased number of agglomerates is formed again, accompanied by a decreased number of detected particles. Due to its potential to redisperse particles from existing hetero-coagulates an aeration of the melt by porous plugs or immersion lances should be the last cleaning step of the melt before casting. In this regard the water model system shows great similarities.

Acknowledgements The authors gratefully acknowledge the German Research Foundation (DFG) for supporting the Collaborative Research Center CRC 920 (Project ID 169148856–subproject B04).

References

1. V.S. Gotsis, G.N. Angelopoulos, D.C. Papamantellos, Process Metall. **72**(5–6), 208–214 (2001). https://doi.org/10.1002/srin.200100107
2. A. Asad, R. Schwarze, Steel Res. Int. **92**(11) (2021). https://doi.org/10.1002/srin.202100122
3. J.N. Israelachvili, *Intermolecular and surface forces* (Academic Press, Elsevier, 2011)
4. D.Y. Kwok, A.W. Neumann, Adv. Coll. Interface. Sci. **81**, 167–249 (1999). https://doi.org/10.1016/S0001-8686(98)00087-6
5. S.R. Gomes de Sousa, A. Leonel, A.J.F. Bombard, Smart Mater. Struct. **29**(5) (2020). https://doi.org/10.1088/1361-665X/ab6abe
6. S. Taniguchi, A. Kikuchi, T. Ise, N. Shoji, ISIJ Int. **36**(1996). https://doi.org/10.2355/isijinternational.36.Suppl_S117
7. F. Wooten, *Optical properies of solids*, Academic Press (1972)
8. X.J. Chen, A.C. Levi, E. Tosatti, Surf. Sci. **251–252**, 641–644 (1990). https://doi.org/10.1016/0039-6028(91)91070-E
9. K.J. Daun, Metall. Mater. Trans. A **47**(7), 3300–3302 (2016). https://doi.org/10.1007/s11661-016-3527-2
10. H. Kobatake, H. Fukuyama, Metall. Mater. Trans. A **47**(7), 3303–3304 (2016). https://doi.org/10.1007/s11661-016-3529-0
11. R. Esquivel-Sirvent, J.V. Escobar, Europhys. Lett. **107**(4) (2014). https://doi.org/10.1209/0295-5075/107/40004
12. K. Sasai, ISIJ Int. **54**(12), 2780–2789 (2014). https://doi.org/10.2355/isijinternational.54.2780
13. P. Knüpfer, J. Fritzsche, T. Leistner et al., Colloids Surf. A **513**, 215–222 (2017). https://doi.org/10.1016/j.colsurfa.2016.10.046
14. L. Ditscherlein, P. Knüpfer, U.A. Peuker, Powder Technol. **357**, 408–416 (2019). https://doi.org/10.1016/j.powtec.2019.08.077
15. K.C. Mills, *Recommended values of thermophysical properties for selected commercial alloys*, Woodhead Publishing (2002)
16. H.J. Butt, C. Brunero, M. Kappl, Surf. Sci. Rep. **59**(1–6), 1–152 (2005). https://doi.org/10.1016/j.surfrep.2005.08.003
17. L. Ditscherlein, U.A. Peuker, Rev. Sci. Instrum.Instrum. **88**(4), 046107 (2017). https://doi.org/10.1063/1.4981531
18. J. Fritzsche, U.A. Peuker, Powder Technol. **289**, 88–94 (2016). https://doi.org/10.1016/j.powtec.2015.11.057
19. O. Laitinen, K. Bauer, J. Niinimäki et al., Powder Technol. **246**, 545–552 (2013). https://doi.org/10.1016/j.powtec.2013.05.051
20. J. Fritzsche, U.A. Peuker, Colloids Surf. A **459**, 166–171 (2014). https://doi.org/10.1016/j.colsurfa.2014.07.002
21. J. Fritzsche, Technische Universität Bergakademie Freiberg (2016)
22. J. Fritzsche, U.A. Peuker, Colloids Surf. A **509**, 457–466 (2016). https://doi.org/10.1016/j.colsurfa.2016.09.051
23. C. Weber, P. Knüpfer, M. Buchmann, et al., Miner. Eng. **167**(2021). https://doi.org/10.1016/j.mineng.2021.106804
24. P. Knüpfer, Dissertation, TU Bergakademie Freiberg (2020)
25. L. Xie, X. Cui, L. Gong et al., Langmuir **36**(12), 2985–3003 (2020). https://doi.org/10.1021/acs.langmuir.9b03573
26. D.Y.C. Chan, E. Klaseboer, R. Manica, Adv. Colloid Interface Sci. **165**(2), 70–90 (2011). https://doi.org/10.1016/j.cis.2010.12.001
27. S.L. Carnie, D.Y.C. Chan, C. Lewis, R. Manica, R.R. Dagastine, Langmuir **21**, 2912–2922 (2005). https://doi.org/10.1021/la0475371
28. J. Ally, M. Kappl, H.J. Butt, Langmuir **28**(30), 11042–11047 (2012). https://doi.org/10.1021/la300539m
29. A.D. Scheludko, D. Nikolov, Colloid Polymer Sci. **253**, 396–403 (1975). https://doi.org/10.1007/BF01382159

30. S. Berbner, F. Löffler, Powder Technol. **78**(3), 273–280 (1994). https://doi.org/10.1016/0032-5910(93)02798-F
31. T. Lai, R. Chen, P. Huang, J. Adhes. Sci. Technol. **29**(2), 133–148 (2015). https://doi.org/10.1080/01694243.2014.977698
32. N.S. Tambe, B. Bhushan, Nanotechnology **15**(11), 1561 (2004). https://doi.org/10.1088/0957-4484/15/11/033
33. M. Shavezipur, W. Gou, C. Carraro et al., J. Microelectromech. Syst. **21**(3), 541–548 (2012). https://doi.org/10.1109/JMEMS.2012.2189363
34. G. Toikka, T. Tran, G.M. Spinks, et al., Vide: Sci. Techn. Appl. (302 Suppl.), 12–16 (2001)
35. R. Polke, Chem. Ing. Tech. **40**(21–22), 1057–1060 (1968). https://doi.org/10.1002/cite.330402106
36. G. Kuczynski, Adv. Coll. Interface. Sci. **3**(3), 275–330 (1972). https://doi.org/10.1016/0001-8686(72)85005-X
37. H. Krupp, R. Polke, K. Unterforsthuber, Chem. Ing. Tec. **41**(5–6), 300–301 (1969). https://doi.org/10.1002/cite.330410515
38. L. Ditscherlein, PhD Thesis, TU Bergakademie Freiberg (2021)
39. J.C. Moreira, N.R. Demarquette, J. Appl. Polym. Sci. **82**(8), 1907–1920 (2001). https://doi.org/10.1002/app.2036
40. P. Vashishta, R.K. Kalia, A. Nakano, et al., J. Appl. Phys. **103**(8) (2008). https://doi.org/10.1063/1.2901171
41. M. Matsui, Phys. Chem. Miner. **23**, 345–353 (1996). https://doi.org/10.1007/BF00199500
42. M. Bouhadja, N. Jakse, A. Pasturel, J. Chem. Phys. **138**(22), 224510 (2013). https://doi.org/10.1063/1.4809523
43. J. Houska, Surf. Coat. Technol. **235**, 333–341 (2013). https://doi.org/10.1016/j.surfcoat.2013.07.062
44. S. Roy, A. Prakash, S. Sandfeld, Condens. Matter—(in revision) (2022). https://doi.org/10.1088/1361-651X/ac8172
45. P. Knüpfer, L. Ditscherlein, U.A. Peuker, Colloids Surf. A Physicochem. Eng. Aspects **530**, 117–123 (2017). https://doi.org/10.1016/j.colsurfa.2017.07.056
46. T.R. Camp, J. Boston Soc. Civ. Eng. **30**, 219–230 (1943)

Chapter 16
Modeling and Evaluation of the Thermo-mechanical Behavior of Filter Materials and Filter Structures

Martin Abendroth⃝, Stephan Roth⃝, Alexander Malik⃝, Andreas Seupel⃝, Meinhard Kuna⃝, and Bjoern Kiefer⃝

16.1 Introduction

Metal melt filtration is a technological process used to clean the melt and to calm the melt flow during casting. The innovative concept of the multifunctional filters studied here centers on complementing the physical removal of impurities (particles) by additional cleaning mechanisms enabled by chemical reactions between melt and filter material as well as active coatings.

Figure 16.1 illustrates the principle of metal melt filtration and also shows the micrograph of an actual filter cross-section. To design and produce multifunctional filters that are able to withstand the very demanding thermo-mechanical conditions to be encountered during such filtration processes is a major challenge. These structures are, for instance, exposed to large temperature rates and gradients, when the hot molten metal hits the filter (thermal shock), which in turn result in thermally-induced stresses that can cause damage or, in extreme cases, complete failure. Additional stresses are caused by the melt flow, since the filters cause a pressure drop in the direction of flow. Even buoyancy related forces can play a role when ceramic filters are immersed in metal melt, due to the significant density differences of these materials. It must also be emphasized that instead of failing in a brittle manner, ceramic filter materials show non-negligible inelastic behavior (plasticity, creep, damage, etc.) at temperature levels (of up to $1.600\,^\circ$C) required for steel filtration.

The foams studied in this research are reticulated foams, which are based on a polyurethane foam coated by a ceramic slurry. The coating is later dried and fired, whereas the polyurethane pyrolizes and sharp edged cavities remain in the strut [2]. The sharp edges are potential locations for the initiation of crack growth. The firing process also influences the thermo-mechanical properties of the foam. Additional

M. Abendroth (✉) · S. Roth · A. Malik · A. Seupel · M. Kuna · B. Kiefer
Institute of Mechanics and Fluid Dynamics, Technische Universität Bergakademie Freiberg,
Lampadiusstr. 4, 09599 Freiberg, Germany
e-mail: Martin.Abendroth@imfd.tu-freiberg.de

© The Author(s) 2024
C. G. Aneziris and H. Biermann (eds.), *Multifunctional Ceramic Filter Systems for Metal Melt Filtration*, Springer Series in Materials Science 337,
https://doi.org/10.1007/978-3-031-40930-1_16

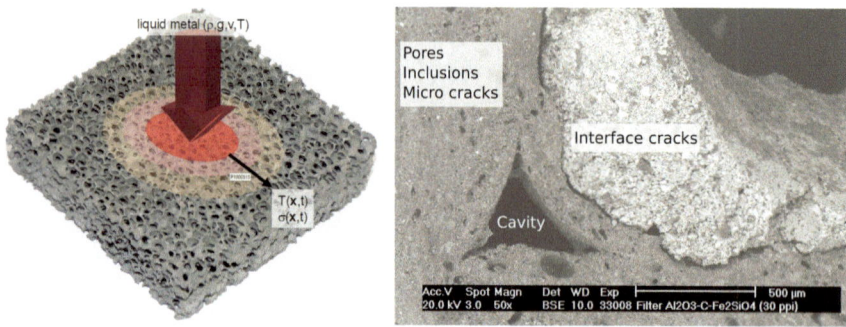

Fig. 16.1 Schematic of liquid metal filtration [1] and a detail of a ceramic filter with a coating

(active) coatings can be applied, which allow specific chemical reactions between melt and filter surface to take place. These reactions can help to reduce certain unwanted elements in the melt.

In order to be able to provide simulation support for the filter design, a thermo-mechanical multiscale modeling approach has been pursued, based on state-of-the-art methods of theoretical and computational mechanics. The goal here was to enable virtual evaluations of the structural integrity and strength of the filters. These are of critical importance, since experimental investigations under realistic process conditions are often very difficult, expensive or even impossible to conduct. To this end, the modeling efforts have focused on three main aspects:

1. The generation of open-cell foam-like Representative Volume Elements (RVEs), with direct control over geometrical features of the foam morphology
2. Multiscale Finite Element Analysis (FEA) of the foam behavior that combines models relevant for different length-scales with scale-bridging concepts (homogenization)
3. More recently, phase-field modeling of an in situ layer formation during the chemical reaction of steel melts and carbon-bonded aluminum oxide filters (reactive filtering phase).

Finite element analysis is generally suitable to model the structural behavior of the foam-like filters on all relevant length-scales. However, such filters consist of a large number of three-dimensionally interconnected struts. Even for macroscopically relatively small filters, the element numbers required for sufficiently accurate meshing under full spatial resolution would thus reach orders of magnitude that could no longer be reasonably handled, even with modern high-performance computers. Homogenization concepts must therefore be applied, if the goal is to predict macroscale properties that are directly influenced by micro- and mesoscale features and behavior, see Fig. 16.2. To this end, the proposed approach is to model the effective thermo-mechanical behavior of the filter material on the microscopic scale (within the struts) via elastoplastic and viscoplastic constitutive models, combined with simple damage approaches. These are established models that are available in

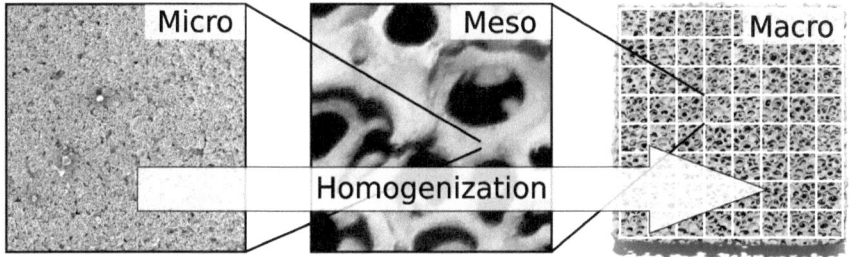

Fig. 16.2 Relevant scales for the three-scale homogenization approach

many commercial FE programs. On the mesoscale, a representative volume element of the foam structure is simulated, in which the detailed geometrical structure is spatially discretized with finite elements. The RVE is then loaded via appropriate boundary conditions and the numerical analysis results in a model for the effective thermo-mechanical behavior of the foam-like mesoscale structure that can be evaluated at every integration point of a macroscale FE simulation.

In general, the effective inelastic behavior of such highly porous structures which the RVEs represent is very complex and may exhibit volumetric strain hardening or softening, and anisotropic non-associated plastic flow behavior, just to name a few phenomena typically encountered. Modeling approaches that try to formulate appropriate flow potentials and flow rules in phenomenological plasticity models are therefore limited in their applicability and accuracy. Classical models include the approach of Deshpande and Fleck [3] and the Cam-Clay model of Roscoe [4], and generalizations suggested in Bigoni-Piccolroaz [5] and Ehlers [6, 7]. There also exists a large body of work regarding the Theory of Porous Media (TPM) that is very relevant to this problem, see, e.g., [8] and the references therein.

The strategy to link the meso- and macro-scale behavior of the ceramic filters followed here, however, is different in nature and can be classified in two categories. Firstly, a novel hybrid approach has been developed, in which the flow potentials and flow rules are replaced by neural networks, while the classical structure of phenomenological elasto-visco-plasticity models known from continuum mechanics is retained. Secondly, direct numerical homogenization via the FE^2 method has been conducted. In this approach, the explicit formulation of a macroscopic constitutive model is typically not necessary. Macroscopic stress-strain relations are instead established by simulating the response of the RVE at each integration point of the macroscale FE problem. Nested, but otherwise separate FE simulation are thus performed on two scales, hence the name FE^2. However, while greatly reducing the degrees of freedom and memory required for the simulation compared to the full spatial resolution for a foam structure of technologically relevant size, this approach is computational still very extensive. Recently, a lot of research has therefore focused on significantly improving the computational efficiency of this method, see [9, 10].

The generated RVEs are, however, not only used for thermo-mechanical FE simulations. The underlying digital geometric models can be exported in different formats. Voxel structures are provided for simulations based on the Lattice-Boltzmann-Method (LBM), where the flow of the liquid melt through the filter can be analyzed [11, 12]. The geometric surface of the generic structures can be triangulated and exported as standard triangle language (STL) data structures, which are commonly used for 3D printing. The printed structures can be analyzed in various experiments to measure actual solid- or fluid-mechanical properties. Very importantly, the models can also be regarded as *digital twins* of real foam structures, which have the advantage that geometrical and topological properties can be easily varied and studied either using numerical or real world experiments.

Finally, recent continuum thermodynamics-based modeling approaches have been devoted to investigating the role of thermo-chemical phenomena as key mechanisms enabling the functionalization of carbon-bonded alumina filters for the purification of steel melt. In particular, the focus is on the formation of an in situ layer of secondary oxides during filtration, which has a high impact on the cleaning efficiency. During this *reactive filtration phase*, a dissolution of alumina from the filter material is assumed in the presence of carbon and liquid iron [13], where the produced oxygen and carbon can diffuse within the melt and nucleate carbon monoxide gas bubbles at impurities yielding a cleaning effect based on flotation [14, 15]. The formation of a dense layer of secondary corundum is proposed as the main reason for the termination of the reactive filtration phase [16], as it is suspected to suppress further carbon supply from the carbon-rich filter to the melt [17]. To capture this, a multi-component/multi-phase phase-field modeling approach is pursued to describe the diffusion, phase transitions, and chemical reactions in the system consisting of filter and melt material. In a first study following this concept, simulation results for the in situ layer formation are presented assuming a simplified filter-melt system.

16.2 Methods

The focus of this section is on the numerical methods that have been established to generate artificial foam structures, to model the effective thermo-mechanical behavior of the foam filters, and to analyze the formation of in situ layers during steel melt filtration with carbon-bonded alumina filters.

16.2.1 *Representative Volume Elements of Foam Structures*

There are several ways to generate RVEs of foam structures. One could use data from computed tomography to map realistic structures. However, in order to reproduce specific geometrical features and to study their influence on the macroscopic behavior, a generic approach is beneficial [1]. The process used here starts with generating a spatially periodic sphere packing. The spheres may be arranged in a lattice structure as shown in Fig. 16.3, or randomly, using algorithms proposed in the liter-

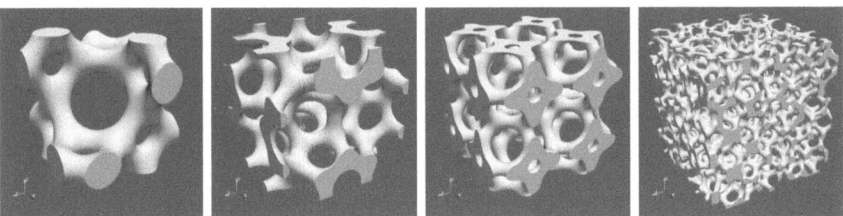

Fig. 16.3 Generic foam structure RVEs with a relative density of $\rho_{rel} = 20\%$ and 2, 8, 16 and 216 pores. The foam with two pores is also known as a Kelvin foam and the one with eight pores as the Wheire-Phelan foam [22]

ature [18–20]. From the sphere packing, a Laguerre tessellation is generated, which contains convex cells around each sphere. But these cells usually exhibit some very small facets, which is not observed in real foams. A relaxation procedure is therefore applied in order to minimize the total foam surface by topological changes and thereby the surface energy [21]. The edges of the relaxed foam structure are mapped into a 3D voxel grid, where each voxel contains the length of struts within itself as a numerical value. This strut density field is smoothed out by a Gaussian filter and finally the surface of the foam structure can be represented by an isocontour (level set). By changing the value of the standard deviation of the Gaussian filter and/or the value for the isocontour, the shape of the struts and the relative density of the generic foam can be adjusted [1].

Subsequently, the generated RVEs can be discretized in different ways with finite elements to perform numerical studies. To analyze mechanical properties such as elasticity, plasticity or creep behavior, the struts are meshed using hexahedral elements as was shown in [1, 23], where sharp edge cavities had to be taken into account for analyzing the fracture mechanical behavior. The same geometries are also used for fluid mechanical simulations, utilizing fine meshed voxel grids (see Fig. 16.4) and assigning either solid or fluid properties to the individual elements [11]. Furthermore, the closed surfaces of the individual RVEs can be exported as STL-files, which may be used to feed 3D printers [24, 25] or being analyzed by software tools like MeshLab to estimate the surface area or topological properties such as the number of cell windows [11].

16.2.2 Hybrid Homogenization Approach

In our notation, homogenized quantities are denoted by an overbar. Homogenized stresses $\bar{\sigma}$ and strains $\bar{\varepsilon}$ are then introduced as the volume-averaged microscopic stress σ and strain ε field, i.e.,

$$\bar{\sigma} := \frac{1}{V} \int_V \sigma \, dV, \quad \bar{\varepsilon} := \frac{1}{V} \int_V \varepsilon \, dV. \tag{16.1}$$

All foam RVEs considered in this research, see the examples in Fig. 16.3, are periodic in all three spatial dimensions. For the mechanical analyses periodic boundary conditions are applied, since this gives the optimum response [26]. The local displacements $\boldsymbol{u}$ at a position $\boldsymbol{x}$ on the RVE boundary ∂V depend on the macroscopic strain $\bar{\boldsymbol{\varepsilon}}$ and a fluctuation $\tilde{\boldsymbol{u}}$

$$\boldsymbol{u} = \bar{\boldsymbol{\varepsilon}}\boldsymbol{x} + \tilde{\boldsymbol{u}} \quad \text{at} \quad \partial V. \tag{16.2}$$

The fluctuations are periodic

$$\tilde{\boldsymbol{u}}(\boldsymbol{x}^-) = \tilde{\boldsymbol{u}}(\boldsymbol{x}^+) \tag{16.3}$$

at homologous points $\boldsymbol{x}^-$ and $\boldsymbol{x}^+$ at the opposing boundaries of the RVE. The fluctuations can be eliminated from (16.2) yielding

$$\boldsymbol{u}(\boldsymbol{x}^+) - \boldsymbol{u}(\boldsymbol{x}^-) = \bar{\boldsymbol{\varepsilon}}(\boldsymbol{x}^+ - \boldsymbol{x}^-) , \tag{16.4}$$

which is implemented as multi point constraints for all sets of homologous points at the RVE boundary. In case of non-periodic meshes a technique developed by Storm et al. [27] may be applied.

Elastic Properties and Failure Limit Surfaces for Foam RVEs

To predict the effective elastic properties and failure limit surfaces local stress fields are determined for six linearly independent strain controlled load cases, denoted by an upper index k in parentheses

$$\bar{\varepsilon}_i^{(k)} = \varepsilon_0 \, \delta_i^{(k)} . \tag{16.5}$$

$\bar{\varepsilon}_i^{(k)}$ denotes the effective strain state for load case k in Voigt notation, whereas ε_0 is a small scalar strain value for which no inelastic response of the RVE is expected. Since $\bar{\varepsilon}_i^{(k)}$ has only a single non-zero component $\delta_i^{(k)}$, we can determine the kth column of the effective stiffness tensor in Voigt notation using

$$\bar{C}_{ik} = \frac{\bar{\sigma}_i^{(k)}}{\bar{\varepsilon}_0} . \tag{16.6}$$

For a failure limit surface the local stress fields $\boldsymbol{\sigma}^{(k)}(\boldsymbol{x})$ for each load case k are computed. Here, $\boldsymbol{x}$ denotes the position within the model. Then, for an arbitrary effective stress state $\bar{\boldsymbol{\sigma}}$ the corresponding strain state is $\bar{\boldsymbol{\varepsilon}} = \bar{\mathbb{C}}^{-1} : \bar{\boldsymbol{\sigma}}$. The resulting local stress field can be computed using the superposition principle

$$\sigma(x) = \sum_{k=1}^{6} \sigma^{(k)}(x)\lambda^{(k)} \quad \text{with} \quad \lambda^{(k)} = \frac{\bar{\varepsilon}_k^{(k)}}{\varepsilon_0} . \tag{16.7}$$

Different failure criteria may be applied, depending on the material. Just to name a few, there are the maximum principal stress criterion, where $\sigma_f(\bar{\sigma}) = \max [\sigma_1(x)] \geq \sigma_c$, or the maximum equivalent stress criterion with $\sigma_f(\bar{\sigma}) = \max [\sigma_{eq}(x)] \geq \sigma_c$, and the Weibull stress criterion for brittle ceramics, where $\sigma_f(\bar{\sigma}) = \sigma_W(\sigma(x)) \geq \sigma_c$. The equivalent stress is defined as

$$\sigma_{eq} = \sqrt{\frac{1}{2} \left[(\sigma_1 - \sigma_2)^2 + (\sigma_2 - \sigma_3)^2 + (\sigma_3 - \sigma_1)^2 \right]} , \tag{16.8}$$

and the Weibull stress using the principle of independent action (PIA)

$$\sigma_W = \left[\sum_{n=1}^{n_{ip}} \frac{V_n}{V_0} \sum_{i=1}^{3} \left\langle \frac{\sigma_i^{(n)} + \left| \sigma_i^{(n)} \right|}{2} \right\rangle^m \right]^{\frac{1}{m}} \tag{16.9}$$

expressed by the sorted principal stresses $\sigma_1 \geq \sigma_2 \geq \sigma_3$.

In general, failure limit surfaces are defined as implicit functions

$$\Phi(\bar{\sigma}) = \sigma_f(\bar{\sigma}) - \sigma_c = 0 , \tag{16.10}$$

where the microscopic loading state depends on the macroscopic stresses $\sigma_f(\bar{\sigma})$. To obtain their shape for a specific failure criterion on the microscale, these functions need to be evaluated for many different macroscopic stress states. The following approach takes samples in stress space in a systematical way. A direction in stress space is defined by

$$N = \frac{I}{\sqrt{3}} \sin(\alpha) + \frac{\hat{N}}{\sqrt{2}} \cos(\alpha), \tag{16.11}$$

with $\alpha \in \left[-\frac{1}{2}\pi, \frac{1}{2}\pi \right]$ denoting the angle between the stress direction and π-plane. The deviatoric part is defined as

$$\hat{N} = \sum_{i=1}^{3} \hat{\lambda}_i M_i, \quad \text{with} \quad M_i = n_i \otimes n_i \quad \text{and,} \quad \hat{\lambda}_i = \begin{bmatrix} \cos(\theta) - \frac{\sin(\theta)}{\sqrt{3}} \\ \frac{2}{\sqrt{3}} \sin(\theta) \\ -\frac{\sin(\theta)}{\sqrt{3}} - \cos(\theta) \end{bmatrix} , \tag{16.12}$$

where n denotes the eigendirections of the stress tensor, $\theta \in \left[-\frac{\pi}{6}, \frac{\pi}{6} \right]$ the Lode angle and I the second order identity tensor. The angles α and θ can be obtained from the corresponding stress tensor $\bar{\sigma}$ and its invariants

$$I_1 = \mathrm{tr}\,(\bar{\sigma})\ , \quad J_2 = \frac{1}{2}\,\bar{s} : \bar{s}, \quad J_3 = \det{(\bar{s})}\ \text{with}\ \bar{s} = \bar{\sigma} - \frac{1}{3}\,I_1\,I \quad (16.13)$$

using

$$\sin\alpha = \frac{I_1}{\sqrt{3}\|\bar{\sigma}\|} \quad \text{and} \quad \sin(3\theta) = \frac{3\sqrt{3}}{2}\,\frac{J_3}{J_2^{(3/2)}}\,. \quad (16.14)$$

This concept of defining a direction N can be used to any second order tensor and will be applied for different purposes in this chapter. Equation (16.11) *a priori* satisfies $\|N\| = 1$ and a *critical* effective stress can be defined $\bar{\sigma}_\mathrm{c} = \lambda_c N$ with λ_c expressed as

$$\lambda_\mathrm{c} = \frac{\sigma_\mathrm{c}}{\sigma_\mathrm{f}(N)}\,. \quad (16.15)$$

Similar approaches have been applied by Zhang et al. [28] and Storm et al. [29] for different realizations of Kelvin foams.

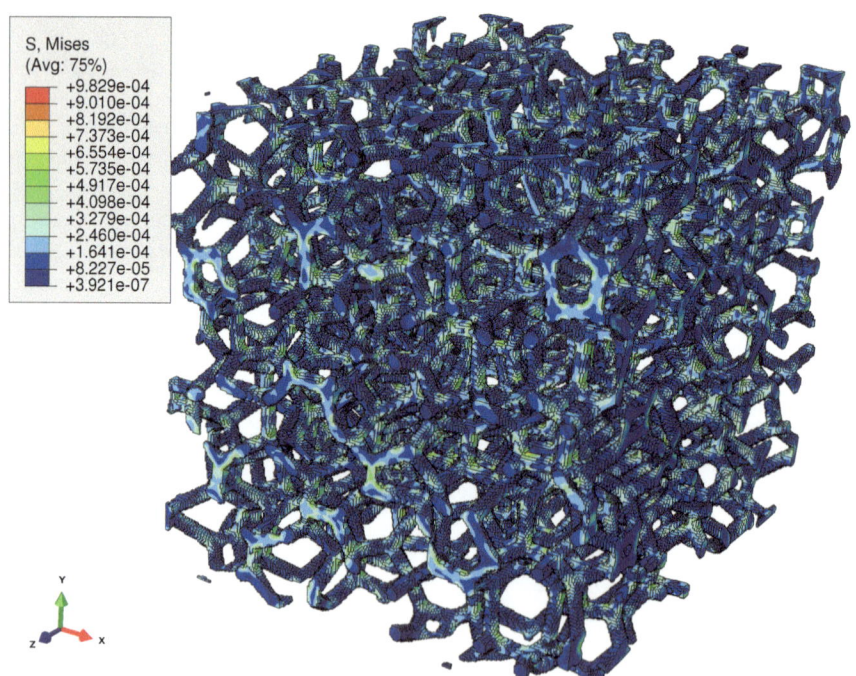

Fig. 16.4 Stress distribution in a F216 foam RVE loaded by an effective shear load case. Here, the FE mesh is made up by cubic voxels for a specific strut geometry and relative density of $\rho_{\mathrm{rel}} = 10\%$

A suitable failure criterion for a fracture mechanical analysis is the coplanar equivalent stress intensity factor (SIF)

$$K_{\text{co}} = \sqrt{K_I^2 + K_{II}^2 + \frac{1}{1-\nu} K_{III}^2}. \tag{16.16}$$

It takes the local stress-state at the crack tip into account caused by superposition of mixed mode loading K_I, K_{II} and K_{III} [30]. For a homogeneous isotropic bulk material the three local SIFs K_i can be related to the J-integral using

$$J = K_i Y_{ij} K_j = \frac{1-\nu^2}{E} \left[K_I^2 + K_{II}^2 \right] + \frac{1+\nu}{E} K_{III}^2. \tag{16.17}$$

Therein, E and ν denote Young's modulus and Poisson's ratio, respectively. Furthermore, Y_{ij} is the Irwin matrix for a homogeneous isotropic material

$$Y_{ij} = \frac{1}{E} \begin{bmatrix} 1-\nu^2 & 0 & 0 \\ 0 & 1-\nu^2 & 0 \\ 0 & 0 & 1+\nu \end{bmatrix}. \tag{16.18}$$

The interaction integral J_j^{int} and the inverted Irwin matrix together with the unit value SIF k_0 are used to separate the SIFs K_i

$$K_i = Y_{ij}^{-1} J_j^{\text{int}} \frac{1}{k_0}. \tag{16.19}$$

The scalar valued interaction integral for mode m is then defined as

$$J_m^{\text{int}} = \lim_{\Gamma \to 0} \int_\Gamma M_{ij}^m n_i q_j \mathrm{d}\Gamma, \tag{16.20}$$

with

$$M_{ij}^m = \sigma_{kl} \varepsilon_{kl}^m \delta_{ij} - \sigma_{ik} u_{k,j}^m - \sigma_{ik}^m u_{k,j}, \tag{16.21}$$

as the superposition of the actual fields of stress σ_{kl}, strain ε_{kl} and displacement gradient $u_{k,i}$ with their corresponding auxiliary fields σ_{kl}^m, ε_{kl}^m and $u_{k,i}^m$, respectively. The auxiliary fields are the *a priori* known near crack tip solutions for pure mode m loading causing a unit value SIF k_0. To evaluate the interaction integrals a finite element model of a Kelvin foam as shown in Fig. 16.5 is utilized. Additional *sub-models* in form of tube like crack models (white mesh in Fig. 16.5 right) are used to compute the interaction integrals along the corresponding crack fronts. The sub-models get as boundary conditions the interpolated displacements of the global Kelvin cell model. More details about this particular modeling approach are given in Settgast et al. [30].

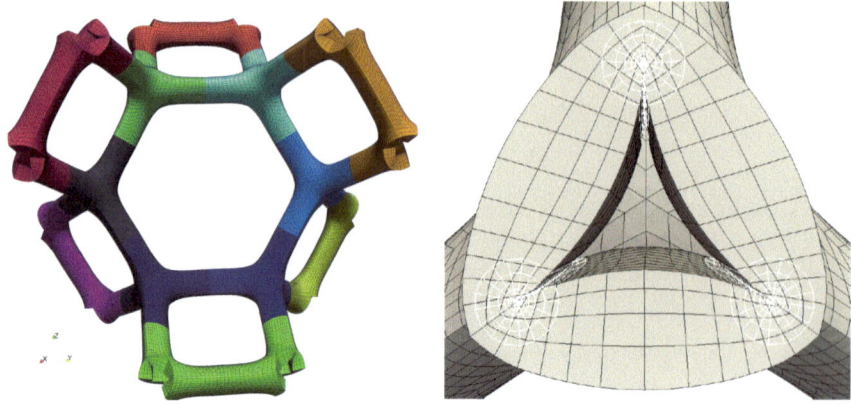

Fig. 16.5 Finite element model of the Kelvin foam including the sharp edged strut cavities and a detail of the mesh (black grid) together with the submodels (white grid) for the calculations of stress intensity factors [30]

Hybrid Constitutive Model Incorporating Neural Networks

In this section a hybrid constitutive model for a macroscopic foam structure is derived. For the sake of simplicity the described model is restricted to elastic-plastic material behavior in a small strain setting. The strain rate tensor is additively split into an elastic and a plastic part

$$\dot{\bar{\varepsilon}} = \dot{\bar{\varepsilon}}^{\text{el}} + \dot{\bar{\varepsilon}}^{\text{pl}} \ . \tag{16.22}$$

Hooke's law is used to relate stress and strain rate

$$\dot{\bar{\sigma}} = \bar{\mathbb{C}} : \dot{\bar{\varepsilon}}^{\text{el}} \ , \tag{16.23}$$

where the components of the effective stiffness tensor $\bar{\mathbb{C}}$ can be determined by at most 6 linear independent load cases as described in detail in [31]. A yield function on the macroscale $\bar{\phi}$ is defined, where the norm of the effective stress is compared with a value, which is predicted by a *neural network* $\text{NN}^{\bar{\phi}}$

$$\bar{\phi} = \|\bar{\sigma}\| - \text{NN}^{\bar{\phi}}\left(\alpha, \theta, \bar{\varepsilon}_q^{\text{pl}}\right) \ . \tag{16.24}$$

The input values for the network are two angles describing the stress direction $N = \bar{\sigma}/\|\bar{\sigma}\|$, where α denotes the angle between stress direction and π-plane and θ the Lode angle, see (16.14). In other words for a given stress direction and accumulated plastic strain $\bar{\varepsilon}_q^{\text{pl}} = \int_0^t \|\dot{\bar{\varepsilon}}^{\text{pl}}\| \, dt$ *internal variable*, the neural network $\text{NN}^{\bar{\phi}}$ predicts the stress amplitude $\|\bar{\sigma}\|$ required to achieve plastic flow.

In general, foams often show non-associated plastic flow, which would require a second *dissipation potential* for the constitutive description, from which the plastic flow direction is derived. Here, it is assumed that the eigendirections of stress and plastic flow are coaxial, which holds for isotropic foams. The plastic strain rate is defined as

$$\dot{\bar{\varepsilon}}^{\mathrm{pl}} = \dot{\bar{\lambda}}\, \bar{\boldsymbol{N}}^{\varepsilon} \quad \text{with} \quad \dot{\bar{\lambda}} = \|\dot{\bar{\boldsymbol{\varepsilon}}}^{\mathrm{pl}}\| \tag{16.25}$$

and the direction of plastic flow is given using (16.11) via two different angles α^{ε} and θ^{ε}, which are predicted by a second neural network

$$\begin{bmatrix} \alpha^{\varepsilon} \\ \theta^{\varepsilon} \end{bmatrix} = \mathrm{NN}^{\varepsilon}\left(\alpha, \theta, \bar{\varepsilon}_{\mathrm{q}}^{\mathrm{pl}}\right), \tag{16.26}$$

which gets the same arguments as $\mathrm{NN}^{\bar{\phi}}$. The Karush-Kuhn-Tucker loading and unloading conditions

$$\bar{\phi} \leq 0, \quad \dot{\bar{\lambda}} \geq 0, \quad \dot{\bar{\lambda}}\bar{\phi} = 0 \tag{16.27}$$

complete the model. Details of the implementation of such a model into FE codes are given in Malik et al. [32] and Settgast et al. [33].

Generation of Training Data and Neural Network Training

The process for the generation of sample data for the neural network training is independent of the choice of the RVE. Using values for $\alpha \in [-\pi/2, \pi/2]$, $\theta \in [-\pi/6, \pi/6]$ and $\lambda \in [0, 1]$ a unit direction in stress or strain space is defined according to Eqs. (16.11) and (16.12). For isotropic foams the eigendirections $\boldsymbol{n}_k$ coincide with the three base vectors. The effective loads for the RVE are defined as

$$\bar{\boldsymbol{\sigma}} = \lambda\, \sigma_{\max} \bar{\boldsymbol{N}} \quad \text{or} \quad \bar{\boldsymbol{\varepsilon}} = \lambda\, \varepsilon_{\max} \bar{\boldsymbol{N}}, \tag{16.28}$$

where $\sigma_{\max}$ and $\varepsilon_{\max}$ denote a maximum amplitude either in stress or strain space. For each loading direction a number of n_λ effective stress and strain tensors are computed using the FEM. The norm of the effective stress tensor is determined as target for the training of $\mathrm{NN}^{\bar{\phi}}$. The effective plastic strain is determined using $\bar{\boldsymbol{\varepsilon}}^{\mathrm{pl}} = \bar{\boldsymbol{\varepsilon}} - \bar{\mathbb{C}}^{-1} : \bar{\boldsymbol{\sigma}}$ and the strain rate is approximated by the finite difference between two subsequent increments indicated by the indices $n - 1$ and n

$$\dot{\bar{\boldsymbol{\varepsilon}}}_n^{\mathrm{pl}} \approx \frac{\Delta \bar{\boldsymbol{\varepsilon}}_n^{\mathrm{pl}}}{\Delta t} = \frac{\bar{\boldsymbol{\varepsilon}}_n^{\mathrm{pl}} - \bar{\boldsymbol{\varepsilon}}_{n-1}^{\mathrm{pl}}}{t_n - t_{n-1}}. \tag{16.29}$$

The internal state variable $\bar{\varepsilon}_q^{pl}$ is summed up incrementally using

$$\bar{\varepsilon}_q^{pl}\Big|_n = \sum_{i=1}^{n} \left\| \Delta\bar{\boldsymbol{\varepsilon}}_i^{pl} \right\| . \tag{16.30}$$

The angles α^ε and θ^ε are derived using (16.14) from the effective plastic strain rate, or α and θ are either given if a stress controlled RVE simulation is performed or derived from the simulated effective stress tensor using (16.14) for a strain controlled simulation. The data base of training samples used by Malik et al. [32] was generated using $n_\alpha = 39$, $n_\theta = 19$ and $n_\lambda = 20$ variations for α, θ and λ. The neural networks are fully connected feed forward neural networks containing a layer of input units (or neurons), one or more hidden layers and a layer of output units. The numbers of units for the input and output layer depend on the problem, whereas the number of hidden layers and units can be chosen freely. Each hidden neuron with index i represents a sigmoidal activation function depending on the input x_i, which is the sum of weighted outputs of the neurons from the previous network layer plus a bias

$$f(x_i) = \frac{1}{1 + e^{-x_i}} \quad \text{with} \quad x_i = \sum_k w_k f(x_k) + b_i . \tag{16.31}$$

The weights w_k and biases b_i are the free parameters of the network, which are determined during a training process. The necessary number of neurons within the hidden layers depend on the complexity of the function which the network is supposed to represent. A good approximation quality can be achieved if the number of free parameters of the neural network corresponds approximately to the square root of the number of training samples. To supervise the training process, the training data are split into two different batches. 90% of the data are used for the network training and the remaining 10% are solely used for the validation of the approximation quality, see Fig. 16.6. The numerical realization of the neural networks, the training and validation process is done with the Python package FFNET [34]. It allows especially the conversion of the neural network into Fortran code, which can be integrated in user material routines (UMAT) for the finite element program Abaqus (Fig. 16.6).

An alternative homogenization approach to model the inelastic behavior of foam structures is the FE^2 method. The concept of FE^2 is to run separate microscale FE simulations of the RVE on each macro material point. Recently, an efficient monolithic formulation was proposed by Lange et al. [10]. The FE^2 method is supposed to produce most accurate homogenized results, however, it is still computational expensive compared to the presented hybrid approach, which resorts to training data that has been computed offline before.

In order to be able to capture size effects of foam structures [35] within the scale transition, higher gradient homogenization schemes have to be considered. Micromorphic theories [36, 37] allow to model these effects by extending the continuum model at the macro scale with additional degrees of freedom. Which of the available

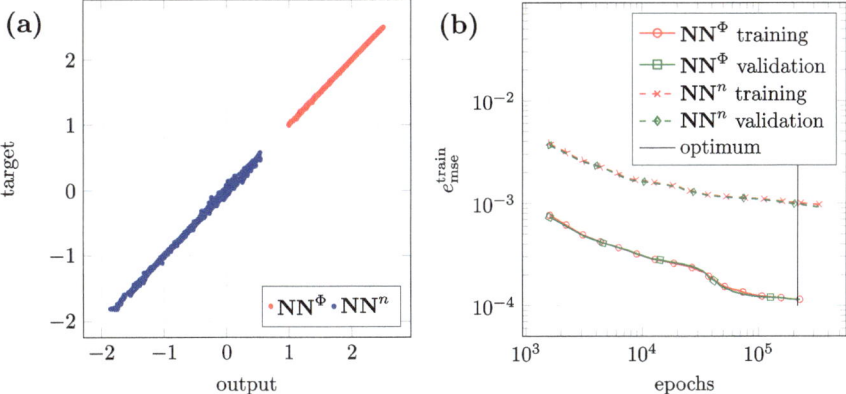

Fig. 16.6 a Accuracy of a trained network. **b** Evolution of training and validation mean square error during the learning process [32]

theories is the most dominant in the considered foam structures is part of ongoing research [38, 39].

16.2.3 Phase-Field Modeling of Multi-component/Multi-phase Systems

The *phase-field method* is a versatile tool to model the microstructure evolution of multi-phase systems, see [40] for an overview. Moreover, phase-field modeling approaches have been applied to various coupled problems such as electrochemistry [41], chemically reactive systems with phase separation [42] and, very recently, to chemo-mechanics focusing on various aspects [43–48]. Available commercial multi-component/multi-phase simulation tools such as DICTRA employ a *sharp interface* description. The intentionally restricted applicability to 1D-problems could be justified so far by the number of practically relevant cases [49]. In addition, DICTRA has access to the mobility database of the Thermo-Calc software necessary to describe the diffusion processes. Furthermore, Thermo-Calc is directly called in order to determine the existing phases [49, 50]. However, the advantage of the *phase-field method* is its straightforward application to 2D and 3D-problems covering the evolution of complex phase-interface topologies in a *diffusive sense*.

Recently, a unifying framework for phase-field modeling of multi-component/multi-phase chemo-mechanics has been proposed by Svendsen et al. [51]. Our phase-field approach is based on a *mixture theory* assuming $i = [1, \ldots, M]$ components, where their individual masses m_i sum up to the total mass m. The constitution of the mixture is characterized by the mass fractions c_i of the components which obey additional sum relations. In addition, *order parameters* v_j are defined as mass fractions to distinguish $j = [1, \ldots, K]$ *phases*, e.g., different aggregate states. The following relations hold

$$\sum_{i=1}^{M} m_i = m, \quad c_i = \frac{m_i}{m}, \quad v_j = \frac{m_j}{m}, \quad \sum_{i=1}^{M} c_i = 1, \quad \sum_{j=1}^{K} v_j = 1. \quad (16.32)$$

The evolution of the component mass fractions and the order parameters are described by coupled, generalized *Cahn-Hilliard* and *Allen-Cahn* type equations [52], respectively. The phase-field model considers the gradients ∇c_i and ∇v_j in order to capture the influence of phase interfaces in a diffusive sense. We derive the balance equations and necessary constitutive relations on the basis of an extended Coleman-Noll procedure for generalized continua as proposed by Hütter [53]. After introducing the conventional mass balance and diffusion equations, a gradient-extended energy balance is exploited in the framework of the *thermodynamics of internal variables* yielding the desired Cahn-Hilliard and Allen-Cahn equations. Thermodynamically consistent relations for the dissipative mechanisms are derived with help of a *dissipation potential approach*. Especially, we focus on the incorporation of chemical conversion processes based on the theory of equilibrium reactions. In Sect. 16.2.3, the phase-field model is re-cast into a *mixed rate variational setting* as proposed by Miehe [43, 54, 55] which enables a proper numerical treatment using finite elements.

Field Equations for Phase-Field Models

The mass balances of the single components can be cast into diffusion equations

$$\rho \, \dot{c}_i = -\nabla \cdot \boldsymbol{h}_i + h_i \quad \forall \, \boldsymbol{x} \in \mathcal{B}, \qquad i = [1, \ldots, M], \qquad (16.33)$$

where ρ is the mass density of the mixture in the current configuration [56]. The global mass conservation yields

$$\dot{\rho} + \rho \, \nabla \cdot \dot{\boldsymbol{u}} = 0 \quad \forall \, \boldsymbol{x} \in \mathcal{B} \qquad (16.34)$$

for the evolution of the density, where $\dot{\boldsymbol{u}}$ denotes the material (barycentric) velocity of the mixture. In order to ensure the balance of total mass we employ restrictions on the flux and source terms $\boldsymbol{h}_i$ and h_i [51], respectively,

$$\boldsymbol{0} = \sum_{i=1}^{M} \boldsymbol{h}_i, \qquad\qquad 0 = \sum_{i=1}^{M} h_i . \qquad (16.35)$$

The energy balance is discussed for the special case of mass transport and phase transition assuming isothermal conditions. The global balance for a spatial region $\mathcal{P}$ convecting with the material body $\mathcal{B}$ reads

$$\int_{\mathcal{P}} \rho \dot{e} \, dv = \int_{\partial \mathcal{P}} \left[\sum_{i=1}^{M} \dot{c}_i \, \Pi_i + \sum_{j=1}^{K} \dot{v}_j \, \Phi_j \right] \cdot \boldsymbol{n} \, da, \tag{16.36}$$

with e being the mass specific internal energy and a flux of generalized power acting on the boundary $\partial \mathcal{P}$ with outward unit normal $\boldsymbol{n}$. Therein, Π_i and Φ_j are power-conjugated microstresses to c_i and v_j, respectively. Applying the divergence theorem to (16.36), we obtain the localized energy balance

$$\dot{e} = \frac{1}{\rho} \sum_{i=1}^{M} \left[\Pi_i \cdot \nabla \dot{c}_i + \nabla \cdot \Pi_i \, \dot{c}_i \right] + \frac{1}{\rho} \sum_{j=1}^{K} \left[\Phi_j \cdot \nabla \dot{v}_j + \nabla \cdot \Phi_j \, \dot{v}_j \right]. \tag{16.37}$$

The generalized flux terms introduce the gradients $\nabla \dot{c}_i$ and $\nabla \dot{v}_j$ into the energy balance. Next, the second law of thermodynamics is considered based on a balance of entropy and Gibbs' fundamental equation [57, 58]. We assume a local entropy balance of the general form

$$\rho \dot{\eta} = -\nabla \cdot \boldsymbol{j} + \pi \quad \forall \boldsymbol{x} \in \mathcal{B}, \tag{16.38}$$

where the restriction $\pi \geq 0$ on the source term states the second law of thermodynamics [57]. The mass specific entropy is denoted as η.

The entropy flux $\boldsymbol{j}$ and the source π are specified following the formalism of the *thermodynamics of internal variables* as comprehensively described by Lebon et al. [58]. To this end, the internal energy e is exchanged by the Helmholtz free energy ψ applying the established Legendre transformation $e = \psi + \vartheta \, \eta$. The time derivative of the latter yields the localized Gibbs fundamental equation

$$\dot{\eta} = \frac{1}{\vartheta} \left[\dot{e} - \dot{\psi} - \eta \, \dot{\vartheta} \right] \quad \forall \boldsymbol{x} \in \mathcal{B}, \tag{16.39}$$

which can be exploited to specify the terms of the entropy balance (16.38) [57, 58]. The free energy depends on the temperature ϑ, and additionally on the deformation gradient which is convenient for a chemo-mechanical extension [43, 51]. However, we introduce the following set of *constitutive state variables*

$$\underline{z} := \left\{ \vartheta, c_i, \nabla c_i, v_j, \nabla v_j \right\}, \tag{16.40}$$

and the free energy function

$$\psi = \hat{\psi} \left(\underline{z} \right) \tag{16.41}$$

for the mass transport and phase transition processes. Note that this set of variables can be extended by kinematic deformation measures in case of chemo-thermo-*mechanical* processes. Here, we finally specify the Gibbs equation (16.39) inserting the time derivative of the free energy (16.41) together with the local energy balance (16.37) and the diffusion equations (16.33) as

$$
\vartheta \, \rho \, \dot{\eta} = - \rho \left[\eta + \partial_\vartheta \hat{\psi} \right] \dot{\vartheta} + \sum_{i=1}^{M} \left[\mathbf{\Pi}_i - \rho \, \partial_{\nabla c_i} \hat{\psi} \right] \cdot \nabla \dot{c}_i
$$

$$
+ \sum_{j=1}^{K} \left[\nabla \cdot \mathbf{\Phi}_j - \rho \, \partial_{v_j} \hat{\psi} \right] \dot{v}_j + \sum_{j=1}^{K} \left[\mathbf{\Phi}_j - \rho \, \partial_{\nabla v_j} \hat{\psi} \right] \cdot \nabla \dot{v}_j
$$

$$
+ \sum_{i=1}^{M} \left[\tfrac{1}{\rho} \nabla \cdot \mathbf{\Pi}_i - \partial_{c_i} \hat{\psi} \right] \left[-\nabla \cdot \mathbf{h}_i + h_i \right] \quad \forall \mathbf{x} \in \mathcal{B} . \tag{16.42}
$$

Following the Coleman-Noll argumentation [53], the terms in brackets in (16.42) having rates of state variables as pre-factors could be chosen so that their dissipative contribution vanishes completely. On the other hand, to discuss the remaining terms in view of (16.38) some dissipative mechanisms could be additionally introduced as discussed by Hütter [53]. We define *energetic* constitutive relations for the entropy and the microstresses

$$
\eta = -\partial_\vartheta \hat{\psi} , \tag{16.43}
$$

$$
\mathbf{\Pi}_i = \rho \, \partial_{\nabla c_i} \hat{\psi} , \qquad\qquad i = [1, \ldots, M] , \tag{16.44}
$$

$$
\mathbf{\Phi}_j = \rho \, \partial_{\nabla v_j} \hat{\psi} , \qquad\qquad j = [1, \ldots, K] , \tag{16.45}
$$

and find *additional balance equations*

$$
\mu_i = \partial_{c_i} \hat{\psi} - \frac{1}{\rho} \nabla \cdot \mathbf{\Pi}_i \quad \forall \mathbf{x} \in \mathcal{B} , \qquad i = [1, \ldots, M] , \tag{16.46}
$$

$$
\Phi_j = \rho \, \partial_{v_j} \hat{\psi} - \nabla \cdot \mathbf{\Phi}_j \quad \forall \mathbf{x} \in \mathcal{B} , \qquad j = [1, \ldots, K] , \tag{16.47}
$$

where μ_i and Φ_j are *generalized chemical potentials* and *dissipative microforces* of the phase transition, respectively. The chemical potential yields the classical definition $\mu_i = \partial_{c_i} \hat{\psi}$ for vanishing gradient terms in the free energy. As mentioned by Hütter [53], the additional balance equations (16.46)–(16.47) appear naturally without an *a priori* introduction of microforce balances as preferred by Gurtin [52]. Furthermore, boundary conditions based on microtractions can be prescribed which includes as a special case the generalized (trivial) boundary conditions utilized by Svendsen et al. [51].

Applying the sum relations (16.32) and (16.35) to the remaining terms in (16.42) yields

$$
\vartheta \, \rho \, \dot{\eta} = - \sum_{j=1}^{K-1} \tilde{\Phi}_j \, \dot{v}_j + \sum_{i=1}^{M-1} \tilde{\mu}_i \, \nabla \cdot \mathbf{h}_i - \sum_{i=1}^{M-1} \tilde{\mu}_i \, h_i , \tag{16.48}
$$

where the *relative* chemical potentials $\tilde{\mu}_i$ and *relative* microforces $\tilde{\Phi}_j$ are introduced

$$\tilde{\mu}_i := \mu_i - \mu_M, \qquad i = [1, \ldots, M-1], \qquad (16.49)$$

$$\tilde{\Phi}_j := \Phi_j - \Phi_K, \qquad j = [1, \ldots, K-1] . \qquad (16.50)$$

The mass flux term in (16.48) can be expressed as a pure convective contribution [57] and a source term which allows to specify

$$\mathcal{D} := \vartheta \, \pi = -\sum_{j=1}^{K-1} \tilde{\Phi}_j \dot{v}_j - \sum_{i=1}^{M-1} \nabla \tilde{\mu}_i \cdot \boldsymbol{h}_i - \sum_{i=1}^{M-1} \tilde{\mu}_i \, h_i \geq 0 . \qquad (16.51)$$

We interpret $\mathcal{D}$ as the dissipation. To ensure $\mathcal{D} \geq 0$, the microforces $\tilde{\Phi}_j$, the mass fluxes $\boldsymbol{h}_i$ and the mass sources/sinks h_i must be chosen accordingly. The latter are related to chemical conversion processes which is addressed in the following.

The source terms h_i are specified for P parallel chemical reactions written in the generalized format

$$\sum_{i=1}^{M} v_i^\alpha X_i \rightleftharpoons 0, \qquad \alpha = [1, \ldots, P] . \qquad (16.52)$$

Therein, X_i denotes the chemical formula of the component i. According to the stoichiometry, a prefactor v_i^α for every component i appears depending on the reaction α. Here, we follow the convention

$$v_i^\alpha = \begin{cases} < 0 & \text{for reactant,} \\ > 0 & \text{for product,} \\ = 0 & \text{for inert component.} \end{cases} \qquad (16.53)$$

Due to the stoichiometric relations and the mass conservation, the mass of components taking part in a chemical reaction α can only change proportional to a single conversion rate density r^α. According to [56] we decided for linear relations

$$h_i = \sum_{\alpha=1}^{P} M_i \, v_i^\alpha r^\alpha, \qquad i = [1, \ldots, M], \qquad (16.54)$$

where M_i is the molar mass of component i.

With the specific source terms (16.54) at hand, the dissipation (16.51) reads

$$\mathcal{D} = -\sum_{j=1}^{K-1} \tilde{\Phi}_j \, \dot{v}_j - \sum_{i=1}^{M-1} \nabla \tilde{\mu}_i \cdot \boldsymbol{h}_i + \sum_{\alpha=1}^{P} A^\alpha \, r^\alpha \geq 0 . \qquad (16.55)$$

Therein the pre-factors of the conversion rate densities r^α can be combined to the *affinity* of the reaction

$$A^\alpha := \hat{A}^\alpha (\tilde{\mu}_i) = -\sum_{i=1}^{M-1} M_i \, v_i^\alpha \, \tilde{\mu}_i, \qquad \alpha = [1, \ldots, P], \qquad (16.56)$$

i.e., its driving force.

In order to ensure positive dissipation, linear Onsager-relations could be chosen between the conjugated variables in (16.55) as established by the *thermodynamics of irreversible processes*, see [56, 57]. In view of the desired variational setting, however, we proceed with a *dissipation potential approach*. For details we refer to the paper of Miehe [54] and the literature therein. In this spirit, a mass specific dissipation potential ϕ with the rate $\underline{\dot{z}}$ as variables and state variables $\underline{z}$ as additional parameters is introduced

$$\phi = \hat{\phi} \left(\underline{\dot{z}}; \underline{z} \right) . \qquad (16.57)$$

The potential ϕ is convex w.r.t. its arguments and zero at $\underline{\dot{z}} = \underline{0}$. A set of generalized *dual* dissipative forces $\underline{\mathbf{Z}}$ is derived as

$$\underline{\mathbf{Z}} := \rho \, \partial_{\underline{\dot{z}}} \hat{\phi} . \qquad (16.58)$$

The part of the dissipation (16.55) proportional to $\underline{\dot{z}}$ can be rewritten

$$\mathcal{D}_{\dot{z}} = \underline{\mathbf{Z}} \cdot \underline{\dot{z}} . \qquad (16.59)$$

The mentioned properties of the dissipation potential ϕ ensure $\mathcal{D}_{\dot{z}} \geq 0$. Moreover, a *dual dissipation potential* ϕ^* can be defined

$$\phi^* = \hat{\phi}^* \left(\underline{\mathbf{Z}}; \underline{z} \right), \qquad \underline{\dot{z}} = \rho \, \partial_{\underline{\mathbf{Z}}} \hat{\phi}^* . \qquad (16.60)$$

using a proper Legendre transformation [54].

In particular we choose

$$\phi = \hat{\phi} \left(\dot{v}_j \right) = \sum_{j=1}^{K-1} \frac{1}{2} \beta_j \, \dot{v}_j^2, \qquad (16.61)$$

$$\phi^* = \hat{\phi}^* \left(\nabla \tilde{\mu}_i, A^\alpha \right) = \sum_{a=1}^{M-1} \sum_{b=1}^{M-1} \frac{1}{2} \nabla \tilde{\mu}_a \cdot \mathbf{M}_{ab} \cdot \nabla \tilde{\mu}_b + \sum_{\alpha}^{P} \frac{1}{2} k^\alpha \left[A^\alpha \right]^2 . \qquad (16.62)$$

The viscosity parameters β_j and the parameters k^α of the conversion rate densities are positive numbers. The Onsager-*mobility tensors* $\boldsymbol{M}_{ab}$ obey $\boldsymbol{M}_{ab} = \boldsymbol{M}_{ba}$. The microforces $\tilde{\Phi}_j$ are derived from ϕ, whereas the fluxes $\boldsymbol{h}_i$ and the conversion rates r^α stem from the dual potentials

$$\tilde{\Phi}_j = -\rho\, \partial_{\dot{v}_j}\hat{\phi} \qquad = -\rho\, \beta_j\, \dot{v}_j, \qquad\qquad j = [1, \ldots, K-1], \qquad (16.63)$$

$$\boldsymbol{h}_i = -\rho\, \partial_{\nabla\tilde{\mu}_i}\hat{\phi}^* \qquad = -\rho \sum_{b=1}^{M-1} \boldsymbol{M}_{ib} \cdot \nabla\tilde{\mu}_b, \quad i = [1, \ldots, M-1], \qquad (16.64)$$

$$r^\alpha = \rho\, \partial_{A^\alpha}\hat{\phi}^* \qquad = \rho\, k^\alpha\, A^\alpha, \qquad\qquad \alpha = [1, \ldots, P] . \qquad (16.65)$$

The source terms finally read

$$h_i = -\rho\, \partial_{\tilde{\mu}_i}\hat{\phi}^* = \rho \sum_\alpha^P M_i\, v_i^\alpha k^\alpha\, A^\alpha, \qquad i = [1, \ldots, M-1] . \qquad (16.66)$$

At this point, the balance equations (16.33), (16.46), (16.47) and the constitutive relations (16.44), (16.45), (16.63)–(16.65) are combined to generate the *field equations*. These form the basis for the variational formulation. For the specific problems to be analyzed we assume that $\rho = \rho_0$, i.e., the density of the mixture keeps its initial value and is additionally constant over the whole domain. Furthermore, we set $\dot{\boldsymbol{u}} = \boldsymbol{0}$. The field equations are written in terms of *variational derivatives* which are defined as

$$\delta_x\hat{f} := \partial_x\hat{f} - \nabla \cdot \partial_{\nabla x}\hat{f}, \qquad f = \hat{f}(x, \nabla x) . \qquad (16.67)$$

Therewith the final field equations read

$$\rho_0\, \dot{c}_i = -\rho_0\, \delta_{\tilde{\mu}_i}\hat{\phi}^* \qquad \forall\, \boldsymbol{x} \in \mathcal{B}, \qquad i = [1, \ldots, M-1], \qquad (16.68)$$

$$\rho_0\, \tilde{\mu}_i = \rho_0\, \delta_{c_i}\hat{\psi} \qquad \forall\, \boldsymbol{x} \in \mathcal{B}, \qquad i = [1, \ldots, M-1], \qquad (16.69)$$

$$-\rho_0\delta_{\dot{v}_j}\hat{\phi} = \rho_0\, \delta_{v_j}\hat{\psi} \qquad \forall\, \boldsymbol{x} \in \mathcal{B}, \qquad j = [1, \ldots, K-1]. \qquad (16.70)$$

Note that $2\,[M-1]$ and $K-1$ equations need to be solved due to the sum relations of the mass fractions. Moreover, the equations (16.70) have the structure of generalized Allen-Cahn equations . Combining (16.69) with (16.68) yields generalized Cahn-Hilliard equations . In the previous set of equations, relative quantities w.r.t. the last component M or last phase K are denoted by a tilde, e.g.,

$$\partial_{c_i}\tilde{\psi} := \partial_{c_i}\hat{\psi} - \partial_{c_M}\hat{\psi}, \qquad i = [1, \ldots, M-1], \qquad (16.71)$$

whose variational derivatives read

$$\delta_{c_i}\tilde{\psi} := \delta_{c_i}\hat{\psi} - \delta_{c_M}\hat{\psi}, \qquad i = [1, \ldots, M-1] . \tag{16.72}$$

Finally, the boundary conditions are specified

$$c_i = \bar{c}_i, \quad \forall \boldsymbol{x} \in \partial\mathcal{B}_c, \qquad -\rho_0\, \partial_{\nabla\tilde{\mu}_i}\hat{\phi}^* \cdot \boldsymbol{n} = \bar{h}_i, \quad \forall \boldsymbol{x} \in \partial\mathcal{B}_h, \tag{16.73}$$

$$\tilde{\mu}_i = \bar{\tilde{\mu}}_i, \quad \forall \boldsymbol{x} \in \partial\mathcal{B}_\mu, \qquad \rho_0\, \partial_{\nabla c_i}\tilde{\psi} \cdot \boldsymbol{n} = \bar{\Pi}_i, \quad \forall \boldsymbol{x} \in \partial\mathcal{B}_\Pi, \tag{16.74}$$

$$v_j = \bar{v}_j, \quad \forall \boldsymbol{x} \in \partial\mathcal{B}_v, \qquad \rho_0\, \partial_{\nabla v_j}\tilde{\psi} \cdot \boldsymbol{n} = \bar{\Phi}_j, \quad \forall \boldsymbol{x} \in \partial\mathcal{B}_\Phi \tag{16.75}$$

with prescribed values at the respective boundaries marked with a bar.

Mixed Rate-Type Variational Setting

Following Miehe [54, 55], the field equations (16.68)–(16.70) appear as the *Euler-Lagrange equations* of a variational problem based on the *mixed rate potential*

$$\Pi^*\left(\dot{c}_i, \tilde{\mu}_i, \dot{v}_j\right) = \frac{d}{dt}\mathcal{E}\left(c_i, v_j\right) + \mathcal{D}^*\left(\dot{c}_i, \tilde{\mu}_i, \dot{v}_j\right) - \mathcal{P}^*_{\mathrm{ext}}\left(\dot{c}_i, \tilde{\mu}_i, \dot{v}_j\right) . \tag{16.76}$$

The *energy storage functional*, the *dissipation functional* and *load functional* of external dead loads read

$$\mathcal{E}\left(c_i, v_j\right) = \int_{\mathcal{B}} \rho_0\, \hat{\psi}\left(c_i, \nabla c_i, v_j, \nabla v_j\right) dv, \tag{16.77}$$

$$\mathcal{D}^*\left(\dot{c}_i, \tilde{\mu}_i, \dot{v}_j\right) = \int_{\mathcal{B}} \left[\rho_0\, \hat{\phi}\left(\dot{v}_j\right) - \sum_{i=1}^{M-1}\rho_0\, \tilde{\mu}_i\, \dot{c}_i - \rho_0\, \hat{\phi}^*\left(\tilde{\mu}_i, \nabla\tilde{\mu}_i\right)\right] dv, \tag{16.78}$$

$$\mathcal{P}^*_{\mathrm{ext}}\left(\dot{c}_i, \tilde{\mu}_i, \dot{v}_j\right) = \int_{\partial\mathcal{B}_h} \sum_{i=1}^{M} \dot{c}_i\, \bar{h}_i\, da + \int_{\partial\mathcal{B}_\Pi} \sum_{i=1}^{M-1} \tilde{\mu}_i\, \bar{\Pi}_i\, da$$

$$+ \int_{\partial\mathcal{B}_\Phi} \sum_{j=1}^{K} \dot{v}_j\, \bar{\Phi}_j\, da . \tag{16.79}$$

The three-field *mixed variational principle* is given by

$$\left\{\dot{c}_i, \tilde{\mu}_i, \dot{v}_j\right\} = \arg\left(\inf_{\dot{c}_i,\dot{v}_j}\, \sup_{\tilde{\mu}_i}\, \Pi^*\left(\dot{c}_i, \tilde{\mu}_i, \dot{v}_j\right)\right) . \tag{16.80}$$

The necessary condition for stationary points reads $\delta\,\Pi^*\left(\dot{c}_i, \dot{v}_j, \tilde{\mu}_i\right) = 0$, where $\delta\,\Pi^*\left(\dot{c}_i, \tilde{\mu}_i, \dot{v}_j\right)$ denotes the first variation of the rate potential. We allow arbitrary variations of the defined field variables, except at boundaries where the field variables are fixed. Note that variations of $\delta \dot{c}_i$ and $\delta \dot{v}_j$ are restricted by the sum relations (16.32). Therefore, we finally find (16.68)–(16.70) as the *Euler-Lagrange equations*.

For the numerical treatment via the finite element method, a time incremental counterpart of the variational principle is derived as explained in detail by Miehe et al. [43, 55]. For discretization in time, an *Euler-backward* scheme is utilized. The spatial discretization is performed by iso-parametric finite elements with low-order, C0-continuous shape functions for all considered primary field variables, which are the mass fractions c_i, the chemical potentials $\tilde{\mu}_i$ and the order parameters v_j. The final non-linear system of algebraic equations to be solved for any time point t is handled with a monolithic Newton-scheme. Due to the variational structure, the considered problem provides an inherently symmetric tangent matrix as pointed out in [43, 55].

16.3 Results and Applications

16.3.1 Application of Foam Models

The generic foam structures described in Sect. 16.2.1 were widely used within the CRC 920 either as geometric models in numerical simulations or as 3D printed structures to conduct specific experiments. Foam RVEs were used to simulate the influence of foam morphology on effective properties (hydraulic turtuosity, viscous and inertial permeability, filtration coefficient) related to metal melt filtration [11]. Asad et al. [59] investigated the immersion process of a ceramic filter in a steel melt. Lehmann et al. [12] used the RVEs as base structures and investigated the influence of specific geometric modifications like additional struts, closed foam windows or streamlined strut cross sections on effective hydraulic properties and filtration performance. 3D printed filter structures based on the RVEs have been used by Wetzig et al. [60] for real world filtration experiments, whereas Bock-Seefeld et al. [25] and Herdering et al. [24, 61] used the RVEs to produce polymer filter templates for customized ceramic foam structures to estimate filtration efficiency, structural filter strength, and integrity. A comparison of mechanical properties between generic and real foam structures performed by Settgast et al. [62] proved the high accuracy of the generic foam models.

16.3.2 Thermo-mechanical Behavior of Foams and Filter Structures

The generation procedure of the foam model allows modifications of the strut shape as explained in Sect. 16.2.1. The influence of the strut shapes on the elastic properties was investigated by Storm et al. [63] on spatial periodic Kelvin cells. The corresponding FE-models were either build from beam or 3D elements. We found that

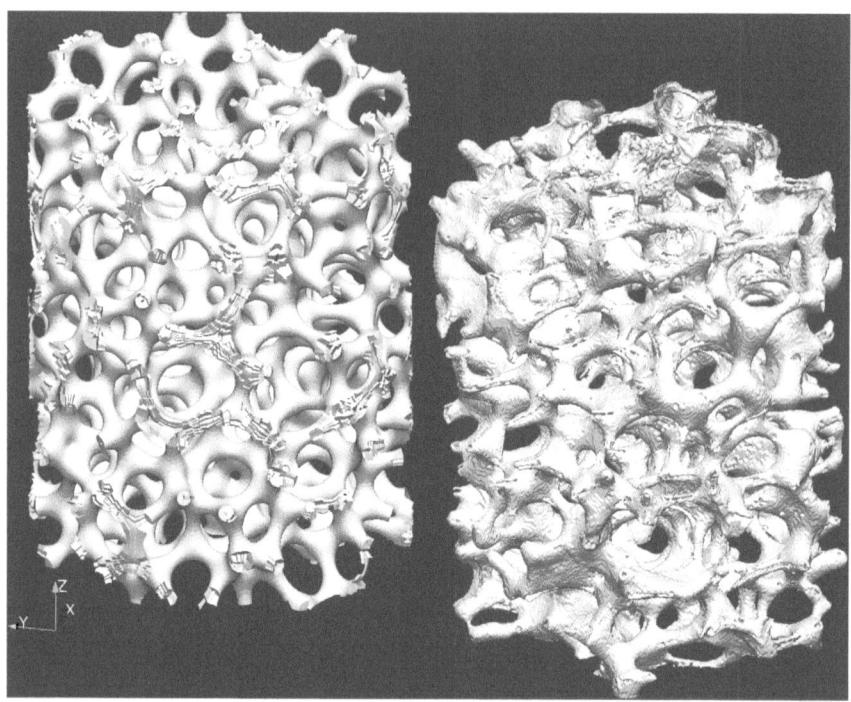

Fig. 16.7 Comparison of a generic foam structure (left) with a relative density of $\varrho = 20\%$, and a CT-scan image (right) of a 10ppi foam sample. For better comparison, all elements of the generated model lying outside the bounding cylinder ($d = 20$ mm, $h = 20$ mm), are removed [1]

beam models systematically underestimate the stiffness of foams, since the additional reinforcement of the foam nodes cannot be represented by beam elements. However, volumetric models follow almost perfectly the Gibson-Ashby relation [64, 65] for the relative elastic modulus depending on the relative density as $\bar{E}/E = [\bar{\rho}/\rho]^2 = \rho_{\text{rel}}^2$.

Furthermore, Storm et al. [66] investigated the influence of the strut shape on the strength of a foam. The cross section radius of the struts was varied according to a hyperbolic equation along their longitudinal axis, where r_e/r_m defines the ratio of the strut radii at the end and the middle of the strut. A ratio $r_e/r_m = 1$ characterizes a strut with constant cross section along its axis. Struts with $r_e/r_m > 1$ have thicker foam nodes and the cross sections become smaller towards the strut centers, which is the typical form in real struts as shown in Fig. 16.7. Interestingly, an optimal ratio can be found for $r_e/r_m \approx 1.4$ maximizing the bending strength of the struts, illustrated in the corresponding Fig. 16.8a. The curvature of struts measured in terms of the ratio between length and curvature radius l_0/r_c has only a minor influence on the strength of foams, which is shown in Fig. 16.8b.

Further results from Storm et al. [66] include the comparison of continuum and hybrid models, where only the foam nodes where meshed using continuum elements, whereas large parts of the struts are modeled using beam elements. The hybrid models

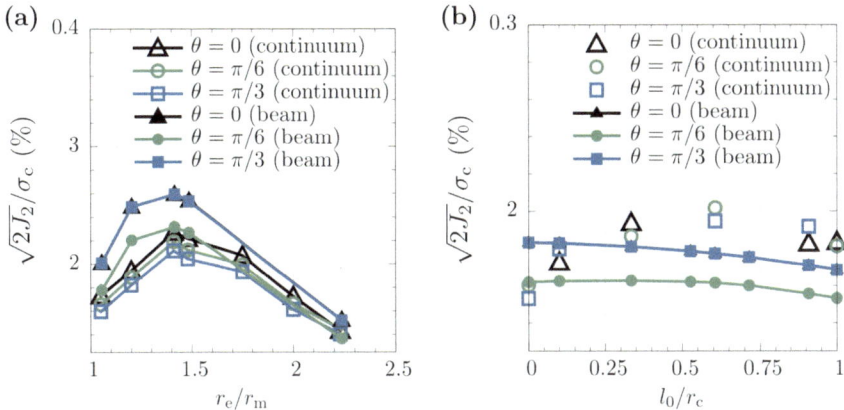

Fig. 16.8 Influence of the **a** material distribution along the strut axis (hyperboloid end to middle radius r_e/r_m) and **b** strut curvature (strut length to curvature radius l_0/r_c) on the norm of critical deviatoric stress in π-plane for continuum and beam models of the Kelvin cell [66]

show an elastic behavior comparable to the continuum models, but the numerical effort required to compute them is considerably reduced.

Small sharp edged cavities remain along the center line of the struts due to the production process. It was found that the influence of the cavities does not have a significant effect on the elastic properties of the foams as long as the relative density remains constant [31].

Failure limit surfaces allow the assessment of strength of filter structures utilizing different failure criteria. Firstly, we determined yield surfaces analytically for Kelvin cells where the struts are assumed to follow the Euler-Bernoulli beam theory. Using this model, results for different failure criteria as von Mises, maximum absolute principal, maximum principal and principal stress criterion with tension compression asymmetry have been calculated [29]. Storm et al. [31] studied the influence of geometrical variations for a Kelvin cell foam topology. These investigations considered the change of cross-section shape of the struts, strut curvature and pore anisotropy. All these features influence the elastic properties significantly. For the generic and almost isotropic foam structure F216 as shown in Fig. 16.3, investigations tackling failure assessment have been conducted using the von Mises criterion, considering different relative densities and strut geometries [1]. Two examples of yield surfaces of the F216 foam RVE are displayed in Figs. 16.9 and 16.10 for a von Mises and Weibull failure criterion, respectively. The macroscopic failure surface for a local von Mises criterion, as shown in Fig. 16.9, is point symmetric with respect to the origin. The shape of the meridian cross section depends on the Lode angle θ. For a Lode angle $\theta = 0$ the maximum J_2-value is located where the hydrostatic stress vanishes ($I_1 = 0$). For negative or positive Lode angles the maximum is shifted towards the negative and positive hydrostatic stress values, respectively. The rightmost diagram shows cross sections for different angles α (cf. (16.11)) projected onto the devia-

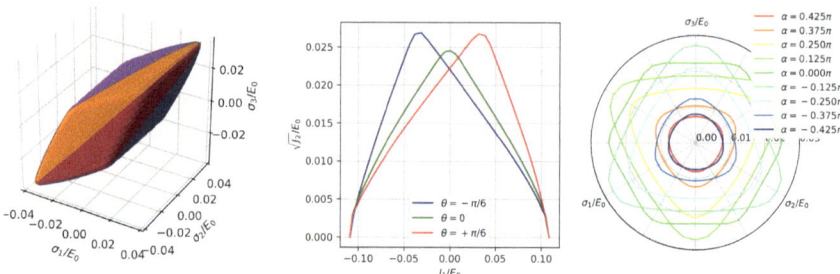

Fig. 16.9 Failure limit surfaces for a F216 foam RVE with a relative density of 20% for a von Mises criterion. (Left) limit surfaces in principal stress space. (Center) meridian cut for different Lode angles θ. (Right) hydrostatic cuts at different angles α

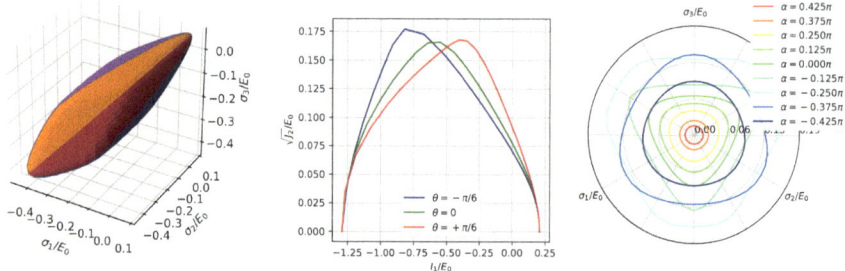

Fig. 16.10 Failure limit surfaces for a F216 foam RVE with a relative density of 20% for a Weibull criterion. (Left) limit surfaces in principal stress space. (Center) meridian cut for different Lode angles θ. (Right) hydrostatic cuts at different angles α

toric plane. For $\alpha = 0$, which implies a pure deviatoric stress state, the cross section exhibits a hexagonal shape, whereas for positive or negative angles the cross section transforms towards a triangular shape but with opposite orientation. To the authors knowledge, for such shapes of failure surfaces no closed form analytical description is available in literature.

The failure surface for the Weibull criterion (cf. Fig. 16.10) given at a failure probability of 63.2% shows a strong asymmetry with respect to the hydrostatic stress axis. The strength in hydrostatic compression is much higher than in hydrostatic tension. The ratio between compressive and tensile strength depends on the relative density of the foam. The meridian cross sections show a similar contour as for the von Mises criterion, but the maximum value for negative Lode angles exceeds the one for positive Lode angles. In general the maxima of the meridian cross sections are observed for negative hydrostatic stress values ($I_1 < 0$) and exceed the ones of the von Mises criteria. The cross section shapes for constant angles α vary from an almost circular shape for $\alpha = 0.425$ to triangular shapes down to $\alpha \approx -0.125$. Around $\alpha \approx -0.25$ an almost hexagonal shape is observed, before for further decreasing α values the shape morphs again towards a triangle, but with opposite orientation compared to those for positive angles α.

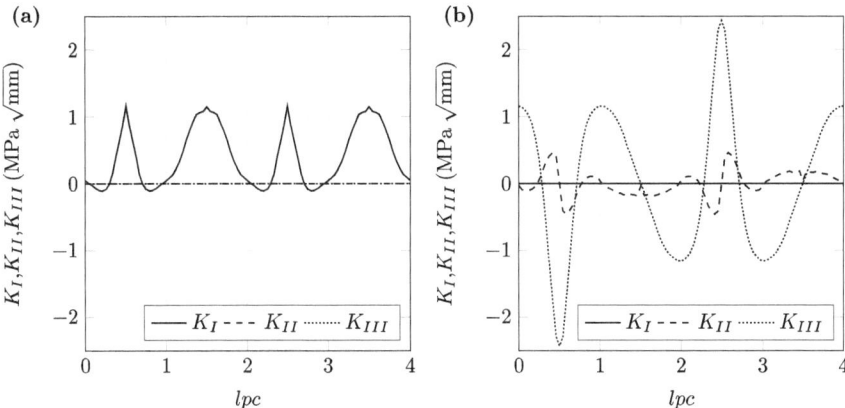

Fig. 16.11 SIFs along the sharp edged cavity within the four-sided strut loop of a Kelvin cell having the normal (001) under effective **a** uniaxial tension $\bar{\sigma}_{11} = 1/16$ MPa and **b** shear loading $\bar{\sigma}_{13} = 1/16$ MPa [30]

The spatial orientation of non-isotropic foam RVEs with respect to the principal directions of the effective stress tensor has an influence on the shape of the corresponding failure surfaces and has been investigated by Zhang and Storm for the special case of orthotropic Kelvin cells [28, 29]. For RVEs containing an increasing number of randomly arranged pores the effective behavior tends towards isotropy and the orientation of the RVE becomes negligible.

The sharp edged cavities remaining from the production process required a fracture mechanical analysis, conducted by Settgast et al. [30]. With help of the interaction integral (16.19), the SIFs are computed for each point along the sharp edges of a Kelvin cell as illustrated in Fig. 16.5. Since the FE analysis is linear elastic, any load case and therewith the corresponding SIFs can be constructed by a scaled superposition of these six independent base load cases. It was found that the fracture limit surfaces always enclose the failure limit surfaces for a von Mises or a maximum principal stress criterion using credible values obtained by experiments for the considered material. From these observations it can be concluded that the local failure is preferably triggered by a critical principal stress on the outer strut surface rather than by the stress concentration at the crack front inside the strut cavities. Figure 16.11 shows exemplary the values of the three SIFs along the local parametric coordinate l_{pc} of a sharp edge around a square shaped foam window for an effective shear loading.

A more critical scenario is thermal shock loading, which occurs at the beginning of the casting process when the melt enters the filter, which may cause the filter to fail even before the melt filtration process starts. Here, the sharp edged cavities are potentially the most critical locations. The occuring stress intensity factors or J-integral values depend strongly on the material properties of the foams bulk material. The dependencies of the maximal J-integral value J^+ along all cavity edges on thermal conductivity h, relative density ρ_{rel} and RVE length l are presented in Fig. 16.12.

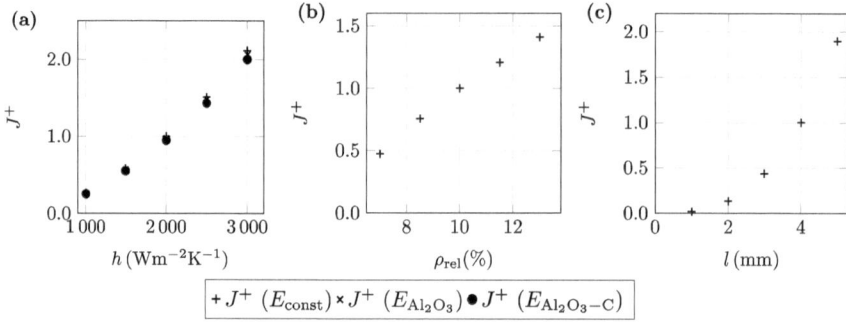

Fig. 16.12 a $J^+(h)$ for $\rho_{\mathrm{rel}} = 10\%$, $l = 4\,\mathrm{mm}$ and different $E(\vartheta)$, **b** $J^+(\rho_{\mathrm{rel}})$ for $h = 2000\,\mathrm{Wm^{-2}K^{-1}}$ and $l = 4\,\mathrm{mm}$ and **c** $J^+(l)$ for $h = 2000\,\mathrm{Wm^{-2}K^{-1}}$ and $\rho_{\mathrm{rel}} = 10\%$ taken from [23]

J^+ depends strongly on the thermal conductivity h with an almost quadratic dependence $J^+ \propto h^2$, because the increasing heat flux leads to higher temperature gradients and therefore higher thermal stresses as well as a higher fracture mechanical loading at the crack front. The influence of the different elastic moduli of the three considered filter materials slightly increases with an increasing heat transfer coefficient. If the relative density ρ_{rel} is increased the size of the sharp-edged cavities does not change. However, the outer surface area A_0 of the foam increases and therewith the heat flux and the fracture mechanical loading at the cracks, with $J^+ \propto \rho_{\mathrm{rel}}^{1/\kappa}$ with the value of the denominator of the exponent $\kappa \gtrsim 1$. The largest impact on J^+ is observed for a changing size of the RVE. It is a combination of two effects. First the crack size changes linearly with the cell size and $J \propto l$. The second effect is that the foam surface area $A_0 \propto l^2$. Both effects result in an almost cubic dependence $J^+ \propto l^3$. The detailed material parameters and the temperature dependent elastic moduli as well as the loading conditions can be found in reference [23].

In some applications, e.g., continuous casting, the ceramic foam filter is subjected to a permanent loading caused by the molten metal for more than a couple of minutes. For that reason, it is important to estimate the long time behavior of the material at high temperatures. In Settgast et al. [67] the creep properties of foams are compared with the creep properties of the bulk material. The following stress-time dependent creep law was assumed

$$\dot{\bar{\varepsilon}}_{\mathrm{eq}}^{\mathrm{cr}} = \left[\frac{\bar{\sigma}_{\mathrm{eq}}}{\bar{A}} \right]^n t^m , \tag{16.81}$$

for the macro scale, wherein the effective equivalent creep strain rate $\dot{\bar{\varepsilon}}_{\mathrm{eq}}^{\mathrm{cr}}$ depends on the effective equivalent stress $\bar{\sigma}_{\mathrm{eq}}$, the structure specific effective stress factor $\bar{A}$, with exponent n and time t with its corresponding exponent m. When applying the same creep law for the bulk material at the micro scale, a good agreement between experimental and simulated creep curves is achieved for uniaxial loading at elevated

Fig. 16.13 Comparison of
creep curves of three tested
and simulated foam samples
for uniaxial loading
($\bar{\sigma} = 0.1$ MPa and
$\vartheta = 1350\,°$C [62]

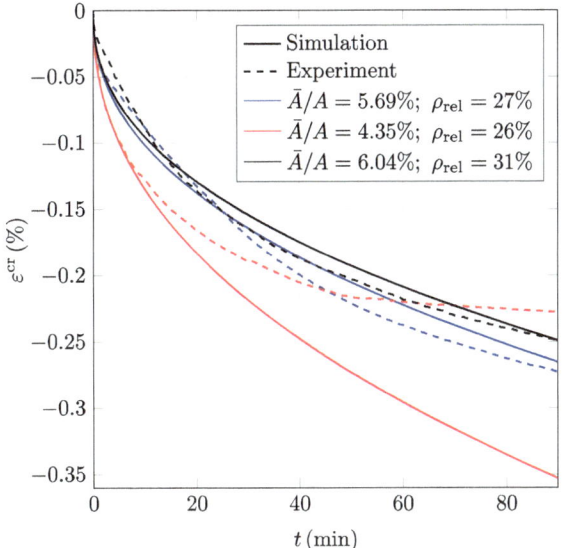

temperature (cf. Fig. 16.13). Interestingly, the exponents n and m are the same for
the local and homogenized constitutive material laws. The ratio between effective
and local stress factor

$$\frac{\bar{A}}{A} \propto \sqrt[n]{\frac{\dot{\bar{\varepsilon}}^{cr}_{eq}}{\dot{\varepsilon}^{cr}_{eq}}} \propto \rho_{rel}^{a/n} \tag{16.82}$$

depends on the structure via exponent a and the relative density ρ_{rel} of the foam. The
exponent a is used to describe the proportionality

$$\frac{\dot{\bar{\varepsilon}}^{cr}_{eq}}{\dot{\varepsilon}^{cr}_{eq}} \propto \left[\frac{1}{\rho_{rel}}\right]^a , \tag{16.83}$$

which is fitted by multiple FE analyses where the relative density is varied. Assuming
a constant relative density the creep rate can be minimized if the strut cross sections
are constant along the strut axis [62, 67].

Utilizing an FE^2 approach in a monolithic setting, proposed by Lange et al. [10],
for which a separate microscale FE simulation of the RVE is run on each macro
material point, the influence of foam morphology on the strength and mechanical
creep behavior can be investigated during realistic filtration scenarios. In [22] a
flow-through filter application (cf. Fig. 16.14) was investigated. As foam RVE the
Wheire-Phelan cell is utilized, because of less computational costs compared to the
F216 foam RVE. The filter causes a pressure drop and therewith mechanical forces
inside the filter. For a constant volumetric flow rate but increasing relative density

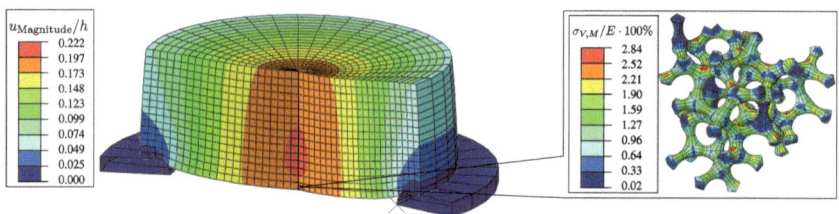

Fig. 16.14 Deformation of a filter during a flow through scenario utilizing the FE^2 homogenization approach [22]

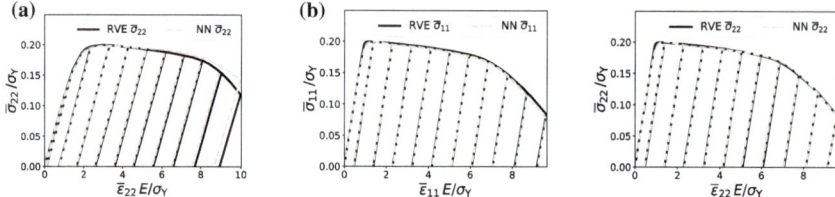

Fig. 16.15 Macroscopic stress-strain curves for **a** uniaxial loading $\bar{\varepsilon}_{22}$, and **b** biaxial loading $\bar{\varepsilon}_{11} = \bar{\varepsilon}_{22}$, where damage is considered. E and σ_Y denote the elastic modulus and yield stress for a local von Mises plasticity model, respectively [33]

the pressure drop increases, which can be described by the Darcy-Forchheimer law. The developed model allows the prediction of elastic and creep deformations for realistic filtration scenarios. The shape of the foam struts has almost no influence on the elastic deflection for a constant relative density. However, with an increasing relative density ρ_{rel} the deflection decreases. An increasing Reynolds number of the flow leads to an increasing deflection, due to the larger pressure drop [11, 22]. In Fig. 16.14 the deformation including creep after one hour of a flow through filtration scenario is illustrated with the underlying FE^2-micro model.

The hybrid approach described in Sect. 16.2.2 employing neural networks was developed to further reduce the computational costs. First, the feasibility of this approach was examined in 2D on a Kelvin cell by Settgast et al. [68]. The model was then extended by a damage formulation [33, 69]. In Fig. 16.15 stress-strain curves are compared from simulations using a fully discretized RVE, with those predicted by the proposed hybrid approach. The hybrid approach can reproduce the RVE simulations almost exactly, even if elastic unloading is taken into account. Also the material degradation due to damage can be reproduced. For these simulations a speed-up factor of 4000 compared to the fully resolved RVE was achieved.

Malik et al. [32] extended the hybrid approach to model also 3D material behavior. As RVE a Wheire-Phelan foam as shown in Fig. 16.3 is used. The corresponding yield surface was identified and its evolution due to strain hardening could be depicted, which is displayed in Fig. 16.16. The comparison of simulations using a fully dis-

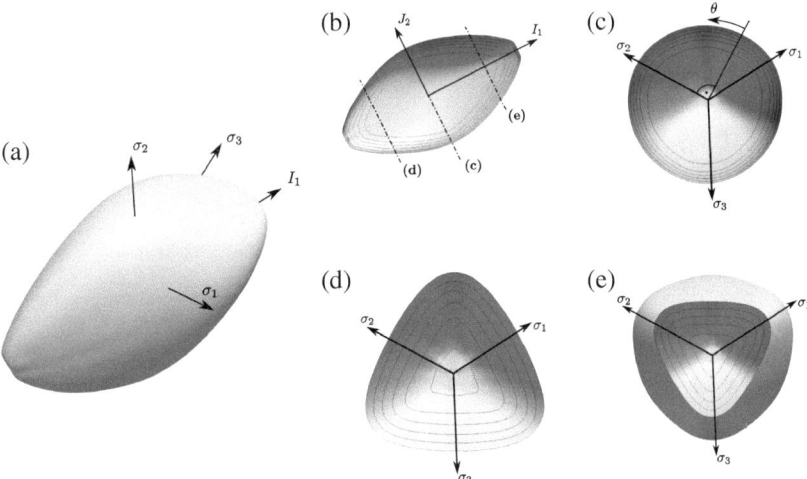

Fig. 16.16 Determined yield surfaces of the investigated Wheire-Phelan RVE for different values of $\bar{\varepsilon}_q^{pl} = 0.005, 0.05, 0.1, 0.15, 0.2, 0.25$ utilizing the approximated yield function NN^Φ in principal stress space. The thin lines indicate the yield surfaces for different values of $\bar{\varepsilon}_q^{pl}$. The symbols σ_1, σ_2 and σ_3 denote the three principal stresses. **a** represents the 3D yield surface, **b** shows the meridian cut with marked deviatoric cutting lines **c**, **d**, and **e** along the hydrostatic axis in I_1^σ-direction without, negative and positive hydrostatic part, respectively taken from [32]

cretized RVE with the results of the hybrid approach was conducted as well. Good agreement was achieved (cf. Fig. 16.17) for different load cases and even for cyclic loading, despite the fact that information on cyclic load cases have not been part of the training data set for the neural networks. It should be emphasized that the flexibility of the neural networks allows the exact reproduction of the short $\bar{\sigma}_{11}$-drop after reaching the maximum load in the elastic regime before hardening, as noticed in the simulations using fully resolved RVEs.

16.3.3 Verification of the Phase-Field Model

In order to verify the FE-implementation we consider a reactive two phase/two component system, where a dissociation reaction

$$A_2 \rightleftharpoons 2\,A \tag{16.84}$$

between the fictitious chemical components A_2 (diatomic, $i = 1$) and A (monoatomic, $i = M = 2$) can occur. The free energy of the mixture is chosen according to a regular solution model with binary interactions

$$\psi = \hat{\psi}_{bul,id}\,(c_i, v_1) + \hat{\psi}_{bul,ex}\,(c_i, v_1) + \hat{\psi}_{int,c}\,(\nabla c_i, v_1) + \hat{\psi}_{int,v}\,(v_1, \nabla v_1) \tag{16.85}$$

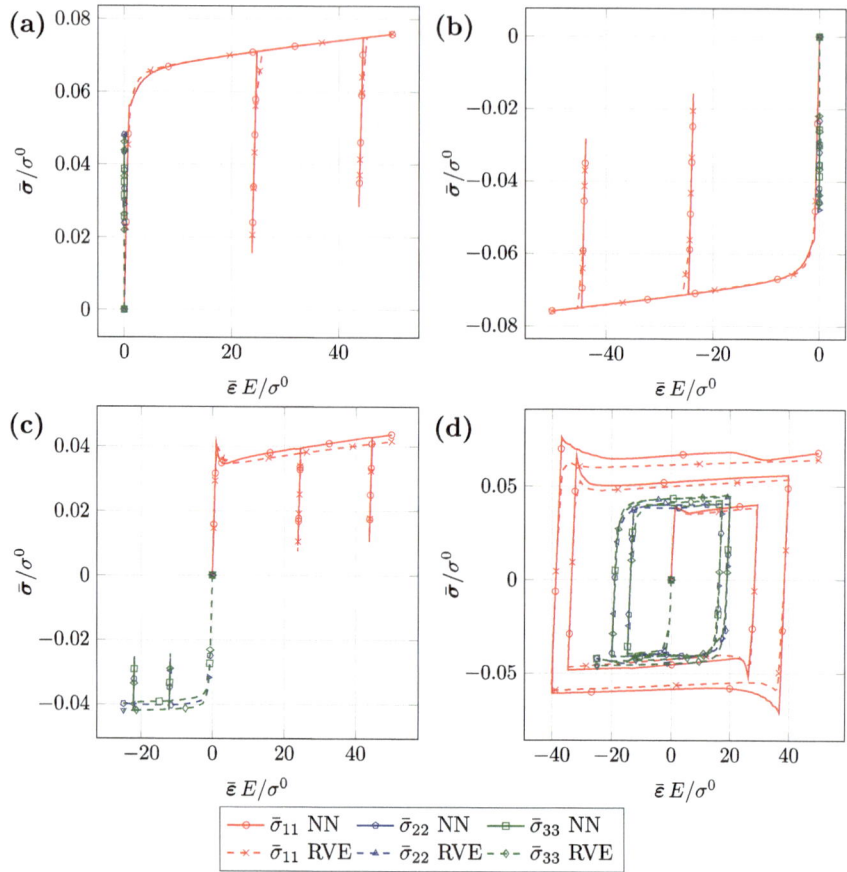

Fig. 16.17 Stress-strain curves of proportional **a** uniaxial tension $\Delta\bar{\varepsilon}_{11} = 0.1$, $\Delta\bar{\varepsilon}_{22} = \Delta\bar{\varepsilon}_{33} = 0$, **b** uniaxial compression $\Delta\bar{\varepsilon}_{11} = -0.1$, $\Delta\bar{\varepsilon}_{22} = \Delta\bar{\varepsilon}_{33} = 0$, **c** deviatoric shear $\Delta\bar{\varepsilon}_{11} = -2\Delta\bar{\varepsilon}_{22} = -2\Delta\bar{\varepsilon}_{33}$ and **d** cyclic deviatoric loading $\Delta\bar{\varepsilon}_{11} = -2\Delta\bar{\varepsilon}_{22} = -2\Delta\bar{\varepsilon}_{33}$ [32]

with the specific contributions

$$\hat{\psi}_{\text{bul,id}}\left(c_i, v_1\right) = \sum_{i=1}^{M} \frac{c_i}{M_i}\left[\left[p\left(v_1\right)\mu_i^{01} + \left[1 - p\left(v_1\right)\right]\mu_i^{02}\right]\right.$$

$$\left. + R\vartheta\ln\left(\frac{c_i}{M_i}\tilde{M}\right)\right], \tag{16.86}$$

$$\hat{\psi}_{\text{bul,ex}}\left(c_i, v_1\right) = \tilde{M}\sum_{i=1}^{M-1}\sum_{k=i+1}^{M}\frac{c_i c_k}{M_i M_k}\left[p\left(v_1\right)L_{ik}^{01} + \left[1 - p\left(v_1\right)\right]L_{ik}^{02}\right], \tag{16.87}$$

$$\hat{\psi}_{\text{int,c}} \left(\boldsymbol{\nabla} c_i, v_1 \right) = \frac{1}{2} \sum_{i=1}^{M} |\boldsymbol{\nabla} c_i|^2 \left[p \left(v_1 \right) \alpha_i^{01} + \left[1 - p \left(v_1 \right) \right] \alpha_i^{02} \right], \tag{16.88}$$

$$\hat{\psi}_{\text{int,v}} \left(v_1, \boldsymbol{\nabla} v_1 \right) = \mathcal{E}^{\Gamma} \left[\frac{6}{L} g(v_1) + \frac{3}{4} L \, |\boldsymbol{\nabla} v_1|^2 \right]. \tag{16.89}$$

Therein, $\psi_{\text{bul,id}}$ and $\psi_{\text{bul,ex}}$ denote the phase dependent energy parts of the ideal solution and excess model, respectively. Moreover, $\psi_{\text{int,c}}$ and $\psi_{\text{int,v}}$ comprise the interface energies between phases characterized either by different component mass fractions c_i or order parameters v_j. The order parameter v_2 is directly substituted by $v_2 = 1 - v_1$ in the given energy. Moreover, v_1 appears via the interpolation function $p(v_1)$ which mixes the chemical potentials of the pure substances in the respective phases. Furthermore, the double well function $g(v_1)$ ensures extrema of the bulk energy at $v_1 = 0$ and $v_1 = 1$

$$p(v_1) = 3v_1^2 - 2v_1^3, \tag{16.90}$$

$$g(v_1) = v_1^2 \left[1 - v_1 \right]^2. \tag{16.91}$$

The molar mass of the mixture

$$\tilde{M} = \left[\sum_{i=1}^{M} \frac{c_i}{M_i} \right]^{-1} \tag{16.92}$$

appears in the logarithmic term of ideal solution and in the excess energy in order to be consistent with a description in molar fractions, as used, e.g., by Bai et al. [47]. The mobility tensors are simplified as $\boldsymbol{M}_{ab} = M_{ab}\boldsymbol{I}$. A constant conversion rate k^1 for reaction (16.84) is considered. The studied model parameters are summarized in Table 16.1.

In Fig. 16.18a, the normalized bulk energies

$$\bar{\psi}_{\text{bul}}^{j} = \frac{\psi_{\text{bul,id}}^{j} + \psi_{\text{bul,ex}}^{j}}{R \, \vartheta} M_2 \tag{16.93}$$

are plotted for $v_1 = 1$ (phase 1) and $v_1 = 0$ (phase 2). By arguments of equilibrium thermodynamics, possible final states of the model system are known for different initial component/phase compositions which can be illustrated with help of Fig. 16.18a. We firstly assume that the mass fraction of A_2 is homogeneous across the domain with $c_1 = 0.92$ at the beginning. The bulk energies can be read off at point 0 in Fig. 16.18a. For a fixed composition, i.e., no chemical reaction, the energy of the system can be minimized, if the system is allowed to decompose into sub-regions with compositions given by the *common tangent* construction (dotted line in Fig. 16.18a) and mass fractions according to the *lever rule* [70]. The energetic state of the system with decomposed phases can be found on the common tangent, i.e., point 2 in Fig. 16.18a. During a simulation, this behavior is only observable if the initial order parameter is

Table 16.1 Model parameters of the verification example (dissociation reaction & phase separation)

Parameter	Meaning	Unit	Value
$R\,\vartheta$	Universal gas constant · temperature	J/mol	10.0
ρ_0	Reference mass density	kg/m^3	1.0
M_1, M_2	Molar mass	kg/mol	2.0, 1.0
μ_1^{01}, μ_1^{02}	Chem. pot. phase 1/phase 2	J/mol	10.0, 15.0
μ_2^{01}, μ_2^{02}	Chem. pot. phase 1/phase 2	J/mol	0.0, −5.0
L_{12}^{01}, L_{12}^{02}	Interaction coeff. phase 1/phase 2	J/mol	25.0, 0.0
M_{11}	Mobility	(kg m^2)/(J s)	1.0
$\alpha_1^{01} = \alpha_1^{02}$	Gradient parameter in phase 1/phase 2	(J m^2)/kg	2.0
$\alpha_2^{01} = \alpha_2^{02}$	Gradient parameter in both phases	(J m^2)/kg	0.0
β_1	Viscosity parameter	(J s)/kg	2.5
k^1	Conversion rate	mol^2/ (J kg s)	$1.0 \cdot 10^{-6}$
v_1^1, v_2^1	Stoichiometric coefficients	–	−1, 2
$\mathcal{E}^\Gamma$	Surface energy	(J m)/kg	5.0
L	Length parameter	m	0.5

not constant $v_1 = 0$ or $v_1 = 1$ throughout the system, i.e., an initial perturbation is necessary. For chemically reactive systems, the mass fractions can evolve, e.g., along the energetic states 1-2-3, until the global minimum of the bulk energy which is point 4 in Fig. 16.18a, with a homogeneous composition and stable phase 2 everywhere. At the intermediate point 3, phase 1 vanishes completely.

For the 2D-simulation, a square domain is considered with edge length $l_0 = 10$ m. As mentioned, the initial mass fraction c_1 of A_2 is homogeneously prescribed:

$$c_1(\boldsymbol{x}, t = 0) = 0.92 \qquad \forall \boldsymbol{x} \in \mathcal{B}. \qquad (16.94)$$

For the order parameter v_1 the following initialization is chosen to realize a perturbation:

$$v_1 = \begin{cases} 0.1 & \text{for } x \leq 5\,\text{m} \\ 0.9 & \text{for } x > 5\,\text{m}. \end{cases} \qquad (16.95)$$

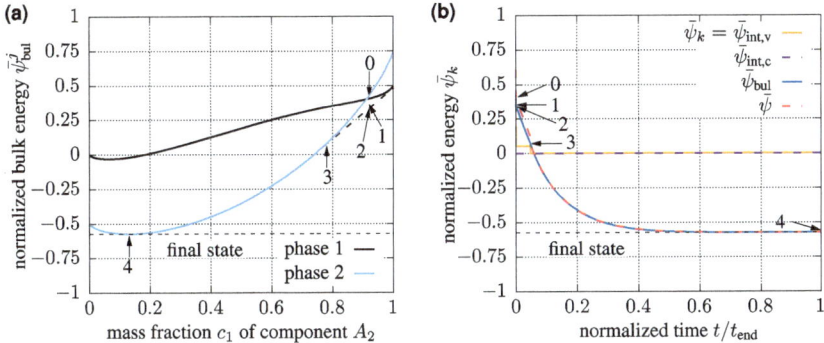

Fig. 16.18 Energy landscapes of the verification example: **a** Bulk energies of the considered phases, and **b** time evolution of system averaged energy contributions. Numbers refer to bulk energies at particular time points

Trivial Neumann boundary conditions are assumed for the mass flux as well as for the microtractions at all boundaries. The domain is discretized by $100 \times 100 = 10^4$ quadrilateral finite elements of equal size.

The process to reach energetic minimum highly depends on the chosen model parameters. For the considered case, phase transition and diffusion proceed much faster than the chemical reaction. In Fig. 16.18b, the evolution of different contributions to the system's whole energy are plotted over time. The specific energies are illustrated as normalized quantities $\bar{\psi}_k$ given by

$$\bar{\psi}_k = M_2 \frac{\int_{\mathcal{B}} \rho_0 \, \hat{\psi}_k(\mathbf{x}) \, \mathrm{d}v}{\rho_0 \, V \, R \, \vartheta}, \tag{16.96}$$

where the index k refers to the terms defined in Eqs. (16.85)–(16.89). The bulk energy $\bar{\psi}_{\mathrm{bul}}$ is introduced as the sum of the ideal solution and excess parts, $\bar{\psi}_{\mathrm{bul,id}}$ and $\bar{\psi}_{\mathrm{bul,ex}}$, respectively. Furthermore, $\bar{\psi}$ is the normalized total free energy.

The numbers 0–4 in Fig. 16.18b belong to specific states of the system's bulk energy and time points which are correspondingly highlighted in the energy landscape Fig. 16.18a. Additionally, the spatial distribution of the order parameter v_1 and the mass fraction c_1 of A_2 are plotted in Fig. 16.20 at these time points. The plots belong to cuts along the x-axis at $y = 0$. These reduced representations contain the full field information of the considered primary variables since the fields just vary along x for the specific problem as shown in Fig. 16.19 for a particular time point.

According to Fig. 16.18b, the total free energy of the system is monotonously minimized during time evolution until the final energetic state 4 which matches the expectation from equilibrium consideration as highlighted in Fig. 16.18a. The following intermediate states of interest should be discussed. From point 0 to 1, the interface energy $\bar{\psi}_{\mathrm{int,v}}$ is quickly decreased, see Fig. 16.18b, in order to relax the initially sharp gradient of v_1 to a typical tanh-shape of the interface [71], compare

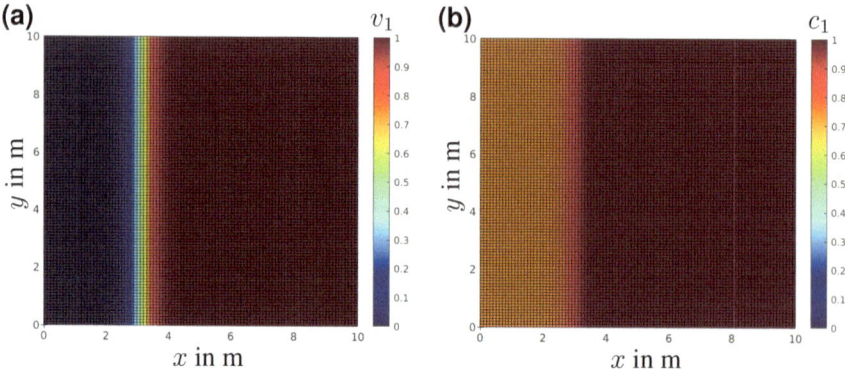

Fig. 16.19 Spatial distribution of **a** the order parameter and **b** the mass fraction c_1 of component A_2 at time point 2 marked in Fig. 16.18

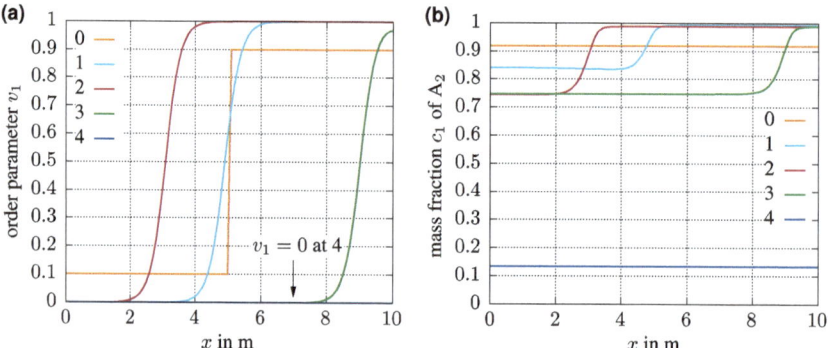

Fig. 16.20 Spatial distribution of **a** the order parameter and **b** the mass fraction c_1 of component A_2 at time points 0–4 marked in Fig. 16.18

Fig. 16.20a. The second gradient energy $\bar{\psi}_{int,c}$ participates, too, but the change is not visible in the diagrams since the chosen gradient parameter seems to have a lower energetic effect. The considered domain consists of two distinct phases at point 1, but the energetic state and the mass fraction c_1 in the phases are still above the common tangent and far away from the intermediate equilibrium constitution according to Fig. 16.20b. Due to the high mobility of diffusion and the slow conversion rate of the chemical reaction, the common tangent is reached at point 2 while the overall composition does not change during this short time span. The interface has moved to $x \approx 3$ mm yielding a larger region of $v_1 = 1$ according to the lever rule. The mass fractions c_1 in the phases located at the left and the right from the interface exhibit the expected equilibrium values at the common tangent. From point 2 to 3 in Fig. 16.18b, the chemical reaction becomes dominant which decreases the overall mass fraction c_1 of A_2 shifting the bulk energy along the common tangent as additionally illustrated in Fig. 16.18a. Correspondingly, the domain of phase 1 shrinks, see Fig. 16.20a, but the mass fractions far from the interface keep the equilibrium values, see Fig. 16.20b. At point 3, the interface starts to decay which is visible as the jump-like change in the interface energy $\bar{\psi}_{\text{int,v}}$ in Fig. 16.18b. The common tangent is left and phase 2 becomes stable during the rest of the purely reactive process. The mass fraction c_1 takes a constant value across the spatial domain and the reaction decreases c_1 to 0.13 as shown in Fig. 16.20b. This state is found at the global energetic minimum marked as point 4 in Fig. 16.18a. Here, the affinity $A^1 \propto \tilde{\mu}_1$ as driving force of the reaction consequently vanishes which defines the (dynamic) *chemical equilibrium*.

The obtained final state of the simulation matches the equilibrium thermodynamics solution, where intermediate evolution steps towards the chemical equilibrium are in qualitative accordance with our expectations explained above. The proposed two phase/two component problem is a versatile benchmark example to verify the numerical treatment relying at least on knowledge and solutions of equilibrium thermodynamics. For the same purpose similar simplified problems have been discussed in recent literature especially for Cahn-Hilliard type problems [47, 55]. Such additional cases have been successfully tested, too, e.g., chemical reaction during spinodal decomposition (Cahn-Hilliard type) for fixed phase 1. However, in this paper we focused on simultaneously active chemical reaction and phase transition for the sake of brevity. We conclude that our FE-implementation of the multi-component/multiphase model yields trustworthy results and can be utilized for more complex problems.

16.4   Conclusions

Along the road tools have been developed to virtually design foam structures. These models can be used in simulations to predict and analyze their thermo-mechanical properties. Using additive manufacturing methods the models can be physically realized and used in experimental investigations regarding filtration phenomena. The conducted numerical investigations provide insights into structure-property relations of filter structures, but can now also contribute statements about the behavior of the entire foam structure in a homogenized manner. Different homogenization approaches can be applied to efficiently predict effective elastic, plastic, creep, fracture, and damage properties of foam structures under loading conditions in metal melt filtration applications. Geometric and topological variations of foams are discussed to improve the thermo-mechanical integrity and the filtration efficieny. Further, thermochemical phenomena of filtration processes, as the formation of in-situ layers can be modeled with developed phase-field models. These new models are able to describe diffusion, phase transition, and chemical reactions in the multi-component/multiphase systems. The in-situ layer formation affects the reactive filtration phase. The knowledge about the kinetics of this effects is very valuable. The developed modeling tools and approaches are not restricted to applications regarding metal melt filtration. They can be applied in many different fields of thermo-chemo-mechanics and research on porous media or meta materials.

Acknowledgements The authors gratefully acknowledge the work and efforts of Dr. -Ing. J. Storm, Dr. -Ing. C. Settgast and Dr. -Ing. Dongshuan Zhang done in the previous funding periods of the CRC 920. This work was funded by the Deutsche Forschungsgemeinschaft (DFG, German Research Foundation)—Project-ID 169148856—CRC 920: Multi-Functional Filters for Metal Melt Filtration—A Contribution towards Zero Defect Materials, subproject B05. Furthermore, the authors acknowledge computing time on the compute cluster of the Faculty of Mathematics and Computer Science of Technische Universität Bergakademie Freiberg, operated by the computing center (URZ) and funded by the Deutsche Forschungsgemeinschaft (DFG, German Research Foundation)—Project-ID 397252409.

References

1. M. Abendroth, E. Werzner, C. Settgast, S. Ray, Adv. Eng. Mater. **19**(9), 1700080 (2017). https://doi.org/10.1002/adem.201700080
2. K. Schwartzwalder, H. Somers, A. Somers. Method of making porous ceramic articles. http://www.google.it/patents/US3090094. US Patent 3,090,094 (1963)
3. V. Deshpande, N. Fleck, J. Mech. Phys. Solids **48**(6–7), 1253 (2000). https://doi.org/10.1016/s0022-5096(99)00082-4
4. K.H. Roscoe, A.N. Schofield, C.P. Wroth, Géotechnique **8**(1), 22 (1958). https://doi.org/10.1680/geot.1958.8.1.22
5. D. Bigoni, A. Piccolroaz, Int. J. Solids Struct. **41**(11–12), 2855 (2004). https://doi.org/10.1016/j.ijsolstr.2003.12.024
6. W. Ehlers, Arch. Appl. Mech. **65**(4), 246 (1995). https://doi.org/10.1007/bf00805464
7. W. Ehlers, O. Avci, Int. J. Numer. Anal. Met. **37**(8), 787 (2012). https://doi.org/10.1002/nag.1121
8. W. Ehlers, J. Bluhm (eds.), *Porous Media—Theory, Experiments and Numerical Applications* (Springer-Verlag, Berlin, Heidelberg, 2002)
9. V.B.C. Tan, K. Raju, H.P. Lee, Comput. Methods Appl. Mech. Eng. **360**, 112694 (2020). https://doi.org/10.1016/j.cma.2019.112694
10. N. Lange, G. Hütter, B. Kiefer, Comput. Methods Appl. Mech. Eng. **382**, 113886 (2021). https://doi.org/10.1016/j.cma.2021.113886
11. E. Werzner, M. Abendroth, C. Demuth, C. Settgast, D. Trimis, H. Kraus, S. Ray, Adv. Eng. Mater. **19**(9), 1700240 (2017). https://doi.org/10.1002/adem.201700240
12. H. Lehmann, E. Werzner, A. Malik, M. Abendroth, S. Ray, B. Jung, Adv. Eng. Mater. **24**(2), 2100878 (2021). https://doi.org/10.1002/adem.202100878
13. T. Zienert, S. Dudczig, O. Fabrichnaya, C.G. Aneziris, Ceram. Int. **41**(2), 2089 (2015). https://doi.org/10.1016/j.ceramint.2014.10.004
14. E. Storti, S. Dudczig, A. Schmidt, G. Schmidt, C.G. Aneziris, Steel Res. Int. **88**(10), 1700142 (2017). https://doi.org/10.1002/srin.201700142
15. A. Asad, M. Haustein, K. Chattopadhyay, C.G. Aneziris, R. Schwarze, JOM **70**(12), 2927 (2018). https://doi.org/10.1007/s11837-018-3117-4
16. A. Schmidt, A. Salomon, S. Dudczig, H. Berek, D. Rafaja, C.G. Aneziris, Adv. Eng. Mater. **19**(9), 1700170 (2017). https://doi.org/10.1002/adem.201700170
17. A. Salomon, M. Motylenko, D. Rafaja, Adv. Eng. Mater. **24**(2), 2100690 (2021). https://doi.org/10.1002/adem.202100690
18. A. Donev, F. Stillinger, S. Torquato, J. Comput. Phys. **202**, 737 (2005). https://doi.org/10.1016/j.cp.2004.08.014
19. A. Donev, F. Stillinger, S. Torquato, J. Comput. Phys. **202**, 765 (2005). https://doi.org/10.1016/j.cp.2004.08.025

20. M. Skoge, A. Donev, F. Stillinger, S. Torquato, Phys. Rev. E **74**, 041127 (2006). https://doi. org/10.1103/PhysRevE.74.041127
21. K. Brakke, Exp. Math. **1**, 141 (1992). https://doi.org/10.2307/1575877
22. N. Lange, M. Abendroth, E. Werzner, G. Hütter, B. Kiefer, Adv. Eng. Mater. **24**(2), 2100784 (2021). https://doi.org/10.1002/adem.202100784
23. C. Settgast, M. Abendroth, M. Kuna, Journal of Multiscale Modelling **07**(04), 1640006 (2016). https://doi.org/10.1142/s1756973716400060
24. A. Herdering, J. Hubálková, M. Abendroth, P. Gehre, C.G. Aneziris, Int. Cer. Rev. **68**(4), 30 (2019). https://doi.org/10.1007/s42411-019-0013-z
25. B. Bock-Seefeld, T. Wetzig, J. Hubálková, G. Schmidt, M. Abendroth, C.G. Aneziris, Adv. Eng. Mater. **24**(2), 2100655 (2021). https://doi.org/10.1002/adem.202100655
26. C. Miehe, A. Koch, Arch. Appl. Mech. **72**, 300 (2002)
27. J. Storm, M. Abendroth, M. Kuna, in *Proceedings of the Unified International Technical Conference on Refractories (UNITECR 2013)* (John Wiley & Sons, Inc., 2014), pp. 897–902. https://doi.org/10.1002/9781118837009.ch153
28. D. Zhang, M. Abendroth, M. Kuna, J. Storm, Int. J. Solids Struct. **75–76**, 1 (2015). https://doi. org/10.1016/j.ijsolstr.2015.04.020
29. J. Storm, M. Abendroth, M. Kuna, Int. J. Mech. Sci. **105**, 70 (2016). https://doi.org/10.1016/ j.ijmecsci.2015.10.014
30. C. Settgast, M. Abendroth, M. Kuna, Arch. Appl. Mech. **86**(1–2), 335 (2016). https://doi.org/ 10.1007/s00419-015-1107-3
31. J. Storm, M. Abendroth, M. Kuna, Mech. Mater. **86**, 1 (2015). https://doi.org/10.1016/j. mechmat.2015.02.012
32. A. Malik, M. Abendroth, G. Hütter, B. Kiefer, Adv. Eng. Mater. **24**(2), 2100641 (2021). https:// doi.org/10.1002/adem.202100641
33. C. Settgast, G. Hütter, M. Kuna, M. Abendroth, Int. J. Plasticity **126**, 102624 (2020). https:// doi.org/10.1016/j.ijplas.2019.11.003
34. M. Wojciechowski, Computer assisted mechanics and engineering sciences **18** (2011)
35. E. Aifantis, Mech. Mater. **35**(3), 259 (2003). https://doi.org/10.1016/S0167-6636(02)00278-8
36. A. Eringen, E. Suhubi, Int. J. Eng. Sci. **2**(2), 189 (1964). https://doi.org/10.1016/0020-7225(64)90004-7
37. R.D. Mindlin, Arch. Ration. Mech. An. **16**(1), 51 (1964). https://doi.org/10.1007/bf00248490
38. G. Hütter, *A Theory for the Homogenisation Towards Micromorphic Media and Its Application to Size Effects and Damage* (Technische Universität Bergakademie Freiberg, Habilitation, 2019)
39. G. Hütter, J. Mech. Phys. Solids **127**, 62 (2019). https://doi.org/10.1016/j.jmps.2019.03.005
40. N. Moelans, B. Blanpain, P. Wollants, Calphad **32**(2), 268 (2008). https://doi.org/10.1016/j. calphad.2007.11.003
41. W. Gathright, M. Jensen, D. Lewis, Electrochem. Commun. **13**, 520 (2011). https://doi.org/10. 1016/j.elecom.2011.02.038
42. S.P. Clavijo, A.F. Sarmiento, L.F.R. Espath, L. Dalcin, A.M.A. Cortes, V.M. Calo, J. Comput. Appl. Math. **350**, 143 (2019). https://doi.org/10.1016/j.cam.2018.10.007
43. C. Miehe, S. Mauthe, H. Ulmer, Int. J. Numer. Meth. Eng. **99**, 131 (2014). https://doi.org/10. 1002/nme.4700
44. C. Cui, R. Ma, E. Martínez-Pañeda, J. Mech. Phys. Solids **147**, 104254 (2021). https://doi.org/ 10.1016/j.jmps.2020.104254
45. H. Attariani, V.I. Levitas, Acta Mater. **220**, 117284 (2021). https://doi.org/10.1016/j.actamat. 2021.117284
46. Y. Bai, D.A. Santos, S. Rezaei, P. Stein, S. Banerjee, B.X. Xu, Int. J. Solids Struct. **228**, 111099 (2021). https://doi.org/10.1016/j.ijsolstr.2021.111099
47. Y. Bai, J.R. Mianroodi, Y. Ma, A. Kwiatkowski da Silva, B. Svendsen, D. Raabe, Acta Mater. **231**, 117899 (2022). https://doi.org/10.1016/j.actamat.2022.117899
48. I. Romero, E.M. Andrés, Á. Ortiz-Toranzo, Comput. Method. Appl. M. **385**, 114013 (2021). https://doi.org/10.1016/j.cma.2021.114013

49. H. Larsson, CALPHAD: Comput Coupling Phase Diagrams Thermochem. **47**, 1 (2014). https://doi.org/10.1016/j.calphad.2014.06.001
50. H. Larsson, L. Höglund, CALPHAD: Comput. Coupling Phase Diagrams Thermochem. **33**, 495 (2009). https://doi.org/10.1016/j.calphad.2009.06.004
51. B. Svendsen, P. Shanthraj, D. Raabe, J. Mech. Phys. Solids **112**, 619 (2018). https://doi.org/10.1016/j.jmps.2017.10.005
52. M.E. Gurtin, Phys. D **92**, 178 (1996). https://doi.org/10.1016/0167-2789(95)00173-5
53. G. Hütter, in *Encyclopedia of Continuum Mechanics*, ed. by H. Altenbach, A. Öchsner (Springer Berlin Heidelberg, 2018), pp. 1–8. https://doi.org/10.1007/978-3-662-53605-657-1
54. C. Miehe, J. Mech. Phys. Solids **59**, 898 (2011). https://doi.org/10.1016/j.jmps.2010.11.001
55. C. Miehe, F.E. Hildebrand, L. Böger, Proc. R. Soc. A **470**(20130641) (2014). https://doi.org/10.1098/rspa.2013.0641
56. I. Müller, Int. J. Solids Struct. **38**, 1105 (2001). https://doi.org/10.1016/S0020-7683(00)00076-7
57. S.R. de Groot, P. Mazur, *Non-Equilibrium Thermodynamics* (Dover Publications Inc, New York, 1984)
58. G. Lebon, D. Jou, J. Casas-Vázquez, *Understanding Non-equilibrium Thermodynamics* (Springer-Verlag, Berlin Heidelberg, 2008)
59. A. Asad, H. Lehmann, B. Jung, R. Schwarze, Adv. Eng. Mater. **24**(2), 2100753 (2022). https://doi.org/10.1002/adem.202270006
60. T. Wetzig, M. Neumann, M. Schwarz, L. Schöttler, M. Abendroth, C.G. Aneziris, Adv. Eng. Mater. **24**(2), 2100777 (2021). https://doi.org/10.1002/adem.202100777
61. A. Herdering, M. Abendroth, P. Gehre, J. Hubálková, C.G. Aneziris, Ceram. Int. **45**(1), 153 (2019). https://doi.org/10.1016/j.ceramint.2018.09.146
62. C. Settgast, Y. Ranglack-Klemm, J. Hubalkova, M. Abendroth, M. Kuna, H. Biermann, J. Eur. Ceram. Soc. **39**(2–3), 610 (2019). https://doi.org/10.1016/j.jeurceramsoc.2018.09.022
63. J. Storm, M. Abendroth, D. Zhang, M. Kuna, Adv. Eng. Mater. **15**(12), 1292 (2013). https://doi.org/10.1002/adem.201300141
64. M. Ashby, Philos. Trans. R. Soc. A: Math. Phys. Eng. Sci. **364**(1838), 15 (2005). https://doi.org/10.1098/rsta.2005.1678
65. L.J. Gibson, M.F. Ashby, J. Zhang, T.C. Triantafillou, Int. J. Mech. Sci. **31**(9), 635 (1989). https://doi.org/10.1016/s0020-7403(89)80001-3
66. J. Storm, M. Abendroth, M. Kuna, Mech. Mater. **137**, 103145 (2019). https://doi.org/10.1016/j.mechmat.2019.103145
67. C. Settgast, J. Solarek, Y. Klemm, M. Abendroth, M. Kuna, H. Biermann, Adv. Eng. Mater. **19**(9), 1700082 (2017). https://doi.org/10.1002/adem.201700082
68. C. Settgast, M. Abendroth, M. Kuna, Mech. Mater. **131**, 1 (2019). https://doi.org/10.1016/j.mechmat.2019.01.015
69. M. Abendroth, G. Hütter, C. Settgast, A. Malik, B. Kiefer, M. Kuna, Technische Mechanik **40**(1), 5 (2020). https://doi.org/10.24352/UB.OVGU-2020-008
70. G. Job, R. Rüffler, *Physical Chemistry from a Different Angle* (Springer International Publishing Switzerland, 2016). https://doi.org/10.1007/978-3-319-15666
71. G.I. Tóth, T. Pusztai, L. Gránásy, Phys. Rev. B **92**(18), 184105 (2015). https://doi.org/10.1103/PhysRevB.92.184105

Chapter 17
Reactive Cleaning and Active Filtration in Continuous Steel Casting

Andy Spitzenberger, Katrin Bauer, and Rüdiger Schwarze

17.1 Introduction

The presence of non-metallic inclusions (NMIs) in steel melt has a significant influence on the properties of the final product. For example, they can cause poor mechanical properties through the formation of clusters, which lead to internal cracks, slivers and blisters [1]. Therefore, the removal of NMIs from the steel melt is an important step to achieve high quality steel products [2].

To be able to remove the NMIs, ceramic foam filters (CFFs) have been used only on the base of pure empirical knowledge in the past [3]. Therefore, the filtration process of CFFs has to be better understood in order to increase the amount of removed NMIs. Two types of filter mechanisms are investigated within Collaborative Research Center 920 (CRC 920): "active" and "reactive" filtration. Active filtration is based on the deposition of NMIs on the filter surface. In contrast, reactive filtration is based on the reaction of oxygen and carbon, which are dissolved in the melt. Carbon monoxide (CO) bubbles are formed by the reaction, which transport the NMIs to the free surface of the melt. Bubble formation can either occur at the filter surface or directly at the inclusions. The carbon, which is involved in the reaction, is dissolved by the melt from the filter.

In the present work, mainly Computational Fluid Dynamics (CFD) is used to investigate the filtration process. A hydrodynamic and magnetohydrodynamic (MHD) model is developed and validated to mimic the flow within the induction crucible furnace (ICF) of the steel casting simulator (SCS) of the CRC 920. Furthermore, the active and reactive cleaning of the melt is considered in the numerical model. The aim of the model is to find out the dominant cleaning mechanism. All models are

A. Spitzenberger (✉) · K. Bauer · R. Schwarze
Institute of Mechanics and Fluid Dynamics, Technische Universität Bergakademie Freiberg, Lampadiusstr. 4, 09599 Freiberg, Germany
e-mail: andy.spitzenberger@imfd.tu-freiberg.de

© The Author(s) 2024
C. G. Aneziris and H. Biermann (eds.), *Multifunctional Ceramic Filter Systems for Metal Melt Filtration*, Springer Series in Materials Science 337,
https://doi.org/10.1007/978-3-031-40930-1_17

implemented within the open-source CFD-library OpenFOAM [4]. Furthermore, the interaction between bubbles and particle was investigated mainly experimentally.

17.2 Modelling

Since there was no solver available in the OpenFOAM library that takes the necessary physics into account, the pimpleFOAM solver was used as a basis for the development of the CFD model. The transient pimpleFOAM solver in its standard form is suitable for Newtonian Fluids and incompressible, turbulent flows.

The mathematical model of the incompressible flow of the melt is given by the conservation equations for mass and momentum:

$$\nabla \cdot \overline{u} = 0, \tag{17.1}$$

$$\frac{\partial \overline{u}}{\partial t} + \nabla \cdot (\overline{uu}) = -\frac{1}{\rho_f} \nabla \overline{p} + \nabla \cdot (\nu_f \nabla \overline{u}) + \nabla \cdot \underline{\tau}^{mod} + \overline{f}_{lor} + S_{filter} + \phi \tag{17.2}$$

Here, $\overline{u}$ and $\overline{p}$ are the Reynolds-averaged or spatially filtered velocity and pressure, respectively. Occurring fluid properties are the density ρ_f and kinematic viscosity ν_f of the steel melt. The Reynolds stress tensor or the subgrid stress tensor is represented by $\underline{\tau}^{mod}$. Additionally implemented terms are the mean Lorentz force $\overline{f}_{lor}$ and the source term S_{filter}, which describes the pressure drop through the CFF. The coupling between Eulerian and Lagrangian phases is included with ϕ. Details describing the latter three terms in (17.2) can be found in Sects. 17.2.2, 17.2.3 and 17.2.4.

17.2.1 Hydrodynamic CFD Model

To validate the hydrodynamic part of the CFD model, a model experiment is used [5]. The experiment is designed to mimic the typical flow in an ICF. The basic setup of the experiment is shown in Fig. 17.1. It mainly consists of a cylindrical vessel (180 mm diameter) filled with a water-glycerin mixture for refractive index matching, an impeller and eight guiding plates. This configuration is intended to produce the double vortex structure, which is typical for the flow in an ICF [6, 7]. The impeller rotates at $n = 400 \, \text{min}^{-1}$ to prevent cavitation and air bubbling. The water-glycerin mixture has a density $\rho_f = 1145 \, \text{kgm}^{-3}$ and a dynamic viscosity $\mu_f = 9.6 \cdot 10^{-3} \, \text{kg s}^{-1} \text{m}^{-1}$. This results in a Reynolds number of approximately $\text{Re} \sim 1500$.

Particle Image Velocimetry (PIV) measurements were performed at this model. Polyamide particles with an average diameter of 22 μm and a density of $\rho_p = 1060 \text{kg/m}^3$ were used as tracers and the central cross-section was illuminated with a

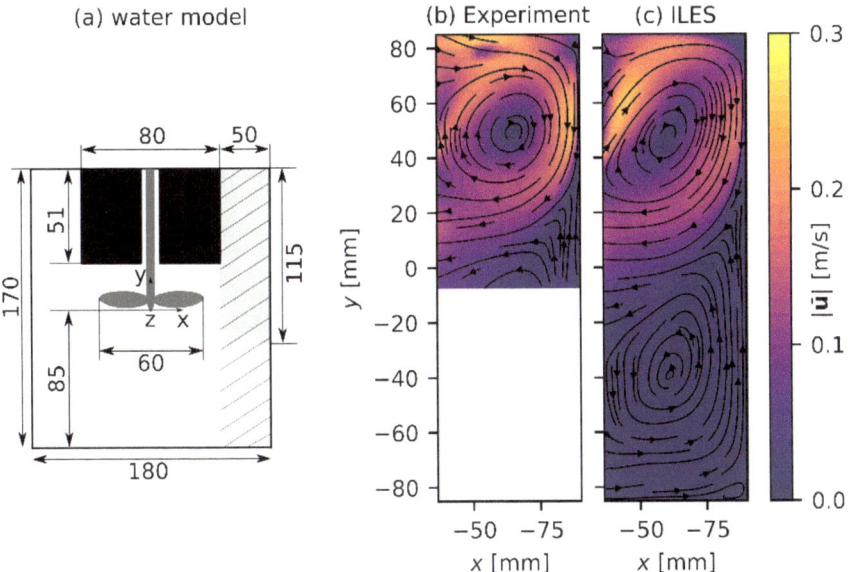

Fig. 17.1 **a** Sketch of the water model with dimensions in mm, gray highlighted area represents the experimental evaluation zone and hatched area the evaluation zone for the simulation. Time-averaged flow field in the stirred vessel for **b** experiment and **c** ILES simulation, reproduced with permission from [5], © 2018 The Minerals, Metals & Materials Society and ASM International

laser light sheet (Raypower 2000, Dantec). Images were acquired using the Phantom V12.1 High Speed camera (400 fps, 2.5 ms exposure time, 1280 × 800 pixels) and a Nikkon Nikkor 35 mm lens. DaVis 8.4.0 software [8] was used to evaluate the velocity fields. Only the gray shaded area in Fig. 17.1a was used for evaluation in order to prevent a distortion of the evaluation by the rotating impeller. During a measurement, a total of 3200 images were recorded over a period of 8 s.

The hydrodynamic CFD model does not include the three additional terms on the right-hand side of the momentum Eq. (17.2). In order to consider the rotation of the impeller in the numerical simulation, the sliding mesh approach was used. Therefore, the pimpleDyMFoam solver was used for the simulation of the flow in the cylinder. For the turbulence modeling the Implicit Large Eddy Simulation [9] (ILES) approach were applied.

The grid created with cfMesh consists of a total of 6 million cells, with the largest cells being 1.1 mm in the fixed zone and 0.6 mm in the rotor zone. In addition, the mesh is refined around the impeller. There are 3 prism layers on the walls to resolve the boundary layer. The CFL number was set to 1. More information on the numerical setup, such as discretization schemes, can be found in Asad et al. [5]. Unless otherwise mentioned, the same discretization schemes are used in all subsequent simulations discussed here.

Figure 17.1b and c gives the mean velocity field in the vessel found in the PIV measurement and the ILES simulations. The velocity field is time averaged over a period of 8s. In both cases, the formation of two toroidal vortices is clearly visible. This pattern is also typical for the flow in an ICF. However, due to the smaller evaluation area in the experiment, the lower vortex in Fig. 17.1b is only partially visible. The shape of the upper vortex differs only slightly between experiment and simulation. Deviations in the flow field occur especially in the area of the guiding plates and the impeller. Due to this acceptable agreement, ILES approach will be used in all the following sections.

17.2.2 Full Magnetohydrodynamic CFD Model

The Lorentz force f_{lor} acting on the fluid, which drives the flow in the ICF, is determined by Maxwell's equations [10]. These are as follows in a simplified form for magnetohydrodynamic (MHD) problems:

$$\nabla \cdot E = 0, \tag{17.3}$$

$$\nabla \cdot B = 0, \tag{17.4}$$

$$\nabla \times B = \mu_0 j, \tag{17.5}$$

$$\nabla \times E = -\frac{\partial B}{\partial t}. \tag{17.6}$$

Here, E is the magnetic induction, B the intensity of the electrical field and μ_0 the magnetic permeability. In addition, Ohm's law in a simplified manner applies to the eddy current density j:

$$j = \sigma_f E, \tag{17.7}$$

with the electrical conductivity of the steel melt σ_f [11]. Combining Maxwell's equations and Ohm's law gives the induction equation for B:

$$\frac{\partial B}{\partial t} = \frac{1}{\mu_0 \sigma_f} \nabla^2 B. \tag{17.8}$$

Finally, the Lorentz force f_{lor} results from:

$$f_{lor} = j \times B. \tag{17.9}$$

The oscillating part of f_{lor} is neglected due to the high frequency of the time harmonic magnetic field f (50–1000 Hz). Because of the inertia of the melt, the flow would not be able to follow the oscillating part [11–13]. In order to calculate the mean part of f_{lor}, the MaxFEM software and a 2D axial symmetrical model of the ICF was used [14, 15]. The resulting field of f_{lor} is interpolated to the 3D CFD mesh in preparation of the CFD simulations using the "KDTree algorithm for nearest neighbor lookup" [16]. Further details regarding the MHD model can be found in the publication of Asad et al. [5].

Validation

The full MHD CFD model of the flow in an ICF is validated using experimental data from Baake et al. [17]. The dimensions and operating conditions of the configuration used by Baake et al. (ICF1) can be found in Fig. 17.2a and Tables 17.1 and 17.2. These also contain the data for the ICF2 investigated in CRC 920, which is discussed in Sect. 17.3. Unlike the ICF2, the ICF1 is run with Wood's metal. The 3D grid of ICF1, created with cfMesh, includes about 4 million cells with a maximum cell size of 2.3 mm. Three prism layers were generated on the walls and top surface. This leads to an average $y^+ = 1$, and therefore the use of wall functions can be omitted [18, 19]. The no-slip boundary condition is applied to the walls. The free surface of the fluid is modeled with a slip-wall, which means that its deformation is neglected. After a settling time of 30 s, the results are averaged over a period of 300 s of flow time. The time step Δt equals 0.001 s and the ILES approach is used for turbulence modeling.

The Lorentz force f_{lor} calculated with MaxFEM software is highest at the side walls of the ICF1, which is caused by the so-called skin effect [5, 20]. The averaged flow field is dominated by two toroidal vortices, with the vortices rotating such that the melt flows along the side wall coming from the bottom and the top. Since the Lorentz force is distributed almost symmetrically, both vortices are also nearly

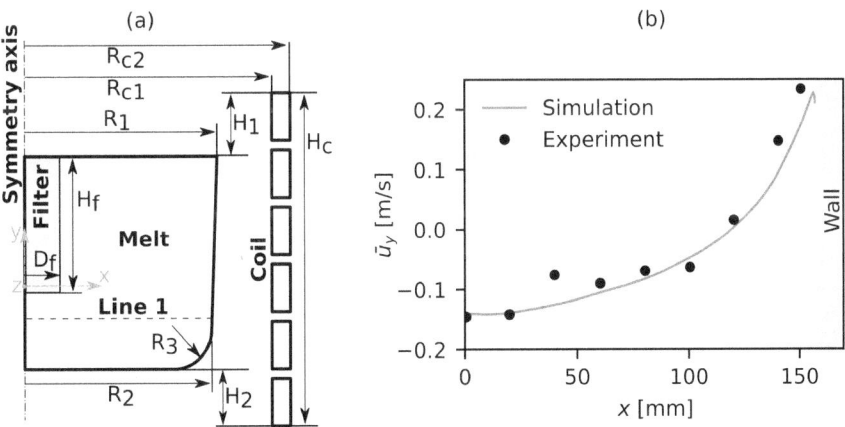

Fig. 17.2 **a** Sketch of the reference crucible furnace (ICF1) and the TU Bergakademie Freiberg furnace (ICF2), see the dimension in Table 17.1 **b** Simulation results of time-averaged vertical flow velocity at a horizontal line compared to experimental data [17] for ICF1, reproduced with permission from [24], © 2016 Elsevier Inc.

Table 17.1 Dimensions in mm of the induction crucible furnaces ICF1 [17] and ICF2, which is investigated in CRC 920, reproduced with permission from [24], © 2016 Elsevier Inc.

	H_1	H_2	H_c	R_{c1}	R_{c2}	R_1	R_2	R_3	H_f	D_f
ICF1	0	0	570	197	207	158	158	–	–	–
ICF2	165	120	400	222.5	240	110	105	20	60	10

Table 17.2 Operation conditions of the crucible furnaces and melt material properties [17, 24–26], reproduced with permission from [24], © 2016 Elsevier Inc.

	ICF1	ICF2
Coil current (rms) I [A]	2000	180
Frequency f [Hz]	400	3400
Number of coil windings n [–]	12	10
Coil length l [m]	0.57	0.4
Magnetic field strength B [T]	0.069	0.01
Fluid	Wood's metal	Steel melt
Density ρ_f [kg m^{-3}]	9400	7000
Dynamic viscosity μ_f [kg m^{-1} s^{-1}]	$4.2 \cdot 10^{-3}$	$6 \cdot 10^{-3}$
Magnetic permeability μ_0 [H m^{-1}]	$1.257 \cdot 10^{-6}$	$1.257 \cdot 10^{-6}$
Electrical conductivity σ_f [S m^{-1}]	$1 \cdot 10^5$	$7 \cdot 10^5$

symmetrical to each other. However, because of typical long-term fluctuations in the flow, a slight asymmetry can be observed despite averaging over 300 s [17, 21–23]. The highest velocities occur at the side walls due to the skin effect.

To validate the full MHD CFD model, the time-averaged vertical velocity $\overline{u}_y$ in experiment and simulation is compared [17]. Figure 17.2b shows its distribution along a horizontal line running from the axis of symmetry through the center of the lower vortex to the side wall. The full MHD CFD model is thus able to reproduce the flow in the ICF1 with good accuracy. Therefore, the model is also applied in the following to the ICF2 investigated in CRC 920. A more detailed description of the model and the validation can be found in Asad et al. [5] and Asad [20].

17.2.3 Ceramic Foam Filter

To model the influence of the ceramic foam filter on the flow in the CFD simulations, it is assumed to be a homogeneous, isotropic, porous medium. The filter causes an additional pressure loss, which is considered in the Navier–Stokes equations (17.2) by the source term

$$S_{\text{filter}} = -\left(\frac{\mu_f}{\kappa_1} + \frac{\rho_f}{\kappa_2}|\bar{u}|\right)\bar{u} \tag{17.10}$$

This relationship corresponds to the Darcy-Forchheimer law, where μ_f corresponds to the dynamic viscosity of the fluid, κ_1 to the permeability and κ_2 to the Forchheimer coefficient. Both, κ_1 and κ_2 were determined using a Direct Numerical Simulation (DNS) by subproject B02 [27]. The values for the filter with 10 pores per inch (ppi), which is used in the following, are $\kappa_1 = 1.96 \cdot 10^{-7} \text{m}^2$ and $\kappa_2 = 2.78 \cdot 10^{-3} \text{m}$. For each inclusion that passes through the filter, a filtration probability

$$\psi = 1 - e^{-\lambda \Delta s} \tag{17.11}$$

is calculated using the law of depth filtration. Here, Δs is the penetration depth and λ is the filtration coefficient, which is a function of the local Reynolds number. The filtration probability is accumulated over several time steps and an effective filtration probability

$$\psi_{\text{eff}}^n = \psi_{\text{eff}}^o + \left(1 - \psi_{\text{eff}}^o\right)\psi^i \tag{17.12}$$

is calculated [27]. The indices n and o denote the current and the previous time step and ψ^i the instantaneous filtration probability according to (17.11). When the inclusion leaves the filter, ψ_{eff} is compared to a random number $\zeta \in [0, 1]$. If $\zeta < \psi_{\text{eff}}$ the inclusion is filtered and therefore removed from the simulation.

17.2.4 Disperse Phase Transport

The discrete phase model (DPM) is used to model the transport of a disperse phase in the melt flow within a Lagrangian framework [28, 29]. The motion of the inclusions and bubbles with mass m_p and velocity v is based on Newtons second law:

$$m_p \frac{dv}{dt} = F_{\text{tot}} \tag{17.13}$$

The total force F_{tot} is the sum of the forces acting on the particle. In the simulations performed, the forces considered include buoyancy force F_B, gravitational force F_G, drag force F_D, virtual mass force F_{VM}, Saffman lift force F_L and electromagnetic pressure force F_{EM}. Since the filtered velocity $\bar{u}$ is approximately the same as the melt velocity u in the case of ILES, the use of a dispersion model is omitted.

Since the particles can be relatively large (see Sect. 17.3), they have an influence on the melt flow. Therefore, the source term

$$\Phi = -\frac{1}{V_{\text{cell}}} \sum F_{\text{tot}} \tag{17.14}$$

is included in the momentum Eq. (17.2) to account for the two-way coupling between particles and fluid. Details on the implementation of the DPM model in OpenFOAM can be found in Asad [20]. Further investigation on the influence of the turbulence model and drag closure on particle motion are provided by Asad et al. [30, 31].

17.3 Numerical Results for Active and Reactive Cleaning

The ICF2 was investigated within the CRC 920 experimentally by Storti et al. [32] and numerically by Asad et al. [27]. However, when comparing the results, large differences were found with respect to the filter performance of the ceramic foam filter. These differences could be due to the fact that only the active filtration was considered in the simulations. Therefore, reactive filtration by CO bubbles will be considered in the following. The bubbles are formed by the reaction of oxygen and carbon dissolved in the melt either at the filter surface or the NMIs.

In the following sections, reactive filtration will be discussed in more detail. Further information can be found in the publications of Asad et al. [33, 34].

17.3.1 Reactive Cleaning—Bubble on Inclusion

Before the start of the simulation, the carbon concentration c in the entire melt is $c = 0$. The carbon required for bubble formation is dissolved from the filter, making it the only source of carbon in the simulation. The transport of the dissolved carbon within the melt is subject to the transport equation

$$\frac{\partial c}{\partial t} + \nabla \cdot (\overline{u}c) = \left[\left(\frac{v_f}{\text{Sc}} + \frac{v_t}{\text{Sc}_t} \right) \nabla c \right] + S_k \tag{17.15}$$

The Schmidt number $\text{Sc} = v_f/D$ is obtained from the diffusion coefficient for carbon in the molten steel $D = 10^{-8} \text{ m}^2 s^{-1}$ [35]. The turbulent Schmidt number is assumed to be $\text{Sc}_t = 1$ and the eddy viscosity v_t is estimated according to Smagorinsky [36, 37]. The sink term S_k considers the reaction and the associated reduction of the carbon concentration.

The reaction between oxygen and carbon is limited to the inclusion surface and therefore cells in which inclusions are present. It is assumed that there is always enough oxygen available in the melt for the reaction. Thus, the oxygen concentration is not considered in the simulations. The reaction reduces the initial amount of carbon $n_0 = cV_{\text{cell}}$ in the cell with volume V_{cell} to

$$n = n_0 - k_r A_p n_p \Delta t. \tag{17.16}$$

Here, $k_r = 14.9 \text{mol m}^{-2}s^{-1}$ is the reaction rate constant, A_p the surface area of the particle and n_p the number of particles in the computational cell. The increase of volume of each bubble at an inclusion is, according to ideal gas law

$$\Delta V_b = \frac{|n - n_0| RT}{p n_p} \tag{17.17}$$

with ideal gas constant R and Temperature T. The corresponding sink term is evaluated as

$$S_k = \frac{n - n_0}{V_{cell} \Delta t}. \tag{17.18}$$

The model assumes, that the CO bubble creates a gas layer around the inclusion, which can grow due to the reaction. Therefore, inclusion and surrounding gas form an equivalent particle with the effective volume $V_{eff} = V_b + V_p$ and mass $m_{eff} = \rho_b V_b + \rho_p V_p$. The CO density in the bubble ρ_b is calculated according to the ideal gas law. The size of the inclusion remains constant throughout the simulation, allowing the equivalent particle to grow only as the CO bubble grows. In addition, effective density ρ_{eff} and effective diameter d_{eff}, which is the diameter of a sphere with the same volume, are calculated. The effective quantities are used to calculate the forces acting on the equivalent particle and the particle motion (17.13). Because of the low void fraction of particles and therefore the low contact probability, the agglomeration of inclusion-bubble aggregates is neglected.

The computational grid for the simulation of the ICF2 was created with cfMesh and has the same properties and boundary conditions as the grid of the ICF1 (see Sect. 17.2.2). ILES was used for turbulence modelling and the same schemes as in Sect. 17.2.1 are applied to the simulation. The time step is $\Delta t = 0.001s$.

First, the flow is simulated without considering the NMIs for 30 s to allow the flow to fully develop. Then $2 \cdot 10^6$ inclusions with a diameter $d_p = 5\,\mu m$ and a density $\rho_p = 3200 \text{kg/m}^3$ are randomly injected close to the top surface. To allow the particles to disperse in the ICF2, the simulation is run again for 30 s. Subsequently, the filter is inserted for 10 s so that carbon can distribute in the melt and bubble formation can begin. The carbon concentration within the filter is $c = 1 \text{mol/m}^3$ and remains constant throughout the immersion phase. After this time, the filter is removed and the magnetic field is deactivated, analogous to the experiments of Dudczig et al. [38] and Storti et al. [32]. Finally, the simulation is run for another 100 s to allow the reaction to continue and the particle-bubble agglomerates to rise to the surface.

Time averaged flow field

The flow field at the vertical midplane and horizontal plane at $y = 0$ shown in Fig. 17.3 is averaged over the immersion time of filter (10 s). It can be seen that a large toroidal vortex dominates the flow in the ICF2 due to the different setup of the

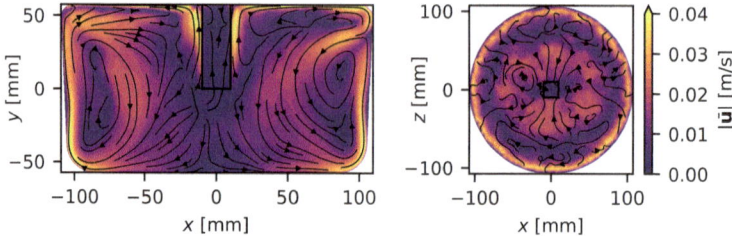

Fig. 17.3 Numerical simulation results of the ICF2 concerning the magnitude of the time-averaged velocity field and the streamlines of the melt flow in the vertical midplane (left) and the horizontal plane at y = 0 (right). Black rectangles illustrate the filter position, reproduced with permission from [20], © 2020 Amjad Asad

magnetic field compared to ICF1. The filter also significantly influences the flow. Inside the filter, the flow velocities are very low due to the high flow resistance. At its surface, on the other hand, the velocity is very high. The rising particle-bubble agglomerates accelerate the melt in this area. The high velocity at the filter surface also ensures improved transport of the carbon away from the filter. Despite time averaging, asymmetric structures and secondary vortices can be seen in the horizontal plane. In addition, when instantaneous flow fields are examined, many turbulent structures are visible, which enhances the mass transport.

Carbon and Particle Distribution

Since the filter is the only source of carbon, the concentration c is highest in it and on its surface. With advancing time, the carbon is better distributed in the ICF2, as can be seen in Fig. 17.4. After removing the filter, the concentration subsequently decreases, since the dissolved carbon is used for bubble formation. Figure 17.5 shows the particle distribution at times $t = 1s$ and $t = 10s$. Only particles with $d_{eff} > d_p$ are included there. Because of the higher carbon concentration directly around the filter, the largest equivalent particles are located in this area. These rise quickly to the surface because of their low effective density. The formation of bubbles is also possible further away from the filter. However, it takes a certain amount of time from the immersion of the filter until particle-bubble agglomerates are also present near the side walls. There, the equivalent particles are smaller because of the lower carbon concentration c. Due to the wetting properties, once NMIs have risen to the free surface, they will remain there and are not redispersed back into the melt. Therefore, as they reach the free surface, the particles are deactivated, preventing them from moving and growing any further.

Due to the ongoing bubble formation, the average particle size increases continuously over the simulation time. In addition, the number of particles with $d_{eff} = 5\,\mu m$ decreases as bubbles form on an increasing number of inclusions.

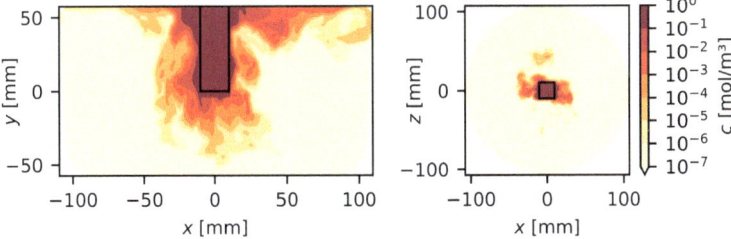

Fig. 17.4 Carbon concentration field in the ICF2 10 s after the immersion of the filter for a constant carbon concentration $c = 1 \, mol/m^3$ in the filter, reproduced with permission from [20], © 2020 Amjad Asad

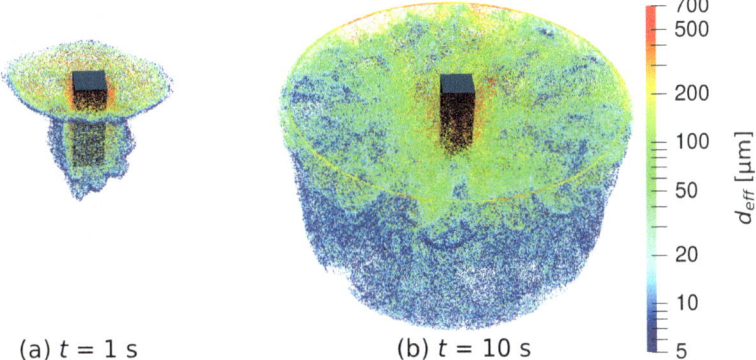

(a) $t = 1$ s   (b) $t = 10$ s

Fig. 17.5 Growth of the equivalent particles (bubbles with attached inclusions) at different times after filter immersion for a constant carbon concentration $c = 1 \, mol/m^3$ in the filter, reproduced with permission from [33], © 2018 The Minerals, Metals & Materials Society

Concentration Decrease

The formation of in-situ layers on the filter wall can cause the carbon concentration c on the filter wall to decrease [39]. In addition to the previously discussed case with a constant concentration in the filter, four cases with decreasing concentration in the filter are investigated. The concentration decrease during the immersion phase is either linear or exponential. The concentration in the filter at the end of the immersion time is either $c_{min} = 0$ or $c_{min} = 0.5 \, mol/m^3$. Considering the decrease in concentration does not significantly affect the flow in ICF2, it will not be discussed in more detail.

With the combination of $c_{min} = 0$ and exponential decrease, there are very small amounts of dissolved carbon in the melt after the filter is removed. This is also reflected in the barely noticeable increase in size of the equivalent particles. With the linear decrease to $c_{min} = 0$, however, there is a higher carbon concentration around the filter. This leads to an improved bubble growth compared to the exponential decrease.

For $c_{min} = 0.5\,\text{mol/m}^3$, the carbon distribution in the melt is very similar for the linear and exponential decrease. The amount of dissolved carbon is significantly larger compared to $c_{min} = 0$, which leads to a larger effective diameter of the equivalent particles.

Melt Cleanliness

The cleanliness of the melt η is defined by the ratio of particles on the surface at the end of the simulation n_{top} to the initial number of particles n_{init}:

$$\eta = \frac{n_{top}}{n_{init}}. \tag{17.19}$$

Figure 17.6 shows the evolution of η over time starting from the immersion of the filter for all cases. It can be seen that the formation of the CO bubbles has a significant effect on the cleanliness of the melt. With increasing time, more and more inclusion-bubble-agglomerates reach the melt surface. At constant carbon concentration in the filter, about 30% of the NMIs have reached the surface and thus are filtered out of the melt 10 s after insertion of the filter. Even after removal of the filter, NMIs continue to be transported to the melt surface. At the end of the simulation, about 63% of the particles have been filtered out of the melt. This result is in the order of the experiments of Storti et al., where 60–95% of the NMIs were removed [32]. Reactive filtration could thus explain the difference between the active filtration simulations of Asad et al. [27] and the experimental results mentioned before.

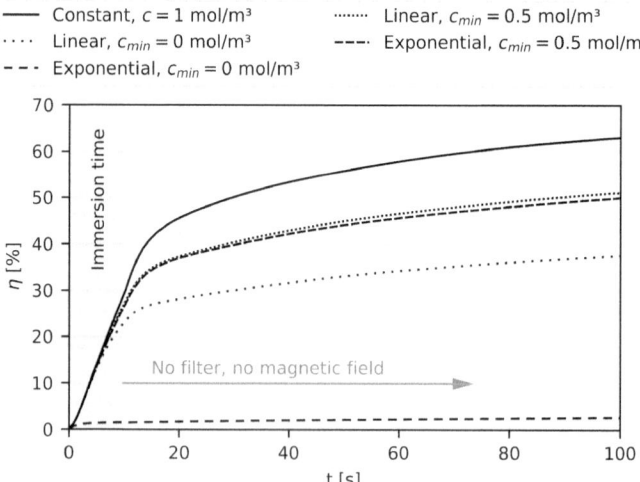

Fig. 17.6 Temporal evolution of the cleaning efficiency η based on different numerical models for the carbon distribution, reproduced with permission from [33], © 2018 The Minerals, Metals & Materials Society

Reducing the C concentration in the filter during the immersion phase to c_{min} provides a reduced cleaning effect. At $c_{min} = 0$, there are large differences between linear and exponential decrease. For the linear decrease, a cleanliness level of 38% is reached at the end of the simulation. With the exponential decrease, almost no cleaning effect is observable. The poorer cleaning effect is closely related to the reduced carbon concentration in the crucible. For $c_{min} = 0.5 \text{mol/m}^3$ there is little difference in the cleanliness of the melt between the two variants, analogous to the carbon concentration field. For both linear and exponential decay, the degree of cleanliness at the end of the simulation is $\eta \approx 50\%$.

17.3.2 Reactive Cleaning—Bubble on Filter

The second possibility for bubble formation besides the formation at inclusions is the formation directly at the filter. Since the filter is considered in the simulation by means of the Darcy Forchheimer law and thus there are no explicit filter walls, the bubble growth at the filter cannot be represented correctly. Therefore, the bubbles are initialized at 32 different positions in the lower half of the filter with a diameter of $1\mu m$. Injection in the lower half of the filter should ensure a long residence time. At each of the points 100 bubbles per second are injected. After injection, the bubble grows to diameter $d_{b,max}$ over a period of $\tau = 10s$. Two types of bubble growth are examined: the bubble volume V_b increases either linearly

$$V_b = V_{b,0} + \left(V_{b,max} - V_{b,0}\right) \cdot \frac{t}{\tau} \qquad (17.20)$$

or exponentially

$$V_b = V_{b,0} \cdot \left(\frac{V_{b,max}}{V_{b,0}}\right)^{t/\tau} \qquad (17.21)$$

$V_{b,0}$ is the initial volume of the bubble and $V_{b,max}$ its maximum volume. The maximum diameters investigated are $d_{b,max} = 100, 300, 500\,\mu m$, which ensures that the bubbles remain spherical [40].

In order to account for the cleaning effect of the bubbles in the simulation, the bubble-inclusion attachment is modeled. The collision probability Ψ between bubbles and inclusions in the same computational cell is calculated as follows [41]:

$$\Psi = \frac{0.25\pi \left(d_p + d_b\right)^2 |U_{rel}| \Delta t n_{min}}{V_{cell}}. \qquad (17.22)$$

Here U_{rel} is the relative velocity between bubble and inclusion and n_{min} is the minimum number of bubbles or inclusions in the cell. The collision probability is compared to a random number ξ to determine a collision occurs. In case of a collision, the corresponding inclusion is removed from the simulation. Two-way coupling between bubbles and melt is considered, but not the collision and agglomeration of NMIs with each other. The filter only influences the flow field and does not actively filter inclusions. The simulation duration is $t = 70s$. Further information can be found in Asad et al. [34].

Bubble Distribution and Melt Cleanliness

The formation of the bubbles directly at the filter again leads to high flow velocities around the filter. Therefore, the flow field is similar to that shown in Fig. 17.3 and will not be discussed in more detail [34].

The distribution of bubbles at time $t = 10\,s$ and for exponential growth with maximum diameter $d_{b,max} = 500\,\mu m$ is shown in Fig. 17.7. The bubbles have a small size at first, causing them to follow the melt flow after their injection. The size of the bubbles increases as the residence time increases. As a result, they experience a greater buoyancy force and begin to rise upward. It can be observed that the bubble ascent starts sooner the larger $d_{b,max}$ is. With linear growth, the bubble size increases faster in the early stages, causing them to rise earlier. At $d_{b,max} = 500\mu m$ and linear increase, the bubbles grow so fast that they cannot distribute in the ICF2 and rise to the surface right at the filter.

The cleanliness of the melt at the end of the simulation (70s) for the different cases is summarized in Table 17.3. A visible cleaning effect occurs only for exponential bubble growth. The best cleanliness of the melt is achieved at exponential growth and $d_{b,max} = 500\,\mu m$. In this case, the bubbles can distribute well due to the initially slow growth. As a result, during the subsequent ascent of the larger bubbles, almost 6% of the inclusions are removed. In the case of linear growth, especially for $d_{b,max} = 500\,\mu m$, the bubbles grow faster in the early stage and thus move upwards

Fig. 17.7 Bubble distribution for bubble formation on the filter surface based on exponential bubble growth with maximum bubble diameter $d_{b,max} = 500\,\mu m$, 10 s after filter immersion, reproduced with permission from [34], © 2019 The Authors. Published by WILEY–VCH Verlag GmbH & Co. KGaA, Weinheim

Table 17.3 Cleanliness of the melt η at $t = 70s$ after filter immersion for different models regarding bubble formation at the filter surface compared to bubble growth on inclusions, reproduced with permission from [34], © 2019 The Authors. Published by WILEY–VCH Verlag GmbH & Co. KGaA, Weinheim

$d_{b,max}[\mu m]$	$\eta[\%]$		
	On filter		On inclusions
	Linear	Exponential	
100	0.79	0.62	61.0
300	0.58	3.48	
500	0.65	5.76	

faster. Therefore, they are not distributed well in the ICF2 and cannot collect many inclusions [34].

Overall, the influence of bubble formation on the filter on the cleanliness level is negligible compared to bubble formation on NMIs. The reason for this could be that small bubbles have a relatively low probability of contact with an inclusion. Larger bubbles, on the other hand, rise quickly to the surface and thus have little time for contact with inclusions.

17.3.3 Combined Reactive and Active Filtration

Simulations of active filtration have shown that the melt cleanliness $\eta = 4\%$ with a single filter (10 ppi) is significantly lower than that of reactive filtration with bubble formation at the NMIs [27, 33]. Therefore, in this section, the cleaning efficiency with combined active and reactive filtration is investigated. For this purpose, one or three filters are placed in the ICF2. The filters act either actively only or actively and reactively. An overview of the different configurations is provided in Table 17.4 and Fig. 17.8. CO bubble formation on the filter is not considered in this case, because of the better cleaning performance of the bubble formation on inclusions.

The combined filtration is simulated for 300 s, with the respective filters acting reactively and releasing carbon to the melt only in the first 10 s. While the filters act reactively, there is a constant concentration of $c = 1 mol/m^3$ in them. Unlike in Sect. 17.3.1, the magnetic field remains active for the entire simulation. The

Table 17.4 Overview of the investigated configurations using the active filtration or reactive cleaning approach, reproduced with permission from [20], © 2020 Amjad Asad

Case	Position (1)	Position (2)	Position (3)
1	–	Active, reactive	–
2	Active, reactive	Active	Active, reactive
3	Active	Active, reactive	Active

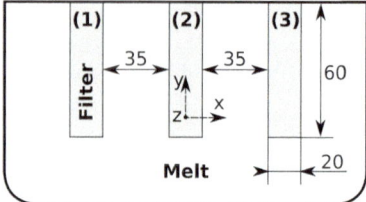

Fig. 17.8 Sketch of the different filter positions for ICF2 with dimensions in mm, reproduced with permission from [24], © 2016 Elsevier Inc.

NMIs have the same properties as in Sect. 17.3.1. Also, the same grid and numerical schemes are used. The time step size is $\Delta t = 0.001$s and ILES is used for the turbulence modeling.

Flow Field

The flow field of case (2) at the vertical midplane and horizontal plane at $y = 0$ in Fig. 17.9 is averaged over the entire simulation duration of 300 s. As expected, the number and position of the filters has a significant influence on the flow field. The toroidal vortex is again dominant. However, it is confined to the space between the side wall and the two outer filters. Compared to Fig. 17.3, the velocity at the filter surface is lower. Since bubble formation only occurs in the first 10 s and the averaging is done over the entire simulation, the mean flow field is barely influenced by the bubble rise. The flow field of case (1) is very similar to that of Fig. 17.3 except for the lower velocity at the filter surface.

The flow fields of cases (2) and (3) are also almost identical with only small differences because of the bubble formation. Due to remaining residual carbon concentration, bubble formation can continue even after the filter is removed from the melt. With a longer simulation time, the differences should disappear, since the flow is then determined by the Lorentz force.

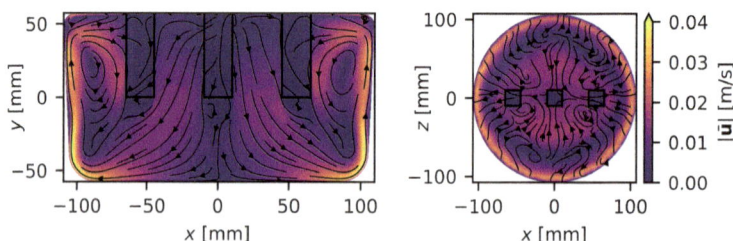

Fig. 17.9 Numerical simulation results of the ICF2 for case (2) (see Table 17.4) concerning the magnitude of the time-averaged velocity field and the streamlines of the melt flow in the vertical midplane (left) and the horizontal plant at $y = 0$ (right). Black rectangles illustrate the filter position, reproduced with permission from [20], © 2020 Amjad Asad

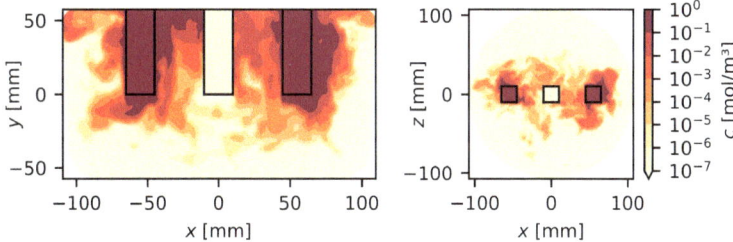

Fig. 17.10 Carbon concentration in the ICF2 in the vertical midplane (left) and the horizontal plane at $y = 0$ (right) at time $t = 10$s for case (2) (see Table 17.4), reproduced with permission from [20], © 2020 Amjad Asad

Carbon Concentration

The carbon concentration field of case (1) was already shown in Fig. 17.4. Figure 17.10 shows the carbon distribution at the end of the reactive phase ($t = 10$s) for case (2). Since there are two reactive filters, the carbon can be better distributed in the crucible. This should be beneficial for the filtration effect due to bubble formation at inclusions. It can also be seen that the carbon practically cannot get into the middle filter. The carbon field of case (3) differs only slightly from case (1) because the additional two filters impede the transport of carbon to the side wall [20].

Melt Cleanliness

The effect of reactive cleaning by bubble formation is again evaluated by the cleanliness of the melt η. Figure 17.11 shows the temporal development of η for all three cases with combined filtration. The effect of active filtration is not considered here. The best results in terms of reactive filtration are provided by case (2), followed by case (1) and case (3). As shown in Sect. 17.3.1, further cleaning of the melt takes place even after the end of the reactive phase. Through carbon distributed in the ICF2, bubbles can still form on inclusions. Because of the two reactive filters in case (2) and therefore better distributed carbon, the cleanliness of the melt at the end of the simulation is $\eta = 83\%$. Case (3) reaches the lowest cleaning effect with $\eta = 63\%$ because of the blocked carbon transport and therefore reduced bubble formation.

Figure 17.12 shows the fraction of inclusions removed at the end of the simulations for both reactive and active filtration. When combining both filtering mechanisms, reactive cleaning again has a stronger effect on the cleanliness of the melt than active filtration. This is true for all three configurations and despite the longer duration of the active filtration. This supports the assumption that the large differences between the experiments of Storti et al. [32] and the active filtration simulations of Asad et al. [27] are caused by the formation of CO bubbles on inclusions. Regarding the combined filter performance, case (2) also provides the best results. In addition, it can be seen that a larger number of active filters also allows more particles to be actively filtered. However, purely active filters can impede reactive filtering, as observed in case (3). Therefore, only slightly more NMIs were filtered in case (3) than in case

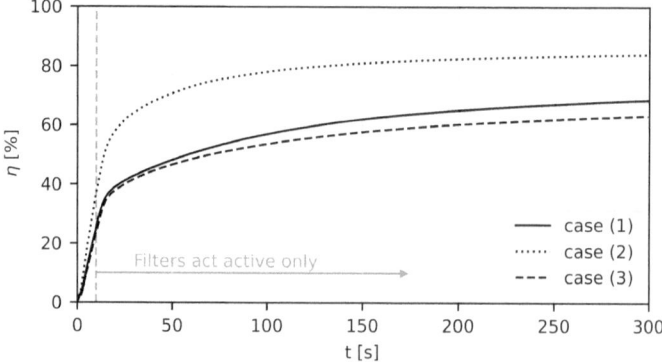

Fig. 17.11 Temporal evolution of the cleaning efficiency concerning the simulation cases of Table 17.4 for the ICF2 based on reactive and active cleaning, reproduced with permission from [20], © 2020 Amjad Asad

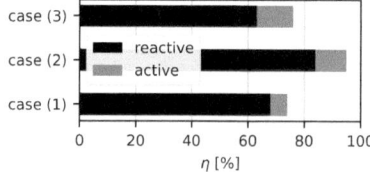

Fig. 17.12 Cleaning efficiency at the end of the simulation ($t = 300\,\text{s}$) concerning the simulation cases of Table 17.4 for the ICF2 based on reactive and active cleaning, reproduced with permission from [20], © 2020 Amjad Asad

(1), although only one filter is present there. Further configurations for combined cleaning can be found in Asad and Schwarze [42].

17.4 Bubble-Particle Interaction

17.4.1 Experimental Setup

To study the interaction between bubbles and particles during reactive cleaning in more detail, a new experimental setup was developed. The setup is shown in Fig. 17.13 and consists of a thin column (1) of dimensions $300 \times 120 \times 30$ mm. The column is filled with pure water in order to avoid additional influences on the bubble formation and motion. The height of the water level is 110 mm. An automatic bubble injector is located centrally at the bottom of the column and is supplied with compressed air. The bubble injector has the advantage of being able to generate reproducible bubbles at an adjustable frequency. Here, the bubble generation frequency was set

Fig. 17.13 Experimental
setup to investigate the
particle-bubble interaction

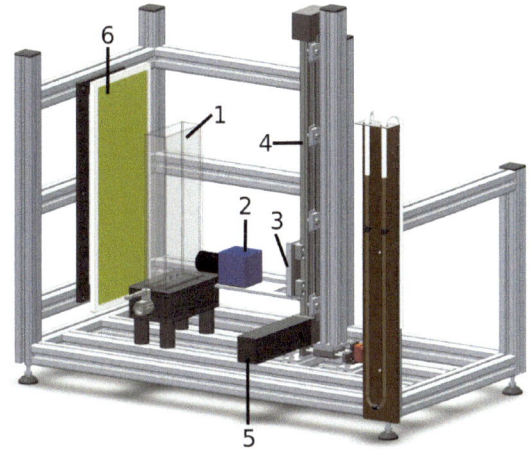

to $f = 2\,\text{Hz}$. The bubble injector device is presented in detail in Ostmann and
Schwarze [43].

Polyester particles with a density $\rho_p \approx 1300\,\text{kg/m}^3$ were dispersed in the water.
The particles are hydrophobic, polydisperse and have a diameter $d_p < 315\,\mu m$.
Moreover, the particles are not spherical.

A MIKROTRON Motion BLITZ EoSens mini2 (Mikrotron GmbH, Unterschleis-
sheim, Germany) high-speed camera (2) was used to capture the ascending bubbles.
The camera is mounted on the guide block (3), which can move vertically along the
linear toothed belt axis (4) (Igus drylin R © ZLW-1040) in order to track the rising
bubble at sufficiently large resolution in a small field of view ($12.8 \times 13.5\,\text{mm}^2$). A
motor (5) with motor control system (Igus drylin R © dryve D1) moves the camera
upwards at approximately $300\,\text{mm/s}$ to match the velocity of the rising bubble. The
image size was 992×1040 pixels which allowed a frame rate of 1200 fps, i.e., an
interframe delay of $833\,\mu s$. In combination with a $100\,\text{mm}$ lens the camera system
provides a resolution $13\,\mu m$ per pixel. The water column was illuminated with a
diffuse LED-based light source (6). A total of 62 image series of rising bubbles were
acquired.

17.4.2   Experimental Results

The Open Source Computer Vision Library (OpenCV) is used to evaluate the images.
First, the bubble is detected in each image of the series and its properties such as
diameter $d_b \approx 2.2\,\text{mm}$, position $\boldsymbol{x}_b$, and velocity $\boldsymbol{v}_b$ are determined. Selected bubble
trajectories are shown in Fig. 17.14a. The trajectories of the bubbles were detected
until the bubble touches an edge of the raw image. Up to a height of 20 mm from their
injection point, the bubbles follow approximately the same trajectory. After that, the

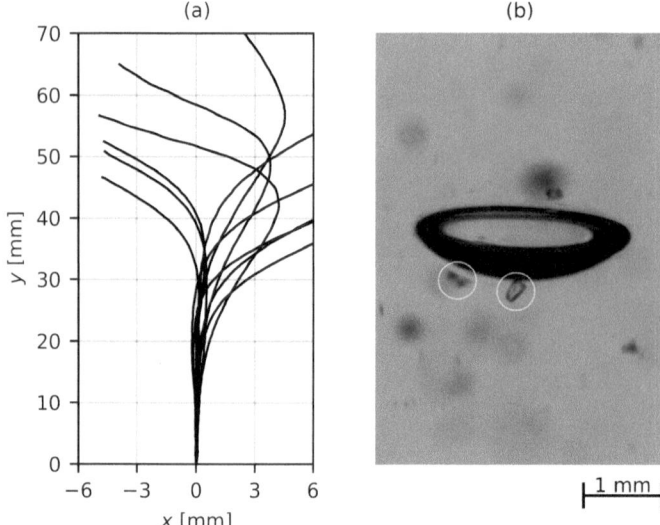

Fig. 17.14 **a** Selected bubble trajectories **b** cropped raw image of a bubble with two attached particles at the downstream side (in circles)

bubbles start to leave the nearly vertical path and the trajectories of the individual bubbles begin to differ significantly, i.e. the particle motion becomes unstable. During the horizontal dispersion, the bubble also tilts in the direction of its motion.

In the next step, the image is cropped to a size of 280 × 440 pixels such that the bubble stays in the center of the image. An example of a cropped image with a bubble in the center can be seen in Fig. 17.14b. The cropped images are used to study the interaction of the rising bubble with the dispersed particles. To evaluate the particle's trajectories, they are tracked using OpenCV's Channel and Spatial Reliability Tracking (CSRT) [44]. For this purpose, the particles to be tracked are manually selected. Thereby, a distinction is made whether the particles are will attach to the bubble or not.

The resulting particle trajectories $x_p(t)$ are shown in Fig. 17.15. The center of the bubble is used as the origin of the coordinate system. The particle velocity v_p is also specified relative to the center of the bubble. All particles which attach to the bubble approach the bubble at $x_{p,0} < r_b$, where $x_{p,0}$ is the first tracked position of a particle and r_b is the bubble radius. It is noticeable that the particles, especially those near $x_p = 0$ and thus the stagnation point, are decelerated before contact with the bubble. As they slide around the bubble, the particles are accelerated by the flow around the bubble until they finally reside on the downstream side of the bubble.

There are also many particles which approach the bubble at $x_{p,0} < r_b$, but do not attach to the bubble (Fig. 17.15b). It should be noted that only those particles are shown which were visible on their entire path around the bubble and were not covered by the bubble. The particles are deflected and accelerated by the flow around the rising bubble. Only the particles closest to $x_p = 0$ are slowed down as they approach the bubble and are influenced by the stagnation point.

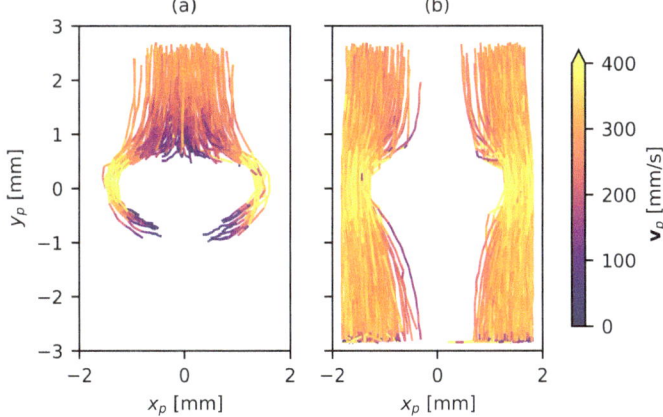

Fig. 17.15 Particle paths around the bubble **a** Particles that attach to the bubble **b** Particles that do not attach to the bubble

In order to assess the dependence of the attachment on the particle position, the attachment probability P_{att} is evaluated as a function of the initial distance of the particles from the symmetry axis of the bubble $|x_{p,0}|$ (Fig. 17.16). For this purpose, nine bins with the width $\Delta x_{p,0} = 0.125$ mm are analyzed. For each bin, the attachment probability P_{att} is calculated as follows:

$$P_{att} = \frac{n_{p,att}}{n_p}. \tag{17.23}$$

Here, n_p is the total number of particles in the bin and $n_{p,att}$ is the number of attaching particles in the bin. A total of 1094 particles were evaluated for this purpose.

The attachment probability decreases with increasing distance from the bubble's symmetry axis. If a particle approaches the bubble along the symmetry axis, the attachment probability is $P_{att} \approx 21\%$. This value decreases for 0.875 mm $< |x_p| < 1$ mm to $P_{att} \approx 2\%$. For $|x_p| > 1$ mm, which is less than the bubble radius $r_b \approx 1.1$ mm, no more attachment of particles was observed.

Fig. 17.16 Attachment Probability P_{att} of the particles in dependence of their relative position to the axis of symmetry of the bubble r_{ax} at the beginning of particle tracking

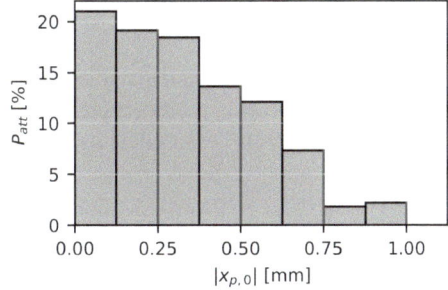

Fig. 17.17 New
experimental test rig for 3D
observations of the
particle-bubble interaction

However, the current experimental setup has disadvantages because the particle
paths can only be determined within the x–y plane. Due to the lack of information
in the z-direction, particles located in front of or behind the bubble are also taken
into account when determining P_{att}. In addition, the evaluation of the distance to the
symmetry axis of the bubble should ideally be done within the x–z plane and not
only in the x-direction. In this case, the attachment probability around the symmetry
axis would be expected to be greater than 21%. Therefore, a new experimental rig
has already been set up, on which 3D observations are possible by implementing
three cameras from three different viewing angles (Fig. 17.17).

17.4.3 Numerical Investigations

In addition to the experimental studies, the bubble rise is also be investigated numerically. For this purpose, the interFoam solver of OpenFOAM, which is based on the
volume of fluid (VOF) method, is used [45]. With the VOF method, the interface is
implicitly represented by the phase fraction α. The VOF-method was also used by
Asad et al. in order to investigate the immersion process of CFFs [46].

The dimensions of the mesh are $60 \times 63 \times 23$ mm and the base mesh consists
of 1.6 million cells with a maximum cell size of 1 mm. Adaptive mesh refinement is

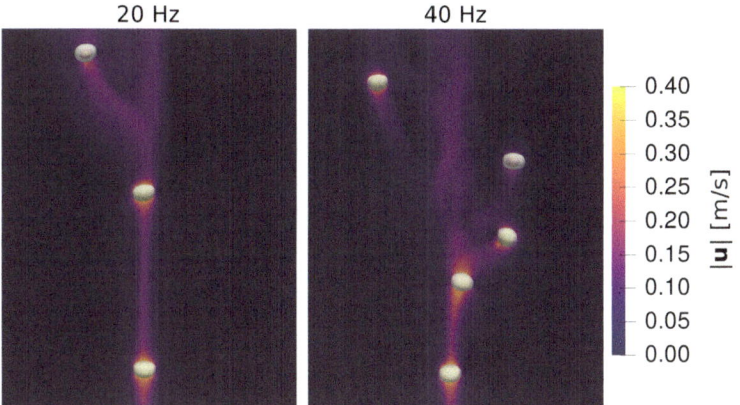

Fig. 17.18 Velocity Field and bubble contours of the VOF-simulations of a rising bubble chain in the vertical midplane for a bubble generation frequency of 20 and 40 Hz

used to capture the interface as accurately as possible. The cells at the phase interface are refined to a size of 62.5 μm.

The bubbles are generated with a frequency $f = 20\,Hz$ or $f = 40\,Hz$ and a diameter $d_b = 2\,mm$ at the bottom of the domain using setFields. The surface tension between air ($\rho_g = 1.22\,\frac{kg}{m^3}$, $v_g = 14.8 \cdot 10^{-6}\,m^2/s$) and water ($\rho_g = 1000\,\frac{kg}{m^3}$, $v_g = 1.0 \cdot 10^{-6}\,m^2/s$) is $\sigma = 0.07\,N/m$. The bottom wall and the side walls are modelled as no slip walls. The simulated time is $t = 1s$.

Figure 17.18 shows the velocity in a cutout of the vertical midplane and the bubble contours ($\alpha = 0.5$). For $f = 20\,Hz$, the bubbles initially rise in a straight line along the same trajectory, as the wake of the bubbles shows. The bubble velocity at the end of the straight-line ascent is approximately 320 mm/s. After about 30 mm of ascent, the bubbles become unstable and leave the straight-line path. At $f = 40\,Hz$, the trajectories of the individual bubbles already start to differ after about 25 mm. The rising velocity of the bubbles is also slightly increased in the case of the higher bubble frequency.

In future simulations, particles will also be taken into account using the Lagrangian approach. In addition, the goal is to implement a model for the particle-bubble interaction and the attachment probability of the hydrophobic particles to the bubble. The implemented model is expected to reflect the experimental results well, so that the new model can be applied to real melt-NMI systems.

17.5 Conclusion

The active and reactive filtration of NMIs from molten steel using CFFs was investigated numerically with OpenFOAM. First, the hydrodynamic model was validated with a water model experiment. The experiment is used to mimic the turbulent and recirculating flow in an ICF. The simulations performed showed that the numerical model was able to reproduce the flow from the experiment well.

Since the flow in an ICF is significantly influenced by the Lorentz force, it was implemented in OpenFOAM. The MHD model was also validated, showing good agreement between data from literature and simulation.

Active filtration of the NMIs is achieved directly through the CFF immersed in the melt. However, the numerical studies showed a low cleaning effect of the active filtration, which was in contrast to the results of Storti et al. [32].

Therefore, a model was implemented in OpenFOAM to account for reactive cleaning. Carbon dissolves from the immersed CFF and distributes itself in the melt via convective and diffusive transport. The carbon can react with dissolved oxygen to form CO bubbles. The bubbles form either directly on the filter or on NMIs distributed in the melt. The bubbles eventually rise to the surface and can carry NMIs with them, thus cleaning the melt. When bubbles are formed on the NMIs, a significant increase in the cleanliness of the melt is observed compared to active filtration. Moreover, the results are in the same order of magnitude as in the experiments of Storti et al. [32]. The formation of bubbles on the filter, on the other hand, had a less significant effect on the cleanliness of the melt.

The combined study of active and reactive filtration also shows the dominance of reactive cleaning. In addition, it was found that the number of filters and their position had a significant effect on the amount of NMIs removed.

The interaction between bubbles and particles in a thin column was also investigated experimentally. The trajectories of the bubbles and particles were evaluated and the attachment probability of the particles was analyzed. In order to extend the investigations, a test rig was set up, which allows 3D measurements. In addition to the experiments, bubble chains were simulated with OpenFOAM. In the future, the simulations will be extended to include particle-bubble interaction models.

Acknowledgements This research was funded by the Deutsche Forschungsgemeinschaft (DFG, German Research Foundation)—Project-ID 169148856-SFB 920, subproject B06. The authors would like to give special thanks to Amjad Asad, whose longtime work in subproject B06 is the basis for the majority of this chapter. Thanks are also due to all those who contributed to the work presented in this chapter: Hannah Aaronson, Pascal Beckstein, Kinnor Chattopadhyay, Cornelius Demuth, Steffan Dudczig, Thomas Gundrum, Martin Haustein, Bernhard Jung, Christoph Kratzsch, Henry Lehmann, Patrick Meergans, Muhammad Ahsan Rauf, Subhashis Ray, Anne Schmidt, Enrico Storti, Michael Triep, Eric Werzner, Thomas Wondrak. We would also like to thank all the SFB 920 members involved as well as our mechanical and electronic workshop for their support. The computations were performed on a Bull Cluster at the Center for Information Services and High-Performance Computing (ZIH) at TU Dresden and the High Performance Compute Cluster at University Computer Centre (URZ) at TU Bergakademie Freiberg.

References

1. L. Zhang, B.G. Thomas, ISIJ Int. **43**, 271–291 (2003). https://doi.org/10.2355/isijinternational. 43.271
2. L. Zhang, Y. Wang, JOM **64**, 1063–1074 (2012). https://doi.org/10.1007/s11837-012-0421-2
3. R.A. Olson III., L.C.B. Martins, Adv. Eng. Mater. **7**, 187 (2005). https://doi.org/10.1002/adem. 200500021
4. OpenFOAM Foundation Ldt., http://www.openfoam.org. Accessed 30 Jun 2022
5. A. Asad, K. Bauer, K. Chattopadhyay, R. Schwarze, Metall. Mater. Trans. B **49**, 1378–1387 (2018). https://doi.org/10.1007/s11663-018-1200-4
6. F. Stefani, T. Gundrum, G. Gerbeth, Phys. Rev. E **70**, 056306 (2004). https://doi.org/10.1103/ PhysRevE.70.056306
7. C. Heinicke, T. Wondrak, Meas. Sci. Technol. **25**, 055302 (2014). https://doi.org/10.1088/0957-0233/25/5/055302
8. DaVis 8.4.0, https://www.lavision.de/. Accessed 30 Jun 2022
9. S. Hickel, Dissertation, Technical University of Munich (2008)
10. P.A. Davidson, *An Introduction to Magnetohydrodynamics*, 1st edn. (Cambridge University Press, Cambridge, 2001)
11. R. J. Moreau, *Magnetohydrodynamics*, 1st edn. (Springer, 1990)
12. J.M. Galpin, Y. Fautrelle, J. Fluid Mech. **239**, 383–408 (1992). https://doi.org/10.1017/S00221 12092004452
13. F. Felten, Y. Fautrelle, Y. Du Terrail, O. Metais, Appl. Math. Model. **28**, 15–27 (2004). https:// doi.org/10.1016/S0307-904X(03)00116-1
14. MaxFEM: Universidade de Santiago de Compostela, https://www.usc.es/en/proxectos/max fem/. Accessed 30 June 2022
15. A. Bermúdez, D. Gómez, P. Salgado, in *Mathematical Models and Numerical Simulation in Electromagnetism*, vol. 74 (Springer Cham, 2014), p. 183
16. S. Maneewongvatana, D. M. Mount, in *Data Structures, Near Neighbor Searches and Methodology*, vol. 59, ed. by M. H. Goldwasser, D. S. Johnson, C. C. McGeoch (American Mathematical Society, 2022), p. 105
17. E. Baake, A. Mühlbauer, A. Jakowitsch, W. Andree, Metall. Mater. Trans. B **26**, 529–536 (1995). https://doi.org/10.1007/BF02653870
18. X. Jiang, C. Lai, *Numerical Techniques for Direct and Large-Eddy Simulations*, 1st edn. (CRC press, 2009)
19. S. Rezaeiravesh, M. Liefvendahl, Phys. Fluids **30**, 055106 (2018). https://doi.org/10.1063/1. 5025131
20. A. Asad, Dissertation, Technische Universität Bergakademie Freiberg (2020)
21. A. Umbrashko, E. Baake, B. Nacke, A. Jakovics, Metall. Mater. Trans. B **37**, 831–838 (2006). https://doi.org/10.1007/s11663-006-0065-0
22. A. Umbrasko, E. Baake, B. Nacke, A. Jakovics, Heat Transf. Res. **39**, 413–421 (2008). https:// doi.org/10.1615/HeatTransRes.v39.i5.50
23. E. Baake, A. Umbrashko, A. Jakovics, AEE **54**, 425 (2005)
24. A. Asad, C. Kratzsch, S. Dudczig, C.G. Aneziris, R. Schwarze, Int. J. Heat Fluid Flow **62**, 299–312 (2016). https://doi.org/10.1016/j.ijheatfluidflow.2016.10.002
25. M. Kirpo. Dissertation, University of Latvia (2009)
26. R. Schwarze, F. Obermeier, Simul. Mater. Sci. Eng. **12**, 985–993 (2004). https://doi.org/10. 1088/0965-0393/12/5/015
27. A. Asad, E. Werzner, C. Demuth, S. Dudczig, A. Schmidt, S. Ray, C.G. Aneziris, R. Schwarze, Adv. Eng. Mater. **19**, 1700085 (2017). https://doi.org/10.1002/adem.201700085
28. L. Cheng, D. Mewes (ed.) *Advances in Multiphase Flow and Heat Transfer* (Bentham Science Publishers, 2012)
29. E. Michaelides, C. T. Crowe, J. D. Schwarzkopf (ed.), *Multiphase Flow Handbook* (CRC Press, 2016)

30. A. Asad, K. Chattopadhyay, R. Schwarze, Metall. Mater. Trans. B **49**, 2270–2277 (2018). https://doi.org/10.1007/s11663-018-1343-3
31. A. Asad, C. Kratzsch, R. Schwarze, Eng. Appl. Comput. Fluid Mech. **11**, 127–141 (2017). https://doi.org/10.1080/19942060.2016.1249410
32. E. Storti, S. Dudczig, A. Schmidt, G. Schmidt, C.G. Aneziris, Steel Res. Int. **88**, 1700142 (2017). https://doi.org/10.1002/srin.201700142
33. A. Asad, M. Haustein, K. Chattopadhyay, C.G. Aneziris, R. Schwarze, JOM **70**, 2927–2933 (2018). https://doi.org/10.1007/s11837-018-3117-4
34. A. Asad, C.G. Aneziris, R. Schwarze, Adv. Eng. Mater. **22**, 1900591 (2020). https://doi.org/10.1002/adem.201900591
35. R.I.L. Guthrie, *Engineering in Process Metallurgy*, 1st edn. (Oxford University Press, Oxford, 1992)
36. D. Mazumdar, J. W. Evans, *Modeling of Steelmaking Processes*, 1st edn. (CRC Press, 2009)
37. J. Smagorinsky, Mon. Weather Rev. **91**, 99–164 (1963). https://doi.org/10.1175/1520-0493(1963)091%3c0099:GCEWTP%3e2.3.CO;2
38. S. Dudczig, C.G. Aneziris, M. Dopita, D. Rafaja, Adv. Eng. Mater. **15**, 1177–1187 (2013). https://doi.org/10.1002/adem.201300121
39. A. Schmidt, A. Salomon, S. Dudczig, H. Berek, D. Rafaja, C.G. Aneziris, Adv. Eng. Mater. **19**, 1700170 (2017). https://doi.org/10.1002/adem.201700170
40. R. Clift, J. Grace, M.E. Weber, *Bubbles, Drops, and Particles*, 1st edn. (Academic Press, London, 1978)
41. M. Sommerfeld, S. Stübing, Powder Technol. **319**, 34–52 (2017). https://doi.org/10.1016/j.powtec.2017.06.016
42. A. Asad, R. Schwarze, Steel Res. Int. **92**, 2100122 (2021). https://doi.org/10.1002/srin.202100122
43. S. Ostmann, R. Schwarze, Rev. Sci. Instrum.Instrum. **89**, 125108 (2018). https://doi.org/10.1063/1.5048708
44. A. Lukezic, T. Vojir, L.C. Zajc, J. Matas, M. Kristan, Discriminative correlation filter with channel and spatial reliability, Paper presented at the *2017 IEEE Conference on Computer Vision and Pattern Recognition*, Honolulu, Hawaii (2017)
45. C.W. Hirt, B.D. Nichols, J. Comput. Phys.Comput. Phys. **39**, 201–225 (1981). https://doi.org/10.1016/0021-9991(81)90145-5
46. A. Asad, H. Lehmann, B. Jung, R. Schwarze, Adv. Eng. Mater. **24**, 2100753 (2021). https://doi.org/10.1002/adem.202100753

Chapter 18
Virtual Prototyping of Metal Melt Filters: A HPC-Based Workflow for Query-Driven Visualization

Henry Lehmann and Bernhard Jung

18.1 Introduction

Recent advancements in additive manufacturing may be a game changer for the design of metal melt filters as the variety of 3D-printable filter geometries is vastly increased in comparison to conventional manufacturing processes. E.g., flow-guiding or surface-increasing elements of different shapes, sizes, and orientations may be added to the pores of foam-like filters, the strut shape may be freely varied, or filters may be designed with controlled variations of pore sizes. In the *Collaborative Research Center 920* (CRC 920), a combination of additive manufacturing with replication is successfully employed for the manufacturing of ceramic filters with hollow struts based on 3D-printed foam templates [1–4]. The particular foam templates for 3D-print are designed based on a geometric modeling approach for conventional open-cell PU foams, which allows for the generation of complete filters with adjusted strut thickness, pore density, and porosity based on periodic elements of hundreds of pores [5]. However, the new variety of filter designs, which can be targeted by 3D-printing technology, implies that a comprehensive exploration of the design space is not feasible using physical prototypes only. In this context, we propose a novel workflow for virtual prototyping of metal melt filters based on the state-of-the-art methods for geometrical modeling, *Computational Fluid Dynamics* (CFD), compression and indexing methods for scientific data, and query-driven visualization.

A major challenge are the large data volumes produced even by single CFD simulations of metal melt flow and even more so for the proposed virtual prototyping approach where many filter designs need to be evaluated inside a *High-Performance Computing* (HPC) environment. The large data problem is particularly severe during the analysis phase, where simulation results are assessed outside of the HPC envi-

H. Lehmann · B. Jung (✉)
Institute for Informatics, Technische Universität Bergakademie Freiberg,
Bernhard-von-Cotta-Straße 2, 09599 Freiberg, Germany
e-mail: jung@informatik.tu-freiberg.de

© The Author(s) 2024
C. G. Aneziris and H. Biermann (eds.), *Multifunctional Ceramic Filter Systems for Metal Melt Filtration*, Springer Series in Materials Science 337,
https://doi.org/10.1007/978-3-031-40930-1_18

ronment using workstations with much lower computing resources. Furthermore, the analysis phase calls for interactive workflows. With data loading times being the main bottleneck, the amount of data loaded into memory should therefore be kept as small as possible.

Motivated by these requirements for interactive data analysis, we developed the LITE-QA (*Lossy In-Situ Tabular Encoding for Query-Driven Analytics*) framework for data management. During CFD simulations in the HPC environment, LITE-QA is used to compress and index the large volumes of simulation data. To support the data analysis phase, the framework adds search engine-like capabilities to data analysis tasks and allows, e.g., for focused visualizations of filter regions that meet the search criteria of the analyst, such as areas with high velocity, vorticity or backflow. Crucially, only very small parts of the simulation data, i.e. those parts that meet the analyst's search criteria, need to be loaded from disk. Figure 18.1 gives an overview of the proposed HPC-based virtual prototyping workflow and the embedded LITE-QA data management framework.

This chapter is organized as follows: Sect. 18.2 gives a brief overview of LITE-QA's data preparation methods (compression and indexing) when run 'in-situ', i.e. integrated into CFD simulations in the HPC environment. Then, Sect. 18.3 shows how the compressed and indexed data can be used in the analysis phase for the creation of query-driven visualizations. In order to demonstrate the feasibility of the proposed approach, Sect. 18.4 presents a virtual prototyping study involving a total of 84 unconventional filter designs. Finally, Sect. 18.5 gives a summary of the proposed

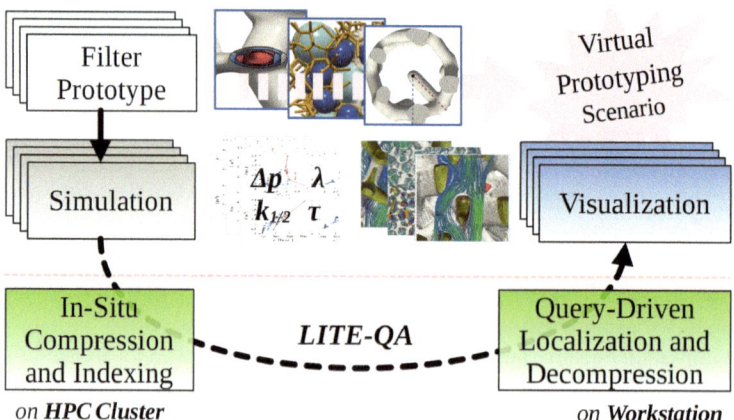

Fig. 18.1 HPC-based virtual prototyping workflow for visualization-assisted evaluation of virtual filter prototypes. CFD simulations of 84 filter prototypes are run in the HPC environment, where the resulting data is compressed and indexed using LITE-QA. The analysis phase outside the HPC environment is supported by LITE-QA's query mechanism for data-efficient, interactive creation of visualizations that provide further insights into the melt flow in the various filter variants. As a result of the virtual prototyping study, filter designs with improved performance and interesting flow characteristics are proposed

virtual prototyping approach and highlights the novel filter designs that performed best in the virtual prototyping study.

18.2   In-Situ Data Compression and Indexing

High-detail numerical simulations have become an increasingly important tool for the development of next-generation metal melt filters. For the CFD simulation of aluminum melts, the *Lattice-Boltzmann Method* (LBM) has been successfully applied at the pore-scale level inside porous filter structures [6–9]. The LBM allows for efficient parallelization in HPC systems, and thus allows the simulation of metal melt flow on high-resolution voxel grids with periodic boundaries [6, 8]. With increasing spatial and temporal resolution, LBM simulations are able to quickly generate large amounts of data in HPC clusters, easily in the order of hundreds of gigabytes or several terabytes and more. Conventional visualization workflows, where all data is stored on disk and loaded into main memory of less powerful workstations, lead to non-interactive workflows for data analysis and visualization or may even be impossible at all. Instead, the present research proposes novel methods for so-called in-situ data preparation, i.e. the reduction and indexing of simulation data while it is still residing in the HPC environment.

18.2.1   Early In-Situ Data Preparation

The simulation of a highly-porous filters requires large grids, i.e. at least 512^3 voxels in order to guarantee numerical stability and resolve fine geometric features like the struts in the required resolution. Due to the voxel-based discretization of the physical domain, the LBM can be setup for simulations of different filters. However, storing only the flow field for post-processing already requires 3×512 MB as 32bit floating point for each of the three components of the velocity vectors for each filter. For performing a local analysis and visualization of the aluminum melt flow inside a filter, a typical data set with five variables, e.g. the flow field and two additional properties, requires ≥ 2.5 GB for each filter and for every additional 40 time steps already additional 100 GB without compression.

Using fast algorithms early in the scientific workflow, the LITE-QA framework for HPC data management, as shown in Fig. 18.2, pursues two data preparation goals:

1. Reduction of storage required for high-resolution data during the running CFD simulations using fast error-bounded lossy compression, e.g. maximum relative error of 1% for decompressed contents, and
2. Creation of a compressed index for providing efficient access to compressed contents, i.e. query-based identification of regions and partial decompression of only the data needed for the visualization task.

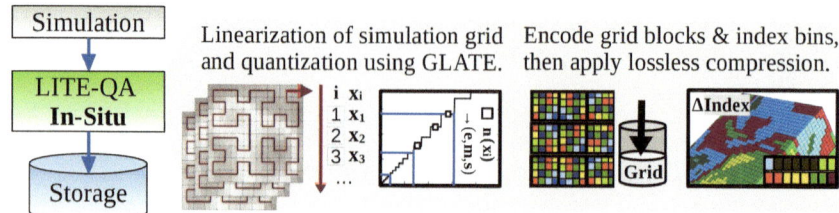

Fig. 18.2 The LITE-QA in-situ processing pipeline achieves high compression rates and is run as part of the simulations inside the HPC environment. The single steps of in-situ processing are grid linearization, data quantization, encoding of grid and index data, and lossless compression of the encodings

Like other in-situ compression methods, e.g. SBD [10], ISABELA [11], ZFP [12], SZ [13] and temporal extensions of them [14, 15], LITE-QA applies lossy compression and reduces the amount of data written out to the file system. LITE-QA applies lossy compression integrated with indexing directly in the simulation processes inside the HPC environment. Thus, temporary storage of large data is avoided already early in the workflow. By using fast algorithms for compression and indexing 'in-situ', i.e. in the HPC environment, the load on storage systems and the bandwidth requirements for network transfer are reduced, while also preparing the data set for later post-processing outside of the HPC environment. By using quantization with a small error of e.g. 1% maximum point-wise error, the full-resolution data sets are stored at a smaller memory footprint, while the full flexibility for post-hoc visualization and analysis is maintained.

18.2.2   Grid Compression and Index Generation

LITE-QA uses fast algorithms for grid compression and the generation of a compressed index, which are based on the sequential and differential encoding employed in the SBD compression algorithm. However, LITE-QA and SBD differ in the data quantization methods used for bounding the point-wise maximum error [16, 17]. Data quantization in SBD is based on a look-up table estimated from the data in the individual subgrids which causes compression artifacts on distributed grids in parallel simulations and on high-resolution temporal data [18]. Instead, LITE-QA compression employs the GLATE (*Grid Linearization and Tabular Encoding*) compressor which uses a step function for quantization, resulting in stable quantization across subgrid boundaries and on high-resolution data.

LITE-QA employs the data encoding of GLATE, i.e. the discrete quantization for linearized numerical data, for the compression of simulation grids in a block-wise manner and the generation of an index based on binning the quantized values. LITE-QA constructs a compact encoding for grid blocks and for index bins [18], which is designed to support partial decompression of simulation results, while being adeuqate for high-degree data reduction using lossless compression techniques on those encodings as backend. For lossless compression of encoded grid blocks and index bins, the bit-packing codec `fastpfor` [19] and the general purpose lossless compressor `zstd`[1] are used.

The temporal extension t-GLATE yields an improved compression rate for temporal data by exploiting the temporal coherence between successive time steps. GLATE establishes a trade-off between data accuracy and compression rate. E.g. for a point-wise maximum error of 1%, the size of the grids is reduced by a factor of five, while data indices are reduced by a factor of three as compared to the uncompressed data assuming the 32bit floating point data type.

For the algorithmic details of LITE-QA, GLATE and t-GLATE we refer to the original publications [16–18]. In the following, representative results on the compression rates achieved for the metal melt filtration simulations conducted in the virtual prototyping study in Sect. 18.4 and, generally, HPC simulations in CRC 920 are presented.

18.2.3 In-Situ Compression Performance

In-situ compression performance is evaluated for an LBM simulation solving an incompressible isothermal flow of liquid aluminum inside a computer-generated monodisperse filter with porosity 90% and Reynolds number $Re = 90$ on a grid of 512^3 voxels. The fluid dynamics is based on the Navier-Stokes equations assuming a superficial velocity of $6 \text{ cm} \cdot \text{s}^{-1}$. The simulation generates a typical data set for visualization, i.e. containing the flow field u, v, w and two additional properties, the velocity magnitude M and a vortex indicator Q. The GLATE quantization for grid and index compression in LITE-QA is operated at 1% maximum point-wise error, which is sufficient for the compression task with respect to local flow visualization.

Non-temporal Compression Performance

Figure 18.3 shows the performance for non-temporal compression of grids using GLATE. Figure 18.3 (1) compares GLATE's compression rates without additional indexing to ZFP, a state-of-the-art lossy floating point compression algorithm inspired by texture compression methods used in graphics hardware [12]. ZFP does not generate a compressed index. Figure 18.3 (2) shows the data size resulting from GLATE compression of the simulation variables with and without additional indexing of the variables u, M, Q. For the compression tests, GLATE restricts the relative error to 1% and ZFP is operated on level 15, which is comparable to a maximum point-wise error of 1% [17].

---

[1] `zstd` is a real-time compression algorithm providing high compression ratios and a very fast decoder, `zstd` is available at `https://facebook.github.io/zstd/`.

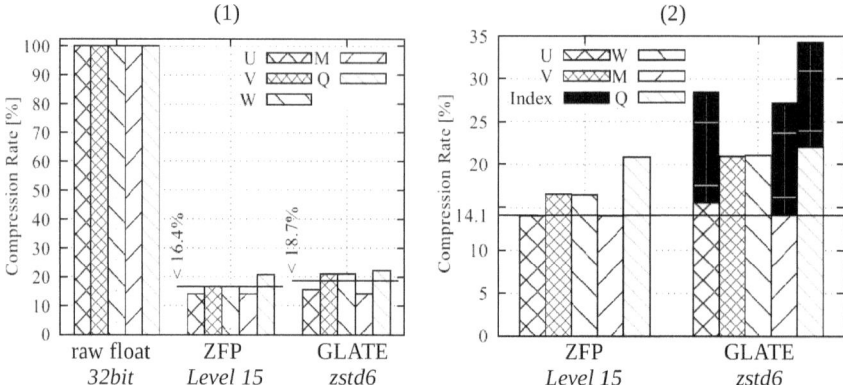

Fig. 18.3 Non-temporal compression performance on grids for variables u, v, w, M, Q and indices for u, M, Q. The data is generated using a LBM simulation of liquid aluminum through a computer-generated filter structure with 90% porosity, superficial velocity $6\,\text{cm} \cdot \text{s}^{-1}$ and Reynolds number of $Re = 90$. (1) GLATE and ZFP achieve at least five fold reduction for compressed grids in average for all variables u, v, w, M, Q. (2) During the more complicated index compression, a three fold reduction on compressed index bins is achieved in average, i.e. only ~12% more storage required as compared to the corresponding compressed grids

On the variables u, v, w, M, Q, GLATE and ZFP achieve an average compression rate of 18.7% and 16.4%. On the index variables u, M, Q, the average compression rate achieved by LITE-QA is 29%. The best index compression is achieved on M with 26%. The index bins for the variables u, M, Q are compressed at a lower rate, as compared to the grid blocks. However, the compressed index requires only ~12% more storage as compared to the corresponding compressed grids.

Temporal Compression Performance

The t-GLATE temporal compression scheme differentiates between so-called key-frames and difference-frames. While key-frames are compressed and decompressed independently, difference-frames reference the previous frame and are compressed with a higher efficiency. The compression efficiency is directly related to the time step size Δt of the simulation and to the amount of difference-frames inserted between key-frames. t-GLATE achieves a trade-off between temporal resolution of exported data, i.e. multiples of Δt, and the resulting compression rate by encoding the data differences to the last exported frame. The approach for difference encoding in t-GLATE falls back to encoding absolute values in the case differences become too large. Therefore, no decline of compression rate is observed, even when the data is exported with a low temporal resolution. t-GLATE is applied to 1024 time steps, which have been exported from the aforementioned simulation of liquid aluminum using multiples of the time step width $\Delta t = 9.13\ \mu\text{s}$ and increasing amounts of difference-frames between key-frames.

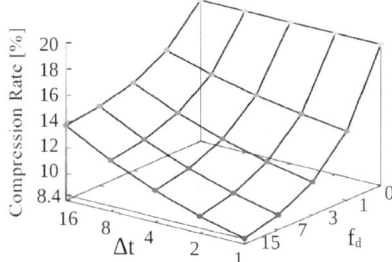

f_d	$1\Delta t$	$2\Delta t$	$4\Delta t$	$8\Delta t$	$16\Delta t$
0	18.7	18.7	18.7	18.7	18.7
1	13.2	13.7	14.3	15.0	16.0
3	10.5	11.1	12.0	13.2	14.6
7	9.1	9.9	10.9	12.3	14.0
15	8.4	9.3	10.4	11.8	13.6

Compression Rate [%]

Fig. 18.4 Temporal compression performance of t-GLATE for a sequence of 1024 time steps of liquid aluminum melt flow through a filter with 90% porosity and superficial velocity $6\ cm \cdot s^{-1}$, Reynolds number $Re = 90$ and time step width $\Delta t = 9.13\ \mu s$. t-GLATE establishes a trade-off between temporal resolution of decompressed data, i.e. multiples 1, 2, 4, 8, 16 of Δt, difference-frames per key-frame $f_d = 0, 1, 3, 7, 15$ and the resulting compression rate

As shown in Fig. 18.4, t-GLATE yields a compression rate of 8.4–18.7% for all time steps with data set variables u, v, w, M, Q in average. Without inserting any difference-frames, the compression rates correspond to non-temporal compression using GLATE and are equivalent to the rates shown in Fig. 18.3. For 15 difference-frames between key-frames, t-GLATE reduces the data set to 8.4% for exporting all 1024 time steps of the simulation run, i.e. the data set is reduced from $2.5\ GB \times 1024 = 2.5\ TB$ to 215 GB.

18.2.4   Summary

In-situ data reduction and indexing methods aim at preparing large-scale simulation data for later analysis outside of the HPC environment. On non-temporal datasets, the GLATE method presented above compresses the LBM simulations to 18.7% in average, comparing to state-of-the-art methods such as ZFP. Further, GLATE was extended to t-GLATE for temporal datasets where compression rates are improved to 8.4% in average, outperforming existing methods. Moreover, LITE-QA combines the GLATE data compression with data indexing needed for query-driven analyses. Whereas uncompressed indices typically require storage amounts of $\geq 100\%$ in addition to the simulation data [20], GLATE compression with integrated indexing reduces the storage requirements for combined simulation and index data to ~27–35%, while guaranteeing a point-wise maximum error of 1%.

18.3 Query-Driven Visualization of Melt Flow

General challenges for the visual exploration of time-dependent scientific data sets and batches of CFD simulations include the non-interactive loading times already for single time steps of a simulation. Furthermore, a purely visual search for interesting features can be tedious and error-prone as flow-relevant features such as high vorticity, high-velocity or backflow occur rather sparsely, often only in a small percentage of all grid cells. Addressing these challenges, query-driven visualizations replace purely visual search in the full dataset with search engine-like capabilities to create visualizations of interesting areas only. The LITE-QA query mechanism, as illustrated in Fig. 18.5, operates on the compressed simulation grid data and index data generated using the HPC data preparation pipeline described in Sect. 18.2.

When searching the flow field for interesting regions with specific characteristics, e.g., areas with significant backflow or very fast flow, it is favorable to access the data in a query-driven manner. Queries make use of the compressed index and are steered using so-called range conditions on simulation variables, e.g., on the velocity magnitude M. Using the index, specific spatial locations are efficiently identified in the flow field where a range condition $a \leq M \leq b$ is true. Given the selected locations from the index, additional data is decompressed and local visualizations of the fluid flow are procedurally generated. Multiple visualizations of an selected region at different time steps can be created to observe the flow evolution. Similarly, comparative visualizations of the same region can be generated for design variants of a filter to gain insights into the effects of geometry modifications.

18.3.1 Querying of Regions in the Flow Field

The visualizations presented in this section use the unsteady simulation data described in Sect. 18.2.3 which contains five variables u, v, w, M, Q at each grid point of the the flow field. Compressed grids and indices were generated in-situ with LITE-QA

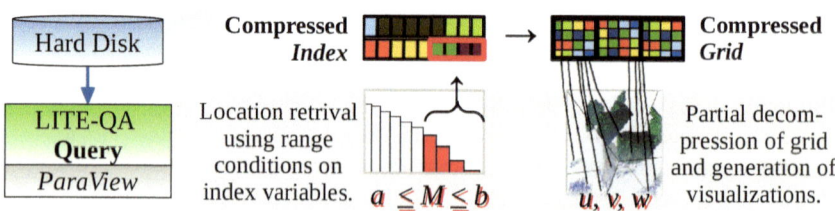

Fig. 18.5 The LITE-QA query processing pipeline integrates a compressed index for realization of a query mechanism used to locate grid cells based on range conditions on variables stored in the index, e.g. $a \leq M \leq b$. Given a set of locations, the flow field u, v, w is decompressed for visualization. The mechanism also minimizes data loading times as only data needed to answer the query is retrieved from disk

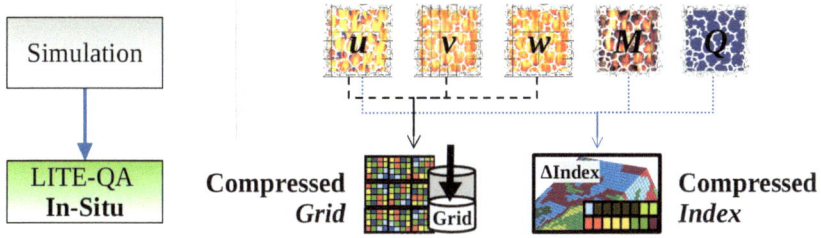

Fig. 18.6 Compressed data set for flow visualization purposes. The data produced by the numerical simulations is stored in a compressed grid for the grid variables u, v, w, and in a compressed index for the index variables u, M, Q accordingly. u is stored as compressed grid and index

during the running simulation, according to the setup as shown in Fig. 18.6. During the data preparation, the flow field u, v, w is stored as a compressed grid, hence u, v, w are called *Grid Variables*, and the variables u, M, Q are stored in a compressed index, hence u, M, Q are called *Index Variables*. The variable u is stored both in the grid and in the index.

Figure 18.7 shows three different regions in the flow field at time step 80. The regions are identified using range conditions on the index variables, i.e. u for identification of backflow, where the stream is redirected against the bulk flow, M for the identification of fast flow and Q for the identification of vortex-like flow. The regions are identified using the following range conditions on the index variables:

$$\begin{aligned}
-\infty \leq u \leq u_0 \text{ for backflow with } u_0 < 0, \\
m_0 \leq M \leq +\infty \text{ for fast flow and} \\
q_0 \leq Q \leq +\infty \text{ for vortex-like flow.}
\end{aligned}$$

In the example, the concrete query parameters were determined as quantiles from value distributions of the index variables u, M, Q, i.e. u_0 as the 5% quantile of the value distribution for values $u < 0$, m_0 as the 85% threshold of the maximum velocity, and q_0 based on the 99% quantile of the value distribution of Q. On the simulation grid composed of 512^3 voxels, the range conditions on Q predictably returns 1% or about 1.34 million grid cells, whereas the amount of cells returned for the queries on u and M depend on the characteristics of their distributions at a given time step.

18.3.2 Implementation of Query Mechanism

The ParaView framework is used as a platform for the implementation of the LITE-QA query mechanism. ParaView is a powerful open-source application for scientific visualization. The central tool for modeling visualizations in ParaView is the so-called visualization pipeline, which arranges algorithms in a network graph and

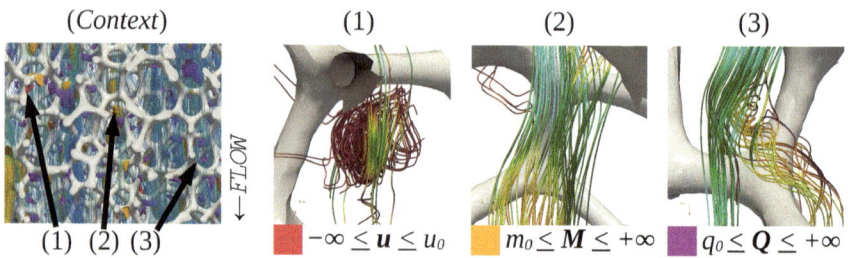

Fig. 18.7 Visualization of regions in the flow field of liquid aluminum melt inside a filter with 90% porosity, superficial velocity 6 cm · s^{-1} and Reynolds number $Re = 90$. Three different regions are located using range conditions on index variables u, M, Q, i.e. (1) backflow in form of a vortex in the slip stream behind a strut in red, (2) fast flow path through a pore window in orange, and (3) fast swirled and vortex-like flow on upstream surface of a strut joint in purple

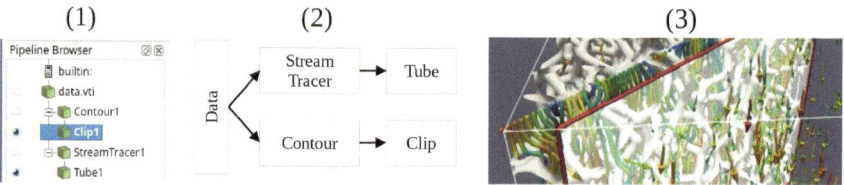

Fig. 18.8 ParaView visualization pipeline used to render the aluminum melt flow through a computer-generated filter structure with 90% porosity. The pipeline (1) as shown in the graphical user interface and (2) displayed as a network flow graph. (3) The pipeline produces the visualization consisting of a clipped iso-contour surface and streamlines as 3D tubes. The root of the pipeline is the data producer reading from a file

defines the data flow by mapping inputs and outputs between them. As shown in Fig. 18.8, algorithms import and transform data, e.g. obtained from simulation results, and perform rendering of 3D objects, e.g. clipped contour surfaces and streamlines with tubes wrapped around them for better visibility.

The LITE-QA query mechanism is implemented using three algorithms for the ParaView visualization pipeline, which execute three different query types modeled after operations provided by the compressed index. The `lqaTable Source` algorithm performs the so-called *Count Query*, the `lqaIndex Source` performs the so-called *Index Query* and `lqaGrid Source` the so-called *Grid Query*. The LITE-QA query algorithms act as data producer. Instead of loading complete uncompressed data sets from files, they use the data index for localization and decompression of only those parts of the simulation grid that are required for the visualization. The queries are typically executed in a hierarchical order, as shown in Fig. 18.9:

1. `lqaTable Source` determines the value distribution of one index variable, e.g. M, and returns a histogram as a `vtkTable`.
2. `lqaIndex Source` evaluates one or more range conditions on index variables, e.g. $a \leq M \leq b$, and returns the grid cell indices I matching the range conditions as a `vtkPolyData` point cloud.

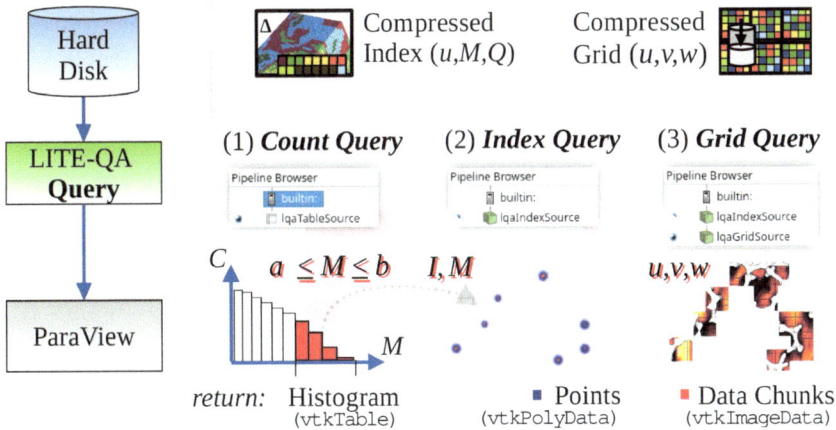

Fig. 18.9 LITE-QA query mechanism operating on compressed grids and compressed indices. Three query types are implemented as data producers for the ParaView visualization pipeline: (1) `lqaTable Source` performs a count query, which returns a histogram C, M as `vtkTable`, (2) `lqaIndex Source` performs an index query, which returns a `vtkPolyData` point cloud representing the indices I of grid cells whose data values M match the range condition $a \leq M \leq b$, and (3) `lqaGrid Source` performs a grid query, which returns a `vtkImageData` containing the decompressed flow field u, v, w at cell indices I

3. `lqaGrid Source` evaluates the range condition according to `lqaIndex Source` and, additionally, decompresses the simulation data at the respective grid cells to return the flow field u, v, w as a `vtkImageData`.

18.3.3 Query-Driven Local Visualization

The generation of the visualization scene is performed using the procedure as shown in Fig. 18.10, where the `lqaIndex Source` and `lqaGrid Source` algorithms are used as a data producer for the visualization pipeline. The range condition for the queries is formulated based on the data distribution of the index variables u, M, Q as explained in Sect. 18.3.1, which are obtained using a preceding count query using `lqaTable Source` as shown in Fig. 18.9. The locations obtained from the index using the range conditions on index variables u, M, Q usually form clusters, which correspond to local phenomena, e.g. backflow, fast preferential flow and vortex-like or swirled flow.

In the example visualization of Fig. 18.10, the point cloud obtained from the index is further decomposed into clusters by using an Euclidean clustering algorithm from the ParaView toolkit. By performing a threshold operation, one specific cluster is selected for visualization based on e.g. the cluster size or a sequential cluster index. Based on the points of the selected cluster a partial decompression of the

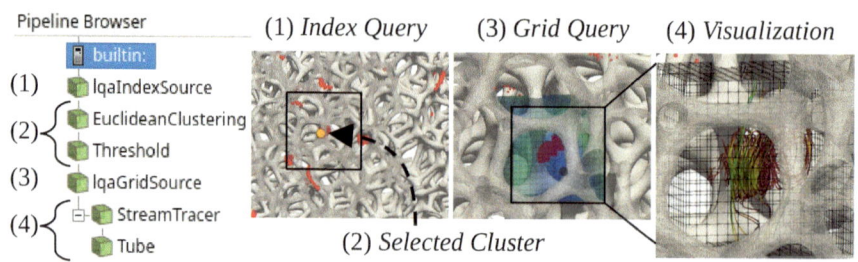

Fig. 18.10 ParaView pipeline for query-driven visualizations based on the LITE-QA framework. (1) determination of region in the flow field using an index query, e.g. $-\infty \leq u \leq u_0$ for backflow, (2) spatial clustering of the resulting point cloud and selection of one cluster, (3) partial decompression of the flow field u, v, w using a grid query on the selected cluster points, and (4) generation of visualization on decompressed data

flow field u, v, w is performed, as described in Sect. 18.3.2. Alternatively, in order to decompress a larger region of the flow field around the selected cluster, the grid query can perform the partial decompression based on a fixed decompression extent, which is defined as an axis aligned bounding box with edge length Δ placed at the geometric center point of the selected cluster. The actual visualization task is performed on the decompressed data, e.g. flow visualization with stream tracer and 3D tubes.

18.3.4 Export of Visualization Scenes

To further support the data analysis, the visualizations of local phenomena can be exported to the web and immersive *Virtual Reality* (VR) environments as shown in Fig. 18.11. Once a visualization has been generated using the procedure described in Sect. 18.3.3, the scene data can be exported to the web and to immersive VR environments using the ExportScene() and SaveData() function from the ParaView Python module respectively. For web export, ParaView generates a standalone HTML version of the scene, which uses WebGL for rendering. For VR, the scene is exported to files and imported into a distributed rendering system based on OpenSceneGraph.

Iterative Generation of Visualizations

Using the LITE-QA query mechanism, visualizations can be exported iteratively in ParaView, e.g. for creating flow animations of selected regions from temporal data sets. Once a region in the flow field has been located in one particular time step of the simulation, the visualization pipeline is evaluated using iterative calls to the UpdatePipeline(time) and SaveScreenshot() function from the ParaView Python module in order to generate frames for an animation at the identified

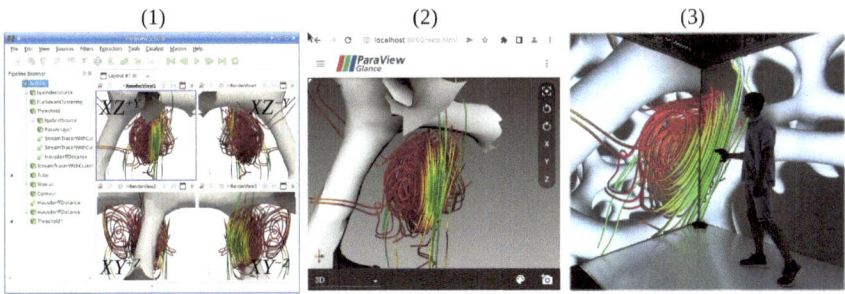

Fig. 18.11 Local visualization of a backflow vortex, which has been located using the LITE-QA query mechanism in (1) the ParaView graphical user interface. The scene is exported to (2) web and (3) immersive VR using the ParaView Python module. An immersive VR application based on OpenSceneGraph is run in the XSITE CAVE (eXtreme definition Spatial Immersion and interacTion Environment) [14, 21], an innovative surround-screen VR environment with an extremely high pixel resolution. The visualization is controlled intuitively using real-time optical tracking and hand-held interaction devices, e.g., adjustment of parameters and temporal and spatial navigation

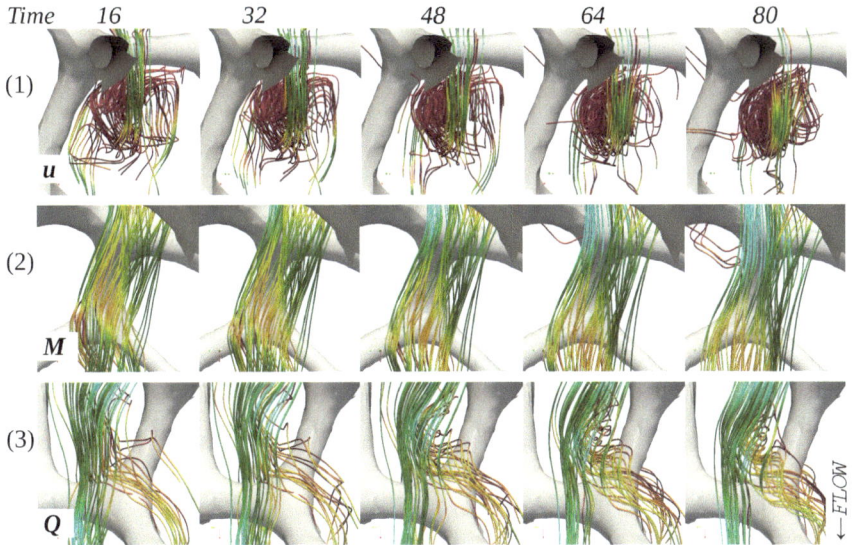

Fig. 18.12 Visualization of local flow evolution over multiple time steps in regions with interesting flow properties as identified by the query-mechanism: (1) strong backflow, i.e. $u \leq u_0 < 0$, (2) fast flow, i.e. $M \geq m_0$, and (3) vortex-like flow, i.e. $Q \geq q_0$

location in a temporal context. Figure 18.12 shows the temporal evolution of the flow of liquid aluminum for the three regions, which are identified in time step 80 using range conditions on the index variables u, M, Q, i.e. a large backflow vortex, fast preferential flow paths and vortex-like or swirled flow.

18.3.5 Summary

Data indexing as described in the previous section enables the efficient identification of regions in the simulation dataset that meet the criteria of user-specified queries. Query-driven visualization utilizes this by creating focused visualizations of large-scale simulations that only include subsets deemed relevant by the analyst such as regions with high vorticity or significant backflow. Query-driven visualization is also computationally very efficient as only the relevant subset of the full dataset needs to be loaded from disk. Besides supporting interactive visualizations of large-scale simulations in desktop, web, and immersive VR environments, the presented approach also supports the automated generation of animations.

While so far the presented methods for in-situ data preparation and query-driven analyses have been presented in the context of single simulations, the next section considers their application for virtual prototyping purposes, where a large number of filter designs is generated, simulated and analyzed.

18.4 Virtual Prototyping Study

In the virtual prototyping scenario presented in this section, a reference filter design is systematically varied in order to create a total of 84 filter variants. Metal melt flow in each of these filters is simulated using the LBM described in [6, 8, 9] and simulation data is compressed and indexed with LITE-QA as described in Sect. 18.2. Evaluation of the candidate filters involves a combination of statistical analyses of global melt flow properties derived directly from the compressed indices, query-driven visualizations that help to gain insights into the local effects of filter variations on the melt flow as well as calculations of global flow characteristics such as filtration efficiency and pressure drop directly obtained from the LBM simulations.

18.4.1 Overview of Investigated Filter Structures

A total of 84 computer-generated filters are screened for the virtual prototyping scenario. The geometry for the filters is obtained from the procedure of representative geometry generation based on a filter skeleton [5]. The filter skeletons are generated from a Laguerre tessellation of periodic sphere packings, which are iteratively adjusted until they reflect the topological properties of real *Ceramic Foam Filters* (CFFs), e.g. the number of faces per cell and the edges per face. The filter geometry is obtained using isotropic Gaussian smoothing on a sharp discretized representation of the strut network embedded into a 3D voxel grid. The reference structures are modeled using a strut aspect ratio of one corresponding to CFFs with circular strut cross section shape as obtained from 3D-printed templates [1, 3, 5]. Based on the modeling procedure, the new filters are generated in the following three groups:

1. (fA) modification of strut shape:

   - f_a—elliptical elongation and flattening of the strut cross section with respect to the bulk flow direction controlled by a strut aspect ratio a,
   - f_{ab}—drop-like strut cross section controlled by a strut aspect ratios a for the upper and b for the lower half of the strut cross section shape, and
   - f_{ba}—reversed drop-like strut cross section shape.

2. (fB) insertion of flow-guiding features:

   - f_w—closing of a total amount of w randomly chosen pore windows,
   - f_α—insertion of finger-like struts on the downstream surface, downward-pointing, inclined by angle α with respect to the bulk flow direction, and
   - f_β—insertion of finger-like struts on the upstream surface, upward-pointing, inclined by angle β.

3. (fC) varying pore size and strut shape within a filter:

   - f_c—continuous thickening of struts from top to bottom, and
   - f_q—systematic arrangements of a total amount of q size-varying pores.

Modifications of Strut Shape and Insertion of Flow-Guiding Features

As shown in Fig. 18.13, all filters in groups (fA) and (fB) are designed by modification of the strut shape and pore geometry of a reference structure f_1^ε with 216 pores. Reference structures are generated for porosities $\varepsilon = 70, 80, 90\%$ which exhibit equal-sized pores with circular strut cross section shape, i.e. f_1^ε corresponds to f_a with aspect ratio $a = 1$ and porosity ε.

The geometric modifications of the strut cross section for f_a is directly integrated into the modeling procedure based on anisotropic Gaussian smoothing [22]. f_{ab} and f_{ba} are generated using a voxel-wise image blending operation, in order to merge flattened and elongated struts into a drop-like strut shape.

For modifications f_w, f_α and f_β, additional flow-guiding features, i.e. closed windows and finger-like struts, are inserted into the strut network prior to the generation of the actual filter surface [22]. The insertion of flow-guiding features reduces the porosity of f_w by 2–5% and for f_α and f_β by approximately 2–2.6%. As a result of the insertion of finger-like struts, the cumulative length of the strut network increases by approximately 14.5%. All filters in groups (fA) and (fB) are generated for porosities $\varepsilon = 70, 80, 90\%$.

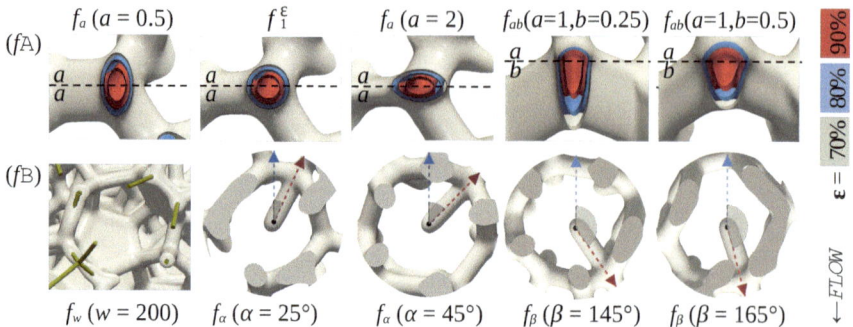

Fig. 18.13 Geometric modifications of the reference structure f_1^ε that has 216 equal-sized pores. Group (fA) with modifications of the strut shape f_a, f_{ab} and f_{ba}. For elliptical deformation f_a, the strut aspect ratio is controlled by parameter a, i.e. elongation for $a < 1$ and flattening for $a > 1$. The reference structure f_1^ε corresponds to $a = 1$. For f_{ab} and f_{ba}, an additional parameter b controls the aspect ratio of the lower part of the strut shape. Filters in group (fB), i.e. f_w, f_α and f_β, have flow-guiding features inserted. For f_w, a total of w windows are randomly closed. For f_α and f_β, downward- and, resp., upward-pointing finger-like struts are inserted inclined by angle α and β with respect to the bulk flow direction. Images for f_α are reproduced with permission [22]

Modifications with Varying Pore Size and Strut Shape

The filters in group (fC) vary pore size and, resp., strut shape along the filter depth. In group f_c, the strut width increases from top to bottom while in group f_q, the pore size is systematically varied in different layouts (see Fig. 18.14). Filters f_c are generated with the filter skeleton of the reference structure f_1^ε. Filters f_q are generated with three new filter skeletons arranging size-varying pores:

1. With $q = 200$ pores, a continuous transition from larger to smaller pores and reverse transition from small to large with 32 larger pores, 72 medium-sized pores and 96 smaller pores. The cumulative strut length decreases by 7.2% with respect to f_1^ε.
2. With $q = 265$ pores, a continuous transition from larger to smaller pores without reverse transition. The cumulative length of the strut network increases by 12.1%.
3. With $q = 320$ pores, an alternating pattern of 32 larger pores and clusters composed of nine smaller pores, which increase the cumulative strut length by 23.2%.

The filters f_q are generated for porosities $\varepsilon = 70, 80, 90\%$, while the filters f_c are generated for porosity $\varepsilon = 85\%$. The porosity value of f_c results from merging two reference structures $f_1^{90\%}$ and $f_1^{80\%}$ using a voxel-wise image blending operation along the filter depth. The procedure accomodates a smooth transition of the strut width and the strut shape across both parts of the filter, i.e. one with 90% porosity for the top and one with 80% porosity for the bottom. In addition to the transition from thin to thick struts, f_c integrates a smooth transition of the strut shape between the top and bottom parts of the filter, where the strut shape at the top is controlled by the aspect ratio $a = c$ and the aspect ratio in the bottom is set to $a = 4$. The filters f_q are

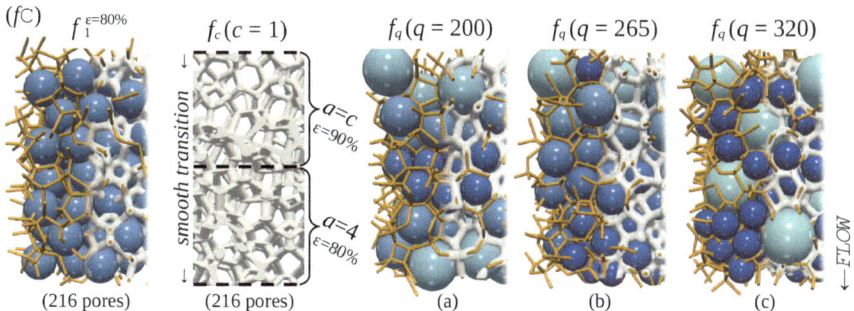

Fig. 18.14 Group (fC) of investigated filters with varying strut thickness f_c and varying pore size f_q. Filters f_c accomodate a continuous transition of thin struts with aspect ratio $a = c$ with $c \leq 2$ at the top to thick struts with fixed aspect ratio $a = 4$ in the bottom part of the filter. The filter f_c is shown for $c = 1$. The filters f_q are modeled with different layouts of $q = 200, 265, 320$ pores with varying size, based on new filter skeletons, i.e. transition of large to small pores (a) with and (b) without reverse transition, and (c) an alternating pattern of large pores and clusters of small pores. The filters f_1^ε and f_q are generated for porosities $\varepsilon = 70, 80, 90\%$ and shown for $\varepsilon = 80\%$. Dark blue spheres indicate smaller pores, whereas turquoise indicates larger pores with respect to f_1^ε

generated with circular strut cross section and uniform strut width for the complete filter.

As a consequence of the designs in group (fC), a variation of the porosity is induced in different spatial regions of the filter domain. For f_c (ix–xi) and f_q (ix) and (x), the porosity decreases along the filter depth in a uniform way according to the increase of strut thickness or the decrease of the pore size. In contrast, for f_q (xi) with $q = 320$ pores, where large pores alternate with clusters of smaller pores, locally increased and decreased porosity is induced on a raster of $4 \times 4 \times 4$ spatial regions in an interlaced manner.

18.4.2   Generation of Simulation Data Sets

A total of 84 simulation data sets are generated in LBM simulations using the parametric modifications as shown in Table 18.1. The 6 modifications in groups (fA) and (fB) are generated with 4 parameterizations and for porosities $\varepsilon = 70, 80, 90\%$ each, yielding 72 concrete filter designs. In group (fC), filters f_q are generated with 3 parameterizations and porosities $\varepsilon = 70, 80, 90\%$ to yield 9 concrete filter designs, while filters f_c are generated with 3 parameterizations and fixed porosity $\varepsilon = 85\%$ for 3 filter designs.

The filters are evaluated for process conditions present during the removal of alumina oxide inclusions from the aluminum melt inside CFFs with 30 PPI as reported for a pilot filtration line [23]. Therefore, the structures are scaled to meet the pore density of the 30 PPI filters [9, 22], resulting in a physical domain of $17.5 \times 17.5 \times$

Table 18.1 Overview of parametric variations for computer-generated filter structures used to generate the simulation data sets for the virtual prototyping scenario. All filters are generated for porosities $\varepsilon = 70, 80, 90\%$ except for f_c with fixed porosity $\varepsilon = 85\%$

Modification		Parameterization			
(fA) *Strut shape*		(i)	(ii)	(iii)	(iv)
– Elliptical struts	f_a	$a = 0.5$	1	2	4
– Drop-like struts	f_{ab}	$a = 1$	1	2	2
		$b = 0.25$	0.5	0.25	0.5
– Drop-like reversed	f_{ba}	$a = 1$	1	2	2
		$b = 0.25$	0.5	0.25	0.5
(fB) *Flow-guiding features*		(v)	(vi)	(vii)	(viii)
– Closed windows	f_w	$w = 50$	100	150	200
– Finger-like downwards	f_α	$\alpha = 15°$	25°	35°	45°
– Finger-like upwards	f_β	$\beta = 135°$	145°	155°	165°
(fC) *Varying geometry*		(ix)	(x)	(xi)	
– Size-varying pores	f_q	$q = 200$	265	320	
– Shape-varying struts	f_c	$c = 0.5$	1	2	

17.5 mm, which is discretized on a grid composed of 512^3 voxels with a spatial resolution $\Delta x = 34.5\ \mu$m for all data sets.

Surface Area of Simulated Filters

The surface area S of the 84 generated filter samples from groups (fA), (fB) and (fC) is shown in Fig. 18.15. The filters f_a, f_{ab}, f_{ba}, f_α, f_β, f_w, f_c and f_q are generated with parameterizations (i–xi) according to Table 18.1. As can be seen, all modifications lead to an increase of the surface area, except f_q (ix), which has a lower surface area than the corresponding reference structure f_1^ε with the same porosity. Filters with drop-like strut shape f_{ab} and f_{ba} in parameterizations (i) and (ii), and filters with finger-like struts f_α and f_β exhibit the largest increase. Among filters with porosity $\varepsilon = 90\%$, filters with closed windows f_w (viii) have the largest surface. Although the filter f_q (xi) has the largest cumulative strut length, its surface area is smaller as compared to f_α and f_β.

Modeling the Fluid Flow

The flow through the open-cell structures is assumed to be periodic with a constant flow rate driven by an imposed pressure gradient in the x-direction and adjusted by a controller, which monitors the flow rate through the inlet and outlet of the filter, while maintaining the prescribed superficial velocity [9]. All process simulations lie within the steady state, where the viscous losses contribute to at most 60% of

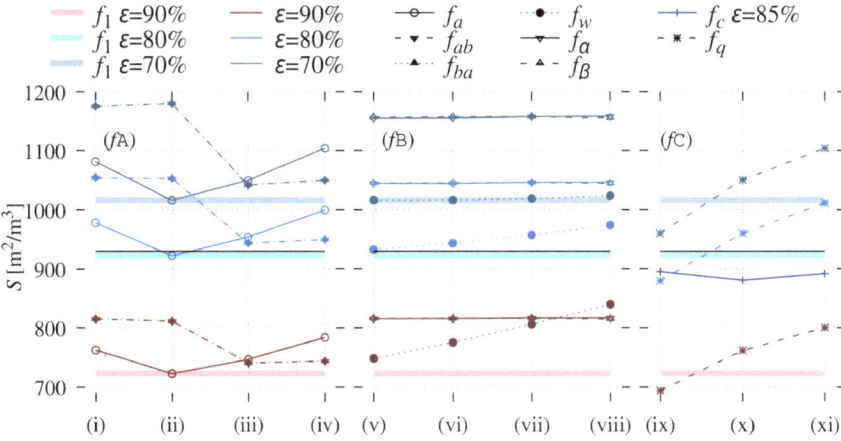

Fig. 18.15 Surface area S of the 84 filter prototypes. The filters are referenced (i–xi) according to Table 18.1. Left (i–iv): group (fA) with modified strut shape f_a, f_{ab} and f_{ba}. Middle (v–viii): group (fB) with with flow-guiding features f_α, f_β and f_w. Right (ix–xi): group (fC) with size-varying pores f_q and shape-varying, thickening struts f_c. The reference structures f_1^ε for porosities $\varepsilon = 70, 80, 90\%$ are shown as horizontal lines. Colors correspond to the porosity of the respective reference structure used for modification, except f_c that has porosity $\varepsilon = 85\%$

Table 18.2 Characteristics of the investigated aluminum filtration process for the reference structure $f_1^{80\%}$, where the average interstitial velocity is given by $u = u_D/\varepsilon$ and the strut width is $d_s = 636\,\mu m$

Dimensionless number	Value	Definition
Reynolds number	1.85×10^1	$Re = \rho u d_s / \mu$
Forchheimer number	1.07×10^0	$Fo = \rho u_D k_1 / (\mu k_2)$
Interception number	3.15×10^{-2}	$d_P^* = d_P / d_s$
Stokes number	1.69×10^{-3}	$St = \rho_P d_P^2 u / (18 \mu d_s)$
Gravitational number	4.80×10^{-2}	$N_G = (\rho_P - \rho) d_P^2 g / 18 \mu u$

the total pressure drop losses [9, 22]. The process is assumed with an aluminum density of $\rho = 2356\ \mathrm{kg \cdot m^{-3}}$ at a temperature of 730°C, superficial velocity of $u_D = 1\ \mathrm{cm \cdot s^{-1}}$, dynamic viscosity $\mu = 1.01 \cdot 10^{-3}\ \mathrm{Pa \cdot s^{-1}}$ and Reynolds number $Re = 18.5$ as defined in Table 18.2.

Application of LITE-QA In-Situ Data Preparation

The data sets are generated using GLATE data compression for the flow field u, v, w, and data indices u, M, Q as described in Sect. 18.3.1. This choice of data indexing allows for the efficient retrieval of flow regions with backflow, high-velocity flow and vortex-like flow. GLATE compression is configured to bound the point-wise maximum error at 1%. As process simulations lie within the steady state it suffices to store only one time step per simulation.

The in-situ data preparation pipeline is applied during the LBM simulations as described in Sect. 18.2. The data sets are compressed to 24.1% in average, where the flow fields u, v, w are compressed to 15.3% in average and the data indices for u, M, Q are compressed to 25.2%. Compared to the unsteady data set used for compression testing, as described in Sect. 18.2.3, the steady solutions are compressed stronger due to the lower Reynolds number [15]. The overall data size for the simulations of 84 computer-generated filters is reduced from $84 \times 2.5\text{GB} = 212.5\text{GB}$ to 51.1 GB including the additionally generated indices, i.e. the flow fields u, v, w account for 19.3 GB and the data indices u, M, Q for 31.8 GB.

18.4.3   Query-Driven Statistical Analyses of Melt Flow

Several statistical analyses of the melt flow can be performed by queries that only rely on the meta-information stored in the header of the LITE-QA compressed indices. As the data amounts stored in the header account for only a very small fraction of the total index size, data loading times are negligible [16].

Particularly, count queries, as described in Sect. 18.3.2, return the data inside the index header for an index variable, including the number of occurrences of each data value in the voxel grid. This data can be used to estimate various statistical quantities such as average, median, minimum, maximum, relative frequencies, and histograms. Following the approach of [24], also the hydraulic tortuosity of the melt flow can be calculated from the velocity distribution, accessible over index variable M, and the velocity distribution in principal flow direction, accessible over index variable u.

Effect of Filter Design on Flow Characteristics

Figure 18.16 shows the tortuosity and maximum velocity for all 84 filter designs. The maximum velocity $M_{\max}$ is presented normalized with respect to the superficial velocity u_D. As can be seen, flattened struts f_a with $a > 1$ and closed windows f_w substantially increase both maximum velocity and flow tortuosity with respect to the underlying reference structure f_1^ε. The only filters which decrease both quantities are the filters with drop-like strut shape f_{ab} and f_{ba}. Different arrangements of size-varying pores f_q only have marginal effects on the tortuosity, however, the maximum velocity increases substantially as the flow is passing through smaller pores with a higher porosity. The insertion of finger-like struts slightly reduces the maximum velocity for f_α and f_β while also the flow tortuosity is decreased indicating a smoothing effect on the flow. The filters f_c, that have porosity $\varepsilon = 85\%$ and implement a continuous variation of strut width and strut shape, lead to the opposite effect, i.e. the maximum velocity and the tortuosity are increased.

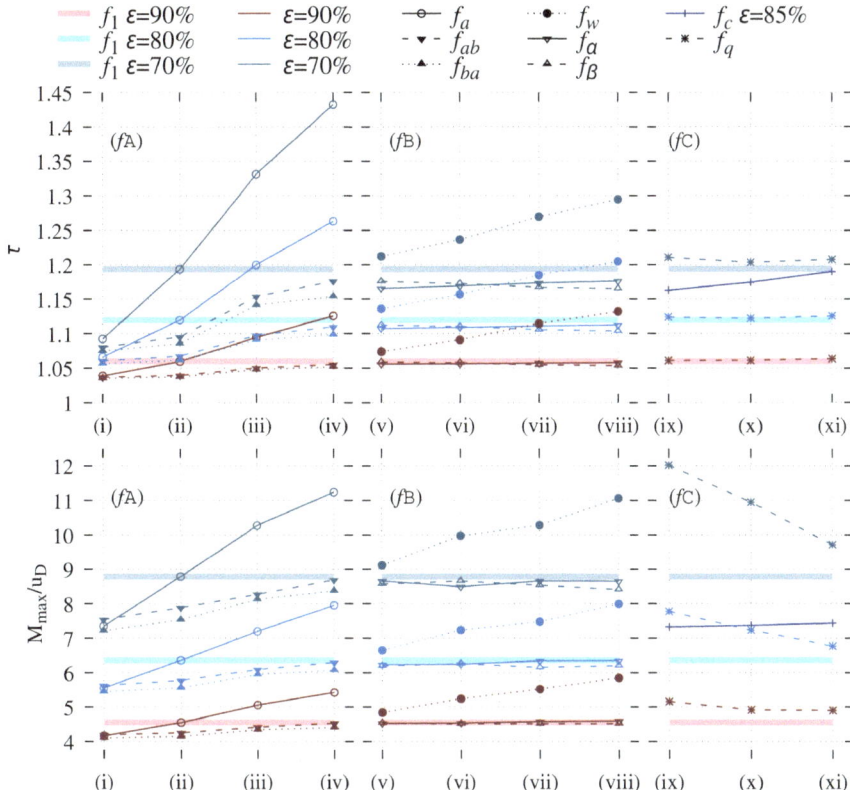

Fig. 18.16 Hydraulic tortuosity τ and normalized maximum velocity $M_{\max}/u_D$ computed using count queries on index variables M and u of the 84 filter prototypes. See Table 18.1 for filter parameterizations (i–xi). The reference structures f_1^ε for porosities $\varepsilon = 70, 80, 90\%$ are shown as horizontal lines. The filter modifications (i–xi) are colored according to the porosity class of the respective reference structure, except f_c that has a porosity of $\varepsilon = 85\%$

Statistical analyses of the flow field reveal global relationships between design modifications and flow characteristics across the complete fluid domain. In order to gain insights, how the design modifications influence the flow characteristics locally, additional visualizations of the melt flow are generated using the LITE-QA query mechanism as described next.

18.4.4 Query-Driven Comparative Visualization of Melt Flow

The LITE-QA query mechanism as described in Sect. 18.3 can be used for the automated creation of melt flow visualizations in specific areas of the filter that exhibit interesting flow characteristics. By visualizing the same filter regions for different filter variants, insights can be gained about the effects of geometric filter modifications on the melt flow.

In particular, comparative visualizations for the flow in specific filters from group $(f\text{A})$, $(f\text{B})$ and $(f\text{C})$ with modified strut shape, inserted flow-guiding features and varying pore geometries are generated. The locations for decompression and visualization are obtained from index queries, which are used to identify specific regions in the flow using range conditions, and from the filter skeleton directly, e.g. the geometric center of a strut or a pore. The visualizations are generated by rendering streamlines and iso-contour surfaces using the ParaView visualization pipeline described in Sect. 18.3.3.

Flow Regions in Computer-Generated Filters and CFF Sample

Figure 18.17 shows visualizations of vortices, which are generated as examples for characteristic flow in two datasets, i.e. the reference structure $f_1^{80\%}$ and a real CFF sample with 216 pores and 77.3% porosity obtained from computer tomography scanning in a previous study [9]. The CFF sample exhibits a pore density, specific surface area and strut width similar to the reference structure. However, the CFF sample also shows several defects, e.g., closed pore windows, cracked struts and deformed pores, due to imperfect manufacturing conditions. For each filter, the LITE-QA query mechanism, as shown in Fig. 18.10, is used to locate vortex regions, i.e. clusters of voxels with high vorticity, inside the flow fields. Five exemplary visualizations are generated for comparing the locations of high-vorticity clusters inside the filter geometries.

For both the computer-generated filter and the real CFF sample, vortices appear close to the upstream and downstream surfaces as well as in pore windows that are oriented in parallel to the bulk flow direction. As the upstream surface is exposed to the direct momentum of the flow, the velocity of vortices is higher as compared to vortices on the downstream surface or in pore windows. On the upstream, the flow crawls up on the surface of struts or closed windows before passing in another direction. On the downstream and in the pore windows, slower vortices appear in the slipstream of struts exhibiting less mass exchange with the bulk flow.

Deformation of Vortex-Like Flow Region

Figure 18.18 shows a vortex on the upstream surface of the reference structure $f_1^{80\%}$ and its deformation due to a different porosity $f_1^{90\%}$, a filter with elliptical strut shape f_a, and a filter with an upward-pointing finger-like strut f_β. After locating the vortex inside the reference structure f_1^ε using an index query, the visualization is regenerated using grid queries for the determined location for the four filter datasets. The camera is placed at three different angles showing the flow deformation due to the shape modifications in front, back and side view.

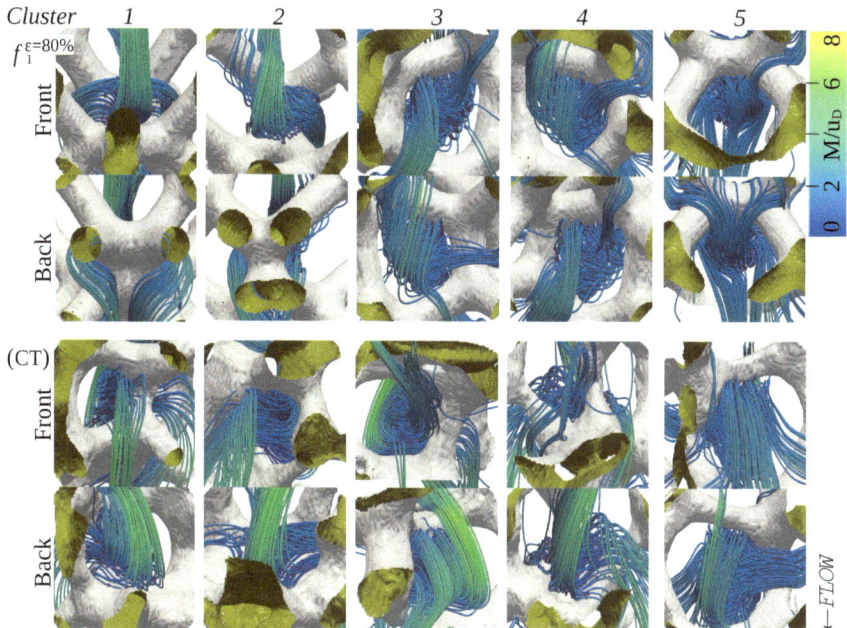

Fig. 18.17 Comparative visualization of vortices in reference structure $f_1^{80\%}$ and (CT) a real CFF sample with porosity of 77.3% obtained from computer tomography scanning. Both filters exhibit the same pore density, specific surface area and strut width. Fast vortices appear close to the upstream surface. Vortices close to the downstream surface and in pore windows are slower

The vortex is found next to a small closed pore window in the reference structure $f_1^{80\%}$, where the flow is redirected and crawling up the strut on the upstream surface. While for f_1^ε with $\varepsilon = 90\%$ the window is open, for $\varepsilon = 80\%$ the window is closed and the vortex region has a larger size. The flow splits in two opposing streams, which merge with the bulk flow while passing two opposing lateral pore windows. The flattened struts in f_a cause the closing of a second pore window, resulting in a stretch of the vortex region along the new upstream surface and an increase of the flow velocity. In contrast, the finger-like struts in f_β result in a decrease of the vortex volume, as the flow is redirected before it enters the vortex region. This local observation may explain the reduced tortuosity as observed in the global statistical analyses of Fig. 18.16.

Flow Around Struts with Modified Shape

Figure 18.19 shows the melt flow around a strut of the reference structure $f_1^{80\%}$ in comparison to filters f_a, f_{ab} and f_{ba} with modified strut shape (elliptical, drop-like, drop-like reversed). A grid query is used to decompress a cubic region of the flow field around the geometric center of a strut which is orthogonal to the bulk flow

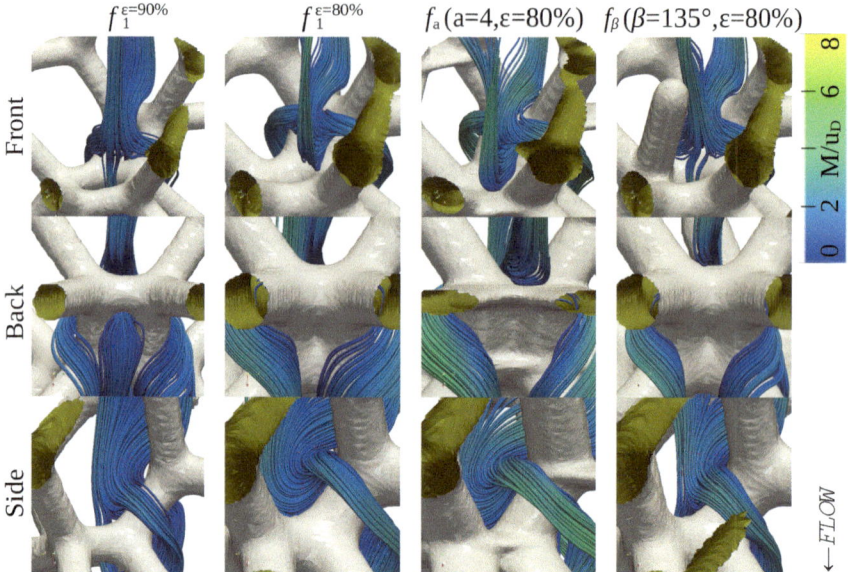

Fig. 18.18 Visualization of a vortex found by a LITE-QA query on the upstream surface of the reference structure $f_1^{80\%}$ and visualizations of the same region in three filter variants. All variants deform the vortex region

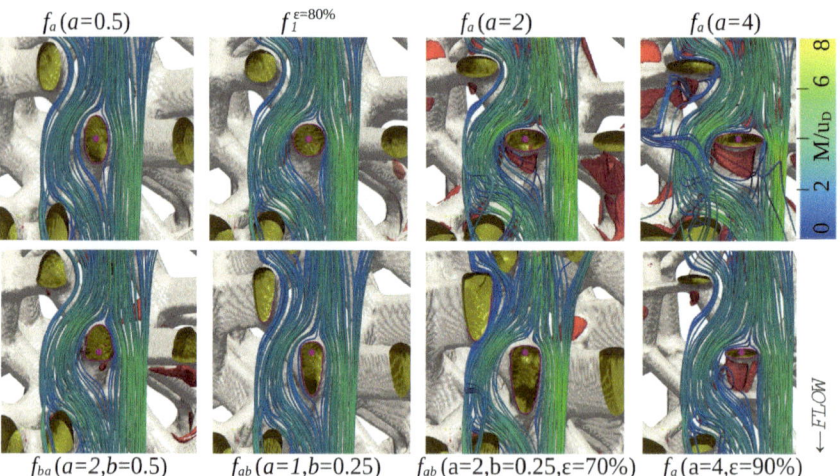

Fig. 18.19 Visualization of the flow around struts with modified shape for filters f_a (elliptical), f_{ab} (drop-like) and f_{ba} (drop-like reversed) in comparison to the reference structure f_1^ε (circular cross section). The iso-contour surfaces in red show fast backflow in the slipstream of the struts. All filters have a porosity of $\varepsilon = 80\%$, unless indicated otherwise

direction. The camera focuses on the strut center and is directed along the strut axis. The iso-contour surfaces in red highlight the fast backflow in the slipstream which is computed as the 5% quantile of the distribution for all values $u < 0$ in each data set.

As can be seen, for flat elliptical struts f_a with $a > 1$, the flow on the upstream surface of the struts exhibits increased velocity, as it is forced around the strut. The flow in the slipstream of the strut is mostly isolated from the bulk flow constituting a backflow region which increases the flow tortuosity. In contrast, for the elongated elliptical shape f_a ($a = 0.5$) and the drop-like strut shapes f_{ab}, the flow is smoothly directed around the strut, decreasing velocity and tortuosity. Interestingly, the reversed drop-like strut shapes f_{ba}, also exhibit decreased velocity and tortuosity. Due to the elongation of the upstream surface without sharp corners, the flow velocity on the upstream surface of f_{ba} is lower as compared to f_1^ε.

Flow Through Pores with Flow-Guiding Features

Figure 18.20 shows the melt flow through a pore of the reference structure f_1^ε in comparison to filters f_a, f_{ab}, f_{ba} with modified strut shape and filters f_α, f_β, f_w that have flow-guiding features inserted. A grid query is used to decompress a cubic region of the flow field around the geometric center of a pore. The camera is rotated towards one pore window and directed orthogonal to the bulk flow. The iso-contour surfaces in orange highlight the fast preferential flow, which is computed for each filter using the 85% threshold of the maximum velocity of $f_1^{80\%}$.

Consistent with the global statistical analyses indicating an increased velocity for flattened elliptical struts f_a with $a > 1$, the local visualization of f_a ($a = 2$) shows a higher-velocity flow (orange iso-contour) at the influx of the pore as compared to the reference structure f_1^ε. However, also consistent with the global analyses, when the porosity is increased from $\varepsilon = 80\%$ to $\varepsilon = 90\%$, even in case of very flat struts f_a ($a = 4$, $\varepsilon = 90\%$) velocity is still smaller as compared to the reference structure (absence of orange iso-contour). Similarly, drop-like strut shapes f_{ab} and f_{ba} decrease the velocity.

For the filters f_α and f_β with finger-like struts, the area of high velocity is deformed and the tortuosity is slightly lowered. The insertion of upward-pointing finger-like struts can cause a redirection of the flow through lateral windows of the pore. For filters f_w with closed windows, statistical analyses of global melt flow indicate an substantial increase of tortuosity and velocity. On the one hand, the flow is forced through a smaller number of pore windows, while on the other hand, regions behind closed windows are formed which are mostly isolated from the bulk flow. The latter effect can be observed for f_w ($w = 200$) in Fig. 18.20.

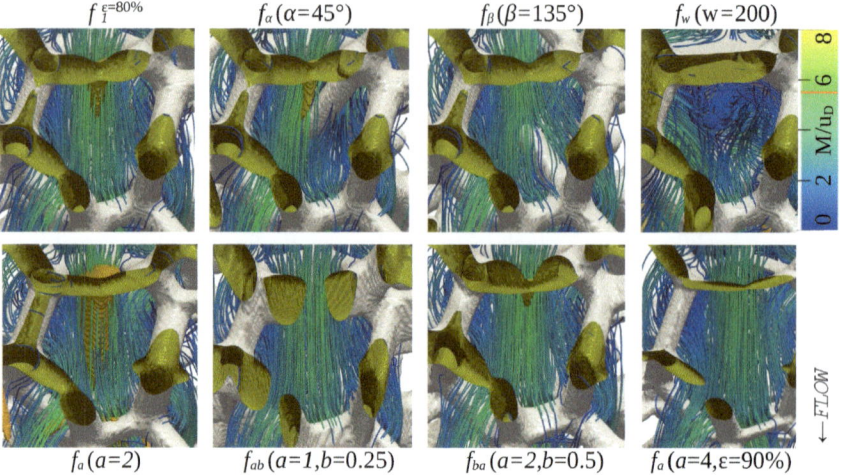

Fig. 18.20 Visualization of the flow through a pore of reference filter f_1^ε, filters f_a, f_{ab}, f_{ba} with modified strut cross section (elliptical, drop-like, drop-like reversed) and filters f_α, f_β, f_w with flow-guiding features (finger-like struts downwards, upwards, closed windows). Regions with high velocity are shown as iso-contour surfaces in orange. All shown filters have porosity $\varepsilon = 80\%$, except f_a $(a = 4)$ with $\varepsilon = 90\%$

Flow Through Size-Varying Pores

Figure 18.21 compares regions with backflow, high-velocity flow and high-vorticity flow for two filters with equal-sized pores and four filters with size-varying pores. The filters with equal-sized pores are the reference structure $f_1^{80\%}$ and the filter f_β $(\beta = 135°)$ with upward-pointing finger-like struts. The three filters f_q are laid out with pores of different diameters across the filter. In filter f_c $(c = 2)$, the effective size of pore cavities decreases from top to bottom due to the continuous transition from thin to thick struts.

A grid query is used to decompress a grid slice with a depth of 128 voxels for one complete filter height and width showing the flow field along the bulk flow direction. Regions with strong backflow, high velocity, and high vorticity in the flow field are found with range queries as described in Sect. 18.3.1 and rendered as iso-contour surfaces. Additionally, streamlines are seeded inside the flow field regions and traced both forwards and backwards along the complete height of the filter.

For filter f_c $(c = 2)$ with a continuous transition from thin to thick struts along the filter depth, the flow is smoother in the top part, as compared to the bottom, where the flow is forced into smaller pores. Almost all backflow, high-velocity and high-vorticity regions are concentrated in the lower half of the filter.

In the filters f_q $(q = 200)$ and f_q $(q = 265)$ with continuous pore size transition along the filter height, the flow is forced into regions of small pores, i.e. middle region of f_q $(q = 200)$ and bottom region of f_q $(q = 265)$, where the fast preferential flow and vortex-like flow concentrates. In contrast, for the filter f_q $(q = 320)$ with an

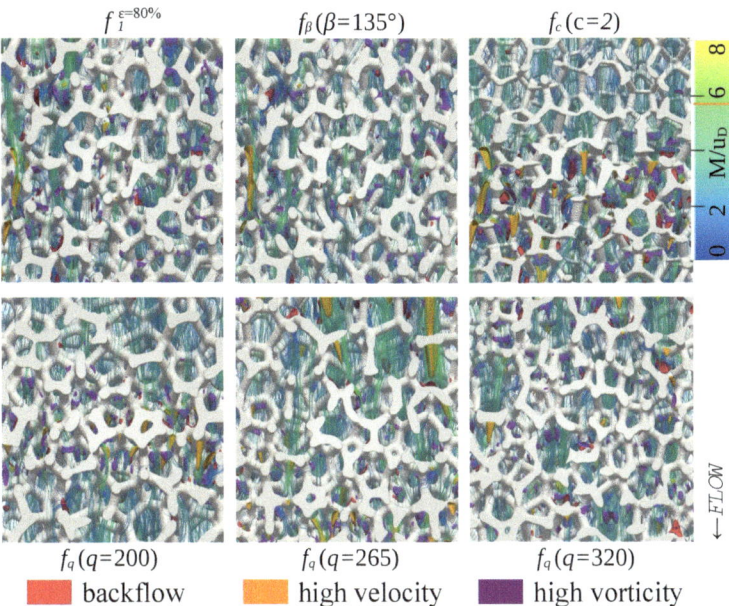

$f_1^{\varepsilon=80\%}$ $f_\beta\,(\beta=135°)$ $f_c\,(c=2)$

$f_q\,(q=200)$ $f_q\,(q=265)$ $f_q\,(q=320)$

🟥 backflow   🟧 high velocity   🟪 high vorticity

Fig. 18.21 Visualizations of regions in the flow field for filters f_q with size-varying pores and f_c with shape-varying struts in comparison to f_β with upward-pointing finger-like struts and the reference structure f_1^ε. All shown filters have porosity $\varepsilon = 80\%$

alternating layout of large and small pores, the regions of fast flow shrink in size as the flow can escape into larger pores more easily. Areas of high vorticity are scattered across the whole filter.

The filter f_c with increasingly thicker struts shows the largest amount of backflow areas among the filters compared in Fig. 18.21 which explains its high tortuosity. In contrast to f_c, the filters f_q do not increase the amount of fast backflow regions. Similar to f_q, the flow in the filters f_β with additional finger-like struts becomes smoother, due to the flow redirection through lateral pore windows as shown in Fig. 18.20, which can effects the size of the backflow regions.

18.4.5   Evaluation of Filtration Performance

The filtration performance of the filters is evaluated based on the filtration coefficient λ and pressure drop p' which are both directly determined from the LBM simulations using the Darcy-Forchheimer law and the conventional depth filtration law.

The Darcy-Forchheimer law describes a simplified relationship between the pressure drop $p' = \mathrm{d}p/\mathrm{d}x$ and the superficial velocity u_D for flow through porous media given by

$$\frac{\mathrm{d}p}{\mathrm{d}x} = -\frac{\mu}{k_1}u_D - \frac{\rho}{k_2}|u_D|u_D.$$

The relationship depends on the Darcy permeability k_1 and the Forchheimer coefficients k_2, which are required to characterize the contribution of the viscous and inertial losses to the overall pressure drop during the process simulations. The parameters are estimated using a computationally efficient approach, which permits comparative assessment with a sufficient accuracy [9].

The filtration coefficient λ is obtained by fitting the simulated inclusion distribution with the conventional depth filtration law given by

$$\frac{\mathrm{d}c}{\mathrm{d}x} = -\lambda c.$$

The inclusion distribution is obtained by tracking 10^5 spherical inclusions with diameter $d_P = 20$ μm and density $\rho_P = 3900$ kg $\cdot$ m^{-3} randomly inserted on the filter inlet. The particle transport, characterized as shown in Table 18.2, is simulated coupled with the LBM under consideration of the drag force for 3 residence times, i.e. an effective filter depth of 52.6 mm [9].

Effect of Filter Design on Filtration Process

Figure 18.22 shows the filtration coefficient λ and pressure drop p' obtained from simulations of all 84 filters. See Table 18.1 for geometric filter modifications and parameterizations (i–xi). An improvement of filtration performance is assessed when the filtration efficiency, measured by λ, increases without substantial impact on the pressure drop p' as compared to the reference structure f_1^ε of corresponding porosity.

From all 84 filters, the filter f_{ab} (iv) with drop-like struts is the only filter which both increases the filtration coefficient λ and at the same time reduces the pressure drop p' with respect to the reference structure f_1^ε with corresponding porosity. Interestingly, the filter f_a (iv) with very flat elliptical strut shapes $a = 4$ and with porosity $\varepsilon = 90\%$ outperforms the reference structure $f_1^{80\%}$ with a higher λ and a lower p'.

All filters yield an increase of λ w.r.t. the respective reference structure f_1^ε with same porosity except f_a (i), i.e. elongated elliptical struts, f_{ba} (i–iv), i.e. inverse drop-like struts, and f_q (ix), i.e. size-varying pores with large-small-large transition. However, a disproportionate increase of the pressure drop p' is observed for f_a (iii–iv), i.e. flat elliptical struts, and f_w (vii–viii), i.e. 150 or more closed windows.

The pressure drop increases only moderately for the filters with finger-like struts f_α and f_β and filters with size-varying pores f_q. The filters f_c with transition from thin to thick struts and porosity $\varepsilon = 85\%$ increase the filtration coefficient λ, while only moderately increasing p' with respect to the reference structure $f_1^{80\%}$.

Concerning the symmetric shape modifications, the following conclusions can be drawn:

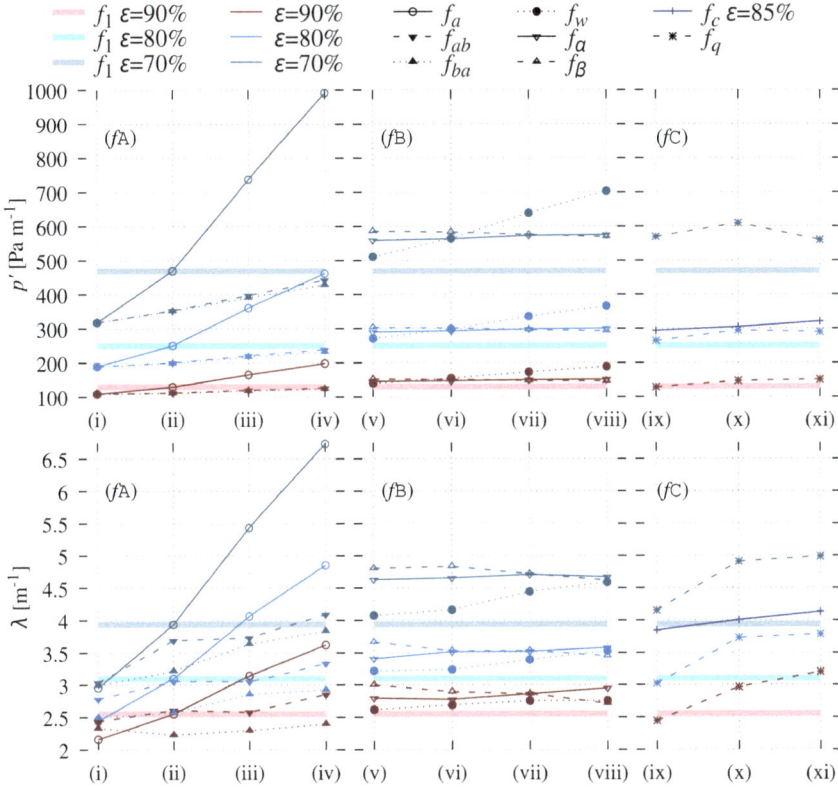

Fig. 18.22 Pressure drop p' and filtration coefficient λ of 84 filter prototypes. See Table 18.1 for geometric filter modifications and parameterizations (i–xi). Group (fA): strut shape modifications f_a, f_{ab}, f_{ba} with parameterizations (i–iv). Group (fB): insertion of flow-guiding features f_α, f_β, f_w with parameterizations (v–viii). Group (fC): varying pore size f_q and strut shape f_c within the filter with parameterizations (ix–xi). The reference structures f_1^ε for porosities $\varepsilon = 70, 80, 90\%$ are shown as horizontal lines. The filters (i–xi) are shown in colors matching the porosity class of the corresponding reference structure, except f_c with fixed porosity $\varepsilon = 85\%$

- filters with flat elliptical struts f_a ($a > 1$) outperform filters with vertically elongated struts f_a ($a < 1$)
- filters with drop-like struts f_{ab} outperform filters with reversed drop-like struts f_{ba}
- filters with upward-pointing finger-like struts f_β outperform filters with downward-pointing finger-like struts f_α.

Selection of Top-Performing Filters

Based on the data shown in Fig. 18.22, a set of top-performing modified filters is determined by comparison to the filtration coefficient λ of the reference structure $f_1^{80\%}$ and the pressure drop p'_{CFF} of the 30 PPI CFF sample shown in Fig. 18.17.

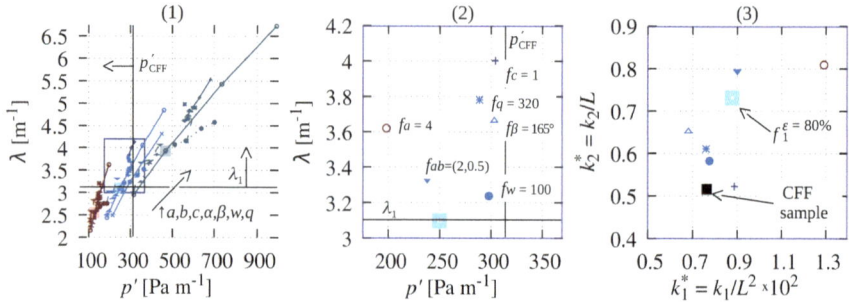

Fig. 18.23 Correlation of pressure drop p' and filtration coefficient λ for (1) all filters and, zooming in into the blue rectangular area and filtering out less well performing parameterizations, (2) the six top-performing filters whose filtration coefficient λ exceeds that of $f_1^{80\%}$ and whose pressure drop is lower than the pressure drop p'_{CFF} of the CFF sample. The six filters are f_c, f_q, f_β, f_a, f_{ab} and f_w with parameterizations as shown in the figure and summarized in Table 18.3. (3) Correlation of dimensionless viscous and inertial permeability k_1^* and k_2^* for the six top-performing filters

Figure 18.23 (1) shows the correlation of the pressure drop p' and the filtration coefficient λ for all filters. The vertical straight line shows the pressure drop of the CFF sample p'_{CFF}.

In order to select a group of filters with increased performance, first, all filters are considered whose filtration coefficient λ is higher as compared to $f_1^{80\%}$ and whose pressure drop p' is lower than the pressure drop p'_{CFF} of the CFF sample. Second, from the considered filters, for each modification one filter is selected, which has the highest filtration coefficient. Figure 18.23 (2) shows six top-performing filters that are also summarized in Table 18.3. The top-performing filters increase λ by 4.5–29% w.r.t. $f_1^{80\%}$. The filters f_a ($a = 4, \varepsilon = 90\%$) and f_{ab} ($a = 2, b = 0.5, \varepsilon = 80\%$) decrease p' by 4.7–20.6%, while the other four top-performing filters increase p' by 15.9–21.9%.

Figure 18.23 (3) shows the viscous and inertial permeability k_1 and k_2 for the top-performing filters. As can be seen, the filters f_a and f_{ab} increase both permeabilities, the filter f_c only decreases the inertial permeability and the remaining filters f_w, f_β and f_q decrease both k_1 and k_2. Notably, the permeabilities of f_β and f_q decrease only slightly although their cumulative strut lengths, due to new struts and many smaller pores, lead to a considerable increase of the surface area by 13.3% and, resp., 9.6% as compared to the reference structure $f_1^{80\%}$.

18.4.6 Discussion

This virtual prototyping study has investigated eight basic modifications of a reference geometry with several parameterizations each, see Table 18.1, yielding 84 virtual filter prototypes in total. For six of the eight modifications, Table 18.3 shows the filter

Table 18.3 Top-performing filters whose filtration coefficient λ exceeds that of f_1^ε and whose pressure drop is lower than the pressure drop p'_{CFF} of the CFF sample. Filters are ranked by λ

Ranked Modification	Filter	$\varepsilon^{[\%]}$	$p'^{[Pa \cdot s^{-1}]}$	$\lambda^{[m^{-1}]}$
1. Shape-varying, thickening struts	$f_c\,(x)^{c=1}$	85	304.2	4.00
2. Size-varying pores, alternating	$f_q\,(xi)^{q=320}$	80	289.0	3.78
3. Finger-like struts upwards	$f_\beta\,(v)^{\beta=135°}$	80	303.2	3.66
4. Flattened strut cross section	$f_a\,(iv)^{a=4}$	90	198.1	3.62
5. Drop-like strut cross section	$f_{ab}\,(iv)^{a=2}_{b=0.5}$	80	237.9	3.33
6. Randomly closed windows	$f_w\,(vi)^{w=100}$	80	298.1	3.24
Reference structure	f_1^ε	80	249.6	3.10

with the best performing parameterization. Two modifications are not included as they perform less well than their counterparts: Upward-pointing finger-like struts out-perform downward-pointing finger-like struts and drop-like strut shapes outperform struts with inverse drop-like shape.

Table 18.3 ranks the modifications according to the filtration coefficient λ. The four best-ranked modifications yield an increase of the filtration coefficient λ by 16.7– 29%, while the other two modifications, i.e. drop-like struts and closed windows, yield substantially lower improvements of 4–6%.

Nonetheless, also the latter modifications show interesting effects when not only the filtration coefficient λ is considered. Drop-like deformations of the strut shape can be parameterized such that pressure drop is substantially lower than in the reference structure without noteworthy decline of λ, e.g. f_{ab} (ii) and (iii) in Fig. 18.22. The deliberate closing of pore windows, which are oriented nearly parallel with respect to the bulk flow, hinders the formation of large slow vortices in the slipstream of struts, as shown in Fig. 18.17. Such observations could inform, e.g., the design of filters with more elaborate window-closing policies as compared to the randomized window closing of this study.

Increased complexity of filter manufacturing is a further issue for some of the modifications. Filters with very thin struts or articulate flat elliptical or even drop-like strut cross sections require a higher fidelity for 3D-printing of the polyurethane templates needed for filter replication. Coating of struts in an evenly manner also becomes more challenging for non-circular strut cross sections.

Positive results regarding filtration performance but without the potential draw-backs concerning their manufacturability are obtained for filter f_β with upward-pointing finger-like struts and the filter f_q with an alternating pattern of small and large pores.

The finger-like struts in f_β in the parametrization shown in Table 18.3 increase filtration efficiency by 18% while only marginally reducing the maximum velocity and the hydraulic tortuosity of the flow. As illustrated in Fig. 18.20, upward-pointing finger-like struts redirect flow through lateral pore windows. This decreases the size of vortices in the slipstream behind struts inside pore windows that are approximately parallel to the bulk flow. Overall, the insertion of upward-pointing finger-like struts substantially increases filtration efficiency, ranked at third place in Table 18.3, while also smoothing the flow.

In f_q of Table 18.3, clusters of nine smaller pores and single larger pores are arranged alternately, inducing a variation of locally increased and decreased porosity in an interlaced manner. As the flow can escape into nearby larger pores, the maximum flow velocity only increases slightly with respect to the reference structure, while the tortuosity remains unchanged. In contrast, in the other investigated layouts of size-varying pores the flow must pass through distinct layers of small pores which causes larger high-velocity regions and higher velocity peaks as shown in Fig. 18.21. The alternating layout of large pores and clusters of smaller pores yields an improvement of λ by 22% as compared to the reference filter. This is the second-best value of all investigated filters, while avoiding potential manufacturing complexities.

The filter f_c, which shows the best improvement of the filtration coefficient λ by 29% within this study, has thin struts in its upper part and further integrates an elliptical flattening. Possibly, other techniques than replication of 3D-printed polyurethane templates are needed to manufacture this filter with high fidelity.

18.5 Conclusions

The advent of additive manufacturing has vastly increased the design space of filters for metal melt filtration. Novel filter geometries can be conceived that vary, e.g., in strut shape and pore size distribution across the filter or that add flow-guiding or surface-increasing features to the pores. Furthermore, all such modifications can be parameterized in numerous ways. In order to assess this large space of possible filter designs before they are actually manufactured, a virtual prototyping workflow based on HPC simulations was presented.

A major challenge are the large data volumes produced even by single CFD simulations of metal melt flow and even more so for virtual prototyping settings where many filter designs need to be evaluated. The large data problem is particularly severe during the analysis phase, where simulation results are assessed outside of the HPC environment, using workstations with much lower computing resources. The LITE-QA framework was developed to address this large data challenge. Integrated into code executed in the HPC environment, the resulting simulation data is compressed and indexed. The data reduction rates of combined data compression and compressed indexing significantly improve on other state-of-the-art methods. As support for the analysis phase, LITE-QA offers query-driven access to the compressed and indexed data. This adds search engine-like capabilities to data analysis tasks and allows, e.g., for focused visualizations of filter regions that meet the search criteria of the analyst, such as areas with high velocity, vorticity or backflow. Crucially, only very small parts of the simulation data, i.e. those parts that meet the analyst's search criteria, need to be loaded from disk. In this way, interactive data analyses are enabled even for very large scientific data sets.

Furthermore, a virtual prototyping study including 84 new filter designs for metal-melt filtration was conducted. The filter designs are created by geometric modification of a reference model that emulates the topology, pore density and strut width of conventional foam filters used for aluminum filtration. Eight basic geometric modifications were investigated that alter the strut shape, add flow-guiding features or vary pore and strut size within a filter. The modification parameters were systematically varied and filter geometries were generated for three porosities. An evaluation of filter performance based on HPC simulations and LITE-QA data management identified the six best-performing filter designs with higher filtration coefficient than the reference filter while only having a moderate effect on melt flow pressure. Local visualizations of regions with interesting flow properties such as backflow, high velocity or high vorticity provide further insights on how the investigated modifications influence the melt flow. The largest improvement of the filtration coefficient by 29% was achieved for a filter with thin struts at the top that increasingly become thicker towards the bottom. The thin struts however require high-fidelity 3D-printing technology. The next-best performing filter designs do not include particularly thin struts or struts of articulate shapes, thus placing lower demands on additive manufacturing processing: A layout where large pores alternate with clusters of small pores with 22% improvement and the insertion of upward-pointing finger-like struts with 18% improvement of the filtration efficiency. These best-performing virtual filter prototypes can be seen as promising candidates for additive manufacturing and further testing as physical prototypes.

Acknowledgements The authors would like to thank the German Research Foundation (DFG) for supporting this investigation carried out by the subproject S02 of the Collaborative Research Centre 920, "Multi-Functional Filters for Metal Melt Filtration—A Contribution towards Zero Defect Materials" (Project-ID 169148856). The authors are also grateful to the Centre for Information Services and High Performance Computing (ZIH) at TU Dresden for providing its facilities for high throughput calculations.

References

1. T. Wetzig, M. Neumann, M. Schwarz, et al., Adv. Eng. Mater. **24**, 2100777 (2022). https://doi.org/10.1002/adem.202100777
2. B. Bock-Seefeld, T. Wetzig, J. Hubálková, et al., Adv. Eng. Mater. **24**, 2100655 (2022). https://doi.org/10.1002/adem.202100655

3. A. Herdering, M. Abendroth, P. Gehre, et al., Ceram. Int. **45**, 153–159 (2019). https://doi.org/ 10.1016/j.ceramint.2018.09.146
4. A. Herdering, J. Hubálková, M. Abendroth, et al., Int. Ceram. Rev. **68**, 30–37 (2019). https:// doi.org/10.1007/s42411-019-0013-z
5. M. Abendroth, E. Werzner, C. Settgast, S. Ray, Adv. Eng. Mater. **19**, 1700080 (2017). https:// doi.org/10.1002/adem.201700080
6. E. Werzner, M.A.A. Mendes, S. Ray, D. Trimis, Adv. Eng. Mater. **15**, 1307–1314 (2013). https://doi.org/10.1002/adem.201300465
7. H. Lehmann, E. Werzner, M.A.A. Mendes, et al., Adv. Eng. Mater. **15**, 1260–1269 (2013). https://doi.org/10.1002/adem.201300129
8. C. Demuth, E. Werzner, M.A.A. Mendes, et al., Adv. Eng. Mater. **19**, 1700238 (2017). https:// doi.org/10.1002/adem.201700238
9. E. Werzner, M. Abendroth, C. Demuth, et al., Adv. Eng. Mater. **19**, 1700240 (2017). https:// doi.org/10.1002/adem.201700240
10. J. Iverson, C. Kamath, G. Karypis, Fast and effective lossy compression algorithms for scientific datasets, in *Euro-par 2012 Parallel Processing* (Springer, 2012), pp. 843–856
11. S. Lakshminarasimhan, N. Shah, S. Ethier, et al., Concur. Comput. Pract. Exp. **25**, 524–540 (2013). https://doi.org/10.1002/cpe.2887
12. P. Lindstrom, Fixed-rate compressed floating-point arrays. IEEE Trans. Visual Comput. Graph. **20**, 2674–2683 (2014)
13. S. Di, F. Cappello, Fast error-bounded lossy HPC data compression with SZ, in *2016 IEEE International Parallel and Distributed Processing Symposium (IPDPS)* (2016), pp. 730–739
14. H. Lehmann, B. Jung, In-situ multi-resolution and temporal data compression for visual exploration of large-scale scientific simulations, in *4th IEEE Symposium on Large Data Analysis and Visualization (LDAV)* (2014), pp. 51–58
15. H. Lehmann, E. Werzner, C. Degenkolb, Optimizing in-situ data compression for large-scale scientific simulations, in *Proceedings of the 24th High Performance Computing Symposium*, Society for Computer Simulation International, San Diego, CA, USA, pp. 5:1–5:8 (2016)
16. H. Lehmann, E. Werzner, C. Demuth, et al., Efficient visualization of large-scale metal melt flow simulations using lossy in-situ tabular encoding for query-driven analytics, in *2018 IEEE International Conference on Computational Science and Engineering (CSE)* (2018), pp. 123–131
17. H. Lehmann, B. Jung, Temporal in-situ compression of scientific floating point data with t-GLATE, in *2018 International Conference on Computational Science and Computational Intelligence (CSCI)* (2018), pp. 1386–1391
18. H. Lehmann, Temporal lossy in-situ compression for computational fluid dynamics simulations. PhD thesis, Faculty for Mathematics & Informatics, Technical University Bergakademie Freiberg, 2018
19. D. Lemire, L. Boytsov, N. Kurz, Softw. Pract. Exp. **46**, 723–749 (2016). https://doi.org/10. 1002/spe.2326
20. K. Wu, S. Ahern, E. W. Bethel, et al., J. Phys. Conf. Ser. **180**, 012053 (2009). https://doi.org/ 10.1088/1742-6596/180/1/012053
21. A. Asad, H. Lehmann, B. Jung, R. Schwarze, Adv. Eng. Mater. **24**, 2100753 (2022). https:// doi.org/10.1002/adem.202100753
22. H. Lehmann, E. Werzner, A. Malik, et al., Adv. Eng. Mater. 2100878 (2021). https://doi.org/ 10.1002/adem.202100878
23. C. Voigt, E. Jäckel, F. Taina, et al., Metall. Mater. Trans. B **48**, 497–505 (2017). https://doi. org/10.1007/s11663-016-0869-5
24. A. Duda, Z. Koza, M. Matyka, Physi. Rev. E **84**, 036319 (2011). https://doi.org/10.1103/ PhysRevE.84.036319

Chapter 19
Registration of Filtration Efficiency of Active or Reactive Filters in Contact with Steel Melt in a Steel Casting Simulator

Steffen Dudczig, Enrico Storti, and Christos G. Aneziris

19.1 Introduction

The mechanical properties of cast steel products are strongly related to the cleanliness of the melt they come from. Both endogenous and exogenous nonmetallic inclusions influence mechanical strength, fatigue resistance and fracture toughness remarkably. The non-metallic inclusions that need to be removed from steel melts are mainly oxides, carbides, nitrides, and sulfides [1]. Different sources of endogenous and exogenous inclusions include reoxidation, slag entrainment, lining erosion, and inclusion agglomeration on linings. Endogenous inclusions include products of deoxidation or inclusions precipitated during cooling and solidification. When aluminum is used as a deoxidation agent, the inclusion population is usually dominated by endogenous alumina particles in a variety of different shapes. In addition, the formation of three-dimensional clusters and their subsequent growth are promoted by to the high interfacial energy of alumina. Small spherical inclusions could account for 90% or more of the total inclusions throughout the secondary metallurgy treatment. However, large inclusions represent from 60% to almost 100% of the oxide volume [2].

Due to the increasing pressure on the steel industry to produce clean, high-quality steel, the improvement of existing steel filtration techniques as well as the development of new materials and systems that can offer higher filtration efficiencies are crucial goals for the steel and refractory industries. Using a so-called steel casting simulator, carbon-bonded alumina components with different types of coatings were tested in the present study. The special apparatus allowed the investigation of filter

S. Dudczig (✉) · E. Storti · C. G. Aneziris
Institute of Ceramic, Refractories and Composite Materials, Technische Universität Bergakademie Freiberg, Agricolastr. 17, 09599 Freiberg, Germany
e-mail: steffen.dudczig@ikfvw.tu-freiberg.de

© The Author(s) 2024
C. G. Aneziris and H. Biermann (eds.), *Multifunctional Ceramic Filter Systems for Metal Melt Filtration*, Springer Series in Materials Science 337,
https://doi.org/10.1007/978-3-031-40930-1_19

surfaces after a simulated filtration with adsorbed particles but without the presence of solid steel all around, as in the case under normal industrial trials. In addition, clean formulations were developed and used in direct contact (one alumina-/spinel-containing crucible per melt) with the melt, in order to reduce the influence of impurities from refractory materials to the minimum. Different active and reactive filter materials were immersed for different immersion times in order to investigate the formation of new layers on the filter surfaces during the contact, to assess the entrapment of inclusions on such layers as well as to evaluate the steel purity after the immersion test. The inclusions remaining in the frozen steel melt were characterized with the aid of an automatic SEM—ASPEX-system, which classifies the inclusions population based on chemistry, as well as position, size and geometry. Finally, a new combined refining process using both reactive and active filters together was explored. After the treatment in the crucible, the remaining cluster of inclusions which were not removed due to buoyancy forces on CO-bubbles at the inclusions (generated by the interaction of steel melt with the carbon bonded reactive filter) were filtered with the aid of active carbon-free filters.

19.2 Experimental Details

This work is divided in two main parts. The first one presents the steel casting simulator in detail, together with the special castable formulations which allowed to reliably test all kinds of different specimens. The second part deals with the experiments performed in the steel casting simulator, in which different ceramic components were put in contact with molten steel.

19.2.1 Steel Casting Simulator

A metal-casting simulator (Systec, Karlstadt, Germany) located since 2010 at the Institute of Ceramics, Refractories and Composite Materials in Freiberg was used. The system was designed to evaluate refractory materials in direct contact with steel melts. The system consists of a water-cooled, gas tight chamber, that is filled with argon to protect all components within (see Fig. 19.1). It is composed of a separated inductive-heated melting unit (150 kVA) with a crucible that can contain up to 100 kg of steel. A revolver system allows to introduce probes to measure the temperature of the melt as well as the dissolved oxygen. In addition, another device is used to immerse prismatic samples (with maximum dimensions of $25 \times 25 \times 125$ mm^3) into the melt and rotate them at 30 rpm. It is also possible to extract melt samples for chemical analysis. The revolver system is equipped with a small chamber, in which an almost oxygen-free atmosphere is obtained through vacuum and argon flowing. After contact with the melt, the sample is extracted and can cool down inside this chamber, which is essential to prevent oxidation of e.g. carbon-bonded specimens.

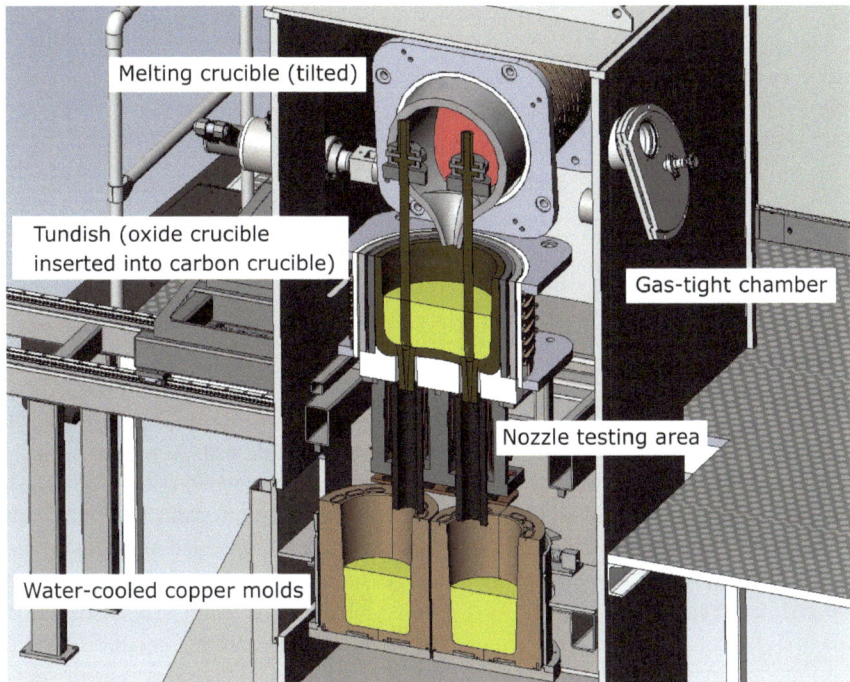

Fig. 19.1 Steel casting simulator setup. Adapted with permission from [1]

Finally, a charging device allows to introduce alloying elements and synthetic slags into the melt.

After conditioning of the molten metal, the crucible is tilted and the melt is cast into an inductive-heated (100 kVA) tundish, which presents three nozzles in the basement (two for tests and one for safety reasons) and a stopper rod system to control the melt flow. After some time, the stoppers are simultaneously lifted and the melt flows through parallel channels. Each consists of a protective carbon rod with a CFC heating unit to achieve a reproducible temperature level and prevent the metal from freezing. From the testing channels, the melt flows into separate water-cooled copper molds for solidification. Both the tundish and the copper molds are constantly weighed, which allows to monitor the melt flow for each of the two testing channels.

Compositions for Steel Casting Simulator

In order to prevent the generation of unwanted inclusions during the experiments with molten metal, special castable formulations were developed in house [3]. The melting crucible is based on alumina (Colisit, Großalmerode, Germany) and the tundish unit consists of a carbon crucible in which an oxide crucible (alumina-mullite castable KM20) is fitted in [1].

Table 19.1 Composition of the refractory castables for production of various components to be used inside the steel casting simulator

Material	KM20	SP1g	SP1
	Amount in wt%		
Tabular alumina	66	61	61
Mullite	20		
Spinel		15	15
Magnesia		6	6
Reactive alumina	11	15	15
Hydratable alumina	3	3	3
Additive (related to tot. solids)	1	1	1
Water (related to tot. solids)	5.5	5.3	6.6

For the tundish unit, a corundum-based refractory castable with 20 wt% of primary mullite and secondary in situ mullite formation proved to be the most suitable (Table). The main component was tabular alumina with a maximum grain size of 5 mm (Almatis GmbH, Ludwigshafen, Germany). Fused mullite was added in the classes 0–0.08, 0–0.5 and 0.7–1.5 mm (Imerys, Paris, France). Hydratable alumina Alphabond 300 (Almatis) in 3 wt% amount was used as binder. The grain size distribution was designed on the base of the Dinger and Funk particle-packing equation, with a distribution modulus of 0.28.

Regarding instead nozzles, adapters, stopper rods and melting crucibles for up to 30 kg of metal, a corundum-based refractory castable with a primary spinel content of 15 wt% was selected. The primary component was still alumina with a maximum grain size of 5 mm (for crucibles, SP1g in Table 19.1) or 3 mm (for adapters, stopper rods and nozzles, SP1 in Table 19.1). The primary spinel (Almatis) was added in the fractions 0–20 μm, 0–90 μm and 0–0.5 mm. The finer fractions were replaced with reactive alumina as well as fine MgO (Refratechnik Steel, Göttingen, Germany) in 6 wt% amount, in order to promote in situ spinel formation and compensate for the shrinkage during sintering.

The preparation of different components for use in the steel casting simulator was performed in agreement with the European Standard DIN EN ISO 1927-5. First, the required amounts of materials were dry mixed for 1 min. After water addition, the mixtures were stirred for additional 5 min. Molds made of steel or plastic with silicone inlays were used to form the castable masses into the desired products. After 24 h, the samples were removed from the molds and dried at 120 °C (with heating rate of 10 K/min) for 12 h. Finally, sintering was performed under air at 1600 °C for 5 h, with heating rate of 1 K/min and an intermediate dwell step of 1 h at 500 °C. Cooling was limited to 3 K/min down to 500 °C, after which the furnace cooled down freely.

19.2.2   Experimental

All experiments performed in the steel casting simulator can be classified as dynamic experiments, i.e. where either the refractory specimen or the melt are in motion. This results in more intensive interactions compared to the static experiments, in order to approach industrial conditions. In the so-called "finger-test", the samples are immersed and rotated for a defined time into the metal (or slag) melt. This is the easiest and fastest type of experiment. More complex experiments can be also performed, by casting the melt from the main crucible into a tundish unit and then through the parallel channels into the copper molds. Before each experiment, the steel casting simulator is evacuated and then filled with argon gas, in order to eliminate oxygen and protect the inductively coupled graphite crucibles required for the heating. In addition, to create defined alumina impurities in situ in the steel melt, 0.5 wt% of an iron oxide mixture was added. The commercially available product (Mineralmühle Leun, Germany) consisted of 75 wt% hematite and 25 wt% magnetite. After setting a temperature of 1650 °C, the oxide mixture was added directly to the melt and an increase of the dissolved oxygen from approximately 10 ppm up to approximately 60 ppm was measured. To create the required endogenous alumina inclusions, 0.05 wt% of pure aluminum metal was added to the melt. Due to this deoxidizing step, the dissolved oxygen content of the melt decreased to the starting values. Through the reaction between oxygen and aluminum, finely dispersed aluminum oxide was formed. After the experiments, the solidified steel as well as the tested samples were thoroughly analyzed.

Casting Through Nozzles

In order to investigate the clogging phenomena, carbon-bonded alumina nozzles based on the composition developed by Emmel were produced by slip casting into plaster molds [1, 4]. The slurry preparation for slip-casting in gypsum molds was performed in a ball mill (plastic container with alumina balls), in which the additives were added and distributed in deionized water in advance. Afterwards the solid parts were added stepwise and mixed for 24 h. After slip-casting and drying for 24 h at 120 °C, the test nozzles were coked at 800 °C under reducing atmosphere in a steel retort under coke grit. Afterwards, the inner surface of the coked nozzles was coated with alumina or mullite slurry, dried at 120 °C and coked again for 2 h at 1400 °C inside an alumina retort under coke grid. Different coatings were applied in a second step to the inner surface of the samples [5]. Before testing, the nozzles were installed melt-tight into the dedicated adapters with a refractory adhesive (Fig. 19.2). A 10 ppi carbon-bonded alumina filter with oxide particles (alumina, spinel, mullite) was placed inside the adapter above the nozzle to add exogenous inclusions to the flowing steel melt. After preparation of the testing area, 100 kg of commercially available 42CrMo4 (AISI 4142) steel were melted inside the crucible and heated up to 1650 °C under inert atmosphere. The oxygen content and the temperature of

Fig. 19.2 Configuration of
the nozzle testing area.
Adapted with permission
from [1]

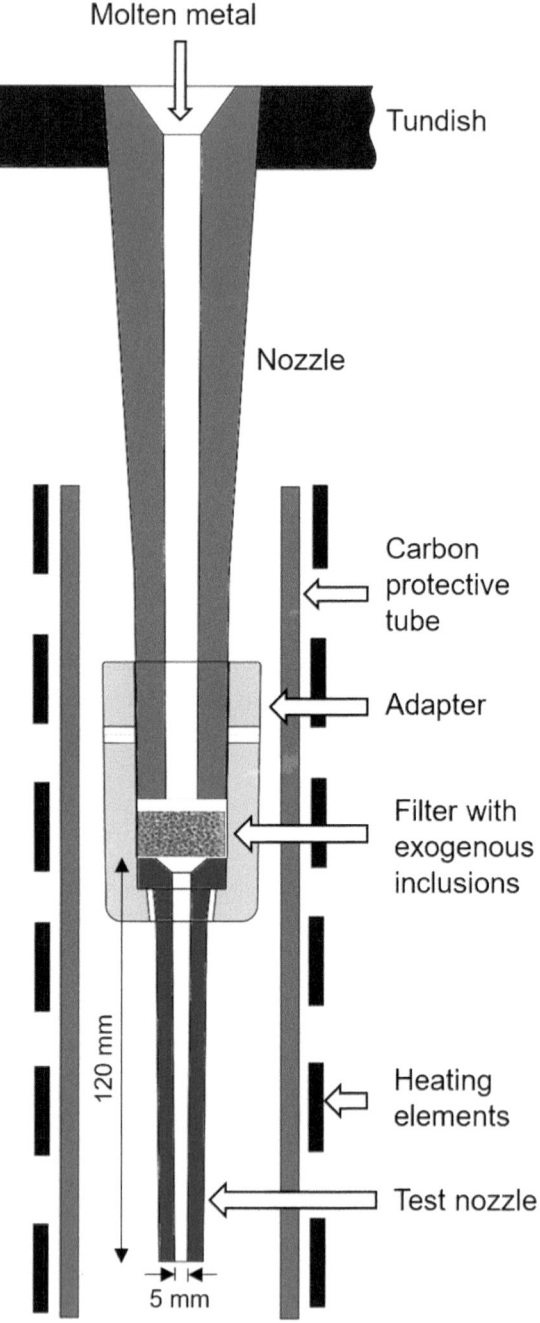

the steel melt were monitored with an oxygen/temperature sensor-system CELOX (Heraeus Electro-Nite, Houthalen, Belgium). After melting, the steel melt was poured into the preheated tundish unit. A short time after pouring (ca. 30 s), the stoppers rods were lifted and the melt flows through the two test nozzles and into the cooled copper molds. The test zone with the test nozzles is preheated by heating elements up to 1600 °C to prevent freezing of the liquid metal and to ensure a constant testing temperature in the nozzle area. The test nozzles before and after steel melt contact as well as the adapter with the inclusion containing reservoir after the steel contact were analyzed with the aid of a microfocus X-ray computed tomograph CT-ALPHA (ProCon X-Ray, Garbsen, Germany). For the visualization of the CT-images the software VGStudio Max 2.1 (Volume Graphics GmbH, Heidelberg, Germany) was used. The microstructural evaluation of the coked surfaces as well as after steel melt contact were carried out by means of scanning electron microscope (SEM XL30, Philips, Germany) in combination with energy dispersive spectrometer (EDS). The samples were partially embedded in resin to stabilize the decomposed layer.

Finger-Tests

Aim of these finger-tests was to assess the thermal shock resistance of the samples at the point of immersion, as well as the chemical interactions with the melt. 10 ppi prismatic ($125 \times 20 \times 20$ mm^3) filter samples were produced using the replica process as described by Emmel and Aneziris [4, 6]. The samples consisted of approximately 30 wt% carbon and 70 wt% alumina after pyrolysis at 800 °C under reducing atmosphere. Several types of ceramic coatings were applied, mainly by cold spraying. Coatings of alumina or carbon-bonded magnesia were applied to these carbon-bonded alumina substrate material, which were then thermally treated at 1400 °C (alumina) or 800 °C (magnesia/carbon) under a reducing atmosphere (a carbon grid). Details of the coatings and of their preparation are given elsewhere [2, 6–8].

Since the melting of 100 kg would have required the use of the whole system (i.e. casting into the tundish and the copper molds) to obtain reasonably small steel blocks, a smaller crucible (for 30–40 kg of steel) was instead installed inside the large one. All finger-tests were performed inside these smaller crucibles, produced in house from the spinel-containing refractory castable (SP1g) described above. Before the immersions, alumina endogenous inclusions were generated through the steel treatment described above. The prismatic samples were subsequently dipped 60 mm deep into the steel melt by entering the system through a sewer port (Fig. 19.3). During the whole experiment, temperature and dissolved oxygen content were monitored after each experimental step. Before and after filter immersion, steel samples were extracted for further analysis. After contact with the melt, the sample was removed and cooled down in a chamber under argon atmosphere to prevent the oxidation of carbon, which could have occurred if the hot sample had had contact with the normal oxygen-containing atmosphere. In addition, the filter sample was not full

Fig. 19.3 Melting unit with revolver system used for the finger tests. Adapted with permission from [2]

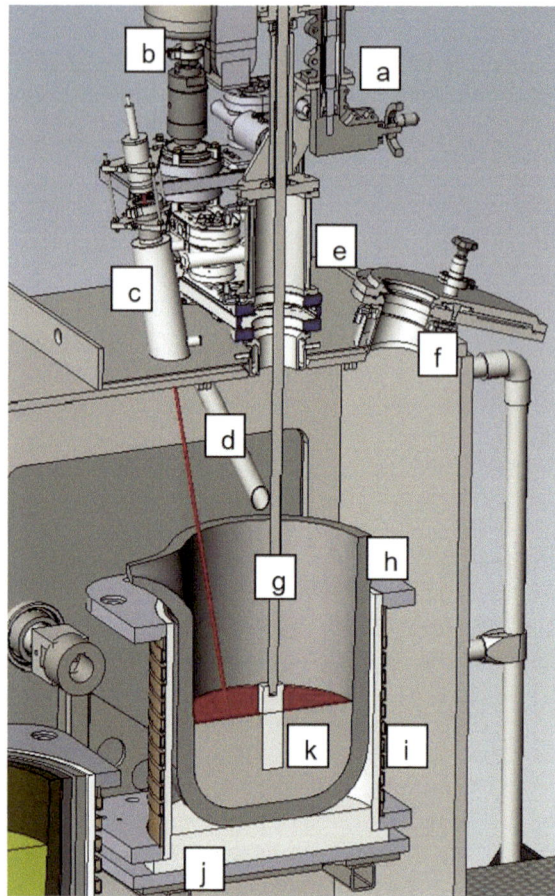

a. Rotatable revolver system
b. Gas-tight reservoir for additives
c. Pyrometer
d. Feeding mechanism
e. Chamber for sample handling
f. Observation window
g. Test sample holder
h. Melting crucible
i. Induction coil
j. Tiltable melting unit
k. Prismatic sample in contact with melt

with steel at the end, allowing for microscopic investigations. The melt was cooled down rapidly by switching off the inductor coil. All obtained steel samples and blocks were analyzed by means of ASPEX and used for other metallographic and mechanical tests.

In order to better understand the reactions that take place at the interface, the thin secondary layer generated on a carbon-bonded alumina filter during immersion in a steel melt was prepared by focused ion beam [9]. The investigations were carried out using the microscopes Versa3D from FEI and Lyra3 from TESCAN, respectively. Both dual beam systems were equipped with 30 kV Ga + ion sources for the FIB (focused ion beam) preparation. Combined Genesis-EDX and TSL-EBSD systems from EDAX Inc. were used for the EBSD investigations. 50 nm thin, transparent lamellas were obtained through a careful preparation. Phase determination was carried out for single points or for defined scans. Topographic information was obtained using the forward scatter detector (FSD), which was positioned below the EBSD screen. Lattice parameters for the identification of involved phases by EBSD were taken from an ICDD database (PDF2008).

The chemical composition of the solidified steel after the trials in the steel casting simulator was first investigated by means of spark emission spectroscopy (Q2 ION, Bruker, USA). Samples of appropriate dimensions were cut from the steel blocks, then polished before the analysis. Each sample was analyzed several times.

ASPEX (FEI, USA) is a special automatic scanning electron microscope (SEM), which is designed for the analysis of metallic specimens: after polishing, areas of approximately 110 mm^2 were scanned for about 8 h. In order to obtain more reliable results, in the latest studies several steel samples were analyzed from each solidified steel block. Inclusions >0.6 μm were detected using a BSE (back-scattered electron) detector based on contrast difference with the steel matrix. Due to the iron content, the average brightness (under EBSD) of the steel matrix is quite high. In comparison, any region with brightness below a certain threshold is counted as inclusion or surface defect, with e.g. Al_2O_3 having a darker tone than MnS due to the different atomic mass of the elements. For each found particle the AFA (automatic feature analysis) included position, geometry, orientation and chemical composition, the latter achieved through EDS (energy-dispersive X-ray spectroscopy). Afterwards, the inclusions were classified by size and by composition according to a classification rules (an example is given in Table 19.2). Rule files are not commonly available and were developed in house using reference specimens of known composition. Ideally, each type of steel requires a new classification rule. In addition, the rule file should be built with regard to the particles of interest. Due to the steel treatment pre-immersion, the most interesting inclusion class was aluminum oxide. For this reason, after eliminating the scratches, dirt and iron oxides due to the preparation process, the class Al_2O_3 is the first in the list. For each particle, if the composition of an inclusion does not fit one rule, the next rule will be checked. The boundaries were selected on the basis of the theoretical composition of each compound. However, the range for each element cannot be too small, otherwise too many particles would not fall under any rule.

Table 19.2 Example of rulefile used to classify the inclusions detected by ASPEX

Class	Restrictions
Alumina	Al > 20 & O > 20 & (Mn + Si + Mg + Ca) < 10 & Mg < 2 & Ca < 2
Galaxite-alumina	Al > 10 & Mn > 2 & Si < 2 & Mg < 2 & Ca < 2
Mn-Si-highAl	Mn > 2 & Si > 2 & Al > 15 & Mg < 2 & Ca < 2
Mn-Si-lowAl	Mn > 2 & Si > 2 & Al < 15 & Al > = 1 & Mg < 2 & Ca < 2
Mn-Si	Mn > 2 & Si > 2 & Al < 1 & Mg < 2 & Ca < 2
MnO-MnS	Mn > 8 & Mn/S > = 1 & (Al + Si + Ti + Cr) < (Mn + S)
CaO-CaS	Ca > 5 & Ca/S > 2
Others	True

Combined Filtration Experiments

As a last approach, combined filtration experiments were performed in the steel casting simulator. Filters with reactive coatings were first immersed into a 100 kg steel melt containing endogenous alumina inclusions. Next, the treated steel was cast through a second (active) filter, which was previously fixed in place through refractory adhesive over a second crucible, in order to physically remove the remaining inclusions from the melt without further reactions. The whole process is presented in Fig. 19.4.

For the reactive part, prismatic filters similar to those used in previous finger-tests were used. For the active filtration, carbon-bonded alumina foam filters composed of 55 wt% Al_2O_3-C and 45 wt% C (STELEX PrO, Foseco International Ltd., Tamworth, UK), a macroporosity of 10 ppi, and a size of $70 \times 70 \times 25$ mm^3 were used as substrates. The Al_2O_3-C filter was flame-spray coated with alumina using a MASTER JET flame-spray unit (Saint-Gobain Coating Solutions, Avignon, France) equipped with alumina-flexicord, which is composed of 99.7 wt% Al_2O_3. The alumina cord was fed to the flame-spray gun and melted at approximately 3120 °C. On impact with the filter substrate, the molten alumina droplets solidified and high-temperature phases were frozen. By lamination of the droplets, a dense coating was formed. All sides of the filter were manually coated for 50 s (the top and bottom faces) or 20 s (the side faces) and applying 4 plain traverses. The macro- and microstructure of a flame-spray coated Al_2O_3-C filter were presented in detail in a recent study by Gehre et al. [10].

After each experiment, 10 steel samples extracted from different regions of the steel cake were analyzed by ASPEX and the results averaged. Numerical simulations indicated that reactive cleaning is beneficial to enhance inclusion removal and that it dominates over active filtration [11]. Finally, computer models suggested the introduction of finger-like struts on the downstream surface of ceramic filters to improve the filtration efficiency [12]. Based on this, 3D-printed polyurethane foams with defined geometry were used as templates for the replica technique. Carbon-bonded

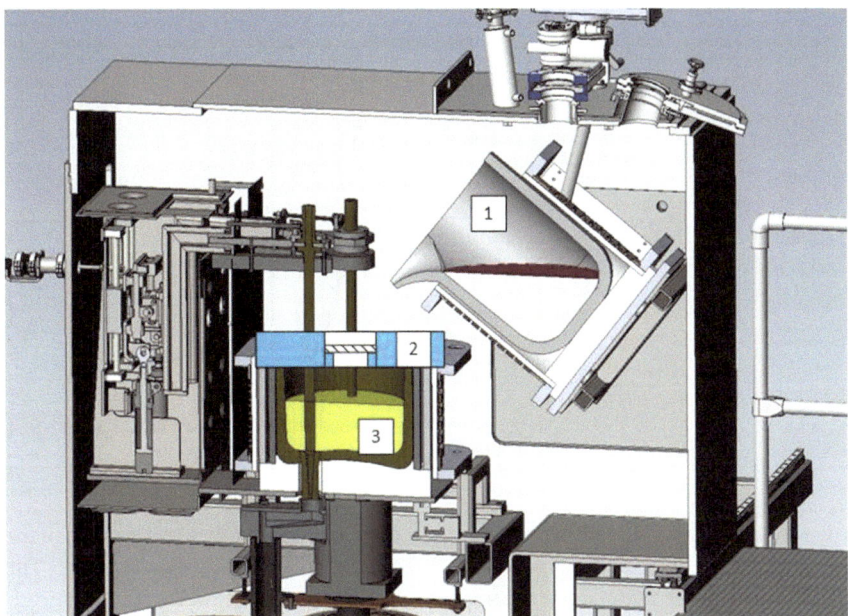

Fig. 19.4 Schematic diagram of the combined filtration test setup. (1) Melting crucible for the immersion of reactive filter; (2) Sample holder with $70 \times 70 \times 25$ mm^3 active filter; (3) Second crucible for the solidification of the steel melt

alumina filters with different coatings were produced from such templates and tested in the steel casting simulator.

19.3   Results and Discussion

19.3.1   Casting Through Nozzles

In spite the fact that both nozzles had approximately the same starting inner diameter, different results of casted steel mass versus time were obtained. After an initial phase of temperature adjustment and filling the test zones, differences in the mass flow could be registered from about 60 s (Fig. 19.5) [1]. The smaller amount of casted steel versus time in the case of the alumina-coated nozzle was caused by a more intense clogging of nonmetallic particles at the inner surface of the tested nozzle, which was macroscopically observed after the test. The clogging layer inside the alumina-coated nozzle consisted of two areas, a dense zone with thickness of about 35 μm and a coral-like zone with thickness in the range 60–80 μm. Investigations by SEM/EDS and EBSD of the clogged particles identified compositions such as spinel and alumina. In case of the mullite-coated nozzle, no remarkable difference of the steel

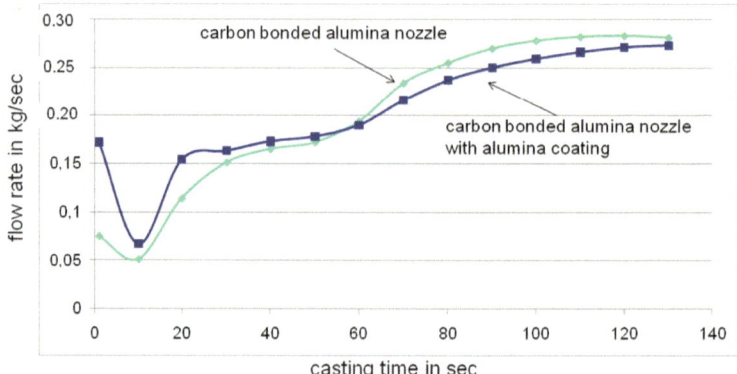

Fig. 19.5 Flow rates of molten steel through carbon-bonded nozzles as function of time. Reproduced with kind permission from [1]

flow with time was detected. Several samples were taken from different regions of the test nozzle. Depending on the position, different morphologies were observed. The coating of a sample taken from the upper part of the nozzle showed a highly porous morphology without any fine particles as present in the coating before the steel contact (Fig. 19.6a). The composition of the coating approached Al_2O_3 with some traces of SiO_2 and some elements from the steel melt such as manganese, iron, and chromium. In contrast, a sample taken from the lowest part of the nozzle (Fig. 19.6b) presented a partial delamination of the former coating from the alumina-carbon substrate and pores with size in the range of 20–50 μm. The former mullite coating consisted of dense parts composed of two phases, darker grains (under BSE) embedded in a light gray matrix, metal particles as well as closed and open pores. An adhesion of exogenous and endogenous particles could not be detected. EDS analysis showed that the gray particles had nearly Al_2O_3 composition with some traces of silicon. The lighter continuous phase consisted of silicon, manganese and aluminum with traces of sodium, titanium, iron, and magnesium. Element mapping demonstrated that the alumina grains were surrounded by silicon and manganese, while iron was mainly present as separate particles. The investigations of samples between 0 and 1 showed a mixture of the previous results, with less pronounced delamination and lower amounts of the Mn-Si-Al-containing phase. The nozzles used in the experiments were scanned by CT before and after the test in the steel casting simulator. Only the presence of small steel particles in the clogging layer allowed to distinguish between the alumina-carbon matrix and the decomposed coating, due to the enrichment of Mn and Fe in the decomposed mullite layer. Delamination and partial destruction of the former coating was clearly visible through CT. In conclusion, a mullite coating on a carbon-bonded alumina component is not suitable for applications at >1520 °C due to the decomposition of mullite. On the other hand, a pronounced clogging for the nozzles with the active alumina coating was observed. Correlating these clogging values to a type of "filtration efficiency" in means of more particles to be deposited

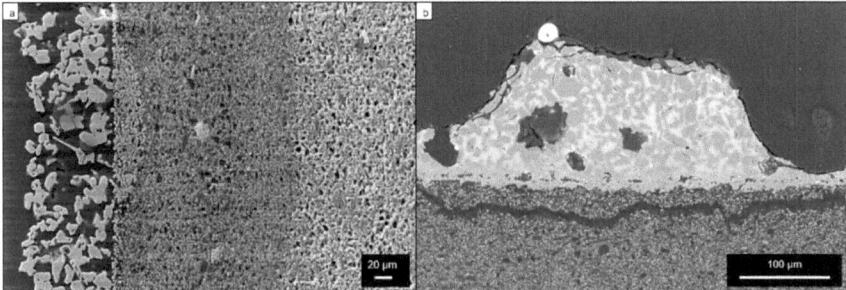

Fig. 19.6 a Highly porous coating (former mullite) on alumina-carbon nozzle (top part) after melt contact. **b** Partially delaminated and decomposed mullite coating (bottom part) after steel contact. Adapted with permission from [5]

on the ceramic collector surface (nozzle or filter), there was a first indication that at least a higher filtration efficiency is expected in case that carbon-bonded filters are coated with an alumina active coating.

19.3.2   Finger-Tests

The finger-tests used carbon-bonded alumina filters with active or reactive coatings. Despite the strong thermal shock during the immersion itself as well as during the rapid cooling process, the filter samples did not show macroscopic damage after the experiment. By comparing the color of the sample surfaces before and after the immersion test, it was observed that no oxidation occurred on the areas without direct melt contact. The change from black to grayish color (Fig. 19.7a and b) in the areas with direct steel contact was caused by several reactions, resulting in decarburization in areas near the surface due to the high solubility of carbon in the steel melt, the interaction between steel, the carbon, and the refractory oxide, and by the deposition of a newly formed phases.

It is worth to mention that the trends for temperature and dissolved oxygen during all performed immersion tests were nearly the same. Table 19.3 presents the values for temperature and dissolved oxygen measured before and after oxidation (the addition of an iron oxide mixture), after deoxidation with aluminum, and after an immersion test for 3 different coated filters [2]. It could be seen that the process of increasing and decreasing the oxygen content by the addition of iron oxide and deoxidation with the aluminum metal was sufficiently reproducible. However, carbon loss in the steel was measured by spark emission spectroscopy after the experiments, in comparison to the delivered steel quality.

Investigations of the filter samples by digital light microscopy delivered interesting results. In general, all filter surfaces were microcrack-free before the immersion

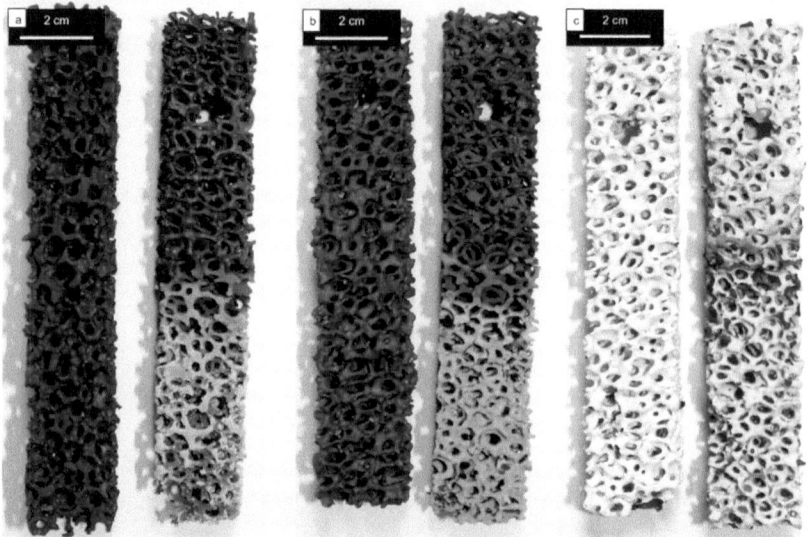

Fig. 19.7 Carbon-bonded alumina filters before and after finger-test. **a** uncoated Al_2O_3-C. **b** MgO-C coated sample. **c** alumina-coated sample. Adapted with permission from [2]

Table 19.3 Measured temperature and dissolved oxygen levels at several steps of the finger-tests. Adapted with permission from [2]

Step	Substrate material	Al_2O_3-C			References
	Coating material	No coating	MgO-C	Al_2O_3	
Before oxidation	T [°C]	1656	1657	1651	1648
	Dissolved O [ppm]	15	13	13	15
After oxidation	T [°C]	1646	1645	1643	1647
	Dissolved O [ppm]	57	55	56	63
After deoxidation	T [°C]	1651	1650	1645	1660
	Dissolved O [ppm]	9	8	8	12
After immersion	T [°C]	1657	1665	1660	–
	Dissolved O [ppm]	22	22	17	–

procedure. After the test, newly formed and loose crystalline phases were found on the surface of uncoated carbon-bonded alumina filters. SEM investigations of carbon-bonded alumina filters after immersion in a steel melt showed agglomerated particles in various shapes, such as platelets and dendrites, which are typical for endogenous inclusions as reported by Dekkers [13, 14]. When testing carbon-bonded alumina filters, the material buildup observed after an immersion test consisted of different structures (from the center of a strut to the surface), reported in several publications [2, 8, 15–19]:

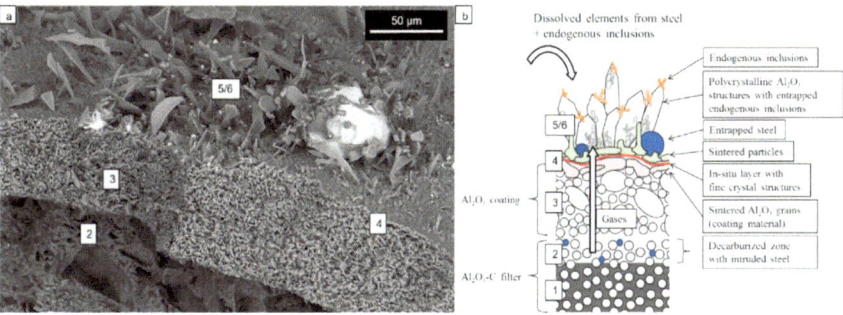

Fig. 19.8 a Surface of alumina-carbon filter with porous alumina coating after finger-test (60 s immersion). **b** Typical layer buildup on carbon-bonded alumina material after steel contact. Adapted with permission from [18]

1. Unaffected carbon-bonded substrate;
2. Decarbonized layer with partially sintered alumina;
3. Oxide porous functional coating–from the production process (optional);
4. Secondary, thin alumina-containing layer resulting from the in situ reaction;
5. Dense collection zone consisting of sintered, polyhedral endogenous alumina inclusions;
6. "Coral-like" collection zone consisting of endogenous alumina particles in complex shapes (Fig. 19.8).

A similar layer buildup was observed throughout all finger-test samples, provided the coating (if present) was porous enough to allow for diffusion of elements from and into the filter material. A thorough discussion of the interactions taking place during melt contact is given in the next section. Zone 1 represents the unreacted carbon-bonded alumina filter material, composed of approximately 70 wt% alumina with typical grain size of 0.5–0.8 μm (d_{50}) and 30 wt% carbon as bonding matrix. Zone 2 consists of alumina grains with typical size in the range 500 nm to 1 μm, partially sintered together into a porous layer with a typical thickness of 5–10 μm. Iron particles were observed at the border between the first 2 zones. The secondary layer (4) has a thickness usually in the 100–400 nm range, sometimes up to 1 μm. It seems to be totally dense even at high magnifications and it reflects the topography of the zone below. This secondary layer is extremely efficient for collecting inclusions with similar chemistry from the melt, and its roughness increases the wetting angle with the steel melt, possibly contributing to the deposition of inclusions too. On top of this layer, the clogging zone can usually be divided into two parts. The first one is dense and consists of relatively small plate-like particles. This is followed by a zone with larger, loose particles in complex shapes.

The first reactive coating consisted of MgO-C, which should generate gaseous Mg when in contact with a steel melt and in turn remove the oxygen by creating secondary MgO [2]. By means of digital light microscopy a dense, crack-rich layer was detected on the filter surface after the immersion test. In addition, the carbon-bonded coating turned to a grayish-white color due to reactions with the steel melt.

Fig. 19.9 **a** Surface of MgO-C coated sample after finger-test. **b** Newly formed spinel whiskers. Adapted with permission from [2]

Under SEM a dense, in situ-formed fine-grained and carbon-free MgO layer with some MgO particle agglomerates on the surface was followed by a porous (3) zone containing some carbon, primary magnesia, and iron droplets. EDX investigations of the elemental composition of the newly formed dense layer as well as the agglomerated particles showed that both consisted primarily of magnesium and oxygen. In addition, newly formed whiskers consisting of $MgAl_2O_4$ were found inside the hollow struts generated from the burnout of the polyurethane foam used as template during the filter production (see Fig. 19.9).

Carbon-bonded calcium aluminates were also tested as highly reactive coatings [20]. After the immersion test, a lot of pores infiltrated with steel (white spots) were found over the CA2-C surface. The rest of the surface consisted of polyhedral particles partially sintered together (see Fig. 19.10). The EDS detector revealed Al, Ca, and O as main constituents of this zone. The CA6–C surface was smoother and did not show any major porosity with entrapped steel (see Fig. 19.11). The difference should be related to the different wettability and the greater potential for the melt to penetrate CA2 than CA6. In addition, gaseous calcium and/or suboxides may be produced from the reaction between the calcium aluminate and the steel melt, with the formation of pores, which would be readily infiltrated by liquid metal. Given the different amount of calcium oxide in the compositions (three times higher in one mole of CA2), this effect should be much stronger in the case of the CA2–C sample than CA6–C. As usual, the applied coating was still detected with a thickness of approximately 40 μm. In some spots, a thin secondary layer was also observed on which endogenous inclusions were collected. Thermodynamic calculations showed that the reduction of CaO from the coatings to Ca vapor should proceed at a slow rate.

Nano-sized materials were also selected for their high specific surface area (thus reactivity) and used in combination with a binder to produce reactive coatings [8, 15–17, 19, 21, 22]. The used water-based slurries had a low solids content, hence very thin coatings were obtained. Due to this, the high reactivity of the nano-sized materials and of the residual carbon from the binders, no residuals from the nano-based coatings were detected on any filter after a finger-test, even for very short (10 s)

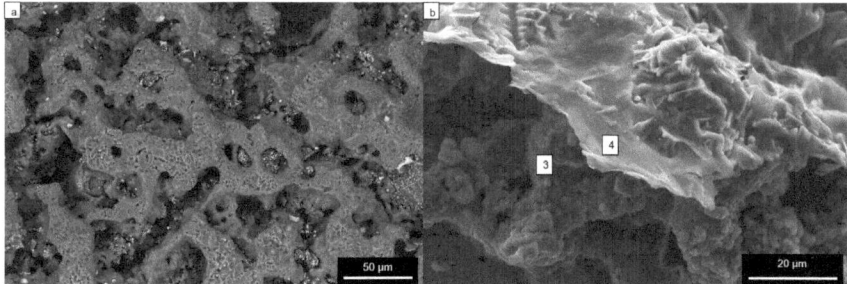

Fig. 19.10 CA2-C filter surface after finger-test. **a** Surface overview. **b** Detail with thin secondary layer. Adapted with permission from [20]

Fig. 19.11 CA6-C filter surface after finger-test. **a** Cross section with coating layer. **b** Detail of thin layer with collected endogenous inclusions. Adapted with permission from [20]

immersion time. For this reason, after contact with a steel melt, the surface of nano-coated filters showed exactly the same features as uncoated carbon-bonded alumina filters (see Fig. 19.12). However, differences in the development of the dense and loose collection zones (5 and 6) were observed, with the nano-coated filters generally showing an increased clogging for the same testing time. It can be assumed that the presence of a nano-based coating before the test resulted in stronger interactions with the steel melt and a faster formation of the secondary (4) layer, which in turn promoted the collection of endogenous inclusions from the melt.

Regarding active coatings, pure alumina was applied by spray coating at first [2, 18]. Such filters were treated at 1400 °C to allow for sintering of the alumina coating. The coating thickness varied between 10 and 100 μm. The determined cold crushing strength of coated filters was $(0.58 \pm 0.08$ N$)/$mm^2, which is about double that of uncoated ones $(0.27 \pm 0.04$ N$)/$mm^2. After the experiment, this coating was still present on the filter surface, although cracks due to thermal shock were detected. They were the more pronounced the longer the immersion time. Frozen steel in the form of droplets was also found as usual. From SEM investigations, the secondary layer (4) with a thickness of a few hundred nm was already detected after a 10 s immersion. Filters immersed for 60 s showed plate-like particles and also

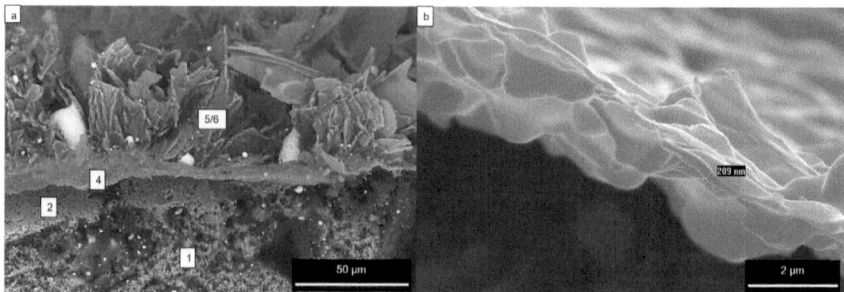

Fig. 19.12 Nano-coated alumina-carbon filter after finger-test (30 s immersion): **a** Cross section view. **b** Detail of the secondary layer (4). Adapted with permission from [16]

fine inclusions in clusters on top of these in some areas (Fig. 19.8). After 120 s, the clusters completely covered the investigated filter surface. EBSD investigations of the thin secondary layer suggested that it consisted of tiny crystals embedded in an amorphous phase. Plate-like particles consisted of trigonal alpha-alumina. Grain orientation within these particles was determined, confirming that they did not grow directly from the coating material but more likely consisted of endogenous inclusions that sintered together due to the high process temperatures. Furthermore, carbon-bonded alumina filters were coated using the flame spray technique, in which a ceramic feedstock is melted at approximately 3160 °C and a dense coating (90 μm) is produced. Very high cooling rates (10^6 K/s) resulted in metastable cubic γ-alumina that was frozen at room temperature. Another difference with cold spraying, thermal spray leads to inhomogeneous coatings due to the difficulty in reaching the filter core. It was observed that the oxygen content of the steel remarkably decreased after immersion of the flame-coated filter, indicating a strong interaction with the melt. Moreover, the alumina coating was extensively covered by a continuous crystalline layer with an average thickness of 45 μm. Compared to a cold-sprayed filter, the microstructure was overall much denser, with plate-like particles covering nearly the whole surface. In this study, no evidence of an in situ formed thin layer was found. It is not surprising that this layer was absent as its formation is promoted by raw material impurities like SiO_2, Na_2O, and K_2O. During the flame spraying process, the raw materials were heated above 3000 °C. At this temperature, these impurities evaporate and hence, are not transported to the filter surface, leading to an alumina coating with high purity.

Besides the investigation of ceramic components after contact with molten metal, the analysis of the solidified steel in terms of remaining inclusions was of critical importance for the subproject C01. Standard chemical analysis by means of spark emission spectroscopy was performed first, followed by automated scanning electron microscopy (ASPEX).

Regarding the steel chemistry after contact with carbon-bonded components, the first experiments involving both uncoated and alumina-coated nozzles already showed some interesting trends [1]. In particular, the Al content decreased in all ingots, suggesting the possible generation of endogenous alumina inclusions. In addition, other elements such as C, Mn, Ni, S and P were also present in lower amounts after the experiments. Further studies involving the use of carbon-bonded alumina filters (Table 19.4) confirmed the trend of aluminum and carbon loss, regardless of the presence and type of coating [2, 15, 18, 19]. MnS inclusions were found in some cases on the filters' surface, as well as regularly in the solidified steel [16, 19, 23, 24]. Sulfides, and particularly MnS, are common impurities in steel and can explain the observed depletion of both elements.

The first ASPEX investigations were used to compare the performance of uncoated carbon-bonded alumina versus nano-coated filters, due to the promising clogging results. One steel melt was used as a reference, in which no filter was immersed. Both uncoated and nano-coated filters showed positive effects: the density of particles with area < 20 μm^2 and especially < 3 μm^2 decreased significantly [19]. On the other hand, the amount of inclusions larger than 50 μm^2 slightly increased. The results were also supported by light microscopy investigations. The nano-coated filter performed significantly better than the uncoated one, especially for a 5 min immersion time. Regarding the chemistry, the majority of inclusions consisted of alumina, followed by MnS (due to the high S content of the steel) and Mn-spinel. Very good filtration efficiency (up to 90% were registered). Despite the good performance of nano-coated filters, these also generated new inclusions (classified as mixed silicates) which were otherwise not detected for the other experiments. In addition, especially after 10 s immersion, the coated filters produced a high amount of coarse particles, which are most detrimental for the fracture toughness of steel. This particular behavior of nano-coated filters was observed in other studies as well [17, 25] (Tables 19.5 and 19.6).

Table 19.4 Steel composition as delivered, of reference melt and after immersion (60 s) of samples. Adapted with permission from [2]

Element	Steel as delivered	Reference melt	No coating	MgO-C	Al$_2$O$_3$
Fe	96.74	96.88	96.87	96.99	97.00
C	**0.400**	**0.372**	**0.359**	**0.365**	**0.366**
Cr	0.995	1.030	1.013	1.063	1.053
Mn	0.781	0.690	0.696	0.692	0.670
Mo	0.197	0.187	0.193	0.182	0.178
Ni	0.200	0.187	0.204	0.155	0.149
Cu	0.290	0.267	0.283	0.204	0.207
Si	0.249	0.196	0.211	0.195	0.207
Al	**0.018**	**0.032**	**0.008**	**0.017**	**0.017**
P	0.011	0.013	0.012	0.011	0.013
S	0.032	0.059	0.052	0.041	0.053
Bal	0.090	0.087	0.189	0.085	0.470

Table 19.5 Density of inclusions found in the solidified steel samples by ASPEX, classified by size. Reproduced with permission from [19]

Inclusion area (μm²)	Inclusions per cm²													Total
	0.1–1	1–3	3–5	5–10	10–20	20–30	30–50	50–80	80–130	130–200	200–500			
No filter	1179	1045	327	417	444	108	40	19	4	1	1			3585
Uncoated 10 s	21	152	152	195	118	23	16	9	3	0	0			689
Nano-coated 10 s	6	110	108	106	79	44	44	58	63	18	8			644
Uncoated 300 s	57	274	129	168	199	103	83	29	14	3	8			1067
Nano-coated 300 s	3	25	35	62	59	58	68	30	11	16	4			371

Table 19.6 Density of inclusions found in the solidified steel samples by ASPEX, classified by chemistry. Reproduced with permission from [19]

Class	Inclusions per cm^2				
	No Filter	Uncoated 10 s	Nano-coated 10 s	Uncoated 300 s	Nano-coated 300 s
Al$_2$O$_3$	942	596	246	488	275
Mn spinel	63	50	6	54	6
Ca aluminate	0	0	1	0	0
Mg spinel	0	0	0	0	0
Al-Mn-Mg-Fe-Ca silicate	0	1	265	1	52
SiO$_2$	45	1	5	8	0
MnO-MnS	2295	13	10	372	15
CaO-CaS	8	6	5	0	1
Others	232	22	106	144	22

Filters with cold-sprayed alumina coatings were tested for different immersion times, from 10 up to 120 s. The proportions of different chemical as well as size classes remained similar for all experiments. Filtration efficiencies in the 50–70% range were registered [18]. Interestingly, with increasing immersion time, there was a trend to more inclusions remaining in the steel, together with a growing amount of SiO$_2$ particles. On the other hand, neither Ca-aluminate, Mg-spinel, Al– Mn–Mg– Fe–Ca–silicate nor CaO–CaS type inclusions were identified in the steel samples. In addition, the density of inclusions > 50 μm slightly increased when immersing alumina-coated filters.

A further study involving both nano-coated and alumina-coated filters showed that the majority of inclusions detected by ASPEX was located in the 0.1–20 μm^2 area region, with a peak at 1–3 μm^2. The use of a filter had always a positive impact on the steel purity, but the effect decreased with increasing immersion time [17]. In addition, particle growth was observed again for the nano-coated filters.

A study involving carbon-bonded calcium aluminate coatings showed the same trends for inclusion size distributions [20]. Most of the detected particles were classified as alumina as expected. Filtration efficiencies close to 100% were registered. However, the steel samples in this case were extracted directly from the steel melt (before and after filter immersion), so the results cannot be directly compared.

19.3.3 Combined Filtration Experiments

SEM investigations on nano-coated and CA2-C coated filters showed very similar features to what reported previously, since the conditions during the immersion were similar to those of simple finger-tests. In case of the active filters instead, at the

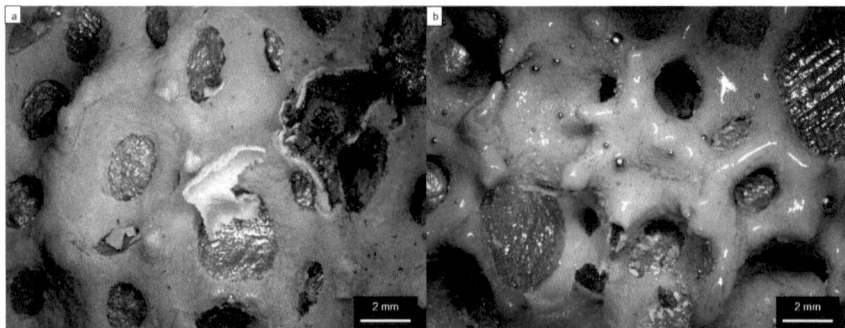

Fig. 19.13 Digital light micrographs of the active filters (bottom side) after combined filtrations with **a** CA2-C and **b** nano-coated filters, respectively

end of the process, some of the steel did not manage to escape the macropores and froze. As a consequence, any remaining melt also solidified directly on top of the filter, which could hardly be analyzed. Generally, the bottom part showed the flame-sprayed alumina coating with collected inclusions and steel droplets. Light microscopy pictures are presented in Fig. 19.13. The alumina coating survived the impact of the molten metal and the thermal shock for the most part, but in some areas it cracked and was lost into the melt or during handling after the experiment. A glassy phase apparently covered the alumina coating after multiple experiments. The bottom surface of the sandwich filter seemed quite rough, with a lot of inclusions over the alumina coating. Since the top portion of the filter consisted of uncoated carbon-bonded alumina, this material reacted on contact with the steel melt and the resulting endogenous inclusions were immediately forced through the bottom (coated) part, in a way that a high number of them was collected on the filter surface.

Under SEM, the surface of the other three samples appeared for the most part very smooth (as suggested by light microscopy investigations), with some interesting formations on top. These usually consisted of steel droplets resting over inclusions with a different composition (see Fig. 19.14b). The EDS signal revealed the presence of glass-forming elements such as Si and P. In particular, about 70% of the smooth layer consisted of SiO_2, with the rest made up of alumina, Na_2O and minor amounts of other oxides. On the other hand, the inclusions most likely consisted of iron phosphate or a similar phase. These observations confirm the presence of a glassy layer on the surface which was speculated before. It was not possible to determine the thickness of such layer in cross section, since it merged with the alumina coating underneath. From multiple images from different filters it was estimated to be at least a few micrometers thick. Since only a thin slag layer is produced during the immersion tests, this layer most likely originated from the impurities contained in the refractory adhesive used to keep the active filter in place. Small grains can be observed within the glassy layer at high magnifications, but the fast cooling of the filter after casting of the steel melt probably did not leave enough time for a complete crystallization. This particular microstructure was observed for the first time on coated carbon-bonded

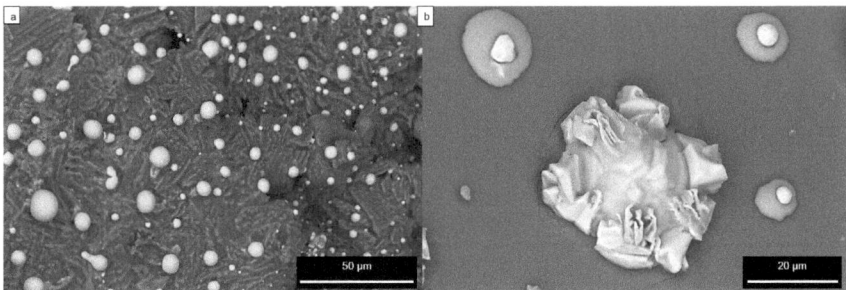

Fig. 19.14 Scanning electron micrographs of the active filters after combined filtrations. **a** Nano-coated filter. **b** No filter immersed

filters in this work. Previous studies only focused on immersion experiments, in which no casting process was involved.

As in simple finger-tests, with combined filtration the amount of carbon decreased considerably for every test (Table 19.7). Interestingly, when performing only the immersion (which is a much faster experiment), the final carbon content in the steel was similar. On the other hand, loss of Cr was much larger in this case. Some Mn was also lost in all experiments, and especially when no filter was immersed. The amounts of other elements such as M, P and S were not affected significantly. Si was lost in all experiments, contributing to the formation of the glassy layer on the active filters (see above) and also to the creation of complex inclusions. Aluminum was strongly depleted whenever a filter was immersed, which is in agreement with previous studies [20]. Most likely this element was involved in the creation of new alumina inclusions during contact between the reactive filter and the steel melt. When no reactive filter was immersed, the Al content was not remarkably affected. The total oxygen content increased for all experiments, and particularly for the combined filtrations.

From the ASPEX size classification, it is clear that all curves tend to follow a log-normal distribution (Fig. 19.15). The combined filtration experiments resulted in larger inclusions on average compared to the reference test. As expected, the majority of inclusions consisted of alumina due to the steel treatment before immersion of the reactive filter, followed by MnS. Some improvement in comparison to simple finger-tests performed under similar conditions were noticed. However, less inclusions were detected in the solidified steel in the case no reactive filter was immersed, i.e. when only casting through the active filter. During the combined filtration experiments, after extraction of the reactive filter, the crucible was immediately tilted and the steel melt cast through the second filter. In this way, any new inclusions which were not already collected at that point were cast together with the steel melt and did not have time to escape to the slag layer via flotation. For this reason, further experiments were performed, with pure alumina filters floating on the steel melt to promote inclusion removal.

Table 19.7 Chemical composition (wt%) of the solidified steel samples obtained by spark emission spectroscopy (O obtained by gas fusion analysis and C by combustion and infrared absorption). "I" indicates the immersion step, "F" indicates the following casting. "CA"—Calcium aluminate coating; "NM"—Nano-based coating; "A"—Flame-sprayed alumina; "S"—Flame-sprayed alumina (bottom half only); "0"—No filter. All coatings applied on pre-coked carbon-bonded alumina filters

Element			Steel sample				
	$I_{CA}\_F_A$	$I_{NM}\_F_A$	$I_0\_F_A$	$I_0\_F_S$	$I_0\_F_0$	I_{CA}	As received
C	0.30	0.29	0.29	0.28	0.30	0.28	0.41
Cr	0.95	0.96	0.95	0.94	0.93	0.84	0.96
Mn	0.65	0.68	0.59	0.59	0.59	0.62	0.77
Mo	0.15	0.16	0.16	0.16	0.16	0.16	0.16
Si	0.18	0.19	0.19	0.21	0.18	0.19	0.27
Al	<0.001	<0.001	0.019	0.019	0.022	<0.001	0.029
Ti	0.0013	0.0011	<0.0018	<0.0021	<0.015	<0.019	0.0044
O	0.0038	0.0037	0.0033	0.0026	0.0026	0.0024	0.0022
Fe	97.54	97.49	97.49	97.68	97.59	97.62	97.36

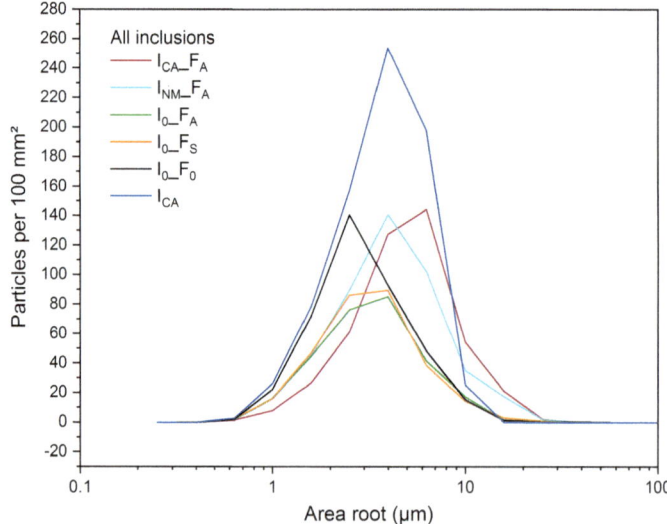

Fig. 19.15 Density of inclusions detected by ASPEX in the steel samples as function of size

19.4 Discussion

19.4.1 Effects on Filter Materials

As mentioned above, one characteristic layer buildup was observed on almost all tested filters, regardless of the presence and type of coating employed. It appears that the in situ formed thin layer can be amorphous, crystalline (α-Al_2O_3) or a mixture of amorphous and crystalline, which depends on the level of impurities of ceramic raw materials, impurities in steel, and alloy elements [18]. Focused ion beam investigations detected only the structure of alpha alumina, despite the variable composition [9]. Other authors detected a vitreous phase on the surface of a decarburized Al_2O_3-C submerged nozzle after contact with steel [26]. They reported that this phase consisted of alumina, silica, and alkali oxides. According to the authors, the aluminum activity in steel increased with the amount of carbon dissolved. The resulting CO gas was then dissolved and provided oxygen for the reoxidation of Al and deposition of alumina. Regarding the clogging layers, plate-like particles clearly consisted of trigonal alpha-alumina. Grain orientation within these particles was determined, confirming that they did not grow directly from the coating material but more likely consisted of endogenous inclusions that sintered together due to the high process temperatures [18]. Small crystals as well as collected inclusions might act as nuclei for the polycrystalline structures observed on the filters' surface.

In presence of a reactive or porous active coating, reaction products and gases can easily move to the filter/steel interface. Consequently, two mass transports probably contribute to the layer buildup observed in many studies: on the one hand, products (e.g., CO, alumina sub-oxides, gaseous Mg and Ca) originating from the carbon-bonded substrate, and, on the other hand, endogenous inclusions as well as dissolved elements from the steel. Regarding the actual reactions taking place, different mechanisms were proposed. Some authors claim that a carbothermic reduction of alumina (with iron as a catalyst) could provide the necessary alumina sub-oxides and carbon monoxide, which migrate through the coating and into the steel melt, remaining available for further reactions [2, 27, 28]. However, thermodynamic studies by Zienert et al. indicated that carbothermic reduction of alumina might not occur under the given experimental conditions [29]. They suggested, instead, a partial dissolution of alumina into the molten steel. No stable equilibrium between alumina and iron liquid is reached because of the constant production of CO gas from the dissolved oxygen and the presence of carbon. Al supersaturation in proximity of the filter surface results in the reprecipitation of alumina. It was observed that the secondary thin layer thickness and the polycrystal size did not increase with immersion time. Aluminum and oxygen showed a concentration gradient through the layer thickness, suggesting that the formation of this structure is at some point limited by diffusion [9]. This would explain why this layer always shows a thickness <1 μm, regardless of the type of functionalization (active or reactive) employed and the immersion time. In addition, no evidence of such a layer was found on filters with dense coatings prepared by flame-spraying. On the other hand, several studies showed that the amount of platelets or particles with complex shapes ("clogging phase") which are collected on this layer slowly increases with the immersion time [17, 19]. The progress of clogging suggests that the interactions at the filter/steel interface still take place, but at a much slower rate. At some point in time, it is possible that dissolution and reprecipitation reach some kind of equilibrium. This would explain why filters delivered a worse performance with longer contact times. This is where the filter behavior changes from "reactive" to "active", according to the definitions proposed by the CRC 920. Endogenous inclusions from the melt are now mainly attracted by the filter surface due to the similar chemistry. The increased roughness and specific surface benefit the filtration process.

19.4.2 Effect on Steel Cleanliness and Filtration Efficiency

From several immersion tests, the Al content as well as the amount of alumina inclusions decreased in all steel samples for short contact times. In addition, other elements such as C, Mn, Ni, S and P were also present in lower amounts after the experiments. With increasing immersion time (60 + s), a higher inclusion density was registered and the filtration efficiency dropped. This was probably due to detachment of particles from the filter surface due to the melt flow. Especially in the case of nano-coatings, a higher amount of large inclusions was detected after the experiments. Since oxides tend to behave as nucleation sites for sulfides, a lower oxygen content (induced by the nano-additives) resulted in larger sulfides and at the same time in a lower inclusion concentration [25]. Apart from the grain growth, the strong reaction with the MWCNTs could also explain the generation of silicates. Carbon and oxygen from the filter material may influence the Si activity in the steel, leading to local saturation and precipitation of silica/silicate inclusions.

In summary, the filters were most efficient during the reactive phase. Multiple studies dealing with the turbulent melt flow driven by the electromagnetic force in an induction furnace were performed within the CRC 920 [30–33]. It was demonstrated that the immersion of a ceramic filter, depending on its permeability and position, has a considerable effect on the flow pattern. Moreover, the model was used to determine the filtration efficiency of ceramic filters. In comparison with the observed performance (30% up to over 95%) from the steel casting simulator experiments, low efficiency values were obtained at first [31]. However, the first simulations did not consider any reactions at the interface, nor the filter rotation. Further studies showed that the formation of CO bubbles on the surface of inclusions increases their rising velocity toward the free surface of the melt. According to the numerical results, about 30% of the starting inclusions are removed after 10 s. This can considerably increase the inclusion removal up to the observed levels and therefore improve the cleanliness of the steel melt [11, 34, 35]. On the other hand, when the bubbling effect of carbon-bonded alumina (due to evolution of CO gas) must be avoided, such as during a continuous casting process, the application of a dense flame-sprayed coating would be beneficial.

19.5 Summary

A special metal-casting simulator allowed to investigate the behavior of nozzles and filters in contact with a steel melt. Appropriate cement-free alumina castable formulations containing mullite or spinel were first developed, in order to produce various components such as nozzles, adapters, stopper rods and crucibles. Next, carbon-bonded alumina samples with or without active and reactive coatings were tested. Exogenous inclusions were introduced into the melt to better characterize the clogging behavior of carbon-bonded alumina nozzles. In comparison with an uncoated nozzle, a more intense clogging of nonmetallic particles at the inner surface of an alumina-coated nozzle resulted in a decreasing melt flow with time. In the case of a mullite coating, delamination and decomposition of the mullite phase were observed, making it unsuitable for applications at >1520 °C. In further approaches, endogenous inclusions were generated in situ by pre-oxidizing and desoxidazing the steel melt. Filters with different active and reactive coatings were immersed for different times in order to investigate the evolution of newly formed phases on the filter surface as well as characterize their cleaning performance based on the analysis of the solidified steel. Despite the different coatings, a general layer buildup was detected on carbon-bonded filters. From the observations, it was proposed that the reactions taking place at the filter/steel interface are at some point limited by diffusion, after which only active clogging proceeds. The inclusions in the frozen steel melt were characterized with the aid of an automatic SEM—ASPEX-system, which identified the chemistry as well as the size and population of the inclusions. Regarding the steel cleanliness, reactive coatings showed promising results especially for very short (10–30 s) immersion time. On the other hand, nano-based coatings promoted the formation of new inclusions with complex chemistry and also the inclusion growth. Computer simulations showed that the formation of CO bubbles (during the reactive phase) on the surface of inclusions increases their rising velocity toward the free surface of the melt, suggesting that about 30% of the starting inclusions are removed after 10 s. In the last approach, investigations of a new combined refining process based on the immersion of reactive filters and a subsequent filtration via carbon-free active filters was investigated, in order to remove the remaining clusters of inclusions. Some improvement in comparison to simple finger-tests performed under similar conditions were noticed.

Acknowledgements The authors gratefully acknowledge the financial support of the German Research Foundation (DFG, Deutsche Forschungsgemeinschaft) for funding this subproject C01 within the frame of the Collaborative Research Center 920, project number 169148856.

We would like to thank our former colleagues Dr.-Ing. Marcus Emmel, Dr.-Ing. habil. Harry Berek and Mr. David Thiele for their experimental support to the subproject. Moreover, we greatly appreciate the contribution of our colleagues at the Chair of Ceramics, in particular Dr.-Ing. Gert Schmidt, Ms. Carolin Ludewig, Dipl.-Ing. Ricardo Fricke, Mr. Lothar Lange, Mr. Udo Venus and Dipl.-Ing. Florian Kerber.

References

1. C.G. Aneziris, S. Dudczig, J. Hubálková, M. Emmel, G. Schmidt, Ceram. Int. **39**(3), 2835–2843 (2013). https://doi.org/10.1016/j.ceramint.2012.09.055
2. S. Dudczig, C.G. Aneziris, M. Emmel, G. Schmidt, J. Hubalkova, H. Berek, Ceram. Int. **40**(10), 16727–16742 (2014). https://doi.org/10.1016/j.ceramint.2014.08.038
3. S. Dudczig, Werkstoffentwicklung von Feuerbetonen für Schlüsselbauteile zur Erfassung von Wechselwirkungen zwischen Stahlschmelzen und Feuerfestmaterialien in einem Stahlguss-simulator, TU Bergakademie Freiberg (2017)
4. M. Emmel, C.G. Aneziris, Ceram. Int. **38**(6), 5165–5173 (2012). https://doi.org/10.1016/j.ceramint.2012.03.022
5. S. Dudczig, C.G. Aneziris, M. Dopita, D. Rafaja, Adv. Eng. Mater. **15**(12), 1177–1187 (2013). https://doi.org/10.1002/adem.201300121
6. M. Emmel, C.G. Aneziris, J. Mater. Res. **28**(17), 2234–2242 (2013). https://doi.org/10.1557/jmr.2013.56
7. M. Emmel, Development of active and reactive carbon bonded filter materials for steel melt filtration, TU Freiberg (2014)
8. E. Storti, Functionalization of carbon-bonded ceramic foam filters with nano-scaled materials for steel melt filtration, TU Bergakademie Freiberg (2018)
9. E. Storti, H. Berek, C.G. Aneziris, Ceram. Int. **44**(12), 14502–14509 (2018). https://doi.org/10.1016/j.ceramint.2018.05.065
10. P. Gehre, A. Schmidt, S. Dudczig, J. Hubálková, C.G. Aneziris, N. Child, I. Delaney, G. Rancoule, D. DeBastiani, J. Am. Ceram. Soc. **101**(7), 3222–3233 (2018). https://doi.org/10.1111/jace.15431
11. A. Asad, R. Schwarze, Steel Res. Int. **92**(11), 1–11 (2021). https://doi.org/10.1002/srin.202100122
12. H. Lehmann, E. Werzner, A. Malik, M. Abendroth, S. Ray, B. Jung, Adv. Eng. Mater. **24**(2), 2100878 (2022). https://doi.org/10.1002/adem.202100878
13. R. Dekkers, B. Blanpain, P. Wollants, F. Haers, C. Vercruyssen, B. Gommers, Ironmak. Steelmak.. Steelmak. **29**(6), 437–444 (2002). https://doi.org/10.1179/030192302225004584
14. R. Dekkers, B. Blanpain, P. Wollants, Metall. Mater. Trans. B **34**(2), 161–171 (2003). https://doi.org/10.1007/s11663-003-0003-3
15. E. Storti, M. Emmel, S. Dudczig, P. Colombo, C.G. Aneziris, J. Eur. Ceram. Soc. **35**(5), 1569–1580 (2015). https://doi.org/10.1016/j.jeurceramsoc.2014.11.026
16. E. Storti, S. Dudczig, G. Schmidt, P. Colombo, C.G. Aneziris, J. Eur. Ceram. Soc. **36**(3), 857–866 (2016). https://doi.org/10.1016/j.jeurceramsoc.2015.10.036
17. E. Storti, S. Dudczig, A. Schmidt, G. Schmidt, C.G. Aneziris, Steel Res. Int. **88**(10), 1700142 (2017). https://doi.org/10.1002/srin.201700142
18. A. Schmidt, A. Salomon, S. Dudczig, H. Berek, D. Rafaja, C.G. Aneziris, Adv. Eng. Mater. **19**(9), 1700170 (2017). https://doi.org/10.1002/adem.201700170
19. E. Storti, S. Dudczig, J. Hubálková, A. Gleinig, A. Weidner, H. Biermann, C.G. Aneziris, Adv. Eng. Mater. **19**(9), 1700153 (2017). https://doi.org/10.1002/adem.201700153
20. E. Storti, M. Farhani, C.G. Aneziris, C. Wöhrmeyer, C. Parr, Steel Res. Int. **88**(11), 1700247 (2017). https://doi.org/10.1002/srin.201700247
21. O. Jankovský, E. Storti, K. Moritz, B. Luchini, A. Jiříčková, C.G. Aneziris, J. Eur. Ceram. Soc. **38**(14), 4732–4738 (2018). https://doi.org/10.1016/j.jeurceramsoc.2018.04.068
22. E. Storti, S. Dudczig, M. Emmel, P. Colombo, C.G. Aneziris, Steel Res. Int. **87**(8), 1030–1037 (2016). https://doi.org/10.1002/srin.201500446
23. A. Weidner, D. Krewerth, B. Witschel, M. Emmel, A. Schmidt, J. Gleinig, O. Volkova, C.G. Aneziris, H. Biermann, Steel Res. Int. **87**(8), 1038–1053 (2016). https://doi.org/10.1002/srin.201500462
24. R. Wagner, A. Schmiedel, S. Dudczig, C.G. Aneziris, O. Volkova, H. Biermann, A. Weidner, Adv. Eng. Mater. **24**(2), 2100640 (2022). https://doi.org/10.1002/adem.202100640

25. S. Henschel, J. Gleinig, T. Lippmann, S. Dudczig, C.G. Aneziris, H. Biermann, L. Krüger, A. Weidner, Adv. Eng. Mater. **19**(9), 1700199 (2017). https://doi.org/10.1002/adem.201700199
26. J. Poirier, Metall. Res. Technol. **112**(4), 410 (2015). https://doi.org/10.1051/metal/2015028
27. R. Khanna, S. Kongkarat, S. Seetharaman, V. Sahajwalla, ISIJ Int. **52**(6), 992–999 (2012). https://doi.org/10.2355/isijinternational.52.992
28. R. Khanna, M. Ikram-Ul Haq, Y. Wang, S. Seetharaman, V. Sahajwalla, Metall. Mater. Trans. B. **42**(4), 677–684 (2011). https://doi.org/10.1007/s11663-011-9520-7
29. T. Zienert, S. Dudczig, O. Fabrichnaya, C.G. Aneziris, Ceram. Int. **41**(2), 2089–2098 (2015). https://doi.org/10.1016/j.ceramint.2014.10.004
30. A. Asad, C. Kratzsch, S. Dudczig, C.G. Aneziris, R. Schwarze, Int. J. Heat Fluid Flow **62**, 299–312 (2016). https://doi.org/10.1016/j.ijheatfluidflow.2016.10.002
31. A. Asad, E. Werzner, C. Demuth, S. Dudczig, A. Schmidt, S. Ray, C.G. Aneziris, R. Schwarze, Adv. Eng. Mater. **19**(9), 1700085 (2017). https://doi.org/10.1002/adem.201700085
32. A. Asad, M. Haustein, K. Chattopadhyay, C.G. Aneziris, R. Schwarze, Jom. **70**(12), 2927–2933 (2018). https://doi.org/10.1007/s11837-018-3117-4
33. A. Asad, K. Bauer, K. Chattopadhyay, R. Schwarze, Metall. Mater. Trans. B Process Metall. Mater. Process. Sci. **49**(3), 1378–1387 (2018). https://doi.org/10.1007/s11663-018-1200-4
34. A. Asad, C.G. Aneziris, R. Schwarze, Numerical investigation of filtration influenced by microscale CO bubbles in steel melt. Adv. Eng. Mater. **22**(2), 1900591 (2020). https://doi.org/10.1002/adem.201900591
35. A. Asad, H. Lehmann, B. Jung, R. Schwarze, Adv. Eng. Mater. **24**(2), 2100753 (2021). https://doi.org/10.1002/adem.202100753

Chapter 20
Decopperization by Utilization of the Filter

Xingwen Wei and Olena Volkova

20.1 Introduction

Steel scrap is a source of metal supplement, and its application should increase due to its advantages of environmental friendly. However, the insufficient recycling process led to the steel scrap usually contain copper as a common tramp element. Copper might be present in the steel scrap unintentionally, as well as it might be added to certain steels intentionally as an alloying element for reaching the required mechanical properties and increasing the corrosion effect [1–4]. The negative impact of copper on the mechanical properties of steel, especially at surface treatment processes under high temperatures has been discussed for a long time [2, 5–8]. The main difficulties for its solution are caused by the factors, including the lower melting point of copper (1083 °C) in comparison with iron (1538 °C) and an unlimited solubility of copper in liquid iron. According to the iron-copper phase diagram, no intermediate compound between iron and copper can be observed. Furthermore, iron has a stronger affinity to oxygen than copper. Thus the formation of copper oxides is difficult in the presence of iron.

Daehn et al. [9] presented how copper contamination would constrain future global steel recycling. Furthermore, Daehn et al. [10] and Sandig et al. [11] listed plenty of removal methods of copper from the steel scrap, including vacuum distillation and active filters. However, nowadays, no efficient method has been proposed for its solution. Table 20.1 summarized a bunch of methods that proposed in literatures.

Expect for the above mentioned methods, Wieliczko et al. [26] and Li et al. [27] stated the excellent efficiency of $ZnAl_2O_4$ spinel on the decopperization process

X. Wei (✉) · O. Volkova
Institute of Iron and Steel Technology, Technische Universität Bergakademie Freiberg, Leipziger Str. 34, 09599 Freiberg, Germany
e-mail: Xingwen.wei@iest.tu-freiberg.de

© The Author(s) 2024
C. G. Aneziris and H. Biermann (eds.), *Multifunctional Ceramic Filter Systems for Metal Melt Filtration*, Springer Series in Materials Science 337,
https://doi.org/10.1007/978-3-031-40930-1_20

Table 20.1 Decopperization methods

Researchers	Methods
Shimpo et al. [12]	Al_2S_3-FeS flux $FeS + Cu = Fe + CuS$
Cohen et al. [13]	Sulfide flux $FeS + Cu = Fe + CuS$
Savov et al. [14]	Evaporation of Cu
Nakazato et al. [15, 16]	Evaporation of Cu from iron silicon alloy and carbon saturated alloy
Yamaguchi et al. [17]	Fe-Pb-C
Zaitsev et al. [18]	Evaporation 100 Pa. 160 tons to 5 h
Wang et al. [19]	$FeS: Na_2S: BaO = 40:20:40$ Increase the S content of molten iron about 0.2%-0.42%
Labaj et al. [20]	The influence of copper content in Fe-Cu alloy on copper evaporation rate
Yamaguchi et al. [21]	Carbon saturated iron via Ag phase into B_2O_3 flux
Labaj et al. [22, 23]	Evaporation kinetic study for 0.509–1.518% copper in Fe-Cu alloy
Hu et al. [24]	FeO-SiO_2-$CaCl_2$ flux $CaCl_2 + Cu_2O + SiO_2 \rightarrow CuCl, Cu_3Cl_3(g) + CaO \cdot SiO_2(s)$
Uchida et al. [25]	$NaCO_3$-FeS flux $FeS + Cu = Fe + CuS$

through the formation and deposition of intermetallic Cu–Zn compound on Al_2O_3-ZnO-C materials, i.e., the carbon may reduce the ZnO oxide to Zn and O, then the released Zn would further react with Cu and forming Cu–Zn compounds that deposit on the surface of Al_2O_3. Moreover, Kim et al. [28] stated the formation of intermetallic compound Cu–Zn during the welding process between Zn coated steel and Cu electrode. The above-mentioned results indicate that $ZnAl_2O_4$ spinel might be a good solution for the decopperization.

In addition, copper anode slime is a by-product that contains the soluble impurities Cu, Se, Au, etc., and insoluble impurities such as Cu_2Se, etc. [29]. The high content of selenium in the anode slime is the source for the main selenium production in the world [30]. Moreover, Kilic et al. [30] reported that Cu_2Se is also a remaining phase of copper in anode slimes after the decopperization process. Instead of Cu_2Se phase, there are many various copper selenide, including Cu_2Se, Cu_3Se_2, CuSe, etc. More types of copper selenides can be found in the work of [31]. In the work of [32], selenium was found to form Cu_2Se and CuSe particles in the liquid iron containing Cu and Se.

The unclarified copper filtration possibility of the $ZnAl_2O_4$ material, the urgency of finding an efficient decopperization method, and the possibility of selenium in the decopperization process led to the current investigation aim to test these possible approaches for the copper removal from the steel scrap. In the present work, $ZnAl_2O_4$ substrates were contacted with the liquid Fe-Cu alloys. At meanwhile, pure alumina substrates were contacted with the liquid Fe-Cu alloys for the comparison. Further-more, selenium was added into the liquid Fe-Cu alloys at 1600 °C to verify its decopperization efficiency.

20.2 Experimental Procedure

20.2.1 The Preparation of Fe-Cu Alloys

A cold crucible induction melter (KIT) (LINN High Therm GmbH) was applied to produce Fe-Cu alloys. The in-detailed description of KIT is reported elsewhere [33]. Armco iron and pure copper (99.999%), and graphite powder were used as primary materials for the sample's preparation.

20.2.2 The Preparation of Preparation of $ZnAl_2O_4$

Pure ZnO (90/RS Carl Jäger, Hilgert, Germany) and Al_2O_3 (Martoxid®MR70, Martinswerk, Bergheim, Germany) were used as the raw materials to prepare the $ZnAl_2O_4$ substrates. The raw materials were mixed in a molar ratio of 48 mol-% Al_2O_3 and 52 mol-% ZnO. The prepared granules were pressed to substrates (diam-eter of 50 mm and a thickness of 5 mm) and further sintered. Afterward, the substrates were ground and polished on one side. The XRD investigations showed the complete conversion of Al_2O_3 and ZnO to $ZnAl_2O_4$ reaction products.

The interaction experiment between Fe-Cu alloys and $ZnAl_2O_4$ substrates were conducted in a hot stage microscope under an argon protective gas (99.999% Ar containing 2–3 ppm oxygen). The Fe-Cu alloy samples with a cylindrical form (height of 8 mm and diameter of 7.5 mm) were prepared, then etched with HCl and H_2O (HCl/H_2O = 1:1) mixed solution to ensure the cleanness of the Fe-Cu alloys. The conducted experiments are summarized in Table 20.2. More detailed experi-mental information including the schematic illustration of the sessile drop method was already reported elsewhere [34].

Table 20.2 The conducted experiments [34]

Fe-Cu (wt%)	$ZnAl_2O_4$ (min)	Al_2O_3 (min)
0.5% Cu, 40 ppm C	90	90
0.5% Cu, 0.5% C	90	90
1% Cu, 40 ppm C	90	90
1% Cu, 0.5% C	90	90
10% Cu, 40 ppm C	90	

20.2.3 *Selenium*

The experiments with the selenium addition were conducted in an induction furnace (MFG40) under an argon protective gas (99.999% Ar containing 2–3 ppm oxygen). The heating rate of the MFG 40 furnace was 20 K/min and the average cooling temperature is around 175 K/min to 300 °C by directly turning off the generator. An in-detailed description of MFG 40 is described elsewhere [35]. Selenium powder (with a purity of 99.9%) and a secondary metallurgical slag were separately added into the liquid Fe-Cu alloy (130 g) through a quartz tube. Fe-Cu alloy with a 1 wt% of copper content was selected for the experiments.

First, around 1 g of selenium powder was added to the liquid Fe-Cu alloys after 5 min holding time at 1600 °C. The holding time was to ensure the fully melting of the alloy samples. Then, if the slag was necessary to be introduced into the liquids, 3 g of a high alumina secondary metallurgical ladle slag was added to the liquid alloy after 5 min of selenium's addition. The selection of the secondary metallurgical ladle slag was taken the consideration of its viscosity and the solids phase distribution. In general, the slags with the lower viscosity has a higher particle removing efficiency. On the other hand, the formation of the copper and selenium particles locate along with the solid particles which have a high melting point was noticed. Therefore, the slag with the chemical composition as shown in Table 20.3 was selected. The phase distribution of the slag and its T_{liq} (°C) $_{calculated}$ were predicted with the aid of FactSage 7.2 (FToxid, FTmisc, liquid slag solution phase and pure solids). The holding time after the slags addition was also settled as 5 min. The experimental procedure is depicted in Eq. (20.1). The chemical analysis of the prepared slag before the experiment was analyzed by applying the X-ray fluorescence spectrometry (XRF; Bruker AXS S8 Tiger, Bruker AXS GmbH, Karlsruhe, Germany), as shown in Table 20.3.

$$\text{Alloy} \xrightarrow{20\,k/\min} \text{liquid alloy at } 1600\,^{\circ}\text{C} \xrightarrow{5\,\min}$$
$$\text{Se addition} \xrightarrow{5\,\min} \text{slag's addition} \xrightarrow{5\,\min} \text{cooling} \qquad (20.1)$$

The chemical composition of the Armco iron and Fe-Cu alloys was measured by the spark spectrometer Foundry-Master UV (Oxford Instruments). Bruker G4 Ikarus and Bruker G8 Galileo combustion analyzers were applied to determine the S, C, and O values. Scanning electron microscopy (SEM) in combination with Energy-Dispersive X-ray Spectroscopy (EDX) were used for morphology and chemical composition analyses (Ultra55, Zeiss NTS GmbH). After the contacting experiments, for $ZnAl_2O_4$ case, the solidified iron sample together with $ZnAl_2O_4$ substrate

Table 20.3 The composition of used slag in wt%

CaO	Al_2O_3	SiO_2	MgO	Fe_2O_3	MnO	T_{liq}, (°C)$_{calculated}$
30.11	42.71	15.08	11.08	0.92	0.09	1705.03

were embedded into resin epoxy and perpendicularly cut for the cross sectional analysis. Solidified slag was separated from Fe-Cu alloy for the SEM/EDX analysis.

20.3 Results and Discussion

20.3.1 $ZnAl_2O_4$ Substrate

Figure 20.1 presents two types of interaction mechanism between Al_2O_3, $ZnAl_2O_4$ substrate, and Fe-Cu alloys. Al_2O_3 substrate and Fe-Cu alloys showed a non-reactive system with (0.5 wt% C) and slight amount of carbon (40 ppm C). The chemical composition of the Fe-Cu alloys was summarized in Table 20.5, copper was found to be remarkably reduced in both cases. It is well known that copper has a great evaporation rate [14]. The copper evaporation mechanism is divided into the following three stages (1). Copper transfer from the bulk of the liquid phase to the interface; (2). Copper evaporation from the liquid metal surface; (3). Transfer of the copper vapours from the interface to the core of the gaseous phase.

As mentioned in previous investigations, the copper evaporation is the reaction of first order [14, 22, 36]. The Eq. (20.2) [14] was used to determine the value of the total mass transfer coefficient based on the measured experimental data and the apparent evaporation rate constant k_{Cu}. k_{Cu} is the main kinetic characteristic of evaporation process.

$$\ln \frac{\%[Cu]}{\%[Cu]_o} = -k_{Cu} \cdot \frac{A}{V} \cdot t \qquad (20.2)$$

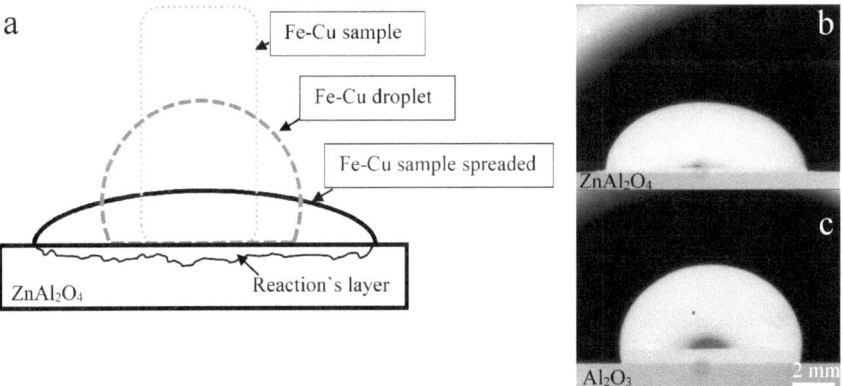

Fig. 20.1 **a** Schematic illustration of Fe-Cu sample melting mechanism on the $ZnAl_2O_4$ substrate. **b** Image of melted Fe-Cu on $ZnAl_2O_4$ substrate. **c** Image of melted Fe-Cu on Al_2O_3 substrate [34]

where %[Cu]—denote content of Cu in the melt (wt%), %[Cu]$_o$—initial content of Cu (wt%), k$_{Cu}$—apparent evaporation rate constant (m · s^{-1}), A—free surface of melt exposed to vacuum (m^2), V—volume of melt (m^3) and t—treatment time (s). As shown in Eq. (20.2) the copper evaporation degree is related to the evaporation free surface. In the present experiments, the liquid Fe-Cu alloy droplet delivered a noticeable free surface for the evaporation process.

On the other hand, a strong reaction was observed between the ZnAl$_2$O$_4$ substrate and Fe-Cu alloys, especially when the carbon content in the Fe-Cu alloys was high. A representative sample after the contact between ZnAl$_2$O$_4$ substrate and liquid Fe-Cu (1 wt%)-C (0.5 wt%) Fe-Cu alloys is presented in Fig. 20.2. The liquid Fe-Cu alloy spread on the ZnAl$_2$O$_4$ substrate while a reaction layer was formed, as pointed in Fig. 20.2 a. Figure 20.2b indicates a cross section analysis between the substrate and solidified Fe-Cu alloy. A copper gradient was detected from the contact interface to the inside of the ZnAl$_2$O$_4$ substrate. Copper content was found to reduce from the interface to the inside of the ZnAl$_2$O$_4$ substrate. This phenomenon indicates a copper diffusion process. Figure 20.2c presents a morphology with a higher magnification, a massive of pores and ZnO oxides were detected. The intermetallic of copper and zinc was not detected after the present investigations. Moreover, the newly formed reaction layer contains a bunch of elements, including Al, Fe, Zn, Mn, Si, and O. However, Cu was not detected in the layer, as presented in Fig. 20.3. Figure 20.4 presents the morphology of the cross-sectional area after the interaction between Fe-10 wt% Cu and ZnAl$_2$O$_4$ substrate. According to the EDX analysis and the elements ratio (see Table 20.4), a significant amount of compounds (containing Cu and Zn) were observed to be scattered in the interacted area. Moreover, copper was also found to diffuse into the substrate.

The oxygen contents in the Fe-Cu alloys were raised after the interaction between liquid Fe-Cu alloy and ZnAl$_2$O$_4$ substrates as shown in Table 20.5. The raised oxygen content in the Fe-Cu alloys was caused by the ZnAl$_2$O$_4$ that reduced by the carbon presented in the alloys. The re-oxidized ZnO powder was then collected in the experimental chamber. On the other hand, the raised oxygen might affect the evaporation of copper. The ways in which oxygen in iron melts may influence the evaporation of the solute elements copper are:

- Adsorbing at the free surface of the melt and hinders evaporation
- Reducing the partial vapour pressure of Cu ($e_{Cu}^o = -0.05$)

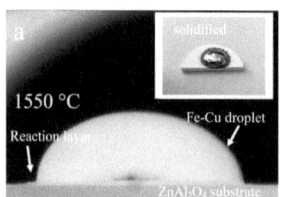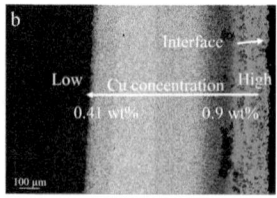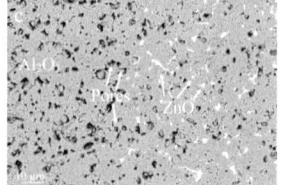

Fig. 20.2 a Images of Fe-Cu droplet on ZnAl$_2$O$_4$ substrate; **b** Copper concentration in the cross section from interface to the substrate; c. Microstructure of cross section of the samples [34]

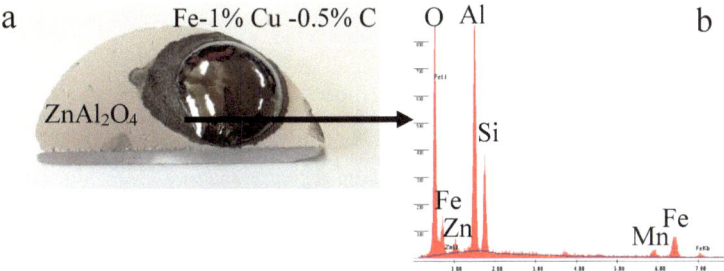

Fig. 20.3 a Image of Fe-1%Cu-0.5%C contact with $ZnAl_2O_4$ substrate; **b** EDX analysis of newly formed reaction layer on the interface of $ZnAl_2O_4$ [34]

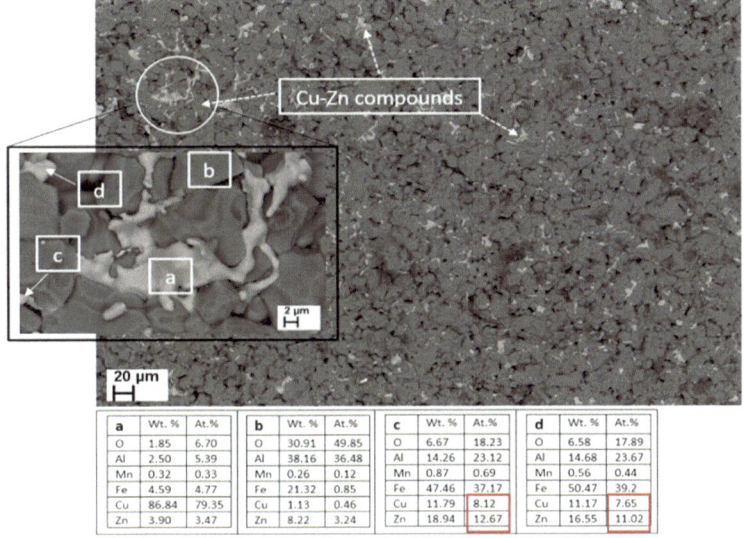

a	Wt. %	At.%		b	Wt. %	At.%		c	Wt. %	At.%		d	Wt. %	At.%
O	1.85	6.70		O	30.91	49.85		O	6.67	18.23		O	6.58	17.89
Al	2.50	5.39		Al	38.16	36.48		Al	14.26	23.12		Al	14.68	23.67
Mn	0.32	0.33		Mn	0.26	0.12		Mn	0.87	0.69		Mn	0.56	0.44
Fe	4.59	4.77		Fe	21.32	0.85		Fe	47.46	37.17		Fe	50.47	39.2
Cu	86.84	79.35		Cu	1.13	0.46		Cu	11.79	8.12		Cu	11.17	7.65
Zn	3.90	3.47		Zn	8.22	3.24		Zn	18.94	12.67		Zn	16.55	11.02

Fig. 20.4 Morphology and EDX results at cross sectional area between Fe-10% Cu alloy and $ZnAl_2O_4$ substrate

Table 20.4 The possible compounds at the cross sectional between Fe-10 wt% Cu and $ZnAl_2O_4$ substrate

Positions	Cu and Zn ratio	Possible compositions
a	$\frac{Cu}{Zn} = \frac{x}{y} = \frac{79.35}{3.47} = 21.14$	Diffusion of Fe-Cu molten alloy
b	$\frac{Cu}{Zn} = \frac{x}{y} = \frac{0.46}{3.24} = 0.142$	$ZnAl_2O_4$
c	$\frac{Cu}{Zn} = \frac{x}{y} = \frac{7.65}{11.02} = 0.694$	Cu_5Zn_8
d	$\frac{Cu}{Zn} = \frac{x}{y} = \frac{8.12}{12.67} = 0.641$	Cu_5Zn_8

Table 20.5 Chemical composition of Fe-Cu alloys after the interaction experiments

Wt%	C	Si	Mn	Al	O	Cu wt%	Cu loss (%)
0.5% Cu Al$_2$O$_3$	40	58	16	36	72	0.382	23.4
0.5% Cu ZnAl$_2$O$_4$	15	138	13	41	423	0.411	17.6
1% Cu Al$_2$O$_3$	40	91	15	55	80	0.718	28.2
1% Cu ZnAl$_2$O$_4$	13	77	13	28	559	0.908	9.2
0.5% C 0.5% Cu Al$_2$O$_3$	0.516 wt	50	27	38	51	0.35	25.0
0.5% C 0.5% Cu ZnAl$_2$O$_4$	14	50	13	38	252	0.321	31.3
0.5% C 1% Cu Al$_2$O$_3$	0.457 wt	50	15	45	62	0.757	24.3
0.5% C 1% Cu ZnAl$_2$O$_4$	18	66	13	46	301	0.722	27.8

- Reacting with the alloy and forms an oxide slag layer which acts a barrier to this evaporation
- Causing Marangoni type of flow at the melt surface which is beneficial to the refining process

Oxygen is a surfactant, it collects on the liquid Fe-Cu alloys surface to balance the surface energy between atmosphere and liquid. In addition, the surface accumulated oxygen blocks the free sites for metal-gas reactions. In other words, the surface located oxygen prevents the copper evaporation to atmosphere. Furthermore, a few ppm of oxygen can significantly reduce the surface tension value [37]. Thereby, the effect of oxygen on the free evaporation rate constant of copper from liquid iron was suggested to be expressed by assuming Langmuir's ideal adsorption isotherm: [34, 38]

$$k_{Cu}^e = k_{Cu}^{eo}(1 - \theta_O) \qquad (20.3)$$

where k_{Cu}^e—free evaporation rate constant of Cu on the surface of liquid iron (m·s^{-1}). k_{Cu}^{eo}—free evaporation rate constant of Cu on the surface of oxygen free liquid iron (m · s^{-1}). θ_O—degree of surface coverage by absorbed oxygen ($\theta_O \leq 1$).

$$K_O^{Cu} \cdot a_O = \frac{\theta_O}{(1 - \theta_O)} \qquad (20.4)$$

where K_O^{Cu}—adsorption equilibrium constant of oxygen on the surface of liquid iron. a_O—activity of oxygen in liquid iron. Combination of Eqs. (20.3) and (20.4) gives the free evaporation rate of Cu on the surface of liquid iron.

$$k_{Cu}^e = \frac{k_{Cu}^{eo}}{(1 + K_O^{Cu} \cdot a_O)} \qquad (20.5)$$

$$\Upsilon_{Fe-O} = \Upsilon_{Fe} - RTln(1 + K_O^{Cu} a_O) \qquad (20.6)$$

where Υ_{Fe} and Υ_{Fe-O} are the surface tension of the iron and oxygen containing iron. The value of K_O^{Cu} was reported as 110 in the Eq. (20.6) by Lee [39].

$$\Upsilon_{Fe-O} = 1.890 - 0.299 \ln(1 + 110 \, a_O) \qquad (20.7)$$

According to the Eqs. (20.3)–(20.7), the high oxygen content might reduce the copper evaporation in the Fe-Cu alloys. With the presence of carbon in the alloys, carbon reduced the oxygen content in the melts and consequently enhanced the copper evaporation.

Based on the archived experimental results, the Cu and Zn compounds were only detected in the case of Fe-10 wt% Cu alloy. However, with the low percent of Cu alloy, the compound was not detected. Therefore, the experimental results indicate that the $ZnAl_2O_4$ based material could not be used as a decopperization approach. In contrary, the $ZnAl_2O_4$ based material may reduce the copper evaporation process by introducing the oxygen into liquid Cu containing alloys.

20.3.2 The Effect of Selenium

The Addition of Selenium

Figure 20.5 presents the detected oxides with the corresponding EDX analysis (see Fig. 20.5(1)). The newly formed oxides were detected on the surface of alumina particle, which is located on the surface of the solidified Fe-Cu alloy after the selenium addition. According to the EDX analysis, the newly formed oxides contain a significant percentage of Cu and Se, it may indicate the formation of copper selenide compound. Moreover, the copper selenides in the present case revealed a hollow form, as arrow pointed in Fig. 20.5b. This phenomenon was probably caused by the high evaporation rate of Se [40]. After the selenium addition, around 7 wt% Cu was reduced in the experiment. The formation of copper selenide was considered as the main reason for the copper reduction.

The Addition of Selenium and Slag

As shown in Fig. 20.6a, solidified slag strongly adhered to the Al_2O_3 crucible. The Fe-Cu alloy surface was fully oxidized. During the experiments, the liquid alloy was shortly explored to the air when Se and slags were added to the liquid alloy. A spherical iron oxide is presented in Fig. 20.6b, with a greater magnification, a massive of Cu and Se containing particles was detected on the iron oxides as shown in Fig. 20.6 c. Furthermore, copper selenides were detected in the side of slags located along with oxides particles, as shown in Fig. 20.6d, see Fig. 20.6(1). Besides this, EDX reveals the particles consisting of Al, Fe, Mg, and O. Based on the EDX analysis, the possible oxide particles were Al_2O_3, $FeAl_2O_4$, and $MgAl_2O_4$. After the addition of Se and slag, around 12 wt% of Cu was reduced.

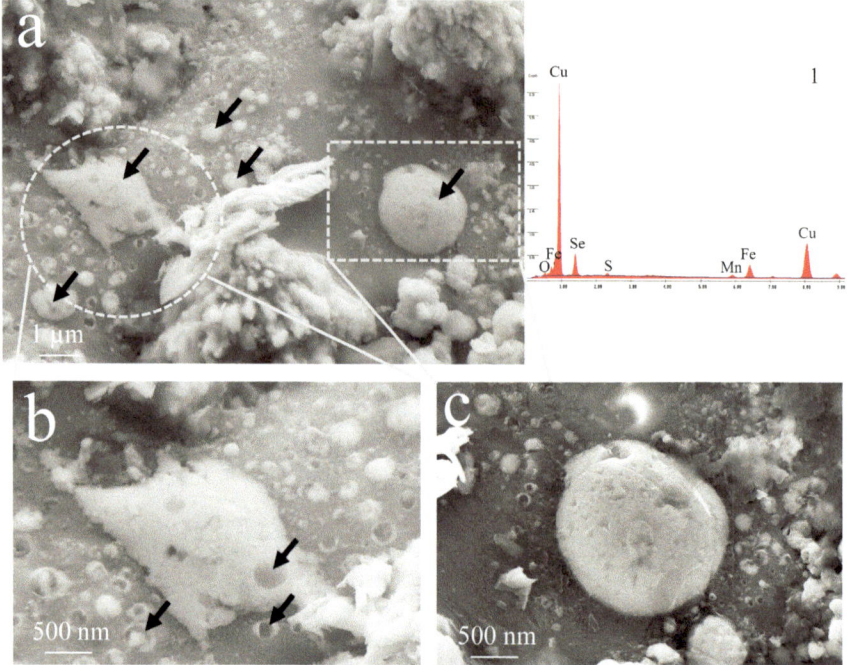

Fig. 20.5 Morphology of oxides particle on the solidified Fe-Cu alloy surface and EDX 1corresponds to the black arrows in figure a [41]

20.4   Conclusion

In the present work, $ZnAl_2O_4$ materials was interacted with the Fe-Cu alloys with and without of presence of carbon to test its decopperization possibility. For the comparison, pure Al_2O_3 has also interacted with the liquid Fe-Cu alloys. The experimental results showed no formation of any compound between copper and zinc. $ZnAl_2O_4$ substrate was found to be strongly reduced by the presence of carbon in the Fe-Cu alloys. On the other hand, pure Al_2O_3 substrate revealed a complete non-reactive system. Copper was mainly reduced through its evaporation process.

With the introducing of selenium into the liquid Fe-Cu alloy at 1600 °C. Copper selenium was detected at the surface of oxide, which is located on the surface of the solidified Fe-Cu alloy. Furthermore, a secondary metallurgical slag was added into the liquid Fe-Cu alloy after 5 min of adding selenium. More copper was found to vanish from the liquid Fe-Cu alloy. Copper selenide particles were later detected along with the oxides which contains complex elements of Al, Fe, Mg, and O. Selenium addition to the liquid Fe-Cu alloy exhibited a decopperization possibility.

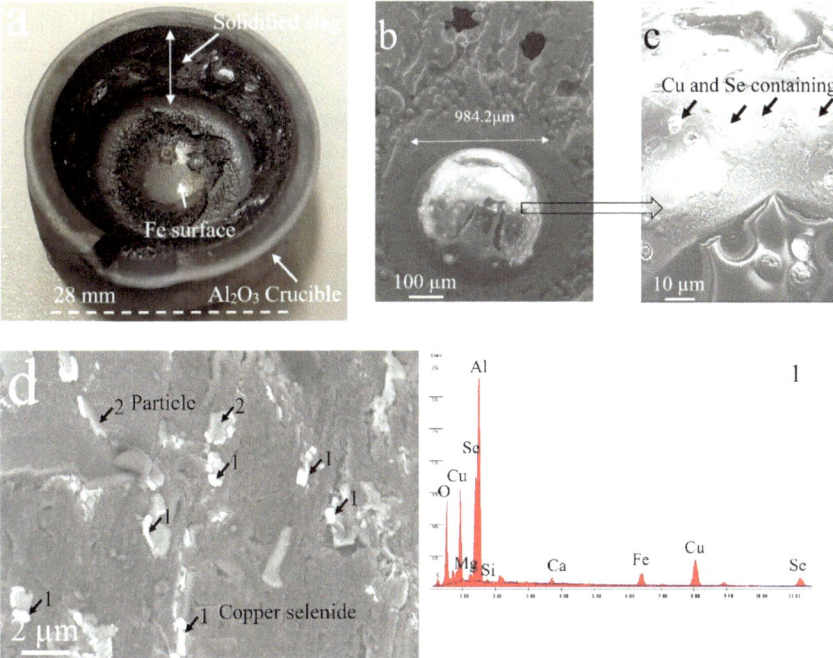

Fig. 20.6 Morphology of oxides particle on the solidified slag surface and EDX 1 corresponds to the black arrows in figure d point 1 [41]

Acknowledgements The investigation was supported by the DFG (German Research Foundation), Project-ID: 169148856-SFB 920, subprojects C01 at the Technical University Bergakademie Freiberg. The authors are grateful for the financial support and helpful discussions. Furthermore, the authors are very grateful for the technical support from Dr.-Ing. Thilo Kreschel for chemical analysis, Ms. Ines Grahl for samples preparation, Dr.-Ing. Armin Franke for morphology analysis, Mr. Marcus Block, and Peter Neuhold for the alloys preparation.

References

1. Z.Y. Cheng, J. Liu, Z.D. Xiang, J. Jia, Y.J. Bi, Intermetallics **120**, 106747 (2020). https://doi. org/10.1016/j.intermet.2020.106747
2. S.-J. Kim, C.G. Lee, T.-H. Lee, C.-S. Oh, ISIJ Int. **42**, 1452–1456 (2002). https://doi.org/10. 2355/isijinternational.42.1452
3. M. Ferhat, A. Benchettara, S. Amara, J. Fundam. Appl. Sci. **6**, 92 (2015). https://doi.org/10. 4314/jfas.v6i1.8
4. Y.-W. Jang, J.-H. Hong, J.-G. Kim, Met. Mater. Int. **15**, 623–629 (2009). https://doi.org/10. 1007/s12540-009-0623-5
5. C. Manque, Tramp elements and steel properties: a progress state of the European project on scrap recycling. Rev. Métallurgie. **95**, 433–442 (1998). https://doi.org/10.1051/metal/199895 040433

6. R. Boom, R. Steffen, Steel Res. **72**, 91–96 (2001). https://doi.org/10.1002/srin.200100090
7. Y. Kondo, Mater. Sci. Forum **696**, 183–188 (2011). https://doi.org/10.4028/www.scientific.net/MSF.696.183
8. K. Shibata, S.-J. Seo, M. Kaga, H. Uchino, A. Sasanuma, K. Asakura, C. Nagasaki, Mater. Trans. **43**, 292–300 (2002). https://doi.org/10.2320/matertrans.43.292
9. K.E. Daehn, A. Cabrera Serrenho, J.M. Allwood, Environ. Sci. Technol. **51**, 6599–6606 (2017). https://doi.org/10.1021/acs.est.7b00997
10. K.E. Daehn, A.C. Serrenho, J. Allwood, Mater. Trans. B. **50**, 1225–1240 (2019). https://doi.org/10.1007/s11663-019-01537-9
11. E.F. Sandig, D. Chebykin, V.V. Prutchykova, O. Fabrychnaya, O. Volkova, Mater. Sci. Forum **959**, 145–160 (2019). https://doi.org/10.4028/www.scientific.net/MSF.959.145
12. R. Shimpo, O. Ogawa, Y. Fukaya, T. Ishikawa, Metall. Mater. Trans. B **28**, 1029–1037 (1997). https://doi.org/10.1007/s11663-997-0057-8
13. A. Cohen, M. Blander, Metall. Mater. Trans. B **29**, 493–495 (1998). https://doi.org/10.1007/s11663-998-0129-4
14. L. Savov, D. Janke, ISIJ Int. **40**, 95–104 (2000). https://doi.org/10.2355/isijinternational.40.95
15. H. Ono-Nakazato, K. Taguchi, Y. Seike, T. Usui, ISIJ Int. **43**, 1691–1697 (2003). https://doi.org/10.2355/isijinternational.43.1691
16. H. Ono-Nakazato, K. Taguchi, T. Usui, ISIJ Int. **43**, 1105–1107 (2003). https://doi.org/10.2355/isijinternational.43.1105
17. K. Yamaguchi, Y. Takeda, Mater. Trans. **44**, 2452–2455 (2003). https://doi.org/10.2320/matertrans.44.2452
18. A.I. Zaitsev, N.E. Zaitseva, E.Kh. Shakhpazov, B.M. Mogutnov, ISIJ Int. **44**, 639–646 (2004). https://doi.org/10.2355/isijinternational.44.639
19. J. Wang, S. Guo, L. Zhou, Q. Li, J. Iron. Steel Res. Int. **16**, 17–21 (2009). https://doi.org/10.1016/S1006-706X(09)60021-2
20. J. Łabaj, J. Lipart, Arch. Mater. Sci. Eng. **52**, 13–17 (2011)
21. K. Yamaguchi, H. Ono, ISIJ Int. **52**, 8 (2012)
22. J. Łabaj, Arch. Metall. Mater. **57** (2012). https://doi.org/10.2478/v10172-012-0005-8
23. ŁABAJ J., O. B., S. G., METALURGIJA. **54**, 265–268 (2011)
24. X. Hu, Z. Yan, P. Jiang, L. Zhu, K. Chou, H. Matsuura, F. Tsukihashi, ISIJ Int. **53**, 920–922 (2013). https://doi.org/10.2355/isijinternational.53.920
25. Y. Uchida, A. Matsui, Y. Kishimoto, Y. Miki, ISIJ Int. **55**, 1549–1557 (2015). https://doi.org/10.2355/isijinternational.ISIJINT-2014-776
26. A.G. Wieliczko, J.N. Griszczenko, M. Krucinski, Hut. Wiadomosci Hut. **65**, 256–259 (1998). ((In Polish))
27. Li. Li, C. Xiang, P. Zhao, C. Wang, H. Tian, S. Li, J. Iron Steel Res. Int. **10**, 5–7 (1998) (In Chinese)
28. Y.G. Kim, I.J. Kim, J.S. Kim, Y.I. Chung, D.Y. Choi, Mater. Trans. **55**, 171–175 (2014). https://doi.org/10.2320/matertrans.M2013244
29. J. Hait, R.K. Jana, S.K. Sanyal, Miner. Process. Extr. Metall. **118**, 240–252 (2009). https://doi.org/10.1179/174328509X431463
30. Y. Kilic, G. Kartal, S. Timur, Int. J. Miner. Process. **124**, 75–82 (2013). https://doi.org/10.1016/j.minpro.2013.04.006
31. C. Wang, J. South. Afr. Inst. Min. Metall. **116**, 593–600 (2016). https://doi.org/10.17159/2411-9717/2016/v116n6a16
32. M. Jenko, J. Fine, D. Mandrino, Surf. Interface Anal. **30**, 350–353 (2000). https://doi.org/10.1002/1096-9918(200008)30
33. T. Shyrokykh, X. Wei, S. Seetharaman, O. Volkova, Metall. Mater. Trans. B **52**, 1472–1483 (2021). https://doi.org/10.1007/s11663-021-02114-9
34. X. Wei, S. Dudczig, D. Chebykin, C.G. Aneziris, O. Volkova, Metals. **11**, 2030 (2021). https://doi.org/10.3390/met11122030
35. O. Kovtun, M. Karbayev, I. Korobeinikov, C. Srishilan, A.K. Shukla, O. Volkova, Steel Res. Int. **20**, 2000607 (2021). https://doi.org/10.1002/srin.202000607

36. T. Matsuo, K.Maya, T. Nishi, K. Shinme, A. Ueno, S. Anezaki, ISIJ. Int. **36**, S62–S65(1996)
37. K. Morohoshi, M. Uchikoshi, M. Isshiki, H. Fukuyama, ISIJ Int. **51**, 1580–1586 (2011). https://doi.org/10.2355/isijinternational.51.1580
38. T. Yoshida, T. Nagasaka, M. Hino, ISIJ Int. **41**, 706–715 (2001). https://doi.org/10.2355/isijinternational.41.706
39. J. Lee, K. Morita, ISIJ Int. **42**, 588–594 (2002)
40. J. Tanabe, Steel Res. Int. **77**, 21–24 (2006). https://doi.org/10.1002/srin.200606125
41. X. Wei, O. Kovtun, A. Yehorov, C.G. Aneziris, O. Volkova, Mater. Lett. **323**, 132543 (2022). https://doi.org/10.1016/j.matlet.2022.132543

Chapter 21
Interactions Between Molten Iron and Carbon Bonded Filter Materials

Xingwen Wei, Enrico Storti, Steffen Dudczig, Olga Fabrichnaya, Christos G. Aneziris, and Olena Volkova

21.1 Introduction

The detrimental effects of the non-metallic inclusion strongly impair the mechanical properties of the final products [1, 2]. In particular, the macro inclusion or the inclusions clusters could lead to the cracks in the final metallic products. Therefore, to improve the cleanness of the metallic products and meet the new market's requirement, it is urgent to minimize the amount of the inclusions in the melts. Ceramic foam filters are considered as a one of the most efficient methods to extract inclusions from the molten liquids though the deposition of inclusions on the ceramic surface. Besides the good attractive nature for the inclusions, the ceramic foam materials should also overcome the great temperature difference in and out of the molten liquids, i.e. the thermal shock resistance.

The addition of carbon could contribute to the improvement of thermal shock resistance of ceramics. Afterward, carbon bonded ceramics filters such as foam (Al_2O_3-C) filters with Al_2O_3 and with carbon bonded MgO coatings for the steel melt filtration were developed [3–5]. Further, to test the filtration efficiency, these filters were immersed into 42CrMo4 steel melts at 1650 °C. Alumina and magnesia layers were detected at the surface of Al_2O_3-C filter and of the Al_2O_3-C filter coated with carbon

X. Wei (✉) · O. Volkova
Institute of Iron and Steel Technology, Technische Universität Bergakademie Freiberg, Leipziger Str. 34, 09599 Freiberg, Germany
e-mail: Xingwen.wei@iest.tu-freiberg.de

E. Storti · S. Dudczig · C. G. Aneziris
Institute of Ceramics, Refractories and Composite Materials, Technische Universität Bergakademie Freiberg, Agricolastraße 17, 09599 Freiberg, Germany

O. Fabrichnaya
Institute of Material Science, Technische Universität Bergakademie Freiberg, Gustav-Zeuner-Str. 5, 09599 Freiberg, Germany

C. G. Aneziris and H. Biermann (eds.), *Multifunctional Ceramic Filter Systems for Metal Melt Filtration*, Springer Series in Materials Science 337,
https://doi.org/10.1007/978-3-031-40930-1_21

bonded MgO. Furthermore, a massive of Al_2O_3 and MgO inclusions were found to accumulate onto the newly formed layers. In contrary, the alumina coated surface remained clean and without the deposition of inclusions [6]. Moreover, the study of [6, 7] also stated the detection of oxides whisker after the carbon bonded ceramic filters materials making the contact with the melts. The composition of whiskers was related to the constitution of the interacted filters materials. The mechanism of the newly formed layers was studied by immersing the carbon bonded filters into the liquids (finger test) [8–10] and interacting with liquid alloy [11]. The formation of the whiskers was investigated in the work of [12]. Besides the above-mentioned carbon boned substrates, calcium aluminates CA2 and CA6 were coated onto the Al_2O_3-C filter and further immersed into the melts for 10 s [13]. After the immersion test, a secondary oxide layer with fine inclusions were observed on filters. Both types of filters delivered satisfying filtration efficiency for non-metallic inclusions.

Based on the literature survey, the presence of carbon was believed to be the reason for the formation of new layers on the filters. The released aluminum for the later formation of the secondary alumina layer was believed to be contributed by the carbothermic reduction of alumina with the presence together of steel and carbon [14]. However, the dissolution mechanism of alumina into melts was later stated in the work of [15] and supported by the calculated simulation. The unclear mechanism that related to the alumina 'reduction' was completely studied by interacting four different types of carbon (10 wt% C) bonded ceramics C-Al_2O_3, Al_2O_3 rich spinel (C-AR78), MgO rich spinel (C-MR66), and $CaO·6Al_2O_3$ (C-CA6) with liquid Armco iron to investigated the possible reactions that induced by the presence of carbon.

In addition, the work of [16] indicated that the inclusions removal efficiency or filtration efficiency is strongly related to the wetting behavior of the inclusions. Therefore, the wettability of the pure ceramics substrates was investigated (1). Al_2O_3 based containing high MgO (MR66), (2). $MgAl_2O_4$ spinel based containing MgO (AR 78), (3). CA 6 ($CaO·6Al_2O_3$), (4). $CaZrO_3$, and (5). Al_2O_3 based containing TiO_2 and ZrO_2 (AZT). Furthermore, after the interaction, the quantity, size, and the composition of the inclusions was analyzed by utilizing an automatic scanning electron microscope combined with an energy dispersive X-Ray spectroscope (EDX) (Aspex, FEI, USA).

21.2 Experimental Procedure

21.2.1 The Preparation of the Substrates

Table 21.1 summarized the raw oxides used for the preparation of the substrates. For the pure ceramic substrates preparation, the additives were dextrin (used as binder for the granulation), Optapix AC95 and PAF35 (Zschimmer & Schwartz— used as temporary binders), Zusoplast G112. On the other hand, for carbon bonded materials carbon black was added to get 10 wt% residual carbons after coking. The

Table 21.1 Raw oxide used for the preparation of the substrates and the related parameters

Substrate material	Raw material(s)	Grain size (d90) (μm)	Roughness (μm)
C-Al$_2$O$_3$	10 wt% + Al$_2$O$_3$	<20	5.5
C-CA6 Bonite	10 wt% + CA6	<20	4.04
C-AR78 spinel	10 wt% + AR78	<20	3.70
C-MR66 spinel	10 wt% + MR66	<90	9.18
MR 66	91.5% Al$_2$O$_3$, 8.5% MgO	<90	2.15
AR 78	99.7% MgAl$_2$O$_4$, 0.3% MgO	<20	2.05
CaZrO$_3$	83.3% CaZrO$_3$, 16.7% ZrO$_2$	<45	3.37
CA 6	94.9% CA 6 (CaO·6Al$_2$O$_3$), 5.0% Al$_2$O$_3$, 0.1% CaO	<20	1.57
AZT	95% Al$_2$O$_3$, 2.5% TiO$_2$ 2.5% ZrO$_2$	<60	8.24

samples were bonded using powder novolac resin and hexametilentetramine as a curing agent. Then, the substrate samples were polished with ultra-fine SiC sandpaper (down to 3 μm particle size). The X-ray diffraction analysis (XRD) was conducted after heat treatment to determine the phase composition of the substrates before the interaction experiments. The instrument (X'Pert PRO MPD X-Ray Diffractometer 3040/60, PANalytical, Germany) operates in Bragg–Brentano geometry with a fixed divergence slit and a rotating sample stage. The parameters were standard Cu Kα radiation ($\lambda = 1.54$ Å), a 2θ-range from 7.5 to 90° with a scan step size of 0.013° and a holding time of 30 s per step. Later, the surface of the substrates confocal laser scanning microscopy (CLS) measured. An area of 5 × 5 mm^2 for each polished sample was scanned. The roughness results obtained by laser microscopy and open porosity was estimated by means of water immersion, according to the European standard DIN EN 993–1. The measured results are summarized in Table 21.1. A more in-detailed description for the preparation of substrates was reported elsewhere [12].

21.2.2 Experimental Setup

The Armco iron, with the chemical composition listed in Table 21.2, was used for the investigations in contact with pure and carbon bonded substrates. The Armco iron samples with a diameter of 7.5 mm and a mass of 2.5 g were prepared and cleaned in an ultrasound device with ethanol and then etched with HCl. The experiments were carried out inside a high temperature microscope using the sessile drop method. Before heating, the chamber of the microscope was evacuated to 10^{-4} atm for the elimination of oxygen. After evacuation, the heating program was started. Thereby, the samples were heated with a heating rate of 40 K/min under Argon atmosphere (Ar 5.0 containing 2–3 ppm oxygen) to the target temperature of 1600 °C for pure

Table 21.2 Composition of Armco iron (Trace elements in ppm)

Fe (wt%)	C	Mn	Si	Al	S	P	O
99.8	40	200	50	<10	40	64	92

ceramics and 1625 °C for carbon bonded ceramics. The Argon gas flow rate was 0.3 L/min. The samples were held at the target temperature for 30 min and 2 h for pure ceramics and carbon bonded ceramics, respectively. The droplet silhouette of molten iron was examined from the video material recorded by an Image Source CCD camera. The setup schema is shown in Fig. 21.1.

To compare with the experiments between liquid Armco iron and carbon bonded substrates, a piece of carbon bonded alumina ($C\text{-}Al_2O_3$) was insert into a platinum container and a quartz glass ampule to ensure a better atmosphere such as (1). Cleanness, (2). Elimination of the influence of gas flow. Furthermore, a gap and orifice were kept between platinum crucible and lid, and on a side of the ampule to ensure the identical atmosphere in crucible and chamber (as shown in Fig. 21.2), which enhanced the reliability of the experiment.

The oxygen content in the experimental atmosphere is controlled by O/CO system, the installed graphite tube for heating process acted as oxygen eliminator by formation of CO and CO_2 gas [17]. The oxygen partial pressure in high temperature microscope was measured during the experiments using a high temperature zirconia oxygen probe, which was installed at the gas outlet. The oxygen partial pressure was calculated according to Eq. (21.1)

$$P(o_2) = 0.2094 \times 10^{\left(\frac{-20.166}{T}\right)EMF} \tag{21.1}$$

where T is Temperature (K) and EMF is sensor signal (mV).

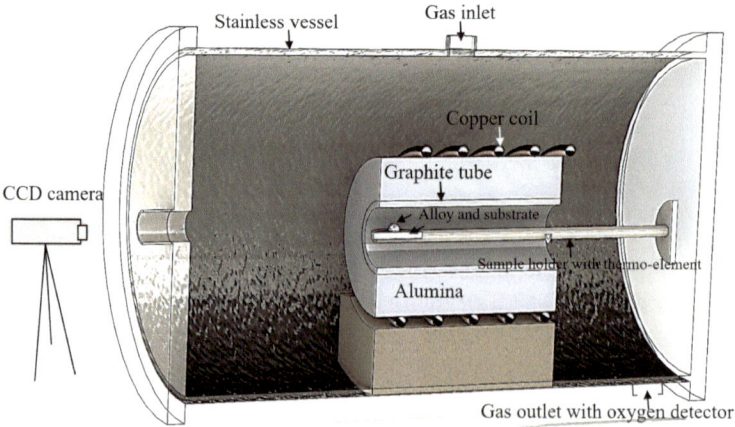

Fig. 21.1 Schematic view of setup, a cross section

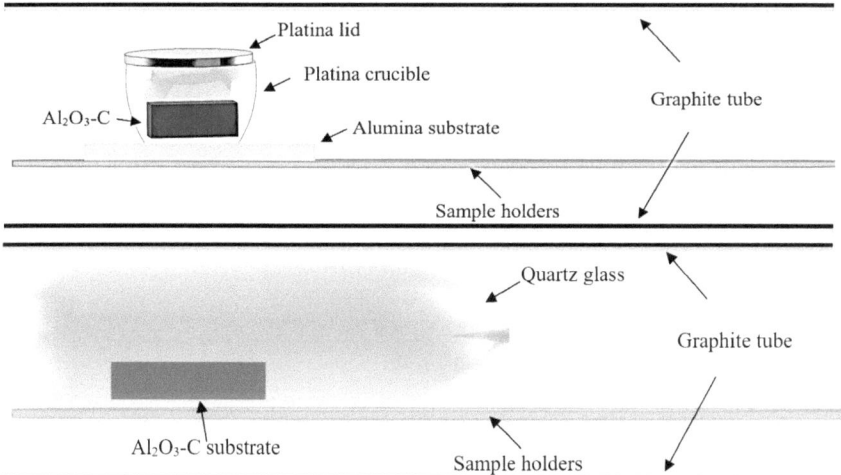

Fig. 21.2 Platinum crucible [12] and quartz glass illustration

For the pure substrates, the oxygen partial pressure was kept constantly at 10^{-21} atm. In the carbon bonded substrates system, the oxygen partial pressure was lower than for the system without carbon due to the carbon oxidation. In the $C-Al_2O_3$ and C-CA 6 cases, the measured oxygen partial pressure is 10^{-24} atm. Furthermore, in the C-AR 78 is $P_{O2} = 10^{-25}$ atm. A lower oxygen content ($P_{O2} = 10^{-26}$ atm at 1625 °C) was detected in the C-MR 66 substrate case.

The spark spectrometer Foundry-Master UV (Oxford Instruments) was employed for the general chemical analysis of the iron samples. Besides, Bruker G4 Ikarus and Bruker G8 Galileo combustion analyzers were utilized to estimate the C, O, and S values. Scanning electron microscopy (SEM) in combination with Energy-Dispersive X-ray Spectroscopy (EDX) were used for morphology and chemical composition analyses (Ultra55, Zeiss NTS GmbH). After the experiments, for pure substrates-iron samples, the solidified iron was pressed into a plate with 1 mm thickness then measured the chemical composition. The cross sectional area between substrates and iron was analyzed by SEM/EDX. On the other hand, carbon bonded substrates and iron samples were entirely embedded in epoxy resin and perpendicularly cut for the cross section SEM/EDX analysis.

21.2.3 Non-metallic Inclusions Determination

After the interaction with pure substrates, the solidified iron droplets were embedded in epoxy resin and cut at the thickest area of the iron droplet from the bottom area and then polished. Afterwards, an automatic scanning electron microscope combined

Table 21.3 Developed Rule-file applied for Aspex-SEM inclusions classification [18]

Class name	Element content (wt%)
Al_2O_3	Al > 20 and O > 20 and Si < 10 and Ti < 10 and (Mn + Mg + Si + Ca + Ti) < 10 and Mn < 10
TiN	Ti > 10 and N > 5
$MnO-Al_2O_3$	Al > 10 and Mn > 2 and (Mg + Si + Ca + Ti) < 10 and Si < 2 and S < 2
$MnO-SiO_2$-high Al_2O_3	Mn > 2 and Si > 2 and Al > 15
$MnO-SiO_2$-low Al_2O_3	Mn > 2 and Si > 2 and 1 < = Al < 15
$MnO-SiO_2-TiO_2$	Mn > 2 and Si > 2 and Al < 1 and Ti > 2
SiO_2	Si/O > = 0.4 and Si/O < 6.2 and Al < 3 and Mg < 3 and Ca < 3 and K < 3 and Mn < 3 and S < 10
SiO_2-MnO	Si/Mn > = 0.7 & Ca < = 0
MnO-MnS	Mn > 8 and (Mn/S) > = 1 and (Al + Si + Ti + Cr) < (Mn + S)
$CaO-Al_2O_3-SiO_2$	Ca > 5 and Al > 5 and Si > 5
CaO-CaS	Ca > 5 and (Ca/S) > 2
Other	Fe < 100

with an energy dispersive X-Ray spectroscope (EDX) Aspex was applied to determine the non-metallic inclusions distribution and their chemical composition. A developed rule file for the inclusions classification is presented in Table 21.3. A more in-detailed description for the inclusions analysis was reported elsewhere [18].

21.3 Results and Discussion

21.3.1 Carbon Bonded Substrates

Figure 21.3 presents the Armco iron samples on the carbon bonded substrate. It can be apparently seen that, the iron samples transferred to non-symmetrical form. Expect for the MR66 case, whiskers were also observed on the surface of the iron sample. In addition, crystals were found to form on the substrate as pointed with arrow. Normally, the iron sample would form a perfect droplet due to surface tension. However, in the present case, the iron samples were covered by the newly formed phase.

Figure 21.4 presents the images of the cooled iron with the contacted carbon bonded substrates. As shown in Fig. 21.4, the carbon bonded substrates were found strongly modified after the interaction between liquid Armco iron and carbon bonded substrates.

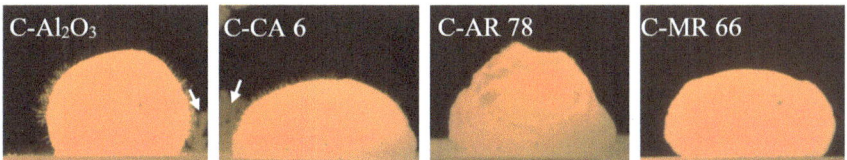

Fig. 21.3 Images of iron on carbon bonded substrates [12]

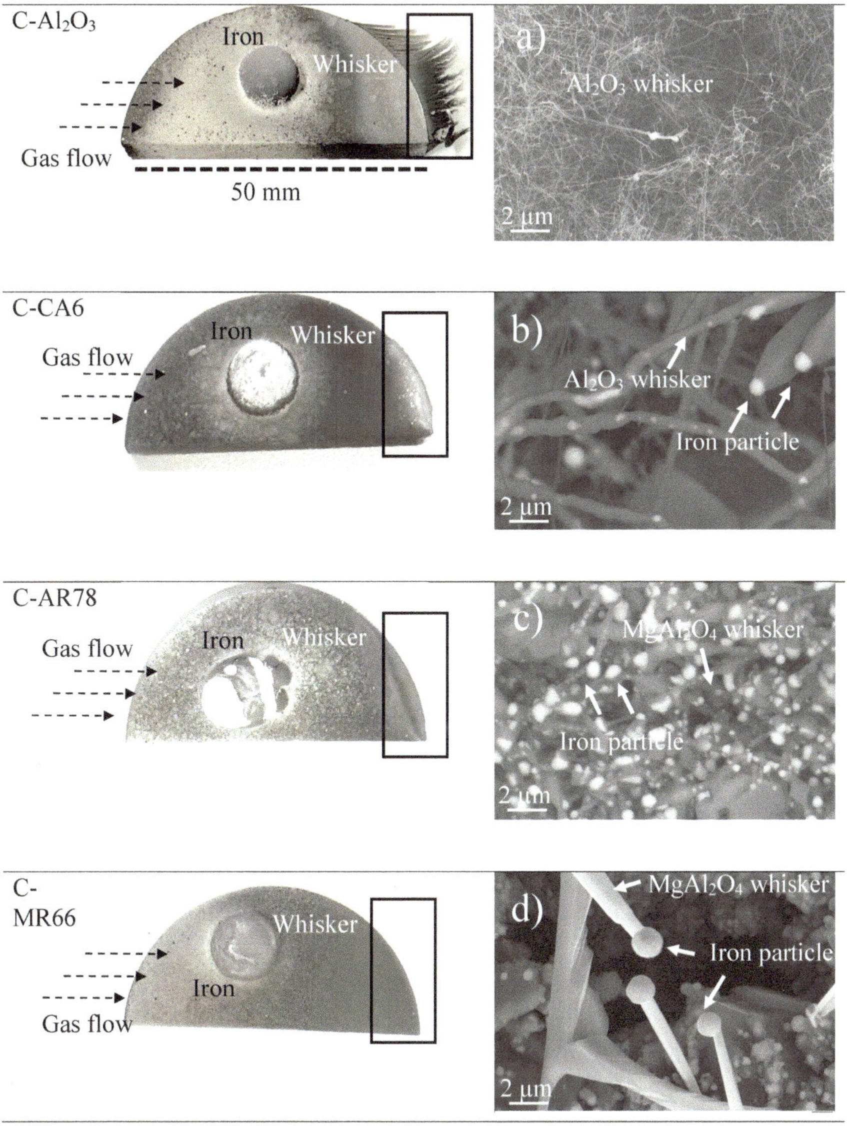

Fig. 21.4 Images of carbon-bonded substrates after interaction with molten iron and morphology of iron capped whiskers. **a** C-Al$_2$O$_3$; **b** C-CA 6; **c** C-AR 78; **d** C-MR 66 [12, 20]

(1) The interfacial color change of the substrates from black to the grey color indicate the great decarburization process. The current used Armco iron sample contains 40 ppm of carbon, this low carbon content indicates a great carbon adsorption tendency of the liquid Armco iron. Furthermore, as already mentioned in the experimental part, the introduced protective gas Argon 5.0 contains also 2–3 ppm oxygen. Therefore, the carbon on interfacial area of the substrates could also been consumed by the oxygen contained in the protective gas.

(2) The formation of oxide layers covering on the iron sample.

(3) The formation of whiskers. The newly formed whiskers were classified as two groups, i.e. with and without the iron particles capping. According to Xie et al. [19], the iron capped whiskers formed by the released elements that dissolved in the melt until its super-saturation, then the whisker starts to grow. Wei et al. [12] stated one more mechanism for the whiskers formation. Here, Wei et al. detected the solid phases Al_4C_3 and Al_4O_4C after the interaction with liquid Armco iron at 1625 °C, as well as the thermodynamic simulation. Afterwards, he stated that the newly formed Al_4C_3 and Al_4O_4C could be oxidized in CO atmosphere and contribute to the alumina whiskers formation.

In Fig. 21.5, the cross sectional analysis of iron and carbon bonded substrates is presented. The newly formed layers covered on the iron sample were revealed as Al_2O_3, $MgAl_2O_4$, $CaO \cdot 2Al_2O_3$, and $MgAl_2O_4$ for C-Al_2O_3, C-AR78, C-CA 6, and C-MR66, respectively. Moreover, iron was found to penetrate into the substrates, especially for C-Al_2O_3 and C-CA 6 substrates. For C-AR78 and C-MR66 substrates, the iron penetration was prevented by the newly formed oxide layers that covered on the iron samples. In the case of C-Al_2O_3 substrate, Al_2O_3 plates was detected accumulated on a newly formed alumina layer. In Fig. 21.5f, a macro non-metallic inclusion $MgAl_2O_4$ with SiO_2 as core in the side of iron was detected. According to the results, the newly formed oxide layers was found to cover on the iron sample and the whiskers formation are related to the composition of the substrates. In other words, it may conclude that the oxide layers consisting elements (Al, Mg, Ca, and O) were released within the interaction process. A proposed mechanism for the elements releasing and the oxides and whiskers formation is presented in Eqs. (21.2) and (21.3).

$$Al_2O_3 / MgAl_2O_4 + MgO \xrightarrow{C,Fe} Mg, Al$$

$$+ CO \xrightarrow{Fe,O} Al_2O_3 / MgAl_2O_4 \text{ oxides layers/} MgAl_2O_4 \text{ whiskers} \tag{21.2}$$

$$CaO \cdot 6Al_2O_3 \xrightarrow{C,Fe} Ca, Al + CO \xrightarrow{Fe,O} CaO \cdot 2Al_2O_3 \text{ oxides layers/} Al_2O_3 \text{ whiskers} \tag{21.3}$$

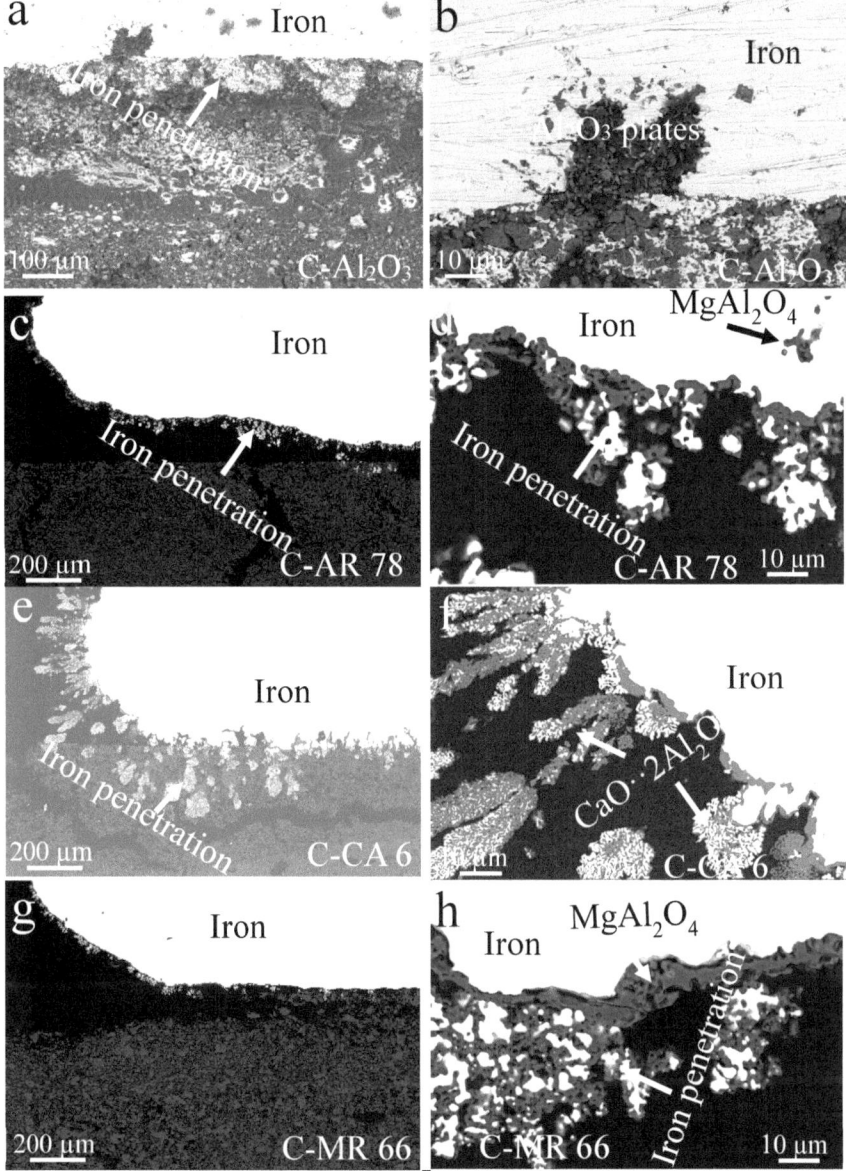

Fig. 21.5 Cross section of iron with the carbon bonded substrates. **a, b** C-Al₂O₃; **c, d** C-AR78; **e, f** C-CA 6; **g, h** C-MR66 [20]

Quartz Glass Ampule and Platina Crucible

The quenched quartz glass ampule was roughly smashed, the microstructure and compositions of the inner/outside surface of ampule and the graphite alumina were carefully analysed with the aid of SEM/EDX as shown in Fig. 21.6.

The idea is that, the for measured whiskers, however were determined as SiO_2 whiskers. But it contains a noticeable Al. Specially the area outside of the ampule. SiO_2 has no effects on producing Al or Al containing gas. The Al observed areas (whiskers) which are far away from the graphite alumina may indicates the reduction process of alumina. In the present case, carbon is the only substance capable for these reduction process.

The investigated experiments without the presence of iron states that the aluminium releasing may occur without the aid of the iron. The shapes of whiskers in both experiments are very different from the results in presence of iron. For the graphite alumina inserted quartz glass ampule system a schematic illustration is shown in Fig. 21.7. Based on the achieved results, the aluminium is apparently released from the C- Al_2O_3 and the most gas products of aluminium could be Al_2O, AlO, and Al, resulting in the detected whiskers crystals on the surface of graphite alumina and in the orifice area, owing to the deoxidation and precipitation reactions. In addition, AlO reacts with CO gas forms alumina as well. The detected Si is owing to the reduction process between SiO_2 and C. In fact, the formation of Si containing gas (SiO) and reoxidation reactions is very complicated. Biernacki and Wotzak [21] have created a mechanism which shown as Eqs. (21.4)–(21.6). The silicon carbide (SiC) will be formed by the reaction of silica (SiO_2) and carbon (C), then SiO_2 reacts with SiC further produce SiO gas. The produced gas SiO will further react with CO and O_2 gas to form SiO_2.

$$SiO_2 + 3C = SiO_{(g)} + 2CO_{(g)} \qquad (21.4)$$

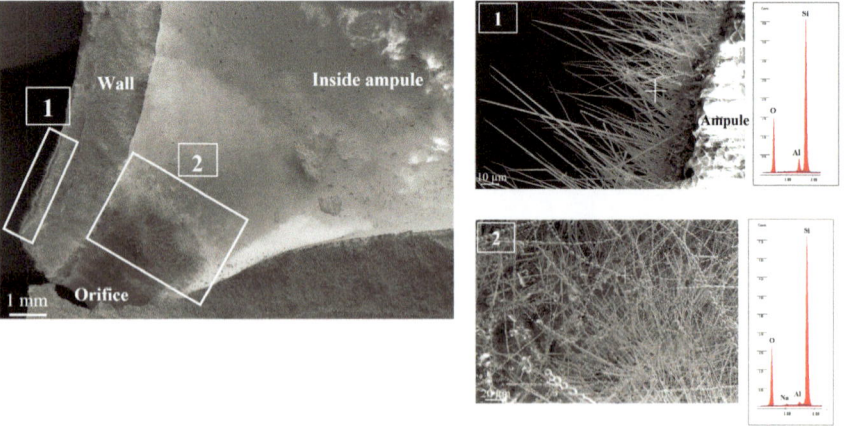

Fig. 21.6 Microstructure and EDX analyses of the quartz glass ampule system on spot 1 and 2

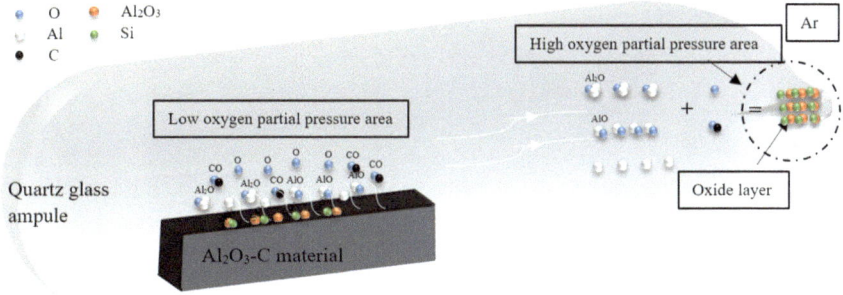

Fig. 21.7 Schematic illustration of the graphite alumina inserted quartz glass ampule system

$$SiO_{2(g)} + 2C = SiC + CO_{(g)} \tag{21.5}$$

$$SiC + 2SiO_2 = 3SiO + CO_{(g)} \tag{21.6}$$

$$SiO + O = SiO_2 \tag{21.7}$$

$$SiO + CO = SiO_2 \tag{21.8}$$

The produced gas products increase the local pressure in ampule system, it led to the released gas products tend to leave the ampule to balance the entire system. On the other side, the raised pressure in ampule blocked the entrance for oxygen as well. Therefore, a significant oxide layer formed on the orifice area. Finally, although the Si can react with alumina, but the detected aluminium located far away from the graphite material which is considered as released with the help of carbon.

In Fig. 21.8 presented the image of the heated carbon bonded alumina in platina crucible. Whiskers on the surface carbon bonded alumina was determined. Figure 21.9 presented the morphology structure of the detected whiskers and its XRD analysed results. The achieved XRD result on the detected dendritic whiskers is presented in Fig. 21.9b, it consisted of alumina.

Based on the detected alumina whiskers on the carbon bonded material, it can be concluded that iron may not have a decisive effect on the carbothermic reduction, since the aluminium release process occurs without the presence of iron. But, on the other hand, the catalytic effect of iron cannot be denied. With the presence of iron, the aluminium releasing process could possibly be enhanced. The explanation for the metallic elements releasing such as Al, Ca, Mg is shown in the later sections with the support of stability of various oxides that existed in present study.

Fig. 21.8 Image of the carbon bonded alumina in platina crucible after heating

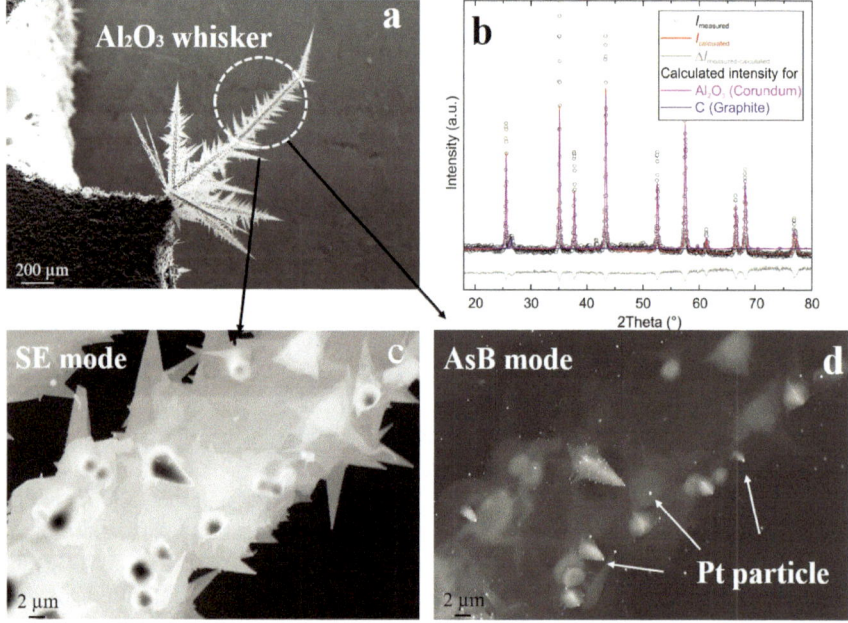

Fig. 21.9 Morphology of the alumina whiskers on the surface of Al_2O_3-C material (**a, c** and **d**). **b** XRD profile of dendritic whiskers [12]

21.3.2 Pure Substrates

The measured contact angle on various substrates were presented in Fig. 21.10. Besides the AZT substrate, the contact angles of all cases exhibited slightly varied value. The contact angle on the AZT substrate increased slightly from the initial time point from 127.3° to 128.4° at 10 min and then decreased to 126.7° at 30 min. Such

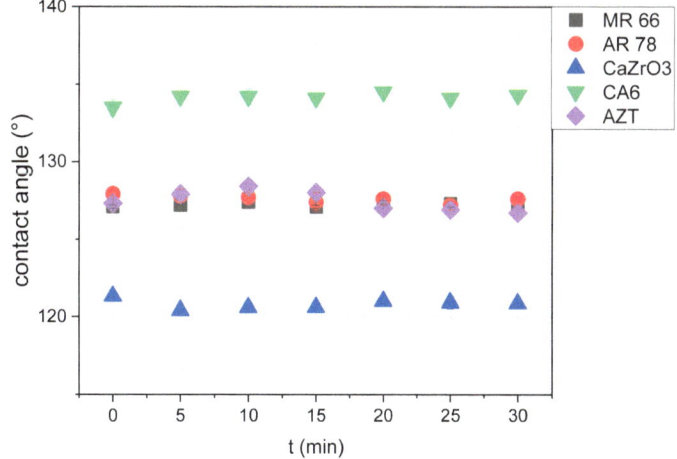

Fig. 21.10   Contact angle of Armco iron on the pure ceramic substrates [18]

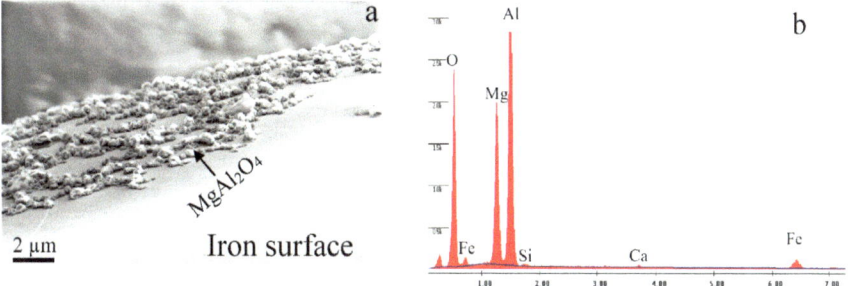

Fig. 21.11   a Spinel formation on iron surface after contacting with MR 66 and **b** corresponding EDX analysis [18]

variation of contact angle was considered caused by the presence of TiO_2. Later, a newly formed phase with consistent of Ti, Si, and Mn in the cross section between AZT substrate and Armco iron was detected. Iron was found to infiltrate into the AZT substrate as well. On the contrary, no reaction was observed with the rested of the substrates, only a percent of SiO_2 was found to deposit on the surface of substrates. Furthermore, $MgAl_2O_4$ spinel on the surface of solidified iron after contacting with MR 66 substrate was determined, as presented in Fig. 21.11.

21.3.3   Inclusions

Figure 21.12 presents the inclusions number with various sizes after making contact with the various substrates. In comparison with the initial iron, the total number of

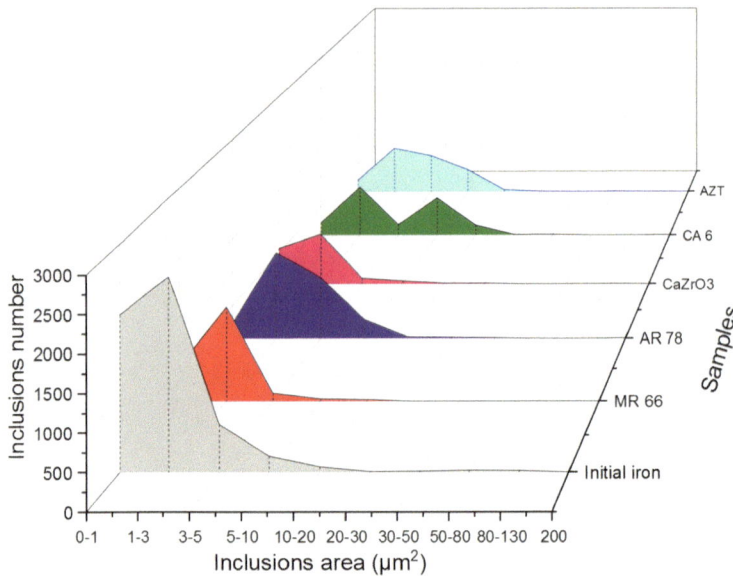

Fig. 21.12 Size distribution of inclusion particles per 100 mm^2 scanned area

inclusions decreased significantly after interaction experiments. The possible explanations for this phenomenon are predicted based on the experimental achieved results: (1). Transportation of inclusions on the sides of the substrates. Deposition of the SiO$_2$ inclusions at the interface of the pure substrates and into the AZT substrate were observed (2). Agglomeration of inclusions. In the initial Armco iron, the inclusions with the size of 1–3 μm^2 showed a peak (see Fig. 21.12 initial iron). However, after the interaction experiment, some peaks showed in the greater size ranges. The greater size of the inclusions indicates the agglomeration process in the liquid iron during the interaction. Three clusters formation mechanisms were proposed in previous investigations, i.e. 1. Collision of single inclusions into a cluster; 2. Single inclusion and cluster; and 3. Cluster to cluster. (3). Evaporation. The evaporation of some elements when the iron sample turns to the liquid state. Mn was found to accumulate on the interface of the solidified iron.

Calculation of the Stability of Oxides

The stability of the oxides MgO, Al$_2$O$_3$, MgAl$_2$O$_4$, CaZrO$_3$, ZrO$_2$, and CA 6 (the oxides formation reactions and the stability calculation method are listed in Table 21.4, free energy ΔG was calculated by utilizing FactSage 7.2 software (FactPS and FToxid), and activity for a_{Me}^x and $a_{Me_xO_y}$ are taken as unity). Figure 21.13 presents the calculated stability of the oxides, the measured oxygen partial pressures were marked in the diagram for the convenience of the comparison. In the pure substrates,

Table 21.4 Used reactions and equations for calculation of oxide stability [18]

$Mg + 2O_2 + 2Al = MgO \cdot Al_2O_3$	(21.9)	$K = \dfrac{a_{Me}^x P_{o2}^y}{a_{Me_X O_Y}}$
$2Mg + O_2 = 2MgO$	(21.10)	
$2\,Al + 3/2O_2 = Al_2O_3$	(21.11)	$In(P_{o2}) = \dfrac{\Delta G}{RT}$
$2Ca + O_2 = 2CaO$	(21.12)	
$2Zr + O_2 = ZrO_2$	(21.13)	
$Ca + Zr + O_2 = CaZrO_3$	(21.14)	
$Ca + 19/2O_2 + 12Al = CaO \cdot 6Al_2O_3$	(21.15)	

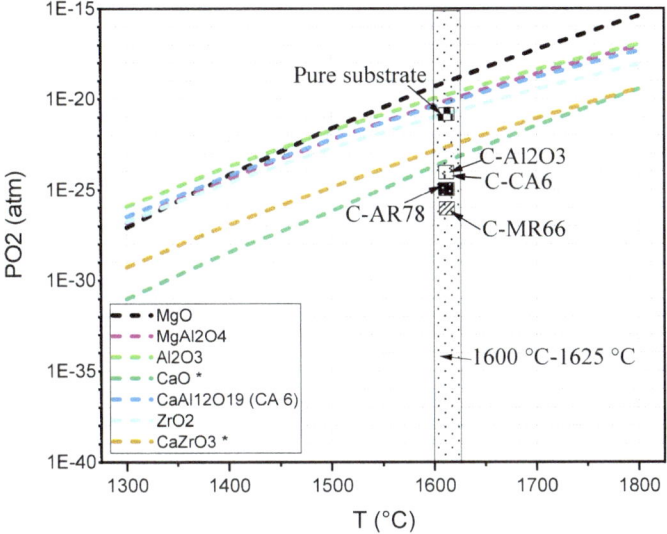

Fig. 21.13 The calculated stability of oxides as a function of temperature and the measured oxygen partial pressure at experimental temperatures

the measured oxygen partial pressure is around 10^{-22} atm at 1600 °C. With this oxygen partial pressure, oxides MgO, Al_2O_3, $MgAl_2O_4$, ZrO_2, and CA 6 could decompose. MgO oxide showed the lowest stability among these oxides. This explained the detected $MgAl_2O_4$ particles on the solidified iron droplet after contacting with MR 66 substrate, as shown in Fig. 21.11. On the other hand, it explains also why the oxygen partial pressure in the case of MR 66 is the lowest. It is due to the presence of the reoxidation of released metallic elements, Mg is easier to be released under the low oxygen partial pressure.

Oxygen partial pressures $P(O_2)$ up to 10^{-25} atm for C-AR 78 substrate, 10^{-26} atm for C-MR 66 substrate, and 10^{-24} atm for C-CA 6 and C-Al_2O_3 substrates and at temperature 1625 °C were measured. It is worth to mention that, the measured oxygen partial pressures were obtained at the gas outlet, i.e. the general oxygen partial pressures in the whole chamber were measured. In fact, the oxygen partial pressures in spherical of carbon bonded substrates could be much lower than the measured values, it means that the decomposition of substrate would take place at low temperatures. It may explain why the iron samples were unshaped in the initial stage of melting process. According to the obtained experimental results and the calculated stability of various oxides, it can be concluded that instead of carbothermic reduction, these oxides would decompose at the present low oxygen partial pressure.

Influence on the Non-metallic Inclusions

The newly formed oxide layers on the iron samples were found to attract the non-metallic inclusions that existed in the iron sample. Inclusions such as SiO_2, CaO, and CaS were found attached to the newly formed oxide layer [20]. Moreover, the SiO_2 was detected as the core of a $MgAl_2O_4$ inclusion. It should be noted that in the initial iron sample SiO_2 was the main inclusion and its size was smaller than 3 μm^2 [18]. By working as a nucleation site, it significantly increased the size of the final inclusion, which is beneficial for the removing process since large inclusions can be easily filtered or removed by buoyancy force. A similar phenomenon was reported by Salomon et al. [7, 11].

On the other hand, the enrichment of metallic elements such as Mg, Al, and Ca in the liquid might result in a higher number of inclusions. However, during the filtration process of molten steel, the diffusion of metallic elements was firstly hindered by carbon at the filter's surface, then by the newly formed oxide layers. Besides this, the filtration time is short, usually in the range of 10–120 s [22]. Therefore, the carbon-bonded filters should not create impurities in the molten iron.

21.4 Conclusion

The present work was aim to analyze the interaction process between various oxides based filter materials with and without the addition of carbon. To complete this, the sessile drop method in a hot stage microscope under protective gas (Argon 5.0) was applied to ensure the cleanness of the experimental chamber. After the interaction experiments between pure substrates and liquid Armco iron, the ASPEX was applied to analyzed the inclusions in the solidified Armco iron including the size, amount, and the type of the inclusions. In general, after the interaction between liquid Armco iron and pure substrates, the most of the substrate remained non-corroded. However, a reaction layer was observed on the AZT substrate. Iron was found to infiltrate into the AZT substrate. Moreover, SiO_2 was found to deposit on the side of all substrates. The contact angle of the liquid Armco iron was also measured and presented. After the interaction experiments, the total number of the inclusions of the all cases decreased. Three factors were pointed out to explain the reduced inclusions number. (1). The increased size of some inclusions indicated the agglomeration of the inclusions;

(2). The transportation of the inclusions to the interface of the substrates; (3). The evaporation of some elements when the iron sample turns to the liquid state.

On the other hand, when carbon was added into the substrates, the interaction process was totally changed. The phenomena such as decarburization of carbon bonded substrates, the formation of whiskers, and oxide layers covered on the iron droplet was observed. In addition, the measured oxygen partial pressures changed with different substrates at the outlet of the hot stage microscope. In order to reveal the influence of iron, a piece of carbon bonded alumina was insert into a quartz glass ampule with an orifice and a platina container with lid and further heated in the hot stage microscope with the same period and temperature. The final results shown that the aluminum was released with the absence of iron. The detected results indicate the newly formed whiskers and oxide layer were caused only by the addition of carbon. Further, the stability of the substrates consisted oxides was calculated. The results showed that the most of the oxides could be decomposed under the low oxygen partial pressures. Moreover, the different oxygen partial pressures were caused also by the various stabilities of the oxides. In summary, with the presence of carbon, the released metallic elements were mainly caused by the low oxygen partial pressures and slightly dissolution.

Acknowledgements The investigation was supported by the DFG (German Research Foundation), Project-ID: 169148856-SFB 920, subprojects A03, C01 at the Technical University Bergakademie Freiberg. The authors are grateful for the financial support and helpful discussions. Furthermore, the authors are very grateful for the technical support by Dr.-Ing. Thilo Kreschel, Ms. Ines Grahl, Dr.-Ing. Armin Franke and Mr. Marcus Block.

References

1. J. Poirier, Metall. Res. Technol. **112**, 410 (2015). https://doi.org/10.1051/metal/2015028
2. A.L.V. da Costa e Silva, J. Mater. Res. Technol. **8**, 2408–2422 (2019). https://doi.org/10.1016/j.jmrt.2019.01.009
3. M. Emmel, C.G. Aneziris, Ceram. Int. **38**, 5165–5173 (2012). https://doi.org/10.1016/j.ceramint.2012.03.022
4. C.G. Aneziris, S. Dudczig, M. Emmel, H. Berek, G. Schmidt, J. Hubalkova, Adv. Eng. Mater. **15**, 46–59 (2013). https://doi.org/10.1002/adem.201200199
5. M. Emmel, C.G. Aneziris, F. Sponza, S. Dudczig, P. Colombo, Ceram. Int. **40**, 13507–13513 (2014). https://doi.org/10.1016/j.ceramint.2014.05.033
6. S. Dudczig, C.G. Aneziris, M. Emmel, G. Schmidt, J. Hubalkova, H. Berek, Ceram. Int. **40**, 16727–16742 (2014). https://doi.org/10.1016/j.ceramint.2014.08.038
7. A. Salomon, M. Dopita, M. Emmel, S. Dudczig, C.G. Aneziris, D. Rafaja, J. Eur. Ceram. Soc. **35**, 795–802 (2015). https://doi.org/10.1016/j.jeurceramsoc.2014.09.033
8. A. Schmidt, J. Fruhstorfer, S. Dudczig, C.G. Aneziris, Adv. Eng. Mater. **22**, 1900647 (2020). https://doi.org/10.1002/adem.201900647
9. T. Zienert, S. Dudczig, P. Malczyk, N. Brachhold, C.G. Aneziris, Adv. Eng. Mater. **22**, 1900811 (2020). https://doi.org/10.1002/adem.201900811
10. E. Storti, H. Berek, C.G. Aneziris, Ceram. Int. **44**, 14502–14509 (2018). https://doi.org/10.1016/j.ceramint.2018.05.065

11. A. Salomon, M. Motylenko, D. Rafaja, Adv. Eng. Mater. **24**, 2100690 (2022) https://doi.org/10.1002/adem.202100690
12. X. Wei, A. Yehorov, E. Storti, S. Dudczig, O. Fabrichnaya, C.G. Aneziris, O. Volkova, Adv. Eng. Mater. **24**, 2100718 (2022). https://doi.org/10.1002/adem.202100718
13. E. Storti, M. Farhani, C.G. Aneziris, C. Wöhrmeyer, C. Parr, Steel Res. Int. **88**, 1700247 (2017). https://doi.org/10.1002/srin.201700247
14. R. Khanna, S. Kongkarat, S. Seetharaman, V. Sahajwalla, ISIJ Int. **52**, 992–999 (2012). https://doi.org/10.2355/isijinternational.52.992
15. T. Zienert, S. Dudczig, O. Fabrichnaya, C.G. Aneziris, Ceram. Int. **41**, 2089–2098 (2015). https://doi.org/10.1016/j.ceramint.2014.10.004
16. C. Voigt, L. Ditscherlein, E. Werzner, T. Zienert, R. Nowak, U. Peuker, N. Sobczak, C.G. Aneziris, Mater. Des. **150**, 75–85 (2018). https://doi.org/10.1016/j.matdes.2018.04.026
17. T. Dubberstein, H.-P. Heller, J. Klostermann, R. Schwarze, J. Brillo, J. Mater. Sci. **50**, 7227–7237 (2015). https://doi.org/10.1007/s10853-015-9277-5
18. X. Wei, S. Dudczig, E. Storti, M. Ilatovskaia, R. Endo, C.G. Aneziris, O. Volkova, J. Eur. Ceram. Soc. **42**, 2535–2544 (2022). https://doi.org/10.1016/j.jeurceramsoc.2022.01.011
19. Z. Xie, F. Ye, J. Wuhan, Univ. Technol.Mater Sci Edn. **24**, 896–902 (2009). https://doi.org/10.1007/s11595-009-6896-1
20. X. Wei, E. Storti, S. Dudczig, A. Yehorov, O. Fabrichnaya, C.G. Aneziris, O. Volkova, J. Eur. Ceram. Soc. **42**, 4676–4685 (2022). https://doi.org/10.1016/j.jeurceramsoc.2022.04.058
21. J.J. Biernacki, G.P. Wotzak, J. Therm. Anal. **35**, 1651–1667 (1989). https://doi.org/10.1007/BF01912940
22. A. Schmidt, A. Salomon, S. Dudczig, H. Berek, D. Rafaja, C.G. Aneziris, Adv. Eng. Mater. **19**, 1700170 (2017). https://doi.org/10.1002/adem.201700170

Chapter 22
High-Temperature Strength and Form Stability of Compact and Cellular Carbon-Bonded Alumina

Horst Biermann, Anja Weidner, and Xian Wu

22.1 Introduction

If not removed timely during the steel production, nonmetallic inclusions (NMIs) will remain in steel products and weaken their mechanical strength, resulting in economic loss or even unexpected disasters. Therefore, the reduction or elimination of NMIs becomes crucial if a high-quality steel supply must be guaranteed. Removal of NMIs by filtration before metal solidification provides an effective way to solve this problem. High-temperature filtration using ceramic foam filters (CFFs) was first applied for the purification of molten aluminum in the 1960s [1]. The rapid development of this technology in aluminum industries began in the 1970s [2, 3]. Its application to other metals followed soon [4]. Due to the much higher melting point of steel (ca. 1536 °C), the filter material for steel filtration should meet very high demands on its thermal and thermomechanical properties. As an engineering material used in functional components such as submerged entry nozzles, stoppers and sliding gates [5], carbon-bonded alumina (Al_2O_3-C) exhibits high thermal conductivity and high thermal shock resistance, which is, therefore, suitable for usage in a high-temperature environment. Its application as filter material in steel metallurgy is intensively investigated within the Collaborative Research Center (CRC) 920 at TU Bergakademie Freiberg, Germany. To characterize and understand the physical properties and (thermo)mechanical behavior, compact Al_2O_3-C specimens as well as foam and spaghetti Al_2O_3-C filters were prepared and studied. Furthermore, the applied pitch binder based on Carbores P was partly or completely replaced by an environmental-friendly binder system based on lactose and tannin. The results can

H. Biermann (✉) · A. Weidner · X. Wu
Institute of Materials Engineering, Technische Universität Bergakademie Freiberg, Gustav-Zeuner-Straße 5, 09599 Freiberg, Germany
e-mail: biermann@ww.tu-freiberg.de

© The Author(s) 2024
C. G. Aneziris and H. Biermann (eds.), *Multifunctional Ceramic Filter Systems for Metal Melt Filtration*, Springer Series in Materials Science 337,
https://doi.org/10.1007/978-3-031-40930-1_22

be used to evaluate the mechanical performance of the materials at room and high temperatures which are required for further optimization of filter design.

22.2 Al_2O_3-C Based on the Carbores P Binder

22.2.1   Manufacture

Compact Al_2O_3-C

The composition of raw materials was based on the formulations reported by Emmel et al. [6]. The main components consisted of α-alumina, Carbores P, carbon black and graphite. The modified pitch Carbores P served not only as the binder but also a carbon source since it remained as carbon residue after pyrolysis. A typical raw material formulation (solid part) is shown in Table 22.1.

Either granules or powders were needed for the later pressing process. Granules were obtained by mixing the raw materials in an automatic mixer equipped with two metallic stirrers for 5 min followed by adding water and glycerol [7]. Subsequently, the granules were dried in an oven to reduce the moisture content. The dried granules were then ready for pressing. In comparison, powders were derived from a slurry produced by mixing the raw materials and additives in water for 24 h in a mill with corundum balls followed by drying, grinding and sieving [8, 9]. The granules or powders were pressed uniaxially or isostatically by a pressing machine under 100–150 MPa using cylindrical or rectangular molds. After this shaping process, a thermal treatment (coking) of the pressed specimens up to 800 or 1400 °C resulted the final products. Compact cylinders were obtained alternatively by drilling from compact bars after coking. Figure 22.1 gives an overview on the pressing routes of preparing cylindrical and rectangular compact Al_2O_3-C specimens.

Cellular Al_2O_3-C

For the manufacture of cellular Al_2O_3-C filters, the above-mentioned pressing routes can certainly not be applied. Al_2O_3-C foam filters—the mostly investigated filter type

Table 22.1   An example of raw material formulation (solid part)

Component	Commercial name	Content (wt%)
α-alumina	Martoxid MR 70	66
Modified pitch	Carbores P	20
Carbon black	Luvomaxx N-991	6
Graphite	AF 96/97	8
Total		**100**

Fig. 22.1 The pressing routes (granule route on the left and slurry route on the right) for production of cylindrical and rectangular compact Al_2O_3-C

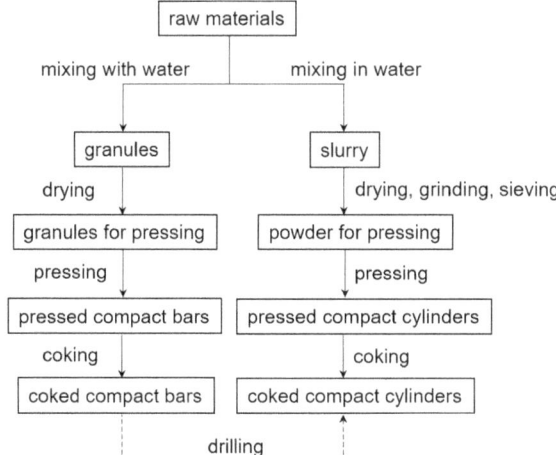

under CRC 920—were prepared through the replica method. Invented independently by Schwartzwalder and Somers as well as Holland in 1963 [1, 10], the replica method is currently widely used for the manufacture of CFFs [6]. To prepare a cylindrical Al_2O_3-C foam filter, a polyurethane foam with a defined geometry and ppi (pores per inch) parameter was used as the template. This was impregnated with the Al_2O_3-C slurry. After drying, the foam surface was sprayed with another slurry with the same composition but different viscosity (i.e. different water content). After drying again, the polymer scaffold was burned out by thermal treatment (up to 800 or 1400 °C). Thus, the replicated ceramic foam was obtained which was either used as the final product or was coated by another formulation of slurry followed by drying and thermal treatment. The corresponding products were assigned as uncoated or coated Al_2O_3-C foam filters, respectively.

Spaghetti filters were manufactured by alginate-based robo gel casting [11]. An alginate-containing Al_2O_3-C slurry was prepared and added into an alkaline earth (calcium or barium) salt solution through a nozzle. The movement of the nozzle was guided and controlled by a robot system. Thus, a stacked spaghetti plate was generated which became stiff after gelation triggered by chelation between the alkaline earth metal ions and the alginate chains. The stiff cast was then removed from the solution and dried. After thermal treatment (up to 800 °C), a coked Al_2O_3-C plate in spaghetti form was obtained. Cylinders for mechanical tests were then cut from the plate. Figure 22.2 gives an overview of the preparation routes to cellular Al_2O_3-C.

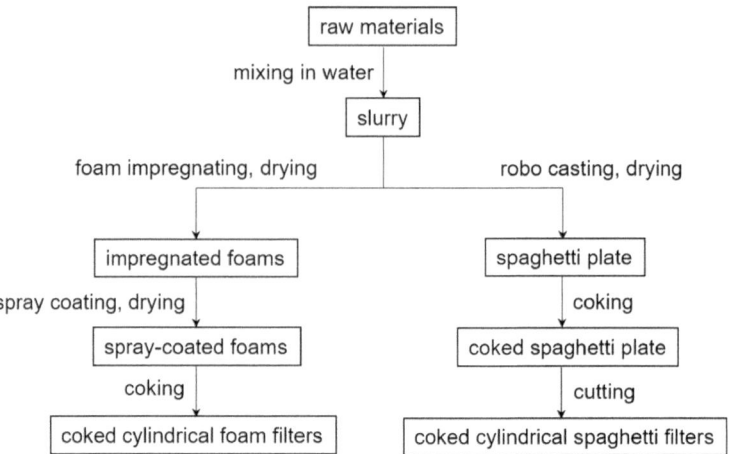

Fig. 22.2 The replica route (left) and the robo gel casting route (right) for production of Al_2O_3-C foam and spaghetti filters

22.2.2 Physical and Microstructural Properties

Compact Al_2O_3-C

Klemm et al. prepared compact Al_2O_3-C initially via the granule route [7]. Cylinders (Ø 50 mm and 50 mm height) and bars (7 mm × 7 mm × 70 mm) were obtained by uniaxial as well as isostatic pressing, followed by coking at 800 °C. To explore the influence of the binder amount, specimens with Carbores P content ranging from 5 to 30 wt% were prepared. It was found that the bulk density of the specimens decreased with the Carbores P content (ca. 1.93 g/cm^3 at 5 wt% Carbores P vs. ca. 1.84 g/cm^3 at 30 wt% Carbores P) (Fig. 22.3a). The corresponding open porosity decreased with increasing Carbores P content which reached a minimum (ca. 29.5%) at 20 wt% Carbores P and increased again at 30 wt% Carbores P (Fig. 22.3b). The degree of volume shrinkage after thermal treatment showed a similar trend with a maximum of ca. 3.6% at 20 wt% Carbores P. Substitution of the coarse-grained Carbores P with a particle size of $d_{50} = 80$ μm by a fine-grained variant ($d_{50} = 5$ μm) at 15 wt% binder content led to an increase in open porosity and decreases in bulk density (Fig. 22.3a and b). Above 15 wt% binder content, uniaxially pressed specimens showed higher open porosity and lower bulk density in comparison to specimens made by isostatic pressing. The volume shrinkage was generally stronger for uniaxially pressed specimens.

Compact Al_2O_3-C cylinders and bars of the same sizes were also prepared via the slurry route [8, 9]. The specimens were shaped by uniaxial pressing and coked at either 800 or 1400 °C. A comparison of two sticks made by the granule and slurry route, respectively, showed that the macrostructure of the former was rougher [9]. Even cracks were found between granules due to the high strength of these. Coking at

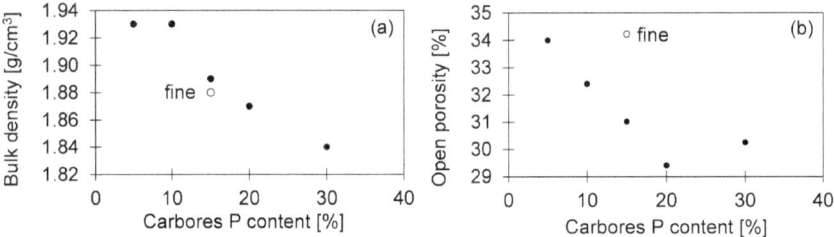

Fig. 22.3 Open porosity **a** and bulk density **b** of compact Al_2O_3-C specimens made by the granule route [7]

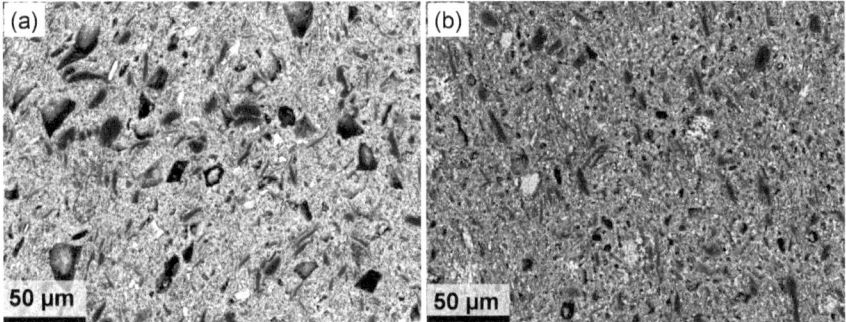

Fig. 22.4 SEM images of compact specimens with 10 wt% binder made by the slurry route: **a** coked at 800 °C; **b** coked at 1400 °C

1400 °C instead of 800 °C led to lower open porosity, stronger shrinkage and higher bulk density since higher temperature favors compaction of the microstructure [8]. A comparison of the scanning electron microscopy (SEM) images of the microstructure of specimens with 10 wt% binder coked at 800 and 1400 °C indicated that the binder residue of the latter had smaller grain sizes (Fig. 22.4).

Cellular Al_2O_3-C

Ranglack-Klemm et al. reported the preparation of carbon nanotubes/alumina nanosheets (CNT-ANS) or alumina nanospheres (CNT-ANB) coated Al_2O_3-C foam filters with 20 wt% Carbores P in cylindrical form (10 ppi, Ø 20 mm and 25 mm height) [12]. Compared with the uncoated Al_2O_3-C foam filters, the coated ones exhibited slightly higher open microporosity and comparable residual carbon content.

Studies were also carried out on Al_2O_3-C-coated as well as Al_2O_3-coated Al_2O_3-C foam filters (20 wt% Carbores P, Ø 20 mm and 25 mm height, 10 ppi) [13]. The results of mercury intrusion porosimetry showed trimodal or multimodal distribution patterns of the specimens with similar open microporosity (ca. 35%). A further

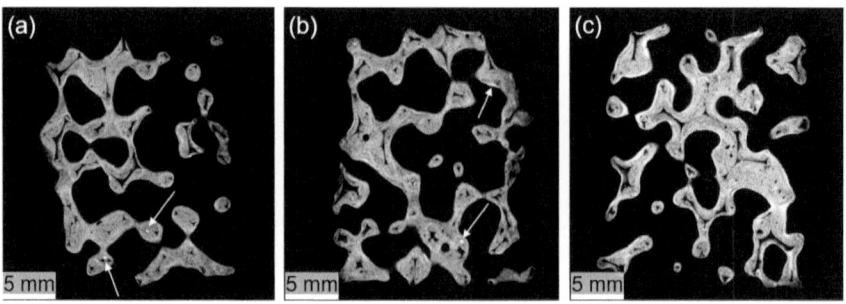

Fig. 22.5 Reconstructed CT images of Al_2O_3-C foam filters: **a** uncoated; **b** coated with Al_2O_3-C; **c** coated with Al_2O_3 [13]

examination by computer tomography (CT) revealed higher macroporosity of the Al_2O_3-coated specimen than that of the Al_2O_3-C-coated one (64 vs. 60%), although the total porosity was almost the same (70%). The CT images also revealed the inhomogeneous distribution of alumina coating (Fig. 22.5). The arrows indicate some bright spots as examples of inhomogeneously distributed alumina.

22.2.3 Mechanical and Thermomechanical Behavior

Mechanical tests were performed on Al_2O_3-C compact specimens (cylinders and bars) as well as Al_2O_3-C foam and spaghetti filters at room and high temperatures. Room temperature tests were carried out under air atmosphere. High-temperature tests were carried out between 700 and 1500 °C. To avoid the oxidation of carbon, it was necessary to run the tests under a protective atmosphere. A specially designed high-temperature testing machine equipped with a sealable chamber and argon supply could meet such a requirement. The specimens were heated by electromagnetic induction in both compression (for cylindrical specimens) and bending (for rectangular specimens) modes. Figure 22.6 shows the machine and the experimental setup for testing cylindrical foam filters. Susceptors made by TZM (titanium-zirconium-molybdenum alloy) were used to enable a homogeneous temperature distribution. More details about the testing machine and testing parameters can be found in the corresponding references [12, 13].

Compact Al_2O_3-C

Compression Behavior

Compression tests at room temperature on uniaxially pressed cylindrical specimens (Ø 50 mm and 50 mm height, coked at 800 °C) made through the granule route showed

(a)  (b)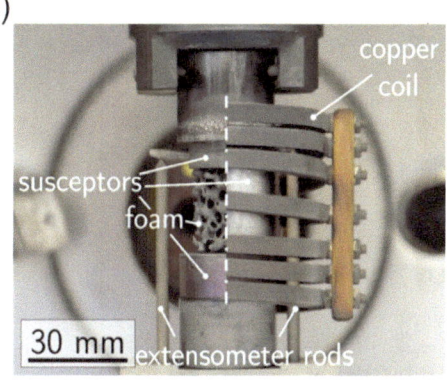

Fig. 22.6 **a** High-temperature testing machine; **b** Experimental setup for testing a foam filter at high temperatures [13]

Fig. 22.7 Cold crushing strength of compact Al$_2$O$_3$-C specimens made by the slurry route coked at 800 and 1400 °C [8]

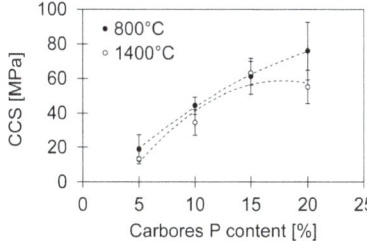

an increase of cold crushing strength (CCS) with the content of Carbores P from 5 wt% (15 MPa) to 20 wt% (44 MPa) and a slight decrease at 30 wt% (42 MPa) [7]. The CCS values of the specimens from the slurry route under otherwise comparable conditions were higher (e.g. 76 MPa vs. 44 MPa at 20 wt% Carbores P) [8]. For the specimens made by the slurry route, those coked at 1400 °C usually exhibited lower CCS than those at 800 °C (Fig. 22.7). At 20 wt% Carbores P, in particular, the difference of CCS reached the highest value of 25 MPa. Therefore, a higher coking temperature did not result in a higher CCS.

Solarek et al. investigated compression behaviors of compact Al$_2$O$_3$-C cylinders (ca. Ø 25 mm and 25 mm height) at room and high temperatures until 1500 °C [14]. The specimens were shaped by uniaxial pressing and coked at 1400 °C. A pronounced hysteresis of the compressive stress–strain curve at room temperature because of plasticization in the carbonaceous matrix was observed (Fig. 22.8). The width of the hysteresis decreased at the second loading.

Creep tests showed an increase in creep rate above 1050 °C (Fig. 22.9) and a power law was proposed to correlate stress (σ) and strain rate at 30 min ($\dot{\varepsilon}_{30min}$) (Eq. 22.1). At 750 and 1050 °C, the stress exponent n was nearly zero. At 1350 °C, n was 2.3. The activation energy of creep above 1150 °C was determined as 263 kJ/mol.

Fig. 22.8 Compressive stress–strain curves of compact Al$_2$O$_3$-C produced by uniaxial pressing at room temperature [14]

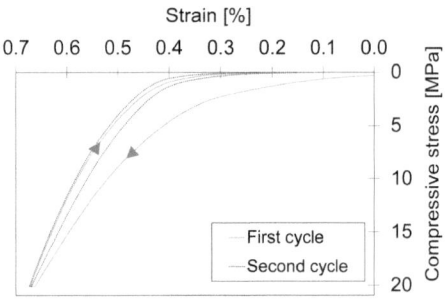

Fig. 22.9 Creep rate of compact Al$_2$O$_3$-C cylinders produced by uniaxial pressing at 750, 1050 and 1350 °C [14]

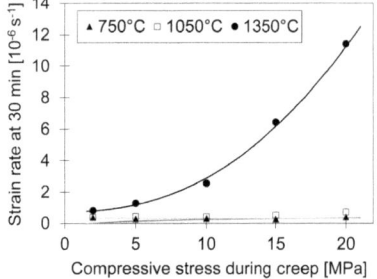

Fig. 22.10 Relaxation stress of uniaxially pressed compact Al$_2$O$_3$-C cylinders after different times at different temperatures [14]

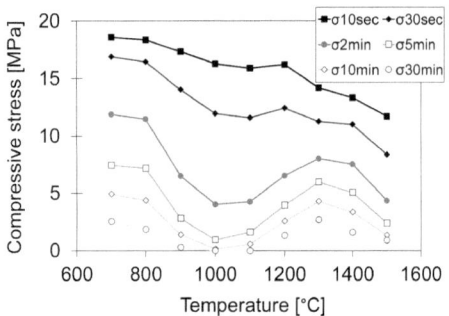

$$\dot{\varepsilon}_{30min} \sim \sigma^n \tag{22.1}$$

Stress relaxation tests showed a minimum remaining stress between 900 and 1200 °C as well as a higher remaining stress between 1300 and 1400 °C (Fig. 22.10). During cyclic stress relaxation, stress relaxation decreased with ongoing cycling. The relatively smaller scatter supported the assumption that high-temperature deformation depends on a plastic process rather than on the size of the major defect [14].

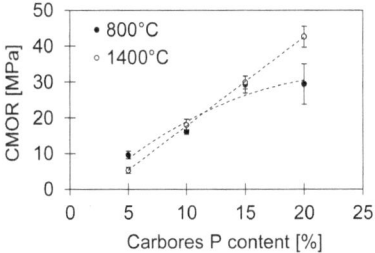

Fig. 22.11 Cold modulus of rupture of compact Al$_2$O$_3$-C specimens made by the slurry route coked at 800 and 1400 °C [8]

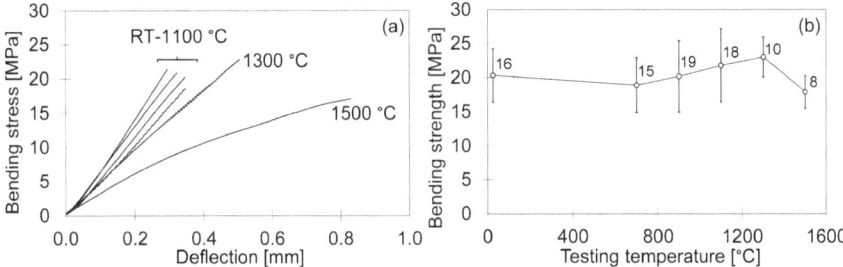

Fig. 22.12 Results of bending tests of compact Al$_2$O$_3$-C bars produced by uniaxial pressing at different temperatures: **a** stress–deflection curves; **b** bending strength versus temperature with the quantity of specimens tested at each temperature [15]

Bending Behavior

Bending tests at room temperature on granule route specimens (7 mm × 7 mm × 70 mm, uniaxially pressed, coked at 800 °C) showed that the cold modulus of rupture (CMOR) values at 20 wt% and 30 wt% Carbores P (17 MPa and 18 MPa) were significantly higher than those at 5–15 wt% Carbores P (<11 MPa) [7]. The corresponding slurry route specimens exhibited higher CMOR which reached 29 MPa at 20 wt% Carbores P [8]. Furthermore, uniaxially pressed specimens showed higher CMOR values than isostatically pressed ones, especially when the Carbores P content was above 15 wt% (Fig. 22.11) [9]. In contrast to the results of CCS values obtained by compression tests, the CMOR value was found higher at 20 wt% Carbores P for specimens coked at 1400 °C than that of 800 °C (42 MPa vs. 28 MPa) [8]. Therefore, a higher coking temperature led to higher bending strength.

Solarek et al. studied the bending behavior of larger compact bars (ca. 25 mm × 25 mm × 145 mm, uniaxially pressed, coked at 1400 °C) at room and high temperatures up to 1500 °C (Fig. 22.12) [15]. A maximum bending strength value was recorded at 1300–1400 °C which could be explained by changes of the material behavior from brittle to ductile leading to reductions in stress concentrations.

In contrast, bending tests on pure graphite bars did not show any softening up to 1500 °C and the strength increased with the temperature [15]. Such differences between graphite and carbon-bonded alumina might result from the higher thermal expansion of alumina in comparison to carbon. Thus, the generated compressive stresses within the graphitic layers of the carbonaceous matrix could promote the glide of the layers already far below 2000 °C. Moreover, they could enhance fracture toughness by closing cracks and increasing resistance against crack growth [15].

Fracture Mechanical Behavior

Compact bars (ca. 25 mm × 25 mm × 145 mm, uniaxially pressed, coked at 1400 °C) were further tested on single edged V-notched beams at room and high temperature [16]. With help of an optical system, the crack mouth opening displacement (CMOD) was measured precisely. Typical force-CMOD-curves of the material are shown in Fig. 22.13.

At room temperature, the force dropped immediately to zero at very small CMOD (30 μm) after a linear-elastic deformation at the beginning. The specimen behaved brittle with instable crack propagation and low work of fracture at room temperature. The fracture toughness K_{Ic} was determined as 0.69 ± 0.07 MPa $\cdot$ m$^{1/2}$. At 1400 °C, the material became ductile and visco-plastic and showed, compared to the result at room temperature, a significant increase of toughness and work of fracture, although a quantitative analysis of the latter was not possible due to large scatter of the CMOD signal. Macroscopic deformation as well as cracks arising from temperature gradients during cooling were observed on specimens after testing at 1400 °C [16].

Cellular Al$_2$O$_3$-C

Solarek et al. performed compression tests on uncoated foam filters (coked at 1400 °C) at 800–1500 °C [15]. Similar to the results of bending tests on compact

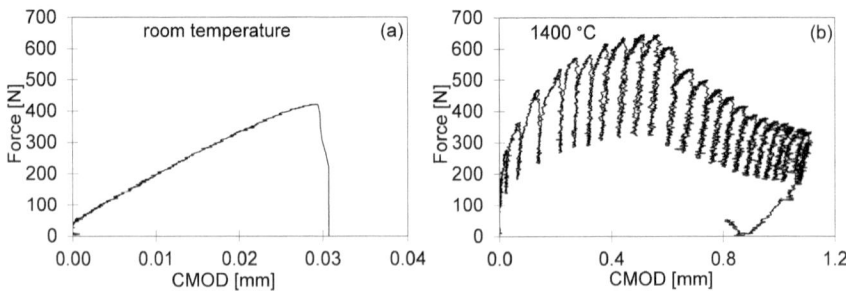

Fig. 22.13 Typical force-CMOD-curves of Al$_2$O$_3$-C: **a** at room temperature; **b** at 1400 °C with unloading steps [16]

bars of the same material, the foam filters exhibited a maximum strength value at 1300–1400 °C, indicating again a brittle-to-ductile transition. Furthermore, the force–displacement curves at temperatures below 1200 °C indicated ruptures of individual foam struts (see the arrows in Fig. 22.14a). At 1400 and 1500 °C, the curves showed fewer loading drops and became smoother, obviously due to significant plastic deformation (Fig. 22.14b). It was supposed that the presence of an electromagnetic field supports the formation of graphitic structures, which was observed using Raman spectroscopy (Fig. 22.15).

Selected specimens after compression tests were examined by SEM in secondary electron mode (Fig. 22.16). Numerous failed foam struts were observed at temperatures not exceeding 1300 °C (Fig. 22.16a). No plastic deformation was visible. Figure 22.16b shows a specimen that was 2 mm shortened after testing at 1500 °C with a partly cracked foam strut at the lower center (Fig. 22.16c).

Ranglack-Klemm et al. investigated the compression behavior of Al_2O_3-C foam filters without and with functional coatings of CNT-ANS or CNT-ANB (see Sect. 2.2.2) at 1100 °C and 1450 °C [12]. The application of functional coatings led to higher compressive strength at 1100 °C with the highest performance achieved by the CNT-ANS coating. The study showed also the brittle character of the filter materials at 1100 °C which turned into viscoplastic behavior with pronounced plastic deformation at 1450 °C (Fig. 22.17).

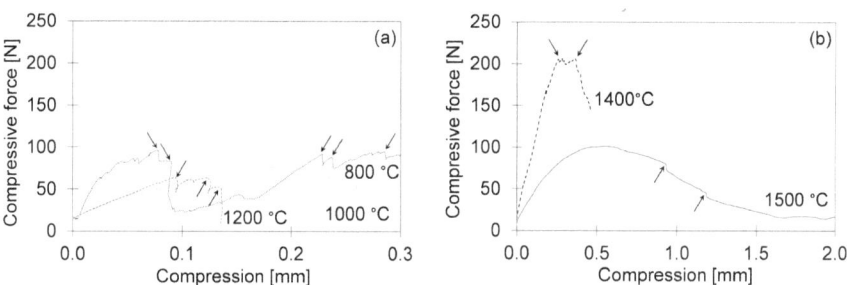

Fig. 22.14 Force-compression curves of Al_2O_3-C foams: **a** at 800, 1000 and 1200 °C; **b** at 1400 and 1500 °C [15]

Fig. 22.15 Raman spectra of Al_2O_3-C foams before (gray) and after compression test at 1500 °C (black) [15]

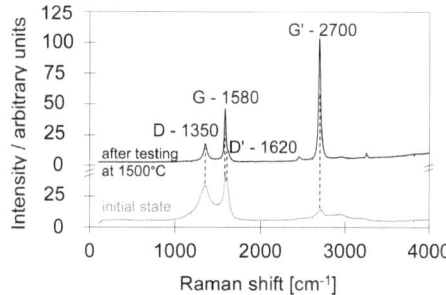

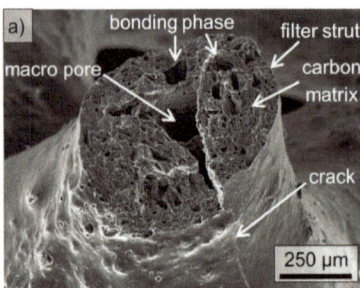
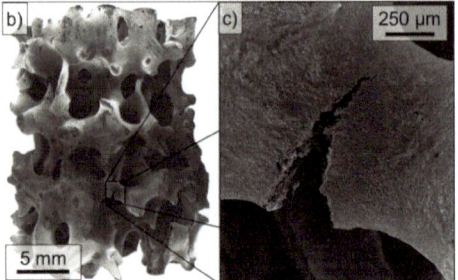

Fig. 22.16 SEM micrographs of Al_2O_3-C foams after compression tests **a** at room temperature and **b, c** at 1500 °C [15]

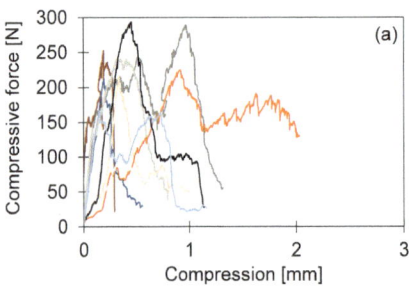
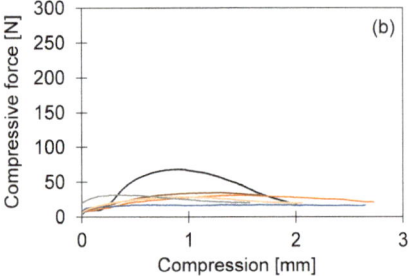

Fig. 22.17 Compressive force-strain curves of Al_2O_3-C foams coated with CNT-ANS tested at **a** 1100 °C and **b** 1450 °C [12]

Compression tests on uncoated, Al_2O_3-C-coated or Al_2O_3-coated Al_2O_3-C foam filters at 1300, 1400 and 1500 °C showed that the coated specimens were more resistant against compression [13]. While brittle behavior was dominant at 1300 °C, the transition from brittle to viscoplastic deformation became apparent at 1400 °C and 1500 °C (Fig. 22.18). Figure 22.19 shows the compressive strength and the corresponding compression values of the uncoated and coated filters tested at 1300, 1400 and 1500 °C. Despite high scatter, the filters coated with Al_2O_3-C (coating 1) exhibit the smallest compressive strength at 1400 °C. This could be explained by the softening of the carbon-containing binder phase which was highly viscous at that temperature, leading to a relatively low compressive strength. At 1500 °C, the viscosity was sufficiently low, so that healing of defects occurred, resulting in higher compressive strength.

According to the SEM analysis on Al_2O_3 coated filters, struts with spalled coating were observed after testing at 1300 °C (Fig. 22.20a). Some arrested cracks and intact coating were observed on a filter after testing at 1400 °C (Fig. 22.20b). Furthermore, a significant viscoplastic deformation led to the buckling of the coating (Fig. 22.20d). A polished section of a filter specimen showed a continuous gap between the filter (I) and the Al_2O_3 coating (II) (Fig. 22.20c). During the thermomechanical testing, this gap caused a wall-slip effect between the coating and substrate.

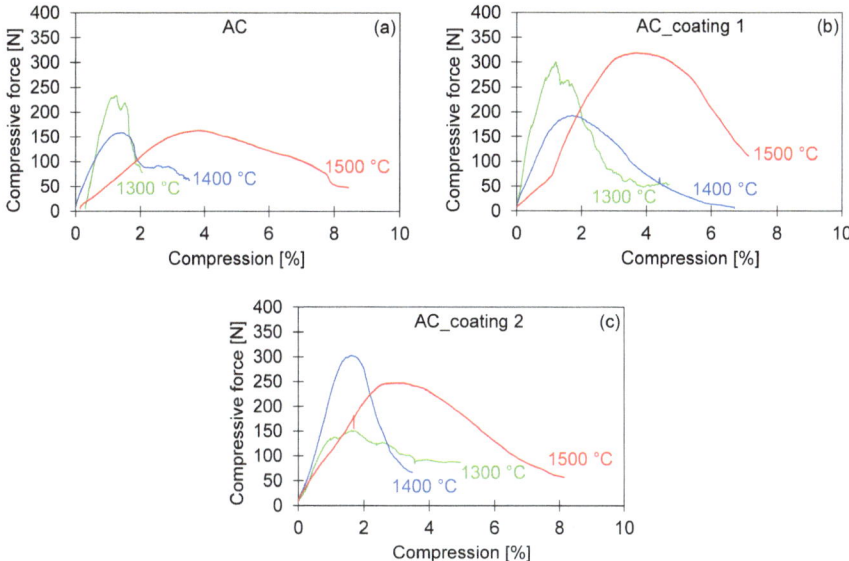

Fig. 22.18 Typical compressive force-strain curves of Al_2O_3-C (AC) foams: **a** uncoated; **b** coated with Al_2O_3-C (coating 1): **c** coated with Al_2O_3 (coating 2) [13]

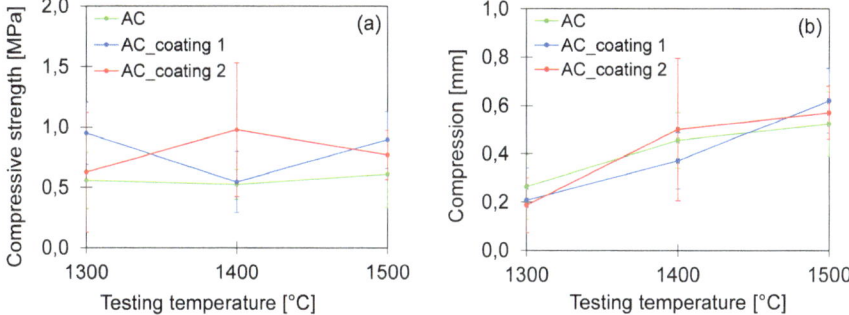

Fig. 22.19 Comparison of the values of compressive strength (**a**) and compression at maximum strength (**b**) tested at different temperatures (AC = uncoated, AC_coating 1 = coated with Al_2O_3-C, AC_coating 2 = coated with Al_2O_3) [13]

To examine the residual thermomechanical properties of Al_2O_3-C foam filters coated with carbon-containing calcium dialuminate (CA_2-C) or calcium hexaaluminate (CA_6-C) after contact with steel melt, the filters were immersed into steel melt under argon atmosphere in a metal casting simulator at 1650 °C for 10 s [17]. Compression tests were then performed on these residual filters at 1100 °C and 1500 °C (Fig. 22.21). Similar to the above-mentioned studies, the filter materials exhibited brittleness at 1100 °C and ductility at 1500 °C with considerable compression strengths at both temperatures. Hence, even after contact with steel melt, the

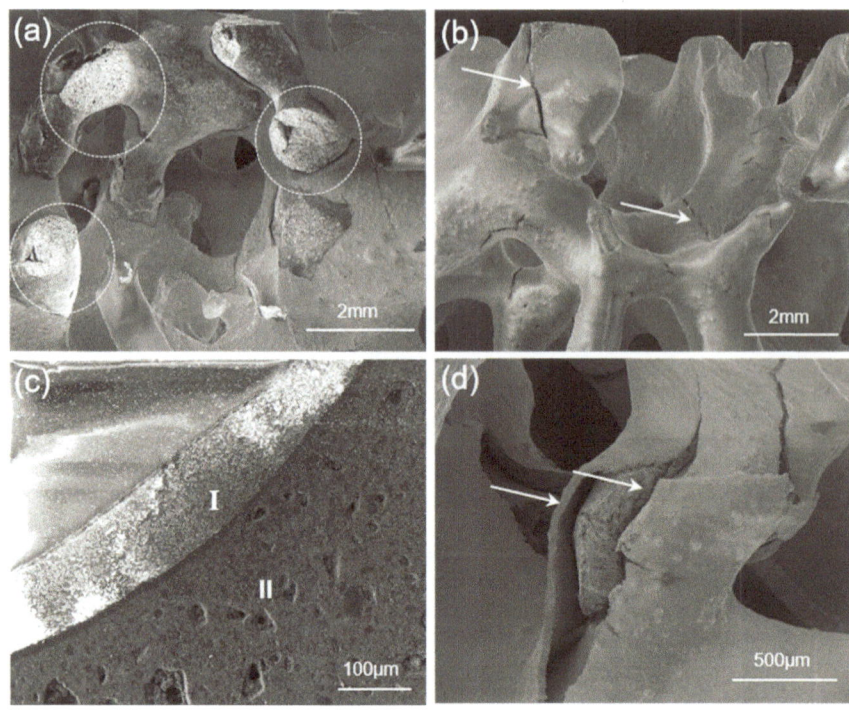

Fig. 22.20 SEM images of Al_2O_3-coated Al_2O_3-C foam filters after testing at different temperatures; **a** 1300 °C; **b** 1400 °C; **c, d** 1500 °C [13]

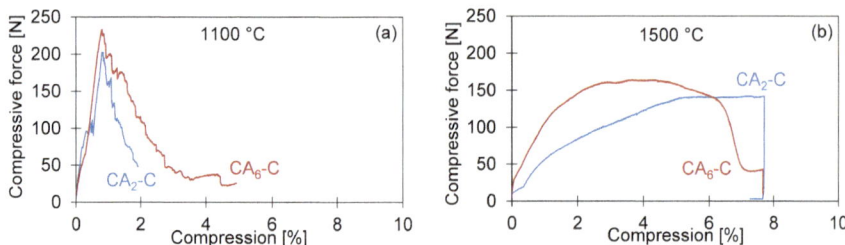

Fig. 22.21 Force-compression curves of residual filters at different temperatures: **a** 1100 °C; **b** 1500 °C [17]. Please note that the tests shown in (**b**) were interrupted shortly before 8% compression due to machine limitations

mechanical behavior of the specimens was not significantly influenced. Also, the pore size distribution and carbon content did not change much after the immersion. Energy dispersive spectroscopy (EDS) analysis of the filter coated with CA_6-C after compression test at 1100 °C showed the remaining coating material (Fig. 22.22) and some steel droplets (Fig. 22.22e).

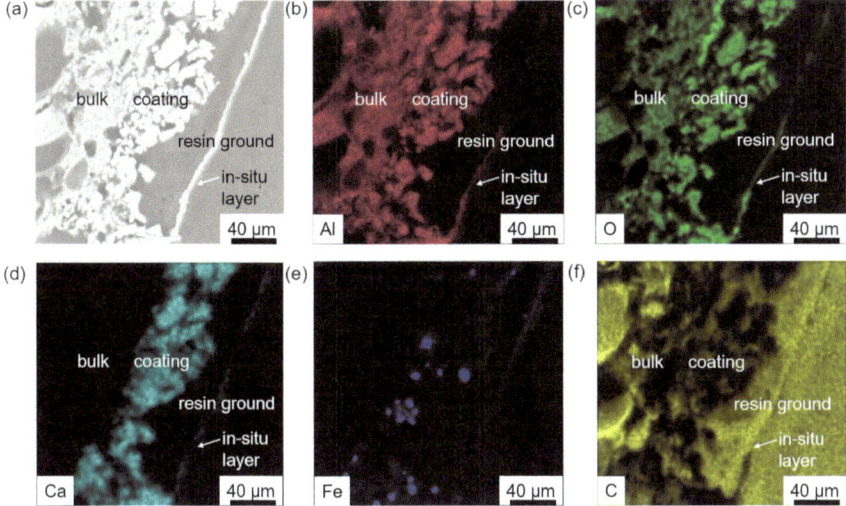

Fig. 22.22 CA$_6$-C-coated foam filter residue after compression test at 1100 °C: **a** back scattered electrons (BSE) micrograph; **b–f**: distribution of elements Al, O, Ca, Fe and C [17]

The compression behavior of Al$_2$O$_3$-C spaghetti filters was studied at room temperature, 800 and 1500 °C (Fig. 22.23) [18]. Ductile deformation occurred only at 1500 °C (Fig. 22.23c). In comparison to the foam filters of similar material composition and size, the spaghetti filters showed significantly higher compression strength at high temperatures. The compressive strength-density diagram of foam and spaghetti filers tested at 800 °C showed a significant difference in the regression slopes, indicating that the higher compressive strength of the spaghetti filters cannot be explained by the higher mass alone (Fig. 22.24). The full struts (not hollow as in the case of foam filters) should have a positive influence on the mechanical strength of the spaghetti filters.

22.3   Al$_2$O$_3$-C Based on the Lactose-Tannin Binder System

22.3.1   The Lactose-Tannin Binder System

Although pitch-based binders have been used for the manufacture of refractories for more than 100 years [19], one crucial drawback is the release of carcinogenic compounds such as benzo[a]pyrene during thermal treatment [20]. In the case of Carbores P, the amount of harmful substances is limited to a lower level. However, their concentration still cannot fully meet the REACH (Registration, Evaluation, Authorisation and Restriction of Chemicals) criterium in Europe [21]. Besides the pitch binders, phenolic resins are also common binders for producing carbon-bonded

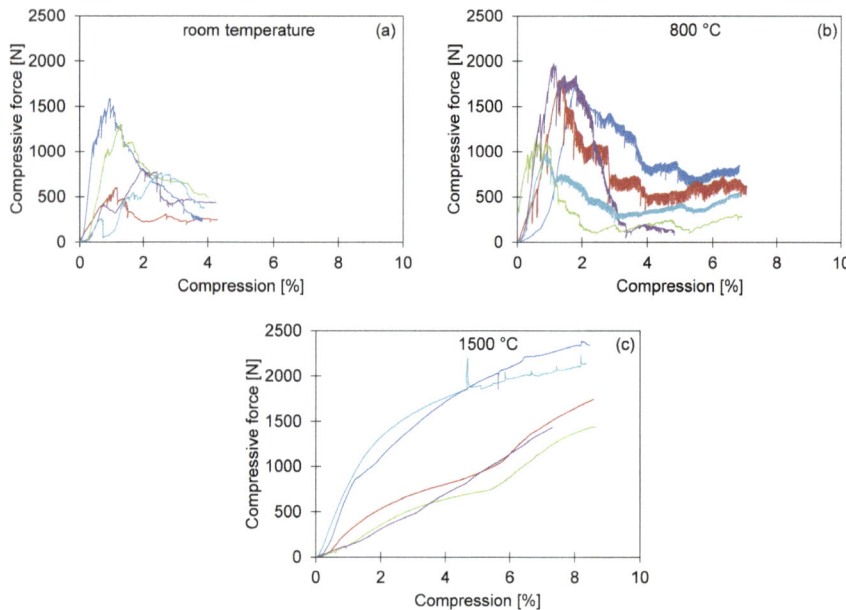

Fig. 22.23 Force-compression curves of Al_2O_3-C spaghetti filters tested at different temperatures: **a** room temperature; **b** 800 °C; **c** 1500 °C [18]

Fig. 22.24 Compressive
strength-density diagram of
Al_2O_3-C foam and spaghetti
filters tested at 800 °C [18]

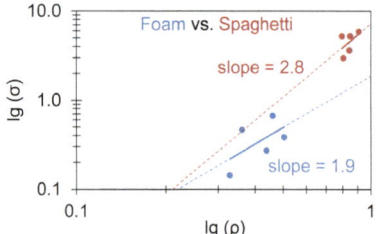

refractories [22]. Unfortunately, the production of theses resins releases free phenol which is hazardous to the environment, too [23]. Therefore, searching for alternative binder systems with environmental-friendly character is meaningful. One promising candidate is a binder system based on a combination of lactose and tannin. First attempts to use the lactose-tannin (L–T) binder system for the manufacture of MgO-C bricks and Al_2O_3-C foam filters showed positive results [24–26].

Lactose (Latin *lac* for milk + chemical suffix *-ose*) is a double sugar existing in dairy products. Cow milk, for example, contains ca. 5 wt% lactose as the second most important component after water. Lactose is a structurally well-defined substance whose chemical structure (combined with one water molecule, i.e. monohydrate) is shown in Fig. 22.25a. Lactose as a binder finds applications in pharmaceutical and food industries as well as for ore processing [27].

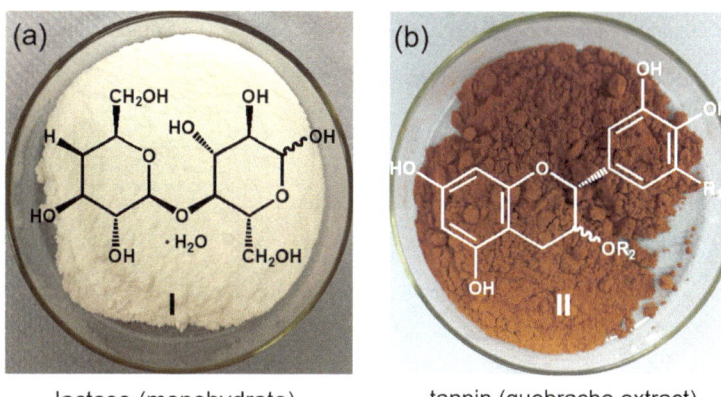

Fig. 22.25 Chemical structures and appearance of lactose (**a**) and tannin (**b**) used in our study

Tannin (from Old High German *tanna* for conifer) is a common component found in many plants such as in tree barks, leaves, stems, fruits, roots, etc. However, tannin is not a substance with a unique molecular structure, but rather a collective name for structurally different molecules which are called vegetable polyphenols. According to Freudenberg (the German chemist who was famous for his contributions to natural organic compounds), tannins are generally divided into two types: hydrolyzable and condensed tannins. Currently, most of the worldwide produced tannins are condensed tannins in which flavonoids (e.g. catechins) are the basic structural unit [28]. The molecular weights of tannins are between 500 and 3000 g/mol. The most important application of tannins is tanning animal skins into leather. The tannin used in our study is a natural condensed tannin extracted from Quebracho trees. Its basic structural fragments are catechins as shown in Fig. 22.25b.

22.3.2 Manufacture

Two strategies were applied for the formulation of the L–T-based binder system: pitch reduction [29] and pitch replacement [30]. In the first case, 80 wt% of Carbores P was replaced by lactose and tannin, i.e. the binder system consisted of 16 wt% L–T and 4 wt% Carbores P (pitch-lean L–T-based binder system). In the second case, Carbores P was completely replaced by lactose and tannin (pitch-free L–T-based binder system). In both cases, the ratio of lactose and tannin was varied systematically. Examples of raw material composition (main solid part) based on the L–T binder system (pitch-lean and pitch-free) are given in Table 22.2.

For the manufacture of L–T-based compact Al_2O_3-C specimens, a slip casting route has been developed [29]. In comparison to the pressing routes, the slip casting route is closer to the replica route for the manufacture of foam filters due to the

Table 22.2 Examples of raw material formulation (main solid part) based on the pitch-lean and pitch-free L–T binder system

Component	Commercial name	Content (wt%)
Example formulation based on the pitch-lean L–T binder system		
α-alumina	Martoxid MR 70	66
Lactose/tannin	Mivolis Milchzucker/Quebracho-Extrakt	16
Modified pitch	Carbores P	4
Carbon black	Luvomaxx N-991	6
Graphite	AF 96/97	8
Total		**100**
Example formulation based on the pitch-free L–T binder system		
α-alumina	Martoxid MR 70	66
Lactose/tannin	Mivolis Milchzucker/Quebracho-Extrakt	20
Carbon black	N-991	6
Graphite	AF 96/97	8
Total		**100**

absence of high pressure during the whole process. Such a slip casting route was, however, not applied to the manufacture of L–T-based rectangular compact Al_2O_3-C bars (ca. 25 mm × 25 mm × 145 mm) since cracks appeared after the casting of the large specimens. Thus, L–T-based rectangular compact Al_2O_3-C specimens were prepared through the slurry route. Figure 22.26 gives an overview of the manufacture of L–T-based compact Al_2O_3-C cylinders and bars.

For L–T-based specimens, a hardening process was necessary before the cast or pressed specimens were sent to coking. During this process, the specimens were heated up to 180 °C in air, while individual tannin molecules were linked with each other through a hardening agent. The most common hardening agent for tannins is hexamethylenetetramine (also called hexamine or hexa) due to its good performance and low toxicity [31, 32]. The coking process was performed at 1000 °C. Coked bars were used directly for mechanical tests. Coked cylinders were machined into desired sizes which were further used as specimens for mechanical tests.

22.3.3 Physical and Microstructural Properties

Cylindrical compact Al_2O_3-C specimens based on the pitch-lean L–T-based binder system with a mass ratio of lactose and tannin ranging from 5:1 to 1:5 were prepared and studied [28]. It was found that the slurry viscosity increased with the tannin content which greatly influenced the quality of the casts. As the result, specimens with L:T = 1:1 and 1:5 could not be machined due to critical cracks appearing after coking, and specimens with L:T = 1:2 and 2:1 were highly porous. High-quality

Fig. 22.26 The slip casting route for L–T-based compact Al₂O₃-C cylinders and the pressing route for L–T-based compact Al₂O₃-C bars

Fig. 22.26 The slip casting route for L–T-based compact Al_2O_3-C cylinders and the pressing route for L–T-based compact Al_2O_3-C bars

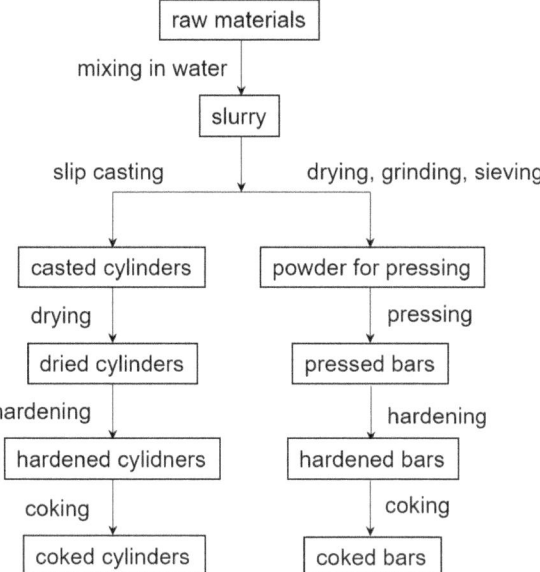

specimens were only producible for L:T = 5:1, 4:1 and 3:1 which showed similar volume shrinkage (5%) and mass reduction (9%) after coking and little difference in residual carbon content (23 wt%). Their bulk density and open porosity values were around 1.7 g·cm^{-3} and >40%, respectively. Furthermore, specimens with L:T = 5:1, 4:1 and 3:1 based on the pitch-free L–T binder system with comparable bulk density and porosity were prepared [30]. According to mercury intrusion porosimetry measurements, the specimens exhibited monomodal pore size distribution [30].

SEM images of the sections of coked cylindrical specimens showed all microstructural features such as alumina particles, pores and isolated carbonaceous bonding aggregates (Fig. 22.27). A microcrack was observed in the specimen with L:T = 3:1 which might be formed during thermal treatment or machining (Fig. 22.27c).

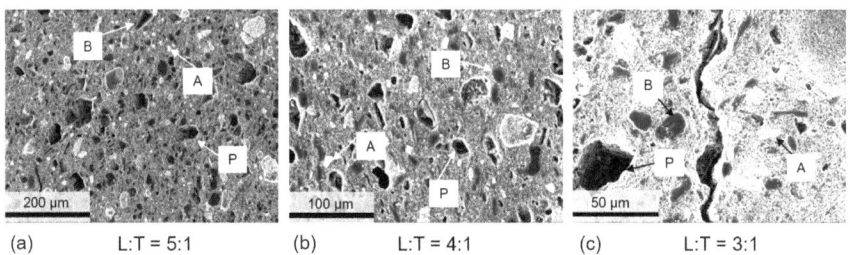

Fig. 22.27 SEM images of pitch-free L–T based compact specimens (A = alumina particle; B = isolated carbonaceous bonding aggregates; P = pore): **a** L:T = 5:1; **b** L:T = 4:1; **c** L:T = 3:1 [30]

22.3.4 Mechanical and Thermomechanical Behavior

Compression tests on cylindrical compact specimens based on the pitch-lean L–T
binder system showed that the CCS values of those with L:T = 5:1 (47 MPa), 4:1
(34 MPa) and 3:1 (28 MPa) were significantly higher than L:T = 2:1 (14 MPa)
and L:T = 1:2 (8 MPa) [29]. They were even higher than or comparable with the
CCS value of the reference specimen (i.e. specimen based on the Carbores P binder
made by slip casting) (32 MPa), although the latter exhibited higher tolerance against
compression since the strain at maximum compressive stress was much higher (3–4
vs. 1–2%) as shown by the CCS test curves (Fig. 22.28). Nevertheless, the splitting
tensile strength values of the L–T specimens (3–5 MPa) were much lower than
that of the reference (11 MPa). This is attributed to the glassy residual carbon of
the pyrolyzed L–T binders which was more fragile against tensile stress. The CCS
curves of the specimens based on the pitch-free L–T binder system showed also
similar character [30].

Thermomechanical tests were focused on cylindrical and rectangular compact
specimens based on the pitch-free L–T binder system with L:T = 5:1, 4:1 and 3:1
[30]. High-temperature bending tests were carried out on bars (ca. 25 mm × 25 mm
× 145 mm) at 700, 900, 1100, 1300 and 1400 °C (Fig. 22.29). The bending strength
ranged from 0.9 to 1.6 MPa and reached generally its maximum at 1100 °C. At
1300 and 1400 °C, the specimens were more deflected. Thus, the brittle-to-ductile

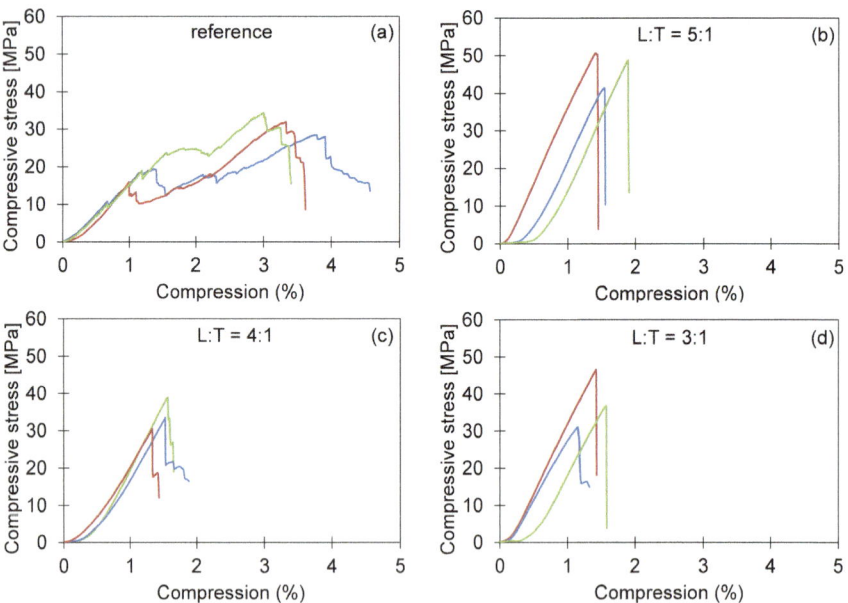

Fig. 22.28 Room temperature compression test curves: **a** pitch-based specimens; **b–d** pitch-lean
L–T based specimens at different L–T ratios [29]

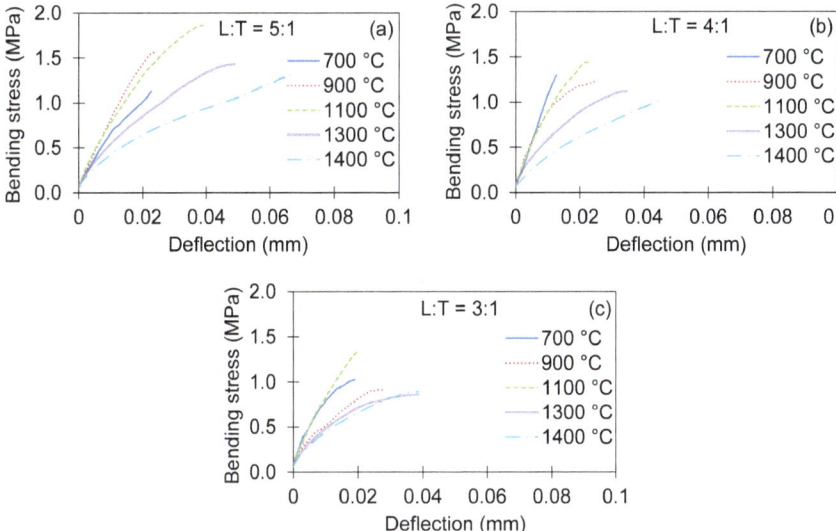

Fig. 22.29 Bending tests on pitch-free L–T-based compact specimens at high temperatures: **a** L:T = 5:1; **b** L:T = 4:1; **c** L:T = 3:1 [30]

transition should take place between 1100 and 1300 °C. L:T = 3:1 exhibited lower bending strength and lower deflection, while L:T = 5:1 showed higher deflection at 1300 and 1400 °C. The bending strengths were significantly weaker than those of the pitch-based specimens (see Fig. 22.12).

Several aspects are considered to be responsible for the observed weakness of the L–T-based bars [30]: (i) the L–T-based bars were more porous in comparison to the pitch-based ones (ca. 45% vs. ca. 37%); (ii) they contained much less residual carbon (ca. 22 wt% vs. ca. 37 wt%); (iii) the coking temperature was lower (1000 vs. 1400 °C); (iv) coking of the L–T binder system generated glassy carbon residue with higher fragility and lower deformability; (v) the hardening process might occur already at room temperature resulting in high hardness of particles which was less favorable for the pressing process.

Compression tests on small cylinders (Ø 8 mm × 12 mm) at the above-mentioned high temperatures showed that the compressive strength was between 22 and 48 MPa (Fig. 22.30). Similar to the bending tests, the maximum strength values were recorded at 1100 °C. The compressive strengths of L:T = 5:1 (23–34 MPa) were lower than those of L:T = 4:1 (38–48 MPa) and L:T = 3:1 (30–44 MPa) at the same test temperatures.

Creep tests on cylinders (Ø 20 mm × 25 mm) were performed at 1100, 1300 and 1400 °C under 10 MPa for 30 min (Fig. 22.31). While the compression values of all specimens were below 0.26% at 1100 °C, the creep effect became much more pronounced at 1300 and 1400 °C, indicating the ductility of the material at these temperatures. The compression of each specimen was even almost doubled as the temperature changed from 1300 to 1400 °C. Compared to the compositions L:T =

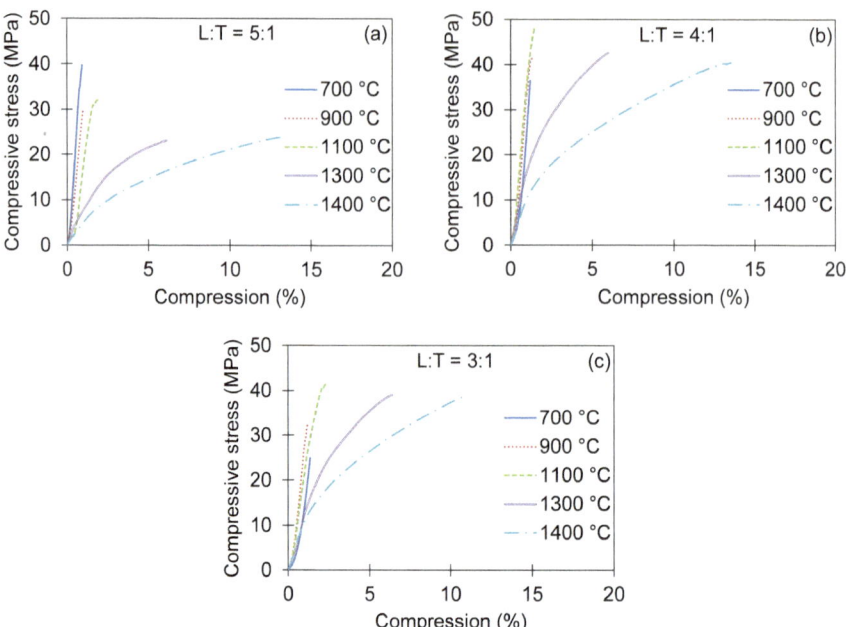

Fig. 22.30 Compression tests on pitch-free L–T-based compact specimens at high temperatures: **a** L:T = 5:1; **b** L:T = 4:1; **c** L:T = 3:1 [30]

4:1 and 3:1, the material L:T = 5:1 showed considerably higher compression at 1300 and 1400 °C.

It was also attempted to perform tests at 1500 °C [30]. As a bending bar was heated to 1500 °C, cracks appeared on the specimens unexpectedly. After cooling down, some solid grey droplets were found on the cracks. Meanwhile, some white substance was observed on test machine components. X-ray diffraction (XRD) analysis identified the grey droplets as a mixture of α-Al_2O_3, Al_2CO and C (graphite or turbostratic carbon), and the white substance as a mixture of Na_2CO_3 and NaOH. According to ball on three balls (B3B) tests as well as Raman spectroscopic investigation carried out by Zielke et al. [33], a progressive graphitization process occurred if the specimens were coked at 800 °C and tested at higher temperatures. We supposed that the sodium ions came from the commercial tannin which are not explicitly shown in its technical data sheet. The formation of the grey droplets was probably due to an overheating triggered by the electromagnetic field for induction heating and enhanced by a transition of the glassy carbon residue (deriving from L–T) into structurally more ordered form. Based on this assumption, hot spots could occur within the specimen through a synergetic interaction between a carbon modification change and an overheating. The appearance of hot spots was also observed e.g. in Field-Assisted Sintering Technology or Spark Plasma Sintering processes where a direct current is applied [34, 35]. A reference heating process up to 1500 °C using

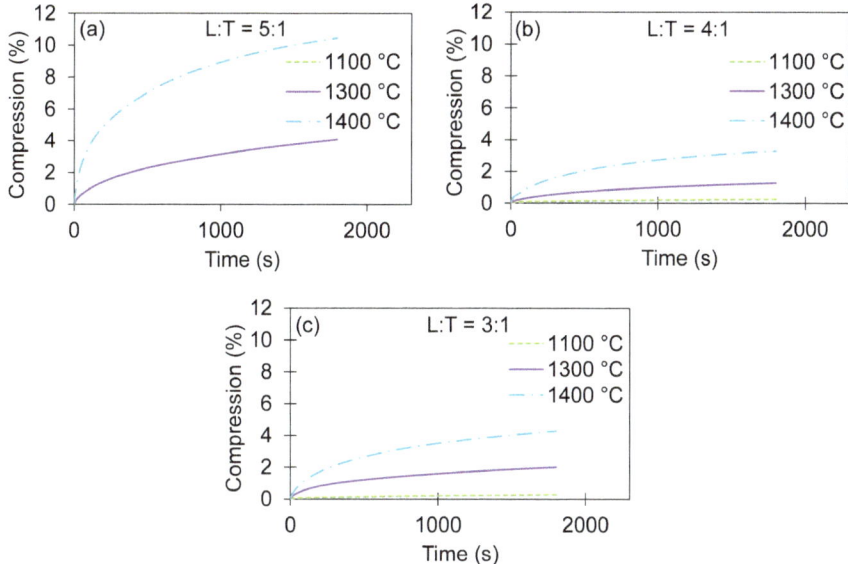

Fig. 22.31 Creep tests (10 MPa, 30 min) on pitch-free L–T-based compact specimens at high temperatures: **a** L:T = 5:1; **b** L:T = 4:1; **c** L:T = 3:1 [30]

conventional muffle oven did not lead to any visible damage on the L–T-based specimens. This supports the above-mentioned assumption. Therefore, mechanical tests on L–T-based specimens at and above 1500 °C should be carried out with resistance heating instead of inductive heating.

22.4 Summary

To characterize and further understand the physical properties and mechanical/thermomechanical behaviors of Al_2O_3-C as filter materials for molten steel, compact cylindrical and rectangular specimens as well as cylindrical foam and spaghetti filters were prepared and investigated. The application of different functional coatings on foam filters was explored. Physical parameters such as bulk density, open porosity and pore size distribution were determined. Compression and bending tests at room and high temperatures up to 1500 °C were performed. Fracture mechanical behavior was also studied. Furthermore, the modified pitch binder Carbores P was partly or completely replaced by an environmental-friendly binder system consisting of lactose and tannin. The composition was optimized by systematically changing the lactose-tannin ratio. A slip casting route was developed for the manufacture of L–T-based compact cylinders. All of the obtained results can be used for further development of the Al_2O_3-C based filter materials.

Acknowledgements The studies were carried out with financial support from the Deutsche Forschungsgemeinschaft (DFG, German Research Foundation) within Collaborative Research Center CRC 920, subproject C02, Project-ID 169148856. The authors gratefully acknowledge the production and testing of some of the materials by Y. Ranglack-Klemm and Dr. J. Solarek, Institute of Materials Engineering, as well as the support from the colleagues at the Institutes of Ceramics, Refractories and Composite Materials as well as Materials Engineering and all other colleagues for their valuable contributions.

References

1. K. Schwartzwalder, A.V. Somers, General motors corporation, US Patent, US3090094, 21 May 1963
2. M.J. Pryor, T.J. Gray, Swiss aluminium limited, US Patent, US3947363, 30 March, 1976
3. J.C. Yarwood, J.E. Dore, R.K. Preuss, Swiss aluminium limited, US Patent, US3962081, 8 June, 1976
4. R.A. Olson III., L.C.B. Martins, Adv. Eng. Mater. **7**, 187 (2005). https://doi.org/10.1002/adem. 200500021
5. V. Roungos, C.G. Aneziris, Refractories Worldforum **3**, 94 (2011)
6. M. Emmel, C.G. Aneziris, Ceram. Int. **38**, 5165 (2012). https://doi.org/10.1016/j.ceramint. 2012.03.022
7. Y. Klemm, H. Biermann, C.G. Aneziris, Ceram. Int. **39**, 6695 (2013). https://doi.org/10.1016/ j.ceramint.2013.01.108
8. Y. Klemm, H. Biermann, C. Aneziris, Adv. Eng. Mater. **15**, 1224 (2013). https://doi.org/10. 1002/adem.201300159
9. Y. Klemm, H. Biermann, C.G. Aneziris, *Proceedings of the UNITECR 2013* (John Wiley & Sons, Inc, Hoboken, 2014)
10. I.J. Holland, US Patent, US3097930, 16 July, 1963
11. T. Wetzig, A. Baaske, S. Karrasch, N. Brachhold, M. Rudolph, C.G. Aneziris, Ceram. Int. **44**, 23024 (2018). https://doi.org/10.1002/adem.201900657
12. Y. Ranglack-Klemm, E. Storti, C. G. Aneziris, H. Biermann, Adv. Eng. Mater. **22**, 1900423 (2020). https://doi.org/10.1002/adem.201900423
13. X. Wu, Y. Ranglack-Klemm, J. Hubálková, J. Solarek, C.G. Aneziris, A. Weidner, H. Biermann, Ceram. Int. **47**, 3920 (2021). https://doi.org/10.1016/j.ceramint.2020.09.255
14. J. Solarek, C. Bachmann, Y. Klemm, C.G. Aneziris, H. Biermann, J. Am. Ceram. Soc. **99**, 1390 (2016). https://doi.org/10.1111/jace.14070
15. J. Solarek, C. Himcinschi, Y. Klemm, C.G. Aneziris, H. Biermann, Carbon **122**, 141 (2017). https://doi.org/10.1016/j.carbon.2017.06.041
16. J. Solarek, Dissertation, TU Bergakademie Freiberg, 2019
17. X. Wu, Y. Ranglack-Klemm, E. Storti, S. Dudczig, C.G. Aneziris, A. Weidner, H. Biermann, Adv. Eng. Mat. **24**, 2100642 (2022). https://doi.org/10.1002/adem.202100642
18. X. Wu, T. Wetzig, C.G. Aneziris, A. Weidner, H. Biermann, Adv. Eng. Mat. **24**, 2100613 (2022). https://doi.org/10.1002/adem.202100613
19. K.K. Kappmeyer, D.H. Hubble, *In High Temperature Oxides Part I – Magnesia, Lime, and Chrome Refractories* (Academic Press, New York, 1970)
20. Coal Tars and Coal-Tar Pitches, in: Report on Carcinogens, 14. Ed., National Institute of Environmental Health and Safety (2016)
21. Verordnung (EG) Nr. 1907/2006 (REACH), Registration, Evaluation, Authorisation and Restriction of Chemicals (2006)
22. E.M.M. Ewais, Carbon based refractories. J. Ceram. Soc. Japan **112**, 517 (2004). https://doi. org/10.2109/jcersj.112.517

23. Phenol, in: Hazardous Substance Fact Sheet, 2. Ed., New Jersey Department of Health (2015)
24. E. Guéguen, C.G. Aneziris, C. Biermann, German Patent, DE102016100083 AI
25. C. Biermann, Dissertation TU Bergakademie Freiberg (2016)
26. C. Himcinschi, C. Biermann, E. Storti, B. Dietrich, G. Wolf, J. Kortus, C.G. Aneziris, J. Eur. Ceram. Soc. **38**, 5580 (2018). https://doi.org/10.1016/j.jeurceramsoc.2018.08.029
27. J.A. Halt, S.K. Kawatra, Mining. Metallurgy & Exploration **31**, 73 (2014). https://doi.org/10.1007/BF03402417
28. A. Arbenz, L. Avérous, Green Chem. **17**, 2626 (2015). https://doi.org/10.1039/c5gc00282f
29. X. Wu, A. Weidner, C.G. Aneziris, H. Biermann, Ceram. Int. **48**, 148 (2022). https://doi.org/10.1016/j.ceramint.2021.09.090
30. X. Wu, A. Weidner, C.G. Aneziris, H. Biermann, Ceram. Int. **49**, 3140 (2023). https://doi.org/10.1016/j.ceramint.2022.12.192
31. G. Tondi, Polymers **9**, 223 (2017). https://doi.org/10.3390/polym9060223
32. A. Pizzi, P. Tekely, J. App. Polym. Sci. **56**, 1645 (1995). https://doi.org/10.1002/app.1995.070561215
33. H. Zielke, T. Wetzig, C. Himcinschi, M. Abendroth, M. Kuna, C.G. Aneziris, Carbon **159**, 324 (2020). https://doi.org/10.1016/j.carbon.2019.12.042
34. Z.A. Munir, U. Anselmi-Tamburini, M. Ohyanagi, J. Mater. Sci. **41**, 763 (2006). https://doi.org/10.1007/s10853-006-6555-2
35. O. Guillon, J. Gonzalez-Julian, B. Dargatz, T. Kessel, G. Schierning, J. Räthel, M. Hermann, Adv. Eng. Mater. **16**, 830 (2014). https://doi.org/10.1002/adem.201300409

Chapter 23
Determination of the Temperature-Dependent Fracture and Damage Properties of Ceramic Filter Materials from Small Scale Specimens

Martin Abendroth, Shahin Takht Firouzeh, Meinhard Kuna, and Bjoern Kiefer

23.1 Introduction

Open cell ceramic foam filters are utilized during metal melt filtration processes, because they offer an active and reactive contribution to clean the metal melt. Additionally, the filters reduce melt turbulence, which also enhances the quality of the cast products due to less casting defects according to a calm melt flow. The integrity of the filters is an important requirement for industrial applications. The filters have to withstand the mechanical and thermal loads without any failure to avoid impurities of the cast product.

Within the CRC 920 new refractory materials are developed [1, 2], which have an improved thermal shock resistance. The various test methods described in this work are not necessarily restricted to certain materials. But to compare the different methods we restrict ourselves to the classes of materials developed within the CRC 920. In particular, this is carbon bonded alumina Al_2O_3-C in different chemical compositions. Here, the experimental methods are used to judge, which composition or heat treatment leads to superior properties of the material with respect to its application.

Small scale specimen testing techniques [3] are especially used if the amount of available material is limited, or if the structures made of the specific material are small. The ceramic bulk materials used for the purpose of metal melt filtration show a distinct size effect. Within large samples different properties may show up depending on the location within the samples, depending on production strategy or heat treatment. Ceramic filters used for metal melt filtration are spatial networks of struts having a diameter between approximately 100 to 2000 μm. Therefore,

M. Abendroth (✉) · S. Takht Firouzeh · M. Kuna · B. Kiefer
Institute of Mechanics and Fluid Dynamics, Technische Universität Bergakademie Freiberg, Lampadiusstr. 4, 09599 Freiberg, Germany
e-mail: Martin.Abendroth@imfd.tu-freiberg.de

© The Author(s) 2024
C. G. Aneziris and H. Biermann (eds.), *Multifunctional Ceramic Filter Systems for Metal Melt Filtration*, Springer Series in Materials Science 337, https://doi.org/10.1007/978-3-031-40930-1_23

specimens having similar dimensions should be used to determine the properties of these filter materials.

The small punch text (SPT) is a promising testing technique for fine-grained ceramic materials [4]. In this test a small disc shaped specimen is placed on ring-like bearing and deformed by a spherical tipped punch. Depending on the material a wide range of properties can be determined [5–7]. The SPT is applicable to determine parameters of simple failure criteria up to parameter sets for complex elastic plastic damage models [8].

The ball on three ball (B3B) test is a special variant of the SPT, where the specimen is supported by three balls instead of a ring-like support. It is especially useful for testing brittle materials or specimens having not perfect plane surfaces [9–11].

The chevron-notched beam (CNB) test is a standardized method to evaluate the fracture toughness of ceramic materials [12–14]. The advantage of this test is that no sharp pre-crack has to be prepared because the crack is forming itself during loading at the beginning of the test. Furthermore, no crack length measurement is required and a stable crack growth can be guaranteed due to a suitable geometry of the notch [15]. The challenge here is to find suitable geometries for sub-sized CNB specimen and their application at high temperatures.

The Brazilian disc test (BDT) utilizes cylindrical specimen, which are compressed along their vertical diameter with increasing load until specimen failure [16]. This test has been developed for brittle rock like materials [17], where grains are rather large compared to the fine grained ceramics within the scope of the CRC 920. The advantage of this test method is that no specimen deformation measurement is necessary. To obtain the material strength only the compression force at failure has to be known. The challenge is to develop a small scale version of this test and its application at high temperatures.

The finite element method is an universal tool to analyze stress and strain fields in structures under arbitrary thermo-mechanical loads. In the context of small scale specimen testing techniques it is especially useful to analyze non-standard test geometries with respect to their suitability to determine strength and fracture toughness of brittle materials under high temperature conditions. From the resulting stress fields simple failure criteria can be easily evaluated. For the analysis of fracture toughness or damage processes special tools have been developed, as the J-integral method [18–20] or cohesive zone approaches [21, 22].

The aim of this contribution is to give an overview about the different experimental and numerical methods to determine the thermo-mechanical properties of refractory ceramics, especially those used for ceramic filters. The results obtained using the different testing techniques will be compared to each other with respect to accuracy and complexity of the test setups.

23.2 Theoretical Foundation

Within this section theoretical approaches are discussed that allow the evaluation of the strength of ceramic filter materials under thermo-mechanical loading. This includes simple classical failure criteria as well as more advanced approaches as fracture and damage mechanics including cohesive zone models.

23.2.1 Failure Criteria

Failure criteria are used to distinguish whether a material can still withstand a certain load or not [23]. The maximum stress criterion assumes that a material fails when the maximum principal stress σ_1 in a material element exceeds the uniaxial tensile strength σ_t or if the absolute value of the minimum principal stress σ_3 is less than the uniaxial compression strength σ_p of the material

$$- \sigma_p < \sigma_3 < \sigma_1 < \sigma_t . \tag{23.1}$$

The Mohr-Coulomb failure criterion describes the critical state of a material with respect to shear as well as normal stress.

$$\frac{M+1}{2} \max [\, |\sigma_1 - \sigma_2| + K(\sigma_1 + \sigma_2), \ |\sigma_1 - \sigma_3| + K(\sigma_1 + \sigma_3),$$
$$|\sigma_2 - \sigma_3| + K(\sigma_2 + \sigma_3) \,] < \sigma_p, \tag{23.2}$$

with

$$M = \frac{\sigma_p}{\sigma_t} \quad \text{and} \quad K = \frac{M-1}{M+1} . \tag{23.3}$$

Especially brittle materials show a scatter of failure stresses. Therefore a statistical description of the failure processes is helpful. The Weibull theory [24] assumes that brittle materials contain randomly dispersed defects. Due to the weakest-link effect, these defects act as crack initiation points, which can cause total failure of the specimen. This theory includes a size effect, because as larger a sample is, as more likely it is to contain a critical defect. The failure probability of the Weibull theory is given as

$$P_f(\sigma) = 1 - \exp\left[-\frac{V_{\text{eff}}}{V_0} \left(\frac{\sigma}{\sigma_0} \right)^m \right], \tag{23.4}$$

where m and σ_0 represent the Weibull modulus and the Weibull reference stress, respectively. The Weibull modulus is a measure to express the amount of scatter, whereas σ_0 corresponds to the strength of the material assuming that the probability of failure is 63.2% considering its volume.

Moreover the size effect is accounted for by the ratio of an effectively loaded volume V_{eff} and a reference volume V_0. The effectively loaded volume is obtained by

$$V_{\text{eff}} = \int_V \left(\frac{\sigma_{\text{eq}}}{\sigma_0} \right)^m dV, \tag{23.5}$$

where σ_{eq} is an appropriate equivalent stress measure. The particular criteria used is based on the principle of independent action (PIA), which assumes that the principal stresses act independently of each other, whereby only positive values are taken into account

$$\sigma_{\text{eq}} = \sqrt[m]{\langle \sigma_I \rangle^m + \langle \sigma_{II} \rangle^m + \langle \sigma_{III} \rangle^m} \quad \text{with} \quad \langle \bullet \rangle = \frac{|\bullet| + \bullet}{2}. \tag{23.6}$$

To estimate the Weibull parameter the maximum likelihood (ML) procedure is the preferred method. Details about the ML method are given by Soltysiak and Zielke et al. [25].

A failure criterion for crack growth is defined according to the theory of linear elastic fracture mechanics. Griffith's theory defines that a critical stress σ_c is needed to propagate a crack with length a

$$\sigma_c = \sqrt{\frac{EG}{\pi a}}, \tag{23.7}$$

where E denotes the Young's modulus of the material and $G = 2\gamma$ an energy which for brittle materials is equal to the specific surface energy γ of the two crack faces. Another suitable criterion is the stress intensity factor (SIF) (for mode I)

$$K_I = \sigma \sqrt{\pi a} \tag{23.8}$$

which can be compared to the fracture toughness value K_{Ic} of the material. Fracture occurs if the stress intensity factor reaches the critical fracture toughness value $K_I = K_{Ic}$. Using finite element simulations SIFs can be determined via the J-integral.

$$K_I = \sqrt{J_I E'} \quad \text{with} \quad E' = \frac{E}{1 - \nu^2} \tag{23.9}$$

for plane strain or $E' = E$ for plane stress conditions, respectively. For the evaluation of the J-integral or stress intensity factors several computational methods are available [19]. The critical mode I fracture toughness is defined as

$$K_{Ic} = Y \sigma_c \sqrt{\pi a}, \tag{23.10}$$

where Y is a dimensionless factor depending on the specimen geometry and loading conditions.

23.2.2 Cohesive Zone Model

The cohesive zone model (CZM) approach is a phenomenological description of damage and its evolution in structural components [22, 26]. CZMs can capture the failure of brittle, ductile or viscoplastic materials. The CZM is a damage model which accounts only for failure by the separation of the material along a given internal surface. CZMs can be combined with arbitrary material models for the adjacent bulk regions. In contrast to continuum damage models the potential crack path must be known from experimental investigations or other numerical analyses. The mechanical behavior of a cohesive zone is described by a traction-separation relation, also called cohesive law.

In general, the cohesive traction separation laws can be described with the following independent model parameters: the cohesive strength t_0, which corresponds to the maximum of the cohesive traction-separation curve, the separation work Γ_0 (i.e. fracture energy), and an internal length s_0, which defines the separation at the maximum traction. For a bi-linear traction separation law as shown in Fig. 23.1 alternatively a separation at total failure s_f could be defined. The specific separation work Γ_0 is denoted as the area under the cohesive traction-separation curve, i.e.,

$$\Gamma_0 = \int_0^\infty t_n ds_n . \tag{23.11}$$

Furthermore, the gradient $E_{cz} = t_0/s_0$ defines the stiffness of the cohesive zone, which describes the initial slope of the curve in the case of a bi-linear law. In the bi-linear cohesive stress-separation curve, damage of the cohesive zone D_{cz} initiates upon reaching the maximum separation stress t_0 with the corresponding separation length s_0. Once the separation length s_f is reached, the damage $D_{cz} = 1$ and the cohesive element is supposed to hold no more traction. It should be mentioned that many different formulations of cohesive laws are available [26], with a large

Fig. 23.1 Bi-linear cohesive law [26]

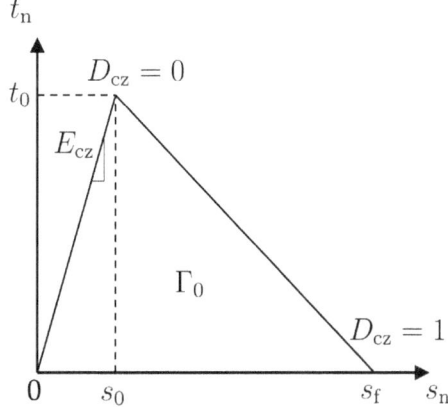

variety of formulations for damage evolution. In the current work the simple bilinear approach is used. The fracture toughness and the separation work are related via Equation (23.9), with the J-Integral replaced by Γ_0.

$$K_{Ic} = \sqrt{\Gamma_0 E'} \qquad (23.12)$$

23.3 Experimental and Simulation Methods

In this section the experimental and simulation methods are described, which are used to determine the thermo-mechanical properties of the materials under consideration.

23.3.1 Material and Specimen Preparation

Carbon-bonded alumina filters, are used for molten steel filtration due to their high thermal shock resistance. As the main component of the Al_2O_3-C, 99.8%-pure alumina (Martoxid MR 70, Martinswerk, Germany, $d_{90} \leq 3.0\,\mu m$) is used. In addition, the carbon content for the slurry is provided from three different sources; modified coal tar pitch powder (Carbores® P, Rütgers, Germany, $d_{90} \leq 0.2\,mm$), fine natural graphite (AF 96/97, Graphit Kopfmühl, Germany, 96.7 wt% carbon, 99.8 wt% $\leq 40\,\mu m$), and carbon black powder (Luvomaxx N-911, Lehmann & Voss & Co., Germany, ≥ 99.0 wt% carbon, > 0.01 wt% ash content, primary particle size of $200 - 500\,nm$). The combination of these sources leads to a final product with 30% carbon content in its final composition. Three different material compositions with a varying amount of Carbores® P were produced. Table 23.1 shows the main chemical composition. Also different specimen manufacturing routes are investigated, where the material block is either pressed or cast into a mold and dried. For the special material variant AC20 the detailed chemical composition is given in Table 23.2.

To prepare the specimens for the applied tests, first, the slurry is poured or pressed into cylindrical or cuboid blocks ($25 \times 25 \times 150\,mm^3$). The green bodies are then coked under reducing conditions at temperatures up to 800 or 1400 °C. The corresponding coking regimes are shown in Fig. 23.2. For the first variant, the heating rate is $1\,K\,min^{-1}$. Starting from 100 °C the temperature is increased, after an increase of 100 °C the temperature is kept constant for 30 min. When the final coking temperature of 800 °C is reached, a holding time of 180 min is specified. The second variant is characterized by a heating rate of $1\,K\,min^{-1}$ to 300 °C. After a holding time of 60 min, the heating rate is increased to $3\,K\,min^{-1}$ until the final temperature of 1400 °C is reached. Subsequently, a holding period of 300 min finalizes the heat treatment. At a temperature of about 235 °C, Cabores® P begins to melt and the non-aromatic hydrocarbons are broken up. With increasing temperature the formation of a liquid-solid mesophase begins, which solidifies at about 500 to 550 °C

Table 23.1 Chemical composition of the investigated Al_2O_3-C materials. The amount of additives is related to the solid content of the material

	Pressed			Casted		
wt%	AC10P	AC15P	AC20P	AC10C	AC15C	AC20C
Al_2O_3	66	66	66	68	67	66
Carbores® P	10	15	20	10	15	20
Graphite	13	10	8	12	10	8
Carbon black	11	9	6	10	8	6
Additives	7.3	7.3	7.3	1.9	1.9	1.9

Table 23.2 Detailed chemical composition of the Al_2O_3-C slurry for an AC20 material

Raw material	% Mass fraction	Additives*	% Mass fraction
Martoxid MR 70	66.0	Castament VP 95 L	0.3
Cabores® P	20.0	Contraspum K 1012	0.1
AF 96/97	7.7	C12C	1.5
Luvomaxx N-991	6.3	*Related to total solid content	

Fig. 23.2 Two different heat treatment processes applied on Al_2O_3-C [27]

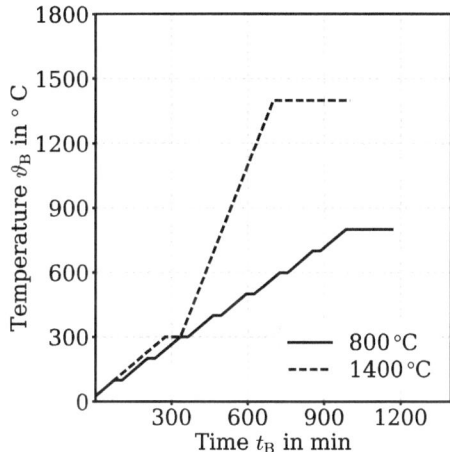

and determines the structure of the later residual coal shock. The modified coal tar pitch therefore acts not only as a carbon source, but also as a binder. The ceramic bonding is thus ensured via carbon content. This is possible because pitches have a good bonding to both oxides and burrs. A shrinkage allowance is necessary because a volume shrinkage of about 6% occurs during the heat treatment. For the SPT and B3B samples, the cylindrical castings are turned to rods with a final diameter of $d = 8$ mm. From the rods discs are cut to the nominal thickness of $t = 0.5$ mm using a cutting device. For the CNB samples, the cuboid bodies are separated and also brought to the final dimensions using an abrasive belt (P100 grit, 162 μm). To cut the notch for the CNB specimens a precision diamond wire saw (Well, Mannheim, Germany) using a wire with diameter $d = 0.3$ mm (Well, Type A3-3) is utilized. The BDT samples are drilled with hollow diamond drilling bits from the cuboid blocks. Various sample diameters from 9.8 mm up to 15.8 mm were obtained using different drilling bits. To obtain the final disc thickness, the cylinders are then cut using a cutting device.

23.3.2 Small Punch Test

The testing device used for the small punch test and its components are integrated into a universal testing machine Inspekt Table 10 kN from Hegewald & Peschke, Germany. It is equipped with a furnace, which allows material testing up to 1200 °C. The furnace has three zones, which are controlled separately. To monitor the specimen temperature an additional thermocouple can be attached through a drilled hole in the apparatus. The loading force is applied via the punch inside the SPT apparatus through an upper linkage, which is directly connected to the load cell of the testing machine. Different load cells can be used depending on the expected measuring range. The measuring ranges of the load cells are 100 N, 500 N, 1 kN and 10 kN. The load cell is located outside the furnace and is actively cooled by a fan to maintain the allowable temperature operating range. Argon as inert gas can be fed into the specimen chamber through the lower pressure linkage to prevent oxidation of the sensitive materials. The components of the SPT apparatus are shown in Fig. 23.3 and are made either of high-temperature steel (1.4841), or of 99,7% pure alumina. The lower housing (c) is attached to the compression linkage and thus forms the connection to the testing machine. The upper housing (b) serves on the one hand to guide the punch (a) and on the other hand to prevent the inert gas from escaping. In this design, the force is applied by a punch and a ceramic loading ball (e) with a diameter $d = 2.5$ mm. The specimen (i) with a diameter $D = 8$ mm and thickness $t = 0.5$ mm is supported by the lower die (h) and centered by a ring (f). Another ring (d) is used to guide the loading ball (f) and the punch. The lower die has an inner hole with a diameter $d = 4$ mm and a chamfer edge with $r = 0.5$ mm × 45°, which results in a support radius $R_a = 2.5$ mm. This multi-part design has the advantage that parts can be changed individually if they are worn out or somehow damaged.

For small deflections of the specimen as it is expected for brittle ceramic materials Börger et al. [9] carried out elastic stress analysis and defined an expression for the failure stress depending on the maximum load F_{max} and geometric measures of the SPT setup.

$$\sigma_f = k \frac{F_{max}}{t^2} = k\sigma_{SP},$$

(23.13)

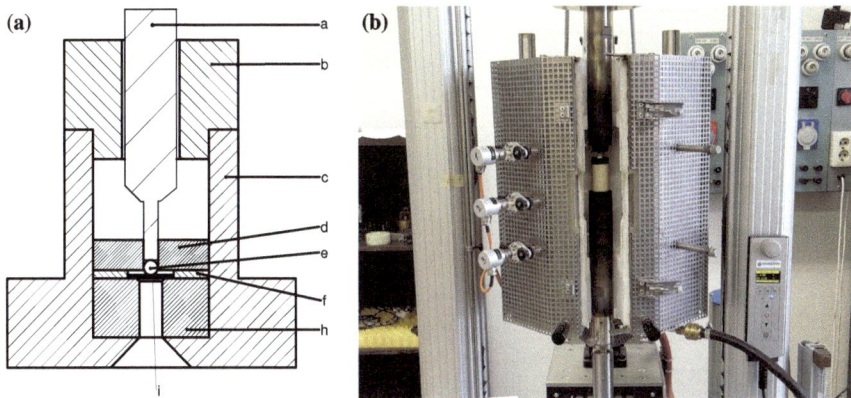

Fig. 23.3 **a** Schematic view of the small punch test apparatus [25]. The specimen radius $R = 4$ mm, specimen thickness $t = 0.5$ mm, support radius $R_a = 4.5$ mm. **b** shows the slightly opened furnace with the SPT apparatus at its center

with

$$k = \frac{3(1 + v)}{4\pi} \left[1 + 2\ln\frac{R_a}{b} + \frac{1 - v}{1 + v} \left(1 - \frac{b^2}{2R_a^2} \right) \frac{R_a^2}{R^2} \right], \quad (23.14)$$

where R denotes the specimen radius, R_a the support radius, t the specimen thickness, v the Poisson's ratio of the specimen material and, b the radius of the contact region between specimen and punch. For the contact radius b in Equation (23.14) different approximations can be found in literature [25]. The dimensionless parameter k can be seen as an empirical function [9], which can be determined by finite element simulations depending on the loading force F and Young's modulus E of the specimen.

23.3.3 Ball on Three Balls Test

The ball on three balls (B3B) test is used to determine the biaxial flexural strength and the material parameters of the cohesive model. In the developed test apparatus, as shown in Fig. 23.4, a disc-shaped specimen with a diameter of $d = 8$ mm and a thickness of $t = 0.5$ mm is supported by three balls and loaded centrically by one ball. The apparatus can be implemented in the same test rig used for SPT for testing temperatures up to $T = 1000\,°C$. A second test rig using a different furnace is used for higher temperatures up to $T = 1400\,°C$. The test apparatus is mounted on the lower linkage of the test rig. The three ceramic support balls (h) are axially aligned by a cage (g). The specimen centering ring (f) ensures that the specimen (i) is correctly positioned. A second centering ring (d) guides the loading ball (e) with a diameter of $d = 2.5$ mm. The punch (a) guided by the upper housing (b) is transmitting the force

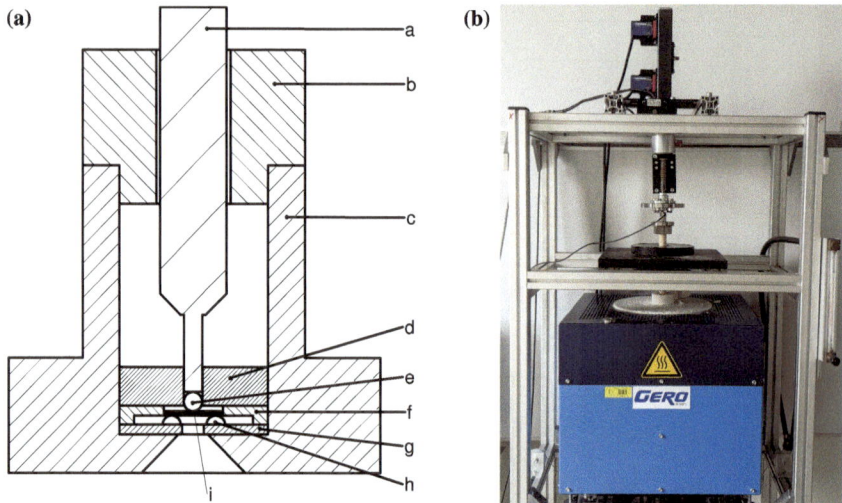

Fig. 23.4 **a** Schematic view of the B3B test setup [28] **b** test rig with the furnace with the B3B apparatus at its center, load cell and the linear motor at top

from the upper pressure plate of the testing machine to the loading ball. Argon as inert gas is introduced through a hole in the lower linkage of the testing machine to prevent decarburization of the specimen during testing at high temperatures. All of the mentioned parts are made of 99.7% pure aluminum oxide. The punch loads the specimen displacement-controlled at a constant displacement rate of $\dot{u} = 0.05$ mm min^{-1} until specimen failure occurs. The applied force is recorded by means of a load cell. As for the SPT the Equations (23.13) and (23.14) can be used to determine the stress at failure [9].

23.3.4 Chevron Notched Beam Test

To determine the fracture toughness K_{Ic} of a brittle material, a four-point bending test with chevron-notched specimens is used. Different configurations of chevron-notched specimens are available in the literature [13, 15, 29]. The triangular-shaped chevron notch is cut into the specimen using a wire saw. It offers the following advantages: First, no initial pre-crack is necessary because a sharp crack develops during loading starting from the tip of the notch. Furthermore, no crack length measurement is required because the maximum force is sufficient to calculate the fracture toughness after failure of the specimen. For the subsequent investigations, the crack length a and the chevron notch parameters a_0 and a_1 are normalized by the specimen height W, resulting accordingly in $\alpha = a/W$ and $\alpha_0 = a_0/W$ and $\alpha_1 = a_1/W$, respectively. Figure 23.5a shows the front view of a specimen with chevron notch

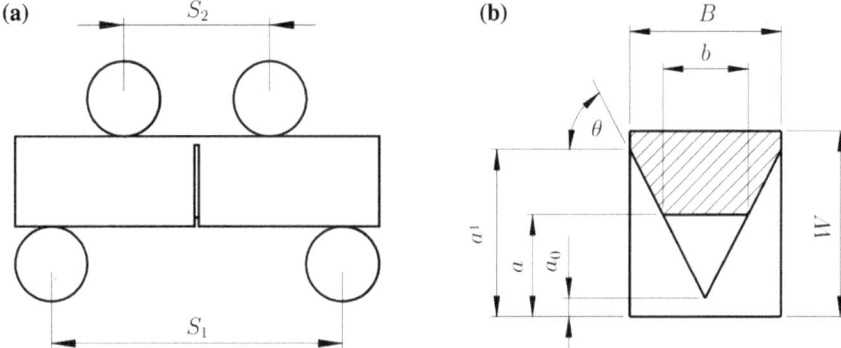

Fig. 23.5 **a** Schematic front view and **b** geometric parameters of a chevron-notched four-point bending specimen [30]

as well as the geometrical quantities from the resulting sectional view, where the section plane corresponds to the notch plane. The outer specimen dimensions are $L = 25$ mm, $B = 5$ mm and $W = 6$ mm. Outer and inner span of the support rollers are given as $S_1 = 20$ mm and $S_2 = 10$ mm, respectively. Figure 23.5b is representative of the crack length a, chevron tip dimension a_0, chevron dimension a_1 and length of crack front b. In the special design used by Zielke et al. [30, 31], one roller of loading and supporting span is replaced by a ball. Both rollers (m) and balls (k) have diameters of $d_s = 5$ mm. A cage structure is assembled to the lower sealing of the testing machine (j). This cage structure was constructed in order to ensure the alignment of the specimen (l) as well as the loading and bearing components. The cage structure is guided by a housing (d) and consists of lower (h), middle (f), upper cage ring (c), and specimen positioning bolts (g). Upper and lower spacers and guiding rods, which are not visible in Fig. 23.6, provide the correct distance between the cage rings. The load from the upper punch (a) is transferred to the inner ball-cylinder pair via loading plate (e) and ball (b).

23.3.5 Brazilian Disc Test

The BDT is a convenient test method for brittle materials. In this test, in-plane compressive forces are applied to the disc-shaped specimens, which cause the formation of in-direct horizontal tensile forces along the vertical center-line of the specimen. The test is considered to be valid, if a central crack is formed along the vertical disc axis and split the disc into two halves. However, excessive stresses are formed along the contact lines as the compressive force increases. Due to the porous structure of the material, high amount of deformation is observed in the vicinity of the contact-line. Therefore, various test configurations were studied at room temperature to obtain a suitable method which yields valid test results. Another criterion regarding the test

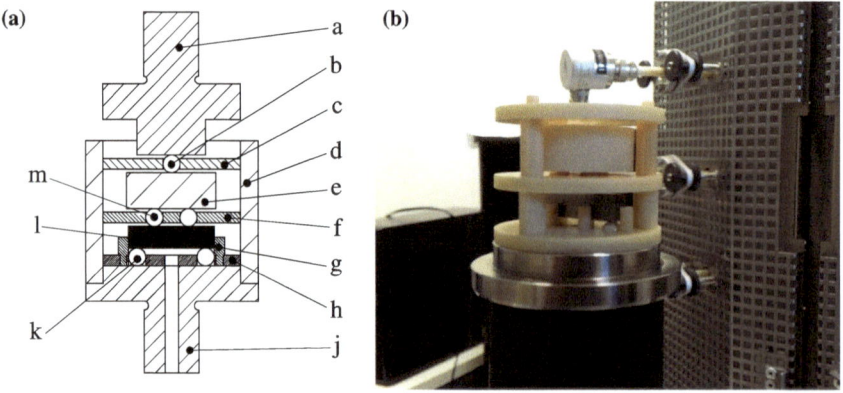

Fig. 23.6 **a** Scheme [30] and **b** close up photograph of the CNB test setup without the outer housing and loading plate

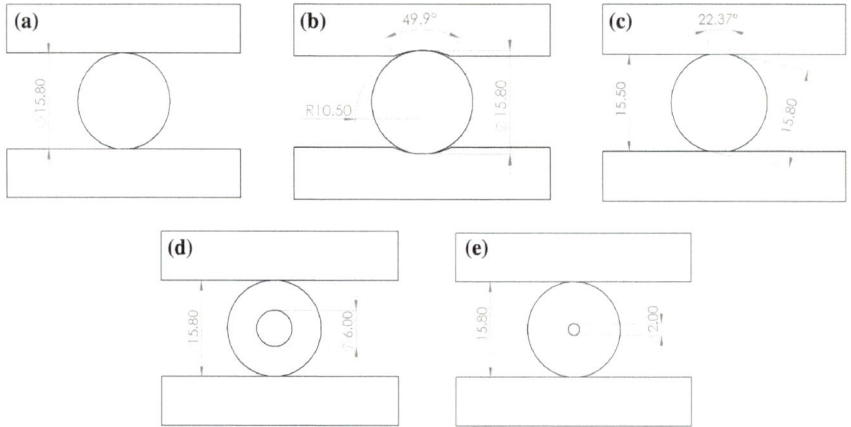

Fig. 23.7 Illustrations of the different BDT test configurations [32] **a** cylindrical specimen, flat punch **b** cylindrical specimen, curved punch **c** flattened specimen, flat punch **d** cylindrical specimen, large hole **e** cylindrical specimen, small hole

variants is the suitability for high temperature testing. The test variants are illustrated in Fig. 23.7. Additionally, to eliminate the false vertical displacement reading caused by the contact area deformation, surface displacement map of the sample was recorded throughout the test using an ARAMIS adjustable digital image correlation (DIC) system. This system consists of two symmetrically focused cameras which can be horizontally adjusted. The surface components are defined by GOM Correlate software which processes the digital images obtained from the cameras. The surface components are marked with the random speckle patterns created by sprayed paint droplet patterns.

Fig. 23.8 Axisymmetric FE model of the SPT to obtain σ_f

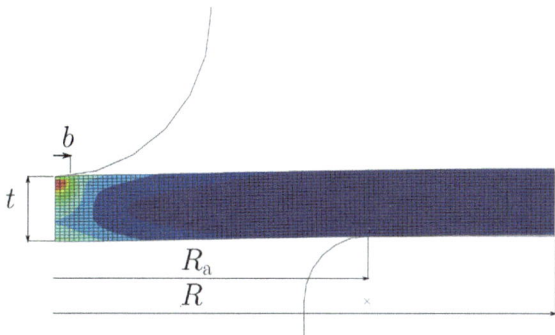

A Shimadzu AGS-10 universal testing machine is used for the room temperature tests. The tests are conducted with a constant punch displacement rate of $0.1\,\mathrm{mm\,min^{-1}}$. To observe the crack initiation sequence a high speed camera (Makro-Vis) is used. Apart from the vertical crack caused by tensile forces at the center, in all of the test variants symmetrical secondary arc-shaped cracks initiating laterally at the contact points were observed. With the help of high speed camera, failure of the disc is recorded for each of the test variants. Ultimately, to apply the BDT at high temperatures, a test configuration is established. The test setup is shown on the right side of Fig. 23.4. Based on the room temperature test configurations, variant from Fig. 23.7e is selected for high temperature testing. To conduct the high temperature testing, the specimen size is further reduced to a diameter of 9.8 mm and a thickness of 6.0 mm with a central hole of 1.5 mm in diameter. To prevent oxidation of the carbon content at higher temperatures, argon gas flow is provided into the test chamber. Applied force vs. punch displacement for each of the specimens is obtained at 1200 °C. Due to the contact area damage and displacement of the test setup, displacement data for high temperature testing is not considered to be accurate. Therefore, to obtain the material parameters, cohesive zone modeling was implemented (Fig. 23.8).

23.3.6 Finite Element Analysis

The finite element method (FEM) is used to analyze all test setups numerically. For the SPT an axisymmetric model is used where the specimen is meshed with axisymmetric 8-node quadratic elements. The receiving die, downholder and punch are modeled using rigid surfaces. Receiving die and downholder are fixed in space. The punch can only move in vertical direction. The contact between the rigid parts and the specimen is realized using the surface to surface interface of Abaqus. A Coulomb friction model with a friction coefficient $\mu = 0.1$ completes the computational setup.

For the B3B test a 3D-model (see Fig. 23.9) is necessary, but the threefold rotational symmetry is taken into account to reduce the computational effort. The specimen is meshed with quadratic 20-node brick elements. The balls are modeled as

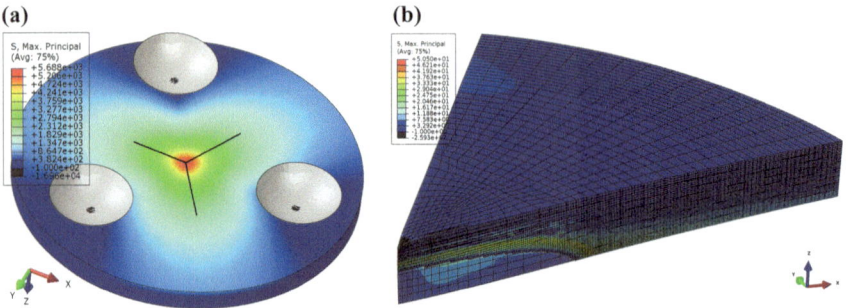

Fig. 23.9 FE models of the B3B test **a** full model viewed from below **b** 1/6 model with a crack in the cohesive zone (support and loading balls not displayed)

rigid surfaces and are fixed in space, except of the loading ball, which can be moved in vertical direction. Also here, the contact between specimen and balls is realized as in the SPT model.

In order to simulate the crack growth in the chevron-notched specimens, an FE model (see Fig. 23.10) with appropriate boundary conditions was developed. 20-node hexahedral continuum elements (C3D20) with quadratic displacement functions are used, with the element edge length being 0.2 mm. An isotropic, linear-elastic material behavior is assigned, which only requires the two material parameters modulus of elasticity E and Poisson's ratio ν. Using the symmetry properties and simplifying the

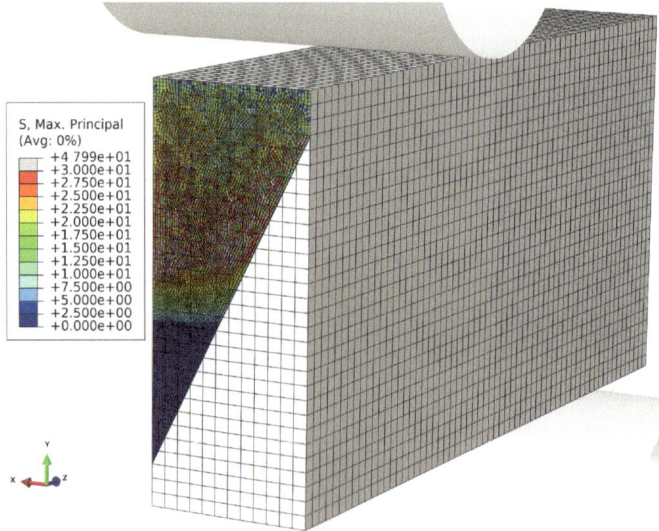

Fig. 23.10 1/4 FE model of the CNB test, where results only for the cohesive elements are displayed. The blue area shows the propagated crack

loading bodies, the model can be reduced to a quarter model with the corresponding symmetry boundary conditions. The support and loading rollers are modeled as rigid bodies and their contact with the sample is simplified as being frictionless. This assumption is permissible due to the rolling bearing of the loading bodies, since these minimize the frictional stresses. Furthermore, the notch radius is defined according to the diameter of the wire saw with $r_K = 0.15$ mm, with 15 nodes along the quadrant. The contact stiffness was defined using a non-linear formulation, where the contact pressure increases exponentially as two contact surfaces approach each other. In this way, roughness and surface waviness of the contact surfaces, which arise due to production, can be taken into account and the run-in characteristics of the measured curves can thus be mapped.

The cohesive elements (COH3D8) have a rectangular base and a linear displacement approach, with element edge lengths ranging from 0.015 to 0.14 mm. Since the cohesive elements are located on a symmetry plane of the model, there are conditions for maintaining the symmetry. The studied parameters of the cohesive model are cohesive stiffness E_{cz}, cohesive strength t_0, work of separation Γ_0, and modulus of elasticity E. Due to its negligible effect the Poisson's ratio was assumed as $\nu = 0.2$.

For the finite element analyses, of the BDT a quarter model of the disc is sufficient. At the vertical symmetry plane cohesive elements are attached with appropriate symmetry conditions. The load is applied using a rigid plane at the top. Figure 23.11 show the models for Var. A and Var. E (cf. Fig. 23.7). The stress analysis shows that for Var. E exists a small stress concentration area at the top of the hole where the crack initiates. For Var. A a rather larger area of tensile stresses along the vertical symmetry plane is observed.

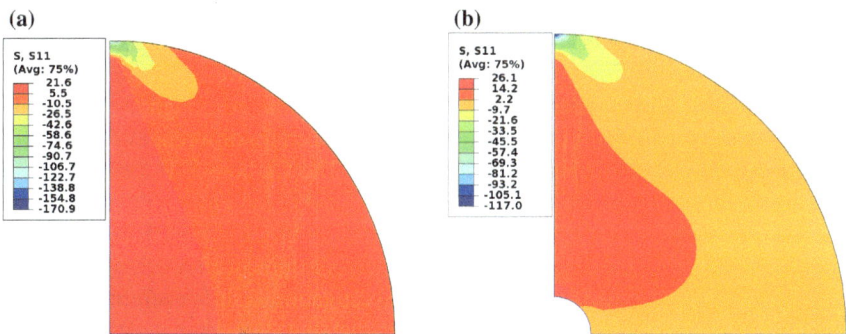

Fig. 23.11 Horizontal component of the stress for Brazilian disc tests **a** round specimen, flat punch **b** cylindrical specimen, small hole

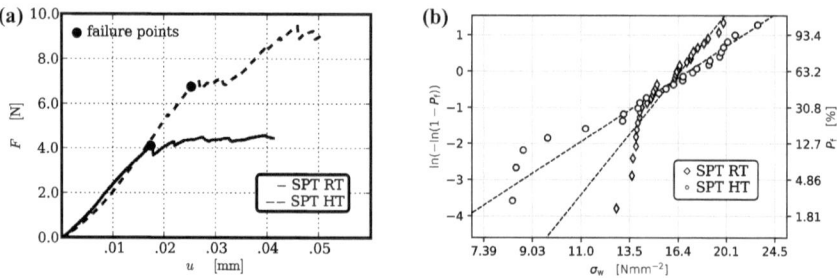

Fig. 23.12 SPT Results for a AC20 material. **a** Typical load-deflection curves at room and high temperatures **b** Weibull plots for 30 specimen tested at room and high temperatures

23.4 Results

23.4.1 Small Punch Test

The SPT was mainly used to obtain the mechanical strength of the different materials. Figure 23.12 shows two typical load vs. deflection curves obtained from SPTs, one measured at room and the other at high temperature (800 °C). At the beginning the curves show an ascending slope as it is expected for a linear elastic material. After a certain point small load drops are observed which are caused by evolving cracks at the lower side of the specimen. The point where the first significant load drop appears defines the point of initial failure and the corresponding failure stress values are obtained using the Equations (23.13) and (23.14) presented in Sect. 23.3.2. The obtained failure stresses have a significant scatter, so a Weibull analysis is performed, the results of which are shown right in Fig. 23.12. The curves represent the Weibull distributions fitted to the corresponding data. Their intersection with the horizontal line corresponding to 63.2% failure probability defines the Weibull reference stress σ_0 and the slope corresponds to the Weibull modulus m. Although the ranges of failure stresses for the different test temperatures overlap, differences for Weibull stress and modulus can be statistically detected. The Carbores® P binder content has a significant influence on the strength of the material system. To investigate this influence, the Carbores® P content of the carbon fraction was varied from 10 wt% to 20 wt%. In the case of the cast material the total carbon content after coking was fixed to 30 wt% for all systems. The result of this investigation is shown left in Fig. 23.13 in the form of a Weibull diagram. It can be seen that, the strength of the cast material system increases with increasing binder content. That is, a larger proportion of coal tar pitch leads to a higher strength of the overall system. However, it can be stated that the binder content has no significant influence on the Weibull modulus. This means that the probability of having a defect with failure-relevant size and location in the microstructure is not influenced by the binder content. This can be linked to the melting of the binder during fabrication.

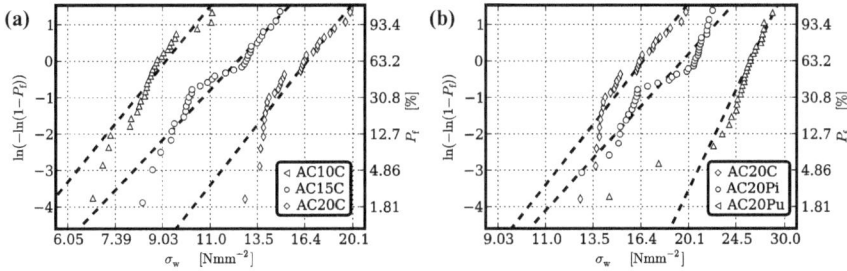

Fig. 23.13 Comparison of the failure probability of SPT specimens **a** influence of the binder content on the strength of the material system [4] **b** influence of the manufacturing route on the strength of the material system [33]

The materials were also evaluated according to three different manufacturing processes. Prior to drying and coking, the slurries were casted (AC20C), isostatically (AC20Pi) or uniaxially (AC20Pu) pressed to $p = 150\,\mathrm{MPa}$. It can be clearly seen on the right side of Fig. 23.13 that the strength of the cast material is lower than the strength of the pressed materials. Uniaxial pressing yields a higher strength of the material with lower scatter than isostatic pressing.

To study the effect of the heat treatment on the material strength, the casted slurries containing 20 wt% Carbores® P were coked at temperatures of 800 °C (AC20C) and 1400 °C (AC20T). Figure 23.14 illustrates the difference of these two heat treatments. The rhombohedral phase of graphite is metastable above 1300 °C. Therefore, a transformation of this phase takes place. In addition, a longer holding time in combination with an elevated temperature leads to an increased graphitization of the coal tar pitch. The higher coking temperature does not exert any significant influence on the Weibull modulus. However, it can be seen that the Weibull stress is lower for the material coked at a higher temperature. This can be explained by the increased crystallization of the carbon. As the degree of crystallization in the carbon increases, the anisotropy of the mechanical and thermal properties also increases. This results

Fig. 23.14 Comparison of the strength of SPT specimens in dependence of the coking temperature of the material [33]

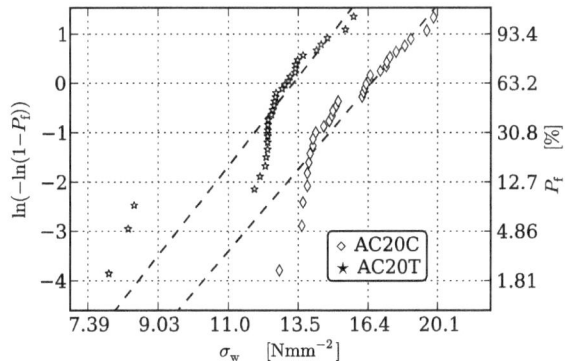

from an increasing regularity in the arrangement of graphene planes. In addition, the removal of impurity atoms between the planes leads to the formation of additional π-bonds, which have a very low binding energy. All these defects facilitate sliding of the graphene planes parallel to the planes. Macroscopically, this can be measured as reduced strength of the material. Based on this investigation, it is clear that the carbon has a significant influence on the strength of the material system. This depends not only on the binder content, but also on the chemistry of the carbon. A lower degree of order of the graphite planes has a positive effect on the strength of the material system.

Soltysiak et al. [34] also investigated the influence of the microstructure on the fracture behavior of carbon bonded alumina. Not all specimen show significant load drops in their load-deflections curves. The failure of some specimen manifests itself just as a decrease in the slope of the load-deflections curves. It was found that the load drops are caused by the breakage of rather large Carbores® P grains near the specimen surface. The absence of large Cabores® P grains is leading to a more continuous fracture of the specimen.

23.4.2 Ball on Three Ball Test

Zielke et al. [35] compared results obtained from SPTs and B3B tests on a AC20C material. The B3B tests were done with different support radii R_a. The failure stresses are obtained similar as to the SPT procedure. For each variant 30 specimen were tested and a Weibull analysis performed. The results are summarized in Table 23.3 and show a good agreement between all different test setups. The support radius has no significant influence on the obtained Weibull parameters.

Further investigations of Zielke et al. [25] using the B3B test were done with respect to different manufacturing routes and test temperatures. The slurry composition corresponds to that of a AC20C material. The results are presented in Fig. 23.15 and Table 23.4. Again the material has greater strength at high temperatures than at room temperature. But there is a significant difference regarding the scatter of the strength data. The Weibull exponent for sprayed specimen is much smaller than for cast specimen. This is because the slip cast specimens show a more homoge-

Table 23.3 Comparison of identified Weibull parameters using B3B tests with varied support radius and SPTs. The values in brackets correspond to 95% confidence interval [35]

	R_a [mm]	σ_0 [MPa]	m
B3B	2.31	21.74 (20.01, 23.67)	4.88 (3.57, 6.56)
B3B	2.89	21.35 (19.66, 23.25)	4.88 (3.57, 6.57)
B3B	3.46	20.56 (19.02, 22.28)	5.18 (3.79, 6.97)
SPT	2.50	19.06 (17.61, 20.57)	5.29 (3.88, 7.13)

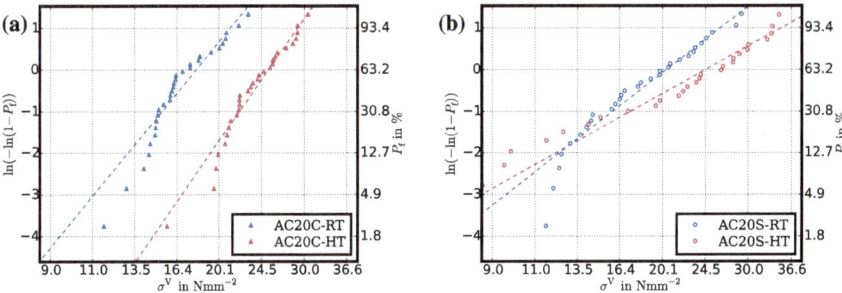

Fig. 23.15 Weibull failure probability plot for different specimen manufacturing processes at different temperatures [25]. **a** Slip cast specimens **b** sprayed specimens

Table 23.4 Comparison of identified Weibull parameters using B3B tests for different manufacturing routes. The values in brackets correspond to 95% confidence interval [25]

Specimen	N	σ_0 [MPa]	m	T [°C]
Slip cast	30	18.04 (16.87, 19.26)	6.18 (4.53, 8.32)	20
Slip cast	30	25.37 (23.97, 26.81)	7.32 (5.36, 9.85)	800
Sprayed	30	20.50 (18.46, 22.69)	3.97 (2.91, 5.35)	20
Sprayed	30	24.61 (21.16, 28.52)	2.85 (2.06, 3.88)	800

neous microstructure than the sprayed ones. Another important aspect regarding the strength of Al_2O_3-C is the coking temperature, which has been investigated by Zielke et al. [27]. In this research the testing facilities have been upgraded with a furnace, which enables test temperatures up to 1600 °C. Figure 23.16 shows the main results. For test temperatures higher than 1200 °C the specimens show no longer neither elastic behavior nor brittle fracture. The deformation at failure increases with higher temperatures and includes some plastic or visco-plastic portions, which results in non-linear rising slopes of the load deflection curves. Figure 23.16b shows the Weibull plot for specimens coked at 800 °C and tested at room temperature. The most interesting effect is displayed in Fig. 23.16c. The material coked at 800 °C has its highest strength also at 800 °C. Similarly, the material coked at 1400 °C has its highest strength at 1400 °C. Therefrom it can be stated that the material strength has its highest values at the coking temperature. At the coking temperature during material production, the material is in a state without micro cracks and an ideal binding of the Al_2O_3 particles to the carbonaceous matrix is established. This state is achieved because of the long holding time. When the specimens are cooled down, micro cracks and pores develop between the two phases due to the different thermal expansion coefficients ($\alpha_C > \alpha_{Al_2O_3}$). These cracks can propagate or branch. This leads to a macroscopic measurable reduced strength. If the testing temperature increases start-

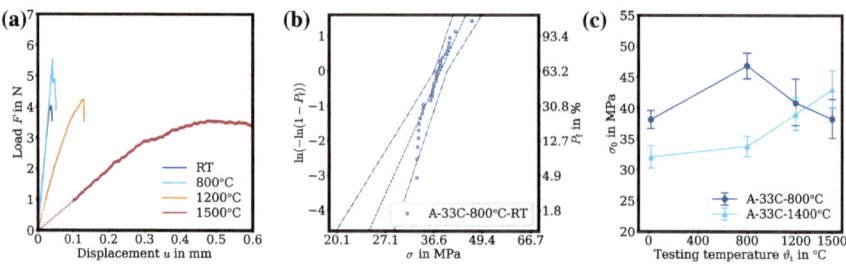

Fig. 23.16 a B3B load-deflection curves at different test temperatures **b** Weibull plot for the specimen coked at 800 °C and tested at room temperature **c** Influence of the coking temperature on the material strength at high temperatures [27]

ing at room temperature, the first small cracks between Al_2O_3 particles and carbon matrix are closed. With a further temperature increase up to the coking temperature, the cracks are closed completely and internal compressive stresses occur leading to a rising strength. If the testing temperature is higher than the coking temperature, the expanding Al_2O_3 particles induce thus tensile stresses in the carbonaceous matrix and new cracks are generated. However, these cracks do not propagate at the particle-matrix interface as before, but within the matrix. The result is a decrease in strength. Fracture toughness was determined by Zielke et al. [28] using simulations of the B3B test utilizing CZM. The underlying FE-model is shown in Fig. 23.9. To identify the cohesive parameters, an optimization approach is utilized. Hereby, an optimal parameter set p^* is found by minimizing the error norm (23.16) between the simulated and measured load deflection curves.

$$p^* = \arg\min\left(\phi(p)\right) \tag{23.15}$$

$$\phi(p) = \frac{1}{N} \sum_{k=1}^{N} \left(\frac{F_{\text{sim}}(u_k, p_i) - F_{\text{exp}}(u_k)}{F_{\text{max}}}\right)^2 \tag{23.16}$$

The parameter set contains the elastic modulus and the unknown cohesive parameters $p = [E, \Gamma_0, t_0]$. The cohesive stiffness E_{cz} has no influence on the result of the simulation, if it is above a certain limit. The displacement range used to compute the error norm goes from zero up to a value slightly larger as the displacement for the maximal force. Figure 23.17 shows the optimization scheme and the comparison between experimental and simulated force displacement curves. The simulation with the cohesive zone model is able to capture the load drop, where the primary cracks appear if the model is properly calibrated. Table 23.5 displays the identified values for three different B3B setups. The material investigated is a AC20 ceramic.

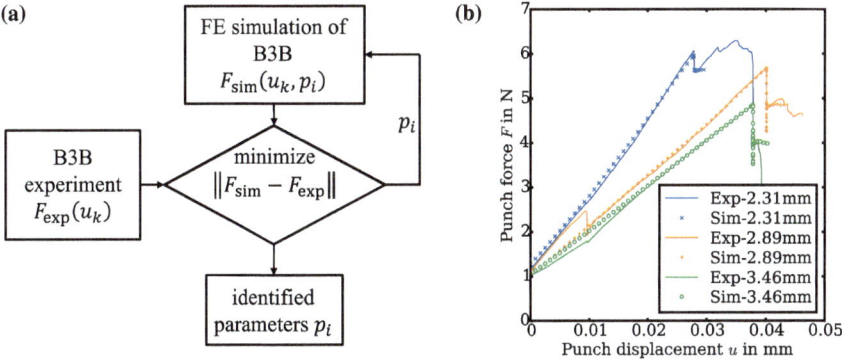

Fig. 23.17 **a** Optimization scheme for the parameter identification by comparison between experimental and **b** simulated force displacement curves [28]

Table 23.5 Identified cohesive parameters and resulting fracture toughness obtained from three B3B tests with varying support radius R_a [28]

R_a [mm]	2.31	2.89	3.46
E [MPa]	5930	6022	8067
Γ_0 [N/mm]	0.043	0.037	0.025
t_0 [MPa]	31.87	36.25	33.88
K_{Ic} MPa$\sqrt{m}$	0.50	0.47	0.45

23.4.3 Chevron-Notched Beam Test

From the CNB test the stress intensity factor can be obtained if energy relations are considered [36]. The available strain energy of the deformed specimen and the necessary energy for crack growth must be equal. The strain energy is a function of applied load F, specimen and crack geometry, which is defined by the dimensionless parameters α, α_0 and α_1. The expression for the stress intensity factor than reads

$$K_I = \frac{F}{B\sqrt{W}} \left[\frac{E\,B}{2(1-\nu^2)} \frac{\alpha_1 - \alpha_0}{\alpha - \alpha_0} \frac{\mathrm{d}C_{\mathrm{CNB}}}{\mathrm{d}\alpha} \right]^{1/2} = \frac{F}{B\sqrt{W}} Y^*. \qquad (23.17)$$

Due to the notch shape the CNB specimen, the length of crack front increases with increasing crack depth α. Regarding the geometry function Y^*, the derivative of the compliance $\mathrm{d}C_{\mathrm{CNB}}/\mathrm{d}\alpha$ of the specimen also increases with crack depth α. But, the notch geometry term $(\alpha_1 - \alpha_0)/(\alpha - \alpha_0)$ decreases with increasing α. These relations lead to the fact that the function $Y^*(\alpha)$ possesses a minimum Y^*_{min}, as shown right in Fig. 23.18. At the crack length $\alpha = \arg\min(Y^*(\alpha))$ the maximum load F_{max} is observed and the critical fracture toughness can be calculated using

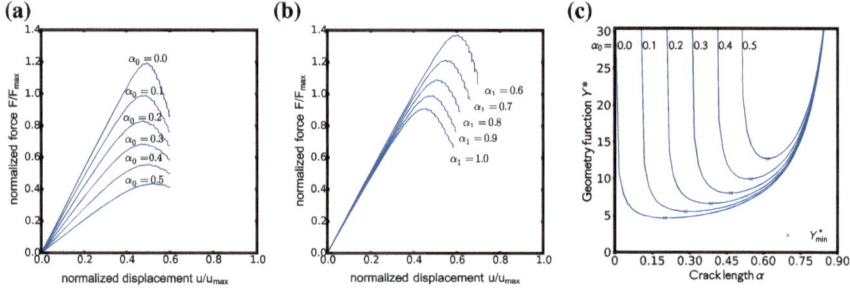

Fig. 23.18 Results of the FE analysis regarding the specimen geometry for the sub-sized CNB specimen [30, 31]. For the left and rightmost diagram $\alpha_1 = 0.9$, for the middle diagram $\alpha_0 = 0.1$

$$K_{Ic} = \frac{F_{max}}{B\sqrt{W}} Y^*_{min} .$$ (23.18)

In the literature [29] empirical formulas can be found for Y^*_{max} based on compliance measurements and FEM simulations, which are applicable for certain ranges of α_0 and α_1. Numerical investigations using a cohesive zone model [31] have shown that for the given notch geometry suitable values for α_0 and α_1 can be found. A test is valid if the crack length at failure is $\alpha_0 < \alpha < \alpha_1$. The probability to get a valid test is best if $\alpha = 0.5(\alpha_1 - \alpha_0)$. Moreover, lower load magnitudes at failure has less impact on the deformation of the test setup. This allows having a testing machine with less stiffness. Therefore, the choice for α_1 is close to 1 and following the right diagram in Fig. 23.18 optimal values for the initial crack length are $\alpha_0 = 0.2 \ldots 0.4$. In the study done by Zielke et al. [31] 15 CNB specimen made of one block of slip casted Al_2O_3-C with $0.18 < \alpha_0 < 0.29$ and $0.89 < \alpha_1 < 0.97$ were tested. 14 of these specimens exhibited a valid load-displacement curve as well as a valid notch geometry and an average fracture toughness value of $K_{Ic} = 0.576 \pm 0.037 \, MPa\sqrt{m}$ was obtained.

23.4.4 Brazilian Disc Test

The variant of the BDT with the small hole is illustrated in Fig. 23.19. The random surface speckle pattern which is used for the DIC measurement is shown in Fig. 23.19a. In Fig. 23.19b shows the specimen after failure, where the central initial and two symmetrical secondary cracks are presented. Figure 23.19c shows snap shots from the high speed cameras and illustrates the growth of the initial crack starting from the hole of the disc.

Using the surface displacements acquired through DIC (see Fig. 23.20a), the displacement output from the simulation is fitted to the experimental values (Fig. 23.20b). For this, elastic parameters are the subject of iteration. Despite differences in sample

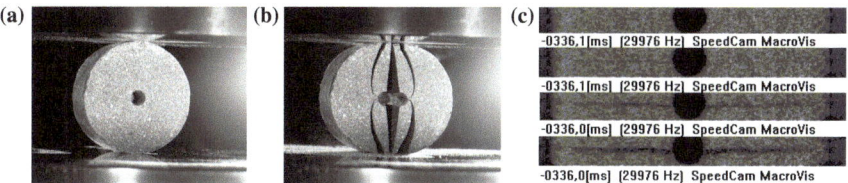

Fig. 23.19 View of a **a** intact **b** failed 16 mm Al$_2$O$_3$-C BDT specimen with a central 2 mm hole and **c** the crack formation (90° tilted)

Table 23.6 Elastic parameters at room temperature identified by each of the test variants [32]

Setup	E [MPa]	ν [–]
Var. A	17.1	0.23
Var. B	15.2	0.28
Var. C	13.8	0.22
Var. D	14.1	0.36
Var. E	14.4	0.25
Average	14.9	0.27

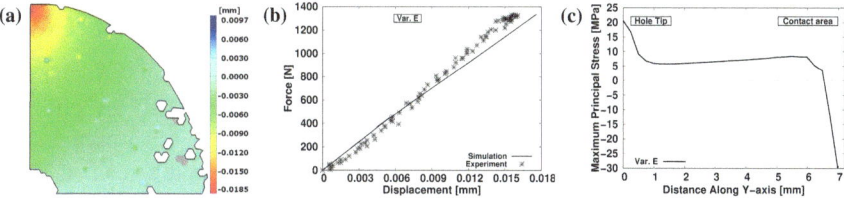

Fig. 23.20 Illustration of the experimental and simulation data for room temperature BDT application **a** Surface displacement map from DIC measurement **b** Force vs. displacement curves from experiment and simulation **c** Variation of the maximum principal stress along the loading diameter of BD

and punch geometries for each of the variants, similar values for the elastic parameters are obtained [32]. The combined analysis DIC and FEM, provides such an accuracy in calculation of the elastic parameters. The average values for $E = 15$ MPa and $\nu = 0.26$ are achieved for Al$_2$O$_3$-C at room temperature (Table 23.6). Moreover to illustrate the formation of tensile stresses along the loading diameter of the specimen, the stress plot of a specimen before failure is given in Fig. 23.20c.

Since DIC is not applicable at high temperature tests, and due to priorly mentioned contact area problems, an accurate reading of the sample deformation is not possible. Therefore, the only reliable test result is the applied force on the specimen. Since the location of the crack formation is well-predicted for specimens with holes, this sample geometry allows the application of cohesive zone modeling. To identify the material parameters at high temperatures measured force data is used in models to investigate cohesive parameters of the material. Using the optimization scheme

Fig. 23.21 Comparison of experimental and with CZM simulated results for BDT at high temperature

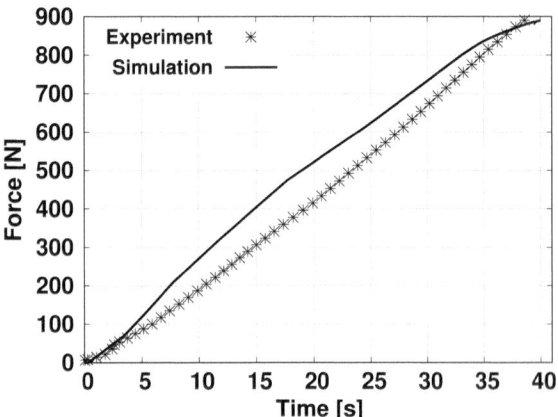

Table 23.7 Identified cohesive parameters and resulting fracture toughness obtained from BDT test at high temperature

E [MPa]	Γ_0 [N/mm]	t_0 [MPa]	K_{Ic} [MPa$\sqrt{m}$]
7350	0.038	26.4	0.55

shown in Fig. 23.17, elastic modulus and cohesive zone parameters could be obtained. Thus, determined high temperature cohesive parameters for Al_2O_3-C are given in Table 23.7. To find out the critical fracture toughness, Equation (23.12) is used. Using the parameters given in Table 23.7, simulation results are in good agreement with the experimental data as shown in Fig. 23.21. The obtained fracture toughness is in good agreement with those obtained from B3B and CNB tests.

23.5 Conclusions

Small scale specimen techniques are suitable methods to determine the temperature-dependent fracture and damage properties of ceramic filter materials. Regarding brittle ceramics the B3B test should be the preferred method, since it provides a proper three point support of the specimen, even if their surface is not perfectly flat. The SPT has a slightly simpler setup but can be applied only if the specimen have a very precise geometry, especially flatness. For both SPT and B3B test simple equations can be used to determine material strength properties from measured critical loads at failure. Due to the large scatter of the experimental data for the brittle materials Al_2O_3 and Al_2O_3-C a larger number of specimens should be tested. The minimum number of required tests is about 25–30 to reach proper confidence intervals. Both tests have been applied successfully to find an optimal chemical material composition as well as suitable production and heat treatment procedures. The experimental costs for both tests are rather low, due to the small amount of material necessary, relatively small testing machines and small furnaces, which can be heated up and cooled quickly and requiring only a small amount of energy.

The CNB test is especially designed to obtain fracture toughness values for brittle ceramics. It yields repeatable experimental results. Although the specimen preparation requires the additional step for cutting the notch, it is not too difficult to obtain. However, the test setup is much more complex than that of SPT or B3B tests and contains multiple small-sized and fragile ceramic parts. The costs for those parts are approximately ten times higher than those of a SPT or B3B tests. The cost of conducting the experiment is comparable to SPT and B3B tests, since the same testing machines and furnaces are usable.

The BDT has been tested as an alternative method to the former ones, because of its very simple test setup. Here, difficulties remain regarding the contact zones between specimen and loading plates, where the high contact stresses can cause local damage near the specimen surface. By varying the geometry of the test it was found, that small holes within the specimen lead to more repeatable results with less scatter. The costs for specimen preparation and testing are comparable to those of SPT and B3B test.

FEM simulations with cohesive zone models enable the identification of fracture toughness values from B3B, CNB or BD tests. Therefore corresponding FE-models are developed, which contain cohesive elements at the potential crack locations. With the help of optimization schemes the cohesive zone parameters can be identified, which are directly related to the fracture toughness of the tested material.

Significant data on material behavior is obtained throughout this study. To sum up, a direct temperature dependency is observed for the strength and fracture toughness of the filter materials. Linear elastic behavior of material at room temperature transforms into viscoplasticity at temperatures above $1200\,°C$. While higher coking temperatures inversely affect the material strength at room temperature, the peak strength of the filter materials are observed at the vicinity of their coking temperature. There is a direct dependency of binder content and strength of filter materials.

Acknowledgements The authors gratefully acknowledge the work and efforts of Dr.-Ing. S. Soltysiak and Dr.-Ing. H. Zielke done in the previous funding periods of the CRC 920. This work was funded by the Deutsche Forschungsgemeinschaft (DFG, German Research Foundation)—Project-ID 169148856—CRC 920: Multi-Functional Filters for Metal Melt Filtration—A Contribution towards Zero Defect Materials, subproject C03. Furthermore, the authors acknowledge computing time on the compute cluster of the Faculty of Mathematics and Computer Science of Technische Universität Bergakademie Freiberg, operated by the computing center (URZ) and funded by the Deutsche Forschungsgemeinschaft (DFG, German Research Foundation)—Project-ID 397252409.

References

1. M. Emmel, C. Aneziris, Ceram. Int. **38**(6), 5165 (2012). https://doi.org/10.1016/j.ceramint.2012.03.022
2. C. Voigt, T. Zienert, P. Schubert, C. Aneziris, J. Hubálková, J. Am. Ceram. Soc. **97**(7), 2046 (2014). https://doi.org/10.1111/jace.12977
3. J. Kazakeviciute, J. Rouse, D. Focatiis, C. Hyde, J. Strain Anal. Eng. Des. **57**(4), 227 (2022). https://doi.org/10.1177/03093247211025208
4. S. Soltysiak, M. Abendroth, M. Kuna, S. Dudczig, Adv. Eng. Mater. **15**(12), 1230 (2013). https://doi.org/10.1002/adem.201300173
5. J. Baik, J. Kameda, O. Buck, Scr. Metall. **17**(12), 1443 (1983). https://doi.org/10.1016/0036-9748(83)90373-3
6. D. Blagoeva, R. Hurst, Mater. Sci. Eng. A **510–511**, 219 (2009). https://doi.org/10.1016/j.msea.2008.05.058
7. J. Torres, A.P. Gordon, J. Mater. Sci. **56**(18), 10707–10744 (2021). https://doi.org/10.1007/s10853-021-05929-8
8. M. Abendroth, M. Kuna, Comput. Mater. Sci. **28**(3–4), 633 (2003). https://doi.org/10.1016/j.commatsci.2003.08.031
9. A. Börger, P. Supancic, R. Danzer, J. Eur. Ceram. Soc. **22**(9–10), 1425 (2002). https://doi.org/10.1016/s0955-2219(01)00458-7
10. R. Danzer, H. Walter, P. Supancic, T. Lube, Z. Wang, A. Börger, J. Eur. Ceram. Soc. **27**(2), 1481 (2007). https://doi.org/10.1016/j.jeurceramsoc.2006.05.034
11. S. Rasche, S. Strobl, M. Kuna, R. Bermejo, T. Lube, Proc. Mater. Sci. **3**, 961 (2014). https://doi.org/10.1016/j.mspro.2014.06.156
12. D. Munz, R. Bubsey, J. Srawley, Int. J. Fract. **16**(4), 359 (1980). https://doi.org/10.1007/bf00018240
13. J. Nakayama, J. Am. Ceram. Soc. **48**(11), 583 (1965). https://doi.org/10.1111/j.1151-2916.1965.tb14677.x
14. H. Tattersall, G. Tappin, J. Mater. Sci. **1**(3), 296 (1966). https://doi.org/10.1007/bf00550177
15. D. Munz, R. Bubsey, J. Shannon, J. Am. Ceram. Soc. **63**(5–6), 300 (1980). https://doi.org/10.1111/j.1151-2916.1980.tb10725.x
16. M. Mellor, I. Hawkes, Eng. Geol. **5**(3), 173 (1971). https://doi.org/10.1016/0013-7952(71)90001-9
17. D. Li, L.N.Y. Wong, Rock Mech. Rock Eng. **46**(2), 269 (2013). https://doi.org/10.1007/s00603-012-0257-7
18. G. Cherepanov, J. Appl. Math. Mech. **31**(3), 503 (1967). https://doi.org/10.1016/0021-8928(67)90034-2
19. M. Kuna, *Finite Elements in Fracture Mechanics* (Springer, Netherlands, 2013). https://doi.org/10.1007/978-94-007-6680-8
20. J. Rice, J. Appl. Mech. **35**(2), 379 (1968). https://doi.org/10.1115/1.3601206
21. K. Park, G.H. Paulino, Appl. Mech. Rev. **64**(6) (2013). https://doi.org/10.1115/1.4023110
22. S. Roth, G. Hütter, M. Kuna, Int. J. Fract. **188**(1), 23 (2014). https://doi.org/10.1007/s10704-014-9942-8
23. H. Altenbach, V.A. Kolupaev, *Classical and Non-Classical Failure Criteria* (Springer Vienna, Vienna, 2015), pp. 1–66. https://doi.org/10.1007/978-3-7091-1835-1_1
24. W. Weibull, *A Statistical Theory of the Strength of Materials*. Handlingar/Ingeniörsvetenskapsakademien (Generalstabens litografiska anstalts förlag, 1939). https://books.google.de/books?id=otVRAQAAIAAJ
25. H. Zielke, A. Schmidt, M. Abendroth, M. Kuna, C.G. Aneziris, Adv. Eng. Mater. **19**(9), 1700083 (2017). https://doi.org/10.1002/adem.201700083
26. S. Roth, M. Kuna, Int. J. Fract. **196**, 147 (2015). https://doi.org/10.1007/s10704-015-0053-y
27. H. Zielke, T. Wetzig, C. Himcinschi, M. Abendroth, M. Kuna, C. Aneziris, Carbon **159**, 324 (2020). https://doi.org/10.1016/j.carbon.2019.12.042

28. H. Zielke, M. Abendroth, M. Kuna, Theoret. Appl. Fract. Mech. **86**, 19 (2016). https://doi.org/10.1016/j.tafmec.2016.09.001
29. D. Munz, J. Shannon, R. Bubsey, Int. J. Fract. **16**(3), R137 (1980). https://doi.org/10.1007/bf00013393
30. H. Zielke, M. Abendroth, M. Kuna, B. Kiefer, Ceram. Int. **44**(12), 13986 (2018). https://doi.org/10.1016/j.ceramint.2018.04.248
31. H. Zielke, M. Abendroth, M. Kuna, Key Eng. Mater. **754**, 71 (2017). https://doi.org/10.4028/www.scientific.net/KEM.754.71
32. S. Takht Firouzeh, M. Abendroth, U. Fischer, C.G. Aneziris, B. Kiefer, Adv. Eng. Mater. **24**(2), 2101081 (2022). https://doi.org/10.1002/adem.202101081
33. S. Soltysiak, M. Abendroth, M. Kuna, Y. Klemm, H. Biermann, Ceram. Int. **40**(7), 9555 (2014). https://doi.org/10.1016/j.ceramint.2014.02.030
34. S. Soltysiak, M. Abendroth, M. Kuna, S. Dudczig, Key Eng. Mater. **592**, 279 (2014). https://doi.org/10.4028/www.scientific.net/KEM.592-593.279
35. H. Zielke, M. Abendroth, M. Kuna, Key Eng. Mater. **713**, 70 (2016). https://doi.org/10.4028/www.scientific.net/KEM.713.70
36. D. Munz, T. Fett (eds.), *Ceramics* (Springer Berlin Heidelberg, 1999). https://doi.org/10.1007/978-3-642-58407-7

Chapter 24
Influence of Internal Defects on the Fatigue Life of Steel and Aluminum Alloys in the VHCF Range

Anja Weidner, Alexander Schmiedel, Mikhail Seleznev, and Horst Biermann

24.1 Introduction

The majority of failure cases of metallic components in mechanical engineering or mobility is caused by cyclic loading and the related fatigue effects. The material behavior under cyclic loading in the range of high and medium stress amplitudes resulting in low and medium service lives (low cycle fatigue, LCF and high cycle fatigue, HCF), respectively, has been studied very intensively over the last 50 years. A very good understanding of the microstructural processes and damage mechanisms ending up in several model approaches (e.g. Antopopolous [1], Essmann et al. [2], Polak [3]) was obtained.

The current scientific challenge of fatigue research concerns the fatigue properties of materials in the range of very high cycle fatigue (VHCF) with number of cycles to failure $N_f > 10^7$. In particular, so-called type II materials are in the focus (see [4]), in particular e.g. high-strength steels, rolling bearing steels or Ni-based alloys. These materials contain discontinuities in the bulk such as nonmetallic inclusions, shrinkage cavities, secondary phases, which are acting as stress concentrators and are causing, therefore, fatigue failure. This results for majority of metallic materials in a multistage fatigue life curve (stress amplitude vs. number of cycles to failure; S–N curve). Whereas fatigue failure is mostly dominated by crack initiation at the surface in the LCF/HCF regime, a shift of the crack initiation site to the interior of the material is observed in the VHCF regime resulting in the well-known fisheye fracture surface (see [5]). The fatigue life in VHCF regime is, therefore, dominated by the crack initiation phase, which covers more than 97% of the fatigue life. According to Murakami

A. Weidner (✉) · A. Schmiedel · M. Seleznev · H. Biermann
Institute of Materials Engineering, Technische Universität Bergakademie Freiberg, Gustav-Zeuner-Straße 5, 09599 Freiberg, Germany
e-mail: weidner@ww.tu-freiberg.de

© The Author(s) 2024   605
C. G. Aneziris and H. Biermann (eds.), *Multifunctional Ceramic Filter Systems for Metal Melt Filtration*, Springer Series in Materials Science 337,
https://doi.org/10.1007/978-3-031-40930-1_24

[5], the fatigue strength depends on the following parameters of nonmetallic inclusions: (i) the size of the inclusion, (ii) the position of the inclusion with respect to the surface, and (iii) the hardness of the bulk material. Regarding the size of the inclusion, Murakami [5] introduced the $\sqrt{area}$ concept, which describes the size of a defect as the square root of the initial defect area projected onto a plane perpendicular to the stress axis, i.e. new defects (small incipient cracks) occurring in the course of the damage are not included in the calculation. Considering the location of nonmetallic inclusions in the tested volume, three different scenarios were introduced: (i) inclusions at the surface, (ii) inclusions below/in contact with the surface, and (iii) inclusions in the bulk. According to these three scenarios, the fatigue strength σ_w can be evaluated considering the largest inclusion's parameters and the hardness of the matrix material as an additional parameter [5]. This empirical model was applied in numerous studies on different materials and described very well the influence of nonmetallic inclusions on the fatigue life. A linear relationship between size of the fisheye and depth of the largest inclusion below the surface was found as well [6, 7]. The formation of a fisheye around internal defects can be accompanied by (i) the formation of a optically dark area (ODA), (ii) a fine granular area (FGA), or (iii) a granular bright facet (GBF), respectively. However, these terms are used quite confusing since often the same fatigue crack zones were labelled with different names (FGA, ODA, GBF) or giving the same name for different zones within the fisheye, which is caused mostly by the fact that subjective visual analysis of optical (ODA, GBF) or scanning electron microscope images is performed [8]. Nevertheless, different mechanisms were presented to describe the FGA around an inclusion: (i) hydrogen embrittlement [9], (ii) long-term repeated contact of the fracture surface under vacuum conditions (interior of the material) [10, 11], and (iii) formation of ultrafine-grained microstructure resulting in a "concavo-convex" pattern smaller than the original microstructure [12].

In particular in high-strength steels, nonmetallic inclusions or even large clusters of smaller inclusions have a detrimental influence on the fatigue behavior resulting in large scatter of the fatigue life [13, 14] or causing severe damage [15]. But also, secondary phases such as intermetallic in aluminum alloys [16] or grains with preferred orientations in Ni-base superalloys [17], Ti- or TiAl-alloys [18] may deteriorate fatigue properties. Therefore, the removal of detrimental nonmetallic inclusions or the prevention of intermetallic are important issues mainly for materials used in safety-relevant components, and, therefore, in the focus of research. Nonmetallic inclusions and/or intermetallic phases form mostly during the solidification process of the material, and are, therefore, irregular and randomly distributed in the solidified material regardless of the manufacturing process [19]. Most detrimental for fatigue properties is the largest nonmetallic inclusion remaining in the volume of the component, since it will act as crack initiating defect [5, 20]. However, it is a demanding and time-consuming process to evaluate the size distribution of inclusions contained in the solidified material by conventional methods such as defect analysis by optical microscopy on microscopic sections out of the metallic components. This method, moreover, provides only two-dimensional information and it is hardly possible to find the largest inclusion. Also, the application of nondestructive methods such as micro computed tomography (μCT) is not suitable for inspection of large metallic components due to the resolution limit. An excellent method for detection of the largest nonmetallic inclusion being present in a certain volume is fatigue testing at low stress amplitudes in the VHCF range.

Metal melt filtration can be used to improve the quality of the melt by reduction of nonmetallic inclusions. Whereas macroscopic exogenous inclusions (>20 μm) being the result of erosion of the refractory materials were directly removed with the slag and/or in the casting system, the removal of endogenous inclusions (<20 μm) is more challenging because they are forming continuously during the casting and solidification process [21]. The primary and secondary nonmetallic inclusions are related to reactions of the liquid melt with deoxidizers; tertiary and quaternary inclusions form below the liquidus temperature of the melt and can be even smaller than 1 μm. Therefore, the removal of endogenous inclusions is a challenging task, which can be hardly implemented by application of conventional ceramic foam filters. Thus, a new generation of active and reactive filters or even a combination of both was developed [22, 23].

The aim of this work is to evaluate the effect of the newly developed filters and filter systems on the fatigue lives of the solidified materials, which was intensively studied for the quenched and tempered cast steel 42CrMo4 as well as for the steel 18CrNiMo7-6, whereas Chap. 25 of this volume is focusing on the influence of the filters on the respective inclusion size distributions as well as their chemistry and their morphology in the solidified steels. In aluminum alloys, the influence of both internal shrinkage cavities (AlSi7Mg) as well as intermetallic phases (AlSi9Cu3) on the fatigue lives was studied.

24.2   Materials and Methods

24.2.1   Materials and Manufacturing

The influence of internal discontinuities on the mechanical behavior under cyclic loading was studied for two steels and two aluminum alloys: (i) quenched and tempered steel 42CrMo4, (ii) hardened steel 18CrNiMo7-6, (iii) AlSi7Mg, and (iv) AlSi9Cu3 produced under different routes.

Quenched and Tempered Steel 42CrMo4

The steel 42CrMo4 was investigated both after melt filtration applied in (i) *industrial sand-casting route* as well as in (ii) laboratory route using a *steel cast simulator*.

The *industrial casting route* was performed at Edelstahlwerke Schemes GmbH (Pirna, Germany). The casting system used for these investigations consisted of three parts: (i) the casting gate (CG), (ii) the casting plate (CP), and (iii) the feeder (F) as

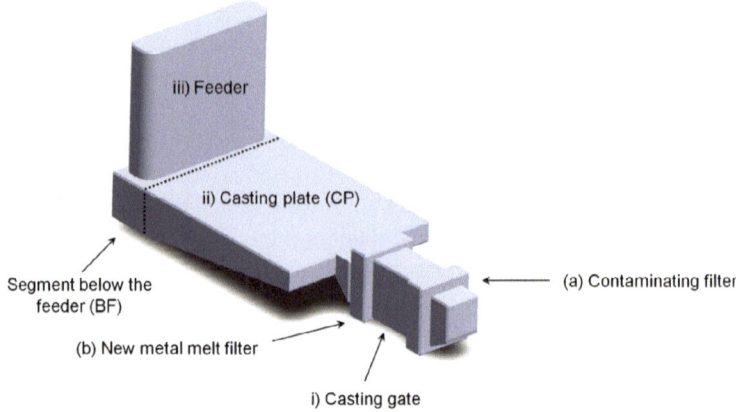

Fig. 24.1 Casting system of industrial sand-casting route for 42CrMo4. Reproduced from [24]

shown in Fig. 24.1. Two filters were applied to the casting gate: (i) a contaminating filter for the artificial impurification of the steel melt with nonmetallic inclusions of the alumina-type (Al_2O_3), and (ii) a cleaning filter for metal melt filtration. For the cleaning filter, two different types of ceramic foam filters were used: (i) active filters based on carbon-bonded alumina (Al_2O_3–C) and (ii) reactive filters based on carbon-bonded magnesia (MgO–C), compare Chap. 1, [23]. Whereas the active filters operated as mechanical filtration of exogenous inclusions as well as of primary and secondary endogenous inclusions by application of a coating offering the same chemical composition as the majority of inclusions in the steel melt, the reactive filters reduce the dissolved oxygen content of the steel melt [23]. The functionalized coatings on the surface of active filters were: (i) alumina (Al_2O_3), (ii) spinel ($MgAl_2O_4$), or (iii) mullite ($3Al_2O_3*2SiO_2$). Fatigue tests were performed for the casting plate (CP) and the region below the feeder (BF). The material of the casting gate as well as the feeder was not investigated by fatigue experiments.

In the laboratory route a *steel cast simulator* (see Chap. 19) was operated at Institut of Ceramis, Refractory and Composite Materials (TU Bergakademie Freiberg, Germany) to investigate the influence of crucible materials, the reactive filtration, the active filtration, and a combination of active and reactive filtration on both the size distribution of nonmetallic inclusions as well as on their chemical composition and morphology. The effect of reactive and/or active filtration on the inclusion distribution as well as on their size and morphology was studied by (i) immersion filtration (reactive), (ii) flow-through filtration (active), and (iii) a combination of (i) and (ii). Whereas carbon-bonded alumina (C–A) filters designed by Emmel et al. [22] were used for immersion filtration, ceramic foam filters of Foseco (STELEX PrO, Foseco International Ltd., Tamworth, UK) were applied for flow-through filtration. Coatings with carbon nano tubes (CNT) together with alumina nano sheets (ANS), calcium aluminate (CA) and an alumina flame-spraying (FS) were applied on both filter types. Details on the application of the coating are described in Chap. 1 and

[25–27]. The working principle of the steel cast simulator is presented in detail in [28, 29]. Before filtration, defined alumina inclusions were generated in the melt by oxidation (addition of 0.5 wt% of iron oxide) and subsequent deoxidation with pure aluminum (0.05 wt%), followed by a specific time–temperature regime [30, 31]. The immersion filters (I) were immersed for 10 s into the steel melt. The flow-through filtration (F) was conducted by tilting the crucible and pouring the steel melt through the filters. In addition, combinations of immersion filtration and flow-through filtration (I_F) were conducted as well. The applied filters and filter combinations had an influence on the chemical composition, in particular on the amount of aluminum and oxygen, of the filtered steel. Table 24.1 summarizes the chemical compositions of steel 42CrMo4 after the treatments in the steel cast simulator. The filter IDs and abbreviations used in Table 24.1 denote reference materials without any filtration (I_0, F_0), alumina flame-spraying (Al_2O_3-f-s, I_A, F_A), alumina-based slurry (AC5, I_{AC5}, F_{AC5}), carbon nano tubes and alumina nano sheets (CNT + ANS, I_{NM}, F_{NM} for nano materials) as well as calcium-hex-aluminate (CA6, I_{CA6}) and calcium-di-aluminate (CA2, I_{CA2}), respectively. Standard methods were used for the determination of the chemical composition such as (i) gas fusion analysis (O), (ii) combustion/infrared absorption analysis (C,S), and optical emission spectroscopy (other elements).

The filter-treated and solidified cast plates and ingots were hot-isostatically pressed (HIP; Bodycote, Germany) in order to reduce closed shrinkage porosity.

Table 24.1 IDs, filter nomenclature and chemical composition (in wt%) of solidified 42CrMo4 batches after immersion and flow-through filtration as well as combination of both

ID	Filter	C	Cr	Mn	Si	S × 10^{-3}	P × 10^{-3}	Al × 10^{-3}	$[O]^{tot}$ × 10^{-3}
I_0[1]	–	0.34	0.95	0.71	0.20	1	21	7	3
$I_0 + F_0$[2]	–	0.30	0.93	0.59	0.18	2	10	2.2	3
I_A	Al_2O_3 f-s	0.33	0.97	0.70	0.21	2	17	3	6
F_A		0.34	0.95	0.64	0.28	1	13	2*	9
I_{AC5}	AC5	0.32	0.96	0.68	0.20	2	16	<1	4
F_{AC5}[3]		0.34	0.95	0.64	0.28	1	13	2[a]	9
I_{NM}	CNT + ANS	0.34	0.97	0.75	0.23	2	23	1	4
F_{NM}		0.32	0.95	0.69	0.18	1	13	2	1
I_{CA6}	CA6	0.33	0.97	0.75	0.22	1	23	2	4
$I_{CA2} + F_A$	CA2 + Al_2O_3 f-s	0.30	0.95	0.65	0.18	2	10	<1	4
$I_{NM} + F_A$	CNT + ANS + Al_2O_3 f-s	0.29	0.96	0.68	0.19	2	10	<1	4

[1] Reference material for immersion filtration
[2] Reference material for combination of immersion and flow-through filtration
[3] Reference material for flow-through filtration
[a] Single measurements below the detection limit of 0.001 wt%. Fe balance

Case Hardening Steel 18CrNiMo7-6

The steel 18CrNiMo7-6 was used for the investigation of the influence of refractory crucible materials on the formation of nonmetallic inclusions under laboratory conditions using the above-mentioned steel cast simulator. The working principle of steel cast simulator is described in more detail in Chap. 19. For this purpose, three different crucible materials were applied: (i) carbon-bonded alumina (A–C), (ii) carbon-bonded alumina–zirconia-titania (AZT–C), and (iii) carbon-bonded alumina with carbon nano tubes and alumina nano sheets (AZT–C–n). Details on manufacturing of the crucible materials can be found in Aneziris et al. and Fruhstorfer et al. [32, 33]. The steel melt was exposed to the crucible materials for 1 h at 1580 °C and, subsequently, solidified. The chemical compositions of the as-delivered steel and the solidified batches after contact with different crucible materials are summarized in Table 24.2. All batches were hot-isostatically pressed (Bodycote, Germany) in order to reduce closed shrinkage porosity. For more details see [34].

Aluminum Alloys AlSi7Mg and AlSi9Cu3

AlSi7Mg (A356) was manufactured in two batches: (i) die casting of cylindrical rods under laboratory conditions (batch 1), and (ii) sand casting (batch 2). Both batches were treated with 0.025 wt% strontium in order to minimize the harmful influence of eutectic. The chemical compositions of both batches determined by optical emission spectroscopy are summarized in Table 24.3. Both batches were hot-isostatically pressed (Bodycote, Germany).

The commercial secondary *AlSi9Cu3* cast alloy was used to study the influence of melt conditioning and filtration on the solidified microstructures and their influence on the mechanical properties, in particular under cyclic loading. A commercially available batch with an increased iron content was used as reference batch (R). The conditioning (C) and the combination of conditioning and filtration (CF) were performed in order to reduce both the iron content as well as the formation of harmful iron-containing intermetallic phases. For this purpose, the melt was doped with the master alloy AlMn20Fe0.3 (Technologica, Germany) and kept at approx. 620 °C and a dwell time of 3.5 h in a ceramic crucible of an electric holding furnace (C).

Table 24.2 Batch nomenclature and chemical composition (in wt%) of solidified 18CrNiMo7-6 steel using different crucible materials. Fe balance. Partly reproduced from [34]

Batch	C	Cr	Ni	Mn	Si	S × 10^{-3}	P × 10^{-3}	Al × 10^{-3}	$[O]^{tot}$ × 10^{-3}
As-delivered	0.19	1.59	1.43	0.24	0.70	27	16	22	25
A-C	0.07	1.44	1.41	0.25	0.63	24	7	<1	32
AZT-C	0.12	1.57	1.39	0.25	0.68	21	13	<1	24
AZT-C-n	0.11	1.55	1.35	0.25	0.61	21	11	<1	20

Table 24.3 Chemical composition (in wt%) of both solidified AlSi7Mg after continuous casting (batch 1) and die casting (batch 2) as well as AlSi9Cu3 for reference batch (R), the conditioned state (C) and conditioned and filtrated batch (CF). Al balance. According to [31, 35]

Batch	Si	Cu	Fe	Mn	Zn	Mg	Sr	Cr
AlSi7Mg								
1	6.83					0.29		
2	6.58					0.28		
AlSi9Cu3								
R	8.70	2.11	1.71	0.09	0.43	0.20	0.03	0.03
C	8.82	2.10	1.60	1.23	0.34	0.22	0.03	0.03
CF	8.73	2.15	0.99	0.47	0.42	0.21	0.01	0.01

Commercial ceramic filters with 30 pores per inch (Hofmann Ceramic, Germany) were used for subsequent filtration (CF). Finally, the melt was poured after degassing into a chill mold giving test bars of 16 mm in diameter and 200 mm in length. The chemical compositions measured by optical emission spectroscopy of reference batch R, batch C after conditioning and batch CF after conditioning and filtration are summarized in Table 24.3. All batches of AlSi9Cu3 were hot-isostatically pressed (Bodycote, Germany) in order to reduce closed shrinkage porosity.

24.2.2 Methods

Characterization of Inclusions

In order to study the effect of nonmetallic inclusions on the fatigue lives of different kinds of materials, the inclusions remaining in the solidified material had to be characterized according to their size, morphology and chemical composition as well as to their size distribution within the solidified materials. Light optical microscopy and scanning electron microscopy were used for detailed quantitative characterization of inclusions. These investigations were corroborated by chemical and electrolytical extraction to study the morphology of inclusions. Detailed quantitative analysis of inclusion size distributions in steel were performed using an automated scanning electron microscope with an integrated EDS detector. These investigations and their results are presented in more detail in Chap. 25.

Fatigue Tests

The fatigue tests were performed at room temperature in resonance mode up to 10^9 cycles at a testing frequency of about $f = 19.5$ kHz by using an ultrasonic fatigue testing machine (University of Natural Resources and Life Sciences, Vienna,

Austria). The testing equipment consisted of an (i) ultrasonic transducer, (ii) amplification horn, and (iii) vibration sensor. For more details see Stanzl-Tschegg et al. [36]. The ultrasonic transducer is generating an ultrasonic wave which is transferred via an amplification horn (titanium) to increase the amplitude of the ultrasonic wave into the specimen. The length of the specimen has to be designed according to the resonant conditions of the investigated material using a modal analysis. Therefore, the Young's modulus, the Poisson ratio and the velocity of sound for longitudinal waves in the material under investigation are needed. The ultrasonic wave propagating through the specimen is reflected on the free end of the specimen generating a standing wave due to the resonant conditions resulting in longitudinal loading of the specimen.

The ultrasonic fatigue test system is operated in vibration control. Thus, the stress amplitudes applied for fatigue tests were calibrated based on chosen vibration amplitudes. For this purpose, two strain gages were glued on opposite sides in the center of the gauge length. In the next step, the vibration amplitudes measured at the vibration sensor, which correlate with the amplitude of the generated ultrasonic wave, was set at three distinct supporting points and the resulting strain values at the strain gages were recorded. Consequently, a linear relation of strain ε vs. vibration amplitude was obtained. The applied stress can then be calculated applying Hooke's law and values of Young's modulus for the tested materials.

Thus, the standing wave causes maximum vibration amplitudes at both ends of the specimen and a zero-vibration node in the center of the gauge length. In contrast, the resulting stress and strain amplitudes are zero at both end faces of the specimen and are at maximum in the center of the gage length as exemplarily shown for a 42CrMo4 specimen in Fig. 24.2a. In the present studies, a parallel gauge length of 9 mm was used instead of hour-glass specimen geometry to increase the cyclically strained material volume.

This technique allowed to fatigue the specimens till 10^9 cycles within a reasonably short time, e.g. few hours to few days depending on the applied stress amplitude and the damping behavior of the investigated material. The damping of the material caused a temperature increase in the gauge length of the specimen of several tens (aluminum) or up to hundreds of Kelvin (austenitic stainless steel). Therefore, pulse/pause mode in combination with additional cooling using compressed air (spot cooler, Eputec, Kaufferingen, Germany) was applied to avoid heating up of the specimen during the fatigue test due to damping of the material. During the pulse of defined length according to the applied stress amplitude the specimen was cyclically strained resulting in a temperature increase, whereas during the pause the gauge length of the specimen cooled down again to room temperature. Figure 24.2b shows the temperature profile within the gauge length of a 42CrMo4 steel in both continuous mode and pulse/pause mode with and without additional cooling. Pulse/pause mode and additional cooling was adopted to keep the temperature increase $\Delta T \leq 10$ K over the entire fatigue test until crack initiation appeared in the specimens. However, the pulse/pause sequences led to effective test frequencies significantly below 20 kHz. All fatigue tests were performed under symmetric push–pull loading conditions (R $= -1$) in laboratory air beside one test series which was carried out at R $= 0.1$. Since

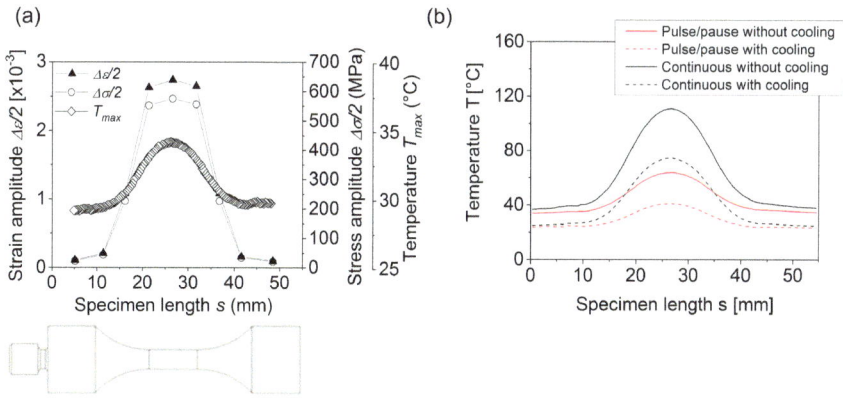

Fig. 24.2 **a** Profile of strain and stress amplitude as well as temperature along a fatigue specimen for steel 42CrMo4 tested at a stress amplitude of $\sigma_a = 400$ MPa. **b** Temperature profile along a fatigue specimen of 42CrMo4 in continuous and pulse (600 ms)/pause (900 ms) mode tested with and without cooling at a stress amplitude of $\sigma_a = 500$ MPa. Reproduced from [37]

the fatigue tests were controlled by the resonant frequency f_{res}, a decrease in f_{res} of about 75 Hz was used as an indicator for the final failure of specimens. The run-out limit of the fatigue tests was set at 10^9 cycles.

In addition to f_{res}, the nonlinearity parameter β_{rel} was recorded in situ during the fatigue tests. The nonlinearity parameter was calculated from the amplitudes A of the first A_1 (20 kHz) and second A_2 (40 kHz) order of harmonic waves of the ultrasonic pulse according to Kumar et al. [38] by Eq. 24.1:

$$\beta_{rel} = (A_2 - 2A_1) - (A_2 - 2A_1)_0 \qquad (24.1)$$

The difference between both amplitudes for current pulses and their difference to the initial undamaged material state (0) is highly sensitive to microstructural processes during cyclic loading (i.e., dislocation patterning, cyclic hardening or softening) as well as to damage processes (crack initiation) [39] and, therefore, an indicator for onset of damage. Thus, β_{rel} should stay more or less constant at zero until crack initiation occurs yielding a sudden change in the nonlinearity parameter.

Specimen Preparation and Thermal Treatment

The specimen geometries used for the materials under investigation are summarized in Fig. 24.3. All fatigue specimens (42CrMo4, 18CrNiMo-7–6, AlSi7Mg, AlSi9Cu3) were manufactured by a CNC lathe from HIPed plates and ingots.

Specimens of 18CrNiMo7-6 were austenitized at 900 °C for 35 min in argon atmosphere and, subsequently, quenched in water resulting in a martensitic microstructure

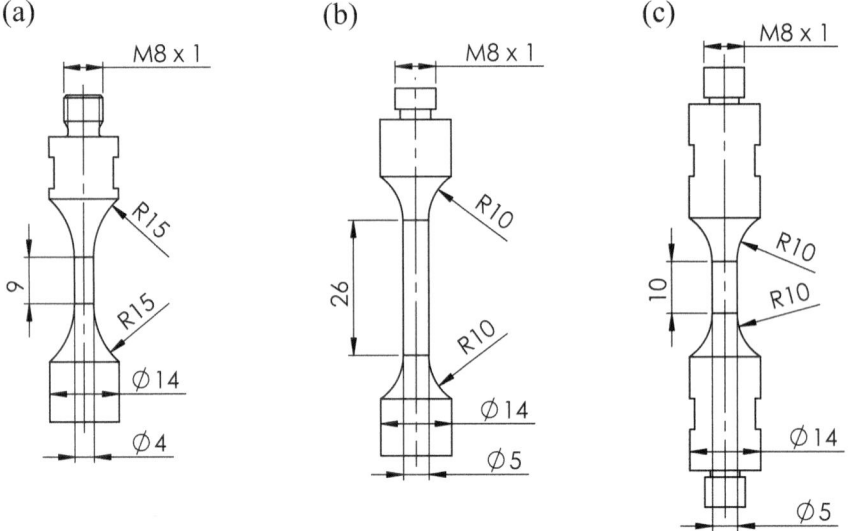

Fig. 24.3 Specimen geometries for ultrasonic fatigue testing. **a** 42CrMo4 and 18CrNiMo7-6.
b AlSi7Mg. **c** AlSi9Cu3 (for tests in combination with in situ acoustic emission measurements).
All dimensions in mm

(see [34]). In order to obtain a high strength which is necessary for the ultrasonic testing, the specimens were not tempered.

The *42CrMo4 specimens* underwent a quenching and tempering treatment: (i) austenitization at 840 °C for 20 min followed by quenching in helium, and (ii) tempering at 560 °C for 1 h in argon atmosphere yielding the typical microstructure of tempered martensite (e.g. [40]).

The gauge lengths of all steel specimens were manually grinded and polished down to 1 μm after the heat treatment. A plasma-nitriding treatment was applied to some 42CrMo4 steel specimens of the industrial sand-casting route and to all 42CrMo4 specimens from the steel cast simulator to avoid crack initiation from the surface. A two-step nitriding process was conducted to achieve a nitriding layer without compound layer with nitriding hardness depths (NHD) of 0.23 and 0.47 mm, respectively. For more details see [41].

Specimens of AlSi7Mg were precipitation hardened according to the peak-aged hardening treatment (T6) by a two-step heat treatment consisting of (i) solution annealing at 560 °C for 0.5 h followed by water quenching, and (ii) ageing at 165 °C for 16 h followed by air cooling. The microstructure of the AlSi7Mg alloy after T6 heat treatment consisted of α-Al solid solution, Al–Si eutectics and a minor content of Mg_2Si precipitates mainly at grain boundaries (see [35]). Batch 1 was tested without any additional grinding and polishing in state after turning resulting with a surface roughness of $R_z = 6.3$ μm, whereas all specimens of batch 2 were carefully grinded and polished down to 1 μm.

Specimens of AlSi9Cu3 were tested in HIP condition without any heat treatment since this alloy is not precipitation hardenable. Again, the gauge lengths of all specimens were carefully grinded and polished down to 1 μm.

Fractography

The fatigue tests were terminated after a decrease of the resonant frequency $\Delta f_{res} = 75$ Hz. Specimens were then cooled in liquid nitrogen and broken by slight mechanical impact taking care not to influence the fatigue fracture surface. All fracture surfaces were analyzed by scanning electron microscopy (SEM). For this purpose, a field-emission SEM (Mira 3 XMU, Tescan, Brno, Czech Republic) was used. Both fracture surfaces were analyzed using secondary electron (SE) contrast, backscattered electron (BSE) contrast and energy dispersive X-ray spectroscopy (EDS) for analyzing the crack initiation defects/nonmetallic inclusions.

From the SEM micrographs in SE contrast the following parameters were determined as shown in Fig. 24.4: (i) the depth of the inclusion/defect S_{inc} (Fig. 24.4a), (ii) the diameter of the fish eye $D_{fisheye}$ (Fig. 24.4a), (iii) the area of the inclusion/defect A_{inc} (Fig. 24.4b), and (iv) the area of the FGA A_{FGA} (Fig. 24.4b)—if present. These parameters were used for the calculation of the fatigue limit σ_w according to Murakami [5] depending on the position of the defect, the core hardness of the material and the $\sqrt{area}$ value of the defect as well as the applied stress amplitude σ_a.

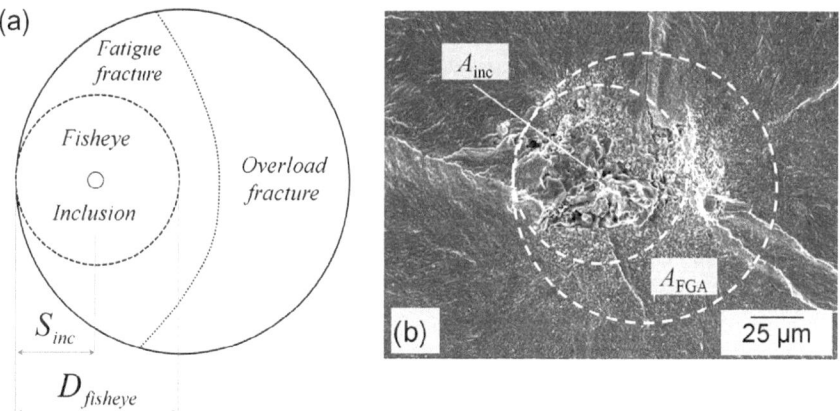

Fig. 24.4 Measured parameters on fracture surfaces. **a** Measurement of the inclusion depth (S_{inc}) and the fisheye diameter $D_{fisheye}$. **b** Determination of the defect area A_{inc} and the area of the fine granular area A_{FGA} (42CrMo4, $\sigma_a = 550$ MPa, $N_f = 1.14 \times 10^8$ cycles). Reproduced from [24]

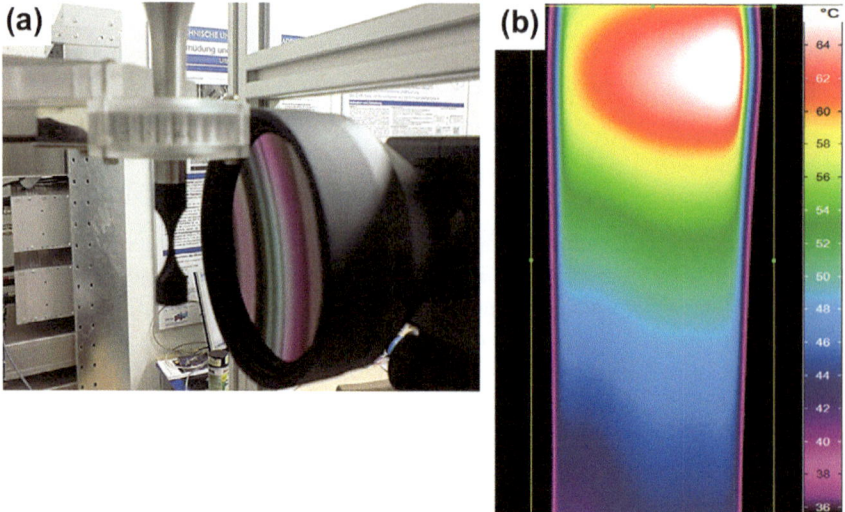

Fig. 24.5 a Experimental setup of the infrared thermocamera Variocam hr (microscopic lens) 50 mm in front of a black-coated specimen mounted in the ultrasonic fatigue testing rig. **b** IR image of the central part of a fatigue specimen close to the end of the fatigue life N_f. Reproduced from [35]

Thermography

During the ultrasonic fatigue tests, thermographic measurements were performed using a long-wave-range infrared (IR) thermocamera VarioCam hr head 600 (InfraTec, Dresden, Germany) operating in the spectral range of 7.5 µm up to 14 µm. The microbolometer-focal plane array detector enables a geometric resolution of 640 × 480 IR pixel. The thermal resolution is <0.03 K at 30 °C. The camera is equipped with a Germanium macro-lens (Zeiss, Jena, Germany) with an anti reflexion coating enabling of field of view (FOV) of 27 × 20 mm at a pixel size of 42 µm. In addition, the camera can be used with an infrared microscope lens enabling higher resolution covering an image field of 16 × 12 mm with a pixel size of 25 µm. For both lenses the focus distance is about 50 mm. The camera is operating at frame rates of up to 50 Hz. Figure 24.5a shows the thermocamera equipped with the microscopic lens in front of the fatigue specimen mounted in the ultrasonic fatigue testing rig. An infrared thermogram of a cyclically strained AlSi7Mg specimen is shown in Fig. 24.5b.

In addition, this thermographic setup was extended by two mirrors made from grinded and polished aluminum sheets (AlMgSi1, EN AW 6082), which were installed behind the fatigue specimen at an angle of 45° each. Details are described in literature [37] and in Chap. 26. This setup allows to obtain a complete image of the specimen's circumference. IR images consist then of three parts (see Fig. 24.6a): (i) direct IR thermograms of the front side (2), and (ii), (iii) IR reflectograms from the back of the specimen (1-left, 3-rigth). The temperature obtained using the Al

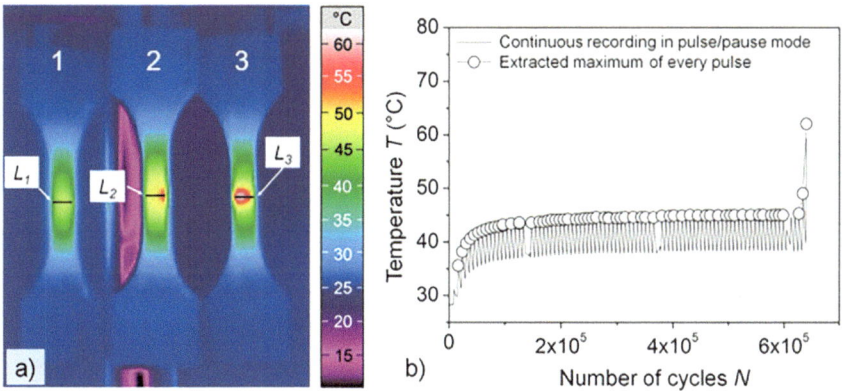

Fig. 24.6 Full-surface view thermographic measurements. **a** Determination of the maximum temperature T_{max} in the direct thermal measurement (L_2) and both reflectograms at the left (L_1) and at the right side (L_3) of the specimen by using horizontal profile lines (L) directly at the crack initiation point. **b** Temperature evolution during pulse/pause cyclic loading and extraction of the maximum temperature of every pulse. Reproduced from [37]

mirrors was about 2 K less than the direct measurement. The measurements were evaluated as follows: Starting from IR-thermograms at fatigue failure N_f of a specimen, the thermal hot spot caused by crack initiation and crack growth was evaluated and indicated by a horizontal line (L_1 to L_3 in Fig. 24.6a). At these lines' L_i the evolution of temperature during the entire fatigue experiment was evaluated as shown in Fig. 24.6b, where the temperature course shows an increase in temperature during pulses and a decrease during the pauses. The maximum temperature $L_{i,max}$ of every pulse was extracted by using an algorithm programmed in Python, compare Fig. 24.6b. The temperature increase was <7 K during the pulse-pause mode in the steady state range of $1 \times 10^5 < N < 1 \times 10^6$ of the fatigue test. At $N < 10^5$, the temperature increased above room temperature due to damping. At $N > 10^6$ crack initiation occurred resulting in a sudden and significant increase in temperature. The direction of view of the thermocamera was marked on the investigated circumference of the specimen enabling, thus, a correlated fractographic analysis of the fracture surface using SEM.

Acoustic Emission

Acoustic emission (AE) measurements were performed in situ during ultrasonic fatigue tests on aluminum alloy AlSi9Cu3. For this purpose, fatigue tests were performed in continuous mode under tensile mean stress ($R = 0.1$) at a stress amplitude of $\sigma_a = 70\,\text{MPa}$. Additional spot cooling was applied to maintain the temperature in the gauge length of the specimens almost constant at $20\,°\text{C} \pm 2\,°\text{C}$. At this stress level, fatigue lives N_f of 1×10^6 cycles up to 1×10^7 cycles were reached resulting in test durations of about 1 min up to 10 min at a test frequency of $f = 19.5$ kHz.

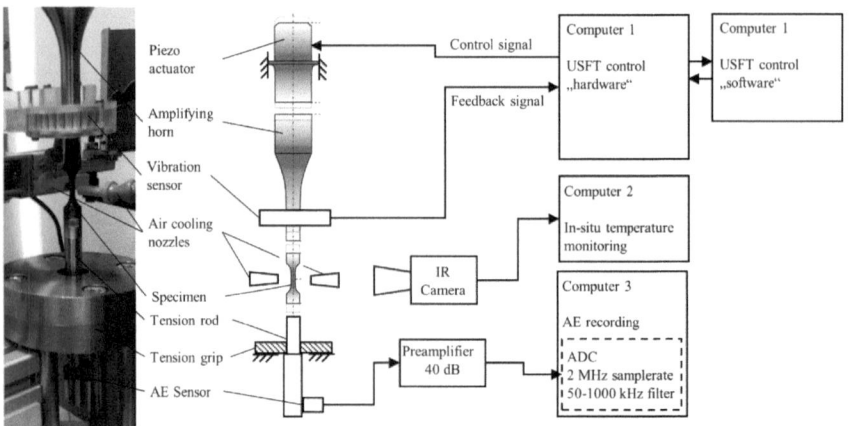

Fig. 24.7 Experimental setup of in situ acoustic emission measurements during ultrasonic fatigue testing under mean tensile stress ($R = 0.1$) on AlSi9Cu3. **a** Photographic image of the resonating load path. **b** Simplified principal scheme and parts description together with control units. Reproduced from [42]

The AE signals were recorded by a broadband Pico-AE sensor (Physical Acoustics Corp., USA), which was attached to the free end of the tensile rod connected to the loaded specimen. The AE waveforms were acquired using the 18-bit PCI-2-based AE system by Physical Acoustics Corp. (USA). The signals of the AE sensor were (i) amplified by 40 dB, (ii) band-pass (50–1000 kHz) filtered, and (iii) recorded continuously without threshold with a sampling frequency of 2 MHz. Due to the high sampling rate and the test duration of several minutes large data sets were expected. Therefore, a series of consecutive continuous AE data streams of 1 min were recorded during each fatigue loading experiment. In total 10 fatigue experiments were performed with AE record. Processing of AE data was conducted via algorithms implemented in MATLAB. Figure 24.7 shows the complete experimental setup for the AE measurements during ultrasonic fatigue testing in combination with infrared thermography. For more details see Seleznev et al. [42].

24.3   Detrimental Effect of Internal Defects and Microstructural Inhomogeneities on Fatigue Life

24.3.1   Industrial Sand-Casting Steel

The applied variants of active (Al_2O_3–C), reactive (MgO–C) and functionalized filters (alumina, spinel or mullite coating) for the melt filtration of the 42CrMo4 steel revealed similar effects on the fatigue behavior in the VHCF regime as shown by Krewerth et al., c.f. Figure 3 in [24]. The only outlier was the Al_2O_3–C filter functionalized with mullite coating showing the lowest fatigue strength caused by dissolution of the mullite coating in contact with the steel melt. In addition, the most common failure mechanism in these investigations was fatigue failure caused by nonmetallic inclusions on or in contact with the surface of the fatigue specimens. Therefore, to focus on the fatigue failures which occurred at internal inclusions all batches were analyzed together excluding the batch with mullite coating. Figure 24.8a shows the fatigue life diagram (stress amplitude vs. number of cycles to failure) for the entity of 42CrMo4 cast plates (CP) including the material below the feeder (BF). Data sets are distinguished in failure due to surface inclusions (filled circles—CP, crosses—BF) and failure at internal inclusions (open circles—CP). In addition, Fig. 24.8b shows the Weibull probability plot in dependence on the location and the size of the crack initiating inclusions.

It is obvious that the material below the feeder contained the largest nonmetallic inclusions located mostly at or in contact with the specimen surface resulting in shortest fatigue lives at all tested stress amplitudes. In contrast, the casting plates contained smaller inclusions. However, internal inclusions causing fatigue failure were in average with 115 μm larger compared to nonmetallic inclusions located at the surface with an average size of 44 μm (compare Fig. 24.8b). Thus, small inclusions at the specimen surface reduced the fatigue lives and were more detrimental than larger internal inclusions which is in good accordance to Murakami [5]. Moreover, both surface and internal inclusions in the casting plates caused significantly higher scatter of fatigue lives (several decades) at the same applied stress amplitudes than the larger inclusions in the material below the feeder, which shows a quite narrow scatter band of fatigue lives (see Fig. 24.8a).

Focusing on crack initiating internal inclusions in the casting plates, four different types of inclusions were found: (i) type 1—large clusters of agglomerates of small Al_2O_3 inclusions (<2 μm) forming a fisheye but no FGA; (ii) type 2—clusters of

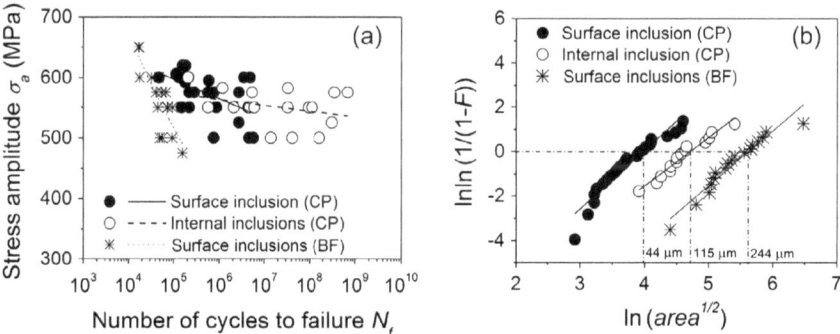

Fig. 24.8 Industrial sand-casting route of steel 42CrMo4—summary of all tested filter variants. **a** Stress versus number of cycles to failure (S–N) plot for different locations of the crack initiation sites for material of the casting plates (CP) and below feeder (BF). **b** Weibull probability plot for the different locations and sizes of the crack initiation inclusions. The average defect sizes are indicated. Reproduced from [24]

nonmetallic inclusions forming both a fisheye and a FGA; (iii) type 3—MnS inclusions, and (iv) type 4—large agglomerates of nonmetallic inclusions. For more details see [24].

The most frequently observed inclusion type is type 2 followed by type 1. Types 3 and 4 were only observed twice. The correlation of inclusion depth S_{inc} and inclusion size $\sqrt{area}$ as well as of fisheye radius $D_{fisheye}/2$ and inclusion depth S_{inc} are shown in Fig. 24.9a, b. It was found also for internal inclusions that smaller inclusions were located closer to the surface (small S_{inc}), whereas larger inclusions were located deeper in the bulk of the specimen volume. Moreover, a linear relationship between inclusion depth and fisheye radius was observed (c.f. [6, 7]). The deeper the inclusion was located below the surface the larger the fisheye could grow.

The size as well as the location below the surface had a significant influence on the fatigue lives as shown in Fig. 24.9c, d. Due to the linear correlation between S_{inc} and fisheye radius $D_{fisheye}/2$, both parameters seem to be interchangeable resulting in similar influences on fatigue lives. The similar trend was found for all four inclusion types. The larger the depth of inclusions or the fisheye radius the larger was the inclusion itself resulting in shorter fatigue lives.

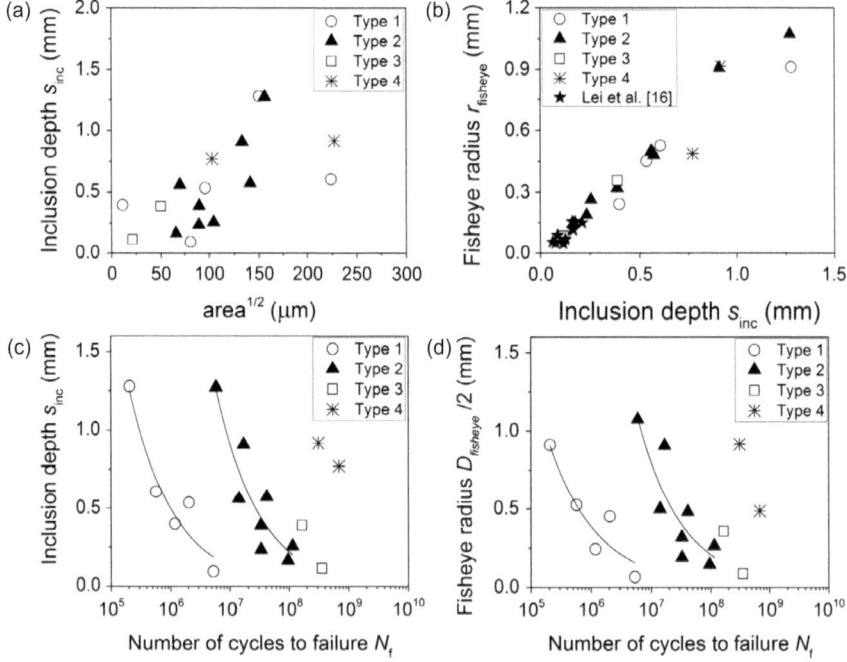

Fig. 24.9 Four different types of fatigue crack initiating inclusions, their geometrical parameters and the influence on the fatigue lives N_f in steel 42CrMo4. **a** Inclusion depth S_{inc} vs. inclusion size $\sqrt{area}$. **b** Linear correlation between fisheye radius $D_{fisheye}/2$ and inclusion depth S_{inc}. **c** Correlation of S_{inc} and N_f. **d** Correlation of $D_{fisheye}/2$ and N_f. In **c** and **d** sizes of individual inclusion types are indicated. Reproduced from [24]

To find the largest inclusion in the studied volume, we had to enhance the fatigue failure caused by internal nonmetallic inclusions. A way to protect the specimens against surface crack initiation was by introduction of compressive residual stresses below the surface. For this purpose, two different strategies were followed: (i) shot peening and (ii) plasma nitriding. However, the shot peening was not efficient since the compressive residual stresses introduced via shot peening were mostly relaxed during ultrasonic fatigue testing. Therefore, results on the influence of plasma nitriding are presented only. Two-step nitriding treatment was performed resulting in NHDs of 0.23 mm and 0.47 mm (see Sect. "Specimen Preparation and Thermal Treatment"). Due to the diffusion of nitrogen, compressive stresses of about -100 MPa and -500 MPa for NHD of 0.23 mm and 0.47, respectively, were measured close to the surfaces [41]. The influence of the nitriding depth on the fatigue behavior is shown in Fig. 24.10a, b for NHD $= 0.23$ mm and 0.47 mm, respectively. In general, all specimens failed at internal inclusions regardless the applied stress amplitude, although the fatigue strength of specimens with NHD of 0.47 mm was significantly higher (650–750 MPa) compared to NHD of 0.23 mm (575–625 MPa). Moreover, majority of specimens with smaller NHD failed directly at the primarily applied stress level (eight out of ten specimens), whereas majority of specimens with NHD 0.47 mm (seven out of ten) reached run-out limit on the primarily applied stress amplitude and the stress amplitude had to be raised several times until fatigue failure occurred (marked by an ellipse in Fig. 24.10b).

Typical fracture surfaces of specimens with different NHD are shown in Fig. 24.11. The fracture surfaces shown in Fig. 24.11a, b, e resemble much on fracture surfaces of nitrided steel specimens cyclically strained in low cycle fatigue (LCF) and high-cycle fatigue (HCF) regime [43]. Clearly, the nitriding layer, the nonmetallic inclusions surrounded by fisheye, and the region of crack growth in the field of compressive residual stresses (wing-shaped region) can be distinguished from the region of globally stable crack growth in the core of the material, and, finally, the region of brittle

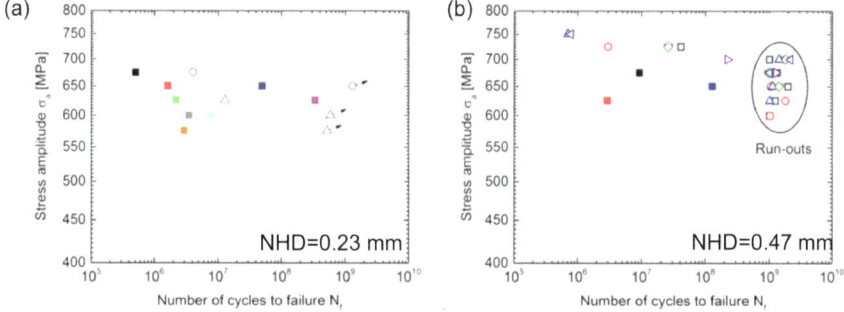

Fig. 24.10 Fatigue behavior of plasma-nitrided steel 42CrMo4. **a** Stress amplitude vs. number of cycles to failure for NHD $= 0.23$ mm and **b** NHD $= 0.47$ mm. Filled symbols correspond to specimens which failed on the primarily applied stress amplitude, whereas open symbols are related to run-out specimens, which were raised in stress amplitude several times until failure occurred. Reproduced from [41]

fracture (due to cooling in liquid nitrogen after end of the fatigue tests). In these cases, the nonmetallic inclusions were located directly in the region of the nitriding layer or in the region influenced by the compressive residual stresses. At the same time, these were specimens, in which the stress amplitudes had to be raised several times in order to reach fatigue failure. However, analyzing the nonlinearity parameter β_{rel} (see Sect. "Fatigue Tests") it turned out that fatigue crack initiation occurred already at the primarily applied stress amplitude σ_{ai} at a certain number of cycles for crack initiation N_i. However, due to the compressive stresses caused by the nitriding layer, the growth of the fatigue crack within the fisheye was hampered until the region of crack growth (wing-shaped regions) extended towards the specimen surface. In contrast, inclusions located close to the center of the cross section far away from the zone of compressive residual stress (Fig. 24.11c, d, f) yielded large fisheyes and failed at the primarily applied stress amplitude σ_a.

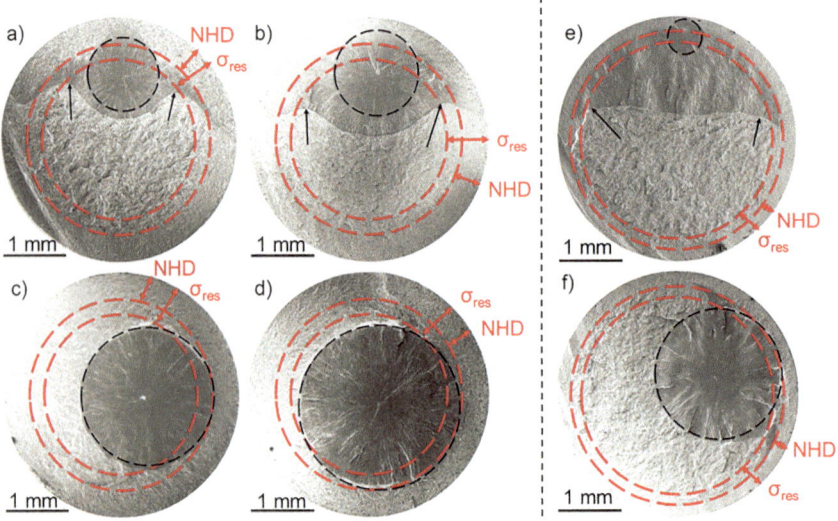

Fig. 24.11 Fracture surfaces of plasma-nitrided 42CrMo4 specimens. **a–d** NHD = 0.47 mm. **e–f** NHD = 0.23 mm. **a** σ_{ai} = 650 MPa, N_i = 448,272,000, S_{inc} = 924 µm. **b** σ_{ai} = 675 MPa, N_i = 9,282,000, S_{inc} = 821 µm. **c** σ_a = 625 MPa, N_f = 2,066,000, S_{inc} = 1549 µm. **d** σ_a = 650 MPa, N_f = 128,932,000, S_{inc} = 1685 µm. **e** σ_{ai} = 575 MPa, N_i = 81,461,000, S_{inc} = 154 µm. **f** σ_a = 600 MPa, N_f = 7,682,000, S_{inc} = 1259 µm. Red dashed circles indicate the NHD and the zone of residual compressive stresses σ_{res}, respectively. Dashed black circles indicate the fisheye around crack-initiating NMIs. Reproduced from [41]

24.3.2 Steel Cast Simulator

Influence of Crucible Material—Steel 18CrNiMo7-6

The melt treatment of the steel 18CrNiMo7-6 in different crucible materials had a significant influence on the chemical composition (compare Table 24.2) of the batches and, consequently, also on the fatigue lives. Beside the desired reduction in aluminium content to reduce the amount of alumina inclusions, the crucible materials caused also a significant loss in carbon, in particular the carbon-bonded alumina (A-C) crucible. Thus, the applied quenching treatment resulted in an ultimate tensile strength of 1200 MPa only. Desired strength values for investigations of the detrimental influence of nonmetallic inclusions should be at least 1400 MPa. The fatigue lives of steel melted in the three crucible materials is shown in Fig. 24.12. Crack initiation occurred either at nonmetallic inclusions located at the surface causing shorter fatigue lives (filled symbols) or at internal nonmetallic inclusions yielding higher number of cycles to failure $N_f > 10^7$ (open symbols). Finally, this resulted in a large scatter of fatigue lives N_f in the range of $10^5 < N_f < 10^9$ cycles.

Significantly different fatigue lives were observed for the three treated steel batches as indicated by the S–N plots in Fig. 24.12. Thus, differences in the fatigue strength at $N = 10^7$ between A–C, AZT–C, and AZT–C–n were about 100 MPa and 200 MPa, respectively, whereby the batches AZT–C–n and A–C exhibited the lowest and highest fatigue strength, respectively. This behavior is related to the different appearance of crack initiating nonmetallic inclusions regarding their chemistry, morphology and size distribution, cf. Chap. 25 or [44]. In addition, crack initiating nonmetallic inclusions were studied on the fracture surfaces. Figure 24.13 shows the relationship between inclusion size ($\sqrt{area}$) and fatigue life N_f (Fig. 24.13a) as well as the correlation between the inclusions' depth and their size (Fig. 24.13b).

It turned out that the AZT–C–n crucible resulted in the largest nonmetallic inclusions, which were located sometimes also in touch with the surface, yielding, however, also high fatigue lives, however, at the lowest stress level. Crack initiating

Fig. 24.12 Stress amplitude vs. number of cycles to failure obtained for steel 18CrNiMo7-6 solidified in different crucible materials: (i) A–C, (ii) AZT–C, and (iii) AZT–C–n. Reproduced from [34]

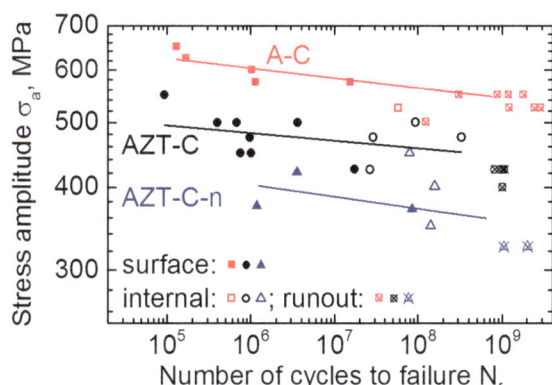

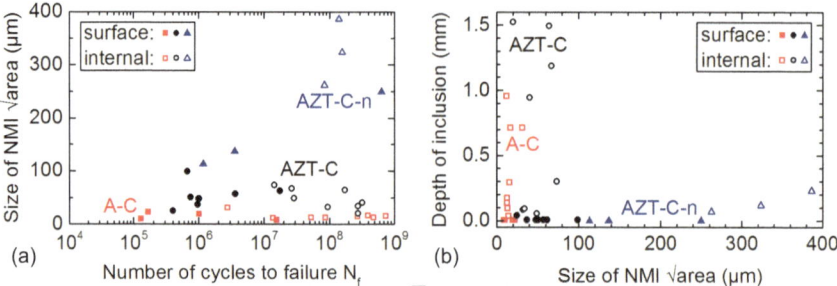

Fig. 24.13 **a** Correlation between crack-initiating inclusion sizes ($\sqrt{\text{area}}$) and numbers of cycles to failure in steel 18CrNiMo7-6. **b** Depth of inclusions versus inclusion sizes ($\sqrt{\text{area}}$). Reproduced from [34]

nonmetallic inclusions of A–C and AZT–C batches, however, had dimensions $\sqrt{\text{area}}$ <100 μm at different inclusion depths resulting in a huge scatter of fatigue lives.

Figure 24.14 shows exemplarily fracture surfaces of batches A–C (Fig. 24.14a), AZT–C (Fig. 24.14b) and AZT–C–n (Fig. 24.14c). The chemical analysis revealed duplex inclusions consisting of Mn–S–Si–O as damage relevant defects for batch A–C, whereas large, dendritic MnS inclusions caused failure in batch AZT–C–n. The change from globular to dendritic MnS was a consequence of the reduced oxygen content in batch AZT–C–n compared to batch A–C (see Table 24.2). For more details on microstructure of duplex inclusions and their formation see [44] and Chap. 25.

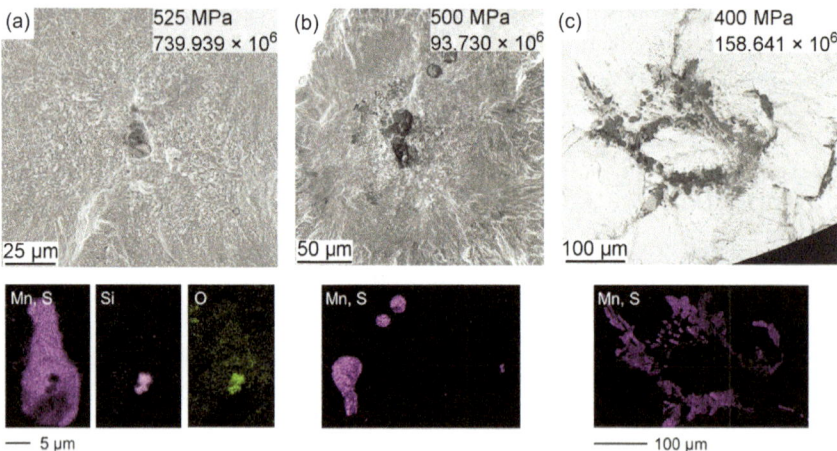

Fig. 24.14 Typical internal crack-initiating nonmetallic inclusions in steel 18CrNiMo7-6. **a** Batch A–C. **b** Batch AZT–C. **c** Batch AZT–C–n. SEM micrographs in SE contrast in combination with EDS element maps of Mn, S, Si, and O, respectively. Reproduced from [34]

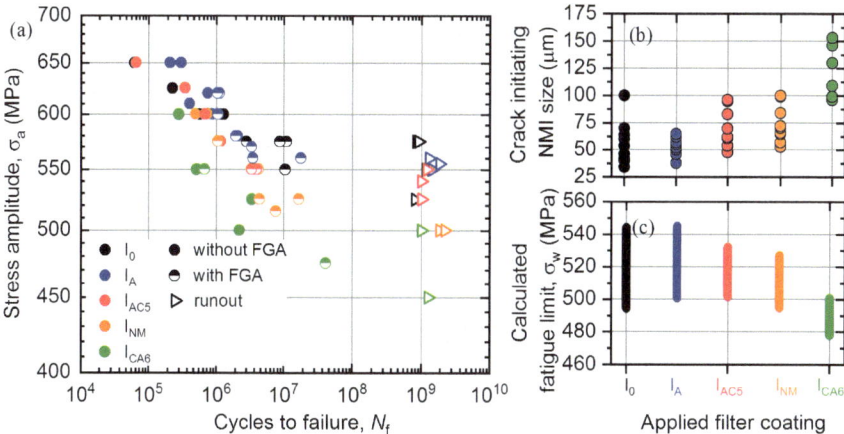

Fig. 24.15 Results of ultrasonic fatigue tests of steel 42CrMo4 after immersion filtration. **a** S–N diagram. **b** Sizes of crack-initiating nonmetallic inclusions. **c** Fatigue limit σ_w calculated according to Matsumoto's approach. For filter IDs see Table 24.1. Partly reproduced from [40]

Immersion Filtration of Steel 42CrMo4

The fatigue behavior of steel 42CrMo4 in reference condition (I_0) and after immersion filtration with different filter coatings (I_A, I_{AC5}, I_{NM}, I_{CA6}, compare Table 24.1) is shown in Fig. 24.15. In the S–N curves fatigue failures by internal nonmetallic inclusions with FGA (filled symbols) and without FGA (half-filled symbols) are distinguished. Run-outs are indicated by open triangles. Again a large scatter of fatigue lives over several decades was observed for all batches at stress amplitudes between 500 and 575 MPa. Batches I_{CA6} and I_{NM} exhibited the lowest fatigue strength values, whereas fatigue lives of I_A, I_{AC5} fall into a scatter band together with I_0. As seen from Fig. 24.15b, the largest crack initiating nonmetallic inclusions with sizes of about 100 µm up to 150 µm were found on the fracture surfaces of batch I_{CA6}. Fatigue failure causing nonmetallic inclusions in all other batches including the reference batch were < 100 µm. Therefore, the calculated fatigue strength σ_w according to Matsomuto's approach [45] yielded the lowest values for batch I_{CA6}. This is in contradiction to the finding that the functionalized coatings on the immersion filters yielded both a significant reduction of the total amount of inclusions as well as a reduction in size of inclusions, cf. Chap. 25.

From Fig. 25.3c in Chap. 25, however, it is visible that the inclusion size distribution of batch I_{CA6} revealed not only a significant reduction of inclusions at all and a shift to smaller inclusions sizes of about 3 µm, but also shows to certain amount an extension to larger inclusion > 100 µm which were also observed on the fracture surfaces.

Figure 24.16 shows an example of a fracture surface containing a plate-like alumina inclusion. Such plate-like alumina inclusions were found to be the major

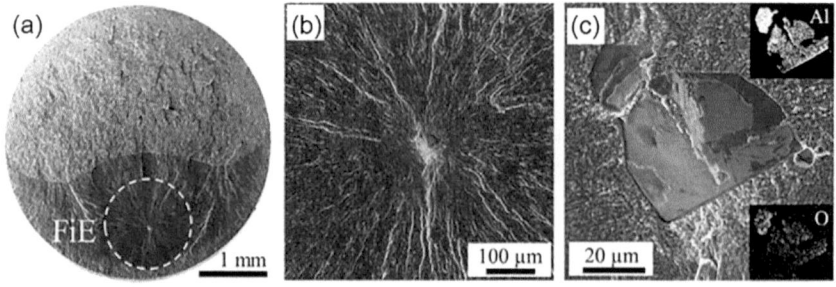

Fig. 24.16 Typical fracture surface of steel 42CrMo4 after immersion filtration. **a** Overview with indicated fisheye. **b** Detailed view on nonmetallic inclusion. **c** Plate-like alumina inclusion together with EDS element maps of aluminum and oxygen. Partly reproduced from [40]

crack initiating defect type in all batches of immersion filtration regardless the coating applied on the filter surfaces.

Flow-Through Filtration of Steel 42CrMo4

The influence of flow-through filtration on the fatigue lives of the steel 42CrMo4 is shown in Fig. 24.17. Beside the reference state—flow-through filtration through a commercial alumina filter (F_{AC5})—two filter coatings applied to AC5 filters were studied, i.e. coating with (i) nano materials (F_{NM}) and (ii) with flame sprayed alumina (F_A). The solidified steel ingots contained huge open and closed shrinkage cavities. Thus, even by HIP these cavities were not fully eliminated. Therefore, the numbers of available specimens from these batches were limited. In addition, fatigue failure of some specimens was caused by shrinkage cavities (see cross symbols in Fig. 24.17). It is obvious that the fatigue strength of the flow-through treated steel batches was significantly lower compared to the reference batch.

Typical fracture surfaces of specimens failed on internal inclusions are shown in Fig. 24.18 for reference batch F_{AC5} without coating (Fig. 24.18a), batch F_{NM} (Fig. 24.18b) and batch F_A (Fig. 24.18c). The reason of the lower fatigue strength of batches F_A and F_{NM} are obviously the size and morphology of crack initiating alumina inclusions compared to batch F_{AC5}.

Whereas the crack initiating inclusions in reference batch F_{AC5} were globular and small, in batches F_A and F_{NM} again large plate-like alumina inclusions were found as crack initiators. Regarding the inclusion size distribution analyzed by Wagner et al. (see Chap. 25) it has to be mentioned, however, that, although batches F_A and F_{NM} revealed much lower number of small inclusions (<5 μm) compared to batch F_{AC5}, all batches contained large nonmetallic inclusions (>60 μm). In addition to the inclusion size, also the inclusion morphology had a significant impact on the fatigue behavior. Thus, the majority of crack initiating internal nonmetallic inclusions had a plate-like morphology varying in size regardless the functionalized coatings used on the immersion filters. It is well known from literature [46] that both the aluminum as

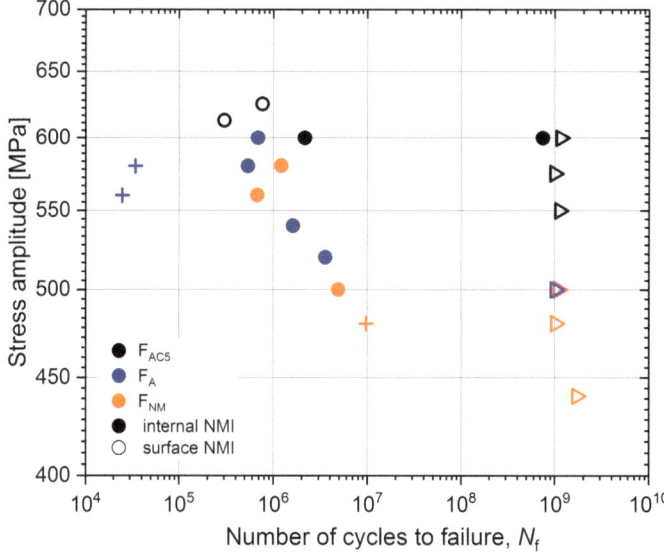

Fig. 24.17 Results of ultrasonic fatigue testing of steel 42CrMo4 after flow-through filtration. Stress amplitude vs. number of cycles to failure. For filter IDs see Table 24.2. Cross symbols represent failure due to shrinkage cavities

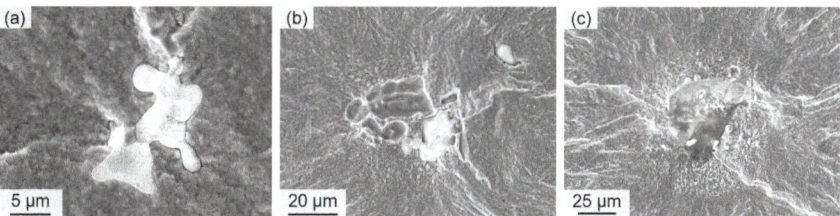

Fig. 24.18 Typical fracture surfaces of 42CrMo4 steel specimens after flow through filtration. **a** Reference state F_{AC5}. **b** AC5 filter coated with nano materials F_{NM}. **c** AC5 filter coated with flame sprayed alumina F_A. For filter IDs see Table 24.2

well as the oxygen content influence the morphology of alumina inclusions. Thus, higher aluminum content (>250 ppm) results in maple-like alumina dendrites [47, 48] independent on the oxygen content. In contrast, low aluminum content (<200 ppm) results at low oxygen content (<80 ppm) in plate-like inclusions or in alumina clusters if the oxygen content is above 80 ppm (see [49, 50] and compare Chap. 25).

Wagner et al. [49] reported that a modified temperature control during the deoxidation of the steel melt in the steel cast simulator resulted in both (i) a higher number of nucleation sites as well as (ii) a supersaturation of aluminum in the steel melt. Whereas (i) caused a shift of the size distribution of alumina inclusions to smaller values (Feret$_{max}$ < 10 μm), dendritic/maple-like alumina inclusions were the result of (ii). Although the plate-like inclusions were avoided by a modified temperature control it turned out that the influence of the inclusions' morphology on the fatigue behavior was small. Since the sizes of both plate-like and the maple-like/dendritic alumina inclusions were comparable, the resulting fatigue strengths were comparable [49]. Nevertheless, the modified temperature regime during the deoxidation of the steel melt was applied for the combination of immersion and flow-through filtration, see the following Sect. "Combination of Immersion and Flow-Through Filtration of Steel 42CrMo4".

Combination of Immersion and Flow-Through Filtration of Steel 42CrMo4

It was described in Sects. "Immersion Filtration of Steel 42CrMo4" and "Flow-Through Filtration of Steel 42CrMo4" that both the immersion filtration as well as the flow-through filtration had no significant positive influence on the fatigue strength of the steel 42CrMo4. However, it is described in Chap. 25 that immersion filters and flow-through filters with different coatings yielded in general both a reduction of the number of remaining nonmetallic inclusions as well as a shift of the inclusion size distribution to smaller sizes. However, it was also found that in all cases the right part of the inclusion size distribution was not positively influenced neither by immersion nor by flow-through filtration. Thus, in all cases large nonmetallic inclusions remained in the solidified steel material causing fatigue failure under cyclic loading. Therefore, a combined filtration was applied consisting of carbon-bonded alumina immersion filters with different coatings (nanomaterial I_{NM} and calciumdialuminate I_{CA2}) and an AC5 flow-through filter with flame-sprayed alumina coating. In addition, the combined filtration treatments were realized together with the modified temperature control during the deoxidation process in the steel cast simulator. The resulting fatigue behavior is shown in Fig. 24.19. It is obvious that there were not that significant differences between the reference batch ($I_0\_F_0$) and the two filter combinations $I_{NM}\_F_A$ and $I_{CA2}\_F_A$ as expected. However, the fatigue strength of $I_{CA2}\_F_A$ was comparable to $I_0\_F_0$, whereas it was lower for $I_{NM}\_F_A$. The reason for this can be seen again in the morphology and size of the nonmetallic inclusions.

Although the inclusions' size distribution discussed in Chap. 25 shows that both combinations of filtration yield again a reduction in the total amount of inclusions, filter combination $I_{NM}\_F_A$ revealed a higher amount of larger inclusions. Figure 24.20 shows typical fatigue crack initiating inclusions found on fracture surfaces of specimens from batch $I_0\_F_0$ (Fig. 24.20a), batch $I_{CA2}\_F_A$ (Fig. 24.20b) and batch I_{NM_FA} (Fig. 24.20c). It is obvious that the reference batch contained the smallest inclusions followed by batch $I_{CA2}\_F_A$, whereas larger inclusions were found on fracture surfaces of batch I_{NM_FA}.

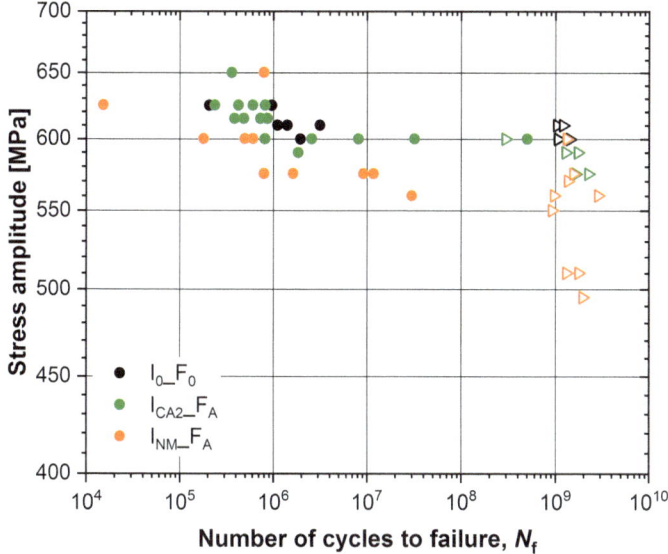

Fig. 24.19 Results of ultrasonic fatigue testing of steel 42CrMo4 after combined immersion and flow-through filtration. Stress amplitude vs. number of cycles to failure. For filter IDs see Table 24.1

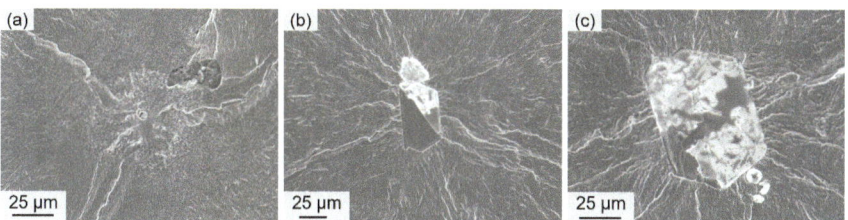

Fig. 24.20 Typical fracture surfaces of 42CrMo4 steel specimens after combined immersion and flow-through filtration. **a** Reference state: no immersion and no flow-through filter—I_0_F_0. **b** Immersion filter coated with calciumdialumina and flow-through filter coated with flame sprayed alumina – I_{CA2}_F_A. **c** Immersion filter coated with nanomaterials and flow-through filter coated with flame sprayed alumina—I_{NM}_F_A. For filter IDs see Table 24.1

Discussion

Three different variants of melt-treated steel 42CrMo4 were studied. Both for immersion and for flow-through filtration different coatings were applied on the surface of carbon-bonded alumina filters. In all cases the crucible material was kept of the same type, knowing well that each batch has its own characteristics. The results of immersion, flow-through and combined immersion/flow-through filtration are summarized in Fig. 24.21. Each filtration type has its own reference data set: I_0—immersion filtration, F_{AC5}—flow through filtration, I_0_F_0 combined filtration. As described above, within the specific filtration type the different coatings on the filter surfaces had no

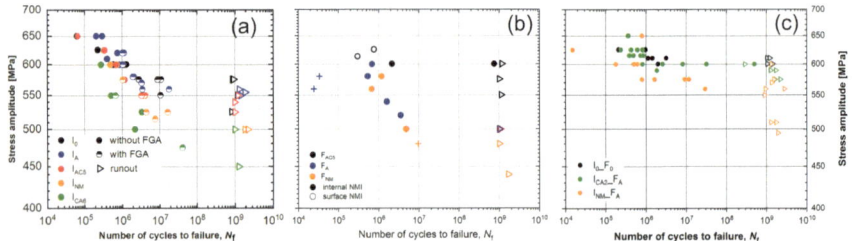

Fig. 24.21 Comparison of fatigue behavior of 42CrMo4 steel after **a** immersion filtration, **b** flow-through filtration, and **c** combined immersion and flow-through filtration. Stress amplitude vs. number of cycles to failure diagrams. For filter IDs see Table 24.1

positive influence on the fatigue behavior. Thus, in all cases the reference material exhibited the best fatigue properties. However, comparing the fatigue results of batches after immersion and the flow-through filtration with results obtained after combined immersion and filtration it turned out that the combination yielded better results compared to each individual treatment. The combination of I_{CA2} and F_A seems to give the best results, i.e. a fatigue strength of 600 MPa.

24.3.3 Aluminum Alloys

AlSi7Mg

The results of the fatigue tests on AlSi7Mg are shown in Fig. 24.22. The S–N diagram in Fig. 24.22a includes the fatigue data obtained both for die casting (batch 1) and sand casting (batch 2). It is obvious that the fatigue lives and strength of die casted (batch 1) and sand casted (batch 2) material are different, which is caused by differences in the microstructure. Batch 1 exhibited a finer microstructure with a smaller secondary dendrite arm spacing (SDAS) of 19 µm compared to the coarser microstructure of batch 2 (SDAS = 26 µm).

Although both batches were HIPed, casting porosity was still present as shown Fig. 24.22b, c. Multiple crack initiating sites were found on the fracture surface: casting pores either (i) in touch with the surface (point 1) or (ii) in the center of the cross section (point 2). In particular batch 1 shows a huge scatter in fatigue lives at a stress amplitude of 110 MPa over several decades caused by the different size of casting porosity. None of the tested specimens failed due to nonmetallic inclusions. However, it was surprising that the shrinkage cavities as a consequence of the casting and solidification process of AlSi7Mg were not fully removed by the HIP process. In addition, Krewerth et al. [35] showed that the fatigue lives of AlSi7Mg alloy were independent on the testing frequency since no differences were found for conventional and ultrasonic resonant testing at 100 Hz and 20 kHz, respectively.

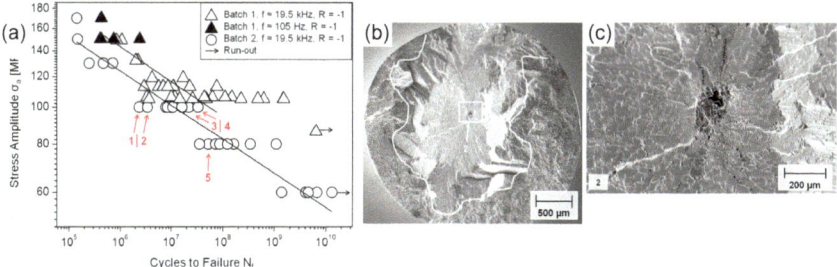

Fig. 24.22 Results of ultrasonic fatigue testing on AlSi7Mg in T6 condition. **a** S–N plot including batch 1 (open triangles) and batch 2 (open circles). Runouts are indicated by arrows. **b, c** Typical fatigue fracture surface of AlSi7Mg (batch 2, $\sigma_a = 100$ MPa, $N_f = 3.45 \times 10^6$), **b** overview, **c** detail of point 2 in (**b**). Numbers and arrows in (**a**) are related to thermographic measurements shown later in Fig. 24.24. Results at f = 105 Hz in detail in [35]. Reproduced from [35]

AlSi9Cu3

Figure 24.23 shows the results of ultrasonic fatigue tests on AlSi9Cu9 in conditioned and filtered state (CF) compared to the reference batch R. As mentioned above, the aim of the conditioning and filtration process was to reduce both the content of iron in the melt and the formation of plate-like β-phase, which would have a detrimental impact on mechanical properties as shown e.g. by Wagner et al. [31] for tensile properties. The microstructures of these two batches were completely different. Whereas batch R with an iron content of 1.7 wt% (see Table 24.3) contained beside α-Al and Al-Si eutectic a large amount of β-phase plates and star-like α_c phase, batch CF with reduced iron content (see Table 24.3) was free of β-phase plates and contained α_c phase in Chinese script morphology. For more details on the microstructure of the two batches see Chap. 25 and literature [31, 51]. Keeping in mind these different microstructures, the fatigue behavior shown in Fig. 24.23 seems to be quite similar for batch CF compared to batch R. However, considering the probability curves (evaluated using Maximum Likelihood method including both failures as well as runouts [52]) for 5%, 50% and 95% of fatigue life, it turned out that the scatter in fatigue lives at a certain stress amplitude was significantly reduced for batch CF (e.g. compare R and CF at $\sigma_a = 100$ MPa). In addition, Fig. 24.23 shows exemplarily two typical fracture surfaces for batch R (c, d) and batch CF (b, e). Specimens of batch R failed mostly from defects located at or in touch with the surface, which were either β-phase plates (Fig. 24.23c) or Al–Fe–Si agglomerates in connection to shrinkage cavity (Fig. 24.23d). In contrast, facture surfaces of specimens of batch CF showed crack initiation occurring at large α_c polyhedra (Fig. 24.23e) located in the interior or in contact with the specimens' surfaces. Some failures in batch CF occurred even with no clearly recognizable crack origin site (Fig. 24.23b).

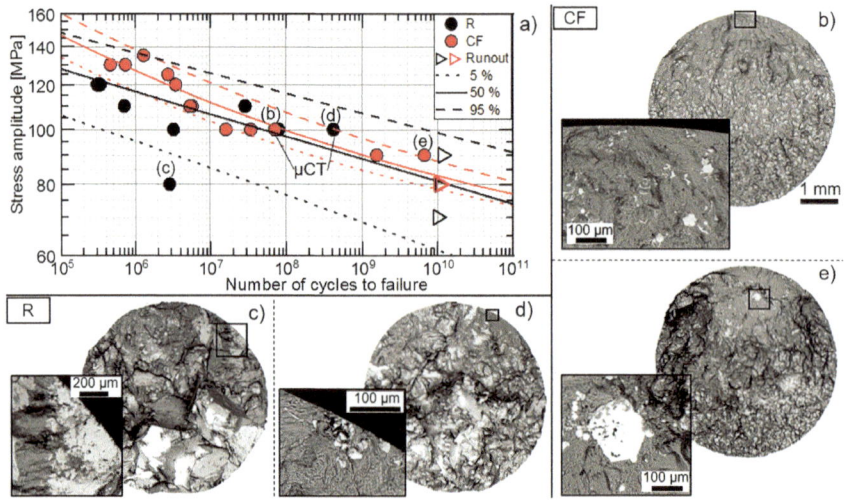

Fig. 24.23 Results of ultrasonic fatigue tests on AlSi9Cu3 alloy. **a** S–N curves of reference batch R (black circles) and batch CF after conditioning and filtration (red circles) together with the 5% (short dashed line), 50% (solid line), and 95% (long dashed line) fatigue life curves (evaluated using Maximum Likelihood method). Open triangles represent runouts. **b–e** Typical fracture surfaces (SEM micrographs in BSE contrast) of specimens from batch R (**c, d**) and CF (**b, e**). Insets in **b–e** show crack initiating defects at higher magnification. Reproduced from [31]

24.4   Determination of Crack Initiation

The above desribed studies on fatigue failure and crack intiation occurring during ultrasonic fatigue testing in the range of high cycle to very high cycle fatigue were corrobareted by additional in situ characterization techniques such as infrared thermography and acoustic emission measurements. Both methods allow to gain additional information regarding the crack initiation time (number of cycles to crack intiation), whereas infrared thermography allows also to study the point of crack intiation in correlation with fractographic analysis of the fracture surfaces.

24.4.1   Thermography

Infrared thermography allows to evaluate the temperature evolution within the specimen during fatigue testing experiments and both the determination of crack initiaton cycle number and location of crack initiation defect. Figure 24.24 shows the overall temperature evolution within the gauge length of five AlSi7Mg specimens tested at $\sigma_a = 100$ MPa (1–4) and $\sigma_a = 80$ MPa (5), respectively. It is obvious that the overall increase in temperature until final fatigue failure depended on the applied stress amplitude. In addition, it was observed that at $\sigma_a = 80$ MPa a saturation of temperature increase was reached after an intial temperaure rise at the beginning of the fatigue test and before final temperature rise due to damage. In contrast, at $\sigma_a = 100$ MPa, no temperature saturation was observed. Instead, a linear temperature increase of about $\Delta T \leq 10$ K occurred. It is obvious from the temperature profiles that about 97% of the fatigue life were spent for crack initiation. Only 3% of the total duration of fatigue experiments were spent for crack propagation, see the significant temperature rise shortly before N_f. The infrared thermograms reveal three different types of fatigue failures: (i) internal crack initiation (4,5), (ii) crack initiation at the surface (3), and (iii) multiple crack initiation (1,2). Obviously multiple crack initiation (1, 2) yielded the shortest fatigue lives compared to internal (4,5) and/or surface (3) crack initiation.

The application of the microscpic lens of the thermocamera with a higher resolution (25 μm/pxl) at a reduced FOV allowed the analysis of the temperature evolution over the circumference of the specimens at the region of the hot-spot causing final failure. To this end, at the point of highest temperature at N_f a horizontal line was placed and the temperature evolution along this line was traced back for the last 3% of fatigue life of the specimen, compare Sect. "Thermography", to study the point of crack initiation in more detail. Finally, these temperature evolutions were correlated with fractographic investiagtions. Figure 24.25 shows an example for correlated fractographic and infrared thermographic investigations on ultrasonically fatigued specimens of aluminum alloy AlSi7Mg. The temperature evolution over the circumference of a specimen cyclically strained a $\sigma_a = 100$ MPa up to $N_f = 2.3 \times 10^6$ cycles is shown in Fig. 24.25a together with five infrared thermograms for different cycles within the last 3% of fatigue life. The overview of fracture surface of this specimen is shown in Fig. 24.25b together with detailed views on the internal defect and crack initiation at subsurface region. The viewing direction of the thermocamera is indicated as well. The temperature profile N_1 revealed a sligth increase in temperature in the center of the profile (I, 2.5 mm) compared to the two sides. However,

Fig. 24.24 Temperature courses measured with infrared thermography during ultrasonic fatigue testing of AlSi7Mg in T6 condition in combination with infrared thermograms at N_f. VarioCam hr head 600 (microscopic lens). Specimens 1–4 were tested $\sigma_a = 100$ MPa, specimen 5 at $\sigma_a = 80$ MPa. Compare Fig. 24.22. Reproduced from [35]

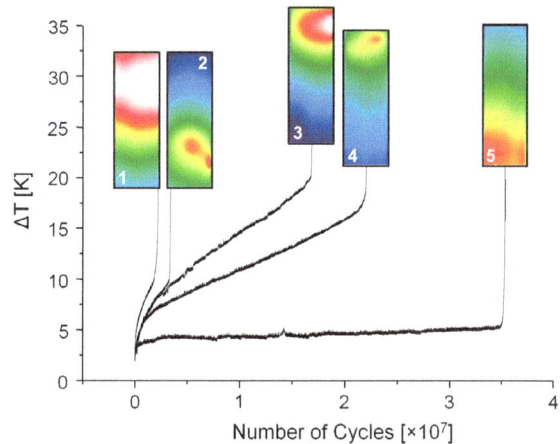

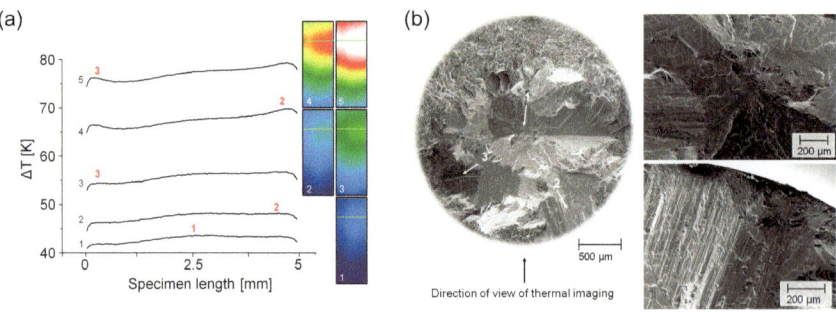

Fig. 24.25 Correlated fractographic and thermographic analysis of damage evolution on cyclically strained AlSi7Mg specimen using VarioCam hr head 600 (microscopic lens). **a** Surface temperature profile of a specimen tested at $\sigma_a = 100$ MPa up to $N_f = 2.3 \times 10^6$. **b** Corresponding fracture surface in overview with indicated crack initiation points 1–3 corresponding to hot spots indicated by red numbers in **a** and detailed view of defects 1 and 3. Viewing direction of thermocamera is indicated by black arrow. Reproduced from [35]

already at N_2 a second hot-spot (II) appeared on the rigth side of the specimen. With increase in number of cylces the temperature at these two hot spots (I, II) was further increasing, whereby the increase at hot-spot II was more pronounced. At N_3 a third hot-spot (III) was generated with less strong temperature increase till fatigue failure (N_5). A comparison with the fracture surface revealed a casting defect in the center of the cross section causing hot-spot I, whereas surface crack initiating was observed on the rigth side of the cross section (hot-spot II).

A third defect was observed on the left side rigth below the specimen surface (hot-spot III). Cracks intiated from I and II merged and caused final fatigue failure.

The infrared thermographic measurements are restricted to measurements of surface temperatures. Thus, cracks initiating on the side of the specimen facing away from the thermocamera will be detected later than defects initiating at the front side due to reasons of heat flow. In order to follow also the temperature evolution on the back side of a specimen facing away from the thermocamera the so-called mirror thermography was developed to study the temperature along the whole circumference of a cylindrical specimen. As described in Sect. "Thermography", two mirrors were placed at angles of $45°$ behind the specimen. For this purpose, the thermocamera was used with macro lens only due to the higher field of view accepting the lower local resolution. This allowed more detailed investigations in terms of correlated fractographic and thermographic analysis, which is even more important for sophisticated materials such as e.g. short-fiber reinforced aluminum matrix composites with more complex fracture surface due to multiple crack initiation points (see [53]). An example of correlated mirror-thermography and fractography is shown in Fig. 24.26 for 42CrMo4 steel specimen tested at $\sigma_a = 500$ MPa up to $N_f = 9.57 \times 10^5$.

The fracture surface with indicated FOV of the thermocamera is shown in Fig. 24.26a, whereas Fig. 24.26b–f shows thermograms at different cycles starting

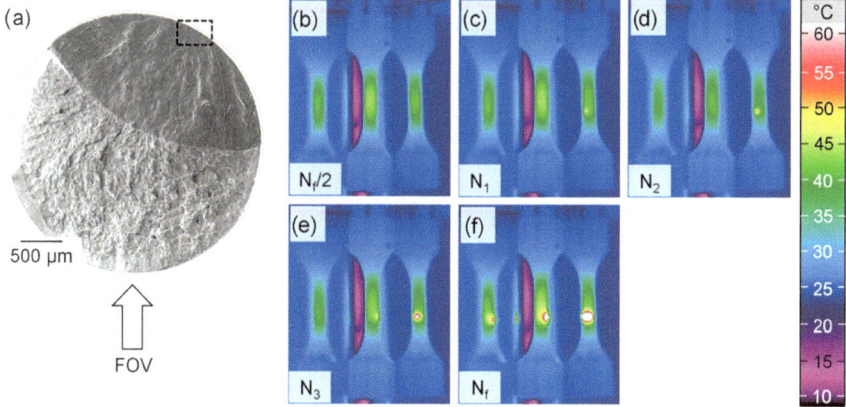

Fig. 24.26 Fracture surface and representative full-surface view thermal images of 42CrMo4 specimen (sand casting) tested at $\sigma_a = 500$ MPa up to $N_f = 9.57 \times 10^5$. VarioCam hr head 600 (macro lens). **a** Fracture surface including the definition of the field of view (FOV) during full-surface view thermographic measurements and the definition of the point of crack initiation (black rectangle). **b–f** Three thermal images of the specimen at **b** $N_{f/2} = 4.81 \times 10^5$, and during final crack growth at **c** $N_1 = 9.35 \times 10^5$, **d** $N_2 = 9.42 \times 10^5$, **e** $N_3 = 9.50 \times 10^5$ and **f** $N_f = 9.57 \times 10^5$. Middle: direct thermogram, left/right: reflectograms. Reproduced from [37]

from $N_{f/2}$ up to N_f. Each of these subfigures contains three thermograms: (i) direct thermogram of the front view (middle), (ii) reflectogram left side, and (iii) reflectogram right side. It should be noted that a slight temperature difference was measured between direct thermogram and reflectograms, which was, however, less than 2 K. It is obvious from Fig. 24.26c that a first temperature hot-spot was observed on the right part of the back side of the specimen (N_1), whereas the direct thermogram of the front side revealed a global warming up only. However, the hot-spot appeared in the direct thermogram of the front view at N_3 only, which was caused by the heat dissipation 8,000 cycles later. This correlated perfect with the fracture surface, where a crack initiation at the specimen surface was observed on the upper right part of the cross section and the fatigue crack did not reach the front of the cross section. The temperature evolution in the region of the first hot-spot can be followed now for all three thermograms until final failure together with the other in situ parameters of the ultrasonic fatigue testing: (i) resonant frequency f_{res} and (ii) non-linearity parameter β_{rel}. Figure 24.27a shows exemplarily the evolution of these parameters over the entire fatigue experiment for the same specimen (42CrMo4, $\sigma_a = 500$ MPa up to $N_f = 9.57 \times 10^5$) as shown in Fig. 24.26, whereby T_{max} indicates the temperature increase in the direct thermogram of the front view. All three parameters show a typical behavior. Thus, the temperature was increasing during the first pulses reaching a saturation. At the same time the resonant frequency f_{res} was decreasing since f_{res} depends significantly on temperature. On the other hand changes in f_{res} reflect also changes in material behavior such as cyclic hardening and/or softening. Therefore,

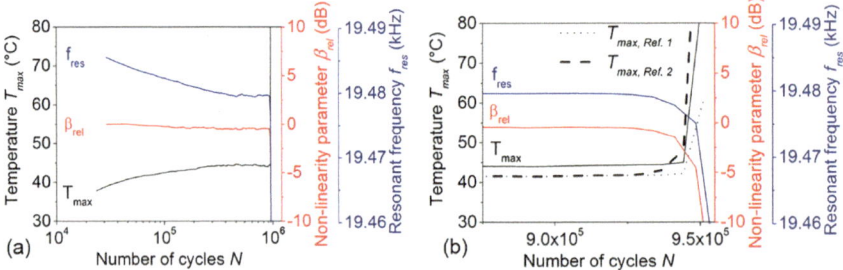

Fig. 24.27 Analysis of the three in situ parameters during ultrasonic fatigue testing: temperature T_{max}, resonant frequency f_{res} and nonlinearity parameter β_{rel}. **a** Evolution of the parameters for 42CrMo4 specimen (sand casting; $\sigma_a = 500$ MPa up to $N_f = 9.57 \times 10^5$) over the entire test duration (same specimen as in Fig. 24.26). **b** Enlarged section of the stage of final crack growth. The indirect thermal measurements in the reflectograms of the mirror faces defined by $T_{max,Ref.1}$ (left reflectogram) and $T_{max,Ref.2}$ (right reflectogram) are included in (**b**). Reproduced from [37]

the non-linearity parameter is more prone for studies on initiation of fatigue damage since it is independent on temperature and reflects only changes in material behavior.

Thus, whereas f_{res} decreased with an increase in T_{max} with ongoing cycling, the non-linearity parameter β_{rel} remained constant at $\beta_{rel} = 0$ over 97% of fatigue life. Only close to N_f a sudden decrease in β_{rel} was observed accompanied by a sudden decrease and increase in f_{res} and T_{max}, respectively. Looking at the last 3% of fatigue life in more detail (Fig. 24.27b), it turned out that the first onset of decrease in f_{res} and β_{rel} was earlier compared to the sudden increase in T_{max}. However, considering now the evolution of temperature of the reflectograms it is obvious that the increase in T_{max} of reflectogram 2 (right side) coincided perfectly with the course of b_{rel} and f_{res}.

The examples shown in Figs. 24.26 and 24.27 are related to surface crack initiation. However, the proposed method works also for internal crack initiation. Figure 24.28 shows an example of steel 42CrMo4 with surface treatment via plasma-nitriding [54]. Thus, the specimen surface was protected against crack initiation due to compressive residual stresses introduced below the surface causing, as a consequence, solely internal crack initiation. Thus, Fig. 24.28a shows the results of the mirror-thermography at N_f, Fig. 24.28b the evolution of f_{res}, T_{max} and Fig. 24.28c the fracture surface. Figure 24.28d is an enlarged view of the last 3% of fatigue life showing the different temperature evolution in the front thermogram (2) and the two reflectograms (1, 3). Clearly the internal crack initiation below the surface and even below the nitriding hardness depth of 0.23 mm is visible on the fracture surface and correlated well with the direct thermogram of the front view.

Following the evolution of temperature in the region of crack initiation in all three thermograms (Fig. 24.28d) shows that in both reflectograms (1, 3) the increase in temperature started later compared to the front view (2). A saturation in temperature was reached over a wide range of fatigue life before final increase of several tens of Kelvin, which coincided with a sudden drop of resonant frequency.

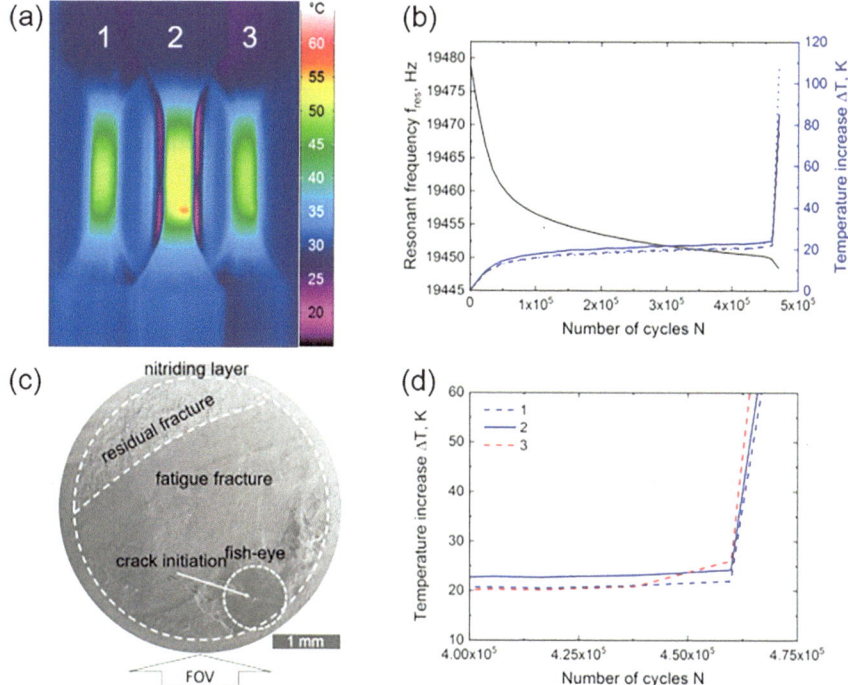

Fig. 24.28 Correlated fractographic and thermographic analysis of 42CrMo4 specimen (sand casting) with a nitriding hardness depth of 0.23 mm tested at $\sigma_a = 625$ MPa up to failure at $N_f = 2{,}182{,}000$. **a** Thermographic images at crack initiation $N = 2{,}161{,}000$. **b** Evolution of resonant frequency and temperature. **c** SEM image of fracture surface crack-initiation nonmetallic inclusion below the nitriding layer

24.4.2 Acoustic Emission Measurements

Whereas infrared thermography is related to the contactless measurement of surface temperatures, the acoustic emissions measured during mechanical loading is a real time method providing volume information and is, therefore, suitable to be applied for studies on damage evolution. The application of AE technique during both monotonic tensile/compressive loading as well as conventional low-frequency cyclic loading [55] is well established, whereas the AE method has been only scarcely applied to ultrasonic fatigue testing [56]. Seleznev et al. [42] adopted the AE measurement technique to the ultrasonic fatigue testing device and developed a method to analyze different modes of fatigue damage during high-frequency testing. Figure 24.29 summarizes the main results of Seleznev et al. obtained during ultrasonic fatigue tests under mean tensile stress (R = 0.1) on aluminum alloy AlSi9Cu3 (see Sect. "AlSi9Cu3"). In total ten fatigue tests including AE and thermographic analysis were performed. Figure 24.29 shows three representative examples. Figure 24.29a–c shows the evolution of the in situ parameters of the ultrasonic fatigue testing

f_{res} and β_{rel}, Fig. 24.29d–f the processed raw AE waveforms of AE records and Fig. 24.29g–i presents the extracted signal power (grey lines) and cumulative power (red lines). In addition, corresponding fracture surfaces of tested specimens are shown in Fig. 24.29j–l. The cumulative AE power evolutions (red lines) shown in Fig. 24.29 g, h can be separated at least into four segments (0-III) with different slopes, which can be correlated to different scenarios of fatigue damage evolution and four stages of crack initiation and propagation (0–IV in Fig. 24.29j, k): (0) crack initiation, (I) mechanically short crack—smooth area, (II) propagation of mechanically short crack, (III) faster crack propagation rates, and (IV) final cleavage fracture. The onset of stage III in Fig. 24.29 g, h, which is correlated to the faster crack growth rates and final failure coincides directly with the pronounced decrease in f_{res} and a sudden increase in β_{rel} in Fig. 24.29a, b. The crack initiation in stage 0 (red dot in Fig. 24.29j, k) is caused by brittle fracture of polyhedral, intermetallic α_c-phase (compare Chap. 25) located either at the surface (Fig. 24.29j) or in the interior (Fig. 24.29h). This is related to the lowest slope of the cumulative power up to roughly half of the total fatigue life. However, this stage (0) was characterized by a relatively high AE activity (Fig. 24.29g), which may be caused by the initiation and arresting of multiple micro cracks before the critical damage was attained and the final crack initiation site was developed [42].

Finally, it can be concluded that the fatigue damage accumulation is highly nonlinear. Moreover, the cumulative AE power can be used as an additional in situ parameter detecting the onset and evolution of fatigue damage. Using this method, is was possible to detect fatigue damage much earlier than the process controlling parameter of resonance frequency f_{res} and the calculated non-linearity parameter β_{rel}. Moreover, Seleznev et al. [42] mentioned that their proposed signal-processing methodology is applicable not only to aluminum-based alloys and the continuous resonant cycling mode but can also be applied to other metallic materials and the pulse-pause testing mode and can be extended to very high cycle fatigue regimes [42].

24.5 Conclusions

This chapter focused on the influence of nonmetallic inclusions in steel or casting defects and intermetallic phases in aluminum alloys on the fatigue behavior. For this purpose, specimens of steel 42CrMo4 were tested after two different routes of steel melt treatment: (i) industrial sand-casting route and (ii) laboratory route using steel cast simulator. In the sand-casting route flow-through filtration using both carbon-bonded alumina and magnesia foam filters with different reactive coatings were applied. In the steel cast simulator both (i) the influence of crucible material as well as (ii) reaction immersion filtration, (iii) flow-through filtration, and (iv) a combination of (ii) and (iii) were studied. Specimens of aluminum alloy AlSi9Cu3 were tested after melt conditioning and filtration. In addition to the studies on fatigue behavior, some methodological developments were implemented such as (i) mirror thermography,

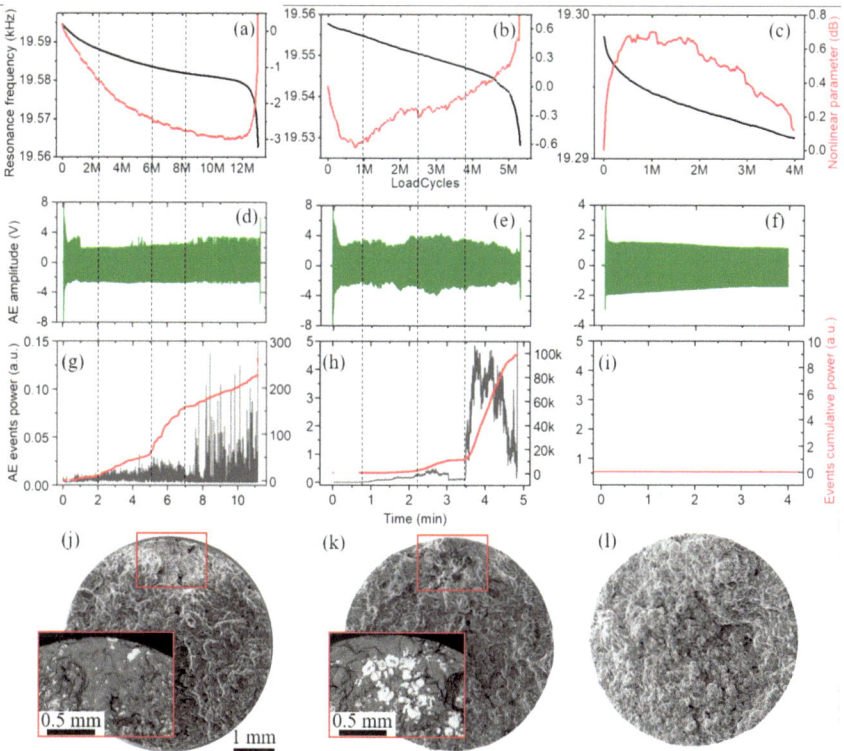

Fig. 24.29 Illustration of the application of AE-based damage monitoring technique during ultrasonic fatigue testing of the AlSi9Cu3 alloy. **a–c** Resonance-based parameters f_{res} and β_{rel} of ultrasonic fatigue testing. **d–f** Raw AE waveforms processed by algorithm described in [42]. **g–i** Resulting signal power extraction (gray curves) and cumulative power (red lines). **j–l** Fracture morphologies obtained by optical microscopy and BSE imaging of the crack initiation sites (insets in **j, k**). **a, d, g, j** represents the case of typical surface crack initiation. **b, e, h, k** represents the case of internal (sub-surface) crack initiation. **c, f, i, l** shows the example where the experiment was interrupted without detected fatigue damage. Reproduced from [42]

(ii) combined analysis of resonant frequency f_{res}, non-linearity parameter β_{rel} and temperature increase due to damage, and (iii) in situ acoustic emission measurements for detection of onset of fatigue failure. The main conclusions can be summarized as follows:

- The carbon-bonded alumina and magnesia with different filter coatings applied in industrial sand casting did not result in a significant influence on the fatigue behavior, except the detrimental effect of the filter with mullite coating. All other applied filters resulted in a huge data set with significant scatter of the fatigue lives over several decades. Moreover, the majority of specimens failed from small inclusions directly located at or in touch with the surface. Therefore, surface treatment in terms of plasma-nitriding was applied, in order to protect the specimen surface and sub-surface regions from crack initiation due to compressive residual stresses introduced below the surface.

- The different applied crucible materials in the steel cast simulator resulted in significant differences in the fatigue life of steel 18CrNiMo18-6 in hardened condition, where A–C crucible material yield the highest fatigue strength in contrast to AZT–C–n yielding the lowest fatigue strength values. This effect is related to the influence of the crucible material on the chemistry, size and morphology of crack initiating nonmetallic inclusions. Thus, steel melted in A–C crucible contained small spherical duplex inclusions, whereas steel melted in an AZT–C–n crucible contained large dendritic MnS.
- The immersion filtration applying carbon-bonded alumina filters with different reactive coatings for 10 s to the steel melt showed slight differences. Thus, coating with calcium-hex-aluminate caused the lowest fatigue strength due to largest remaining nonmetallic inclusions. In contrast, flame sprayed alumina coating (I_A) yield comparable fatigue strength as the reference state (I_0).
- The flow-through filtration did not show any positive influence on the fatigue behavior compared to reference state (F_{AC5}). All treated batches contained some amount of large open and closed shrinkage cavities. No significant difference in fatigue lives was found for flame-sprayed alumina coating (F_A) or coating with nano materials (F_{NM}).
- Both immersion filtration as well as flow-through filtration revealed large plate-like alumina inclusions as crack initiating defects. Since the morphology of alumina inclusions significantly depended on the Al:O ratio, a modified temperature control during deoxidation process was proposed resulting in more maple like or dendritic morphology of alumina. However, their sizes were not significantly reduced, why the fatigue strength was not significantly influenced.
- A combination of immersion filtration and flow-through filtration was applied together with the modified temperature to steel 42CrMo4 resulting in the best fatigue strength for a combination of immersion filter coated with calcium-di-aluminate and a flow through filter coated with flame-sprayed alumina.
- The melt conditioning and filtration applied to AlSi9Cu3 revealed that the harmful plate-like β intermetallic can be avoided and, therefore, slightly enhanced fatigue strength was reached compared to the reference state.
- The correlated infrared thermograhic and fractographic analysis allowed together with the analysis of the resonant-based parameters f_{res} and β_{rel} for a better understanding of the onset of fatigue crack growth caused by nonmetallic inclusions. In particular, the full-surface view imaging gives valuable insight into crack initiation time and site.
- In situ acoustic emission measurements applied during ultrasonic fatigue tests are a suitable tool to gain more knowledge and understanding on different stages of fatigue failure.

Acknowledgements This work was funded by the Deutsche Forschungsgemeinschaft (DFG, German Research Foundation)—Project-ID 169148856—SFB 920, subproject C04. The steels were provided by Institute of Ceramics, Refractory and Composite Materials or by Edelstahlwerke Schmees, Al was provided by Foundry Institute. The authors gratefully acknowledge the research work of Dr.-Ing. Dominik Krewerth, Tim Lippmann, Johannes Gleinig and Ruben Wagner (all Institute of Materials Engineering) providing a major contribution to the present chapter. The quenching and tempering treatment as well as the chemical analysis of the investigated materials by Dr. Thilo Kreschel (Institute of Iron and Steel Technology) is gratefully acknowledged. The surface protection by nitriding by company Vacutherm (Brand-Erbisdorf) is acknowledged. In addition, the authors would like to thank Mrs. Birgit Witschel (Institute of Materials Science) for extensive metallographic work and several students for preparation/polishing of gage lengths of fatigue specimens.

References

1. J.G. Antonopoulos, L.M. Brown, A.T. Winter, Philos. Mag. **34**(4), 549 (1976). https://doi.org/10.1080/14786437608223793
2. U. Essmann, U. Gösele, H. Mughrabi, Philos. Mag. A **44**(2), 405 (1981). https://doi.org/10.1080/01418618108239541
3. J. Polák, Mater. Sci. Eng. **92**, 71 (1987). https://doi.org/10.1016/0025-5416(87)90157-1
4. H. Mughrabi, Int. J. Fatigue **28**(11), 1501 (2006). https://doi.org/10.1016/j.ijfatigue.2005.05.018
5. Y. Murakami, T. Nomoto, T. Ueda, Fatigue Fract. Eng. Mater. Struct. **22**(7), 581 (1999). https://doi.org/10.1046/j.1460-2695.1999.00187.x
6. K. Shiozawa, L. Lu, Fatigue Fract. Eng. Mater. Struct. **25**(8–9), 813 (2002). https://doi.org/10.1046/j.1460-2695.2002.00567.x
7. Z. Lei, Y. Hong, J. Xie, C. Sun, A. Zhao, Mater. Sci. Eng. A **558**, 234 (2012). https://doi.org/10.1016/j.msea.2012.07.118
8. M. Seleznev, E. Merson, A. Weidner, H. Biermann, Int. J. Fatigue **126**, 258 (2019). https://doi.org/10.1016/j.ijfatigue.2019.05.011
9. Y. Murakami, Metal fatigue: effects of small defects and nonmetallic inclusions. Elsevier Ltd. (2002)
10. T. Sakai, Y. Sato, Y. Nagano, M. Takeda, N. Oguma, Int. J. Fatigue **28**(11), 1547 (2006). https://doi.org/10.1016/j.ijfatigue.2005.04.018
11. K. Shiozawa, Y. Morii, S. Nishino, L. Lu, Int. J. Fatigue **28**(11), 1521 (2006). https://doi.org/10.1016/j.ijfatigue.2005.08.015
12. T. Nakamura, H. Oguma, Y. Shinohara, Proc. Eng. **2**(1), 2121 (2010). https://doi.org/10.1016/j.proeng.2010.03.228
13. Y. Murakami, S. Beretta, Extremes **2**, 123 (1999). https://doi.org/10.1023/A:1009976418553
14. Y. Furuya, T. Matsuoka, T. Abe, K. Yamaguchi, Scr. Mater. **46**(2), 157 (2002). https://doi.org/10.1016/S1359-6462(01)01213-1
15. C. Klinger, D. Bettge, Eng. Fail. Anal. **35**, 66 (2013). https://doi.org/10.1016/j.engfailanal.2012.11.008
16. D. Schwerdt, B. Pyttel, C. Berge, Int. J. Fatigue **33**(1), 33 (2011). https://doi.org/10.1016/j.ijfatigue.2010.05.008
17. C. Stöcker, M. Zimmermann, H.-J. Christ, Int. J. Fatigue **33**(1), 2 (2011). https://doi.org/10.1016/j.ijfatigue.2010.04.008
18. J. Günther, D. Krewerth, T. Lippmann, S. Leuders, T. Tröster, A. Weidner, H. Biermann, T. Niendorf, Int. J. Fatigue **94**(2), 236 (2017). https://doi.org/10.1016/j.ijfatigue.2016.05.018
19. E. Bayraktar, I. Garcias, C. Bathias, Int. J. Fatigue **28**(11), 1590 (2006). https://doi.org/10.1016/j.ijfatigue.2005.09.019

20. Y. Furuya, S. Matsuoka, T. Abe, Metall. Mater. Trans. **34**, 2517 (2003). https://doi.org/10.1007/s11661-003-0011-6
21. L. Zhang, JOM **65**, 1138 (2013). https://doi.org/10.1007/s11837-013-0688-y
22. M. Emmel, C.G. Aneziris, Ceram. Int. **38**(6), 5165 (2012). https://doi.org/10.1016/j.ceramint.2012.03.022
23. M. Emmel, C.G. Aneziris, G. Schmidt, D. Krewerth, H. Biermann, Adv. Eng. Mater. **15**(12), 1188 (2013). https://doi.org/10.1002/adem.201300118
24. D. Krewerth, T. Lippmann, A. Weidner, H. Biermann, Int. J. Fatigue **84**, 40 (2016). https://doi.org/10.1016/j.ijfatigue.2015.11.001
25. E. Storti, M. Emmel, S. Dudczig, P. Colombo, C.G. Aneziris, J. Eur. Ceram. Soc. **35**(5), 1569 (2015). https://doi.org/10.1016/j.jeurceramsoc.2014.11.026
26. E. Storti, M. Farhani, C.G. Aneziris, C. Wöhrmeyer, C. Parr, Steel Res. Int. **88**(11), 1700247 (2017). https://doi.org/10.1002/srin.201700247
27. A. Schmidt, A. Salomon, S. Dudczig, H. Berek, D. Rafaja, C.G. Aneziris, Adv. Eng. Mater. **19**(9), 1700170 (2017). https://doi.org/10.1002/adem.201700170
28. J. Fruhstorfer, L. Schöttler, S. Dudczig, G. Schmidt, P. Gehre, C.G. Aneziris, J. Am. Ceram. Soc. **36**(5), 1299 (2016). https://doi.org/10.1016/j.jeurceramsoc.2015.11.038
29. J. Fruhstorfer, S. Dudczig, M. Rudolph, G. Schmidt, N. Brachhold, L. Schöttler, D. Rafaja, C.G. Aneziris, Metall. Mater. Trans. B **49**, 1499 (2018). https://doi.org/10.1007/s11663-018-1216-9
30. S. Dudczig, C.G. Aneziris, M. Emmel, G. Schmidt, J. Hubalkova, H. Berek, Ceram. Int. **40B**(10), 16727 (2014). https://doi.org/10.1016/j.ceramint.2014.08.038
31. R. Wagner, M. Seleznev, H. Fischer, R. Ditscherlein, H. Becker, B.G. Dietrich, A. Keßler, T. Leißner, G. Wolf, A. Leineweber, U.A. Peuker, H. Biermann, A. Weidner, Mater. Charact. **174**, 111039 (2021). https://doi.org/10.1016/j.matchar.2021.111039
32. C.G. Aneziris, S. Dudczig, J. Hubalkova, M. Emmel, G. Schmidt, Ceram. Int. **39**(3), 2835 (2013). https://doi.org/10.1016/j.ceramint.2012.09.055
33. J. Fruhstorfer, S. Dudczig, P. Gehre, G. Schmidt, N. Brachhold, L. Schöttler, C.G. Aneziris, Steel Res. Int. **87**(8), 1014 (2016). https://doi.org/10.1002/srin.201600023
34. S. Henschel, J. Gleinig, T. Lippmann, S. Dudczig, C.G. Aneziris, H. Biermann, L. Krüger, A. Weidner, Adv. Eng. Mater. **19**(9), 1700199 (2017). https://doi.org/10.1002/adem.201700199
35. D. Krewerth, A. Weidner, H. Biermann, D. Krewerth, A. Weidner, H. Biermann, Ultrasonics **53**, 1441 (2013). https://doi.org/10.1016/j.ultras.2013.03.001
36. S. Stanzl-Tschegg, Int. J. Fatigue **60**, 2 (2014). https://doi.org/10.1016/j.ijfatigue.2012.11.016
37. D. Krewerth, T. Lippmann, A. Weidner, H. Biermann, Int. J. Fatigue **80**, 459 (2015). https://doi.org/10.1016/j.ijfatigue.2015.07.013
38. A. Kumar, C.J. Torbet, J. Wayne Jones, T.M. Pollock, J. Appl. Phys. **106**, 024904 (2009). https://doi.org/10.1063/1.3169520
39. A. Kumar, C.J. Torbet, T.M. Pollock, J.W. Jones, Acta Mater. **58**(6), 2143 (2010). https://doi.org/10.1016/j.actamat.2009.11.055
40. M. Seleznev, S. Henschel, E. Storti, C.G. Aneziris, L. Krüger, A. Weidner, H. Biermann, Adv. Eng. Mater. **22**(2), 1900540 (2019). https://doi.org/10.1002/adem.201900540
41. A. Weidner, T. Lippmann, H. Biermann, J. Mater. Res. **32**, 4305 (2017). https://doi.org/10.1557/jmr.2017.308
42. M. Seleznev, A. Weidner, H. Biermann, A. Vinogradov, Int. J. Fatigue **142**, 105918 (2021). https://doi.org/10.1016/j.ijfatigue.2020.105918
43. H-J. Spies, A. Dalke, in *Comprehensive Materials Processing*, 1st edn., ed. by M.S.J. Hashimi (Elsevier Books, Amsterdam, 2014), p. 439
44. J. Gleinig, A. Weidner, J. Fruhstorfer, C.G. Aneziris, O. Volkova, H. Biermann, Metall. Mater. Trans. B **50**, 337 (2019). https://doi.org/10.1007/s11663-018-1431-4
45. Y. Murakami, *Metal Fatigue Effects of Small Defects and Nonmetallic Inclusions* (Elsevier, Oxford, UK, 2002)
46. W. Tiekink, R. Boom, A. Overbosch, R. Kooter, S. Sridhar, Ironmaking Steelmaking **37**(7), 488 (2010). https://doi.org/10.1179/030192310X12700328925822

47. E. Steinmetz, C. Andreae, Steel Res. **62**(2), 54 (1991). https://doi.org/10.1002/srin.199101250
48. R. Dekkers, B. Blanpain, P. Wollants, F. Haers, C. Vercruyssen, B. Gommers, Ironmaking Steelmaking **29**(6), 437 (2002). https://doi.org/10.1179/030192302225004584
49. R. Wagner, A. Schmiedel, S. Dudczig, C.G. Aneziris, O. Volkova, H. Biermann, A. Weidner, Adv. Eng. Mater. **24**(2), 2100640 (2021). https://doi.org/10.1002/adem.202100640
50. R. Wagner, R. Lehnert, E. Storti, L. Ditscherlein, C. Schröder, S. Dudczig, U. Peuker, O. Volkova, C.G. Aneziris, H. Biermann, A. Weidner, Mater. Charact. **193**, 112257 (2022). https://doi.org/10.1016/j.matchar.2022.112257
51. H. Becker, A. Thum, B. Distl, M.J. Kriegel, A. Leineweber, Metall. and Mater. Trans. A. **49**, 6375 (2018). https://doi.org/10.1007/s11661-018-4930-7
52. F. Pascual, W. Meeker, Analysis of fatigue data with runouts based on a model with nonconstant standard deviation and a fatigue limit parameter. J. Test. Eval. **25**(3), 292 (1997)
53. A. Illgen, A. Weidner, H. Biermann, Int. J. Fatigue **113**, 299 (2018). https://doi.org/10.1016/j.ijfatigue.2018.04.025
54. T. Lippmann, A. Weidner, H. Biermann, Thermographische Untersuchung des VHCF-Schädigungsverhaltens am Beispiel des Vergütungsstahls G42CrMo4. In: Fortschritte in der Werkstoffprüfung für Forschung und Praxis: Tagung Werkstoffprüfung (2016) Verlag Stahleisen GmbH, ISBN: 9783514008304, 179–184
55. A. Vinogradov, V. Patlan, S. Hashimoto, Philos. Mag. A **81**(6), 1427 (2001). https://doi.org/10.1080/01418610108214356
56. M. Shiwa, Y. Furuya, H. Yamawaki, K. Ito, M. Enoki, Mater. Trans. **51**(8), 1404 (2010). https://doi.org/10.2320/matertrans.M2010074

Chapter 25
Analysis of Detrimental Inclusions in Steel and Aluminum

Anja Weidner, Ruben Wagner, Mikhail Seleznev, and Horst Biermann

25.1 Introduction

Inclusions are one of the most important issues for metallic materials due to their detrimental effects during production and in application under mechanical load. In steels, exogenous inclusions arise from interactions of the steel melt with its surroundings and are rare in modern steels [1]. Endogenous inclusions, on the other hand, stem from deoxidation treatment or precipitate due to decreasing solubility with decrease in temperature [2]. Accordingly, most of the inclusions in steels are oxides, followed by sulfides and nitrides. Depending on their size, morphology and distribution, they have different effects on the production and mechanical properties. In particular, for thin foil production, large alumina inclusions or alumina clusters are harmful [3, 4]. In service, quasi-static mechanical properties mainly depend on the overall distribution of inclusions, while the maximum sized inclusion may cause fatigue failure [1, 5].

In (recycled) aluminum alloys, the presence of iron plays a decisive role, since it can form primary intermetallic phases solidifying prior to Al-grains of the solid solution [6]. The size of these intermetallic phases can be of the order of millimeters. Among the iron-rich intermetallic phases in Al-Si alloys, the β phase $Al_{4.5}FeSi$ is probably the most detrimental intermetallic phase due to its plate-shaped morphology. This phase causes reduced permeability and feeding, leading to an

A. Weidner (✉) · R. Wagner · M. Seleznev · H. Biermann
Institute of Materials Engineering, Technische Universität Bergakademie Freiberg, Gustav-Zeuner-Str. 5, 09599 Freiberg, Germany
e-mail: Weidner@ww.tu-freiberg.de

M. Seleznev
e-mail: Seleznev@iwt.tu-freiberg.de

H. Biermann
e-mail: Biermann@iwt.tu-freiberg.de

© The Author(s) 2024
C. G. Aneziris and H. Biermann (eds.), *Multifunctional Ceramic Filter Systems for Metal Melt Filtration*, Springer Series in Materials Science 337,
https://doi.org/10.1007/978-3-031-40930-1_25

increase in shrinkage porosity [7, 8]. In general, primary intermetallic phases affect the mechanical properties of Al alloys [9–11].

Due to the detrimental character of inclusions in both, steels and aluminum alloys, they have been investigated for decades, firstly to avoid them, or secondly to tailor them in such a way that they have a less harmful effect on mechanical properties. In steels, typically secondary metallurgy, filtration and other techniques are used during melt treatment [12–14]. In aluminum alloys, doping of the melt, varying cooling rates and filtration techniques are applied to tailor the morphology of intermetallic phases and to reduce the iron content in the melt [6, 11, 15–17].

Analyzing inclusions, independent from the matrix material, is crucial for the materials quality control as well as for estimating the materials properties. A distinction is made between qualitative and quantitative inclusion analysis. The latter is suitable for determining the size distribution of nonmetallic inclusions in steel by means of either light optical microscopy (LOM) or automated scanning electron microscopy (SEM) with integrated energy-dispersive X-ray spectroscopy (EDS) detector determining the chemical composition of each single inclusion [18]. The size and distribution of intermetallic phases in aluminum alloys can be analyzed by X-ray microtomography (μCT, also XMT) [19].

In order to contribute to the understanding of nonmetallic inclusions (NMIs) in steel and intermetallic phases in aluminum alloys, the current chapter presents results of inclusion analyses of different materials. Therefore, melt conditioning and filtration of a secondary AlSi9Cu3 alloy was performed to reduce the iron content and, thus, change the amount and morphology of intermetallic phases. Furthermore, based on the case hardening steel 18CrNiMo7-6, different crucible materials were investigated with a special focus on the resulting type, population and size distribution of nonmetallic inclusions. The influence of different filtration treatments on the NMIs was further examined on the quenched and tempered steel 42CrMo4. Immersion filters, flow-through filters and a combination of both were used for reactive and active filtration of the 42CrMo4 steel. The resulting NMIs were further characterized by nanoindentation to evaluate their harmfulness under mechanical load.

25.2 Methods and Materials

25.2.1 Materials

Different materials were used to study the influences of the crucible material, the filtration system as well as conditioning and filtration of the metal melt on the formation of inclusions: (i) the case hardening steel 18CrNiMo7-6 with different crucible materials, (ii) immersion and/or flow-through metal melt filtration of the quenched and tempered steel 42CrMo4, and (iii) metal melt conditioning and filtration of aluminum alloy AlSi9Cu3.

Case Hardening Steel 18CrNiMo7-6

The case hardening steel 18CrNiMo7-6 (AISI 4317) is usually applied for gear parts and is, therefore, well suited for investigations on nonmetallic inclusions due to their effect on the fatigue lifetime of powertrain components. Within a steel casting simulator, described in detail in Chap. 19 and in literature [20], various refractory crucible materials were used to study their reaction with steel melt without steel killing or any other treatments. The steel melt was exposed to crucibles of carbon-bonded alumina (A-C), carbon-bonded alumina–zirconia-titania (AZT-C) and AZT-C with carbon nanotubes and alumina nanosheets (AZT-C-n) for 1 h at 1580 °C. The solidified steel was subsequently hot-isostatically pressed (HIP; Bodycote, Germany), austenitized under argon atmosphere (35 min at 900 °C) and subsequently quenched in water.

Quenched and Tempered Steel 42CrMo4

For investigations of the effect of reactive and/or active steel melt filtration (cf Chap. 19) applied as immersion or flow-through filters on the formation of nonmetallic inclusions, the steel 42CrMo4 (AISI 4142) was used. Since this steel is utilized for highly stressed components whose strength can be affected by nonmetallic inclusions, it is well suited for studies on their formation and distribution. Within a steel casting simulator, creation of defined alumina inclusions was performed by oxidation with an iron oxide (0.5 wt%) and subsequent deoxidation with pure aluminum (0.05 wt%) following a decidedly time–temperature regime [21, 22]. A-C foam filters were immersed into the steel melt for several seconds. Carbon nanotubes (CNT) together with alumina nano sheets (ANS), calcium aluminate (CA) as well as an alumina flame-spraying were used as coatings on the immersion filters and flow-through filters, respectively. For flow-through filtration, the crucible was tilted and steel melt was poured through carbon-bonded Al_2O_3 ceramic foam filters (STELEX PrO, Foseco International Ltd., Tamworth, UK). For details on the coatings design, production and properties the reader is referred to the literature [20, 23, 24].

Aluminum Alloy AlSi9Cu3

The aluminum–silicon high pressure die casting alloy AlSi9Cu3 (EN AC 46,000) is widely used as secondary aluminum alloy with an elevated iron content up to 1.3 wt% (EN 1706:2020). The contained iron causes formation of intermetallic inclusions, solidifying prior to Al-grains. In particular, plates of the β phase affect mechanical properties of the alloy and must, therefore, be avoided. Melt conditioning and filtration caused both the formation of less harmful intermetallic inclusions as well as the reduction of the iron content. A master alloy AlMn20Fe0.3 (Technologica, Germany) was used for doping the aluminum melt [25]. The melt was then held in an electric furnace at 620 °C for 3.5 h. For the subsequent filtration process into a melting furnace, filters with 30 pores per inch (Hofmann Ceramic, Germany) were used. After degassing of the melt and reaching the intended casting temperature, the melt was poured into a chill mold, producing test bars of 200 mm length and a diameter of 16 mm. Experimental details can be found elsewhere [19]. HIP was performed for all cylinders.

25.2.2 Characterization of Inclusions

Metallographic sections from different positions of the respective material in HIPed condition were cut, ground and polished with diamond paste down to 1 µm for inclusion analysis. Additionally, individual sections in as-solidified condition were investigated. Detailed characterization of inclusions was performed with a field emission scanning electron microscope (SEM, MIRA 3 XMU, Tescan, Czech Republic) using secondary electron (SE) as well as backscattered electron (BSE) contrast. Chemical and crystallographic information were gained by energy-dispersive X-ray spectroscopy (EDS) and electron backscatter diffraction (EBSD, EDAX/Ametek, USA). Depending on the material, the used contrast and the desired interaction volume, acceleration voltages between 10 and 30 kV were applied at the SEM. Light optical microscopy (LOM) was performed by means of GX51 microscope with a built-in camera XC10 (Olympus). Inclusion detection was based on the gray scale contrast difference between the matrix and the inclusions.

Classification of Inclusions by Automated SEM ASPEX

For quantitative analysis of nonmetallic inclusions in steel samples, an automated SEM (ASPEX PSEM eXpress, FEI, USA) with an integrated EDS detector was used. ASPEX recognizes nonmetallic inclusions based on grayscale thresholding under BSE contrast and measures their chemical compositions by means of EDS. Depending on the chemical composition of the inclusions, they were assigned to chemical classes according to a rule file. The design of the rule file strongly depends on the material and the contained inclusions requiring extensive qualitative and quantitative investigations. The rule files used will be referred to in the appropriate sections. Depiction of the inclusions size distributions were either provided in terms of the square root of the projected area of inclusions ($\sqrt{\text{area}}$) according to Murakami [26] or in terms of $Feret_{max}$ in order not to underestimate the real dimension of inclusions in 3D.

TEM-Investigations of Inclusions

FIB lamellae from nonmetallic inclusions with approx. dimensions of $10 \times 4 \ \mu m^2$ and thickness of approx. 100 nm were cut using an FEI Helios NanoLab 600i. A transmission electron microscope (TEM, JEM-2200FS, JEOL, Japan) was used to investigate the inclusion phases, their interface to the matrix, adherent submicron particles and the respective chemical composition by high resolution TEM, selected area electron diffraction (SAED) and electron energy loss spectroscopy (EELS).

Nanoindentation of Inclusions

Nanoindentation offers the opportunity to receive micromechanical properties of features like nonmetallic inclusions in micrometer range. Evaluating the inclusions hardness can help to estimate their harmfulness under mechanical stress.

The indentation of a Berkovich indenter is recorded during testing in a load–displacement curve, which is used afterwards for evaluating the indentation hardness according to the Oliver-Pharr method [27]. Besides the hardness, the indentation modulus of individual phases and the maximum shear stress can be determined [28]. In this work, a picoindenter (PI 87, Hysitron, USA) was used, which was mounted on the SEM stage. The combination of nanoindentation with SEM enables pinpoint indentations of inclusions of only 5 μm diameter. However, the nonmetallic character of the inclusions as well as the resolution limit of the SEM itself limit the depiction of hardness indents in nonmetallic inclusions. Therefore, atomic force microscopy (AFM XE-100, Park Systems, South Korea) was used to measure the indents' surface and volume as well as their surroundings. This provided information about the material behavior during the indentation process with regard to the true contact area between indenter and surface depending on pile-up or sink-in of the surrounding material.

X-ray Microtomography of AlSi9Cu3

X-ray microtomography was applied to study microstructural details, phases and pores in the aluminum alloy. A CT Xradia 510 Versa (Carl Zeiss Microscopy GmbH, Deutschland) was used to scan specimens of AlSi9Cu3 with a power of 7 W at 80 kV before and after ultrasonic fatigue testing (USFT). Three fields of view were stitched to an overall scan field of 5.8×13.8 mm^2 with a voxel size of 5.7 μm [19]. Image analysis was performed with the Fiji distribution [29] of the software ImageJ [30]. For pixel-based segmentations the plugin Trainable Weka Segmentation 3D was used [31].

25.2.3 Extraction of Nonmetallic Inclusions

In order to be able to discriminate various filtration methods, the morphology of inclusions needs to be investigated. Depending on the chemical composition of NMIs

and the resulting corrosion properties, chemical as well as electrolytic extraction were
used as described in the following.

Chemical Extraction

Chemical extraction of nonmetallic inclusions from case hardening steel
18CrNiMo7-6 was performed with hydrochloric acid, mixed 1:1 with deionized
water, at 70 °C in an ultrasonic bath. After dissolution of approx. 0.1 g of steel in
100 ml of the acid, 200 ml deionized water were added for dilution. By the help of a
vacuum pump, the nonmetallic inclusions were filtered by polycarbonate membrane
filters with an open pore size of 0.2 μm. The washed and dried filters were covered
by a thin carbon layer providing conductivity [32]. This procedure was applied to
study oxide inclusions.

Electrolytic Extraction

Compared to chemical extraction, the more time-consuming electrolytic extraction
offers investigation of less chemically stable inclusions like MnS. Therefore, poten-
tiostatic electrolytic extraction of 42CrMo4 steel was operated at 2–3 V and 30–
60 mA. With an electrolyte consisting of 10% acetylacetone, 1% tetramethylam-
monium chloride, and methanol, a dissolution rate of 0.3 up to 0.8 mg min^{-1} was
realized [22, 33]. Filtration of remaining nonmetallic inclusions was performed as
described above.

25.3 Influence of Crucible Material on Nonmetallic Inclusions in 18CrNiMo7-6

In this section, investigations on the effect of different crucible materials on the
NMIs in the case hardening steel 18CrNiMo7-6 are presented. Therefore, melting
crucibles of carbon-bonded alumina (A-C), carbon-bonded alumina-titania-zirconia
(AZT-C), and AZT-C with carbon nanotubes and alumina nanosheets (AZT-C-n)
were used for steel treatment in a steel casting simulator, as described in Sect. "Case
Hardening Steel 18CrNiMo7-6". Each of the batches included 20–25 kg of steel. For
detailed investigations the reader is referred to a related study [32]. The chemical
compositions of the respective batches are denoted in Table 25.1.

The greatest influence of the crucible materials on the chemical compositions
of the batches of 18CrNiMo7-6 steel compared to the as-delivered condition was
found with respect to the contents of Al, O, and C. Due to decarburization during the
holding time of one hour, the carbon content was reduced in all batches. Moreover, a
slight decrease in the concentration of oxygen was observed for both AZT batches.

Table 25.1 Chemical composition (wt%) of as-delivered steel and of batches after treatment in different crucibles. Fe balance. Reprinted from [32, 34] by permissions from Springer Nature and Wiley, respectively

Batch	C	Cr	Ni	Mo	Mn	S	P	Al	$[O]^{total}$
As-delivered	0.19	1.59	1.43	0.24	0.70	0.027	0.016	0.022	0.0025
A-C	0.07	1.44	1.41	0.25	0.63	0.024	0.007	<0.001	0.0032
AZT-C	0.12	1.57	1.39	0.25	0.68	0.021	0.013	<0.001	0.0024
AZT-C-n	0.11	1.55	1.35	0.25	0.61	0.021	0.011	<0.001	0.0020

The most significant change was observed with respect to the aluminum content, which was reduced to values below the detection limit in all treated batches [32, 34].

The size distributions of all NMIs detected by ASPEX are presented in Fig. 25.1a. The batch with the A-C crucible revealed the highest proportion of NMIs of all three batches, and exhibited a log-normal distribution. Instead, the steels derived from both of the AZT crucibles contained fewer inclusions, distributed in a bimodal manner. While the first peak was accounted to small inclusions <1 μm, the second peak was attributed to larger inclusions with approx. 20 μm √area. Although, batch AZT-C had a slightly higher number of inclusions compared to AZT-C-n, the insert in Fig. 25.1a shows that for sizes >80 μm, this trend is reversed [32].

The two main types of NMIs, detected on metallographic sections of all three batches were plain sulfides and duplex inclusions consisting of an oxide core and a sulfide shell. A typical duplex inclusion is displayed in Fig. 25.1b together with the results of a combined EBSD-EDS analysis. According to the EDS maps, this inclusion consisted of an oxide core and a sulfidic shell. The former was identified

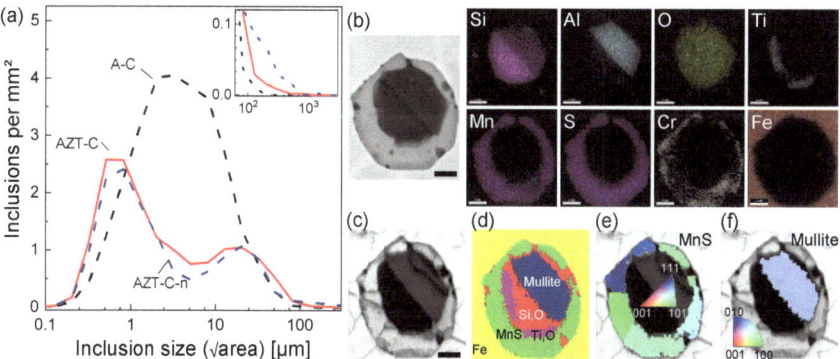

Fig. 25.1 a Size distribution of NMIs determined by ASPEX for three steel batches treated in different crucibles. Analyses of a duplex inclusion by **b** SE imaging and related EDS-maps, **c** EBSD band contrast, **d** phase map, and inverse pole figures for **e** MnS and **f** mullite phase. Scale bar corresponds to 2 μm. Reproduced by permission from Springer Nature [32]

as mullite (Fig. 25.1d, f) and some remaining oxide parts, and the latter as MnS (Fig. 25.1e).

Based on investigations of further inclusions, it was determined that the oxide cores were mostly amorphous and belonged to the ternary system SiO_2-Al_2O_3-MnO. Small amounts of TiO_x were also found at the interface between the core and the shell. The shell itself was composed of polycrystalline MnS (Fig. 25.1e) in most cases, with small silica particles incorporated. Further detailed investigations on the duplex as well as sulfidic inclusions can be found elsewhere [32].

Since the steel batch from the A-C crucible revealed a significantly higher amount of duplex inclusions, it can be assumed that a higher total oxygen content leads to an increased number of duplex inclusions. At constant levels of Mn and S in all three batches, the number of sulfides remained constant, but their size distribution was affected by the number of oxides, as they act as nucleation sites [32].

25.4 Nonmetallic Inclusions in 42CrMo4 After Metal Melt Filtration

Reactive and active filters can be applied by immersion filtration or by flow-through filtration. In this section, studies of (i) immersion filtration, (ii) flow-through filtration and (iii) a combination of immersion and flow-through filtration and, therefore, a combination of reactive and active filtration are presented. In the following, immersed filters will be abbreviated with "I" and flow-through filters with "F", followed by the respective filtration coating description (subscript).

The impact of these filtration systems on the inclusions' morphology and size distribution was investigated on metallographic sections. As fatigue crack initiation takes place at the largest inclusion, fatigue testing was performed to reliably reveal the largest inclusion within a cyclically strained volume. Therefore, fatigue crack surfaces were investigated after ultrasonic fatigue testing (see Sect. "Influence of Temperature During Steel Melt Treatment on Nonmetallic Inclusions").

25.4.1 Immersion of Filters into the Steel Melt

Immersion tests with carbon-bonded ceramic foam filters with different functional coatings were performed in a steel casting simulator. Therefore, approx. 40 kg of 42CrMo4 steel were melted and the ceramic foam filters were dipped into the steel melt for 10 s while rotating along its own central axis. Experimental details can be found elsewhere [5, 20]. The functional coatings of the respective filters along with the chemical compositions of the solidified steel batches after filter immersion are denoted in Table 25.2.

Table 25.2 Nomenclature of filter coatings types and chemical composition (wt%) of respective solidified steel batches after filter immersion tests. Fe balance. Reproduced from [5] with permission from Wiley

Filter coating	Batch abbreviation	C	Cr	Mn	Si	S	P	Al	$[O]^{total}$
No filter	I_0 (Ref.)	0.34	0.95	0.71	0.20	0.001	0.021	0.007	0.003
Al_2O_3-f-s	I_A	0.33	0.97	0.70	0.21	0.002	0.017	0.003	0.006
AC5	I_{AC5}	0.32	0.96	0.68	0.20	0.002	0.016	<0.001	0.004
CNT + ANS	I_{NM}	0.34	0.97	0.75	0.23	0.002	0.023	0.001	0.004
CA6	I_{CA6}	0.33	0.97	0.75	0.22	0.001	0.023	0.002	0.004

The nomenclature given in Table 25.2 denotes coatings based on alumina flame-spraying (Al_2O_3-f-s, I_A), alumina-based slurry with carbon (AC5, I_{AC5}), carbon nanotubes and alumina nanosheets (CNT + ANS, I_{NM} for Nano Materials) as well as calcium hex-aluminate (CA6, I_{CA6}). Detailed description on the coatings design, production and properties are given elsewhere [20, 23, 24]. Compared to the reference I_0, the amount of aluminum in the steel batches after immersion tests was significantly reduced. Among the filtered batches, batch I_A represented the batch with the highest contents of aluminum and oxygen. All other chemical elements mentioned in Table 25.2 were only insignificantly affected by the immersion treatment.

For the sake of investigating the inclusions' morphologies and chemical compositions, chemical extraction as well as EDS measurements were carried out. Figure 25.2 shows different inclusions' morphologies (upper row, SEM imaging) and chemical compositions (lower row, EDS measurements).

The morphology of the inclusions may be divided into polyhedra (Fig. 25.2c, d), globules (Fig. 25.2b) and plates (Fig. 25.2a). After chemical extraction of NMIs from their steel matrix, most of the extracted NMIs were found to have a diameter of less than 10 μm. Only the category of plate-shaped alumina inclusions comprised larger particles (Fig. 25.2a). Nearly all detected inclusions were oxides. Since MnS is dissolved in the extraction acid, it was only found as precipitation on the surface of multiphase inclusions on metallographic sections (Fig. 25.2d). Besides these multiphase inclusions, also homogeneous inclusions appeared (Fig. 25.2a).

The results of quantitative analyses of the inclusions within the respective steel batches by ASPEX and LOM are shown in Fig. 25.3 together with the largest inclusions, observed on fatigue fracture surfaces.

The ASPEX analyses confirmed that the inclusions were almost exclusively oxides. In addition to the major portion of pure alumina, other mixed oxides occurred (Fig. 25.3a). Due to the immersion treatment, the total number of inclusions per unit area increased in batches I_A and I_{AC5}, but decreased in batches I_{NM} and I_{CA6}. Independent from the filter coatings, the amount of alumina inclusions was reduced in all batches compared to the reference batch. At the same time, all coated immersion filters increased the proportion of other oxides, as given in detail in Fig. 25.3b.

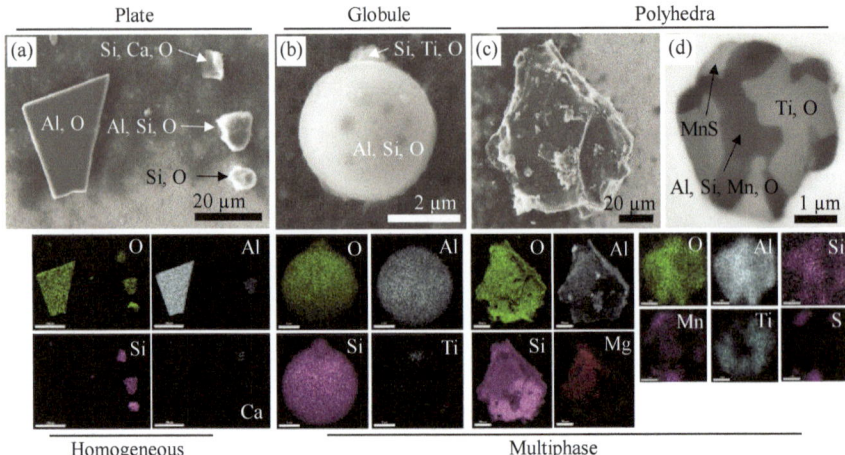

Fig. 25.2 Nonmetallic inclusions after **a–c** chemical extraction and **d** sectioning of 42CrMo4 steel. Different morphology types are **a** plates, **b** globules and **c–d** complex-shaped polyhedra. The inclusions may be divided by their chemical composition into **a** homogeneous and **b–d** multiphase inclusions according to the EDS maps below each inclusion. Reproduced from [5] with permission from Wiley

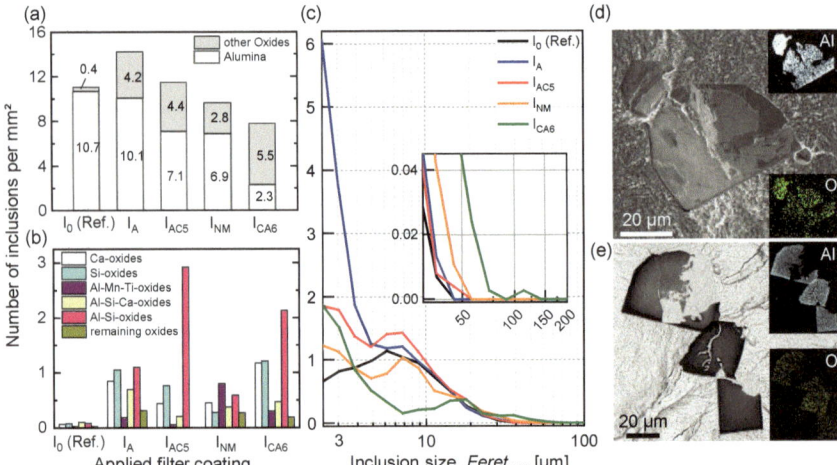

Fig. 25.3 **a–c** Analysis of inclusions from polished sections and **d, e** fatigue fracture surfaces. **a, b** Inclusions' distribution detected by ASPEX and **c** LOM. **a** Distinguishing between the number of alumina inclusions and 'other oxides' and **b** the number of each oxide type in the respective steel batches detected by ASPEX analysis. **c** Size distribution of all inclusions by LOM analysis; inset focuses on the largest inclusions found on sections. **d–e** Plate-like alumina inclusions found on fatigue fracture surfaces together with EDS-mappings revealing their chemical compositions. Partly reproduced by permissions from Wiley [5] and Elsevier [35, 36]

Regarding the total amount of (oxide) inclusions and the reduction of alumina inclusions, the filter I_{CA6} seems to be the most promising.

The size distribution of NMIs determined by LOM analysis shown in Fig. 25.3c revealed significant changes due to the coatings on the immersion filters. The distributions of NMIs in batches I_{AC5} (red line) and I_{NM} (orange line) follow the general trend of the reference batch I_0 (Ref., black line), dropping to zero at approx. 40–50 µm $Feret_{max}$. The largest NMIs of batch I_A (blue line) found on sections had also a size of approx. 40 µm, whereas a significant increase was observed for NMIs smaller than 6 µm. The most interesting influence of the coated immersion filter was observed in batch I_{CA6}. While small inclusions with a size range of 4–20 µm occurred much less frequently, the size distribution of large NMIs showed almost double increase in $Feret_{max}$ at 0.01 mm^{-2} areal density from 50 to around 100 µm [5]. These large NMIs appeared as plate-like alumina inclusions and were found on fatigue fracture surfaces (Fig. 25.3d, e) [35, 36].

25.4.2 Flow-Through Filtration of Steel Melt

Flow-through filtration of approx. 100 kg steel melt with carbon-bonded ceramic foam filters with different functional coatings were performed in the steel casting simulator. The functional coatings on the respective filters along with the chemical compositions of the solidified steel batches after filtration are listed in Table 25.3.

According to the abbreviations for filter coatings introduced in Sect. "Immersion of Filters into the Steel Melt", these were also applied here for flow-through filters (F). For detailed information on filter coatings, the reader is referred to related studies [20, 23, 24].

For analysis of inclusions from polished sections of these three batches, a total area of more than 2500 mm^2 was analyzed by ASPEX. By means of LOM, in addition to the ASPEX specimens, further specimens with an inspected area of almost 5500 mm^2 were analyzed. The inclusions' size distributions as well as the proportions of chemical classes within the respective steel batches are shown in Fig. 25.4.

Table 25.3 Nomenclature of filter coatings types and chemical composition (wt%) of respective solidified steel batches after flow-through filtration. Fe balance

Filter coating	Batch abbreviation	C	Cr	Mn	Si	S	P	Al	$[O]^{total}$
AC5 (Ref.)	F_{AC5}	0.34	0.95	0.64	0.28	0.001	0.013	0.002*	0.009
CNT + ANS	F_{NM}	0.32	0.95	0.69	0.18	0.001	0.013	0.020	0.001
Al$_2$O$_3$-f-s	F_A	0.34	0.94	0.62	0.22	0.001	0.013	0.015	0.002

* Single measurements below the detection limit of 0.001 wt%

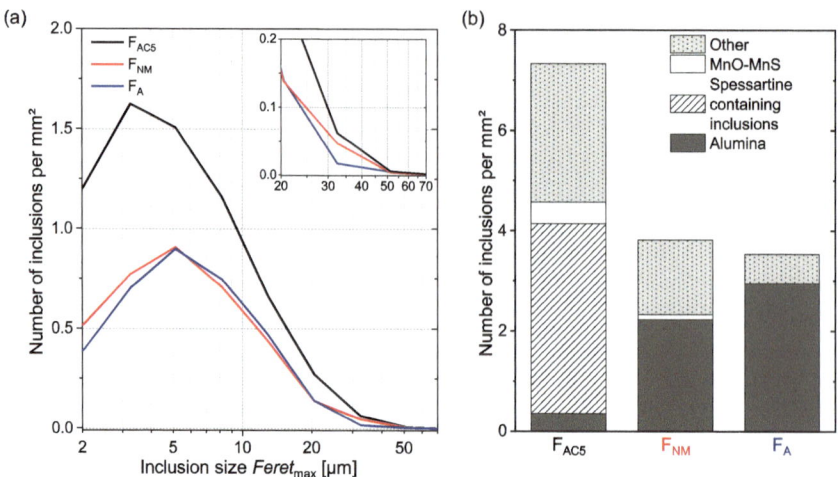

Fig. 25.4 Analysis of inclusions in 42CrMo4 steel after flow-through filtration from polished sections. **a** Size distribution of all inclusions by ASPEX analysis; inset focuses on the largest inclusions found on sections. **b** Inclusion density and chemical classes within the respective batches

The sizes of NMIs were estimated by $Feret_{max}$ instead of commonly used $\sqrt{area}$ [26] due to the fact that plate-like alumina inclusions were found on fatigue fracture surfaces. These plates appear as needles on 2D sections and might be underestimated if the size of NMIs is approximated by $\sqrt{area}$. The size distributions in Fig. 25.4a show that F_{AC5} revealed more NMIs than batches F_{NM} and F_A. However, the insert at the top of Fig. 25.4a points out that large NMIs appeared in all batches in the size class of max. 63 μm ($Feret_{max}$). Plate-like alumina inclusions found on fatigue fracture surfaces of batches F_{NM} and F_A (not shown here) even revealed inclusions with 147 μm and 112 μm ($Feret_{max}$), respectively. While these largest inclusions are relevant to fatigue loading, static material properties depend on the cross-section weakened by NMIs [5]. The area percentages covered by NMIs were at 0.023%, 0.009% and 0.007% for batches F_{AC5}, F_{NM} and F_A, respectively. This gives rise to the assumption that F_{AC5} might reveal the worst mechanical properties under quasi-static loading. To distinguish the chemical classes of NMIs, a rule file according to Table 25.4 was used (only the main parts are given).

The analysis of chemical classes of NMIs in Fig. 25.4b shows great differences between the batches after flow-through filtration. Batch F_A contained the highest amount of alumina inclusions but the lowest overall number of NMIs per area. In batch F_{NM} less alumina inclusions but a higher amount of 'other' inclusions containing small amounts of calcium aluminates and different oxides occurred. The greatest change in the chemical classes of the inclusions was observed in batch F_{AC5}. Here, the amount of alumina inclusions was drastically reduced, which is also reflected in the greatly reduced overall aluminum content in the solidified steel (cf. Table 25.3). Instead, due to the reactive filtration new multiphase inclusions were formed, characterized by inclusions with a matrix of spessartine $Mn_3Al_2Si_3O_{12}$,

Table 25.4 Rule file used to distinguish chemical classes of NMIs in solidified steel after flow-through filtration

Chemical class	Rule, element content in at%
Alumina (Al_2O_3)	Al > 5 and O > 5 and Mn < 5 and S < 20 and Ca < 5
Spessartine containing inclusions	Mn < 33.4 and (Al + Mn + Si) > 14.5 and Al > 1 and Si > 1 and Mn > 1 and S < 5
MnO-MnS	Mn > 6 and S > 5 and Al < 20 and Si < 10 and Ca < 5
Other	True

embedding primarily small alumina inclusions [37]. This spessartine matrix is known for its low melting temperature of approx. 1200 °C [38, 39]. The total number of inclusions per area increased significantly in this batch. Although the spessartine containing inclusions are usually smaller than the alumina inclusions, they occasionally caused fatigue failure. Figure 25.5a shows a spessartine multiphase inclusion on a fatigue fracture surface of batch F_{AC5} with corresponding EDS mappings.

Both, the multiphase NMI on the fatigue fracture surface (Fig. 25.5a), as well as the multiphase NMI found on a section (Fig. 25.5b) are characterized by alumina embedded in a matrix of spessartine as indicated by associated EDS-mappings. As can be derived from Fig. 25.4b, batches F_{NM} and F_A mainly contained alumina inclusions. On the surface of these alumina inclusions MnS precipitated (black arrows in Fig. 25.5c, d), which is well known from the literature [14, 32]. Besides MnS,

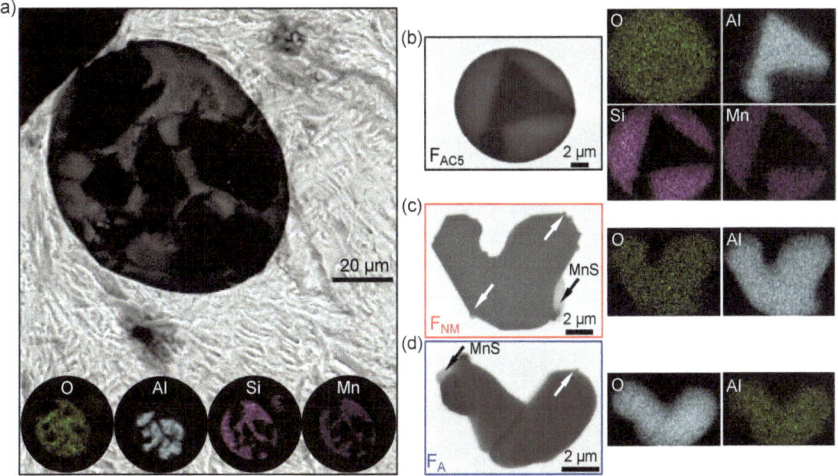

Fig. 25.5 Nonmetallic inclusions on fatigue fracture surface **a** and **b–d** on sections. **a, b** Multiphase nonmetallic inclusions from batch F_{AC5}, consisting of an alumina core and a spessartine matrix, indicated by EDS-maps. **c, d** NMIs from sections of batches F_{NM} and F_A, respectively, marked by colored frames, together with EDS maps. Black arrows indicate MnS precipitates; white arrows indicate BN precipitates

small satellites precipitated on alumina inclusions (white arrows in Fig. 25.5c, d), which were identified as BN by EELS. Detailed TEM-investigations on NMIs of batch F_{NM} are shown in Fig. 25.6.

In Fig. 25.6, an alumina inclusion is visible. The diffraction pattern from SAED analysis of this particular alumina inclusion did not show a typical diffraction pattern from α-Al_2O_3 (corundum), but a metastable condition (Fig. 25.6c). However, the common structures of the metastable alumina phases δ and θ occurring along the transformation path from metastable γ-Al_2O_3 to thermodynamically stable α-Al_2O_3, which are described in Chap. 6.3, did not fit to this diffraction pattern. Therefore, this metastable alumina inclusion appears to have a defective intermediate state that did not completely transform to α-Al_2O_3 during steel melt treatment. Alumina inclusions of batch F_A were also identified as metastable Al_2O_3. The BN shown here was detected in all three batches, identified as h-BN by means of EELS, shown

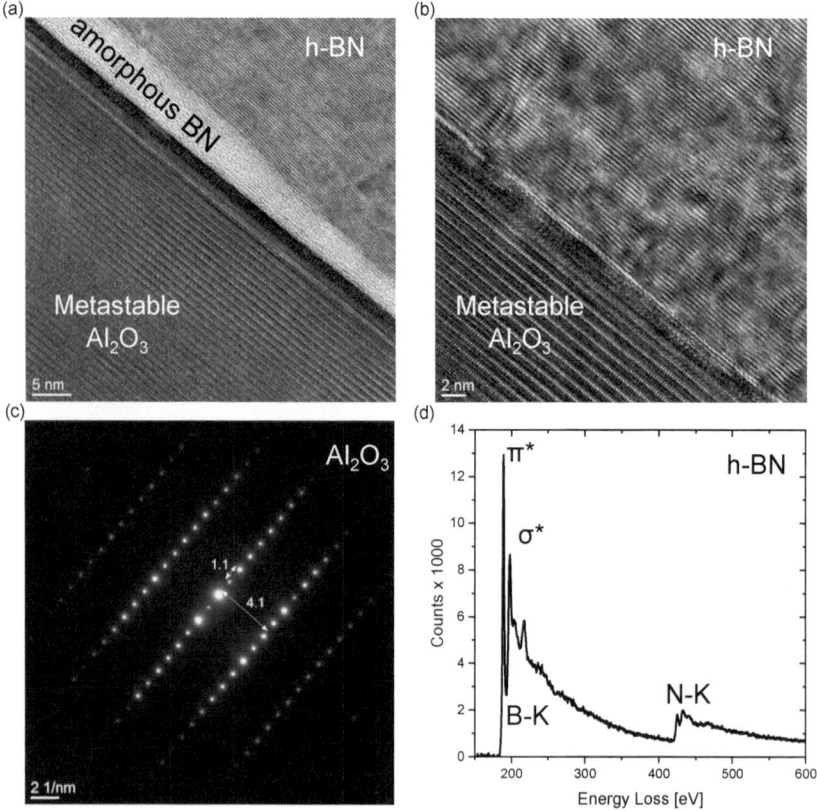

Fig. 25.6 HR-TEM investigations on a nonmetallic inclusion of batch F_{NM}. **a** h-BN precipitated on metastable Al_2O_3 with an amorphous BN interlayer, **b** direct interface between metastable Al_2O_3 and h-BN with parallel lattice planes between both phases. **c** SAED diffraction pattern of metastable alumina inclusion shown in **a** and **b**; **d** EELS spectrum of h-BN shown in **a** and **b**

in Fig. 25.6d [40]. At the same inclusion, Al_2O_3-BN interfaces with and without an interlayer were found. The interlayer consisted of amorphous BN (Fig. 25.6a). In case of a direct interface between metastable Al_2O_3 and h-BN, parallel lattice planes between both phases occurred (Fig. 25.6b). The above mentioned spessartine phase occurring in batch F_{AC5} was found to be amorphous. These investigations show that (i) alumina acts as nucleation site for MnS and BN and, (ii) the metastable alumina inclusions can most probably be assigned to ternary endogenous inclusions, which formed below the liquidus temperature of steel and, therefore, did not finally transform into α-Al_2O_3 during solidification within the crucible.

25.4.3 Influence of Temperature During Steel Melt Treatment on Nonmetallic Inclusions

Due to the above mentioned steel melt filtration, plate-like alumina inclusions were formed, which deteriorated the materials fatigue strength [5, 35, 36]. To avoid the formation of plate-like alumina inclusions, the temperature during steel melt treatment within the steel casting simulator was reduced from 1650 °C (for flow-through filtration, Sect. "Flow-Through Filtration of Steel Melt") to 1600 °C. For experimental details the reader is referred to a related study [22]. The batch with the reduced steel melt temperature during deoxidation is abbreviated as I_0(LT) for 'low temperature' and compared to I_0(Ref.) from Table 25.2. The chemical compositions of both batches are presented in Table 25.5.

In as-received condition, significant differences in the chemical composition of the two batches were present only with respect to the aluminum contents of 350 ppm and 290 ppm in I_0(Ref.) and I_0(LT), respectively. After the treatment, the solidified steel exhibited slightly different total oxygen contents of 30 ppm and 24 ppm in I_0(Ref.) and I_0(LT), respectively. Due to the steel melt treatment within the steel casting

Table 25.5 Chemical composition (wt%) of the solidified 42CrMo4 of batch I_0(Ref.) with a high temperature of 1630 °C during deoxidation and batch I_0(LT) with a lower temperature of 1600 °C during deoxidation in 'as-received' and 'after treatment' condition, respectively. Fe balance. Reproduced from [22] with permission from Wiley

Batch abbreviation	C	Cr	Mn	Si	S	P	Al	$[O]^{total}$
I_0(Ref.) as received	0.41	0.96	0.79	0.24	0.001	0.014	0.035	–
I_0(Ref.) after treatment	0.34	0.95	0.71	0.20	0.001	0.021	0.007	0.0030
I_0(LT) as received	0.41	0.96	0.77	0.27	0.002	0.014	0.029	0.0022
I_0(LT) after treatment	0.31	0.95	0.68	0.18	0.001	0.010	0.026	0.0024

simulator, the contents of carbon dropped below the values of the standard EN ISO 683-2 [41], which is already known from another study [34]. All other elements were almost independent from the temperature level during steel melt treatment [22].

The inclusion size distributions of both batches together with a bar chart revealing the NMIs' sizes obtained on fatigue fracture surfaces are shown in Fig. 25.7.

In Fig. 25.7a, a shift of the inclusion size distribution of I_0(LT) towards smaller sizes is visible. This shift is about 2 µm at the maxima of the two probability distributions, while the basic shapes of the size distributions are similar. The inset at the top of Fig. 25.7a focusing on the largest inclusions shows that for both batches the largest NMIs occurred in the size class until about 40 µm $Feret_{max}$. Since information about the largest NMIs are not easy accessible via 2D sections, the size distributions are supplemented by the largest NMIs observed on fatigue fracture surfaces, representing the largest NMI within the respective cyclically loaded volume. The bar chart reveals that for both batches, the NMIs found on fatigue fracture surfaces significantly exceeded the maximum sizes of the NMIs observed on sections. 70% of NMIs observed on fatigue fracture surfaces of batch I_0(Ref.) revealed a plate-like morphology, exemplarily shown in Fig. 25.7b. Instead, in batch I_0(LT) all NMIs on fatigue fracture surfaces appeared maple-like or dendritic [42] (Fig. 25.7c). Thus, it was possible to avoid plate-like alumina inclusions with the reduced temperature

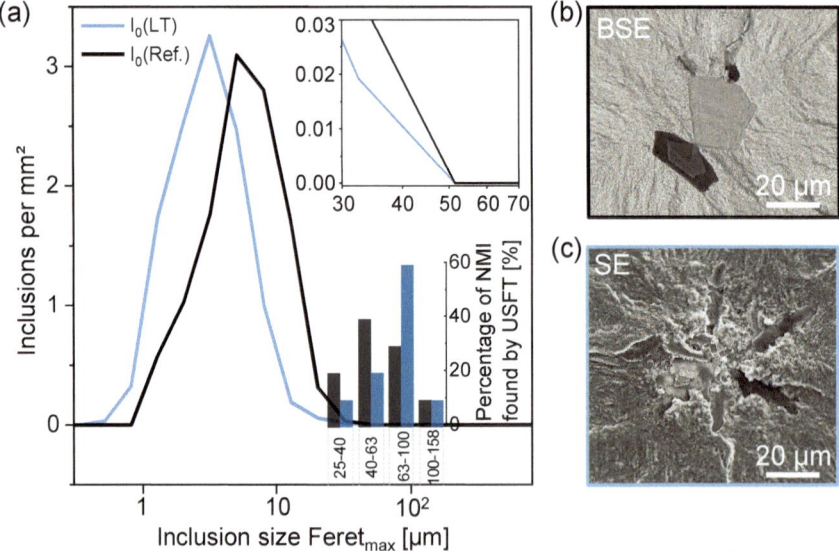

Fig. 25.7 **a** Inclusion size distribution determined by ASPEX from polished sections of batches I_0(Ref.) (cf. Sect. 25.4.1) and I_0(LT); bar chart reveals percentage portions of inclusions of the respective size classes found on fatigue fracture surfaces. **b** and **c** Typical crack-inducing alumina inclusions on fatigue fracture surfaces. **b** Plate-like NMI of batch I_0(Ref.) in BSE contrast, **c** maple-like alumina inclusion of batch I_0(LT) in SE contrast. Reproduced from [22] with permission from Wiley

during the steel melt treatment. However, the size of the maple-like or dendritic alumina inclusions still requires steel melt filtration, since the ultrasonic fatigue experiments showed that large inclusions with sizes of more than 100 μm were present [22].

25.4.4 Combined Immersion Filtration and Flow-Through Filtration

In previous sections, the resulting NMIs after immersion filtration (Sect. "Immersion of Filters into the Steel Melt") and flow-through filtration (Sect. "Flow-Through Filtration of Steel Melt") were described. Depending on the filter coatings, both types of treatments achieved a significant reduction in aluminum levels and a partly change of the inclusions' chemistry. However, the steel batches still contained large detrimental plate-like alumina inclusions. Therefore, the temperature regime during steel melt treatment was adjusted to avoid the formation of plate-like alumina inclusions (Sect. "Influence of Temperature During Steel Melt Treatment on Nonmetallic Inclusions").

In this section, the influence of the combination of immersion filtration, flow-through filtration and an adjusted temperature regime during steel melt treatment on the NMIs will be presented. For this purpose, two different coatings on immersion filters were combined with an alumina flame-spray coating on flow-through filters (cf. Section 25.4.1). These two kinds of combined filtrations will be compared to two different batches of flow-through filtration and a reference batch without any filtration, as denoted in Table 25.6.

The chemical compositions of the individual batches showed the largest differences in terms of the aluminum content in the solidified steel. The immersion filtration of batches $I_{CA}\_F_A$ and $I_{NM}\_F_A$ caused a significant reduction of the aluminum content below the detection limit, while the other batches ranged at approx. 20 ppm. The content of oxygen, on the other hand, showed that the combined filtration caused slightly higher values of 38 ppm and 37 ppm for batches $I_{CA}\_F_A$ and $I_{NM}\_F_A$, respectively. Batch $I_0\_F_A$ revealed 33 ppm, while $I_0\_F_S$ and $I_0\_F_0$ remained at 26 ppm, which is comparable to the oxygen content of 22 ppm of the as-delivered condition of batch $I_0(LT)$ (cf. Sect. "Influence of Temperature During Steel Melt Treatment on Nonmetallic Inclusions"). All other elements showed no significant differences between the batches.

Due to the fact that some of the batches revealed clusters of alumina inclusions, both on fatigue fracture surfaces as well as on metallographic sections, an experimentally and stochastically based method for prediction of endurance properties and fatigue life was developed [43]. Therefore, 91 polished sections of a batch with pronounced cluster formation with a total area of more than 10,000 mm^2 as well as 40 fatigue fracture surfaces failed due to internal crack initiation were analyzed. Detailed description of the procedure was presented elsewhere [43]. In the approach

Table 25.6 Nomenclature of filter coatings types and chemical composition (wt%) of respective steel batches after immersion filtration and flow-through filtration. Fe balance. Partly reproduced from [37] with permission from Elsevier

Immersed filter	Flow-through filter	Batch abbreviation	C	Cr	Mn	Si	S	P	Al	$[O]^{total}$
CA2-coated	Al_2O_3-f-s	$I_{CA}\_F_A$	0.30	0.95	0.65	0.18	0.002	0.01	<0.001*	0.0038
Nano-coated	Al_2O_3-f-s	$I_{NM}\_F_A$	0.29	0.96	0.68	0.19	0.002	0.01	<0.001*	0.0037
N/A	Al_2O_3-f-s	$I_0\_F_A$	0.29	0.95	0.59	0.19	0.002	0.01	0.019	0.0033
N/A	Carbon-bonded + Al_2O_3-f-s	$I_0\_F_S$	0.28	0.94	0.59	0.21	0.002	0.01	0.019	0.0026
N/A	N/A	$I_0\_F_0$	0.30	0.93	0.59	0.18	0.002	0.01	0.022	0.0026

* Single measurements below the detection limit of 0.001 wt%

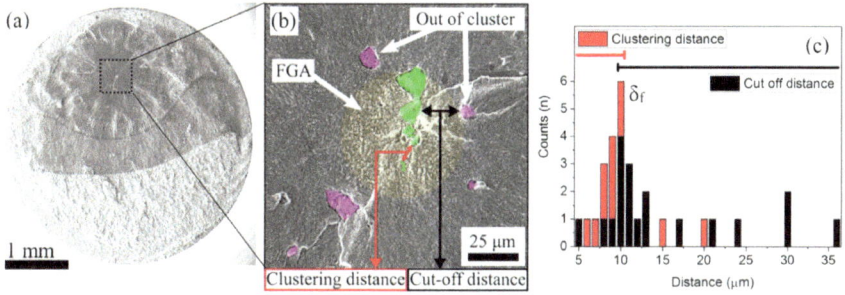

Fig. 25.8 a Estimating the critical distance δ_f between single inclusions from a fatigue fracture surface. **b** Fine-granular area (FGA) in the center of a fisheye with NMIs. **c** Summary histogram of measured distances of all fatigue fracture surfaces. Reproduced from [43] with permission from Wiley

the critical distance δ determined from sections (δ_s) and from fatigue fracture surfaces (δ_f) was used. Figure 25.8 shows the principle procedure estimating a critical distance δ_f.

As shown in Fig. 25.8b, NMIs located inside the FGA were considered as part of a crack-initiating cluster. The cut-off distance was determined from the distance of NMIs outside the FGA to the nearest NMI inside the cluster NMI. The histogram in Fig. 25.8c summarizes the analysis of measured distances between NMIs of all fatigue fracture surfaces. Accordingly, the clustering distance was mostly less than 10 μm, whereas the cut-off distance mostly exceeded 10 μm. Thus, the critical distance δ_f was around 10 μm, which is in good agreement with the formation of so-called intermediate aggregates of NMIs of 5–10 μm size found by Yin et al. [44]. The critical distance between NMIs gained from 2D sections δ_s is given in Fig. 25.9.

Compared to the exemplary curve for complete spatial randomness (blue dashed line), the point of inflection of the experimental distribution (black line) appears at a distance of 13 μm, which is referred to the critical distance δ_s. Applying this procedure to all 91 sections, the summarized distribution (insert in Fig. 25.9) revealed a peak at around 10 μm, which is in excellent agreement with the fractographic approach. The approximation of the inclusions' shape by a circle or an ellipse showed no significant difference.

According to this method, NMIs with a distance less than or equal to 10 μm were counted as one inclusion in the subsequent analyses, in order to meet their detrimental effect under mechanical loading. This procedure is exemplarily shown for a cluster of alumina inclusions on a fatigue fracture surface of batch $I_{CA}\_F_A$ in Fig. 25.10.

Analysis of inclusions from polished sections of the above-mentioned batches for combined filtration was carried out by ASPEX on a total area of almost 4,000 mm². For LOM analyses, further specimens were investigated in addition to the ASPEX specimens, resulting in a total area of more than 7,000 mm². The inclusion size distributions as well as the proportions of chemical classes within the respective steel batches are shown in Fig. 25.11.

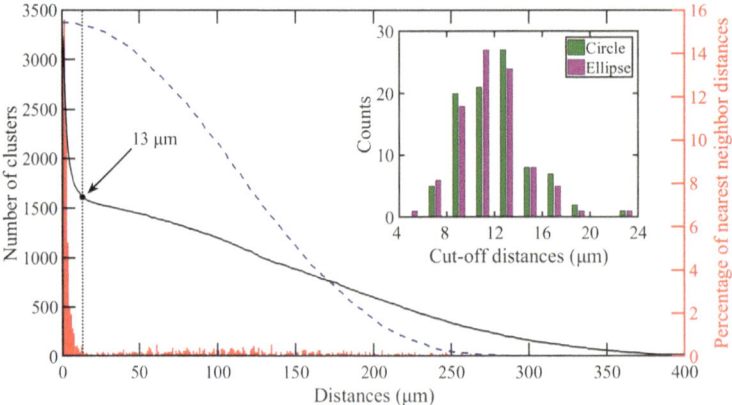

Fig. 25.9 Number of clusters for the given maximum pairwise distance within clusters (black line), inflecting at 13 μm. Exemplary curve for complete spatial randomness (blue dashed line). Histogram of nearest neighbor distances of NMIs of one specimen (red bars). Summary of points of inflection for all 91 specimens (inset). Reproduced from [43] with permission from Wiley

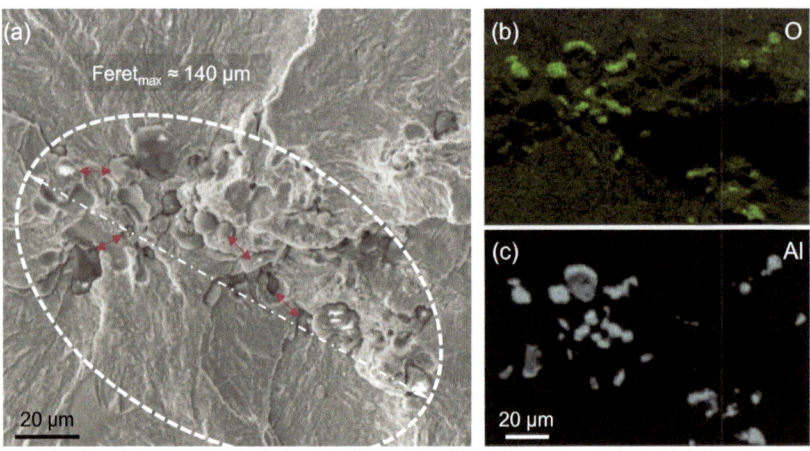

Fig. 25.10 **a** Cluster of alumina inclusions in batch I_{CA}_F_A acting as fatigue crack initiation site with EDS maps of **b** O and **c** Al

The size distribution of NMIs from batches without immersion filtration, namely I_0_F_A, I_0_F_S and I_0_F_0 revealed a similar curve starting from 5 μm (*Feret*$_{max}$) to larger NMIs. The batches with immersion as well as flow-through filtration I_{CA}_F_A, I_{NM}_F_A show a curve, which is slightly shifted towards larger sizes. However, considering cluster formation of small NMIs, these clusters occasionally cause a broadening of the size distributions up to more than 100 μm *Feret*$_{max}$ as can be seen from the inset at the top of Fig. 25.11a. This inset also reveals that the curve of

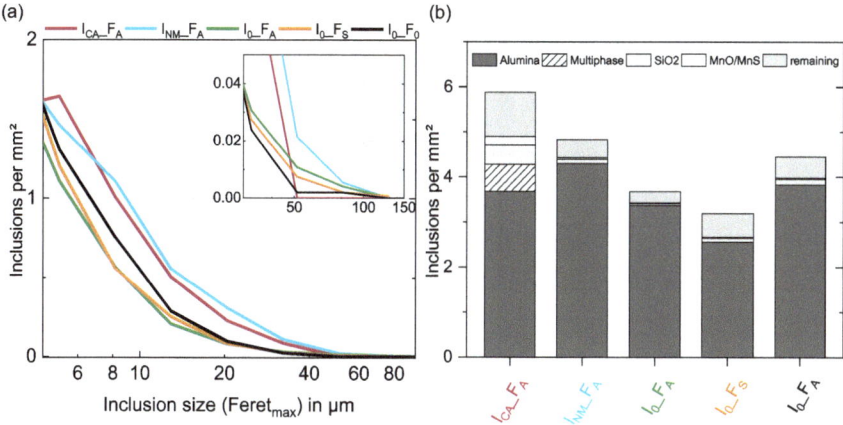

Fig. 25.11 Analysis of inclusions in 42CrMo4 steel after combined immersion filtration and flow-through filtration from polished sections. **a** Size distribution of all inclusions from LOM analyses in HIP condition considering clusters of NMIs; inset focuses on the largest inclusions found on sections. **b** Inclusion density and chemical classes within the respective batches from ASPEX analyses in HIP condition

$I_{CA}\_F_A$ (red) might be shifted to larger NMIs, but does not reveal NMIs larger than 40 µm, whereas in batches $I_0\_F_A$ (green) and $I_0\_F_S$ (orange) NMIs in the range of 100 µm were observed. However, it must be mentioned that not the identical sample area per batch was examined. For example, a larger total area analyzed makes the occurrence of very large NMIs at the edges of the distribution more likely.

Figure 25.11b shows the total number of NMIs per area as well as the chemical classes according to a rule file, which can be found elsewhere [37]. Compared to the unfiltrated batch $I_0\_F_0$, $I_0\_F_S$ followed by $I_0\_F_A$ revealed the lowest total amount of inclusions per area, whereas these batches had the largest NMIs, visible in the insert of the size distribution in Fig. 25.11a. Batches $I_{CA}\_F_A$ and $I_{NM}\_F_A$ were found to have the highest inclusions density. It is noticeable that multiphase inclusions occur in batch $I_{CA}\_F_A$ alone. Therefore, the immersion filter with calcium aluminate coating seems to cause the formation of those multiphase NMIs.

For evaluating the morphology of the NMIs, electrolytic extraction was conducted. Exemplary NMIs for each batch are shown in Fig. 25.12.

As expected from the chemical classes examined in Fig. 25.11b, most of the NMIs found after electrolytic extraction were alumina inclusions. The morphology of NMIs ranged from compact to flat to dendritic inclusions. The formation of dendritic alumina inclusions indicated a local aluminum supersaturation [22, 45]. The two batches with the lowest aluminum content in the solidified steel, $I_{CA}\_F_A$ and $I_{NM}\_F_A$, showed no dendritic inclusions. This will be discussed below.

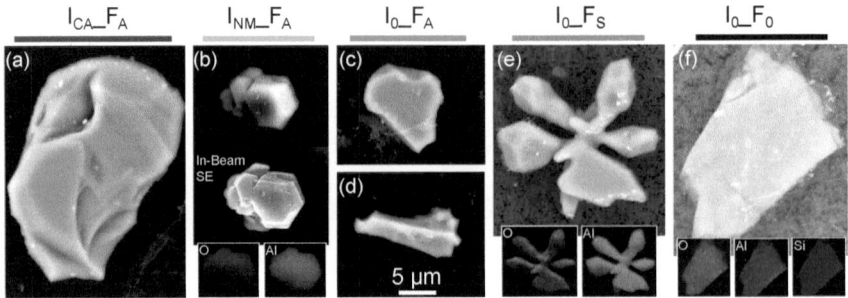

Fig. 25.12 Exemplary nonmetallic inclusions from electrolytic extraction of batches $I_{CA}\_F_A$, $I_{NM}\_F_A$, $I_0\_F_A$, $I_0\_F_S$ and $I_0\_F_0$. **a–e** Alumina inclusions of different morphologies. **f** Nonmetallic inclusion containing Al, Si and O

25.4.5 Discussion

Various filtration methods were investigated on the steel 42CrMo4. In the following, their influence on the size distribution, morphology and chemistry of NMIs will be discussed. In addition, the mechanical properties of NMIs studied by nanoindentation are discussed in terms of their harmfulness under mechanical load. When comparing the different batches, it must always be taken into account that there were differences in terms of the temperature during steel melt treatment (Sect. "Influence of Temperature During Steel Melt Treatment on Nonmetallic Inclusions"), the reference materials and the amount of material processed. In addition, clustering was accounted for the analysis of the size distribution of NMIs in the batches of combined filtration. These differences may have some effect on the NMIs, which must be considered in a quantitative comparison of the different batches.

Influence of Steel Melt Filtration on Nonmetallic Inclusions

Since alumina inclusions were the main inclusions occurring in the studied batches of 42CrMo4 after filtration, the mechanical properties of the quenched and tempered steel depended mainly on their size distribution and morphology.

The total density of NMIs affected the tensile deformability and energy dissipation of 42CrMo4 [5] as well as the fracture toughness under quasi-static loading, see Chap. 27. Thus, regardless of the morphology of the NMIs, the portion of the material weakened by the total amount of NMIs is responsible for tensile ductility and energy dissipation. Within the batches after immersion filtration (Sect. "Immersion of Filters into the Steel Melt"), batches I_{CA6} and I_{NM} revealed the lowest total amounts of NMIs of approx. 7.8 NMIs/mm^2 and 9.7 NMIs/mm^2, respectively, while batch I_A had the highest density of approx. 14.3 NMIs/mm^2. In contrast, when the latter filter coating (alumina flame-sprayed) was applied to flow-through filters, it achieved lower densities of NMIs of approx. 3.6 NMIs/mm^2 (F_A) and 3.7 NMIs/mm^2 ($I_0\_F_A$). The coating with nanomaterials also generated a significantly lower density of NMIs of 3.8 NMIs/mm^2 when used as a flow-through filter F_{NM}.

Combining the nanomaterials coating and the Al_2O_3 flame-spray coating (I_{NM_FA}) led to density of 4.8 NMIs/mm^2 (cf. Fig. 25.11). The combination of a CA2-coating on an immersion filter with an Al_2O_3 flame-sprayed flow-through filter (I_{CA_FA}) caused a comparatively high density of 5.9 NMIs/mm^2 and the formation of additional NMIs, which was already observed for batch I_{CA6}. However, the formation of spessartine multiphase inclusions slightly reduced the total amount of alumina inclusions. The flow-through filtration consisting of a carbon-bonded filter and a filter with Al_2O_3 flame-spray coating ($I_0\_F_S$) led to a very low density of 3.2 NMIs/mm^2.

While such low values are favorable for tensile deformability as well as energy dissipation under quasi-static loading, the morphology and size distributions of NMIs play a crucial role under fatigue loading. Therefore, for the latter batch $I_0\_F_S$ it must be considered, that NMIs with more than 100 μm *Feret*$_{max}$ were observed. This also applies to batch $I_0\_F_A$.

In order to improve the prediction of NMIs' morphology and their behavior under fatigue loading, a schematic representation of alumina inclusions depending on the contents of oxygen and aluminum in the solidified steel adapted from Tiekink et al. [45] is shown in Fig. 25.13. Therefore, NMIs causing fatigue crack initiation of the respective batches were evaluated regarding their morphology.

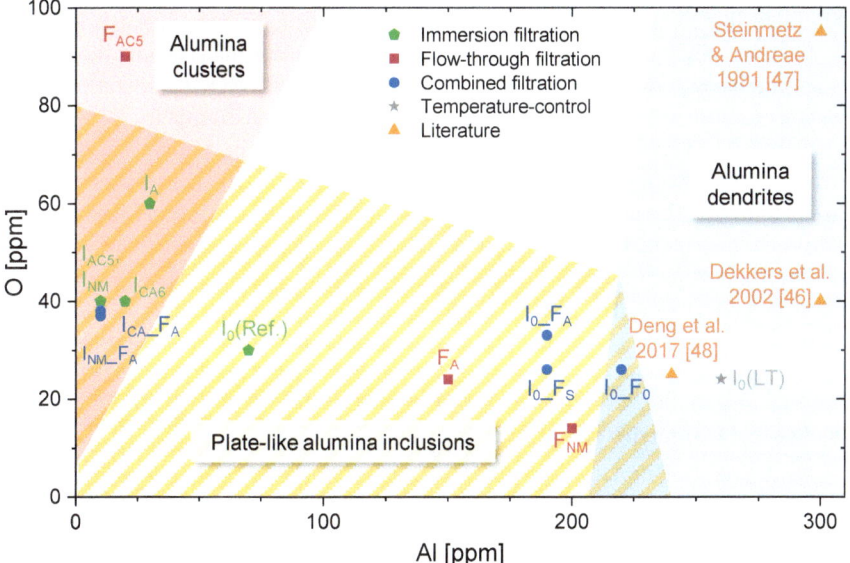

Fig. 25.13 Morphologies of alumina inclusions after treatment with and without metal steel filtration according to Tiekink et al. [45]. Inclusions from the literature are mentioned with their respective reference

The figure shows three different regions of alumina morphologies: (i) alumina clusters (red), (ii) plate-like alumina inclusions (yellow striped), and (iii) alumina dendrites (blue). In overlapping areas, batches exhibited different morphologies on fatigue fracture surfaces.

All batches with immersion filtration only (green pentagons) were characterized by a low aluminum content and revealed exclusively plate-like alumina inclusions on fatigue fracture surfaces. Batches filtered with flow-through filters only (red: F_A, F_{NM}; blue: I_0_F_A, I_0_F_S), ranked in a similar Al:O ratio, had a significantly higher aluminum content than batches with immersion filtration and also showed exclusively plate-like alumina inclusions. On fatigue fracture surfaces of the reference batch I_0_F_0, different morphologies were found, which appeared compact or dendritic, while batch I_0(LT) with a lower deoxidation temperature and an even higher amount of aluminum revealed dendritic or maple-like alumina inclusions only. This is in agreement with dendrite alumina morphologies reported by Dekkers et al. [46], Steinmetz and Andreae [47], and Deng et al. [48].

Compared to all other batches, F_{AC5} showed the highest oxygen content and a low aluminum content. NMIs, which initiated fatigue fracture of batch F_{AC5} had a compact morphology and were mostly spessartine containing inclusions due to the drastically reduced amount of alumina inclusions in this batch (cf. Fig. 25.4b).

The two batches with a combined filter system exhibited special features. Although they have oxygen and aluminum contents similar to batches with only plate-like inclusions, both clusters and plate-like alumina inclusions occurred here. In batch I_{CA}_F_A, 75% of the NMIs causing fatigue crack initiation were clusters and 25% plate-like alumina inclusions. On fatigue fracture surfaces of batch I_{NM}_F_A, 37.5% of the NMIs were alumina clusters and 62.7% plate-like alumina inclusions.

The analyses of NMIs in batches of filtered 42CrMo4 showed that metallographic characterization of NMIs is a powerful tool to estimate the steels quality. Considering the chemical composition of the as-received material and the filtrated material can already help estimating the morphology of largest inclusions, which are crucial under fatigue loading. The total density of NMIs per area, on the other hand, can be used to estimate the weakened portion of the material and, thus, the tensile deformability and energy dissipation. Nevertheless, the determination of mechanical properties is necessary to assess the quality of the steel, since metallographic analyses and chemical compositions cannot completely represent the steel properties [34].

Mechanical Properties of Nonmetallic Inclusions in 42CrMo4

Besides the size distribution and the total density of NMIs per area, their mechanical properties are of special interest for evaluating their harmfulness as stress concentrators in steel. In particular, the mechanical properties of spessartine containing multiphase inclusions, appearing in some batches after metal melt filtration (Figs. 25.4, 25.11), were rarely studied in literature. For the investigation of the mechanical properties of NMIs by nanoindentation, NMIs were divided into different chemical classes: Al_2O_3, Al_2O_3 embedded in a matrix of spessartine, multiphase inclusions containing alumina und titanium oxide, spessartine inclusions and MnS. Representative load–displacement curves of the respective classes of NMIs are depicted in Fig. 25.14 [37].

The load–displacement curves revealed large differences between the single chemical classes of NMIs. Under displacement-controlled nanoindentation up to 100 nm displacement, mean loads of about 5–7 mN were required for Al_2O_3 and Al_2O_3 in spessartine. Despite their ceramic character, multiphase and pure spessartine inclusions were already close to the mean F_{max} of the ferritic-pearlitic steel matrix. The residual indentation depths h_r point out the strong plastic behavior of MnS and the steel matrix, compared to Al_2O_3 and Al_2O_3 in spessartine. Load–displacement curves in Al_2O_3 showed an abrupt elastic–plastic transition (pop-in), highlighted by

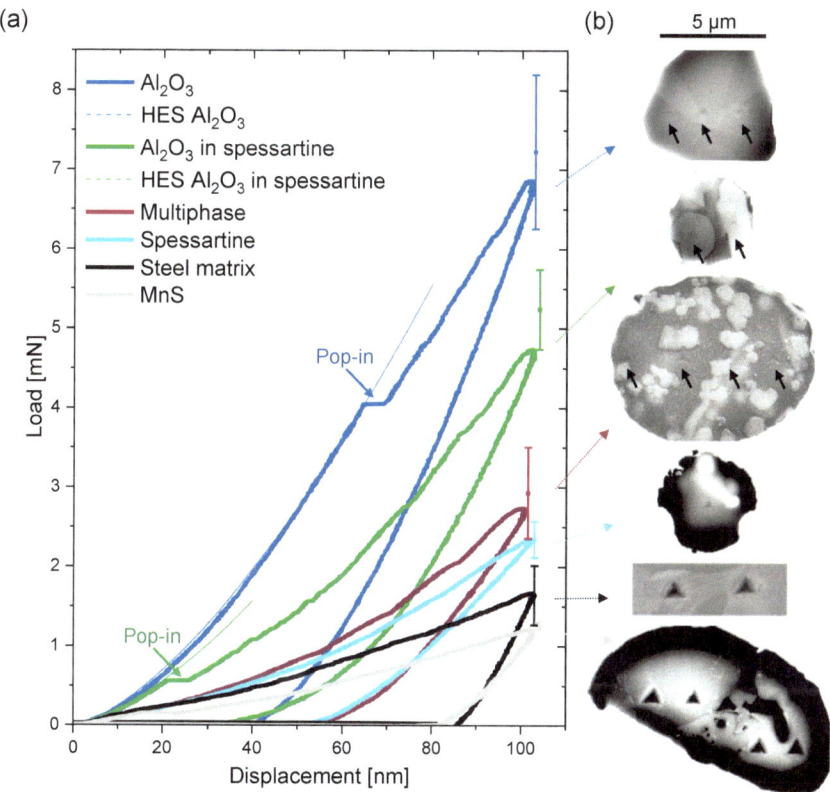

Fig. 25.14 **a** Representative load–displacement curves and **b** exemplary images of different classes of NMIs in 42CrMo4 steel: Al_2O_3, Al_2O_3 in a matrix of spessartine, multiphase inclusions, spessartine inclusions and MnS compared to the steel matrix of 42CrMo4. Black arrows mark indents if hardly visible. Hertzian elastic solution (HES) for Al_2O_3 and Al_2O_3 in spessartine fitted with a tip radius of 400 nm indicate the elastic–plastic transition upon pop-in events. Error bars on each curve show the mean and standard deviation of the maximum force at 100 nm displacement for the respective class. Reproduced from [37] with permission from Elsevier

the Hertzian elastic solution (HES), which is well known from literature [49, 50]. For detailed investigations, the reader is referred to [37].

Figure 25.15 shows, that with increasing amount of spessartine (Al_2O_3 in spessartine < spessartine multiphase < spessartine), the hardness of the respective NMIs decreased and, therefore, came closer to the hardness of the steel matrix (Fig. 25.15, blue triangles). In the same order, the values of the Young's modulus also decrease (red circles). In contrast, the residual indentation depths increase with increasing content of spessartine. Accordingly, the investigated spessartine multiphase inclusions promise not only a higher amount of plasticity compared to alumina, but also the ability to collect smaller detrimental oxide inclusions like titanium oxide and alumina. Furthermore, the low-melting character of spessartine containing inclusions, investigated in detail in a related study, makes them suitable to contribute to the prevention of nozzle clogging after metal melt filtration [37].

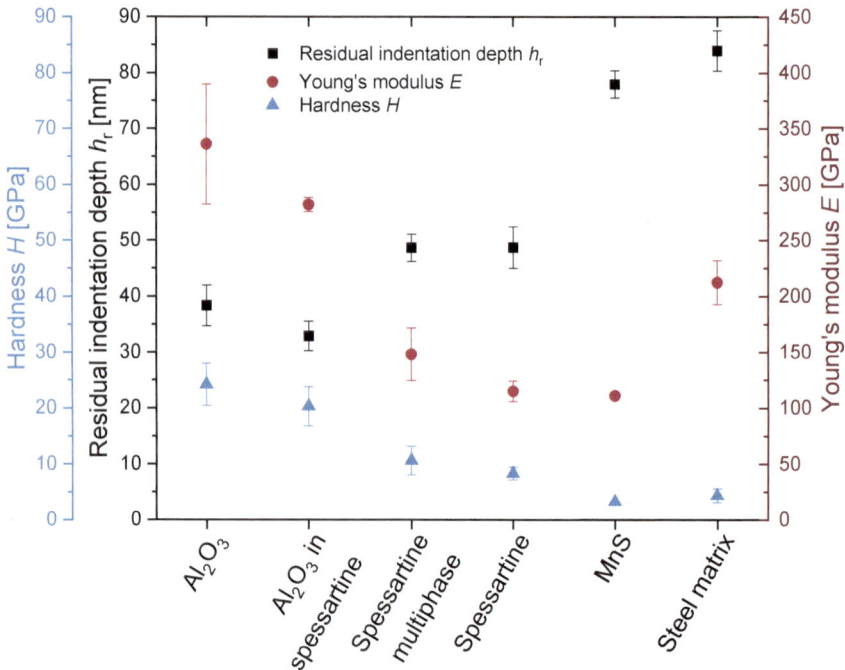

Fig. 25.15 Mechanical properties of different classes of NMIs in 42CrMo4 steel as well as the steel matrix gained by nanoindentation: Hardness H (blue triangles), residual indentation depth h_r (black rectangles) and Young's modulus E (red circles). Standard deviation of class Al_2O_3 in spessartine as an orientation due to only two available indents. Reproduced from [37] with permission from Elsevier

25.5 Effect of Melt Conditioning and Filtration on Intermetallic Phases in AlSi9Cu3

The aim of conditioning and filtration (hereinafter referred to as CF) of AlSi9Cu3 was to reduce the amount of iron within the final casting. Due to the conditioning of the aluminum melt with the master alloy AlMn20Fe0.3 and the subsequent filtration process the initial iron content of 1.60 wt% was reduced to 0.99 wt%. A reference batch (R) with an iron content of 1.71 wt% served as comparison. Combined EBSD-EDS measurements were used to identify the phases contained in batches R and CF (Fig. 25.16).

Figure 25.16a shows batch R with the Al matrix and Al-Si eutectic next to plates of the β phase and the star-like α_c phase, which were indexed by Kikuchi patterns. Particularly in the eutectic, areas with low pattern quality can be incorrectly indexed. The phases of batch CF were indexed based on their chemical composition. Compared to batch R, no β phase plates occurred, but α_c phase in Chinese script morphology. Due to added strontium, Al–Si eutectic was modified in both batches. After reduction of the iron content by almost 40% to 0.99 wt% in batch CF, the remaining iron was contained in the α_c phase in Chinese script or star-like morphology [6, 19].

The mentioned plates of the β phase in batch R sometimes reached 1 mm in length [51] and, therefore, drastically reduced the mechanical properties of the casting [19]. Based on the investigation of fracture surfaces of tensile specimens from both batches R and CF, it becomes clear how strongly the β phase plates affected the mechanical properties under uniaxial tension (Fig. 25.17a).

In batch R, fracture takes places exclusively along β phase plates. The fact that the plates were split and appeared at the corresponding positions of the counterparts underlines their brittle behavior under uniaxial load. Besides, the white arrow in

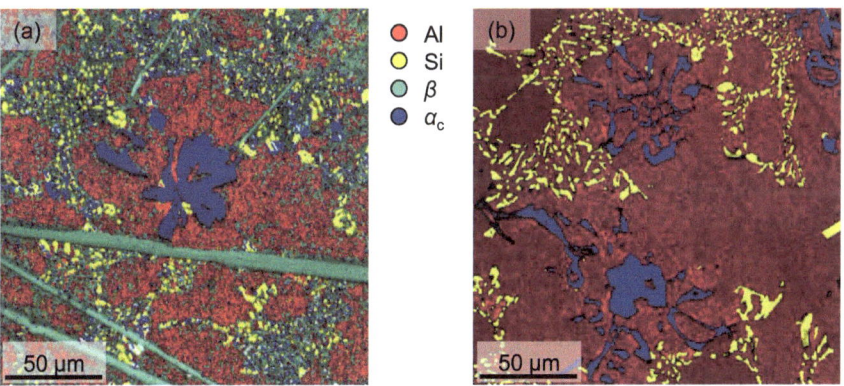

Fig. 25.16 EBSD micrographs of representative microstructures. **a** Reference state with phase indexing by Kikuchi patterns. **b** Melt conditioned state (CF) with phase indexing by EDS composition. Color code indicated in the middle. Reproduced from [19] with permission from Elsevier

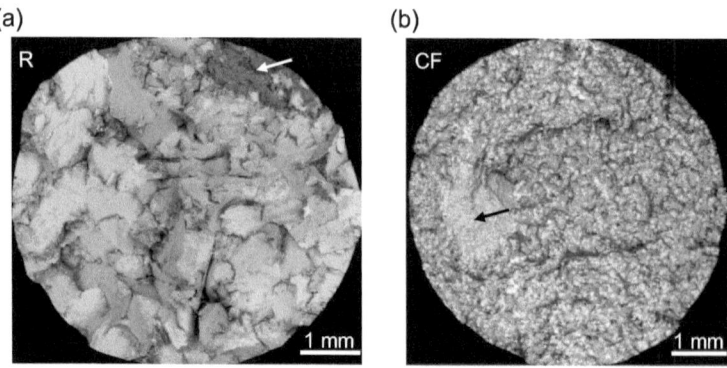

Fig. 25.17 Fracture surfaces of AlSi9Cu3 after tensile tests, **a** batch R and **b** batch CF. White arrow in **a** indicates an oxide skin. Black arrow in **b** indicates pronounced occurrence of iron. Partly reproduced from [19] with permission from Elsevier

Fig. 25.17a marks an oxide skin containing Al, Mg and O stemming from the casting process [52]. In contrast, ductile features dominated the fracture surface of tensile test specimens of batch CF (Fig. 25.17b). The black arrow indicates a pronounced occurrence of iron. Batch R achieved a mean ultimate tensile strength (UTS) of 118 ± 8 MPa and failed completely brittle. Batch CF, instead, achieved an UTS of 238 ± 2 MPa and a remarkable elongation at fracture (A) of about 2% exceeding the required minimum value of the standard EN 1706 of $A < 1\%$. Under uniaxial tensile load, the α_c phase in Chinese script morphology, which was formed due to the melt conditioning, seems to play a minor role. More detailed investigations as well as the stress–strain curves of batches R and CF can be found elsewhere [19].

The harmful behavior of β phase plates under uniaxial tensile load is distinctively pronounced and well investigated [11, 53]. However, there is a lack of knowledge about the behavior of intermetallic phases in aluminum under fatigue loading. To address this lack of information, X-ray microtomography (μCT) scans before and after ultrasonic fatigue testing were performed. Phase segmentation based on a gray scale threshold value was not applicable for the current material, because the gray scales of respective phases were too close to be segmented. Thus, from the μCT data sets, the fatigue crack was segmented by machine learning (ML) techniques using Trainable Weka Segmentation (TWS) [54]. This procedure of fatigue crack segmentation is schematically shown in Fig. 25.18.

Starting from slices of an identical position within the USFT specimen before (*pre* USFT, Fig. 25.18a) and after (*post* USFT, Fig. 25.18b) fatigue testing, ML algorithms were trained based on the slice *post* USFT in several iteration steps until a satisfactory (*good*) segmentation of the phases was achieved. The result of the segmentation was provided for each image component (phases, fatigue crack) as a probability gray scale map (Fig. 25.18c–f). The lighter a pixel the higher the probability that it belongs to the respective phase. These maps were then binarized and superimposed with any other condition. In Fig. 25.18, the fatigue crack is exemplarily binarized and superimposed

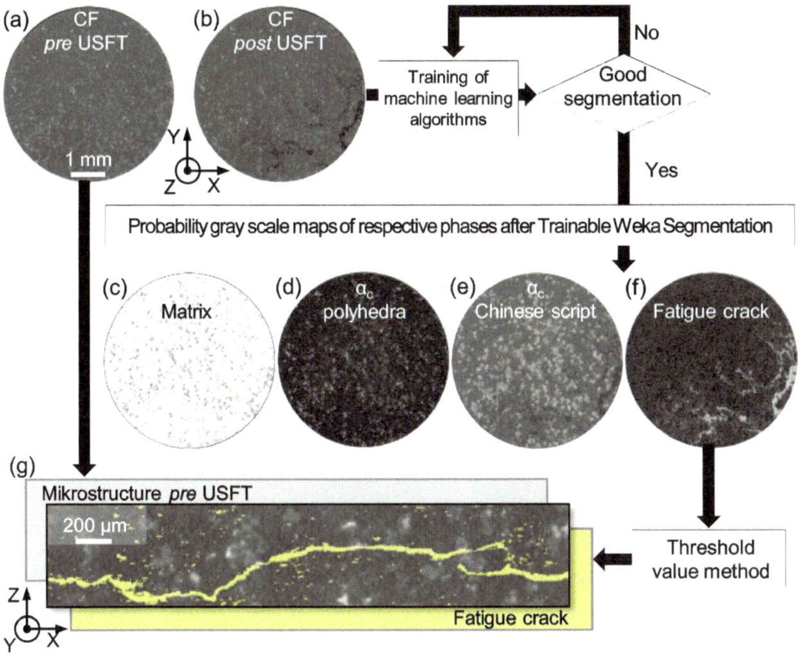

Fig. 25.18 Scheme for superposition of the segmented fatigue crack with the initial microstructure *pre* USFT. **a** µCT image in initial state *pre* USFT. **b** µCT image *post* USFT. Image portions segmented by ML algorithms of TWS as probability maps in gray scales: **c** Al matrix, **d** polyhedral α_c phase, **e** α_c phase in Chinese script morphology, **f** fatigue crack. **g** Superposition of the fatigue crack (yellow) with microstructure *pre* USFT. Reproduced from [55]

with the microstructure of the *pre* USFT condition (Fig. 25.18g). By the help of this superposition it became visible, that the fatigue crack path in batch CF mainly proceeded through the Al matrix and occasionally through the α_c phase in Chinese script morphology. The polyhedral α_c phase, conversely, was mostly bypassed.

In batch R, the high amount of brittle β phase plates yielded locally fast crack growth [56]. This led to a large scatter in the S–N curve [19]. Thus, the β phase was found to be the most detrimental phase under fatigue loading and caused a wide spread in the fatigue lifetime. However, the influence was less pronounced than expected in view of the behavior under tensile stress.

To further enhance the understanding of the behavior of the β phase under mechanical loading, in particular fatigue, the µCT data sets were analyzed by means of digital volume correlation (DVC) [57]. By these investigations the strain localizations around the fatigue cracks in batches R and CF were evaluated. It was found that the radius of strain localization around the fatigue crack was larger in batch R containing the brittle β phase than in batch CF. This is due to the random orientation of β phase plates acting as crack guidance, which, therefore, promoted crack deflection and increased the area of strain localization. The facilitation of the fatigue crack propagation by the brittle behavior of the β phase might, therefore, be compensated by a significant crack deflection along the loading axis reducing the harmful character of the β phase under fatigue loading.

25.6 Conclusions

This chapter comprised the analysis of inclusions in steel as well as in an Al-Si alloy depending on the melt treatment, filtration technique and the crucible material. (i) Different crucible materials were applied to the case hardening steel 18CrNiMo7-6. (ii) Immersion and/or flow-through metal melt filtration were utilized for the quenched and tempered steel 42CrMo4. (iii) Metal melt conditioning and filtration were performed for the Al-Si alloy AlSi9Cu3. The following general conclusions can be drawn regarding the influence of these treatments on the inclusions in the respective material:

- In batches of steel 18CrNiMo7-6, treated within crucibles from A-C, AZT-C and AZT-C-n, sulfides were detected as well as duplex inclusions with a core of different oxides and a sulfidic shell.
- The total oxygen content in steel 18CrNiMo7-6 significantly affected the size distribution of NMIs. While batch A-C with a low oxygen content revealed a log-normal size distribution, batches AZT-C and AZT-C-n with higher amounts of oxygen showed a bimodal inclusion size distribution.
- Immersion as well as flow-through filtration of 42CrMo4 steel were found to significantly affect the NMIs. Depending on the filter chemistry (reactive/active), the total contents of oxygen and aluminum were changed. After immersion treatment, all batches had a low Al:O ratio, which caused formation of plate-like alumina inclusions. Batches filtered with flow-through filters had a significantly higher aluminum content than batches with immersion filtration and showed exclusively plate-like alumina inclusions, too. Only in two reference batches without filtration, the aluminum supersaturation was high enough to form dendrite alumina inclusions.
- A lower temperature during deoxidation of 42CrMo4 steel was found to cause a higher number of nucleation sites for alumina inclusions, which shifts the inclusion size distribution towards smaller sizes.
- In 42CrMo4 steel, batch F_{AC5} as well as batches in which a filter coating of calcium aluminate was applied ($I_{CA\_}F_A$, I_{CA6}) revealed the formation of multiphase inclusions. These multiphase inclusions entrap smaller particles of alumina and titanium oxide in a matrix of spessartine $Mn_3Al_2Si_3O_{12}$. Although the total amount of NMIs increased due to multiphase inclusions, their effects on mechanical properties of the steel were likely to be negligible due to their small size, round-shaped morphology, and mechanical properties as verified by nanoindentation.

- If clusters of nonmetallic inclusions are formed in steel, they must be evaluated as a coherent inclusion, taking into account a critical distance, in order to correctly assess their harmfulness under mechanical stress.
- Melt conditioning with Mn and subsequent melt filtration of an Al-Si-alloy was found to significantly reduce the iron content and caused the formation of less detrimental α_c phase instead of plate-like β phase.
- For intermetallic phases in aluminum alloys, phase segmentation from μCT data sets by machine learning algorithms of Trainable Weka Segmentation enabled investigations on the behavior of individual phases under mechanical loading. Although this technique was applied to analyze a fatigue crack path, it is not limited to the analysis of this type of mechanical loading.

Acknowledgements This work was funded by the Deutsche Forschungsgemeinschaft (DFG, German Research Foundation)—Project-ID 169148856—SFB 920, subproject C04. The authors gratefully acknowledge the work and research of Dr.-Ing. Dominik Krewerth, Alexander Schmiedel, Johannes Gleinig and Tim Lippmann providing a major contribution to the present chapter. Dr-Ing. Mykhaylo Motylenko and Ralf Ditscherlein are acknowledged for performing TEM analyses and μCT investigations, respectively. In addition, the authors would like to thank Mrs. Birgit Witschel for her extensive metallographic work and Mrs. Astrid Leuteritz for FIB preparation.

References

1. A.L.V. Da Costa e Silva, J. Mater. Res. Technol. **7**(3), 283 (2018). https://doi.org/10.1016/j.jmrt.2018.04.003
2. L. Zhang, B.G. Thomas (eds.), *Inclusions in Continuous Casting of Steel.* XXIV National Steelmaking Symposium, Morelia, Mich, Mexico, 2003
3. R.P. Batista, A.A. Martins, A.L.V.d. Costa e Silva, J. Min. Metall. B Metall. **53**(3), 357 (2017). https://doi.org/10.2298/JMMB170730047P
4. L.D. Way, Mater. Sci. Technol. **17**(10), 1175 (2001). https://doi.org/10.1179/026708301101509142
5. M. Seleznev, S. Henschel, E. Storti, C.G. Aneziris, L. Krüger, A. Weidner, H. Biermann, Adv. Eng. Mater. **50**(22), 1900540 (2020). https://doi.org/10.1002/adem.201900540
6. H. Becker, T. Bergh, P.E. Vullum, A. Leineweber, Y. Li, Materialia **5**, 100198 (2019). https://doi.org/10.1016/j.mtla.2018.100198
7. L. Liu, A.M.A. Mohamed, A.M. Samuel, F.H. Samuel, H.W. Doty, S. Valtierra, Metall. Mater. Trans. A **40**(10), 2457 (2009). https://doi.org/10.1007/s11661-009-9944-8
8. J.A. Taylor, Procedia Mater. Sci. **1**, 19 (2012). https://doi.org/10.1016/j.mspro.2012.06.004
9. R.R. McCullough, J.B. Jordon, P.G. Allison, T. Rushing, L. Garcia, Int. J. Fatigue **119**, 52 (2019). https://doi.org/10.1016/j.ijfatigue.2018.09.023
10. S. Seifeddine, S. Johansson, I.L. Svensson, Mater. Sci. Eng. A **490**(1), 385 (2008). https://doi.org/10.1016/j.msea.2008.01.056
11. S. Ferraro, G. Timelli, Metall. Mater. Trans. B **46**(2), 1022 (2015). https://doi.org/10.1007/s11663-014-0260-3
12. K. Raiber, P. Hammerschmid, D. Janke, ISIJ Int. **35**(4), 380 (1995). https://doi.org/10.2355/isijinternational.35.380
13. S. Ali, R. Mutharasan, D. Apelian, Metall. Trans. B **16**(4), 725 (1985). https://doi.org/10.1007/BF02667509

14. S. Ogibayashi, *Advances in Technology of Oxide Metallurgy*. Nippon Steel (1994)
15. H. Becker, A. Thum, B. Distl, M.J. Kriegel, A. Leineweber, Metall. Mater. Trans. A **49**(12), 6375 (2018). https://doi.org/10.1007/s11661-018-4930-7
16. H.L. de Moraes, J.R.d. Oliveira, D.C.R. Espinosa, J.A.S. Tenório, Mater. Trans., JIM **47**(7), 1731 (2006). https://doi.org/10.2320/matertrans.47.1731
17. S. Ferraro, A. Bjurenstedt, S. Seifeddine, Metall. Mater. Trans. A **46**(8), 3713 (2015). https://doi.org/10.1007/s11661-015-2942-0
18. B.G. Bartosiaki, J.A.M. Pereira, W.V. Bielefeldt, A.C.F. Vilela, J. Mater. Res. Technol. **4**(3), 235 (2015). https://doi.org/10.1016/j.jmrt.2015.01.008
19. R. Wagner, M. Seleznev, H. Fischer, R. Ditscherlein, H. Becker, B.G. Dietrich, A. Keßler, T. Leißner, G. Wolf, A. Leineweber, U.A. Peuker, H. Biermann, A. Weidner, Mater. Charact. **174**, 111039 (2021). https://doi.org/10.1016/j.matchar.2021.111039
20. C.G. Aneziris, S. Dudczig, M. Emmel, H. Berek, G. Schmidt, J. Hubalkova, Adv. Eng. Mater. **15**, 46 (2013). https://doi.org/10.1002/adem.201200199
21. S. Dudczig, C.G. Aneziris, M. Emmel, G. Schmidt, J. Hubalkova, H. Berek, Ceram. Int. **40**(10), 16727 (2014). https://doi.org/10.1016/j.ceramint.2014.08.038
22. R. Wagner, A. Schmiedel, S. Dudczig, C.G. Aneziris, O. Volkova, H. Biermann, A. Weidner, Adv. Eng. Mater. **24**(2), 2100640 (2022). https://doi.org/10.1002/adem.202100640
23. E. Storti, M. Farhani, C.G. Aneziris, C. Wöhrmeyer, C. Parr, Steel Res. Int. **88**(11), 1700247 (2017). https://doi.org/10.1002/srin.201700247
24. A. Schmidt, A. Salomon, S. Dudczig, H. Berek, D. Rafaja, C.G. Aneziris, Adv. Eng. Mater. **19**(9), 1700170 (2017). https://doi.org/10.1002/adem.201700170
25. B.G. Dietrich, H. Becker, M. Smolka, A. Keßler, A. Leineweber, G. Wolf, Adv. Eng. Mater. **19**(9), 1700161 (2017). https://doi.org/10.1002/adem.201700161
26. Y. Murakami, *Metal Fatigue* (Elsevier, Oxford, 2002)
27. W.C. Oliver, G.M. Pharr, J. Mater. Res. **7**(6), 1564 (1992). https://doi.org/10.1557/JMR.1992.1564
28. R. Lehnert, A. Weidner, M. Motylenko, H. Biermann, Adv. Eng. Mater. **21**(5), 1800801 (2019). https://doi.org/10.1002/adem.201800801
29. J. Schindelin, I. Arganda-Carreras, E. Frise, V. Kaynig, M. Longair, T. Pietzsch, S. Preibisch, C. Rueden, S. Saalfeld, B. Schmid, J.-Y. Tinevez, D.J. White, V. Hartenstein, K. Eliceiri, P. Tomancak, A. Cardona, Nat. Methods **9**(7), 676 (2012). https://doi.org/10.1038/nmeth.2019
30. C.A. Schneider, W.S. Rasband, K.W. Eliceiri, Nat. Methods **9**(7) (2012). https://doi.org/10.1038/nmeth.2089
31. I. Arganda-Carreras, V. Kaynig, C. Rueden, J. Schindelin, A. Cardona, H.S. Seung, *Trainable_ Segmentation: Release V3.1.2* (Zenodo2016)
32. J. Gleinig, A. Weidner, J. Fruhstorfer, C.G. Aneziris, O. Volkova, H. Biermann, Metall. Mater. Trans. B **50**, 337 (2019). https://doi.org/10.1007/s11663-018-1431-4
33. X. Zhang, L. Zhang, W. Yang, Y. Zhang, Y. Ren, Y. Dong, Metall. Res. Technol. **114**(1), 113 (2017). https://doi.org/10.1051/metal/2016056
34. S. Henschel, J. Gleinig, T. Lippmann, S. Dudczig, C.G. Aneziris, H. Biermann, L. Krüger, A. Weidner, Adv. Eng. Mater. **19**(9), 1700199 (2017). https://doi.org/10.1002/adem.201700199
35. M. Seleznev, E. Merson, A. Weidner, H. Biermann, Int. J. Fatigue **126**, 258 (2019). https://doi.org/10.1016/j.ijfatigue.2019.05.011
36. M. Seleznev, J. Gleinig, K.Y. Wong, A. Weidner, H. Biermann, Procedia Struct. Integrity **13**, 2071 (2018). https://doi.org/10.1016/j.prostr.2018.12.206
37. R. Wagner, R. Lehnert, E. Storti, L. Ditscherlein, C. Schröder, S. Dudczig, U.A. Peuker, O. Volkova, C.G. Aneziris, H. Biermann, A. Weidner, Mater. Charact. **193**, 112257 (2022). https://doi.org/10.1016/j.matchar.2022.112257
38. R.B. Snow, J. Am. Ceram. Soc. **26**, 11 (1943). https://doi.org/10.1111/j.1151-2916.1943.tb15177.x
39. H. Zhang, C. Liu, Q. Lin, B. Wang, X. Liu, Q. Fang, Metall. Mater. Trans. B **50**(1), 459 (2019). https://doi.org/10.1007/s11663-018-1451-0

40. J.Y. Huang, H. Yasuda, H. Mori, J. Am. Ceram. Soc. **83**(2), 403 (2000). https://doi.org/10.1111/j.1151-2916.2000.tb01204.x
41. DIN Deutsches Institut für Normung e.V., *Heat-Treatable Steels, Alloy Steels and Free-Cutting Steels,* 77th edn. Beuth Verlag, Berlin DIN EN ISO 683-2 (2018)
42. K. Wasai, K. Mukai, A. Miyanaga, ISIJ Int. **42**(5), 459 (2002). https://doi.org/10.2355/isijinternational.42.459
43. M. Seleznev, K.Y. Wong, D. Stoyan, A. Weidner, H. Biermann, Steel Res. Int. **89**(11), 1800216 (2018). https://doi.org/10.1002/srin.201800216
44. H. Yin, H. Shibata, T. Emi, M. Suzuki, ISIJ Int. **37**(10), 936 (1997). https://doi.org/10.2355/isijinternational.37.936
45. W. Tiekink, R. Boom, A. Overbosch, R. Kooter, S. Sridhar, Ironmaking Steelmaking **37**(7), 488 (2010). https://doi.org/10.1179/030192310X12700328925822
46. R. Dekkers, B. Blanpain, P. Wollants, F. Haers, C. Vercruyssen, B. Gommers, Ironmaking Steelmaking **29**(6), 437 (2002). https://doi.org/10.1179/030192302225004584
47. E. Steinmetz, C. Andreae, Steel Res. **62**(2), 54 (1991). https://doi.org/10.1002/srin.199101250
48. X. Deng, C. Ji, Y. Cui, Z. Tian, X. Yin, X. Shao, Y. Yang, A. McLean, Ironmaking Steelmaking **44**(10), 739 (2017). https://doi.org/10.1080/03019233.2017.1368958
49. C. Lu, Y.-W. Mai, P.L. Tam, Y.G. Shen, Philos. Mag. Lett. **887**(6), 409 (2007). https://doi.org/10.1080/09500830701203156
50. W.G. Mao, Y.G. Shen, C. Lu, Scr. Mater. **65**(2), 54 (2011). https://doi.org/10.1016/j.scriptamat.2011.03.022
51. R. Wagner (2020). https://doi.org/10.25532/OPARA-87
52. L. Lattanzi, A. Fabrizi, A. Fortini, M. Merlin, G. Timelli, Procedia Struct. Integrity **7**, 505 (2017). https://doi.org/10.1016/j.prostr.2017.11.119
53. J.A. Taylor, 35th Australian Foundry Institute National Conference, **148** (2004)
54. I. Arganda-Carreras, V. Kaynig, C. Rueden, K.W. Eliceiri, J. Schindelin, A. Cardona, H. Sebastian Seung, Bioinformatics **33**(15), 2424 (2017). https://doi.org/10.1093/bioinformatics/btx180
55. R. Wagner, A. Weidner, M. Seleznev, H. Fischer, R. Ditscherlein, A. Keßler, T. Leißner, G. Wolf, U.A. Peuker, H. Biermann, in *Tagung Werkstoffprüfung. Werkstoffe und Bauteile auf dem Prüfstand.* Berlin, online, 2020 (Deutscher Verband für Materialforschung und –prüfung e.V.), 129. https://doi.org/10.48447/WP-2020-054
56. S. Seifeddine, I.L. Svensson, Metallurgical Science and Technology 27–1 (2009)
57. R. Wagner, E. Noack, R. Ditscherlein, T. Leißner, U.A. Peuker, H. Biermann, A. Weidner, in *Werkstoffe und Bauteile auf dem Prüfstand.* Tagung Werkstoffprüfung, Aachen, online, 2021 (Stahlistitut VDEh, Düsseldorf, 2021), 283

Chapter 26
A Numerical Investigation of Heat Generation Due to Dissipation in Ultrasonic Fatigue Testing of 42CrMo4 Steel Employing Thermography Data

Michael Koster, Alexander Schmiedel, Ruben Wagner, Anja Weidner, Horst Biermann, Michael Budnitzki, and Stefan Sandfeld

26.1 Introduction

In ultrasonic fatigue testing non-metallic inclusions in steels can cause fracture under cyclic loading, where crack initiation and propagation in the bulk of samples induce a local temperature increase, see e.g. [1–3]. Further, heating can be observed along the gauge length of samples without failure. There exist studies to characterize dissipation from a quantitative and qualitative point of view, based on thermomechanics, using thermography data from ultrasonic fatigue testing, see e.g. [4–7]. In Boulanger et al. [7] and the references therein, the heat equation is used as a basis, which includes terms specifying different sources of heating of the material, capturing intrinsic dissipation, thermoelastic coupling, microstructural coupling and an external heat source term accounts for heat exchange with the environment. By introducing certain

M. Koster · M. Budnitzki (✉) · S. Sandfeld
Institute for Advanced Simulation (IAS-9: Materials Data Science and Informatics), Forschungszentrum Jülich GmbH, 52428 Jülich, Germany
e-mail: m.budnitzki@fz-juelich.de

S. Sandfeld
Jülich-Aachen Research Alliance (JARA-CSD), Forschungszentrum Jülich GmbH, 52428 Jülich, Germany

Chair of Materials Data Science and Materials Informatics, Faculty 5–Georesources and Materials Engineering, RWTH Aachen University, 52056 Aachen, Germany

A. Schmiedel · R. Wagner · A. Weidner · H. Biermann
Institute of Materials Engineering, Technische Universität Bergakademie Freiberg, Gustav-Zeuner-Straße 5, 09599 Freiberg, Germany

© The Author(s) 2024   679
C. G. Aneziris and H. Biermann (eds.), *Multifunctional Ceramic Filter Systems for Metal Melt Filtration*, Springer Series in Materials Science 337,
https://doi.org/10.1007/978-3-031-40930-1_26

assumptions these contributions can be condensed to just one collective heat source, see [7].

In all of these analyses the evaluation of the heat equation is based on a reduction of the spatial dimension, treating boundary and initial conditions via introduction of constants, to be determined using experimental data, and assessing dissipation in a simplified manner. As a diffusion-type equation may be subject to boundary conditions of Dirichlet, Neumann and Robin types, see e.g. [8, 9], the intricacies of these formulations are avoided in the investigations introduced before. In contrast, in Yang et al. [6] natural convection and radiation are introduced in the 1D heat equation, based on [10]. In this procedure boundary conditions are applied in terms of parameters that are fitted to experimental temperature data. Furthermore, a finite element computation of pure heat transfer is performed, accounting for heating induced by dissipation, applying a constant inward heat flux at the gauge length portion of the model geometry.

In the present study a fully-coupled linear thermoelastic model is used to capture a complete pulse phase of an ultrasonic fatigue experiment, excluding fracture and major plastic effects. Numerical simulations have been performed on the JURECA general-purpose supercomputer [11]. The experiment comprises 15 pulse–pause phases in total, each of which exhibits approximately 1800 load cycles. Each load cycle is modeled applying sinusoidal excitation. Thermal boundary conditions are derived from experimental data. Further, from the difference of a computational reference data set and experimental temperature data the intensity and geometry of a volumetric heat source are deduced, applying a fitting procedure. This allows for a detailed assessment of the evolution of heat generation, interpreted as dissipation in the bulk of the sample.

26.2 Experimental Setup

The heating in steel as a consequence of dissipative effects due to high frequency and fully reversed loading (R = -1) was studied, using an ultrasonic fatigue testing equipment (UFTE, University of Natural Resources and Life Sciences, Vienna, Austria) for very high cycle fatigue (VHCF) experiments. The UFTE, as shown in Fig. 26.1, is operating by resonant vibration of the sample at a frequency of about 20 kHz. An ultrasonic transducer is generating an ultrasonic wave which is transferred via an amplification horn (titanium) to increase the amplitude of the ultrasonic wave in the fatigue sample. The length of the sample was designed via a modal analysis of the resonant conditions of the investigated material. Young's modulus E, Poisson's ratio v and the velocity of sound for longitudinal waves c_L in the material under investigation are needed. The ultrasonic wave propagating through the sample is reflected on the free end of the sample generating a standing wave due to the resonant conditions resulting in longitudinal fatigue of the sample. Thus, the resulting stress amplitudes are zero at both end faces of the sample and have a maximum value at the centre of the gauge length, illustrated in Fig. 26.1. This technique allows for the application of up

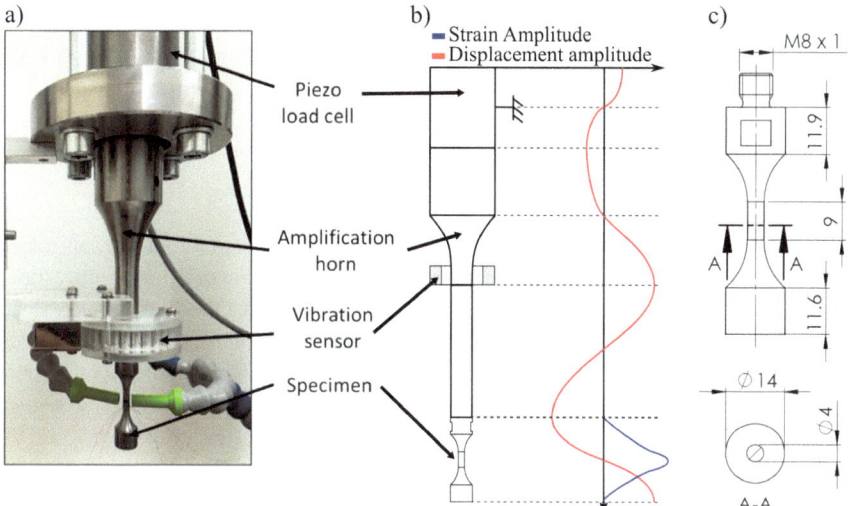

Fig. 26.1 Ultrasonic fatigue testing setup with **a** overview, **b** corresponding stress and displacement amplitude distribution along the entire loading path and **c** sample geometry

to 10^9 cycles within a reasonably short time. The temperature of the sample increases during resonant vibration due to material damping (dissipation), the magnitude of which depends on the applied stress amplitude and material behaviour. Therefore, a favourable pulse/pause ratio has to be chosen to limit the temperature increase to 10 K over the entire experiment, i.e. vibration of the sample and rest alternate during the pulse and the pause stage, respectively. In this study, the influence of compressed air cooling does not have to be considered and, therefore, a pulse/pause ratio was chosen, which led to an effective test frequency of about $f_{eff} = 1$ kHz.

The test material was 42CrMo4 steel in quenched and tempered condition, see [12]. The microstructure after quenching and tempering treatment consists of tempered martensite. The sample geometry designed according to the resonant condition at 19.3 kHz using material parameters $E = 210$ GPa, $\upsilon = 0.285$ and $c_L = 5179$ m/s is shown in Fig. 26.1 with a constant gauge length. A parallel gauge length of 9 mm and a diameter of 4 mm were used.

Ultrasonic fatigue testing is a vibration-controlled method. Thus, the stress amplitudes applied for fatigue tests have to be calibrated based on the chosen vibration amplitudes. For this purpose, two strain gages were glued on opposite sides at the centre of the gauge length. In the next step, the vibration amplitude, which correlates with the amplitude of the generated ultrasonic wave of the ultrasonic fatigue testing equipment, was set at three distinct supporting points and the resulting strains ε at the strain gages were recorded. In addition, displacement measurements Δx were performed at the free end of the samples using a fibre optic sensor (MTI 2100 Fotonic-Sensor, MTI Instruments Inc., New York, USA). Consequently, a linear relation of strain ε and displacement Δx, respectively, vs. vibration amplitude VA was obtained.

Then, the applied stress σ can be calculated from strain ε by Hooke's law using the value of Young's modulus of $E = 210$ GPa. Finally, both linear relations, i.e. ε vs. VA and Δx vs. VA, were merged into the following linear equation with $[\sigma] = [E\,\varepsilon] = [B] = $ MPa, $[\Delta x] = \mu$m and $[A] = $ MPa/μm:

$$E\varepsilon = A\Delta x - B .$$

Here, $A = 9.42$ MPa/μm and $B = 1.39$ MPa were used for cylindrical samples.

Infrared thermographic measurements were conducted in situ during ultrasonic fatigue loading. To this end, a long wave range (7 to 14 μm) thermal camera Vario hr head (Infratec Dresden) with a 640 × 240 pixel focal plane array detector was used, enabling a lateral resolution of 25 μm at a thermal resolution of 0.03 K. Thermography measurements at a sampling rate of 50 Hz were performed at higher stress levels in order to observe measurable heat output during ultrasonic load. To ensure a defined value of the coefficient of emission of 0.96, black lacquer (Dupli-Color® SUPERTHERM) was applied to the samples. Since the thermal camera just captured the front view of the sample, a special mirror thermography was adopted to capture the entire circumference of the sample, see [1]. For this purpose, two mirror-polished aluminium plates were placed at an angle of 45° behind the sample, shown on the left in Fig. 26.2. Thus, three thermograms were recorded at the same time, i.e. the real thermogram of the front view of the camera and two reflectograms of the infrared radiation of the two mirrors behind the sample. The temperature deviation between the front thermogram and reflectograms was less than 2 K. Finally, a sample was subject to the desired stress amplitude for a relatively short period, i.e. 100 ms (see Sect. 26.3.2) and the thermogram. The considerations, therefore, refer to heat dissipation during the period of the ultrasonic pulse rather than from the increase of temperature due to the crack growth and final fracture, as shown in the reflectograms in Fig. 26.2.

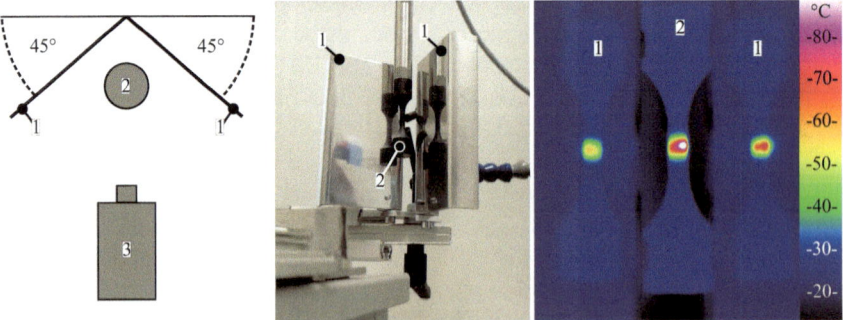

Fig. 26.2 In situ infrared mirror thermography setup and associated infrared thermograms. The mirrors as well as their reflections are marked as (1). The sample is marked as (2) and the infrared camera is marked as (3). As an example, the reflectogram of a 42CrMo4 sample showing a hot spot due to an internal defect was used

26.3 Modeling Approach

The goal of this study is to use experimental information combined with a well-established constitutive theory to mimic temperature evolution observed in an ultrasonic fatigue testing setup of 42CrMo4 steel samples, discussed in Sect. 26.2. Section 26.3.1 provides the model's equations as well as the corresponding weak form and time discretization. Section 26.3.2 deals with the kind of data obtained in experiments and with a way to explore this data in order to gain some insight into the nature of the temperature evolution. In Sect. 26.3.3 a closer look at the data and an in-depth explanation of the methods applied are presented. To model material behavior, different quantities are derived from thermography data and employed via initial and boundary conditions. The results of finite element computations of the temperature evolution during pulse and pulse-decay phases are compared with the experimental outcome.

26.3.1 Mechanical Model

The mechanical framework under consideration is linear thermoelasticity. Here, just a very brief outline of the underlying relations is given. For a comprehensive derivation see e.g. [13]. The free energy density is of the form

$$\rho \psi = \mu \operatorname{tr}\left(\hat{\boldsymbol{\varepsilon}}^2\right) + \frac{1}{2} K (\operatorname{tr} \boldsymbol{\varepsilon})^2 - 3K\alpha \vartheta \operatorname{tr} \boldsymbol{\varepsilon} - \frac{1}{2}\rho c \vartheta^2, \tag{26.1}$$

where ρ is the mass density, μ denotes the shear modulus and K is the bulk modulus of the material. The parameter α is referred to as the coefficient of thermal expansion, c is called the specific heat capacity. The symbol $\boldsymbol{\varepsilon}$ denotes the small strain tensor and $\hat{\boldsymbol{\varepsilon}}$ represents its deviatoric part. The difference between the local temperature θ and the constant reference temperature θ_0 is introduced via the variable $\vartheta = \theta - \theta_0$. The elastic energy contribution is represented by the first two terms on the right-hand side of (26.1). The third term is a coupling term relating elasticity and temperature and the last term in the free energy is exclusively associated with temperature. Based on the free energy the corresponding constitutive relations for entropy and stress can be derived. Using the Gibbs relation and introducing the Fourier model of isotropic heat conduction

$$\mathbf{q} = -\lambda \nabla \vartheta,$$

where λ is the coefficient of heat conduction, the heat equation can be written as

$$\rho c_d \dot{\vartheta} + \theta_0 3K\alpha \operatorname{tr} \dot{\boldsymbol{\varepsilon}} = \lambda \nabla \cdot \nabla \vartheta + \rho r. \tag{26.2}$$

The mechanical contribution is defined via Cauchy's equation of equilibrium

$$\nabla \cdot \boldsymbol{\sigma} + \rho \mathbf{b} = 0, \tag{26.3}$$

where $\boldsymbol{\sigma}$ denotes the stress tensor and $\mathbf{b}$ is the body force vector per unit mass. From the right-hand side of (26.3) it can be seen that inertia effects will not be considered in what follows, see e.g. [4, 7]. To implement this model within the framework of the finite element library FEniCS, see [14, 15], the weak form of Eqs. (26.2) and (26.3) is needed. Therefore, the equations are multiplied by test functions and integration is performed over the problem domain. For Eq. (26.2) the associated test functions are denoted $\delta\vartheta$ and the ϑ are referred to as the trial functions. Then, performing integration by parts on terms containing second-order spatial derivatives and applying the divergence theorem yields the variational statement with boundary terms, associated with Neumann and Robin boundary conditions. As (26.2) contains time derivatives of the solution variables, a time discretization procedure has to be applied; here the backward Euler scheme is used, see e.g. [16]. The fully coupled bilinear form is given by

$$\begin{aligned}a((\vartheta, \mathbf{u}), (\delta\vartheta, \delta\mathbf{u})) &= \int_\Omega \rho\, c_d\, \vartheta\, \delta\vartheta\, dV + \int_\Omega \theta_0\, 3K\alpha \operatorname{tr}\boldsymbol{\varepsilon}\, \delta\vartheta\, dV \\ &+ \sum_i \int_{\partial\Omega_R^i} \Delta t(\cdot)\vartheta^{(\star)}\, \delta\vartheta\, da + \int_\Omega \Delta t\, \lambda\nabla\vartheta \cdot \nabla\delta\vartheta\, dV \\ &+ \int_\Omega \left[2\mu\,\hat{\boldsymbol{e}} + K \operatorname{tr}(\boldsymbol{\varepsilon})\mathbf{I} - 3K\alpha\,\vartheta\mathbf{I}\right] : \delta\boldsymbol{\varepsilon}\, dV\end{aligned}$$

and the fully coupled linear form can be expressed as

$$\begin{aligned}L(\delta\vartheta, \delta\mathbf{u}) &= \int_\Omega \rho\, c_d\, \vartheta^n\, \delta\vartheta\, dV + \int_\Omega \theta_0\, 3K\alpha \operatorname{tr}\boldsymbol{\varepsilon}^n\, \delta\vartheta\, dV \\ &- \sum_i \int_{\partial\Omega_N^i} \Delta t\, q^{n+1}\, \delta\vartheta\, da + \sum_i \int_{\partial\Omega_R^i} \Delta t(\cdot)\vartheta_\infty^{(\star)}\, \delta\vartheta\, da \\ &+ \int_\Omega \Delta t\, \rho\, r^{n+1}\, \delta\vartheta\, dV + \int_\Omega \rho\mathbf{b} \cdot \delta\mathbf{u}\, dV + \int_{\partial\Omega} \mathbf{t} \cdot \delta\mathbf{u}\, da,\end{aligned} \tag{26.4}$$

where superscript $n + 1$ denotes the current time step, n refers to the previous time step and Δt is the step size. Superscripts have been omitted for the sake of readability on trial functions. Subscripts N and R are used in the boundary terms to indicate the i terms accounting for Neumann and Robin boundary conditions, respectively, where $(\cdot) = \{h, \varepsilon\sigma\}$ and $(\star) = \{1, 4\}$ have been defined to introduce one general Robin boundary condition term, accounting for convection and radiation. The problem to be solved using the finite element method can then be expressed as

$$a((\vartheta, \mathbf{u}), (\delta\vartheta, \delta\mathbf{u})) = L(\delta\vartheta, \delta\mathbf{u}). \tag{26.5}$$

26.3.2 Experimental Data

The experimental data consist of a time series of arrays of pixel values, captured by a thermal camera, see Fig. 26.3.

It is assumed that in the experiment under investigation the temperature at the sample's surface at fixed positions along the gauge length's axis is approximately equal to the mean temperature on the associated cross sections, see e.g. [4]. Hence, temperature distribution in the sample must be the same in the front and rear views. Therefore, just a small rectangular part of each of the thermal maps, comprising the gauge length of the sample's front view, is considered. The mean temperature on cross sections is computed as the mean value of pixel values taken along a horizontal row in the sub-arrays defined for each frame.

The mean temperature values of the gauge length can be plotted versus time for each instant, illustrated in Fig. 26.3b. It can be seen that there are 15 pulse–pause phases, each having the same qualitative behavior, where a steep temperature increase during pulse operation is followed by a temperature decrease during pause mode. Each pulse phase is comprised of a pulse and pulse-decay period with a decay of oscillations. In pulse there is an excitation period of $t_{\text{pulse}} = 100\,\text{ms}$, after that there is a pulse-decay of $t_{\text{pulse-decay}} = 200\,\text{ms}$. Each pause phase has a duration of $t_{\text{pause}} = 2000\,\text{ms}$. In addition, it can be observed that the maximum temperature increases with every excitation phase, similar to what can be found in loading of Stage I type, see e.g. [16].

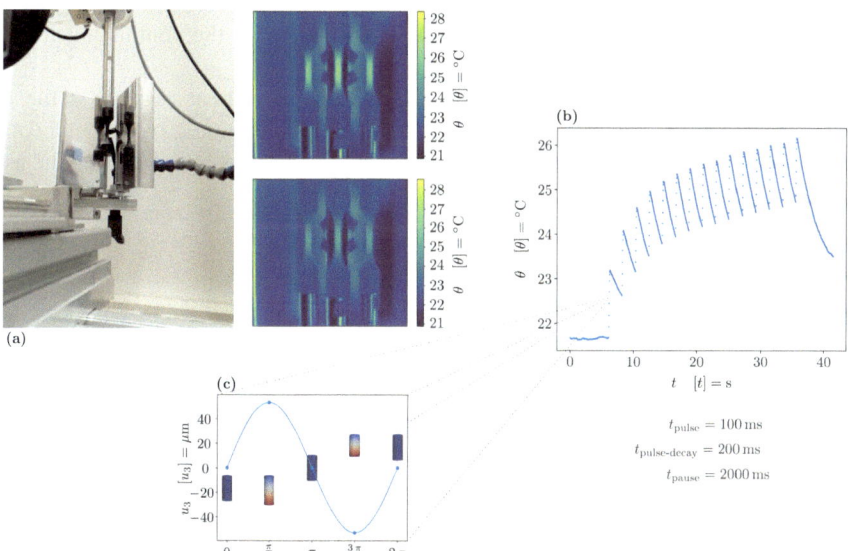

Fig. 26.3 Experimental setup with thermography plots of pulse and pause phases (**a**), time evolution of the mean temperature of the sample's gauge length (**b**) and the displacement evolution throughout a single pulse (**c**)

Table 26.1 Material parameters for 42CrMo4

Parameter	Value and units	Source
Density ρ	$7.8 \cdot 10^{-9}\,\mathrm{N\,s^2/mm^4}$	Measurement
Young's modulus E	$212 \cdot 10^3\,\mathrm{N/mm^2}$	Richter [17]
Poisson's ratio ν	0.285	Richter [17]
Specific heat capacity c_d	$453 \cdot 10^6\,\mathrm{mm^2/s^2\,K}$	Measurement
Thermal conductivity λ	$39.915\,\mathrm{N/s\,K}$	Measurement
Thermal expansion α	$11.5 \cdot 10^{-6}\,\mathrm{1/K}$	Richter [17]
Reference temperature θ_0	$293.15\,\mathrm{K}$	
Ambient temperature θ_∞	$293.15\,\mathrm{K}$	
Emissivity ε	0.96	Krewerth et al. [2]
Stefan-Boltzmann constant σ	$5.67 \cdot 10^{-11}\,\mathrm{N/s\,mm\,K^4}$	Baehr et al. [8]

After inspection of the overall temperature evolution of the sample, the third pulse–pause phase is selected as the basis for the following investigation. Therefore, the experimental data of interest will be taken from this phase. In what follows, a 1D temperature difference representation along the sample's gauge length will be used, obtained by defining equidistant cut planes orthogonal to the sample's axis. The mean temperature difference values on each of these slices is plotted against coordinate x_3. As it is presumed that temperature on the sample's cross sections is near-constant, this is a means of reduction of the 3D surface temperature at defined positions along the gauge length.

The material parameters for 42CrMo4 steel required in the continuum model outlined in Sect. 26.3.1 are given in Table 26.1.

As the temperature differences encountered in the experiment are small, a linear model formulation with constant parameters is used. In addition, experimental data can be employed in the numerical computation in terms of initial and boundary conditions.

26.3.3 Methods and Results

For the simulation only the gauge length part of the sample is considered, where a cylinder geometry with radius $r = 2\,\mathrm{mm}$ and height $h = 9\,\mathrm{mm}$ is assumed. The cylinder axis is parallel to the x_3-direction and the bottom base is at the origin of coordinates. Hence, boundary conditions may be applied at the top or the bottom base or on the lateral surface.

The mechanical boundary conditions are prescribed via displacements, based on experimental meta data. In the experiment the top surface is fixed and the bottom surface experiences a sinusoidal excitation in pulse operation. Therefore, Dirichlet boundary conditions for the bottom base of the cylinder geometry for a single pulse–pause phase can be formulated as

$$
u_b^D(t) = \begin{cases} \left[0, 0, -a \sin\left(\frac{\pi}{2} \frac{t}{\Delta t}\right)\right]^{\mathrm{T}} & \text{if } t \in \left[0, \tilde{t}\right] \\ 0 & \text{if } t \in \left(\tilde{t}, 2.3\right]. \end{cases} \tag{26.6}
$$

The starting point is set to a reference time of $t = 0$ s. In (26.6) and in what follows subscript b and superscript D are used to denote the cylinder's bottom surface at $x_3 = 0$ and Dirichlet boundary conditions, respectively. The end of excitation is not immediately obvious as in pulse-decay there still is excitation of the sample, but a value of $\tilde{t} \approx 0.18$ s can be identified as the time, where heating of the sample stops. An illustration of the deformation, according to Eq. (26.6), for a single loading cycle is given in Fig. 26.3, where the color plots of the sample indicate the magnitude of displacement in the x_3-direction. The amplitude a can be found from the experimental setup, given in Sect. 26.2. In the experiment there are approximately 1800 full load cycles per pulse phase followed by some pulse-decay. Assuming four time steps in each load cycle, this gives a total number of $n = 7200$ time steps, giving a step size of $\Delta t = \tilde{t}/n = 2.5 \cdot 10^{-5}$ s for an even distribution of cycles on $\left[0, \tilde{t}\right]$ with a constant time step value. Here, decay is not considered explicitly since there are no experimental details available. It is observed that in tension a decrease in temperature in the bulk of the sample occurred, whereas during compression the temperature increased and at the end of each loading cycle. No influence of displacements on temperature was seen, cf. [7]. At the top surface of the geometry, displacements are set to

$$
u_t^D = 0,
$$

where here and in what follows subscript t refers to the top surface of the cylinder geometry at $x_3 = 9$ mm. For the mechanical contribution to the coupled problem no additional types of boundary conditions will be explicitly prescribed and no body forces will be considered. Thus, the associated terms in the linear form (26.4) vanish.

It is not immediately obvious which kind of thermal boundary conditions can mimic material behavior as found in the experiment and how the respective quantities in the boundary expressions can be derived from thermography data. As indicated above, the sample's geometry has been modeled as a cylinder, characterizing the gauge length of the sample. Hence, thermal boundary conditions at the bottom and top of the real sample cannot be immediately employed in the computation. What is more, the top of the sample is connected to the amplification horn and the bottom of the sample is free, as pointed out in Sect. 26.2. This means that heat transfer across the top and bottom bases of the cylinder geometry is based on different mechanisms than in the experiment. In what follows, the third combined pulse and pulse–decay portion of the experiment will be analyzed to inspect the underlying heat generation and transfer mechanisms. The interval of interest starts where temperature difference increases, associated with pulse operation. Next, the different aspects of modeling will be examined in detail. For the simulation the point of departure is when the lowest temperature difference value is attained in the second cooling phase, being defined as reference time $t_0 = 0$.

Next, temperature boundary conditions will be defined, based on experimental data. Heat conduction in the bulk can be modeled using Dirichlet or Neumann boundary conditions at the bottom and the top of the cylinder geometry. For interaction with the environment, Robin conditions may be introduced at the lateral surface. It can be observed that the 1D experimental temperature profile along the gauge length is parabolic, where the maximum value can be found at the center of the sample, increasing continuously in the course of the experiment. Heating of the sample's gauge length must be due to dissipation during excitation, captured in a phenomenological fashion in the computation. From the temperature distribution along the gauge length it can be deduced that there is no inward heat flux from the sample's bottom and top boundaries. In addition, it is difficult to quantify the heat loss via outward heat flow, convection or radiation. Since temperature differences are small, it will be assumed negligible. Hence, neither Neumann nor Robin boundary conditions will be considered during pulse operation. The definition of Dirichlet boundary conditions, based on thermography data, is the more accessible alternative. To use experimental information the mean values of horizontally aligned pixel values at the desired boundaries are computed from gauge length temperature arrays for all points in time captured during pulse and part of pulse–decay operation. To use experimental boundary data in the computation, a function fit must be performed for the bottom and top boundary, respectively. The discrete mean values over time follow a sigmoid type distribution. Hence, a logistic function fit seems to be a natural choice, see e.g. [18]. The corresponding functions, used for the bottom and top boundaries, are given as

$$\vartheta_b^D(t) = \frac{L_1}{1 + e^{-k_1(t - t_1^0)}} + \vartheta_b^{\min}; \; \vartheta_t^D(t) = \frac{L_2}{1 + e^{-k_2(t - t_2^0)}} + \vartheta_t^{\min}, \qquad (26.7)$$

where the L_i are associated with the maximum values, the k_i represent the logistic growth rates and the t_i^0 are the midpoints of the curves. The plots of functions (26.7), shown in Fig. 26.4, are shifted upwards by values $\vartheta_b^{\min}$ and $\vartheta_t^{\min}$, respectively.

In addition, an initial state of the system at the reference time must be defined, where for displacements this corresponds to the undeformed configuration of the sample, i.e. $u^0 = u(x, t = 0) = 0$ holds, where here and what follows the supersript 0 indicates an initial condition. The initial temperature state is prescribed as the 1D distribution along the gauge length at $t = 0$ s. The polynomial

$$\vartheta_{\text{pulse}}^0 = \vartheta_{\text{pulse}}(x_3, t = 0) = \sum_{i=0}^{n} a_i \, x_3^i \qquad (26.8)$$

is employed to fit the initial temperature difference distribution, used in the simulation, see Fig. 26.5.

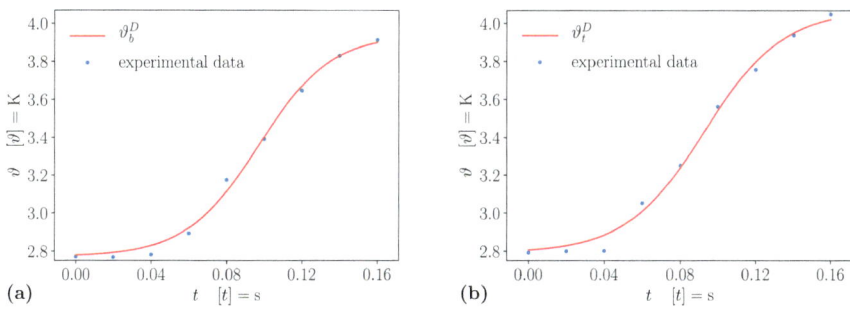

Fig. 26.4 Dirichlet temperature difference logistic fit for the **a** bottom and **b** top of the sample

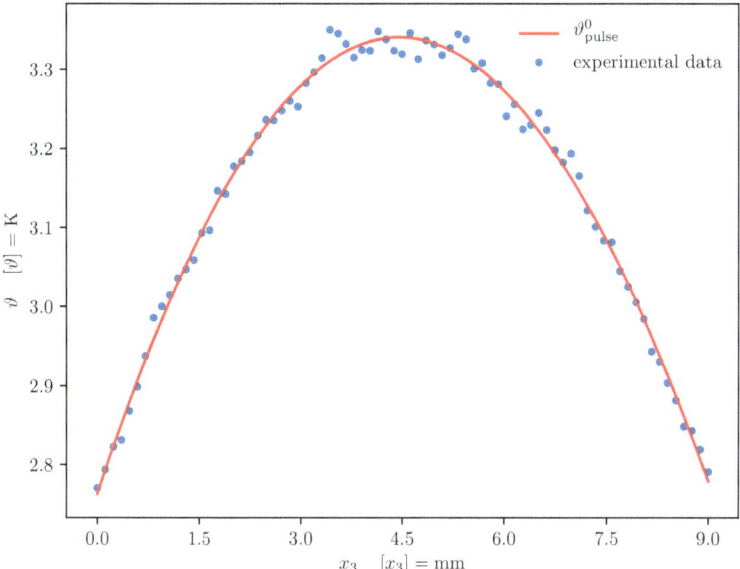

Fig. 26.5 Initial temperature difference polynomial fit for the pulse phase

Based on the reasoning in Blanche et al. [4], it is assumed that the temperature evolution observed in the experiment can be reproduced as the result of the superposition of proper boundary conditions and a volumetric heat source in the bulk of the cylinder geometry. The idea is as follows: In ultrasonic fatigue testing sample heating may have diverse characteristics. In the three-stage model in [19] the first stage is characterized by a near-linear temperature increase. After that, a plateau-like temperature evolution can be found, followed by a steep increase in temperature, observed shortly before failure. If there is failure due to non-metallic inclusions in the bulk of the sample this is accompanied by a large local temperature rise, appearing as a hot spot on the sample's surface, see e.g. [1, 2]. In the following, just heating attributed to dissipation without damage, related to Stage I, described before, will be

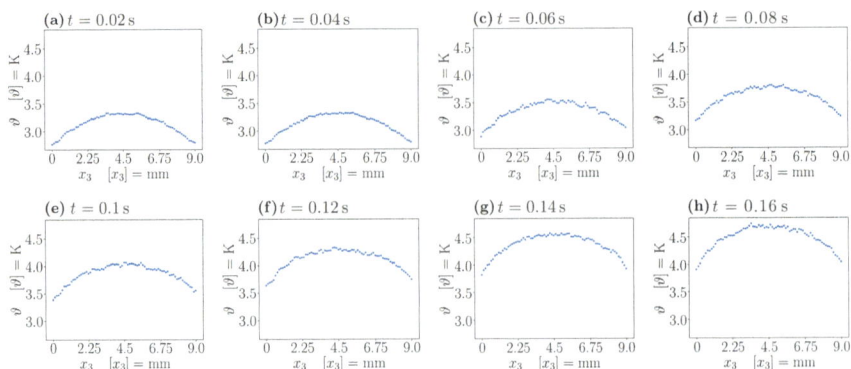

Fig. 26.6 Experimental 1D temperature difference profile evolution in the pulse phase

inspected, where the experiments discussed in Sect. 26.2 have been designed to not exhibit major plastic effects, fracture or damage.

Mimicking heating in the bulk of the sample is be done using a phenomenological approach, defining a heat source function. In the simulation the evolution of the temperature distribution in the sample is the outcome of the setup with initial and boundary conditions and the heat source in Eq. (26.4). The difference of 1D temperature profiles based on a reference computation and the parabolic 1D experimental distributions is used to deduce the associated geometry along the x_3 direction and intensity of the volumetric heat source. The experimental data series is shown in Fig. 26.6.

The reference computation is performed without prescribing any additional heating in Eq. (26.4). Thermal initial and Dirichlet boundary conditions are used as specified in Eqs. (26.8) and (26.7). Computation is done in a 3D finite element setting, where in a postprocessing step results are transformed to a 1D representation, computing mean temperature values on cross sections of the sample geometry at fixed positions, coinciding with positions of cut planes used for the experimental data. The resulting series of data points is then plotted along the x_3-axis of the cylinder. The computational reference data series is given in Fig. 26.7. It can be seen immediately that some kind of heat supply in the bulk of the sample is needed to turn these profiles into parabolic shape, as found in the experimental data.

Hence, the difference of discrete experimental and computational reference data, provided in Fig. 26.8, is used as a basis to deduce the geometry and intensity of a volumetric heat source. At the beginning of the time series, the distribution of points is noisy, but turns into a geometry with a symmetric plateau about the center of the gauge length and a sharp fall-off towards the sides. Presuming the geometry of the scatter plots shown below approximately correspond to the geometry of a heat source in the bulk of the sample in the x_3-direction, the super Gaussian formulation

$$f(x_3) = a \exp\left[-\left(x_3 - x_3^0/b\right)^{2n_1}\right] \tag{26.9}$$

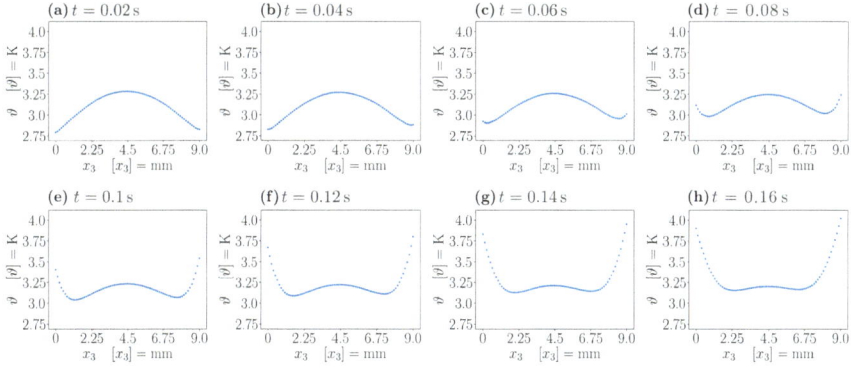

Fig. 26.7 Computational 1D reference temperature difference profile evolution

is used to model the flat-top temperature profiles, see e.g. [20]. The associated function fit is illustrated in Fig. 26.8. Function (26.9) is defined in terms of parameter a, denoting the maximum value of the associated curve, x_3^0 is the center point, b represents the value of half the width of the flat-top and the order of the super Gaussian is prescribed via $2n_1$, where parameters a and b will be determined using curve fitting. It turns out that a value of $n_1 = 4$ is a proper choice to fit the data, using (26.9), shown in Fig. 26.8.

To get some insight into the evolution of parameters a and b in the course of the pulse phase the discrete parameter values are plotted versus time in Fig. 26.9.

To identify a formulation for the intensity of the heat source, the evolution of parameter a is inspected. At the beginning of the pulse phase the heat generated in the bulk of the sample is close to zero, followed by a near-linear increase of a for $0.04\,\text{s} < t < 0.14\,\text{s}$. Eventually, there is still some increase in intensity, but at a reduced rate. From the experimental setup these characteristics seem reasonable,

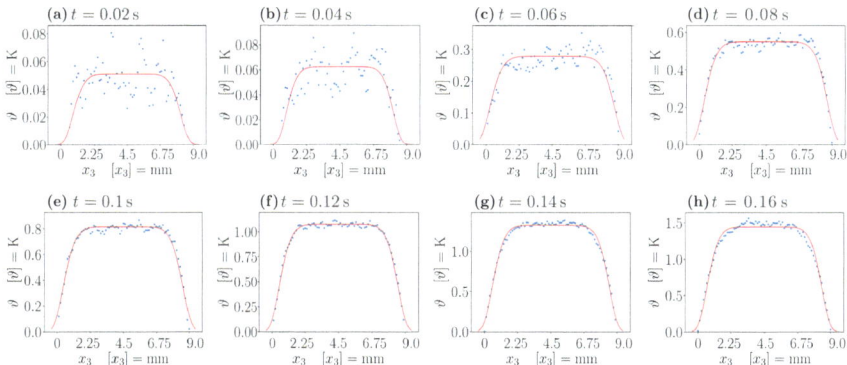

Fig. 26.8 1D experimental-computational temperature difference data and associated super Gaussian fit

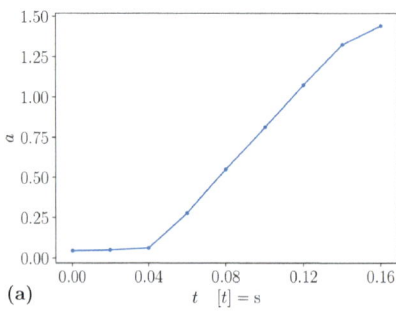
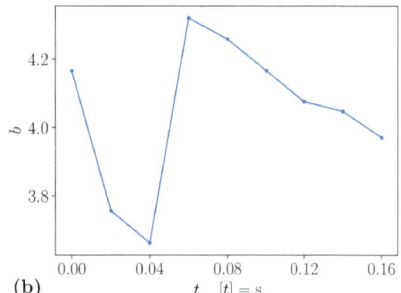

Fig. 26.9 Evolution of parameters a and b

since at the beginning of the experiment, a standing wave is induced by the testing device, where it takes some time to produce oscillations at the desired frequency. This complies with the very low amount of heating observed at the beginning of the pulse. Then, there is a steady excitation of the sample, continuously heating up the material. The delayed decrease in the rate of evolution of a after $t = 0.14$ s can be attributed to pulse-decay. Assuming a piecewise-linear time dependence of a on intervals $[t_i, t_{i+1})$, there is a correlation with the respective heat source strength, corresponding to a piecewise-constant intensity of the heat source function introduced in (26.9). To employ this relation, a normalized representation of the slopes γ_i, defined on intervals $[t_i, t_{i+1})$ for $i = 0, \ldots, 8$, is chosen, given as $\tilde{\gamma}_i = \gamma_i / \gamma_{max}$, using the maximum slope value γ_{max}. As data fitting has been performed in terms of temperature difference values, a constant factor must be introduced to relate the internal rate of heating per unit mass to the resulting temperature difference in the sample, given by

$$\check{a}_i = \tilde{\gamma}_i \, 7.3 \cdot 10^9, \text{ on } [t_i, t_{i+1}), i = 0, \ldots, 8.$$

Then, $\rho r \approx 5.8 \cdot 10^4$ W/m^3 for the maximum $\check{a}_i$, where $[\check{a}_i] = (\text{kg mm}^3)/(\text{N s}^5)$. Using [7] as a reference, this appears reasonable, compared with the value of 2.0×10^5 W/m^3 stated, where it must be borne in mind that the temperature increase encountered in this study has a value of approximately 11 K. Therefore, the piecewise constant representation

$$\check{a}(t) = \sum_{i=0}^{8} \check{a}_i \, \chi_{[t_i, t_{i+1})}$$

is introduced for the heat source intensity, with indicator functions $\chi_{(t_i, t_{i+1})}$, defined via

$$\chi_{(t_i, t_{i+1})}(t) = \begin{cases} 1 \text{ if } t \in [t_i, t_{i+1}) \\ 0 \text{ if } t \notin [t_i, t_{i+1}). \end{cases}$$

Since parameter b is immediately related to the geometry of the volumetric heat source along the cylinder axis, the interpretation of its evolution is different. The right part of Fig. 26.9 reveals the variation of b over time, where for simplicity a linear time dependence is assumed, giving

$$b(t) = \alpha_i + \beta_i(t - t_i) \text{ on} [t_i, t_{i+1}), i = 0, \dots, 8. \quad (26.10)$$

In function (26.10) the α_i denote the discrete b_i and the β_i represent the values of slope on intervals $[t_i, t_{i+1}]$. The derived relations constitute the part of the volumetric heat source definition that can be obtained from experimental thermography data. To get a spatial formulation, the 1D super Gaussian function of coordinate x_3 must be combined with the geometry of the heat source on cross sections at fixed values of x_3. Assuming that heat generation in the bulk of the sample on cut planes is constant in the interior and has a sharp fall-off towards the lateral surface a super Gaussian formulation may be employed, again, where this kind of circular distribution can be defined as

$$g(x_1, x_2) = \exp\left\{-\left[(x_1/r_c)^2 + (x_2/r_c)^2\right]^{2n_2}\right\},$$

with $r_c = 2\,\text{mm}$ the radius of the cylinder geometry and $n_2 = 8$. Eventually, combining the above results yields

$$r(\mathbf{x}, t) = \check{a}(t) \exp\left\{-\left\{\left[(x_1/r_c)^2 + (x_2/r_c)^2\right]^{2n_2} + \left(x_3 - x_3^0/b(t)\right)^{2n_1}\right\}\right\}, \quad (26.11)$$

representing the spatial and temporal evolution of heat in the bulk of the sample, induced by dissipation, see e.g. [8].

Then, using (26.11) in (26.4) with initial and boundary conditions as in the reference computation, it turns out that the resulting computational 1D temperature difference distribution along the gauge length resembles a parabolic profile and is in good agreement with the experimental result, shown in Fig. 26.10.

26.4   Summary and Outlook

In this study experimental thermography data are used in the initial and boundary conditions of the fully-coupled linear thermoelasticity model. Based on experimental data and metadata the construction of a volumetric heat source, associated with dissipation, is provided in detail. It is shown that the outcome is in good agreement with experimental findings for the temperature evolution in pulse operation of an ultrasonic fatigue experiment. The rate of heat and geometry associated with the heat source definition are based on fitting. It was found that the resulting intensity can be interpreted in an intuitive way. For the evolution of the width of the distribution in the x_3-direction the interpretation turned out to be more intricate. The distribution

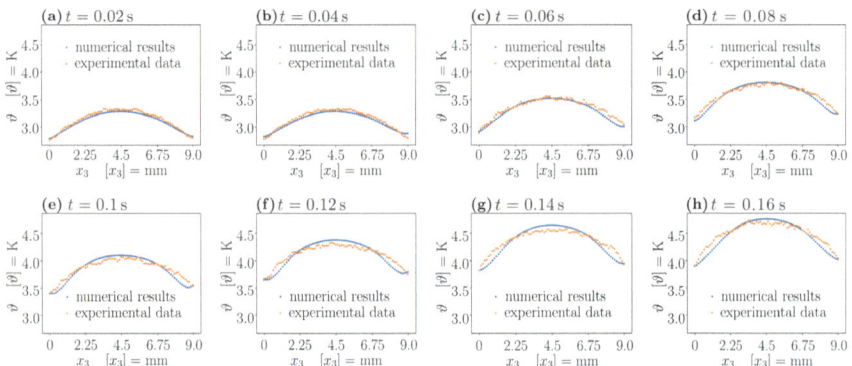

Fig. 26.10 Comparison of experimental and computational 1D temperature difference profiles for the pulse phase

of experimental-computational temperature difference values is pretty noisy up to a time of $t = 0.06\,\text{s}$, explaining a quite strong variation in parameter b at the beginning of the pulse. What is more, heat generation might be more susceptible to a decay in amplitude of excitation, illuminating the decrease of b after $t = 0.1\,\text{s}$.

The investigations presented before are meant as a preliminary study, discussing dissipation from a phenomenological point of view. To allow for a physical interpretation of dissipation, the equations derived and implemented in the FEniCS finite element library can be enhanced to capture thermoplasticity and damage. Information on the size of non-metallic inclusions, see [21], will be considered to define a proper resolution of the numerical mesh. It is envisaged that more complex material behavior can be inspected in the following analyses based on experimental thermography data exhibiting characteristics of damage.

Acknowledgements This work was supported financially by the German Research Foundation (DFG) within the framework of the Collaborative Research Center "Multi-Functional Filters for Metal Melt Filtration" (CRC 920, project number 169148856, subproject C04). The authors gratefully acknowledge the computing time granted by the JARA Vergabegremium and provided on the JARA Partition part of the supercomputer JURECA at Forschungszentrum Jülich. Special thanks go to Dr. Rhena Wulf (Institute of Thermal Engineering, TU Bergakademie Freiberg).

References

1. D. Krewerth, T. Lippmann, A. Weidner, H. Biermann, Int. J. Fatigue **80**, 459–467 (2015). https://doi.org/10.1016/j.ijfatigue.2015.07.013
2. D. Krewerth, A. Weidner, H. Biermann, Ultrasonics **53**, 1441–1449 (2013). https://doi.org/10.1016/j.ultras.2013.03.001
3. D. Wagner, N. Ranc, C. Bathias, P.C. Paris, Fatigue Fract. Eng. Mater. Struct. **33**, 12–21 (2009). https://doi.org/10.1111/j.1460-2695.2009.01410.x

4. A. Blanche, A. Chrysochoos, N. Ranc, V. Favier, Exp. Mech. **55**, 699–709 (2015). https://doi.org/10.1007/s11340-014-9857-3
5. C. Doudard, S. Calloch, F. Hild, S. Roux, Mech. Mater. **42**, 55–62 (2010). https://doi.org/10.1016/j.mechmat.2009.09.005
6. W. Yang, X. Guo, Q. Guo, Int. J. Fatigue **138**, 105717 (2020). https://doi.org/10.1016/j.ijfatigue.2020.105717
7. T. Boulanger, A. Chrysochoos, C. Mabru, A. Galtier, Int. J. Fatigue **26**(3), 221–229 (2004). https://doi.org/10.1016/S0142-1123(03)00171-3
8. H.D. Baehr, K. Stephan, *Heat and Mass Transfer* (Springer, Berlin, 2011)
9. S.J. Farlow, *Partial Differential Equations for Scientists and Engineers* (Dover Publications, New York, 1993)
10. A. Chrysochoos, H. Louche, Int. J. Eng. Sci. **38**(16), 1759–1788 (2000). https://doi.org/10.1016/S0020-7225(00)00002-1
11. Jülich Supercomputing Centre. JURECA: Data centric and booster modules implementing the modular supercomputing architecture at Jülich supercomputing centre. J. Larg. Scale Res. Facil. **7**, A182 (2018). https://doi.org/10.17815/jlsrf-7-182
12. A. Schmiedel, T. Kirste, R. Morgenstern, A. Weidner, H. Biermann, Int. J. Fatigue **152**, 106437 (2021). https://doi.org/10.1016/j.ijfatigue.2021.106437
13. P. Haupt, *Continuum Mechanics and Theory of Materials* (Springer, New York, 2002)
14. A. Logg, K.-A. Mardal, G. Wells, *Automated Solution of Differential Equations by the Finite Element Method* (Springer, New York, 2012)
15. M.S. Alnaes et al., The FEniCS project version 1.5. Arch. Numer. Softw. 3 (2015)
16. H.P. Langtangen, A. Logg et al., *Solving PDEs in Python: The FEniCS Tutorial I* (Springer, New York, 2016)
17. F. Richter, *The Physical Properties of Steels: The 100 Steels Programme Part I: Tables and Figures* (Verlag Stahleisen, Düsseldorf, 1983)
18. H. Belyadi, A. Haghighat, *Machine Learning Guide for Oil and Gas Using Python: A Step-by-Step Breakdown with Data, Algorithms, Codes, and Applications* (Gulf Professional Publishing, Cambridge, 2021)
19. Z. Teng, H. Wu, C. Boller, P. Starke, Fatigue Fract. Eng. Mater. Struct. **43**, 2854–2866 (2020). https://doi.org/10.1111/ffe.13303
20. K. Gillen-Christandl, G.D. Gillen, M.J. Piotrowicz, M. Saffman, Comparison of Gaussian and super Gaussian laser beams for addressing atomic qubits. Appl. Phys. B **122**, 131 (2016). https://doi.org/10.1007/s00340-016-6407-y
21. R. Wagner, A. Schmiedel, S. Dudczig, C.G. Aneziris, O. Volkova, H. Biermann, A. Weidner, Adv. Eng. Mater. **24**(2), 2100640 (2021). https://doi.org/10.1002/adem.202100640

Chapter 27
Effect of Non-metallic Inclusions on the Temperature and Strain-Rate-Dependent Strength, Deformation and Toughness Behavior of High-Strength Quenched and Tempered Steel

Kevin Koch, Sebastian Henschel, and Lutz Krüger

27.1 Introduction

Non-metallic inclusions occur in various steps during steelmaking, including desulfurization and deoxidation [1]. However, the reaction products can only be transferred partially into the slag. The aim of the Collaborative Research Center (CRC) 920 "Multifunctional Filters for Molten Metal Filtration—A Contribution to Zero Defect Materials" consists of removing the remaining particles and dissolved components by providing active and reactive surfaces to trap impurities and thus to increase the purity of molten metals [2]. Since a complete removal of these impurities is not possible, the remaining impurities in the casting component occur as non-metallic inclusions [3].

Non-metallic inclusions represent stress concentration points in a mechanically loaded component. Under these conditions, the risk of cleavage fracture increases. This can be critical even in ductile materials. However, most ductile materials tend to show ductile fracture which consists of void nucleation, void growth and void coalescence [4]. With more stress concentration points, these stages of ductile fracture happen earlier, resulting in lower strength, deformability and toughness of the material. As a result, the functional properties, especially those of safety components are restricted. This includes safety against brittle fracture at low temperatures and crack resistance to stable crack propagation. In addition, failure occurs due to

K. Koch · S. Henschel · L. Krüger (✉)
Institute of Materials Engineering, Technische Universität Bergakademie Freiberg, Gustav-Zeuner-Straße 5, 09599 Freiberg, Germany
e-mail: krueger@ww.tu-freiberg.de

© The Author(s) 2024
C. G. Aneziris and H. Biermann (eds.), *Multifunctional Ceramic Filter Systems for Metal Melt Filtration*, Springer Series in Materials Science 337,
https://doi.org/10.1007/978-3-031-40930-1_27

697

inhomogeneous distribution of non-metallic inclusions. This leads to a scatter of the mechanical properties deformability and toughness that complicates a mathematical analysis [5–7].

During operation, components are often exposed to changing temperatures. High temperatures can increase the deformability and ductility, while low temperatures favor brittle material behavior [8, 9]. Additionally, sudden force initiation is particularly critical for safety components, since brittle fracture occurs particularly in hazardous situations (e.g. vehicle crash). Because such components are subject to previous operational stress, crack-like imperfections and their crack resistance behavior under sudden stress must also be known [10]. When considering the toughness behavior of high-strength steels, both temperature and loading rate effects have to be taken into account [11].

Various methods already exist for determining the resistance to crack initiation at high and very high stress rates, e.g. [12, 13]. These were only applied for materials with elastic–plastic material behavior. In addition to already established methods, a relatively new method for fracture mechanics testing under high stress rates is presented. This method uses the principle of the split Hopkinson pressure bar (Kolsky Bar) for detailed analysis of forces and displacements during sudden loading. Other non-contact methods (laser interferometry, high-speed photography) were applied to support this analysis.

Material damage can be examined by using various in situ methods, e. g. optical microscopy, thermography and acoustic emissions [14]. The latter was applied for investigations under high loading rates to determine the onset of crack initiation and to characterize deformation processes.

The experimental determination of the material behavior as a function of the inclusion characteristics leads to a deeper understanding of the influence of different non-metallic inclusions, especially manganese sulfides and aluminum oxides. The relationship between the inclusion characteristics and the temperature and loading-rate-dependent strength, deformation and toughness behavior are discussed.

27.2 Materials and Methods

Materials Quenched and tempered 42CrMo4 steel is used for various applications, e.g. in automotive engineering. Due to its favorable strength and toughness properties, it is suited for drive and transmission components such as crankshafts, gears and piston rods.

The investigated 42CrMo4 steel was processed by sand casting. An alumina-coated ceramic foam filter in the gating system was used to intentionally contaminate steel melt with non-metallic inclusions. A second filter was then used to purify the steel melt. Table 27.1 gives an overview of the filters used for contamination and purification. Two batches of filtered steel castings were investigated. In both cases, carbon-bonded alumina filters were used for purification. An uncoated filter was used

Table 27.1 Characteristics of the multifunctional filters

Batch	Contamination	Purification
C1	Al_2O_3	Al_2O_3-C
C2	Al_2O_3	Al_2O_3-C + $Al_2O_3 \cdot SiO_2$

for the processing of batch C1. For C2, a filter with mullite coating was used. More detailed information on the process is given by Emmel et al. [15].

For comparison, two other batches of 42CrMo4 were used as a reference. Batch R represents a hot-rolled material. RS was processed by continuous casting and shows a higher sulfur content compared to the other batches. Details on the chemical composition of the investigated steels are given in Table 27.2. In order to remove shrinkage cavities, the material underwent hot isostatic pressing [16]. The machined samples were austenitized at 840 °C for 20 min in a vacuum atmosphere, quenched in a He stream and tempered at 560 °C for 60 min in a N_2 atmosphere.

After heat treatment, the microstructure of the steels consists of martensite, see Fig. 27.1. Non-metallic inclusions of Al_2O_3 as well as MnS were observed. The characteristics of the non-metallic inclusions are shown in Table 27.3.

Another three batches of material were processed by field assisted sintering technology (FAST). For this purpose, 42CrMo4 steel powder was used. Different amounts of Al_2O_3 powder were added to simulate non-metallic inclusions. Further information on the process is given in Koch et al. [17]. Table 27.4 shows the chemical composition of the three different steel batches.

The microstructures of the steels processed by FAST are shown in Fig. 27.2. It can be observed that the added alumina particles are located along the boundaries of the steel particles. With an increasing amount of Al_2O_3, the distance between the non-metallic particles decrease. Within the areas which are occupied by the steel particles, no alumina particles were observed, see Fig. 27.2c.

After machining, the samples were austenitized at 840 °C for 20 min in a vacuum atmosphere, quenched at 15 bars in a He stream and tempered at 450 °C for 60 min in an N_2 atmosphere.

Tensile Testing The strength and deformation behavior was examined in tensile tests. For this purpose, quasi-static and dynamic strain rates as well as different temperatures were used. Table 27.5 gives an overview of the test program. Quasi-static tensile tests at strain rates of 10^0 s^{-1} and less were performed with a universal testing machine Zwick 1476.

Table 27.2 Chemical composition of investigated 42CrMo4 steels

Batch	C	Si	Al	S	P
R	0.44	0.04	0.04	0.006	0.012
RS	0.41	0.25	0.02	0.031	0.012
C1	0.41	0.53	0.09	0.007	0.011
C2	0.43	0.54	0.09	0.007	0.011

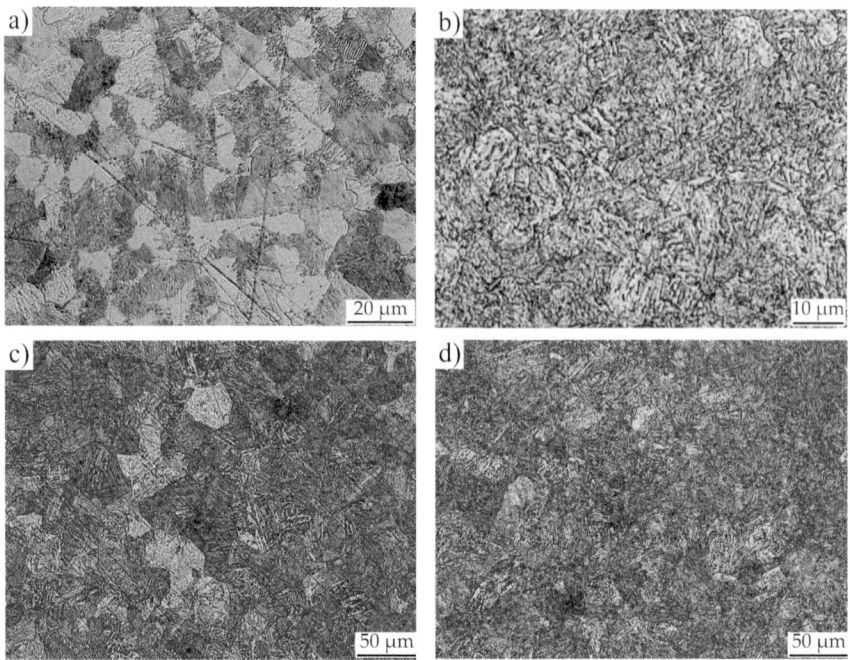

Fig. 27.1 Microstructure of the batches R (**a**), RS (**b**), C1 (**c**), C2 (**d**). Etched with Vilella's reagent

Table 27.3 Characteristics of non-metallic inclusions

Batch	Inclusion size [μm]	Inclusions/area [mm^{-2}]	Inclusion content [10^{-3} Vol.%]	Mean distance [μm]
R	4.24	5.1	8.0	125
RS	2.81	62.2	42	53
C1	2.00	13.1	5.0	88
C2	1.98	13.5	5.5	87

Table 27.4 Chemical composition of investigated sintered 42CrMo4 steel

Batch	C	Si	Al	S	P
S0	0.36	0.21	0.018	0.011	0.015
S1	0.37	0.23	0.460	0.011	0.014
S2	0.33	0.21	3.540	0.012	0.010

For strain rates at approx. 10^{-3} s^{-1}, the specimen geometry (ISO 6892 [18]) shown in Fig. 27.3a was used. In this case, the elongation of the specimen was determined by a clip-on extensometer. For tests at strain rates of 10^0 s^{-1} (ISO 26203–2 [19]) and more, specially-shaped specimens were used, see Fig. 27.3b. The force as well

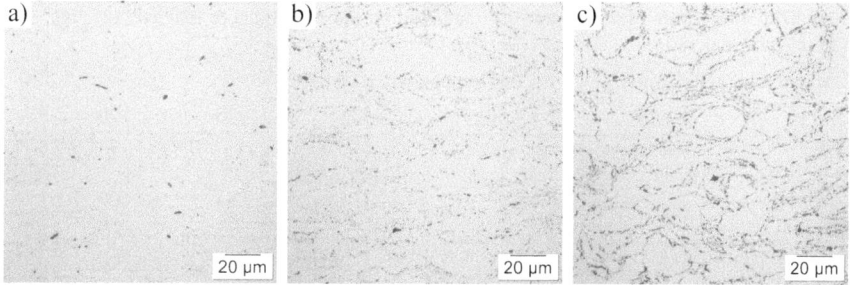

Fig. 27.2 Microstructure of the batches S0 (**a**), S1 (**b**) and S2 (**c**) unetched [17]

Table 27.5 Investigated test temperatures and strain rates for tensile testing, (X): only selected materials

Strain rate [1/s]	Test temperature [°C]			
	−60	−40	−20	20
$0.4–1\cdot10^{-3}$	(X)	X	X	X
$1\cdot10^{0}$		X		X
$1\cdot10^{1}$				X
$1\cdot10^{2}$				X
$1\cdot10^{3}$		(X)		(X)

as the elongation in these tests were measured with strain gauges on the specimen. Additionally, tests at strain rates of 10^{2} s^{-1} and 10^{3} s^{-1} were performed in a drop weight testing machine and a rotating wheel test machine, respectively.

Fracture Toughness Testing The aim of the fracture toughness tests was to determine the relationship between the crack resistance J and the stable crack propagation Δa. The material resistance to crack initiation is derived from the crack resistance curve. It can also be derived from the crack tip blunting (stretch zone width, SZW). The fracture mechanics tests were carried out over a wide range of loading rates and temperatures, see Table 27.6.

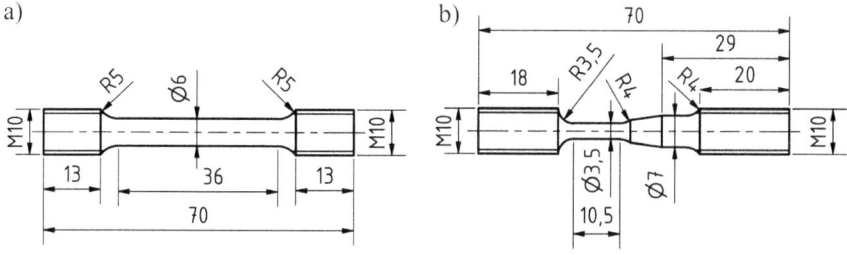

Fig. 27.3 Tensile test specimen used for **a** quasi-static tests and **b** dynamic tests [20]

Table 27.6 Investigated test temperatures and loading rates for fracture toughness testing, (X): only selected materials

Loading rate [MPa $\sqrt{m}$/s]	Test temperature [°C]			
	−60	−40	−20	20
10^0	(X)	X	(X)	X
10^5	(X)	X		X
10^6		(X)		(X)

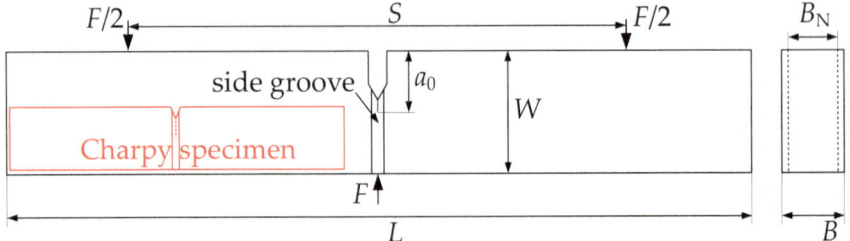

Fig. 27.4 Cracked SENB specimen, $W/B = 2$; $a_0/W = 0.5$; $B_N/B = 0.8$ with position of the Charpy specimens which are made from the broken samples

Quasi-static tests were carried out with a servo-hydraulic universal testing machine MTS 810 to determine the crack resistance curve. Single-edge notch bending (SENB) specimens ($B = 10$ mm) according to ASTM E 1820 [21] or ISO 12135 [22] were used, see Fig. 27.4. The length L of the specimen was 120 mm. Hence, two smaller Charpy-type bending specimens could be made from the broken samples.

The stress intensity factor K_I and the current crack length ($a = a_0 + \Delta a$) was determined from the elastic compliance according to ISO 12135. The J-integral is calculated by

$$ J = \frac{K^2(1 - v^2)}{E} + \frac{1.9 U_{\text{pl}}}{B_N(W - a_0)} \cdot \left(1 - \frac{\Delta a}{2(W - a_0)}\right) \qquad (27.1) $$

Fracture toughness tests at loading rates of about 10^5 MPa$\sqrt{m}$s^{-1} were performed in an instrumented pendulum impact test machine (PSd 300, WPM Leipzig). The test setup is shown in Fig. 27.5. The force was measured by an instrumented tup. The displacement of the specimen was measured in force direction by a laser doppler interferometer (Polytec OFV-525). Further information on this technique is given in Henschel et al. [23].

The dynamic fracture toughness tests have been performed with the low-blow method to create different amounts of stable crack growth. The impact energy was varied by setting the deflection angle of the pendulum prior to each test. Investigations were carried out at impact energies $E_0 = 2.3 - 3.5$ J which correspond to impact velocities $v_0 = 0.48 - 0.59$ m/s. The J-Integral was calculated from the force–displacement curve similar to quasi-static testing:

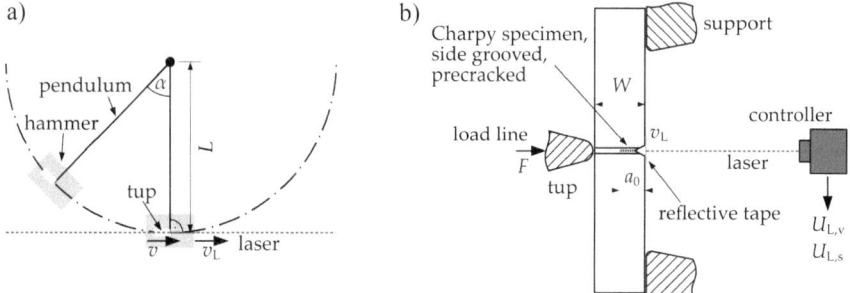

Fig. 27.5 Schematic test setup with laser measurement

$$J_d = \frac{K^2(1 - v^2)}{E} + \frac{2W_{pl}}{B_N(W - a_0)} \cdot \left(1 - \frac{\Delta a}{2(W - a_0)}\right) \qquad (27.2)$$

With

$$W_{pl} = \int_0^s F \, ds' - \frac{F^2 C_0}{2} \qquad (27.3)$$

Here, W_{pl} is the plastic portion of deformation energy and C_0 is the initial compliance.

Loading rates of about $10^6 \, \text{MPa}\sqrt{\text{m}}\text{s}^{-1}$ were achieved by applying the split Hopkinson pressure bar (Kolsky bar) technique. The setup shown in Fig. 27.6 consists of four different bars that are axially aligned.

The striker bar is propelled by compressed air and hits the incident bar and introduces a nearly rectangular pressure pulse of known amplitude $\varepsilon_{I,max}$:

$$\varepsilon_{I,max} = \frac{v_{St}}{2c_B} \qquad (27.4)$$

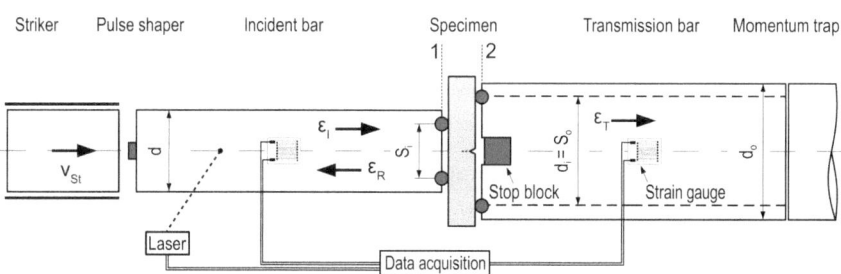

Fig. 27.6 Setup of the split Hopkinson pressure bar used for fracture toughness tests. The length of both incident and transmitted bars is 1.5 m

Here, v_{St} and c_B are the striker bar velocity and the bulk sound velocity of the bars, respectively. Equation (27.4) is only valid, if striker and incident bars consist of the same material and have the same diameter.

This pressure pulse propagates along the incident bar and is partially transmitted into the specimen (and the transmitted bar) and partially reflected as a tensile pulse. The pulse in the transmitted bar propagates further in the momentum trap that eventually separates from the transmitted bar, which leads to a single loading of the specimen. The elastic deformation of the bars that are made of high-strength aluminum (AA7075) is measured by strain gauges at the center of the bars. At the incident bar, a laser interferometer was used to measure the particle velocity. Since particle velocity and strain in the bar are proportional, an additional measurement site was established. This measurement site was applied if relatively long incident pulses were used. Details are found in Henschel et al. [24]. The signal acquisition (20 MSample/s) was triggered by two light barriers near the impact site of the striker bar. Furthermore, the light barriers were also used to calculate v_{St}.

From the incident, reflected and transmitted pulses (ε_I, ε_R and ε_T, respectively), the forces and displacements at the interfaces 1 and 2 were calculated. For the sake of brevity, the equations are not shown here, but can be found in Henschel [20].

The analysis of the force equilibrium is a crucial step. If the axial forces F_1 and F_2 at the interfaces 1 and 2 are approximately equal, inertial effects can be neglected. Consequently, there is a proportionality between the acting force $F = F_1 = F_2$ and the stress intensity factor K_I at the crack tip. Hence, equations for quasi-static calculation of K and J [24] can also be applied here. On the other hand, if there are significant inertia effects, the forces F_1 and F_2 are treated separately. In order to calculate the time-dependent stress-intensity factors, two-point loading from each side can be assumed. This method is described in detail in Henschel [20, 25]. The pressure pulse can be shaped by using small pieces of deformable material at the impact site, see Fig. 27.6. Consequently, this affects the rise time of the incident pulse.

Acoustic Emission Analysis The relation between different mechanisms of material damage and the acoustic emission (AE) signal was studied at different temperatures and loading rates. The main focus was to investigate ductile fracture at ambient temperature. However, tests at $-40\,°C$ were aimed to examine the transition from ductile to brittle behavior. To this end, quasi-static tensile tests as well as quasi-static and dynamic fracture toughness tests were carried out with one or more acoustic transducers mounted to the specimen to record acoustic emissions. Details on the test setup are given in Kietov et al. [26].

The AE signal was recorded continuously in order to identify possible events of material damage. In order to analyze the signal, first transient signals were identified by the threshold technique. Secondly, the continuous low-amplitude signal was analyzed with respect to signal energy and power. Usually, transient signals are associated with crack growth and fracture, while continuous signals correspond to plastic deformation or dislocation motion, see Fig. 27.7. Further information on the test method and evaluation is given in Kietov et al. [26, 27].

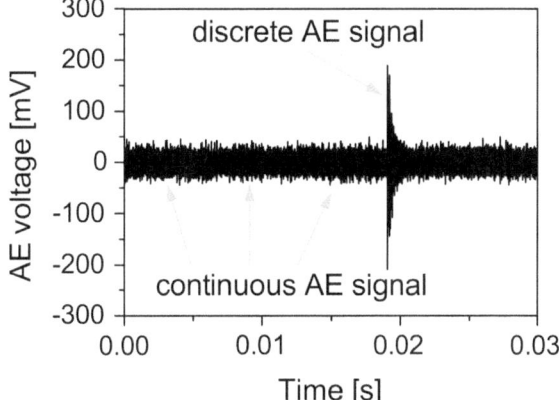

Fig. 27.7 Example of continuous and discrete AE signals [27], with permission

Mixed-Mode Testing In order to study the fracture toughness under mixed-mode loading, compact tension shear specimens (CTS) were used. Details can be found in Henschel et al. [28]. The specimen dimensions were based on recommendations of Richard [29] and scaled for a thickness of 8 mm. The heat treatment consisted of austenitization (840 °C, 20 min, vacuum), quenching in a He stream at 15 bars and tempering (450 °C, 1 h, N_2). The stress intensity factors K_I and K_{II} were determined by finite-element analysis. Equations used for the calculation of K_I and K_{II} are given in Henschel et al. [28]. The calculated stress intensity factors were compared to experimentally measured stress intensity factors. To this end, strain gauges were applied to analyze the stress field around the crack tip, according to Sarangi et al. [30].

The aim of the investigations is to determine the mixed-mode fracture toughness for a high-strength steel as well as the effect of the loading angle on the crack growth.

27.3   Temperature and Strain-Rate Dependent Material Behavior of High-Strength Steels

27.3.1   Strength and Deformability

The strength and deformation behavior was investigated in uniaxial tensile tests for various temperatures and strain rates. Figure 27.8 shows the effect of the temperature for the steels R and RS. The yield strength R_{eL} increases at low temperatures. This can be explained by thermally activated dislocation motion. The plastic elongation at fracture A also increases with decreasing temperature. This is primarily due to an increasing uniform strain A_g, while necking is only slightly affected by temperature. Therefore, the ratio A_g/A increases with decreasing temperature. The reduction of area Z decreases slightly with decreasing temperature which indicates an embrittlement.

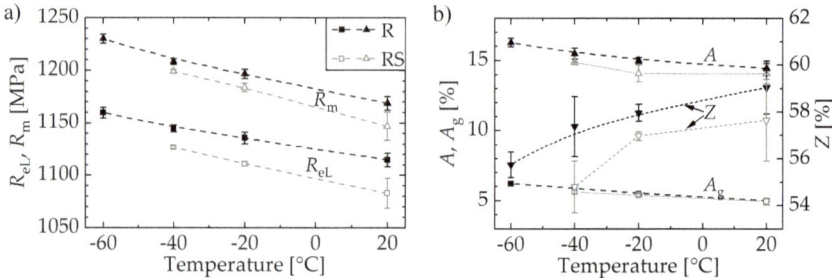

Fig. 27.8 Strength and deformation characteristics of R and RS depending on the temperature

The higher tensile strength R_m of R compared to RS can be explained by the higher content of C. However, both steels show a similar strain hardening behavior.

Figure 27.9 shows the results of the tested steel castings. The strength values of C1 and C2 showed no significant differences. However, slight differences in tensile strength and yield strength can be explained by differences in the chemical composition.

The deformability shows significant scatter. However, the impurification with alumina particles only slightly affects the properties which were investigated in tensile tests. Compared to batches R and RS, the steel castings also show an increasing uniform strain at low temperatures. The elongation at fracture is unaffected by temperature, which can be seen for batch C1. On the other hand, C2 showed decreasing elongation at fracture at low temperatures.

The reduction of area decreased at very low temperatures. The embrittlement at low temperatures is affected by non-metallic inclusions. Figure 27.10 shows the fracture surfaces of the steels C1 and C2. The agglomerate in C2 is identified as the main reason for the relatively low deformability, see the fracture path lines in Fig. 27.10b. Due to low temperature and the notch effect of the agglomerate, a mixture of ductile fracture and cleavage fracture is observed.

Figure 27.11 shows the relationship between the size of the largest damage-relevant agglomerate and the elongation at fracture. It can be seen that an increasing

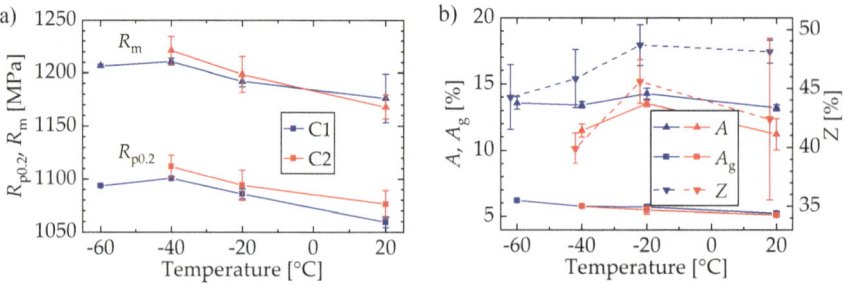

Fig. 27.9 Strength and deformation characteristics of C1 and C2 depending on the temperature

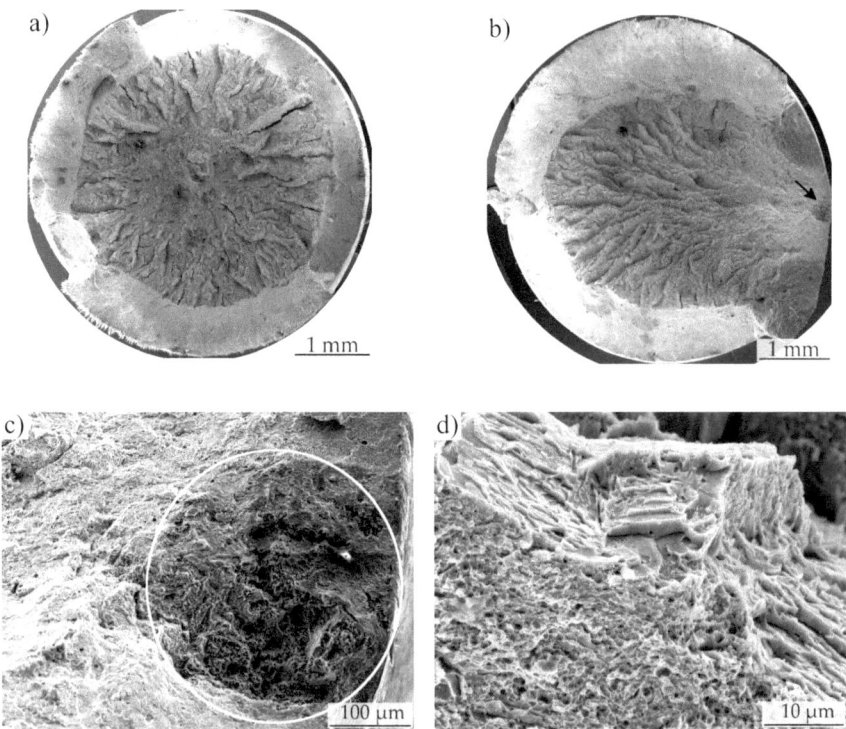

Fig. 27.10   Fracture surfaces of tensile test specimens: **a** C1, **b–d** C2. $T = -40\,°C$

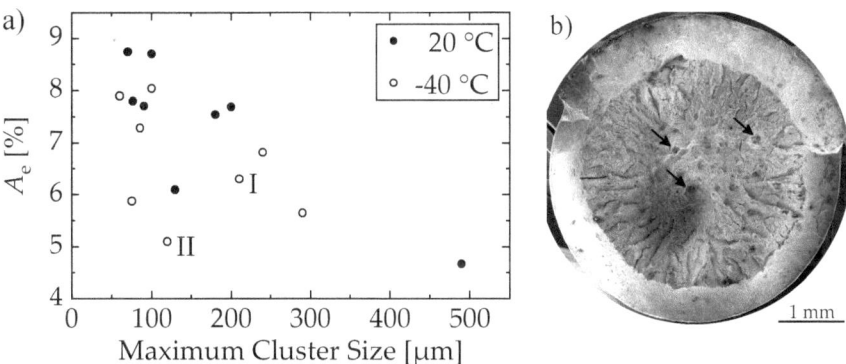

Fig. 27.11   Relation between elongation during necking A_e and size of the damage relevant inclusion cluster according to [31]. I: specimen in Fig. 27.10b, II: specimen in Fig. 27.11b

agglomerate size reduces the elongation at fracture. Furthermore, the amount of agglomerates and their arrangement in the specimen volume have an effect. Figure 27.11b shows a fracture surface with several small agglomerates.

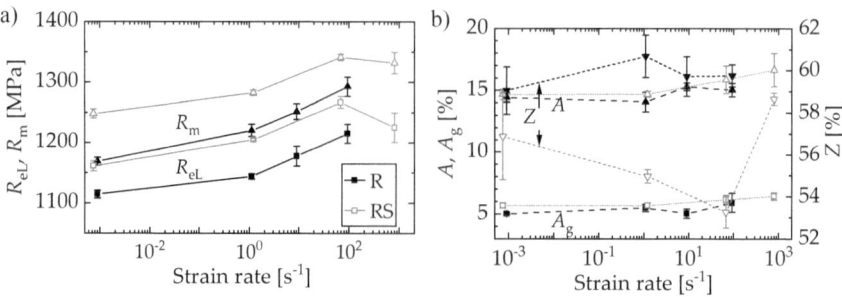

Fig. 27.12 Strength and deformation of R and RS depending on the strain rate, $T = 20\ °C$

Due to small spatial distance, a relatively low strain during necking was observed (mark II). On the other hand, another sample shows a significantly greater elongation during necking, although the inclusion agglomerate is almost twice as large in diameter (mark I and Fig. 27.10b and c). However, there is no other agglomerate in the weakest cross-section. An effect of the temperature was not observed.

The effect of the strain rate on the strength of R and RS is shown in Fig. 27.12a. Analogously to a decreasing temperature, an increasing strain rate causes an increase of yield strength and tensile strength. Due to adiabatic heating, the increasing strain rate between 10^2 and $10^3\ s^{-1}$ did not affect the tensile strength. This leads to softening of the material which is superposed by the strain hardening. Figure 27.12b shows that the deformation parameters of steel R are not depending on the strain rate. In contrast, the reduction of area of the steel RS shows a strong decrease up to strain rates of $10^2\ s^{-1}$. The increase of Z by a further increase of strain rate is associated with the increase in temperature and correlates with an increasing ductility.

The effect of the strain rate on the strength and deformability parameters are shown in Fig. 27.13. The steel castings show a similar behavior compared to the steels R and RS, as a positive strain rate effect occurs. The effect of the strain rate on yield and tensile strength is more significant at $T = -40\ °C$ than at 20 °C. However, the steel C2 is an exception to this. The deformability does not show a distinct dependence. At $T = 20\ °C$, the strain at failure and the reduction of area decrease, in particular for strain rates greater than $10^2\ s^{-1}$. Additionally, the elongation at fracture is not temperature dependent at $T = -40\ °C$. Furthermore, the behavior of the uniform elongation correlates with embrittlement at the highest strain rate.

In Henschel et al. [32], C1 and C2 were compared regarding the strength and deformation behavior. It was found that C2 has a higher ductility at quasi-static loading in the temperature range from −40 to 20 °C. Taking the deformability parameters of the other steel castings into account as well as the strain rate dependency, it can be assumed that structural differences within the cast plates are the most probable cause for apparent differences in mechanical behavior. In previous studies, it was additionally found, that the crucible material [33] and the usage of an additional immersion filter [34] may have a significant effect on the spatial inclusion distribution and, consequently, on the mechanical properties.

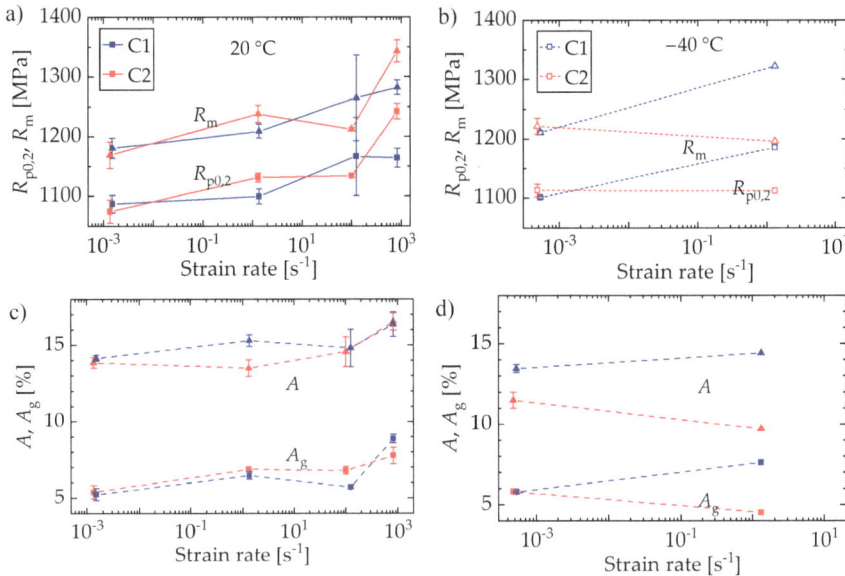

Fig. 27.13 Strength and deformability of C1 and C2 depending on the strain rate for $T = 20$ °C (left) and -40 °C (right)

The results of the tensile tests of the sintered materials are shown in Fig. 27.14. The tensile strength and strain at failure decrease with an increasing content of alumina particles. It should be mentioned that the sintered material showed more brittle behavior in general compared to the previously mentioned 42CrMo4 batches. Batch S3 shows no plastic deformation, neither at quasi-static loading nor at dynamic loading. Hence, failure occurs before the yield strength is reached. Microscopic examinations underneath the fracture surfaces showed secondary cracks along the former steel particle boundaries, see Fig. 27.14. In this area, the ceramic particles are arranged with small distances to each other. Due to the increased notch effect, crack deflection was favored. The batch S0 already shows low strain at failure without the addition of aluminum oxide, compared to the 42CrMo4 batches mentioned before. This can be attributed to oxide layers on the steel particles. These layers are formed e.g. during powder processing. This leads to an internal notch effect in the sintered material, which results in a reduced deformability. This effect is intensified by the addition of the ceramic particles.

In order to examine the material damage by means of acoustic emissions (AE), the voltage signal of a piezo-electric transducer mounted to the specimen was recorded continuously during tensile tests. Only steel R was investigated for this purpose. For the analysis, damage relevant signals had first to be discriminated from the electrical and mechanical background noise that was also recorded. Transient signals were identified with a threshold-based technique. However, continuous signals show very low amplitudes and are more difficult to distinguish from background noise.

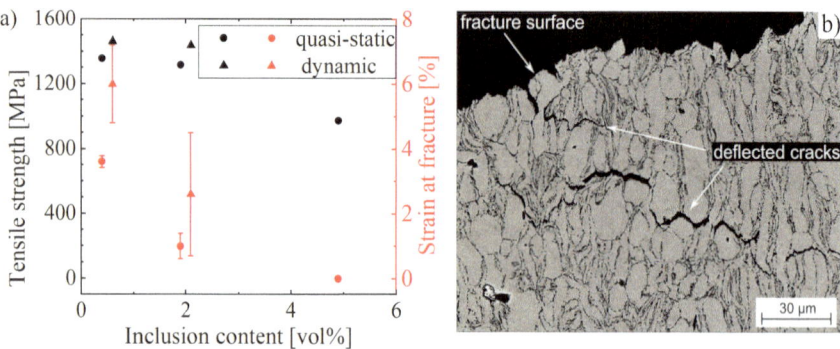

Fig. 27.14 a Strength and deformation behavior of the sintered materials at $T = 20\,°C$, **b** microscopic structure of S3 perpendicular to the fracture surface [17]

Therefore, the signal was analyzed with respect to the frequency of the AE signals with a statistical approach. This initial test was performed under quasi-static loading, without any plastic deformation. Detailed information is given in Kietov et al. [27]. The results are shown in Fig. 27.15. It could be observed that transient signals occur mostly in a median frequency range between 250 and 600 kHz. However, continuous signals showed median frequencies between around 180 to 240 kHz. Both signals could be very well discriminated from the background noise which showed lower median frequencies.

The actual tensile tests with AE recording were carried out up to failure as mentioned earlier in this chapter. Figure 27.16 shows the stress–strain diagram and the corresponding AE signal for steel R. It can be observed from the rapid increase of the transient AE rate that the majority of transient signals occurred during elastic deformation.

At the beginning of the yield plateau (Lüders strain), the transient AE activity started to decrease. Shortly before fracture, another rapid increase of transient signals was observed. The continuous AE signals showed a different behavior. Until the

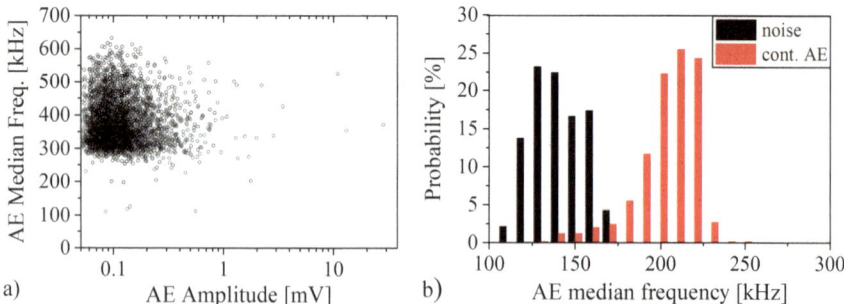

Fig. 27.15 a Median frequency and amplitude of the transient AE signals, **b** median frequency of the continuous AE signals compared to noise during tensile tests [27], with permission

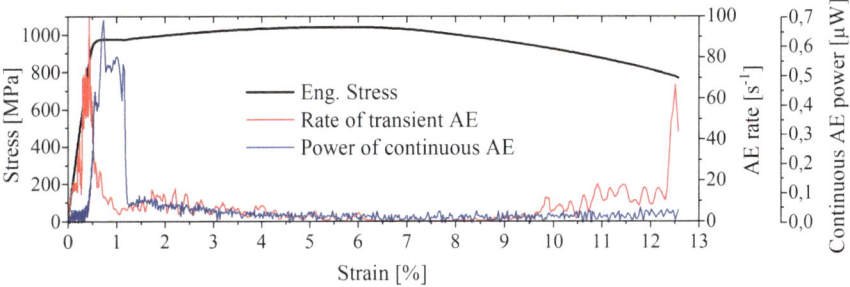

Fig. 27.16 Stress–Strain-diagram and corresponding AE signal (transient and continuous) for steel R, $T = 20\,°C$, strain rate $10^{-3}\,s^{-1}$ [27], with permission

beginning of yielding, the level of the continuous AE remains below the noise. During the yielding plateau a significant increase of the power of the continuous AE was observed. This was attributed to the high amount of dislocation motion which occurs at the first deformation of the material. It only happened at the yield plateau and no other further increase happened after the continuous AE dropped back to noise level. Effects that correspond to damage in context of non-metallic inclusions were not observed during these tensile tests.

27.3.2 Fracture Toughness

Figure 27.17 shows the J-Δa curves for the steels R, RS, C1 and C2. The difference between the steels R and RS for $\Delta a < 0.2$ mm is insignificant. However, at higher Δa, $dJ/d\Delta a$ and J are significantly higher for steel R. In Fig. 27.17b one can see that the cast steels with the added non-metallic inclusions show both a lower toughness level as well as a lower slope of the J-Δa curve. Minor differences are observed between the industrially produced and contaminated steels C1 and C2.

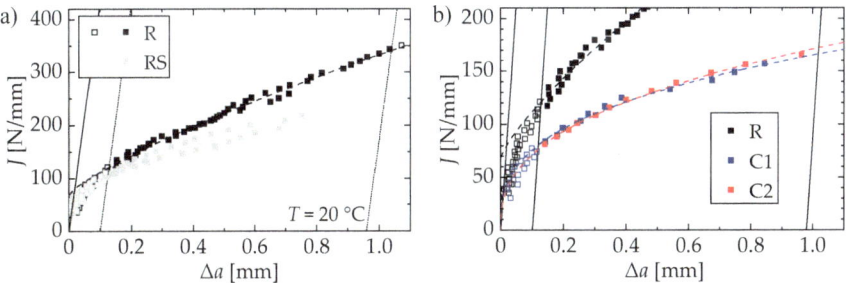

Fig. 27.17 J-Δa curves at $T = 20\,°C$: **a** steels R and RS, **b** steels R, C1 and C2

The J_{iBL} and $J_{0.2BL}$ values derived from the J-Δa curves are shown in Fig. 27.18. In Fig. 27.18a it can be seen that the material resistances to both physical and engineering crack initiation (J_{iBL} or $J_{0.2BL}$) do not differ significantly for the steels R and RS. Likewise, Fig. 27.18b shows no significant differences between the two cast steels. On the other hand, the material resistances J_{iBL} and $J_{0.2BL}$ of the cast steels are significantly lower than the references R and RS. At $T = 20$ °C, an average of 30 N/mm is determined for J_{iBL} for the cast steels, while approx. 70–80 N/mm are measured for R and RS, respectively. The differences between the physical crack initiation (J_{iBL}) determined from the J-Δa curve and the crack initiation (J_{iSZW}) determined from the fracture surface are discussed later.

Figure 27.19 shows the effect of temperature on the J-Δa curves for the steels R and RS. The values of J_{iBL}, J_{SZW} and $J_{0.2BL}$ have already been shown in Fig. 27.18. No significant relationship between crack resistance and temperature was observed. Multiple cases of pop-in behavior or completely unstable failure were observed at all temperatures. No J-Δa curve was obtained under these conditions. Nevertheless, the data points of the steel R for $T = -60$ °C are shown in Fig. 27.19. In addition, the characteristic values of crack initiation were estimated and shown in brackets in Fig. 27.18. In case of an unstable crack propagation after $\Delta a > 0.2$ mm, only the toughness J_u can be determined according to ISO 12135. However, this neither describes the crack initiation nor is it independent of the thickness.

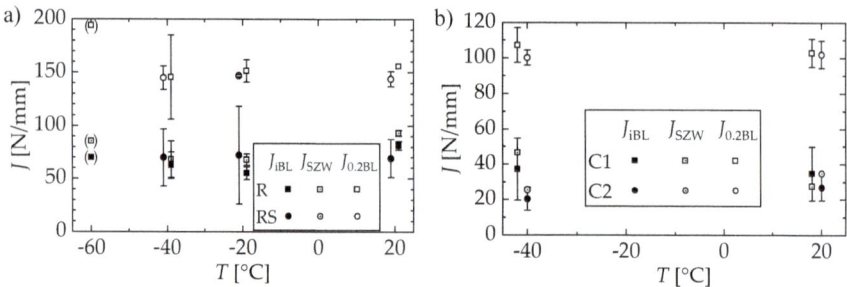

Fig. 27.18 Effect of temperature on the characteristics of crack initiation J_{iBL}, J_{SZW} and $J_{0.2BL}$: **a** steels R and RS, **b** steels C1 and C2

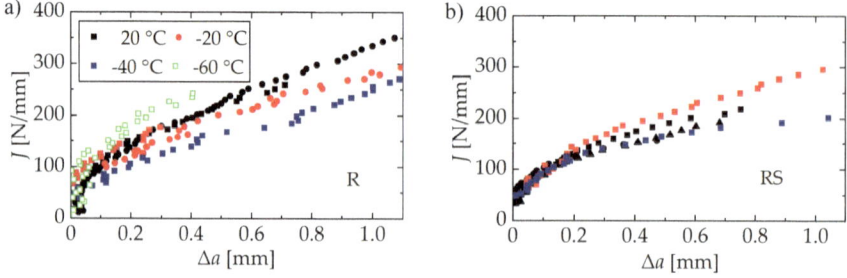

Fig. 27.19 Effect of temperature on the J-Δa curve: **a** steel R, **b** steel RS

Analogously to the steels R and RS, C1 and C2 show no significant temperature dependence of crack initiation, see Fig. 27.20. Large scatter was observed at low temperatures, which can be explained by the cleavage fracture surfaces and the associated higher scatter to be expected in the ductile–brittle transition region [35]. This scatter is not due to pop-ins. The fracture toughness is affected by the strength and deformability of the material. It was shown, that the strength of all investigated steels increases with decreasing temperature. An attempt to model the resistance to crack initiation at quasi-static loading rates was previously shown in Henschel and Krüger [36]. It was observed, that the distance of the particles within an agglomerate determines the toughness behavior.

The increase is about 30 MPa or 3% in the temperature range from 20 to −40 °C. On the other hand, there was no clear connection between temperature and deformability. Steels R and RS showed a slight increase in A and a slight decrease in Z with decreasing temperature (about +6% and −4%, respectively). No significant effect of temperature on the deformability parameters A and Z was observed for the cast steels. Due to small differences and scatter of the deformability, the fracture toughness is independent of the temperature.

The effect of the loading rate on fracture toughness is described and discussed below, including the temperature effect at high loading rates. The $J_d−\Delta a$ curves for the impact tests in the pendulum impact test machine are shown in Figs. 27.21 and 27.22. The results for 20 °C are represented by full symbols and those for −40 °C by open symbols. The crack resistance curve was determined by the $J−\Delta a$ values from the multiple specimen method (MSM) using the low-blow technique. Additionally, the normalization method (NM) according to ISO 26843 [37] enables the $J−\Delta a$ values to be estimated up to the end of the respective test. This corresponds to the turning point in the low-blow test. A complete $J−\Delta a$ curve is thus calculated from the force–displacement curve by using the NM.

A comparison of the measured correlations between J and Δa and those determined using the normalization method according to ISO 26843 shows that J is overestimated and Δa is underestimated, see Fig. 27.21. Lucon [38] also points to this circumstance using the example of two reactor pressure vessel steels (20MnMoNi5-5 and ASTM A533B Cl.1).

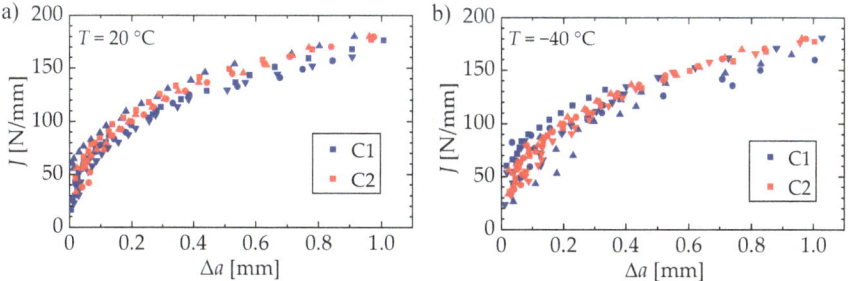

Fig. 27.20 $J−\Delta a$ curves for the cast steels C1 and C2: **a** 20 °C, **b** −40 °C

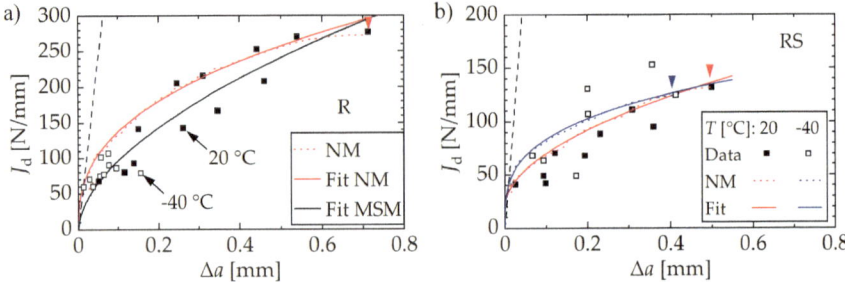

Fig. 27.21 Dynamic $J_d - \Delta a$ curves for steels **a** R and **b** RS at different temperatures. The Displacement was calculated by double integration

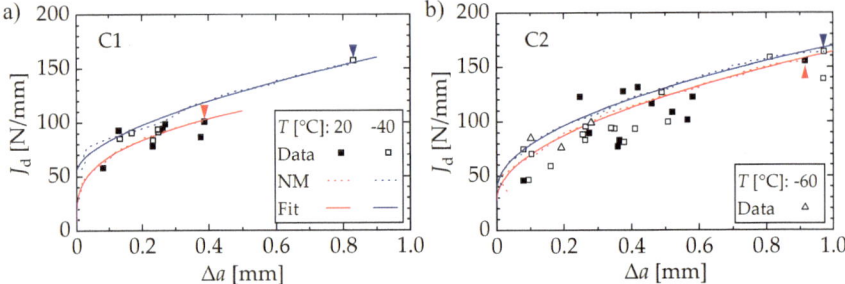

Fig. 27.22 Dynamic $J_d - \Delta a$ curves for steels **a** C1 and **b** C2 at different temperatures

Analogously to quasi-static tests, no temperature dependency of the crack resistance curves was determined under dynamic loading. Due to the multiple specimen method and slightly different crack lengths, the $J_d - \Delta a$ curves show relatively high scatter, see Fig. 27.21a. Different crack lengths can be caused by tolerances in the production of B and W.

The characteristic parameters of crack initiation (J_{iBL}, J_{SZW} and $J_{0.2BL}$) were determined based on the crack resistance curves, see Fig. 27.23. On the one hand, the crack resistance curve determined using several low-blow tests was evaluated for this purpose. On the other hand, the $J - \Delta a$ curve determined using the normalization method analyzed. It was already established in Fig. 27.23 that the crack resistance curve determined using the normalization method shows relatively large values of J_d for small Δa. This circumstance also affects J_{iBL} and J_{iSZW}. The technical crack initiation parameter $J_{0.2BL}$, on the other hand, is hardly affected by the evaluation method. Furthermore, it can be seen that the difference between J_{iBL} and J_{iSZW} is smaller on average when using the normalization method than when using the multiple specimen method.

A slight temperature effect of the crack initiation parameters between $-40\ ^\circ C$ and $20\ ^\circ C$ was observed for the steels RS, C1 and C2. With decreasing temperature, the fracture toughness increases. Occasionally, cleavage fracture was detected. The

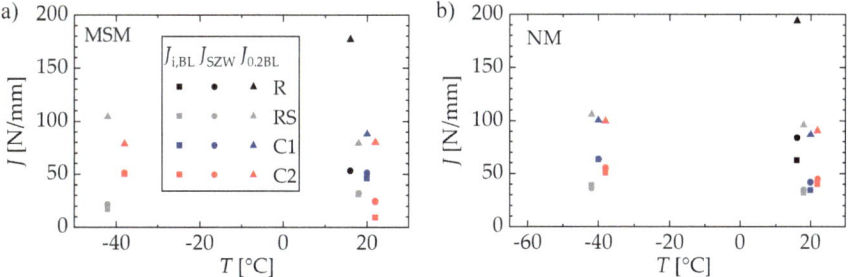

Fig. 27.23 Crack resistance parameters estimated from $J - \Delta a$ curves: **a** multiple specimen method (MSM), **b** normalization method (NM)

investigated temperature interval is therefore at the lower end of the upper-shelf toughness. The results of dynamic tests at very high loading rates (10^6 MPa$\sqrt{}$ms^{-1}) are shown in Fig. 27.24.

For steel R a significant temperature effect can be observed. Regardless of the loading rate, a decreasing temperature leads to a decrease in fracture toughness of about 20 MPa$\sqrt{}$m. In the case of quasi-static loading, this correlates to a 23% reduction in toughness compared to room temperature. At very high loading rates, a significantly greater decrease in toughness of 44% is observed. The steels R and RS show a decreasing toughness with an increasing strain rate. However, the toughness loss for loading rates $>10^5$ MPa$\sqrt{}$ms^{-1} is less pronounced for steel RS.

The microscopic damage analysis shows ductile failure. Figure 27.25a shows that oxidic inclusions are detached from the metallic matrix even at low levels of stress. With a favorable spatial arrangement of the non-metallic inclusions, shear bands form between the inclusions or inclusion clusters. This phenomenon can be recognized by the steep flanks on the fracture surfaces see Fig. 27.26a.

The resistance to physical crack initiation is based on the blunting capacity of the crack tip. The blunting depends to a large extent on the distribution of the non-metallic inclusions in terms of size and position in front of the crack tip [39], see Fig. 27.26.

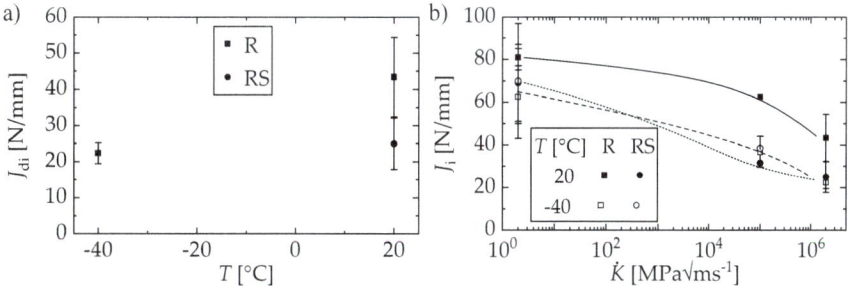

Fig. 27.24 Effect of temperature and loading rate on fracture toughness at crack initiation: **a** loading rate $2 \cdot 10^6$ MPa$\sqrt{}$ms^{-1}, **b** loading rate effect, $J_i = J_{iBL}$

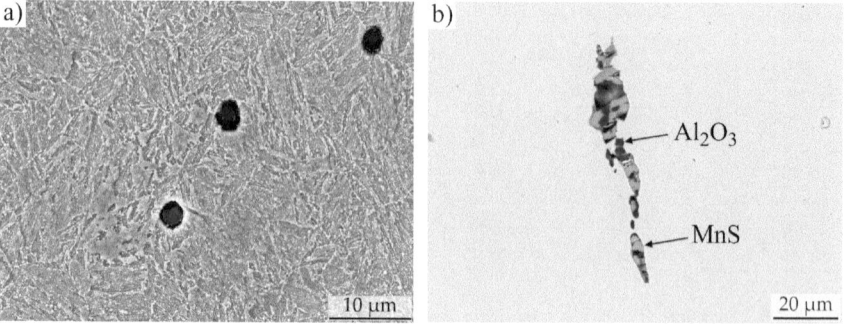

Fig. 27.25 Microscopic damage evolution on steel R under pure elastic stress: **a** Detachment of oxidic inclusions ($\sigma = 700$ MPa, strain rate 10^{-3} s^{-1}), **b** fracture and partial detachment of multiphase inclusions ($\varepsilon = 16\%$, strain rate $9 \cdot 10^2$ s^{-1}) [20]

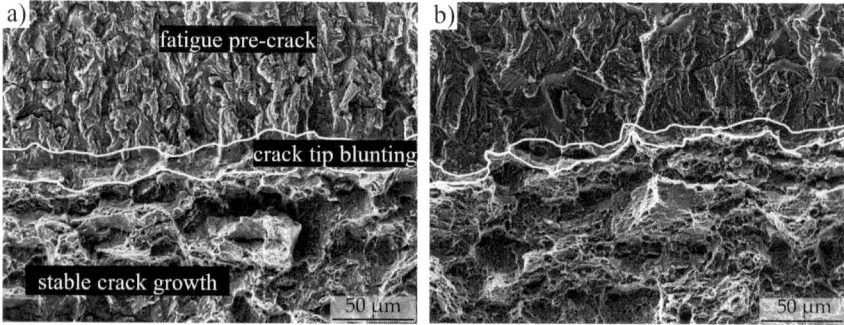

Fig. 27.26 Crack tip blunting behavior: **a** steel R, **b** steel C1

During quasi-static fracture toughness testing, the acoustic emissions have been analyzed for the steel R. The acoustic emissions were processed the same way as described for tensile testing. Figure 27.27 shows a load–displacement curve for a specimen tested at $T = 20$ °C. The number of transient AE increases almost linear over time. However, the continuous AE showed a different behavior, which is presented by a significant change of the power level.

Figure 27.28a shows the energy of the transient AE during fracture toughness testing. The emitted energy of the transient AE remains relatively low at the beginning, but becomes significantly stronger at a certain point. According to Roy [40, 41], this behavior could be attributed to large amounts of stable crack growth for this type of steels. By analyzing of the continuous AE, the size of the plastic zone could be estimated. This was done by calculating the volumetric AE energy density from the continuous AE power. The specific equations for this calculation are given in Kietov et al. [27]. Figure 27.28b shows the comparision of the plastic zone sizes estimated by the continuous AE and by the stress intensity factor. It shows good coincidence until the onset of crack growth which happened at 0.7 to 0.8 mm of displacement.

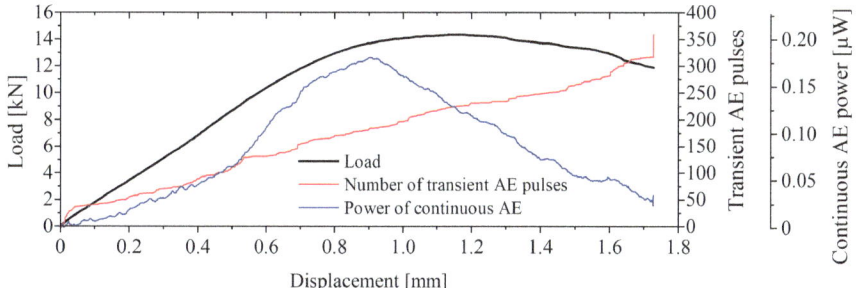

Fig. 27.27 Analysis of acoustic emission during quasi-static loading [27], with permission

After the onset of crack growth, both estimated plastic zone sizes begin to deviate, since the plastic zone size is understimated by K_I. Hence, the analysis of the AE gives a more realistic measure of the plastic zone. Furthermore, the beginning of the deviation of the plastic zone size can be used to identify the onset of crack growth. Consequently, the acoustic emission analysis shows its advantages during the study of both unstable fracture [42], and stable fracture.

The results of mixed-mode fracture toughness tests on steel RS will be discussed in the following. A total of four different loading angles was tested in order to investigate the effect on the fracture toughness and crack growth. The loading angle $\alpha = 0°$ resembles pure mode I loading. Hence, the ratio K_I/K_{II} decreases with increasing α. It could be observed that all tested loading angles showed nearly linear-elastic behavior. The maximum force until fracture increases with increasing α. Detailed information on the force measuring technique with strain gauges as well as the displacement measuring technique is given in Henschel et al. [28].

The results for K_{IQ} and K_{IIQ} at different loading angles are presented in Fig. 27.29a. The mode I fracture toughness decreased with increasing loading angle, while the mode II fracture toughness increased. This behavior can be described with the limiting curve shown in Fig. 27.29b. Within the area of the curve, the material will not fail. Flat as well as slant areas could be observed on the fracture surfaces,

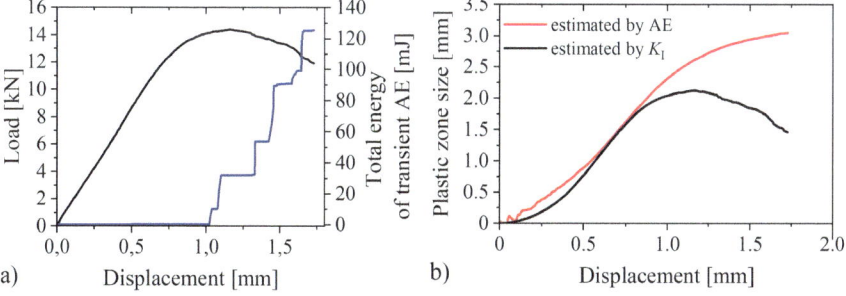

Fig. 27.28 a Energy of transient AE used to determine $J_{AE, Et}$ during quasi-static loading, **b** evolution of the plastic zone size [27], with permission

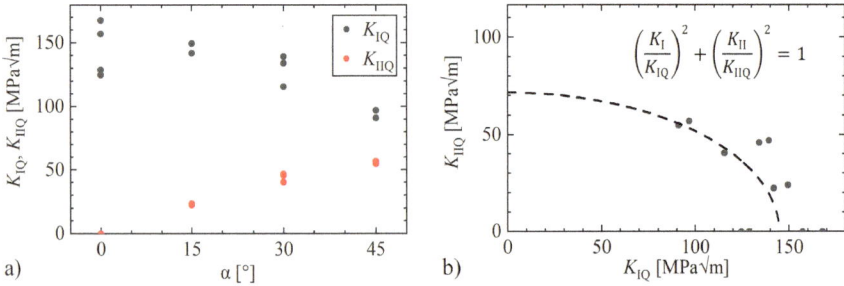

Fig. 27.29 Mixed mode fracture toughness (**a**) K_{IQ} and K_{IIQ} as a function of α, approximate criterion [28], with permission

Fig. 27.30 Fracture surfaces: both flat and slant fracture regions were observed, [28], with permission

see Fig. 27.30. This indicates a mixture of plane strain and plane stress on the specimen. Crack tip blunting was observed at the tip of the fatigue pre-crack. The crack extension predominantly consisted of unstable ductile fracture, which was promoted by non-metallic inclusions.

27.4   Conclusions

This chapter contributes to the understanding of the effect of non-metallic inclusions on the mechanical material behavior of quenched and tempered 42CrMo4 steel. The strength, deformability and fracture toughness behavior were examined at different temperatures and loading rates under uniaxial loading. The most important findings are presented below.

- All examined steels showed increasing strength with decreasing temperature and increasing strain rate. The relationship between deformability and strain rate is not clear. Due to the strain rate, adiabatic heating occurs, which is superimposed on the strain rate-related hardening. Non-metallic inclusions did not have a significant effect on the strength and the deformability.

- Higher inclusion contents lead to a decreasing fracture toughness under quasi-static loading. However, the effect was not observed at very low temperatures or very high loading rates. The non-metallic inclusions lead to a deflection of the crack front. Inclusion agglomerates are particularly critical.
- The material damage could be examined based on the AE analysis. Different AE events enabled different damage mechanisms to be identified. Hence, the onset of crack growth and the size of the plastic zone at the crack tip could be estimated.

Acknowledgements The authors thank the German Research Foundation (DFG) for the financial support of the investigations in the Collaborative Research Center 920—Project-ID 169148856, subproject C05. The acoustic emission analysis of Wladimir Kietov and the experimental support of Birgit Witschel, Georg Maiberg, Sascha Graf, Lars Reichert and Florian Posselt are gratefully acknowledged.

References

1. L. Zhang, B.G. Thomas, Metall. Mater. Trans. B **37**, 733 (2006). https://doi.org/10.1007/s11663-006-0057-0
2. C.G. Aneziris, U. Fischer, M. Emmel, J. Hubálková, H. Berek, Keram. Z. **64**, 124 (2012)
3. H.V. Atkinson, G. Shi, Prog. Mater. Sci. **48**(5), 457 (2003). https://doi.org/10.1016/S0079-6425(02)00014-2
4. R.D. Thomson, J.W. Hancock, Int. J. Fract. **26**, 99 (1984). https://doi.org/10.1007/BF01157547
5. K. Wallin, P. Nevasmaa, A. Laukkanen, T. Planman, Eng. Fract. Mech. **71**(16–17), 2329 (2004). https://doi.org/10.1016/j.engfracmech.2004.01.010
6. W.C. Leslie, Trans. Iron Steel Soc. AIME **2**, 1 (1983)
7. W.M. Garrison Jr., N.R. Moody, J. Phys. Chem. Solids **48**(11), 1035 (1987). https://doi.org/10.1016/0022-3697(87)90118-1
8. Y.-C. Lin, M.-S. Chen, J. Zhang, Mat. Sci. Eng. A **499**(1–2), 88 (2009). https://doi.org/10.1016/j.msea.2007.11.119
9. G. Quan, Y. Tong, G. Luo, J. Zhou, Comp. Mater. Sci. **50**(1), 167 (2010). https://doi.org/10.1016/j.commatsci.2010.07.021
10. R. Chaouadi, J.L. Puzzolante, Int. J. Pres. Ves. Pip. **85**(11), 752 (2008). https://doi.org/10.1016/j.ijpvp.2008.08.004
11. Y.C. Chi, S. Lee, K. Cho, J. Duffy, Mat. Sci. Eng. A **114**, 105 (1989). https://doi.org/10.1016/0921-5093(89)90850-2
12. H.-W. Viehrig, J. Böhmert, J. Džugan, in *From Charpy to Present Impact Testing*. ed. by D. François, A. Pineau (Elsevier, Amsterdam, 2002), p.245
13. F. Jiang, K.S. Vecchio, Rev. Sci. Instrum. **78**, 063903 (2007). https://doi.org/10.1063/1.2746630
14. A. Weidner, *Deformation Processes in TRIP/TWIP Steels* (2020). https://doi.org/10.1007/978-3-030-37149-4
15. M. Emmel, C.G. Aneziris, G. Schmidt, D. Krewerth, H. Biermann, Adv. Eng. Mater. **15**(12), 1188 (2013). https://doi.org/10.1002/adem.201300118
16. Bodycote plc, *Hot Isostatic Pressing–Simple Equations for Better Materials* (2014)
17. K. Koch, V. Kietov, S. Henschel, L. Krüger, EPJ Web Conf. **250**, 03001 (2021). https://doi.org/10.1051/epjconf/202125003001
18. International Organization for Standardization, *Metallic Materials–Tensile Testing–Part 1: Method of test at room temperature* (ISO 6892–1) (2019)

19. International Organization for Standardization, *Metallic Materials — Tensile Testing at High Strain Rates — Part 2: Servo-Hydraulic and Other Test Systems* (ISO 26203–2) (2011)
20. S. Henschel, Dissertation. TU Bergakademie Freiberg (2018). https://nbn-resolving.org/urn: nbn:de:bsz:105-qucosa2-331498
21. American Society for Testing and Materials (ASTM), *Standard Test Method for Measurement of Fracture Toughness*. American Society for Testing and Materials, West Conshohocken, PA (ASTM E 1820) (2015)
22. International Organization for Standardization, *Metallic Materials–Unified Method of Test for the Determination of Quasistatic Fracture Toughness*. International Organization for Standardization, Geneva, Switzerland (ISO 12135) (2016)
23. S. Henschel, L. Krüger, Mater. Test. **57**(10), 837 (2015). https://doi.org/10.3139/120.110785
24. S. Henschel, L. Krüger, Eng. Fract. Mech. **133**(S1), 62 (2015). https://doi.org/10.1016/j.engfracmech.2015.05.020
25. S. Henschel, L. Krüger, Int. J. Fract. **201**, 235 (2016). https://doi.org/10.1007/s10704-016-0127-5
26. V. Kietov, S. Henschel, L. Krüger, Eng. Fract. Mech. **188**, 58 (2018). https://doi.org/10.1016/j.engfracmech.2017.07.009
27. V. Kietov, S. Henschel, L. Krüger, Eng. Fract. Mech. **210**, 320 (2019). https://doi.org/10.1016/j.engfracmech.2018.06.035
28. S. Henschel, F. Posselt, S. Dudczig, T. Wetzig, C.G. Aneziris, L. Krüger, Proc. Struct. Int. **28**, 1369 (2020). https://doi.org/10.1016/j.prostr.2020.10.108
29. H.A. Richard, *Bruchvorhersagen bei überlagerter Normal- und Schubbeanspruchung von Rissen* (VDI, Düsseldorf, 1985)
30. H. Sarangi, K. Murthy, D. Chakraborty, Eng. Fract. Mech. **88**, 63 (2012). https://doi.org/10.1016/j.engfracmech.2012.04.006
31. S. Henschel, L. Krüger, MATEC Web Conf. **12**, 04007 (2014). https://doi.org/10.1051/matecconf/20141204007
32. S. Henschel, D. Krewerth, F. Ballani, A. Weidner, L. Krüger, H. Biermann, M. Emmel, C.G. Aneziris, Adv. Eng. Mater. **15**(12), 1216 (2013). https://doi.org/10.1002/adem.201300125
33. S. Henschel, J. Gleinig, T. Lippmann, S. Dudczig, C.G. Aneziris, H. Biermann, L. Krüger, A. Weidner, Adv. Eng. Mater. **19**(9), 1700199 (2017). https://doi.org/10.1002/adem.201700199
34. M. Seleznev, S. Henschel, E. Storti, C.G. Aneziris, L. Krüger, A. Weidner, H. Biermann, Adv. Eng. Mater. **22**(2), 1900540 (2020). https://doi.org/10.1002/adem.201900540
35. A. Krabiell, Dissertation. RWTH Aachen (1982)
36. S. Henschel, L. Krüger, Steel Res. Int. **87**(1), 29 (2016). https://doi.org/10.1002/srin.201400567
37. International Organization for Standardization, *Metallic Materials–Measurement of Fracture Toughness at Impact Loading Rates using Precracked Charpy-Type Test Pieces*. International Organization for Standardization, Geneva, Switzerland (ISO 26843) (2015)
38. E. Lucon, J. ASTM Int. **8**(10), 103644 (2011). https://doi.org/10.1520/JAI103644
39. S. Henschel, L. Krüger, Fract. Struct. Integrity **9**(34), 326 (2015). https://doi.org/10.3221/IGF-ESIS.34.35
40. H. Roy, N. Parida, S. Sivaprasad, S. Tarafder, K.K. Ray, Mat. Sci. Eng. A **486**(1–2), 562 (2008). https://doi.org/10.1016/j.msea.2007.09.036
41. H. Roy, H.N. Bar, S. Sivaprasad, S. Tarafder, K.K. Ray, Int. J. Pres. Ves. Pip. **87**(10), 543 (2010). https://doi.org/10.1016/j.ijpvp.2010.08.016
42. S. Henschel, V. Kietov, F. Deirmina, M. Pellizzari, L. Krüger, Mat. Sci. Eng. A **709**, 152 (2018). https://doi.org/10.1016/j.msea.2017.10.053

Chapter 28
Influence of Filter Structure and Casting System on Filtration Efficiency in Aluminum Mold Casting

Benedict Baumann, Andreas Keßler, Claudia Dommaschk, and Gotthard Wolf

28.1 Introduction

In 1974, the company Swiss Aluminium Limited developed the first ceramic foam filters for the filtration of liquid metal to increase the purity of the molten metal [1]. Only one year later, another patent was registered, which improved the ceramic foam filter concerning its properties [2]. Since then, ceramic foam filters have been continuously improved as well as modified and are now available on the market in a variety of different materials, porosities, sizes, and geometries. In the cause of the energy-intensive raw aluminum production, the use of secondary raw materials becomes more and more important. Through this the melts get more contaminated with impurities; the further development of foam ceramic filters is essential for the foundry industry. Through this, the SFB 920 started with the development of intelligent filter materials and filter systems. The SFB 920 is focusing on the further development of filter materials for ceramic foam filters for the foundry industry. Thus, a major contribution to Zero Defect Materials has already been made for molten metal filtration.

B. Baumann (✉) · A. Keßler · C. Dommaschk · G. Wolf
Foundry Institute, Technische Universität Bergakademie Freiberg, Bernhard-von-Cotta-Str. 4, 09599 Freiberg, Germany
e-mail: Benedict.Baumann@gi.tu-freiberg.de

A. Keßler
e-mail: Andreas.Kessler@gi.tu-freiberg.de

C. Dommaschk
e-mail: Claudia.Dommaschk@gi.tu-freiberg.de

G. Wolf
e-mail: Gotthard.Wolf@gi.tu-freiberg.de

© The Author(s) 2024
C. G. Aneziris and H. Biermann (eds.), *Multifunctional Ceramic Filter Systems for Metal Melt Filtration*, Springer Series in Materials Science 337,
https://doi.org/10.1007/978-3-031-40930-1_28

In foundry practice, those high-performance filters are used for the separation of exo- and endogenous non-metallic impurities that reduce the mechanical and casting properties of parts. In addition to cleaning the melt, the filters also ensure laminar flow in the gating system and thus prevent new oxide formation in the mold cavity. Thus, filters in the casting system offer the possibility of sustainably cleaning the melt during the casting process, i.e. directly before the mold is filled. The implementation of the ceramic foam filter in the gating system does not follow any clear rules and in foundry practice is usually guided by the available space on the pattern plate and, in the best case, by the design recommendations of the filter manufacturer. In conventional foundry simulations, the filter is only considered as a flow resistance for the melt. A prediction of the filtration effect is hardly possible with the simulation programs. Simulations have also been used in foundry research mainly to study the flow behavior of the melt through the filter. For example, Barkhudarov and Hirt [3] simulated the formation of impurities as a result of turbulent mold filling. Zadeh and Campbell [4] compared simulation results from MagmaSoft and Flow-3D with results from real casting trials. They could not find any correlation between the results, because the simulations do not take into account the formation of oxide surfaces on the melt front and the cooling of the melt through the filter. A more microscopic approach was taken by Acosta et al. [5–7] and Werzner et al. [8, 9] by simulating the flow of the melt in the separate pores and examined the effect of the flow on the deposition of the impurities on the pore walls. The simulations were limited to separate pores only and not to the total filter system. Thus, the effect of the filter position in the casting system on the filtration efficiency of the ceramic foam filters has not yet been adequately investigated on a macroscopic level.

The present work investigates the effect of filter position on the filtration efficiency of ceramic foam filters. For this purpose, real filter structures are scanned by micro-computed tomography. This scan is then loaded as an STL file into the simulation program Flow-3D. Thus, it is possible to investigate the four most common filter positions concerning their filtration efficiency. Furthermore, it is investigated how far the length of the filter and the roughness of the filter surface influence the particle filtration. The simulation results are then compared with results from real casting trials.

28.2 Simulation

The CFD simulation program Flow-3D from Flow Science is used for the simulations. The program is able to represent the filter as an independent structure in the casting system as well as the implementation of particles in a defined number and size. In the following, the further parameters of the simulations will be explained in more detail.

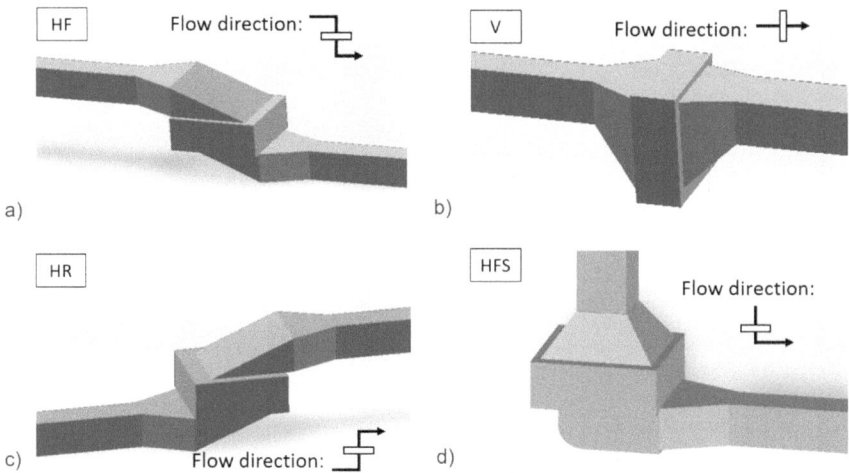

Fig. 28.1 Examined filter positions **a** horizontally falling (HF); **b** vertically (V); **c** horizontally rising (HR); **d** horizontally falling sprue (HFS) [11]

28.2.1 Geometry

For the simulations, four different filter chamber geometries respectively filter positions are investigated which can be seen in Fig. 28.1. The design of the filter chamber geometry was based on the design guidelines of Campbell [10]. Campbell set up the guidelines about a maximum possible laminarization of the melt as well as concerning foundry-specific framework conditions, but not with the focus on achieving the highest possible separation efficiency. The simulated filter chamber geometries are thus also used in practice and can be described as follows:

- (a) Filter position HF: The filter lies **h**orizontally in the runner and is flowed through from top to bottom (**f**alling).
- (b) Filter position V: The filter stands **v**ertically in the runner.
- (c) Filter position HR: The filter lies **h**orizontally in the runner and is flowed through from bottom to top (**r**ising).
- (d) Filter position HFS: The filter is located **h**orizontally below the **s**prue and the flow is from top to bottom (**f**alling).

28.2.2 Filter

Real filter geometries are used for the simulations with Flow-3D. To generate the filter geometries, commercially available 20 and 30 ppi ceramic foam filters with dimensions of $50 \times 50 \times 22$ mm are scanned using micro-CT. The scanned data set is then converted into an STL file (see Fig. 28.2) and can be used directly in

Fig. 28.2 Data set of a
ceramic foam filter geometry
scanned by micro-CT and
converted into an STL file
[11]

Flow-3D as geometry. To define the geometry as a filter in the simulation, the filter is
defined with certain parameters. These parameters are the surface roughness of the
filter and the drag coefficient. The drag coefficient defines how well or poorly the
particles adhere to the filter when they come into contact with it. For the simulations,
the drag coefficient is defined with the parameter 1, which means that every particle
that touches the filter sticks to the filter wall and is thus separated from the melt. The
surface roughness is used to define the filter material. For aluminum casting alloys,
filters made of alumina are used in practice. Fankhänel et al. [12] measured the
roughness of various filter materials and found, among other results, that the surface
roughness of alumina filters is 1.7 µm. For the simulations, the surface roughness is
thus defined as 1.7 µm. Furthermore, a simulation series is carried out with a surface
roughness of 7.3 µm. This value corresponds to the surface roughness of mullite
filters and represents a high contrast to the surface roughness of the alumina filters
to investigate a possible influence of the roughness of the filters on the separation
efficiency. To investigate the possible influence of the length of the filter on the
filtration efficiency, the filter position HF is also simulated with a double-length 30
ppi filter in the dimensions 50 × 50 × 44 mm and surface roughness of 1.7 µm. The
STL file of the 30 ppi filter is multiplied and arranged in series so that the two filters
are connected in a row and thus produce a longer filter.

28.2.3   Melt

An AlSi7Mg0.3 is selected as the aluminum casting alloy for the simulation. To minimize the computation time for the simulations, heat transfer processes between melt and mold wall as well as melt and filter are not considered.

28.2.4   Particle

Voigt et al. [13] and Le Brun et al. [14] investigated the separation efficiency of ceramic foam filters for continuous aluminum casting in practical casting tests at Constellium. A defined impurity content was set in the melt and the number, as well as size of the particles before and after the filter, were measured using LiMCA. Powdered alumina (Al_2O_3) and powdered spinel ($MgAl_2O_4$) were added to the melt to ensure a high impurity content in the melt. From over 140 individual measurements, an average particle count of 17,500 particles was determined, which is also the number of particles that flow through the filter per simulation. It should be noted that this is the number of particles in an intended contaminated melt. The impurity content does not represent the impurity content in industrial foundry practice. Table 28.1 shows the two particle types used in the simulation with their density, size, and number.

As with the tests at Constellium, two different types of particles are defined for the simulation to investigate the deposition efficiency of different types of particles. On the one hand, the deposition of aluminum oxide is analyzed, cause it is the most frequently occurring non-metallic inclusion in aluminum melts, and on the other hand, the deposition of spinel particles are examined, which can form in aluminum melts due to the production process. The spinel particles were intentionally chosen to be larger because they can grow in the melt and are therefore larger than the usually finely distributed oxide skins. Apart from the density, size, and distribution of the particles, no other parameters such as the shape or surface properties of the particles can be defined. Furthermore, there are no interactions between the particles themselves, and between particles and mold walls as well as particles and melt.

Table 28.1 Overview of the particles used in the simulations [11]

Type	Density [g/cm^3]	Size [μm]	Percentage [%]	Quantity [number]
Al_2O_3 (Alumina)	3.95	25	45	7875
		35	30	5250
		45	7.5	1313
$MgAl_2O_4$ (Spinel)	3.5	45	7.5	1312
		55	5	875
		80	5	875

Since the particles are much smaller than the pore diameter of the filters and there are no interactions between the individual particles, only deep-bed filtration can be represented with the simulation, but not cake or sieve filtration.

28.2.5 Further Boundary Conditions

Figure 28.3 shows the simulation process using the example of the HF filter layer with a 20 ppi filter. As can be seen, the particles are arranged in a bulk in front of the filter. To reduce the computing time of the simulations, the system is described as semi-stationary, i.e. at the beginning of the simulation the mold cavity is already filled with melt. The driving force for the melt flow is the metallostatic pressure respectively the geometry of the sprue. The metallostatic pressure can be regarded as approximately the same for all filter positions. When the melt is set in motion at the start of the simulation, the particles are also carried along by the melt flow and thus pass through the filter. The simulation runtime is limited to the duration of the filtration process. The simulation is completed when all free particles not bound in the filter have left the filter. To reduce the computing time even further, the entire filter is not examined, but only a 10 mm thick section, so that the examined area has a size of 50 × 10 × 22 mm.

To investigate in which areas of the filter particles were deposited, the filters are subdivided into 3 segments for evaluation. Figure 28.4 shows an example of the subsections for the filter layer HF and V.

Table 28.2 shows all 13 simulations performed with the respective parameters, where the number and type of particles are always identical.

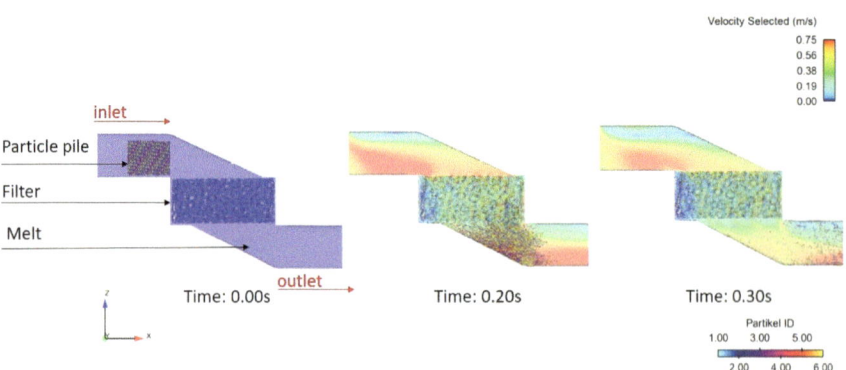

Fig. 28.3 Example of a simulation with the filter position HF for a 20 ppi filter [11]

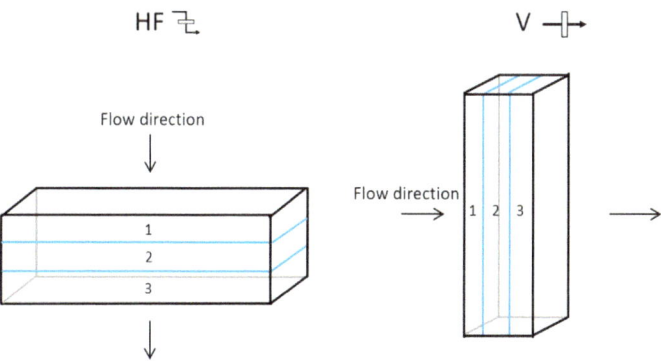

Fig. 28.4 Subdivision of the filters into individual segments

Table 28.2 Summary of all simulations performed with the different variables

Simulation number	Variables			
	Position filter chamber	Filter length [mm]	Filter porosity [ppi]	Surface roughness (μm)
1	HF ⫪	22	20	1,7
2	HF ⫪	22	30	1,7
3	V ⊣⊢	22	20	1,7
4	V ⊣⊢	22	30	1,7
5	HR ⌐	22	20	1,7
6	HR ⌐	22	30	1,7
7	HFS ·∟	22	20	1,7
8	HFS ·∟	22	30	1,7
9	HF ⫪	22	20	7,3
10	HF ⫪	22	30	7,3
11	V ⊣⊢	22	20	7,3
12	V ⊣⊢	22	30	7,3
13	HF ⫪	44	30	1,7

28.3 Results of the Simulation

To compare the different filter positions, the filtrations efficiency E was calculated for each one. The filtration efficiency E is defined through the number of particles before or in the filter and the number of particles after the filter. Where N_0 stands for the number of particles before or in the filter and N_1 for the number of particles after the filter.

$$E = \left(1 - \frac{N_0 - N_1}{N_0}\right) \times 100[\%] \tag{28.1}$$

Table 28.3 Filtration efficiencies determined in the simulations [11]

Filter position	20 ppi efficiency (%)	30 ppi efficiency (%)	Percentage increase (%)
HF ⏄	27.9	36.8	31.9
V ⊣⊢	22.7	33.2	46.3
HR ⌐	23.4	29.5	26.1
HFS ⌐	20.8	33.2	59.6

Table 28.3 gives an overview of the filtration efficiencies of the individual filter positions for the simulation of the 20 and 30 ppi filters with a surface roughness of 1.7 μm. As can be seen, the filter position in which the filter lies horizontally in the runner and is flowed through from top to bottom (HF) has the highest filtration efficiency for both types of filters 20 and 30 ppi. For all filter positions, a clear increase in filtration efficiency of at least 26% can be observed when using a 30 ppi filter compared to a 20 ppi filter. Especially for the filter positions where the melt flows directly into the filter (V and HFS), a significant increase in filtration efficiency of at least 46% can be observed.

28.3.1 Influence of Particle Size on Filtration Efficiency

To determine a possible influence of the simulated particle sizes (see Table 28.1) on the separation performance, the filtration efficiency of the individual particle sizes was determined for each filter position. The results for the 20 and 30 ppi filters are shown in Figs. 28.5 and 28.6. It can be also seen that the HF filter position has the highest filtration efficiency across all particle size classes for the 20 and 30 ppi filters. For the 20 ppi filters, there is generally no concrete difference in the filtration efficiency of the individual particle sizes from 25 to 80 μm. Even with the 30 ppi filters, there is hardly any significant difference between the particle sizes of 25 μm up to 55 μm. Only from a particle size of 80 μm is an increase in filtration efficiency noticeable. The differences between the individual particle size classes are too small compared to the pore diameter of the filter and so no significant difference can be proved. It is expected that the filtration efficiency increases with increasing particle size or decreasing pore size.

28.3.2 Influence of Particle Type on Filtration Efficiency

The particle size of 45 μm was used to study the influence of the particle type on the filtration efficiency. The same number of alumina particles and spinel particles were defined for each simulation (see Table 28.1). Table 28.4 shows the filtration efficiencies of the 45 μm particles at the HF and HFS filter positions for the 20

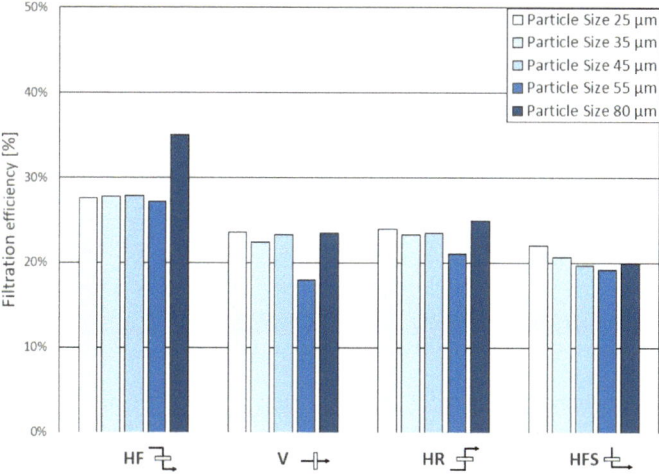

Fig. 28.5 Filtration efficiency of the different filter positions for the various particle sizes for the simulations with a 20 ppi filter [11]

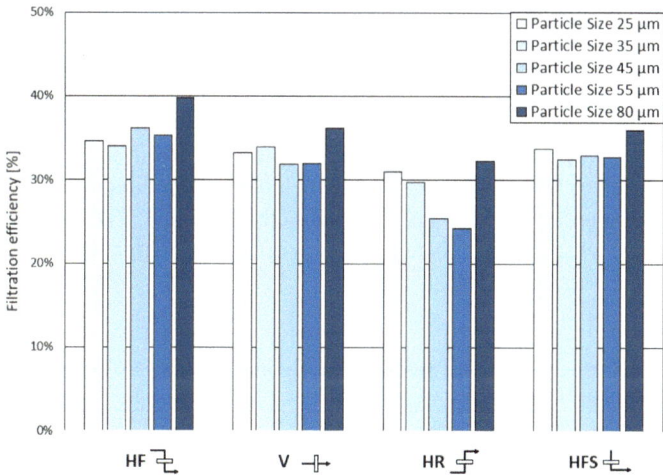

Fig. 28.6 Filtration efficiency of the different filter positions for the various particle sizes for the simulations with a 30 ppi filter [11]

and 30 ppi filters, respectively. As can be seen, the filtration efficiencies do not differ significantly. The largest change is for the HF filter position (30 ppi) with a difference of 3.4% points. The two different particle types were separated with an average difference of approximately 1% point. Thus, the influence of the particle type in the simulation is negligible. Probably the difference in density between the two particles is too small to cause a difference in filtration efficiency.

Table 28.4 Filtration efficiencies of the 45 μm alumina and spinel particles [11]

Filter position	Filter porosity [ppi]	Alumina particles 45 μm	Spinel particles 45 μm
		Filtration efficiency [%]	Filtration efficiency [%]
HF ⊐⌐	20	27.9	27.7
HF ⊐⌐	30	37.9	34.5
HFS ⌐⌐	20	19.8	19.4
HFS ⌐⌐	30	32.8	32.9

28.3.3 Deposition of Particles in the Filter

To investigate where the particles are deposited in the filter, the simulated filters were divided into 3 segments as shown in Fig. 28.4. Table 28.5 shows the respective proportion of particles deposited in one of the three areas for the filter positions HF and V for 20 and 30 ppi filters respectively.

As can be seen from the table, the difference between 20 and 30 ppi filters at filter position HF is between 0.1 and 0.4% points. For filter position V, the distinctions are also only 0.6 to 1.8% points, so no significant difference can be detected between 20 and 30 ppi filters. However, it can be seen from the table that the particles in filter position HF are deposited mainly in the first third of the filter (approx. 44%) and that the sedimentation of the particles decreases with increasing filter length. In comparison, the particles in filter layer V are deposited to a likewise high degree in all 3 segments. One possible reason for this could be found in the flow direction of the filters. Whereas with filter layer HF the melt has to be redirected before it passes the filter, with filter layer V the melt flows directly through it. The redirection slows down the melt, which leads to better filtration efficiency [15] and thus explains the increased separation efficiency.

Table 28.5 Percentage of deposited particles in each filter plane, as well as the difference (Δ) between 20 and 30 ppi [11]

Filter position	Filter porosity [ppi]	Filter plane					
		1		2		3	
		Fraction of filtered particles [%]	Δ	Fraction of filtered particles [%]	Δ	Fraction of filtered particles [%]	Δ
HF ⊐⌐	20	44.2	0.4	31.4	0.1	24.4	0.3
HF ⊐⌐	30	43.8		31.5		24.7	
V ⊣⊢	20	35.0	0.6	36.0	1.8	29.0	1.3
V ⊣⊢	30	35.6		34.2		30.2	

Table 28.6 Comparison of filtration efficiency with a different surface roughness of the filters

Filter			Filtration efficiency [%]	
Position	Surface roughness [μm]	Filter porosity [ppi]	Total	Δ
HF ⌐⌐	1,7	20	27,9	−3,1
HF ⌐⌐	7,3	20	24,8	
HF ⌐⌐	1,7	30	36,8	−2,5
HF ⌐⌐	7,3	30	34,3	
V —⊩→	1,7	20	22,7	−1,4
V —⊩→	7,3	20	21,3	
V —⊩→	1,7	30	33,2	−0,7
V —⊩→	7,3	30	32,5	

28.3.4 Influence of the Surface Roughness of the Filter on the Filtration Efficiency

To investigate a possible influence of the surface roughness of the filter on the filtration efficiency, the filter positions HF and V were simulated with a surface roughness of 1.7 μm on the one hand and with a surface roughness of 7.3 μm on the other hand for 20 and 30 ppi filters (see Table 28.2). The calculated filtration efficiencies are shown in Table 28.6.

As can be seen from the table, the filtration efficiencies are always lower for the filters with higher surface roughness. It should be noted that the difference in filtration efficiency for filter position HF, with a max. of 3.1% points, is significantly higher than for filter position V, which has a difference of 1.4% points. Since, due to the redirection, the melt flows through the filter more slowly at filter position HF than at filter position V, it is therefore possible that the surface roughness of the filter has a more significant effect on the filtration efficiency at slower flow rates than at faster flow rates.

28.3.5 Influence of Filter Length on Filtration Efficiency

To investigate the influence of the filter length on the filtration efficiency, a 30 ppi filter with a double filter length of 44 mm was simulated at a surface roughness of 1.7 μm for the filter layer HF. Table 28.7 shows the results of the simulation using the double filter length as well as the single filter length with otherwise identical parameters.

As can be seen, the difference between the two filtration efficiencies is 13.7% points, which corresponds to an increase of approx. 37%. Therefore, a filter twice as long is much more efficient in filtering particles. In this simulation, the filter is again

Table 28.7 Comparison of filtration efficiencies with different filter lengths

Filter			Filtration efficiency [%]	
Position	Filter length [mm]	Filter porosity [ppi]	Total	Δ
HF ⌐⌐	22	30	36,8	+13,7
HF ⌐⌐	44	30	50,5	

Table 28.8 Percentage of deposited particles in each filter plane, as well as the difference between 22 and 44 mm filter length

Filter position	Filter length [mm]	Filter plane		
		1	2	3
		Fraction of filtered particles [%]	Fraction of filtered particles [%]	Fraction of filtered particles [%]
HF ⌐⌐	22	43,8	31,5	24,7
HF ⌐⌐	44	47,0	30,1	22,8

divided into 3 segments to investigate in which areas the particles are deposited in the filter (see Table 28.8).

Just as with the normal filter length of 22 mm, almost half of the particles are located in the first third of the filter when using a filter length of 44 mm.

28.4   Summary of the Simulations

Table 28.9 shows all simulations with the variables: Filter position, filter length, porosity, and surface roughness of the filter as well as the corresponding filtration efficiencies. For each filter position, the use of a 30 ppi filter increases the filtration efficiency by 26% up to 59.6% which corresponds to an average increase of about 42.4% compared to a 20 ppi filter. The filter position HF has the highest filtration efficiency for both the 20 ppi and 30 ppi filters. Increasing the surface roughness of the filters from 1.7 to 7.3 μm leads to a marginal reduction in filtration efficiency, and the filter position HF achieves a higher filtration efficiency than the filter position V again. By doubling the filter length from 22 to 44 mm, the filtration efficiency of the filter position HF was increased by 37% to 50.5%, which provides the highest filtration efficiency. It should be noted, however, that most particles are deposited in the first third of the filter and that in practice a longer filter can lead to increased cooling of the melt and thus to freezing of the melt in the filter.

Table 28.9 Summary of all simulations with the corresponding variables as well as the filtrate efficiencies

Simulation number	Variables				Filtration efficiency [%]
	Filter position	Filter length [mm]	Filter porosity [ppi]	Surface roughness (μm)	
1	HF ⌐L	22	20	1,7	27,9
2	HF ⌐L	22	30	1,7	36,8
3	V ─╟→	22	20	1,7	22,7
4	V ─╟→	22	30	1,7	33,2
5	HR ⌐	22	20	1,7	23,4
6	HR ⌐	22	30	1,7	29,5
7	HFS ⊥	22	20	1,7	20,8
8	HFS ⊥	22	30	1,7	33,2
9	HF ⌐L	22	20	7,3	24,8
10	HF ⌐L	22	30	7,3	34,3
11	V ─╟→	22	20	7,3	21,3
12	V ─╟→	22	30	7,3	32,5
13	HF ⌐L	44	30	1,7	50,5

28.5   Casting Trials

The simulations have shown that the deposition of particles in the filter depends on the filter position. To verify this, casting trials are carried out by adding foreign particles to a melt and then casting it in four molds with the filter positions used in the simulations. The castings were generated with both 20 and 30 ppi filters. The casting trials aim to analyze the deposition behaviors of the particles in the filter. Since in reality deep-bed filtration, as well as sieve and cake filtration, are effective, a comparison of the filtration efficiency between simulation and casting trials is not useful.

28.5.1   Geometry

As in the simulations, the filter positions HF, V, HR, and HFS (see Fig. 28.1) are also examined in the casting trials. The corresponding models differ only in the design of the filter chamber and are otherwise identical from the casting technology point of view. The main focus of the models is that design-related influencing factors are minimized and the results can only be assigned to the different filter chamber geometries. The molds are all made of furan resin-bonded molding material.

Fig. 28.7 Micrograph of the Duralcan® alloy with 15 wt% Al$_2$O$_3$

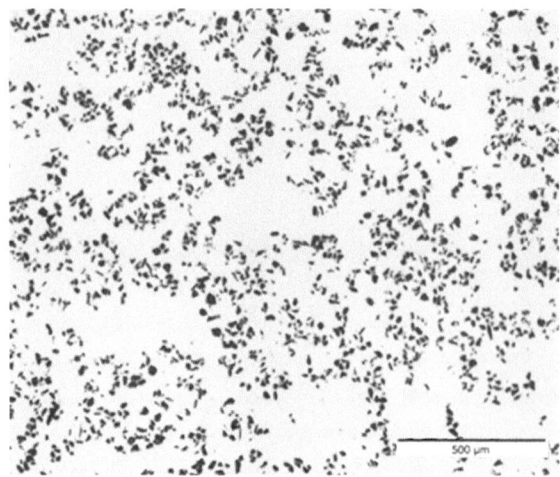

28.5.2 Filter

For the casting trials, only filters made of alumina with dimensions of 50 × 50 × 22 mm from Hofmann Ceramic GmbH (Breitscheid, Germany) with porosities of 20 ppi and 30 ppi were used. The filters from Hofmann Ceramic GmbH were also used for the micro-CT scan and thus for the simulations.

28.5.3 Melt with Foreign Particles

As in the simulations, an AlSi7Mg0.3 is used as the base alloy for the casting trials. To investigate the behavior of the particles during the filtration process, Duralcan® is added to the melt. Duralcan® is a metal matrix composite (MMC) that is normally used for "high-end" applications such as the frame of the space shuttle [16]. Dural-can® consists of an aluminum matrix reinforced with Al$_2$O$_3$ particles. To enrich the melt with 20 µm of foreign particles, Duraclan® was added. Figure 28.7 shows a micrograph of the Duralcan® master alloy used, which is reinforced with 15% Al$_2$O$_3$ particles.

28.5.4 Execution of the Casting Trials

The melt is prepared in a resistance-heated 20 kW crucible furnace with a capacity of 30 kg of molten aluminum. Per batch, 27 kg of ingot material of the alloy AlSi7Mg0.3 is melted and degassed for 20 min with an impeller. At a temperature of 750 °C, 3 wt.% Duralcan® is added, which corresponds to 0.8 kg Duralcan® on the total melting quantity. Before the metal is transferred from the crucible to the ladle, it is homogenized manually by stirring with a graphite rod. This ensures that the Al_2O_3 particles are homogeneously distributed in the melt and their sedimentation is prevented. The alloy was poured with a casting temperature of 730 °C. From one crucible, 8 molds with the same filter chamber geometry are cast. After solidification and cooling of the metal, the casting is removed from the mold and the filter chamber with the filter cast in it is prepared for metallographic examinations.

28.5.5 Metallographic Examinations of the Filter

Like the simulations, the filters are also divided into three segments (see Fig. 28.4). Individual images of each zone are made by using a 3D microscope with 200 × magnification. This was fitted into a panoramic image and processed by image analysis. The size, morphology, and appearance of the particles are known from previous investigations and therefore a simple identification of the particles is enabled. Figure 28.8 shows an example of the metallographic evaluation of a 20 ppi filter with the filter position HF. The red-colored areas are the filtered Al_2O_3 particles from the Duralcan® alloy.

28.6 Results of the Casting Trials

As in the simulations, the filters were also divided into 3 sections for the casting trials (see Fig. 28.4) to investigate where the particles were deposited in the filter. Figure 28.9 shows the assembled photomicrographs of the filter positions examined for the 20 and 30 ppi filters and the percentage of particles deposited in each plane. The color scale helps to identify especially effective zones. As can be seen in the figure, in all filter positions where the flow is vertical (HFS, HF, HR), approx. 50

Fig. 28.8 Example of filter evaluation using filter position HF with a 20 ppi filter [11]

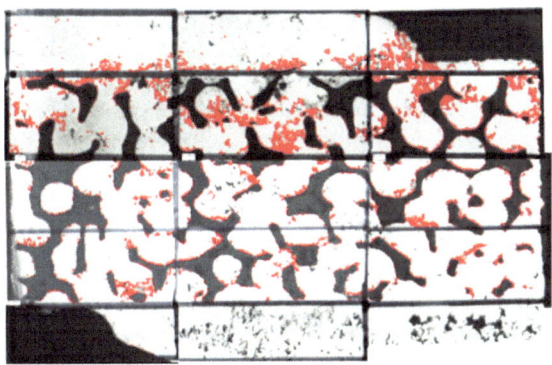

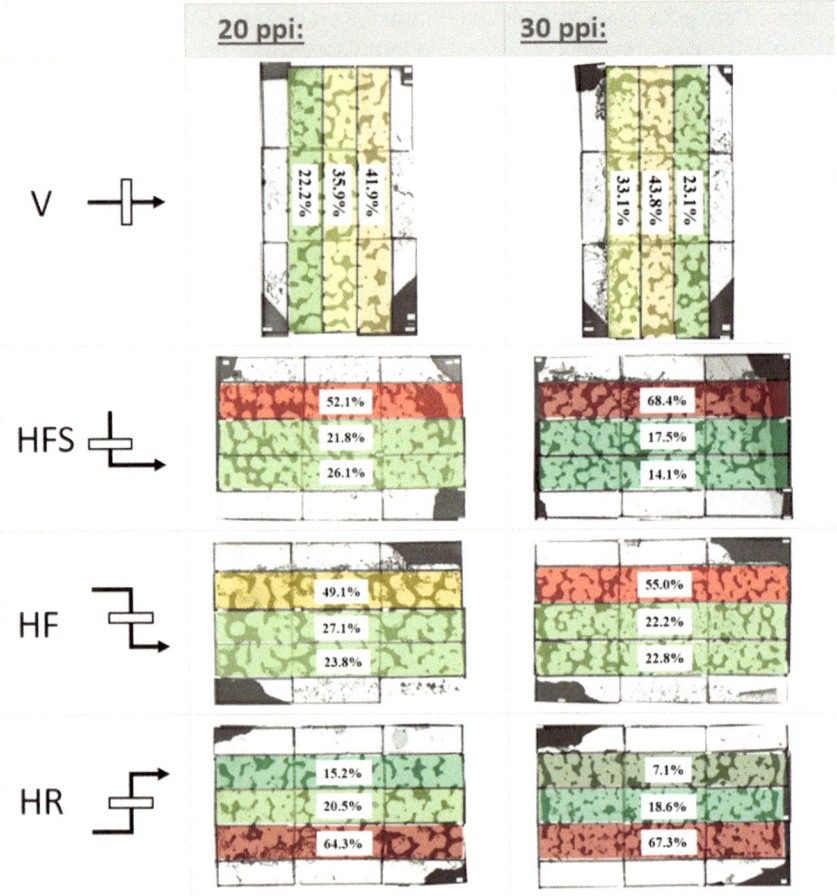

Fig. 28.9 Percentage of deposited particles in each filter plane for the cast filter positions V, HFS, HF, and HR for the 20 and 30 ppi filters

to almost 70% of all particles are separated in the first third of the filter. In almost all cases, the filtration effect decreases with increasing filter length, so that in the last third of the filter only 7 to 26% of the particles are separated. The deposition of particles behaves differently in filter position V, where the filter stands horizontally in the mold. In this filter position, the deposition of particles in the individual filter levels is more uniform and ranges between 22 and 44%. This could be due to the increased flow rate compared to the other filter positions, which ensures that the particles are carried further into the depth of the filter. In the filter position HFS, the filtration efficiency is at its highest in the first third of the filter, despite the increased melt velocity. This can be due to the fact that, in this filter position, the melt hits the filter directly from the gate and is therefore strongly turbulent, which leads to the formation of oxides. These oxides are separated by sieve and cake filtration upstream of the filter so that the 20 μm Al_2O_3 particles are separated primarily by the effect of cake filtration in the first third of the filter. It was also observed in the simulations that the particles are deposited more uniformly over the total filter length in filter position V and that most of the particles are deposited in the first third of the filter in filter position HF.

28.7 Conclusion

In the present work, the influence of the filter position on the filtration efficiency of 20 ppi and 30 ppi ceramic foam filters was investigated for the first time by means of numerical simulations. Part of the simulation results was subsequently evaluated with real casting trials. Both the simulations and the casting trials were able to show similar separation behavior of the particles in the filter layers. This demonstrates that the numerical simulation of particle filtration from a metallic melt gives predictions in reasonable agreement with experimental measurements.

The following conclusions are drawn from the simulations:

- The highest filtration efficiency is achieved when the filter is placed horizontally in the runner and the flow is from top to bottom.
- In all filter positions where the filter lies horizontally in the runner, deposition of inclusions occurs primarily in the first third of the filter. If the filter is positioned vertically in the runner, deposition tends to occur more uniformly along the length of the filter. This could also be proven with real casting trials.
- The simulations have shown that the roughness of the filter surface has a minor influence on the filtration efficiency. The filtration efficiency decreases slightly when the roughness of the filter surface increases.
- By doubling the filter length from 22 to 44 mm, a 37% increase in filtration efficiency can be achieved for Al_2O_3 particles with a size of 20 μm. Nevertheless, 47% of the particles are deposited in the first 15 mm of the 44 mm long filter, so that the use of longer filters is not advised from a foundry technology point of view [17].

Acknowledgements The authors would like to acknowledge the German Research Foundation (Deutsche Forschungsgemeinschaft, DFG) for funding this work and project as a part of the Collaborative Research Centre 920 "Multi-Functional Filters for Metal Melt Filtration – A Contribution towards Zero Defect Materials" (Project ID: 169148856 – CRC 920, Subproject S03).

Particular acknowledgment is expressed to the commitment of the staff members of the Foundry Department, especially to Mr. Klaus Eigenfeld, Mrs. Eva Hoppach, and Mr. Björn G. Dietrich.

References

1. J.P. Michael, T.J. Gray, U.S. Patent 3,893,917, 8 July 1975
2. J.C. Yarwood, J.E. Dore, R.K. Preuss, U.S. Patent 3,962,081, 8 June 1976
3. M.R. Barkhudarov, C.W. Hirt, Tracking defects. (Flow Science, Inc.) https://www.flow3d.com/wp-content/uploads/2014/08/Tracking-Defects.pdf. Accessed 03 June 2022
4. A. Zadeh, J. Campbell, AFS transactions. 02–020 (2002)
5. F.A. Acosta G., A.H. Castillejos E., Metall. Mater. Trans. B **31**, 491–502 (2000). https://doi.org/10.1007/s11663-000-0155-3
6. F.A. Acosta G., A.H. Castillejos E., Metall. Mater. Trans. B **31**, 503–514 (2000). https://doi.org/10.1007/s11663-000-0156-2
7. F.A. Acosta G., A.H. Castillejos E., J.M. Almanza R., A. Flores V., Metall. Mater. Trans. B **26**, 159–171 (1995). https://doi.org/10.1007/BF02648988
8. C. Demuth, E. Werzner, M.A.A. Mendes, H. Krause, D. Trimis, S. Ray, Adv. Eng. Mater. 19 9, (2017). https://doi.org/10.1002/adem.201700238
9. E. Werzner, M. Abendroth, C. Demuth, C. Settgast, D. Trimis, H. Krause, S. Ray, Adv. Eng. Mater. 19 9, (2017). https://doi.org/10.1002/adem.201700240
10. J. Campbell (ed.), *Complete Casting Handbook. Metal Casting Processes, Metallurgy, Techniques and Design*, (Butterworth-Heinemann Ltd, Oxford, 2015)
11. B. Baumann, A. Kessler, E. Hoppach, G. Wolf, M. Szucki, O. Hilger, Arch. Foundry Eng. Vo. **21**(3), 70–80 (2021). https://doi.org/10.24425/afe.2021.138668
12. B. Fankhänel, M. Stelter, C. Voigt, C.G. Aneziris, Adv. Eng. Mater. 19 9, (2017). https://doi.org/10.1002/adem.201700084
13. C. Voigt, E. Jäckel, F. Taina, T. Zienert, A. Salomon, G. Wolf, C.G. Aneziris, P. Le Brun, Metall. Mater. Trans. B. **48**(1), 497–505 (2017). https://doi.org/10.1007/s11663-016-0869-5
14. P. Le Brun, F. Taina, C. Voigt, E. Jäckel, C.G. Aneziris, in E. Williams (Eds.), Light Metals 785–789 (2016). https://doi.org/10.1007/978-3-319-48251-4_133
15. H. Duval, C. Rivière, É. Laé, P. Le Brun, J. Guillot, Metall. Mater. Trans. B. **40**(2), 233–246 (2009). https://doi.org/10.1007/s11663-008-9222-y
16. O. Beffort, Metal matrix composites: properties, applications and machining (in German), Paper presented at the 6. Internationales IWF-Kolloquium, ETH Zürich, Egerkingen, Schweiz, 18–19 April 2002
17. E. Jäckel, Dissertation. Technische Universität Bergakademie Freiberg (2019)

Chapter 29
Cleanness of Molten Steel—Active and Reactive, Exchangeable Filter Systems for the Continuous Casting of Steel

Tony Wetzig, Andreas Baaske, Sven Karrasch, Steffen Dudczig, and Christos G. Aneziris

29.1 Introduction

Steel remains one of the most versatile and most important materials in the global economy. The ongoing innovation in steelmaking and casting processes lays the foundation for major advances in new technologies across all industries. In the face of climate change, the improvement of steel plant efficiency makes an important contribution to the reduction of emissions and global energy consumption.

In this regard, the control of non-metallic inclusions in cast steel melts plays a major role for improving the performance and reliability of steel products as well as to reduce tool wear and scrap rates during their production [1].

Avoiding unnecessary inclusion formation and removing inclusions by exploiting their physical (buoyancy) characteristics are a top priority. In modern steel plants,

T. Wetzig (✉) · S. Dudczig · C. G. Aneziris
Institute of Ceramics, Refractories and Composite Materials, Technische Universität
Bergakademie Freiberg, Agricolastraße 17, 09599 Freiberg, Germany
e-mail: tony.wetzig@ikfvw.tu-freiberg.de

S. Dudczig
e-mail: steffen.dudczig@ikfvw.tu-freiberg.de

C. G. Aneziris
e-mail: aneziris@ikfvw.tu-freiberg.de

A. Baaske · S. Karrasch
Thyssenkrupp Steel Europe AG, Kaiser-Wilhelm-Straße 100, 47166 Duisburg, Germany
e-mail: andreas.baaske@thyssenkrupp.com

S. Karrasch
e-mail: sven.karrasch@thyssenkrupp.com

C. G. Aneziris and H. Biermann (eds.), *Multifunctional Ceramic Filter Systems for Metal Melt Filtration*, Springer Series in Materials Science 337,
https://doi.org/10.1007/978-3-031-40930-1_29

this can be achieved by applying advanced secondary metallurgy technologies [2, 3], flow control measures [4–9] as well as high-performance slags and refractories [1, 3, 10, 11]. Unfortunately, these methods are rather ineffective regarding contained micro inclusions or tertiary and quaternary inclusions, which form in late stages of the casting and cooling procedure [1]. The agglomeration of these small inclusions can result in the delayed formation of clusters with critical size [12–14].

An effective way to deal with these types of inclusions is the steel melt filtration by means of reactive filter materials such as carbon-bonded alumina, which possesses the necessary refractoriness and thermal shock resistance as well as the capability to interact with the molten steel and contained non-metallic micro inclusions by means of reactive filtration mechanisms [15–19].

In the foundry industry, metal melt filtration already is the state of the art. During the batch process, the defined quantity of melt is forced through the open pores of the filters installed in the gating system or feeders [20–24]. In steelmaking, the dominant casting process is the so-called continuous casting. Implementation of filters in this continuous process poses several challenges incorporating long casting sequences comprising hundreds or thousands of tons of cast steel in the presence of corrosive slags and atmosphere. Filters installed in the so-called tundish, a buffer vessel between the ladle and the molds, need to maintain their filtration performance and structural integrity over much longer time periods. Thus, the risk of premature clogging or fracture is increased to an unacceptable level in most cases using conventional filter types [4]. The application of multi-hole dam filters or baffle filters provides a compromise between steel melt filtration, flow control and reliability [2, 4, 6, 7, 25–27]. However, due to their simple structure, the removal of micro inclusions with diameters smaller than 20 µm is rather inefficient [7, 26].

As a consequence, a novel approach to filtration in continuous casting is necessary. By immersing filters from the top into the tundish, the filters can be replaced at any time during the ongoing casting procedure [28]. The concept was investigated based on two different filter types.

First, cylindrical carbon-bonded alumina foam filters were developed based on the replication technique [29, 30]. The replication of the polyurethane foam structure was investigated using different coating routines. Additionally, a novel robo gel-casting process for the manufacturing of so-called "spaghetti" filters based on alginate was investigated [31]. These gel-cast cellular structures have the advantage that no sacrificial template is needed resulting in a filter with full filter strut cross-section, higher mechanical strength and an adjustable cellular macrostructure.

After mechanical and structural investigations to find the most suitable manufacturing parameters for both foam replication and gel casting, prototypes of both filter types were evaluated a steel casting simulator. Following the lab-scale casting investigations, industrial tests were performed at the continuous casting plant of thyssenkrupp Steel Europe in Duisburg, Germany. The findings delivered the proof of principle for the new steel melt filtration concept in continuous casting in agreement with the models of the CRC 920.

29.2 Carbon-Bonded Alumina Foam Filters

For the manufacturing of the foam filters for industrial continuous casting of steel, cylindrical polyurethane foams with 10 pores per inch acted as templates. The diameter and the height of the templates was 200 mm respectively and a central bore with a diameter of 40 mm was applied for fixing the filter during the immersion procedure. Optionally, eight additional circularly arranged bores with 40 mm diameter were implemented in the foam structure. These "macro-channels" aimed for additional melt flow conditioning during application [28].

Both foam geometries, either with macro-channels (8c) or without macro-channels (0c) were investigated in industrial casting trials. The structurally weaker 8c geometry was used as a basis for preliminary lab-scale investigations regarding the structural and mechanical properties of the filters as a function of the manufacturing process. Figure 29.1 shows the polyurethane templates and the final carbon-bonded filters for both geometries.

The filter material selected for the investigation was carbon-bonded alumina, mainly due to its high resistance against thermal shock, corrosion and erosion [32], its favorable priming properties in contact with molten steel [16] as well as its reactive filtration effects on the melt cleanliness [33]. The dry base material composition according to Emmel et al. [15] contained 66 wt.% fine-grained alumina, 20 wt.% modified coal tar pitch, 7.7 wt.% graphite and 6.3 wt.% carbon black. The dry mixture was transformed into a slurry by adding water, suitable dispersing and defoaming agents as well as lignin sulfonate as wetting and temporary binding agent. The solid content was adjusted depending on the intended coating procedure.

The primary coating of the templates was performed by impregnation to wet the foam with the ceramic material and subsequent centrifugation in order to remove excessive slurry from the pore structure. For that purpose, a high-viscosity slurry with 81.1% solid content was prepared using an intensive mixer. With the aid of additional coatings, the filter mass and the mean strut thickness can be increased in order to increase the filter strength at the cost of filter porosity. Thus, three different coating routines were investigated:

- impregnation/centrifugation (C),
- impregnation/centrifugation + surface spray coating (CS) and
- impregnation/centrifugation + dip coating + surface spray coating (CDS).

The solid content of the base slurry was adjusted to a solid content of 60 wt.% for dip coating and 70 wt.% for spray coating. Both slurries were prepared by ball mixing on a drum roller mixer for 24 h.

First, key rheological properties of the used slurries were investigated. The yield stress was analyzed by means of stress-controlled oscillation tests and the flow curve was determined in controlled shear rate tests. The key findings are shown in Table 29.1.

All slurries possessed shear-thinning behavior, which is considered favorable for the replication technique. In this way, the removal of excessive slurry is supported

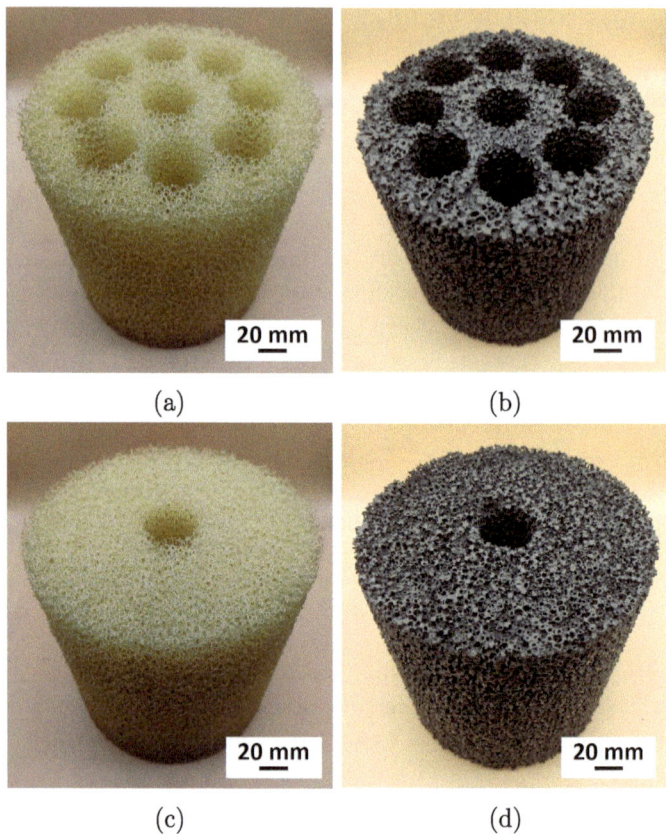

Fig. 29.1 Polyurethane foams and fired carbon-bonded alumina filters with 8c geometry (**a**, **b**) and 0c geometry (**c**, **d**) [34]

Table 29.1 Rheological features of investigated slurries for the manufacturing of carbon-bonded alumina foam filters [29]

Slurry application	Dip coating	Spray coating	Centrifugation
Solid content/wt.%	60.0	70.0	81.1
Yield stress/Pa	0.9	1.7	81.3
Dynamic viscosity ($\dot{\gamma} = 100\ s^{-1}$)			
– Upward ramp/mPa s	26.7	117.4	5292.0
– Downward ramp/mPa s	23.8	129.3	8370.4
Non-Newtonian behavior	Shear-thinning	Shear-thinning	Shear-thinning
Time-dependent behavior	Slightly thixotropic	Slightly rheopectic	Strongly rheopectic

by a reduction in viscosity due to the applied shear stresses during centrifugation or drainage after dip coating. The centrifugation slurry with a high solid content of 81.1 wt.% furthermore exhibited a considerable yield point which promoted the proper adhesion of the primary coating on the smooth polyurethane surface. Rheopectic effects were observed for high solid contents as well. Most likely, this effect originated from the carbon black content with nanometer-sized primary particles, which tend to exhibit the time-dependent build-up of structural particle networks [45]. The rheopectic behavior should be considered while processing especially the paste-like centrifugation slurry in order to avoid time-dependent changes of the coating quality.

After the rheological study, batches of filters were manufactured following the aforementioned coating routines C, CS and CDS. The filters were dried again after each individual coating step for at least 24 h at room temperature. Finally, the samples were fired at 800 °C in reducing atmosphere with an average heating rate of 0.8 K min^{-1}. During the firing process, the mesophase pitch binder was pyrolyzed to form the carbonaceous matrix [35, 36].

After manufacturing of the filters using the three different coating routines, the porosity and the bulk density of the filters were analyzed. To avoid sampling errors due to inevitable property gradients, the measurements were performed on the whole component in each case. The total filter component volume was calculated from the filter dimensions (excluding bores) according to DIN EN 1094–4. The apparent mass of the sample during the immersion in water was measured according to DIN EN 993–1. The true density of the ground bulk material was analyzed by means of a helium pycnometer.

From this data, the total, the infiltrable, the non-infiltrable porosity as well as the bulk density were calculated. The infiltrable filter porosity, which is available for steel melt filtration, determines the filter capacity and is an important parameter. Non-infiltrable porosity does not contribute to the filtration process and is undesired for the most part. The results of the investigation are listed in Table 29.2.

The true density of the carbon-bonded alumina filter material was 2.96 g cm^{-3} and identical for all filters. The total porosity and the bulk density are directly proportional to the filter mass and thus depended on the number of applied coating steps. As parts of

Table 29.2 Structural properties of 8c filters manufactured by different coating routines [29]

Filter type	8c-C	8c-CS	8c-CDS
Total component volume/dm^3	6.08 ± 0.10	6.05 ± 0.04	5.99 ± 0.05
Effective filter volume/dm^3	4.09 ± 0.09	3.94 ± 0.06	4.0 ± 0.07
Filter mass/g	1129.2 ± 19.2	1394.1 ± 16.1	1760.7 ± 6.7
Effective bulk density/g cm^{-3}	0.28 ± 0.01	0.35 ± 0.01	0.44 ± 0.01
Total porosity ε_t/%	90.4 ± 0.3	88.1 ± 0.3	85.1 ± 0.3
Infiltrable porosity ε_{inf}/%	84.2 ± 0.5	80.6 ± 0.7	77.4 ± 0.5
Non-infiltrable porosity ε_{nin}/%	6.2 ± 0.2	7.4 ± 0.4	7.8 ± 0.3
Ratio $\varepsilon_{inf}/\varepsilon_t$/%	93.1 ± 0.3	91.5 ± 0.5	90.8 ± 0.4

the total porosity, the infiltrable share shrank with increasing filter mass, whereas the non-infiltrable porosity increased, e.g. due to the formation of encapsulated filter pore cells and closed material porosity. Since the infiltrable porosity directly affects the filter capacity and the permeability, procedure C was most favorable. Nonetheless, all three filter types showed consistently high porosity and were thus considered suitable for further investigations.

From the data, the filtration capacity can be estimated according to Raiber et al. [24]. The total mass of steel melt, which can be treated by one filter, depends highly on the initial inclusion content of the melt batch as well as the effective filter pore utilization. Exemplary, an 8c-CDS filter with an infiltrable porosity of 77.4% can filter approximately 112 tons of steel melt, assuming a pore utilization of 20% and a volume of removed alumina inclusions equivalent to total oxygen reduction of 10 ppm. The actual pore utilization is difficult to estimate and depends e.g. on the permeability over time and the relative density of the formed clogging layers. The initial inclusion content of the steel melt can also vary strongly across batches. As a consequence, the investigation of exchangeable filter systems is necessary in order to address the high variance of resulting filter life times.

As mentioned before, the mechanical properties as well as the flow dynamics, i.e. permeability and tortuosity, are strongly influenced by the local strut geometry and mean strut diameter. By means of digital light microscopy, the mean strut diameter was investigated at different positions of the filter structure. The macro-channels of the 8c filter geometry separate the filter in a core structure and a shell structure. Segments of both parts were sampled for all three filter types and investigated on both sides regarding the minimum diameter of surface-near individual struts. The results are presented in Fig. 29.2.

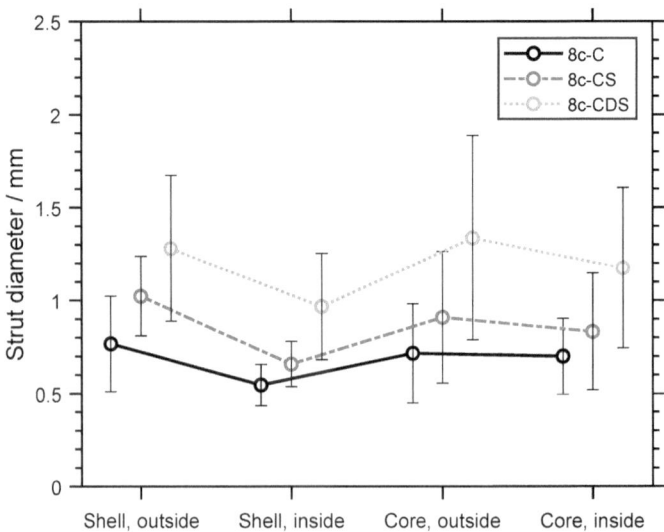

Fig. 29.2 Strut diameter distribution of 8c filters manufactured by different coating routines [29]

As expected, every additional coating increased the mean and the standard deviation of the strut diameter at all radial positions. In this way, also local inhomogeneity originating from the primary coating can be amplified by additional coatings due to enhanced adhesion. The strut diameter distribution as a function of radial position also showed some characteristic features. It was very similar for all filter types, so it can be assumed that the primary coating by impregnation and centrifugation was crucial. The shear-thinning behavior of the applied slurry in combination with the centrifugal force increasing with increased distance to the center of rotation explains why there was a tendency to a lower mean strut diameter in the shell structure of the filter compared to the core structure. However, the values at the outer shell and core positions were elevated due to the filter geometry and slurry surface tension, i.e. additional energy is necessary to remove slurry at the transition to the macro channels (core, outside) and at the filters' lateral surface (shell, outside) resulting in local slurry accumulation. For CS and CDS, the results indicate that the surface spray coating amplified this local inhomogeneity. On the one hand, the local strut thickness increase has negative effects on the filter porosity and permeability. On the other hand, the core structure and at the lateral surface can be subject to critical mechanical stresses during the installation and the application of the filter, e.g. regarding the central mounting principle and abrasive effects. Thus, the local reinforcement could be beneficial.

Figure 29.3 shows representative light micrographs of the shell/inside position for all three filter types. For CS and CDS samples, the spray coating significantly increased the surface roughness due to the fine dispersion of the material and the high air/slurry ratio. According to the Wenzel model [35], this should lead to a higher apparent contact angle in poor wetting state and thus an overall higher adhesion force for inclusions trapped on the filter surface [24]. Furthermore, there was a shift from a tapered to a more cylindrical strut morphology with additional coatings, which was expected to have further implications for the mechanical properties of the filters.

The mechanical properties of the carbon-bonded alumina were already investigated in many studies regarding smaller filter samples [15, 36] and compact refractory samples [37, 38] at room temperature as well as at high temperatures [39]. However, the large dimensions and special geometry of the filter design for continuous casting as well as the adapted manufacturing process was expected to have a crucial impact on

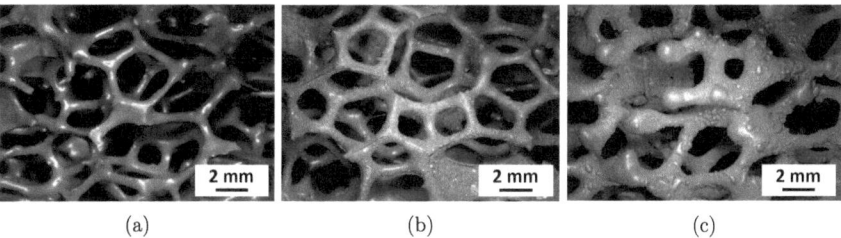

(a)   (b)   (c)

Fig. 29.3 Digital light micrographs showing the strut morphology of filters manufactured by different coating routines: 8c-C (**a**), 8c-CS (**b**) and 8c-CDS (**c**) [29]

Fig. 29.4 Distinct fracture patterns of 8c-C (I), 8c-CS (II) and 8c-CDS (III) filters after compressive mechanical testing [29]

the mechanical performance of the cellular structure. To investigate the mechanical properties of the filters, compressive mechanical tests were performed.

Three distinct types of fracture patterns were observed in general (see Fig. 29.4):

- Local fracture in large segments (C);
- Intact core structure but fracture of the shell structure (CS);
- Mainly intact structure with stepwise fracture in the top section (CDS).

The mean strut diameter and its characteristic distribution as a function of the distance from the center as well as the morphology of the strut pose possible origins of the distinct behavior. The higher slope in the progression of the force–displacement curve for the CDS filters indicated an elevated static Young's modulus. According to D'Angelo et al. [40], a higher mean strut thickness results in an increased stiffness of the foam structure. Regarding the strut shape, strong tapering causes local stress concentrations under load. A higher mean strut thickness can help to compensate the impact of the strut shape. The effect of the pore size distribution and the brittleness on the fracture behavior of foam materials was also observed by Hubálková et al. [41] It was concluded that the higher strut thickness and reduced strut tapering in CDS filters led to a more homogenous stress distribution with less local stress concentration. In contrast, C filters with thin, tapered struts exhibited fracture in large segments which is in agreement with the before-mentioned assumptions. Double-coated foams (CS) exhibited a special fracture pattern due to the reinforcement by a single spray coating. The spray coating did not substantially increase the apparent Young's modulus, thus the effect was assumed to originate from the local increase of the strut thickness near the lateral surface of the filter component. As a consequence, fracture of large segments like in filter C was avoided until the filter shell broke. Despite the apparently strong impact of the different strut properties, it has to be kept in mind that the mechanical testing of cellular foam structures depends on many factors and that the stress distribution within the foam strongly differs from that observed in dense materials.

From the force–displacement data and the reference cross-section area of the cylindrical filters including or excluding the macro-channels, the compressive strength was calculated, see Table 29.3. As expected, the filter strength was positively correlated to the filter mass and thus to the number of coatings. In absolute terms, the CS filters showed only a slight increase in filter strength compared to C filters. The aforementioned positive effect of the spray coating on the fracture behavior coincided with a subjectively higher resistance against attrition or mechanical impact. Nonetheless, the spray coating alone did not suffice to reinforce the whole filter volume and to increase the overall mechanical strength of the component. Only the application of an additional dip coating in the framework of procedure CDS, yielded the necessary significant increase of the filter strength.

Based on previous investigations of the carbon-bonded material, it can be assumed that the high-temperature mechanical performance correlates with the room-temperature mechanical strength. Solarek et al. [39] showed that the mechanical strength of carbon-bonded alumina even increases at high temperatures. One drawback is the potential plastic deformation at these temperatures. For that reason, the mechanical load, which can be expected during the immersion of the filters in the continuous casting tundish, was estimated. Considering the comparatively low flow rate in the center of the industrial tundish (below 0.1 m s^{-1} [5, 42, 43]), the main contribution to the mechanical load should come from buoyancy effects. Thereby, the most critical conditions are present during the immersion process. During the initial phase of immersion, the filter is wrapped in a sacrificial material protecting the component against slag contact. In that condition, the immersed component acts as a full cylinder until the sacrificial material decomposes. In that state, the buoyancy-related compressive stress on the bottom plane of the cylinder in a steel melt (density of 7 g cm^{-3}) was estimated at approximately 0.013 MPa, a value sufficiently below the room-temperature compressive strength. Furthermore, the estimated stress value is way below the stress at which high-temperature plastic deformation was first observed by Solarek et al. (above 0.1 MPa [39]). It also has to be considered that after the initial immersion phase, the filters inevitably will undergo further microstructural transformation since the melt temperature ($>1550 \text{ }^{\circ}\text{C}$) is significantly higher than the filters' firing temperature ($800 \text{ }^{\circ}\text{C}$) [44]. In decarburized zones, the residual alumina portion of the material will experience densification by sintering. These effects may result in a strength increase over time.

Table 29.3 Compressive strength of tested carbon-bonded alumina filters (data extracted from [29])

Compressive strength	8-C	8-CS	8-CDS
Incl. macro-channels/MPa	0.06 ± 0.00	0.08 ± 0.01	0.23 ± 0.02
Excl. macro-channels/MPa	0.10 ± 0.00	0.12 ± 0.02	0.35 ± 0.03

The macrostructure of one CDS filter was exemplarily analyzed by means of micro-computed tomography (μCT) in order to visualize radial property gradients and inhomogeneity. The resolution was restricted to 262.4 μm due to the large size of the filter (for more detailed information about the μCT setup check Hubálková et al. [41]). With the aid of the software MAVI (Modular Algorithms for Volume Images, Fraunhofer ITWM), the porosity was analyzed in segments with the dimensions of $60 \times 60 \times 120$ voxels. The 48 analyzed segments were located at 8 positions within the filter shell and 8 positions within the filter core, each at 3 different height levels.

The mean porosity across all segments amounted to 59.7% with a standard deviation of 5.0%. This value is lower than the porosity determined by the Archimedes principle due to the limited resolution and the binarization of the data. Figure 29.5 shows the cross-sectional mapping of the interpolated porosity data in 3 different x–y planes of the filter. The lowest porosity in the test was observed in the bottom-core region. The reason was local cluster formation due to slurry accumulation. The primary coating led to radial inhomogeneity which gets amplified by the dip coating, especially at the bottom of the filter due to the gravitational drainage of excess slurry. Despite local inhomogeneity, the porosity ranged between 55 and 65% in the majority of regions.

Based on the findings, 8c-CDS filters showed the best compromise between mechanical performance and sufficiently high porosity. Therefore, the CDS procedure was chosen as a basis for tests in contact with molten steel. Prior to industrial tests, the 8c-CDS filter prototype was evaluated in a lab-scale casting test. The filter was placed in the tundish vessel of a steel casting simulator (Systec GmbH, Karlstadt, Germany) and preheated while 101.1 kg of 42CrMo4 (1.7225) steel were remelted by inductive heating in an alumina-spinel crucible at 1650 °C under argon atmosphere. The exact buildup and specifications of the device as well as the associated peripheral equipment are described by Dudczig et al. [45]. After the melting and preheating procedure, the melt was cast on top of the filter by tilting the crucible. After the impact on the filter, the melt flowed into copper molds through nozzles at the bottom of the tundish vessel.

Figure 29.6 shows the sample after the steel casting simulator test. No critical deformation or signs of fracture were observed indicating sufficient high-temperature

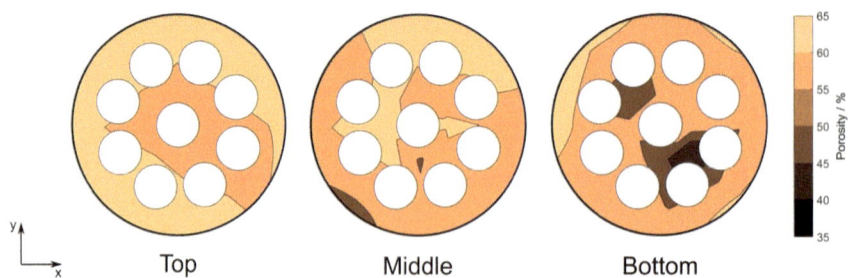

Fig. 29.5 Porosity distribution of an 8c-CDS filter derived from reconstructed μCT volume image analysis [29]

50 mm

(a)   (b)   (c)

Fig. 29.6 8c-CDS filter tested in contact with molten steel inside the tundish of the steel casting simulator (**a**); filter top (**b**) and lateral surface (**c**) including solidified steel residues and in-situ formed layers [29, 34]

mechanical strength and thermal shock resistance. A μCT analysis confirmed this observation and revealed that the filter diameter shrank by 2.2% due to ongoing sintering effects.

Filter regions with prolonged melt contact, i.e. in the impact zone or below the formed melt bath level (h ≈ 50 mm), showed a color change (see Fig. 29.6) due to local decarburization by carbon dissolution and CO bubble formation as well as the in-situ formation of alumina layers. SEM investigations revealed that in-situ layer formation was more pronounced in greater distance to the outlet nozzles. The structures resembled the alumina layers originating from carbothermic reactions as frequently observed in previous investigations of the Collaborative Research Center 920 [16–18, 36]. The location-dependent difference in the present investigation can be explained by the melt flow conditions within the melt bath. High melt flow rates near the nozzles most likely resulted in the continuous degradation of the local carbon concentration by steady removal of dissolved carbon at the melt-filter interface. With higher distance to the nozzle, lower flow rates allow for local increase of the carbon concentration in the melt facilitating the reactive filtration by carbothermic reactions. Thus, the comparatively low flow rates in the center of an industrial tundish for continuous casting were expected to be beneficial for the filtration process.

Based on the findings from lab-scale investigations, the filter prototypes manufactured by the triple coating procedure CDS were considered ready for industrial testing at thyssenkrupp Steel Europe (Bruckhausen, Germany). Nonetheless, it had to be considered that the application in an industrial continuous casting plant yielded more severe casting conditions than the lab-scale procedure. Due to the lack of the protective argon atmosphere, the filter could not be preheated resulting in more severe thermal shock. Furthermore, the inevitable contact to the slag and oxygen-rich atmosphere at high temperatures can lead to corrosion and oxidation of the material during the immersion procedure. The potential failure of the filters under these conditions does not only depend on the filter material and the manufacturing process but also on the filter geometry. The macro-channels of the 8c filter geometry are beneficial for the manufacturing of the filters and may support filtration through melt flow conditioning. Unfortunately, the thin foam bridges between them also represent weak spots regarding damage by thermal shock or corrosive attack. As a consequence, one CDS filter without macro-channels (0c-CDS) was tested besides 8c-CDS filters. One 8c-CDS filter was furthermore functionalized with an additional alumina coating (8c-CDSA) according to Schmidt et al. [17]. The resulting filter configurations are shown in Fig. 29.7. Each filter was attached to a isostatically pressed carbon-bonded alumina shaft using either high-alumina chromium-free mortar (HA; >94% Al_2O_3) or a core adhesive based on clay and water glass (CW). To support the filters with macro-channels mechanically, slip-cast disks with fitting cross-section were placed between the foam and the shaft in the case of sample 8c-CDS and 8c-CDSA. All of the filter components were wrapped in glass fiber fabric sleeves in order to protect the material against initial slag contact.

The continuous casting procedure of the plant was not changed substantially. The underlying casting sequences comprised five ladle batches of about 380 tons steel melt each. The first batch was cast as usual without filter to fill the tundish and apply the slag. The used slag cover consisted of a combination of low-melting calcium aluminate slag and a top layer of silica-rich powder for thermal insulation. The shaft of the filter component was connected to a slewing crane, which is commonly used for immersing measuring probes into the steel melt within the tundish.

The filter was immersed into the melt during the first ladle exchange of the casting sequence, i.e. after finishing the casting of the first melt batch. For this purpose, the ladle shroud was removed and the ladle turret was rotated. The turret was stopped after a rotation of 90 °C and the filter was positioned over the tundish and immersed into the melt by means of the slewing crane. Subsequently, the ladle exchange was completed by positioning the new ladle and the ladle shroud, opening the slide gates and refilling the tundish to a melt mass of about 80 tons. During the process, the filter experienced inevitable slag contact due to the penetration of the slag cover and the temporary decrease of the tundish melt bath level. After refilling the tundish,

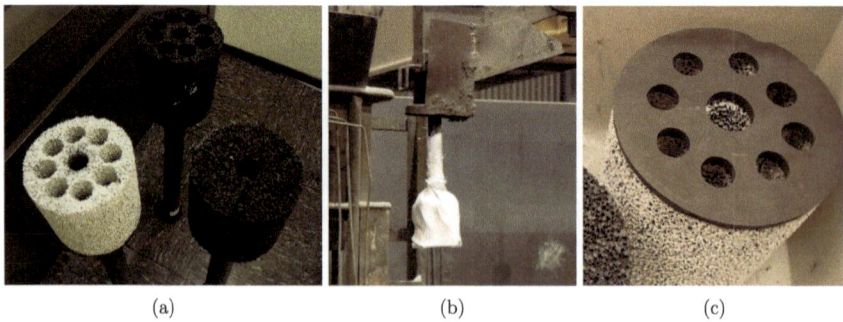

(a)   (b)   (c)

Fig. 29.7 Different carbon-bonded alumina filter samples mounted to alumina-carbon shafts (**a**), exemplary filter with glass fiber fabric sleeve as the protection against the initial slag contact prior to immersion in the tundish by means of a pivoting crane (**b**), and alumina-carbon disk as support against buoyancy-related mechanical stresses [30]

the melt bath depth amounted to approximately 1000 mm and the immersed filter was located about 200 mm below the melt-slag interface. The melt temperature was between 1550 and 1560 °C. The filter remained in the melt for the casting of one melt batch (m ≈ 380 t, t ≈ 45 min) until the next ladle exchange. The filter removal was performed by following the immersion procedure in inverse order. In total, three casting tests were performed using an ultra-low carbon interstitial-free steel grade (sequence A) and a low carbon steel grade (sequence B). The corresponding test parameters are listed in Table 29.4.

The appearances of all three samples after removal from the melt are depicted in Fig. 29.8. In all cases, a disk of solidified slag stayed attached to the shaft indicating the depth of immersing. Directly below the former melt-slag-refractory interface, a pronounced corrosion zone was observed. That implies that the carbon-bonded shaft material was corroded by partial dissolution in the melt, oxidation as well as reaction with slag elements. The effects were promoted by erosion and transport mechanisms due to Marangoni stirring [46].

Table 29.4 Industrial filter immersion test parameters [30]

Test	Sample type	Al$_2$O$_3$-C disk	Glass fiber fabric sleeve	Adhesive	Casting sequence (batch)
Pretrial	8c-CDS	No	No	CW	–
1	8c-CDS	Yes	Yes	HA	A (2nd of 5)
2	8c-CDSA	Yes	No	CW	A (4th of 5)
3	0c-CDS	No	Yes	HA	B (2nd of 5)

(a)   (b)   (c)

Fig. 29.8 Filter samples after immersion in the industrial continuous casting tundish; **a** 8c-CDS, **b** 8c-CDSA, **c** 0c-CDS [30]

The tested filter configurations behaved very differently in contact with the melt, slag and atmosphere. Sample 8c-CDSA was completely disintegrated due to insufficient adhesion to the shaft. The core adhesive, which is usually applied for the mating of dense refractories did not provide the necessary adhesion, corrosion resistance and refractoriness to fixate the filter component under these severe conditions. As a consequence, the impact of the functional alumina coating could not be investigated in the presented study. The high-alumina mortar used for the samples 8c-CDS and 0c-CDS ensured proper bonding between the shaft and the filter material over the whole testing duration, indicating that the right choice of the adhesive is a major influencing factor. However, 8c-CDS was damaged nonetheless, i.e. the bridges between the macro-channels failed and the outer shell of the filter was destroyed. Thus, if macro channels are desired, the distance between the macro-channels in the foam templates should be widened by reducing the number or the diameter of the macro-channels. Only filter 0c-CDS showed the necessary thermal shock resistance, mechanical strength and corrosion resistance to survive the test without major damage or deformation.

The tested 0c-CDS filter was analyzed by micro-computed tomography and visual inspection, see Fig. 29.9. According to two-dimensional reconstructed slice images, the core of the filter and the shaft exhibited no cracks. Furthermore, single steel droplets were observed in the whole filter structure indicating good melt intrusion and priming during the process. However, the pores were not completely filled with steel meaning that also the outflow of steel melt during the filter removal was ensured [4, 24]. By visual inspection, the filter could be divided into three major contact zones with distinct characteristics: (I) the filter top with attached slag, (II) the filter's lateral surface with dense layers and (III) the filter bottom which was covered in structures resembling inclusion clusters.

Strut samples from all three contact zones were analyzed in detail in order to validate the models and findings of previous lab scale investigations regarding occurring filtration mechanisms [17–19, 29, 33, 47]. Samples of filter struts from the three identified regions were investigated by digital light microscopy to examine differences in filter structure and appearance, by X-Ray diffraction (XRD) to analyze their overall phase composition as well as by scanning electron microscopy (SEM), energy-dispersive X-Ray spectroscopy (EDS) and electron back scatter diffraction (EBSD) to investigate changes in the microstructure, newly formed phases or other evidence for chemical reactions and filtration mechanisms.

To get an overview, the struts were examined by means of digital light microscopy first (see Fig. 29.10). Struts from zone I, the filter top with pronounced slag contact especially during the filter removal, showed signs of oxidation and corrosion. Whether the oxidation occurred before or after the removal is not fully clarified. Samples from the lateral filter surface (zone II) were covered in dense, crystalline layers. The core strut material was fully intact indicating that no significant corrosion or oxidation by slag contact occurred. The lateral filter surface was subject to the melt flow during the whole casting test, implying that the layer formations consisted of micro inclusions removed by reactive filtration mechanisms similar to effects observed in lab-scale investigations [17, 18, 47]. Struts from zone III at the filter

(a)   (b)

Fig. 29.9 Reconstructed μCT cross-sectional image (**a**) and photograph (**b**) of the 0c-CDS filter tested in industrial filter immersion tests exhibiting (I) a top slag contact zone, (II) a lateral contact zone with dense clogging layers and (III) a bottom contact zone with inclusion clusters [30]

bottom exhibited a similar base structure including intact carbon-bonded alumina filter material in the strut core sealed with a dense crystalline layer. However, on top of the layer formation, carpets of dendritic structures were observed. The first assumption was that these structures consisted of clusters of inclusions removed by direct collision with the filter material followed by adhesion and sintering (active filtration). The presence of the clusters at the filter bottom suggests that they were mainly transported by buoyancy.

In the next step, material from all three zones was ground and analyzed for their phase composition by means of XRD. A pure slag sample was used in the case of zone I and $2\,CaO \cdot Al_2O_3 \cdot SiO_2$ (gehlenite, C_2AS) was found as main constituent with small amounts of $2\,CaO \cdot SiO_2$ (belite, C_2S). These phases are typical regarding the combined use of a calcium aluminate slag and a silica-rich insulating powder. In zones II and III, corundum and graphite were identified and indicated the presence of intact

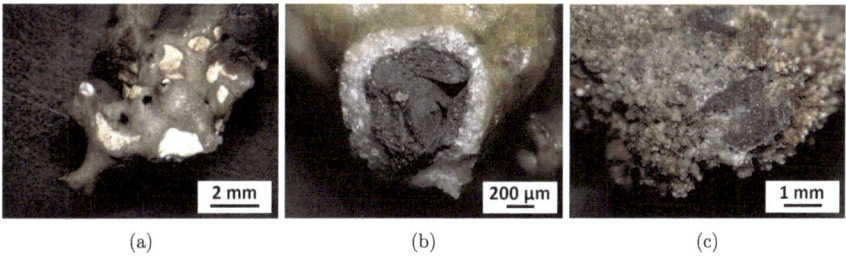

(a)   (b)   (c)

Fig. 29.10 Digital light micrographs of strut samples from **a** contact zone I, **b** contact zone II and **c** contact zone III of the 0c-CDS filter after the industrial immersion test [30]

filter base material. The calcium aluminates $CaO \cdot 2\ Al_2O_3$ (CA_2) and $CaO \cdot Al_2O_3$ (CA) were also found in samples from both zones. The presence of calcium-rich phases indicated that slag contact occurred although no silicon-containing phases were detected. The main difference of the zone 3 phase composition was the identification of calcium aluminates with higher Al_2O_3 content, namely calcium hexaaluminate $CaO \cdot 6\ Al_2O_3$ (CA_6) and calcium β-alumina x $CaO \cdot 11\ Al_2O_3$ (C_xA_{11}, x ≈ 1), which indicated that either less slag was involved in the formation of the phases or the calcium aluminate composition was enriched by Al_2O_3 due to the higher removal rate of new inclusions.

For understanding the processes which took place during the immersion test, analyzing the dense layers formed in contact zone 2 was of special interest. SEM, EDS and EBSD investigations were performed for this purpose. By means of SEM, an intermediate transition layer was detected between the filter base material and the dense layers around the strut. Figure 29.11 shows the SEM analyses of the interface between the filter material and the intermediate layer as well as the interface between the intermediate layer and the dense top layer. The thickness of the transition layer varied locally and amounted to approximately 50 μm on average. EDS mapping for eight essential elements and EBSD-based image quality (IQ) and phase mapping analyses revealed significant differences in the chemical composition and microstructure of the different regions of interest (ROI).

At the interface between the base material (ROI 1) and the transition zone (ROI 2), see Fig. 29.11a–c, similar aluminum, carbon and oxygen concentrations were detected in the whole analyzed area. It can be concluded that the transition layer zone originally was part of the filter material. A unique feature of the transition zone was its increased calcium content and the presence of minor amounts of magnesium and silicon indicating slag contact. Apparently, the thin transition layer represented corroded filter material originating from slag penetration despite the usage of the protective glass fiber sleeve. EBSD image quality (IQ) and phase mapping revealed that larger CA_2 crystals formed in the transition layer, which were clearly distinguishable from the fine-grained alumina crystals of the filter material. The larger crystal size most likely originated from the filling of larger pores in the filter material with slag and the subsequent chemical reaction of the slag and the alumina. The corrosion was not harmful since it was restricted to a rather thin layer. In fact, the newly formed calcium aluminate phase could be beneficial for the filtration according to Storti et al. [19], who observed a particularly high filtration efficiency for filters functionalized with calcium aluminate.

The same analysis procedure was carried out at the interface between the transition zone (ROI 2) and the dense top layer (ROI 3), see Fig. 29.11d–f. According to SEM micrographs, the dense top layer possessed a much more homogenous microstructure than the transition layer. Furthermore, the EDS mapping revealed significant differences in the chemical composition. The intensity of aluminum and oxygen were clearly increased in the dense layer while the calcium concentration was similar in the whole investigated area. Traces of carbon, iron, silicon or magnesium were only found in few spots of the dense layer. From the mapping, it can be concluded that the dense layer did not originate from the carbon-bonded alumina base material and

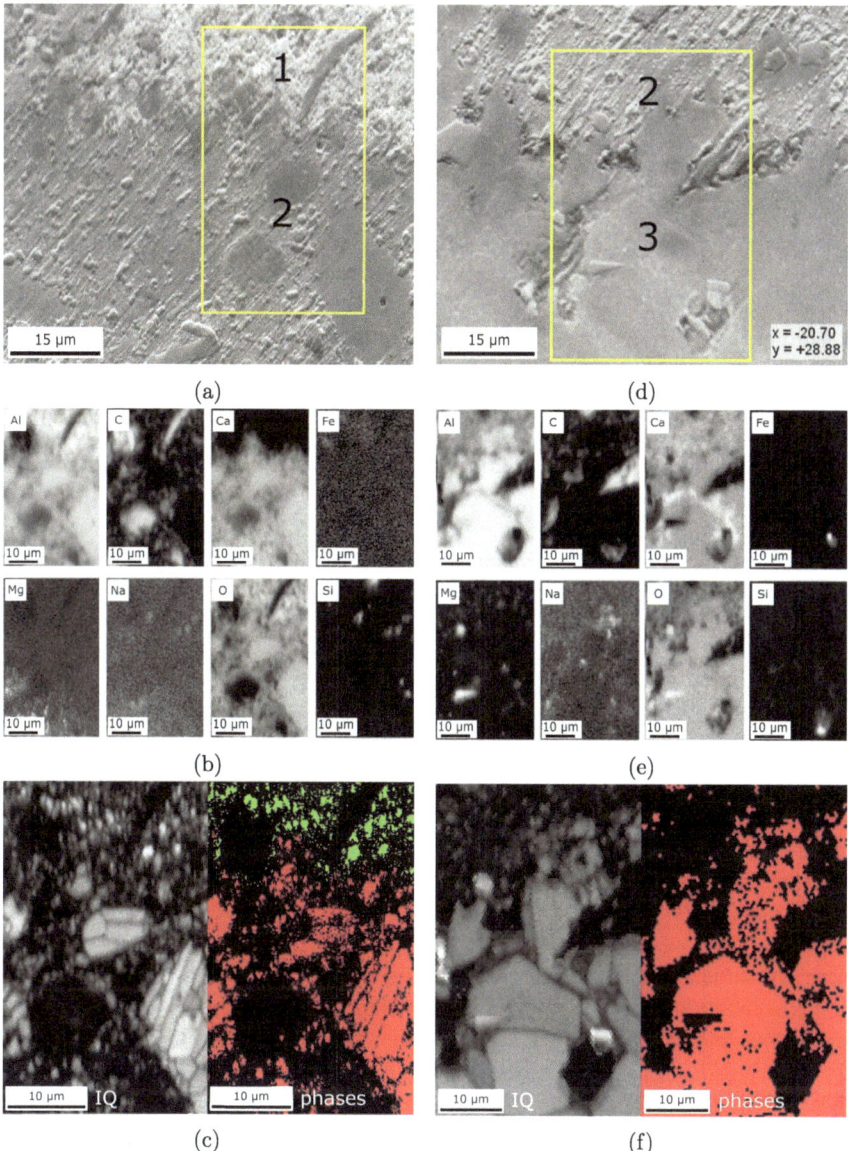

Fig. 29.11 SEM micrograph (**a**, **d**), EDS mapping (**b**, **e**) as well as EBSD image quality (IQ) and phase mapping (**c**, **f**) of the interface between the base material (1) and the transition layer (2) and the interface between the transition layer (2) and the clogging layer (3) [30]

that the impact of slag was lower than for the formation of the transition layer. On the other hand, the increased alumina concentration suggested that the dense layers originating from alumina micro inclusions which reacted with the calcium-enriched transition layer to form CA_2 near the interface. The crystal growth was promoted by sintering effects at temperatures over 1550 °C.

The different regions of interest were also examined by means of EBSD and EDS spot analyses, see Fig. 29.12. As in the EBSD phase mapping, trigonal α-alumina was identified in ROI 1 confirming that unaltered filter material was present. The transition layer (ROI 2) showed very weak electron backscatter patterns without clear phase classification (spinels and calcium aluminates at few locations). According to the EDS spectrum, the transition layer contained both typical slag elements (calcium, magnesium, sodium and silicon) as well as typical filter material constituents (aluminum and carbon). This finding supports the hypothesis of superficial corrosion of the filter material by slag. In the dense top layer (ROI 3), no sodium, magnesia or silicon were found, indicating that the slag had only indirect impact on the formation. Furthermore, no carbon was detected. As a consequence, it can be assumed that removal or precipitation of alumina or aluminates from the steel melt occurred. From the first EDS spot analyses, it already became evident that the Al/Ca ratio was elevated in the dense clogging layer. To confirm this observation, a series of spot analyses was performed in pure slag samples (32 spots) and the dense clogging layer (12 spots). The Al/Ca EDS intensity ratio amounted to 1.5 ± 0.1 in the slag and 5.9 ± 0.1 in the clogging layer.

Finally, also strut fragments from zone III, the filter bottom, were analyzed via SEM and EDS, see Fig. 29.13. As for the lateral filter surface (zone II), the intact carbon-bonded alumina filter struts were covered in dense layers. However, pronounced inclusion clusters were identified on top of these layers across the whole sample. The morphology of the single inclusions was either coral-like and plate-like and resembled the typical inclusion morphology reported for low-carbon alumina-killed steels in the literature [1, 18]. In order to clarify differences in the chemical composition between those types of clusters, EDS analyses were performed. The main constituents were aluminum and oxygen with traces of sodium as well as calcium, magnesium, potassium and iron. Coral-like structures exhibited slightly higher trace element intensities and lower aluminum intensities. This could explain the identification of β-Al_2O_3 in the previous XRD investigation. The common location of the observed coral-like structures on top of plate-like structures implies that they were formed due to contact to slag during the filter removal. Following that assumption, the subjacent carpet of plate-like structures most likely marked the actual filter surface with attached inclusion clusters removed from the steel melt by active filtration.

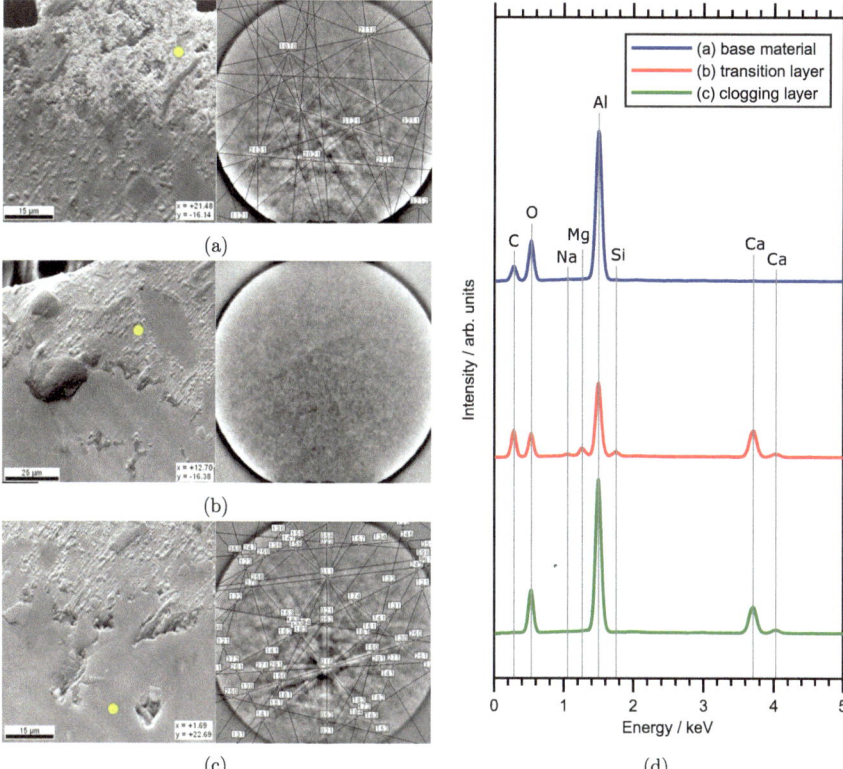

Fig. 29.12 SEM micrographs and EBSPs of the unaltered filter base material (**a**), intermediate transition layers (**b**) and outer clogging layers (**c**) as well as comparative EDS spectra (**d**) of strut samples from zone II [30]

29.3   Gel-Cast Alginate "Spaghetti" Filters

Replica foam filters are the state of the art for steel melt filtration, however, due to their typical hollow strut network structure, their mechanical strength is decreased [48, 49]. For low-strength materials like carbon-bonded alumina, the resulting structural weakness may lead to damage during transport, handling, installation, and application [29, 30]. Another drawback of replica filters is the special set of characteristics required from the used polymer templates, e.g. regarding their cost efficiency, open-celled structure, toughness, flexibility, degradation behavior during firing and availability, which limits the number of economically viable template materials [48].

In order to address those issues, an alternative technique for the manufacturing of cellular ceramics was investigated. The process is based on the in-situ gel casting of an alginate-containing ceramic suspension. The 3D cellular filter structures are obtained by stacking continuous strands of gelled material in a periodical fashion without use

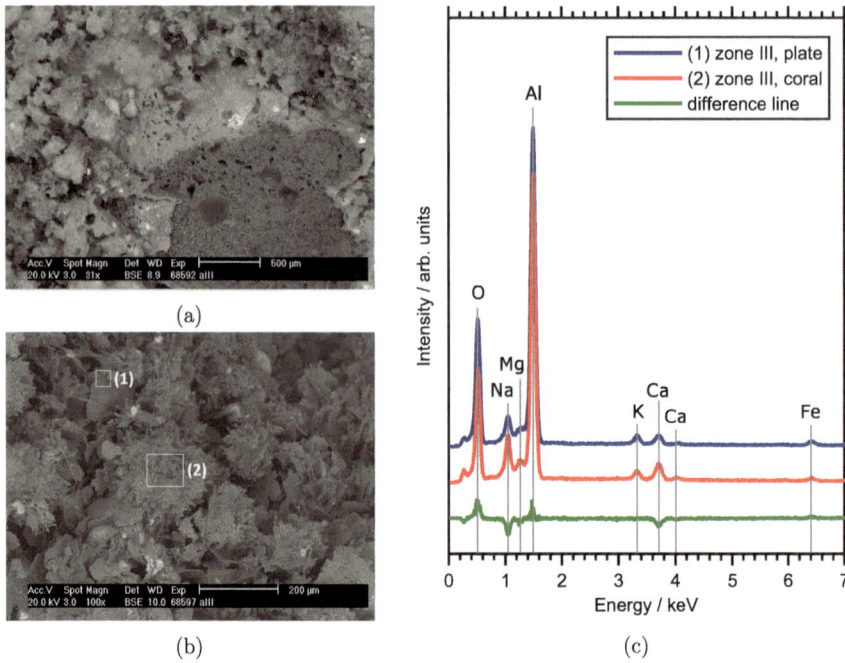

Fig. 29.13 SEM micrographs of the zone III filter strut cross-section (**a**) and detected inclusion clusters (**b**) as well as representative EDS spectra (**c**) of clusters with with plate-like (1) or coral-like (2) morphology [30]

of any sacrificial template material [50]. The technique exploits the natural gelation capability of so-called alginates, which are water-dispersible polysaccharides derived from brown seaweeds [51, 52] and consist of two different monomers: β-D-mannuronate (M) and α-L-guluronate (G). For the gel-casting process, the diaxially linked blocks within the polysaccharide chains consisting of two neighboring G monomers (GG blocks) are of special interest due to their characteristic structure, which enables the gel transition of alginate sols. Providing mobile alkaline earth ions like Ca^{2+} or Ba^{2+} enables the bonding of two GG blocks from different polysaccharide chain segments by a combination of the chelation of the metal ion by hydroxyl oxygen atoms and of intermolecular ionic bonds. This process leads to 3D cross-linking. The resulting structures, which are responsible for the stiffness of the alginate gel, are often referred to as "egg box" model [51]. If the local alkaline earth concentration is increased suddenly, the gel transition takes place more or less instantly [53].

In the novel robo gel-casting process, the alginate-containing ceramic slurry is pumped into a container with $CaCl_2$ solution to trigger the gelation, similar to processes known from the food or pharma industry such as the encapsulation of substances in alginate beads [54]. To obtain the periodic 3D structure, the nozzle is moved by means of a controllable robot system. After the gel-casting process, the additively manufactured structure is kept in the $CaCl_2$ solution for several minutes to ensure complete gelation, followed by drying, coating and firing [50]. Potential disadvantages of the technique arise from the low necessary slurry viscosity and thus limitations regarding the dimensional stability and the achievable geometry of the structure in comparison to other robocasting processes. However, the technique allows for the manufacturing of large dimensions at high printing speeds (up to 100 mm s^{-1} with the presented setup). As a consequence, the gel-casting process is also applicable for large-scale cellular structures from refractory materials with coarser particle size.

For the present investigation, the same carbon-bonded alumina base composition as for the before-mentioned foam filters was applied. For the gel-casting slurry, an alginate content of 0.65% and a solid content of 55.6% exhibited the best compromise between workability, stability and gel strength in preliminary tests. The resulting composition was applied to manufacture prismatic samples for lab-scale tests in contact with molten steel.

The gel-cast prismatic filters were investigated in immersion tests under argon atmosphere using a 42CrMo4 steel melt at temperatures above 1600 °C (see Fig. 29.14). The test procedure allows for the analysis of the thermal shock perfor- mance under practically relevant conditions. Additionally, the tested filter sample can be removed to drain the melt and cooled slowly under inert conditions in order to avoid oxidation. In this way, the former filter-melt interface is exposed in a damage-free manner and can be analyzed directly and without alteration in subsequent investiga- tions. In the present case, no critical thermal shock damage, corrosion or creep defor- mation was observed. As for conventional carbon-bonded filters, the filter surface underwent a change in appearance in contact with molten steel (see Fig. 29.14a) and thus was analyzed by means of digital light microscopy, SEM and EDS. The investigations revealed two distinct types of surface structures: carpets of shard-like translucent structures (see Fig. 29.14b), which are in agreement with in-situ formed layers observed in previous steel casting simulator tests, e.g. Storti et al. [36], and vermicular structures of unknown nature (see Fig. 29.14c and d).

In order to analyze the qualitative element composition of the detected structures, EDS investigations were performed. For all structures, minor traces of Mg and Si were detected, which most likely arose from alginate residues or the steel melt. Shard- like structures consisted mainly of Al and O with traces of Ca (from the cross-linking solution) indicating the presence of alumina and calcium aluminates. The latter are reported to have a potentially favorable impact on steel melt cleanliness [19]. In the

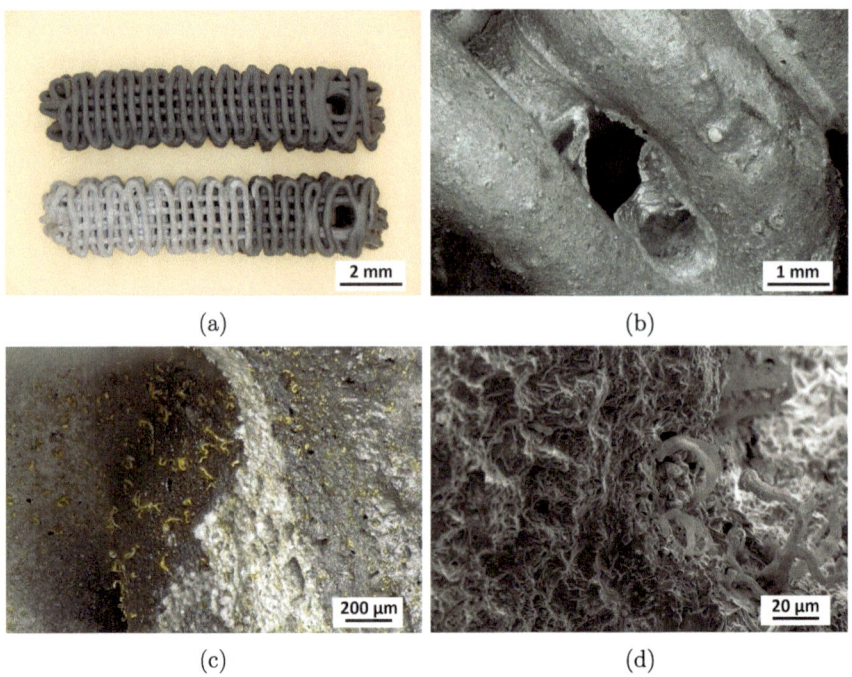

(a)   (b)

(c)   (d)

Fig. 29.14 Steel casting simulator immersion test [31]; Prismatic gel-cast filters before and after immersion in molten steel (**a**), digital light micrographs of translucent shard-like structures (**b**) and vermicular structures (**c**) on the filter surface and SEM micrograph of the resulting filter surface topology (**d**)

case of vermicular structures, Al, Fe and traces of Cl were detected. The structures were formed due to the presence of $CaCl_2$ from the gel-casting procedure and its local reaction with the melt and filter material.

An additional steel casting simulator test with a flat filter (75×75 mm^2, 16 filter layers, 5 mm strut spacing) was performed in an impingement setup. Approximately 100 kg of steel melt were cast on the filter and flowed through in into the tundish vessel (see Fig. 29.15). As for the finger test, the gel-cast filter survived without critical damage or deformation. Digital light microscopy analyses revealed inclusion clusters and in-situ formed layers on the filter surface. As a consequence of the forced melt passage, the detected structures were more pronounced than those observed after short-term finger immersion tests.

(a)   (b)   (c)

Fig. 29.15 Steel casting simulator impingement test [34]; casting procedure (**a**) as well as digital light micrographs of the filter macrostructure after melt contact (**b**) and removed dendritic inclusions (**c**)

After the evaluation of the process, a larger gantry robot system with a working space of $500 \times 500 \times 200 \, m^3$ and a tube with diameter of 3.2 mm was applied for the purpose of upscaling. A polymer nozzle with a diameter of 2 mm was used to ensure the required resolution and reproducible strut thickness. Thanks to the modification of the tube-nozzle-system the pumping process was simplified allowing for higher slurry viscosity. Therefore, the slurry solid content was increased to 70 wt.% to reduce the dry shrinkage of the material and to increase the dimensional stability of the gel-cast structure. A lower sodium alginate content of 0.5 wt.% was chosen to account for the strong decrease of the water volume ratio. The linear shrinkage of single struts amounted to 11.9% for the given composition. While a spray coating can be sufficient to interconnect the struts of small gel-cast filters, a dip coating was more suitable to ensure proper coating of the whole volume of upscaled gel-cast filters. The dip-coating slurry composition and preparation were adopted from the CDS foam filter manufacturing procedure.

Figure 29.16 shows the upscaled gel-cast filters at different process steps. The dry shrinkage showed some anisotropy and amounted to about 5% in length and 9% in width. The reason for this behavior is the high number of contact points between the alternating strut layers, which hinder the individual strut shrinkage in longitudinal direction. The shrinkage in the strut diameter, can partially compensate the resulting mismatch. Nonetheless, residual stresses and slight deformation could not be avoided with the present composition.

Cylindrical samples were drilled from the fired filter for high-temperature mechanical testing [55]. The compressive strength of distinct samples was measured at room temperature, at 800 °C and at 1500 °C under controlled argon atmosphere. As described by Solarek et al. [39] for foam filters of the same material, pseudoplastic behavior was observed at 1500 °C and resulted in premature reaching of the maximum sample displacement. Potential reasons for this behavior are the continuation of sintering and pyrolysis effects, the sliding of graphite layers or high-temperature creep within the oxide portion. The room temperature strength of the samples amounted to 3.12 ± 1.31 MPa. A strength increase to 4.60 ± 1.23 was observed at 800 °C due to crack closure and compressive stresses within the material originating from the thermal expansion mismatch between alumina and the carbon

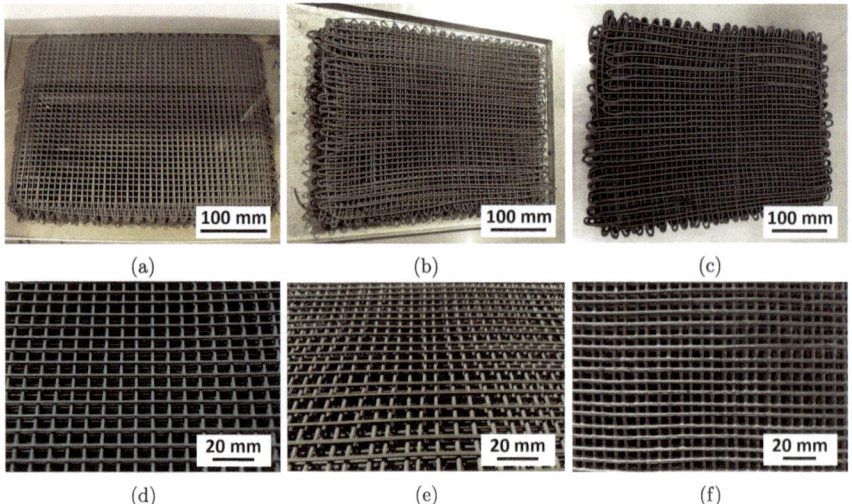

(a)   (b)   (c)

(d)   (e)   (f)

Fig. 29.16 Gel-cast spaghetti filters upscaled for industrial application in continuous steel casting directly after gel casting (**a, d**), after drying (**b, e**) and after coating and firing (**c, f**) [34]

matrix. Similar behavior was detected by Solarek et al. [39] for carbon-bonded foam filters and by Zielke et al. [56] for carbon-bonded bulk material. Compared to tested foam filters, the compressive strength of the gel-cast filters was approximately an order of magnitude higher. However, that was only achieved at the expense of filter porosity. The mean total porosity of the gel-cast samples amounted to only 72% compared to approximately 91% for the aforementioned CDS foam filters.

The upscaled gel-cast filters were tested in an industrial continuous casting plant. The plates were immersed into the tundish using the same procedure as for the trials with foam filters. Figure 29.17 shows the filter after the test. Due to a crane malfunction, the filter was excessively exposed to the slag and air at high temperatures leading to excessive corrosion and oxidation. Fortunately, central parts of the filter were protected from the slag. By means of digital light microscopy, in-situ formed layers were detected. The structures were very similar to those observed in the framework of the foam filter investigation. Thus, it can be concluded that the same filtration mechanisms occurred.

The present investigation delivered first proofs of principle for the new filtration concept by filter immersion using carbon-bonded alumina foam filters and gel-cast filters. Nonetheless, further investigations are necessary to improve the gel-casting technology and immersion procedure of filter plates and to research the filtration efficiency of foam filters and gel-cast filters by the new filter immersion concept in direct comparison.

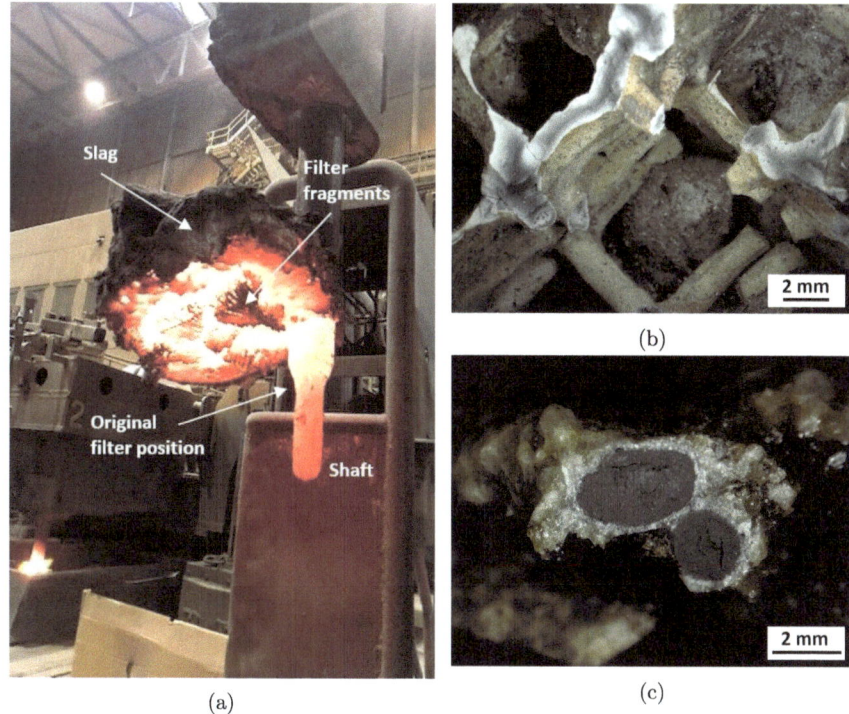

Fig. 29.17 Industrial immersion test in continuous steel casting [34]; filter residues after excessive slag and air exposure at high temperatures (**a**) as well as digital light micrographs of oxidized filter struts (**b**) and intact filter struts with unaltered carbon-bonded alumina matrix surrounded by dense clogging layers (**c**)

Acknowledgements The authors would like to acknowledge the German Research Foundation (DFG) in terms of the Collaborative Research Center 920-Project-ID 169148856, subproject T01, for the financial support of this research. Furthermore, special thanks go to J. Hubálková and B. Luchini for performing and supporting the computed tomography analyses, M. Rudolph for conducting the XRD investigation as well as N. Brachhold and G. Schmidt for performing the SEM, EDS and EBSD analyses.

References

1. L. Zhang, B.G. Thomas, Inclusions in continuous casting of steel, in *Pro-ceedings of XXIV National Steelmaking Symposium*, Morelia, Mich, Mexico, Nov 26–28 2003, 138–183 (2003)
2. C.M. Lee et al., Rev. Met. Paris **90**(4), 501–506 (1993). https://doi.org/10.1051/metal/199390 040501
3. J. Poirier, Metall. Res. Technol. **112**(4), 410 (2015). https://doi.org/10.1051/metal/2015028
4. J. Mancini, J. van der Stel, Rev. Met. Paris **89**(3), 269–278 (1992). https://doi.org/10.1051/ metal/199289030269

5. L. Zhang, S. Taniguchi, K. Cai, Metall Materi. Trans. B **31**(2), 253–266 (2000). https://doi.org/10.1007/s11663-000-0044-9
6. L. Zhang, B.G. Thomas, Inclusion investigation during clean steel production at baosteel, in *Proceeding of Iron and Steel Society Technology Conference*, Warrendale, PA: ISS-AIME, Apr 27–30 2003, 141–156 (2003)
7. O.B. Isaev, Metallurgist **53**(11–12), 672–678 (2009). https://doi.org/10.1007/s11015-010-9231-3
8. Y. Miki, B.G. Thomas, Metall. Materi. Trans. B **30**(4), 639–654 (1999). https://doi.org/10.1007/s11663-999-0025-6
9. L. Zhang, J. Aoki, B.G. Thomas, Metall. Materi. Trans. B **37**(3), 361–379 (2006). https://doi.org/10.1007/s11663-006-0021-z
10. C. Liu, F. Huang, X. Wang, Metall. Materi. Trans. B **47**(2), 999–1009 (2016). https://doi.org/10.1007/s11663-016-0592-2
11. S.K. Michelic, U.D. Salgado, C. Bernhard, I.O.P. Conf, Ser. Mater. Sci. Eng. **143**, 12010 (2016). https://doi.org/10.1088/1757-899X/143/1/012010
12. L. Zhang, JOM **65**(9), 1138–1144 (2013). https://doi.org/10.1007/s11837-013-0688-y
13. H. Ling, L. Zhang, JOM **65**(9), 1155–1163 (2013). https://doi.org/10.1007/s11837-013-0689-x
14. L. Zhang, W. Pluschkell, Ironmak. Steelmak. **30**(2), 106–110 (2013). https://doi.org/10.1179/030192303225001766
15. M. Emmel, C.G. Aneziris, Ceram. Int. **38**(6), 5165–5173 (2012). https://doi.org/10.1016/j.ceramint.2012.03.022
16. S. Dudczig et al., Ceram. Int. **40**(10), 16727–16742 (2014). https://doi.org/10.1016/j.ceramint.2014.08.038
17. A. Schmidt et al., Adv. Eng. Mater. **19**(9), 1700170 (2017). https://doi.org/10.1002/adem.201700170
18. E. Storti et al., Steel Res. Int., **88**(10), 1700142 (2017). https://doi.org/10.1002/srin.201700142
19. E. Storti et al., Steel Res. Int., **88**(11), 1700247 (2017). https://doi.org/10.1002/srin.201700247
20. K. Uemura et al., ISIJ Int. **32**(1), 150–156 (1992). https://doi.org/10.2355/isijinternational.32.150
21. D. Apelian, R. Mutharasan, S. Ali, J. Mater. Sci. **20**(10), 3501–3514 (1985). https://doi.org/10.1007/BF01113756
22. V.N. Antsiferov, Powder Metall. Met. Ceram. **42**(9/10), 474–476 (2003). https://doi.org/10.1023/B:PMMC.0000013219.43559.46
23. D. Janke, Grundlegende Untersuchungen zur Optimierung der Filtration von Stahlschmelzen - Abschlußbericht, Luxemburg: Amt für amtliche Veröff. der Europ. Gemeinschaften (1996)
24. K. Raiber, P. Hammerschmid, D. Janke, ISIJ Int. **35**(4), 380–388 (1995). https://doi.org/10.2355/isijinternational.35.380
25. A.S. Kondrat'ev et al., Refractories, **31**(7–8), 384–391 (1990). https://doi.org/10.1007/BF01281545
26. K. Janiszewski, Archiv. Metall. Mater. **58**(2) (2013). https://doi.org/10.2478/amm-2013-0029
27. K. Janiszewski, B. Panic, Metalurgija **53**(3), 339–342 (2014)
28. A. Baaske et al., Ceramic filters and filter systems for continuous metal melt filtration, US Patent US20170292173A1 (2017)
29. T. Wetzig et al., Ceram. Int. **44**(15), 18143–18155 (2018). https://doi.org/10.1016/j.ceramint.2018.07.022
30. T. Wetzig et al., Ceram. Int. **44**(18), 23024–23034 (2018). https://doi.org/10.1016/j.ceramint.2018.09.105
31. T. Wetzig et al., Adv. Eng. Mater. **22**(2), 1900657 (2020). https://doi.org/10.1002/adem.201900657
32. J. Fruhstorfer et al., Steel Res. Int., **87**(8), 1014–1023 (2016). https://doi.org/10.1002/srin.201600023
33. J. Fruhstorfer et al., Metall. Mater. Trans. B **49**(3), 1499–1521 (2018). https://doi.org/10.1007/s11663-018-1216-9

34. T. Wetzig, New approaches for steel melt filtration in continuous casting of steel. Dissertation, TU Bergakademie Freiberg (2022)
35. C.G. Aneziris, M. Hampel, Int. J. Appl. Ceram. Technol. **5**(5), 469–479 (2008). https://doi.org/10.1111/j.1744-7402.2008.02223.x
36. E. Storti et al., J. Eur. Ceram. Soc. **35**(5), 1569–1580 (2015). https://doi.org/10.1016/j.jeurceramsoc.2014.11.026
37. Y. Klemm, H. Biermann, C.G. Aneziris, Ceram. Int. **39**(6), 6695–6702 (2013). https://doi.org/10.1016/j.ceramint.2013.01.108
38. Y. Klemm, H. Biermann, C.G. Aneziris, Adv. Eng. Mater. **15**(12), 1224–1229 (2013). https://doi.org/10.1002/adem.201300159
39. J. Solarek et al., Carbon **122**, 141–149 (2017). https://doi.org/10.1016/j.carbon.2017.06.041
40. C. D'Angelo, A. Ortona, P. Colombo, Acta Mater. **60**(19), 6692–6702 (2012). https://doi.org/10.1016/j.actamat.2012.08.039
41. J. Hubálková et al., Adv. Eng. Mater. **19**(9), 1700286 (2017). https://doi.org/10.1002/adem.201700286
42. R.D. Morales et al., ISIJ Int. **39**(5), 455–462 (1999). https://doi.org/10.2355/isijinternational.39.455
43. A.K. Sinha, Y. Sahai, ISIJ Int. **33**(5), 556–566 (1993). https://doi.org/10.2355/isijinternational.33.556
44. C. Röder et al., J. Raman Spectrosc. **45**(1), 128–132 (2014). https://doi.org/10.1002/jrs.4426
45. S. Dudczig et al., Adv. Eng. Mater. **15**(12), 1177–1187 (2013). https://doi.org/10.1002/adem.201300121
46. A.F. Dick et al., ISIJ Int. **37**(2), 102–108 (1997). https://doi.org/10.2355/isijinternational.37.102
47. T. Zienert et al., Ceram. Int. **41**(2), 2089–2098 (2015). https://doi.org/10.1016/j.ceramint.2014.10.004
48. P. Colombo, Philosophical transactions. Ser. A Math. Phys. Eng. Sci. **364**(1838), 109–124 (2006). https://doi.org/10.1098/rsta.2005.1683
49. U.F. Vogt et al., J. Eur. Ceram. Soc. **30**(15), 3005–3011 (2010). https://doi.org/10.1016/j.jeurceramsoc.2010.06.003
50. C.G. Aneziris et al., *A process for producing carbon-containing ceramic components*. German Patent DE102015221853A1 (2017)
51. P. Gacesa, Carbohyd. Polym. **8**(3), 161–182 (1988). https://doi.org/10.1016/0144-8617(88)90001-X
52. M. Fertah et al., Arab. J. Chem. **10**, 3707–3714 (2017). https://doi.org/10.1016/j.arabjc.2014.05.003
53. M. Oppelt et al., Metall. Materi. Trans. B **45**(6), 2000–2008 (2014). https://doi.org/10.1007/s11663-014-0151-7
54. A.H. King, Flavor encapsulation with alginates. in: Flavor encapsulation. (American Chemical Society, Washington, DC, 1988). vol 370, pp. 122–125
55. X. Wu et al., Adv. Eng. Mater. **24**(2), 2100613 (2021). https://doi.org/10.1002/adem.202100613
56. H. Zielke et al., Carbon **159**, 324–332 (2020). https://doi.org/10.1016/j.carbon.2019.12.042

Chapter 30
Numerical Simulation of Continuous Steel Casting Regarding the Enhancement of the Cleanliness of Molten Steel

Sebastian Neumann, Amjad Asad, and Rüdiger Schwarze

30.1 Introduction

The material research in the field of steelmaking based on continuous casting is further focusing the improvement of quality and performance. Nonmetallic inclusions (NMI) e.g. Al_2O_3, SiO_2, $MgAl_2O_4$ develop due to unavoidable formation processes such as reoxidation, deoxidation, corrosion or erosion of the refractory [1–5]. Micro inclusions do not float, they rather stay in the melt and end up in the casting mold. This lowers, based on the NMI quantity, the quality of the steel product [1, 2].

For this reason, the research strongly focuses the cleaning of the steel melt during continuous casting by reducing the amount of NMI in the melt. Besides the improvement of buoyancy-driven inclusion removal by e.g. flow control [2, 6–10], bubble flotation [8, 11] and secondary metallurgy [7], is the melt filtration another promising procedure. Especially the application of ceramic filters achieves high cleaning efficiencies [12–19].

Active filtration, the deposition of the NMI on the filter wall, shows moderate decrease of inclusions in the melt [14–16]. However, the cleaning efficiency can be increased with alumina active coatings [18].

High cleaning efficiencies and an in-situ layer formation on the filter surfaces are observed using alumina-coated carbon-bonded filters [19]. The associated process is a reactive cleaning process. In the melt occurs a reaction of the present oxygen (O) and the dissolved carbon (C) from the filter coating. In consequence, CO bubbles

S. Neumann (✉) · A. Asad · R. Schwarze
Institute of Mechanics and Fluid Dynamics, Technische Universität Bergakademie Freiberg, Lampadiusstraße 4, 09599 Freiberg, Germany
e-mail: Sebastian.Neumann@imfd.tu-freiberg.de

© The Author(s) 2024
C. G. Aneziris and H. Biermann (eds.), *Multifunctional Ceramic Filter Systems for Metal Melt Filtration*, Springer Series in Materials Science 337,
https://doi.org/10.1007/978-3-031-40930-1_30

develop at the NMI surface. Finally, the inclusions float to the slag, driven by the bubble rise [19–21].

Investigations of reactive cleaning in a batch filtration system revealed high cleaning efficiencies of approx. 95% after a short-term period of 10 s [22]. This very promising approach was further investigated using an exchangeable carbon-bonded alumina-coated foam filter in an industrial scale continuous steel casting tundish. "Very low" cleaning efficiencies are reported, which are related to a low surface ratio of the filter to the tundish cross section. However, there were indications for reactive cleaning (clogging layer) and active filtration (inclusion cluster) [23]. Industrial scale experiments are rare, expensive and hard to postprocess. Therefore, the importance of the application of numerical simulation in the field of continuous casting processes rises constantly. A lot of literature discusses physical and mathematical models, validations as well as experiments or numerical simulations of melt flows [24–47]. Furthermore, several studies are published on the design of the tundish for flow controlling purposes and inclusion behavior [25–27, 29–32]. This underlines the importance of the research in this field.

This chapter investigates the active and reactive cleaning of NMI using computational fluid dynamics (CFD). The research concentrates on a prototype laboratory scale one-strand tundish and an industrial prototype two-strand tundish. The numerical model, discussed by Neumann et al. [38, 44] and Asad et al. [24, 39, 45, 46] is adopted for the multiphase flow in the tundish using active filtration and reactive cleaning.

30.2 Numerical Model

The numerical model has been previously presented in Chap. 17, by Asad et al. [24, 39, 45, 46] and Neumann et al. [38, 44]. In short, the Eulerian–Lagrangian model includes the steel melt flow, NMI transport and their deposition on the filter (active filtration), the transport of dissolved carbon, the CO reaction and bubble generation at the NMI surface as well as the inclusion flotation (reactive cleaning).

The Newtonian steel melt flow is assumed to be incompressible and isothermal. The turbulent flow field is calculated with the implicit large eddy simulation (ILES) approach (see Chap. 17, Asad et al. [38, 40, 41, 44–46]). The spatially filtered mass and momentum conservation equations are described in Eqs. 17.1 and 17.2. A special characteristic using ILES is the usage of a specific numerical convection term discretization scheme to mimic the subgrid-scale Reynolds stress tensor of Eq. 17.2. Explicit modeling of the subgrid-scale turbulence in not needed. The continuous and disperse phases are directly coupled, see Eqs. 17.2 and 17.14. The pressure drop in the porous ceramic foam filter is modeled by the Darcy-Forchheimer law. It is accounted by the source term see Eq. 17.10, in the momentum conservation equation.

The reader is referred to Asad et al. [24, 39, 45, 46] and Neumann et. al [38, 44] for the geometric properties of the pore size of 10 pores per inch (ppi) applied for the ceramic foam filter. The disperse phase transport is discussed in Chap. 17 using the Lagrangian framework, see Eq. 17.13.

The modeled cleaning strategies are the active filtration, the removal of inclusions out of the melt by deposition of the NMIs on the filter surface, and reactive cleaning, the reaction of dissolved carbon and oxygen in the melt, which creates bubbles at the NMI and lead to their flotation and cleaning out of the melt. Active filtration is defined by an effective filtration probability (see Chap. 17, Eqs. 17.11 and 17.12) for the NMI deposition on the filter surface.

Reactive cleaning of exchangeable carbon compound and an alumina coated filter [22, 23] is modeled by a species transport equation for the concentration of the dissolved carbon, see Chap. 17 Eq. 17.15. The model for the bubble formation on each inclusion uses the ideal gas law Eq. 17.17. A sink term model defines the reduction of carbon due to the reaction with the oxygen present in the melt (see Eq. 17.18), see also Kim et al. [48]. The growth of the bubble is managed with a coupled Lagrangian approach. Inclusion and bubble are calculated together using an equivalent spherical particle with effective volume/mass depending on the local carbon concentration, ref. Asad et al. [24, 39, 45, 46], Neumann et al. [38, 44] and Chap. 17.

30.3 Simulation Setup

The numerical models presented before are applied to numerical simulations of two different tundish designs, see Fig. 30.1. The laboratory scale prototype tundish (Fig. 30.1a), published by Neumann et al. [38], is a one-strand tundish with a volume of approx. 62,5 L. In Fig. 30.1b, the industrial scale prototype two-strand tundish is shown with a volume of approx. 8000 L, see Neumann et al. [44].

In general, the numerical simulation of both designs focuses on active filtration as well as reactive cleaning. Therefore, an advanced version of the standard pimple-Foam solver including the presented numerical models is used based on the open source computational fluid dynamics library OpenFOAM in version 5.0. Eulerian–Lagrangian simulations are performed for the continuous phase of the melt and the dispersed phase of the NMI or the NMI and bubble aggregate represented by equivalent particles. Incompressible and isothermal conditions are assumed. Table 30.1 shows the material parameters used for steel melt and the nonmetallic inclusions. Table 30.2 summarizes the most important simulation settings for the hexahedral mesh, the operation conditions, numerical time step and inclusion load in the melt. For meshing details is the reader referred to the specific publications of Neumann et al. [38, 44]. The turbulent flow regimes for both designs are in comparable magnitudes by using the corresponding operation conditions. The industrial scale tundish simulation is adopted to the inlet parameters of the real plant experiment of Wetzig et al. [23].

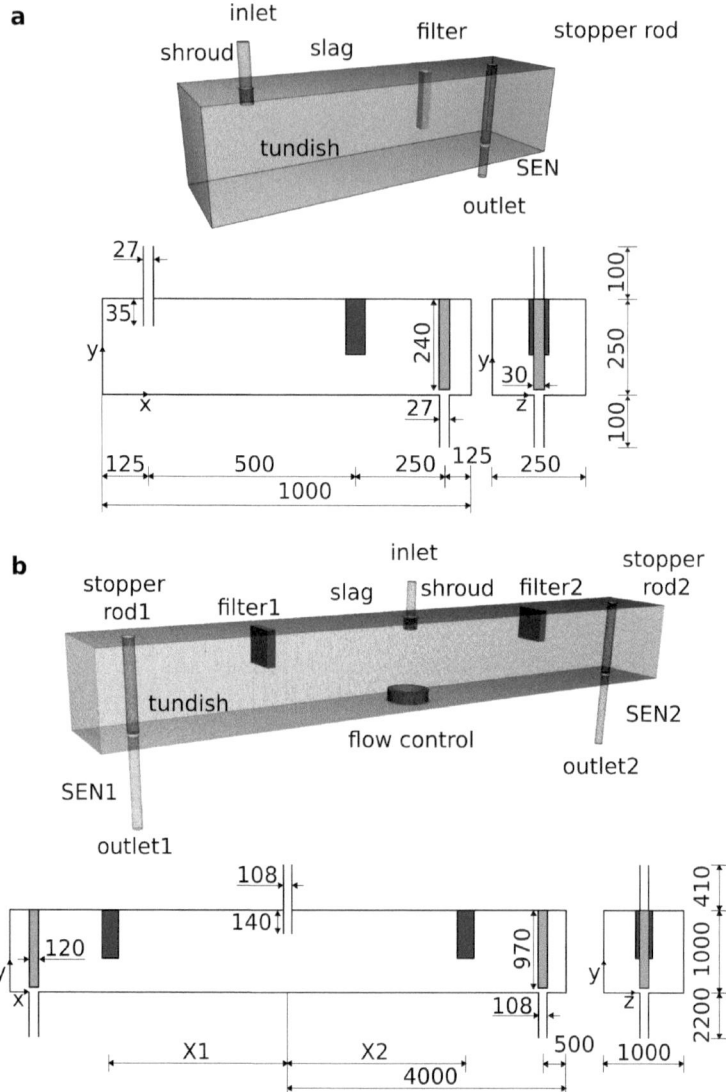

Fig. 30.1 Sketch of the one-strand prototype tundish **a** and two-strand industrial tundish **b** with dimensions (in mm), boundary names and simulation mesh, adapted with permission from [38] copyright 2019 The Minerals, Metals & Materials Society and ASM International, and with permission from [44], 2019 The Authors. Published by WILEY–VCH Verlag GmbH & Co. KGaA, Weinheim

Table 30.1 Material parameter of steel melt and nonmetallic inclusions, adapted with permission from [38] copyright 2019 The Minerals, Metals & Materials Society and ASM International, and with permission from [44], 2019 The Authors. Published by WILEY–VCH Verlag GmbH & Co. KGaA, Weinheim

	Variable	Units	Value
Liquid			
Density	ϱ_f	$\left[\text{kg} \cdot \text{m}^{-3}\right]$	7000
Dynamic viscosity	μ_f	$\left[\text{kg} \cdot \text{s}^{-1} \cdot \text{m}^{-1}\right]$	$6 \cdot 10^{-3}$
Inclusion			
Density	ϱ_i	$\left[\text{kg} \cdot \text{m}^{-3}\right]$	3000
Diameter	d_i	μm	20

Table 30.2 Numerical simulation settings from [38, 44], adapted with permission from [38] copyright 2019 The Minerals, Metals & Materials Society

Tundish	Units	Laboratory scale	Industrial scale
Cell size	[mm]	4	16
Mesh size	[−]	~$1.4 \cdot 10^6$	~$2.1 \cdot 10^6$
Time step	[s]	$2 \cdot 10^{-4}$	$2 \cdot 10^{-3}$
Mass flow	$\left[\text{kg} \cdot \text{s}^{-1}\right]$	12	140
Inlet velocity U_i	$\left[\text{m} \cdot \text{s}^{-1}\right]$	3, 00	2, 19
Reynolds number inlet	[−]	$1 \cdot 10^5$	$2.7 \cdot 10^5$
Reynolds number bulk	[−]	$8 \cdot 10^3$	$2.3 \cdot 10^4$
Number of particle	[−]	~$1.2 \cdot 10^6$	~$2.1 \cdot 10^6$

The discretization schemes are second-order accurate for spatial and temporal discretization. An exception is the spatial discretization of the carbon concentration, which is utilized with a first-order upwind scheme [38, 44].

Neumann et al. [38] analyzed the sensitivity of the simulation model regarding active filtration and reactive cleaning for the one-strand laboratory scale tundish. Independent results for the mesh sensitivity are published [38, 44]. The active filtration is not influenced by the mesh size. However, applying different cell sizes using reactive cleaning leads to deviations of up to 10% between coarsest and finest mesh [38]. The simulations of the industrial two-strand tundish are made using the coarse grid with the aim to reduce the simulation effort and costs [44].

Neumann et al. [38] also reported a sensitivity study regarding time-step dependence. A maximal deviation of approx. 10% is observed for reactive cleaning using high and low time steps. For the industrial tundish simulations, larger time steps have been used due to computational limitations [44].

The reader is referred to Neumann et al. [38] for further sensitivity studies regarding the number of inclusions and several operation conditions.

The boundary conditions are equivalent for both designs. At the inlet, a constant velocity is set. Walls are defined with no slip conditions. The slag has slip conditions for pressure and velocity. At the outlet, an inlet–outlet condition is used. Further information can be found in [38, 44].

The initial conditions for the simulations are equally prepared for both designs. A Reynolds-Averaged-Navier–Stokes (RANS) simulation and a subsequent Unsteady-RANS (URANS) build the basis to get statistical steady-state conditions. Furthermore, a single-phase ILES simulation provides the desired flow field for the following multiphase ILES simulation of the melt and the homogeneously dispersed NMI in the tundish. The detailed time durations and setups can be found in the particular publication of Neumann et al. [38, 44].

30.4   Laboratory One-Strand Tundish

Neumann et al. [38] investigated the laboratory one-strand tundish regarding active filtration and reactive cleaning. Exchangeable carbon-bonded alumina foam filters are inserted in the steel melt during the simulations with different dimensions, see Table 30.3.

In general, the global tundish flow structure is less influenced by the foam filter after insertion. The time-averaged velocity field in longitudinal mid-plane is presented in Fig. 30.2a. It shows, the well-known flow structure with the developing inlet jet, a large recirculation zone in the tundish center and the outgoing flow. Close to the filter the velocity is slightly decelerated based on the pressure drop in the filter, modeled by the Darcy-Forchheimer law.

Figure 30.2b reveals the carbon concentration in the tundish 4s after the insertion of the carbon-bonded foam filter. Carbon is quickly distributed within the melt due to the high turbulent diffusion. The carbon concentration in the flow field varies based on the distance to the filter and the number of reactions due to the NMI distribution. Inclusions close to the filter rapidly float to the slag based on the high concentration of carbon.

Table 30.3 Filter size and position laboratory tundish, adapted with permission from [38] copyright 2019 The Minerals, Metals & Materials Society

Filter	Units	AF1 RC1	AF2 RC2	AF3 RC3	AF4 RC4
Length	[mm]	25	35	50	35
Width	[mm]	25	35	50	35
Height	[mm]	157	157	157	80
Volume scale	[−]	Basis	2 : 1	4 : 1	1 : 1
V_F/V_T	[%]	0.16	0.31	0.63	0.16

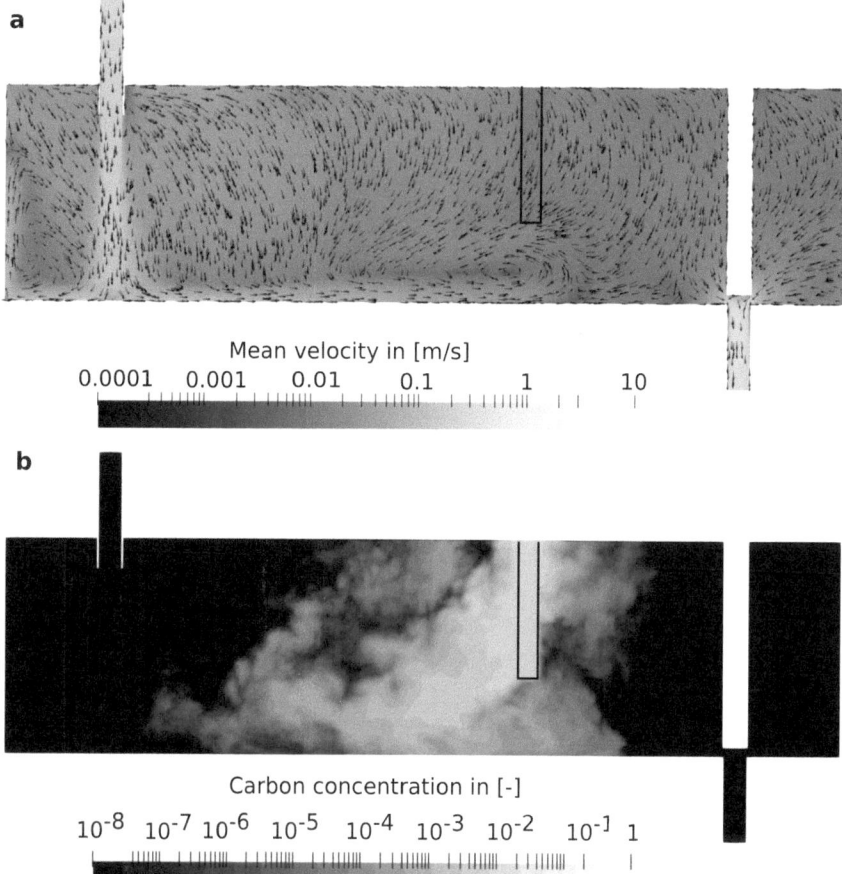

Fig. 30.2 Longitudinal tundish mid plane of the one-strand laboratory-scale prototype tundish with filter inserted a time-averaged velocity field b carbon concentration during reactive cleaning, adapted with permission from [38] copyright 2019 The Minerals, Metals & Materials Society

Figure 30.3a compares active filtration (AF) and reactive cleaning (RC) regarding the cleaning efficiency of the NMI after $8s$ of filter immersion. The cleaning efficiency is represented by the deposition of NMI's on the filter or sticking at the slag. In summary, active filtration, the selective deposition of inclusions at the filter surface, shows no significant contribution to the cleaning of the tundish melt.

Not even different dimensions of the filter design (AF1–AF4, RC1–RC4), see Table 30.3 and Fig. 30.3a, have shown any influence on active filtration efficiency. Most NMI remain in the tundish or escape the domain through the submerged exit nozzle.

However, reactive cleaning shows a great potential as cleaning strategy in tundish systems. Figure 30.3a shows a maximum cleaning efficiency of 44% for filter type 3 after a time of $8s$ after immersion into the tundish. This is not surprising because

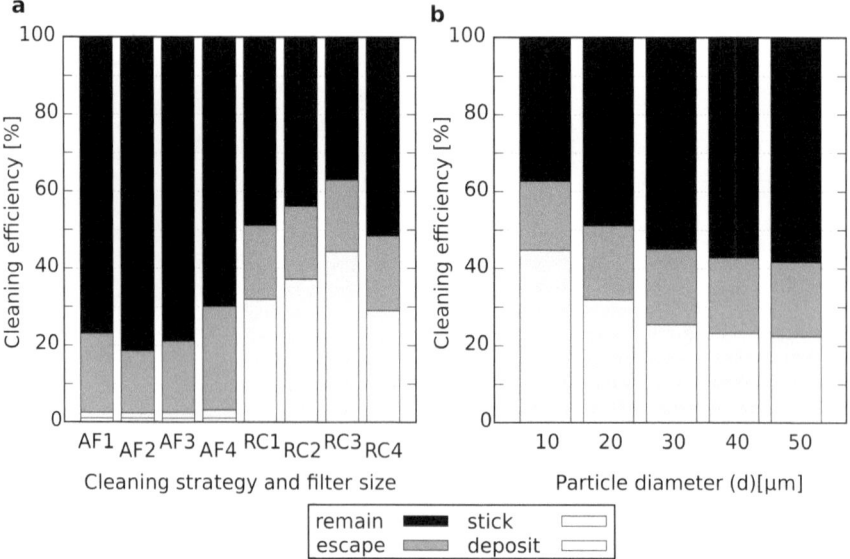

Fig. 30.3 Cleaning efficiencies after 8*s* of filter immersion considering a active filtration and reactive cleaning for various filter sizes, and b during reactive cleaning for different particle diameters, adapted with permission from [38] copyright 2019 The Minerals, Metals & Materials Society

of the highest filter volume of filter RC3, which consequently leads to a high carbon concentration in the tundish melt. Therefore, a multiplicity of reactions generating bubbles at the NMI surfaces result in a high rate of rising particles to the slag.

Furthermore, the filter design has also an impact on the reactive cleaning efficiency. The filters RC1 and RC4 with comparable volume, show different cleaning efficiencies. Carbon seems to be better distributed within the melt, when a compact filter with lower aspect ratio is used. Consequently, the cleaning efficiency is higher in this case (see Fig. 30.3a).

The reactive cleaning efficiency considering the particle diameter 8*s* after immersion of the filter is depicted in Fig. 30.3b. With increasing inclusion diameter, the cleaning performance using filter RC1 decreases due to the higher particle mass of the inclusions. Of course, the effect of the generated CO bubbles is reduced in the lift force. However, high cleaning efficiencies in range of 20−45% are achieved.

30.5 Industrial Two-Strand Tundish

The intention to investigate an industrial prototype two-strand tundish is to learn more about the cleaning strategies in plant applications. Exchangeable carbon-bonded alumina-coated foam filters could be a promising application to clean the steel melt in the tundish using active filtration and reactive cleaning. A great advantage is the less constructive intervention with simultaneously high cleaning efficiencies using exchangeable filter systems.

Compared to the laboratory scale tundish, the dimensions as well as the capacities are much higher for the industrial tundish. The flow structure in a two-strand tundish may differ to the one-strand design. Moreover, the findings of Neumann et al. [38] reveal a distinct dependence on the cleaning efficiency with the filter design. Therefore, filter design (shape) as well as the flow guidance (flow controller) to the filter is investigated to enhance the removal of the NMI [44], see drawings in Fig. 30.4.

Flow controlling units positioned on the ground of the tundish centered under the inlet shroud with different inclination angles (45°, 60°, 90°) and a turbostop are investigated regarding cleaning efficiencies of active as well as reactive cleaning (Fig. 30.4a). Furthermore, a quadratic filter (cuboid), adopted from the dimensions of the filter of Wetzig et al. [23], and a paddle-shaped filter with doubled width and height as well as halved length compared to the cuboid filter are researched, see Fig. 30.4b. Another probable influencing parameter to the cleaning efficiency is the positioning of the filter in the tundish melt. Analyzed locations can be found in Fig. 30.1b and Table 30.4.

Additionally, the cleaning duration is considered for different periods. Short-term ($10s$), mid-term ($100s$) and long-term ($400s$, one tundish filling sequence) investigations for active filtration and reactive cleaning after filter immersion are performed. Detailed information about the setup of the simulations can be found in Neumann et al. [44].

Neumann et al. [44] also published long-term simulation results without cleaning activities. The intention of this simulation was to count those NMI, which enter the cuboid or paddle filter volume. From this information, promising combinations of filter design, position and flow controlling unit are deduced. Finally, the highest counting rates occur for the filter positions close to the inlet nozzle (e.g. P1, P4) using flow controller FC1, FC3 or no flow controller. Additionally, the paddle filter shows higher counting rates than the cuboid.

Figure 30.5 displays the simulation results for active filtration and reactive cleaning at the filter position P1 regarding short-term (10 s) and mid-term (100 s) investigations and different flow controlling units.

Obviously, active filtration (Fig. 30.5a) shows poor cleaning efficiency compared to reactive cleaning (Fig. 30.5b). The application of flow controlling units has an influence on the results of both cleaning strategies, but the highest cleaning efficiencies occur using no flow controller (noFC). Active filtration in Fig. 30.5a shows a distinct dependency on the filter design. The paddle shape enhances slightly the NMI cleaning.

However, based on the promising findings of the long-term ($400s$) simulations without cleaning activities of Neumann et al. [44], higher cleaning rates were expected. But altogether, the results are comparable to the findings of the laboratory one-strand tundish, see Fig. 30.3a.

Figure 30.5b presents the results using reactive cleaning strategy. A slight dependence of the cleaning efficiency on the filter shape is recognized for the short-term ($10s$) simulations. The paddle-shaped filter shows higher cleaning efficiencies for

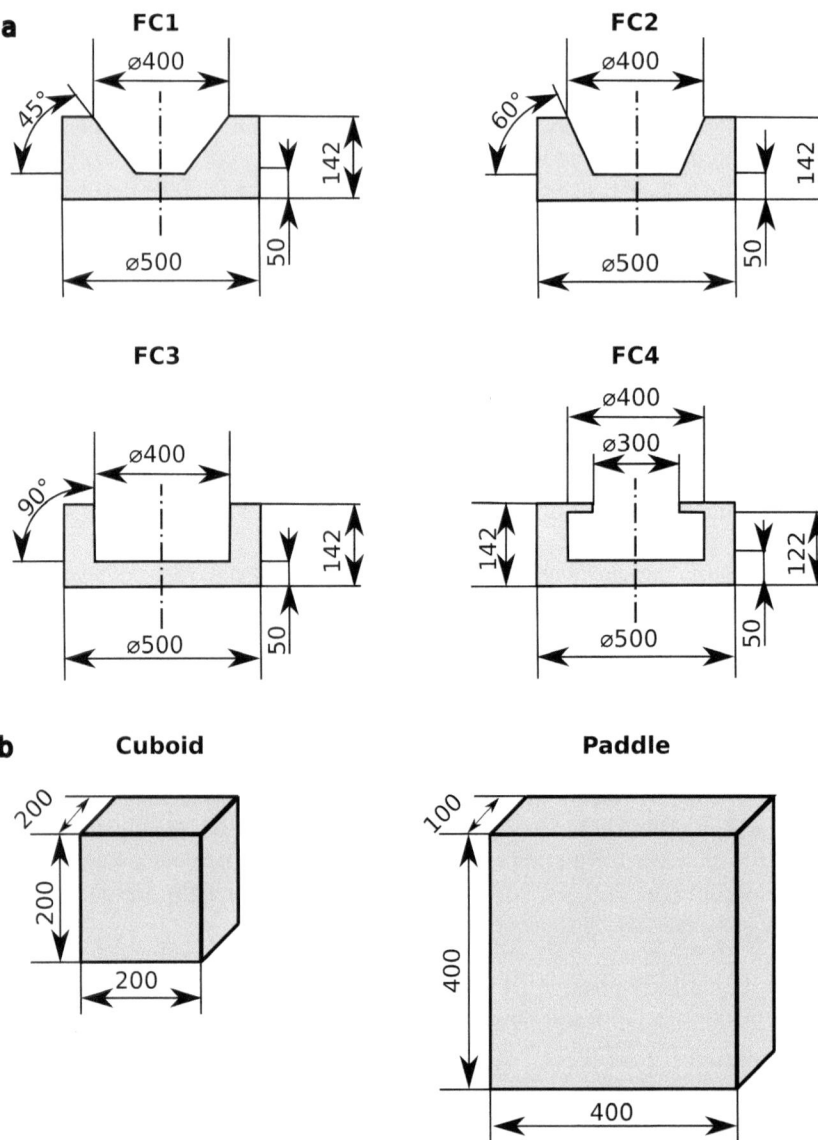

Fig. 30.4 Drawings of the researched a flow controller (FC) and b filters (F) (dimensions in mm), adapted with permission from [44], 2019 The Authors. Published by WILEY–VCH Verlag GmbH & Co. KGaA, Weinheim

Table 30.4 Filter positions for the industrial tundish, adapted with permission from [44], 2019 The Authors. Published by WILEY–VCH Verlag GmbH & Co. KGaA, Weinheim

Point	P1	P2	P3
X1 [m]	−0.75	−1.50	−2.25
Point	P4	P5	P6
X2 [m]	1.00	2.00	3.00

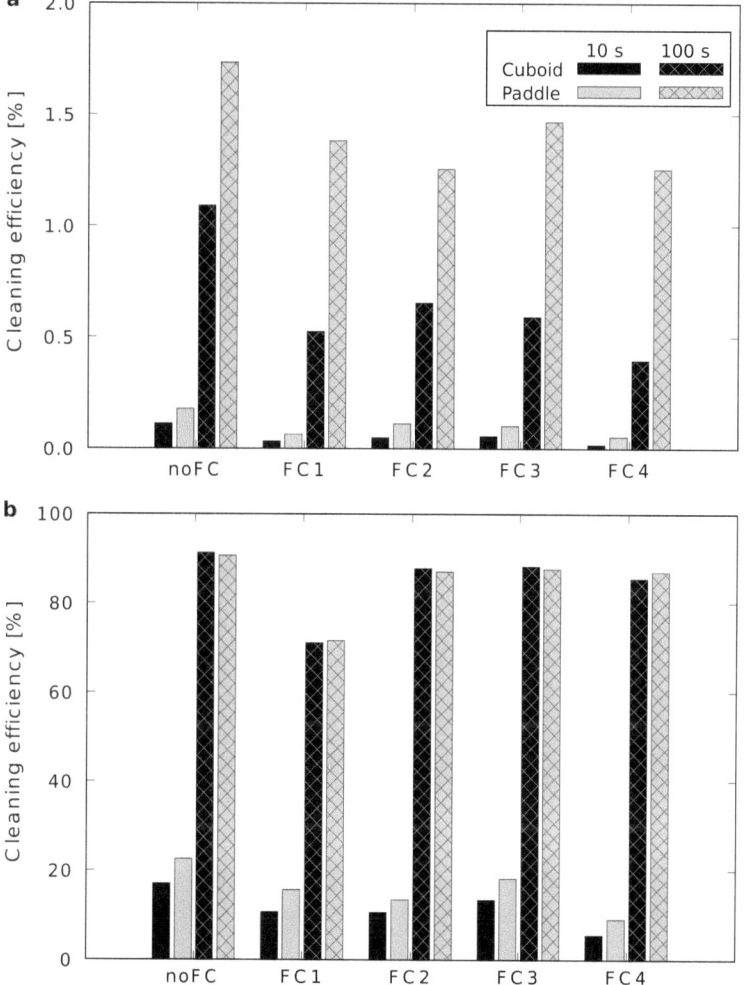

Fig. 30.5 Cleaning efficiencies regarding a active filtration and b reactive cleaning considering short term (10s) and mid-term (100s) operation as well as different flow controller (FC) and cuboid and paddle filter shape at position P1 for the industrial-scale tundish, adapted with permission from [44], 2019 The Authors. Published by WILEY–VCH Verlag GmbH & Co. KGaA, Weinheim

all flow controlling units. However, for the mid-term ($100s$) investigations the shape differences are neglectable.

The most important result is that with increasing duration of the reactive filter application, the cleaning efficiency massively increases. This is a consequence of the current model for the carbon concentration, which constantly supplies carbon to the melt based on the assumption of a durable filter. A further outcome of this is, that the current filter size (Fig. 30.4) is to small and larger filters will lead to higher cleaning efficiencies within a shorter period of time.

Figure 30.6 shows the simulation results regarding the different cleaning strategies, cuboid and paddle filter shape, various cleaning durations as well as different filter positions. Again, active filtration shows only small cleaning effectiveness, see Fig. 30.6a. Filter positioning as well as filter design have neglectable influence to the results. Higher cleaning durations improve the cleaning efficiency, but they are still low.

In contrast, reactive cleaning shows high cleaning efficiencies, see Fig. 30.6b. However, filter position and filter shape have less influence. The simulation results show again the time dependence of the reactive cleaning after filter insertion. The mid-term ($100s$) observations show cleaning efficiencies of approx. 90%.

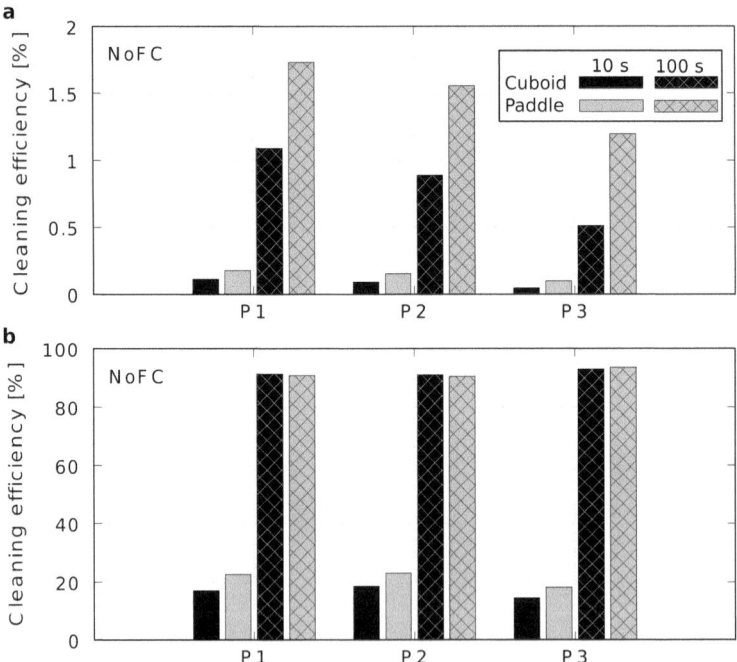

Fig. 30.6 Cleaning efficiencies regarding a active filtration and b reactive cleaning considering short-term ($10s$) and mid-term ($100s$) operation without flow controller (noFC) for the cuboid and paddle filter shape at filter positions P1-P3 for the industrial scale tundish, adapted with permission from [44], 2019 The Authors. Published by Wiley–VCH Verlag GmbH & Co. KGaA, Weinheim

Figure 30.7 depicts the cleaning efficiencies with regard to active filtration and reactive cleaning considering long-term ($400s$) operation as well as different flow controller (noFC, FC2) at filter positions P1 and P2 for the cuboid and paddle shaped filter. Reactive cleaning shows in Fig. 30.7b still high cleaning efficiencies of about 90% independent from flow controlling unit, filter shape or position. However, the long-term observations also increase the degree of purity by active filtration markedly, Fig. 30.7a. Due to the high residence time of the NMI, the alumina-coated, carbon-bonded ceramic hybrid foam filter develops active cleaning efficiencies of approx. $20 - 30\%$, based on the assumption of constant cleaning effectiveness. Clear trends for the application of flow controlling unit, filter position or shape are not observable. However, there are better or worse combinations. If no flow controller is applied, the cuboid filter in position P1 shows the best results. In contrary, the paddle-shaped filter in position P2 together with the flow controlling unit FC2 is the most effective active filtration combination.

Finally, all investigations regarding reactive cleaning (see Figs. 30.5, 30.6 and 30.7) are almost independent of the filter volume. The paddle filter has twice the volume of the cuboid, but distinct differences concerning the cleaning efficiencies are not observable. This finding is in contrast to the results of the laboratory scale tundish (see Fig. 30.3b), where the filter volume influences the carbon concentration in the melt, which leads to higher cleaning efficiencies. This is an interesting fact, since the relative ratio of the filter volume to the tundish volume is for both simulation setups in the same range of about $0.2 - 0.5\%$. Of course, the absolute dimensions of the filter applied in the industrial-scale tundish are in maximum $100 - 200$ times larger than the filter used in the laboratory-scale tundish. But in summary, the relative volume ratio seems not to be an expressive indication variable for the choice of the filter size.

30.6 Conclusions

This chapter describes the application of exchangeable alumina-coated, carbon-bonded ceramic hybrid foam filter in tundish melt flows during continuous casting. A laboratory and an industrial-scale tundish are investigated. With regards to the cleaning efficiency, two different cleaning strategies are compared.

Filter using the active filtration approach directly remove the non-metallic inclusions out of the melt by the deposition of the particles onto the filter surface. In contrast, reactive cleaning is based on a reaction of carbon and oxygen generating carbon monoxide bubbles directly at the inclusions. With increasing bubble volume, inclusion and bubble float to the slag are finally cleaned out of the steel melt.

A numerical model which considers the melt flow, particle motion and includes the different cleaning strategies is developed for the open-source computational fluid dynamics library OpenFOAM.

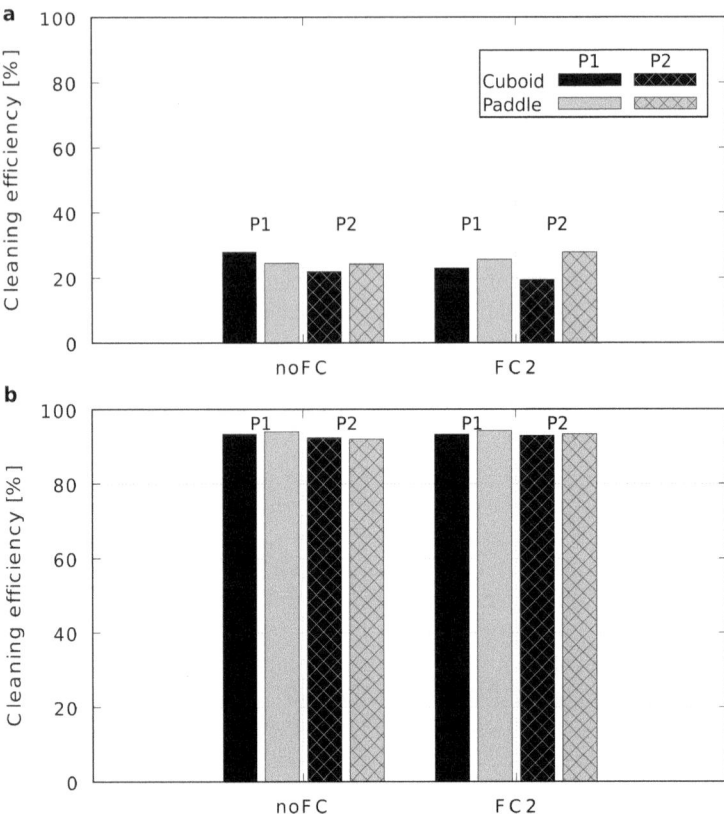

Fig. 30.7 Cleaning efficiencies regarding **a** active filtration and **b** reactive cleaning considering long-term (400 s) operation as well as different flow controller (noFC, FC2) for the cuboid and paddle shaped filter at filter positions P1 and P2 for the industrial-scale tundish, adapted with permission from [44], 2019 The Authors. Published by WILEY–VCH Verlag GmbH & Co. KGaA, Weinheim

Parameter studies concerning different filter parameter (design, size, position, cleaning strategy), time periods, various flow controlling units and particle sizes are performed.

Active filtration shows in contrast to reactive cleaning much lower cleaning efficiencies for both tundish types. There are dependencies with regards to the combination of filter shape, position and flow controlling unit. However, long-term observations show reasonable cleaning efficiencies up to about 30% for active filtration.

In contrast, reactive cleaning has a high efficiency close to 100% and quickly removes the non-metallic inclusion. This cleaning strategy is almost independent from the flow controlling unit or the filter design or position. However, the filter volume has an influence to the cleaning efficiency for the laboratory scale tundish, but the investigations for the industrial scale tundish do not show such trends.

Future work should focus on the validation of the models in industrial experiments under plant conditions. Here, NMI concentrations in the incoming (shroud) and outgoing (submerged entry nozzle) melt flow should be detected in both, plant experiments and numerical simulations. The research should also inspect the influence of the dissolved carbon from the ceramic foam filter, in more detail.

Acknowledgements The authors gratefully acknowledge the German Research Foundation (DFG) for supporting the Collaborative Research Center 920, subprojects: T01, B06. The research work was funded by the Deutsche Forschungsgemeinschaft (DFG, German Research Foundation)—Project-ID 169148856—SFB 920. The authors thank Tom Kasper for his work in SFB 920 transfer project T01. The computations were performed on a Bull Cluster at the Center of Information Services and High Performance Computing (ZIH) at TU Dresden and the High Performance Cluster at the University Computer Centre (URZ) at the TU Bergakademie Freiberg.

References

1. L. Zhang, B.G. Thomas, ISIJ Int. **43**, 271–291 (2003). https://doi.org/10.2355/isijinternational.43.271
2. L. Zhang, B.G. Thomas, in *Proceedings of XXIV National Steelmaking Symposium*, Morelia, Mich, Mexico, 26–28 November 2003
3. L. Zhang, in *TMS Annual Meeting*, Orlando, 2007, ed. by A. Gokhale, J. Li, T. Okabe
4. R. Dekkers, B. Blanpain, P. Wollants, F. Haers, B. Gommers, C. Vercruyssen, Steel Res. Int. **74**, 351–355 (2003). https://doi.org/10.1002/srin.200300197
5. W. Yang, H. Duan, L. Zhang, Y. Ren, Jom-J. Min. Met. Mat. S. **65**, 1173–1180 (2013). https://doi.org/10.1007/s11837-013-0687-z
6. L. Zhang, S. Taniguchi, K. Cai, Metall. Mater. Trans. B **31**, 253–266 (2000). https://doi.org/10.1007/s11663-000-0044-9
7. C.M. Lee, I.S. Choi, B.G. Bak, J.M. Lee, Rev. Met. Paris **90**, 501–506 (1993). https://doi.org/10.1051/metal/199390040501
8. J. Mancini, J. van der Stel, Rev. Met. Paris **89**, 269–278 (1992). https://doi.org/10.1051/metal/199289030269
9. O.B. Isaev, Metallurgist **53**, 672–678 (2009). https://doi.org/10.1007/s11015-010-9231-3
10. Y. Miki, B.G. Thomas, Metall. Mater. Trans. B **30**, 639–654 (1999). https://doi.org/10.1007/s11663-999-0025-6
11. L. Zhang, Steel Res. Int. **77**, 158–169 (2006). https://doi.org/10.1002/srin.200606370
12. A.S. Kondrat'ev, V.N. Popov, L.M. Aksel'rod, M.R. Baranovskii, S.A. Suvorov, N.B. Tebuev, Refractories **31**, 384–391 (1990). https://doi.org/10.1007/BF01281545
13. K. Janiszewski, Arch. Metall. Mater. **58**, 513–521 (2013). https://doi.org/10.2478/amm-2013-0029
14. K. Janiszewski, B. Panic, Metalurgija **53**, 339 (2014)
15. K. Janiszewski, Steel Res. Int. **84**, 288–296 (2013). https://doi.org/10.1002/srin.201200077
16. K. Janiszewski, B. Gajdzik, K. Gryc, L. Socha, A. Bogdał, SSP **226**, 189–192 (2015). https://doi.org/10.4028/www.scientific.net/SSP.226.189
17. K.-I. Uemura, M. Takahashi, S. Koyama, M. Nitta, ISIJ Int. **32**, 150–156 (1992)
18. C.G. Aneziris, S. Dudczig, J. Hubálková, M. Emmel, G. Schmidt, Ceram. Int. **39**, 2835–2843 (2013). https://doi.org/10.1016/j.ceramint.2012.09.055
19. A. Schmidt, A. Salomon, S. Dudczig, H. Berek, D. Rafaja, C.G. Aneziris, Adv. Eng. Mater. **19**, 1700170 (2017). https://doi.org/10.1002/adem.201700170

20. M. Schlautmann, B. Kleimt, A. Kubbe, R. Teworte, D. Rzehak, D. Senk, A. Jaklic, M. Klinar, Stahl Eisen **131**, 57–65 (2011)
21. D. Rzehak, Dissertation (in German), RWTH Aachen (2013)
22. E. Storti, S. Dudczig, A. Schmidt, G. Schmidt, C.G. Aneziris, Steel Res. Int. **88**, 1700142 (2017). https://doi.org/10.1002/srin.201700142
23. T. Wetzig, A. Baaske, S. Karrasch, N. Brachhold, M. Rudolph, C.G. Aneziris, Ceram. Int. **44**, 23024–23034 (2018). https://doi.org/10.1016/j.ceramint.2018.09.105
24. A. Asad, M. Haustein, K. Chattopadhyay, C.G. Aneziris, R. Schwarze, JOM **70**, 2927–2933 (2018). https://doi.org/10.1007/s11837-018-3117-4
25. A. Cwudziński, Steel Res. Int. **85**, 623–631 (2014). https://doi.org/10.1002/srin.201300079
26. A. Cwudziński, Steel Res. Int. **85**, 902–917 (2014). https://doi.org/10.1002/srin.201300284
27. R. Schwarze, D. Haubold, C. Kratzsch, Ironmaking Steelmaking **42**, 148–153 (2015). https://doi.org/10.1179/1743281214Y.0000000221
28. A. Vakhrushev, M. Wu, A. Ludwig, G. Nitzl, Y. Tang, G. Hackl, R. Wincor, Steel Res. Int. **88**, 1600276 (2017). https://doi.org/10.1002/srin.201600276
29. Q. Yue, C. Zhang, X.Z. Wang, Metalurgija **56**, 126–130 (2017)
30. M. Warzecha, T. Merder, P. Warzecha, Arch. Metall. Mater. **60**, 215–220 (2015). https://doi.org/10.1515/amm-2015-0034
31. P. Warzecha, A.M. Hutny, M. Warzecha, T. Merder, Arch. Metall. Mater. **61**, 2071–2078 (2016). https://doi.org/10.1515/amm-2016-0333
32. Y. Jin, X. Dong, F. Yang, C. Cheng, Y. Li, W. Wang, Metals **8**, 611 (2018). https://doi.org/10.3390/met8080611
33. J. Huang, Z. Yuan, S. Shi, B. Wang, C. Liu, Metals **9**, 239 (2019). https://doi.org/10.3390/met9020239
34. D. Mazumdar, R.I.L. Guthrie, ISIJ Int. **39**, 524–547 (1999). https://doi.org/10.2355/isijinternational.39.524
35. B.G. Thomas, Metall. Mater. Trans. B **33**, 795–812 (2002). https://doi.org/10.1007/s11663-002-0063-9
36. K. Chattopadhyay, M. Isac, R.I.L. Guthrie, ISIJ Int. **50**, 331–348 (2010). https://doi.org/10.2355/isijinternational.50.331
37. M. Vynnycky, Metals **8**, 928 (2018). https://doi.org/10.3390/met8110928
38. S. Neumann, A. Asad, T. Kasper, R. Schwarze, Metall. Mater. Trans. B **50**, 2334–2342 (2019). https://doi.org/10.1007/s11663-019-01637-6
39. A. Asad, E. Werzner, C. Demuth, S. Dudczig, A. Schmidt, S. Ray, C.G. Aneziris, R. Schwarze, Adv. Eng. Mater. **19**, 1700085 (2017). https://doi.org/10.1002/adem.201700085
40. A. Asad, K. Bauer, K. Chattopadhyay, R. Schwarze, Metall. Mater. Trans. B. Sci. **49**, 1378–1387 (2018). https://doi.org/10.1007/s11663-018-1200-4
41. A. Asad, K. Chattopadhyay, R. Schwarze, Metall. Mater. Trans. B Process Metall. Mater. **49**, 2270–2277 (2018). https://doi.org/10.1007/s11663-018-1343-3
42. A. Asad, C. Kratzsch, S. Dudczig, C.G. Aneziris, R. Schwarze, Int. J. Heat Fluid Flow **62**, 299–312 (2016). https://doi.org/10.1016/j.ijheatfluidflow.2016.10.002
43. R. Schwarze, J. Klostermann, C. Brücker, Int. J. Heat Fluid Flow **29**, 1688–1698 (2008). https://doi.org/10.1016/j.ijheatfluidflow.2008.08.001
44. S. Neumann, A. Asad, R. Schwarze, Adv. Eng. Mater. **22**, 1900658 (2020). https://doi.org/10.1002/adem.201900658
45. A. Asad, C.G. Aneziris, R. Schwarze, Adv. Eng. Mater. **22**, 1900591 (2020). https://doi.org/10.1002/adem.201900591
46. A. Asad, R. Schwarze, Steel Res. Int. **92**, 2100122 (2021). https://doi.org/10.1002/srin.202100122
47. A. Asad, H. Lehmann, B. Jung, R. Schwarze, Adv. Eng. Mater. **24**, 2100753 (2021). https://doi.org/10.1002/adem.202100753
48. H.-S. Kim, J.G. Kim, Y. Sasaki, ISIJ Int. **50**, 678–685 (2010). https://doi.org/10.2355/isijinternational.50.678

Chapter 31
Precipitation of Iron-Containing Intermetallic Phases from Aluminum Alloys by Metal Melt Filtration

Johannes Paul Schoß, Andreas Keßler, Claudia Dommaschk, Michal Szucki, and Gotthard Wolf

31.1 Introduction

Iron (Fe) is considered as an accompanying impurity in aluminum (Al) casting alloys. However, due to the formation of the pre-eutectic, Fe-containing β-AlFeSi phase, which exhibits a plate-like morphology, its presence is detrimental to the mechanical and casting properties [1, 2]. Since an increasing Fe content in the melt resulted in the diffusion-related formation of coarse plates, the effective feeding with residual melt is hindered [3, 4]. In addition, shrinkage porosity and notching stresses are increased, which significantly reduce the ductility of the castings as well [4–6]. Fe impurities become particularly enriched in secondary Al alloys due to insufficient scrap separation in the global aluminum recycling process [7–9]. The usage of secondary Al alloys for castings is much more beneficial from an economic and environmental aspect than the production of primary aluminum. Compared to the primary production (in the fused-salt electrolysis process), secondary alloys require only 5% of the energy and CO_2 emissions [10]. Moreover, this contributes to the energy-intensive electrolytic reduction of Al_2O_3. Currently, the reduction of Fe in secondary Al alloys is performed by dilution with high-cost pure aluminum ($\geq 50\%$). Therefore, investigations are conducted to reduce the Fe content in a commercial secondary AlSi9Cu3 alloy using a two-step procedure.

First, Fe-rich intermetallics are generated in the melt using melt conditioning by adding alloying elements such as manganese (Mn), chromium (Cr), nickel (Ni), beryllium (Be), cobalt (Co), etc., resulting in a modification of the morphology into a compact and filterable α-intermetallic (or also called sludge phase) [9, 11–13]. With

J. P. Schoß (✉) · A. Keßler · C. Dommaschk · M. Szucki · G. Wolf
Foundry Institute, Technische Universität Bergakademie Freiberg, Bernhard-Von-Cotta-Str. 4, 09599 Freiberg, Germany
e-mail: Johannes.Schoss@gi.tu-freiberg.de

© The Author(s) 2024
C. G. Aneziris and H. Biermann (eds.), *Multifunctional Ceramic Filter Systems for Metal Melt Filtration*, Springer Series in Materials Science 337,
https://doi.org/10.1007/978-3-031-40930-1_31

increasing the initial Fe, Mn, and Cr content, the formation temperature of the Fe-rich intermetallics shifts to higher temperatures, forming a primary phase, i.e., before the formation of the dendrite network (α-Al solid solution) [14–17]. Consequently, the melt temperature must be set below the formation temperature of the Fe-rich inter-metallic phase and above the formation temperature of Al-dendrites in order to obtain the Fe-containing intermetallics. As a result, the residual melt exhibits an Fe-reduced content. In the second step, the intermetallic phases can be finally separated by an appropriate filtration technique using multifunctional and/or specially coated ceramic foam filters (CFF), which are developed in cooperation with other subprojects in the Collaborative Research Center 920 (CRC 920).

In order to implement the two-step procedure for a potential filtration technology, the influences on the formation of the Fe-rich intermetallics has to be investigated. In particular, the formation characteristics of intermetallics depend on the tempera-ture, time, cooling rate, and initial chemical composition [18–21]. For this purpose, differential scanning calorimetry (DSC) cooling curves are analyzed to examine the different influences (e.g., temperature, chemical composition, etc.) on the formation of intermetallic phases. Based on these insights, the operating range can be defined to initiate the filtration process. Furthermore, the potential for Fe removal is to be evaluated via experimental filtration trials using a specially developed laboratory filtration apparatus. The efficiency of Fe removal is determined using variations in time and temperature. Additionally, the influence of multifunctional filter coatings and materials is also considered. Besides different filter porosities (20, 30 pores per inch - ppi), foam ceramic filters (CFF) made of alumina, C-bonded alumina filters, and alumina CFF with rough alumina coating are used. Finally, the transfer project T03 will implement an appropriate filtration process technology for an industrial application in the light metal foundry. Hence, the setup and evaluation in industrial scale trials are presented. Industrially manufactured CFF as well as appropriate filter structures from the CRC 920 will be tested.

31.2 Experimental Section

31.2.1 Castings and Materials

Commercial secondary AlSi9Cu3(Fe) Al alloy (EN AC-46000, VAR 226 D) from Nemak (Dillingen, Germany), master alloys AlMn20[1] and AlCr20[1] (Technologica, Germany), and steel scrap (Schönheider Guss, Germany) were used to obtain various alloy compositions. Table 31.1 shows the element contents of the used base materials.

---

[1] Numbers refers to chemical composition by weight%.

Table 31.1 Chemical compositions (weight%) of the base materials in the optical emission spectrometer (OES) SPECTROMAXx (Ametek, USA), where the element contents are within the tolerances of the standardizations (DIN EN 1706 and DIN EN 10268) [22]

Alloy	Al [%]	Si [%]	Fe [%]	Cu [%]	Mn [%]	Mg [%]	Cr [%]	Ni [%]	Zn [%]	Others [%]
AlSi9Cu3(Fe)	87.3	8.63	0.973	2.14	0.094	0.251	0.039	0.049	0.408	0.116
AlMn20[1]	77.6	0.086	0.384	0.006	21.32	0.026	0.138	0.013	0.014	0.413
AlCr20[1]	78.3	0.086	0.810	0.005	0.076	0.047	19.80	0.028	0.058	0.790

Alloy	Fe [%]	C [%]	Si [%]	Mn [%]	P [%]	S [%]	Cr [%]	Ni [%]	Al [%]	Others [%]
HC180P	99.0	0.008	0.153	0.355	0.074	0.007	0.037	0.040	0.159	0.167

Table 31.2 Chemical compositions (weight%) of various series of an AlSi9Cu3(Fe) alloy with weight fraction ratio of w_{Mn}/w_{Fe} and sludge factor (SF) according to Eq. 31.1 in optical emission spectrometer (OES) SPECTROMAXx (Ametek, USA) [22]

Alloy	Al [%]	Si [%]	Fe [%]	Cu [%]	Mn [%]	Mg [%]	Cr [%]	Ni [%]	Zn [%]	w_{Mn}/w_{Fe} ratio	Sludge factor (SF)
Leg A	87.1	8.87	0.954	2.14	0.106	0.178	0.039	0.054	0.424	0.11	1.28
Leg B	86.0	8.87	1.31	2.19	0.363	0.193	0.274	0.071	0.616	0.28	2.86
Leg C	85.4	8.97	1.58	2.30	0.603	0.159	0.181	0.040	0.596	0.39	3.30
Leg D	84.9	9.24	1.61	2.18	1.01	0.127	0.259	0.042	0.515	0.63	4.41
Leg E	85.5	8.23	1.58	1.92	1.47	0.177	0.235	0.043	0.412	0.93	5.23
Leg I	85.4	8.99	0.829	3.13	0.346	0.212	0.040	0.071	0.772	0.42	1.64
Leg II	85.2	8.96	0.991	3.19	0.342	0.214	0.038	0.070	0.777	0.35	1.79
Leg III	85.1	9.01	1.18	3.13	0.338	0.210	0.038	0.068	0.771	0.29	1.97
Leg IV	84.9	8.85	1.22	3.09	0.662	0.202	0.040	0.068	0.754	0.54	2.66
Leg V	85.0	8.67	1.20	3.06	0.860	0.201	0.040	0.067	0.743	0.72	3.04
Leg VI	84.7	8.70	1.20	3.02	1.20	0.193	0.041	0.066	0.729	1.00	3.72
Leg VIII	84.4	8.46	1.27	2.94	1.17	0.186	0.628	0.070	0.715	0.92	5.49
Leg IX	84.1	8.43	1.28	3.04	1.17	0.178	0.888	0.069	0.711	0.91	6.28
Leg X	84.1	8.20	1.29	2.99	1.20	0.171	1.18	0.067	0.690	0.93	7.23

The different chemical compositions are constituted to investigate the formation of primary Fe-rich intermetallics (sludge phases). They are shown in Table 31.2. Their preparation was realized in an electric resistance furnace (Nabertherm, Germany) with a clay-bonded ISO graphite crucible size A 60 (Mammut-Wetro, Germany) and a capacity of ≈25 kg Al. The AlSi9Cu3(Fe) base alloy was melted first at 820 °C. Then, the Fe, Mn, and Cr contents were adjusted by adding steel scrap (HC180P) and the master alloys (AlMn20 un AlCr20) by liquid metal transfer. Due to their higher melting temperature, the master alloys were externally melted to temperatures above 1000 °C, as reported in previous studies [21, 22]. For each chemical composition, the chemical analysis was examined first. After that, samples were poured into a preheated metallic die, from which specimens were taken for metallography and DSC measurements. Finally, the prepared melt was transferred to a specially designed laboratory filtration apparatus for experimental filtration trials using a ladle with ≈3 kg capacity.

31.2.2 Analytical Methods

The chemical composition values correspond to the average of three individual sparks in the optical emission spectrometer (OES) SPECTROMAXx (Ametek, USA). The microstructural examinations were conducted on polished cross-sections prepared for the final stage with colloidal silica suspension OP-U using a semiautomatic grinding and polishing machine TegraPol-31 (Struers, Germany). For obtaining a better contrast of Fe-rich intermetallics, a hot sulfuric acid solution (80 °C, 1 min, 30% H_2SO_4) was used, leading to a dark visualization of these phases compared to the less-affected silicon [22]. As a result, the area fractions of intermetallic phases were considerably easier to determine using image analysis software Stream Motion (Olympus, Japan). Based on a polyhedral and Chinese script-like morphology, detection examples using image analysis software and hot sulfuric acid etching are shown in Fig. 31.1.

For experimental filtration trials, panoramic images were taken first on the polished cross-sections and then after etching by a Keyence VHX 2000 digital microscope (Keyence, Japan). Afterwards, the area fractions were obtained via the software Stream Motion. Supplementary microstructural investigations were performed by scanning electron microscope (SEM; VEGA3 W-REM, Tescan, Germany) using secondary electron (SE) and backscattered electron (BSE) images. Finally, measurements (mappings and spot analysis) for elemental distribution were transacted in SEM by energy-dispersive X-ray spectroscopy (EDS; XFlash 610 M detector, Bruker Nano, Germany).

Fig. 31.1 Illustration to determine the area fractions of Fe-rich intermetallic phases (representative of a polyhedral –left– and Chinese script-like morphology –right–) based on exemplary cross-section in light microscopy: **a, b** original polished micrographs, **c, d** after etching (80 °C, 1 min, 30% H_2SO_4) to obtain a better contrast, and **e, f** detected Fe-rich intermetallics according to a gray scale value in red [22]

For determining the formation temperature of Fe-rich intermetallics and solidification temperature of Al-dendrites, the differential scanning calorimetry (DSC) measurements were realized in an STA 449 F5 Jupiter heat flow DSC with a Proteus software (Netzsch, Germany). The DSC device operates with three individually adjustable mass flow controllers (MFC), where a constant flow rate of argon gas of 40 mL min^{-1} was applied for the DSC cooling curves. As a reference, an empty crucible was used for each DSC curve. The investigated specimens (slices: diameter of 3.0 mm, height of 2.0 mm, and weight of $\approx$40 mg) were always subjected to the same temperature regime: heating with a constant rate of 20 K min^{-1}, holding the temperature of 800 °C for 30 min, and then varying the cooling rate between 0.5 and 40 K min^{-1} [22]. Consequently, the peak temperature and an extrapolated onset temperature of the Fe-rich intermetallic phases were determined from the DSC cooling curves. Thereby, the onset temperature was extrapolated using the tangent intersection method taken on the inflection point and the interpolated baseline. The baseline is the curve section where no sample reaction occurred before and after the detecting peak. Accordingly, the DSC cooling curves must be fitted horizontally and smoothed (elimination of background noise). The described procedure has been used in preliminary studies [14, 22].

31.2.3 Setup of the Laboratory Filtration Apparatus

The experimental trials were conducted on a specially designed laboratory filtration apparatus (Fig. 31.2). Predefined thermal conditions provided by the heating element (CeraSys 1300; Berghütten, Germany) allow high repeatability and good comparability between the batches in the filtration process. The heating module has an electrical power of 2.1 kW, which is limited due to the dimensions of the ferritic heat conductor (CrFeAl139) and the output voltage ($\approx$30 V) of the used converter (transformer). The base material (secondary Al alloy) can be melted or retained liquid in an A 3 graphite crucible (Mammut-Wetro, Germany), which is inserted into the heating element and placed on a base plate with a pedestal made of refractory material. The graphite crucible is fixed between the filter box half (stainless steel) and the base plate via four turnbuckles. A gasket made of Alsitra KP refractory paper (Rath, Germany) is integrated between the filter box half and the graphite crucible. Thereafter, a 30 ppi ceramic foam filter (CFF) with dimensions 50 × 50 × 22 mm (Hofmann Ceramic, Germany) is located in the cavity or chamber of the filter box half. To prevent bypass formation during filtration, another gasket (refractory paper) is added between the two halves of the filter box. The other half is mounted inversely with hexagonal cylinder head screws and has a channel with a receptacle for a screw-in flange to which a vacuum pump is connected. The vacuum can be applied before and/or during filtration and achieve a pressure of $\approx$500 mbar. The filtration process is initiated by tilting the apparatus (180°) with the aid of pivot bolts and structural abutments, transferring the conditioned melt from one graphite crucible to another. The opposite graphite crucible A 3 is clamped using a closing lever push rod clamp FL-121/45.

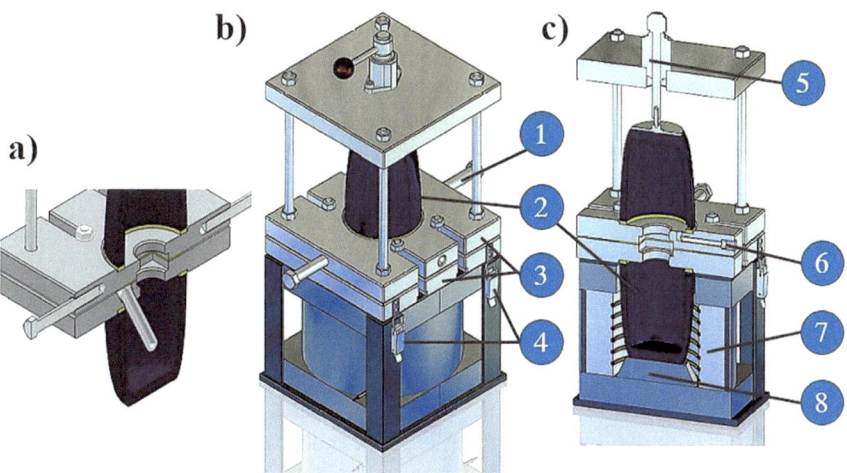

Fig. 31.2 Laboratory filtration apparatus in **a** isometric sectional view with silicon nitride protection tube to record the bath temperature, **b** isometric illustration of the complete device, and **c** dimetric sectional illustration including markings (numbers 1–8) of the component and material designation [22]

Pos.	Designation of components
1	Pivot bolt: M16 with hexagon head screw (stainless steel)
2	Graphite crucible of form A 3 for 1 kg Aluminum
3	Filter box halves with 240 x 238 x 30 mm (stainless steel)
4	Turnbuckles GN832-55-Ni (stainless steel)
5	Closing lever: Push rod clamp FL-121/45
6	Mounting for screw-in flange M16 DN 16 KF (stainless steel)
7	Heating element: CeraSys™ 1300 with an electric power of 2.1 kW
8	Base-plate of refractory material with the dimension of 230 x 228 x 40 mm

The operating principle and heat balance assessment was already described in preliminary study [22]. Accordingly, excessive heat loss was reduced using an appropriate pressure-resistant, fiber-reinforced insulation material (OxOx, Schunk Carbon Technology, Germany), resulting in a decrease of preheating temperature within the filter box halves to the range of 220–260 °C. The temperature range corresponds to the preheating used in permanent die casting. The fiber-reinforced aluminum oxide ceramic was inserted as a crucible cover in a machined groove of the metallic filter box half above the A 3 crucible. The thermal balance of the device was monitored by four thermocouples placed at different positions as described in the literature [22]. The melt measurements were obtained using a thermocouple in a protective tube made of silicon nitride (StarCeram N 7000; Kyocera Fineceramics, Germany). Silicon nitride has a comparatively high thermal conductivity of ≈ 20 W mK^{-1}, allowing relatively instantaneous measurement of the temperature signatures in the melt bath. The predefined thermal conditions simultaneously led to a predetermined cooling rate during the shutdown of the apparatus. The cooling rate was recorded in the molten range (730–600 °C), i.e., before the formation of the Al dendrite network, and consisted of 3.8 K min^{-1} [22]. Hence, the influences on forming primary solidifying Fe-rich intermetallics are investigated in DSC cooling curves at 3.8 K min^{-1} in the following for subsequent filtration trials.

31.3   Microstructure and DSC Cooling Curves

The formation temperature of the primary Fe-rich intermetallic phases is examined in DSC cooling curves with different cooling rates in two series based on an AlSi9Cu3(Fe) alloy with varying initial Fe, Mn, and Cr contents. In particular, Mn and Cr lead to the phase transition from a harmful plate-like β-Al$_{4.5}$FeSi morphology into a less detrimental cubic α-intermetallic [1, 13, 19, 20]. These α-intermetallics exhibit a Chinese-script, a star-like, and a coarse polyhedral morphology [18–20]. The literature widely describes the effect of phase transformation of plate-shaped β-Al$_{4.5}$FeSi [1, 20, 21, 23, 24]. Mn and Cr, inter alia (such as Ni, Be, Co, etc.), represent the most effective elements for influencing the morphology of Fe-rich intermetallics [13, 17, 19, 25, 26]. Thus, the formation of Fe-rich α-intermetallic (sludge phase) can be calculated via an empirical equation as sludge factor (SF) regarding the initial weight fraction w_M of the elements M = Fe, Mn, and Cr [12, 14, 17, 20, 25].

$$SF = 100 \ (w_{Fe} + 2w_{Mn} + 3w_{Cr}) \tag{31.1}$$

The different weight fractions of elements Fe, Mn, and Cr are represented as the variation of the initial chemical composition by the sludge factor (SF). For this purpose, microstructural investigations are conducted starting from the first series of alloy compositions (Leg A–E, Table 31.2) in the as-cast state. The compositions are partly beyond the tolerances of the technical standard DIN EN 1706, resulting in initial Fe contents referring to previous studies by Wagner et al. [1]. Since Mackay [27] investigated the influence of an increasing Fe content on the formation temperature of the plate-like β-Al$_{4.5}$FeSi phase, a pre-dendritic formation of the Fe-rich β-intermetallics was detected starting from ≈1.2 wt% Fe via thermal analysis cooling curves (at 6 and 30 K min^{-1}) in an A356 alloy (corresponds to AlSi7Mg0.3) and A413 alloy (corresponds to AlSi12(Fe)). Based on these findings, the initial Fe content was successively increased in the experimental series to ≈1.6 wt% (Leg A–E) and ≈1.2 wt% Fe (Leg I–X), respectively. A similar correlation of primary formation of Fe-rich intermetallics between initial Fe content and formation temperature was established by Shabestari [28] on an A413 alloy. The empirical correlation of formation temperature of primary solidifying intermetallics (sludge) and initial weight fraction of Fe is consequently provided [22, 28].

$$T_{Sludge} = \left(645.7 + 34.2 \ (100 \ w_{Fe})^2\right)°C \tag{31.2}$$

According to Eq. 31.2, an increase in Fe content from 0.95 to 1.31 wt.% (Leg A to B) would raise the formation temperature from 676.6 to 704.4 °C. However, the objective is to create sludge particles (α-intermetallics) as a primary solidifying phase in a filterable size ($>100 \ \mu m^2$), thus, the elements Mn and Cr are also added to equal proportions to Fe to modify the morphology. Therefore, the filtration efficiency for Fe removal would be improved, and the operating range can be extended to initiate an appropriate filtration process.

Low contents of Cr (>0.04 wt%) and Mn (>0.3 wt%) lead to a transformation of the β-intermetallic into a cubic α-intermetallic [18]. Consequently, in the first series (Leg A–E), the Mn content is successively increased from 0.1 wt% to ≈ 1.5 wt%. Then, the Cr weight fraction increases slightly to ≈ 0.2 wt% and remains constant. Due to the increasing Fe, Mn, and Cr contents, the w_{Mn}/w_{Fe} ratio and sludge factor SF also increase, corresponding to Eq. 31.1. In the AlSi9Cu3(Fe) base alloy (Leg A), their values amount 0.11 (w_{Mn}/w_{Fe}) and 1.28 (SF). From Leg C to Leg E, the ratio reaches 0.39 and 1.0, with a rising SF of 3.30, and finally 5.23. The cross-sectional micrographs of these specimens in gravity die casting are shown in Fig. 31.3.

The secondary AlSi9Cu3(Fe) alloy (Leg A) consists the typical microstructure of α-Al dendrites, eutectic Al-Si phase, a compact Cu-containing phase (Al_2Cu), and the intermetallic Fe-containing phase (β-$Al_{4.5}FeSi$, marked in Fig. 31.3a). By increasing the initial Mn content (to ≈ 1.5 wt%, Leg E), the phase transition occurred as described, leading to a primary solidified Fe-rich α-intermetallic (marked with number 1 in Fig. 31.3b). Further phase components are also observed, wherein the eutectic Al-Si phase gradually recedes into the interdendritic regions of the α-Al solid solution due to the higher volume-related formation of Fe-rich intermetallics. These α-intermetallics exhibit the mentioned star-like and/or coarse polyhedral morphology.

The reason for this is the increasing Mn content, which equally increases the w_{Mn}/w_{Fe} ratio. In Leg E, no needle-like or plate-like β-morphology of intermetallic phases is observed. Therefore, Fig. 31.4 shows SEM images in backscattered electron (BSE) contrast for selected intermetallic morphologies. In sample Leg C, the evidence

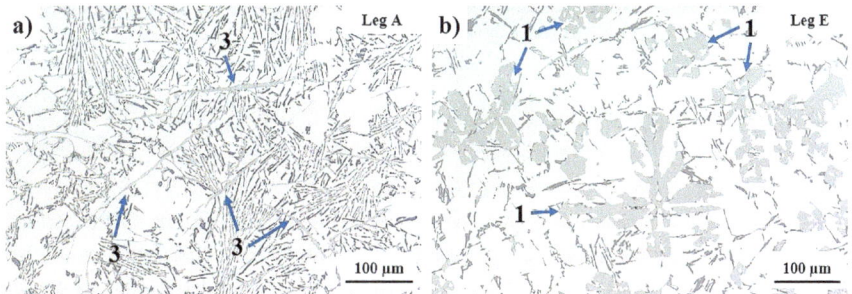

Fig. 31.3 Optical micrographs of polished cross-sections of two exemplary alloys in the as-cast state for **a** Leg A and **b** Leg E (compared to Table 31.2). The marked arrows with numbers 1 and 3 represent the different morphology of Fe-rich intermetallics according to the literature [22]

of plate-like β-intermetallics can be seen adjacent to star-like α-intermetallics (Fig. 31.4a). However, this indicates an incomplete phase transition in Leg C resulting in a w_{Mn}/w_{Fe} ratio of 0.39. In Leg D, no plate-like (or needle-like in depth etching) β-intermetallics occur in optical micrographs or SEM images, confirming the critical ratio (w_{Mn}/w_{Fe} of 0.5) referred to the literature [11, 22]. In addition, a further particular morphology, i.e., an altered Chinese script-like morphology, is significant in Leg C (Fig. 31.4b). Theoretically, the coral-like morphology (as Chinese script-like intermetallics) corresponds to a pre-eutectic or post-dendritic signature in cooling curves. Hence, their volume-related extension is limited within the interdendritic regions, resulting in a mainly branched form of these particles. Similar morphology has been reported by Fabrizi and Timelli [25], indicating an abrupt transition in the growth mechanism initiated by a strong supercooling effect or local concentration difference. According to Cao and Campbell [29, 30], the influence of strontium (Sr) also affects the growth mechanism of the Fe-rich intermetallics, thus, a coral-like morphology may be formed instead of a finely branched Chinese script-like structure. Due to the chemical composition of secondary commercial alloys, this influence is distinctive, as they are mainly provided in the permanently refined state (Sr ≥ 200 ppm) [22, 25, 29–31].

Figure 31.4f depicts a section of primary solidifying Fe-rich α-intermetallics after deep etching with hot sodium hydroxide solution. Thereby, the growth mechanism of the α-intermetallic particles can be demonstrated. This is similar to the growth mechanism of Al dendrites. Starting from centric nuclei, which are usually formed as regular hollowed dodecahedrons [17, 18, 23, 25], the continuous growth of primary dendrite arms occurs in the {110} plane along the < 100 > direction. Subsequently, secondary and ternary dendrite arms form on the primary arms. In the 2D optical images, the coalescing ternary arms of the Fe-rich intermetallic particles appear as a window shaped (hollow cube) structure. As diffusion progresses, closed coarse polyhedral intermetallic particles are finally obtained. From Fig. 31.4f, the progressive growth of the secondary arms, which emerge from the center, can be observed. According to the literature [17–19, 25], the cooling rate and growth mechanism are directly related. However, despite that, a correlation between initial chemical composition and the growth mechanism and/or morphology can be established in the following from DSC cooling curves.

The formation temperature of stable and metastable phase compounds was examined based on the table shown in Fig. 31.5. In addition to the distinctive phase compounds of Al dendrites, eutectic Al-Si phase, and Cu-containing complexes (peaks 2–7), which are typical for this type of alloy, the formation of the primary α-intermetallics (sludge, peak 1) was also observed [14–16]. However, all reaction peaks except the Mg_2Si phase (peak 5) have been detected in the DSC cooling curves. Regarding the formation temperature, a tendency to higher temperatures of the Fe-rich α-intermetallics with increasing initial content of Fe, Mn, and Cr can be observed in the cooling curves at a constant cooling rate of 3.8 K min^{-1} (Fig. 31.5). This tendency to form primary Fe-rich intermetallics (onset temperature) is observed for any cooling rate in all alloy compositions according to the literature data [14–16, 22]. However, no formation temperature of primary solidified α-intermetallics is

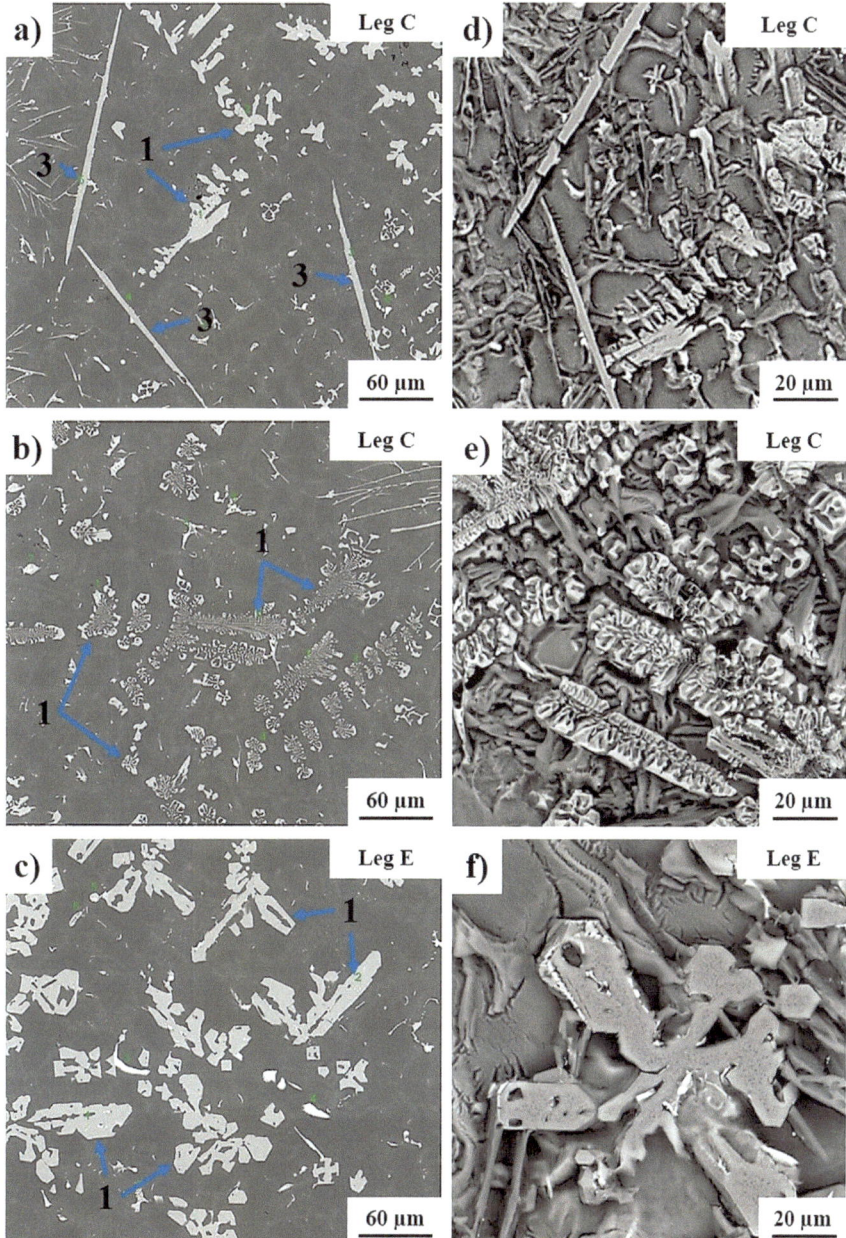

Fig. 31.4 SEM images in BSE contrast (**a–c**) from polished cross-sections of exemplary alloy compositions of the first series (Leg A–E) in the as-cast state, representing the Fe-rich intermetallics in plate-like β-$Al_{4.5}FeSi$ morphology (no. 3), Chinese-script or coral-like, as well as the star-like morphology of α-intermetallics (no. 1). SEM images of characteristically morphological structures of Fe-rich intermetallics (**d–f**) are represented after deep-etching with hot sodium hydroxide solution (65 °C, 1 min, NaOH)

detectable in the base alloy (Leg A). Starting from Leg B, the formation temperature increases from 677.2 to 695.3 °C (Leg D) and reaches a maximum of 704.7 °C with the further addition of Fe, Mn, and Cr (Leg E). The peak temperatures for the α-Al dendrites and the Al-Si eutectic remained nearly constant at ≈585 °C and 565 °C in all alloys, respectively. The Cu-containing phases (Al_2Cu and $Al_5Mg_8Si_2Cu_2$) behaved similarly at ≈495 °C. Due to measurement noise, the distinction of these phases was complicated, hence they are simply summarized under the designation 6 - 7.

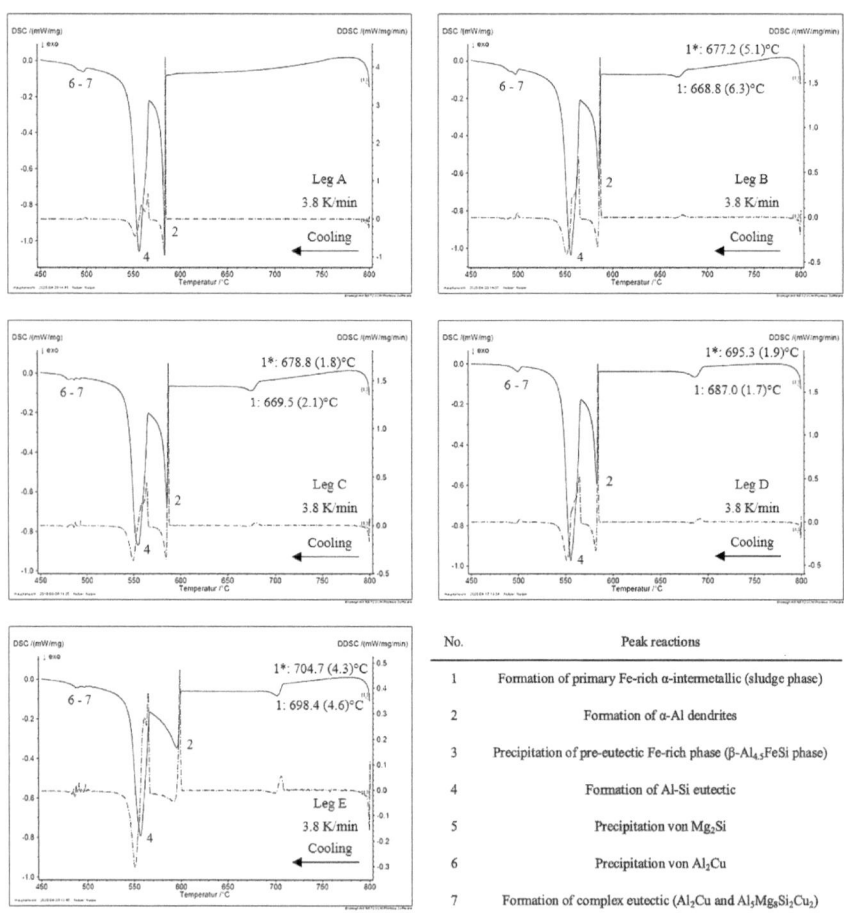

Fig. 31.5 DSC cooling curves with the first derivative (DDSC) of the first series of alloys (Leg A–E) at constant cooling rate (3.8 K min^{-1}). The average of three individual measurements for maximum peak (1) and onset temperature (1*) is presented, including the corresponding standard deviation (in parentheses). The designations 1–7 correlate with the peak reactions and/or phase formation of the AlSi9Cu3(Fe) alloy type, according to the literature data [22]

In contrast, the β-$Al_{4.5}FeSi$ phase was revealed in the base Al alloy and is expected to form a pre-eutectic phase at about 583.8 °C [14, 22]. However, the micrographs also showed their evidence in other alloys (including Leg C). Furthermore, the pre-eutectic β-$Al_{4.5}FeSi$ phase is not clearly displayed as individual peaks in the DSC curves. The β-intermetallics can be merely detected as a gradient alteration in the first derivative of the curve. Likewise, this is evident as an inflexion point in peak 4 within the DSC cooling curves. However, this implies that no onset or peak temperature can be determined for the β-$Al_{4.5}FeSi$ phase. Consequently, the pre-eutectic β-phase and the eutectic Al-Si phase are being consolidated as peak 4 [22].

In preliminary studies, the deferral in peak and onset temperatures was also observed depending on different cooling rates (0.5–40 K min^{-1}), particularly for the first series of alloy compositions (Leg A–E). Thereby, the onset temperature dropped by about 26 K from $\approx$687 °C towards 661 °C in Leg C. However, the formation temperature of the Fe-rich intermetallics decreased simultaneously in Leg D and E by about 24 K and 8 K, respectively, as reported in the literature [14, 16, 22].

Since the specific enthalpy ΔH theoretically reflects the area fractions of the respective phase formation in the microstructure, microstructural investigations were examined on the half-width cross-sections of the DSC samples. The microstructure was significantly affected by the cooling rate. With increasing cooling rate from 3.8 to 40 K min^{-1} the Al dendrites, the eutectic Al-Si phase, as well as the lamellar and/or complex microstructure of Cu-containing phases became considerably finer dispersed [14, 15, 22].

Similarly, the primary solidifying Fe-rich α-intermetallics are affected regarding the growth mechanism by the cooling rate. As the cooling rate increases from 0.5 over 3.8 to 40 K min^{-1}, the morphology is impaired from coarse, polyhedral α-intermetallic particles towards an open, star-like, and/or window-like morphology, which has been described previously [14, 15, 22]. Moreover, the number of primary α-particles increases with increasing cooling rate due to a significantly larger super-cooling during solidification. Consequently, the time interval in the solidification range of the Fe-rich α-intermetallics is shorter, interrupting the growth mechanism towards closed intermetallic particles and deferring it to the post-dendritic and/or pre-eutectic solidification range. This leads to a Chinese script-like or coral-like manifestation of the α-intermetallics in the interdendritic regions due to a strong supercooling effect.

The cooling rate and thus the occurring morphology of Fe-rich intermetallics is predetermined depending on the casting process (e.g., $\approx$10 K min^{-1}, representing gravity sand casting conditions [21]). However, due to the thermal stability of the filtration apparatus used, the cooling rate was predefined as 3.8 K min^{-1}. In prelimi-nary study, compact polyhedral α-intermetallics have been detected by EDS measure-ments in scanning electron microscopy (SEM) at this cooling rate [22]. Therefore, a second series with increasing weight fractions of Cr (Leg I–X, Table 31.2) is conducted to determine the formation range of primary solidifying α-intermetallics by further DSC measurements. Herein, the successive addition of Cr to equal propor-tions as Fe and Mn (to $\approx$1.2 wt%) has been carried out. According to Eq. 31.1, the sludge factor (SF) also increases, enabling the determination of the formation range of Fe-rich intermetallic particles by the extrapolated onset temperatures in the DSC cooling curves for an applicable filtration process.

Figure 31.6 depicts the formation (or operating) range by two regression curves (dashed lines) obtained for the Fe-rich α-intermetallics (sludge phase) and the Al dendrites. Corresponding to the chemical compositions of the second series (Leg I–X), the increase of Fe content from initially 0.8 to ≈1.2 wt% occurs first. Equation 31.2 indicates that the formation temperature would thereby rise from ≈669 °C to 693 °C. However, the empirical calculation is not consistent with the onset temperatures of the DSC cooling curves. Consequently, increasing the initial Fe content merely resulted in an insufficient widening of the formation range for Fe-rich α-intermetallics. Similarly, the operating range for potential removal of the Fe-rich phases would be constrained. Further increasing of the initial Mn content from initially ≈0.3 to 1.2 wt% while maintaining a constant Cr (≈0.04 wt%) extended the formation range (by about 50 K). Finally, a subsequent increase of the Cr content to also 1.2 wt% significantly promoted on the formation temperature of the Fe-rich α-intermetallics to above 700 °C. Since, according to Eq. 31.1, Cr has the greatest impact (factor 3) on sludge formation, the formation range is widened to initiate an appropriate filtration process.

The formation temperature values from the second series (Leg I–X) are also compared to those from the first series (designated as Leg A–E). These values coincide with the regression curves (dashed lines). Strikingly, the higher effect of Cr on the formation temperature can be emphasized by comparing samples from Leg B and III. The averaged formation temperature for Fe-rich α-intermetallics is ≈677 °C (Leg B) and is considerably higher than that of Leg III, despite that the chemical compositions merely differ by 0.1 wt% Fe and 0.2 wt% Cr. Hence, Cr and Mn are recommended to add in equal proportions in the conditioning step to ensure a reliable operating range for a metal melt filtration process.

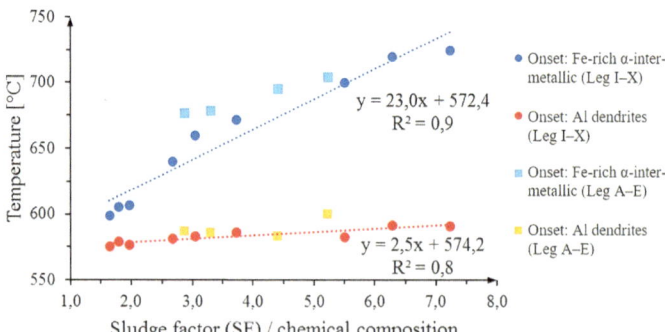

Fig. 31.6 Determination of the formation range (operating range) via two regression lines (incl. equations) obtained from the onset temperature in DSC cooling curves for the Fe-rich α-intermetallics (sludge phase) and the Al dendrites at a cooling rate of 3.8 K min^{-1}. The chemical compositions of the experimental series are designated as Leg A–E and Leg I–X

31.4 Experimental Filtration Trials in a Laboratory Filtration Apparatus

Samples before and after filtration were examined by image analysis using the previously described laboratory filtration apparatus to appraise the removal potential of Fe-rich intermetallic phases. For this purpose, an examination field was established for a predefined alloy composition (exemplary on Leg D) with the variation of time and temperature (620 °C, 655 °C, and 685 °C). After finalizing a predetermined time and temperature, the filtration process was initiated by tilting the apparatus (180°). Consequently, the residual Fe-depleted melt can be purified from the Fe-rich particles by a 30 ppi foam ceramic filter (CFF). The parameters for investigation were preselected based on the literature [14, 15, 22]. Ashtari et al. [2], e.g., reported on the sludge formation after short times of retention of ≈15 min in an A356 alloy. Therefore, the minimum dwell time was selected for 20 min. Conversely, studies on the time-dependent formation of Fe-rich intermetallics (sludge) by Dietrich et al. [21] were conducted for 4 h, 24 h, and 96 h. However, the sludge formation stagnated after a certain time, and thus a time exceeding of 4 h is not required for the formation of coarse polyhedral α-intermetallics. Accordingly, the maximum time was set at 120 min, considering that extended dwell times will not promote further phase growth [21, 22]. Hence, the examination was conducted for 20, 70, and 120 min.

Regarding the temperature, they were obtained from the onset and peak temperatures of the respective DSC cooling curves of the alloy. The peak temperature was preferred due to the reliable formation of intermetallic particles. According to Fig. 31.6, the formation temperature of the Al dendrites increases moderately along the regression line and equation. Consequently, 620 °C was selected as the lower filterable limit to ensure the range of 20–30 K above their formation temperature as a safety margin. Moreover, CalPhaD simulation results (JMatPro) revealed the highest filterable quantity of α-intermetallics at this temperature [22]. This is evident in the microscopic images of the half-widths taken from selective samples in quenching experiments.

Figure 31.7 displays the micrographs for the base AlSi9Cu3(Fe) alloy (Leg A), the conditioned alloy (Leg D) used for the examination field, as well as the dimensions of the quenched sample stick. Following the sampling principle using glass tubes and a Peleus ball, the molten metal with expected primarily solidified particles was pulled into the cavity of the glass tube via negative pressure. The thin glass tubes had an inner diameter of ≈5 mm. The extracted material was then quenched in water bath (at room temperature) in order to preserve the microstructure state. The manifestation of the Fe-rich intermetallic particles in terms of size and distribution during extraction at 620 °C in Leg D provides evidence for their formation as a primary solidifying phase. Etching (hot sulfuric acid solution) leads to higher contrast (blackening) of the intermetallics compared to the remaining microstructure. The remaining microstructure (including Al dendrites, Al-Si eutectic, etc.) is finely formed and dispersed due to the relatively high quenching cooling rate (see Leg A). In contrast, the intermetallic α-particles are present in the desired size (>100 μm) and occurred predominantly in section A/3 of the stick specimen due to the suction effect. Consequently, the coarse polyhedral α-intermetallics must have formed previously during temperature conduction in the laboratory apparatus.

Concluding from the evidence for the primary solidification of Fe-rich intermetallics, evaluation of samples before and after filtration by optical microscopy was performed for each parameter combination pursuant to the literature [22]. For each section, the porosity was determined first on the polished samples and then compared to the total fraction, consisting of pore and etched Fe-rich intermetallics fractions. Samples for the determination of these fractions are provided in Fig. 31.8.

Figure 31.8 presents exemplarily micrographs of section III before filtration (685 °C and 120 min) and after filtration (685 °C and 120 min, as well as 620 °C and 20 min). The division of sections and the determination of area fractions have been reported in preliminary investigations [21, 22]. Porosity localization differs between the samples before and after filtration due to procedural reasons via 180°

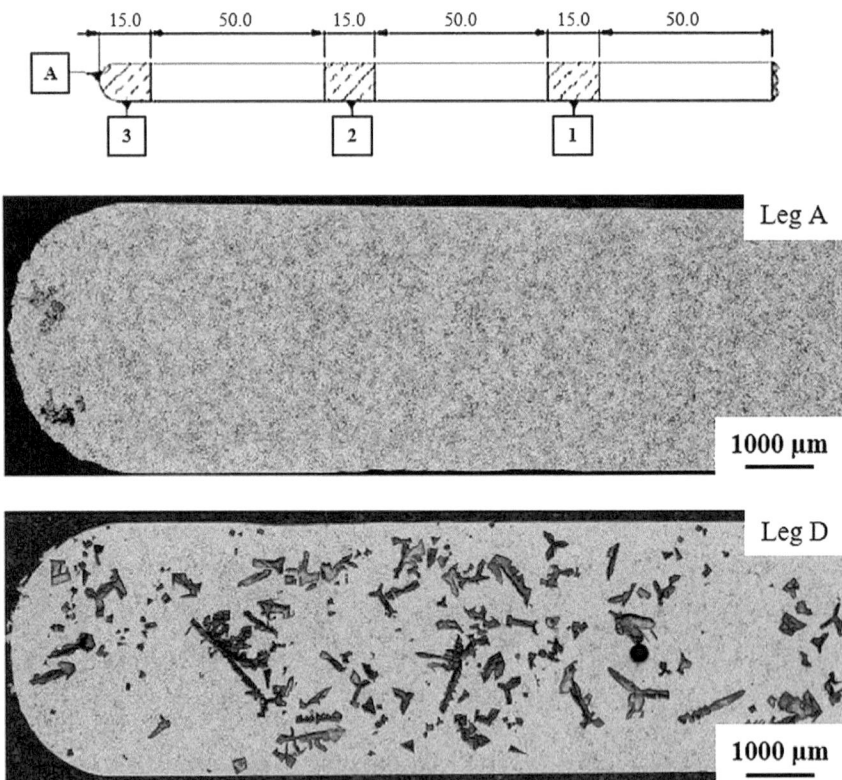

Fig. 31.7 Optical micrographs of etched cross-sections of the half-widths of sampling sticks from section A/3 taken from the sampling scheme presented above, exemplified on two alloy designations (Leg A and E). The specimens were extracted from the laboratory apparatus at 620 °C and at height of ≈20 mm above the crucible bottom

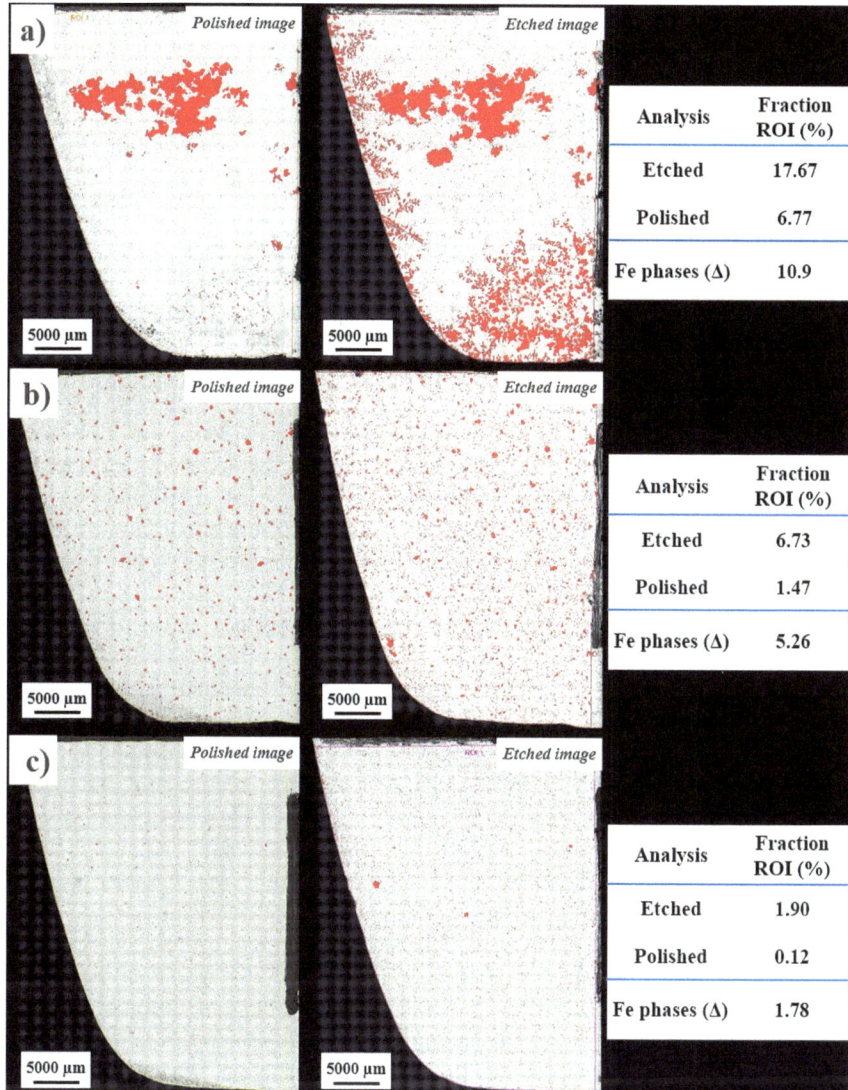

Fig. 31.8 Examination of exemplary samples to determine the porosity fraction (polished samples) and the total fraction (etched samples: porosity and Fe-rich phases) based on section III for the parameters **a** 685 °C and 120 min before filtration, **b** after filtration, and **c** 620 °C and 20 min after filtration. Values correspond to the literature data [22]

tilting. The area fraction of the Fe-rich intermetallics is calculated by the difference of the total minus porosity fractions. They amount to 10.9% before, as well as 5.26%, and 1.78% after filtration, respectively. Therein, the inhomogeneous distribution of the Fe-containing intermetallics is striking. Before filtration, the α-intermetallics are distributed along the edges of the crucible wall (Fig. 31.8a), while after filtration, the intermetallic particles are more homogeneous and finely distributed. The manifestation of particle distribution is explained by the pronounced primary solidification of the intermetallic phases, as already shown in the DSC cooling curves (Fig. 31.5). In this case, the α-intermetallic preferentially forms heterogeneously on pre-existing surfaces [22, 29, 31, 32]. Consequently, the samples exhibit an exogenous, shell-like solidification type before the filtration process is initiated. The samples after filtration show isolated compact intermetallic particles (with an area size of $\approx 10.6 \ \mu m^2$). During further cooling in the solidification range, the particles may have formed depending on their filtration or process temperature. This evidence was also shown in preliminary investigations based on the summary of the detected area fractions of the sections as an upper (I and II), as well as lower segment (III and IV) [22]. The reduction in area fraction in samples after filtration from about 5.0% (685 °C) to finally $\approx 1.5\%$ (620 °C) was due to the temperature-dependent solubility of Fe in molten aluminum [33].

The comparison of Fig. 31.8b to c shows notably reduced and detected intermetallic particles after etching at 620 °C, although the dwell time was shortened by 100 min prior to the initiation of the filtration process. The quantitative consideration (area fractions) is consistent with the qualitative results of chemical composition after the filtration (Table 31.3), showing no further reduction in α-intermetallic particles regardless of the dwell times (20, 70, and 120 min) [22]. Accordingly, the greater influence on the formation of α-intermetallics is inferred by temperature than by time. These assertions are consistent with the literature, indicating that intermetallic particles can form continuously throughout the entire solidification range until the onset of the formation of the the α-Al dendrites [17, 23, 24].

Fe reduction was initially achieved on the remaining crucible material, which had not been used for microstructural examinations. However, the results of the optical emission spectrometer (OES) measurements required to be more differentiated due to the distribution of Fe-rich intermetallic particles. In consequence, the sample materials were subsequently remelted in the crucible induction furnace at above 800 °C to provide a reliable evaluation of the chemical composition. The results of Fe, Mn, and Cr examination in the OES of the remelted sample material before and after filtration according to the scope of this study are listed in Table 31.3 [22].

Particularly, the chemical compositions of all samples before filtration are similar to the reference alloy Leg D (1.6 wt% Fe, 1.0 wt% Mn, and 0.26 wt% Cr). Otherwise, the element contents of Fe, Mn, and Cr of the filtered samples frequently decline below the limiting contents of the standard AlSi9Cu3(Fe) alloy composition (DIN EN 1706). Comparing the parameter combinations of 685 °C and 120 min with 620 °C and 120 min, the Fe content decreases from 1.130% to 0.808% and remains at ≈ 0.8 wt% Fe, even when the holding time is shortened to 70 or 20 min. Thus, the

Table 31.3 Experimentally determined chemical compositions (weight%) after remelting of the sample material in the examination field with a variation of time and temperature on an exemplary predefined alloy composition (Leg D) in the optical emission spectrometer OES SPECTROMAXx (Ametek, USA). Values correspond to the literature data [22]

Alloy: Leg D	Parameter	Al [%]	Si [%]	Fe [%]	Cu [%]	Mn [%]	Mg [%]	Cr [%]	Ni [%]	Zn [%]
Before filtration	685 °C/ 120 min	85.5	9.05	1.640	2.22	0.959	0.123	0.259	0.054	0.643
	655 °C/70 min	84.9	9.14	1.640	2.18	0.935	0.119	0.272	0.077	0.629
	620 °C/20 min	84.9	9.08	1.610	2.16	0.974	0.118	0.277	0.058	0.644
	620 °C/70 min	84.9	9.05	1.700	2.17	0.988	0.099	0.286	0.040	0.622
	620 °C/ 120 min	84.9	9.09	1.660	2.18	0.985	0.106	0.283	0.050	0.628
After filtration	685 °C/ 120 min	86.0	9.02	1.130	2.24	0.584	0.113	0.091	0.041	0.654
	655 °C/70 min	86.0	9.09	1.050	2.27	0.533	0.113	0.079	0.042	0.657
	620 °C/20 min	86.5	8.99	0.855	2.24	0.390	0.097	0.051	0.088	0.659
	620 °C/70 min	86.7	8.92	0.785	2.30	0.294	0.060	0.031	0.074	0.662
	620 °C/ 120 min	86.7	8.97	0.808	2.26	0.340	0.083	0.036	0.041	0.669

Fe content reaches the lowest values of 0.855%, 0.785% and 0.808 wt%, which represent the maximum reduction to about 50% in average (at 620 °C) [22].

With increasing the process or holding temperature, the Fe reduction decreases significantly to about 35% (655 °C) and 30% (685 °C), respectively. The highest reduction of Mn and Cr is likewise achieved at 620 °C with ≈66% and 86% [22]. Consequently, the two-step process consisting of conditioning and filtration allows a noticeable reduction of the Fe, Mn, and Cr elements. A relatively high proportion of Fe can be bound due to the formation of intermetallics and separated via the laboratory filtration apparatus. Moreover, the maximum Fe reduction at 620 °C correlates to the highest quantity of α-intermetallics in the two-phase range, according to the JMatPro® simulation results. Based on these findings from the examination field, additional filtration trials were performed in the second experimental series (Leg I–X) regarding the most evidentially effective parameter combination (620 °C and 20 min) for Fe removal.

Figure 31.9 depicts the remaining elemental contents of Fe, Mn, and Cr after filtration using a 30 ppi alumina CFF in the second series (Leg I–X, Table 31.2). The initial concentrations are proportionated up to 1.2 wt%. The remaining sample materials were also remelted at 800 °C to analyze the influence of the varying initial

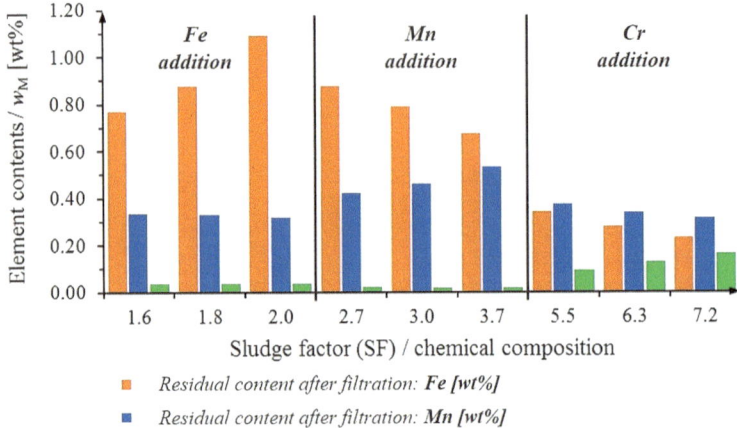

Fig. 31.9 Determination of elemental contents of Fe, Mn, and Cr after filtration of the Fe-rich intermetallics (sludge) based on the two-step procedure for the alloy compositions of the second series (Leg I–X, Table 31.2). The successively increasing addition of the respective element to the initial chemical composition is labeled with respect to the residual element contents

chemical composition in OES on Fe removal. The respective addition of elements is marked in Fig. 31.9. With increasing the Fe, Mn, and Cr content, the initial sludge factor varied simultaneously, which explains that the residual contents in the filtered samples were plotted over the sludge factor. According to Eq. 31.1, the increase of initial Fe resulted in a rising SF from 1.64 to 1.97. Consequently, the Fe concentration in the residual content also ascended to over 1.0 wt%.

The further addition of Mn (to 1.2 wt%) increases the SF from 2.66 to 3.72. Moreover, the solidification range extends to form Fe-rich intermetallics (Fig. 31.6). The residual Fe content in the filtered samples is thus reduced from 0.879 to 0.677 wt%. However, the residual Mn content in the filtrate increases similarly from 0.421 to 0.532 wt% and exceeds the tolerance level specified in the standard. With the further proportional addition of Cr (up to 1.2 wt%), the content of both elements decreased. As a result, the Fe content declines to below 0.4 wt%, and the Mn content falls to less than 0.35 wt% below the tolerance limits. Logically, the addition of Cr leads to an increase in the residual Cr content, which still remains within the standard's limits (DIN EN 1706).

Therefore, the promising results obtained in the laboratory filtration apparatus including the findings regarding the formation characteristics of Fe-rich inter-metallics, need to be implemented for an industrial application. For this purpose, additional industrial scale filtration technology is presented in the following.

31.5 Development of a Filtration Process Technology on an Industrial Scale

During the operation of the new filtration process technology, the temperature management was adjusted according to the most effective parameter combination for a crucible induction furnace with a maximum capacity of ≈150 kg of liquid aluminum. For this purpose, the chemical composition was selected according to the optimum operating range determined previously. Accordingly, the conditioned melt is maintained at 620 °C for 20 min and then superheated to ≈700 °C in the crucible induction furnace with a heating rate of 12 K min^{-1}. Exceeding the superheating temperature should be avoided in order to achieve the desired formation range of Fe-rich α-intermetallics (Fig. 31.6).

The modular design of the newly developed industrial filtration apparatus consists of the modules inlet, outlet, and filtration unit (filter box chamber) and comprises a total capacity of about 48 kg. The inlet and outlet modules can be demounted at any time, and the unit can be integrated into an existing continuous launder system. However, the principle schematic shown in Fig. 31.10a enables a batch operation mode. Considering a compromise of using both institutional filters, produced in cooperation with the subprojects A01, A02, and A07, and industrially manufactured CFF (trapezoidal for continuous casting), the smallest 7-inch filter chamber size was elected. Thus, alumina-based CFF (e.g., Sivex filters; Pyrotek, Czech Republic) [34], C-bonded alumina filters, and alternative filter materials and coatings can be used. In this context, different filter materials and coatings were evaluated in advance using the laboratory filtration apparatus in order to evaluate suitable filter types under industrial conditions with higher melt volume (up to 100 kg). For this purpose, the C-bonded Al_2O_3 CFFs are used, which were developed in the third funding period of the Collaborative Research Center 920. They are manufactured in dimensions 178 × 178 × 50 mm (trapezoidal) by the Institute of Ceramics, Refractories, and Composite Materials and tested on an industrial scale. The fabricated CFF is thus inserted into the filtration unit's filter chamber, including the gasket. Predefined filter positions of 3° ensure improved flow conditions on the inlet side. Both, standard 6.5 mm thick fiber and 3.5 mm thick expandable material are used as gasket types. The inlet and outlet modules, as well as the completely assembled filtration device are presented in Fig. 31.10.

The attachment modules of the inlet and outlet shown in Figs. 31.10b and c were manufactured as negative models from a milled part made of 24 mm thick plywood (birch) in ribbed construction. Castable hydraulic-setting concrete refractory (Rath, Germany) was used for the modular parts. In accordance with the manufacturer's instructions, fabrication was conducted with the aid of a vibratory table and a prescribed drying process. As insulation material, microporous insulation panels WDS (Rath, Germany) with a thickness of 50 mm were used and integrated with the modular parts into a welded steel structure with flanged surfaces (S235JR). The first modification occurred on the outlet side, where a pouring lip was implemented instead of plug construction. In order to ensure a continuous and complete wetting

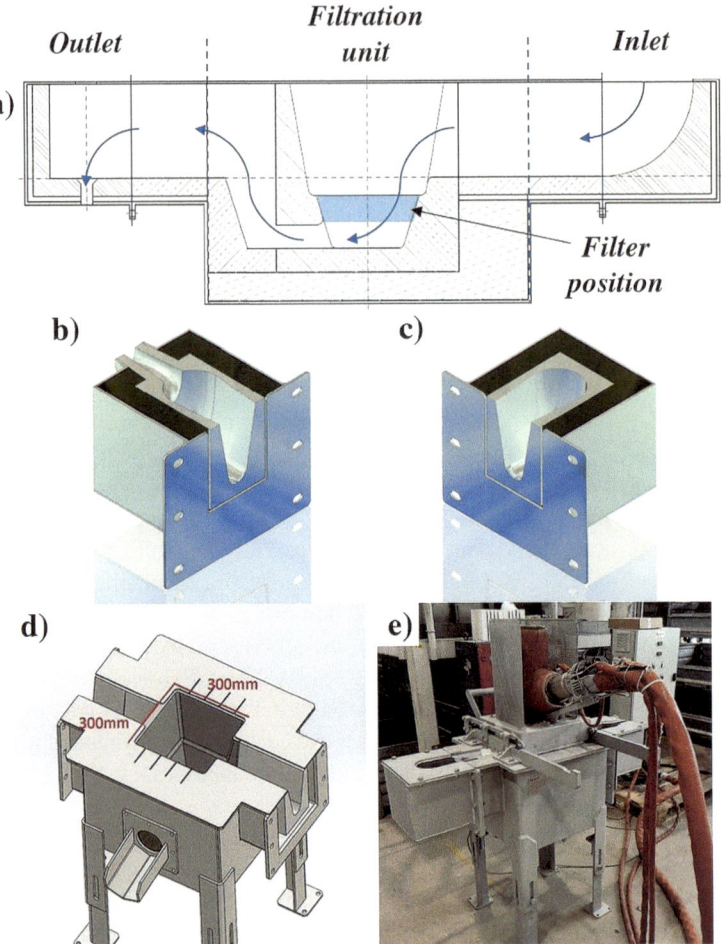

Fig. 31.10 The modular setup of the newly developed filtration technology for investigations on metal melt filtration of Fe-rich intermetallics on an industrial scale. **a** Schematic illustration of the setup, including **b** isometric projection of the outlet module and **c** inlet module, ensuring a consistent metallostatic level (priming height) of ≈160 mm above the 7″ CFF by modifying a pouring lip instead of tapping plug. **d** dimetric projection of the filtration unit (7-inch filter box chamber) and **e** illustration of the completely assembled filtration apparatus with a capacity of ≈50 kg of molten aluminum

of the filter surface, this was necessary to provide a consistent metallostatic height above the filter position during the filtration process. The casting lip operates as overflow with a constructive height of 130 mm above the launder bottom and additionally 30 mm height to the upper filter chamber rim, resulting in excess pouring metal (>48 kg) being collected in cavities at the outlet side. The height of 160 mm is based on the priming height of the manufacturer's recommendation: Pyrotek filtration best practices [35], allowing filter grades or ppi for CFF to be used up to 30 ppi. Figure 31.10d presents the dimetric projection of the filtration unit. Therein, a side-mounted draining stone is seen, which allows to empty the entire filtration device after the successful filtration process. In addition, the completely assembled filtration apparatus is demonstrated in Fig. 31.10e. In particular, the heater (Leister LE 10000HT) mounted on the lid of the filter box requires an electrical output power of $\approx$15 kW, which provides an output temperature of $\approx$780 °C to preheat the filter chamber and CFF to over 500 °C. However, the essential modification of preheating is particularly formative due to the C-bonded Al_2O_3 CFF since conventionally used gas burners would not be compatible with this filter material.

Figure 31.11a constitutes a representative cross-section of the 20 ppi C-bonded Al_2O_3 CFF used for a conditioned AlSi9Cu3 alloy with 0.866 wt% Fe, 0.781 wt% Mn, and 0.718 wt% Cr (SF: 4.58). This reveals a distinctive filter cake of intermetallic particles over the entire filter area. After weighing the CFF before and after the filtration process, i.e., the unloaded and loaded filter with particles, the result indicates a weight increase from the initial 920 g to 4032 g. Consequently, the CFF separated a significantly large amount of particles through the filter cake. The pouring temperature in the ladle was 690 °C. Measuring points in the launder system at the inlet and outlet using the silicon nitride protection tubes provided a temperature of $\approx$671 °C in the filtration device. This can be attributed to the transfilling process. The measurements during filtration imply that the process temperature for SF of 4.58 must have been below the required formation temperature of $\approx$678 °C, according to Fig. 31.6. Therefore, distinct loading or filter cake was observed on the filter surface with compact polyhedral intermetallic particles of >100 μm (Fig. 31.11b). Furthermore, additional oxide films can be detected, which are retained within the filter cake (Fig. 31.11c). The elemental distribution in the EDS mapping from a section of Fig. 31.11b indicates the manifestation of Fe-rich intermetallics.

Figure 31.12 displays the SEM images in BSE contrast, as well as the distribution of the elements Al, Si, C, O, Fe, Mn, Cr, and Cu in the investigated section. Herein, Al, Si, and Cu represent the origin AlSi9Cu3 matrix alloy. The Fe-rich intermetallics retained in the filter cake are evident by the Fe, Mn, and Cr elements. The elemental distribution analysis includes only 1.96 wt%, 0.62 wt%, and 0.41 wt% of these elements, respectively, related to the entire section. The elements C and O are characterized by higher weight percentages, namely 39.79 and 7.33 wt%, and represent the filter material used for the CFF. Due to the new filtration process technology, a considerably high portion of Fe was separated, resulting from forming primary solidifying Fe-rich intermetallics. The elemental contents for Fe, Mn, and Cr were reduced in consequence of the filtration process by about 41%, 45%, and 61%, to values of 0.514 wt% (Fe), 0.432 wt% (Mn), and 0.283 wt% (Cr), respectively.

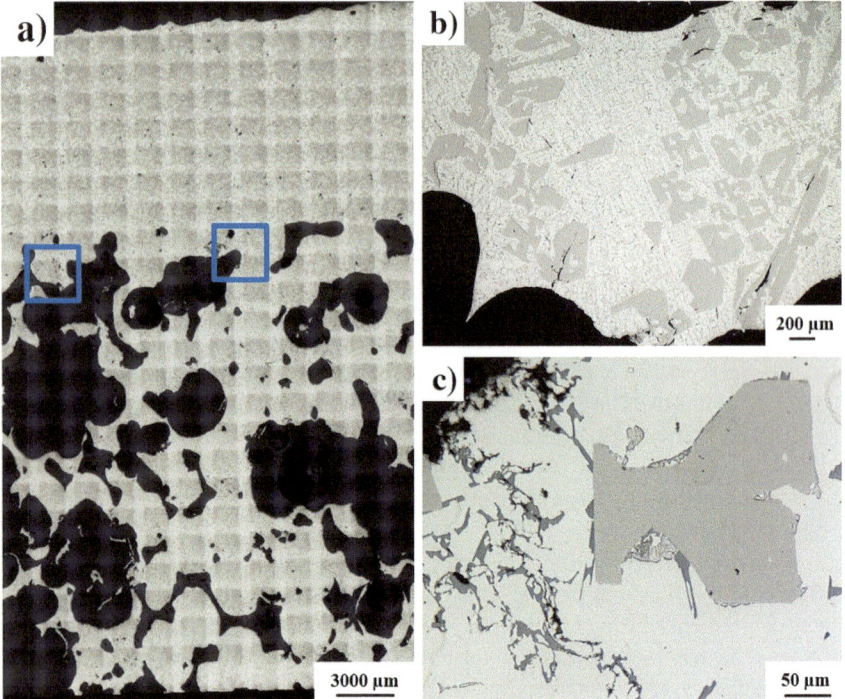

Fig. 31.11 Optical micrographs of the 20 ppi C-bonded Al$_2$O$_3$ filter of the conditioned Al alloy poured at 670 °C with **a** panoramic, and **b** individual images at 100×, and **c** 1000× magnification. The intermetallic particles are visible over the entire area in the filter cake

31.6 Conclusions

In the transfer project T03, the Fe content of recycled and secondary aluminum alloys should be reduced to less than 0.4 wt% for an industrial application at Nemak Dillingen. Based on findings from the second funding period, the formation characteristics of Fe-rich intermetallics were investigated in collaboration with other subprojects. Subsequently, DSC cooling curves were analyzed to determine the formation temperatures of Fe-rich intermetallics under semi-technical and practical conditions. Consequently, the operating range for a corresponding industrial filtration process was defined by regression lines for the formation of Fe-rich intermetallics. The microstructural scope of this study was adopted to determine the most effective filtration parameters for achieving the greatest potential for separation of Fe-rich intermetallic particles. Subsequent investigation with the most effective parameter combination and variation of the initial chemical composition resulted in the highest Fe reduction to below 0.4 wt% (project target value). Furthermore, an appropriate filtration process technology has been developed, to separate Fe-rich particles on an

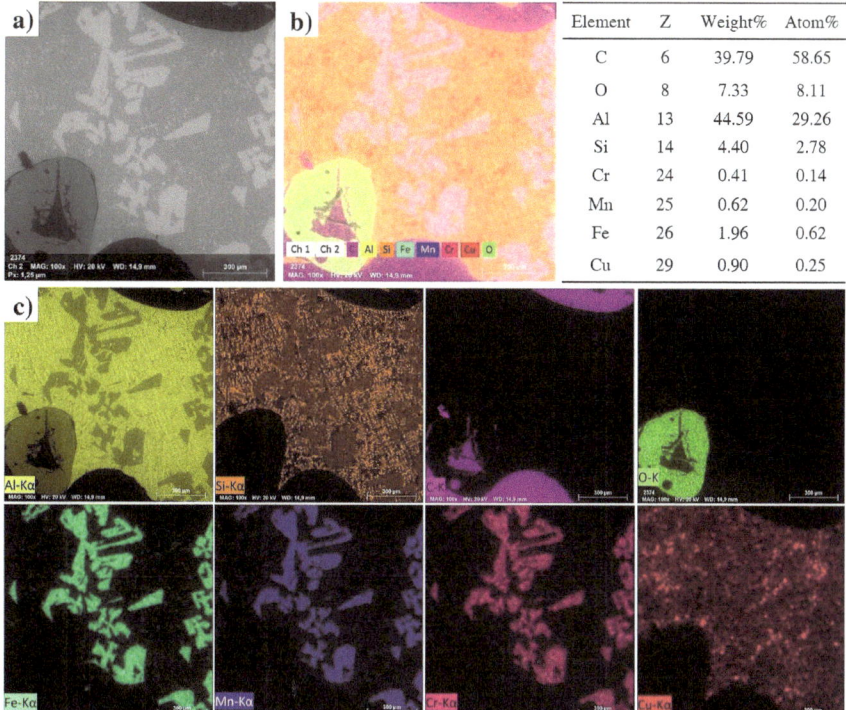

Element	Z	Weight%	Atom%
C	6	39.79	58.65
O	8	7.33	8.11
Al	13	44.59	29.26
Si	14	4.40	2.78
Cr	24	0.41	0.14
Mn	25	0.62	0.20
Fe	26	1.96	0.62
Cu	29	0.90	0.25

Fig. 31.12 SEM image in BSE contrasts: **a** the polished section from Fig. 31.11b, and **b** illustration of the elemental distribution in the cross-section including the detected elements, atomic number, weight%, and atomic% listed in the right table. **c** Individual element distribution (EDS mapping) of Al, Si, C, O, Fe, Mn, Cr, and Cu. SEM images were acquired with a magnification of 100x, a working distance (WD) of 14.9 mm, and a scale of 300 μm

industrial scale. However, due to the Fe reduction of ≈40–50% to 0.514 wt%, additional optimization of the filtration technology is required. Nevertheless, the CFF analysis indicates significant deposition of Fe-rich intermetallics via cake filtration, where the evidence of Fe-rich particles was revealed in EDS measurements. Further optimization options include external commissioning of auxiliary heating elements (launder heater: CeraSys 1150) with further adjustment of the filtration temperature to a lower level, as well as additional investigations of multifunctional filter coatings and materials with respect to Fe removal efficiencies.

Acknowledgements The authors would like to acknowledge the German Research Foundation (Deutsche Forschungsgemeinschaft, DFG) for funding this work and project as a part of the Collaborative Research Centre 920 "Multi-Functional Filters for Metal Melt Filtration - A Contribution towards Zero Defect Materials" (Project ID: 169148856 - CRC 920, Subproject T03). Particular acknowledgment is expressed to the commitment of the staff members of the Foundry Department, especially to Mr. Björn G. Dietrich, Mr. Michal Smolka, as well as Ms. Natalia Mrowka, Ms. Maria Raczek, and Prof. Michal Szucki for their assistance in the realization of this work. Furthermore, especially acknowledge to the company of Nemak Dillingen for their support and assistance of these studies.

Data Availability The data that support the findings of this study are available from the corresponding authors upon reasonable request.

References

1. R. Wagner, M. Seleznev, H. Fischer, R. Ditscherlein, H. Becker, B.G. Dietrich, A. Keßler, T. Leißner, G. Wolf, A. Leineweber, U.A. Peuker, H. Biermann, A. Weidner, Mater Charact **174**, 111039 (2021). https://doi.org/10.1016/j.matchar.2021.111039
2. P. Ashtari, K. Tetley-Gerard, K. Sadayappan, Can. Metall. Q. **51**, 75 (2012). https://doi.org/10.1179/1879139511Y.0000000026
3. C.M. Dinnis, J.A. Taylor, A.K. Dahle, Mater. Sci. Eng. A **425**, 286 (2006). https://doi.org/10.1016/j.msea.2006.03.045
4. E. Taghaddos, M.M. Hejazi, R. Taghiabadi, S.G. Shabestari, J. Alloys Compd. **468**, 539 (2009). https://doi.org/10.1016/j.jallcom.2008.01.079
5. L. Ceschini, A. Morri, S. Toschi, A. Bjurenstedt, S. Seifeddine, Metals **8**, 268 (2018). https://doi.org/10.3390/met8040268
6. M.V. Kral, Mater. Lett. **59**, 2271 (2005). https://doi.org/10.1016/j.matlet.2004.05.091
7. G. Gaustad, E. Olivetti, R. Kirchain, Resour. Conserv. Recycl.. Conserv. Recycl. **58**, 79 (2012). https://doi.org/10.1016/j.resconrec.2011.10.010
8. B. Prillhofer, H. Antrekowitsch, BHM, **152**(3), 53 (2007)
9. L. Zhang, J. Gao, L.N.W. Damoah, D.G. Robertson, Miner. Process. Extr. Metall. Rev. **33**, 99 (2012). https://doi.org/10.1080/08827508.2010.542211
10. Umweltbundesamt, factsheet-aluminium_fi_barrierefrei.pdf [Online], (2019). http://www.umweltbundesamt.de/sites/default/files/medien/3521/dokumente/factsheet-aluminium_fi_bar rierefrei.pdf. Accessed September 2020
11. D. Bolibruchová, M. Žihalová, Mater. Technol. **49**, 681 (2015)
12. H.L. de Moraes, J.R. de Oliveira, D.C. Romano Espinosa, J.A. Soares Tenório, Mater. Trans. **47**, 1731 (2006). https://doi.org/10.2320/matertrans.47.1731
13. G. Gustafsson, T. Thorvaldsson, G.L. Dunlop, Metall. Trans. A **17A**, 45 (1986). https://doi.org/10.1007/BF02644441
14. S. Ferraro, A. Bjurenstedt, S. Seifeddine, Metall. Mater. Trans. A **46A**, 3713 (2015). https://doi.org/10.1007/s11661-015-2942-0
15. G. Timelli, S. Capuzz, A. Fabrizi, D. Caliari, Mater. Sci. Forum **828–829**, 415 (2015). https://doi.org/10.4028/www.scientific.net/MSF.828-829.415
16. G. Timelli, S. Capuzzi, A. Fabrizi, J. Therm. Anal. Calorim.Calorim. **123**, 249 (2016). https://doi.org/10.1007/s10973-015-4952-y
17. T. Gao, K. Hu, L. Wang, B. Zhang, X. Liu, Results Phys. **7**, 1051 (2017). https://doi.org/10.1016/j.rinp.2017.02.040
18. H. Becker, A. Leineweber, Mater Charact **141**, 406 (2018). https://doi.org/10.1016/j.matchar.2018.05.013
19. H. Becker, A. Thum, B. Distl, M.J. Kriegel, A. Leineweber, Metall. Mater. Trans. A **49A**, 6375 (2018). https://doi.org/10.1007/s11661-018-4930-7
20. H. Becker, T. Bergh, P.E. Vullum, A. Leineweber, Y. Li, Acta Mater. **5**, 100198 (2019). https://doi.org/10.1016/j.mtla.2018.100198
21. B.G. Dietrich, H. Becker, M. Smolka, A. Keßler, A. Leineweber, G. Wolf, Adv. Eng. Mater. **19**, 1700161 (2017). https://doi.org/10.1002/adem.201700161
22. J.P. Schoß, H. Becker, A. Keßler, A. Leineweber, G. Wolf, Adv. Eng. Mater. **24**(2), 2100695 (2022). https://doi.org/10.1002/adem.202100695

23. T. Gao, Y. Wu, C. Li, X. Liu, Mater. Lett. **110**, 191 (2013). https://doi.org/10.1016/j.matlet.2013.08.039
24. S. Ferraro, A. Fabrizi, G. Timelli, Mater. Chem. Phys. **153**, 168 (2015). https://doi.org/10.1016/j.matchemphys.2014.12.050
25. A. Fabrizi, G. Timelli, IOP Conf. Ser. Mater. Sci. Eng. **117**, 012017 (2016). https://doi.org/10.1088/1757-899X/117/1/012017
26. E. Cinkilic, C.D. Ridgeway, X. Yan, A.A. Luo, Metall. Mater. Trans. A **50A**, 5945 (2019). https://doi.org/10.1007/s11661-019-05469-6
27. R.I. Mackay, Master thesis (McGill University, Canada, 1996)
28. S.G. Shabestari, Mater. Sci. Eng. A **383**, 289 (2004). https://doi.org/10.1016/j.msea.2004.06.022
29. X. Cao, J. Campbell, Metall. Mater. Trans. A **34A**, 1409 (2003). https://doi.org/10.1007/s11661-003-0253-3
30. X. Cao, J. Campbell, Mater. Sci. Technol. **20**, 514 (2004). https://doi.org/10.1179/026708304225012026
31. X. Cao, N. Saunders, J. Campell, J. Mater. Sci. **39**(7), 2303 (2004). https://doi.org/10.1023/B:JMSC.0000019991.70334.5f
32. W. Khalifa, F.H. Samuel, J.E. Gruzleski, H.W. Doty, S. Valtierra, Metall. Mater. Trans. A **36A**, 1017 (2005)
33. J.A. Taylor, Proc. Mater. Sci. **1**, 19 (2012). https://doi.org/10.1016/j.mspro.2012.06.004
34. Pyrotek Inc., 1697-SIVEX-Low-Phosphate-LP-EN.pdf (2017). https://www.pyrotek.com/DeliverFile/04d3de7b541afba64daed5b57991e9d3. Accessed March 2021
35. Pyrotek Inc., Pyrotek Filtration Best practices.pdf (2017). https://www.pyrotek.com/DeliverFile/e8949c2cdcf637e617f6ea2fcdb28d09. Accessed February 2021

Chapter 32
Functionalized Feeders, Hollowware, Spider Bricks and Starter Casting Tubes for Increasing the Purity in Steel Casting Processes

Tony Wetzig, Matthias Schwarz, Leandro Schöttler, Patrick Gehre, and Christos G. Aneziris

32.1 Introduction

Steel ingot casting lost most of its market share regarding steel manufacturing to the more cost-efficient continuous casting processes. However, ingot casting remains a crucial technology, especially for the casting of specialty steels or small batches. Independent of the applied casting technology, the steel melt cleanliness is essential to ensure high performance of the end product and the efficiency of the steel plant [1]. During casting, micro inclusions with low natural buoyancy may remain in the melt, enter the mold and coagulate to detrimental inclusion clusters [2–6]. A popular approach to remove residual non-metallic inclusions prior to casting is the steel melt filtration [7–17].

T. Wetzig (✉) · P. Gehre · C. G. Aneziris
Institute of Ceramics, Refractories and Composite Materials, Technische Universität Bergakademie Freiberg, Agricolastraße 17, 09599 Freiberg, Germany
e-mail: tony.wetzig@ikfvw.tu-freiberg.de

P. Gehre
e-mail: patrick.gehre@ikfvw.tu-freiberg.de

C. G. Aneziris
e-mail: aneziris@ikfvw.tu-freiberg.de

M. Schwarz · L. Schöttler
Deutsche Edelstahlwerke Specialty Steel GmbH & Co. KG, Obere Kaiserstraße, 57078 Siegen, Germany
e-mail: matthias.schwarz@dew-stahl.com

L. Schöttler
e-mail: leandro.schoettler@dew-stahl.com

© The Author(s) 2024
C. G. Aneziris and H. Biermann (eds.), *Multifunctional Ceramic Filter Systems for Metal Melt Filtration*, Springer Series in Materials Science 337,
https://doi.org/10.1007/978-3-031-40930-1_32

The proper choice of refractories and filter materials in contact with the molten steel is crucial to controlling the final inclusion content of the melt. In bottom-teeming ingot casting, the parts are not preheated and thus must exhibit sufficient thermal shock resistance in order to avoid damage and contamination of the melt [18]. Also the erosion and corrosion resistance have to be considered.

In this regard, the surface of materials applied near the mold should be chemically inert in contact with the steel melt to avoid late gas bubble formation and melt contamination [19]. Common filter materials for steel castings with high refractoriness, high-temperature strength, thermal shock resistance and corrosion resistance are zirconia [20] and carbon-bonded alumina [16, 17, 21–24]. Flame-sprayed surface coatings possess further potential for steel melt filtration due to their favorable thermomechanical behavior [25].

The presented study comprised the research, manufacturing and application of 3 different types of components:

1. Carbon-bonded alumina hybrid filters with tailored geometry and flame-sprayed alumina coating for active filtration in the runner of a bottom-teeming ingot casting of steel.
2. Pressure slip-cast high-alumina feeders and hollowware [26, 27] with carbon-bonded alumina functional coating for reactive "filtration" in bottom-teeming ingot casting of steel.
3. Extruded filter starter tubes for the steel melt filtration in the tundish of a continuous casting plant.

The former two approaches can be effectively brought together in a combined filter system. At first, the melt is cast into the feeder and comes into contact with the carbon-bonded functionalized coating of the feeders. In the process, carbothermic reactions as known from reactive filtration mechanisms by reactive carbon-bonded alumina filters result in the removal of inclusions and dissolved aluminum and oxygen from the steel melt by CO bubble formation and floatation as well as in-situ layer formation and supported inclusion coagulation [22]. The melt flows through the hollowware into the pressure-slipcast spider brick with improved corrosion and erosion resistance avoiding the new formation of inclusions and gets distributed into the horizontal runner system. In the openings of the spider brick and/or in the end pieces of the runner, filters with oxidic surface clean the melt by active filtration. The filters are designed by a hybrid approach. The filter geometry is tailored to fit the runner geometry by means of computer-aided design (CAD) of 5ppi foams and subsequent selective laser sintering of polymer templates [28]. To obtain a filter with high thermal shock resistance and sufficient mechanical strength, the templates are replicated with the aid carbon-bonded alumina slurries [21, 29]. In order to avoid reactive interactions with the melt, e.g. CO bubble formation, close to the mold, the carbon-bonded filter is coated with alumina by flame spraying [25, 30].

At the time of writing this chapter, the project in cooperation with the company Deutsche Edelstahlwerke Specialty Steels GmbH (DEW) is still in progress. As a consequence, the focus of the chapter will be on the research of the applied filters for bottom-teeming ingot casting. Nonetheless, a short overview and outlook are given for the pressure slip casting, functionalization and application of feeders and the research of starter casting tubes.

32.2 Flame-Sprayed Hybrid Filters for Bottom-Teeming Ingot Casting

The positioning of choice for filters in the bottom-teeming ingot casting system was the runner end piece due to the underlying protection against movement of the filters. The inner diameter of the horizontal runner was 60 mm. In the end piece of the runner (see Fig. 32.1b), the melt is deflected in a 90° angle through a vertical opening with smaller diameter leading to the molds. The tight fit of the filter was achieved by modeling of the foam via CAD. The resulting foam geometry was a cylinder (h = 75 mm, d = 55 mm) with a circumferentially rounded off edge (R_c = 20 mm) on one side (see Fig. 32.1a).

In order to reduce the pressure drop during casting and ensure some margin for the additive manufacturing of the template, a pore density of 5 pores per inch (ppi) was used for the CAD model instead of 10 ppi, which is more common for steel melt filtration. The structure was generated by a multi-stage geometric modeling process according to Abendroth et al. [32]. In the procedure, the typical pore structure of reticulated open-cell foams is simulated and adjusted regarding the desired strut thickness and pore density. By means of the software SolidWorks 2017 (Dassault Systèmes SolidWorks Corp., France), the final designs were provided in .stl-format. Besides the final near-net-shape cupola geometry, also prismatic templates (75 × 75 × 25 mm³) and cylindrical foams (d = h = 50 mm) were designed for preliminary tests.

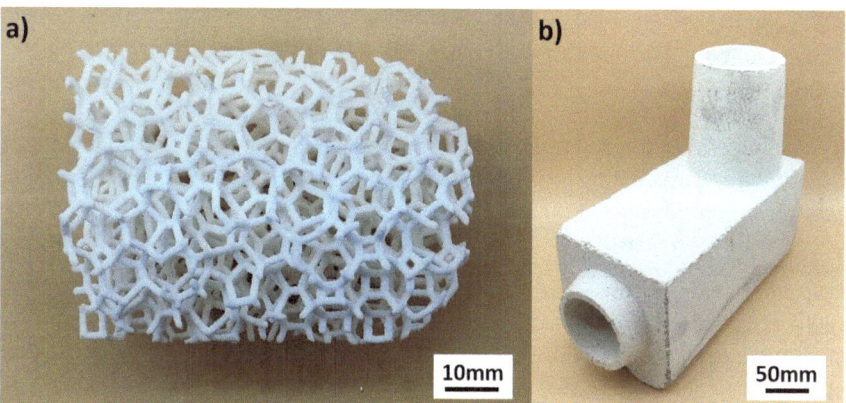

Fig. 32.1 **a** Printed template (3D) via SLS with tailored near-net-shape geometry for the application in the runner end piece of the industrial bottom-teeming ingot casting setup; **b** runner end piece for industrial bottom-teeming ingot casting [31]

By means of the open-source software slic3r.org v.1.29 (Alessandro Ranellucci, Italy), the CAD data were converted from the .stl-format to .gcode-format for the 3D printer Sharebot SnowWhite (Sharebot S.r.l., Italy). The additive manufacturing of the foam templates was based on selective laser sintering (SLS) using thermoplastic polyurethane (TPU) Luvosint TPU X92A-2 WT (Lehmann & Voss & Co. KG, Germany) and printing parameters according to Herdering et al. [28, 31]. Printed prismatic foams were cut into miniature cube samples ($22 \times 22 \times 22$ mm^3) prior to the filter manufacturing by the replica technique. For the lab-scale investigations, the custom 5 ppi structure by means of additive manufacturing (3d) was compared to reference filters based on commercial 10 ppi PU foams (ref).

The applied base material was adopted from Emmel et al. [21] and consisted of 66 wt.% fine-grained alumina, 20 wt.% modified coal-tar pitch (carbonaceous binder), 7.7 wt.% graphite (carbonaceous filler), and 6.3 wt.% carbon black (carbonaceous filler). Ammonium lignin sulfonate (wetting agent, binder), a dispersing agent and defoamer were added to facilitate the creation of stable water-based suspensions. The solid content of the slurry was chosen based on the underlying coating technique and required rheological profile.

Slurries for primary coatings, requiring a high viscosity (solid content of 81.1 wt.%) were prepared in an intensive mixer. Slurries for secondary coatings (solid content of 60.0 or 70.0 wt.%) were prepared by ball mixing on a drum roller for 24 h. Besides slurry-based coating, also vacuum re-infiltration with low-viscosity, liquid, modified coal-tar pitch was investigated. The following techniques were tested and applied within this study:

- impregnation of the foam template with slurry (solid content of 81.1 wt.%) and removal of excess slurry by means of rollers (R),
- impregnation of the foam template with slurry (solid content of 81.1 wt.%) and removal of excess material by means of centrifugation (C),
- dip coating by immersing the foam sample into a low-viscosity slurry (solid content of 60.0 wt.%) and subsequent drainage (D),
- spray coating with slurry (solid content of 70.0 wt.%) using spray gun (S),
- pre-coating of the foam template by immersing the sample into the liquid, modified coal-tar pitch and subsequent drainage (P),
- vacuum re-infiltration of fired filters by immersing them into liquid, modified coal-tar pitch and evacuating the samples for 30 min at 0.1 bar followed by re-inflation with air and subsequent drainage (I).

Single coatings provide insufficient mechanical strength of the resulting filter and each of the coating techniques exhibited specific advantages and disadvantages. Thus, multi-step coating routines had to be applied by combining the listed techniques in an efficient way. In the following, combining the labels of chained single coating techniques in chronological order marks the label of the resulting coating routine and associated samples. The investigated coating routines were RS, CS, CD, CDS, PRS, RSI and CDSI. In each case, the samples were dried for at least 24 h at room temperature after each coating step. After the final coating was applied and dried, the samples were fired at 800 °C with a heating rate of 0.5 K/s and a dwell time of 3 h.

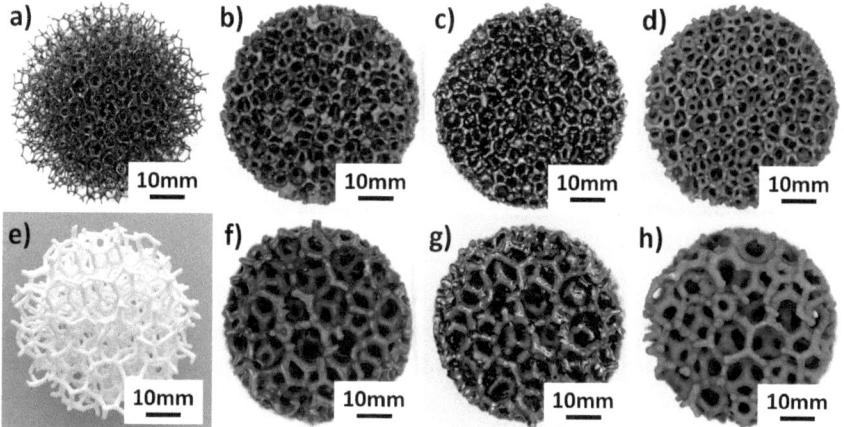

Fig. 32.2 Sample cross-section of CDSI filters with cylindrical geometry in different processing stages: **a** ref foam template; **b** fired CDS-ref filter; **c** CDSI-ref filter before second firing; **d** CDSI-ref filter after second firing; **e** 3D foam template; **f** fired CDS-3d filter; **g** CDSI-3d filter before second firing; **h** CDSI-3d filter after second firing [31]

Each firing was performed in petcoke-filled retorts to ensure a reducing atmosphere and avoid oxidation. The routines including a re-infiltration step (RSI and CDSI) had to be fired twice, i.e. once before and once after re-infiltration.

For a better understanding, Fig. 32.2 shows cylindrical CDSI-ref and CDSI-3d samples at the different processing stages and the sample preparation of CDSI-3d is exemplarily described in the following:

1. manufacturing of 5 ppi TPU foam templates (3d) by SLS (see Fig. 32.2e),
2. impregnation of the templates, subsequent centrifugation (C) and drying,
3. dip coating (D) of the samples and subsequent drying,
4. applying a spray coating on the samples (S) and subsequent drying,
5. firing of the samples (see Fig. 32.2f),
6. vacuum re-infiltration of the fired samples with liquid pitch and subsequent drying (see Fig. 32.2g),
7. second firing of the samples (see Fig. 32.2h).

For a first evaluation of the procedures, miniature cube samples were manufactured and investigated. First, one cube sample per coating routine was cut to $12 \times 12 \times 12$ mm^3 and analyzed in a mercury intrusion porosimeter Autopore 5 (Micromeritics, USA). The used penetrometers had a cup volume of 15 cm^3. The stem volume (capillary volume) was 1.13 cm^3. The measured pore size distribution is referred to as pore entry diameter (d_p) distribution in the following due to the path dependency of the measurement, i.e. smaller pores may "block" larger pores until sufficient pressure for infiltration is achieved resulting in distortions of the results. Based on the final intrusion volume, the open porosity (ε_o) of the strut material was calculated.

The pore entry diameter distribution of different 3d and ref samples is shown in Fig. 32.3. Please note, that the analysis is restricted to material porosity. Both the functional filter pores and any cavities left behind after the pyrolysis of the polymer templates are too large to be detected by mercury intrusion. The reference samples were very similar, whereas the 3d samples showed very versatile pore entry diameter distributions. The most differences were observed for entry pores above 500 nm. RSI-3d was the only sample with significant shift for pores smaller than 500 nm, most likely due to the re-infiltration and the secondary pyrolysis of the pitch resulting in a very distinct microstructure. CS-3d, CDS-3d and PRS-3d were most similar to the reference samples. For better comparison, Table 32.1 lists the measured open porosity and the pore entry size ($d_{p,q}$) at core percentiles (q) for each filter type. The $d_{p,10}$ and $d_{p,25}$ values exhibited the largest differences. The $d_{p,10}$ value of almost all 3d samples was lower than those of references. This could be explained by sharp edges in the triangular strut cavities of the 10 ppi ref foams which were not completely infiltrated yet at the start of the measurement. The round cavities within 3d struts do not show this geometry effect. As an exception, RS-3d showed the highest $d_{p,10}$ value in the test which could originate from cracks in the material. RS-3d, CD-3d and RSI-3d exhibited high $d_{p,25}$ values and a high share of entry pores in the range of 10^4 to 10^5 nm. Although the exact reason is not clear, an increased number of cracks, elevated surface roughness or air inclusions could be potential origins of this behavior.

PRS-3d and RSI-3d exhibited the lowest implied open material porosity in the test. It can be concluded that both the pre-coating and the re-infiltration with liquid pitch reduced the porosity of the carbon-bonded alumina. The samples CD-3d and RS-3d showed the largest porosity in the test.

Ten fired samples of each filter type were used to measure the compressive strength (σ_c). Prior to testing, the mass (m) and the sample dimensions were determined to calculate the bulk density (ρ_b) of the miniature cube samples. The associated results

Fig. 32.3 Cumulative pore volume as a function of pore entry diameter (d_p) of equivalent cylindrical pores for miniature cube samples measured by means of mercury intrusion [31]

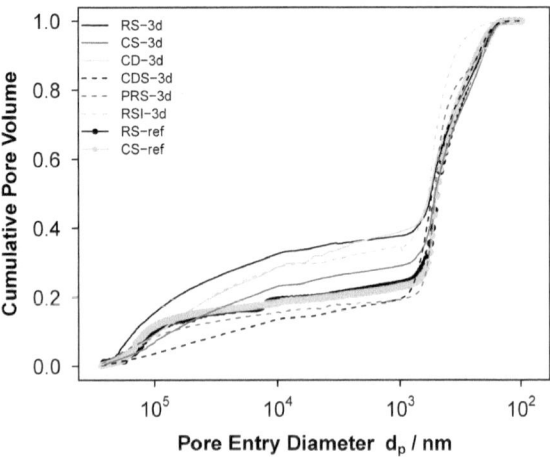

Table 32.1 Open porosity (ε_0) and pore entry diameters ($d_{p,q}$) at distinct percentiles (q) for miniature cube samples measured via mercury intrusion porosimetry [31]

Filter type	ε_0 [%]	$d_{p,10}$ [nm]	$d_{p,25}$ [nm]	$d_{p,50}$ [nm]	$d_{p,75}$ [nm]	$d_{p,90}$ [nm]
RS-3d	47.1	145 375	35 782	550	336	215
CS-3d	42.2	72 061	4 948	504	302	204
CD-3d	48.8	90 831	16 804	566	335	209
CDS-3d	40.8	23 067	754	540	346	227
PRS-3d	31.2	75 083	628	506	399	223
RSI-3d	34.1	85 893	18 619	573	455	346
RS-ref	40.6	104 971	786	490	329	226
CS-ref	45.4	116 340	697	491	339	230

are listed in Table 32.2. Naturally, the number of coatings correlated with the filter mass and bulk density for 3d samples. The increased strut thickness has a direct impact on the mechanical strength of the material. However, this effect is overlaid by microstructural effects, e.g. varying surface quality or the presence of cracks and air inclusions.

In order to visualize potential deviations from an ideal mechanical behavior, the mechanical strength of the filters was visualized as a function of the bulk density (see Fig. 32.4). Following the models of Gibson and Ashby [33], the logarithmized relative strength of the foam material should be proportional to its logarithmized relative density. Since the base material was identical for all samples, the relation can be illustrated by using the compressive strength and bulk density instead. If a filter batch exhibits strong deviation from the ideal linear relationship, this is an indication for the presence of significant differences in the microstructure.

Linear regression of the mean filter strength data on the mean filter bulk density across all tested 3d filter types yielded a good linear model fit ($R^2 = 0.972$). It can be concluded that most of the differences in filter strength using distinct coating routines originated from the underlying bulk density of the filters. The determined

Table 32.2 Mean mass (m), bulk density (ρ_b), and compressive strength (σ_c) data for fired miniature cube sample batches [31]

Filter type	m [g]	ρ_b [gcm^{-3}]	σ_c [MPa]
RS-3d	5.38 ± 0.57	0.29	0.20 ± 0.04
CS-3d	4.89 ± 0.54	0.28	0.23 ± 0.05
CD-3d	3.12 ± 0.48	0.20	0.08 ± 0.02
CDS-3d	6.10 ± 0.57	0.36	0.33 ± 0.06
PRS-3d	5.67 ± 0.59	0.27	0.18 ± 0.06
RSI-3d	6.14 ± 0.62	0.33	0.36 ± 0.08
RS-ref	5.85 ± 0.75	0.37	0.41 ± 0.08
CS-ref	4.69 ± 0.50	0.32	0.31 ± 0.05

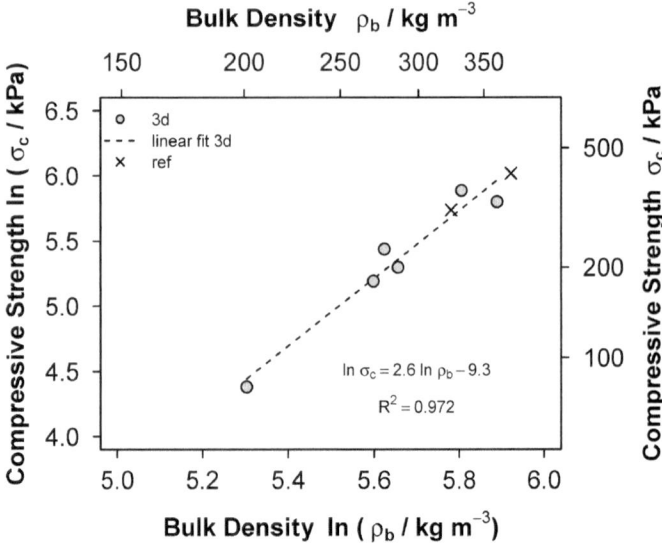

Fig. 32.4 Compressive strength (σ_c) of miniature cube samples as a function of the bulk density (ρ_b) [31]

slope of the regression line was 2.6 and much higher than the value of 1.5 described in the original Gibson-Ashby model for bending-dominated cellular materials [33]. The authors assumed that the hollow strut cavities in the replicated foams were the origin of this behavior. Contrary to the ideal model, a certain minimum foam bulk density is necessary to obtain a stable foam. Above that critical value, an increasing coating thickness not only increases the mechanical strength in proportion to the bulk density but it also reduces the share of the cavities within the strut volume resulting in overcompensation and a higher slope of the model.

CS-3d and RSI-3d samples exhibited a tendency towards mechanical strength exceeding the values predicted by the linear regression model. The data of reference samples was in line with that of 3d samples despite the fact that they were not included in the linear regression model. The mechanical strength of the reference samples (0.3 to 0.4 N/mm^2) was in the standard range of mechanical strength observed for similar carbon-bonded alumina filters for steel melt filtration from the literature [21].

On the basis of the miniature cube investigations, the range of suitable processing routes was reevaluated. Routine CD was excluded for further tests due to insufficient coating thickness leading to low bulk density and a mechanical strength far below the reference standard. Furthermore, the PRS procedure was not pursued since the efforts involving the use of liquid pitch were not justified by significant benefits. Instead, the re-infiltration procedure was pursued also for CDS base filters (CDSI).

Using the resulting set of five coating routines, cylindrical ref and 3d samples with a geometry close to that of the final components were produced. To better understand the effect of the upscaled filter geometry on the achievable coating thickness, the average foam mass was investigated after each manufacturing step (see Table 32.3). Regarding the variance of the filter mass, rolling was the most unreliable primary coating procedure for cylinder samples. Centrifugation showed very low variance in comparison but also resulted in lower filter mass after the first coating. It can be concluded that rolling provided inefficient removal of excess slurry, resulting in coating thickness inhomogeneity. Spray coating showed higher variance than dip coating, especially for ref samples due to their higher pore density of 10 ppi impeding the coating of the core of the cylindrical filter component.

The shrinkage behavior of the cylindrical samples was investigated as well (see Table 32.4). Due to the different mass of 3d foam templates (m $\approx$ 7.0 g) and ref foam templates (m $\approx$ 3.5 g), the volatile mass released during the firing procedure varied and should be considered in the heating program in order to avoid damage. Basic routines without re-infiltration, i.e. RS, CS and CDS, altogether resulted in an average total mass loss of 18.3% for 3d samples and 13.0% for ref samples. Accounting for the mass of the foam templates, all of the mentioned filter types showed a total mass loss of approximately 5% after firing, i.e. the used polymer template had no significant impact on the pyrolysis of the carbon-bonded material. In the case of re-infiltrated samples (RSI, CDSI), the mass loss was solely associated with the pyrolysis of the re-infiltration pitch. Due to the high amount of volatile solvents, the mass loss after firing was rather high. However, it showed low variance and was comparable for both 3d and ref samples, indicating reproducibility.

For cylindrical ref and 3d filters manufactured by RS, CS and CDS routines rather isotropic shrinkage was observed in radial and longitudinal direction at between 2 and 4%. The resulting volume shrinkage ranged between 6 and 10%, with the highest values for CDS filters. The authors assumed that the use of dip-coating slurries with low solid content resulted in higher pre-firing porosity. According to mercury intrusion analyses, the open material porosity without re-infiltration was lowest for

Table 32.3 Average mass of cylindrical filter samples at distinct processing stages [31]

Filter type	Dry filter mass [g], after			
(count)	First coating [a]	Second coating	Third coating	Firing
RS-3d (3)	30.80 $\pm$ 5.13	54.95 $\pm$ 2.15	–	45.75 $\pm$ 1.94
CS-3d (3)	22.23 $\pm$ 0.32	41.74 $\pm$ 2.33	–	33.03 $\pm$ 2.19
CDS-3d (3)	22.13 $\pm$ 0.40	31.67 $\pm$ 0.51	52.90 $\pm$ 2.01	43.74 $\pm$ 2.07
RSI-3d (3)	60.43 $\pm$ 2.56	–	–	51.45 $\pm$ 2.06
CDSI-3d (3)	57.33 $\pm$ 0.67	–	–	49.31 $\pm$ 0.41
RS-ref (3)	19.43 $\pm$ 3.78	40.74 $\pm$ 4.04	–	35.25 $\pm$ 4.12
CS-ref (3)	15.60 $\pm$ 0.35	38.10 $\pm$ 1.37	–	32.75 $\pm$ 1.11
CDS-ref (3)	16.47 $\pm$ 0.81	28.83 $\pm$ 1.18	52.26 $\pm$ 3.55	46.23 $\pm$ 3.54
RSI-ref (1)	48.50	–	–	41.94
CDSI-ref (1)	59.90	–	–	50.50

[a] For RSI and CDSI samples, the first coating is referring to the re-infiltration of pre-fired filters

Table 32.4 Mean shrinkage of cylindrical filter samples with respect to mass, height, diameter, and volume after firing [31]

Filter type	Percentage change after firing in				
–	Mass [%] incl. PU	Mass [%] excl. PU	Height [%]	Diameter [%]	Volume [%]
RS-3d	−16.7 ± 0.3	−4.88 ± 0.31	−2.2 ± 1.1	−2.2 ± 1.2	−6.5 ± 1.7
CS-3d	−20.9 ± 0.8	−5.27 ± 0.11	−2.1 ± 0.9	−2.4 ± 0.4	−6.7 ± 0.5
CDS-3d	−17.3 ± 0.8	−5.03 ± 0.59	−3.2 ± 1.3	−3.7 ± 1.2	−10.1 ± 2.8
RSI-3d [a]	−14.9 ± 0.2	–	−1.1 ± 1.8	−1.0 ± 0.8	−3.1 ± 3.4
CDSI-3d[a]	−14.0 ± 1.4	–	−0.1 ± 0.2	−1.5 ± 0.6	−3.2 ± 1.2
RS-ref	−13.6 ± 1.7	−5.63 ± 1.03	−2.5 ± 0.9	−2.1 ± 0.2	−6.5 ± 1.0
CS-ref	−14.0 ± 0.3	−5.53 ± 0.66	−2.6 ± 1.2	−3.0 ± 0.9	−8.4 ± 0.9
CDS-ref	−11.6 ± 0.9	−5.43 ± 0.68	−2.3 ± 0.8	−3.1 ± 0.4	−5.9 ± 4.2
RSI-ref[a]	−13.5	–	+1.3	+2.0	+5.4
CDSI-ref[a]	−15.7	–	−2.7	+0.6	−1.5

[a] For RSI and CDSI samples, the data are referring to the secondary firing after the reinfiltration only

CDS-3d, indicating stronger densification and potentially higher shrinkage. Differences in the microstructure of distinct coating layers and their effects on the pyrolysis dynamics could not be excluded. The dimensional shrinkage was comparatively low for secondary firing after re-infiltration. The RSI-ref sample even exhibited positive volume changes. Most likely, bloating effects occurred due to the gas release during the pyrolysis.

As for the investigation of miniature samples, the dimensions and the mass of the cylindrical samples were used to determine their bulk density prior to measuring their compressive strength (see Table 32.5). CS-ref and RS-ref exhibited similar bulk density and mechanical strength data, indicating that rolling and centrifugation primary coating both were suitable for ref samples. In contrast, RS-3d showed extremely low strength at rather high bulk density confirming the detrimental effect of coating inhomogeneity observed in the filter mass analysis. Furthermore, the samples showed bad surface quality and cracks before the testing. Due to the low pore density of the 3d templates in combination with the upscaled cylinder geometry, the rolling technique was not suitable for the reliable removal of excess slurry.

The CDS procedure as well as re-infiltration led to increased bulk density and compressive strength. Despite primary coating by rolling, RSI-3d showed very high strength, however, also with the highest standard deviation in the test series indicating low reproducibility. The highest strength values were observed for the RSI-ref (0.50 MPa) sample and the CDSI-ref (0.95 MPa) sample. Further investigations are necessary since only one sample was tested in each case, however, the tendency implies high potential for pitch re-infiltration, especially for conventional foams.

As for miniature cube sample data, the logarithmized bulk density and compressive strength of cylindrical filters were plotted in Fig. 32.5. Separate linear regressions

Table 32.5 Mean bulk density (ρ_b) and compressive strength (σ_c) data for cylindrical filter batches [31]

Filter type	ρ_b [gcm^{-3}]	σ_c [MPa]
RS-3d [a]	0.36 ± 0.03	0.04 ± 0.01
CS-3d	0.26 ± 0.02	0.12 ± 0.01
CDS-3d	0.34 ± 0.01	0.29 ± 0.02
RSI-3d	0.40 ± 0.02	0.47 ± 0.17
CDSI-3d	0.40 ± 0.01	0.52 ± 0.01
RS-ref	0.27 ± 0.03	0.11 ± 0.04
CS-ref	0.26 ± 0.01	0.11 ± 0.01
CDS-ref	0.36 ± 0.02	0.32 ± 0.07
RSI-ref	0.32	0.50
CDSI-ref	0.40	0.95

[a] RS-3d samples showed local damage at the component edges prior to testing σ_c might be underestimated

were performed for 3d filters (excluding outlier RS-3d) and ref filters. The fit for ref samples ($R2 = 0.920$) was impaired by the strong positive outliers represented by the re-infiltrated samples RSI-ref and CDSI-ref. No pronounced deviation was observed for the 3d data resulting in a very good fit of the associated regression ($R^2 = 0.998$) and indicating that the re-infiltration was not as effective as for ref samples. Nonetheless, the re-infiltration can be used to improve the filter strength if needed without pronounced reduction of the functional filter porosity but by increasing the strut density instead. The model slopes were 4.7 for the ref regression and 3.3 for the 3d regression, indicating even higher side effects than for miniature cube samples. Due to the larger cylinder geometry, more inhomogeneity effects, e.g. inhibited penetration of the spray coating, had to be compensated besides the impact of the strut cavities.

Finally, foam templates with cupola geometry for the industrial casting were printed via SLS and replicated using the CDS routine. The fired filters flame-sprayed with commercial alumina (Pure Alumina Thermal Spray Flexicord, Saint-Gobain, France) by means of a MasterJet flame spray gun (Saint-Gobain, France). For detailed information about the spray process parameters, see Neumann et al. [30].

During the flame spraying of the CDS-3d prototype filters, the spalling of the flame-spray coating was observed (see Fig. 32.6). The insufficient adhesion originated from local inhomogeneity of the base material (varying strut thickness) and low surface roughness of the struts. To address these problems, a second batch based on CS was manufactured (to reduce inhomogeneity by dip coating) and the spray coating was performed with higher air-to-slurry ratio (to increase surface roughness). With these modifications, the quality of the flame-sprayed coating was improved substantially and the flame-spraying process was reproducible for the whole batch (5 filters).

The batch of flame-sprayed CS-3d prototype filters was sent to the industrial project partner Deutsche Edelstahlwerke Specialty Steels GmbH & Co. KG (Siegen,

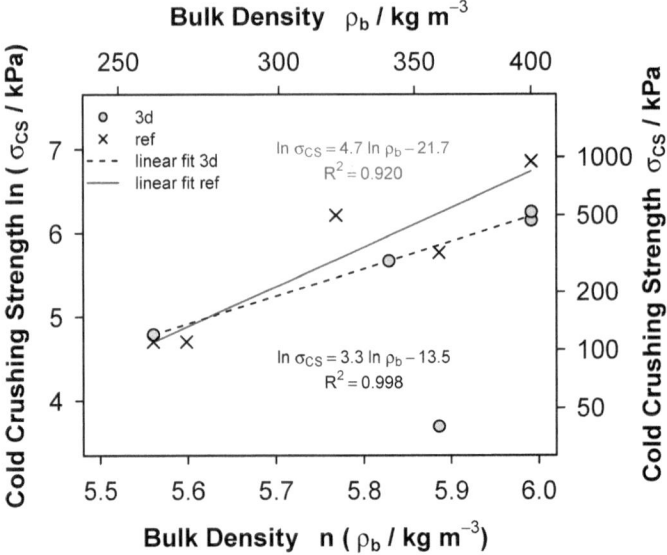

Fig. 32.5 Compressive strength (σ_c) of cylindrical filters as a function of bulk density (ρ_b) [31]

Germany) and tested in bottom-teeming ingot casting. Unfortunately, the melt froze in the filter due to insufficient priming (see Fig. 32.7). The interface between the solidified steel and the filter was investigated by means of digital light microscopy. No signs of oxidation or microstructural damage were observed in the filter indicating that the filter material survived the severe thermomechanical stresses. Furthermore, the flame-spray coating was still intact.

Future casting trials and analyses are necessary. In order to reduce the risk of melt freezing, higher casting temperatures, lower filter pore density and a shortened filter geometry are considered. Furthermore, the filtration efficiency will be analyzed and the flow within the casting system will be simulated.

32.3 Pressure Slip-Cast Hollowware and Extruded Starter Casting Tubes

As an outlook, this chapter shall give a short overview of developments regarding the manufacturing and investigation of pressure slip-cast hollowware components for steel ingot casting by bottom-teeming and extruded cellular starter tubes for continuous casting.

The conception of pressure slip casting molds for feeders and hollowware components in cooperation with the company DORST Technologies was successful. With the aid of a coarse-grained alumina slip composition containing an ecofriendly binder system based on konjak flour and welan gum, which was already successfully tested

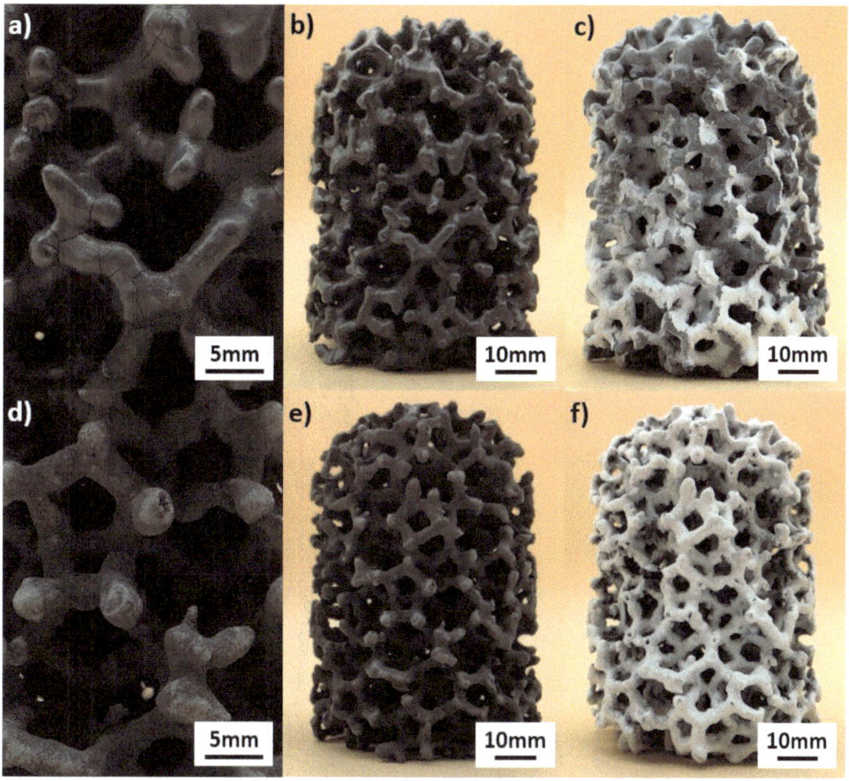

Fig. 32.6 Flame-spayed prototype filters with tailored near-net-shape geometry: **a** smooth surface morphology of CDS-3d filters; **b** CDS-3d filters after firing; **c** CDS-3d filters after flame spraying; **d** rough surface morphology of modified CS-3d filters; **e** modified CS-3d filters after firing; **f** modified CS-3d filters after flame spraying [31]

for the pressure slip casting of spiderbricks [26, 27], the manufacturing of first feeder prototypes was achieved (see Fig. 32.8). The high surface quality in combination with functional coatings is expected to have a positive impact on the melt cleanliness in comparison to traditional refractory systems.

In ongoing investigations, the functionalization with carbon-bonded reactive coatings, the sintering behavior, the lab-scale behavior in contact with steel melts and finally the industrial application in ingot casting and its effect on the steel cleanliness will be addressed.

Regarding the development of cellular starter casting tubes for continuous casting tundishes, comprehensive preliminary tests for the extrusion of distinct refractory materials were performed. To determine the composition with the best compromise regarding its resistance against thermal shock, mechanical loads and oxidation at

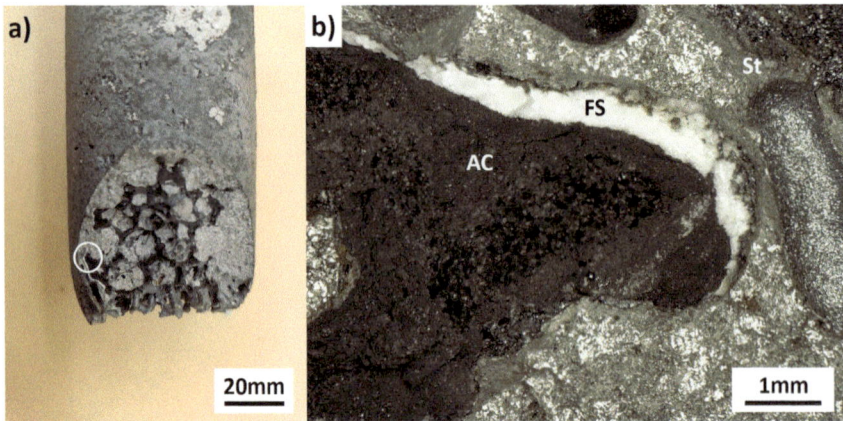

Fig. 32.7 Flame-sprayed CS-3d filter tested in industrial bottom-teeming ingot casting: **a** solidified steel with exposed filter; **b** digital light micrograph showing the intact carbon-bonded alumina filter material (AC), the flame-sprayed alumina coating (FS) and the solidified steel (St) [31]

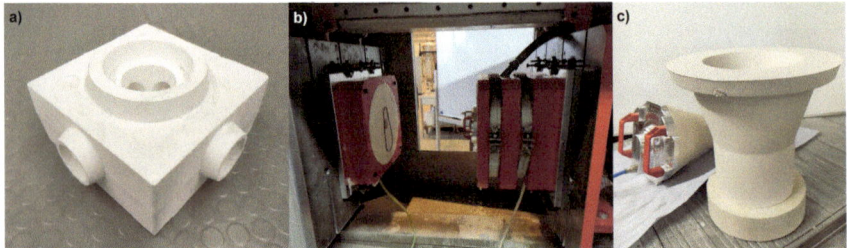

Fig. 32.8 Pressure-slip casting of refractory hollowware: **a** pressure slip-cast spider brick, **b** mold for pressure slip casting of feeders, **c** pressure slip-cast feeder prototype

high temperatures, alumina-based materials with or without addition of coarser fractions, carbonaceous material and zirconia were extruded with the aid of cellulose-based plasticizers and systematically analyzed. In the next step, first starter casting tube prototypes were manufactured (see Fig. 32.9). In following investigations, the prototypes will be tested in lab-scale and industrial steel casting trials.

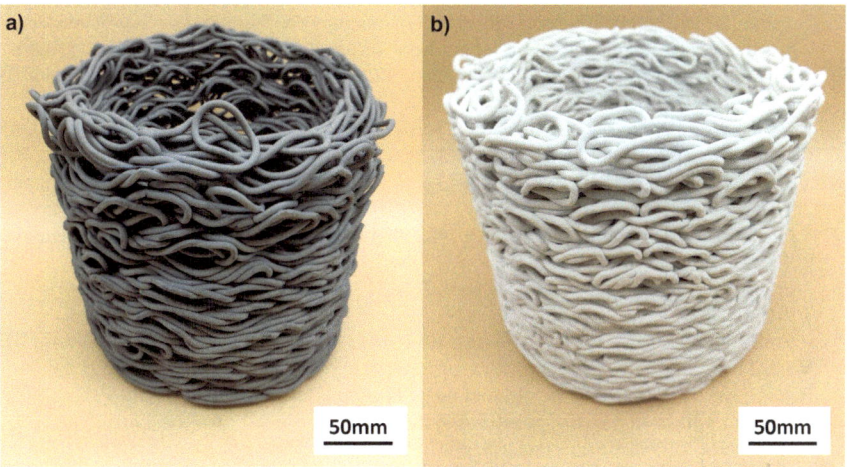

Fig. 32.9 Extruded carbon-bonded (**a**) and carbon-free (**b**) spaghetti starter casting tubes

Acknowledgements The authors would like to acknowledge the German Research Foundation (DFG) in terms of the Collaborative Research Center 920—Project-ID 169148856, subproject T04, for the financial support of this research. Furthermore, special thanks go to M. Abendroth for modeling the foam structures and N. Franke for supporting the pressure slip casting.

References

1. L. Zhang, B.G. Thomas, Inclusions in continuous casting of steel, in *Proceedings of XXIV National Steelmaking Symposium*, Morelia, Mich, Mexico, 26–28 Nov 2003, pp. 138–183
2. R. Dekkers, B. Blanpain, P. Wollants, Metall. Mater. Trans. B Mater. Trans B **34**(2), 161–171 (2003). https://doi.org/10.1007/s11663-003-0003-3
3. R. Dekkers et al., Ironmaking Steelmaking **29**(6), 437–444 (2013). https://doi.org/10.1179/030 192302225004584
4. L. Zhang, JOM **65**(9), 1138–1144 (2013). https://doi.org/10.1007/s11837-013-0688-y
5. L. Zheng et al., ISIJ Int. **56**(6), 926–935 (2016). https://doi.org/10.2355/isijinternational.ISI JINT-2015-561
6. H. Ling, L. Zhang, H. Li, Metall. Mater. Trans. B **47**(5), 2991–3012 (2016). https://doi.org/10. 1007/s11663-016-0743-5
7. C.M. Lee et al., Rev. Met. Paris **90**(4), 501–506 (1993). https://doi.org/10.1051/metal/199390 040501
8. D. Apelian, R. Mutharasan, S. Ali, J. Mater. Sci. **20**(10), 3501–3514 (1985). https://doi.org/ 10.1007/BF01113756
9. K. Yamada et al., ISIJ Int. **27**(11), 873–877 (1987). https://doi.org/10.2355/isijinternationall 966.27.873
10. A.S. Kondrat'ev et al., Refractories, **31**(7–8), 384–391 (1990). https://doi.org/10.1007/BF0128 1545
11. K. Uemura et al., ISIJ Int. **32**(1), 150–156 (1992). https://doi.org/10.2355/isijinternational. 32.150

12. D. Janke, Grundlegende Untersuchungen zur Optimierung der Filtration von Stahlschmelzen—Abschlußbericht, Luxemburg: Amt für amtliche Veröff. der Europ. Gemeinschaften (1996)
13. V.N. Antsiferov, Powder Metall. Met. Ceram. **42**(9/10), 474–476 (2003). https://doi.org/10.1023/B:PMMC.0000013219.43559.46
14. K. Janiszewski, Arch. Metall. Mater. **58**(2) (2013). https://doi.org/10.2478/amm-2013-0029
15. K. Janiszewski, B. Panic, Metalurgija **53**(3), 339–342 (2014)
16. T. Wetzig et al., Ceram. Int. **44**(15), 18143–18155 (2018). https://doi.org/10.1016/j.ceramint.2018.07.022
17. T. Wetzig et al., Ceram. Int. **44**(18), 23024–23034 (2018). https://doi.org/10.1016/j.ceramint.2018.09.105
18. J. Fruhstorfer et al., J. Ceram. Sci. Technol. **7**(2), 173–182 (2016). https://doi.org/10.4416/JCST2016-00010
19. J. Fruhstorfer et al., Steel Res. Int. **87**(8), 1014–1023 (2016). https://doi.org/10.1002/srin.201600023
20. R.A. Olson III, L.C.B. Martins, Liquid metal filtration, in: *Cellular Ceramics* (Wiley-VCH, Weinheim, Chichester, 2010), pp. 403–415
21. M. Emmel, C.G. Aneziris, Ceram. Int. **38**(6), 5165–5173 (2012). https://doi.org/10.1016/j.ceramint.2012.03.022
22. S. Dudczig et al., Ceram. Int. **40**(10), 16727–16742 (2014). https://doi.org/10.1016/j.ceramint.2014.08.038
23. E. Storti et al., J. Eur. Ceram. Soc. **35**(5), 1569–1580 (2015). https://doi.org/10.1016/j.jeurceramsoc.2014.11.026
24. T. Wetzig et al., Adv. Eng. Mater. **22**(2), 1900657 (2020). https://doi.org/10.1002/adem.201900657
25. A. Herdering et al., Ceram. Int. **45**(1), 153–159 (2019). https://doi.org/10.1016/j.ceramint.2018.09.146
26. U. Klippel, C.G. Aneziris, A.J. Metzger, Adv. Eng. Mater. **13**(1–2), 68–76 (2011). https://doi.org/10.1002/adem.201000215
27. N. Gerlach, *Funktionale Feuerfestbauteile auf Basis des Druckschlickergussverfahrens*. Dissertation, TU Bergakademie Freiberg (2018)
28. A. Herdering et al., Interceram. - Int. Ceram. Rev. **68**(4), 30–37 (2019). https://doi.org/10.1007/s42411-019-0013-z
29. K. Schwartzwalder, AV. Somers, Method of making porous ceramic articles (1963)
30. M. Neumann et al., Ceram. Int. **45**(7), 8761–8766 (2019). https://doi.org/10.1016/j.ceramint.2019.01.200
31. T. Wetzig et al., Adv. Eng. Mater. **24**(2), 2100777 (2022). https://doi.org/10.1002/adem.202100777
32. M. Abendroth et al., Adv. Eng. Mater. **19**(9), 1700080 (2017). https://doi.org/10.1002/adem.201700080
33. M.F. Ashby, Cellular solids—scaling of properties, in: Cellular Ceramics (Wiley-VCH, Weinheim, Chichester, 2010), pp. 3–17

Chapter 33
Increasing Cleanliness of Al-melts by Additon of Ceramic Fibers

Daniel Hoppach and Urs A. Peuker

33.1 Particle Separation in Ceramic Foam Filters in a Water Model System

This chapter deals with the increase in filtration efficiency by additional usage of ceramic fibers. The results shown here are generated in a water model system, which is already introduced earlier in subproject B01 and B04 [1–5]. Based on the properties of water, mainly it's dynamic viscosity, it can be suited as a model melt when additionally the poor wetting is considered. Since most of the impurities found in Al-melts are alumina based, they are also applied to the model melt. Besides experiments in water model system tensile test of castings are taken to determine the mechanical properties. Hence, the impact of ceramic fibers on a sand casted AlSi10Mg-alloy is described.

Usually, CFF's are applied to remove particulate non-metallic inclusions as they lead to a decrease in mechanical properties of the casting [6]. A crude classification of these ceramic foam filters with porosities typically above 70% takes place with the number of pores counted per square inch (ppi-value). The higher the ppi-value the smaller the pores. The strut thickness decreases, but as the solid volume increases the total solids volume and hence the ceramic foam surface increases [5]. Thereby the contact probability between particles in the melt and the filter strut increases and additionally more surface is available for particle separation. It results from this correlation a higher filtration efficiency (Fig. 33.1).

The filtration efficiency η in water model system is given by the following Eq. 33.1:

D. Hoppach · U. A. Peuker (✉)
Institute of Mechanical Process Engineering and Mineral Processing, Technische Universität Bergakademie Freiberg, Agricolastraße 1, 09599 Freiberg, Germany
e-mail: urs.peuker@mvtat.tu-freiberg.de

D. Hoppach
e-mail: daniel.hoppach@mvtat.tu-freiberg.de

© The Author(s) 2024
C. G. Aneziris and H. Biermann (eds.), *Multifunctional Ceramic Filter Systems for Metal Melt Filtration*, Springer Series in Materials Science 337,
https://doi.org/10.1007/978-3-031-40930-1_33

Fig. 33.1 Integral filtration
efficiencies η_I in water
model system with 10 and
30 ppi CFF. The impurities
are made of Al_2O_3 with an
initial particle size in
between 20 and 100 μm. The
approaching velocity is 3.2
cm/s [5]

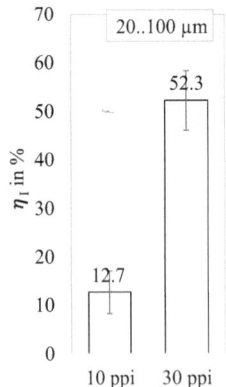

$$\eta = \frac{m_{P,CFF}}{m_{P,CFF} + m_{P,FC}} \times 100 \ in\% \qquad (33.1)$$

with the mass of particles separated in the CFF $m_{P,CFF}$ and the mass of particles in the filtrate, respectively separated particles in the filter cloth $m_{P,FC}$ subsequent to the CFF in the water model system. The filtration efficiency in the water model system is, as already stated dependent on the structure parameters of the CFF, as well as the approaching velocity of the melt, the particle system and the particle size distribution [5]. Investigations in Al-melts using LiMCA-system and / or micrographic analysis also depict these dependencies [7].

But, depending on the casting method the application of ceramic foam filters is limited to their ppi-value, as the pressure gradient is directly connected to the pore size. If the resistance of the CFF is too high for the melt to flow through the pores a freezing might occur as the melt cools down below liquidus temperature and leads to a casting stop (see Fig. 33.2).

Knowing that the ppi-value of the CFF needs to match the casting method in terms of process stability it follows from the above conclusions that small impurities in the single digit μm-range are difficult to be removed with CFF, as transport mechanism to the CFF strut are weak (Fig. 33.3). Thus in the case of sand casting the employed CFF commonly exhibit a ppi-value of around 10.

33.2 Ceramic Fibers for Enhanced Filtration Efficiency and Cleanliness of Al-melts

The restricted application of CFF due to their ppi-value in respect of process stability during casting leads to a lack of impurity removal in the single digit μm-range which needs to be balanced. A promising approach to remove impurities in this order of magnitude is the application of ceramic fibers. The employed ceramic fibers CeraFib 99 from CeraFib GmbH are made of corundum with 99% of Al_2O_3 and 1% of oxidic

Fig. 33.2 Solidified Aluminum melt during impingement on CFF in sand casting, which resulted in a casting stop

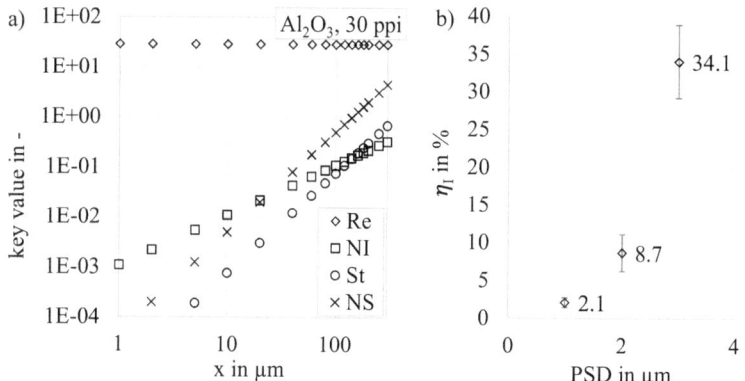

Fig. 33.3 **a** Key figure values in dependence on particle size in RT system when a 30 ppi CFF is used for separation of Al_2O_3-impurities. Approaching velocity is 3.2 cm/s. Strut thickness of 912 µm, obtained from CT-analysis, is used for calculation of key values, **b** integral filtration efficiencies η_I of 30 ppi CFF in dependency of added Al_2O_3-impurity PSD in RT system. Approach velocity is 3.2 cm/s

additions [8]. These fibers feature a diameter of around 10–12 µm (Fig. 33.4) and exhibit a continuous operation resistance temperature of 1250 °C which is above common Al-melt casting temperatures (720–750 °C). Due to these characteristics the application of these fibers for removing impurities in Al-melts is given.

The implementation of small amounts of these ceramic fibers in the water model system with vertical flow through on top of the CFF's already has a positive effect on the filtration efficiency (Fig. 33.5). Independent on the approaching velocity already

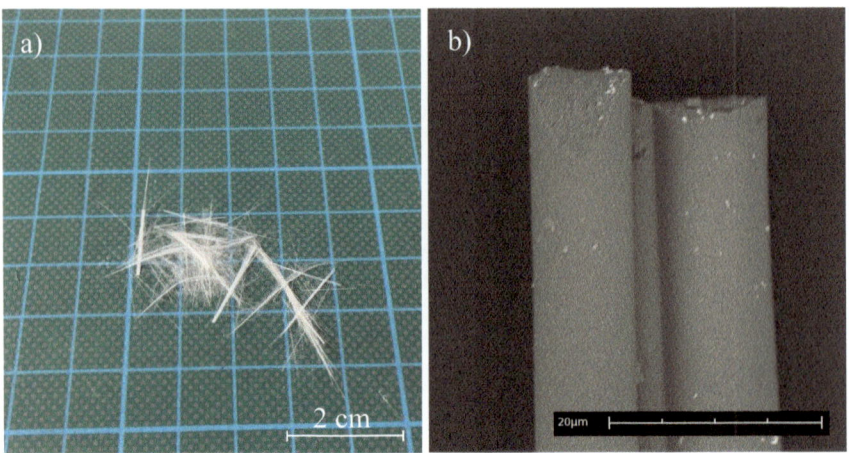

Fig. 33.4 **a** CeraFib 99 fibers with a mass of 0.02 g cut into segments of 15 to 20 mm in length, **b** SEM-analysis of a bunch of three fibers with a diameter of around 10–12 μm

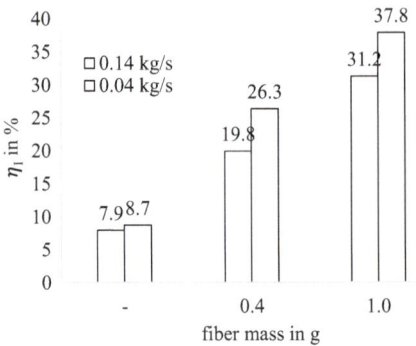

Fig. 33.5 Local filtration efficiencies η_L with the fibers placed on top of a 30 ppi carrier CFF (0.1 to 1.0 g). The approach velocity is 3.2 respectively 11.2 cm/s. Increase in fiber mass leads to rising filtration efficiency. The particle system used is Al_2O_3 with a PSD of 2 to 20 μm

minor amounts of 0.4 g of fibers added on top of a 30 ppi CFF with the dimensions of $50 \times 50 \times 20$ mm^3 lead to an increase of the integral filtration efficiency. Higher amounts of added fiber mass lead to a further increase in integral filtration efficiency. In this experimental setup the CFF acts as a carrier and prevents the fibers from entering in the filtrate, which is equal to a contamination of the casting. Therefore a fiber length of at least 15 to 20 mm ensures a detention when 10 and 30 ppi carrier CFF are used. The average pore diameter of these CFF is 4.92 mm respectively 2.30 mm. Computer tomography was used to gather this information. However, fiber mass can not be arbitrary raised, due to a lift in pressure drop as well which often goes along with a freezing of the melt.

Two options of ceramic fiber addition with potential industrial application is introduced subsequently. The insertion of fibers into the impured model melt before the CFF derived from precoat filtration provides one means of fiber addition. Here, the fibers can start separating the impurities prior to the actual casting. During the casting process these fibers with separated impurities are washed up on the carrier CFF and can continue with the cleaning of the melt.

Placing the ceramic fibers in between two CFF´s is the second possibility. An optimal positioning of the fibers is ensured when the CFF´s exhibit a hollow space as it is shown in Fig. 33.6 and is consecutively stated as encapsulated arrangement. Furthermore, this adjustment guarantees the whole melt to pass through the fibers and raises the contact probability between impurities and fibers.

Water model experiments with the encapsulated arrangement of 30 ppi CFF and varying fiber masses were carried out to determine the pressure drop $\Delta P/\Delta L$. Two different mass fluxes of 0.15 and 0.5 kgs^{-1}, which corresponds to approaching velocities u of 12 and 40 cms^{-1} respectively, were adjusted. The results indicate the relation between pressure drop and fiber mass but also approach velocity. However, in the case of the water model system no freezing of the model melt occurs and the evaluation of an upper limit for fiber mass needs to be evaluated for sand casting with Al-melt as well (Fig. 33.7).

Sand casting trials with an AlSi10Mg alloy were carried out at Formguss Dresden GmbH to determine the maximum amount of fiber mass that can be added in between two 10 ppi CFF ($35 \times 35 \times 22$ mm^3). By placing 0.06 g of ceramic fibers in between the CFF´s the time for casting a mould (Fig. 33.8) of 4.5 kg increases from 14 to around 20 s, which means that the mass flux decrease from 0.32 kgs^{-1} to 0.23 kgs^{-1}. This an indication for a raised pressure drop. The casting temperature was about

Fig. 33.6 Encapsulated CFF arrangement with hollow space to place the fibers

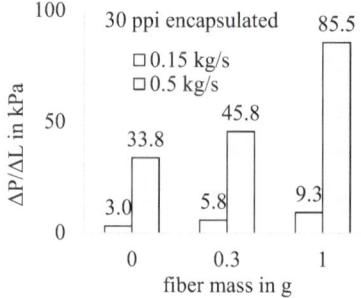

Fig. 33.7 Pressure drop in water model system with ceramic fibers encapsulated in two 30 ppi CFF ($50 \times 50 \times 22$ mm^3) at different mass fluxes

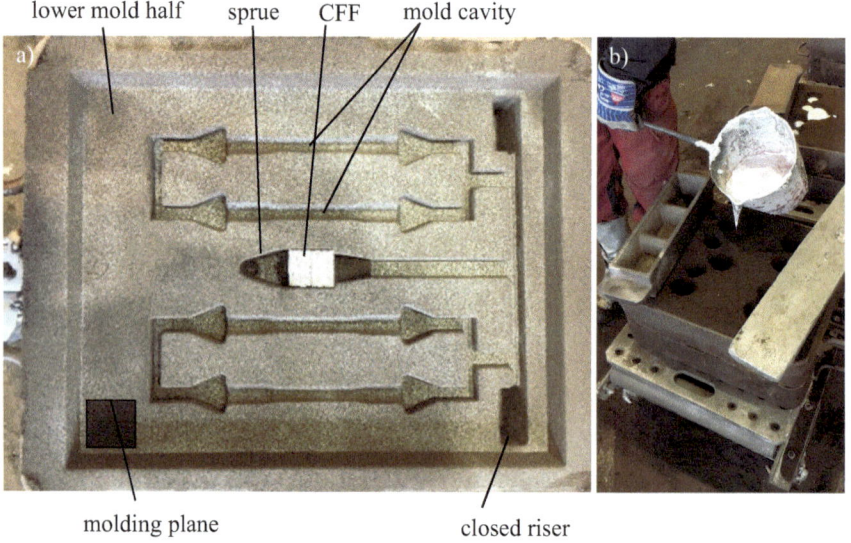

Fig. 33.8 **a** Lower mold half of the sand mold to cast four cylindrical samples with a diameter of 15 mm which can be used for tensile test evaluation, **b** form filling of the casting mold with a ladle

720 °C. It can be concluded for sand casting with ceramic fibers and two 10 ppi CFF a maximum amount of 0.06 g of fibers can be added (Fig. 33.9).

Hence, further evaluation of filtration efficiency in water model system is done with 0.05 g of ceramic fibers. The addition of this amount of fibers seems to be few, but looking at the surface of the fibers with a diameter of 10 to 12 μm it becomes clear that they significantly increase the surface provided for possible particle separation (Table 33.1).

Moreover, the small diameter, compared to a 10 ppi CFF strut of around 1.6 mm, gathered with computer tomography (Zeiss Xradia 510 Versa) at the Institute of Mechanical Process Engineering and Mineral Processing, leads to a high contact

Fig. 33.9 Detailed view on the fibers placed in between the 10 ppi CFF in the casting mold

Table 33.1 Characteristic dimensions of fibers with different diameters compared to 10 and 30 ppi CFF ($50 \times 50 \times 22$ mm^3). For a better comparison between fibers and CFF´s the strut thickness was used to calculate a length which corresponds to a total strut length of the whole CFF. Strut length of the 10 and 30 ppi CFF are 1.614 respectively 0.912 mm

Fiber/CFF	Density	Mass	Length	Volume	Surface	Spec. surface
–	g/cm^3	g	cm	cm^3	cm^2	1/cm
10–12 μm	3.95	0.05	13,491	0.013	47	3637
300 μm	1.15	0.13	62	0.113	15	134
10 ppi	3.95	21.44 ± 0.92	272	9.7	237	24
30 ppi		21.95 ± 1.013	851	12.1	520	43

probability between impurities and fibers. Whereas for the 10 ppi CFF with a depth of 20 mm the particles in the melt have a relatively low contact probability as only around 8 to 10 struts are passed. A 10 ppi CFF ($50 \times 50 \times 22$ mm^3) further exhibits a medium pore size of around 4.9 mm. Having a material volume of 9.7 cm^3 the surface provided for particle separation is 237 cm^2. In comparison to that, a 30 ppi CFF is characterised through a strut thickness of 0.9 mm and a pore size of 0.23 cm. The volume yields to 12.1 cm^3 and results in a suface area of 520 cm^2.

The addition of 0.05 g of fibers with a diameter of 10–12 μm accompanies with additional surface of 47 cm^2, which is about 20% of surface for potential particle separation, but a volume of only 0.013 cm^3. A characterization of fibers and CFF´s is summarized in Table 33.1.

Experiments in water model system with 30 ppi CFF in encapsulated arrangement ($50 \times 50x40$mm^3) and use of different fiber diameters from 10–12 μm and 300 μm illustrate the positive effect of fiber addition. There, 0.05 g of the fibers with a diameter of 10–12 μm is placed in between the CFF. In the case of the fibers with a diameter

Fig. 33.10 Water model results of integral filtration efficiencies in dependency of fiber diameter and approach velocity

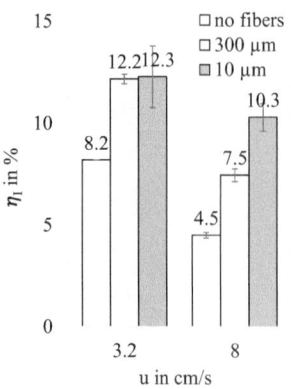

of 300 μm made of nylon thread and a density of 1.15 g/cm³ a mass of 0.13 g is applied. Hence, an almost tenfold higher volume of 0.0113 cm³ of fibers is added but the provided surface is only 15 cm² and thus less than one third. Nylon thread is employed to investigate the influence of fiber diameter on the filtration efficiency in water based model system. It can not be used for Al-melts. For this experiment, two different approaching velocities of 3.2 and 8 cm/s were choosen. The impurities are represented by Al₂O₃-particles with a particle size distribution of 2–20 μm. The results of the integral filtration efficiencies and the corresponding local separatetd masses can be seen in Fig. 33.10 respectively Fig. 33.11a and b.

Approaching two 30 ppi CFF with 3.2 cm/s in the water based model system reaches an integral filtration efficiency of 8.2%. An increase in approach velocity

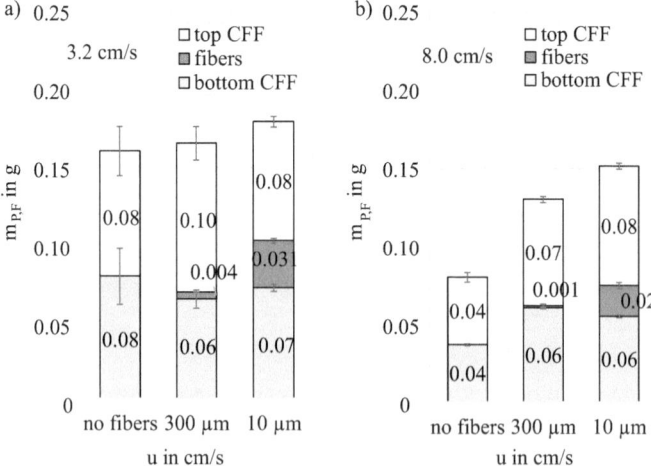

Fig. 33.11 Water model results of local separated masses in dependency of fiber diameter and approach velocity, **a** 3.2 cm/s and **b** 8.0 cm/s

decreases the integral filtration efficiency to 4.5%. The usage of fibers indepen-dent on the fiber diameter leads to an increase in integral filtration efficiency but at higher approaching velocities the influence of the fiber diameter becomes visible. Smaller fiber diameter have the advantage of providing more surface and raising the contact probability and thus increase the integral filtration efficiency. Thus, the lower the diameter the higher the integral filtration efficiency and fiber addition is more beneficial even when only small amounts are provided.

A more detailed view on the contribution of fibers is gained when analyzing the local separated particle masses (Fig. 33.11). For both approaching velocities, a lower fiber diameter indicates a stronger affection on the particle separation. However, in the case of a higher approaching velocity of around 8 cm/s an impact on the particle separation in the up- and downstream CFF is visible. Compared to the separated particle masses in the CFF alignment with no fiber use the particle separation in the up- and downstream CFF is increased. In the case of 10–12 μm fiber use the upstream CFF exhibits with 0.08 g even a doubled mass of separated particles.

As already mentioned, a fiber addition derived from precoat filtration is another possibility and can be directly compared with the encapsulated arrangement in water based model system. A application of precoat filtration in the semi-automated pilot plant which operates with a constant flow rate by adjusting the pressure in the vessel is not possible due to a time consuming cleaning subsequent to each filtration trial. Some fibers remain in the vessel without being washed up at the upper side of the approached CFF and gravimetric analysis is complicated due to complex dismantling of the storage vessel. Thus, these trials are carried out in a small-scale water based experimental setup with easy access to the storage vessel. But the experimental setup, as well as the vertical flow through of the CFF´s is similar to the semi-automated pilot plant. For these experiments the already well known Al_2O_3-fibers with a diameter of 10–12 μm are used by adding 0.05 g in the stirred storage vessel of the with 2 to 20 μm Al_2O_3-particles impured model melt. The approach velocity for these trials was set to 10 cm/s.

The results can be seen in Fig. 33.12. A comparison between no fiber addition, precoat fiber addition and encapsulated alignment is done.

Advantageously, the precoat fiber addition increases the residence time of the fibers in the water based model melt and hence should increase the contact probability too. Impurities already can adhere to the fibers in the storage vessel and are washed up at the CFF surface. However, as already mentioned some fibers may remain in the storage vessel. This behaviour is also observed when these fibers are added to an AlSi10Mg melt, where fiber addition becomes challenging due to the high surface tension of the Al-melt. In contrast to the precoat filtration, a placing of the fibers in between two CFF lowers residence time which is then directly related to the approach velocity.

However, the usage of 30 ppi CFF with a depth of 40 mm leads to a total separation of 0.02 g of particles. Thereby the upstream filter separates more particles than the downstream filter (0.0113 to 0.0084 g) which is already observed in B01 and literature as well [9]. Insertion of fibers in the storage vessel leads to an additional separation of 0.0037 g of impurities. A varying amount of particle loaded fibers still remains

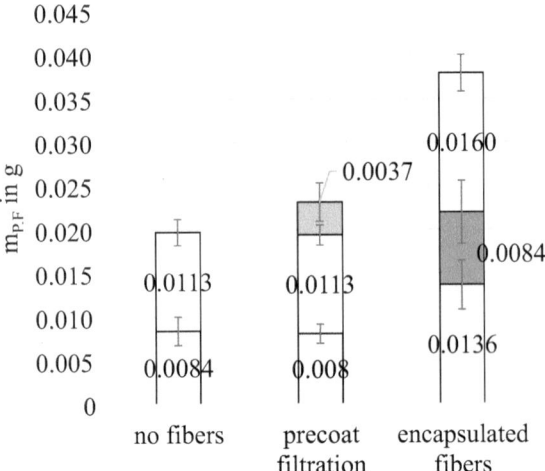

Fig. 33.12 Local separated particle masses $m_{P,F}$ in the case of two CFF ($50 \times 50 \times 22$ mm^3) with no fiber use and fiber use as precoat filtration or in encapsulated alignment

in the storage vessel and had to be taken out for evaluation of the mass of separated particles. With precoat filtration there is no affectation on the amount of separated particles in the up- and downstream CFF, where separation of 0.0113 and 0.0084 g respectively took place.

Encapsulated fiber addition leads to a total amount of 0.0375 g for this arrangement. Thereby the separated mass of particles is much higher and the advantageous placing of the fibers leads to more than a doubled amount of separated particles in the fibers (0.0084 g), compared to precoat filtration. One reason might be that this arrangement assures the entire water based model melt with all the impurities to pass through the encapsulated alignment.

The positive affectation on the particle separation of the surrounding CFF is obvious, as in the up- and downstream CFF´s the particle separation is raised to 50% compared to the precoat filtration. This finding in small scale water based experimental setup assists the results obtained with approaching velocities of 8 cm/s and illustrates the capability of encapsulated fiber addition, especially at higher approaching velocities on particle separation. This promising approach lead to a patent application for metal melt filtration [10]. SEM-analysis of loaded fibers with a diameter of 10–12 μm in water based model system proved evidence of particle separation in single-digit μm range (Fig. 33.13).

A transfer of precoat fiber addition to a sand casting with AlSi10Mg alloy needs to clear the addition of the fibers into the Al-melt. Therefore, several fiber addition strategies were testet in a hand ladle with ca. 2 kg of Al-melt. A simple adding on top of the melt with a subsequent mixing is not productive as the fibers form a bunch, surrounded by an oxide layer. A foregoing removal of the present oxide layer shows no positive effect, as the oxide layer forms immediately again. Placing the fibers on the bottom of the hand ladle and filling it with Al-melt also does not lead to a homogeneous dispersion of the fibers, as the fibers will remain at the interface of

Fig. 33.13 Separated Al$_2$O$_3$ particles on Al$_2$O$_3$ fibers with a diameter of 10–12 μm in water based model system [5]

the melt. Wrapping the fibers in Al-foils before addition is also non-satisfying as the Al-foil melts immediately and the fibers are hold back at the interface and can not penetrate the melt. Further investigations were done with casting trials of cubes with the dimensions of 50 × 50 × 50 mm^3. Besides the simple placing of the fibers on the bottom (Fig. 33.14a) more options were tested where half of the cube was filled with Al-melt. Then, the fibers were added on top of the still liquid melt and doused with further Al-melt. But the fibers were not wetted. Besides this liquid/liquid strategy the fibers were added on a semi-solid / mushy Al-melt which prevents the fibers from floating when further Al-melt is added (Fig. 33.14b). This method turned out to be the only possible investigated way of fiber addition. These cubes with a known amount of fibers then can be added to a crucible for precoat filtration. A subsequent casting of this fiber treated Al-melt into ingots was performed to estimate where these fibers are located and if impurities might adhere. Therefore, the ingots were cut into several specimen and the cross-sectional area was observed. But no fibers were found as they mostly remain in the hand ladle with other backlogs, such as oxide skins. A practical transfer of ceramic fiber precoat addition into an Al-melt casting process seems to be difficult in terms of fiber dispersion due to the high surface tension of the melt (Fig. 33.15).

Besides precoat fiber addition the promising results of the water model experiments are transferred into a AlSi10Mg sand casting system with the experimental setup already shown in Fig. 33.8a. According to Formguss Dresden GmbH the alloy is refined with strontium. A mass of 0.03 and 0.06 g of the fibers were placed in between two 10 ppi CFF (35 × 35 × 22mm^3).

AlSi10Mg alloy is a casting alloy, which is characterized by a dendritic solidification of the aluminium rich α-solid solution. As already stated, the impact of the

Fig. 33.14 **a** Fiber addition on the bottom of a sand mold with subsequent casting. The fibers are not wetted and remain at the bottom, **b** Semi-liquid/liquid fiber addition, where the fibers are poured in the casting cube

Fig. 33.15 **a** Ingot casting with **b** Intersections to examine the presence of ceramic fibers in cross-sectional areas

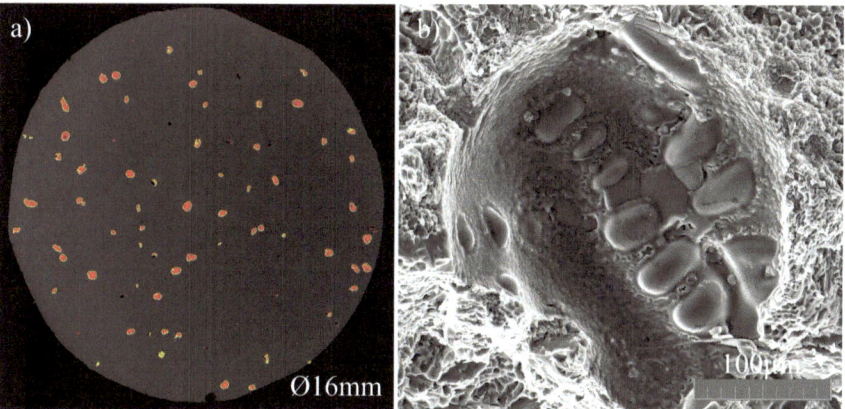

Fig. 33.16   a SEM image from intersection of a casted tensile test specimen with a diameter of 16 mm. Present gas inclusions are marked in red color. The pore area in this example is determined to 1.4% and decreases the mechanical properties, **b** detail of a gas pore on a fracture surface of a tensile test specimen with dendrite arms from the solidified α- Al reaching into the pore

ceramic fibers on the particle separation mainly in single digit μm-range should be observed. SEM analysis indicates the presence of gaseous inclusions with diameter of more than 100 μm (Fig. 33.16). Thereby, these pores superimpose the decreasing effect of the present particular inclusions below 10 μm on the mechanical properties and need to be removed with a hot isostatic pressing (HIP). The HIP was performed at bodycote GmbH and the application of a high temperature and a confining pressure closes the pores. Thus, the porosity is reduced to 0.1% and the impact of the fibers on the smallest particular inclusions in the casting can be evaluated.

Subsequent to the HIP a T6 heat treatment, which consists of a solution annealing (525 °C, 6 h), a quenching in water and an artificial ageing (167 °C, 7 h), was performed. The heat treatment increases tensile strength as well as yield strength and decreases the elongation at fracture. Thus, the sensitivity in respect of present impurities increases when tensile testing is performed.

In order to clarify the experimental proceeding consisting of melt refining, HIP and T6 heat treatment it has to be pointed out that the mechanical properties of the test specimen are already shifted to elevated values as the largest defects and impurities are removed. Nevertheless, a positive influence by adding fibers is still visible (Fig. 33.17).

The mechanical properties of the as-cast specimen are 93.5 MPa respectively 139.0 MPa for yield strength and tensile strength. Elongation at fracture is about 2% and Brinell hardness number is 57.0 ± 1.9 HBW10. A HIP and T6 heat treatment increases the Brinell hardness to values of 99.0 ± 3.7 HBW10 due to fine dispersed precipitations.

It should be noted, that the application of a 10 ppi CFF is not apparently ineffective. But the effect of a HIP and T6 heat treatment is much higher, as they lead to a strong raise in the yield strength to 220.5 ± 8.2 MPa and tensile strength increases to 260.0 ± 8.7 MPa. Nevertheless, a fiber application of 0.03 g still increases the yield and

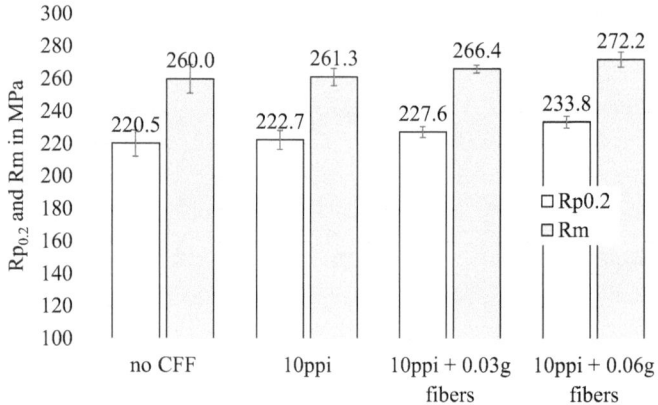

Fig. 33.17 Yield and tensile strength of the tested HIP and T6 heat treated specimen. Elongation at fracture does not change in between the conditions and is $1.7 \pm 0.2\%$

tensile strength to values of 227.6 ± 3.4 MPa respectively 266.4 ± 2.4 MPa. Further increase of fiber application to 0.06 g again increases the yield and tensile strength to 233.8 ± 3.6 MPa respectively 272.2 ± 4.6 MPa and proves the effectiveness of ceramic fibers on the particle separation.

It also needs to be mentioned, that the application of 10 ppi CFF and fibers leads to a significant decrease in the iron content of up to 50% determined with spark spectrometry (Figs. 33.18 and 33.19). It can be assumed that the CFF as well as the fibers remove iron rich phases from the Al-melt, which has a positive effect on the mechanical properties. Structural analysis with SEM and light microscope shows the presence of intermetallic β-Al_7FeSi_2 phase which is known to be hard and brittle and thus unwanted. Although the presence of non-metallic inclusions, such as $MgSi_2O$, is also reduced by applying CFF and fibers (Fig. 33.19). Thereby, Mg-silicates are common non-metallic inclusions in AlSi-alloys [11]. The employed 10 ppi CFF reduces the proportion of $MgSi_2O$ already to 32% and fiber use of 0.03 g leads to a proportion of 14%.

The analysis of the ceramic fibers after a sand casting is performed. There, the fiber radius can be estimated with 5.4 ± 0.14 μm and is consistent with the manufacturer

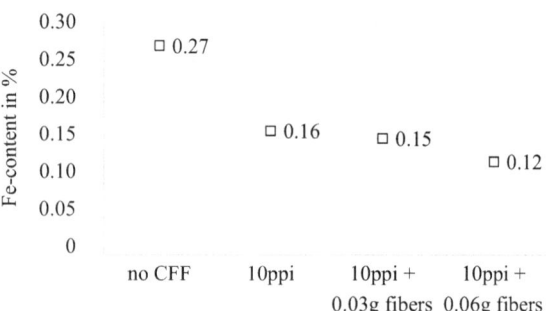

Fig. 33.18 Reduced Fe-content when the Al-melt is treated with CFF and fibers

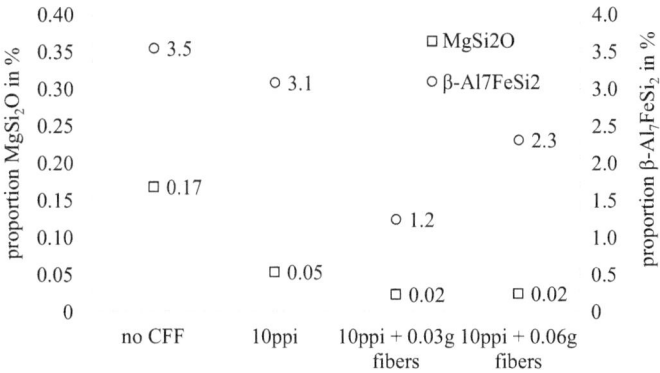

Fig. 33.19 Reducing content of non-metallic inclusions ($MgSi_2O$) and intermetallic phase (β-Al_7FeSi_2)

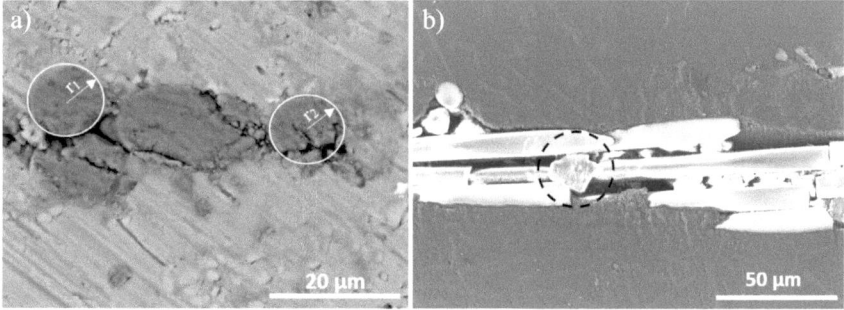

Fig. 33.20 **a** Detailed view of ceramic fibers after usage in an encapsulated arrangement with 10 ppi CFF for the estimation of the fiber radius, **b** microsection of ceramic fibers with entrapped non-metallic inclusion (encircled)

specification where a diameter in between 10 and 12 μm is stated (Fig. 33.20a). The fibers are molded in the AlSi10Mg-alloy. A detailed view of a microsection with ceramic fibers proves the separation of non-metallic inclusions with a diameter close to the single digit μm-range (Fig. 33.20b). The chemical composition of this Mg–silicate was determined with EDX-measurement to 58.5 at% oxygen, 21.5 at% magnesium and 14.1 at% silicon. This detection, as well as the verifiable decrease of the non-metallic inclusion $MgSi_2O$ shown in Fig. 33.19 lead to the conclusion of a positive influence of the fibers on the melt filtration and hence the mechanical properties.

A more detailed study of the particle separation on the ceramic fibers in water based model system was done with a cone beam computer tomograph Xradia 510 Versa from Zeiss with a resolution of 0.376 μm voxel size[12]. The approaching velocity was 3.2 cm/s. The impurities are represented by Al_2O_3 with a initial PSD of 2–20 μm. The segmentation was done in the image processing software ImageJ and the data evaluation was done with the software VG STUDIO MAX 3.3 (Fig. 33.21).

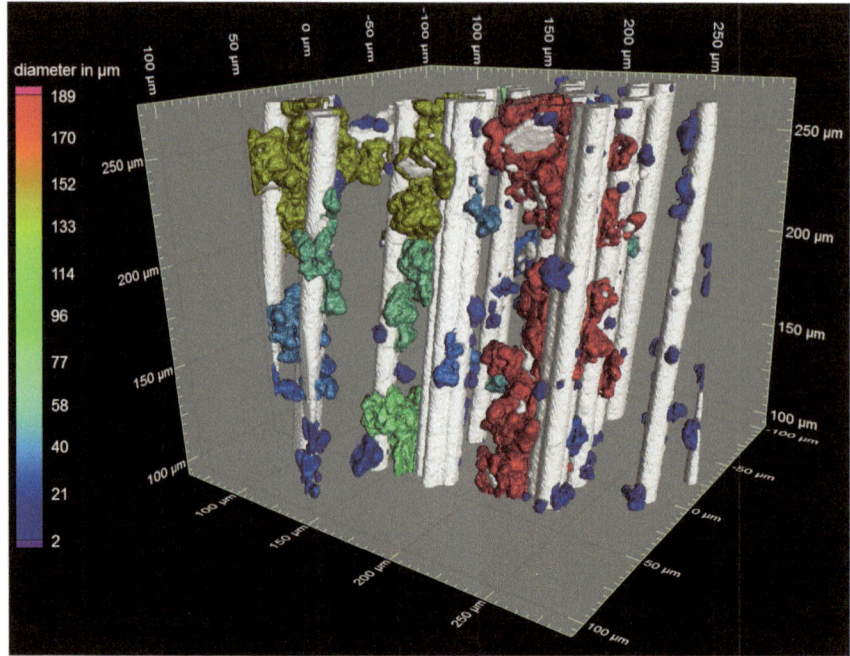

Fig. 33.21 Examined section of the loaded ceramic fibers in water based model system (ca. 200 × 200 × 300 μm³). The separatet coherent particles are coloured according to their diameter [12]

For a better understanding of the data, it is important to note that the particle separation in Fig. 33.21 represents the end of the filtration experiment in the water based model system. With ongoing time of filtration, particle separation takes place and more and more particles are entrapped. Thus, it is not possible to conclude from this snapshot on the particle size of the impurities when they separate. But the evaluation of the number of particles according to their diameter proves evidence of the assumed particle separation in the single digit μm-range (Fig. 33.22). It becomes visible, that the majority of separated particles is in the single digit μm-range. 128 particles exhibit a radius of 5 μm or below. More 47 particles are in the range of 6 to 10 μm. Of course, the volume represented by these impurities is dwindling small compared to the largest separated coherent particles. But as the initial state of the particle diameter just before the separation can not be evaluated the data needs to be read as a snapshot at the time where the filtration experiment stopped.

An evaluation of this data set in VG STUDIO MAX 3.3 with the sphere method tool indicates that also the clustered coherent separated particles exhibit a diameter in the single digit μm-range with a modal value of 5.3 μm (Fig. 33.23). Thereby, detected particles range from 0.5 to 8.0 μm. During particle separation on the fibers it can be assumed that the particles start to separate on single fibers as it can be seen in Fig. 33.23 on the right side and with further separation the particles grow into the fiber interspace until they form bridges and coalesce to a large cluster.

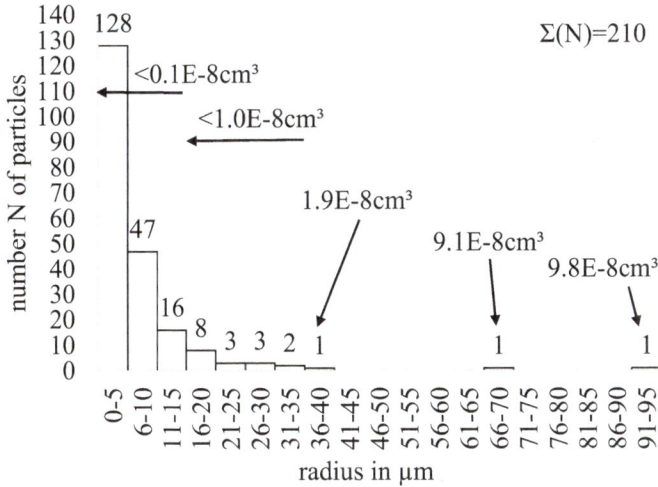

Fig. 33.22 Number of Al$_2$O$_3$-particles detected in the examined section separated on the ceramic fibers [12]

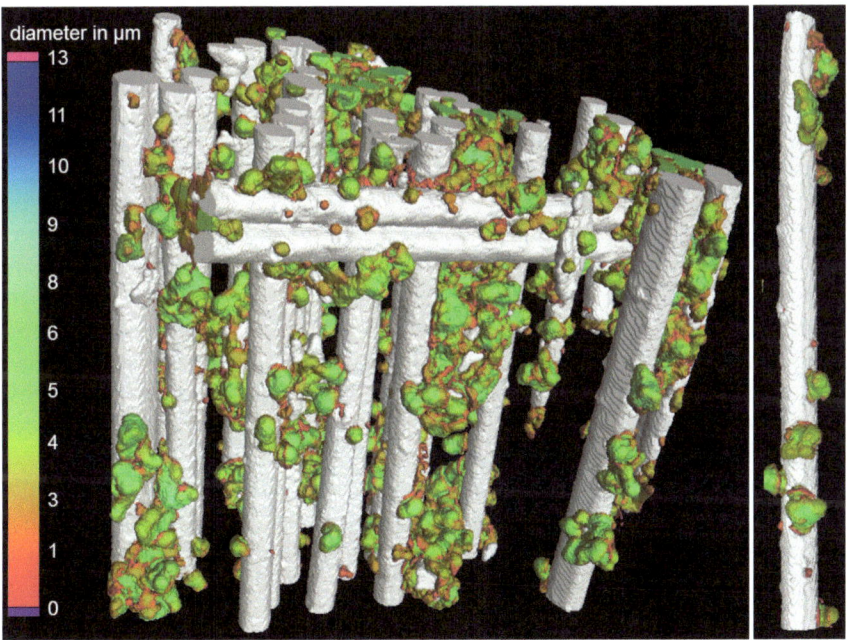

Fig. 33.23 Graphic representation with VG STUDIO MAX 3.3 of the single particle diameter of the Al$_2$O$_3$-impurities evaluated with the sphere method tool in water based model system and illustration of particle separation in single digit μm-range on a single fiber [12]

33.3 Conclusion

The presented results obtained in water model system setup as well as in sand casting with an AlSi10Mg-alloy indicate a high potential of fiber addition on the particle separation for enhanced melt cleanliness and thus better mechanical properties. Two different possibilities of fiber addition were studied, the precoat fiber addition and the encapsulated fiber addition. The water based model melt results clearly indicate the advantageous fiber addition in encapsulated arrangement where they develop their full potential even with positive effect on the up- and downstream CFF with local enhanced particle separation. In contrast, the precoat fiber application may appear at first glance beneficial as the fibers are inserted into the storage vessel in the water model system respectively the crucible. But a satisfactory fiber addition into the Al-melt could not be achieved. Furthermore, the precoating of all fibers can not be fully ensured on the CFF as some fiber remain in the crucible.

A transfer of encapsulated fiber addition in sand casting of an AlSi10Mg-alloy confirms the findings in water based model system as the mechanical properties, in particular the yield and tensile strength are increased. The evaluation of loaded fibers in water based model system with high-resolution computer tomography proves the assumption of significant particle separation and provides information regarding the particle deposition to the fibers. Thereby, only small amounts of fibers in the range of 0.05 g are needed. The addition of the ceramic fibers can be simply made by placing them in between two CFF [10] and thus requires no extra investment of time or knowledge and could be immediately implemented in manufacturing process.

It should be noted that the increase of melt cleanliness can be achieved with a great many of actions through CFF application. In this context it is worth mentioning the investigations in B01 where, among others, the impact of graduated CFF structures with varying ppi-values or the CFF surface roughness was studied demonstrating the positive influence on the filtration efficiency [5, 13]. Also, alternative CFF arrangements were evaluated. The replacement of a CFF volume through such as a loose filling of diced CFF with equal volume is proved to increase the filtration efficiency in water model system [14]. A sand casting of this CFF alignment lead to an increase in tensile strength from 266.7 ± 5.3 to 269.8 ± 6.2 MPa. Hence, a patent application was generated (Fig. 33.24) [12].

Concluding, the sum of all single actions will lead to an increase in filtration efficiency and thus melt cleanliness with elevated mechanical properties of the casting. Here, encapsulated ceramic fiber addition represents one promising measure that can be easily implemented in casting routine of foundry industry.

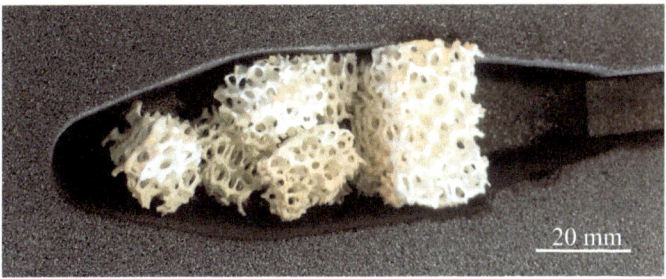

Fig. 33.24 Diced CFF ($16 \times 16 \times 20mm^3$) placed in a sand mold where a CFF with the dimensions of $35 \times 35 \times 20mm^3$ keeps the dices in place

Acknowledgements The authors would like to thank Ralf Schünemann, Steffen Scholz, and Thomas Hantusch for their support in the planning and setting up of the filtration plant. Furthermore, we thank Ralf Ditscherlein and Erik Löwer for the CT measurements and Ruben Wagner, as well as Philipp Schramm, for material characterization. The authors also would like to thank Formguss Dresden GmbH for the realization of the casting trials. The authors gratefully acknowledge the German Research Foundation (DFG) for supporting the Collaborative Research Center CRC 920 (Project ID 169148856–subproject T05).

References

1. J. Fritzsche, Dissertation, TU Bergakademie Freiberg (2016)
2. F. Heuzeroth, Dissertation, TU Bergakademie Freiberg (2016)
3. P. Knüpfer, Dissertation, TU Bergakademie Freiberg (2020)
4. L. Ditscherlein, Dissertation, TU Bergakademie Freiberg (2021)
5. D. Hoppach, Dissertation, TU Bergakademie Freiberg (2022)
6. S. Bell, B. Davis, A. Javaid, et al., in *Enhanced Recycling, Action Plan 2000 on Climate Change, Minerals and Metals Program* (2006), Vol. Report No. 2003-20(CF), p. 15
7. R. Guthrie, M. Li, Metall. Mater. Trans. B **32**, 1081–1093 (2001) https://doi.org/10.1007/s11663-001-0096-5
8. CeraFib GmbH, https://www.cerafib.de. Accessed 10 Dec 2021
9. C. Tian, R. Guthrie, Metall. Mater. Trans. B **26**, 537–546 (1995). https://doi.org/10.1007/BF02653871
10. D. Hoppach U.A. Peuker, Patent No. DE 10 2018 126 326
11. H. Ye, J. Mater. Eng. Perf. **12**, 288–297 (2003). https://doi.org/10.1361/105994903770343132
12. D. Hoppach, U.A. Peuker, Using ceramic fibres for enhancing filtration efficiency in al-melts based on a room-temperature model system. Paper presented at the *FILTECH*, Köln, 8–10 March 2022
13. L. Ditscherlein, A. Schmidt, E. Storti et al., Adv. Eng. Mater. **19**(9), 1700088 (2017). https://doi.org/10.1002/adem.201700088
14. D. Hoppach, U.A. Peuker, Patent No. DE 10 2019,117,513

Chapter 34
Fatigue Lives and Damage Mechanisms at Elevated Temperatures of Steel 42CrMo4 in the HCF and VHCF Regime

Alexander Schmiedel, Thomas Kirste, Roman Morgenstern, Anja Weidner, and Horst Biermann

34.1 Introduction

Metallic materials for engineering applications processed in industrial manufacturing often contain different types of defects such as non-metallic inclusions (NMI) or structural defects such as microshrinkages, which is one of the important issues of industrial materials. As a consequence, the detrimental effect of such defects and, therefore, their impact on the fatigue strength have to be considered when these materials are parts of engineering designs. Furthermore, the fatigue strength is also influenced by the type of the fatigue load as well as environmental conditions, i.e. temperature.

For a deeper understanding of the fatigue failure inducing damage mechanisms, the knowledge of the processes of crack initiation as well as crack propagation is essential. In general, crack initiation occurs from the samples' surface at lifetimes up

A. Schmiedel (✉) · A. Weidner · H. Biermann
Institute of Materials Engineering, Technische Universität Bergakademie Freiberg, Gustav-Zeuner-Str. 5, 09599 Freiberg, Germany
e-mail: A.Schmiedel@iwt.tu-freiberg.de

A. Weidner
e-mail: Weidner@ww.tu-freiberg.de

H. Biermann
e-mail: Biermann@ww.tu-freiberg.de

T. Kirste · R. Morgenstern
Federal-Mogul Nürnberg GmbH, a Tenneco Group Company, Nopitschstraße 67, 90441 Nürnberg, Germany
e-mail: Thomas.Kirste@tenneco.com

R. Morgenstern
e-mail: Roman.Morgenstern@tenneco.com

C. G. Aneziris and H. Biermann (eds.), *Multifunctional Ceramic Filter Systems for Metal Melt Filtration*, Springer Series in Materials Science 337,
https://doi.org/10.1007/978-3-031-40930-1_34

853

to 10^7 [1–5]. At a higher number of cycles, i.e. at lower stress amplitudes in the HCF and VHCF regime, the crack initiation site shifts from the surface to the interior of the sample due to the more harmful effect of internal defects than the surface roughness. Furthermore, the detrimental effect of these defects such as NMIs or pores depends also on the type, the size and the position of the defects within the sample. Relevant contributions concerning the understanding of the damage mechanism leading to fatigue failure were published since the 1980s by Miller [6–9]. The author described short crack initiation and propagation based on microstructural fracture mechanics and elastic–plastic fracture mechanics since it was shown that linear-elastic fracture mechanics was not sufficient to describe very small cracks [10]. Therefore, Miller suggested three regimes of crack growth starting after nucleation with microstructurally short cracks followed by physically small cracks and, finally, long cracks. Of particular interest, regarding the damage mechanism in the HCF and VHCF regimes, are short cracks, i.e. microstructurally and physically short cracks. Microstructurally short cracks have the dimension of the longest microstructural barrier, e.g. martensite plates and propagate under shear stress to the next microstructural barrier, e.g. grain boundaries [11]. It is readily conceivable that the crack arrest at such barriers and a considerable amount of cycles to failure were consumed under these conditions regarding crack growth. However, growth rates of short cracks can vary in a wide range and are strongly influenced by the microstructure and the environment.

The temperature has a significant impact on the material's crack resistance. For applications subjected to various temperatures, the influence of the temperature on the fatigue strength has to be considered in a wide range of cycles, i.e. in the HCF and VHCF regime. Fatigue experiments in the HCF regime using relatively low test frequencies are well-established. Fatigue tests in the VHCF range using the ultrasonic resonance technique at ambient temperatures have been established as state of the art procedure to determine the fatigue properties of various materials [12–19]. However, only a few experimental evaluations of the fatigue behaviour in the VHCF regime using the ultrasonic resonance technique at elevated temperatures were published [20–29].

In this chapter, the fatigue behaviour of two batches of the 42CrMo4 steel (forged and cast) were investigated by performing fatigue experiments at a stress ratio R = −1 up to 10^{10} cycles and at test temperatures of 295 K (room temperature, RT) 473 and 773 K. The fatigue tests were carried out at test frequencies of 90 Hz and 20 kHz in the HCF and VCHF regimes, respectively. Four types of defects were determined as relevant crack initiation sites and, in consideration of the individual defect sizes, the fatigue lives were examined using a short crack model [30, 31]. Based on these investigations, the temperature-dependent damage mechanism leading to fatigue failure is discussed.

34.2 Materials and Methods

34.2.1 Materials

The study was performed on two different batches of 42CrMo4 steel, i.e. the cast state (0.42 wt% C, 0.81 wt% Cr, 0.19 wt% Mo, 0.96 wt% Mn, 0.42 wt% Si), which is referred to as CS in the following, as well as the wrought state (0.44 wt% C, 0.95 wt% Cr, 0.18 wt% Mo, 0.78 wt% Mn, 0.32 wt% Si), which is referred as WS in the following. Both batches were subjected to the quenching and tempering heat treatment procedure which is referred to as HT2, with the following characteristics: (i) austenitization at 1123 K for 1 h under inert gas atmosphere and quenching in oil and (ii) tempering at 773 K for 2 h. Furthermore, batch CS was subject to a further heat treatment, which is referred to as HT1 consisting of (i) austenitization at 1153 K for 2 h and quenching in oil followed by (ii) tempering at 858 K for 1 h. Each heat treatment led to a martensitic microstructure. The values of Vickers hardness at elevated temperatures (473 K and 773 K) were derived from the ultimate tensile strength (UTS) according to DIN EN ISO 18265. At RT, the Vickers hardness was determined by using a standard hardness tester (LECO M-400-G3, St. Joseph, USA). The values of Young's modulus were determined using dynamic resonant measurements in the range from RT up to 773 K.

34.2.2 Microstructure

The materials microstructures were characterized by scanning electron microscopy (SEM) using a field-emission SEM (MIRA 3 XMU, TESCAN, Czech Republic) equipped with an electron backscattered diffraction (EBSD) system (EDAX/Ametek, USA) and OIM (TSL). Therefore, secondary electron (SE) contrast and EBSD were applied. EBSD measurements were performed to determine grain sizes and crystallographic orientation of the different material states. Therefore, metallographical samples were prepared and vibration polished (SiO_2 suspension with 0.02 μm size) for several hours.

Metallographic cross sections were investigated by an optical microscope (Zeiss Observer Z1, Zeiss, Jena, Germany) and the defect distributions were determined by analysis software (AxioVision, Zeiss, Jena, Germany) using grey scale correlation. A total area of 3650 mm^2 of material CS was investigated at 200 × magnification and a total area of 114 mm^2 of material WS was investigated at 100 × magnification. Various defects were classified according to the technical guideline "Evaluation of inclusions in special steels based on their surface areas" (SEP 1571) [32].

34.2.3 Fatigue Tests

34.3 General Aspects of HCF and VHCF Tests

Fatigue tests were performed at three test temperatures (RT, 473 K, and 773 K) at a stress ratio of R = −1 up to 10^{10} cycles. Some of the fatigue tests were performed using a resonance pulsator (Amsler, Neftenbach, Switzerland) at a test frequency of about 90 Hz. Fatigue data determined by this approach are referred to as '90 Hz' in the following. Above 10^8 cycles, the tests were stopped and considered as run-outs. The other fatigue tests were performed using ultrasonic fatigue testing equipment (UFTE) developed and produced by the University of Natural Resources and Life Sciences (Vienna, Austria) at a test frequency of about 20 kHz. Fatigue data determined by this approach are referred to as 20 kHz in the following. The physical background of this ultrasonic testing method was presented in detail by Stanzl-Tschegg [12].

Depending on the test method, different sample geometries were required. A CNC lathe was used to produce the samples used for 90 Hz fatigue tests. They had a diameter of 5 mm and a parallel gauge length of 20 mm. The surface roughness of the samples was machined to $Rz = 2$ μm. Fatigue tests of the 90 Hz samples at elevated temperatures were realized by using a chamber furnace around the sample and the grip system to ensure isothermal conditions. The temperature was controlled by thermocouples of type K attached at the sample's gauge length. The sample geometries used for 20 kHz fatigue tests were determined by FEM modal analysis (Ansys, Inc.) individually for each testing temperature to ensure the correct length for the desired resonance frequency of about 20 kHz. Although the total length of the samples depends on the temperature, the gauge length was always constant with a diameter of 4 mm and a parallel length of 9 mm. The samples were also machined using a CNC lathe with a surface roughness of $Rz = 3$ μm. An additional cooling system (EPUTEC Drucklufttechnik, Kaufering, Germany) was applied to prevent overheating at RT fatigue tests. The fatigue tests at elevated temperatures were performed by using an induction heating system (Systemtechnik Skorna, Typ S230, Sulzbach-Rosenberg, Germany). The temperature was measured by an infrared pyrometer (IMPAC IGA 140/23, LumaSense Technologies, Frankfurt, Germany). Therefore, the samples were painted with black lacquer (Dupli-Color® SUPERTHERM), which is heat-resistant up to 1073 K. In addition, a compressed air cooling system was used at elevated temperature fatigue testing to ensure a homogeneous temperature profile and maximum temperature increase during the tests ΔT to less than 10 K. Also the method a defined ratio of constrained ultrasonic pulse durations followed by a pause, which is in the following referred to as pulse/pause ratio, was used to limit sample heating during the fatigue tests. The fatigue testing system for elevated temperatures and the calibration procedure are given in detail in the following section as well as in previous work [27].

Details of the fatigue testing system for elevated temperature testing and the related calibration procedure

The ultrasonic fatigue testing system for elevated temperature testing was used to perform tests up to 773 K. The ultrasonic fatigue testing equipment for RT tests was extended by an induction heating system, working with a water-cooled copper coil and a high-frequency generator. In detail, the main parts of the testing system are shown in Fig. 34.1 and are referred to as ultrasonic transducer (see (1) Fig. 34.1a) to introduce the ultrasonic wave into the load path, the booster horn (see (2) Fig. 34.1a) to increase the amplitude of the ultrasonic wave, the λ-rod (see (7) Fig. 34.1b) to transmit the wave into the sample (see (8) Fig. 34.1b), and the vibration sensor (see (3) Fig. 34.1a) to measure the feedback signal of the stationary wave. In addition, the following components were included to perform the fatigue tests at elevated temperatures: the compressed air cooling system (see (4) Fig. 34.1a) aligned to different positions on the load path, the infrared pyrometer (see (5) Fig. 34.1a) focused on the centre of the sample's gauge length (see the red laser dot in Fig. 34.1b), watercooled copper coil of the induction heating system (see (6) Fig. 34.1a) and the fibre optic sensor (MTI 2100 Fotonic-Sensor, MTI Instruments Inc., New York, USA) (see (9) Fig. 34.1b).

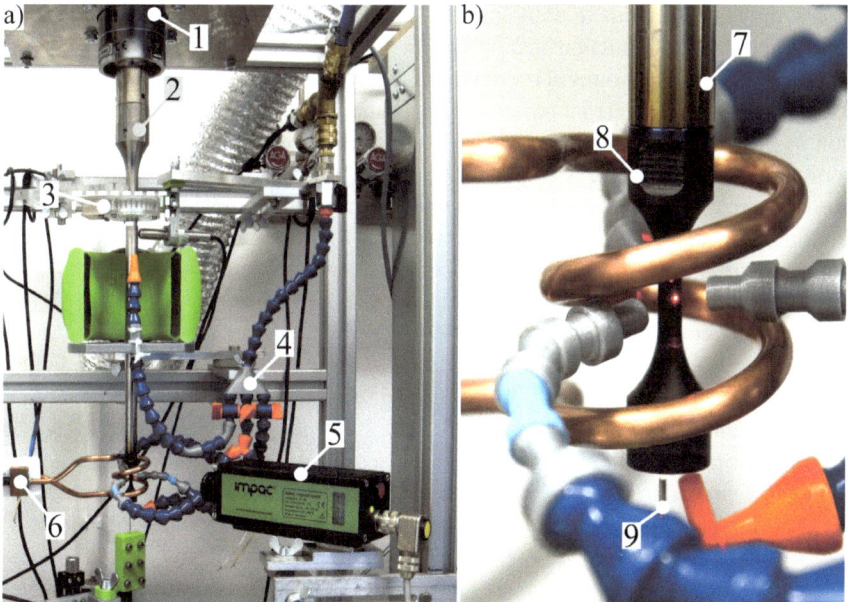

Fig. 34.1 Ultrasonic fatigue testing system for elevated temperatures. **a** Overview. **b** In detail: (1) ultrasonic transducer, (2) booster horn, (3) vibration sensor, (4) compressed air cooling system, (5) infrared pyrometer, (6) copper coil of induction heating system, (7) λ-rod, (8) sample, (9) fibre optic sensor. Reproduced from [27]

In order to perform the fatigue tests at a specific temperature, it is important that the temperature distribution is as homogeneous as possible within the sample's gauge length. However, various environmental parameters affect the distribution of the temperature, i.e. (i) material damping as a function of stress amplitude, (ii) energy input of the induction heating system, (iii) ambient temperature, (iv) thermal conduction of the entire load path, (v) thermal convection primarily influenced by the compressed air cooling system, and (vi) thermal radiation. Therefore, the compressed air cooling performance can be adjusted for each position of the load path (see orange and grey nozzles in Fig. 34.1a and b). A schematic illustration of the heat in- and output due to conduction, convection and induction is given in Fig. 34.2a. Heat input generated by the induction heating system is referred to as q_I and depends on generator power settings as well as the size and the number of windings of the copper coil. The controller of the induction heating system, which was a Labview (National Instruments, Austin, USA) based application, used the signal of the infrared pyrometer as the controlled variable, which is determined at the centre of the sample (see red dot in Fig. 34.1b). Further heat input is generated by the damping of the material and is referred to as q_D. The amount of heat input over time due to material damping depends, in general, on the tested material as well as the applied stress amplitude and the pulse/pause ratio. Therefore, a sufficient cooling system and a reasonable pulse/pause ratio are necessary at RT experiments. At elevated temperature experiments, the heat input due to damping can be used to keep the test temperature while less power from the induction heating system is required. However, material damping leads to higher temperatures at the centre of the sample. Therefore, sufficient cooling is required also at elevated temperature experiments, i.e. heat output due to convection q_K. The optimum cooling setting is ensured by individual flow rates of each cooling point on the load path (see orange and grey nozzles in Fig. 34.1a and b). Furthermore, heat output occurs due to conduction q_K over the whole load path, i.e. from the threaded side of the sample into the λ-rod up to the booster horn as well as from thermal radiation q_{TR}. A homogeneous temperature profile can be achieved by well-balanced settings of the parameters determining heat in- and output. Homogeneous temperature distribution was verified using a thermocamera (VarioCAM hr head 780, INFRATEC, Dresden, Germany) An example of the temperature field is shown in the thermograms of samples during fatigue tests at 473 K and 773 K in Fig. 34.2b and c, respectively.

The calibration of the UFTE for each series of ultrasonic fatigue samples is a mandatory procedure to perform fatigue tests at the desired stress amplitude. In general, this procedure provides the correlation between vibration amplitude set at the UFTE and the resulting strain in the centre of the sample's gauge length. This was achieved at RT by using strain gages (KFG-1-120-C1, Kyowa Electronic Instruments, Tokyo, Japan) clued in the middle of the gauge length. The values of these strain gages were correlated with the vibration amplitude by data extrapolation on the base of three specified amplitudes. The available vibration amplitude during the calibration is limited by the maximum stress that can be applied to the strain gages. Therefore, an extrapolation of the data is necessary due to the usually higher strain than the strain gages can tolerate during the fatigue experiment. At elevated temperatures,

a)   b)   c)

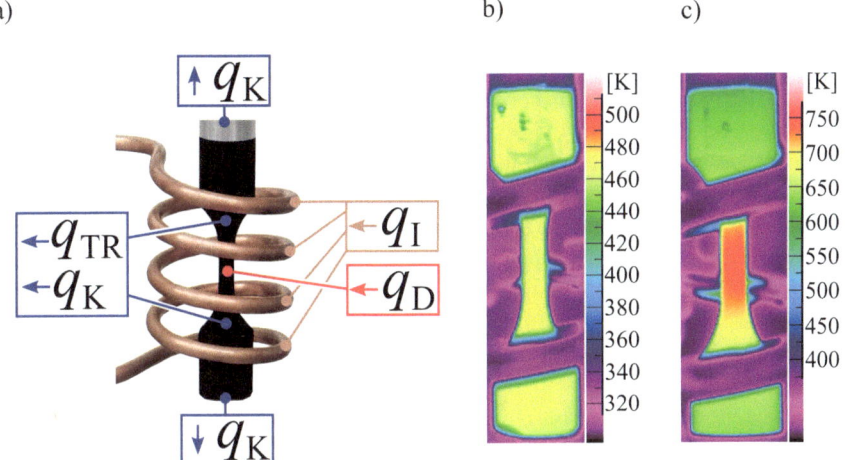

Fig. 34.2 Thermal condition of a sample during the fatigue experiment. **a** Schematic illustration of heat in- and output. Thermogram of a sample tested at **b** 473 K and **c** 773 K. Reproduced from [27]

strain gages are not applicable especially in the case of an induction heating system due to the induction of a voltage to the strain gage, which makes strain measurement by strain gages almost impossible while induction heating. Thus, the displacement of the sample during the fatigue test was considered as an additional parameter to determine the correlation between strain in the centre of the sample and the vibration amplitude of the UFTE. In particular, it is indispensable to perform calibration at elevated temperatures for materials with a significant change of Young's modulus as a function of temperature, e.g. the investigated steel 42CrMo4.

A fibre optical sensor has been used to measure the displacement at the unthreaded end of the sample as shown in (9) of Fig. 34.1b. Optical displacement measurement is very well applicable for fatigue tests at elevated temperatures and was already reported by several authors [21, 22, 24]. A calibration method for elevated temperature was applied, where strain and displacement measurements were performed simultaneously at RT and solely displacement measurements were conducted at elevated temperatures. The stress amplitude was, therefore, calculated as a function of temperature according to Eq. (34.1). The equation includes the temperature-dependent parameters displacement Δx_T and Young's modulus E_T, and the parameter κ, which is temperature independent

$$\sigma_a(T) = \frac{\Delta x_T}{\kappa} \cdot E_T \tag{34.1}$$

The parameter κ is calculated according to Eq. (34.2) and includes the ratio between samples displacement Δx_{RT} to strain ε_{RT} at RT as well as the elongation of the samples gauge length l_0 due to thermal expansion α owing to temperature increase ΔT. Consequently, the value of κ has the dimension of a length.

$$\kappa = \frac{\Delta x_{RT}}{\varepsilon_{RT}} + (l_0 \cdot \alpha \cdot \Delta T) \qquad (34.2)$$

This approach delivers the correlation between stress amplitude σ_a and displacement Δx as a function of temperature. Therefore, the vibration amplitude of the UFTE has to be adjusted to the temperature dependent displacement of the sample for each test temperature. As shown in Fig. 34.3, there is a linear relation between vibration amplitude, e.g. three supporting points at 100, 200 and 300, and displacement with different slopes depending on the temperature. Therefore, it is possible to extrapolate the vibration amplitude (VA) of the UFTE for specific displacements and, consequently, for specific stress amplitudes since the correlation between stress and displacement is given in Eq. (34.1).

In the case of higher temperatures, surface oxidation affects the displacement measurement. Since the displacement measurement via fotonic sensor is very sensitive concerning the reflected light from the surface, Ni-coating of the surface was performed for calibration samples, which is sufficient up to 673 K. The slope to 773 K (see (*) in Fig. 34.3) was determined by extrapolation using linear regression. (see grey dashed line in Fig. 34.3). Consequently, the correlation between VA and displacement can also be determined at 773 K even if no direct measurement of displacement is possible at this temperature. The accuracy of the calibration procedure depends on precise strain measurements at RT using strain gages, displacement measurements at RT and elevated temperatures using fibre optic sensor and the correct value of Young's modulus at each test temperature.

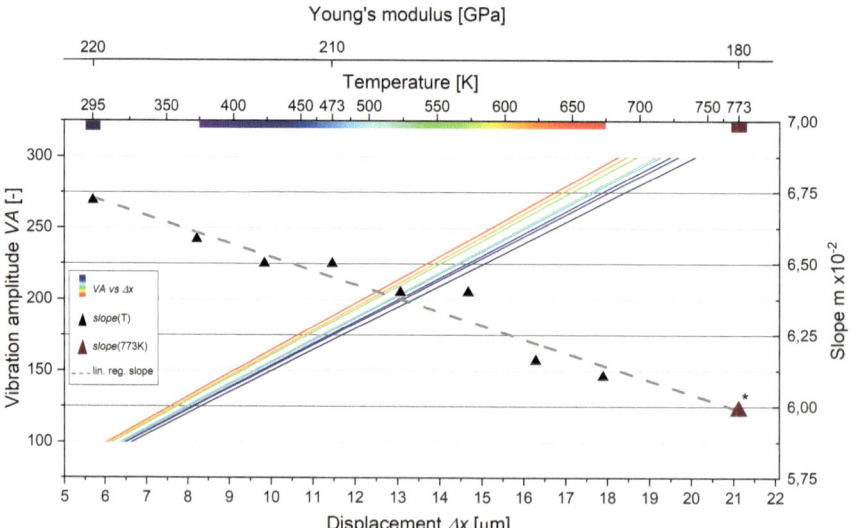

Fig. 34.3 Exemplarily calibration results of material WS(HT2) for 773 K fatigue tests. Vibration amplitude versus displacement curves (coloured lines) and slope vs temperature (black triangles and dashed line). (*) Slope at 773 K calculated by extrapolation from lower temperatures

34.3.1 Fractography

The fractographic analysis of fracture surfaces of failed samples is substantially important in the understanding of the mechanisms leading to fatigue failure. Thus, an optical microscope (Zeiss Discovery V20, Jena, Germany) and SEM were utilized to perform fracture surface analysis. The following typical features on the fracture surface were determined: (i) the fatigue failure initiating defect, (ii) the size of the defect approximated as an ellipse given as $\sqrt{\text{area}}$, (iii) the diameter of the fisheye (FiE) in case of presence, the diameter of the fine granular area (FGA) in case of presence, and (v) the shortest distance between the fatigue failure initiating defect and the sample's surface. Reduced surface roughness is typical for the area within the FiE compared to the region of the fracture surface outside the FiE. Furthermore, the formation of an FGA inside the FiE can occur due to an internal defect, which is a frequently observed phenomenon in the VHCF range, especially for high strength steels [33]. The occurrence of an FGA depends on the size of the defect, i.e. only relatively low values of defect-related SIF (about < 5 MPa$\sqrt{\text{m}}$) lead to FGA formation. The FGA is characterized by a fine granular layer due to the coalescence of micro-debonding [1, 34].

A decrease of the resonance frequency of approximately 75 Hz is used to declare samples tested at 20 kHz as failed, whereas the number of cycles to failure was determined after the samples broke during the fatigue tests for samples tested at 90 Hz. For the fractographical investigations, the samples of the stopped tests were broken. Therefore, they were cooled in liquid nitrogen and opened, finally, by slight mechanical impacts without any influence on characteristic features of fatigue failure. Samples without a decrease in the resonance frequency reaching more than 10^9 cycles (20 kHz) or reaching more than 10^8 cycles (90 Hz) were declared as runouts. Fracture surfaces of all samples tested at 20 kHz or 90 Hz were investigated by SEM or optical microscopy, respectively.

Concerning the presence of different types of defects, i.e. different chemical compositions, energy-dispersive X-ray spectroscopy (EDX) was performed.

34.3.2 Fatigue Lives

The fatigue strength of the investigated batches of the steel 42CrMo4 was influenced by various material properties, e.g. tensile strength, hardness, ductility and present defects. The damage mechanisms leading to fatigue failure depended on these material properties as well as the apparent microstructure. Temperature as an environmental influence affects these material properties as well. Since crack growth was in question as the dominant mechanism leading to fatigue failure instead of crack initiation, the influence of temperature has to be considered. Therefore, a model for short crack growth using elastic–plastic fracture mechanics (EPFM) based on considerations of Miller [6–9] was applied to the experimental data.

The applied crack growth model is based on the model introduced by Chapetti [30, 31]. It considered the difference between stress intensity factor (SIF) range ΔK and SIF range threshold ΔK_{th} as an effective parameter, which was first introduced by Zheng and Hirt [35] as one of the various generalisations of Paris' law already published [36–38] in the past (see Eq. (34.3)). To correlate the model of short crack growth to the data of the fatigue experiments performed with the investigated batches of steel 42CrMo4 in the HCF and VHCF range, Eq. (34.4) was derived from Eq. (34.3) to calculate the number of cycles to failure $N_{f.num}$. Here, the difference between the SIF range and SIF-threshold range $(\Delta K - \Delta K_{th})$ as well as the material coefficients C and m have to be determined. ΔK was calculated as a function of the stress range and the crack length a (see Eq. (34.5)). The value of Y depends on the applied type of crack and load. Finally, the integration over the crack length a has to be carried out to calculate the number of cycles to failure. The application of the short crack growth model by Chapetti [30, 31] on the fatigue data of the investigated steel 42CrMo4 in the HCF and VHCF range was published in detail in earlier work [27, 28].

$$\frac{da}{dN} = C*(\Delta K - \Delta K_{th})^{m} \tag{34.3}$$

$$N_{f.num} = \int_{a_0}^{a_f} \frac{1}{C*(\Delta K - \Delta K_{th})^{m}} da \tag{34.4}$$

$$\Delta K = Y \Delta \sigma \sqrt{\pi a} \tag{34.5}$$

ΔK_{th} for long crack propagation consists of two parts: (i) intrinsic ΔK_{dR} and (ii) extrinsic SIF range ΔK_c. ΔK_{dR} is determined by the plain fatigue limit of the material as well as the distance to a microstructural barrier, e.g. grain boundary, which is also sufficient to arrest short crack growth [39] according to Eq. (34.5). The distance to a microstructural barrier corresponds to the size distribution of the microstructure, i.e. the distance is equal to the size of a martensite plate regarding the investigated steel 42CrMo4. ΔK_c is determined by various values, e.g. a material constant on the difference between the thresholds of crack growth for long cracks ΔK_{thR} and the microstructure depending on the shortest barrier, and increases with the crack length a up to saturation. In the following, the unit system for crack propagation and SIF is [$m/cycle$] and [$MPa\sqrt{m}$], respectively.

The crack growth starts with the initial length of a_0 up to the length until failure a_f. Both values have to be derived from the fracture surfaces of each sample (see Fig. 34.4), i.e. a_0 and a_f are related to the defects size and the FiE's size, respectively. Therefore, the values of a_0 and a_f were derived from the area of the defects, i.e. approximated as an ellipse, and finally calculated according to Eq. (34.6). It is assumed that the majority of fatigue life is consumed when the crack reaches the size of the FiE. Only a few cycles are still needed for crack growth outside the FiE up to final failure. In cases, where no FiE was observed, the crack length until failure a_f was assumed as the shortest distance between the defect and the sample's surface.

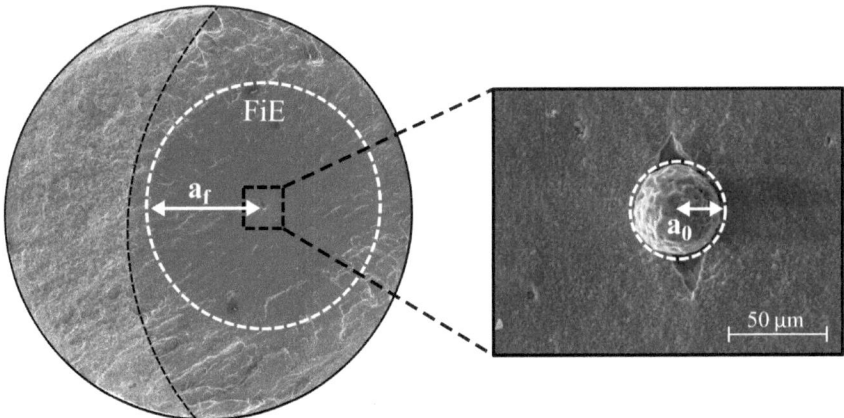

Fig. 34.4 Schematic illustration of the determination of initial crack length a_0 and fatigue crack length a_f using fracture surface analysis. Reproduced from [28]

$$a = \sqrt{\frac{area}{\pi}} \tag{34.6}$$

This procedure allowed the calculation of the number of cycles to failure $N_{f.num}$, based on the results and observations of the fatigue experiments and the fracture surface analysis. The remaining parameters C, m and ΔK_{thR} to calculate the number of cycles to failure $N_{f.num}$ (see Eq. (34.4)), which were referred to as free parameters of the short crack growth model in the following, were determined by calculating the highest correlation between experimental and the calculated number of cycles to failure N_f and $N_{f.num}$, respectively. Therefore, various sets of these free parameters were applied.

34.4 Results

34.4.1 Materials Characterization

The cast (CS) and the wrought state (WS) of the steel 42CrMo4 were investigated in the present study. Batch CS was subject to the heat treatments HT1 and HT2 whereas batch WS was subject to HT2 (see details on heat treatment in Sect. 34.2.1 Materials). Characteristic properties of all investigated materials were determined at RT, 473 K and 773 K, i.e. yield strength at 0.2% plastic strain ($R_{p0.2}$), ultimate tensile strength (UTS), elongation at fracture (A_f) and Vickers hardness (HV10), as given in Table 34.1

Microstructures of all three materials were investigated using EBSD. The crystallographic orientation distributions and the grain sizes were determined. All materials

Table 34.1 Characteristic properties of steel 42CrMo4 at 295 K, 473 K and 773 K according to [27, 28]; * hardness values at 473 K and 773 K were derived from UTS according to DIN EN ISO 18265

	CS(HT1)			CS(HT2)			WS(HT2)		
Temperature [K]	295	473	773	295	473	773	295	473	773
$R_{p0.2}$ [MPa]	844	733	556	1179	1006	597	1163	1006	700
UTS [MPa]	1000	927	663	1251	1188	793	1260	1206	760
A_f [%]	12	11	22	10	9	18	6	5	13
HV10	320	290*	205*	420	370*	250*	425	376*	237*

showed a microstructure characterized by tempered martensite (see Fig. 34.5) and no pronounced texture. The values of the grain size were evaluated as 5.5 μm ± 4.3 μm, 1.9 μm ± 1.5 μm and 1.5 μm ± 0.9 μm, for CS(HT1), CS(HT2) and WS(HT2), respectively.

The classification of defects is based on SEP 1571 [32]. An overview of types of defects and their characteristic features are given in Table 34.2. This classification considers the following characteristic features of the defects: (i) the deformability, (ii) the form, (iii) the distribution, and (iv) the typical representative. Here, NMIs were designated as type A, BC and D defects with type BC as a combination of type B (Al_2O_3, MgO) and C (SiO_2). Microshrinkes are defined in this work as defect type MS. NMIs were present in the wrought material WS(HT2) and microshrinkages as typical defects for cast materials were present in both materials of the cast batch CS(HT1) and CS(HT2). NMIs have to be assumed as present also in the cast materials CS(HT1 and HT2) due to the manufacturing process. However, the detrimental effect of cast defects such as microshrinkages is significantly higher due to their size and morphology. Consequently, microshrinkages were the dominant defects leading to fatigue failure in all cases of the investigated cast batch. The corresponding defects observed by fracture surface analysis are shown in Fig. 34.7.

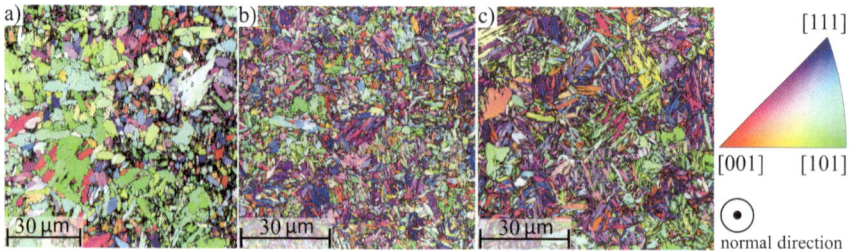

Fig. 34.5 EBSD scan of investigated materials with inverse pole figure according to the surface normal. **a** CS(HT1). **b** CS(HT2), **c** WS(HT2). Reproduced from [28]

Table 34.2 Overview of types of defects and their characteristic features examined in this work (based on SEP 1571 [32]) according to [28]

Type	A	BC	D	MS
Representative	MnS	Al_2O_3, MgO, SiO_2	Al_2O_3, MgO, CaS	Microshrinkage
Deformability	Yes	Al_2O_3, MgO no	Oxide no	Yes
		SiO_2 yes	Sulfide yes	
Form	Elongated	Crumbled or elongated	Globular	Random
Distribution	Stringer	Stringer	Random	Random

34.4.2 Fatigue Testing

The fatigue data of the investigated batches of steel 42CrMo4 tested at RT, 473 K and 773 K are shown in Fig. 34.6. Experiments performed at 90 Hz and 20 kHz testing frequencies are marked by triangles and squares, respectively. The regression lines were calculated concerning the combined 90 Hz and 20 kHz fatigue data at each test temperature. The maximum likelihood method based on the approach by Pascual and Meeker [40] was used to calculate the regression line for 50% failure probability as well as the regression lines for 5% and 95% failure probability, which are marked by coloured areas with respect to the test temperatures. The influence of the test temperature on the fatigue strength of each material is clearly seen in Fig. 34.6, i.e. the fatigue strength decreased with an increase in temperature and showed the highest values at RT and lowest at 773 K. For all materials, the decrease in fatigue strength from RT to 473 K was moderate compared to the decrease to 773 K. Table 34.3 shows characteristic values of the SN-curves, i.e. calculated fatigue strength at 10^6 and 10^9 cycles for each test temperature as well as the slope parameter k^* (see Eq. (34.7)) and the scatter T_N.

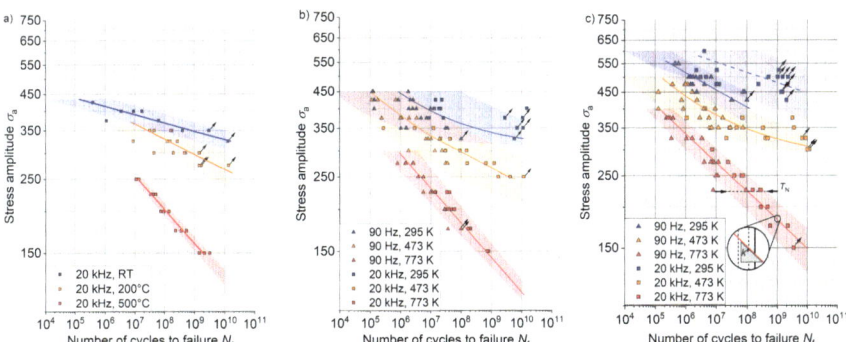

Fig. 34.6 Stress versus number of cycles to failure (SN) curves of steel 42CrMo4 at RT (blue), 473 K (orange) and 773 K (red) including runouts (black arrows). **a** Material CS(HT1). **b** Material CS(HT2). **c** Material WS(HT2). Reproduced from [28]

Table 34.3 Characteristic parameters of SN-curves of the cast (CS(HT1), CS(HT2)) and the wrought (WS(HT2)) steel 42CrMo4 at RT, 473 K and 773 K. Values of material WS(HT2) at RT based on data of 90 Hz experiments (blue solid line in Fig. 34.6c) are marked as [a], values of material WS(HT2) at RT based on data of 20 kHz experiments (blue dashed line in Fig. 34.6c) are marked as [b]. Values of the fatigue strength of material CS(HT1) at 10^6 cycles are beyond the calculated SN-curve and, therefore, marked as not available (NA). Data of CS(HT2) and WS(HT2) according to [28] and Data CS(HT1) according to [27]

Material state	CS(HT1)			CS(HT2)			WS(HT2)		
Temperature [K]	295	473	773	295	473	773	295	473	773
Fatigue strength $\sigma_{a.1}$ at 10^6 [MPa]	415	NA	NA	445	380	325	515[a]	435	335
Fatigue strength $\sigma_{a.2}$ at 10^9 [MPa]	346	295	160	340	265	145	480[b]	320	180
Slope parameter $k^* = \Delta\log(N_f = 10^9)/\Delta\log(\sigma_{a.2})$	57	30	11	48	21	10	32[b]	40	11
$T_N = 1$: ($N_{Ps = 5\%}/N_{Ps = 95\%}$)	>1000	79.7	2.5	>1000	>1000	7.9	>1000	640	18

The cast material CS(HT2) showed slightly higher fatigue strength compared to CS(HT1). It has to be noted that for CS(HT1) only fatigue data of 20 kHz test frequency were available. For CS(HT1) a decrease in calculated fatigue strength at 10^9 cycles was observed from RT to 473 K and 773 K by 15% and 54%, respectively. For CS(HT2) a decrease in calculated fatigue strength at 10^9 cycles was observed from RT to 473 K and 773 K by 22% and 57%, respectively. Also, the slope parameter k^* and the scatter T_N showed their lowest value at 773 K, i.e. $k^* = 11$ and $T_N = 2.5$ for CS(HT1) and $k^* = 10$ and $T_N = 7.9$ for CS(HT2).

Batch WS showed superior fatigue strength at each test temperature compared to both materials of CS batch. The scatter T_N of the fatigue data was in all cases significantly higher at RT and 473 K compared to 773 K. This is due to the low slope of the calculated SN-curve at lower temperatures as shown by the lowest values of slope parameter k^* at 773 K. Only a small number of runouts was observed at a test temperature of 773 K, whereas runouts occurred more frequently at temperatures below 773 K.

$$k^* = \frac{\Delta\log(N_f = 10^9)}{\Delta\log(\sigma_{a.2})} \tag{34.7}$$

The wrought material WS(HT2) shows a significant difference of the fatigue data obtained at 90 Hz or 20 kHz at RT. Thus, two separate SN-curves were calculated, whereby the 20 kHz SN-curve is approximately 100 MPa above the 90 Hz SN-curve. At 473 K and 773 K, one SN-curve was calculated for each temperature including both 90 Hz and 20 kHz fatigue data. According to the calculated values of the SN-curves at each test temperature (see Table 34.3), it is shown that the fatigue strength at 10^9 cycles decreased from RT to 473 K and 773 K by 33% and 62%, respectively. Furthermore, the slope parameter k* and the scatter T_N decreased significantly with an increase in test temperature with its lowest value of $k* = 11$ and $T_N = 18$ at 773 K.

34.4.3 Fractography and Defect Distribution

The fracture surfaces of all failed samples were observed using SEM and optical microscopy. The types of defects leading to fatigue failure, based on the classification by SEP 1571 [32] (see Table 34.2), were determined. An overview of defects observed by SEM is given in Fig. 34.7a-d. EDS was used to confirm the observed defects according to the classification. Therefore, 204 fracture surfaces were analysed by SE- or light microscope. It revealed that the following defects were observed only in material WS(HT2): (i) globular alumina (type D), (ii) alumina-cluster (type BC), and (iii) elongated manganese-sulphide (type A). Microshrinkages (type MS) were only observed in the cast materials CS(HT1) and CS(HT2), which is related to the manufacturing process. In most cases, no FGA was observed for both CS and WS batches. The $\sqrt{\text{area}}$ concept of Murakami [41] was applied to express the size of the defects.

No dependencies were found for the size of the defects and the shortest distance to the sample surface with respect to the number of cycles to failure. However, there is a dependency on the size of the FiE and the distance of the defect to the sample surface (see highlighted data points coloured in red in Fig. 34.8) at 773 K. Fracture surfaces of samples tested at lower temperatures showed no clear dependency and, in general, frequently lower values of FiE size (cf. 295 K and 473 K to 773 K in overview and detailed view of Fig. 34.8).

The distribution of defect types with respect to the classification according to the technical guideline SEP 1571 [32] (see Table 34.2) and the test temperature are given in Fig. 34.9. The sizes of observed defects of type MS (microshrinkages) tended to increase with an increase in test temperature up to 773 K, however, even with a high scatter of the $\sqrt{\text{area}}$ parameter. Defects of type A (elongated manganese-sulphide) showed the lowest mean values of $\sqrt{\text{area}}$ parameter at each temperature. However, the

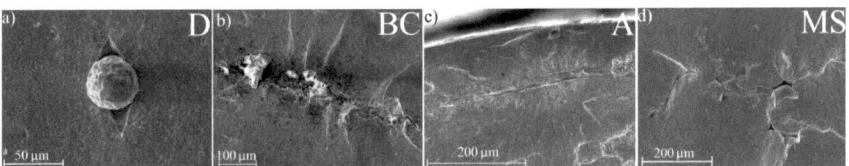

Fig. 34.7 Defects observed by SEM present in wrought material WS, (**a–c**) and cast material CS, (**d**). **a** Alumina (globular). **b** Alumina-cluster. **c** MnS (elongated). **d** Microshrinkage. Reproduced from [28]

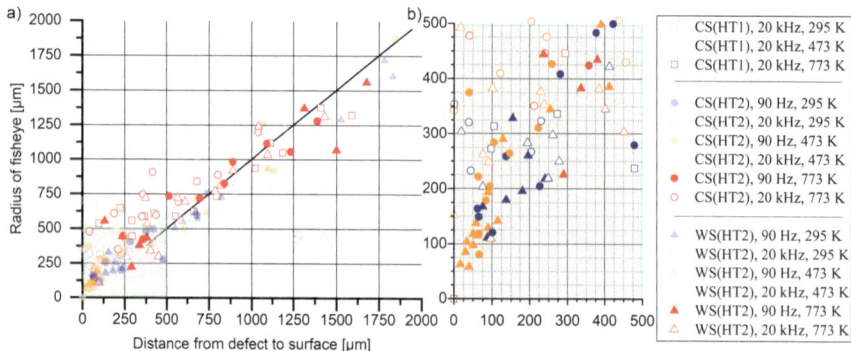

Fig. 34.8 Fisheyes radius vs shortest distance of defect to the surface for all materials of both batches CS and WS at RT, 473 K and 773 K. Data points for 295 K and 473 K tests partially illustrated transparently

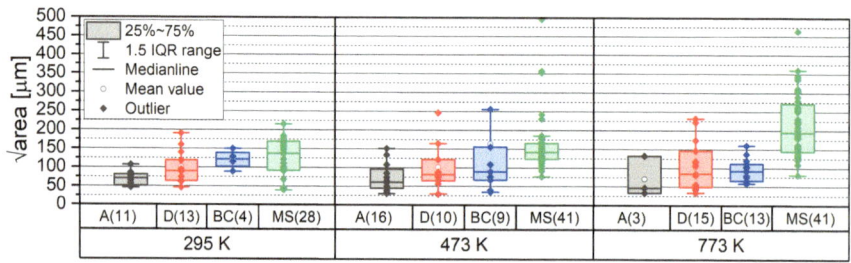

Fig. 34.9 Distribution of defect size ($\sqrt{area}$ parameter by Murakami [42]) observed by fracture surface analysis with respect of defect classification according to Table 34.2 and test temperature. The amount of each type of defect is given in brackets. A: MnS (elongated), BC: Alumina-cluster, D: Alumina (globular), MS: Microshrinkage. Partly reproduced from [28]

majority of type A observed for RT (11) and 473 K (16) and only a few (3) at 773 K. In contrast, defects of type BC (large alumina-cluster) occurred only rarely at RT (4) and considerably more at 473 K (9) and 773 K (13). Majority of fatigue failure relevant defects were alumina containing defects, i.e. types D (globular alumina) and BC. These defects showed the lowest and the highest mean values of $\sqrt{area}$ parameter in material WS(HT2), respectively. Previous investigations [28] showed that the largest defects within the test volume of the samples cause fatigue failure with the highest probability whether NMIs or microshrinkages.

34.4.4 Fatigue Life Prediction

The crack growth model according to Chapetti [30, 31] was applied to the investigated materials CS(HT1 and HT2) and WS(HT2). The results of fracture surface analysis, i.e. defect size, FiE size and shortest distance to the surface, as well as the fatigue data, i.e. the number of cycles to failure and the applied stress amplitude, were utilized for the calculations. According to Eq. (34.8), the numbers of cycles to failure $N_{\text{f.num}}$ were calculated, which required the determination of the intrinsic and extrinsic SIF range and enables also the calculation of k according to Eq. (34.9). The calculation of the intrinsic SIF range ΔK_{dR} required assumptions concerning the size of the microstructural barrier b and the plain fatigue limit, i.e. the grain size and the particular stress amplitude at 10^9 cycles were used for this purpose, respectively. The intrinsic SIF range then increases with increasing crack length a and the extrinsic SIF range can be derived from the ratio between the current intrinsic SIF range and the threshold value for long crack propagation ΔK_{thR}.

$$N_{\text{f.num}}(C, m, \Delta K_{\text{thR}}) = \int_{a_0}^{a_f} \frac{1}{C*\left(\Delta \sigma \sqrt{\pi a} - \left(\Delta K_{\text{dR}} + (\Delta K_{\text{thR}} - \Delta K_{\text{dR}})\left(1 - e^{-k(a-b)}\right)\right)\right)^m} \, da \quad (34.8)$$

$$k = \frac{1}{4b} \frac{\Delta K_{\text{dR}}}{\Delta K_{\text{thR}} - \Delta K_{\text{dR}}} \quad (34.9)$$

Multiple sets of different combinations of the free parameters C, m and ΔK_{thR} were applied to Eq. (34.8) to find the best possible approximation of $N_{\text{f.num}}$ to the experimental values of N_f, for each material at each test temperature. They were varied in range and spacing, i.e. $10^{-20} < C < 10^{-11}$ [MPa$\sqrt{\text{m}}$] (logarithmically spaced, values per decade $= 10$), $0.5 < \Delta K_{\text{thR}} < 13$ [MPa$\sqrt{\text{m}}$] (linearly spaced, step size $\leq$ 0.2), and $2 < m < 12$ (linearly spaced, step size ≤ 0.2). The number of variations of the parameters C, m and ΔK_{thR} were adapted with respect to expectable values to minimize the number of calculation loops. The individual initial (a_0) and the final (a_f) crack lengths were derived from the area (measured as a polygon) of the defect and the FiE according to Eq. (34.6) as shown in Fig. 34.4, respectively. The application of the short crack growth model, i.e. the calculation of the numbers of cycles to failure, showed a reasonably good agreement with the experimentally determined number of cycles to failure at 773 K using 90 Hz and 20 kHz fatigue tests and absolutely no agreement at lower temperatures. The fitting parameters of the model using the fatigue data and results of the fracture surface analysis of samples tested at 773 K are given in Table 34.4.

Table 34.4 Fitting parameters of the applied short crack growth model by Chapetti [30, 31] for fatigue tests performed at a temperature of 773 K and a stress ratio of R $= -1$ according to [27, 28]

Material	ΔK_{dR}	ΔK_{thR}	σ_a (N $= 10^9$)	b	m	C	R^2
	MPa$\sqrt{\text{m}}$	MPa$\sqrt{\text{m}}$	MPa	μm	–	MPa$\sqrt{\text{m}}$	–
CS(HT1)	1.3	3.8	160	5.5	3.0	4.6×10^{-14}	0.92
CS(HT2)	0.7	2.3	145	1.9	4.2	3.0×10^{-13}	0.96
WS(HT2)	0.8	1.7	180	1.5	4.6	2.2×10^{-13}	0.92

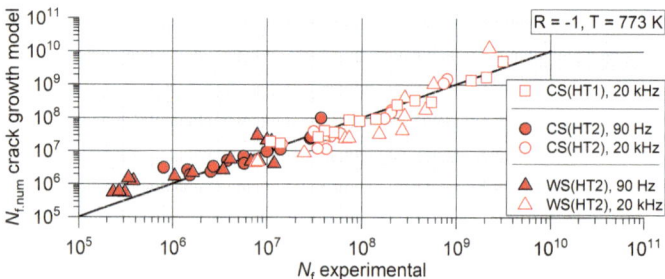

Fig. 34.10 Correlation of calculated and experimentally obtained number of cycles to failure using individual initial (a_0) and final crack lengths (a_f) derived from fracture surface analysis. The black line indicates the coincidence of $N_{f.num}$ and N_f

A significant correlation of calculated and experimentally determined values of the number of cycles to failure was observed only at 773 K, i.e. $R^2 = 0.92$, 0.96 and 0.92 for materials CS(HT1), CS(HT2) and WS(HT2) respectively. However, no dependency between $N_{f.num}$ and N_f was achieved through the application of the short crack growth model to fatigue tests at lower temperatures of RT and 473 K (Fig. 34.10).

34.5 Discussion

34.5.1 Application of Test Frequency on Fatigue Livetimes of Quenched and Tempered Steel 42CrMo4

Fatigue tests were performed at 90 Hz and 20 kHz test frequencies at temperatures of RT, 473 K and 773 K on wrought and cast batches of steel 42CrMo4, which has been subjected to two different quenching and tempering treatments. As a consequence, materials that received HT2 showed, in general, higher values of tensile strength and hardness due to smaller grain sizes compared to the material that received HT1. As already discussed by Murakami [41] on tempered steel, there is a linear relationship between hardness and the fatigue limit up to approximately a hardness of 400HV. Steels with higher hardness are characterized by an increased sensitivity to inclusions and small defects, which can decrease the fatigue limit [41]. The influence of different types of defects present in the investigated wrought and the cast batches are discussed in detail in the following.

Since two different fatigue testing systems were used operating at 90 Hz and 20 kHz test frequencies, a frequency effect, which would manifest in significantly different SN-curves has to be considered. The magnitude of frequency effects can be influenced by e.g. the crystal lattice structure, surface quality, atmosphere, sample geometry and strain rate during the fatigue tests [3, 43–45]. No influence of testing

Table 34.5 Percentage decrease of hardness HV10, UTS and fatigue strength at 10^9 cycles for elevated temperatures in relation to the values at RT. Values at RT are given in the corresponding unit and values at elevated temperatures in percentage relation to RT values

	CS(HT1)			CS(HT2)			WS(HT2)		
Temperature [K]	295	473	773	295	473	773	295	473	773
HV10 [%, %]	**320**	−10	−36	**420**	−12	−40	**425**	−12	−44
UTS [MPa, %, %]	**1000**	−7	−34	**1251**	−5	−37	**1260**	−4	−40
Fatigue strength $\sigma_{a.2}$ at 10^9 [MPa, %, %]	**346**	−15	−54	**340**	−12	−57	**480**	−33	−62

frequencies was observed for batches CS at all three temperatures as well as batch WS at 473 K and 773 K. Batch WS showed only at RT a difference in the SN-curves for 90 Hz and 20 kHz fatigue tests. Therefore, the reason for the behaviour of material WS(HT2) at RT with a high amount of runouts at a wide range of stress amplitudes is not clear yet. An increased sensitivity to crack initiation at inclusions, such as the NMIs present in the wrought material, can be assumed since the hardness was above 400HV as already discussed by Murakami [41]. Further influencing factors for a frequency effect, e.g. different surface roughnesses or notches, can be excluded since the quality of all samples was almost identical.

The results of the fatigue tests showed a decrease in fatigue strength due to an increase in the test temperature for each investigated material, i.e. WS(HT2), CS(HT1) and CS(HT2) as shown in Fig. 34.6 and Table 34.3. This is in agreement with considerations of Murakami [41] and Chapetti [46] regarding the temperature dependence of UTS as well as Vickers hardness. The percentage decrease of hardness HV10, UTS and the resulting decrease in fatigue strength at 10^9 cycles for elevated temperatures in relation to the values at RT is given in Table 34.5.

Material CS(HT2) showed significantly higher values of tensile strength and hardness (see Table 34.1) compared to material CS(HT1). However, no increase in fatigue strength was observed due to the different heat treatments as expected from the literature [5]. The wrought material WS(HT2) showed the highest fatigue strength at each temperature compared to both materials CS(HT1) and CS(HT2) of the cast batch. This superior fatigue strength can not be explained by the quasi-static properties of the material, e.g. the values of hardness or tensile strength since they are comparable for HT2 treatment of CS and WS batches. Therefore, the influence of the defects on the fatigue strengths has to be considered in detail.

34.5.2 The Detrimental Effect of Internal Defects on the Fatigue Strength

A classification of different defects based on the technical guideline SEP 1571 [32] and the results of the fracture surface analysis was established (see Table 34.1).

Defects of type MS (microshrinkages) were observed only in materials CS(HT1) and CS(HT2). NMIs, i.e. defects of type D (globular alumina), BC (alumina-cluster) and A (elongated manganese-sulphide), were observed only in material WS(HT2). The distribution of defects leading to fatigue failure at the different test temperatures (see Fig. 34.9) revealed the following results: (i) elongated manganese-sulphides were more frequently fatigue crack initiating at RT and 473 K compared to only a few samples that failed at 773 K due to this type of defect, (ii) alumina-clusters were more frequently crack initiating at 773 K compared to a decreasing number of samples which failed at RT and 473 K, and (iii) globular aluminas caused fatigue failure by similar frequencies at all three temperatures. Thus, it appears that defects containing alumina, i.e. type D (globular alumina) and BC (alumina-cluster), tend to be more detrimental with increasing temperature compared to the defect type A containing manganese-sulphides, which tends to be more detrimental at lower temperatures. The defects present in all materials CS(HT1), CS(HT2) and WS(HT2) exceed the threshold value for FGA formation of approximately 4.2–5 MPa$\sqrt{m}$ [3, 47–51] due to their size, which explains the absence of an FGA in most of the investigated fracture surfaces. Regarding the superior fatigue strength of material WS(HT2) compared to material CS(HT2) with equivalent characteristic properties (cf. $R_{p0.2}$, UTS, A_f, HV10 in Table 34.1 or grain sizes of 1.9 μm ± 1.5 μm for CS(HT2)) and 1.5 μm ± 0.9 μm for WS(HT2), the fundamental difference concerning the detrimental effect of the defects, i.e. NMIs and microshrinkages, has to be considered. Thus, these defect types will be discussed in more detail in the following.

Within the group of NMIs, different chemical compositions and morphologies lead to different properties of these defects. In general, the manganese-sulphide containing defects of type A are characterized by relatively low values of hardness compared to the surrounding steel matrix as well as a significant decrease in hardness at elevated temperatures [52, 53]. Manganese-sulphide NMIs reveal a value of thermal expansion at RT of about $\alpha_{th} \approx 18.1 \times 10^{-6}$ K^{-1} [52]. Contrary to this, alumina containing NMIs of types D and BC are characterized by relatively high values of hardness compared to the surrounding steel matrix, low sensitivity of the hardness due to increasing temperatures and only a slight change of ductility at elevated temperatures of 773 K [52, 54, 55]. Alumina containing NMIs reveal a value of thermal expansion at RT of about $\alpha_{th} \approx 8.3 \times 10^{-6}$ K^{-1} [52]. NMIs with a different value of thermal expansion than the surrounding matrix, i.e. low alloy steel ($\alpha_{th} \approx 12 \times 10^{-6}$ K^{-1} at RT) lead to additional internal stress fields due to heating or cooling of the material. According to Ånmark et al. [52], NMIs with higher values of thermal expansion than the matrix results in additional stress in the matrix around the NMI (cf. defect type A) when heating. On the other hand, NMIs with lower values of thermal expansion than the matrix result in additional stress in the NMI (cf. defect type D and BC) when heating. The absolute value of thermal expansion mismatch between steel matrix and each type of NMI is comparable. However, manganese-sulphide NMIs as root cause for the fatigue failure were observed more frequently at RT and 473 K but only a few at 773 K (see Fig. 34.9). In contrast, alumina containing NMIs of types D and BC were observed as root cause for the fatigue failure approximately evenly at each temperature. This leads to the assumption, that

NMIs hardness is more significant concerning the detrimental effect of the defects since manganese-sulphide containing defects are characterized by significantly lower values of hardness compared to alumina containing NMIs. Therefore, defect type A (elongated manganese-sulphide) is expected less detrimental due to lower hardness and significant hardness reduction at elevated temperatures compared to defect types D (globular alumina) and BC (alumina-cluster) with higher values of hardness and low influence of the temperature on the hardness.

In the cast materials, the fatigue relevant defect is type MS (microshrinkage). The fracture surface analysis revealed an increasing defect size (as $\sqrt{area}$ parameter) with increasing test temperature, i.e. median line at 138 μm at RT, 142 μm at 473 K and 197 μm at 773 K (see Fig. 34.9). Thus, it is recognized that the tolerable defect size increased at elevated test temperatures from RT up to 773 K. Furthermore, the notch effect has to be considered as an influencing factor on the fatigue strength of materials CS(HT1) and CS(HT2) due to the size and morphology of the cast defects. The notch effect is less detrimental at elevated temperatures (773 K) compared to lower temperatures (RT, 473 K) [38, 56–58]. However, the morphology of the cast defects concerning multiple spots of stress concentration and, therefore, the origin of crack nucleation highly depends on the morphology of the defects [59, 60].

34.5.3 Fatigue Life and Damage Mechanism at Various Temperatures

All investigated materials showed a dependency of a decrease in fatigue strength at an increase in temperature. The highest fatigue strengths of each tested material were always observed at RT and the lowest fatigue strengths at 773 K. Based on the results of the fatigue tests, the fracture surface analysis, the material characterization and the application of the short crack growth model, two main aspects were considered with respect to the temperature dependency of the observed fatigue lives: (i) change of quasi-static materials' properties, and (ii) change of the damage mechanism leading to fatigue failure from RT to 773 K.

As already shown in Table 34.1, Vickers hardness and tensile strength of all batches of the steel 42CrMo4 decreased with an increase in temperature. However, there was only a minor decrease in these values from RT to 473 K, but a significant decrease at 773 K. Therefore, the temperature-dependent quasi-static materials' properties have to be related to the decrease in fatigue strengths of the investigated materials, which is in agreement with the considerations of Chapetti [46] and Murakami [41]. Comparing materials CS(HT2) and WS(HT2), values of UTS, tensile strength, hardness and elongation to fracture (see Table 34.1) are comparable but the results of the fatigue tests were different. At 10^9 cycles to failure, the fatigue strength of material CS(HT2) was lower than WS(HT2) at each test temperature, i.e. a reduction of the fatigue strength of material CS(HT2) compared to material WS(HT2) by 29% at RT, 17% and 473 K and 19% and 773 K. The higher fatigue strength of material WS(HT2) can be addressed to the more or less detrimental effect of different types of defects. The batch WS failed only on NMIs and the batch CS failed only on microshrinkages. The present NMIs (see Table 34.2) were apparently less detrimental compared to the microshrinkages as already discussed above. Concerning the size distribution of defects leading to fatigue failure (see Fig. 34.9) as observed by fracture surface analysis, the lower fatigue strength of samples including microshrinkages is justified due to their significantly larger size than the NMIs. This dependency of fatigue strength on the size (as $\sqrt{\text{area}}$ parameter) of the defects was already reported by Murakami [42] as linear related in a double logarithmic scale for annealed medium carbon steels.

As a result of the fracture surface analysis, a dependency of the size of the FiE and the distance of the defects to the sample's surface was found (see Fig. 34.8). This dependency was significant for samples tested at elevated temperatures of 773 K and indistinct at lower temperatures. Therefore, it was assumed that the fatigue life of the investigated batches of 42CrMo4 at 773 K was dominated by crack growth and not as typical at RT by crack initiation. This assumption was confirmed by the applicability of the short crack model by Chapetti [30, 31] using for each sample tested at 773 K the fatigue data and individual defect characteristics. Based on the distinct regimes of crack growth introduced by Miller [6–9] as well as the microstructural investigations and results of the short crack growth model, we assume a three-stage damage mechanism.

Crack nucleation occurs in stage I, i.e. a crack initiated preferably at the interface of the defect and the steel matrix. In the case of batches CS, only microshrinkages were crack initiation sites. In the case of batch WS, NMIs yielded crack initiation, whereas, the frequency of elongated manganese-sulphide as fatigue relevant defect decreased with increasing test temperatures and alumina containing NMIs were observed at all temperatures approximately at the same frequency (cf. type of defects in Fig. 34.9). In this stage, the crack propagation after initiation was characterized by the localisation within an individual grain and an incremental increase of the crack length due to the release of a small number of dislocations at the crack tip during one loading cycle. Under these conditions, the microstructurally short cracks according to Miller [9] were assumed to develop simultaneously at multiple preferred locations in the interior of the sample. The growth rate of those cracks is expected to decelerate or they arrested when reaching the microstructural barrier, e.g. grain boundary. Moreover, it is assumed that crack nucleation and growth of microstructurally short cracks occurred more or less instantaneously at 773 K if the stress amplitude was high enough. Therefore, the crack resistance at 773 K was significantly lower compared to RT and 473 K, which is in agreement with lower fatigue strength at 773 K and reduced scatter since overcoming the microstructural barrier is favoured at 773 K (see fatigue strength and scatter in Fig. 34.6).

By overcoming the microstructural barrier, stage II of damage is reached. It is assumed that stage II crack growth occurred only inside the FiE. The threshold value to overcome the barrier ΔK_{dR} was calculated separately for each material at RT, 473 K and 773 K (see Table 34.4). The lowest values of ΔK_{dR} were obtained at 773 K. Depending on the microstructural barrier, the neighbour grain orientation or the stress level, crack arrest can occur, which is more likely at RT and 473 K than at 773 K. Due to the frequent occurrence of crack arrest at RT and 473 K, the short crack model is not applicable at these temperatures. Furthermore, the significantly lower values of slope parameter $k*$ at 773 K fatigue tests indicate higher crack growth rates at this temperature compared to RT and 473 K in this stage.

Crack growth outside the FiE is referred as to stage III of damage and is associated with long crack growth. Most of the cycles to failure were consumed in stages I and II. Stage III is related to long crack propagation, which is characterized by a significant increase in growth rate. Therefore, only a few cycles were consumed in this stage up to the termination of the fatigue test.

Finally, it is concluded that the fatigue life at RT and 473 K is characterized by crack initiation as the dominating mechanism, whereas crack growth is dominating the fatigue life at 773 K.

34.6 Conclusions

In this chapter wrought (WS) and cast (CS) batches of the quenched and tempered steel 42CrMo4 with two different heat treatments (HT1 and HT2), i.e. HT1 with lower tensile strength and Vickers hardness compared to HT2, of 42CrMo4 steel were studied. The batches were investigated regarding the fatigue behaviour in the HCF and VHCF regimes at various temperatures of RT, 473 K and 773 K performing fatigue testing at 90 Hz and 20 kHz frequency. The main results of the investigations are the following:

1. As a result of the manufacturing process, fatigue failure was caused by microshrinkages and NMIs for both materials CS(HT1) and CS(HT2) of the cast batch and material WS(HT2) of the wrought batch, respectively. Material WS(HT2) showed the highest fatigue strength at each test temperature, whereas, the fatigue strength of all investigated materials decreased by an increase in temperature. The lowest fatigue strength was observed at a test temperature of 773 K.
2. The detrimental effect of the present defects on the fatigue strengths of all investigated materials was analysed. Both groups of defects, i.e. NMIs and microshrinkages, showed no dependency on size or distance to the surface and the number of cycles to failure. However, it can be reasonably assumed that the largest defects of each type within both groups lead to fatigue failure.
3. Based on the individual fatigue data and the fracture surface analysis of each sample, i.e. size of defect and FiE, a short crack growth model was applied. The calculated number of cycles to failure showed a distinct correlation with the experimentally evaluated values only at 773 K test temperature. Therefore, it was assumed that crack initiation is the relevant mechanism of fatigue, dominanting at RT and 473 K, whereas, crack growth is the fatigue life determining mechanism at 773 K.

Acknowledgements The authors gratefully acknowledge the German Research Foundation (DFG) for supporting the Collaborative Research Centre CRC 920 - Project-ID 169148856, subproject T02 and the project partner Federal-Mogul GmbH Nürnberg a Tenneco group company, Thomas Kirste and Dr.-Ing. Roman Morgenstern, Diane Hübgen (Institute of Materials Science), and Dr.-Ing. Sebastian Henkel (Institute of Materials Engeenierng).

References

1. T. Sakai, Y. Sato, N. Oguma, FFEMS **25**, 765 (2002). https://doi.org/10.1046/j.1460-2695.2002.00574.x
2. Y. Murakami, T. Nomoto, T. Ueda, FFEMS **22**, 581 (1999). https://doi.org/10.1046/j.1460-2695.1999.00187.x
3. S. Stanzl-Tschegg, B. Schönbauer, Proc. Eng. **2**, 1547 (2010). https://doi.org/10.1016/j.proeng.2010.03.167
4. K. Shiozawa, Y. Morii, S. Nishino, L. Lu, Int. J. Fatigue **28**, 1521 (2006). https://doi.org/10.1016/j.ijfatigue.2005.08.015
5. D. Jeddi, T. Palin-Luc, Fatigue Fract. Eng. Mater. Struct. **41**, 969 (2018). https://doi.org/10.1111/ffe.12779
6. K.J. Miller, FFEMS **5**, 223 (1982). https://doi.org/10.1111/j.1460-2695.1982.tb01250.x
7. K.J. Miller, FFEMS **10**, 75 (1987). https://doi.org/10.1111/j.1460-2695.1987.tb01150.x
8. K.J. Miller, FFEMS **10**, 93 (1987). https://doi.org/10.1111/j.1460-2695.1987.tb01153.x
9. K.J. Miller, Fatigue Frac. Eng. Mat. Struct. **16**, 931 (1993). https://doi.org/10.1111/j.1460-2695.1993.tb00129.x
10. D. L. Davidson, in *Modelling Problems in Crack Tip Mechanics*, ed. by J.T. Pindera, B.R. Krasnowski (Springer Netherlands, Dordrecht, 1984), pp. 217–228. https://doi.org/10.1007/978-94-009-6198-2_13
11. E.R. de los Rios, Z. Tang, K.J. Miller, FFEMS **7**, 97 (1984). https://doi.org/10.1111/j.1460-2695.1984.tb00408.x
12. S. Stanzl-Tschegg, Int. J. Fatigue **60**, 2 (2014). https://doi.org/10.1016/j.ijfatigue.2012.11.016
13. C.S. Bandara, S.C. Siriwardane, U.I. Dissanayake, R. Dissanayake, Eng. Fail. Anal. **45**, 421 (2014). https://doi.org/10.1016/j.engfailanal.2014.07.015
14. T. Sakai, A. Nakagawa, N. Oguma, Y. Nakamura, A. Ueno, S. Kikuchi, A. Sakaida, Int. J. Fatigue **93**, 339 (2016). https://doi.org/10.1016/j.ijfatigue.2016.05.029
15. C. Bathias, Int. J. Fatigue **28**, 1438 (2006). https://doi.org/10.1016/j.ijfatigue.2005.09.020
16. D. Krewerth, A. Weidner, H. Biermann, Ultrasonics **53**, 1441 (2013). https://doi.org/10.1016/j.ultras.2013.03.001
17. D. Krewerth, T. Lippmann, A. Weidner, H. Biermann, Int. J. Fatigue **80**, 459 (2015). https://doi.org/10.1016/j.ijfatigue.2015.07.013
18. S. Henschel, D. Krewerth, F. Ballani, A. Weidner, L. Krüger, H. Biermann, M. Emmel, C.G. Aneziris, Adv. Eng. Mater. **15**, 1216 (2013). https://doi.org/10.1002/adem.201300125
19. D. Krewerth, A. Weidner, H. Biermann, Adv. Eng. Mater. **15**, (2013). https://doi.org/10.1002/adem.201300124
20. A. Shyam, C. J. Torbet, S. K. Jha, J. M. Larsen, M. J. Caton, C. J. Szczepanski, T. M. Pollock, J. W. Jones, in *Superalloys 2004 (Tenth International Symposium)* (2004), pp. 259–268
21. J.Z. Yi, C.J. Torbet, Q. Feng, T.M. Pollock, J.W. Jones, Mater. Sci. Eng. A **443**, 142 (2007). https://doi.org/10.1016/j.msea.2006.08.028
22. X. Zhu, A. Shyam, J. Jones, H. Mayer, J. Lasecki, J. Allison, Int. J. Fatigue **28**, 1566 (2006). https://doi.org/10.1016/j.ijfatigue.2005.04.016
23. Y. Furuya, K. Kobayashi, M. Hayakawa, M. Sakamoto, Y. Koizumi, H. Harada, Mater. Lett. **69**, 1 (2012). https://doi.org/10.1016/j.matlet.2011.11.066

24. D. Wagner, F.J. Cavalieri, C. Bathias, N. Ranc, Propul. Power Res. **1**, 29 (2012). https://doi.org/10.1016/j.jppr.2012.10.008
25. A. Cervellon, J. Cormier, F. Mauget, Z. Hervier, Int. J. Fatigue **104**, 251 (2017). https://doi.org/10.1016/j.ijfatigue.2017.07.021
26. A. Amanov, Y.-S. Pyun, J.-H. Kim, C.-M. Suh, I.-S. Cho, H.-D. Kim, Q. Wang, M.K. Khan, Fatigue Fract. Eng. Mater. Struct. **38**, 1266 (2015). https://doi.org/10.1111/ffe.12330
27. A. Schmiedel, S. Henkel, T. Kirste, R. Morgenstern, A. Weidner, H. Biermann, Fatigue Fract. Eng. Mater. Struct. **43**, 2455 (2020). https://doi.org/10.1111/ffe.13316
28. A. Schmiedel, T. Kirste, R. Morgenstern, A. Weidner, H. Biermann, Int. J. Fatigue **152**, 106437 (2021). https://doi.org/10.1016/j.ijfatigue.2021.106437
29. A. Schmiedel, C. Burkhardt, S. Henkel, A. Weidner, H. Biermann, Metals **11**, (2021). https://doi.org/10.3390/met11111682
30. M.D. Chapetti, T. Tagawa, T. Miyata, Mater. Sci. Eng. A **356**, 227 (2003). https://doi.org/10.1016/S0921-5093(03)00135-7
31. M.D. Chapetti, T. Tagawa, T. Miyata, Mater. Sci. Eng. A **356**, 236 (2003). https://doi.org/10.1016/S0921-5093(03)00136-9
32. SEP 1571 Teil 1:2017–08, (n.d.)
33. Y. Hong, X. Liu, Z. Lei, C. Sun, Int. J. Fatigue **89**, 108 (2016). https://doi.org/10.1016/j.ijfatigue.2015.11.029
34. T. Sakai, H. Harada, N. Oguma, in *Proceedings of ECF-16, CD-ROM* (2006)
35. Z. Xiulin, M. A. Hirt, Eng. Fract. Mech. **18**, 965 (1983). https://doi.org/10.1016/0013-7944(83)90070-X
36. K. Hussain, Eng. Fract. Mech. **58**, 327 (1997). https://doi.org/10.1016/S0013-7944(97)00102-1
37. D. Raabe, *Computational Materials Science* (Wiley-VCH, Weinheim, 1998)
38. D. Radaj, M. Vormwald, *Ermüdungsfestigkeit*, 3., neubearbeitete und erweiterte Auflage (Springer-Verlag Berlin Heidelberg, Berlin, Heidelberg, 2007). https://doi.org/10.1007/978-3-540-71459-0
39. M. Chapetti, Int. J. Fatigue **25**, 1319 (2003). https://doi.org/10.1016/S0142-1123(03)00065-3
40. F. Pascual, W. Meeker, J. Test. Eval. **25**, 292 (1997). https://doi.org/10.1520/JTE11341J
41. Y. Murakami, JSME Int. J. Ser. 1, Solid Mech. Strength Mater. **32**, 167 (1989). https://doi.org/10.1299/jsmea1988.32.2_167
42. Y. Murakami, *Metal Fatigue: Effects of Small Defects and Nonmetallic Inclusions* (Elsevier, Waltham, MA, 2002). https://doi.org/10.1016/B978-0-08-044064-4.X5000-2
43. N. Schneider, J. Bödecker, C. Berger, M. Oechsner, Int. J. Fatigue **93**, 224 (2016). https://doi.org/10.1016/j.ijfatigue.2016.05.013
44. W. Peng, Y. Zhang, B. Qiu, H. Xue, AASRI Procedia **2**, 127 (2012). https://doi.org/10.1016/j.aasri.2012.09.024
45. B. Guennec, A. Ueno, T. Sakai, M. Takanashi, Y. Itabashi, Int. J. Fatigue **66**, 29 (2014). https://doi.org/10.1016/j.ijfatigue.2014.03.005
46. M. Chapetti, Procedia Struct. Integrity **7**, 229 (2017). https://doi.org/10.1016/j.prostr.2017.11.082
47. Y. Hong, C. Sun, Theoret. Appl. Fract. Mech. **92**, 331 (2017). https://doi.org/10.1016/j.tafmec.2017.05.002
48. T. Karsch, H. Bomas, H.-W. Zoch, S. Mändl, Int. J. Fatigue **60**, 74 (2014). https://doi.org/10.1016/j.ijfatigue.2013.09.006
49. M. Seleznev, S. Henschel, E. Storti, C. G. Aneziris, L. Krüger, A. Weidner, H. Biermann, Adv. Eng. Mater. n/a, 1900540 (2019). https://doi.org/10.1002/adem.201900540
50. Y. Hu, C. Sun, Y. Hong, Fatigue Fract. Eng. Mater. Struct. **41**, 1717 (2018). https://doi.org/10.1111/ffe.12811
51. M. Seleznev, E. Merson, A. Weidner, H. Biermann, Int. J. Fatigue **126**, (2019). https://doi.org/10.1016/j.ijfatigue.2019.05.011
52. N. Ånmark, A. Karasev, P.G. Jönsson, Materials (Basel, Switzerland)**8**, 751 (2015). https://doi.org/10.3390/ma8020751

53. F. Matsuno, S. Nishikida, H. Ikesaki, Trans. Iron Steel Instit. Jpn. **25**, 989 (1985). https://doi.org/10.2355/isijinternational1966.25.989
54. A.L.V. da Costa e Silva, J. Mater. Res. Technol. **8**, 2408 (2019). https://doi.org/10.1016/j.jmrt.2019.01.009
55. P.A. Thornton, J. Mater. Sci. **6**, 347 (1971). https://doi.org/10.1007/BF02403103
56. P. Lukáš, L. Kunz, B. Weiss, R. Stickler, Fatigue Fract. Eng. Mater. Struct. **9**, 195 (1986). https://doi.org/10.1111/j.1460-2695.1986.tb00446.x
57. P. Lukás, L. Kunz, B. Weiss, R. Stickler, Fatigue Fract. Eng. Mater. Struct. **12**, 175 (1989). https://doi.org/10.1111/j.1460-2695.1989.tb00525.x
58. P. Gallo, F. Berto, Frattura Ed Integrità Strutturale **34**, 180 (2015). https://doi.org/10.3221/IGF-ESIS.34.19
59. M. Wicke, M. Luetje, I. Bacaicoa, A. Brueckner-Foit, Procedia Struct. Integrity **2**, 2643 (2016). https://doi.org/10.1016/j.prostr.2016.06.330
60. I. Serrano-Munoz, J.-Y. Buffiere, C. Verdu, Y. Gaillard, P. Mu, Y. Nadot, Int. J. Fatigue **82**, 361 (2016). https://doi.org/10.1016/j.ijfatigue.2015.07.032

Index

© The Editor(s) (if applicable) and The Author(s) 2024
C. G. Aneziris and H. Biermann (eds.), *Multifunctional Ceramic Filter Systems for Metal Melt Filtration*, Springer Series in Materials Science 337,
https://doi.org/10.1007/978-3-031-40930-1